Old and New Questions in Physics, Cosmology, Philosophy, and Theoretical Biology

Essays in Honor of Wolfgang Yourgrau

W. Yourgrau

Wolfgang Yourgrau in the late 1950s

Old and New Questions in Physics, Cosmology, Philosophy, and Theoretical Biology

Essays in Honor of Wolfgang Yourgrau

Edited by
Alwyn van der Merwe
University of Denver, Denver, Colorado

Plenum Press • New York and London

Library of Congress Cataloging in Publication Data

Main entry under title:

Old and new questions in physics, cosmology, philosophy, and theoretical biology.

Includes bibliographical references and index.

1. Physics—Addresses, essays, lectures. 2. Cosmology—Addresses, essays, lectures. 3.Biology—Addresses, essays, lectures. 4. Science—Philosophy—Addresses, essays, lectures. 5. Yourgrau, Wolfgang—Addresses, essays, lectures. 6. Physicists—United States—Biography—Addresses, essays, lectures. I. Yourgrau, Wolfgang. II. Van der Merwe, Alwyn.

QC71.043 1983 500 83-2173

ISBN 0-306-40962-3

A Division of Plenum Publishing Corporation
233 Spring Street, New York, N.Y. 10013

Printed in the United States of America

Contributors

T. Anders • *Universität München, Munich, GFR*
Francisco J. Ayala • *University of California, Davis, USA*
A. O. Barut • *University of Colorado, Boulder, USA*
Heinz Beran • *Austrian Radio Corporation (Vienna) and British Broadcasting Corporation (London)*
Peter G. Bergmann • *Syracuse University, New York, and New York University, New York, USA*
Jay Boyd Best • *Colorado State University, Fort Collins, USA*
D. Bohm • *Birkbeck College, University of London, England*
Sir Hermann Bondi • *Natural Environment Research Council, Swindon, England*
Prince Louis de Broglie • *Institut Henri Poincaré, Paris, France*
O. Costa de Beauregard • *Institut Henri Poincaré, Paris, France*
Nicola Cufaro-Petroni • *Istituto di Fisica, Bari, Italy*
H. Dehnen • *Universität Konstanz, GFR*
Stanley P. Gudder • *University of Denver, Colorado, USA*
D. ter Haar • *University of Oxford, England*
Jean Heidmann • *Observatoire de Paris, Meudon, France*
B. J. Hiley • *Birkbeck College, University of London, England*
H. Hönl • *Universität Freiburg, GFR*
L. P. Horwitz • *Tel Aviv University, Israel*
C. J. Isham • *Imperial College of Science and Technology, London, England*
D. Ivanenko • *Moscow State University, USSR*
Toivo Jaakkola • *University of Helsinki, Finland*
W. Jachmann • *Universität München, Munich, GFR*
Stanley L. Jaki • *Seton Hall University, South Orange, New Jersey, USA*
Alfred Kastler • *Université Paris VI, France*
Morris Lazerowitz • *Smith College, Northampton, Massachusetts, USA*
R. Bruce Lindsay • *Brown University, Providence, Rhode Island, USA*
G. Ludwig • *Philipps-Universität, Marburg, GFR*
Stanley Mandelstam • *University of California, Berkeley, USA*
Jean-Paul Marchand • *University of Denver, Colorado, USA*
Henry Margenau • *Yale University, New Haven, Connecticut, USA*

ANDRÉ MERCIER • *Universität Bern, Switzerland*
ARNE NAESS • *University of Oslo, Norway*
JOHN NORTH • *University of Oxford, England*
D. J. O'CONNOR • *University of Exeter, England*
JAMES L. PARK • *Washington State University, Pullman, USA*
A. R. PEACOCKE • *University of Cambridge, England*
JEAN-CLAUDE PECKER • *Collège de France, Paris, France*
HERBERT PIETSCHMANN • *Universität Wien, Austria*
C. PIRON • *Université de Genève, Switzerland*
SIR KARL POPPER • *London School of Economics and Political Science, England*
ANTONIO F. RAÑADA • *Universidad Complutense de Madrid, Spain*
ERASMO RECAMI • *University of Catania, Italy*
R. ROMPE • *GDR Physical Society, Berlin, GDR*
H. SALECKER • *Universität München, Munich, GFR*
RICHARD SCHLEGEL • *Michigan State University, East Lansing, USA*
F. SELLERI • *Università di Bari, Italy*
ROMAN U. SEXL • *Universität Wien, Austria*
RALPH F. SIMMONS, JR. • *Washington State University, Pullman, USA*
HENRYK SKOLIMOWSKI • *University of Michigan, Ann Arbor, USA*
MICHAEL J. STEPHEN • *Rutgers University, Piscataway, New Jersey, USA*
E. C. G. SUDARSHAN • *University of Texas, Austin, USA*
M. SURDIN • *Centre des Faibles Radioactivités, Gif-sur-Yvette, France*
J. L. SYNGE • *Dublin Institute for Advanced Studies, Ireland*
ALBERT SZENT-GYÖRGYI • *National Foundation for Cancer Research, Woods Hole, Massachusetts, USA*
K. TAHIR SHAH • *International Centre for Theoretical Physics, Trieste, Italy*
HERMAN TENNESSEN • *University of Alberta, Edmonton, Canada*
HÅKAN TÖRNEBOHM • *University of Gothenburg, Sweden*
H.-J. TREDER • *Akademie der Wissenschaften der DDR, Berlin, GDR*
ALWYN VAN DER MERWE • *University of Denver, Colorado, USA*
J.-P. VIGIER • *Institut Henri Poincaré, Paris, France*
K. WADAN • *Universität München, Munich, GFR*
A. WEHRL • *Universität Wien, Austria*
HARALD WERGELAND • *University of Trondheim, Norway*
J. O. WISDOM • *Wilmont House, Castlebridge, County Wexford, Ireland*
JAMES F. WOODWARD • *California State University, Fullerton, USA*
H. D. ZEH • *Universität Heidelberg, GFR*

Dedication

Remembering Wolfgang Yourgrau

Sir Karl Popper, FRS
London School of Economics and Political Science
London, England

It was approximately in 1957 or 1958 that I received a letter, and not long afterwards a visit, from Wolfgang Yourgrau. The name was known to me from two sources. I had come across one or two of his articles, which I had found interesting and which incited my curiosity: I remember that they were written in Jerusalem and in South Africa. And I had received a letter from Erwin Schrödinger in which he referred to Yourgrau, and recommended him to me.

As usual in such cases, I had unconsciously formed some kind of picture of the man in advance of his visit; and as usual, this theory or anticipation became conscious to me by the refutation of this prejudice when the man arrived in person. I remember particularly well what it was that struck me as completely different from my expectation: it was, physically, the quickness and energy of the movements of the man, creating a certain mercurial kind of impression; and spiritually, the human warmth of a most unusual and intense personality.

Very soon we called each other by our first names; and very soon we became real friends.

There was a tremendous range of subjects that were of common interest to us. But although we hardly ever agreed on anything (except on our common hostility to right wing totalitarians and our love and admiration for Einstein, de Broglie, and Schrödinger), we both felt personally very close. I do not think that our disagreements ever affected in the least our mutual good will. Wolfgang and I were of course essentially different. He was a gifted organizer, and a man who had a great number of friends; and he lived—or so it appeared to me—in the world of science; while I knew very few people.

It was through his instrumentality that my wife and I stayed in 1966 for a considerable time in Denver, and during this period I met not only Wolfgang's wife Thella and their three sons, but also some of his great friends, among them George Gamow, outstanding among the great physicists of our time by his incredible fertile imagination and by his intuitive daring.

This was in 1966.

Another of Wolfgang's great friends whom I met in Denver was Henry Margenau who soon became cofounder with Wolfgang of the *Foundations of Physics*, now edited in the spirit of its founders and very well looked after by Alwyn van der Merwe. And there was, in the field of human science, Allen Breck, who edited with Wolfgang a volume on *Physics, Logic, and History*, a report on a most interesting International Colloquium organized by Wolfgang which took place while we were in Denver in 1966.

After 1966 Wolfgang visited me a few times in England. But in the last years before his death, all our plans for a meeting were frustrated, mainly by ill health. The last of these planned yet frustrated meetings was meant to take place only a few months before Wolfgang died.

He was a personality who will for ever remain alive in the memory of all who knew him.

Karl Popper

Preface

Simply to say that this is a collection of essays in honor of the late Wolfgang Yourgrau (1908–1979) is to explain, at least for—the obviously many—"insiders," the unusually wide-ranging title of the present volume.

In a Foreword to the Proceedings of the First International Colloquium (focusing on logic, physical reality, and history), held at the University of Denver in May of 1966 under their leadership, Wolfgang Yourgrau and Allen Breck wrote, in an oblique reference to C. P. Snow: "Indeed there are not two or three or four cultures: there is only one culture; our generation has lost its awareness of this. . . . Historians, logicians, physicists—all are banded in one common enterprise, namely in their desire to weave an enlightened fabric of human knowledge."

Augment, if you will, the foregoing categories of scholars with biologists, philosophers, cosmologists, and theologians—all of whom, in addition to historians, Wolfgang Yourgrau, by dint of his inextinguishable enthusiasm and charismatic qualities, assembled in Denver for the Second and Third International Colloquia (in 1967 and 1974, respectively)—and a few other besides, and one arrives at a statement of the *credo* which Yourgrau not only professed, but consistently exemplified throughout his adult life. It is, therefore, entirely appropriate that the present rather monumental tome of 59 essays, contributed by 66 of Wolfgang Yourgrau's colleagues and admirers, should in their depth and scope reflect the diversity of his interests and scholarly writings. (The bulk of the latter, incidentally, is listed on pp. 30–34.)

In recognition of Yourgrau's aversion towards any kind of strict compartmentalization of knowledge, we resisted the temptation to extensively subdivide the collection of essays, zealously grouping together only those papers that "belong" together. Instead we settled for their largely alphabetical ordering by author. It did, however, seem advisable, if only for the guidance of the reader, to make a single division of the volume into two parts of roughly comparable size.

Part A, constituting what one might call the "technical" half of the book, contains substantive papers on elementary particles, the foundations of quantum theory, general relativity, and cosmology. These investigations should engage the attention primarily of schooled theoretical and mathematical physicists, cosmologists, and philosophers of science.

In Part B, the discussion turns to less arcane, if equally absorbing, intellectual matters. In fact, a veritable smorgasbord of topics are submitted for cogitation; they range, *inter alia*, from astonishing findings on the sentient behavior of primitive life forms, through fresh perspectives on some of the perennial questions within the province of ethics and theology, to incisive studies in the method, history, and philosophy of science. Since the understanding of these essays is not bedeviled to any problematic extent by mathematics and technical jargon, they should be readily accessible and stimulating to any intelligent educated layman, no less than to the specialist.

University of Denver Alwyn van der Merwe

Contents

Introduction 1
Alwyn van der Merwe

Part A: Current Problems in Physics and Cosmology

1 • On the Connections Between Pico- and Mega-Cosmos (According To the Unification and Geometrization Suggested by Einstein, Eddington, and Schrödinger 37
H.-J. Treder

2 • The Zitterbewegung and the Einstein A Coefficient of Spontaneous Emission 53
A. O. Barut

3 • Supersymmetry Groups 59
Peter G. Bergmann

4 • Relativistic Phase Space Arising Out of the Dirac Algebra 67
D. Bohm and B. J. Hiley

5 • Why Mourn the Passing of Determinacy 77
Hermann Bondi

6 • On the True Ideas Underlying Wave Mechanics 83
Louis de Broglie

7 • *CPT* Invariance as Basic for Interpreting Quantum Mechanics 87
O. Costa de Beauregard

8 • The Hilbert Space Axiom in Quantum Mechanics 109
Stanley P. Gudder

9 • The Cyclotron–Soliton Model of Terrestrial Kilometric Radiation ... 129
D. ter Haar

10 • Clumpy Irregular Galaxies: A Progress Report ... 133
Jean Heidmann

11 • The Aporias of Cosmology and the Attempts at Overcoming Them by Nonstandard Models ... 145
H. Hönl and H. Dehnen

12 • On Relativistic Quantum Theory ... 169
L. P. Horwitz

13 • Vacuum Tunneling in Static Space-Times ... 189
C. J. Isham

14 • On the Extensions of General Relativity ... 213
D. Ivanenko

15 • Status and Perspectives of Cosmology ... 223
Toivo Jaakkola

16 • The Connection between the Objective Description of Macrosystems and Quantum Mechanics of "Many Particles" ... 243
G. Ludwig

17 • The Current Theory of Strong Interactions and the Problem of Quark Confinement ... 265
Stanley Mandelstam

18 • Statistical Inference by Minimal Bures Distances ... 275
Jean-Paul Marchand

19 • The Knots of Quantum Thermodynamics ... 289
James L. Park and Ralph F. Simmons, Jr.

20 • About the Facts of Cosmological Significance ... 309
Jean-Claude Pecker

21 • Stochastic Interpretation of Relativistic Quantum Equations 325
N. Cufaro-Petroni and J.-P. Vigier

22 • New Quantum Mechanics ... 345
C. Piron

23 • ON NONLINEAR CLASSICAL DIRAC FIELDS AND QUANTUM PHYSICS 363
ANTONIO F. RAÑADA

24 • ELEMENTARY PARTICLES AS MICROUNIVERSES 377
ERASMO RECAMI

25 • PRECISION MATHEMATICS AND APPROXIMATION MATHEMATICS IN PHYSICS 397
R. ROMPE AND H.-J. TREDER

26 • ELECTROWEAK INTERACTION IN POLARIZED ELECTRON–POSITRON AND ELECTRON–ELECTRON SCATTERING 411
T. ANDERS, W. JACHMANN, H. SALECKER, AND K. WADAN

27 • INFINITY AND THE NATURAL WORLD 421
RICHARD SCHLEGEL

28 • INDUCED SUPERFLUORESCENCE AND THE NATURE OF THE WAVE–PARTICLE DUALITY 433
F. SELLERI AND J.-P. VIGIER

29 • BREAKDOWN OF PREDICTABILITY: AN INVESTIGATION INTO THE NATURE OF SINGULARITIES 437
K. TAHIR SHAH

30 • WAVE FUNCTIONS IN ONE-DIMENSIONAL DISORDERED SYSTEMS WITH LONG-RANGE HOPPING 451
MICHAEL J. STEPHEN

31 • PERCEPTION OF QUANTUM SYSTEMS 457
E. C. G. SUDARSHAN

32 • STOCHASTIC ELECTRODYNAMICS: AN OVERVIEW 469
M. SURDIN

33 • ON THE INCREASE OF ENTROPY IN THE CARLEMAN MODEL II 495
A. WEHRL

34 • THE KLEIN PARADOX REVISITED 503
HARALD WERGELAND

35 • EINSTEIN NONLOCALITY, SPACE-TIME STRUCTURE, AND THERMODYANICS 517
H. D. ZEH

Part B. Issues in Biology, Philosophy, and the History of Science

1 • Biology and Physics: Reflections on Reductionism 525
Francisco J. Ayala

2 • The Age of Analysis Reassessed 535
Heinz Beran

3 • Transphyletic Animal Similarities and Predictive Toxicology 549
Jay Boyd Best

4 • The Wronging of Wright 593
Stanley L. Jaki

5 • On the Historical Development of the Indistinguishability Concept for Microparticles 607
Alfred Kastler

6 • Philosophical Semantics 625
Morris Lazerowitz

7 • Social Exemplifications of Physical Principles 647
R. Bruce Lindsay

8 • Seneca's Ethics Viewed from a Modern Standpoint 659
Henry Margenau

9 • Physics and Philosophy 671
André Mercier

10 • Einstein, Spinoza, and God 683
Arne Naess

11 • The Satellites of Jupiter, from Galileo to Bradley 689
John North

12 • Emergent Properties 719
D. J. O'Connor

13 • The Theory of Relativity and Our World View 733
A. R. Peacocke

14 • Science beyond Science 753
Herbert Pietschmann

15 • God as Cosmic Guest-Worker 765
Roman U. Sexl

16 • A Model of Reality as Mind 769
Henryk Skolimowski

17 • A Matter of Chance 789
J. L. Synge

18 • On the Nature and Origin of Life 805
Albert Szent-Györgyi

19 • Science, History, Philosophy, and World Views 809
Herman Tennessen

20 • Queries about Inquiries 847
Håkan Törnebohm

21 • The Nature of the Physical World 873
J. O. Wisdom

22 • Early Attempts at a Unitary Understanding of Nature 885
James F. Woodward

Subject Index 909

Name Index 913

INTRODUCTION*

For the Record: Wolfgang Yourgrau (1908–1979)

ALWYN VAN DER MERWE
Department of Physics
University of Denver
Denver, Colorado 80208

Homo sum: humani nihil a me alienum puto—Terence,
Heauton Timoroumenos, Act 1

Much has already been written elsewhere[1,2] about Wolfgang Yourgrau—in whose memory, and as an expression of their friendship and esteem, more than sixty scholars, active in the forefront of their respective specialities and, significantly, representing diverse nationalities and backgrounds, have graciously volunteered to contribute to this *Festschrift*.

All the same, the existing public record is incomplete in some essential respects. I am, therefore, grateful for the opportunity afforded by the present publication to rectify matters by augmenting the literature concerning Yourgrau in basically two partially overlapping areas.

Firstly, I intend to provide additional perspective on some historical events in which Yourgrau's activities impinged on the public domain. Secondly, I should like, especially in consideration of Wolfgang's family and those individuals whose world lines crossed his, to furnish some psychological understanding of Yourgrau's complex personality, a few aspects of which, no doubt due to their veiled origin, remained somewhat of an enigma even to his close friends.

I fully realize that the road to the latter objective is replete with pitfalls. To "know thyself" is hard enough; to fathom thy friend might prove impossible! Nonetheless, a balanced, if not objective, recounting of events will be attempted. And if in this pursuit I should fail, I am confident that "Dieu me pardonnera—c'est Son metier"—to invoke a saying Wolfgang himself was fond of quoting.

The last circa 30 years of his career, Wolfgang Yourgrau spent in academic surroundings (in the U.S.A. and overseas), engaged in what he always most wanted to do: teaching and the pursuit of his own writing and research. Colleagues and university administrators who encountered him under those circumstances divided generally speaking neatly into two categories: Aligned on one side was the (not so loyal) opposition, consisting of those who resented Yourgrau's outspokenness, self-assertion, and not insignificant ego. On the other side stood those who apparently held, with Goethe, that "Die Irrthümer des Menschen machen ihn eigentlich liebenswürdig" and were more than willing to overlook Yourgrau's "inconsequential" failings in recognition of his preponderant virtues.

Students and colleagues of the latter persuasion most likely had a fond perception of Yourgrau—until the closing phase of his life, when incessant pain and the ravages of ill health tragically distorted the picture—as an unusually interesting composite of several parts: A gregarious individual, of muscular physique, attentive to sartorial detail, and very cosmopolitan in outlook and behavior, whose company was always in demand—not least on account of his endearing talent for infusing humor and *Gemütlichkeit* into nearly every situation and for leaving his visitors or hosts with a mild feeling of . . . well, euphoria; a polymath, with a special predilection for theoretical physics; an academic who placed work and scholarship on a pedestal and could marshal an imposing mental discipline to strictly apportion his indefatigible energy between work and play, without allowing the one compartment to interfere with the other.

These were some of the behavioral traits which, by reason of their high visibility, determined to a large degree Yourgrau's public image. But of far greater significance in interpersonal relationships, and doubtlessly fateful for the course of his early life, were two deeper-lying traits, ones very indicative of Yourgrau's strength of character: first, his uncompromising commitment to a few matters of principle and, second, his action-oriented humanitarian concern in the concrete for disadvantaged members of society and any fellow being who had been wronged in any way.

The saying, "Man is the sum of his experiences," contains in the case of Wolfgang (Heinrich Joachim) Yourgrau more than an element of truth. Born on November 16, 1908 in either Berlin or Katowice (Upper Silesia)—apparently nobody knows which for sure—the young Wolfgang's life almost immediately became beset by contradictions and trauma that were to leave an indelible scar on his developing psyche. At the root of the problem were the peculiarities surrounding his parents:

First, there was his mother, born as Selma Federmann in the industrial German city of Katowice—which transferred to Polish rule only in 1921. Consistently regarding herself as Jewish, she nonetheless was, at the time of her marriage, not a serious practitioner of Judaism. More pertinently, her primary identification was not with Jewry but with Germany, in which country the Federmanns—a predominantly Jewish family with major Catholic and Protestant admixtures through intermarriage—have been rooted for centuries. "My mother," according to Wolfgang Yourgrau, "would have said 'I am Jewish, but first of all I am a German.' In this attitude she was no exception. For, do not forget that German Jews, on the whole, were dreadfully nationalistic in outlook."[3]

Consider next the father, Josef (?) Yourgrau: A devout Catholic and patriotic Belgian, a doctor of organic chemistry domiciled in Brussels. Fluent in several languages, besides his native French, his knowledge of German was nevertheless only marginal. In fact, he harbored an aversion to everything German—a sentiment that was to become intensified under the German occupation of Belgium during the First World War.

Not surprisingly, the marriage between these partners was of short duration, ending in divorce (in the German city of Hanover) when Wolfgang was only one year old. Although he was put in custody of his mother, Wolfgang's father retained the right to have his son visit him in Brussels on a regular basis.

From that moment on, and throughout childhood and adolescence, Wolfgang remained torn, in mind and soul, between his parents and between the disparate worlds they represented, his search for identity allegorically accentuated by his frequent railroad trips from Berlin to Brussels and back.*

Particularly agonizing was the question of national allegiance that arose in this context. The empathy Wolfgang felt towards his father, coupled with their exclusive use of French as a medium of communication, translated into an admiration for the "Belgian character and the Belgian way of life."[3] On the other hand, he was no less attached to his doting German-speaking mother—although "never quite forgiving her for the divorce."[3] Under her imposing influence and through the inculcation provided by his German education, from Kindergarten through University, Yourgrau's German citizenship evolved from a *de jure* fact to its acceptance also in reality. (The process of accommodation ultimately ripened into a profound respect for the best in German literature, music, and science—which, during the emotional debates of World War II, Yourgrau was staunchly to defend in the face of opposition that sought to equate Germany with Hitler.†)

World War I, which trapped Wolfgang with his mother in Berlin, was a period of poignant anxiety, in the first place because it saw Belgium—which wished to

*It was on one of these trips, in the early thirties, that Yourgrau suffered a disastrous accident. In his own words,

> A fire broke out on the night train to Berlin. For miles and miles the train [raced] through Belgium and . . . Germany like a burning torch. Passengers suffered carbon monoxide poisoning and various bodily injuries. I . . . jumped, almost unconscious, through the window and finally [ended up] in the *Gewerbekrankenhaus* in Berlin, where I was treated . . . for 4½ months.[4]

Serious as it was, this misfortune produced one bonus for the student Yourgrau—in the form of solicitous regular visits by Albert Einstein and his wife for the duration of his confinement.[4]

†In an allusion to that period, Yourgrau recalled in 1975:

> We made a very big distinction between Germans and Nazis. The Germans are the people who produced Beethoven and Mozart and some of the greatest geniuses of all times . . . the greatest writers and composers, Jews and Gentiles. . . . That Germany used to be called the most cultured country, next to France, in the world. You don't shrug off the culture of centuries because it was interrupted by an interval of Nazis, tragic and far-reaching as it was.[3]

remain neutral—overrun by German forces, striking out for France. "When Germany celebrated victories," Yourgrau later recalled,[3] "I felt awful because I knew that my father," who was serving with the Belgian army, "was on the losing side."

The military conflict also gave Wolfgang a first-hand look at the suffering that accompanies war. As early as the end of 1915, Berlin was experiencing critical food shortages. Wolfgang's mother, alarmed by his severe state of malnutrition—which gave rise to tuberculosis of his left lung—arranged for the boy to go to a location near the French border, where her brother, a medical doctor, was in charge of an army hospital. Food in the hospital was better and more plentiful than in Berlin. But the other aspects of life were not an improvement. As told later by Yourgrau,

> The hospital was short of manpower, so I helped out as much as I could. Every night we went to the railroad station to collect wounded German soldiers arriving from the western front. I was taught, at the age of eight—and I still have nightmares as a result—how to treat people who were screaming from pain and dying and I learnt how to assist my uncle in administering palliative morphine injections to soldiers who were beyond repair.[3]

Having completed his *Abitur* at the Werner-Siemens Realgymnasium of Berlin, Yourgrau, at the age of 17, entered the von Humboldt University of that city. Ahead lay "wonderful years full of excitement." With celebrated physicists such as Max von Laue, Albert Einstein, and Erwin Schrödinger—not to mention luminaries in other areas—gracing its faculty, Berlin University was at that time one of the leading institutions of learning in the world. Yourgrau reveled in this heady atmosphere, sampling widely from the rich intellectual menu the curriculum had to offer—from abstruse lectures delivered by Einstein to seminars led by Wolfgang Köhler (of *Gestalt* psychology fame). Coming from a family (on his mother's side) where every second "uncle" was a medical doctor, Yourgrau early acquired a lasting interest in medicine. At the University of Berlin, this interest—perhaps through the example provided by Schrödinger, whom he in a sense idolized—led Yourgrau also into the area of biological science. Philosophy and then theoretical physics became, however, his main subjects. Serving first as a tutor in philosophy and subsequently as an assistant to Schrödinger, he completed his studies in 1932, earning the Dr. phil. degree magna cum laude, with a dissertation entitled "Investigations on Color Blindness—A Physical Inquiry."

Outside the walls of Berlin University, Yourgrau concurrently underwent a different, but no less decisive, educational process.

The late 1920s, as Yourgrau was embarking on his university studies, witnessed the apparent recovery of the Weimar Republic from the economic and political crises of the postwar era. It was a time of rising hopes, but not for long: Soon after 1929 a worldwide economic depression ushered in mass unemployment and business failures, which in turn caused social and political tensions to increase at an ominous pace. Perhaps nowhere in Germany were the unsettling effects of these developments more visible than in its capital. Paradoxically, however, they coincided with what has been described as Berlin's "Golden Age," when this metropolis came to occupy a leadership role not only in science (as epitomized by von Humboldt University), but also in the worlds of theater, filmmaking, literature, music, and other creative pursuits. Berlin during those years was alive with an enormous intellectual excitement and glorying in the presence of artists, writers, and scholars who flocked from all parts of Europe and further afield to share in the city's flourishing intellectual and creative life.

Yourgrau was enthralled and swept along by this ferment, becoming very much part of the prevailing *Zeitgeist*. A well-trained and rather accomplished cellist himself,* he and his mother, both "passionately addicted to chamber music,"[4] rarely failed to take front-row seats at the concerts conducted by artists such as the highly acclaimed Michael Taube. As a member of the Berlin PEN-Club he also had the opportunity to meet leading playwrights, editors, essayists, and novelists of the day and, on occasion, to become actively involved in the direction and production of plays. It was in those surroundings that Yourgrau made the acquaintance of the famed novelist and dramatist Arnold Zweig, who was later to become his collaborator in a journalistic venture (see below).

Into the crowd of creative and "beautiful" people Yourgrau fitted with ease, not only by temperament, but also because the means were at hand. In fact, his father and mother, both at that time being wealthy, vied with materialistic tokens for the affection of "their only child, their *Bube*," to the point where, Yourgrau recalled, "I lived like a foppish snob in an elegant three-room apartment, had a man servant (!), owned stylish clothes and [drove] a Fiat."[4]

Characteristically, all this did not blunt Yourgrau's perception for the other face of Berlin: A world of sadness and decay, of joblessness and despair—existing side by side with genius and decadence, with the glitter of packed opera houses and concert halls. His determination—despite his own pampered lifestyle, or perhaps because of it—to help rectify in some measure this glaring social imbalance, proved in the end to be his tragic undoing.

A year before entering university, while still a gymnasium student, Yourgrau—sensitized to suffering by his early experience in the army hospital and outraged by the poverty and misery he saw among some of his school mates—joined the *Sozialistische Arbeiterpartei* (SAP), a minor left-wing offshoot of the German Social Democratic Party (SPD).†

The SAP, while shying away from the Communist Party (KPD) on its left, was increasingly to fight the rising Nazi Party on its far right—in concert with the SPD

*On occasion Yourgrau took over from the regular cellist "in a quartet that met on Friday evenings in the Pragerstrasse," and which included Einstein—"whose admirable knowledge of musicology," Yourgrau was moved to observe, "must not be confounded with his proficiency as a violinist."[4]

†By an interesting turn of the wheel, Yourgrau, partly at the suggestion of Walter Berendsohn (see later), was in 1970 to be the only U.S. citizen (among five individuals) to nominate Willy Brandt—then leader of the SPD and Chancellor of the German Federal Republic, but already in 1933 active in his party's affairs—for the 1971 Nobel Peace Prize.[5]

and KPD. "I was member No. 7 of the SAP, whose leader was Max Seydewitz,"* Yourgrau in 1975 explained.[3]

> In my college years I went out on the stump for our party, holding speeches, and recruiting new members. Our party was striving after a socialist state of the kind you have in Britain today, in which the coal, steel, and other important industries would be nationalized. The people who joined my party were on the whole not super intellectuals. Some of those did voluntarily join our ranks, but they were not the people I addressed. I spoke to the havenots, the unemployed, the powerless—the very people the Nazis were first successfully to convert to their side. (Only much later Hitler was to gain support among the wealthy—who were, after all, wary of the word "Socialist" in the title of his party.)

It was understandable under these circumstances that identification of German Jewry—"which was among the wealthiest and most educated in the world"—with the aims of the SAP remained confined to a small fraction of idealists of the type that Yourgrau represented.

Increasingly, Yourgrau found himself on a collision course with the Nazi Party, which after 1929 was rapidly gaining mass support and becoming more openly anti-Semitic. Hitler and his supporters, more than social ills, now are perceived as the real enemy. "I often heard Hitler speak," Yourgrau later recalled.[3] Everything he said and wrote was evil . . . far beyond mere political opposition. It was individuals he wanted to destroy, the opposition: Excise it, kill it, put it into concentration camps." Taking time off from his studies, with the permission of a sympathetic University Rector, Yourgrau traveled for months from city to city, making speeches against Hitler and drumming up support for SAP politicians who sought to run for office in the upcoming *Reichstag* elections.

Adolf Hitler legally took office as *Reichskanzler* on 30 January 1933. In the elections of March 5, the Nazis and Nationalists won a bare majority of *Reichstag* seats. On March 23, the Enabling Act—opposed only by the SPD and the disbarred KPD—conferred on Hitler dictatorial powers. The Third Reich was launched.

*Max Seydewitz, born 19 December 1892; founder and first Chairman of the SAP; in exile in Norway and Sweden, 1933–1945; Prime Minister of Saxony, 1947–1952. A sentimental reunion took place between Yourgrau and the aging Seydewitz in Dresden, on the occasion of the Einstein celebrations held in the GDR in March of 1979. Reporting on this meeting, Seydewitz wrote[6]

> Ich habe ihn bei seinem Besuch bei mir wiedererkannt, und alles, was er mir über sein Leben in der Zeit vor der Errichtung der Naziherrschaft erzählte, stimmte voll und ganz mit meiner Erinnerung an ihn überein. Er war damals ein Student aus gutsituiertem bürgerlichen Hause, der sich aus innerer Überzeugung der klassenbewussten Arbeiterbewegung angeschlossen hatte. Alles was er mir über sein politisches Engagement aus der Zeit bis Anfang 1933 erzählte, entsprach der Wahrheit, was ich an vielen Einzelheiten seiner Erzählung festellte, die nur er und ich aus der damaligen Zeit wissen konnte. Was er nach 1933 getan hat, kann ich nicht wissen, weil sich unsere Wege bis 1979 nicht wieder gekreuzt haben. Sicher aber ist, dass er aus Deutschland als Verfolgter des Naziregimes flüchten musste und von seinem früheren Vermögen alles oder zumindest das Meiste verloren hat.

For Yourgrau, the moment of truth arrived a week later, on April 1, *boykott Tag,** in an event that was forever to remain etched on his mind.

Citing basically from Yourgrau's oral memoirs[3] and a letter[7]† he wrote to Walter Berendsohn‡ in July, 1947, this is what happened:

> I was sitting in a *Lokal* on Kaiserallee, opposite Café Josty, where I used to have breakfast, when a group of troopers from *Sturmabteilung* Wilmersdorf entered the establishment. I heard one of the men say, "That's him." I continued pretending to peruse the *Völkischer Beobachter*—the Nazi daily which at that time was almost compulsory reading—but I knew what was coming. Among the group was a lad whom my mother, out of pity, once took into our home because he was jobless and peniless—a member of the *Lumpen Proletariat.* We fed and clothed him. First he joined the Communists, now he was with the Nazis. As the group was approaching my table, I couldn't help noticing that he was wearing my sweater and riding boots.
>
> Passing my table, one trooper turned around and swung his fist at me, breaking my left cheekbone—which, incidentally, never completely healed. I fell flat on the floor, bleeding like a pig. They thought I was finished. But they did not reckon with the fact that I was an amateur boxer—who represented my University in the middle-weight class, besides being very active in tennis and hockey. My training was that when you go down, you mustn't stay down. So, almost as an automatic reflex, I picked myself up, and in a well-aimed professional blow, knocked in the front teeth of my assailant. He fell through the glass door of the *Lokal,* cut all over his hands and face by the shattered glass, and landed on the sidewalk on all fours.
>
> He shouted back, "That Socialist swine, he tried to kill me!" So, from inside the *Lokal,* his companions got up, grabbed the marble-top tables of the place and battered me with these objects and their clubs until I passed out unconscious on the floor.
>
> When I came to, a *Schupo* (policeman)—who knew me from some lectures that I gave at an evening school for disadvantaged people—was at my side. He advised me, "Do not stay around. They will return to kill you. You must disappear!" (Unlike the feared SA, or "Brown Shirts," the German police were not—yet—an instrument of the Nazis.) At his behest, a medic, who accompanied my attackers but then stayed behind, applied a makeshift bandage to my wounds, whispering in my ear that he was unspeakably sorry about it all. I was really moved, despite everything. But when he had completed his Samaritan deed, he opened his hand and demanded one *Reichsmark*—that being what the bandage had cost him! I paid the Mark and was once again instructed how singular were the workings of a Nazi mind.

*Day on which persecution of Jews began in Germany, with national boycott of all Jewish businesses and professions by SA pickets.

†Quoted with kind permission of Dr. Werner Berthold acting on behalf of the Deutsche Bibliothek, Frankfurt am Main, GFR.

‡Walter Berendsohn, born 10 September 1884 in Hamburg, Germany; professor, specializing in Scandinavian literature at the University of Hamburg, 1926–1933; left Germany in 1933; in 1947, at the time of Yourgrau's correspondence, guest professor of literary history at the University of Stockholm.

Hastily leaving the *Lokal,* I roamed around in a daze for about half an hour, trying to figure out where to go. Certainly I could not return to my apartment, since that is where the stormtroopers would be looking for me. Then I heard someone call out to me from a passing chauffeur-driven car—an elegant *Horch,* the favorite vehicle of the German upper class. It was a female acquaintance of earlier days, but now, as I was about to learn, a gynecologist and a big shot in the Nazi circles. Appalled by my appearance, and having quickly sized up the situation from my brief explanation, she matter-of-factly told me to come with her. I was again bleeding profusely and ached all over. Feeling lousy and defeated, I could not care anymore. I got in and slumped down.

She took me to the suburb of Grünau, where, it turned out, she was running a sanatorium, to which female patients of hers (mostly abortion cases) were sent for recuperation. Here I spent quite a long time recovering from my nearly fatal injuries. The possibility of infection—these were the days before antibiotics—was our biggest worry. But my earlier physical conditioning and the good care showered on me by nurses and cooks pulled me through. Aside from the chauffeur, I was the only male in this *Naziheim* for lonely and often victimized patients. When I had sufficiently recovered, I was pressured evenings to play the piano, accompanying these unfortunate ladies in their singing of sentimental songs. I also had to take them for boat rides on the nearby river lake. In short, it was an atmosphere that defies description. I had, of course, every reason to be grateful to the physician, who perhaps saved my life. We separated, after about three months, in friendship, but painfully avoiding any political discussions.

She dropped me off in the middle of town, where I mingled with the bustling crowds. The next several months, I led a dreamlike existence: No two nights sleeping in the same location, always on the guard, but still involved in the publishing of pamphlets from a basement of Kaiserallee.

In late December, a young *Schupo,* forging the signature of his superior, furnished me with a *Sichtvermerk,* thus finally enabling me to escape from Germany.

With money my father had supplied, I first went to Poland; every other country refused me entry, apparently because I was taken for a Communist. Here, in Poland, I addressed antifascist meetings and joined a troupe of politically conscious German actors. My job was that of "dramatic producer and stage manager." One day, after a performance, I was arrested, handcuffed and summarily thrown in jail. I was released only after the Polish representative at the League of Nations directly intervened with Pilsudski* on my behalf. Moreover, I had to agree to leave Poland within 24 hours.

And so I took off for Riga, the capital of Latvia, where I unwisely gave a lecture against fascism in a workers' club. I received friendly advice to make my exit from Latvia within three days.

At the end of my rope, I decided once more to enter Poland, where my wife† —an ophthalmic surgeon and a former assistant in the eye department of Neuköllner *Krankenhaus*—stayed behind in order to undergo an operation.

I succeeded in persuading the British Consul in Riga to issue me a visa, although I was already on the blacklist. In my distress I also turned to Robert Weltsch (editor, *Jüdische Weltrundschau*), who arranged the issuing to me of a certificate allowing me to settle in the far-off British Mandate of Palestine.

*Józef Pilsudski, Polish General and Socialist leader, a virtual dictator from 1926 to his death in 1935.
†The reference here is to Yourgrau's first wife.

> Earlier, I had a conversation with Friedrich Wolf* in Warsaw, where he was staging *Der gelben Fleck.* He unconditionally wanted me to accompany him to Moscow. This would have suited my wife, qualified to practice medicine and a Russian by birth, very well indeed. I thanked Wolf for his offer, but preferred not to go to Russia. With a view of joining my father, I then traveled to Brussels. But the Belgians, already furious about my Prussian citizenship, demanded that I first do military duty—in the Belgian Congo—for several years before they would grant me Belgian citizenship. Thus, in the end, Palestine appeared to me to be the best solution after all.

Stepping ashore in the port city of Haifa in the British Mandate of Palestine some time during 1934 or later, Yourgrau was standing on the threshold of an era that had further tests for his mettle in store.

Appointed first as a lecturer with the educational and cultural division of the *Histadrut* (Jewish Federation of Labor), Yourgrau lectured before various audiences on political themes and, later, also on Islamic religion and philosophy—subjects, which, because of the proximity of Arab population groups, increasingly evoked his scholarly curiosity.

At the same time he contributed essays and book reviews in the Hebrew press and assisted his Zionist friend Robert Weltsch on his *Mitteilungsblatt.*

From the *Histadrut,* Yourgrau went to an equivalent position with the HOGOA *(Hitachdut Oley Germania ve Oley Austria),* a society for immigrants from Germany and Austria, where he remained occupied for several years with lectures and seminars on political and cultural themes.

The move to HOGOA marked Yourgrau's increasing involvement with the vexing problems surrounding the socioeconomic and political assimilation of the new, mainly German-speaking, immigrants—which counted about 50,000 around 1942—into the *Yishuv,* i.e., the Jewish community of Palestine. These problems—the nature and origins of which Hans-Albert Walter, for example, has carefully analyzed in his monumental work on German Exile literature[8]—led, around 1941–1942, to the formation of the *Aliya Chadasha* (literally, New Immigration), a partly cultural, partly political organization from the ranks of the HOGOA. Yourgrau was one of the founders of this party and, in fact, directed its first campaign in the city of Haifa.

On another internal question, that of the political position of the Arab population within Palestine, Yourgrau came to hold a very definite, if aberrant opinion. Speaking to a reporter in 1965, Yourgrau explained.

> While . . . in Jerusalem, I belonged to a political group which was led by the great American Dr. Magnes† and the French philosopher Martin Buber. Our group was against the dissection of the Jewish state from the Arab sphere. . . . We tried

*Friedrich Wolf (1888–1953), German physician, diplomat, poet, and revolutionary dramatist; obliged to leave Germany in 1933, he spent the war years in the USSR.

†Judah L. Magnes (1877–1948), U.S.-born rabbi and Zionist leader; Chancellor (1925–1935) and first President (1935–1948) of the Hebrew University in Jerusalem.

> arrive at a Judeo-Arab state in which both the ethnic groups would be equally represented. We did not want the Jews to be surrounded by a hostile Arab sea. But politically we lost.[9]*

Viewing himself, in common with many others in the German *Aliyah*, as an exile from Europe, rather than a settler in Palestine, Yourgrau's interest in domestic issues was, however, overshadowed by his worries about the unfolding and world-shaking foreign developments. Nazism, its devastation of the cultural values in which his family had been rooted for centuries and then, starting in September of 1939, the daily progress of World War II remained foremost in his mind. He yearned for a chance to meaningfully continue his fight against fascism and make his own contribution to the war against the Axis powers—which now (1941–1942) was heroically being waged on the high seas, the Russian front, and in the deserts of North Africa (by the British Eighth Army).

The opportunity to translate this longing into action, at least on the ideological level, and at the same time to promote the free interchange of opinions on other pertinent issues, came in the spring of 1942 with the launching of a periodical called *Orient*.†

Especially since some confusion surrounded the birth of this publication, it might be wise to quote Yourgrau literally on this score from his earlier-mentioned letter to Berendsohn[7]:

> In April 1942 I was offered the license for a weekly newspaper in the German language in Haifa. A married couple owned the license and tried to turn it into a profit. Since the (British mandate) Government rarely granted new licenses, the offer seemed very attractive, although I had not earlier thought of publishing a newspaper. I immediately took up contact with Arnold [Zweig],‡ an old friend of mine . . . , and asked him for advice. Arnold was not very keen to get himself involved in any adventure. But when I explained that all he had to do was to supply one article per week, and that he would neither materially nor editorially be responsible for the paper, he reckoned he would give it a try. . . . And so I bought the license and became the sole publisher and editor of the paper. Not having any money, I had to borrow from friends and acquaintances. Because the license was not in my name (licenses in Palestine could not officially be "bought"

*To the extent that this implies an opposition to the State of Israel, historical veracity demands that we mention here the evolution of Yourgrau's views on this point in the light of subsequent events: "I was against its establishment, as I also told [Chaim] Weizmann and [David] Ben-Gurion," Yourgrau declared in 1975.[3] "But now that Israel has demonstrated its potential . . . its ability to survive surrounded by an Arab sea, I must take off my hat in admiration. I am convinced Israel is here to stay. It should never be allowed to disappear."

†Complete sets of issues of *Orient* are available, among others, in the Deutsche Bibliothek, Frankfurt am Main (GFR) and at the Akademie der Kunste der DDR in East Berlin. An extensive dissertation on *Orient* appears in the cited work by H.-A. Walter.[8] Reference 10 contains an editorial from *Orient*, while a complete bibliography of the journal can be found in Ref. 11. NOTE (added in proof): *Orient* has now appeared as Vols. 14 and 15 of *Exilliterator*, H.-A. Walter and W. Berthold, eds. (Gerstenberg Verlag, Hildesheim, 1982).

‡Arnold Zweig (1887–1968), German-born (Silesia) novelist and dramatist; characterized by E. Loewy[10] as a "socialist and humanistic Zionist"; denationalized, 1933; in exile in Palestine, 1933–1948; 1958 recipient of the International Lenin Peace prize.

§A. Roth and L. Briszk.

> or "sold"), there appeared on the back of the periodical two obscure names,§ those of the individuals who gave me the license but who had neither *de jure* nor *de facto* even the slightest to do with the weekly. I repeat: I was the sole publisher and editor* and, comically, also acted vis-à-vis the Government in this dual capacity—although the license was not in my name! I, however, informed the Government of all this and received from the then Information Officer (Director of the public relations office) the official blessing. . . . The title "Orient" I had to retain, although I had no intention of publishing a journal for Orientalists. But the license was tied to this title. The periodical had little to do with the Orient and was given from the very start a pronounced antifascist character. I had *no* connection with *any* party, *any* government, or *any* movement as usually constituted. I wanted to build up an independent weekly periodical.

The premier issue of *Orient* (under its new management)† came off the press in Haifa on 10 April 1942. In the first of what was to become his regular lead articles,‡ entitled *Auftakt*, Yourgrau outlined the program of his weekly as follows:

> *Orient* is the name of this journal. With this title we associate a program: It stands not only for a geographical concept, but for an intellectual signpost, as well. The yearning for Palestine elevated the European Jews above their local basis, it broadened their intellectual horizon in all areas with the addition of a new realization. Confronted with reality at the end of the Zionist goal, we witness the return of the pendulum: The horizon becomes narrow, the idealist becomes bourgeois, the idea degenerates into a soil patriotism, into provincialism, the blinkers obscure the intellectual, cultural, and political fields of view.
>
> We live in the Mid-East, Arab nations are our neighbors, religions with alien ethical and moral content impinge upon us, and India casts its giant shadow over our small land, Palestine. Very real is the danger for acquiring a distorted optics. It is the dream of the suppressed to attain unlimited power; it is their tragic fate to lose sight of moderation, to interpret freedom falsely. There where the instincts of the masses are daily aroused, where the street influences the political leadership, in these times of aimlessness and contradiction—under such circumstances there will always be but a few that are capable of advocating reason and objective appraisal of political events. This Palestine is our destiny. Few have striven for it, many have opposed it. A homeland one can, however, not decree, it can only be built through love and reliance and ingenuity. Along with this must go the courage to fight for the goal, at any price, with responsibility and determination, and often risking unpopularity. Palestine is an integral part of the Mid-East. This insight must someday gain acceptance if history is not to condemn us as dreamers and visionaries who failed to grasp a big opportunity.
>
> The present journal will be an independent one. The deeper reason for us not having a public opinion resides in the circumstance that . . . the involvement of public interest smothers every attempt at intellectual independence. Political parties, organizations, and scientific institutions all strive in jealous competition to bring about the near totalitarian control of public life. Conversations with

*This point is also emphasized in correspondence exchanged in 1942 between Arnold Zweig and Christopher Holme,[12] in which the former wrote: " . . . benutze ich die Gelegenheit gern, zu versichern, dass die gesamte redaktionelle, verlegerische und sogar drucktechnische Arbeit am *Orient* ausschliesslich in den Händen von Herrn Dr. Yourgrau lag und liegt."

†Formally designated as No. 2 of Vol. 3, due to the legalities explained above.

‡Two of which, amusingly, appeared under the pseudonym "Tuggelin," a name Yourgrau was later to give to one of his sons.

young people reveal most clearly the extent to which the process of intellectual standardization has progressed. This gives rise to frightening perspectives for the character formation of young people—and not only of them. Under such conditions the requirements of a cultural ideal can never be fulfilled. Our journal will provide the platform for all those who are ready to make pertinent contributions to the problems of this land and to the large questions of our time.

Our journal appears in the German language. Concerning this fact we ourselves can deliver various comments and the sharpest polemics. This newspaper is meant to reach the reader for whom the command of the Hebrew language will for the duration of the War remain an unattainable goal. Our appeal extends only to this circle. Intellectual discussions with *these* friends who are endeavoring to appreciate our viewpoint interest us. Against undistinguished reactions of a few intolerant fellow citizens we shall be immune.

This journal is not an organ for emigrant literature. Our platform stands at the disposal of old Zionist interests in no way different than it does to the immigrant who, even in this country, is struggling for clarification of his own political convictions. One can never eliminate thoughts and ideologies simply by declaring them taboo. We are determined that only good manners and honorable discussions will serve as criteria on this score.

Every Fascist stirring, every attempt to restrict the right to freedom of opinion—which is part of mankind's eternal heritage—against this we declare an uncompromising war!

We dispose in our community over a large number of pedagogues, writers, scientists, and political economists; experience teaches us anew each day that we can engage only a fraction of these intellectuals in the production process. Added to this material deprivation, there are intellectual and psychological stresses for which no escape valve exists. For the here involved legion of known and unknown friends in need we intend to be a mouthpiece, an expression for their legitimate demands and wishes. The more differentiable a person is, the more defenseless he becomes against the harshness and merciless methods of the struggle for existence. There must come about something like a community of intellectuals, otherwise we shall be responsible for, and accessories to, the decline and downfall of precious human beings.

Reminiscing about the "line" the journal took, Yourgrau later had this to say[7]:

We were, no doubt, very much to the left, without though fostering political affiliations. . . . while Arnold Zweig concentrated on the European situation, I attempted to analyze the situation in Palestine and draw attention to the dangers of a narrow nationalism and a self-righteous attitude. My model was the magnificent Ossietzki*—and nothing flattered me more than when readers assured me in their correspondence that I was a worthy disciple of this accomplished journalist."

[Yourgrau quickly added, "Be assured I have in this connection no illusion. I [was] too much of an honest admirer of Ossietzki, to succumb to such megalomania."]

*Carl von Ossietzki (1888–1938), German journalist and pacifist; became editor in 1927 of *Die Weltbühne*, a liberal political weekly; herein unmasked the secret preparations for war by the leaders of the *Reichswehr*; awarded the 1935 Nobel Prize for Peace—which was interpreted as an expression of worldwide censure of Nazism.

Apart from Zweig, who was responsible for themes of a cultural and political nature, regular contributors of name to *Orient* were: Else Lasker-Schüler* (who often contributed poems), Franz Goldstein (who took care of literary criticism), Louis Fürnberg,† and the Dutchman Pierre van Paassen. For the rest, Yourgrau recalled,[7] "I allowed medical doctors, students, young journalists, one or two talented poets, some lawyers and earlier jurists, etc. to say what they felt like saying."

The consistently antifascist attitude of *Orient* was particularly evident in its steadfast debunking—even after the mass murder of the Jews became known—of some of the simplifications and generalizations concerning Germany that characterized part of the public opinion in Palestine, as elsewhere, at that time. To *Orient* it remained axiomatic that the Nazi regime, not the Germans, was the enemy, inasmuch as it viewed the latter as the victims, rather than the accomplices, of the former.

Although it was Arnold Zweig who initiated the debate in *Orient* on the "Hitler-is-not-Germany" thesis (in a series of articles entitled *Antigermanismus*), it is Yourgrau's subsequent editorial on the same matter which, because of his unique background, is of special interest.

To quote H.-A. Walter,[8]‡

> Yourgrau's presentation, taking the form of an inquiry into the diverse theories of fascism, refuted, first of all, the *Volkscharakter* thesis, with the comment that fascism, as everyone knows, originated in Italy, but that, nonetheless, the Italian nation had quite different "attributes" from the German nation. Thereupon he embraced the view that fascism was a product of the German *Geistesgeschichte*. The latter he conceded, contained, among others, philosophical systems and models, "whose authors one could regard as intellectual fathers of German totalitarianism"; the Nazis did, however, by no means succeed because a Fichte, Klages, Nietzsche, or Spengler existed. As the prime intellectual breeding ground of fascism Yourgrau identified rather the *Ungleichzeitigkeit* (Bloch) that since the Enlightenment had become increasingly visible in all walks of life—i.e., the gap between individual awareness and the ever more rapidly unfolding politicoeconomic, technical, and so forth, developments—which it had failed to digest. The uncertainty arising from this gap had (according to Yourgrau) penetrated deeply into the consciousness of everyone, but the German nation had been affected by it more than others.

Yourgrau went on to amplify the latter assertion and concluded, like Zweig, *but* in a more generalized form:

> In the light of all this, I feel justified in claiming that the phenomenon of fascism can be demonstrated as latent in more or less recognizable fashion in nearly all nations and peoples on earth. . . . Accordingly, to defeat Hitler is not yet to liquidate fascism. (*Orient*, 29 January 1943.)

*Else Lasker-Schüler (1869–1945), German-born lyric poetess and dramatist; one of the most important representatives of expressionism; in exile in Jerusalem from 1937 until her death.[10]

†Louis Fürnberg (1909–1957), Czechoslovakian-born lyric poet, narrative writer, and journalist; exile in Palestine, 1941–1946.[10]

‡We thank Dr. Uwe Schweikert, representing J. B. Metzlersche Verlagsbuchhandlung, Stuttgart, for kind permission to quote from Ref. 8 as well as Ref. 10.

The conviction embodied here is one which Yourgrau was frequently to reiterate in later years.

The subscribers to *Orient* typically were:

> young immigrants from Western Europe (German, Austrian, and Czechoslovakian), students, aging German academics, artists, and young workers who disapproved of the conditions inside the *Yishuv* and who did not feel themselves spiritually at home in the *Histadrut*.[7]

The journal suffered, amongst others, from a permanent lack of funding, and much improvisation was required to keep it alive: "I never knew, not even a week ahead of time, where I would get the money for the printer," Yourgrau declared.[7] "For this reason some issues were half printed, half stenciled. . . . I acquired paper, helped with the setting and printing, took care of dispatching, and often personally delivered copies to the newsstands or even to the subscribers—in other words, it was all very laborious and very romantic."

Despite this lack of resources, *Orient* was not destined to remain obscure. Public reaction to its appearance was intense and swift in developing.

"Already the first edition," as Yourgrau told it,[7]

> provoked opposition. Paradoxically, not only the fanatical Hebraists and old Zionists opposed us, but also wide sectors of the new European immigrants. The organization of German-speaking immigrants, the HOGOA, as well as the political party of the *Aliyah Chadasha*, which I helped to found, personally resented my paper. People started a boycott, wrote letters, and wildly agitated against us. They called us Communists, traitors to Zionism, and cafe literati. We were reproached for being a demoralizing influence, uprooted intellectuals, and typical *Weltbühne* people; in short, there was no party or organization in the country who accepted us.

In retrospect, at least, the furor was to be expected. First there was the obvious observation, already noted by Yourgrau[7]: "The Zionists lacked a positive attitude towards my undertaking; to the extreme Socialists I was not radical enough; to the language fanatics publication through the German medium was hated."* As for the "paradoxical" opposition to *Orient* from the side of the German-speaking "establishment," the following remarks of H.-A. Walter are enlightening:

> The *Orient* developed its political conceptions—and its attitude towards the assimilation problem [for the German immigrants]—from a radical democratic perspective. In this it went a significant step beyond the liberalism of the *Aliya Chadasha*, without positively disavowing it. The *Aliya Chadasha*, to be sure, after an initial silence, disowned the *Orient*. In view of the above, this was only to be expected. For people of the kind the German *Aliya* were on the whole, who basically had little interest in politics and were hardly active in political organizations, for people whose lifestyle in Germany had existed above all in being

*Cf., however, H.-A. Walter,[8] who rules out language as a motive—on the grounds that two German dailies and one German weekly which preceded *Orient* were immune against attacks.

> inconspicuous and keeping the peace, for such people *Orient* had to appear as too "radical," i.e., too political and too principle-oriented, also as too "loud" and too "ostentatious," for them to regard it as their mouthpiece. This was the case the more because the periodical tied the concerns of the *Aliya* to other fundamental questions and viewed them paradigmatically and in a much broader context than the *Aliya* and their representatives cared for.[8]

The antagonism from the side of the HOGOA went so far that Yourgrau was forced to relinquish his position as secretary of the *Aliya Chadasha*:

> The executive Committee of the party in Haifa issued me an ultimatum: Either I continue to organize and represent the party, promptly resigning the editorship (or at least writing under a pseudonym), or I discontinue my activities on behalf of the party. I perceived this ultimatum as improper and tactless and forthwith resigned my political office.[7]

The boycott campaign against advertisers, printers, and distributors associated with *Orient*—and which was effective in keeping the circulation figure for the journal below a thousand—was now building towards a climax. In Haifa the manuscript copy for *Orient* was being passed around from one printer to another because of threats. When one night the bookbinding shop was put on fire, Yourgrau decided to leave Haifa and continue operations in Jerusalem.

"Here in Jerusalem," Yourgrau later recalled,[7]

> I found a well qualified printer, who neatly brought out our badly roughed-up mouthpiece. One evening—it was February 2, 8:30 p.m., 1943—I received a call informing me that our printing shop was destroyed by a bomb. Twenty minutes later I found a scene which was reminiscent of a medieval destruction technique. Several bombs must have been placed at various points inside the shop. The detonation was so violent that the people in the neighborhood mistook it for an aerial bombardment by the Nazis and tumbled into the air raid cellar.
>
> When I immediately after this militant attempt at persuasion wanted to publish the paper with another printer, I found to my dismay, that not only Jewish but also Arab printers had been warned by "unknown" that they were liable to suffer the same outcome as the printer who had been destroyed. After long negotiations, I finally found a Christian-Arab printer who was prepared to print at least the cover of my closing issue, naturally for an enormously high price, which we had no choice but to pay. The contents of the issue we could only stencil—that was all. I was warned not, under any circumstances, after the bomb attempt, to publish another issue. I felt however morally compelled, despite the clearly existing danger, to let our public know why publication was suspended. . . . The Hebrew press, interestingly, failed to report the destruction of our printing premises and took, so to speak, no notice of us. We brought out our first issue in April 1942 and our last in April 1943.

We shall not join in the speculation (cf. Refs. 8 and 11) concerning the identity of those responsible for the bomb attack that put *Orient* out of existence—Yourgrau himself is reticent on this point. Neither is this the place to pass judgment on the pragmatic wisdom or otherwise of injecting the "radical" debates of *Orient* into a society whose survival was at that time being imperiled by internal stresses and geopolitical designs. (Yourgrau in retrospect referred to *Orient* as "that bold and very unwise venture of mine."[13]) What is pertinent, though, from a personality point of view, is the demonstration which the crises surrounding *Orient* provided (as did, a decade earlier, the life threatening events under incipient National Socialist rule) of Yourgrau's uncommon courage and stubborn adherence to the principle of freedom of expression, regardless of the price (which traits in interpersonal affairs had their counterpart in the tendency to speak his mind without fear of intimidation).

Yourgrau himself, in a backward look, said it most poignantly:

> I knew from the beginning that I was letting myself in for an adventure à la Don Quixote. It was clear to me from the start that I would incur tremendous debts, that I would be persecuted, and that I would in the end be all alone. All this came true, but it befell me not unexpectedly. Now you will ask me, quite justifiably: Why this all? Was it worth it? Perhaps yes. Perhaps I would today repeat the same experience, despite debts, ostracism, social boycott, loss of friends, etc. I believe that there should be from time to time such crazy types, who freely take upon themselves a Don Quixote existence, provided that they do it with eyes open, fully aware and without illusions![7]

In a lighter vein, we cannot help reflecting, from the background of personal observation, that other personality traits were in all likelihood involved in a venture such as *Orient*. Yourgrau was an action-oriented individual and an incurable optimist who wasted no time brooding on a problem or concern when a dynamic solution seemed possible. And no matter what the odds against success were, Yourgrau would approach every new day with a (nondenominational) prayer and the full expectation that a miracle would occur. The beauty of it all was that, not infrequently, it actually did!

After this "bitter experience"[7] with *Orient*, Yourgrau, now living in Jerusalem, deep in debt and without a job, was offered a position as Editor of its German Language Radio by the Public Information Office of the British Government of Palestine—"with whom," Yourgrau recalled,[7] "I was not always on the best footing," as witnessed by the fact that "one issue of *Orient* was banned by the (British) censor who was displeased by my approach to England's Middle Eastern politics."

Christopher Holme, head of the Public Information Office from 1941 to 1946, who in May of 1943 signed Yourgrau's letter of appointment, recalled more than 37 years later his first encounter and subsequent relationship with Yourgrau in the following terms:

> Wolfgang came to see me officially, with an application for newsprint for his journal . . . —the Public Information Office also managed the newsprint rationing scheme for the Palestine press. *Orient* was duly allocated a small ration. Later I remember his press being blown up . . . no doubt because they objected to his views (liberal) or language (German) or both.
>
> I remember a smallish ebullient figure whose conversation took me straight back to the Europe from which many of us felt so cut off in those war years. (I don't mean England—that was a different kind of severance!) A visit from Your-

> grau would have been an enlivening break in official routine. We often talked physics, in which as an arts man I had always a frustrated interest, and also no doubt mathematics, in which I took a serious spare-time course with a Hebrew University lecturer during the last two years of the war. But these were by no means our only topics. I still have two little books that he gave me, one on physics (Nernst), the other on politics (Richard Wagner!)[14]*

Delighted by the opportunity to resume his ideological role in the war against the Axis Powers—which was now increasingly being fought in the Mediterranean theater—Yourgrau threw himself body and soul into the German-language broadcasts which emanated daily from Jerusalem to listeners in the Middle-East and the Balkan States. "I led this department for nearly one year," Yourgrau wrote in 1947,[7]

> prepared the texts for our news bulletins, comments and plays; I spoke personally nearly every night and molded my broadcasts into a highly aggressive antifascist weapon. Hermann Vallentin from Berlin and Kathinka Küster (a frequent and valued contributor to *Orient*) assisted me successfully as announcer and script-writer, respectively.

Indeed, Yourgrau's services earned him the high praise of Christopher Holme, who wrote in a note to Yourgrau after the conclusion of his period of duty:

> I wish to convey my thanks to you and your colleague Mrs. Ginsberg [Kathinka Küster] for your unfailing devotion to duty during the time you have been with us. In my opinion, the original thought and energy which you put into your work as head of our German news service constituted an important contribution to the Allied cause.[15]

To John Peet,† a close friend of Yourgrau's and his immediate superior at Jerusalem Radio, we are indebted for an evocative portrait of Yourgrau during his stint of service with the Palestine Information Office. We can do no better than to quote *in extenso* from his communication to us:[16]‡

> In mid-1943, I finally met Wolfgang Yourgrau (after having learnt of him through reading of *Orient*). With considerable help from Arnold Zweig, whom I got to know in the meantime, I received a post as radio news editor in the official Palestine Information Office (run ultimately by the British colonial service), which provided the news bulletins for Jerusalem Radio, the Palestine Government station. Before I moved to Jerusalem, having been released from the British Forces, Arnold Zweig told me I would be meeting a friend of his who also worked in the Information Office. And on my first day there I met him. I remember him as a busy, bustling man, untidily dressed in a fairly shabby grey suit splashed with

*Quotation with kind permission of the correspondent.

†John Peet, born in London, 1915; 1939–1943, served in British Forces in Palestine; 1943–1945, News Editor, Jerusalem Radio; 1945–1950 correspondent of Reuters News Agency; moved to East Germany in 1950; there edited the English-language periodical Democratic German Report for 24 years; today a member of a far-flung team of specialized translators working on a definitive 50-volume English edition of the collected works of Marx and Engels (published jointly in Moscow, London, and New York).

‡With kind permission of the correspondent.

ash* from the cigar permanently in his mouth, his pockets bulging with pieces of paper, and always with more paper in his hands.

For the next two years I met him practically every day. We worked closely together, particularly after I had been promoted to head of the radio news department. Wolfgang was responsible for the daily news bulletins in German which went out from Jerusalem. I had to write the key news bulletins in English, and these were then circulated to the other language departments (Hebrew, Arabic, French, Greek, Polish, and German), which translated and shortened or reassorted them, according to the different listener groups.

And almost every day, as soon as a new btach of carbon-copies was distributed to the language groups, Wolfgang would dart into the room, puffing heavily on his cigar, and start interpreting the news of the day back to me, suggesting improvements in wording and presentation or providing often very acute interpretations. "Now this meeting of the Big Three in Teheran," he would start off, and then launch into a discourse about what it meant, what it might mean, what it could lead to; often he strengthened the discourses with fairly obscure historical parallels drawn from his wide reading. The discourses at first came in heavily-accented "Katzenjammer Kid" English; but when he had finally assured himself that I spoke German at least as well as he spoke English, he was relieved to fall back on German.

We soon discovered that we agreed on many subjects, and when we did not agree, at least we knew and understood what the other was talking about; and our professional relationship soon developed into personal friendship. He was, I think, pleased to discover, amongst the rather insular and limited British colonial officers who staffed the office, somebody who not only spoke fluent German, but who also had a fair grasp of European political realities and who could often cap his quotations when he started off some discussion by saying "You will remember that Marx said . . . ."

When the flow of news was slack, he would sit around in my office and talk, and drink many cups of Turkish coffee, which was hawked around the building from office to office all day long. Very often we had a quick lunch together, and talked—about the Palestine problem, which was always present, about how the war was going, and particularly about how the world would look after the war.

Free time was scant, but several times he took me to visit a friend of his, a rather impressive German-Jewish red-head,† who served us coffee and cakes and then, seated at the piano, sang us songs by Bertolt Brecht, Kurt Tucholsky, and Wedekind. It was a curiously nostalgic replant in Jerusalem of liberal-literary pre-Hitler Berlin.

One thing I remember very clearly was that, although Yourgrau could be very grim and determined, he loved to laugh and produce witticisms, at which he would laugh immoderately himself. He was annoyed at the relative weakness of his English, which made it difficult to produce cracks which everybody in the newsroom could enjoy. One of his forms of humour was to produce, with the utmost solemnity, a profound theory about almost anything, which culminated in an absurdity.

Details of his private life in Jerusalem, if I ever knew them, elude me today.

*Very much out of character for the "modern" Yourgrau, at least, who was strikingly fastidious in matters of dress; perhaps in 1943 a transient outward manifestation of his new occupation.

†Apparently Kathinka Küster.

> He was certainly not affluent, but his regular employment in the Information Office provided him with enough to live on. He was extremely and deeply concerned about the miserable living conditions of some German-Jewish intellectuals in Jerusalem who were penniless, in particular his long-time friend Else Lasker-Schüler. E. L.-S. figures today in all literary histories as an outstanding, or the outstanding, German impressionist poet; but in Jerusalem of 1944 she was a helpless, tiny, bewildered old lady, existing in a garret on the charity of a handful of friends. Wolfgang (who might well have known E. L.-S. from pre-Hitler Berlin, where she had been a noted figure in left-wing literary circles), visited her regularly, bringing food and a little cash. In several cases, where more money was needed in an emergency—doctor's bills, I believe—he asked me for help, which I was able to give. Once he took me to visit her. I suggested buying flowers, but he urged me to take some tea and sugar instead. She died, aged 76, in January 1945, and was buried at the foot of the Mount of Olives. Wolfgang was one of the very few mourners.*
>
> Otherwise I have no memory of Wolfgang's acquaintances in Jerusalem. Although he was a very politically oriented person, he appeared to have little connection with political life at the time I knew him, apart from a small and fairly ineffectual group around Dr. Magnes of the Hebrew University, which propagated a two-nation progressive Palestine state, in opposition to the main-line Zionist aim of a purely Jewish state disregarding the Arabs.
>
> One further personal memory. Although Wolfgang was only six years older than I, and my subordinate in the office, I always regarded him as far older (which he was not), and far wiser (which he probably was).

Challenging as the position at the Information Office was, Yourgrau was looking for a more active involvement in the war effort: "At the beginning of the War I reported to the British Army, desiring to join up," Yourgrau later wrote.[7]

> I was, however turned down because an officer's rank— which I wanted for obvious reasons— was not in those days immediately given to German nationals, even if they were in possession of a Palestinian pass. Consequently, I stayed with my radio, published occasionally lead articles in Hebrew and English papers, wrote book reviews—and waited for my chance.

The awaited chance came in the middle of April 1944 with Yourgrau's induction into the American Office of Strategic Services (OSS).† At that time, things were beginning to look up for the Allies: The total defeat of the Axis Army in North Africa was approaching its first anniversary and the subsequent allied invasion of the lower Italian peninsula had resulted in the formal surrender of Italy— in September of 1943.

*This vignette provides an anything-but-isolated example of Yourgrau's uncommon ability to empathize. It evokes a remark by Arnold Zweig in a letter to Walter Berendsohn: "Yourgrau was the only person in Palestine who asked himself in late 1941 on what I actually could live—boycotted as I was by the censorship, also the secret Haganahistic, of contact with the international newspapers, also those of England."[17] All this will have a familiar ring to many of Yourgrau's students and colleagues of later years.

†A United States intelligence agency established in 1942 under the jurisdiction of the Joint Chiefs of Staff. Its mission was to gather information about adversary nations and sabotage their morale and war potential. The OSS was abolished in 1945, but many of its functions were later assumed by the present Central Intelligence Agency (CIA).

But Rome and the rest of Italy was still under German control. And so, of course, were the Balkan States, not to mention the whole of Western Europe.

Responsibility for American clandestine operations in the Balkans lay with the Middle Eastern Headquarters of the OSS in Cairo—at least until it passed (in late Spring of 1944) to the Allied Command in Italy. In March 1944, the Cairo office (then headed by Col. John Toulmin) was urgently searching for suitable individuals to man its Morale Operations (MO) Branch.

The purpose of Morale Operations, according to the author R. H. Smith, an authority on the OSS,

> was to influence enemy thinking by means of "black propaganda" that would appear as though it had come from within the enemy's own ranks. MO therefore distributed forged newspapers and military orders, operated clandestine transmitters that purported to be broadcasting from within enemy territory, and began rumor campaigns.[18]

The Acting Chief of MO in Cairo, John Fistere, not being able to find qualified Germans in the Egyptian capital, traveled to Jerusalem to see whether suitable candidates for the MO vacancies could be found among the German refugees of that city. First he paid a visit to Lt. Col. Henry Hunloke, the chief security officer of Palestine. Reporting subsequently to his superior, John Fistere thus described what happened: "I told him what I wanted . . . thoroughly secure and . . . thoroughly competent persons. He combed his files and memory" and came up with "two outstanding bodies [sic] . . . Mrs. Ginsberg Küster and a Doctor Yourgrau."*[19]

After assuring himself that Yourgrau and Küster had the unqualified endorsement of their superior at the British Information Office (Christopher Holme) and also of the chief of censorship for Palestine, John Fistere proposed both individuals for service with the OSS—adding, in an interesting footnote to his report, that Yourgrau's conduct in connection with *Orient* was "an indication of the character of the man and should be viewed as a credit mark in his security investigation."†

Yourgrau (and Küster) accepted the employment offer subsequently made by OSS and served with its MO section from April 15 to the end of October 1944.

*It transpired[3] that Yourgrau was first brought to the attention of the Pentagon by the celebrated novelist Thomas Mann (1875–1955) who, after 1938, was living in the USA. Yourgrau knew Mann through his son Klaus and daughter Erika (the actress wife of the poet W. H. Auden).

†Readers who might be familiar with the remarkable career of Kathinka Ginsberg (née Küster) would be interested in what John Fistere concluded on her behalf:

> I believe it is very indicative of this body's character that she returned, under partial cover, to Berlin in 1939 or 1940, ostensibly to clean up some personal affairs, but actually to arrange for the evacuation of several hundred children from Germany. She is a woman of considerable poise and dignity, obviously well-bred, and if I am a judge, of great strength of character.[20]

Reminiscing about those six months, Yourgrau later characterized his activities as follows:[3]

> We had a team of six or eight people [notably including Kathinka Küster] with myself as editor and chief. We established broadcasting stations which carried my comments in German to the German occupation forces. Speaking like a Nazi officer, I told them that everything was lost. We tried to undermine their morale—that was the whole purpose. So, for example, I was smuggled into Turkey, then officially neutral [declaring war on Germany and Japan only in February of 1945], and established a transmitter on one of the Turkish islands. This was quite a strong station, whose signals went out to the Balkans for 6 or 8 hours every day.
>
> We also arranged to have agents dropped behind enemy lines. Their task was to blow up installations such as bridges and help organize resistance movements and supply them with money, clothes, and food.
>
> Then, for months, we prepared fake editions of the weekly *Schutzstaffel* (SS) newspaper, copies of which were then dropped by the hundreds of thousands over occupied territory. The editorials, advertisements, etc., from the first line to the last, were all written in typical Nazi idiom. Our printing presses were German and we took care to use German newsprint (bought via Switzerland). Thus, every precaution was taken to make our product appear to any of its finders like the genuine article from Berlin. But, of course, into each edition were weaved subtle demoralizing items, such as human interest stories surrounding casualty figures, which we theorized would have their effect on Nazi morale in the long run.

An achievement of the Cairo OSS-MO Branch that should be recorded is the role it played in the abduction of Major-General Karl Kreipe,* Divisional Commander of the *Wehrmacht* on the Island of Crete.†

The abduction of Kreipe was carried out by two daring young Englishmen, William Stanley Moss and Patrick Leigh-Fermor,‡ assisted by a handful of Cretan guerrillas. It ranks as *the* capture of World War II; never before or afterwards did the Allies kidnap a Nazi of such a high rank. The exciting details of the capture itself are told in a book[21] and an article[22] written by Moss and Leigh-Fermor, respectively

The rationale behind the capture of Kreipe, according to Leigh-Fermor, was as follows:[22]

> The original plan was to capture General Müller, commander of the XXII Bremen Panzer Grenadier Division based on Heráklion. The general was hated and feared for the appalling harshness of his rule—labour gangs, mass shooting of hostages, reprisal destruction of villages, torture and executions by the Gestapo. However, late in March General Müller was replaced by General Kreipe who

*His promotion to Lieutenant-General came through on the very day after his capture!

†The British and Greek forces on Crete succumbed to a German airborne invasion in 1941 and the island remained occupied until late in 1944.

‡Both of whom received the British DSO for their exploit.

> became our quarry. He had commanded divisions on the Leningrad and Kuban sectors and was decorated with the Knight's Cross. Although the capture of Kreipe would not give the Cretans the same satisfaction as the capture of Müller, the German forces on Crete would suffer a severe blow to their self-confidence and prestige, and the action would have a strong tonic effect among the Cretans: it would be a triumph for the resistance movement which had kept the island so effectively, and improbably, united.

Unfortunately, the record is not explicit about the role, if any, that the Cairo OSS played in the planning that preceded and accompanied this operation.* The evidence is, however, clear that its MO Branch was deeply involved in the subsequent intelligence and propaganda exploitation of this event.

General Kreipe was waylaid on the evening of April 26, 1944 while returning from his headquarters (at Ano Archanes) to his villa five miles away. Imprisoned in the back seat of his staff car, he was driven past 22 control posts, receiving full salutes from the guards, and through the busy center of Heráklion, before the car was abandoned somewhere along the north coast. For the next eighteen days, while the German garrison scoured the rugged island with plane and patrol, Kreipe and his captors made their perilous way across to the snow-capped Mt. Ida to an evacuation point on the southwest coast of Crete. There the party successfully rendezvoused with a waiting ship on the night of May 14/15.

General Kreipe was taken by boat to Mersa Matruh and then driven to Cairo. There, for months, he was shared between British and American intelligence agencies. "He spent two months with us while I interrogated him," Yourgrau recalled.[3] Kreipe, the thirteenth child and the breadwinner in a family of fifteen headed by an indigent pastor, was a proud man who stood on his dignity and was treated accordingly. "He was an old-fashioned German officer, a nationalist, who actually despised Hitler, but would never betray his country or his commander in chief. It was a question of honor and tradition," Yourgrau observed.

In a (semi-?) monthly report,[23] John Fistere, Acting Chief of Cairo OSS-MO provided further details on the activities of his unit in the wake of Kreipe's abduction:

> We concocted, in collaboration with [illegible], a campaign of rumors and stickers to help persuade the German garrisons that General Kreipe . . . actually escaped with British connivance. . . . The reports of agents from Crete . . . and from many other sections of Greece indicate that nearly all garrisons believe that Kreipe really did run out on them. . . . An interesting part of the plan was a letter presumed to have been written by another General by the name of Franz Krech,† who about the same time was killed by Greek *andartes* in the Peloponnesus. . . .

*It might be germane to the paucity of our information on this point to mention that OSS relations with the British Secret Service in Cairo were characterized by an atmosphere of mutual distrust, rivalry, and concealment.[18]

†This letter, directed at German soldiers, starts off with a declaration that the (presumed) writer has made the agonizing decision to defect to the other side. The writer continues with a detailed motivation for his decision and ends with an appeal to German solidiers "not to throw their lives away senselessly" for the sake of "a handful of criminals," but rather "to live for a new Germany."

> Each BLO in Greece had a copy of this letter and so did several of our own men. We also sent copies of it to Istanbul to be shown around. . . . We had no report on the success of this part of the campaign until we received the monitoring of the Russian Moscow official radio in Greek for June 30th. A copy of that is also enclosed.*

John Fistere concluded, "It is not often that an MO campaign bears such identifiable fruits"—retrospectively a statement in which scholars of worldwide OSS triumphs and disasters would heartily concur.

Yourgrau's employment with OSS was terminated at the end of October 1944—at which time the German occupation forces had been driven from practically all of the Balkan States and MO operations in the Middle Eastern theater were concluded. One of his last acts was an extensive and thoughtful memorandum entitled, "Suggestions for MO in Post-War Germany and Austria."†

In evaluating Yourgrau's activities with OSS, Col. Harry S. Aldrich, who was to replace Col. John Toulmin during early 1945 as head of the OSS Middle East Headquarters, expressed himself in 1944 as follows:

> Dr. Wolfgang Yourgrau was employed . . . as political advisor on German affairs. As part of his work, he conducted research, made analyses, and engaged in creative activities. We were impressed with his knowledge and grasp of the subject, and feel that he made substantial, although immeasurable, contributions towards the defeat of the enemy.[24]

On Yourgrau's return to Palestine, the British Government—anticipating that the fall of Hitler's "Fortress Europe" was then only a matter of time—appointed him as editor in charge of the victory issues of both *Hagalal* and *Al Muntada*, the regularly published official Hebrew and Arabic newspapers.‡ "For months," according to Yourgrau,[7]

> I prepared these issues, which were offered for sale on the day of the Victory parade [V-E Day, May 8, 1945]. The Hebraic number contained contributions of Arnold Zweig, Franz Goldstein, Martin Buber, Hugo Bergmann, Ernst Simon, Fritz Schiff, Field Marshal Jan Smuts, myself, and others. The Arabic one contained an appeal by Prof. Hitti of Princeton and other oustanding Arabic scholars. Bernhard Zondek wrote in a marvelous essay, in both newspapers, on the development of medicine during the War, [and] Richard Koebner published a very interesting article, in the Hebraic issue, on the relation between the English and the Jews . . . The Victory issues sold out in four hours, so that literally not a single further copy could be bought.

*Presumably, then, this stratagem to lower the morale of the *Wehrmacht* was a success, too. A copy of the enclosure in question is not at our disposal.

†Communication within OSS-MO had its personal and light moments, too. "Tell your son that I am ready to send him stamps occasionally," Yourgrau wrote, *inter alia*, in a memo to one of his superiors. Then, probably impelled by an irresistible urge, he added: "Stamp collecting, by the way, is one of the few vices you and I never indulged in." We mention this as perhaps a typical example of Yourgrau's disarming sense of humor and his inability to be awed by authority.

‡A third government newspaper, the *Forum*, regularly appeared in English.

At War's end, Yourgrau felt he had earned some repose. Forsaking *Hagalgal*, where for a while he had been engaged in the writing of lead articles, Yourgrau in October 1945 joined the School of Higher Studies as a lecturer of logic, ethics, and history of philosophy.

The School of Higher Studies (in the "German Colony" of Old Jerusalem), a creation of the British Council, was a prestigious branch of the University of London—which awarded all the School's degrees, from baccalaureates to doctorates. Yourgrau's appointment to this institute marked the start of an era in which he would redirect his phenomenal energy into the resumption of his academic career and the pursuit of his preempted original lines of research. He soon became Head of the School's department of logic and scientific method and subsequently Acting Dean of its Faculty of Arts and Sciences.

"I have, therefore, after a stormy period of twelve years, replete with experiences of a diverse nature, found a home, so to speak," Yourgrau wrote Berendsohn in 1947.[7] "I am now . . . a philosopher, not any more a journalist, editor, or traveling lecturer. . . . You can imagine how satisfied and nearly happy I am again to occupy an academic position."

Happiness in *academia* was matched by domestic bliss, which Yourgrau shared with his second wife, South-African born Thella Garber since their marriage in the middle of 1947. In some ways alike but in other respects quite the opposite of Yourgrau, the regal yet strikingly vivacious Miss Garber, a most talented violinist and a stern upholder of the King's English, was as much *sui generis* in her own set of priorities as Yourgrau was in his. And in the make-up of her personality she, too, was the product of her antecedents.

While Wolfgang Yourgrau's psyche bore in many respects the imprints of his mother's deeply caring but dominating personality, Thella Garber's view of herself and life was shaped to the core by that of her self-denying and deeply idealistic father. A Russian Jew who immigrated with his new bride to South Africa around the turn of the century, Joel Garber (born 1874) was a devoted scholar since childhood of the *Tanach* and the *Talmud*, who all his life "served the Zionist cause with every fibre of his body."[25]

Deriving a respectable income from the ownership and management of a furniture store in the flourishing town of Ermelo (Transvaal), he spent his leisure hours studying English and Hebrew and writing Hebrew poetry and prose—many examples of which got published. A deep respect for learning and scholarship led him to demand, and get, nothing less than excellence in academic achievements from his five sons—*all* of whom became medical doctors of considerable repute—and his two daughters.

This was the family background of young Miss Garber who, at the outbreak of the War, happened to be visiting Jerusalem, after capping her music studies with a Licentiate from London's Royal College of Music. She stayed on in this city, serving with the British Army well beyond the end of the War—narrowly escaping death ("forewarned by precognition") in the IZL's July 1946 bomb attack on the British

Government offices in Jerusalem's King David Hotel. She met Yourgrau on V-E Day—an auspicious beginning to a period of rising expectations.

The euphoria on the domestic and academic fronts was, however, of short duration. Once again Yourgrau's dreams of peace and an academic career were about to be dashed by political developments beyond his control.

In February of 1947, Great Britain, despairing of successfully dealing with the Arab opposition to the creation of a Jewish state in Palestine, turned the whole problem over to the United Nations. On 29 November 1947, the General Assembly of this body, led by the USA and the USSR, adopted with a two-thirds majority a resolution to divide Palestine into a Jewish state, an Arab state, and a small internationally administered zone including Jerusalem. Great Britain abstained from voting and the Arabs announced their intention to resist partition. As the British began to withdraw in early 1948, Arabs and Jews started to prepare for war.

Arab rioters took to the streets of Jerusalem already on the morrow of the U.N. vote, and violence directed at Jews intensified and spread to all parts of the country from that day onward. While the British police stood idly by, the defense of the *Yishuv* became the sole responsibility of the *Haganah*.

The effects of encounters between the *Haganah* and armed Arab bands were felt also in the German Colony, where the Yourgraus had their home. "We lived in a house in the Arab quarter," Yourgrau recalled,[3]

> because these houses were cheaper and had very thick insulating walls. [Moreover, the School of Higher Studies was within walking distance.] When relations between the Jews and Arabs turned unpleasant, we were caught in the middle, together with the British who surrounded us. Some nights we spent more time under than in bed in order to avoid stray bullets entering through our windows. The War was over, I thought, but here we were still in the line of fire—while all I wanted was to enjoy a happy married life. My college having closed down in any event, I decided, on advice of the British High Commissioner [Sir Arthur Cunningham] that it was time to leave. We were driven to the Tel-Aviv Airport in a bullet-proof car he put at our disposal. While my wife boarded a plane for Johannesburg, I flew off to Nicosia (Cyprus) on another mission.

Yourgrau undertook his trip to Cyprus at the beginning of 1948 for the purpose of determining, at the request of the British Colonial Office, whether a branch of the University of London (similar to the just then disbanded School of Higher Studies) should be opened on that island.

The scene he found in Nicosia and the rest of Cyprus was a familiar one, with the British administrator again caught between two antagonistic ethnic groups, this time an overwhelming Greek majority (clamoring for union with Greece) and a Turkish minority (favoring partition of the island). One aim the British hoped to achieve with the proposed unversity college was, according to Yourgrau,[3]

> to indirectly smooth out the intransigent rage existing between Greeks and Turks. So [Yourgrau continued] I stayed on as a guest of the Governor and for two months interviewed teachers, judges, doctors, and anyone else who enjoyed

> a status within the two population groups, on the feasibility of the proposed institution. My subsequent report on this matter concluded in effect: God help Britain if she should ever attempt to establish a university on Cyprus. It would be burnt down in one day! I handed over my recommendations to the authorities and took the very next plane to South Africa—ahead of the outbreak of violence and widespread terrorism [against the British].

In South Africa, where he went by the end of March 1948 at the urging of his wife, Yourgrau at last found the respite from any further personal involvement in violence and political strife, which by then he felt was his due. His marriage to Thella Garber was blessed within a year by the birth of twin sons, Bertrand and Tuggelin, and some time later by a third son, Palle. His wife's cerebral and outgoing family, fascinated by Yourgrau's sense of humor and otherwise kindred personality, pressed him to its bosom. Yourgrau experienced, perhaps for the first time of his life, a comfortable sense of belonging and an atmosphere conducive to an exclusive preoccupation with scholarship.

For more than a decade Yourgrau taught, with spellbinding flamboyance and obvious enthusiasm, at the Universities of Cape Town, Witwatersrand (Johannesburg), and Natal (Pietermaritzburg), at the same time concentrating on his writing and research. Although his lectures initially and primarily dealt with traditional philosophy and the philosophy of science, the center of Yourgrau's scholarly interests gradually shifted towards theoretical physics. The motivating force behind this development was not purely intellectual. Speaking later of his post-Palestine trend away from the humanities—which received its major impetus from the publication in 1955 of the classic *Variational Principles in Dynamics and Quantum Theory* (coauthored with Stanley Mandelstam) and another boost from *Treatise of Irreversible and Statistical Thermophysics* (in collaboration with Gough Raw and myself)—Yourgrau confided, "I wanted in my own mind to get as far away as possible from the disturbing experiences of my earlier life."

In the late 1950s, with his fiftieth birthday approaching, Yourgrau was undergoing a midlife intellectual crisis. Sitting at the southern tip of Africa, he felt acutely isolated from interaction with leading thinkers in philosophy and physical science and judged the development of his own potential thwarted as a result. Having come to know and like Americans as a result of his association with them during the War, and perceiving the United States as the place where the intellectual action was, Yourgrau's thoughts once again turned to emigration. He took up correspondence with the physicist Robert Oppenheimer (then Director of the Institute for Advanced Study at Princeton) and other U.S. scholars, soliciting their advice on employment opportunities in the United States. The assurances that Oppenheimer was able to give, together with the award of a Bollingen Fellowship (on the recommendations of Oppenheimer and Pascual Jordan) and a temporary job offer at the University of Minnesota, encouraged Yourgrau to take the crucial step: Selling everything his family owned, apart from personal belongings and an inseparable plethora of books,* Yourgrau, his spouse and their three sons† set sail for the New World.‡

The first few years of life in the United States for Yourgrau were times of great intellectual excitement. But it was also a period of grave economic uncertainties and related crises. That the Yourgrau family maintained its sanity under these stresses was in no small measure a tribute to the *materfamilias*, Thella Yourgrau, Wolfgang Yourgrau's saintly complement, who with dignity and an unshakeable faith in the finally successful outcome of events, remained a much-needed source of morale in every crisis.

Arriving in the United States in the beginning of 1959, Yourgrau served first as Research Professor at the Minnesota Center for the Philosophy of Science (headed by Herbert Feigl) and then as Cowling Professor in logic and philosophy of science at nearby Carleton College. Subsequently (1960–1963) he became professor and chairman in the history and philosophy of science department at Smith College in Northampton, Massachusetts. In 1963 he accepted a similar but permanent position at the University of Denver.

J'y suis, j'y reste! Although, in the next decade and beyond, Yourgrau was to travel widely, both inside the United States and abroad for the purpose of meeting scholars and friends whose company he enjoyed, no less than for reasons of lecturing and research—it was to America and Denver that he always returned. He also became a U.S. citizen at the first opportunity. The man long in search of a national identity, but also the believer in Socialism and the iconoclast of patriotic shibboleths, had, in the cradle of free enterprise, the land frequently extolled not only by politicians and chauvinists as "the greatest country on earth," finally and paradoxically found peace and a spiritual home. Well, "in a way,"[3] at least!

Remarkably, the bulk of Yourgrau's books and papers on theoretical physics (with an emphasis on relativity, quantum theory, and thermophysics) saw the light after he settled in the United States, during the physically declining years of his life.§

But Yourgrau's interest in other areas of learning also then remained alive, as witness his active participation in the three International Colloquia, which he and

*Yourgrau's impressive library of more than 6,000 books followed him to the United States. This entire collection, comprising many rare and valuable books and reflecting Yourgrau's wide spectrum of interests, is presently housed at the Kanazawa Institute of Technology, in the port city of Kanazawa, Japan. Items identified by the Yourgrau signature and originally belonging to the collection of rare editions and valuable documents that made up his library in Jerusalem, were left by the British Council in trusteeship with the Hebrew University of Jerusalem after proclamation of the State of Israel.

†Yourgrau's daughter from his first marriage, Maya, it should be mentioned, was living with her mother in Israel.

‡The boat trip first took them to England, where Yourgrau paid a visit to another one of his correspondents, the aging philosopher Bertrand Russell, whom Yourgrau admired to the extent that he had given the name Bertrand to one of his sons. It was also on this trip that Yourgrau for the first time met (the then still unknighted) Karl Popper, who was to become a lifelong friend.

§A bibliography appears at the conclusion of this Introduction.

Allen Breck organized at the University of Denver between 1966 and 1974 with the purpose of bringing a multidisciplinary approach to bear on issues in physics, logic, history, biology, philosophy, cosmology, and theology (see Bibliography).

Commenting on the scope of Yourgrau's publications, Henry Margenau had this to say:

> Their enormous range has made him known to a vast number of scholars; but their impression was perhaps less deep, even among physicists, than an equally monolithic effort in a single speciality would have been. He grasped for, and achieved, an understanding in many fields, and this today is a rare accomplishment.[1]

Among Yourgrau's lasting contributions to physics must also be counted the international monthly, *Foundations of Physics*, which he founded together with Henry Margenau in 1969 and coedited (first with Margenau and then myself) until his death. Actually, the conception of this journal dates back to the first meetings between its founders, at the time when Yourgrau was still teaching at Smith (and Margenau at Yale). As Margenau recalls it, "During this period we also had our first discussions on the desirability of a journal that would bring together the problematic issues of physics and philosophy."[1]

Yourgrau also was a cofounder (in 1967) of the *International Journal of Theoretical Physics* and the recipient of numerous distinctions and honors, among which he especially valued the Swiss Einstein Medal, awarded to him in 1970.

Physically ravaged by a debilitating illness and unremitting pain, Yourgrau quietly expired on 18 July 1979 in the midst of a losing race against time to complete, among others, his contribution to a *Festschrift* commemorating the centenary of Einstein's birth. In view of his near reverence for and close identification with Einstein, both as a human being and a scientist, it is difficult to suppress the feeling that Yourgrau was consoled in his last moments by the thought that he lasted long enough to celebrate, if imperfectly, this historic milestone.

Finally, as someone who had viewed Wolfgang Yourgrau from close by for a quarter of a century, I might be forgiven for repeating here (from an earlier eulogy) a brief personal epitaph: Wolfgang Yourgrau was a deeply emotional man. Those surrounding him he could drive to quiet and not-so-quiet desperation at times. But mostly it was his sense of optimism, his positive thinking, and his contagious sense of humor that prevailed. Above all stood his helpfulness—which made people search him out for counsel and companionship when they had nowhere else to go. He was unforgiving of the failings and lapses in the loyalty of his friends. This was no accident: It simply reflected the standards he set himself in these matters. He would go to extreme lengths to fulfill the confidence people placed in him. A familiar phrase around his home or office was: "You can depend on me." And so, by God, one could, as many of his students and colleagues would testify. His interest in the personal problems of his students, his colleagues, and sometimes even strangers was always genuine and nearly always a prelude to an extension of an offer to help. Indeed, in Wolfgang Yourgrau, this author, in common with other admirers, colleagues, and generations of his students had the privilege of viewing not only an exceptionally courageous man and a scholar with a profound respect for learning and man's intellectual heritage, but also an even rarer occurrence: the marriage within one individual of a true concern for mankind in the abstract with empathy for his fellows in the concrete.

Acknowledgments

I should like to express my deep appreciation to the following individuals who, in some way or other, directly or indirectly, have contributed to my preparation of this essay: Professor Bernhard Abrahamsson, Professor Robert Amme, Dr. Werner Berthold, Professor Allen Breck, Dr. Gerda Breit, Mr. Joseph Coors, Mr. William Coors, Professor and Mrs. Burton Feldman, Mr. Harry Fitzwater, Professor and Mrs. Aaron Goldman, Mrs. Betty Greenwell, Mrs. Guilty Gross, Mr. Christopher Holme, Col. Patrick Leigh-Fermor, Professor John Livingston, Dr. Ernst Loewy, Dr. Virginia Mitchell, Dr. Gerhard Nellhaus, Professor Stoffel Nienaber, Mr. John Peet, Congresswoman Patricia Schroeder, Mr. George Wendell, Mrs. Grete Wohl, Dr. Uwe Schweikert, Mr. Bertrand Yourgrau, Mr. Tuggelin Yourgrau, and Professor Palle Yourgrau.

I owe a special debt of gratitude to Mrs. Thella Yourgrau, without whose constant encouragement and generous cooperation this biography in its present form would not have materialized.

Finally, our sincerest thanks are due to the contributors to this *Festschrift* for the extraordinary patience they have shown during the time that obstacles delaying its publication had to be overcome, and to the staff of Plenum Publishing Corporation (especially Mr. Steven Pisano) for making the special, intensive effort finally required to complete the publication process on time.

Secretarial expenses incurred in this project were partially covered by a faculty research grant from the University of Denver. This assistance is herewith gratefully acknowledged.

References and Notes

1. H. Margenau, *Found. Phys.* **10**, 3 (1980).
2. A. van der Merwe, *Nature* **284**, 289 (1980); *Physics Today*, April 1980, p. 69.
3. Classified *Oral Memoir*, May 1975. Generally speaking, quotations from this source are in compressed or paraphrased form.
4. W. Yourgrau, "Einstein—And the Vanity of Academia," in *Albert Einstein: His Influence on Physics, Philosophy, and Politics*, P. C. Aichelburg and R. U. Sexl, editors (Vieweg, Braunschweig, 1979).
5. *New York Times*, 21 October 1971.

6. Letter from Max Seydewitz to Dr. Hans Knoch, dated 17 September 1979.
7. Letter (in German) from Wolfgang Yourgrau to Professor Walter Berendsohn, dated 26 July 1947. Original in the "*Sammlung* Berendsohn," Deutsche Bibliothek, Frankfurt am Main.
8. H.-A. Walter, *Deutsche Exilliteratur 1933–1950, Band 4: Exilpresse* (J. B. Metzlersche Verlagsbuchhandlung, Stuttgart, 1978), especially the chapter entitled "Orient," pp. 679–733 and 814–821.
9. *The Denver Clarion*, 16 February 1965.
10. E. Loewy, *Literarische und Politische Texte aus dem Deutschen Exil* 1933–1945 (J. B. Metzlersche Verlagsbuchhandling, Stuttgart, 1979).
11. *Orient. Haifa* 1942–1943. *Bibliografie einer Zeitschrift*, compiled by V. Riedel with a Foreword by R. Hirsch (Aufbau-Verlag, Berlin and Weimar, 1973).
12. Letter from Arnold Zweig to Christopher Holme, dated 23 September 1942.
13. Letter from Wolfgang Yourgrau to Dr. Ernst Loewy, dated 23 June 1977.
14. Letter from Christopher Holme to this author, dated 20 August 1980.
15. Letter from Christopher Holme to Wolfgang Yourgrau, dated 11 November 1944.
16. Letter from John Peet to this author, dated 12 November 1979.
17. Letter (in German) from Arnold Zweig to Walter Berendsohn, dated 13 September 1946. Original in the "*Sammlung* Berendsohn," Deutsche Bibliothek, Frankfurt am Main; also quoted in Ref. 8.
18. R. H. Smith, *OSS: The Secret History of America's First Central Intelligence Agency* (University of California Press, Berkeley, 1972).
19. Letter from John Fistere to Major C. T. S. Keep, dated 19 March 1944. Original in files of the CIA, Washington, D.C.
20. Letter from John Fistere to Major C. T. S. Keep, dated 19 March 1944. Original in files of the CIA, Washington, D.C.
21. W. S. Moss, *Ill Met by Moonlight* (The Macmillan Co., New York, 1950).
22. P. Leigh-Fermor, "How to Steal a General," in *History of the Second World War* (Purnell and Sons in cooperation with the Imperial War Museum, London, 1966–1968), B. L. Hart (editor-in-chief), Vol. 5, No. 7.
23. Letter from John Fistere to Lt. Col. Kenneth Mann, dated 3 July 1944. Original in files of the CIA, Washington, D.C.
24. Letter from Col. Harry S. Aldrich to Christopher Holme, dated 30 October 1944. Original in files of the CIA, Washington, D.C.
25. *Record and South African Jewish Chronicle*, December 1959.

A List of Publications by Wolfgang Yourgrau

"Treatise on Miguel de Unamuno," *Hagalgal* (Palestine), 1936.

"Analysis of the Political Situation in the Middle East," *Die Neue Weltbühne* (France), 1938.

"Auftakt" (programmatic foreword to *Orient*), *Orient* (Palestine), 10 April 1942.

"Gandhi oder Nehru?," *Orient*, 17 April 1942.

". . . Geprellt!" (on the situation of West-European immigrants within Palestinian political parties), *Orient*, 24 April 1942.

"Bilanz . . ." (on the labor movement and Zionism), *Orient*, 1 May 1942.

"Antworten" (replies to critical letters of readers to *Orient*), *Orient*, 1 May 1942.

"Marche Macabre: Wien" (on Viennese intellectuals of the 20th century), *Orient*, 8 May 1942.

"Am Vorabend" (on Winston Churchill and World War II), *Orient*, 15 May 1942.

"Lehrerstreik" (on Palestinian school affairs), *Orient*, 22 May 1942.

"In Sachen 'Orient'," *Orient*, 29 May 1942.

"Die Woche" (on *Orient*), *Orient*, 5 June 1942.

"Erdölversorgung und Kriegsentscheidung" (on the German oil supply), *Orient*, 12 June 1942.

"Notiz der Woche" (on the British–Soviet Mutual Assistance Pact of 26 May 1942 and on terrorism against the German-language newspaper "BNN," Tel Aviv), *Orient*, 19 June 1942.

"Erdölversorgung und Kriegsentscheidung (II)" (on the oil potential of the USSR), *Orient*, 19 June 1942.

"Notiz der Woche" (on the Jewish antifascist resistance), *Orient*, 26 June 1942.

"Notiz der Woche" (on World War II, antifascist resistance, and Palestinian politics), *Orient*, 3 July 1942.

“Der Umbruch” (on changes in political thinking during World War II), *Orient*, 10 July 1942.
“Notiz der Woche” (on Turkish politics and the Turkish presidents Refik Saydam and Schükrü Sarajoglu), *Orient*, 17 July 1942.
“Notiz der Woche” (on Indian politics in 1942), *Orient*, 24 July 1942.
“Appell an den Leser” (on *Orient*), *Orient*, 31 July 1942.
“Die ‘überlegene’ Rasse” (on the Japanese drug traffic), *Orient*, 31 July 1942.
“Notiz der Woche” (on Palestinian politics), *Orient*, 7 August 1942.
Abschied von Gandhi” (on Mohandas Karamchand Gandhi), *Orient*, 14 August 1942.
“Notiz der Woche” (on the establishment of a Second Front), *Orient*, 21 August 1942.
“Notiz der Woche” (on *Orient*), *Orient*, 28 August 1942.
“Notiz der Woche” (on Zionism and Palestinian politics), *Orient*, 11 September 1942.
“Notiz der Woche” (on Palestinian politics in 1942), *Orient*, 18 September 1942.
“Antwort an einen Gegner (I)” (on *Orient*), *Orient*, 18 September 1942.
“Kleines Jubiläum” (on the half-year existence of *Orient*), *Orient*, 25 September 1942.
“Heimat oder Asyl” (on German emigrants in Palestine), *Orient*, 2 October 1942.
“Einordnung und frühes Leid (I)” (on Zionism), *Orient*, 16 October 1942.
“Einordnung und frühes Leid (II)” (on Zionism), *Orient*, 23 October 1942.
“Einordnung und frühes Leid (III)” (on German emigrants in Palestine), *Orient*, 30 October 1942.
“Renovatio, Restitutio, Restauratio!” (on World War II), *Orient*, 20 November 1942.
“Alija Chadascha” (on West-European immigrants in Palestine), *Orient*, 4 December 1942.
“Der ungläubige Thomas” (on British social politics), *Orient*, 11 December 1942.
“Not—ohne Ende?” (on the Jewish question), *Orient*, 18 December 1942.
“ . . . Alleinseligmachend?” (on Zionism), *Orient*, 25 December 1942.
“Georg Landauer: Zwischen zwei Revolutionen,” *Orient*, 25 December 1942.
“Ausklang und Einkehr: 1942 . . . 1943” (on World War II), *Orient*, 1 January 1943.
“Aktion—und Aktionen” (on World War II and the antifascist resistance in 1943), *Orient*, 8 January 1943.
“Man rückt ab” (on Zionism in Palestine), *Orient*, 15 January 1943.
“Am 30. 1. 1933, als ein gewisser Herr Hitler . . .” (on Adolf Hitler and facism), *Orient*, 29 January 1943.
“Nach einer Bombe” (on the bomb attack on the *Orient* printing press), *Orient*, 7 April 1943.
“Thomas Mann: a Cultural-Sociological Study,” *Hagalgal*, 1946.
“Logical Positivism and its Metaphysical Criticism,” *Hagalgal*, 1946.
“The Natural Philosophy of Max Planck,” *Hagalgal*, 1947.
“The Symbolism of E. Cassirer and A. N. Whitehead,” *Hagalgal*, 1947.
“On Martin Heidegger—a Study on Existentialism,” *Hagalgal*, 1947.
“The Great Albert Einstein,” *Jewish Affairs* (South Africa), 1949.
Virgil the Universal (T. Haarhof), a book review, *Forum* (South Africa), 1949.
“Goethe,” *Suid-Afrika* (South Africa), 1949.
“Reflections on the Natural Philosophy of Goethe,” *Philosophy* (Britain), 1951.
“On Recent Trends in Natural Philosophy,” *South African Journal of Science* **47**, 169 (1951).
Science in South Africa (published by the Council for Scientific and Industrial Research, Pretoria, South Africa), a book review, *Archives Internationales d’Histoire des Sciences* **14**, 197 (1951).
“Collegium Naturalis Philosophiae,” *Archives Internationales d’Histoire des Sciences* **15**, 374 (1951).
Scientific Autobiography and Other Papers (M. Planck), a book review, *South African Journal of Science*, 1951.
“General System Thoery and its Implications for Natural Philosophy,” *Scientia* (Italy), 1952.
Philosophy of Nature (M. Schlick), a book review, *South African Journal of Science*, 1952.
Variational Principles in Dynamics and Quantum Theory, with S. Mandelstam (Sir Isaac Pitman and Sons, London, 1955; W. B. Saunders, Philadelphia, 1968; Dover Publications, New York, 1979), 1st, 2nd, and 3rd editions.
“Some Notes on the Epistemology of Saint Thomas,” *Theoria* (South Africa), 1955.
“‘Acceleration’ of Chemical Reactions,” with C. J. G. Raw, *Nature* **178**, 809 (1956).
“Variational Principles and Chemical Reactions,” with C. J. G. Raw, *Il Nuovo Cimento* **5**, 472 (1957).
“Are ‘First Principles’ First Principles?,” *Bulletin of Philosophy, University of Cape Town* (South Africa), 1957.
“On the Nature of Mathematical Constants,” with D. Livingston, *Methodos* (Italy), 1958.

"Time Variation of Chemical Affinity," with C. J. G. Raw, *Nature* **181**, 480 (1958).
"Symmetry" (on H. Weyl), *Philosophy and Phenomenological Research* (USA) **18**, March 1958.
"The Cognitive Value of Nietzsche's Philosophy," *Theoria* (South Africa) **10**, 1 (1958).
Philosophy of Education, a book review, *The Australasian Journal of Philosophy* (Australia), 1959.
Chronic Radiation Hazards (G. J. Neary *et al.*), a book review, *Nuclear Physics* **12**, 223 (1959).
"Philosophy of Science and Analysis—A Reply to M. Lazerowitz," *Philosophy* (Britain) **35**, 1 (1960).
"A Double-Iteration Property of Boolean Functions," with C. Lyngholm, *Notre Dame Journal of Formal Logic* **1**, 111 (1960).
The Basic Concepts of Mathematics (K. Menger), a book review, *The Journal of Symbolic Logic* (USA) **25**, June 1960.
"Some Problems Concerning Fundamental Constants in Physics," in *Current Issues in the Philosophy of Science*, H. Feigl and G. Maxwell, eds. (Holt, Rinehart, and Winston, New York, 1961).
Max Planck Festschrift (B. von Kockel et al.), a book review, *Physics Today*, 1961.
"Challenge to Dualism," a review of *From Dualism to Unity in Quantum Physics* (A. Landé), *The British Journal for the Philosophy of Science* **12**, 158 (1961).
"On the Logical Status of So-Called Thought Experiments," in *Proceedings of the International Congress of History and Philosophy of Science*, Ithaca, New York (Hermann, Paris, 1962).
"On the Reality of Elementary Particles," in *The Critical Approach to Science and Philosophy* (in Honor of Karl R. Popper), M. Bunge, ed. (The Free Press of Glencoe, London, 1964).
Treatise on Irreversible and Statistical Thermophysics—An Introduction to Nonclassical Thermodynamics, with A. van der Merwe and C. J. G. Raw (Macmillan, New York, 1966; Dover Publications, New York, 1982).
"Language, Spatial Concepts, and Physics," in *Mind, Matter, and Method: Essays in Philosophy and Science in Honor of Herbert Feigl*, P. K. Feyerabend and G. Maxwell, eds. (University of Minnesota Press, Minneapolis, 1966).
"On Invariance Principles," in *Proceedings of the Einstein Symposium*, German Academy of Science, Berlin, 1965 (Akademie-Verlag, Berlin, 1966).
"On the New Physics and Modern Literature," *Denver Quarterly* (USA), 1966.
"Report on the International Colloquium I in Logic, Physical Reality, and History at the University of Denver, May 1966," with A. D. Breck and E. A. Lindell, *Denver Quarterly* (USA), 1966.
"On Models and Thought Experiments in Quantum Theory," *Monatsberichte der Deutschen Akademie der Wissenschaften zu Berlin* **9**, 865 (1967).
"On Paradoxes in Physics," in *Proceedings of the International Colloquium in the Philosophy of Science*, London, 1965 (North-Holland Publishing Co., Amsterdam, 1968).
"On Verification in General Relativity Theory—An Appraisal in Light of the Solar Quadrupole Moment Hypothesis," in *Proceedings of the XI International Congress of the History of Sciences*, Warsaw, 1965 (Polish Academy of Sciences, 1968).
"A Budget of Paradoxes in Physics," in *Problems of the Philosophy of Science*, I. Lakatos and A. Musgrave, eds. (North-Holland Publishing Co., Amsterdam, 1968).
"A New, Formalized Version of the Verifiability Principle," with C. Works, *Ratio* (Britain) **10**, 54 (1968).
"Did Ernst Mach 'Miss the Target'?," with A. van der Merwe, *Synthese* **18**, 234 (1968).
Max Planck als Philosoph (H. Kretzschmar), a book review, *Science* (U.S.A.), 1968.
"Note on Duality in Propositional Calculus," with C. Works, *Notre Dame Journal of Formal Logic* **9**, 284 (1968).
"Entropy Balance in Photosynthesis," with A. van der Merwe, *Proceedings of the National Academy of Sciences* **59**, 734 (1968).
"Gödel and Physical Theory," *Mind* (Britain) **78**, 77 (1969).
"Verification of Proof—an Undecided Issue?," in *Boston Studies in the Philosophy of Science*, R. S. Cohen and M. W. Wartofsky, eds. Vol. V. (D. Reidel, Dordrecht, 1969).
Bioelectronics (A. Szent-Györgyi), a book review, *Nature* (1969).
Physics, Logic, and History (Proceedings of the International Colloquium I, University of Denver, 1966), coedited with A. D. Breck (Plenum Press, New York, 1970).
"Variable Stars and the Frequency Dependence of the Speed of Light in a Gravitational Field," with J. F. Woodward, *Annalen der Physik* **25**, 334 (1970).
"*Assumption and Myth in Physical Theory* (H. Bondi), a book review, *Inquiry* (Norway), 1970.

"The Cosmos of George Gamow," *New Scientist* (Britain), 1970.

"A Paradox in the Interaction of the Gravitational and Electromagnetic Fields?," with J. F. Woodward, *Nature* **226,** 619 (1970).

"Donors and Acceptors," *Nature* **226** (1970).

Perspectives in Quantum Theory—Essays in Honor of Alfred Landé, coedited with A. van der Merwe (The M.I.T. Press, Cambridge, 1971; Dover Publications, New York, 1979).

"A Paradox in the Interaction of the Gravitational and Electromagnetic Fields?: A Reply to Reinhardt," with J. F. Woodward, *Nature* **229,** 36 (1971).

"Alfred Landé and the Development of Quantum Theory," with A. van der Merwe, in *Perspectives in Quantum Theory,* W. Yourgrau and A. van der Merwe, eds. (M.I.T. Press, Cambridge, 1971).

"Nietzsche and Modern Physics," *Denver Quarterly* (USA), 1971.

"Tired Light and the 'Missing Mass' Problem," with J. F. Woodward, *Acta Physica Hungarica* **30,** 323 (1971).

Biology, History, and Natural Philosophy (Proceedings of the International Colloquium II, University of Denver, 1967), coedited with A. D. Breck (Plenum Press, New York, 1972).

"Vacuum Dispersion, Photon Rest Mass, and the Speed of Light in a Gravitational Field," with J. F. Woodward, *Annalen der Physik* **28,** 88 (1972).

"The Incompatibility of Mach's Principle and the Principle of Equivalence in Current Gravitational Theory," with J. F. Woodward, *The British Journal for the Philosophy of Science* **23,** 111 (1972).

"Entropy (Positive and Negative), Information, and Statistical Thermodynamics," with A. J. van der Merwe, in *Cosmology, Fusion and Other Matters,* F. Reines, ed. (University of Colorado Press, Boulder, Colorado, 1972).

"Marginal Notes on Schrödinger," in *Biology, History, and Natural Philosophy,* A. D. Breck and W. Yourgrau, eds. (Plenum Press, New York, 1972).

"Further Evidence for an Anomalous Interaction Between the Electromagnetic and Gravitational Fields," with J. F. Woodward, *Il Nuovo Cimento* **9B,** 440 (1972).

"Difficulties Concerning a Finite Photon Rest Mass," with J. F. Woodward, *Nature* **241,** 338 (1973).

Mach's Principle, the Equivalence Principle, and Gravitation: A Rejoinder to Newburgh," with J. F. Woodward, *British Journal for the Philosophy of Science* **24,** 264 (1973).

"Observed Deflection of Light by the Sun as a Function of Solar Distance," with P. Merat, J.-C. Pecker, and J.-P. Vigier, *Astronomy and Astrophysics* **32,** 471 (1974).

"Tired Light, Lorentz Covariance, and Conservation Principles," with J. F. Woodward, *Acta Physica Hungarica* **37,** 283 (1974).

"Gravitational Attenuation and the Internal Heating of Planetary Bodies," with R. J. Crowley and J. F. Woodward, *Astronomische Nachrichten* **295,** 203 (1974).

"Mach's Principle and the Rest Mass of the Graviton," with J. F. Woodward and R. J. Crowley *Physical Review D* **11,** 1371 (1975).

"Mach's Principle: Micro- or Macrophysical?," with J. F. Woodward, *British Journal for the Philosophy of Science* **26,** 137 (1975).

"Mach's Principle, Gravitation, and Electromagnetism: The Electromagnetic Inertia of Charged Elementary Particles," with J. F. Woodward, *Astronomische Nachrichten* **296,** 111 (1975).

"Formalism versus Interpretation—An Issue Revisited," *Methodology and Science* (The Netherlands) 8 (1975).

"Cosmological Implications of Anomalous Redshifts—A Possible Working Hypothesis," with T. Jaakkola, M. Moles, J.-P. Vigier, and J.-C. Pecker, *Foundations of Physics* **5,** 257 (1975).

"Some New Aspects of Relativity: Remarks on Keswani's Paper," with J. F. Woodward, *British Journal for the Philosophy of Science* **26,** 133 (1975).

Cosmology, History, and Theology (Proceedings of the International Colloquium III, University of Denver, 1974), coedited with A. D. Breck (Plenum Press, New York, 1976).

"Some Methodological Problems in Quantum Physics," with L. S. Mayants and A. J. van der Merwe, *Annalen der Physik* **33,** 21 (1976).

"Reflections on Philosophy of Science," in *Vistas in Physical Reality,* E. Laszlo and E. B. Sellon, eds. (Plenum Press, New York, 1976).

"Cosmology and Logic—An Intractable Issue," in *Essays in Memory of Imre Lakatos,* R. S. Cohen, P. K. Feyerabend, and M. W. Wartofsky, eds. (D. Reidel Publishing Co., Dordrecht, 1976).

"Alfred Landé," *Physics Today* **29**, (5), 82 (1976).
"Formalism versus Interpretation—An Issue Revisited," in *Yearbook of The Institute of General Semantics*, M. Kendig, ed. (Institute of General Semantics, Lakeville, Connecticut, 1976).
"On Some Cosmological Theories and Constants," in *Cosmology, History, and Theology*, W. Yourgrau and A. D. Breck, eds. (Plenum Press, New York, 1976).
"La constante de Hubble mise en question," with L. Nottale, J.-C. Pecker, and J.-P. Vigier, *La Recherche* 7(68), 529 (1976).
"Statistical Inference and Quantum Mechanical Measurement," with R. W. Benoist and J.-P. Marchand, *Foundations of Physics* **7**, 827 (1977).
"On Gravitational Shock-Waves," with H.-J. Treder, *International Journal of Theoretical Physics* **16**, 233 (1977).
"Dirac's Classical Theory of Electrons in the Unitary Field Theory of Einstein with Tele-Parallelism," with H.-J. Treder, *Physics Letters* **68A**, 415 (1978).
"On General-Relativistic and Gauge Field Theories," with H.-J. Treder, *Foundations of Physics* **8**, 695 (1978).
"Gravitational Field Equations of Fourth Order and Supersymmetry," with H. von Borzeszkowski and H.-J. Treder, *Annalen der Physik* **35**, 471 (1978).
"Addendum to 'Statistical Inference and Quantum Mechanical Measurement'," with R. W. Benoist and J.-P. Marchand, *Foundations of Physics* **8**, 117 (1978).
"On Self-Gravitation," with H.-J. Treder, *Physics Letters* **64A**, 25 (1978).
On General Relativity, with A. Mercier and H.-J. Treder (Akademie-Verlag, Berlin, 1979).
"Classical Spontaneous Breakdown of Symmetry and Induction of Inertia," with D.-E. Liebscher, *Annalen der Physik* **36**, 20 (1979).
"On the Increase of Entropy in the Carleman Model," with A. Wehrl, *Physics Letters* **72A**, 13 (1979).
"Einstein—And the Vanity of Academia," in *Albert Einstein—His Influence on Physics, Philosophy and Politics*, P. C. Aichelburg and R. U. Sexl, eds. (Vieweg, Braunschweig and Wiesbaden, 1979).
"Electro-Gravitational Induction: A Re-examination of a Classical Conjecture," with H.-J. Treder and J. F. Woodward, *Gerlands Beiträge zur Geophysik* (GDR) **88**, 11 (1979).
"Equivalence principle, gravitational collapse, and the classical particle problem," with H.-H. von Borzeszkowski, *J. Phys. A. Math. Gen.* (Britain) **12**, 361 (1979).
"On Singularities in Electrodynamics and Gravitational Theory," with H.-H von Borzeszkowski and H.-J. Treder, *Astronomische Nachrichten* **300**, 57 (1979).
Fundamental Principles of General Relativity Theories—Local and Global Aspects of Gravitation and Cosmology, with H.-J. Treder, H.-H. von Borzeszkowski, and A. van der Merwe (Plenum Press, New York, (1980).
"Simultaneous Measurement, Phase-Space Distributions, and Quantum State Determination," with J. L. Park and W. Band, *Annalen der Physik* **37**, 1 (1980).
"On the Mode of Propagation of Light," with L. Jánossy and A. van der Merwe, *Astronomische Nachrichten* **301**, 1 (1980).

PART A

CURRENT PROBLEMS IN PHYSICS AND COSMOLOGY

1

On the Connections between Pico- and Mega-Cosmos

(According to the Unification and Geometrization Suggested by Einstein, Eddington, and Schrödinger)

H.-J. Treder
Akademie der Wissenschaften der DDR
Berlin, GDR

1. Unification and Geometrization

Wolfgang Yourgrau has always—and most intensively during the last years—entered into the great physical question asked by his teachers Einstein and Schrödinger: How can one found the different types of micro- and macrophysical interactions and their orders of magnitude within the framework of a unified field theory seeing its pattern in general relativity? To answer this question, Wolfgang considered the ideas of Helmholtz, Planck, Hilbert, Einstein, Weyl, Eddington, Schrödinger, and Heisenberg, assuming a Hamiltonian principle with geometrically founded symmetries, universal constants, and meaningful numerical factors as the key that would unlock the door to classical and quantum aspects of physics.

Yourgrau dedicated numerous papers to these problems of symmetries, universal constants, and the dimensionless numbers which can be constructed from them. He was especially interested in Eddington's cosmological number, which contains an empirical hint at the fundamental relations between elementary particles and cosmology. It was a primary intuition of general relativity to unify Mach's principle and the determination of dynamics by the cosmic masses (the Mach–Einstein doctrine of the relativity of inertia) in such a way that there actually would follow conditions on microphysics from the megaphysics of relativistic cosmology.

In long discussions with Wolfgang, I could convince him that Mach's principle and the Mach–Einstein doctrine are not only epistemological and cosmological postulates, but principles which can be formulated in a rigorous mathematical and physical fashion and contain the foundation of Eddington's number. Wolfgang contributed his later ideas concerning this subject to both of the last books he coauthored: to *On General Relativity*,[1] written together with André Mercier and H.-J. Treder, and to a book by W. Yourgrau, Alwyn van der Merwe, Horst-H. von Borzeszkowski, and H.-J. Treder published posthumously.[2]

Unfortunately, other ideas and plans had to stay outlines.* I hope that the following paragraphs on the Einstein and Schrödinger program of a geometrization of physics and on the Einstein–Eddington problem of a relation between micro- and macrocosmos will also represent the ideas that Wolfgang and I have repeatedly discussed during the last years.

2. Hermitian Relativity and Chromodynamics

All proposals that have been considered to unify general relativity theory (GRT) with the theories of quantum fields and elementary particles and to show up new microscopic aspects of gravitation, which would facilitate the renormalization of quantum-field singularities, involve certain supplementary geometrical structures representing a hypothetical generalization of the properties of the Einstein vacuum. The same applies to attempts to solve internal problems of GRT.

Several "post-Einsteinian" gravitational theories were mathematically established in analogy to the unitary geometrical field theories developed by Einstein and his colleagues for the purpose of unifying both classical macroscopic fields, gravitation, and electromagnetism. These theories enrich also the four-dimensional Riemann–Einstein spaces of GRT by additional geometrical objects and often attempt to put them in correspondence with electromagnetism. On the other hand, Schrödinger had mentioned in 1944 that other, e.g., mesonlike, fields also could appear in a unitary geometrical field theory; and Pauli mentioned in 1935 that the only universal coupling existing besides the gravitational one is the universal Fermi coupling of the radioactive weak interaction. For this reason, some authors believe more in a unification of the theories of gravitation and radioactivity. This would lead to a theory of a universal geometrical field, where the Newton–Einstein gravitation would represent the long-range component and radioactivity the ultrashort-range component of the universal field.

However, in such theories it is always a problem to overcome the difficulties resulting from the differences in the orders of magnitude. A generalized gravitational theory is first of all a theory with an extremely weak coupling between field and source. New symmetries may enlarge qualitative features of the field, but they do not

*Here should be cited some of the post-1960 papers in which Wolfgang Yourgrau, mostly in coauthorship with Liebscher, von Borzeszkowski, Woodward, and myself, considered topics concerning the unification of physics, the geometrical interpretation of its symmetries, and the relations between micro- and macrocosmos; see pp. 30–34.

increase the order of magnitude of the coupling constant. For instance, from the so-called "Einstein–Cartan theories of gravitation" there follows, in addition to Einstein's gravitation, some kind of universal "Fermi coupling"; but the coupling constant

$$\propto \frac{8\pi G h}{c^3}\, hc \approx 10^{-81}\ \text{erg cm}^3$$

is too small by a factor of 10^{30}. The theories of "supergravitation," with a super-gauge group resulting from a generalization of the linearized Einstein group, provide particle masses of the gauge-field quanta which are the Eddington number 10^{40} smaller than the actual masses μ, namely,

$$\propto \frac{G\mu^2}{hc} = \frac{G\mu}{c^2}\frac{\mu c}{h}$$

But the approaches to generalized gravitational theories starting from GRT are also designed to overcome the internal difficulties of Einstein's equations of gravitation: They are to eliminate the occurrence of potential-theoretical singularities as well as dynamical singularities, e.g., the gravitational collapse and the "big bang" singularities. To evade the first state of singularities, i.e., to guarantee stability and self-consistency of particle models, one has to consider modifications of Einstein's equations which become significant for large values of curvature (or energy density). However, the dynamical singularities can possibly be removed by another coupling between gravitational and matter fields which becomes effective for gravitational radii GM/c^2 large in comparison with the geometrical dimension. This brings about the new effect of the "absorption of gravitation." (Such an effect results from the tetrad theory of gravitation, which shows up possibilities for testing the absorption of gravitation experimentally.)

The unitary field theories, which enrich the Riemannian space-time V_4 of GRT by the Einstein–Cartan teleparallelism, replace, according to Einstein, an "amorphous body" (the "Bezugs-mollusken" of GRT) by a crystal lattice. This procedure reduces, of course, the covariance properties of the V_4 defined by GRT. The appearance of the reference mollusk corresponds to Einstein's "general relativity," i.e., to the invariance under arbitrary Lorentz rotations of the reference systems. Indeed, the tetrads h_ν^A describing these systems appear in the Riemannian metric $g_{\mu\nu}$ ($\mu, \nu = 0, 1, 2, 3$) of the V_4 only via the Lorentz-invariant combination $g_{\mu\nu} = h_\mu^A h_{A\nu}$ ($A = 0, 1, 2, 3$). On the other hand, the teleparallelized V_4 contains all the 16 components h_ν^A explicitly as four Lorentz vectors. Therefore, a tetrad field theory permits only "rigid" Lorentz rotations and represents accordingly only a theory with "special relativity," in the sense of special relativity theory.

The physical interpretation of geometrical entities can also be connected with a restriction on the transformations which are mathematically admissible. The "physicalization" of geometry leads to a preference for certain mathematical representations. The five-dimensional theory of relativity provides a well-known example of this fact. Its physical interpretation requires us to use the representations $X^4 = f(X^\nu)$ of the space-time V_4 in V_5 and thus restricts the general five-dimensional coordinate transformations to the combined Einstein–Maxwell group:

$$\overline{X}^\nu = \overline{X}^\nu(X^\lambda), \qquad \overline{X}^4 = X^4 + \phi(X^\lambda) \qquad (\nu, \lambda = 0, 1, 2, 3)$$

On the other hand, Einstein's idea of a reference-tetrad theory as geometrical unification of gravito- and electrodynamics may be realized by abandoning the Maxwell gauge group and gauging the vector potential as in Dirac's theory of electrons:

$$A_\nu A^\nu = m^2c^4/e^2$$

(e is the elementary charge and m the particle mass.)

On abandoning the Lorentz covariance of the reference tetrads h_ν^A and the gauge invariance of the electromagnetic field, it becomes possible to identify the vector potential A_ν with the timelike tetrad components h_ν^0 by assuming $h_\nu^0 = (e/mc^2)A_\nu$. (Maxwell's electrodynamics is then replaced by Dirac's.)[3]

The difference of this teleparallelized Riemann geometry and the Riemann geometry of GRT is negligible as long as the ratio of gravitational and electric forces characterized by Eddington's number $\propto Gm^2/e^2$ can be regarded as sufficiently small.

The unitary geometrical field theories should unify quantum and general relativity theories. According to Einstein, all matter fields result only from the differences between the unitary field equations and the Einstein–Maxwell equations of GRT in regions of large field strength. Only the classical fields, gravitation and electromagnetism, then would represent genuine fields, and all the other fields of quantum field theory would be considered as their nonclassical "tails."

3. Planck's Units and Eddington's Numbers

The history of Einstein's Hermitian generalization of GRT, in quest of a unitary world geometry, is noteworthy. Einstein proposed this theory already in 1925 (after his analysis of Eddington's "purely affine" field theory); but he withdrew it soon afterwards because the Hermitian symmetry of this theory implies the existence of antimatter, i.e., a physical consequence in which Einstein could not believe (1925).[4] The Hermitian geometry (with real coordinates) necessarily involves a CPT theorem.

Only in 1945 does Einstein return to his Hermitian-symmetric relativity theory (HRT), his opinion now being shared by Schrödinger. The first problem arising in such a theory refers to the "super-gauge group." In order to find a physical interpretation of the Hermitian (thus, nonsymmetric) world geometry, the equations of HRT must be invariant with respect to substitutions

$$\Gamma^\alpha_{\lambda\mu} \rightarrow \Gamma^\alpha_{\lambda\mu} + \delta^\alpha_\lambda \phi_\mu$$

of the affine connection.

In GRT this problem is solved in a trivial manner; in HRT the necessary existence of this A group led Einstein to the realization that the imaginary part $g_{\underset{\vee}{\mu\nu}} = -g_{\underset{\vee}{\nu\mu}}$ of the metric corresponds to a vector potential.

The Hermitian symmetric fundamental tensor of HRT satisfies the following relations:

$$
\begin{aligned}
g_{\mu\nu} &= g_{\underline{\mu\nu}} + g_{\underset{\vee}{\mu\nu}} = g^{*}{}_{\nu\mu}, \qquad g_{\mu\alpha}g^{\nu\alpha} = \delta^{\nu}_{\mu} \\
g^{\mu\nu} &= g^{\underline{\mu\nu}} + g^{\underset{\vee}{\mu\nu}} = g^{*\nu\mu}, \qquad g = \det g_{\mu\nu} = g^{*} \\
g_{\underline{\mu\nu}} &= \tfrac{1}{2}(g_{\mu\nu} + g_{\nu\mu}), \qquad g_{\underset{\vee}{\mu\nu}} = \tfrac{1}{2}(g_{\mu\nu} - g_{\nu\mu})
\end{aligned}
\tag{1}
$$

The Hermitian field equations read

$$g_{\mu\nu,\lambda} - g_{\alpha\nu}\Gamma^{\alpha}_{\mu\lambda} - g_{\mu\alpha}\Gamma^{\alpha}_{\lambda\nu} = g^{*}{}_{\mu\nu,\lambda} - g^{*}{}_{\alpha\nu}\Gamma^{*\alpha}{}_{\mu\lambda} - g^{*}{}_{\mu\alpha}\Gamma^{*\alpha}{}_{\lambda\nu} = 0, \qquad \Gamma^{\alpha}_{\mu\nu} = \Gamma^{*\alpha}{}_{\nu\mu} \tag{2}$$

$$[(-g)^{1/2}g^{\underset{\vee}{\mu\nu}}]_{,\nu} = 0 \leftrightarrow \Gamma^{\lambda}_{\alpha\lambda} = \Gamma^{*\lambda}{}_{\alpha\lambda} \tag{3}$$

$$R_{\underline{\mu\nu}} = \tfrac{1}{2}(R_{\mu\nu} + R_{\nu\mu}) = 0, \qquad R_{\mu\nu} = R^{*}{}_{\nu\mu} \tag{4}$$

and

$$R_{\underset{\vee}{\mu\nu},\lambda} + R_{\underset{\vee}{\nu\lambda},\mu} + R_{\underset{\vee}{\lambda\mu},\nu} = 0, \qquad R_{\underset{\vee}{\mu\nu}} = \tfrac{1}{2}(R_{\mu\nu} - R_{\nu\mu}) \tag{5}$$

where (2) determines the "affine connection" $\Gamma^{\alpha}_{\mu\nu}$ uniquely. These equations are not independent because the $R_{\mu\nu}$ satisfy the generalized Bianchi identities[5]

$$[(-g)^{1/2}R_{\underline{\mu\alpha}}g^{\underline{\alpha\nu}}]_{,\nu} - \tfrac{1}{2}(-g)^{1/2}g^{\underline{\mu\nu}}R_{\underline{\mu\nu},\alpha} = \tfrac{1}{2}(-g)^{1/2}g^{\underset{\vee}{\mu\nu}}(R_{\underset{\vee}{\mu\nu},\alpha} + R_{\underset{\vee}{\nu\alpha},\mu} + R_{\underset{\vee}{\alpha\mu},\nu}) \tag{6}$$

providing the complete classical dynamics.

Equation (3) says that the theory is super-gauge invariant under transformations of the A group; therefore it must be satisfied identically. From this fact, the existence of a real vector potential follows:

$$g^{\underset{\vee}{\mu\nu}} = \frac{i}{(-g)^{1/2}}\,\eta^{\mu\nu\sigma\tau}A_{\tau,\sigma} \tag{7}$$

Equations (4) can be interpreted as gravitational equations with field matter $T_{\mu\nu} = T_{\nu\mu}$. The real part of $g_{\mu\nu}$ represents the gravitational potential $a_{\mu\nu}$, and $F_{\mu\nu}$ corresponds to a real bivector:

$$g_{\mu\nu} = a_{\mu\nu} + iF_{\mu\nu} = a_{\mu\nu} - iF_{\nu\mu} = g^{*}{}_{\nu\mu}, \qquad a_{\mu\nu} = g_{\underline{\mu\nu}}, \qquad iF_{\mu\nu} = g_{\underset{\vee}{\mu\nu}} \tag{8}$$

Then Eq. (4) may be written as "Riemann's equations"

$$R_{\underline{\mu\nu}} - G_{\mu\nu} = \frac{8\pi G}{c^2}\left(T_{\mu\nu} - \frac{1}{2}a_{\mu\nu}T\right) \tag{9a}$$

where $G_{\mu\nu} = G_{\nu\mu}$ is the Ricci tensor of GRT.[6–8]

Finally, Eqs. (5) define by their singularities pointlike field sources:

$$R_{\underset{\vee}{\mu\nu},\lambda} + R_{\underset{\vee}{\nu\lambda},\mu} + R_{\underset{\vee}{\lambda\mu},\nu} = -\frac{i}{2}(-g)^{1/2}\eta_{\mu\nu\sigma\tau}\sum_{A}^{n}\varepsilon_A \dot{X}^{\tau}_{A}\delta(A) \tag{9b}$$

Einstein perceived in 1949 that identifying A_ν with the electromagnetic potential, i.e., A_0 with the Coulomb potential, the Hermitian symmetry results in a wrong sign of the electrostatic forces $\propto 1/r^2$ (attraction instead of repulsion between similar charges!). Since 1950, Einstein considered therefore only asymmetric real fields $g_{\mu\nu} \neq g_{\nu\mu}$. However, for an asymmetric field theory Pauli's criticism is justified that this generalization of GRT does not have a genuine geometrical meaning.

Moreover, Infeld could show in 1950 that the Einstein equations (2)–(5), approximately integrated for quasistatic fields according to the Einstein–Infeld–Hoffmann method, do not provide Coulomb-like forces between "point charges." (By mistake one even supposed that the antisymmetric $g_{\underset{\vee}{\mu\nu}}$ was a "force-free" field.)

However, it was shown by Treder in 1957 that Einstein's Hermitian equations lead in fact to additional nongravitational forces between two "point charges" ε_A and ε_B; these forces have the form

$$\mathcal{F}_{AB} = \frac{c^4}{4G}\,\epsilon_A\epsilon_B\,\frac{\mathbf{r}_{AB}}{r_{AB}}$$

for simple delta-like distributions of charge, and for general point particles they read

$$\mathcal{F}_{AB} = \frac{c^4}{2G}\left(-\frac{[\alpha_A\epsilon_B + \alpha_B\epsilon_A]}{r_{AB}^2} + \frac{\epsilon_A\epsilon_B}{2}\right)\frac{\mathbf{r}_{AB}}{r_{AB}}$$

But the HRT had already physically "passed away" at this time, and such forces were only considered as its epitaph.

As a matter of fact, Einstein's HRT comprises, via its real part, gravitation describing vacuum and the macrocosmos and, by its imaginary part, the "picocosmos" of the substructure of elementary particles (quarks) and, accordingly, the structure of hadrons as "substances."[9]

As Eddington mentioned, one can obviate Einstein's objection to Weyl's geometry, criticizing that it violates microscopic reversibility, if one continues the Riemannian geometry of GRT into the complex part of the manifold. While a real, physically significant generalization of GRT destroys fundamental principles of physics, a complex generalization of geometry introduces, according to Eddington's arguments, substantial matter into the theory.

Such a generalization of GRT referred to the requirement of quantum mechanics that complex wave fields be considered. The Hermitian symmetry of quantum mechanics was also assumed to be valid in HRT: Only Hermitian geometrical objects can be interpreted physically (in real space-time coordinates).

However, Einstein's HRT does not imply, as Einstein assumed, firstly gravitation and electromagnetism with the zero rest mass gravitons and photons and, secondly, electrons as field sources whose rest masses are given by the field energy. This theory unifies actually gravitodynamics and chromodynamics, i.e., the gravitational field, determined by the real part of the metric, and the "gluon field", represented by the dual of the imaginary part of the metric. Accordingly, the field quanta without rest mass are the gravitons and the gluons. Both types of particles should not occur as normal free particles: In GRT the gravitons are quasiparticles, showing that there are limitations on measurements resulting from the quantum structure of the substantial measuring devices. Free gluons cannot exist by "confinements" of chromodynamics either.

The substantial particles of HRT, carrying "color charges"

$$\eta_A = (c^2/G^{1/2})\varepsilon_A$$

and rest masses determined by the field, are identical with the "quarks." But those cannot exist as isolated particles. Indeed, from the gravitational part of Einstein's HRT equation it follows that such isolated color charges would possess an unbounded field mass, which corresponds to a diverging gravitational potential of the isolated quark.

Only n-quark systems with a vanishing total color charge

$$\sum_A^n \eta_A = 0$$

have bounded rest masses which are determined asymptotically by their gravitodynamics and amount to

$$\sim \frac{1}{c^2} \sum_A^n \eta_A^2 L$$

where L is the linear dimension of the quark system. The attraction between the quarks,

$$\frac{\eta_A \eta_B}{4} = \frac{c^4}{4G} \epsilon_A \epsilon_B$$

does not depend on distance. The latter is an integrability condition of the gravitational part (9a) of the HRT equations, where the imaginary field

$$g_{\underset{\vee}{\mu\nu}} = \tfrac{1}{2}(g_{\mu\nu} - g_{\nu\mu}) = iF_{\mu\nu}$$

is the source of gravitation.

The comparison of the order of magnitude of the imaginary part $g_{\underset{\vee}{\mu\nu}}$ of the Hermitian metric $g_{\mu\nu} = g^*_{\nu\mu}$ with the real part

$$a_{\mu\nu} = \tfrac{1}{2}(g_{\mu\nu} - g_{\nu\mu}) \approx \eta_{\mu\nu}$$

i.e., with the quasi-Minkowski value of the Riemannian metric of GRT, justifies the expectations of Eddington and Einstein. The value of $(-g_{\underset{\vee}{\mu\nu}} g^{\underset{\vee}{\mu\nu}})^{1/2}$ is determined by the root of Eddington's number, i.e., by the ratio of gravitation and strong interaction (in subatomic regions).

Indeed, using

$$g_{\underset{\vee}{ik}} \approx i\eta_{ikl}\epsilon \frac{X^l}{r}$$

from

$$|\delta| \propto \frac{c^4}{G}\epsilon^2 \propto \eta^2$$

it follows for hadronic matter that

$$m \propto \eta^2 L/c^2$$

where L is a subatomic length having the order of magnitude of the Compton wavelength, $L \propto h/mc$. Accordingly, one obtains

$$m \propto \eta^2 \frac{L}{c^2} \propto \eta^2 \frac{h}{mc^3} \rightarrow m \sim \eta \left(\frac{h}{c^3}\right)^{1/2}, \qquad \eta \sim m \left(\frac{c^3}{h}\right)^{1/2}$$

and

$$\epsilon = \left(\frac{G}{c^4}\right)^{1/2} \eta \propto \left(\frac{G}{hc}\right)^{1/2} m = \left(\frac{Gm^2}{hc}\right)^{1/2} \approx 10^{-20}$$

(if m equals the meson mass). Thus, the real part of the metric has the order of magnitude "one" and the imaginary part amounts to $\approx 10^{-20}$, i.e., to the root of the Eddington number 10^{-40}.

Einstein's HRT provides for the confinement of the quarks. An isolated quark having a diverging gravitational potential does not match the cosmos. Only quark systems with

$$\sum_{A}^{n} \eta_A = 0$$

possess finite heavy masses $\sim m$.

Finally, the Mach–Einstein doctrine of the relativity of inertia asserts that the inertial masses of particles are due to their membership in the cosmic system. According to this gravitodynamical interpretation of GRT, any inertial mass is a homogeneous function of the (quasi-Newtonian) gravitational potential of the totality of all the other N cosmic particles (see below).

HRT does not include electrodynamic and weak interactions. The field quanta of the Maxwell field, i.e., Einstein's photons without rest mass, are not quasiparticles but genuine free particles. Within the framework of the gauge-field theory, electrodynamics and radioactivity have to be unified in a unitary field theory with photons, of vanishing rest mass, and heavy mesons as field quanta. The potential is then a combination of Coulomb and Yukawa potentials, namely, the Bopp–Podolsky potential

$$\propto \frac{e}{r}\left[1 - \exp\left(\frac{r}{\Lambda_\mu}\right)\right]$$

This potential is everywhere regular and admits accordingly the existence of pointlike particles without inertial structure representing field sources and having the elementary charge e. Their classical field masses are of the order of magnitude

$$\not\approx \frac{e^2}{\Lambda_\mu c^2} \approx \frac{e^2}{hc}\mu$$

where

$$\Lambda_\mu = \frac{h}{\mu c}$$

is the effective range of the weak interaction, determining the radius of action of the pointlike particles. These pointlike sources are leptons with nonvanishing rest masses.

Einstein's dualism "ether and (substantial) matter" is doubled: In HRT, the real part, gravitation, faces the imaginary part, the gluon field, i.e., the macrocosmos faces the picocosmos. The electromagnetism considered by Einstein, Eddington, and Weyl as the "substance" is not included in HRT; it represents the far-reaching component of the unified theory of electrodynamics and radioactivity.

HRT has to identify the pointlike field sources of nonvanishing field mass with quarks. Quarks cannot exist freely, but must necessarily form structured systems with rest mass and finite dimension L. The quanta of HRT, gravitons and gluons, are quasiparticles without rest mass, reflecting the occurrence of limitations on measurement for space-time geometry.

The unified theory of electrodynamic and weak interactions defines the field sources by the pointlike leptons (with vanishing or nonvanishing rest masses). Their quanta without rest mass (W mesons) are also free pointlike particles.

It is a decisive feature of HRT that the masses of particle systems and their laws of motion are, as in GRT, determined gravitodynamically; GRT and HRT involve classical mechanics as limiting cases. All the other field theories operate in the framework of special relativity; here the principles of mechanics must be required as additional axioms.

Further, it may be an important point that we have no possibility for a classical model of the "confinements" in the framework of special relativity. In special-relativistic field theories, field equations of the fourth order must be of the type $k^2\Box\phi - \Box^2\phi = Q\delta(x)$ with $k \neq 0$, and in the limit $k \to 0$ all forces between deltalike charges (and the field masses of these charges) vanish. Therefore, there are no forces of the type $|\delta_{AB}| = \text{const}$ between point charges in special-relativistic field theories.

4. Mach–Einstein Doctrine

A field theory is strictly local; its principles of relativity refer to a purely local representation: The motion is always considered as motion with respect to the infinitesimal environment. Accordingly, in order to attempt a global description of physics, one needs supplementary global principles, called "telescopic principles" by Planck.[2] In a field theory, the reference systems of dynamics are additional functions of integration. They thus are not determined by microscopic assertions and results of field physics, but given via an additional principle, e.g., a cosmological one, making assertions about physics "at infinity."

The macroscopic dynamics is therefore essentially determined by boundary conditions in the sense of the first Newtonian law supplementary to the field equations. A field physics is a pure microscopic, and thus not complete, physics.

The relativistic field theory realizes its principles of relativity and covariance by offering an infinite variety of universes as mathematically possible and virtually contained in the variety of possible world models.

Thc actual cosmos specifies one and only one reference system as its "proper system."

Indeed, in the actual cosmos investigated by astronomy there is an empirical equivalence of three reference systems which are introduced by different methods. This coincidence implies a cancellation of the relativity principles: the inertial system of celestial mechanics determined dynamically, the astronomical fundamental system given by the galaxies, and the black background radiation defining the same class of reference systems. (The isotropy of the background radiation even leads to Newton's absolute reference system.)

From the viewpoint of a field theory these reference systems defined macroscopically are equivalent to a purely accidental choice of boundary and initial conditions. The cosmological world models of GRT specifying these reference systems are, from the viewpoint of GRT, certain special solutions of Einstein's equations with matter. Accordingly, the number of heavy particles in the Einstein cosmos, Eddington's cosmological number

$$N \sim \left(\frac{hc}{G\mu^2}\right)^2 \approx 10^{78}, \qquad \mu = \text{mass of baryons}$$

is also a purely accidental constant of integration.[1]

However, the cosmos is unique. We do not want to investigate some universes that are mathematically conceivable—only the one actually existing; but the latter does not result from the axioms of the field theory in a unique manner. On the one hand, the microscopic principles do too much; they provide not only the actual cosmos but an infinite variety of models, as well. On the other hand, they are not able to determine macroscopic physics at all: They furnish only necessary but no sufficient conditions. The microscopic character of field physics imposes a special supposition on the structure of physical laws. Local physics must be supplemented by telescopic principles: In general, integral principles uniquely lead to differential equations, while the inverse operation is ambiguous.

Newton's concept of an absolute space already implies such telescopic principles; Huygens sought to refer dynamics to the fixed stars by a principle of the reciprocity of motions.

Mach's principle (its realization is one of the great problems of GRT) requires the formulation of a relativistic dynamics which contains Newton's absolute dynamics as the limiting case where the motion is referred to the totality of cosmic masses.

However, a telescopic theory (in the sense of Planck) cannot be formulated either in the three-dimensional space of classical physics or in the four-dimensional space-time of the strictly local theory of relativity. From the telescopic viewpoint, every degree of freedom of a physical system provides generally one dimension of the configuration space V_{3N}.

Assuming, according to Hertz, a universe of N atomic particles of equal mass represented by mass points, the space to be used in telescopic physics is given by the $3N$-dimensional configuration space V_{3N} considered by Hertz. In general, there exist connections and interactions which are anholonomic, etc., i.e., nonintegrable. A general and complete separation of the Hertzian $3N$-dimensional space, i.e., its representation as a product of the N three-dimensional spaces of the single particles A,

$$V_{3N} = \prod_{A}^{N} \underset{(A)}{V_3}$$

is generally only possible if the connections between the particles are holonomic and accordingly capable of being represented in the form of Newtonian forces.

If the N particles are identical, this is assumed in classical physics, especially by Hertz (Boltzmann statistics!); the N factor spaces V_3 may also be identified. Thus, for holonomic connections, the three-dimensional absolute Newtonian space is again obtained. There is an obvious analogy to Schrödinger's wave mechanics: It is also formulated in a configuration space. But, unlike our mechanics, in quantum mechanics the particles are not identical, such that the partial spaces may not be identified. (Just this is reflected by the Einstein–Podolsky–Rosen paradox.) In the Hertzian configuration space, for the "total universe" $[N \sim (hc/G\mu^2)^2]$ it is now possible to formulate the dynamics with respect to the cosmic particles themselves and, accordingly, to use in dynamics only reciprocal functions of the particles, as was required by Huygens, Leibniz, and Poincaré. Following Mach and Einstein, one then has to postulate that the inertia of the N point masses A is completely induced by their mutual gravitation; inertia has to be described by a homogeneous scalar function of the local gravitational potential. Its main part stems from the average gravitational potential Φ of the universe.

According to this Mach–Einstein doctrine, Einstein's equivalence of inertia and gravitation means a reduction of inertia to the quasi-Newtonian gravitational interaction. The metric of the Hertzian configuration space V_{3N} of the N particles is then also a homogeneous function of the gravitational interaction acting between these particles. Thus, Eddington's particle number becomes a fundamental constant of dynamics.

The relations between each particle A and the cosmos, representing the embodiment of all particles, is determined by the Mach–Einstein doctrine, which can be formulated as an anholonomic Hertz condition. It requires that for each particle A the following relation always be satisfied:

$$\mathbf{p}_A = -\sum_{B}^{N}{}' \mathbf{p}_B = -\mathbf{P} \qquad (\mathbf{p}_A = \{p_i\}_A; i = 1, 2, 3)$$

where $\mathbf{p}_A$ is the momentum of the particle A and $\mathbf{P}$ is the total momentum of the remaining universe. These "recoils" do not have any influence on the motion of the particle A only if the total mass M and, accordingly, the particle number N of the universe tends to infinity.

On the other hand, in the Einstein cosmos which represents the relativistic model of a dynamics fulfilling Mach's principle, one obtains for the average gravitational potential[2,10]

$$\phi = -G\frac{M}{R} = -G\sum_{A}^{N}\frac{m_A}{r_A} = -\frac{4\pi}{3}G\rho R^2 = -\frac{c^2}{3}$$

(R is the radius of the Einstein cosmos and ρ the average cosmic density.) The bounded total cosmic mass M and the bounded average gravitational potential lead to post-Newtonian correction terms of classical gravitodynamics, that is to the Einstein–Mach effects of inertial induction by moving masses (a sort of Thirring–Lense effect which is known from GRT) and to the Einstein effects of relativistic celestial mechanics. All Einstein effects accordingly depend on the ratio of the local Newtonian potential

$$\varphi_A = -\frac{Gm_A}{r_A}$$

and the average cosmic potential,

$$\frac{\varphi}{\phi} = \frac{m}{M}\frac{R}{r} = \frac{3G}{c^2}\frac{m}{r}$$

The metric of V_{3N} and the metric g_{ik} of V_3 become, by reason of the Mach–Einstein doctrine, a homogeneous function of the Newtonian gravitational potentials φ_A of the N cosmic masses m_a. By contrast, according to Einstein's strictly local general principle of relativity, the metric $g_{\mu\nu}$ of the space-time V_4 is an inhomogeneous function of local gravitation. However, both approaches lead to local space curvatures of V_3 and generally also to additional (three-dimensional) vector potentials g_{0i}. In GRT, those correspond to the so-called Thirring–Lense effect. In the telescopic gravitational theories, they are an expression of Galilei's and Huygens' reciprocity of motions and are reflected by the Thirring effect. Actually, the Einstein effects result in a locally exact version, as correction terms of a Newtonian celestial mechanics, if the constants G and c are assumed to be equal to the values determined experimentally and the dynamics is telescopically referred to the Einstein cosmos with

$$\phi = -c^2/3$$

An unboundedly large cosmological potential, as implied by Newton's cosmology, would ascribe to the constant G/c^2 an unbounded effective value. It is evident that, in such an infinite cosmos, Newton's classical gravitodynamics would be valid instead of GRT.

5. Machian Universes

Neglecting the local Einsteinian effects of gravitation and assuming the Mach–Einstein gauge condition

$$\frac{3G}{c^2} \sum_B{}' \frac{m_B}{r_{AB}} = 1$$

the general expression for the canonical momentum of a particle A with the heavy mass m_A,

$$\mathfrak{p}_A = \frac{3G}{c^2} m_A \sum_B{}' m_B \frac{\dot{\mathfrak{r}}_{AB}}{r_{AB}}$$

provides a formulation of the reciprocity of motion:

$$\mathfrak{p}_A = m_A \left(\dot{\mathfrak{r}}_A - \frac{3G}{c^2} \sum_B{}' \frac{m_B \dot{\mathfrak{r}}_B}{r_{AB}} \right)$$

The canonical momentum of the particle A referred to a specified reference system depends not only on the velocity $\dot{\mathfrak{r}}_A$, but also on the negative velocities $-\dot{\mathfrak{r}}_B$ of all the other particles of the universe; the latter part contains the weight factor

$$\frac{3G}{c^2} \frac{m_B}{r_{AB}}$$

This dependence solves the paradoxes of relativity formulated by Huygens, Neumann, and Mach: One cannot distinguish physically if the particle A moves at a speed $\dot{\mathfrak{r}}_A = \dot{\mathfrak{r}}$ with respect to the universe or if the universe moves with the opposite speed $\dot{\mathfrak{r}}_B = -\dot{\mathfrak{r}}$ with respect to the particle. In the first case, one has

$$\mathfrak{p}_A = m_A \dot{\mathfrak{r}}_A = m_A \dot{\mathfrak{r}}$$

and, in the second,

$$\mathfrak{p}_A = -m_A \frac{3G}{c^2} \sum_B{}' \frac{m_B \dot{\mathfrak{r}}_B}{r_{AB}} = m_A \dot{\mathfrak{r}} \frac{3G}{c^2} \sum_B{}' \frac{m_B}{r_{AB}} = m_A \dot{\mathfrak{r}}$$

This fact again makes the "telescopic" character of an inertia-free gravitodynamics evident. All particles contribute to the momentum of a particle A. Accordingly, the configuration space is not separable into the corresponding three-dimensional particle spaces.

In this telescopic dynamics, the inertial forces cause this nonseparability due to the relativity of motions. In arbitrary reference systems, the local momenta $m_A\dot{\mathbf{r}}_A$ and the telescopic terms

$$-\frac{3G}{c^2}m_A\sum_B{}'\frac{m_B\dot{\mathbf{r}}_B}{r_{AB}}$$

have the same meaning for the canonical momentum $\mathbf{p}_A$ of a particle.

For a universe expanding in accordance with the Hubble law,

$$\dot{\mathbf{r}}_{AB} = H\mathbf{r}_{AB} = \frac{\dot{R}}{R}\mathbf{r}_{AB}$$

we obtain the following approximate expression for the momentum of a particle:

$$\mathbf{p}_A = m_A(\dot{\mathbf{r}}_A + H\mathbf{r}_A) \qquad (H = \text{Hubble constant} = \dot{R}/R)$$

which results from the relativity of inertia.

Because of $\dot{\phi} = -H\phi$, the Mach–Einstein gauge provides

$$\dot{m}_A = -Hm_A$$

such that, up to nonlinear terms, the acceleration is given by the Newtonian expression

$$\dot{\mathbf{p}}_A = m_A\ddot{\mathbf{r}}_A$$

In the next (post-Newtonian) limit, i.e., in the Einsteinian approximation, the momentum $\mathbf{p}_A$ of a particle is also a function of the local variation $\delta\phi$ of the gravitational potential ϕ:

$$\mathbf{p}_A = m_A\left[\left(1+\frac{\delta\phi}{\phi}\right)\dot{\mathbf{r}}_A + H\mathbf{r}_A\right]$$

For the Kepler problem (and, in GRT, for the Schwarzschild problem) of the motion of a test particle A around a local gravitational center of mass m', $\delta\phi$ is equal to the local Newtonian potential φ:

$$\delta\phi = \varphi = -Gm'/r$$

Accordingly, the post-Newtonian definition of the particle momentum $\mathbf{p}_A$ provides, for a static Einstein cosmos (with $H = 0$), just Einstein's approximate formulation of Mach's relativity of inertia in GRT:

$$\mathbf{p}_A = m_A\left(1+\frac{3G}{c^2}\frac{m'}{r_A}\right)\dot{\mathbf{r}}_A$$

Using this relation, the equations of motion

$$\dot{\mathbf{p}}_A = -\frac{Gm_Am'}{r_A^3}\mathbf{r}_A$$

immediately lead to the general-relativistic Einstein effects of celestial mechanics. These equations of motion correspond to the effective Riemannian metric of V_3

$$g_{ik} = \delta_{ik}\left(1 + \frac{3Gm'}{c^2 r}\right) \qquad (i, k = 1, 2, 3)$$

References and Notes

1. A. Mercier, H.-J Treder, and W. Yourgrau, *On General Relativity* (Akademie-Verlag, Berlin, 1979).
2. H.-J. Treder, H. -H. v. Borzeszkowski, A. van der Merwe, and W. Yourgrau, *Fundamental Principles of General Relativity Theories: Local and Global Aspects of Gravitation and Cosmology* (Plenum, New York, 1980).
3. H.-J. Treder and W. Yourgrau, *Phys. Lett.* **68A,** 415 (1978).
4. H.-J. Treder, *Die Geometrisierung der Physik—zum 50. Jahrestag der Einsteinschen Akademie* (communication, Berlin, 1975).
5. A. Einstein, *The Meaning of Relativity,* 3rd edition (Princeton University Press, Princeton, New Jersey, 1950).
6. L. Infeld, *Nature* **166,** 1075 (1950).
7. H.-J. Treder, *Wiss. Z. Humboldt-Univ. Berlin Math. Naturwiss Reihe* **4,** 9 (1955).
8. V. Hlavatý, *Geometry of Einstein's Unified Field Theory* (Noordhoff, Groningen, 1957).
9. H.-J. Treder, *Ann. Phys.* (Leipzig) **19,** 370 (1950); **36** (1980).
10. H.-J. Treder, *Die Relativität der Trägheit* (Akademie-Verlag, Berlin, 1972); H.-J. Treder, *Über Prinzipien der Dynamik bei Einstein, Hertz, Mach, und Poincaré* (Akademie-Verlag, Berlin, 1974).

2

The Zitterbewegung and the Einstein *A* Coefficient of Spontaneous Emission

A. O. Barut
Department of Physics
University of Colorado
Boulder, Colorado 80309

1. The Creation of a "Photon"

The process of emission and absorption of a photon by the electron is *the* ultimate basic elementary process of quantum electrodynamics (QED). The QED does not further analyze the details and any possible spatial and temporal structure of this process. Questions concerning the duration of the emission process, the direction of the emission, in fact, the shape of the photon itself, can only be posed and answered in a probabilistic sense and according to the rules of the quantum theory of measurement.

The emission (and absorption) process may not be as simple as we have come to be accustomed to. One must first of all not always use naively the perturbation theory picture of an ($ee\gamma$) vertex, characterized by a coupling constant α (which is an input of the theory), for all real and virtual emission processes. Furthermore, a single physical photon is quite different from the photon concept of QED-perturbation theory, i.e., the photon line that enters the Feynman diagrams (or, a plane wave ϵe^{ikx} with $k^2 = 0$). We recall that Einstein has introduced his light quantum as a *localized* energy quantum, distributed discontinuously in space, and this as a "heuristic point of view concerning the creation and conversion of light."[1] Recent experiments on individual photons and electrons may be interpreted by assuming the photon to be indeed a *localized* object (not a plane wave) which has a spatial extension of the order of the coherence length, but whose energy may be much more localized, perhaps pointlike, within the wave train.[2] The shape of the photon can be deformed; its frequency spectrum need not consist of a single value (e.g., a photon emitted by a transition from a state which is a superposition of two neighboring states to another sharp ground state). If we call the actual emitted lump of light quantum a single photon, as indeed we should, then it surely carries some information about the way it was produced (it may even contain nonlinear terms in the electric and magnetic fields),[3] hence a photon may not have the kind of individuality that we tend to attribute to it as a *particle of mass zero*.

I believe the structure of the photon, and further, even the internal structure of the electron, have important consequences on the interpretation and foundation of quantum theory and the theory of measurement. For example, in a double-slit experiment, a single photon can "interfere with itself"[4] when two parts of its wave go through two different slits and then recombine; yet the localized energy will go through one of the slits only.

Perhaps a nonperturbative treatment of the emission process in QED will give the properties of the physical localized photon. But this has to be done in some other way than by summing up the Feynman diagrams. Beyond this, it has been further suggested recently that we look in detail and closely at the "act of creation" of a "photon" from an accelerated electron (say in an excited state in an atom), rather than taking the emission vertex as a primary input.[5] This goes really beyond QED, and the purpose is to aim to calculate the shape, energy, and other characteristics of the emitted photon.

The process of emission (or absorption) of a photon has a correspondence-principle-limit in the classical radiation formula. However, the quantum emission seems to be qualitatively different from the classical radiation in that the emitted energy of the radiation is proportional to the frequency of radiation, $E = \hbar\omega$, instead of the third power of frequency. The latter dependence comes from the radiation of an accelerated point charge. However, for the relativistic Dirac electron, the motion of the charge does not coincide with the motion of the center of mass of the electron: the charge performs a so-called "*Zitterbewegung*" with the eigenvalues of velocity equal to c (the velocity of light) around the center of mass. Although the charge, when considered to be moving with the center of mass, radiates an energy proportional to ω^3, the radiation from the *Zitterbewegung* is quite different. The former can be thought to be the average over the radiation from the *Zitterbewegung*. Using the properties of the *Zitterbewegung* of the relativistic Dirac electron, and applying the radiation formula to the *Zitterbewegung* and some heuristic technical assumptions, it was indeed possible to derive the linear relationship $E = \hbar\omega$, and the value of the Planck constant $\hbar$ in terms of the values of e and c (or, the value of $\alpha = e^2/\hbar c$).[5] Of course, any argument going beyond an established theory in physics can at first be only heuristic. We must therefore see more precisely in which direction the theory is being generalized by also rederiving known results from the new point of view. For this purpose we consider here the spontaneous emission of a photon from an excited state using the radiation formula, again applied to the *Zitterbewegung*, and discuss the details of the remark[6] that Einstein's A coefficient can be calculated by this method. Possible generalizations will also be given.

2. A Theory of Spontaneous Emission

There are two standard derivations of the rate of spontaneous emission, hence of the A coefficient: (1) The original Einstein derivation using thermodynamic equilibrium and the Planck distribution law; (2) the derivation in quantum electrodynamics, in which an interaction of the form $H_{\text{int}} = \mathbf{p} \cdot \mathbf{A}$ or $\mathbf{d} \cdot \mathbf{E}$ ($\mathbf{d}$ = dipole moment operator) is used between an atomic system and the quantized radiation, and in first-order time-dependent perturbation theory, the matrix element between the state $|\psi_2, N = 0\rangle$ and the state $|\psi_1, N = 1\rangle$ is evaluated. Here N refers to the number of "quanta" in the radiation field.

Now for the relativistic Dirac electron, the velocity of the charge is given by

$$\dot{\mathbf{x}} = \frac{i}{\hbar}[H,\mathbf{x}] = c\alpha \tag{1}$$

with eigenvalues of each component being $\pm c$. The components of the velocity do not commute, a peculiar motion which Schrödinger appropriately called *Zitterbewegung.*[7] The coordinate x in the Dirac equation will be identified with the position of the charge because in the minimal coupling to the electromagnetic field, the field couples locally to the charge at the point x. One can introduce another average coordinate x_A, the center of mass of the electron, whose velocity, according to the laws of relativistic mechanics, is given by

$$\dot{\mathbf{x}}_A = c^2H^{-1}\mathbf{p} \tag{2}$$

where p is the momentum of the Dirac electron and m its rest mass.

In fact, Eq. (1) can be integrated, and one can write[8]

$$\mathbf{x}(t) = \mathbf{x}_A(t) + \xi(t)$$
$$\xi(t) = \frac{1}{2}i\hbar c \exp\left(2\frac{i}{\hbar}Ht\right)[\alpha(0) - cH^{-1}\mathbf{p}]H^{-1}$$

as the solution of quantum Heisenberg equations. We note that the frequency and amplitude of the *Zitterbewegung* for a free electron are of the order of $2mc^2/h^2$ and $h/2mc$, respectively.

The idea is now to apply the radiation formula to the acceleration of the charge $\ddot{x}$, and relate it to that of $\ddot{x}_A$. The radiated power from the charge is given by[9]

$$\frac{dE}{dt} = \frac{2}{3}\frac{e^2}{c^3}\overline{\ddot{\mathbf{x}}^2} \tag{3}$$

where the bar denotes a time average. Because x is an operator, we can obtain in two ways a number from Eq. (3): either by taking the eigenvalue of the operator, or by taking a matrix element of the operator.

We assume that underlying the quantum description the charge performs a stochastic motion around the center-of-mass motion with instantaneous velocity c. Then,

Eq. (3) evaluated with the eigenvalues of the operator $\ddot{\mathrm{x}}$ is identified with the actual power of quantum emission. If we give to this microscopic motion of $\dot{\mathrm{x}}$, during acceleration, a frequency ω, and an amplitude c, we can derive[6] for the energy E emitted during one period the Planck relation $E = \hbar\omega$.

On the other hand, quantum mechanics describes the behavior of the electron by a probability wave function $\psi(x)$ (instead of an actual microscopic motion). In this case we can evaluate the matrix element of (3) between any two states

$$\left\langle \psi_2 \middle| \frac{dE}{dt} \middle| \psi_1 \right\rangle = \frac{2e^2}{3c^3} \langle \psi_2 | \ddot{\mathrm{x}}^2 | \psi_1 \rangle \tag{4}$$

Therefore the ratio

$$A = \frac{\left\langle \psi_2 \middle| \dfrac{dE}{dt} \middle| \psi_1 \right\rangle}{\displaystyle\int_0^T \left(\frac{dE}{dt} \right) dt} = \frac{1}{\hbar\omega} \left\langle \psi_2 \middle| \frac{dE}{dt} \middle| \psi_1 \right\rangle \tag{5}$$

is just the Einstein A coefficient, the probability of emission of the energy quantum $\hbar\omega$, or the rate of spontaneous emission (or inverse lifetime).

We now discuss the evaluation of (4) using (1). Expanding the states ψ_1, and ψ_2,

$$\psi_i = \sum_n a_n^{(i)} | n \rangle$$

where $|n\rangle$ are states of definite energy E_n, and using

$$\alpha(t) = e^{-iHt/\hbar} \alpha(0) e^{iHt/\hbar} \tag{6}$$

we first evaluate

$$\begin{aligned}
\langle \psi_2 | \ddot{\mathrm{x}} | \psi_1 \rangle &= c \langle \psi_2 | \dot{\alpha}(t) | \psi_1 \rangle \\
&= \frac{i}{\hbar} c \langle \psi_2 | [H, \alpha] | \psi_1 \rangle \\
&= \frac{ic}{\hbar} \sum_{n,m} a_n^{(2)*} a_m^{(1)} (E_n - E_m) \langle n | \alpha(t) | m \rangle \\
&= c \sum_{n,m} a_n^{(2)*} a_m^{(1)} \frac{i}{\hbar} (E_n - E_m) \exp \left[-\frac{i}{\hbar} (E_n - E_m) t \right] \langle n | \alpha(0) | m \rangle
\end{aligned}$$

Next we have to evaluate $|\langle \psi_2 | \ddot{x} | \psi_1 \rangle|^2$. In general there will be a lot of terms, but for simplicity we consider the monochromatic case, and assume that there are two definite eigenstates of energy and $(1/\hbar)(E_2 - E_1) \equiv \omega$. In this case

$$|\langle \psi_2 | \ddot{\mathrm{x}} | \psi_1 \rangle|^2 = c^2 \omega^2 |\langle 2 | \alpha(0) | 1 \rangle|^2 \tag{7}$$

The nonrelativistic limit of the matrix element of $\alpha(0)$ is given by

$$\langle \psi_2 | \alpha(0) | \psi_1 \rangle \rightarrow \frac{1}{m} \psi_2 \left(\vec{p} + \frac{e}{c} \vec{A} \right) \psi_1 \tag{8}$$

The matrix element of the power of emitted radiation becomes

$$\left\langle \psi_2 \left| \frac{dE}{dt} \right| \psi_1 \right\rangle \rightarrow \frac{2}{3}\frac{e^2}{c^3}\omega^2\frac{1}{m^2}\left|\left\langle \psi_2 \left| \vec{p} + \frac{e}{c}\vec{A} \right| \psi_1 \right\rangle\right|^2$$

Neglecting the **A** term, passing from the matrix element of $\vec{p}$ to that of $\vec{r}$, $\langle 2|\vec{p}|1\rangle = m\omega\langle 2|\vec{r}|1\rangle$, and using the fact that the radiation has two degrees of polarization, we obtain

$$\left\langle \psi_2 \left| \frac{dE}{dt} \right| \psi_1 \right\rangle \cong \frac{4}{3}\frac{e^2}{c^3}\omega^4|\langle 2|\vec{r}|1\rangle|^2 \tag{9}$$

Consequently, from Eq. (5), the A coefficient is

$$A \cong \frac{4}{3}\frac{e^2}{c^3}\frac{\omega^3}{\hbar}|\vec{r}_{12}|^2 \tag{10}$$

3. Further Problems

The present theory of spontaneous emission opens up a number of new directions for further study:

(1) The study of the properties of the Lienard–Wiechert potentials $A_\mu(x)$ using the eigenvalues of the velocity $\dot{x} = c$ of the *Zitterbewegung* and the corresponding Lienard–Wiechert potential due to the Möller transition current $\langle\psi_2|\vec{\alpha}|\psi_1\rangle$ parallel exactly our comparison of two radiation formulas. This would answer some of the questions concerning the nature of the "photon" discussed in Section 1.

(2) The use of more general states (instead of the stationary states), and more general relativistic expressions for the spontaneous emission (instead of the nonrelativistic approximation).

(3) Assuming that the (classical) electron does indeed perform, because of its self-field, a kind of (stochastic) motion around its center of mass, something like the Dirac electron, we may try to reconcile classical and quantum theories.[10] The true understanding of quantum theory can only come if we can calculate the numerical value of h (or α). Hence the assumptions made in our derivation must be further studied. Furthermore the appearance of the wave function from the averaging over the "Zitterbewegung," the limit from the quantum radiation energy relationship, $E = \hbar\omega$, to the classical radiation energy relationship $E \sim \omega^3$, must be investigated.

The self-energy effects can be evaluated in classical electrodynamics in closed form and they lead to nonlinear dynamical equations. Such nonlinear equations can indeed show stochastic or chaotic behavior, in addition to certain periodic, limit-cycle (i.e., or stationary) behavior. It seems that the wave-mechanical description includes in a probabilistic sense some of these nonlinear self-energy effects. From this point of view it does not look too strange that one can derive the emission of a photon and Planck's constant from the classical radiation formula applied to *Zitterbewegung*.

References and Notes

1. A. Einstein, *Ann. Phys. (Leipzig)*, 4th Ser. **17,** 132–148 (1905).
2. See, e.g., H. Paul, *Wiss. Fortschr.* **29,** 62 (1979).
3. M. Lax and M. Zwanziger, *Phys. Rev.* A **7,** 750–771 (1973).
4. P. A. M. Dirac, *Quantum Mechanics*, 4th Ed. (Clarendon Press, Oxford, 1958).
5. A. O. Barut, *Z. Naturforsch.* **33a,** 993–994 (1978).
6. A. O. Barut, *Czech. J. Phys.* **B29,** 3–5 (1979).
7. E. Schrödinger, "Über die kraftefreie Bewegung in der relativistischen Quantenmechanik," *Sitzungsber. Preuss. Akad. Wiss.* **XXIV,** 418–428 (1930).
8. A. O. Barut and A. J. Bracken, *Phys. Rev. D* **23,** 2454 (1981) and *D* **24,** 3333 (1981).
9. W. Heitler, *The Quantum Theory of Radiation*, 3rd. ed. (Oxford Univ. Press, Oxford, 1954), pp. 25, 179.
10. A. O. Barut, in *Quantum Theory and the Structures of Time and Space*, Vol. III, L. Castell and C. F. von Weizsäcker, editors (C. Hanser Verlag, Münich, 1979), pp. 175–182.

3

SUPERSYMMETRY GROUPS*

PETER G. BERGMANN
Physics Department
Syracuse University
Syracuse, New York 13210

Physics Department
New York University
New York, New York 10003

1. INTRODUCTION

In recent years a new type of symmetry, known as "supersymmetry," has assumed considerable interest, because it appears to provide a channel for unifying the external and the internal symmetries. Whereas the external symmetries relate to our ideas of space and time and are associated with the notions of energy, linear momentum, and angular momentum (including spin), the internal symmetries are associated with perceived other properties of elementary particles, their electric charge, their isospin, strangeness, lepton and baryon number, color, charm, etc. Among elementary particles there are to be found families that combine bosons and fermions; accordingly, supersymmetry includes transformations that lead from bosonic to fermionic states, and vice versa. The term "supergravity" denotes theories in which supersymmetry is gauged and in which the external (space-time) symmetry is that of diffeomorphisms (rather than the Poincaré group), characteristic of general relativity.

Most of the work to date has been based on the introduction of hypercomplex numbers forming a Grassmann algebra, usually over the field of complex numbers. A somewhat different proposal has been put forward recently,[1] which, as will be shown in this paper, uses a different type of symmetry group.

In order to understand better the relationship between conventional supersymmetry and our modification, it appeared useful to look at standard supersymmetry and its groups in a wider context. Section 2 is concerned with a generalization of supersymmetry, which includes the treatments based on the Grassmann algebras as a special case.

*Research supported in part by National Science Foundation grant No. 78-06721.

Gürsey and Marchildon[2] constructed supersymmetry groups by considering linear vector spaces whose members possess components that are themselves Grassmann elements, and linear mappings of such linear vector spaces on themselves, including those that mix the odd and even Grassmann components. They restrict the mappings to those that preserve unchanged a certain bilinear form which acts on pairs of vectors and has values within the domain of even Grassmann numbers. These mappings form a group. Gürsey's and Marchildon's procedure, it turns out, can be applied to more general hypercomplex number systems.

Remarkably, the generalized problem has nontrivial solutions, which appear identical in form to those exhibited by Gürsey and Marchildon. That the resulting groups are in fact not isomorphic to each other is disguised by the notation, which is the one in almost universal use today.

Grassmann algebras are homomorphic to the algebra of ordinary (complex or real) numbers, and this fact forms the topic of Section 3. The homomorphism is significant from a formal point of view because it is a characteristic of "supergroups" viewed as abstract Lie groups. Physically the homomorphism implies that any "superfield" contains an ordinary field, whose dynamical behavior is unaffected by the presence of the pure Grassman fields.

Section 4 discusses further our recent proposal,[1] which in essence introduces a nonassociative algebra for the system of "numbers" to be employed. If one follows as closely as possible the procedures used by Gürsey and Marchildon, one constructs a linear vector space whose even components are ordinary numbers and whose odd components are members of the nonassociative algebra. The bilinear form that resembles the one used in Section 2 differs from the latter in that its values are ordinary numbers. The group of vector transformations that leave this form invariant is isomorphic to an (ungraded) unitary group, and therefore simple.

2. Graded Systems with Hypercomplex Numbers

Grassmann algebras are typical of hypercomplex number systems that admit a grading, but they are not the only hypercomplex systems with that property. In what follows we shall consider vectors and matrices whose entries are graded hypercomplex numbers not necessarily Grassmann. We shall require also the availability of an operation that is the analog of the complex conjugate of a complex number, and of the Hermitian adjoint of matrices; we shall call that operation *adjunction*. We shall first define the properties required of our hypercomplex numbers, give two examples that are not Grassmann, and proceed next to vectors and matrices. The section will conclude with the construction of the generators of the symmetry groups.

Of our hypercomplex numbers we shall require that they form a hypercomplex associative algebra, and that each member of the system can be decomposed additively and uniquely into an "even" and an "odd" part. Both even and odd hypercomplex numbers are to be subject to "adjunction." The adjoint of an even number is to be even, the adjoint of an odd number odd. The adjoint of an ordinary (complex) number is to be its conjugate complex, and adjunction is to commute with addition. With respect to multiplication (which in general is noncommutative) adjunction is to obey the rule

$$(AB)^\dagger = \pm B^\dagger A^\dagger \tag{1}$$

The minus sign is to hold if and only if both factors are odd.

These postulates all apply to Grassmann numbers, but they are somewhat more general. They are compatible with Clifford algebras (such as the creation and annihilation operators of a fermionic field). They, for instance, apply to a complex-four-dimensional linear vector space spanned by the four Pauli spin matrices, with σ_x and σ_y taken to be odd numbers and σ_z and 1 to be even, and provided the adjoints of these four basic vectors are to be defined as follows:

$$\begin{aligned} \sigma_x^\dagger &= i\sigma_x, \qquad \sigma_y^\dagger = i\sigma_y \\ \sigma_z^\dagger &= \sigma_z, \qquad 1^\dagger = 1 \end{aligned} \tag{2}$$

The second example is a graded complex-eight-dimensional algebra. The four odd basis vectors $\tau_0, \tau_1, \ldots, \tau_3$ and the four even basis vectors $1, \sigma_1, \ldots, \sigma_3$ are all to be self-adjoint. The multiplication table is as follows, with all relations to be cyclic with respect to the three indices 1, 2, 3:

$$\begin{aligned} &\tau_0\tau_i = \tau_i\tau_0 = i\sigma_i, \qquad i = 1, 2, 3 \\ &\tau_1\tau_2 = -\tau_2\tau_1 = \sigma_3, \qquad \text{etc.} \\ &\tau_1^2 = \tau_2^2 = \tau_3^2 = \tau_0^2 = i \\ &\tau_0\sigma_i = \sigma_i\tau_0 = \tau_i \\ &\sigma_1\tau_1 = \tau_1\sigma_1 = \sigma_2\tau_2 = \text{[illegible]} = -\tau_0 \\ &\sigma_1\sigma_2 = -\sigma_2\sigma_1 = -i\sigma_3, \qquad \text{etc.} \\ &\sigma_1\tau_2 = \tau_1\sigma_2 = -\tau_2\sigma_1 = -\sigma_2\tau_1 = -i\tau_3, \qquad \text{etc.} \\ &\sigma_1^2 = \sigma_2^2 = \sigma_3^2 = 1 \end{aligned} \tag{3}$$

This algebra can be built up from the three odd elements τ_i in that the remaining linearly independent elements are all (multiple) products of these three.

Especially the last example demonstrates that both among the odd and the even elements there can be some that commute, others that anticommute. We shall now proceed with the general formalism, which presupposes no more than the condition (1) (along with associativity, etc.) and construct vectors and matrices.

A vector U is to possess even and odd components, with the notation

$$U = \begin{pmatrix} u \\ \psi \end{pmatrix} \qquad (u \text{ even}, \psi \text{ odd}) \tag{4}$$

The adjoint vector $U\dagger$ is to be defined as a row vector,

$$U^\dagger = (u^\dagger, \psi^\dagger) \tag{5}$$

A matrix, likewise, can be decomposed into even and odd submatrices:

$$\mathcal{M} = \begin{pmatrix} M, & \rho^{\dagger} \\ \sigma, & m \end{pmatrix} \tag{6}$$

M and m are even and square submatrices (i.e., matrices whose entries are all even), whereas $\rho^{\dagger}$ and σ are odd and rectangular. The adjoint of $\mathcal{M}$ is to be defined as

$$\mathcal{M}^{\dagger} = \begin{pmatrix} M^{\dagger}, & \sigma^{\dagger} \\ -\rho, & m^{\dagger} \end{pmatrix} \tag{7}$$

With this definition we have the two rules

$$\begin{aligned} (\mathcal{M}U)^{\dagger} &= U^{\dagger}\mathcal{M}^{\dagger} \\ (\mathcal{M}\mathcal{N})^{\dagger} &= \mathcal{N}^{\dagger}\mathcal{M}^{\dagger} \end{aligned} \tag{8}$$

Next, a bilinear form (U, V) is to be defined as follows:

$$(U, V) \equiv U^{\dagger}\mathcal{H}V \equiv (u^{\dagger}, \psi^{\dagger}) \begin{pmatrix} \eta, & 0 \\ 0, & \epsilon \end{pmatrix} \begin{pmatrix} v \\ \chi \end{pmatrix} \tag{9}$$

where the even submatrices η and ϵ are self-adjoint and anti-self-adjoint, respectively:

$$\eta^{\dagger} = \eta, \qquad \epsilon^{\dagger} = -\epsilon \tag{10}$$

The value of the form (U, V) is an even hypercomplex number whose adjoint is (V, U). (U, U) is self-adjoint.

The form (U, V) will be invariant under linear transformations of the vector space U by means of matrices of type (6) if these matrices satisfy the condition

$$\mathcal{M}^{\dagger}\mathcal{H}\mathcal{M} = \mathcal{H} \tag{11}$$

or

$$\begin{aligned} M^{\dagger}\eta M + \sigma^{\dagger}\epsilon\sigma &= \eta \\ m^{\dagger}\epsilon\sigma &= \rho\eta M \\ m^{\dagger}\epsilon m - \rho\eta\rho^{\dagger} &= \epsilon \end{aligned} \tag{12}$$

The solutions of this set of conditions form a group, which is the analog of the group of unitary matrices in the algebra of complex numbers.

The remarkable fact is that there are nontrivial solutions, in spite of the fact that in terms of complex components a large number of bilinear forms must be kept invariant. These solutions can be obtained, following, e.g., Gürsey and Marchildon,[2] by exponentiating the solutions of the infinitesimal problem corresponding to the conditions (12),

$$\begin{aligned} M^{\dagger}\eta + \eta M &= 0 \\ \epsilon\sigma &= \rho\eta \\ m^{\dagger}\epsilon + \epsilon m &= 0 \end{aligned} \tag{13}$$

If η is nonsingular, then ρ is uniquely determined by σ, which in turn is subject to no further condition. ηM is anti-self-adjoint, and ϵm self-adjoint, with no further conditions on M or m.

If the dimensionality of the even and the odd parts of the linear vector space spanned by vectors of type (4) are K and k, respectively, then there are $\frac{1}{2}K(K - 1)$ "linearly independent" generators of type M, $\frac{1}{2}k(K + 1)$ generators of type m, and kK generators of type σ. The term "linearly independent" is to be understood with respect to the hypercomplex number system adopted. The (scalar) coefficients of the even generators M and m must be even hypercomplex numbers, those of the odd generators σ odd. For a Grassmann algebra with a set of base elements $e_1, \ldots, e_n$, for instance, there are 2^{n-1} distinct elements of each kind that are linearly independent with respect to their complex (not hypercomplex) coefficients. Interpreted as an ordinary Lie group, a supergroup over a Grassmann algebra has a complex dimensionality equal to

$$D = 2^{n-2}[(K + k)^2 + (k - K)] \tag{14}$$

The next section will be concerned with the structure of groups over Grassmann algebras.

3. Structure of Supersymmetry Groups

The appearance of equations such as (12) and (13) might make it appear that the structure of supersymmetry groups is independent of the hypercomplex algebra employed, but this appearance is deceptive. Even if it is agreed that the underlying algebra is Grassmann, the details of a supersymmetry group depend not only on the number of even and odd components of the linear vector space (1), which are exhibited by notations such as $SU(m/n)$, but also by the dimensionality of the chosen Grassmann algebra. In this respect, supersymmetry groups resemble standard Lie groups, in which also the real and the complex groups differ from each other, though the algebraic relations among the generators appear alike.

Grassmann algebras are homomorphic to algebras of ordinary numbers, and this homomorphism carries over to the corresponding supersymmetry groups. Any element of a Grassmann algebra can be decomposed into its odd and its even parts, and the latter decomposes further into its "pure Grassmann" (nilpotent) part and the remainder, which is an ordinary number. For a Grassmann element a denote that ordinary number by a_0. Then we have for addition and for multiplication the following rules, which establish the homomorphism:

$$\begin{aligned}(a + b)_0 &\equiv a_0 + b_0 \\ (ab)_0 &\equiv a_0 b_0 \equiv (ba)_0\end{aligned} \tag{15}$$

This homomorphism can be extended to the supersymmetry group that is represented by the matrices (6), (11). Replace each matrix $\mathcal{M}$ by the complex part of M, to be denoted by M_0. Likewise replace the matrix $\mathcal{H}$, (9), by the complex part of η, denoted by η_0. The matrices M_0 form an (ordinary, ungraded) unitary group, which leaves η_0 invariant. This group is homomorphic to the original supergroup; the invariant subgroup consists of all those matrices (6), (11) whose complex parts M_0 are the K-dimensional unit matrix.

This homomorphism is not a general property of supersymmetry groups, but only of those based on Grassmann algebras, the ones ordinarily dealt with in the recent literature. The basis for the homomorphism is the fact that products of pure Grassmann elements are always pure Grassmann, in other words, that under no circumstances can products of odd or even pure Grassmann elements contribute to the part M_0 of a product matrix. Likewise, it is only the M_0 part of any matrix (6) that acts on the complex part u_0,

$$U_0 = \begin{pmatrix} u_0 \\ 0 \end{pmatrix} \tag{16}$$

of a vector (4).

In quantum field theory two fermions can combine to produce a boson, as in the annihilation of an electron–positron pair. Likewise the product of a fermion creation and annihilation operator can contribute to occupation numbers, total energy, etc. All of this is possible only because creation and annihilation operators together form a Clifford algebra, not a Grassmann algebra, and this is one reason why Grassmann algebras may not be completely satisfactory in the mathematical formulation of fermion fields in physics.

Given the infinitesimal generators of the whole supersymmetry group with the conditions (13), the total number of entries that can be chosen freely are, as mentioned previously, $\frac{1}{2}K(K+1)$ even and $\frac{1}{2}k(k-1)$ odd hypercomplex numbers. The hypercomplex entries of commutators of any of these generators again obey the conditions (13), as verified by a straightforward computation. This calculation does not require explicit reference to the multiplication table of the chosen hypercomplex algebra; it suffices to know that the algebra is closed and associative.

The details of the multiplication table do, however, enter when the commutator of two generators is to be represented as a particular combination of the linearly independent basis generators, in other words when the structure constants of the supersymmetry group are to be determined. That is why, even with a given hypercomplex number system, two generators with the same single nonvanishing matrix entry are not equivalent if the respective nonvanishing entries are different (i.e., linearly independent) hypercomplex numbers.

4. Modified Algebras and Related Symmetries

In Ref. 1 a different "number" system was introduced, in which multiplication is not associative. The numbers of this modified system are still graded into two classes, "even" and "odd," but all even numbers are ordinary complex numbers. Vectors can be cast into the form (4). There will also be an adjunction, with the understanding that the product rule for "numbers" is the same as in Eq. (1). With the help of a bilinear form of the type (9) we associate with any two vectors of type (4) a complex number, which goes over into its conjugate complex if the two factors U, V are interchanged.

A vector of type (4) can be represented in terms of the complex components of the constituent "numbers," a representation that reveals the true (complex) dimensionality of the linear vector space employed. In terms of such a representation the form (9) is an ordinary finite-dimensional Hermitian form. The linear substitutions that preserve the form of $\mathcal{H}$ will form one of the standard unitary groups, and thus be simple. Though these substitutions can be represented in the form of matrices having the general form (6), the submatrix m must be thought of as a square array of entries which in turn are not simply complex numbers but involve linear substitutions within the linear vector space of odd numbers.

The appearance of entries that are not elements of the algebra of numbers is both possible and unavoidable because of the fact that the algebras considered in this section are nonassociative, in contrast to those of Section 2. In an associative algebra the product of any two hypercomplex numbers is again a hypercomplex number. Thus, though even in an associative algebra there are linear substitutions available that cannot be represented by matrices of type (6), the product of two matrices of type (6) will be a matrix of the same type. If two such matrices satisfy the conditions (12), their product will satisfy the same conditions.

In a nonassociative number system it is possible to consider matrices acting on vectors of type (4) whose entries are "numbers." But the result of two such linear transformations carried out consecutively need not be a matrix of the same kind, i.e., its entries need not be "numbers." The generalization is slight: The entries of the submatrix m can be even (i.e., complex) numbers or dyadic products of two odd numbers, the first factor being a column vector, the second a row vector.

Specifically, consider two infinitesimal matrices that satisfy the conditons (13), with both M and m zero. Their commutator will be of the form

$$\begin{aligned} M &= \eta^{-1}(\sigma_2^\dagger \epsilon \sigma_1 - \sigma_1^\dagger \epsilon \sigma_2) \\ m &= (\sigma_2 \eta^{-1} \sigma_1^\dagger - \sigma_1 \eta^{-1} \sigma_2^\dagger)\epsilon \end{aligned} \tag{17}$$

The entries of the matrix M are complex numbers. Those of the matrix m are submatrices, representing "dyadic" products of the "vectors" of type σ ("column vectors") by those of type $\sigma^\dagger$ ("row vectors"). Quite generally the entries of that matrix cannot be restricted to complex numbers (even though m is "even") if the Lie algebra of the generators is to be closed.

The fact that the symmetry group of this section is simple proves that it cannot be isomorphic to the symmetry groups associated with Grassmann algebras, which are not simple (as demonstrated in Section 3). The invariance groups of this section are interesting in that the entries of the submatrices M are complex numbers. As the generators of type M form an obvious subgroup (though not an invariant subgroup), the elements of that subgroup can be interpreted as space-time symmetries more realistically than the corresponding transformations among the "even" but hypercomplex components of a Grassmann algebra.

5. Conclusion

This paper has been devoted primarily to a discussion of supersymmetry groups as abstract groups. These fall broadly into two classes, one associated with associative hypercomplex number systems, whose "even" members have a much richer structure than ordinary complex numbers, the other based on nonassociative "numbers" constructed so that its even elements are simply complex numbers. Supersymmetry groups are ordinary Lie groups, not always simple; but in both classes a shorthand notation such as $SU(m/n)$ (with fixed values of m, n) contains a collection of abstract groups that are not isomorphic to each other.

The linear spaces that are mapped on themselves by matrices of type (7), (12) can be decomposed into invariant (nonlinear) subspaces that are the analogs of mass shells and light cones. In the case of Grassmann algebras such an invariant subspace may consist of all vectors for which the real part of the norm has a fixed value, spacelike, timelike, or lightlike. For algebras of the kind discussed in Section 4 this classification is the only one available. For Grassmann algebras a finer classification also exists, one in which all components of the norm are fixed and taken into account. For other hypercomplex number systems, this more elaborate classification may be the only one to exist.

Given the huge variety of supersymmetry groups it is not too difficult to find such groups that contain as subgroups symmetries believed to play a role in elementary particle physics. Any further selection would seem to depend on constructing complete physical theories that incorporate whatever supersymmetry is being considered and on studying their dynamical consequences.

Acknowledgments

The author has profited greatly from discussions with A. Komar, with P. Van Nieuwenhuizen, and with Y. Ne'eman. He wishes to acknowledge the stimulation he received at a recent course at the Ettore Majorana Centre for Scientific Culture at Erice, Sicily.

References and Notes

1. P. Bergmann and A. Komar, *Int. J. Theor. Phys.* **17**, 387 (1978).
2. F. Gürsey and L. Marchildon, *J. Math. Phys.* **19**, 942 (1978).

4

Relativistic Phase Space Arising out of the Dirac Algebra

D. Bohm and B. J. Hiley
Department of Physics
Birkbeck College
(University of London)
Malet Street
London WC1E 7HX, England

1. Introduction

The spinor plays a fundamental role in the quantum theory (especially in its relativistic form) but its overall structure and physical significance are not clear.

Various attempts have indeed been made to interpret the meaning of the spinor. Of these, the most elementary is the use of stereographic projection.[1] From this, one may come to the interpretation through a vector plus a flag,[2] and also through a set of triads.[3] Next, there is a different but closely related approach, in terms of the chords of circles.[4]

What appears, at least at first sight, to be quite another idea is to start with a Clifford algebra, and to interpret the coefficients as antisymmetric tensors[5]

$$\phi = A_0 + A_\mu \gamma^\mu + A_{[\mu\nu]}\gamma^\mu\gamma^\nu + A_{[\mu\nu\lambda]}\gamma^\mu\gamma^\nu\gamma^\lambda + A_{[\mu\nu\lambda\alpha]}\gamma^\mu\gamma^\nu\gamma^\lambda\gamma^\alpha \tag{1}$$

Still another way of dealing with such questions, but along the same general lines is Schönberg's connection of spinors with phase space.[6]

In this paper, we shall start by combining the Clifford algebra with Schönberg's phase space interpretation, showing the interrelationship of these two, and developing the whole theory in a systematic way. In doing this, we will use the algebraic approach in such a way as to bring out the connection between quantum theory and classical theory.[7] This will enable us directly to comprehend the meaning of Dirac's equation in the classical limit.

We begin by showing how the Dirac algebra can be expressed as the direct product of two dual Grassmann algebras, which can in turn be written as products of fermionic creation and anihilation operators. Each antisymmetric tensor of a given rank is then the coefficient of a term in the algebra corresponding to the excitation of a number of "vector fermions" equal in number to the rank of the tensor. From this, a connection between the fermionic operators and boundary operators, similar to that of Kähler,[8] will be derived.

Finally, starting with the Dirac equation, containing electromagnetic potentials, we obtain a Liouville equation, in which spin is seen to correspond to a rotary transformation on a set of constants of the motion, thus showing how the notion of spin goes beyond ordinary classical concepts as expressed in terms of the Liouville equation.

2. Brief Summary of the Connection between Nonrelativistic Quantum Algebra and an Algebra on Phase Space

In a previous paper[7] we have developed in some detail the connection of the nonrelativistic quantum algebra with an algebra on phase space. Briefly, we start by proposing that the density matrix, $\rho(x', x)$, or more generally a certain non-Hermitian extension of it, called the characteristic matrix, $\xi(x',x)$, be taken as a basic concept while the state vector is seen as an abstraction from this. We then make the Wigner–Moyal transformation[7]

$$F(X, P) = \frac{1}{2\pi} \int \xi\left(X - \frac{\eta}{2}, X + \frac{\eta}{2}\right) e^{-iP\eta}\, d\eta \tag{2}$$

where

$$X = \frac{x + x'}{2}, \qquad \eta = x - x'$$

and we obtain the generalized Liouville equation

$$\frac{\partial F}{\partial t} + LF = 0 \tag{3}$$

where L is the generalized Liouville operator, which reduces exactly to the well-known classical form

$$LF = \frac{\partial H}{\partial P}\frac{\partial F}{\partial X} - \frac{\partial H}{\partial X}\frac{\partial F}{\partial P} \tag{4}$$

for the special case of the harmonic oscillator and approximately for the general hamiltonian, in the classical limit (of high temperatures).

When $\xi(x', x)$ is a Hermitian operator, F is real, but not in general nonnegative.

It is this which has prevented a direct interpretation of F as a probability density. Instead, we have proposed, for the general (non-Hermitian) $\xi(x', x)$, that $F = U + iV$ be regarded as a pair of related constants of the motion (which may be negative as well as positive).

An essential step in our work has been to treat $\xi(x', x)$ in two ways. First of all, it is a matrix operator, in a vector space, indexed by x. To simplify the discussion, let us approximate the range, x, as discrete and finite, so that it has n "states." Then there will be n^2 independent elements in $\xi(x', x)$. What we now do is to regard these as components of a higher vector space, having n^2 elements. (As n approaches infinity, this becomes effectively the phase space). We call these vectors, ξ_α. We then introduce new matrix operator, $O'_{\alpha,\alpha'}$, having n^4 elements, which evidently are the basis of a matrix algebra on the phase space. When we make the Wigner–Moyal transformation we are regarding $\xi(x', x)$ as a vector in this higher-dimensional space, on which the Liouville equation acts as an operator. The interchange between operator and vector interpretation for $\xi(x', x)$ is essential for expressing clearly the relationship between classical and quantum laws of motion.

In our paper, we have developed this interpretation in some detail, proposing a new kind of nonnegative probability function and also a way of extending the theory to cover certain new domains. All of this may ultimately be significant in the relativistic theory, but for the purposes of the present paper what is important will be just the connection between quantum mechanics and Liouville's equation, whose solutions are to be considered as constants of the motion, with the aid of the double interpretation of $\xi(x', x)$ as an operator in the configuration space and as a vector in the phase space.

3. Connection between Clifford Algebras and Grassmann Algebras

Before going in detail into the discussion of relativistic phase space, it is necessary to express in a simple and systematic way the connection between Clifford algebras and Grassmann algebras.

As indicated in Section 2, we begin with the density matrix, which must, however, now be enriched with spinor indices, so that we write

$$\rho = \rho_{ij}(x'^\mu, x^\mu) \tag{5}$$

where i and j are the usual set of four Dirac indices. We then extend the above to the non-Hermitian matrix

$$\xi = \xi_{ij}(x'^\mu, x^\mu) \tag{6}$$

and we note that ξ, like ρ, satisfies the two Dirac equations

$$\gamma^\mu \frac{\partial \xi}{\partial x'^\mu} = 0, \qquad \frac{\partial \xi}{\partial x^\mu} \gamma^\mu = 0 \tag{7}$$

ξ can be expressed as a Clifford algebra, essentially the same as (1) except that the coefficients are, in general, complex. That is to say, we are treating ξ as a Dirac matrix. But, as suggested in Section 2, we are also going to regard ξ as a vector V in a higher-dimensional vector space.

We now note that when γ^μ operates on ξ from the left, this will commute with γ^μ operating on ξ from the right. To distinguish these two operators when they act on the higher vector space, we shall designate them, respectively, as

$$\vec{\gamma}^\mu \quad \text{and} \quad \overleftarrow{\gamma}^\mu$$

with

$$[\vec{\gamma}^\mu, \overleftarrow{\gamma}^\nu] = 0 \tag{8}$$

Of course, we also have the usual anticommution rules

$$\{\vec{\gamma}^\mu, \vec{\gamma}^\nu\} = \{\overleftarrow{\gamma}^\mu, \overleftarrow{\gamma}^\nu\} = 2g^{\mu\nu}$$

What we now aim to do is to arrive at Schönberg's representation of the Clifford algebra in terms of fermionic creation and annihilation operators (but from an opposite point of departure). The first step is to obtain a set of eight operators, similar to the $\vec{\gamma}^\mu$ and $\overleftarrow{\gamma}^\mu$ [which, however, all anticommute in spite of Eq. (8), which defines two commuting sets of four operators]. To obtain this, we consider

$$\vec{\gamma}^5 = i\vec{\gamma}^1\vec{\gamma}^2\vec{\gamma}^3\vec{\gamma}^4 \quad \text{and} \quad \overleftarrow{\gamma}^5 = i\overleftarrow{\gamma}^4\overleftarrow{\gamma}^3\overleftarrow{\gamma}^2\overleftarrow{\gamma}^1$$

the product $\vec{\gamma}_5\overleftarrow{\gamma}_5$ anticommutes with all the $\vec{\gamma}^\mu$ and the $\overleftarrow{\gamma}^\mu$. We then define

$$\overset{+}{\gamma}{}^\mu = \vec{\gamma}^\mu, \qquad \overset{-}{\gamma}{}^\mu = \vec{\gamma}^5\overleftarrow{\gamma}^5\overleftarrow{\gamma}^\mu \tag{9}$$

and obtain

$$\{\overset{+}{\gamma}{}^\mu, \overset{-}{\gamma}{}^\nu\} = 0, \qquad \{\overset{+}{\gamma}{}^\mu, \overset{+}{\gamma}{}^\nu\} = 2g^{\mu\nu}, \qquad \{\overset{-}{\gamma}{}^\mu, \overset{-}{\gamma}{}^\nu\} = -2g^{\mu\nu} \tag{10}$$

We now go on to define the operators

$$a^{+\mu} = \frac{\overset{+}{\gamma}{}^\mu + \overset{-}{\gamma}{}^\mu}{2}, \qquad a^\mu = \frac{\overset{+}{\gamma}{}^\mu - \overset{-}{\gamma}{}^\mu}{2} \tag{11}$$

These satisfy the anticommutation relationships

$$\begin{aligned} \{a^{+\mu}, a^{+\nu}\} &= \{a^\mu, a^\nu\} = 0 \\ \{a^{+\mu}, a^\nu\} &= g^{\mu\nu} \end{aligned} \tag{12}$$

We have thus arrived at Schönberg's set of creation and annihilation operators, which we have, however, derived from the two supplementary Clifford algebras, generated, respectively, from $\overleftarrow{\gamma}^\mu$ and $\vec{\gamma}^\mu$. [These are, however, to be distinguished from the original Clifford algebra given by Eq. (1).] Clearly the a^μ define a Grassmann algebra with

16 elements, as do the $a^{+\mu}$, so that together they cover the same 256 elements that are covered by the product of the two supplementary Clifford algebras.

The infinitesimal element of the Lorentz group is given by

$$L_g = \frac{\epsilon^{\mu\nu}}{4}(\vec{\gamma}^\mu\vec{\gamma}^\nu + \overleftarrow{\gamma}^\mu\overleftarrow{\gamma}^\nu) = \epsilon_{\mu\nu}a^{+\mu}a^\nu \tag{13}$$

From the above, we readily verify that $a^{+\mu}$ and a^ν do indeed transform as vectors.

When we use the "fermionic" operators $a^{+\mu}$ and a^ν, it is convenient to define an "empty" state vector, V_e, satisfying

$$a^\mu V_e = 0 \qquad \text{for all } \mu \tag{14}$$

Let us now go back to the representations (1) in terms of the original Clifford algebra, and consider the unit scalar

$$S = (1)_{ij}$$

We have

$$a^\mu S = \left(\frac{\overset{+}{\gamma}{}^\mu - \overset{-}{\gamma}{}^\mu}{2}\right) S = \tfrac{1}{2}(\gamma^\mu 1 - \gamma^5 1 \gamma^5 \gamma^\mu) = 0$$

where we have written $\vec{\gamma}^\mu \cdot 1 = \gamma^\mu$ and $\vec{\gamma}^5 \cdot 1 = \gamma^5$; $1 \cdot \overleftarrow{\gamma}^5 = \gamma^5$, and $1 \cdot \overleftarrow{\gamma}^\mu = \gamma^\mu$. This means that the scalar S is equivalent to V_e, the "empty" state of the "fermionic" representation.

We have also

$$a^{+\mu}S = \frac{\gamma^\mu 1 + \gamma^5 1 \gamma^5 \gamma^\mu}{2} = \gamma^\mu$$

and similarly we can easily show that products, such as $a^{+\mu}a^{+\nu}S$ with $\mu \neq \nu$, also reduce to the corresponding products, $\vec{\gamma}^\mu\vec{\gamma}\gamma^\nu \cdot 1 = \gamma^\mu\gamma^\nu$. This means that we can write for the general state vector

$$V = [A_0 + a^{+\mu}A_\mu + a^{+\mu}a^{+\nu}A_{[\mu\nu]} + a^{+\mu}a^{+\nu}a^{+\alpha}A_{[\mu\nu\alpha]} + a^{+\mu}a^{+\nu}a^{+\alpha}a^{+\beta}A_{[\mu\nu\alpha\beta]}]V_e \tag{15}$$

This expresses the connection between the Clifford algebra and Schönberg's use of a Grassmann algebra.

In the Clifford algebra, the operator $i\gamma^5 = -\gamma^1\gamma^2\gamma^3\gamma^4$ produces what has been called a duality transformation. That is,

$$-i\gamma^5\phi = \tilde{A}_0 + \tilde{A}_\mu\gamma^\mu + A_{[\mu\nu]}\gamma^\mu\gamma^\nu + \tilde{A}_{[\mu\nu\lambda]}\gamma^\mu\gamma^\nu\gamma^\lambda + \tilde{A}_{[\mu\nu\lambda\alpha]}\gamma^\mu\gamma^\nu\gamma^\lambda\gamma^\alpha \tag{16}$$

where the $\tilde{A}$ are the duals, as usually defined. For example,

$$\tilde{A}_\mu = A_{[\nu\alpha\beta]}\frac{\epsilon^{\nu\alpha\beta\gamma}}{3!}g_{\gamma\mu}$$

Noting that γ^5 is operating from the right, we obtain, in the "fermionic" representation,

$$\gamma^5 = \vec{\gamma}^5 = i\vec{\gamma}^1\vec{\gamma}^2\vec{\gamma}^3\vec{\gamma}^4$$

Let us choose any of the $\vec{\gamma}^\mu$, and express it as

$$\vec{\gamma}^\mu = \overset{+}{\gamma}^\mu = a^{+\mu} + a^\mu$$

We then obtain

$$\overset{+}{\gamma}^\mu a^\alpha \overset{+}{\gamma}^\mu = (a^{+\mu} + a^\mu)a^\alpha(a^{+\mu} + a^\mu)$$

when $\alpha = \mu$, we obtain $a^{+\mu}$, and when $\alpha \neq \mu$, we obtain $-a^\alpha$. So, the operator γ^μ transforms a^μ into $a^{+\mu}$ and (as can easily be shown) $a^{+\mu}$ into a^μ. Their product, $i\vec{\gamma}^5$, thus interchanges all creation and annihilation operators. So the duality operation transforms the "empty" state vector, V_e into the "full" state V_f, and as can easily be verified, it changes an arbitrary state vector, V, into $\tilde{V}$.

Let us now take as an example the Dirac equation operating from the left on the original Clifford algebra

$$\left(\gamma^\mu \frac{\partial}{\partial x^\mu}\right)\phi = \left(\vec{\gamma}^\mu \frac{\partial}{\partial x^\mu}\right)\phi = 0$$

Writing once again

$$\vec{\gamma}^\mu = \overset{+}{\gamma}^\mu = a^{+\mu} + a^\mu$$

we obtain

$$(a^{+\mu} + a^\mu)\frac{\partial}{\partial x^\mu} V = 0$$

where V is the state vector corresponding to ϕ. Consider a typical term of V, such as $(a^{+\alpha}A_\alpha)V_e$. Then $a^{+\mu}a^{+\alpha}(\partial A_\alpha/\partial x^\mu)V_e$ corresponds to the exterior derivative of A_α. This is the boundary operator[9] [which evidently satisfies $(a^{+\mu}\,\partial/\partial x^\mu)(a^{+\lambda}\,\partial/\partial x^\lambda) = 0$, because of antisymmetry of all products of $a^{+\mu}$]. Similarly $a^\mu(\partial/\partial x^\mu)\cdot A_\alpha a^{+\alpha}V_e = g^{\mu\alpha}(\partial A_\alpha/\partial x^\mu)$ V_e, which corresponds to the coboundary operator.

If D represents the duality operation (given by $i\vec{\gamma}^5$) we can write $a^\mu = Da^{+\mu}D^{-1}$, and obtain

$$\left(\vec{\gamma}^\mu \frac{\partial}{\partial x^\mu}\right)V = a^\mu \frac{\partial}{\partial x^\mu} + Da^\mu \frac{\partial}{\partial x^\mu} D^{-1} \tag{17}$$

and this is just the expression for the Dirac equation that Kähler[8] has derived, in another way.

The Dirac operator can thus be regarded from the point of view of algebraic topology as the sum of a boundary operator and a coboundary operator, while the terms of the Grassmann algebra correspond to a set of complexes. We have, however, been able to derive the result fairly directly, because the relationship of antisymmetric tensors of different rank is more simply expressed in terms of the Grassmann algebra rather than in terms of the Clifford algebra.

The result relates to a preliminary connection between quantum theory and cohomology theory that we have made earlier.[10] But now, we can develop this connection further by relating the homology operations to the Clifford and Grassmann algebras which are basic to relativistic quantum theory. We expect to go into this question in later papers.

4. The Dirac Equation and Relativistic Phase Space

We return to the two Dirac equations (7) which are now written in terms of the vector space, V:

$$\overrightarrow{\gamma}^{\mu}\frac{\partial V}{\partial x'^{\mu}} = 0, \qquad \overleftarrow{\gamma}^{\mu}\frac{\partial V}{\partial x^{\mu}} = 0 \tag{18}$$

As in Eq. (2), we introduce $X = (x'^{\mu} + x^{\mu})/2$, $\eta^{\mu} = x^{\mu} - x'^{\mu}$ along with a relativistic generalization of the Wigner–Moyal transformation

$$\overline{V}(X^{\mu}, P^{\mu}) = \frac{1}{2\pi}\int V(x'^{\mu}, x^{\mu})e^{-iP_{\mu}\eta^{\mu}}\, d^4\eta \tag{19}$$

We multiply the second of the equations (18) by $\gamma_5\gamma_5$ and we get

$$\overset{+}{\gamma}^{\mu}\frac{\partial V}{\partial x'^{\mu}} = 0, \qquad \overset{-}{\gamma}^{\mu}\frac{\partial V}{\partial x^{\mu}} = 0$$

From (11) we write the above as

$$(a^{+\mu} + a^{\mu})\frac{\partial V}{\partial x'^{\mu}} = 0, \qquad (a^{+\mu} - a^{\mu})\frac{\partial V}{\partial x^{\mu}} = 0$$

By adding and subtracting the above equations we obtain

$$\frac{a^{+\mu}}{2}\frac{\partial V}{\partial X^{\mu}} - a^{\mu}\frac{\partial V}{\partial \eta^{\mu}} = 0 \tag{20a}$$

$$\frac{a^{\mu}}{2}\frac{\partial V}{\partial X^{\mu}} - a^{+\mu}\frac{\partial V}{\partial \eta^{\mu}} = 0 \tag{20b}$$

Under the Wigner–Moyal transformation (20a) becomes

$$\left(\frac{a^{+\mu}}{2}\frac{\partial}{\partial X^{\mu}} - ia^{\mu}P_{\mu}\right)\overline{V} = 0$$

By applying the above operators twice we are led to

$$g^{\mu\nu}P_\nu \frac{\partial \overline{V}}{\partial X^\mu} = 0 \tag{21}$$

This is the (classical) Liouville equation for a free particle.

We have thus seen that certain combinations of the Dirac operators appearing in Eq. (20a), when applied twice, imply the Liouville equation. If, however, we had added (20a) and (20b), and then applied the resulting operator twice we would have obtained the Klein–Gordon equation which, of course, contains further quantum-mechanical implications, beyond those of the classical Liouville equation. Some of these implications have been discussed in Ref. 7, but for the purposes of this paper, such questions can be set aside.

As pointed out in Section 2 the coefficients of the $\overline{V}$, as obtianed by the Wigner–Moyal transformation on (15), are now to be interpreted as constants of the motion. The Dirac equation applied to the original Clifford algebra thus implies a whole set of antisymmetric tensors that are related constants of the motion, beyond the single pair (real and imaginary parts of a complex function) which appears in the nonrelativistic theory. The appearance of such a set implies some new physical concepts, which we shall discuss, at least in part, further on in this section.

Let us now consider the effect of electromagnetic potentials, $A_\mu(x^\nu)$. The Dirac equations (18) become

$$\begin{aligned} \overrightarrow{\gamma}^\mu \left[\frac{\partial}{\partial x'^\mu} - i\frac{e}{c}A_\mu(x'^\nu)\right] V = 0 \\ \overleftarrow{\gamma}^\mu \left[\frac{\partial}{\partial x^\mu} + i\frac{e}{c}A_\mu(x^\nu)\right] V = 0 \end{aligned} \tag{22}$$

Expanding

$$A_\mu(x^\nu) \cong A_\mu(X^\nu) + \frac{\eta^\alpha}{2}\frac{A_\mu}{\partial X^\alpha}(X^\nu)$$

and

$$A_\mu(x'^\nu) \cong A_\mu(X^\nu) - \frac{\eta^\alpha}{2}\frac{\partial A_\mu}{\partial X^\alpha}(X^\nu)$$

and following the same substitutions that led to Eq. (21), we obtain for the generalized Liouville operator,

$$\begin{aligned} L_q = g^{\mu\nu}\left(P_\mu + \frac{e}{c}A_\mu\right)\frac{\partial}{\partial X^\mu} + g^{\mu\nu}\left(P_\mu + \frac{e}{c}A_\mu\right)\frac{e}{c}\frac{\partial A_\nu}{\partial X^\sigma}\frac{\partial}{\partial p_\sigma} \\ - \frac{e}{c}\frac{a^{+\mu}a^\nu}{4}\left(\frac{\partial A_\mu}{\partial X^\nu} - \frac{\partial A_\nu}{\partial X^\mu}\right) - i\frac{a^{+\mu}a^{+\nu}}{4}\frac{e}{c}\frac{\partial}{\partial X^\sigma}\left(\frac{\partial A_\mu}{\partial X^\nu} - \frac{\partial A_\nu}{\partial X^\mu}\right)\frac{\partial}{\partial P_\sigma} \end{aligned} \tag{23}$$

The first term in the above represents the modified contribution to the Liouville equation coming from the motion of the trajectory through space. The second represents the contribution of the electromagnetic force. The third corresponds to the effects of "spin" of the Dirac electron. That is to say, $\overline{V}$ changes, not only because of changes of position and momentum in a trajectory, but also because the electromagnetic fields, $F_{\mu\nu}$, generate what is equivalent to a rotation (Lorentz) transformation among the components of the V. These latter thus cease to be constants of the motion. Spin is therefore a new kind of movement, which involves a change of what were originally constants of the motion. This means that the solutions of Liouville's equation now undergo a transformation that does not arise from the "particle" motions, so that they have begun to take on an independent physical meaning. Spin is thus not a property of trajectories but of the field, $\overline{V}(X^{\mu}, P^{\nu})$ (which is constituted of antisymmetric tensors).

Finally, there is the fourth term in (23). This represents a kind of "force" on the trajectories due to the electromagnetic field (because it involves $\partial/\partial P_{\sigma}$). But it also involves fields, V, through the operators $a^{+\mu}a^{+\nu}$. We may regard it as describing the effect of spin on the trajectories. (This term will be unimportant if the field functions, $\overline{V}$, are spread out over a large region in phase space.)

5. Conclusion

We have developed an interpretation of Dirac's equation in terms of an original Clifford algebra [given in Eq. (1)], which systematically incorporates all the antisymmetric tensors, the Grassmann algebra and its dual, and a pair of supplementary (right and left) Clifford algebras. From there, we have derived a Liouville equation. By thus putting Dirac's equation into a "language" broad enough to relate it to classical mechanics, we can see in some detail how spin goes beyond ordinary classical concepts. These latter attribute all the electronic motion to a particle trajectory, and imply that the solutions of Liouville's equation are a mere "shadow" of the trajectories, which are regarded as descriptions of the basic reality. We see that, on the contrary, the solutions of Liouville's equation are fields having certain independence of motion, and that the trajectories depend on these fields in an irreducible way. So the two together are needed for a complete account, and this is beyond the classical way of thinking.

A related result has been obtained nonrelativistically in our previous paper,[4] where we show that the concept of a self-determined trajectory has a limited applicability and that these limits are determined by the solutions of Liouville's equations, which indeed also have a meaning going beyond that of being mere descriptions of trajectories.

References and Notes

1. F. A. M. Frescura and B. J. Hiley, *Am. J. Phys.* **49,** 152 (1980).
2. R. Penrose, *Battelle Rencontres,* C. M. DeWitt and J. A. Wheeler, editors (Benjamin, New York, 1968), p. 121.

3. H. A. Kramers, *Quantum Mechanics* (Dover, New York, 1964).
4. D. Bohm, *Perspectives in Quantum Theory. Essays in Honour of Alfred Landé*, W. Yourgrau and A. van der Merwe, editors (MIT, Cambridge, Massachusetts, 1971), p. 78.
5. E. M. Corson, *Introduction to Tensors, Spinors, and Relativistic Wave Equations* (Blackie, London, 1955).
6. M. Schönberg, *Nuovo Cimento Supp.* **6,** 356 (1957).
7. D. Bohm and B. J. Hiley, *Foundations of Physics* **11,** 179 (1981).
8. E. Kähler, *Deutsch. Akad. Wiss. Berlin Kl. Math. Phys. Tech.* No. 1, 5 (1961).
9. For a defintion of these terms, see P. J. Hilton and S. Wylie, *Homology Theory* (Cambridge University Press, Cambridge, 1960).
10. D. Bohm, B. J. Hiley, and A. Stuart, *Int. J. Theor. Phys.* **3,** 171 (1970).

5

Why Mourn the Passing of Determinacy?

Hermann Bondi
Natural Environment Research Council
Swindon SN2 1EU, England

1. Introduction

Well over half a century after Heisenberg's formulation of the uncertainty principle, there still exist some physicists who are uncomfortable with indeterminacy, and numerous philosophically inclined nonphysicists who think of indeterminacy as a bad dream that one may hope to awake from some day. Einstein's well-known views reinforce these attitudes.

It will be my thesis in this article that "determinacy" always was an absurd concept incompatible with the very possibility of science, and based on mathematical misunderstandings. The demonstration of the absurdity of a deterministic picture does not, however, point uniquely to Heisenberg uncertainty and to Born statistics as the only possible resolution of the difficulty; it merely makes one breathe a sigh of relief that such a straightforward escape from the nonsense of determinisms should be so readily available.

2. Dependence on Initial Conditions

First, a few remarks about the mathematical point. In a theory in physics such as dynamics or electromagnetism (and I will confine my remarks to physics, being too unfamiliar with other fields of science) one has a set of differential equations linking the initial, supposedly known situation with subsequent later situations of the system in a rigorous and perfectly definite way. It was this picture, characterized by Newtonian dynamics applied to simple systems, that generated belief in determinacy. This was strengthened by the later developments of analytical dynamics, notably the Hamilton–Jacobi method which sought to mirror the behavior of any dynamical system through suitable transformations in the uniform rotation of a sphere under no forces.

While for a long time it was thought that only mathematical complexities prevented this method being applicable to *any* dynamical system, much later it was shown, as Prigogine has reminded us, that this transformation could only be carried out for a very limited set of simple systems. The importance of this fact lies in that it shows how complex even in dynamics the link between the initial and the later situations can be. This is so crucial, because science is empirical. Knowledge of the initial situation can never be perfect, but only approximate. Unless a narrow range of initial situations implies a modest range of later outcomes, the exactness of the link is irrelevant. If a small error in the description of the initial situation may totally alter the subsequent behavior of the system, then it makes no sense to speak of a deterministic description. The image of determinism only arises from the systems that are very simple, relatively rare in nature, but a common subject of study. Only if the later configuration depends very stably on the initial conditions can one speak of a deterministic connection. But such stable connections arise only rarely in fact, the systems much studied forming an unrepresentative set.

3. Unpredictability is Essential

But quite generally no discussion makes sense in which the empirical character of science is not fully allowed for at every stage. A crucial means of acquiring empirical knowledge is through experiments devised to test a hypothesis. An experiment always involves a great deal of complexity. But the test of the hypothesis needs to emerge clearly from all this welter of detail. Thus the making irrelevant of other extraneous influences is a central issue in any such empirical test of a hypothesis. An absurdly simple example occurs when in a strange room I want to find out which switch controls which light. I operate one switch and observe that a light comes on. Next I deal with several other switches before I return to the first one. I may then observe that whenever I turn it on the respective light comes on and whenever I turn it off this light goes off, as a test of the hypothesis that this switch, and only this switch, controls this light. Of course at any one stage, alternative hypotheses are imaginable, viz., that my light always goes off first 3 seconds and then 8 seconds after some other light comes on (if this, say, was the actual sequence of events). Only by operating my switch at *arbitrary* times, *unrelated* to any other operations, can I begin to exclude such hypotheses. However, the more complications I allow in the switching gear, the less difficult will it be to maintain such alternative hypotheses through any given cycle of events. Their inability to predict successfully may count against them, but so would a bulb failure count against my main hypothesis. The essential means to rob the alternatives of credibility must, however, be my *unpredictable* choice of the moment of switching.

As a Marx Brothers fan, I might perhaps relate the mirror scene in "Duck Soup," where Groucho, spying in disguise, descends a stairway in a to him strange house, and sees, through what may be a large mirror or an opening, a like figure descending a stairway. Is it another spy, perhaps a confrere, on the same errand, or is it his own mirror image? To test, he retreats to one side and, after due thought crosses the mirror hopping in some absurd pose and, lo and behold, the other figure does the same. Of course this *does not prove* that it is a mirror, for after all his confrere might have had just the same idea. So he needs to test again and thinks up some other ridiculous pose for exposing himself to the "mirror," with the same result. Only on the fourth crossing of the opening does his "image" differ from him and they scatter in confusion.

The essence of this story is that even if you think you will cross the "mirror" in what you regard as an unpredictable pose at an unpredictable moment, somebody of a similar mentality might hit on just the same idea. Unpredictability is of the essence of any good experimental arrangement for excluding extraneous factors. Even if we look at the simple example of planetary orbits in the solar system, then on consideration we find that we regard them as such a good test of Newton's theory of gravitation only because we view the particular sets of masses, orbital radii, etc., of the individual planets as in some sense *fortuitous*. Otherwise, it would be easy to argue that only just the actual orbits occupied are conics (modified somewhat by mutual interaction) and that all other potential orbits are more complex.

4. What is "Fortuitous"?

It is therefore central to our disentangling the essential facets of an experiment from the supposedly incidental ones that we bring in some factors which we regard as purely fortuitous relative to the substance of the experiment, like the time of switching on the apparatus, the location of the laboratory, what the experimenters had eaten for breakfast, etc. But how certain can we be that these are truly fortuitous and do not have an unexpected direct relation to the intended experimental test of some hypothesis? One way is by believing that what we are testing is very much simpler than the complexity of the mental processes of the experimenters that control the moment of switching on, their breakfast foods, etc. But the more complex the subject of our study, the less we can be certain that there is not some interplay. The experimental result is, as has been so well established, not an isolated "fact" but an amalgam of background knowledge, response of the equipment, interpretation, etc. The tighter we believe the causal links are, the more difficult or impossible it becomes to disentangle them. Yet it is always essential in science to separate the substantive from extraneous material that, however substantive in its own right, may be regarded as fortuitous in relation to what we are trying to find out.

A good example is the solar system. Kepler and Newton were successful in describing and analyzing the motions of the planets and satellites only because they divided the knowledge at their disposal arbitrarily (and the more arbitrary we view the division to be, the greater must be our admiration for the genius displayed) between one set (where is each body when) and another set (why are the bodies of just such size, why do they have this number of satellites, what determines their mean distances from the Sun, what is the constitution of each planet, etc.). The first problem they solved completely; the second, which is essentially that of the origin of the solar system, remains a problem to this day. Medawar's description of science as "the art of the soluble" can hardly have a finer example than this choice by Kepler and by Newton to confine their attention to the then soluble part of the problem.

If one takes a truly deterministic outlook, then nothing is fortuitous. Jupiter having the mass it has, and having the period of revolution it has, cannot be viewed as less substantive or more incidental than the elliptical shape of its orbit. But progress in science is only possible if we do not attempt to explain everything at once. Thus, in a truly deterministic universe, science would become infeasible through inability to regard anything as fortuitous. Since no two events have identical antecedents, each must be regarded as *sui generis*, and laws become impossible to formulate.

At this stage it may be argued that the fortuitous element needs to be introduced and is introduced through the human experimenter. When he switches on his apparatus would be regarded as irrelevant to the substance of his experiment. But with a fully deterministic outlook every human action has also its identifiable causes and the same argument holds as before. There are only two ways out for the determinist: (i) an insistence that certain phenomena are inherently so simple that they can be assumed to be unrelated to the complexities of human decision making; (ii) that, although the inanimate world is deterministic, human beings (or all animals, or all living matter) are not governed by deterministic laws.

Escape route (i) would limit science to inherently simple domains and in any case the definition of simplicity is wholly arbitrary.

Escape route (ii) is a throwback to *vitalism* and as such is surely unacceptable today.

5. How Much Indeterminacy?

We see therefore that some measure of indeterminacy is necessary for physics to be possible. But, it will be objected, in an utterly indeterminate universe there are no regularities and thus physics again is impossible. Hence physics can only be possible in a world in which there is a limited, but nonzero, amount of indeterminacy. I will argue that our universe is just of this kind.

It is frequently thought that indeterminacy is confined to the atomic scale and below. Clearly there is indeterminacy on that scale and, as I have pointed out elsewhere, its most primitive form, the indistinguishability (i.e., identical nature) of like atoms, was already implied by Dalton's atomic hypothesis. What indeterminacy does for us is to limit the extent to which causal links can be traced backwards, i.e., it defines events that must truly be regarded as *fortuitous*.

Most familiar amongst such fortuitous events is the moment of disintegration of a radioactive nucleus. Because this moment is fortuitous and only because it is fortuitous can we regard the life of the daughter nucleus as unaffected by the actual length of life of the parent nucleus, and so can define an independent half-life of the daughter species of nuclei. The statistical law of radioactive decay is only so simple because there are no antecedent causes. When we recall our human limitations and their consequence, viz., that only simple laws can be discovered, then the significance of this observation becomes clear.

It may be argued that however sound these arguments may be, indeterminacy is only relevant on the very small, the atomic, scale, and is irrelevant to macroscopic physics. But such an argument implies a denial of the existence of *amplifiers* of indeterminacy, amplifiers that have macroscopic consequences of microscopic events.

It is easy to engineer such amplifiers; indeed, they are common. One only has to turn up the volume of a radio set without input to hear the random arrival of electrons in the first amplifying stage as true noise, or to see the similarly generated dance of dots on a TV screen when there is no input which equally are random through the operation of Heisenberg indeterminacy. When a fluorescent light is switched on, the delay before it begins to shine similarly has a major component in the random arrival of ionizing particles from cosmic ray collison products, where not only the original arrival of a cosmic ray particle is random, but the direction of motion of the collison products is truly Heisenberg indeterminate.

While these macroscopic effects of indeterminacy are mere by-products, in the random number generators essential in advanced computing they are the engineered aim of the device.

6. Instabilities

Let us move on from these artifacts to natural systems, whose scientific description through models frequently displays *instabilities*, i.e., situations the aftermaths of which cannot be forecast by the model. Of course this only implies that the model is not sufficiently refined and that the aftermath is determined by forces or constraints excluded from consideration by the simplification involved in the construction of the model. More complex models may resolve some such uncertainties by including factors negligible except in relation to these instabilities. One then arrives at a progression of models. A simple example is the sense of rotation of bath water flowing out through the plug hole. Once such a rotation has been established, it persists. What determines its sense? In the first instance the motions of the water just before the plug was pulled out may have had a circulatory component. The sense of this will persist and indeed it will be amplified by the concentration of flow at the hole. But what if we made sure there were no motions in the water antecedent to opening the outflow (and making sure that the removal of the closure did not itself generate a sense of rotation)? Then it is the irregularities and deviations from perfect symmetry of the bathtub. But what if we make sure the tub is very symmetrical? If its accuracy is better than about one part in a million (the ratio of the rotation period of water at the throat of the hole to the length of the day) then the vertical component of the Earth's angular velocity will determine the sense of rotation. But what if our bath, with utterly still water in a highly symmetrical tub, is situated on the equator? Will the water refuse to run out? Surely not, but less accessible causes will determine the sense of rotation.

sible causes will determine the sense of rotation. Eventually it will be molecular indeterminacy that fixes the sense of rotation randomly.

Similarly we can analyze instability in natural processes, such as the huge heat engine of our atmosphere that gives us the weather. (Note that in this case, the rotation of the Earth, so minor a matter for the bathtub, is dominant.) The more the science of meteorology advances, the more clearly it will first define and then resolve instabilities. But is it altogether fanciful to expect that eventually it will reveal instabilities whose outcome is so evenly balanced that the Heisenberg indeterminacy of atomic events will be at the root of the actual outcome?

The discussion of amplifiers of indeterminancy has so far been restricted to inanimate matter. But we do regard our own actions as somewhat unpredictable, and this unpredictability was vital in isolating the essential features of an experimental setup from the supposedly irrelevant ones, such as the moment of switching on the apparatus. But how truly unpredictable are we humans? Will the advances of behavioral psychology makes us more and more predictable? There will be a definite barrier to this if we ourselves are amplifiers of atomic indeterminacy. Is this to be expected? If we can show that there would be an evolutionary advantage in having an amplifier of indeterminacy as a component of the central nervous system then it is likely that this will exist. For amplification is not strange to living organism; indeed, each of them is the amplification of the single cell as which it started life at conception. Moreover, minute electric currents control the systems.

It is easy to think of cases where such an amplifier would have evolutionary advantage. Any small organism, with a confined central nervous system, cannot hope to beat bigger ones preying on it with a fixed and predetermined sequence of moves since the predators can learn what this is, many of them having at their disposal much bigger brains. But however big the predator's brain, he cannot foretell truly *random* moves of his prey. However small an organism, being guided by atomic indeterminacy should be a feasible task for its computing capacity. Whether the jerky pattern of a house fly's movements is of this kind, I cannot tell, but the possibility certainly exists. If this were so, would we deny to our mental equipment the possibility of including as a component a random generator which a house fly possesses?

7. Preconditions for Physics

The upshot of all these arguments is thus twofold: An input of fortuitous, uncaused events is necessary to disentangle the causal nexus of the world and so to render physics possible. Secondly, amplifiers of indeterminacy, known and hypothesized, ensure that Heisenberg indeterminancy is not totally confined to the very small. Physics can only exist in a world full, but not completely full, of causal links. Thus the very possibility of physics illuminates the nature of our world, and shows that belief in complete determinacy is as incompatible with science as is belief in a world governed by hordes of unpredictably acting demons.

6

On the True Ideas Underlying Wave Mechanics*

Louis de Broglie
Institut Henri Poincaré
Paris, France

I presented the first principles of wave mechanics in three Notes that appeared in *Comptes rendus* of September–October 1923 and later gave a more elaborate version in my Doctoral Thesis submitted on the 25th of November, 1924. My essential idea was to generalize for all particles the coexistence of waves and particles which had been discovered by Einstein in 1905 for the case of light and photons. In conformity with the clear ideas of classical physics, I tried to imagine a real physical wave which transported minute and localized objects through space in the course of time. Two ways of doing so then occurred to me. The first is not taught anymore in the ordinary classroom, but today I consider it by far the more profound. It was sketched in my 1923 Notes and further developed in the first chapter of my Thesis. It was based on the difference between the relativistic transformations of the frequencies of a wave and of a clock. Assuming that a particle possesses an internal vibration, which causes it to resemble a little clock, I supposed that this clock moved in its wave in such a manner that its internal vibration remained always in phase with the vibration of the wave. This is the postulate of "phase coincidence." These hypotheses appeared to me necessary because the relation $W = h\nu$, when applied to a particle, implies the existence of an internal frequency ν of the particle, while it is known since the work of Planck and Einstein that ν is *also* the frequency of the wave carrying the particle. Hence, the particle appears to be incorporated in the wave, such that it occupies a tiny region where the wave amplitude is very large. From this one can derive the well-known formula $p = h/\lambda$. In the second chapter of my Thesis, I then showed that, in the case where the wave propagation is studied in the approximation of geometric

*A translation, by J.-P. Marchand and A. van der Merwe, from the French version of this article, which originally appeared in *C. R. Acad. Sci. Paris, Series B*, 277 (16 July 1973).

optics, one is led to the identification of Fermat's principle with Maupertuis' principle of least action and hence again to the formula $p = h/\lambda$.

Let us try to emphasize the differences existing between these two modes of reasoning. The first, viz., the postulate of phase coincidence, is of an essentially relativistic nature, considering that it is based on the difference between two relativistically transforming formulas; whereas the second, the identification of the principles of Fermat and Maupertuis, has nothing essentially relativistic about it, since these two principles are equally valid in classical and relativistic theories. Another difference separating the two methods is that the first reasoning is valid for all wave propagations, whereas the second makes sense only for those propagations which take place in the approximation of geometric optics.

Following the appearance of my Thesis, my ideas have often been misinterpreted by those saying that, according to me, the electron *was* the wave, which, of course, did away with the particle itself. It seems that adoption of this idea motivated Schrödinger in his very elegant 1926 papers to write down for the first time the propagation equation of a wave, which he called the Ψ wave. This he did for the electron, but only in the Newtonian approximation and without taking the spin into account. In this manner, Schrödinger was able to compute exactly the wave processes corresponding to the quantified states of an atomic system, as conceived classically in the works of Bohr and his followers. It is quite certain that Schrödinger then viewed his Ψ wave as a physical wave. But he abandoned completely any idea of localizing the particle in this wave, so that the picture which he formed of the atom and, more generally, of the Ψ waves, made no provision for localized particles. This had very grave consequences and made Schrödinger's use of configuration space paradoxical in the case of particle systems. Soon afterwards, Born introduced the normalization of the Ψ wave, by which he modified arbitrarily the amplitude of the wave and hence deprived it of all physical reality. The normalized Ψ wave was thus transformed into a simple probabilistic representation which, while leading to a large number of exact predictions, provided no understandable picture of the coexistence between waves and particles.

Schrödinger's work has the merit of making one see clearly that wave mechanics, when applied to atomic systems, gives rise to problems where the approximation of geometric optics is no longer valid. It follows from this that Fermat's principle ceases to be applicable and no longer allows us to define a ray which could be interpreted as a particle trajectory. If, therefore, one refuses to invoke the principle of phase coincidence, one is led to the conclusion that it is impossible to ascribe a trajectory to the particle within its wave and to the claim that the particle can have only isolated localizations *without intermediate positions*. But such a conception raises serious difficulties, in particular the one pointed out by Einstein at the Solvay Congress of 1927. It can be summarized as follows: Consider a source that emits a spherical wave transporting a particle. After a moment, the particle manifests its presence at a point of the spherical wave by its localized effect on a detector. It is evident that the emission of the particle by the source is the *cause* of its arrival at the detector. But the causal link between the two phenomena implies the existence of a trajectory, and to deny this existence is to renounce causality and to deprive oneself of any understanding.

Let me now make an important observation. Since normalization, which arbitrarily modifies the wave's amplitude, does however not alter its phase, customary quantum mechanics is able to define the same frequency and the same wavelength as my theory, and it is this feature which makes the former a powerful theory leading to a large number of precise results. But, in opposition to what is usually done, quantum mechanics does not have the right to postulate $W = h\nu$ and $p = h/\lambda$, because the energy W and the momentum p of a particle are properties which are associated with the concept of a localized object that moves through space along a trajectory. The reason I was able to establish these formulas was that I advanced the hypothesis of a particle localized inside its wave.

When, in 1928, I assumed the functions of a lecturer, I presented the ideas then current in quantum mechanics, and for many years I abstained from developing my own original ideas.

But in the last 20 years I have again convinced myself that one should return to the idea that a particle is a very small object that is localized and moves along a trajectory. As I have shown in a series of increasingly thorough studies,[1] this idea can be implemented, while conserving the statistical significance of the normalized Ψ wave, with the aid of my concept of a particle guided by its wave, if one completes this notion by a so-called hidden thermodynamics, whose development opens up very novel perspectives. A particularly important consequence of this thermodynamics is that the principle of least action appears now as an aspect of the second principle of thermodynamics.[2]

It is important to note the astonishing fact that, in light and particle optics, one can predict with extreme precision an enormous number of phenomena on the sole ground of wave propagation, and without invoking the corpuscular structure of the energy carried by the waves. In the case of interference and diffraction phenomena, the statistical postulate of Born is sufficient to explain the observational data. But whereas in the usual quantum theory one has to assume this postulate arbitrarily, I am able to supply a logical basis for it. There exists, however, a case where the postulate of phase coincidence appears to provide an explanation which the usual theory is unable to give; it obtains when one considers the action of a Hertzian wave with frequency ν on an oscillating circuit or any other device adjusted to this frequency. It is, in fact, natural to think that some of the photons carried by the wave transmit the energy to the oscillating circuit in the form of sudden impulses, which compensate the damping. But the energy thus transmitted to the circuit cannot sustain its natural oscillation unless these impulses have a rhythm corresponding to the natural frequency of the circuit, which is that of the wave. This seems to prove that the incident photons possess an internal oscillation frequency equal to that of the wave, which in turn confirms the principle of phase coincidence, while the ordinary theory provides no analogous idea.

In conclusion, I think that my original ideas, which I have amended and further developed over the last years, permit us to understand the real nature of the coexistence between waves and particles, whereas customary quantum mechanics and its extensions furnish only an exact statistical view of quantum phenomena, without revealing their true structure. The principle of phase coincidence teaches us, in fact, that there exists a corpuscular dynamics, having the character of a dynamics with *variable proper mass*, which underlies all wave propagation phenomena, even if these are considered outside the approximation leading to geometric optics. And I think this is a point which the usual quantum mechanics has not allowed one to see.

As at my age I cannot hope to continue my personal work for a long time, I would like to express the hope that some young researchers will dedicate themselves to the development, in the direction I have indicated these last years, of the ideas that half a century ago gave rise to the birth of wave mechanics in France.

References and Notes

1. L. de Broglie, *La réinterprétation de la Mécanique ondulatoire* (Gauthier-Villars, Paris, 1971).
2. L. de Broglie, *La Thermodynamique de la particule isolée* (Gauthier-Villars, Paris, 1964).

7

CPT Invariance as Basic for Interpreting Quantum Mechanics

O. Costa de Beauregard
Institut Henri Poincaré
Paris, France

1. Summary

This paper is an expanded version of one recently published in *Foundations of Physics* and is a continuation of previous works devoted to the EPR correlation.

The leading idea remains that the EPR correlation (either in its well-known form of nonseparability of future measurements, or in its less known time-reversed form of nonseparability of past preparations) displays the intrinsic time symmetry existing in almost all physical theories at the elementary level. But, as explicit Lorentz invariance has been an essential requirement in both the formalization and the conceptualization of my papers, the noninvariant concept of *T* symmetry has to yield in favor of the invariant concept of *PT* symmetry, or even (as *C* symmetry is not universally valid) to that of *CPT* invariance.

A distinction is then drawn between "macro" special relativity, defined by invariance under the orthochronous Lorentz group and submission to the retarded causality concept, and "micro" special relativity, defined by invariance under the full Lorentz group and including *CPT* symmetry. The *CPT* theorem clearly implies that "micro special relativity" *is* relativity theory at the quantal level. It is thus of fundamental significance not only in the search of interaction Lagrangians etc., but *also in the basic interpretation of quantum mechanics, including the understanding of the* EPR *correlation.*

While the experimental existence of the EPR correlations is manifestly incompatible with macro relativity, *it is fully consistent with micro relativity*. It goes without saying that going from a retarded concept of causality to one that is *CPT* invariant has very radical consequences, which are briefly discussed.

2. Introduction

A preceding paper[1] emphasized that intrinsic time symmetry—a property shared by classical theories of dynamics, wave propagation, probability, and information, and by the quantal theories of particles and (largely[2]) of fields—is of critical importance in the fundamental problem of the interpretation of the quantum theory. It was stressed there that the wave collapse,[3] that is, the *stochastic event*, or *transition* of quantum mechanics shares the *time symmetry* of all elementary phenomena. An example given was the formula[4]

$$\langle x|a\rangle = \langle x|x'\rangle \langle x'|a\rangle \tag{1}$$

solving the Cauchy problem[5] for the Klein–Gordon, or the spinning waves equations of Dirac, Petiau–Duffin–Kemmer, etc. In the relativistically covariant formalism of first quantization[6] this formula yields the expansion of the wave function $\langle x|a\rangle$ at any point-instant x in terms of the complete set of orthogonal Jordan–Pauli propagators $\langle x|x'\rangle$ with apexes x' on an arbitrary spacelike surface σ, the coefficients of the expansion being the values $\langle x'|a\rangle$ of the wave function on σ. Orthogonality of two Jordan–Pauli propagators with spacelike separation of their apexes x' and x'' follows from the formula

$$\langle x'|x''\rangle = \langle x'|x\rangle \langle x|x''\rangle \tag{2}$$

as the Jordan–Pauli propagator is zero outside the light cone.

In formulas (1) and (2) the operation $|x\rangle\langle x|$ is an invariant integral over a spacelike surface σ, in the form of the flux of the (conservative) Gordon or Dirac style 4-current.

Formula (1) shows that the Jordan–Pauli propagator is the eigenfunction in the covariant position measurement problem,[7] formulated as "Do we find at the pseudoinstant σ the particle crossing a given element of σ?" (the corresponding probability density being the flux of the Gordon or Dirac style current, respectively). The point is that the Jordan–Pauli propagator is nonzero inside *both* the future and the past, so that the stochastic event associated with the position measurement is a *time symmetric* "collapse-and-anticollapse" (Fig. 1).

Paradoxical as this concept may seem, it *is* the one appropriate for discussing the Einstein[8] or EPR[9] correlation, where (so to speak) the "dice are cast" at two (fuzzy) point-instants L and N with a spacelike separation, and are nevertheless correlated through their common past origin at C! In association with a spinning wave equation, formula (1) expresses a *position and polarization* measurement such as those performed in the Einstein[8] or EPR[9] correlation experiments.[10] It displays a retroactive effect, which is exactly the (paradoxical) element needed for explaining the Einstein correlation. It is *this* point that will be tested in the experiment in preparation by Aspect.[11]

My 1974 article[1] has been followed by a series of technical papers[12] devoted to the Einstein correlation, together with the interpretation of quantum mechanics. In

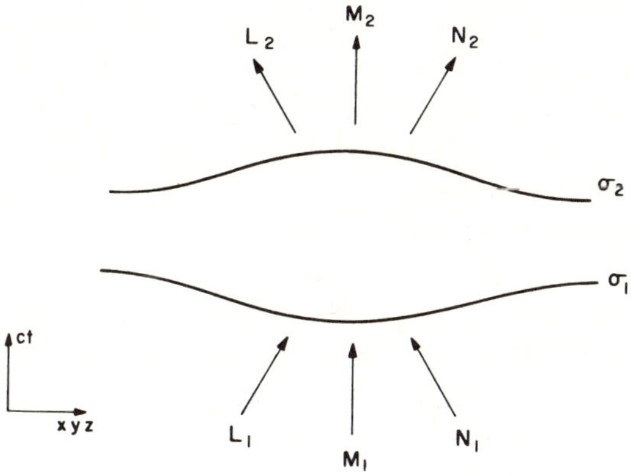

FIGURE 1. *S*-matrix formalism (interaction picture). *CPT* invariance entails reciprocity between preparations L_1, M_1, N_1, and measurements L_2, M_2, N_2. The *CPT*-invariant collapse-and-anticollapse concept consists in one element $\langle \Phi_i | \psi_j \rangle$ of the transition matrix being selected.

these, relativistic covariance and time symmetry are systematically upheld in both the mathematical formalism and the wording. From this conjunction stems the *necessity of explicitly taking a new step.*

Time symmetry was understood in these papers[1,12] in the classical sense it had in, say, the celebrated Loschmidt and Zermelo paradoxes of statistical mechanics. But time symmetry is not a Lorentz invariant concept, so what is then really meant is "*PT* symmetry" in the sense of reversal of all four axes of space-time.[13] But, as Lee and Yang have taught us, *PT* invariance is not the end of the story. As the physics of elementary particles is not always *C* invariant, we must go to *CPT* invariance—which, after general relativity and the Heisenberg uncertainty relations, ties one more connection between physics and geometry.

So, in the straight line of my previous article,[1] the present one aims at emphasizing that the *CPT invariance of relativistic quantum mechanics*—both in its first[6] and second quantized[14] formalizations—has paramount consequences not only at the technical level of choosing interaction Lagrangians etc., but *also at the fundamental level of understanding the interpretation of quantum mechanics.*

At this point a preliminary clarification is needed. Following Einstein[8] himself, some distinguished physicists have expressed more or less correlated feelings according to which the EPR[9] correlation implies either an open conflict[15] between the relativity and the quantum theories, or at least a cold war situation where a careful diplomacy in the wording can prevent a conflagration.[16] Following Stapp,[17] these theorists favor in one way or another the strategy of controlled retreat; they introduce the concept of a hidden ether,[18] conceptually providing an absolute (fluid) space and time reference frame, or else, the concept[19] of particles interacting through a Dirac:[20] relativistic gas, and propagating in their own interior spacelike signals, moving forward with respect to the proper time.

In my opinion such *ad hoc* proposals are extremely artificial and doomed to failure. The history of physics has repeatedly shown that breakthroughs are gained by means of boldly consistent interpretations of an operational mathematical formalism. As examples I recall Maxwell's prediction that the squared velocity of light waves is the ratio of the electric and magnetic permeabilities of the vacuum, Einstein's lecture of the group property of the Lorentz–Poincaré formulas, and Anderson's experimental confirmation of the Dirac–Weyl antiparticle concept. *Blitzkrieg* and "fleeing forward," rather than strategic retreat, wins the battles of science. Also, as will be explained in Section 4, I feel that the important experimental results[21] of Wilson, Lowe, and Butt and of Bruno, d'Agostino, and Maroni are very hard to justify by means of the sort of theories just mentioned.

But then, why is it that some authors[8,15,16] feel that there is a conflict between relativity and quantum mechanics at the level of Einstein correlations, while I feel that there is none at all? The reason is this: Although, as elementary particle physicists, they know well, and apply technically, the *CPT*-invariance principle, these authors, when discussing the interpretation of quantum mechanics, fall back on Einstein's 1905 recipe for defining the special relativity theory: invariance through the orthochronous Lorentz group, and submission to the principle of retarded causality (Einstein's prohibition to telegraph into the past). There is absolutely no doubt that special relativity understood in *this* way is incompatible with the physical existence of both the Einstein correlation[10,21] and of the time-reversed Einstein correlation.[22]

But on the other hand (and in full conformity with Eberhard's[23] statement that a treatment of the Einstein correlation which is *both relativistic and quantal* implies a rejection of the retarded causality concept) I have shown in detail[12] that *special relativity understood as invariance through the full Lorentz group, including CPT invariance, is fully consistent with the Einstein correlation.* More precisely, the physical existence of the correlation, together with assumed Lorentz invariance, *entails* invariance through the *full* Lorentz group, including *CPT* invariance. And this should certainly not come as a surprise, since the latter invariance is *the relativistic invariance of quantum field theory.*[24]

One is thus led to distinguish between the *macro special relativity*, defined by invariance through the orthochronous Lorentz group and the retarded causality concept, and the quantal, or *micro special relativity*, defined by invariance through the full Lorentz group and *CPT* symmetry. I find it surprising that quantal theorists do adhere to micro special relativity in their technical work, and to macro special relativity when aiming at the interpretation of quantum mechanics. My choice is the opposite: to use *micro special relativity for interpreting quantum mechanics.* And this is *Blitzkrieg.*

Another significant point in the relativistic treatment of the Einstein correlation is the antinomy between two philosophies: the "advancing time" philosophy, as expressed in an often quoted sentence of Isaac Newton, and the "four-dimensional extended space-time" philosophy, as expressed in an often quoted sentence of Hermann Minkowski. In the realm of quantum mechanics the Newtonian time concept (according to which, in Bergson's words, "the universe dies and is reborn at every instant *t*") of universal time has reappeared in Schrödinger's famous 1926 articles, while the Minkowskian four-dimensional geometrical spirit pervades Feynman's celebrated 1949 papers. As a compromise, there is the Tomonaga–Schwinger technique of the "advancing arbitrary spacelike σ," which Dyson has proved to be mathematically equivalent to Feynman's.

It so happens that those authors favoring, in the interpretation of quantum mechanics, Einstein's 1905 macro relativity, also favor the Tomonaga–Schwinger advancing σ philosophy—apparently because it looks more akin to the macroscopic, anthropocentric, causality concept. My *Blitzkrieg* strategy deliberately makes the opposite choice of a Feynman-style philosophy, where causality is conceived in terms of four-dimensional space-time connections—which are *CPT* invariant. At this point one author should be quoted, as having prominently recommended, in special relativity, such an approach, in an extremely elegant mathematical formalization: A. D. Fokker.[25] I am (and especially in the Einstein correlation problem) a definite follower of Fokker's "direct-action-at-a-distance philosophy," and a proponent of the time symmetric causality concept that is implied in it. I explain in Section 4 why I deem it definitely superior to the Tomonaga–Schwinger approach for dealing with the Einstein correlation.

Section 5 discusses a technical point. It is in fact a rehabilitation of the Racah[26] time reversal, which is shown to be much more consistent with the *geometrical*, static, concept of time reversal than is the Wigner[27] motion reversal. While there certainly is an affinity between the Tomonaga–Schwinger "advancing σ" philosophy and Wigner's concept of motion reversal, it will be shown that, on the other hand, there is a definite affinity between Feynman's "space-time extended interactions" philosophy and Racah's concept of time reversal.

Summarizing, the aim of the present article, following previous ones,[1,12] is to promote, as the appropriate paradigm of relativistic quantum mechanics (and, inside of it, of the Einstein correlation), the Lorentz and *CPT*-invariant framework of "micro relativity," expressed in the Minkowski and Feynman style of an extended space-time geometry. Philosophically speaking, this entails (as clearly recognized by Eberhard[23]), the replacement of the classical retarded causality concept by a *CPT-invariant microcausality concept*. From this *necessarily* follow some dramatic consequences, which are briefly discussed in Section 6.

3. Lorentz Plus *CPT* Invariance in the Preparation and Measurement Problem

As previously said, the T operation is here understood as geometrical reversal of the time axis—just as the P operation is understood as geometrical reversal of the three space axes. Thus, our T operation (which we will term the Racah[25] time reversal) is different from the Wigner[26] motion reversal $\mathcal{T}$ operation generally used in quantum physics—whereas $P \equiv \mathcal{P}$. The PT operation reverses all four components of 4-vectors, including of course energy-momenta (whereas the $\mathcal{T}$ operation reverses the momenta and not the energies). The PT operation leaves unchanged the 6-component angular momenta, and thus reverses helicities.

The PT operation just defined exchanges particles and antiparticles as defined à la Stuckelberg–Feynman. Therefore, in this scheme, we set by definition $C^{-1} \equiv PT$, whence $CPT = 1$ (which is a reinterpretation of the $\mathcal{CPT}$ invariance theorem[24]). Obviously, C or PT noninvariance entails chirality noninvariance. Considered as a passive transformation the PT symmetry will reverse all the Feynman arrows, so that, in order to restore the original situation, we must reverse the signposts, that is, exchange particles and antiparticles. So, CPT invariance is truly built into the Feynman scheme (and the C, P, T philosophy is more in harmony with it than is the $\mathcal{C}$, $\mathcal{P}$, $\mathcal{T}$ philosophy).

Now, considered as an active transformation, a CPT symmetry performed on a Feynman diagram (Fig. 1) will exchange[28] the preparations L_1, M_1, N_1, ... and the measurements L_2, M_2, N_2, ... , so that a major consequence of my basic assumption is *intrinsic symmetry between the preparation and measurement concepts*—a feeling certainly not shared by B. d'Espagnat[15]—but does a scientific revolution[29] go on without "sound and fury"?

Thus CPT invariance (just as $\mathcal{CPT}$ invariance) entails a "detailed balance theorem" of the form $(A + B + \cdots \to P + Q + \cdots) \sim (\bar{P} + \bar{Q} + \cdots \to \bar{A} + \bar{B} + \cdots)$, but something more can be said. If C invariance does not hold, $C(PT)$ invariance does, and *this* is exactly the four-dimensional analog of the $\mathcal{CP}$ invariance concept which has large, but not universal, validity. The feeling of uneasiness created by the $\mathcal{CP}$ violation in K meson decay may thus be alleviated by going to the symmetries of four-dimensional geometry.

Again considered as an active transformation, a CPT symmetry will exchange nonseparability of future measurements upon subsystems that have interacted[10,21] into nonseparability of past preparations that will interact.[22] I have used[12] this argument as *a priori* supporting conservation of the correlation formula between linear polarization states at very large spatial distances,[21] and also in Aspect's[11] experiment where the polarizers are turned while the photons are flying, because common sense believes the time-reversed statements to be obvious. The latter could of course be tested[12] in anticascade experiments, which have become routine since the advent of the dye laser.[30] So, if one does not question retarded causality in the time-reversed, anticascade experiments, the CPT invariance principle renders useless the Wilson–Lowe–Butt, Bruno–d'Agostino–Maroni, and Aspect experiments. But, on the other hand, as Born's probabilistic interpretation of quantum mechanics in the form of a "wavelike probability calculus" forbids[12] that the correlated measurements in an EPR experiment (or the correlated preparations in a time-reversed EPR experiment) do confer *properties belonging* to the subsystems, the retarded causality concept may then seem less obvious than it was with the classical calculus of probabilities. Then the CPT-invariance principle may be used to link experimental results of the direct and the reversed EPR correlation experiments.

My *CPT*-invariance basic philosophy certainly requires that I define and use a time symmetric concept of the wave collapse. This, as recalled in the Introduction, I have largely done in my series of papers.[12] In the context of the *S*-matrix formalism the following *CPT*-invariant concept of the wave collapse holds: given the Schwinger–Feynman transition amplitude $\langle\Phi|\Psi\rangle$ between a specified set of preparations $|\Phi\rangle$ and a specified set of measurements $|\Psi\rangle$, the single stochastic event, or *collapse*, consists in that *one element* $\langle\Phi_i|\Psi_j\rangle$ *of the transition amplitude matrix is actualized.*

Then one can either consider a predictive problem, by collecting the various outcomes of a given preparation, or a "blind"[31] retrodictive problem, by collecting the possible preparations of a given measurement result. For computing the corresponding sets of probabilities, it is convenient to project, respectively, $|\Phi\rangle$ upon $|\Psi\rangle$, which is termed *collapsing* the wave function, or to project $|\Psi\rangle$ upon $|\Phi\rangle$, which I call *anticollapsing* the wave function. These are two asymmetric (but symmetric to each other) ways of thinking of, and computing, the Hermitian scalar product—which is intrinsically symmetric in $|\Phi\rangle$ and $|\Psi\rangle$.

Clearly, the predictive procedure uses retarded waves, while the retrodictive procedure uses advanced waves—a fact recognized by Fock,[32] Watanabe,[31] and others.[33] Born's "wavelike probability calculus" essentially links together *lawlike symmetry* and *factlike asymmetry* of, on the one hand, *predictive and retrodictive probabilities* and of, on the other hand, *retarded and advanced waves.*

A very important point should now be stressed, that will be expanded upon in Section 4. The (wavelike) probability calculus used in quantum mechanics essentially is a *conditional probability* calculus: it holds "if" such preparation and "if" such measurement, *as specified in the formula*, is performed.

Therefore, the *S*-matrix transition amplitude $\langle\Phi|\Psi\rangle$ (Fig. 1) essentially holds *if* each and every one of the ingoing preparations L_1, M_1, N_1 . . . , and each and every one of the outgoing measurements L_2, M_2, N_2 . . . , *as stated in the formula*, are performed.

It is thus clear that the probability concept inherent in quantum mechanics (here, in relativistic quantum mechanics) is *neither objective nor subjective because it is both.* Probability (whence, also, *information*) thus appears as *the hinge around which mind and matter are interacting*—a point to be further expanded upon in Section 6, in the context of *CPT* invariance.

So, finally, the *paradoxical fact* is that the so-called *local* preparations L_1, M_1, N_1, . . . and measurements, L_2, M_2, N_2, . . . are *nonseparable.*[34] I have argued[12] that *the only conceivable connection between them is through the Feynman zigzags* (Fig. 1) *via the interaction region C*, as implied by the very mathematics, and also required by the physics—as these zigzags are *the only occupied channels.*

Thus, *the Einstein*[8] or *EPR*[9] *paradox is born from the union of two earlier paradoxes: intrinsic time symmetry* in the Loschmidt and Zermelo fashion, and *wavelike probability calculus* à la Born. The Einstein or EPR correlation is tied *via* the Feynman zigzags, either as a nonseparability of future measurements via their common past preparation[10,21] (Einstein correlation proper), or as a nonseparability of past preparations via their common future measurement[22] (time-reversed Einstein correlation).

CPT invariance and wavelike calculus of probabilities are built into the Feynman treatment of the S matrix. They show up together as the paradoxical Einstein correlation—the wizard of which is the Feynman zigzag.

The deterministically evolving state vector is a concept that is (1) not testable, (2) theoretically ambiguous, (3) unnecessary for computing transition amplitudes.

It is not testable. Considering the transition amplitude $\langle \Phi_1 | U_{12}^{-1} | \Psi_2 \rangle$ between a preparation $|\Phi_1\rangle$ and a subsequent measurement $|\Psi_2\rangle$, we are definitely not allowed to tamper with the system in between (without, according to Bohr, changing the very nature of the problem).

It is theoretically ambiguous. Denoting σ an arbitrary Tomonaga–Schwinger spacelike surface lying between the pseudotimes σ_1 of the preparation and σ_2 of the measurement, the state of the system at σ is taken as $|U_{1\sigma}\Phi_1\rangle$ if we believe in retarded waves and as $|U_{\sigma 2}^{-1}\Psi_\epsilon\rangle$ if we believe in advanced waves, without any possibility of a Popper "falsification."

It is unnecessary. For computing the transition amplitude $\langle \Phi_1 | U_{12}^{-1} | \Phi_2 \rangle$, the evolution equation of the Tomonaga–Schwinger unitary operator:

$$U^{-1}[\delta/\delta\sigma(x)]U = itU^{-1}HU$$

is necessary and sufficient.

The philosophy consonant with these remarks is that the system is in a definite state only just after a preparation (state $|\Phi_1\rangle$) and just before a measurement (state $|\Phi_2\rangle$). In between it is *transiting* between $|\Phi_1\rangle$ and $|\Phi_2\rangle$, and, thus, feeling the retarded influence of the preparation at σ_1 and the advanced influence of the measurement at σ_2.

This CPT-invariant conceptualization clearly yields a straightforward interpretation of the "delayed choice experiments."[11,37,38]

4. Physics of the Collapse-and-Retrocollapse Concept

When intrinsic time symmetry is taken care of together with the Copenhagen concept that it is the act of observation which collapses the wave function, unexpected insights are gained concerning the interaction between mind and matter at the elementary level.

Consider first (Fig. 2) the Heisenberg microscope thought experiment, which has been so "instrumental" for promoting the new paradigm. In the object plane, an electron E scatters a photon, which is observed at I in the retina of the observer, where the wave collapse admittedly takes place.[35] It is from *that* observation that the position E of the electron in the object plane is inferred, so that the *collapse* at I truly entails

FIGURE 2. The Heisenberg microscope thought experiment: collapse at *I* of the photon wave, retrocollapse at *E* of the electron wave.

a *retrocollapse* at *E*. This is the very argument I[36] had used in 1953 when promoting my interpretation[12] of the EPR correlation.

Concerning the Heisenberg microscope, C. von Weiszäcker[37] has made a significant remark. The observer can choose to bring into focus either the image plane, in which case he will "measure" the electron's position, or the microscope's focal plane, in which case he will "measure" the electron's momentum. Moreover, if the microscope is very long, he can in principle make his choice *after* the scattering process has taken place. This is a significant remark, which adds strength and precision to the retrocollapse concept.

Von Weiszäcker's type of argument has recently been reemphasized by J. A. Wheeler[38] in his "time delayed quantal measurements" thought experiments. These lead him to conclude that "no measurement is a measurement as long as it is not concluded," which is operationally all right. The point, however, is, that relativistically speaking, the whole procedure should be pictured in space-time, where the "future," whatever it *is*, cannot be else than what it will have been. Therefore the "choice" that is made at the macrolevel, and the (subsequently) observed "collapse," does in some sense retrocollapse the (prior) observed effect.

Of course we are definitely *not* speaking of a "mechanical disturbance" produced by the setting of the macroscopic apparatus, and "retroacting" upon the so-called "quantal object." Rather, we are speaking of a psychophysical interaction between the observer and the observed, to which we will come back in Section VII. This point is closely connected with the one made by Bohr[39] in his answer to Einstein–Podolsky–Rosen[9]—except that Bohr, like Wheeler, sticks to operationalism and does not discuss intrinsic time symmetry.

Incidentally, there *is* a time delayed quantal experiment now in progress: the one by A. Aspect.[11]

Coming back to my[36] 1953 argument, let us see how the retrocollapse concept ties in with the distant Einstein correlations.

In his intervention[8] at the 1927 Fifth Solvay Council, which truly started the whole affair, what Einstein said was this. Consider (Fig. 3) a plane wave falling normally upon a diffracting screen with an aperture *C*. An associated particle "falling inside the hole" will later blacken some grain *L* of a semispheric photographic plate with center at *C*. Then, asked Einstein, what sort of a correlation ties any of the nonblackened grains *N* to the one *L* that is blackened? Of course there would be no paradox at all if it could be said, and thought, that, so to speak, "the dice are cast at *C*." But the statistical interpretation of quantum mechanics, just produced by Born in 1926, has it, for quite compelling reasons,[12] that it is not at *C*, but at *L* and/or *N*, that "the dice are cast." Whence the paradox. Einstein significantly added that such a

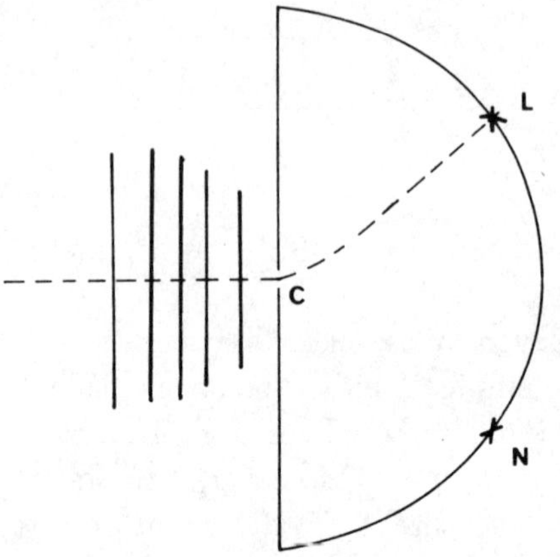

FIGURE 3. Einstein's 1927 thought experiment: how is the correlation tied between any nonblackened grain *N* and the blackened grain *L*?

"spacelike telegraphing" seemed to contradict the most fundamental assumption of his relativity theory—and this, as already said, is part of the present discussion. Incidentally, Renninger[40] has later produced a "negative" version of the argument.

In 1947 this tricky problem was much discussed by L. de Broglie and his coworkers. Having given some thought to the question one day I said this: According to Born, "the dice are cast" where and when a measurement is made, that is, in the present case, at *L* and/or *N*. Now there *is* one (and only one) information channel connecting *L* and *N*: the zigzag going backwards through *C*. Moreover, all known elementary physical processes are time symmetric; so this one, which undoubtedly is "elementary," should be time symmetric. This means that *the causality concept is time symmetric at the elementary level.*

de Broglie positively took that proposal as a mad one; in 1953 only did he accept to forward my "Note aux Comptes Rendus."[36] However, as it is by now quite obvious that the EPR correlation is in itself "something crazy," all the proposals made for solving it are, in some way or another, "crazy."[41] And so mine[1,12] is now freely discussed.[23]

Another typical example of both the Einstein correlation and the retrocollapse concept is provided by the observation by two or more observers *L* and *N* of the same quantal transition *C*—for example (Fig. 4) scintillations of α particles upon a ZnS

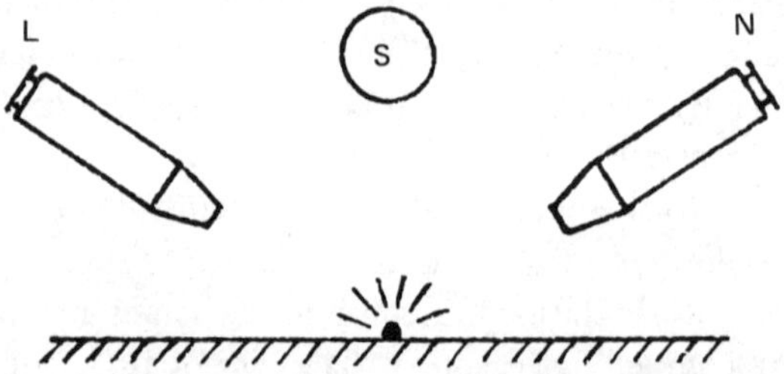

FIGURE 4. Scintillations of α particles upon a ZnS screen as observed by two (Einstein correlated) observers *L* and *N*.

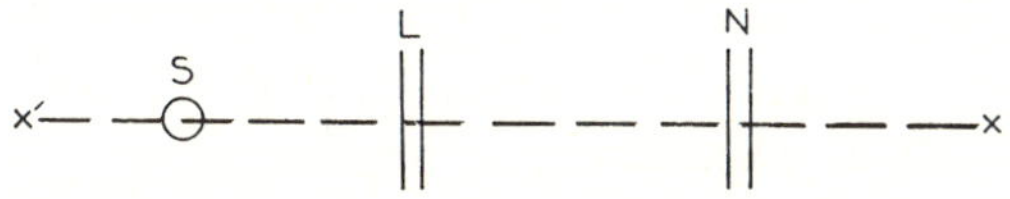

FIGURE 5. Successive collapses of polarization states of photons.

screen. The correlated collapses occur at *L* and *N* and, as both observers do agree upon what they see, they are producing the *same* retrocollapse at *C*—in their common past. It thus turns out that *they are not independent* from each other—as they feel they are—because, *by logical necessity*, they are cooperating, or competing.

Finally, the collapse-and-retrocollapse concept should be discussed in itself. Consider for example (Fig. 5) a light beam emanating from a source *S* and crossing in succession two linear polarizers *L* and *N*, the directions of which are *A* and *B*. Where and when does the transition of a photon's polarization from *A* to *B* occur?

The classical answer was, when the photons issuing from *L* impinge upon *N*. This is, however, a *macroscopical prejudice*, impressed upon our minds by our familiarity with the *factlike* preponderance of retarded over advanced waves.

From the mathematical formalism *per se* a very different concept follows. The transition occurs "somewhere in between" *L* and *N*, and consists in $\cos^2(A - B)$ of the photons jumping from the retarded wave that has passed *L* onto the advanced wave that will pass *N*.

Consider also (Fig. 6) those photons emanating from a source *S* and passing in succession two small apertures *A* and *B* inside screens *L* and *N*. *Mutatis mutandis* the discourse is the same as before and, as very explicit pictures can be drawn in this case (Figs. 5a and 5b), no more comment will be made.

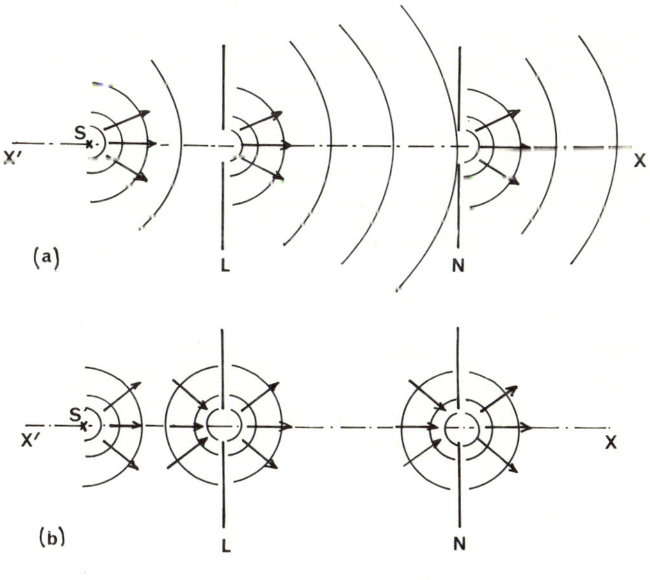

FIGURE 6. Successive collapses of positions of photons: (1) classical time-asymmetric concept; (b) proposed new time-symmetric concept.

One need not point out that the thought experiment just discussed is the three-dimensional image of the four-dimensional one discussed in the Introduction [Formulas (1) and (2)], with the abscissa x playing the role of the time t (Fig. 1).

5. Discarding a Frequent Misconception

As an example consider a pair of photons a and b flying oppositely upon an axis x in the laboratory frame, prepared in one of the two spin-zero states

$$\sqrt{2}\,\psi = L_aL_b \pm R_aR_b = \begin{cases} Y_aY_b + Z_aZ_b \\ i[Y_aZ_b - Z_aY_b] \end{cases} \tag{3}$$

and suppose that a (circular or linear) polarization measurement is performed on photon a alone. It is true that it does collapse the a substate into either L_a or R_a, or Y_a or Z_a. But it would be utterly wrong to then conclude, by carelessly reading formulas (1), that it does also collapse *physically* the b substate into, respectively, L_b or R_b, or Y_b or Z_b, or Z_b or Y_b.

This is easily shown by the following *reductio ad absurdum*. We can measure simultaneously in the laboratory frame the *circular polarization* on photon a and the *linear polarization* on photon b, the corresponding transition amplitude being

$$2\Psi = L_aY_b + iL_aZ_b \pm R_aY_b \mp iR_aZ_b \tag{4}$$

The question is of course: *which* of the two measurements collapses the *other* substate?

Moreover, the two measurements need not even be simultaneous, because *the correlation formula is symmetric in a and b*, and thus *invariant with respect to the time ordering of the measurements.*

In the case of photons, the measurements performed upon a and b always define a spacelike vector LN, so that the time ordering $t_a - t_b$ is *relative* in an (x, t) Lorentz transformation. And, as such a Lorentz transformation conserves the (circular or linear) polarization states, it is completely meaningless to conceive of a wave collapse occurring "first" at L (or N), and then affecting "instantly" the other photon, b (or a).

One further step should even be taken. Not only the relative distances $x_L - x_C$ and $x_N - x_C$ between the detectors L and N and the source C, but also (Fig. 2) the relative velocities along x, $v_L - v_C$ and $v_N - v_C$, are arbitrary, the polarization states being invariant with respect to v_C, v_L, v_N. Therefore, *it is absolutely excluded that a preferred reference frame exists for formalizing and understanding the Einstein correlations* (as some authors have suggested).

So, finally, what is the answer to the question raised by the *reductio ad absurdum* argument? It is (as previously stated): *a word "if" is attached to each and every symbol L, R, Y or Z present in the formulas* (3) and (4). It is under *this* assumption that a collapse of photon a into L_a, or R_a, or Y_a, or Z_a, does collapse the photon b into,

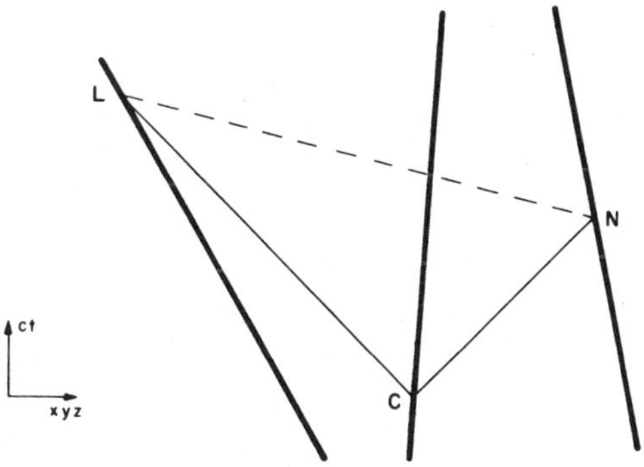

FIGURE 7. The source C (cascading atoms) and the detecting devices L and N can have arbitrary velocities inside an (x, ct) plane, and the time ordering of the detecting events is relative in an (x, ct) Lorentz transformation. Therefore it is excluded that a preferred reference frame exists for describing the Einstein correlation.

respectively, L_b, or R_b, or Y_b, (or Z_b), or Z_b (or Y_b): *if* the *strictly correlated* measurement is performed on photon b.

And *what* if no measurement at all is performed on photon b? Well, as a measurement performed on photon a alone is *definitely not* a proper measurement performed upon the a and b system, then, *nothing at all can be said of b*.

In the S-matrix context, when two measurements L and N are said to be performed upon subsystems a and b, the verb "to be" is understood in the atemporal sense appropriate to Minkowski's four-dimensional geometry. If, instead, we choose everyday language, the appropriate wording becomes[42]: *If* (say) a circular polarization measurement performed upon a yields (or has yielded, or will yield) (say) the L_a state, then, *if* a circular polarization measurement is (or has been, or will be) performed upon b, it yields (or has yielded, or will yield) the L_b state.

Things being so, the superiority of the Feynman philosophy over the Tomonaga–Schwinger philosophy consists in the following:

In the nonrelativistic Schrödinger formalism, which the covariant Tomonaga–Schwinger approach mimics to a certain extent, one of the two collapses, say L, occurs first, and is thought of as collapsing the overall $|\Phi\rangle$ into, say, $|\Psi\rangle$. Then, the "second" measurement, N, collapses $|\Psi\rangle$. It finally turns out that the correlation formula is symmetric in L and N. The difference between the Schrödinger and the Tomonaga–Schwinger treatments is that, in the first one, the time ordering of the collapses is absolute, whereas, in the latter, it is relative to the σ family and, therefore, reversible (if the LN vector is spacelike).

It is thus clear that the asymmetry in the reasoning and in the computation is spurious, and just as irrelevant as the ordering of credit and debt entries inside an account recording, where it does not affect the overall balance.

The point is that, thanks to its use of propagators in the space-time picture,[43] or of the eigenfunctions of the 4-frequency picture,[12] *the Feynman method directly displays the L and N symmetry of the correlation formula.*

Therefore, notwithstanding its remoteness from common sense (or, should we rather say, *because* of it), the Fokker[25] and Feynman philosophy is the most appropriate in our problem.[44] And this, of course, brings us back to Eberhard's[23] statement, that a radical change in the causality concept is required by the Einstein correlations, in relativistic quantum mechanics.

6. Racah versus Wigner Time Reversal

While the well-known Wigner[27] motion reversal operation $\mathcal{T}$ is obviously quite consonant with the Schrödinger advancing time, and the Tomonaga-Schwinger advancing σ, philosophy, the Racah[26] time reversal operation T is generally discarded with little comment as being nonphysical. It can be consistently used, however, as recognized by Watanabe[45] and by Jauch and Rohrlich.[46] We intend to show here that (as defined in the framework of the Dirac electron theory) the Racah time reversal T *exactly* is the geometrical reversal of the time axis which is appropriate in the four-dimensional space-time geometry, and is thus *naturally akin to the Feynman zigzags philosophy*.

Consider, in units such that $c = 1$ and $\hbar = 1$, ($i, j, k, l = 1, 2, 3, 4$; $x^4 = it$) the Dirac equation

$$\{\gamma^i(i\partial_i - eA_i) + m\}\psi = 0 \tag{5}$$

and define $\gamma^{ij\cdots} \equiv \gamma^i\gamma^{j\cdots}$ if all indexes are different, and zero otherwise. Obviously, the operations ($\gamma^5 \equiv \gamma^{1234}$)

$$\psi' = \gamma^i\psi, \qquad \psi'' = \gamma^{jkl}\psi, \qquad \psi''' = \gamma^{1234}\psi = \gamma^5\psi \tag{6}$$

respectively, reverse axes (j, k, l), (i), and (1, 2, 3, 4). Thus, in particular, $P\psi = \gamma^4\psi$ (as is well known), $T\psi = \gamma^{123}\psi$ (Racah), $PT\psi = \gamma^5\psi$.

Under the PT operation, e and m are (by definition) invariant; the 4-current $i\bar{\psi}\gamma_i\psi$ behaves as a true vector—it changes sign.

The electric charge Q, defined by

$$24Q = \epsilon_{ijkl}\iiint j^i d_1x^j d_2x^k d_3x^l \tag{7}$$

where ϵ_{ijkl} denotes the Levi-Cività object, d_1x^j, d_2x^k, d_3x^l three 4-vectors collinear neither to each other nor to the 4-current j^i, obviously behaves as a true scalar[47]—which is consistent with the invariance of e.

It is obvious on (5) that the change in sign of $\gamma^i(i\partial_i - eA_i)$ in a PT symmetry is equivalent to changing the sign of the mass term m. Therefore, by requiring that $CPT \equiv 1$, we can define the C operation as changing the sign of the mass m, which (as e is invariant by definition) changes the sign of e/m.

As the 6-component angular momentum $x^ip^j - x^jp^i$ is invariant by the PT symmetry, the helicity is changed. Therefore, C noninvariance essentially means helicity noninvariance.

On the whole, this CPT system of concepts is geometrically very symmetric, and quite consonant with the Feynman zigzags philosophy. There is no doubt that it can be used consistently. One potential advantage in it is that *all three operations* C, P,

T *are unitary*, with eigenvalues ± 1; for example, the Jordan–Pauli propagator has parities $P = +1$, $T = -1$, $PT \equiv C = -1$.

The electron-positron pair of equations ($i = 1, 2, 3, 4$)

$$(\gamma_i k^i \pm ik)\psi = 0$$

(directly displaying Feynman's concept of antiparticles) is quite reminiscent of the Weyl neutrino–antineutrino pair of equations

$$(\sigma \cdot \mathbf{k} \pm E)\varphi = 0$$

The change in sign of the scalar product "4-velocity by 4-momentum" displays $C(PT)$ invariance just as that of "3-spin by 3-momentum" displays CP invariance.

Incidentally, Recami and co-workers[48] have before me used this Racah style conceptualization of CPT invariance.

Note Added in Proof

A reexamination of the $\mathcal{CPT}$ theorem has led me to the following conclusions: Direct inspection of a pair of relativistically adjoint wave equations (e.g., Dirac's for spin ½) shows that charge conjugation $\mathcal{C}$ is the product of rest mass reversal C by exchange Z (symbolically, $\psi \rightleftharpoons \bar{\psi}$) of the two equations. Similarly, the (non-covariant) Wigner motion reversal $\mathcal{C}$ is the product of the Racah time reversal T by Z. So, via the formulas

$$\mathcal{C} = CZ \qquad \mathcal{P} = P \qquad \mathcal{T} = TZ \tag{8}$$

the $\mathcal{CPT}$ theorem is rewritten as

$$\mathcal{CPT} = CPT = 1 \tag{9}$$

The latter (but not the former) expression displays *manifest Lorentz invariance*.

Strictly speaking, the operation PT is not geometrical spacetime reversal, because it changes the sign of the scalar product $\gamma^i(i\partial_i - eA_i)$ of (loosely speaking) the 4-velocity and the momentum-energy operators. Truly, $\mathcal{PT} = PTZ$ is covariant motion reversal.

As shown in Fig. 8, *particle–antiparticle exchange* and *covariant motion reversal* are the *passive* and the *active* aspects of the same mathematical operation. Also, *true geometrical reversal of all four spacetime axes* is the identity operation in formula (9).

My forthcoming paper in *Foundations of Physics* explains all this in detail.

7. The Most Ominous Consequence of CPT Invariance: Reciprocity of the Conversion Negentropy $\rightleftharpoons$ Information

In his famous book *Symmetries and Reflections* E. Wigner[49] expresses a very bold—but very logical—assumption. He argues, from both the general principle that to every "action" there corresponds a "reaction," and from the specific interpretation

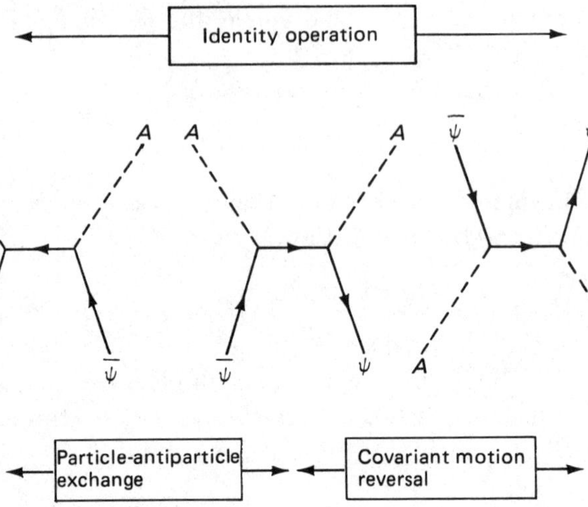

FIGURE 8. Particle–antiparticle exchange and covariant motion reversal as the passive and the active aspects of the same mathematical operation. Identity operation as geometrical reversal of all four space time axes. The example is, of course, the $e^+ e^- \rightleftharpoons 2\gamma$ transition.

scheme of quantum mechanics, that *there should exist a reaction of the observer upon the observed.*[50] He goes even so far as to add: "Every phenomenon is unexpected and most unlikely until it has been discovered—and some of them remain unreasonable for a long time after they have been discovered." He says also that he knows of no philosopher having already uttered such a statement. There is at least one, however (and this may come as a surprise): Descartes.[51]

The general principle of CPT invariance upheld in this paper obviously entails the conclusion already drawn by Wigner. If, following the Copenhagen–Göttingen school, the stochastic event of *wave collapse* is attributed to an act of consciousness of the observer, then, *via* the principle of *CPT* invariance, it is (as already said) a *collapse-and-anticollapse*, so that the act of cognizance (the by-product of which is the emission of a retarded wave) is coupled to an act of will (the by-product of which is the absorption of an advanced wave). As an example of this intrinsically time symmetric phenomenon we have discussed[12] the covariant position measurement process. As previously said, asymmetry shows up *only* at the macro level, and it is, in Mehlberg's[52] words, *factlike* and *not lawlike*.

What we are saying is nothing else than a very familiar statement in information theory. The information concept has two faces: gain in knowledge and organizing power, which show up, respectively in (say) the reception and the emission of a message. And this was already known to Aristotle (and, more explicitly, to Thomas Aquinas). To say that information as knowledge is a trivial concept, and information as organization a somewhat esoteric concept, is nothing other than a statement of the Second Law—which is factlike. It is, I believe, the deepest of all expressions of the Second Law.

But now, very much as there is a lawlike symmetry and a factlike asymmetry between particles and antiparticles, *and* that, knowing this, it is not impossible to unveil the hidden, negative, face of energies, in the form of antiparticles, so it must be possible to unveil also the hidden, advanced, face of the information waves of quantum mechanics. All the more so that there is a partial binding between these two lawlike symmetries and factlike asymmetries. It is well known that, by definition, the Feynman propagator links positive energies with the retardation, and negative energies with advance concept:

$$D_F = D_{\text{ret}} + D_- = D_{\text{adv}} + D_+$$

Also, the Jordan–Pauli propagator has the two expressions:

$$D \equiv D_+ - D_- = D_{\text{ret}} - D_{\text{adv}}$$

It goes without saying that accepting the appearance of advanced waves at the macro level means acceptance of the phenomena termed *precognition* and *psychokinesis* in parapsychology—two phenomena not distinguishable from each other at this level of the discussion. Other physicists have made, in a slightly different context, an essentially identical statement.[53]

H. Schmidt[54] has very carefully discussed the phenomenology of a small contribution of advanced waves against a strong background of retarded waves, and has shown that it is no more incompatible with the routine working of macro physics than is a small contribution of antiparticles against a strong background of particles. Schmidt's contribution is significant, because it disposes of sweeping statements of the kind "As one cannot kill one's grandfather in his cradle, the (macroscopic) advanced waves concept is *a priori* ridiculous." Was it not Lavoisier who had stated that "As there are no stones inside the sky, no stones can fall from the sky?"

Schmidt had good reasons when writing the quoted paper. He has published, in respected journals, the positive results of experiments in psychokinesis, and not only of "real time" ones,[55] but also of retroactive ones.[56] The latter, of course, are *extremely* paradoxical. What I intend to show now is that *they are necessarily implied for consistency of the Copenhagen–Göttingen interpretation of quantum mechanics.*

If it is true that it is the act of observation which collapses the wave function, *what* if the result of a measurement is recorded automatically and read only later? Of course it must be read at some time, otherwise nothing can be said.

Schmidt's[56] retropsychokinetic experiments provide an answer to this question, and I cannot imagine any other one. Fantastic as it looks, this answer is not contrary to facts,[54] and even, says Schmidt,[56] it *is* fact.

Of course, it would be quite significant to couple an EPR correlation experiment with a psychokinetic experiment, and thus see if psychokinesis performed on one beam would affect the other one. An attempt has been made in this direction, with a negative result,[57] but, as an essential ingredient was missing—psychokinesis on one beam—the experiment in itself was nonsignificant.

It has been repeatedly stated[15] that the Einstein spacelike correlation cannot be used for telegraphing outside the light cone. This is undoubtedly true if one tries to use it in connection with some gadget—because of the (factlike) Second Law. But, if we can overcome the Second Law, by making use of the reciprocity of the conversion *negentropy* $\rightleftharpoons$ *information* through psychokinesis, telegraphing outside the light cone via Feynman zigzags becomes possible.

8. "Wavicles" as Information Channels

Ever since the discovery of the wave–particle dualism (Einstein's photons and de Broglie's matter waves) the "complementarity" between the two concepts has remained somewhat enigmatic. At the recent International Conference "Science et Conscience" held in Cordoba, F. Capra[58] has expressed the opinion that, rather than "objects endowed with properties" (which they definitely cannot be) elementary particles should be conceived as "space-time channels carrying information between preparations and/or measurements"—an idea quite akin to the "space-time telegraph" I am speaking of.

Such a channel is, in fact, the specific wave equation obeyed by the said "elementary particle"; and the information is expressed as its occupation number.

Perhaps one can then wonder if, in this sort of "yes or no" questioning of Nature so well formulated in terms of projection operators, there is not something more than mere mathematical symbolism: are we perhaps questioning the switchboard operator?

This, of course, amounts to asking if there is not some sort of rudimentary "life" or "consciousness" attached to the elementary particles—*not at all* an unreasonable question, because *where* should one put a sharp boundary between life and nonlife, when going down from animal or plant to the particles, through those viruses which are crystallizable molecules? Let it be emphasized that such a question is neutral between materialism and spiritualism, being not committed as to which—matter or psyche—is the "epiphenomenon" of the other.

Two authors at least, Witmer[59] and Cochran,[60] have raised such a question in physics journals. I refer to them, adding only the remark that, in such a perspective, "psychokinesis" would be understood as "suggestion."

9. Phenomenology of *CPT* Invariance

Duhem,[29] and quite a few others after him, have convincingly shown that there is no such thing as a crucial experiment. Acceptance of a new paradigm therefore builds up through an accumulation of observed facts in conformity with it.

As previously said,[12] one main aspect of the paradigm I am proposing is that the link of the distant Einstein correlation is a Feynman zigzag, with the connection either in the past or in the future. I review here some of the facts which are explained in a quite straightforward way by this hypothesis, but some of which do not seem to be so easily justifiable in other schemes.

1. Nonseparability of two distant stochastic measurements[10,21] or of two distant preparations,[22] physically connected through their common past or future, respectively.

2. Intrinsic symmetry between these two concepts, as experimentally verifiable.

3. Theoretically predicted, and experimentally verified,[21] invariance of the correlation formula through arbitrary displacements of the measurement events along the respective beams. This really amounts to a geographical exploration of the streams forming the Feynman zigzag which channels the correlation.[44]

4. Aspect's[11] experiment (if it turns out as supporting quantum mechanics).

5. The Einstein correlation plus psychokinesis experiment, as defined at the end of Section 6 (if it yields a positive answer).

6. Schmidt's[56] retropsychokinesis experiments.

10. Conclusion

My formalization and interpretation of the 1927 Einstein[8] and 1935 EPR[9] correlation strictly adheres to the mathematics of relativistic quantum mechanics, either in its second quantized Tomonaga–Schwinger–Feynman–Dyson expression, or in the first quantized expression I[6] have proposed. It consists in a straightforward reading of the (very operational) mathematics: no more, but no less, than "the scriptures." In this, I believe, I am carefully listening to some illustrious teaching: Copernicus and Einstein (not to speak of others) have unveiled the sense of the scriptures by wording a discourse exactly fitting an operational mathematical recipe: heliocentric kinematics, or Lorentz–Poincaré kinematics. It then turned out that a large collection of (sometimes paradoxical) phenomena, other than those already taken care of, but also implicit in the formalism, were true.

My guess is that the various consequences of Lorentz plus *CPT* invariance (as tied in with Born's wavelike principle of adding partial amplitudes) pursued in the most systematic and uncompromising way, will open new vistas in the problem of the interaction between mind and matter. These, implied in the very concept of probability, are enunciated with much more acuity, through the use of Born's principle, in the quantal measurement problem.

References and Notes

1. O. Costa de Beauregard, "Time Symmetry and Interpretation of Quantum Mechanics," *Found. Phys.* 6, 539 (1976).
2. The following discussions will show that *CPT* invariance is the covariant (and very legitimate) heir of the classical *T* symmetry.
3. *State vector collapse* is taken by some as a more rigorous wording. We express here this concept by the shorter and more intuitive wording of *wave collapse*.
4. We are using Dirac's notation together with his remark that an expansion $\psi^a(x) = \Sigma c_i^a \phi_i(x)$ can be written (summation sign omitted) $\langle a|x\rangle = \langle a|i\rangle \langle i|x\rangle$, where ψ and the ϕ's are interpreted as transition amplitudes. See P. A. M. Dirac, *The Principles of Quantum Mechanics* (Oxford Clarendon Press, 1948), p. 79, and A. Landé, *New Foundations of Quantum Mechanics* (Cambridge University Press, Cambridge, 1965), p. 83.
5. J. Schwinger, *Phys. Rev.* 74, 1439 (1948). See p. 1451.

6. O. Costa de Beauregard, *Précis de mécanique quantique relativiste* (Dunod, Paris, 1967). See also *Synthese* **35**, 129 (1977), p. 143, reprinted in *Hans Reichenbach, Logical Empiricist*, W. C. Salmon, editor (D. Reidel, Dordrecht, 1979), p. 341–366.
7. As, in the formalism of Ref. 6, the Jordan–Pauli propagator is the Fourier transform of the Fourier nucleus, the position operator associated with the Klein–Gordon equation is the 4-vector x *modulo that it ends on* σ (that is, 3 degrees of freedom and not 4; for example, the components of $\mathbf{x}$). This statement does *not* contradict the more complicated expression of the Newton–Wigner position operator, where by definition only positive frequencies are accepted, because my formalism essentially requires *both* the positive and the negative frequencies on an equal footing (private correspondence with R. F. O'Connell).
8. A. Einstein, in *Rapports et Discussions du 5e Conseil Solvay* (Gauthier Villars, Paris, 1928), p. 253–256. In this early intuitive discussion of the paradox, Einstein incidentally referred to an apparent conflict with his special relativity theory.
9. A. Einstein, B. Podolsky, and N. Rosen, *Phys. Rev.* **47**, 777 (1935). The analysis is in terms of the nonrelativistic Schrödinger formalism; the sting of the paradox is still more painful in relativistic quantum mechanics.
10. S. J. Freedman and J. F. Clauser, *Phys. Rev. Lett.* **28**, 938 (1972); J. F. Clauser, *Ibid.* **36**, 1223 (1976); E. S. Fry and R. C. Thompson, *Ibid.* **37**, 465 (1976).
11. A. Aspect, *Phys. Lett.* **54A**, 117 (1975); *Phys. Rev. D* **14**, 1944 (1976).
12. O. Costa de Beauregard, *Nuovo Cimento* **42B**, 41 (1977) and **51B**, 267 (1979); *Lett. Nuovo Cimento* **19**, 113 (1977) and **25**, 91 (1979); *Phys. Lett.* **60A**, 93 (1977) and **67A**, 171 (1978).
13. The Wigner motion reversal is *not* defined as geometrical reversal of the time axis, whereas the Racah time reversal is. This point will be discussed in the fourth section. In fact, it was already *PT* symmetry that Loschmidt and Zermelo were stressing in the framework of the Galileo–Newton mechanics.
14. S. I. Tomonaga, *Prog. Theor. Phys.* **1**, 27 (1946); J. Schwinger, *Phys. Rev.* **74**, 1439 (1948); F. J. Dyson, *Phys. Rev.* **75**, 486 (1949); R. P. Feynman, *Phys. Rev.* **76**, 749, 769 (1949).
15. B. D'Espagnat, *Conceptual Foundations of Quantum Mechanics*, 2nd Ed. (Benjamin, New York, 1976), pp. 90, 119, 238, 265, and 281.
16. J. F. Clauser and A. Shimony, *Rep. Prog. Phys.* **41**, 1881 (1978); see p. 1920.
17. H. P. Stapp, *Found. Phys.* **7**, 313 (1977).
18. B. D'Espagnat and J. S. Bell, private communications.
19. N. Cufaro Petroni and J. P. Vigier, *Lett. Nuovo Cimento* **25**, 151 (1979).
20. P. A. M. Dirac, Nature **168**, 906 (1951).
21. A. R. Wilson, J. Lowe, and D. K. Butt, *J. Phys. G* **2** 613 (1976); M. Bruno, M. d'Agostino, and C. Maroni, *Nuovo Cimento* **40B**, 143 (1977).
22. R. L. Pflegor and L. Mandel, *Phys. Rev.* **159**, 1084 (1967); *Jn. Opt. Soc. Am.* **58**, 946 (1968).
23. P. Eberhard, *Nuovo Cimento* **46B**, 392 (1978).
24. G. Lüders, *K. Dansk Videns. Selsk.* **28**, 5 (1954) and *Ann. Phys.* (Leipzig) **2**, 1 (1957); W. Pauli, *Niels Bohr and the development of physics*, W. Pauli, L. Rosenfeld, and V. Weisskopf, editors (Pergamon, New York, 1955), p. 30.
25. A. D. Fokker, *Time and Space, Weight and Inertia* (Pergamon, New York, 1965).
26. G. Racah, *Nuovo Cimento* **14**, 322 (1937).
27. E. Wigner, *Gott. Nachr.* **31**, 546 (1932).
28. It should be noted that the Feynman propagator $D_{F(-)} \equiv \overline{D} - ½(D_+ + D_-)$ is *PT* invariant, as is also the anti-Feynman propagator $D_{F(+)} \equiv \overline{D} + ½(D_+ + D_-)$ obtained by sidestepping the two poles the other way around. As is well known, use of the Feynman propagator ensures automatically an exponential decay if used in prediction (but, symmetrically, an exponential build-up if used in retrodiction). The opposite would follow from use of the anti-Feynman propagator. See Ref. 46, p. 408.
29. T. S. Kuhn, *The Structure of Scientific Revolutions* (University of Chicago Press, Chicago, 1962; 2nd ed, 1970). The essence of Kuhn's philosophy is already found in P. Duhem, *The Aim and Structure of Physical Theory*, Part II, Chaps. 4 and 6 (Princeton Univ. Press) translated after the French 1913 edition by P. P. Wiener.
30. See P. F. Liao and G. C. Bjorklund, *Phys. Rev. Lett.* **36**, 584 (1976), for an anticascade experiment with polarizers.

31. S. Watanabe, *Rev. Mod. Phys.* **27,** 179 (1955).
32. V. Fock, *Dokl. Akad. Nauk SSSR* **60,** 1157 (1948).
33. O. Costa de Beauregard, *Cah. Phys.* **2,** 317 (1958); Y. Aharonov, P. G. Bergmann, and Y. Lebowitz, *Phys. Rev.* **134B,** 1410 (1964); F. J. Belinfante, *Measurements and Time Reversal in Objective Quantum Theory* (Pergamon, New York, 1975); P. C. W. Davies, *The physics of time asymmetry* (Surrey Univ. Press, 1974).
34. This is d'Espagnat's (Ref. 15) very appropriate wording, following Einstein, Podolsky, and Rosen (Ref. 9).
35. Of course, absolutely speaking, this is not the end of the story, as the optical nerve and brain area belong to the picture; and *where* should we stop and say "here and now I have seen the photon"? J. von Neumann, *Mathematische Grundlagen der Quantum Mechanik* (Springer, Berlin, 1932), Chap. VI.
36. O. Costa de Beauregard, *C. R. Acad. Sci. Paris* **236,** 1632 (1953).
37. C. von Weiszäcker, *Z. Phys.* **70,** 114 (1931).
38. J. A. Wheeler, *Mathematical Foundations of Quantum Theory,* J. A. Marlow, editor (Academic Press, New York, 1977), p. 9.
39. N. Bohr, *Phys. Rev.* **48,** 696 (1935).
40. W. Renninger, *Physik* **158,** 417 (1960); *Phys. Z.* **136,** 251 (1963).
41. "Your theory is crazy, but not crazy enough to be true" (N. Bohr). "For any speculation which does not at first glance look crazy, there is no chance" (F. J. Dyson).
42. O. Costa de Beauregard, *Lett. Nuovo Cimento* **26,** 135 (1979).
43. R. Payen and J. M. Vigoureux, *Lett. Nuovo Cimento* **20,** 263 (1977).
44. Suppose some archaeologists discover that similar cultures have developed along two converging rivers running through barren country. Will they assume that there has been some direct interaction between any site on one river and any site on the other? Or rather, that the influence has traveled up and down, *via* navigation on the rivers?
45. S. Watanabe, *Phys. Rev.* **84,** 1008 (1951).
46. J. M. Jauch and F. Rohrlich, *The Theory of photons and electrons,* (Addison-Wesley, Reading, Massachusetts, 1955); see pp. 88–96.
47. Jauch and Rohrlich, Ref. 46, make the surprising statement that it does *not;* this is because they do not reverse (as they should) the sign of the 3-volume element.
48. R. Mignani and E. Recami, *Lett. Nuovo Cim.* **11,** 421 (1974); E. Recami and G. Zumino, *Nuovo Cim.* **33A,** 205 (1976)
49. E. Wigner, *Symmetries and Reflections* (M.I.T. Press, Cambridge, Massachusetts, 1967), pp. 171–184.
50. In almost all textbooks on quantum mechanics it is stated that there is a reaction of the measuring apparatus upon the observed system. It is very queer that a necessary consequence of this statement is not drawn, namely, the reaction of *the observer* upon the so-called observed system—because, where is the severance between the observer and the measuring apparatus?
51. R. Descartes, *Lettres* (Adam—Tannery eds.) I, 222 (letter 525) and III, 663 (letter 302).
52. H. Mehlberg, *Current Issues in the Philosophy of Science,* H. Feigl and G. Maxwell, editors (Holt, Rinchart, Winston, New York, 1961), p. 105.
53. E. H. Walker, in *Quantum Physics and Parapsychology,* L. Oteri, editor (Parapsychology Foundation, 1975), p. 1; R. D. Mattuck and E. H. Walker, *The Iceland Papers,* A. Puharitch, editor. Foreword by B. Josephson (Essential Research Associates, Amherst, Massachusetts, 1979), p. 111.
54. H. Schmidt, *Found. Phys.* **8,** 464 (1978).
55. H. Schmidt, *Bull. Am. Phys. Soc.* **24,** 38 (1978).
56. H. Schmidt, *Proc. Intern. Conf. Cybernetics and Society* (IEEE, New York, 1977), p. 535.
57. J. Hall, C. Kim, B. McElroy, and A. Shimony, *Found. Phys.* **7,** 759 (1977).
58. F. Capra, Le tao de la physique, in *Science et Conscience,* Stock, Paris (1980), pp. 43–55 (see especially pp. 45–46).
59. E. E. Witmer, *Am. J. Phys.* **35,** 40 (1967).
60. A. A. Cochran, *Found. Phys.* **1,** 235 (1971).

8

The Hilbert Space Axiom in Quantum Mechanics

Stanley P. Gudder
Department of Mathematics and Computer Science
University of Denver
Denver, Colorado 80208

1. Introduction

The Hilbert space axiom is the basic postulate of conventional quantum mechanics. In essence it maintains that the observables for an isolated quantum mechanical system $\mathcal{S}$ can be represented by self-adjoint operators on a complex Hilbert space H; the states for $\mathcal{S}$ can be represented by density operators on H; and the dynamics can be represented by a one-parameter unitary group on H. The power of this axiom is indisputable, but what is its justification? Where does the Hilbert space come from? And if it is necessary to use Hilbert spaces, why must they be complex?

Justification for the Hilbert space axiom has been a source of investigation since the early days of axiomatic quantum mechanics. One of the first papers with this subject in mind was due to G. Birkhoff and J. von Neumann,[1] and the research continues today. Some progress has been made toward solving this problem but it has not been completely resolved to the satisfaction of the majority of investigators.

In this article we shall survey some of the different directions that have been taken in attacking this problem. We begin with a precise statement of the Hilbert space axiom due to Dirac and von Neumann. We then consider the quantum logic approach to this question. A hidden variables model is next described which shows the possibility of another quantum world. Piron's theorem and the number field are then discussed. We also consider the algebraic and operational approaches to the problem.

2. Dirac–von Neumann Axioms

In this section we give the axioms of traditional quantum mechanics due to Dirac and von Neumann. Although these two investigators formulated quantum mechanics in similar ways,[6,39] we shall find it convenient to follow von Neumann. The first three axioms concern quantum statics.

(A1) The observables of an isolated quantum system $\mathcal{S}$ can be represented by self-adjoint operators on a complex Hilbert space H.

(A2) The states of $\mathcal{S}$ can be represented by positive trace class operators of trace one on H (that is, by *density operators*).

(A3) The probability that an observable represented by A has a value in a Borel set $\Delta \subseteq R$ when $\mathcal{S}$ is in a state represented by W is $\mathrm{tr}[P^A(\Delta)W]$, where $P^A(\cdot)$ is the resolution of identity for A.

A general density operator represents a so-called *mixed state*. The *pure states* are states that cannot be written as a nontrivial convex combination of two other states. It is easy to show that a pure state is represented by a one-dimensional projection P_ψ, where P_ψ is the projection onto the subspace spanned by $0 \neq \psi \in H$. In terms of a pure state P_ψ, Axiom (A3) becomes $\mathrm{tr}[P^A(\Delta)W] = \langle P^A(\Delta)\psi, \psi\rangle$. The last axiom is the axiom of quantum dynamics.

(A4) If $\mathcal{S}$ is in a pure state represented by a unit vector $\psi \in H$ at time $t = 0$, then the state of $\mathcal{S}$ at time t is represented by the unit vector $\psi_t = [\exp(-itH_0/\hbar)]\psi$, where H_0 represents the energy observable.

It turns out that with a few physical assumptions, one can "derive" the other three axioms from either (A1) or (A2). Also, (A1) is equivalent (in a sense) to the following axiom.

(A1′) The true–false experiments of $\mathcal{S}$ are represented by orthogonal projections on H.

True–false experiments correspond to apparatuses which give two possible results such as a counter or spin-1/2 measurement. Such experiments are sometimes called propositions or questions. A proposition corresponds to an observable that cannot be decomposed into simpler observables. The reason for the equivalence of (A1) and (A1′) is the following. If (A1) holds, then a yes–no experiment corresponds to a self-adjoint operator A. By the spectral theorem,[7,25] A has the form $A = \int \lambda P^A(d\lambda)$, so if A is indecomposable we must have $A = \lambda_0 P_0$ for some $\lambda_0 \in R$ and orthogonal projection P_0. We may assume that A has the two values 0 and 1, so $A = P_0$. Conversely, suppose (A1′) holds. If $\hat{A}$ is an observable and Δ a real Borel set, then corresponding to the pair $(\hat{A}, \Delta)$ we have the proposition "$\hat{A}$ has a value in Δ." We denote the corresponding orthogonal projection by $P^A(\Delta)$. It is not hard to justify that $\Delta \to P^A(\Delta)$ should be a projection-valued measure, so $\hat{A}$ corresponds to the self-adjoint operator $A = \int \lambda P^A(d\lambda)$. We shall have more to say about this later when we treat the general concept of an observable.

One of the objects of the quantum logic approach to quantum mechanics is to justify axiom (A1′). Let us suppose that (A1′) holds. A state $\hat{s}$ of the system $\mathcal{S}$ determines the probability $\hat{s}(\hat{a})$ that any proposition $\hat{a}$ is true when $\mathcal{S}$ is in the state $\hat{s}$. It follows that $\hat{s}$ corresponds to a probability measure on the set of orthogonal projections on H. Axiom (A2) now follows from Gleason's theorem.[11,38] In fact, according to this theorem there exists a unique density operator W such that $\hat{s}(\hat{a}) = \operatorname{tr}(aW)$, where a is the projection corresponding to $\hat{a}$. If $\hat{A}$ is an observable with corresponding self-adjoint operator $A = \int \lambda P^A(d\lambda)$, it follows from our previous discussion that $P^A(\Delta)$ is the proposition "$\hat{A}$ has a value in Δ." Applying (A2), the probability that $\hat{A}$ has a value in Δ when $\mathcal{S}$ is in the state $\hat{s}$ is $\operatorname{tr}[P^A(\Delta)W]$, so (A3) holds. Now suppose $\mathcal{S}$ is in the pure state represented by a unit vector $\psi \in H$ at time $t = 0$. For an isolated system it is reasonable to assume that the state $\psi_t = \alpha(t)\psi$ at time $t > 0$ is also pure. In order to preserve probability, ψ_t must be a unit vector. It follows from Wigner's theorem[38] that $\alpha(t)$ extends to a unitary or antiunitary operator on H. Moreover, it is clear that $t \to \alpha(t)$ is a one-parameter group of automorphisms on H and it is reasonable to assume that $\alpha(t)$ is strongly continuous in t. It follows from Stone's theorem[7,37] that $\alpha(t) = \exp(-itH_0/\hbar)$ for a self-adjoint operator H_0. Hence, (A4) holds.

One of the objects of the algebraic approach to quantum mechanics is to justify (A1). As we have seen, once (A1) holds, we can "derive" (A1′) and from the above, (A2), (A3), and (A4) hold.

In the operational approach, one of the objects is to justify (A2). Suppose (A2) holds and $\hat{A}$ is a bounded observable for $\mathcal{S}$. If $\hat{s}$ is a state for $\mathcal{S}$, then $\hat{s}$ determines the expectation $\hat{s}(\hat{A})$ of $\hat{A}$ when $\mathcal{S}$ is prepared according to $\hat{s}$. It is then reasonable that $\hat{s} \to \hat{s}(\hat{A})$ preserves convex combinations of states. Now represent the set of states in terms of the set of density operators $\mathcal{D}$ as in (A2) and equip $\mathcal{D}$ with the trace norm. Then bounded observables can be represented by affine functionals on $\mathcal{D}$ which can be extended to linear functionals on the Banach space $\mathcal{W}$ of self-adjoint trace class operators with the trace norm. We make the mild assumption that these representing linear functionals are continuous on $\mathcal{W}$ and are thus in the dual $\mathcal{W}'$ of $\mathcal{W}$. But $\mathcal{W}'$ is isomorphic to the set of bounded self-adjoint operators on H[5,7] and hence (A1) holds. Again, the other axioms now follow.

We thus see that the Axioms (A1), (A1′), and (A2) are physically equivalent and that the other axioms follow from any one of these three. For this reason we shall regard any one of these three axioms as the "Hilbert space axiom" for quantum mechanics. In the sequel we shall examine various attempts at justifying this axiom.

3. The Quantum Logic Approach

In the quantum logic approach the primitive axiomatic elements are the experimental propositions (or questions, or events).[1,18,25,27,30,38] Let $L = \{a, b, c, \ldots\}$ be the set of propositions for a quantum system $\mathcal{S}$. For $a, b \in L$, if b is true (i.e., is experimentally verified) whenever a is true, we write $a \leq b$. It is clear that $a \leq a$, and $a \leq b$, $b \leq c$ imply that $a \leq c$. If $a \leq b$ and $b \leq a$, then a and b are experimentally indistinguishable so we then assume that $a = b$. In this way $(L, \leq)$ is a partially ordered set (poset). We assume that there exists an absurd proposition 0 and a trivial proposition 1 satisfying $0 \leq a \leq 1$ for all $a \in L$. Since it would be hard to justify at this point, we do not assume that L is a lattice. That is, we do not assume that the greatest lower bound $a \wedge b$ or the least upper bound $a \vee b$ exist for all $a, b \in L$.

For $a \in L$ we denote the proposition which is true if and only if a is false by a'. It is clear that $a'' = a$, $a \leq b$ implies that $b' \leq a'$, and $a \vee a' = 1$ for every $a, b \in L$. Thus $(L, \leq, ')$ is an orthocomplemented poset. If $a \leq b'$ we write $a \perp b$ and say that a and b are *orthogonal*. If a_i, $i = 1, 2, \ldots$, are mutually orthogonal we assume that $\vee a_i$ exists in L. Hence $(L, \leq, \cdot)$ is σ-orthocomplete.

We say that $a, b \in L$ are *compatible* (Written $a \to b$) if there exist mutually orthogonal elements $a_1, b_1, c \in L$ such that $a = a_1 \vee c$, $b = b_2 \vee c$. Compatible propositions correspond to noninterfering experiments. As we shall see, the main justification for the above definition is that a and b are compatible if and only if they can be measured by a single observable. The prototype example of $a \leq b$ is two counters with volumes of sensitivity $\Delta_1 \subseteq \Delta_2 \subseteq R^3$. We can then think of dividing the larger counter into two parts $\Delta_2 = \Delta_1 \cup (\Delta_2 \cap \Delta_1^c)$, where Δ_1^c is the complement of Δ_1. Letting $\Delta_3 = \Delta_2 \cap \Delta_1^c$ we see that Δ_1, Δ_3, ϕ are mutually disjoint sets and $\Delta_1 = \phi \cup \Delta_1$, $\Delta_2 = \Delta_3 \cup \Delta_1$. If a_3 denotes the counter with volume of sensitivity Δ_3 we see that a_1 a_3 and 0 are mutually orthogonal propositions and $a = 0 \vee a$, $b = a_3 \vee a$ so a and b are compatible. Motivated by these considerations we postulate

$$a \leq b \text{ implies } a \leftrightarrow b \tag{1}$$

An orthocomplemented poset satisfying (1) is called an *orthomodular* poset. A σ-orthocomplete orthomodular poset is called a *logic*.

A *state* on a σ-orthocomplete orthocomplemented poset L is a map $m: L \to [0, 1]$ satisfying $m(1) = 1$ and $m(\vee a_i) = \Sigma m(a_i)$ if the a_i are mutually orthogonal, $i = 1, 2, \ldots$. We interpret $m(a)$ as the probability that the proposition a is true when the system is the state m. If L has a separating set of states, then (1) holds automatically; that is, L is a logic. Indeed, suppose $a \leq b$. Since $a \leq (b')'$, we have $a \perp b'$ and hence $a \vee b'$ exists. By De Morgan's law (which holds in this framework), $(a \vee b')' = a' \wedge b$ exists. Since $a \perp (a' \wedge b)$, $a \vee (b \wedge a')$ exists. If m is a state, then

$$\begin{aligned} m[a \vee (b \wedge a')] &= m(a) + m(b \wedge a') = m(a) + m[b' \vee a)'] \\ &= m(a) + 1 - m(b' \vee a) = m(a) + 1 - m(a) - m(b') \\ &= m(b) \end{aligned}$$

Hence $b = a \vee (b \wedge a')$ and $a = a \vee 0$ so $a \leftrightarrow b$.

An *observable* on a logic L is a σ-homomorphism $x: B(R) \to L$ where $B(R)$ is the Borel σ algebra on R. That is, $x(R) = 1$, $x(\Delta_1) \perp x(\Delta_2)$ if $\Delta_1 \cap \Delta_2 = \phi$, $x(\cup \Delta_i) = \vee$

$x(\Delta_i)$ if $\Delta_i \cap \Delta_j = \phi, i \neq j = 1, 2, \ldots$. We interpret $x(\Delta)$ as the proposition that the observable x has a value in the set Δ. With this interpretation, the above properties of x are easily justified. Two observables x and y are *compatible* (Written $x \leftrightarrow y$) if $x(E \leftrightarrow y(F)$ for every $E, F \in B(R)$. Compatible observables correspond to noninterfering experiments. The proof of the following theorem is straightforward (cf. Refs. 18, 38).

Theorem 1. (a) The range of an observable on L is a Boolean sub-σ-algebra of L. (b) Two propositions are compatible if and only if they are in the range of a single observable.

If x is an observable and m is a state on the logic L, it follows from the above interpretation that $m[x(\Delta)]$ is the probability that x has a value in $\Delta \in B(R)$ when the system is in the state m. We are then justified in calling the probability measure $\Delta \rightarrow m[x(\Delta)]$ the *distribution* of x in the state m. The *expectation* of x in the state m becomes $E_m(x) = \int \lambda m[x(d\lambda)]$ if the integral exists.

We now give some examples of logics. Let Ω be a nonempty set and let $L(\Omega)$ be a σ algebra of subsets of Ω. Under the usual set-theoretic order and complementation, $L(\Omega)$ becomes a logic. In $L(\Omega)$ any two elements are compatible and $L(\Omega)$ corresponds to the logic of classical mechanics. In this case the states reduce to probability measures on $L(\Omega)$. Moreover, it can be shown[38] that corresponding to an observable $x{:}B(R) \rightarrow L(\Omega)$ there exists a unique measurable function $f{:}\Omega \rightarrow R$ such that $x(\Delta) = f^{-1}(\Delta)$ for all $\Delta \in B(R)$. Hence, observables reduce to measurable functions in this framework. If f is an observable (measurable function) and μ is a state (probability measure), the expectation reduces to the usual integral $E_\mu(f) = \int f\, d\mu$.

For another example, let H be a complex Hilbert space and let $L(H)$ be the set of orthogonal projections on H. Under the usual order and orthocomplementation, $L(H)$ becomes a logic. This is the logic of traditional quantum mechanics. In this case, two projections are compatible if and only if they commute.[10,38] As a result of the spectral theorem, observables can be identified with self-adjoint operators on H. Applying Gleason's theorem, any state m on $L(H)$ has the form $m(P) = \mathrm{tr}(PW)$, $P \in L(H)$, where W is a unique density operator. The expectation of an observable (self-adjoint operator) A in the state m becomes $E_m(A) = \mathrm{tr}(AW)$.

For a third example, let Ω be a nonempty set and let C be a collection of subsets of Ω satisfying (1) $\Omega \in C$; (2) $\cup A_i \in C$ for any sequence of mutually disjoint $A_i \in C$; (3) $A \in C$ implies the complement $A^c \in C$. We call C a *σ class* of subsets of Ω. It is not hard to check that a σ class under the usual set-theoretic order and complementation is a logic. Also, it is clear that a σ class C is a generalization of a logic $L(\Omega)$ of classical mechanics. If $A, B \in C$, one can show[15,17] that $A \leftrightarrow B$ if and only if $A \cap B \in C$. Hence, unlike a classical logic, there can be interfering (noncompatible) propositions in a σ class. We thus see that σ classes share an important property with the traditional logic $L(H)$ of quantum mechanics. The states again reduce to probability measures on (Ω, C) and the observables reduce to measurable functions on (Ω, C). If f is a measurable function and μ a probability measure on (Ω, C), the expectation reduces to the usual integral $E_\mu(f) = \int f\, d\mu$. In the next section we shall give the relationship between this example and a hidden variables theory.

We now give a concrete example of a σ class which might have physical importance. Let $\lambda > 0$ and let $\Omega = [0, n\lambda]$, where $n > 1$ is an integer. If C is the collection of Lebesgue measurable subsets of Ω whose Lebesgue measures μ are integer multiples of λ, then C is a σ class but not a σ algebra. This example can be used for an "elementary length" theory in which λ is the elementary length.[14,20] Since there are many incompatible elements in C, this example exhibits the interference effects characteristic of quantum mechanics.

4. Hidden Variables and the Quantum Miracle

A state m on a logic L is *dispersion-free* if m has only the values 0 and 1. A dispersion-free state provides a deterministic description since every proposition is either true or false with certainty when the system is prepared in such a state. A set of states S is *order-determining* if $m(a) \leq m(b)$ for every $m \in S$ implies that $a \leq b$. The traditional Hilbertian logic $L(H)$, dim $H > 2$, has no dispersion-free states.[18] However, a classical logic $L(\Omega)$ has an order-determining set of dispersion-free states. Indeed, the set of probability measures concentrated at the points of Ω provide such a set. In the same way a σ class has an order-determining set of dispersion-free states.

We say that a logic L *admits hidden variables* if L has an order-determining set of dispersion-free states. The important point is that there are logics other than the classical ones which admit hidden variables and these retain the interference effects characteristic of quantum mechanics.

We have seen that a σ class admits hidden variables. Conversely, suppose a logic L admits hidden variables and let S be an order-determining set of dispersion-free states on L. Let h be the map from L into the collection of subsets of S defined by

$$h(a) = \{m \in S : m(a) = 1\}$$

and let C be the range of h. It can be shown that C is a σ class in S and that $h: L \rightarrow C$ is an isomorphism.[15,18] We thus have the following result.

Theorem 2. A logic L admits hidden variables if and only if L is isomorphic to a σ class C. In this case, C can be taken to be a σ class of subsets of the set of dispersion-free states on L.

How does this fit in with the nonexistence proofs for hidden variables? Most of these proofs show that if a model admits hidden variables, then the model must be classical in the sense that all elements of the model are mutually compatible. This is not the case in the above theorem. The reason for the difference is that in the nonexistence proofs more restrictions are placed on L and on the definition of a state than we have made. For example, in Ref. 25 it is assumed that L is a lattice and that a state

m satisfies the condition: if $m(a) = 1$ and $m(b) = 1$, then $m(a \wedge b) = 1$. This last condition has been criticized in many sources (for example, Ref. 2).

We now consider the quantum miracle. Suppose x and y are observables for which $x + y$ can be constructed in a meaningful way (for example, this can be done in our previous examples of logics). Let m be a state and find $E_m(x)$, $E_m(y)$, and $E_m(x + y)$. If $x \leftrightarrow y$, then it is physically evident that we should have

$$E_m(x + y) = E_m(x) + E_m(y) \tag{2}$$

However, if $x \not\leftrightarrow y$, then there seems to be no physical reason for (2) to hold. In fact, it would be surprising if expectations were additive for every state. Indeed, if x and y are highly interfering observables, then the observable $x + y$ has little relation with the observables x and y. Why then should (2) hold? The quantum miracle is that in traditional quantum mechanics (2) always holds when both sides of the equation exist. In fact, from Gleason's theorem we have for self-adjoint operators A, B that

$$E_m(A + B) = \mathrm{tr}[(A + B)W] = \mathrm{tr}AW + \mathrm{tr}BW = E_m(A) + E_m(B)$$

In the author's opinion, the additivity of expectations is a consequence of the very strong Hilbert space axiom and is not physically inevitable.

We now show that if the logic admits hidden variables, then expectations need not be additive. The following example is due to J. Zerbe.[40] Let Z denote the integers and let $\Omega = \{(i, j): i, j \in Z\}$. Let $R_m = \{(i, j): i = m\}$, $C_n = \{(i, j): j = n\}$, and $D_t = \{(i, j): i + j = t\}$. Let Σ_R, Σ_C, Σ_D be the σ algebras generated by $\{R_m: m \in Z\}$, $\{C_n: n \in Z\}$, $\{D_t: t \in Z\}$, respectively. Since $R_m \cap C_n \neq \phi$, $R_m \cap D_t = \phi$, $C_n \cap D_t \neq \phi$ for every m, n, t it follows that $C = \Sigma_R \cup \Sigma_C \cup \Sigma_D$ is a σ class. Define the probability measure μ on C by $\mu(R_0) = \mu(C_0) = \mu(D_1) = 1$, $\mu(R_m) = \mu(C_n) = \mu(D_t) = 0$ for m, $n \neq 0$, $t \neq 1$. Define the measurable functions f, g by $f(i, j) = i$, $g(i, j) = j$. Then $(f + g)(i, j) = i + j$ is measurable. Now $E_\mu(f) = E_\mu(g) = 0$ but $E_\mu(f + g) = 1$.

We now give another example. As already noted, if $\dim H > 2$, then $L(H)$ has no dispersion-free states and hence does not admit hidden variables. We shall now show that if $\dim H = 2$, then $L(H)$ admits hidden variables and the expectation functional need not be additive. Since such logics are used to describe spin-½ systems, this gives a physically applicable Hilbertian logic that admits hidden variables. Let H be the two-dimensional Hilbert space $\mathbb{C}^2$. The nontrivial projections in $L(H)$ are one-dimensional. Given two different one-dimensional projections $a, b \in L(H)$, a dispersion-free state m can be constructed satisfying $m(a) = m(b') = 1$, $m(a') = m(b) = 0$. It follows that $L(H)$ admits hidden variables.

Let a, b be the following one-dimensional projections:

$$a = \begin{pmatrix} 1 & 0 \\ 0 & 0 \end{pmatrix}, \qquad b = \begin{pmatrix} ½ & ½ \\ ½ & ½ \end{pmatrix}$$

and let A, B be the spin observables

$$A = \begin{pmatrix} 1 & 0 \\ 0 & -1 \end{pmatrix} = a - a', \qquad B = \begin{pmatrix} 0 & 1 \\ 1 & 0 \end{pmatrix} = b - b'$$

The sum

$$A + B = \begin{pmatrix} 1 & 1 \\ 1 & -1 \end{pmatrix}$$

has eigenvalues $\pm\sqrt{2}$ and hence has the form $A + B = \sqrt{2}d - \sqrt{2}d'$, where d, d' are one-dimensional projections. Since $A + B$ does not commute with A or B, all the above one-dimensional projections are distinct. Define a state m on $L(H)$ by $m(0) = 0$, $m(I) = 1$, $m(a) = 1$, $m(a') = 0$, and $m(c) = 1/2$ for every $c \neq 0, 1, a, a'$. The expectations become

$$\begin{aligned} E_m(A) &= m(a) - m(a') = 1 \\ E_m(B) &= m(b) - m(b') = 0 \\ E_m(A + B) &= \sqrt{2}m(d) - \sqrt{2}m(d') = 0 \end{aligned}$$

Hence, $E_m(A + B) \neq E_m(A) + E_m(B)$ and E_m is not additive.

5. Piron's Theorem

In this section we give Piron's method[29,30] for justifying the Hilbert space axiom (also, see Refs. 38, 41). This method employs the quantum logic framework for quantum mechanics and thus attempts to justify Axiom (A1′). That is, we seek to show that L is isomorphic to a sublogic of $L(H)$ for some complex Hilbert space H. It can be shown that this cannot be done for an arbitrary logic L,[12,13] so additional assumptions must be placed on L. We do not have the space to give all the physical justifications for the additional assumptions (some of which are quite controversial) but leave it to the reader to examine these in the literature.[25,30] However, we shall give a precise mathematical description of the required axioms and their consequences.

We begin with the relevant definitions. A poset P is a *complete lattice* if the least upper bound and the greatest lower bound of any collection of elements of P exist. We say that b *covers* a if $a < b$ and $a < c < b$ implies that $c = a$ or $c = b$. An *atom* is an element that covers 0. That is, the atoms are the minimal nonzero elements of P. We say that P is *atomic* if for every $0 \neq b \in P$ there exists an atom $a \leq b$. We say that P is *semimodular* if a is an atom such that $a \wedge b = 0$, then $a \vee b$ covers b.

Now let $\mathcal{S}$ be a quantum system and let L be the logic of experimental propositions for $\mathcal{S}$. Our additional axioms are that L is a complete, atomic, semimodular lattice. Thus L becomes a complete, atomic, semimodular, orthomodular lattice. We call such an L a *projective logic*.

The *direct union* of a collection of projective logics L_α (denoted $\vee L_\alpha$) is the Cartesian product ΠL_α where we define $x \leq y$ if $x_\alpha \leq y_\alpha$ for every α and $(x')_\alpha = x'_\alpha$, $x, y \in \Pi L_\alpha$. It is easy to check that $\vee L_\alpha$ is again a projective logic. We say that a projective logic L is *irreducible* if L is not the direct union of two projective logics each containing more than one element. The *center* of a logic L is

$$C(L) = \{a \in L : a \leftrightarrow b \text{ for all } b \in L\}$$

Lemma 1.[30] A projective logic is irreducible if and only if $C(L) = \{0, 1\}$.

Theorem 3.[30] Any projective logic is the direct union of irreducible projective logics.

Because of Theorem 3, it is sufficient to study irreducible projective logics. The main result, roughly speaking, is that an irreducible projective logic is isomorphic to the logic of all closed subspaces of a vector space over a division ring. To state this result precisely, we need some definitions.

An *involution* on a division ring D is a map $^*: D \to D$ satisfying $(\alpha + \beta)^* = \alpha^* + \beta^*$, $(\alpha\beta)^* = \beta^*\alpha^*$, $\alpha^{**} = \alpha$, for all $\alpha, \beta \in D$. Let $\mathbf{V}$ be a vector space over the division ring D with involution *. A map $\phi: \mathbf{V}x\mathbf{V} \to D$ is a *definite hermitian form* if $\phi(x + \alpha y, z) = \phi(x, z) + \alpha\phi(y, z)$, $\phi(x, y)^* = \phi(y, x)$, and $\phi(x, x) = 0$ implies $x = 0$. For $M \subseteq \mathbf{V}$ we write

$$M^\perp = \{y : \phi(x, y) = 0 \text{ for all } x \in M\}$$

Clearly, $M^\perp$ is a subspace in $\mathbf{V}$, $M \cap M^\perp = \{0\}$ and $M \subseteq M^\perp$. If $M = M^{\perp\perp}$ we call M a *ϕ-closed* subspace. The pair $(\mathbf{V}, \phi)$ is called *Hilbertian* if for any ϕ-closed subspace M of $\mathbf{V}$ we have $\mathbf{V} = M + M^\perp$.

Theorem 4.[38] Let $(\mathbf{V}, \phi)$ be a Hilbertian vector space and let $L(\mathbf{V}, \phi)$ be the set of ϕ-closed subspaces of $\mathbf{V}$ partially ordered by inclusion. Then $L(\mathbf{V}, \phi)$ is an irreducible projective logic.

The converse of Theorem 4 is Piron's theorem. To state this theorem we need an additional definition. For an element b of a logic L, a *0–b chain* is a sequence $\{a_i\}$ in L such that $0 < a_1 < a_2 < \cdots < b$. The *length* of a 0–$b$ chain is the number of elements of the chain. The *rank* of b is the supremum of the lengths of the 0–b chains.

Theorem 5.[38] Let L be an irreducible projective logic with an element of rank at least three. Then there exists a Hilbertian vector space $(\mathbf{V}, \phi)$ over a division ring D such that L is isomorphic to $L(\mathbf{V}, \phi)$.

The division ring D in Theorem 5 is uniquely determined by L (up to an isomorphism).[38] Unfortunately, to further determine D, more assumptions on L must be made and we shall treat this in the next section. Three important possibilities for D are the reals R, the complexes $\mathbb{C}$, and the quaternions Q. Also, we have made no mention of the completeness of the Hilbertian vector space $(\mathbf{V}, \phi)$ in Theorem 5. In the case of R, $\mathbb{C}$, or Q we have the following result.

Theorem 6.[38] Let $(\mathbf{V}, \phi)$ be a Hilbertian vector space over one of the division rings R, $\mathbb{C}$, or Q. Then $(\mathbf{V}, \phi)$ is complete; that is, $(\mathbf{V}, \phi)$ is a Hilbert space.

6. The Division Ring

We have seen from Piron's theorem that an irreducible projective logic L with an element of rank at least three is isomorphic to the lattice $L(\mathbf{V}, \phi, D)$ of ϕ-closed subspaces (and hence orthogonal projections) of a Hilbertian vector space $\mathbf{V}$ over a division ring D. For this reason we shall assume, in this section, that L has the form $L(\mathbf{V}, \phi, D)$ as above, and investigate physical assumptions which ensure that D is R, $\mathbb{C}$, or Q.

Our first assumption is that the pure states of L are concentrated on atoms. That is, if m is pure, there exists a unique atom $a(m)$ such that $m(b) = 1$ implies $a(m) \leq b$ and $m \to a(m)$ is injective. This is a physically reasonable condition which holds in the examples of logics in Section 3.

An observable x on a measurable space (Ω, Σ) is defined in the same way as an observable based on $B(R)$. If (Ω, Σ, μ) is a measure space, we say that an observable x on (Ω, Σ, μ) is *smooth* if $\mu(E) = 0$ whenever $x(E) = 0$.

We now show that any reasonable physical system admits a smooth observable. Most physical systems contain the following mathematical structure. There is a locally compact, second countable, Hausdorff space S, which we call a *physical space* and a locally compact, second countable, Hausdorff group G, which we call a *symmetry group* acting as a continuous, transitive, transformation group on S.[16] Furthermore, there is a representation $g \to W_g$ of G on the group of automorphisms *aut*(L) of L which is continuous in the sense that $g \to m(W_g a)$ is a continuous function for every state m and every $a \in L$. The Borel σ algebra of S being denoted by $B(S)$, most physical systems admit an observable $x : b(S) \to L$ that satisfies $x(g\Delta) = W_g x(\Delta)$ for all $\Delta \in B(S)$, $g \in G$. We call such observables *covariant*. One example of a covariant observable is the position observable, but there are many others. A nontrivial σ-finite measure μ on $B(S)$ is *quasiinvariant* if $\mu(\Delta) = 0$ if and only if $\mu(g\,\Delta) = 0$ for all $\Delta \in B(S)$, $g \in G$. It can be shown that there exist quasiinvariant measures on $B(S)$.[16] See Ref. 22 for the proof of the following.

Theorem 7. Let S be a physical space, G a symmetry group on S, $g \to W_g$ a continuous representation of G on aut(L), and μ a quasiinvariant measure. If $x : B(S) \to L$ is a covariant observable, then there is a set $F \in B(S)$ with $\mu(F) \neq 0$ such that $x : B(S) \to L$ is smooth.

Let x be an observable on a measure space (S, B, μ). We now consider a requirement in which x is maximum in a certain sense. Denoting the range of x by $R(x)$, maximality relative to a set of states M means (roughly) that $R(x)$ is so large that states on $R(x)$ can be uniquely extended to states in M, that superpositions of states on $R(x)$ correspond to superpositions in M, and that compatibility of states on $R(x)$ correspond to compatibility in M. We now make the above vague statements mathematically precise. If m is a state with domain $R(x)$, then clearly m defines a probability measure $\Delta \to m[x(\Delta)]$ on B. For this reason we call such states *probability measures* on $R(x)$. We first assume that there is a set of pure states M such that every probability measure ν on $R(x)$ has a unique extension $\bar{\nu} \in M$. If m_1, m_2 are distinct pure states, we say that a pure state m_3 is a *superposition* of m_1 and m_2 if $a(m_3) < a(m_1) \vee a(m_2)$. We must now define what is meant by a superposition of probability measures on $R(x)$. Suppose ν_1, ν_2 are distinct probability measures on $R(x)$ that are absolutely continuous relative to μ. If there exist real numbers α, β such that

$$(d\nu_3/d\mu)^{1/2} = \alpha\,(d\nu_1/d\mu)^{1/2} + \beta(d\nu_2/d\mu)^{1/2}$$

we say that ν_3 is a *superposition* of ν_1 and ν_2. Our next assumption is that if ν_3 is a superposition of ν_1 and ν_2, then $\bar{\nu}_3$ is a superposition of $\bar{\nu}_1$ and $\bar{\nu}_2$. Finally, we say that two distinct probability measures ν_1 and ν_2 that are absolutely continuous relative to μ are *compatible* if

$$[(\tfrac{1}{2}d\nu_1/d\mu)^{1/2} + (\tfrac{1}{2}d\nu_2/d\mu)^{1/2}]^2$$

is a density function for a probability measure, and we assume that if ν_1 and ν_2 are compatible, then so are $a(\bar{\nu}_1)$ and $a(\bar{\nu}_2)$. The motivation for this definition of compatibility stems from the fact that if ν_1 and ν_2 are compatible, then

$$\int[(\tfrac{1}{2}d\nu_1/d\mu)^{1/2} + (\tfrac{1}{2}d\nu_2/d\mu)^{1/2}]d\mu = 1 + \tfrac{1}{2}\int(d\nu_1/d\mu)^{1/2}\,(d\nu_2/d\mu)^{1/2}\,d\mu$$

which implies that $(d\nu_1/d\mu)^{1/2}$ and $(d\nu_2/d\mu)^{1/2}$ are orthogonal normalized functions in $L_2(S, \mu)$. Now unit vectors in $L_2(S, \mu)$ correspond to atoms and distinct atoms in a proposition system are compatible if and only if they are orthogonal.

Actually we need more than the three conditions of the previous paragraph. We need the three conditions to hold for signed probability measures.[22] This is easily formulated since a signed probability measure can be uniquely represented in terms of two probability measures.[22] We thus say that an observable x on (S, D, μ) is *maximal* relative to a set of pure states M if the three conditions of the previous paragraph generalized to signed probability measures hold. It can be shown that the usual position observable for a nonrelativistic particle is smooth and maximal.[22] There are also examples of smooth, maximal observables with finite spectrum.[22] The main result of this section is the following.

Theorem 8.[22] If $L(V, \phi, D)$ admits a smooth, maximal observable, then D contains the real field R as a subfield.

If D contains R in its center and is a finite extension of R, then applying a theorem of Frobenius,[22] D is R, $\mathbb{C}$, or Q. Other physical criteria must be used to determine which of these three is the correct one.

7. Algebraic Approach

In this section we consider the algebraic approach to quantum mechanics[8,23,35,36] and show what it has to contribute toward justifying the Hilbert space axiom.

Let $\mathcal{A}$ be a complex algebra. An *involution* on $\mathcal{A}$ is a map $(:\mathcal{A} \to \mathcal{A}$ satisfying

:$A^{**} = A$, $(\alpha A + \beta B)^* = \bar{\alpha}A^* + \bar{\beta}B^*$, and $(AB)^* = B^*A^*$ for all $\alpha, \beta \in \mathbb{C}$, $A, B \in \mathcal{A}$. An algebra with involution * is called a **-algebra*. We call $\mathcal{A}$ a *normed algebra* if $\mathcal{A}$ is equipped with a norm $\|\cdot\|$ satisfying $\|AB\| \leq \|A\|\|B\|$ for all $A, B \in \mathcal{A}$. A *Banach *-algebra* is a complete normed *-algebra satisfying $\|A^*\| = \|A\|$ for all $A \in \mathcal{A}$. A *C* algebra* is a complete normed *-algebra satisfying $\|A^*A\| = \|A\|^2$ for all $A \in \mathcal{A}$. A C* algebra is a Banach *-algebra. Indeed, if $\mathcal{A}$ is a C* algebra, then for every $A \in \mathcal{A}$ we have

$$\|A\|^2 = \|A^*A\| \leq \|A^*\|\|A\|$$

Hence, $\|A\| \leq \|A^*\|$ and replacing A by A^* gives $\|A^*\| \leq \|A\|$ so $\mathcal{A}$ is a Banach *-algebra. We shall always assume that our *-algebras contain an identity I.

Let $\mathcal{A}$ be a *-algebra. An element $A \in \mathcal{A}$ is *self-adjoint* if $A = A^*$; A is *positive* if $A = B^*B$ for some $B \in \mathcal{A}$. In the algebraic approach to quantum mechanics, an attempt is made to justify the Hilbert space axiom in the form (A1). Thus, one tries to show that the observables of a quantum system can be represented by self-adjoint operators on a complex Hilbert space. For simplicity, only bounded observables are usually considered, although some work has been done for unbounded observables.[19,33] The first step is to assume that the set of bounded observables can be represented as self-adjoint elements of a Banach *-algebra (or C*-algebra) $\mathcal{A}$. Although the physical justification for this assumption is not complete, some progress has been made.[8,36]

The next step is to represent $\mathcal{A}$ as a C* algebra of operators on a complex Hilbert space H. If $\mathcal{A}$ is a Banach *-algebra, a *representation* of $\mathcal{A}$ is a *-homomorphism Π of $\mathcal{A}$ into a subset of the set $\mathcal{L}(H)$ of bounded linear operators on H. That is, $\Pi: \mathcal{A} \to \mathcal{L}(H)$ preserves addition, multiplication, scalar multiplication, and involution. A representation $\Pi: \mathcal{A} \to \mathcal{L}(H)$ is *continuous* if $\|\Pi(A)\| \leq M\|A\|$ for all $A \in \mathcal{A}$ and some fixed $M \geq 0$. It is an interesting fact that representations, which are required to preserve only the algebraic operations, are automatically continuous. We thus obtain topological properties free.

Theorem 9.[8,28] Every representation $\Pi: \mathcal{A} \to \mathcal{L}(H)$ of a Banach *-algebra $\mathcal{A}$ is continuous. In fact, $\|\Pi(A)\| \leq \|A\|$ for all $A \in \mathcal{A}$.

A representation $\Pi: \mathcal{A} \to \mathcal{L}(H)$ is *cyclic* if there exists a vector $\phi \in H$ such that $\{\Pi(A)\phi : A \in \mathcal{A}\}$ is dense in H. We then call ϕ a *cyclic vector*. It can be shown[28] that any representation can be decomposed into a direct sum of cyclic representations.

A *positive linear functional* on a Banach *-algebra $\mathcal{A}$ is a linear functional $\omega: \mathcal{A} \to \mathbb{C}$ which satisfies $\omega(A^*A) \geq 0$ for all $A \in \mathcal{A}$. A *state* is a positive linear functional ω for which $\omega(I) = 1$. The states represent expectation functionals corresponding to physical states. The main result of this section completes the algebraic justification of the Hilbert space axiom. It is called the GNS construction (after Gelfand, Naimark, and Segal, Ref. 28, 35).

Theorem 10. Let $\mathcal{A}$ be a Banach *-algebra. To every cyclic representation Π of $\mathcal{A}$ with cyclic vector ϕ, there corresponds a positive linear functional $\omega(A) = \langle\Pi$

$(A)\phi, \phi\rangle$. Conversely, to every positive linear functional ω on $\mathcal{A}$ there corresponds a cyclic representation Π with cyclic vector ϕ such that $\omega(A) = \langle \Pi(A)\phi, \phi\rangle$ for all $A \in \mathcal{A}$.

In the above theorem, ω is a state if and only if the corresponding vector ϕ is a unit vector. In short, corresponding to a state ω of a physical system $\mathcal{S}$, the bounded observables of $\mathcal{S}$ can be represented by bounded self-adjoint operators on a Hilbert space H and the state ω can be represented as the usual expectation functional $A \to \langle A\phi, \phi\rangle$.

8. Operations and Baer *-Semigroups

In one of its formulations, the operational approach to quantum mechanics can be used to attempt to justify the Hilbert space axiom in the form (A2).[1,5,18] However, we shall consider a different formulation of this approach and follow J. Pool.[31,32] This formulation is closer to the quantum logic approach and will be more convenient for our purposes.

Pool begins with a triple (E, S, P), where E corresponds to the set of quantum events, S corresponds to the set of states, and for all $p \in E$, $\alpha \in S$, $P(p, \alpha) \in [0, 1]$ gives the probability that the event p occurs when the system is in the state α. For $p \in E$, define $S_1(p) = \{\alpha \in S: P(p, \alpha) = 1\}$, $S_0(p) = \{\alpha \in S: P(p, \alpha) = 0\}$. We call (E, S, P) an *event-state structure* if axioms (ES1)–(ES7) below are satisfied. Pool[31] has given physical justification for the axioms. In any case, the reader can easily supply them.

(ES1) If $p, q \in E$ and $S_1(p) = S_1(q)$, then $p = q$.
(ES2) There exists an element $1 \in E$ such that $S_1(1) = S$.
(ES3) If $p, q \in E$ and $S_1(p) \subseteq S_1(q)$, then $S_0(q) \subseteq S_0(p)$.
(ES4) If $p \in E$, then there exists a $p' \in E$ such that $S_1(p') = S_0(p)$, $S_0(p') = S_1(p)$.
(ES5) If $p_1, p_2, \ldots \subset E$ and $S_1(p_i) \subseteq S_0(p_j)$ for $i \neq j$, then there exists a $p \in E$ such that $S_1(p_i) \subseteq S_1(p)$ for all i; if $S_1(p_i) \subseteq S_1(q)$ for all i, then $S_1(p) \subseteq S_1(q)$; and $P(p, \alpha) = \Sigma(p_i, \alpha)$ for all $\alpha \in S$.
(ES6) If $\alpha, \beta \in S$ and $P(p, \alpha) = P(p, \beta)$ for all $p \in E$, then $\alpha = \beta$.
(ES7) If $\alpha_1, \alpha_2, \ldots \in S$, $t_1, t_2, \ldots \in [0, 1]$ with $\Sigma t_i = 1$, then there exists an $\alpha \in S$ such that $P(p, \alpha) = \Sigma t_i P(p, \alpha_i)$ for all $p \in E$.

We call the elements of E *events* and the elements of S *states*. If $p, q \in E$ we define $p \leq q$ when $S_1(p) \subseteq S_1(q)$ and $p \perp q$ when $S_1(p) \subseteq S_1(q') = S_0(q)$. For $\alpha \in S$, define $\mu_\alpha: E \to [0, 1]$ by $\mu_\alpha(p) = P(p, \alpha)$ and let $\hat{S} = \{\mu_\alpha: \alpha \in S\}$.

Theorem 11. [Pool] If (E, S, P) is an event-state structure, then $(E, \leq, ')$ is a σ-orthocomplete orthomodular poset. If $\alpha \in S$, then $\mu_\alpha(1) = 1$ and $\mu_\alpha(\vee p_i) = \Sigma\mu_\alpha(p_i)$ for any sequence of mutually orthogonal $p_i \in E$.

Let (E, S, P) be an event-state structure, and let Σ denote the set of maps $x: D_x \subseteq S \to S$. We define $1 \in \Sigma$ to be the identity map on S and $0 \in \Sigma$ is defined by $D_0 = \phi$. If $x, y \in \Sigma$, then $x \circ y \in \Sigma$ is defined by $D_{x \circ y} = \{\alpha \in D_y : y\alpha \in D_x\}$ and $(x \circ y)\alpha = x(y\alpha)$ for all $\alpha \in D_{x \circ y}$. An *event-state-operation structure* is a 4-tuple (E, S, P, Ω) where (E, S, P) is an event-state structure and $\Omega: E \to \Sigma$, $\Omega: p \to \Omega_p$ satisfies

(O1) If $p \in E$, then $D_{\Omega_p} \equiv D_p = \{\alpha \in S: P(p, \alpha) \neq 0\}$.
(O2) If $p \in E$, $\alpha \in D_p$ and $P(p, \alpha) = 1$, then $\Omega_p\alpha = \alpha$.
(O3) If $p \in E$ and $\alpha \in D_p$, then $P(p, \Omega_p\alpha) = 1$.
(O4) If $p_1, \ldots, p_n, q_1, \ldots, q_m \in E$ and $\Omega_{p1} \circ \ldots \circ\Omega_{pn}$ $= \Omega_{q1} \circ \ldots \circ\Omega_{qm}$, then $\Omega_{pn} \circ\Omega_{pn-1} \circ \ldots \circ\Omega_{p1}$ $= \Omega_{qm} \circ\Omega_{qm-1} \circ \ldots \circ\Omega_{q1}$.
(O5) If $x = \Omega_{p1}\circ \ldots \circ\Omega_{pn}$, then there exists a $q_x \in E$ such that $S_1(q_x) = D_x^c$ (the complement of D_x).
(O6) If $p, q \in E$, $q \leq p$ and $\alpha \in D_p$, then $P(q, \Omega_p\alpha) = P(q, \alpha)/P(p, \alpha)$.
(O7) If $p, q \in E$ are compatible and $\alpha \in D_p$, then $P(q, \Omega_p\alpha) = P(p \wedge q, \Omega_p\alpha)$.

For $p \in E$, $\alpha \in D_p$ we call $\Omega_p\alpha$ the *state conditioned on the event p and the state α*. The elements of the set

$$S_\Omega = \{\Omega_{p1} \circ \ldots \circ\Omega_{pn} : p_1, \ldots, p_n \in E\} \subseteq \Sigma$$

are called *operations*. We define the map $^*: S_\Omega \to S_\Omega$ as follows: if $x = \Omega_{p1} \circ \ldots \circ\Omega_{pn} \in S_\Omega$ then $x^* = \Omega_{pn} \circ \ldots \circ\Omega_{p1}$. It is not hard to show that q_x in (O5) is unique. For $x \in S_\Omega$, we define $x' = \Omega_{qx}$, where q_x is this unique event.

Let S be a semigroup. An *involution* * on S is a map from S to S satisfying $s^{**} = s$ and $(st)^* = t^*s^*$ for every $s, t \in S$. An element $e \in S$ is a *projection* if $e^2 = e = e^*$. A *Baer *-semigroup* is a semigroup S with 0, 1 and an involution * such that for every $s \in S$ there exists a projection $s' \in S$ such that $st = 0$ if and only if $s't = t$. It can be shown that the above projection s' is unique. A *closed projection* is a projection e which satisfies $e = e''$. Denote the set of closed projections in S by $P'(S)$, and for $e, f \in P'(S)$ define $e \leq f$ if $ef = fe = e$. D. Foulis has shown[9,10] that $[P'(S), \leq, ']$ is an orthomodular lattice. We say that S is a *coordinatizing* Baer *-semigroup for an orthomodular lattice L if L is isomorphic to $P'(S)$.

Theorem 12. [Pool] Let (E, S, P, Ω) be an event-state-operation structure. Then $(E, \leq, ')$ is a σ-orthocomplete orthomodular lattice and $S_\Omega, \circ, {}^*, ')$ is a coordinatizing Baer *-semigroup for $(E, \leq, ')$.

9. Strong Consistency

Let L be a σ-orthocomplete orthomodular lattice which we interpret as the logic corresponding to a quantum system. D. Foulis[9,10] has shown that L can always be coordinated by a Baer *-semigroup. Unfortunately, S need not be unique up to an isomorphism. However, as we have seen in the previous section, one of these coordinatizing Baer *-semigroups has an important physical meaning, namely, as the set of operations on the state of space of L. We now justify the Hilbert space axiom by characterizing those Baer *-semigroups that can be injectively represented in a complex Hilbert space H. Because L is coordinatized by S, this results in an isomorphism of L with a sublogic of $L(H)$.

To illustrate our result, let H be a complex Hilbert space and let $\mathcal{L}(H)$ be the set of bounded linear operators on H. For $A \in \mathcal{L}(H)$, let A' denote the projection onto the null space of A. Let $S \subseteq \mathcal{L}(H)$ be a semigroup containing I and A^*, A' whenever $A \in S$, and suppose $\|A\| \leq 1$ for all $A \in S$. Then S is an example of a Baer *-semigroup. For $x \in H$, define $\phi(A) = \langle Ax, x\rangle$, $A \in S$. Then $\phi(0) = 0$, and using Schwarz's inequality we see that ϕ is bounded. Moreover, if $\{\lambda_1, \ldots, \lambda_n\} \subseteq \mathbb{C}$, $\{A_1, \ldots, A_n\} \subseteq S$ we see that

$$\Sigma\bar{\lambda}_i\lambda_j\phi(A_i^*A_j) = \Sigma\bar{\lambda}_i\lambda_j\langle A_jx, A_ix\rangle = \|\Sigma\lambda_jA_jx\|^2 \geq 0$$

We shall call such functions "*-definite." The *-definite function ϕ has an additional important property. If $\phi(A^*A) = 0$, then $Ax = 0$, which implies that $A'x = x$. Hence, $\phi(A') = \phi(I)$. We shall call *-definite functions with this property "consistent." We thus see that S possesses a separating set of consistent *-definite functions. Our main result shows that a slightly stronger condition characterizes Baer *-semigroups which are isomorphic to an $S \subseteq \mathcal{L}(H)$.

Now let S be an arbitrary Baer *-semigroup. A *Baer *-representation* ρ of S in H is a map $\rho: S \to \mathcal{L}(H)$ satisfying

(R1) $\rho(1) = I$;
(R2) $\rho(st) = \rho(s)\rho(t)$ for all $s, t \in S$;
(R3) $\rho(s^*) = \rho(s)^*$ for all $s \in S$;
(R4) $\rho(s') = \rho(s)'$ for all $s \in S$;
(R5) $\|\rho(s)\| \leq 1$ for all $s \in S$.

A Baer *-representation $\rho: S \to \mathcal{L}(H)$ is *cyclic* if there exists an $x \in H$ such that the linear hull of $\{\rho(s)x : s \in S\}$ is dense in H. It is not difficult to show that any Baer *-representation is a direct sum of cyclic ones. For this reason, we shall mainly restrict our attention to cyclic Baer *-representations.

A bounded complex-valued function ϕ on S is **-definite*[3,26] if $\phi(0) = 0$ and for any finite collections $\{\lambda_1, \ldots, \lambda_n\} \subseteq \mathbb{C}$, $\{s_1, \ldots, s_n\} \subseteq S$, we have

$$\sum_{i,j=1}^{n} \bar{\lambda}_i\lambda_j\phi(s_i^*s_j) \geq 0$$

Notice that if ϕ is *-definite and $t \in S$, then the function $\phi_t(s) = \phi(t^*st)$ is also *-definite. We say that a *-definite function ϕ is *consistent* if $\phi(s^*s) = 0$ implies that

$\phi(s') = \phi(1)$. In a Baer *-semigroup, $s^*s = 0$ implies that $s' = 1$. Hence, a consistent *-definite function is one that respects the above implication. For example, if ϕ is faithful [$\phi(s^*s) = 0$ implies that $s = 0$], then ϕ_t is consistent for all $t \in S$. Indeed,

$$\begin{aligned}\phi_t(s^*s) = 0 &\Rightarrow \phi[(st)^*st] = 0 \Rightarrow st = 0 \Rightarrow s't = t\\ &\Rightarrow \phi_t(s') = \phi(t^*s't) = \phi(t^*t) = \phi_t(1)\end{aligned}$$

In the next paragraph we shall define ϕ to be strongly consistent if limits of certain *-definite functions derived from ϕ are consistent.

Let ϕ be a *-definite function on S and let $\{c_1, \ldots, c_m\} \subseteq \mathbb{C}$, $\{t_1, \ldots, t_m\} \subseteq S$. Define the function $\phi_1 \equiv \phi\{c_i, t_i\}$ by

$$\phi_1(s) = \sum_{k,l} \bar{c}_k c_l\, \phi(t_k^* s t_l)$$

It is not hard to show that ϕ_1 is again *-definite. Now let $\{c_1^k, \ldots, c_{n(k)}^k\} \subseteq \mathbb{C}$, $\{t_1^k, \ldots, t_{n(k)}^k\} \subseteq S$, $k = 1, 2, \ldots$, be a sequence of finite sets and let $\phi_k \equiv \phi\{c_i^k, t_i^k\}$. We say that the sequence ϕ_k *converges* if

$$\lim_{k\to\infty} \phi_k(1) = \operatorname{Re} \lim_{k,l\to\infty} \sum_{i,j} \bar{c}_i^l c_j^k \phi(t_i^{l*} t_j^k)$$

Lemma 2.[21] If ϕ_k converges, then $\psi(s) \equiv \lim_{k\to\infty} \phi_k(s)$ exists for all $s \in S$ and ψ is *-definite.

We call ψ in Lemma 2 a *limit function* for ϕ. We say that ϕ is *strongly consistent* if every limit function for ϕ is consistent.

Theorem 13.[21] (a) Let $\rho: S \to \mathcal{L}(H)$ be a Baer *-representation, $x \in H$, and $\phi(s) = \langle \rho(s)x, x\rangle$, $s \in S$. Then ϕ is a strongly consistent *-definite function. Conversely, if ϕ is a strongly consistent *-definite function on S, then there exists a Baer *-representation $\rho: S \to \mathcal{L}(H)$ which is cylic with cyclic vector $x \in H$ such that $\phi(s) = \langle \rho(s)x, x\rangle$ for all $s \in S$. (b) S admits an injective Baer *-representation if and only if S possesses a separating set of strongly consistent *-definite function.

10. Connection to the Algebraic Approach

In this section we give a physical significance for *-definite functions using a connection to the algebraic approach. Let S be an arbitrary *-semigroup with 0 and 1, and let $\mathcal{A}$ be a Banach *-algebra with unit I. We say that $\mathcal{A}$ is an *enveloping* Banach *-algebra for S if there is an injection $\rho: S \to \mathcal{A}$ satisfying (R1), (R2), (R3), (R5) of Section 9 and

$$\text{(R6)}\ \mathcal{A} = \left\{\sum_{i=1}^{n} c_i\, \rho(s_i) : c_i \in \mathbb{C},\ s_i \in S,\ n \in N\right\}^{\wedge}$$

where ^ denotes the closure.

We now show that any *-semigroup S with 0 and 1 possesses an enveloping Banach *-algebra. Let $l_1(S)$ be the set of functions $f: S \to \mathbb{C}$ satisfying $f(s) = 0$ except for countably many $s \in S$ and $\|f\| \equiv \Sigma_{s \neq 0} |f(s)| < \infty$. Moreover, we identify any two functions f and g if $f(s) = g(s)$ for every $s \neq 0$. Under pointwise sums and scalar products it is clear that $l_1(S)$ is a complex normed linear space. For $f, g \in l_1(S)$ we define

$$(fg)(s) = \Sigma\{f(r)g(t): r, t \in S, rt = s\}$$

For $f \in l_1(S)$ we define the involution f^* of f by $f^*(s) = \overline{f(s^*)}$. Let $\delta: S \to l_1(S)$ be the map $[\delta(s)](t) = \delta_{s,t}$ (the Kronecker delta). The proof of the next lemma uses results from Refs. 24, 34.

Lemma 3. Under the map $\delta: S \to l_1(S)$, $l_1(S)$ is an enveloping Banach *-algebra for S. Moreover, every element of $l_1(S)$ has the form $f = \Sigma c_i\delta(s_i)$ where $\|f\| = \Sigma|c_i| < \infty$.

Lemma 4.[21] Let S be a *-semigroup with 0, 1 and let $\mathcal{A}$ be an enveloping Banach *-algebra for S with enveloping map $\delta: S \to \mathcal{A}$. If ϕ is a *-definite function on S, then there exists a unique positive linear functional ω on $\mathcal{A}$ such that $\omega[\delta(s)] = \phi(s)$ for all $s \in S$. Conversely, if ω is a positive linear functional on $\mathcal{A}$, then $\phi(\cdot) = \omega[\delta(\cdot)]$ is a *-definite function on S. If $\phi(1) = 1$, then ω is a state.

In the algebraic approach to quantum mechanics, the bounded observables of a quantum system are represented by self-adjoint elements of a Banach *-algebra (usually a C *-algebra) $\mathcal{A}$ and the expectation functionals are represented by states on $\mathcal{A}$. The above lemma gives a physical interpretation for the *-definite functions on S. The normalized *-definite functions on S are the restrictions to S of a state on any enveloping Banach *-algebra for S. The consistency condition has an important physical significance. To see this let S be a Baer *-semigroup and let $\mathcal{A}$ be an enveloping Banach *-algebra with enveloping map $\delta: S \to \mathcal{A}$. Let ϕ be a normalized consistent *-definite function on S and let ω be the corresponding unique state on $\mathcal{A}$. In some approaches to quantum mechanics[4,5,18] the positive elements of S (i.e., elements of the form s^*s) are interpreted as quantum mechanical "effects." Then $\omega[\delta(s^*s)] = \phi(s^*s)$ is interpreted as the probability that the effect s^*s is observed when the system is in the state ω. The probabilistic interpretation is mathematically reasonable since a state ω is continuous with norm 1[8,28] and hence

$$0 \leq \omega[\delta(s^*s)] \leq \|\delta(s^*s)\| \leq 1$$

Now if s^*s is an effect, $(s^*s)'$ corresponds to the event which occurs if and only if the effect s^*s is not observed. It follows that if the effect s^*s has zero probability of being observed in the state ω, then $(s^*s)'$ occurs with certainty in the state ω. Since $(s^*s)' = s'$,[9] we should have $\omega[\delta(s^*s)] = 0$ implies that $\omega[\delta(s')] = 1$. In terms of ϕ this becomes $\phi(s^*s) = 0$ implies $\phi(s') = 1$. This is precisely the consistency condition.

References and Notes

1. G. Birkhoff and J. von Neumann, "The Logic of Quantum Mechanics," *Ann. Math.* **37,** 823–843 (1936).
2. D. Bohm and J. Bub, "A Refutation of the Proof of Jauch and Piron that Hidden Variables Can Be Excluded in Quantum Mechanics," *Rev. Mod. Phys.* **38,** 470–475 (1966).
3. A. Clifford and G. Preston, *The Algebraic Theory of Semigroups,* Vol. 1 (Amer. Math. Soc. Providence, Rhode Island, 1961).
4. E. B. Davies, *Quantum Theory of Open Systems* (Academic Press, New York, 1976).
5. E. B. Davies and J. Lewis, "An Operational Approach to Quantum Probability," *Commun. Math. Phys.* **17,** 239–260 (1970).
6. P. Dirac, *The Principles of Quantum Mechanics* (Clarendon, Oxford, England, 1930).
7. N. Dunford and J. Schwartz, *Linear Operators,* Parts 1 and 2 (Wiley Interscience, New York, 1958 and 1963).
8. G. Emch, *Algebraic Methods in Statistical Mechanics and Quantum Field Theory* (Wiley Interscience, New York, 1972).
9. D. Foulis, "Baer *-Semigroups," *Proc. Am. Math. Soc.* **11,** 648–654 (1960).
10. D. Foulis, "A Note on Orthomodular Lattices," *Portugal Math.* **21,** 65–72 (1962).
11. A. Gleason, "Measures on Closed Subspaces of a Hilbert Space," *J. Rat. Mech. Anal.* **6,** 885–893 (1957).
12. R. Greechie, "An Orthomodular Lattice with a Full Set of States not Embeddable in Hilbert Space," *Carib. J. Sci. Math.* **2,** 15–26 (1969).
13. R. Greechie, "Orthomodular Lattices Admitting no States," *J. Comb. Theory* **10,** 119–132 (1971).
14. S. Gudder, "Elementary Length Topologies in Physics," *SIAM J. Appl. Math.* **16,** 1011–1019 (1968).
15. S. Gudder, "Quantum Probability Spaces," *Proc. Am. Math. Soc.* **21,** 296–302 (1969).
16. S. Gudder, "Quantum Logics, Physical Space, Position Observables and Symmetries," *Rep. Math. Phys.* **3,** 193–202 (1972).
17. S. Gudder, "Generalized Measure Theory," *Found. Phys.* **3,** 399–411 (1973).
18. S. Gudder, *Stochastic Methods in Quantum Mechanics* (North-Holland, New York, 1979).
19. S. Gudder and R. Hudson, "A Noncommutative Probability Theory," *Trans. Am. Math. Soc.* **245,** 1–41 (1978).
20. S. Gudder and J.-P. Marchand, "A Coarse-Grained Measure Theory," *Bull. Acad. Polon. Sci.* **28,** 557–564 (1980).
21. S. Gudder and J. Michel, "Representations of Baer *-Semigroups" *Proc. Amer. Math. Soc.* **81,** 157–163 (1981).
22. S. Gudder and C. Piron, "Observables and the Field in Quantum Mechanics," *J. Math. Phys.* **12,** 1583–1588 (1971).
23. R. Haag and D. Kastler, "An Algebraic Approach to Quantum Field Theory," *J. Math. Phys.* **5,** 848–861 (1964).
24. E. Hewitt and H. Zuckerman, "The l_1-Algebra of a Commutative Semigroup," *Trans. Am. Math. Soc.* **83,** 70–97 (1956).
25. J. Jauch, *Foundations of Quantum Mechanics* (Addison-Wesley, Reading, Massachusetts, 1968).
26. R. Lindahl and P. Maserick, "Positive-Definite Functions on Involution Semigroups," *Duke Math. J.* **39,** 771–782 (1971).
27. G. Mackey, *The Mathematical Foundations of Quantum Mechanics* (Benjamin, New York, 1963).
28. M. Naimark, *Normed Algebras* (Noordhoff, Gröningen, The Netherlands 1972).
29. C. Piron, "Axiomatique quantique," *Helv. Phys. Acta.* **37,** 439–468 (1964).
30. C. Piron, *Foundations of Quantum Mechanics* (Benjamin, New York, 1976).
31. J. Pool, "Baer *-Semigroups and the Logic of Quantum Mechanics," *Commun. Math. Phys.* **9,** 118–141 (1968).
32. J. Pool, "Semi-Modularity and the Logic of Quantum Mechanics," *Commun. Math. Phys.* **9,** 212–228 (1968).
33. R. Powers, "Self-Adjoint Algebras of Unbounded Operators," *Commun. Math. Phys.* **21,** 85–124 (1971).

34. G. Rüttiman, "On the Logical Structure of Quantum Mechanics," *Found. Phys.* **1,** 173–182 (1972).
35. I. Segal, "Postulates for General Quantum Mechanics," *Ann. Math.* **48,** 930–948 (1947).
36. I. Segal, *Mathematical Problems of Relativistic Physics* (Amer. Math. Soc., Providence, Rhode Island, 1963).
37. M. Stone, "On One-Parameter Unitary Groups in Hilbert Space," *Ann. Math.* **33,** 643–648 (1932).
38. V. Varadarajan, *Geometry of Quantum Theory,* Vol. 1 (Van Nostrand Reinhold, Princeton, New Jersey, 1968).
39. J. von Neumann, *Mathematical Foundations of Quantum Mechanics* (Princeton Univ. Press, Princeton, New Jersey, 1955).
40. J. Zerbe, "Generalized Measure Theory," Dissertation, Univ. of Denver, Denver, Colorado (1979).
41. N. Zierler, "Axioms for Non-relativistic Quantum Mechanics," *Pac. J. Math.* **11,** 1151–1169 (1961).

9

THE CYCLOTRON–SOLITON MODEL OF TERRESTRIAL KILOMETRIC RADIATION

D. TER HAAR
Department of Theoretical Physics
University of Oxford
Oxford OX1 3NP, England

Recently, Istomin, Petviashvili, and Pokhotelov,[1] referred to in what follows as IPP (we must warn the reader that there are a number of minor errors in this paper), have developed a theory to explain the terrestrial or auroral kilometric radiation (AKR). We refer to IPP and a paper by Gurnett[2] for a survey of the experimental data. The IPP model has been analyzed in considerable detail by Cole and Pokhotelov,[3] who found that the model could explain many features of the AKR quantitatively and especially the fact that the power in the AKR is roughly 10% of the power in the auroral electron beams which excite the plasma turbulence in the region where the AKR originates. The IPP model uses the fact that the electron beams will excite short-wavelength electrostatic cyclotron waves which will be concentrated in solitonlike structures—called cyclotron solitons. Energy will be pumped into these solitons until the electric field in the soliton reaches a certain value. The oscillating charges in the structure will then produce electromagnetic radiation at twice the gyrofrequency which in the region where the AKR is produced corresponds to kilometric radiation. The soliton structure is found by using an ad hoc procedure employed widely for various situations, for instance, by Petviashvili[4,5]: the dispersion relation of the relevant waves is interpreted as an eikonal equation from which then the partial differential equation for the soliton envelope is constructed. IPP quote the book by Mikhailovskii[6] as the source of their dispersion equation,

$$\omega = n\omega_B\{1 + \alpha(k_\perp r_H)^{-3} + [(k_2 r_H)^2(k_\perp r_H)^3]/\alpha n^2\} \tag{1}$$

where ω_B is the gyrofrequency, n the number of the relevant harmonic, $\alpha = (\omega_p/\omega_B)^2/(2\pi)^{1/2}$, ω_p the Langmuir frequency, k_z and $k_\perp$ the longitudinal and transverse wave numbers, and r_H the electron Larmor radius; they state that Eq. (1) is valid when

$k_\perp r_H \gg 1$, although they seem also to have assumed that they are considering nearly transverse waves, such that $(k_z r_H)^2(k_\perp r_H)^6/\alpha^2 \lesssim 1$, or possibly even $\ll 1$.

One next considers the diamagnetic current due to the cyclotron waves; this reduces the main magnetic field, B_0, by an amount [see IPP, Eq. (2.4)] δB given by the equation

$$\frac{\delta B}{B_0} = \frac{\delta\omega_B}{\omega_B} = -\frac{n^2}{4}\frac{\omega_B^2}{(\omega - n\omega_B)^2}\frac{\alpha}{(k_\perp r_H)^3}\frac{|\nabla\phi|^2}{B_0^2} \tag{2}$$

where ϕ is the electric potential.

The Petviashvili procedure now proceeds as follows. Consider one of the Maxwell equations in this case

$$\mathrm{div}\epsilon\nabla\phi = 0 \tag{3}$$

or, operating on this equation with div^{-1}:

$$\hat{\epsilon}\nabla\phi = 0 \tag{4}$$

where now $\hat{\epsilon}$ is interpreted as an operator, acting upon $\nabla\phi$, which in accordance with Eq. (1) we can write in the form

$$\hat{\epsilon} = \frac{\alpha n^2\omega_B^2}{\omega(\omega - n\omega_B)}\left[\frac{\omega(\omega - n\omega_B)}{\alpha n^2\omega_B^2}(k_\perp r_H)^3 - 1 - \frac{(k_z r_H)^2(k_\perp r_H)^6}{\alpha^2 n^2}\right] \tag{5}$$

with ω and $\mathbf{k}$ being interpreted as the operators $i\partial/\partial t$ and $-i\nabla$.

Writing Eq. (4) in the form [cf. Eq. (5)]

$$\left[\frac{\omega(\omega - n\omega_B)}{\alpha n^2\omega_B^2}(k_\perp r_H)^3 - 1 - \frac{(k_z r_H)^2(k_\perp r_H)^6}{\alpha^2 n^2}\right]\nabla\phi = 0 \tag{6}$$

and replacing $\omega - n\omega_B$ by $\omega - n\omega_B - n\delta\omega_B$, with $\delta\omega_B$ given by Eq. (2) we get

$$\left[\frac{(k_\perp r_H)^3}{\alpha}\frac{\omega(\omega - n\omega_B)}{n\omega_B^2} - 1 - \frac{(k_z r_H)^2(k_\perp r_H)^6}{\alpha^2 n^2}\right]\nabla\phi = \frac{(k_\perp r_H)^3}{\alpha}\frac{\delta B}{B_0}\nabla\phi = -\frac{n^2}{4}\frac{\omega_B^2}{(\omega - n\omega_B)^2}\frac{|\nabla\phi|^2}{B_0^2} \tag{7}$$

Looking for a solution of ϕ in the form

$$\phi = \tilde{\phi}\exp\left[-in\omega_B t(1 - \tfrac{1}{2}A^2)\right] \tag{8}$$

we find the following equation for $\tilde{\phi}$:

$$B_0^2 A^4 \hat{L}\nabla\tilde{\phi} = |\nabla\tilde{\phi}|^2\nabla\tilde{\phi} \tag{9}$$

where $\hat{L}$ is the operator defined by Eq. (2.6) of IPP. If we next introduce dimensionless cylindrical coordinates ρ and ζ through the relations

$$\rho = \left(\frac{2\alpha}{A^2}\right)^{1/3}\frac{r}{r_H}, \qquad \zeta = \tfrac{1}{2}nA^2\frac{z}{r_H} \tag{10}$$

and, assuming axial symmetry, write

$$\tilde{\phi} = r_H \frac{H_0 A^{8/3}}{(2\alpha)^{1/3}} F(\rho, \zeta) \tag{11}$$

we find Eq. (2.8) of IPP which has soliton solutions with longitudinal and transverse dimensions L_z and $L_\perp$ given by Eqs. (2.10) of IPP:

$$L_\perp \sim r_H A^{2/3}/\alpha^{1/3}, \qquad L_z \sim z_H/nA^2 \tag{12}$$

IPP show that energy will be pumped into the cyclotron solitons until $L_z\, \omega_B \sim v_b$, where v_b is the beam velocity, and this leads to the equalities

$$v_T/v_b \sim A^2 \sim E/B_0 \tag{13}$$

Through the fusion of cyclotron waves into electromagnetic waves the solitons will produce the AKR, and IPP show that the power P produced in this way by a single soliton is given by the equation

$$P \sim 250 c r_H^2 B_0^2 \left(\frac{v_T}{c}\right)^4 \left(\frac{v_T}{v_b}\right)^{10/3} \tag{14}$$

We have spelled out the derivation of Eq. (2.8) of IPP in considerably more detail than in the original paper.

If $v_T/v_b \sim v_b/c \sim 0.1$, P reaches a value of $\sim 10^{-2}$ W per soliton, while Eqs. (12) lead to transverse and longitudinal dimensions of 2 and 20 m, respectively. The auroral region of the ionosphere is $\sim$ 600 km wide and 15000 km long so that close packing could lead to $\sim 10^{15}$ solitons at any one time, easily producing the observed power of 10^9 W.

However, it is questionable whether the mechanism can work. One way to look at this is by considering the performance coefficient ϵ, which is the ratio of the emitted power P to the beam energy incident per unit time upon the soliton $\frac{1}{2} m v_b^2 n_b v_b \pi L_\perp^2$. We have

$$\epsilon \sim 10^4 \frac{n_0}{n_b} \frac{v_T}{c} \left(\frac{v_T}{v_c}\right)^{17/3} \tag{15}$$

IPP point out that one can put an upper bound on ϵ from the requirement that the energy transferred by the beam to the soliton should exceed the radiated energy. From this requirement it follows that [note that there is a misprint in Eq. (4.11) of IPP]

$$\epsilon_{\max} \sim 10 \frac{v_T}{c} \left(\frac{c}{v_b}\right)^3 \left(\frac{v_T}{v_b}\right)^{11/3} \tag{16}$$

which for $v_T/v_b \sim v_b/c \sim 0.1$ just about reaches the required value 0.1. It is tight, but might conceivably just be sufficient. However, if one uses the values given by Cole and Pokhotelov,[3] namely, $v_T/v_b \lesssim 0.03$, $v_b/c \sim 0.1$, one finds $\epsilon_{\max} \sim 10^{-3}$ to 10^{-4} and the cyclotron-soliton theory of the AKR will no longer be able to explain quantitatively the observed radiation intensity and the relation between the emitted radiation energy and the energy of the auroral electron beams.

Acknowledgments

I am grateful to Professor K. Cole for putting at my disposal a preprint of his paper with Dr. Pokhotelov, and to Professor Ya. N. Istomin for informing me that the dispersion relation used by IPP is based on Eq. (7.18) of Mikhailovskii's book.

References and Notes

1. Ya. N. Istomin, V. I. Petviashvili, and O. A. Pokhotelov, *Sov. J. Plasma Phys.* **4**, 76 (1978).
2. D. A. Gurnett, *J. Geophys. Res.* **79**, 4227 (1974).
3. K. D. Cole and O. A. Pokhotelov, "Cyclotron Solitons—Source of Earth's Kilometric Radiation, *Plasma Phys.* **22**, 595 (1980).
4. V. I. Petviashvili, *Sov. J. Plasma Phys.* **1**, 15 (1975).
5. V. I. Petviashvili, *Sov. J. Plasma Phys.* **2**, 247 (1976).
6. A. B. Mikhailovskii, *Theory of Plasma Instabilities*, Vol. I (Plenum, New York, 1973).

10

Clumpy Irregular Galaxies
A Progress Report

Jean Heidmann
Observatoire de Paris
Meudon, France

1. The Stage

Our planet Earth revolves around the Sun, 150 million kilometers away. Our Sun is just one of the 100 billions or so stars which swarm across our Galaxy, a vast system with diameter around 100 000 light-years.* Our Galaxy is one of the billion other galaxies which our most powerful telescopes are able to detect in the depths of space, at distances up to several billions light-years.

It is believed that all this vast universe came into existence 10 to 15 billion years ago, from a kind of tremendous Fiat Lux, the famous Big Bang, from which it sprang into an expansion phase which we experience now as the running away of all galaxies. It is believed also that galaxies formed out of the primeval gas pervading the primitive universe in the first one or two hundred million years of its existence. This formation proceeded through the condensation of clouds of gas into stars, in which gravitational energy, converted into heat, ignited nuclear reactions, providing the energy radiated by these celestial bodies.

It is known, however, that our Sun is a star only five billion years old. The picture is that it condensed quite recently, in the midst of our Galaxy, from gas which was reinjected into interstellar space by first-generation stars which, when reaching a state of nuclear fuel exhaustion, lost their equilibrium and blew away a sizable part of their atmosphere. That this can happen is presently being witnessed: It is the supernova phenomenon, in which a star suddenly becomes as bright as a whole galaxy for a month or two. What is practically witnessed, too, is the slow birth of new stars, mainly

*1 light-year = 9.4×10^{17} cm.

by sophisticated infrared and radio-millimetric wave observations, in the core of highly condensed interstellar gas clouds.

This brief preamble has set a stage: one of tremendous scale, spanning aeons of time, on which a tumultuous processing of huge amounts of matter involves enormous quantities of energy. And this is not all that is happening: Quasars, active nuclei of peculiar galaxies, are places in the universe where equally important or even activities on a much larger scale are occurring; with this huge difference, however: that we do not understand what is going on, how the "machine" works, whether or not still unknown physical processes, phenomena, or laws are playing a dominant role. It is from this perspective that "clumpy" irregular galaxies appear on the stage. The present chapter, devoted to these objects, will not be a report on a settled subject, giving the final status of a particular topic. It is rather a progress report on a question which is still being investigated, and the outcome of which is still unknown. Maybe, in the coming years, future work will show that very strange phenomena are occurring in clumpy irregular galaxies or, in a more matter-of-fact way, that it boils down to quite classical processes. However, in order not to discourage the reader, let us say that it is practically safe to state at present that in these galaxies stars are currently forming on a scale unknown before, which is a minimum datum worth working on.

2. Markarian Galaxies

Back in the early seventies, Markarian galaxies started to be much talked about. Radio astronomy had already led to the discovery of galaxies emitting powerful fluxes of radio waves; these emissions come from the nuclei, i.e., central parts, of galaxies, or from two galaxy-size plasma blobs symmetrically disposed with respect to the parent galaxy.

These observations indicated that nuclei of galaxies were able to eject beams of matter and energy, from a region with quite small dimensions, in an unexpectedly efficient way. The acute problems raised by such processes culminated in the discovery of quasars, objects still more powerful and more compact. Two decenniums later the mystery surrounding them is still so deep that some theoreticians have recourse to black holes or, more safely, to "black boxes" to deal with this phenomenon.

Ambartsumian was the first to suggest that nuclei of galaxies were able to eject matter on such a scale as to give birth, progressively, to whole galaxies—a view completely different from the classical one, according to which galaxies condensed from the primeval gas filling the universe. Investigation of galaxies with active nuclei could then be rewarding. Thus, Ambartsumian's observatory in Byurakan, Armenia, started to look for them. The chosen criterion of activity was the presence of excess near-ultraviolet light in the galactic bodies, which was to be detected by a Schmidt telescope with an objective prism, the largest in the world. So it happened that Markarian discovered his galaxies, now numbering about a thousand.

3. THE 21-CM LINE

Quite a number of workers started to investigate the "active" galaxies by various means. My colleagues Lucette Bottinelli and Lucienne Gouguenheim and I detected them for the first time in the 21-cm line of neutral hydrogen, with the aid of the large radiotelescope at Nancay, France, in 1971. This radio emission comes from the interstellar gas which pervades the galaxies and contains precious information. The observed central frequency of the line gives the radial velocity of the object and hence its distance—through Hubble's law of expansion. The total intensity gives the amount of neutral hydrogen, of primary importance for the questions of star formation. And the width of the line gives the values of the internal velocities, either turbulent or rotational, which are directly linked, through size and the virial theorem, to the total mass of the system. It is thus that our attention was caught by three of them, Mkn 7, 8, and 325, which at the same time: (1) had very large internal velocities, 250–500 km s^{-1}; (2) were very bright, of absolute magnitudes* -20 to -22; (3) were quite large, up to 80 000 light-years across; (4) looked irregularly shaped. This was peculiar, because irregular galaxies were not known to be either that bright or large or that much in motion.

4. LARGE TELESCOPE PHOTOGRAPHS

The photographs we had then were those from the *Palomar Sky Survey*, a photographic atlas covering all sky, but with a small scale. In order to see better, we obtained individual photographs on a larger scale, with large telescopes. Thus, starting in 1973, in collaboration with Caterina Casini, from Milano Institute of Astronomy,

*Absolute magnitude = 5 − 2.5 log L, where L is intrinsic luminosity in Sun units; -20 corresponds to 10^{10} Suns.

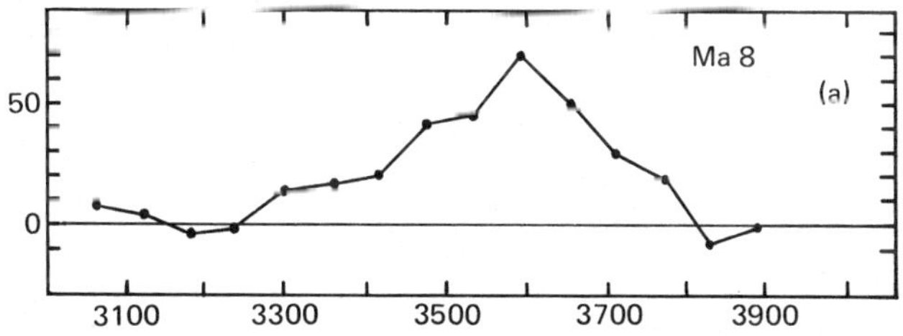

FIGURE 1. 21-cm line profile of Mkn 8 in 10^{-26} erg cm^{-2} s^{-1} Hz^{-1} versus radial velocity (through Doppler shift) with respect to the Sun in km s^{-1}. The central velocity corresponds to a distance of 150,000 light-years; the line flux is produced by 10^{10} solar masses of neutral hydrogen and the width of the line is 400 km s^{-1}.

we used the Haute Provence (France) and Asiago (Italy) 2-m telescopes, and even the Caucasus 6-m Soviet telescope, in collaboration with Igor D. Karachentsev.

These pictures of the peculiar irregular galaxies revealed a quite peculiar morphology: half a dozen or so bright "clumps" loosely scattered in a common and fainter diffuse envelope—while classical irregular galaxies usually appear as a patchy amorphous stretch of light, with one or two dominant, but not exceptionally bright, knots. We found another example, Mkn 296; while Mkn 297, investigated by Duflot, du Puy, and de Veny, could be assimulated to the same class.

In order to find more of the same objects, we took large-scale photographs of a selection from Vorontsov-Vel'yaminov's *Atlas of Peculiar Galaxies*, in collaboration with Cesare Barbieri at Asiago and Brent R. Tully in Hawaii and found 2 or 3 possible candidates.

Thus these first investigations point to a class of quite rare galaxies characterized by an excess near-ultraviolet emission, a clumpy irregular morphology, and by lumi-

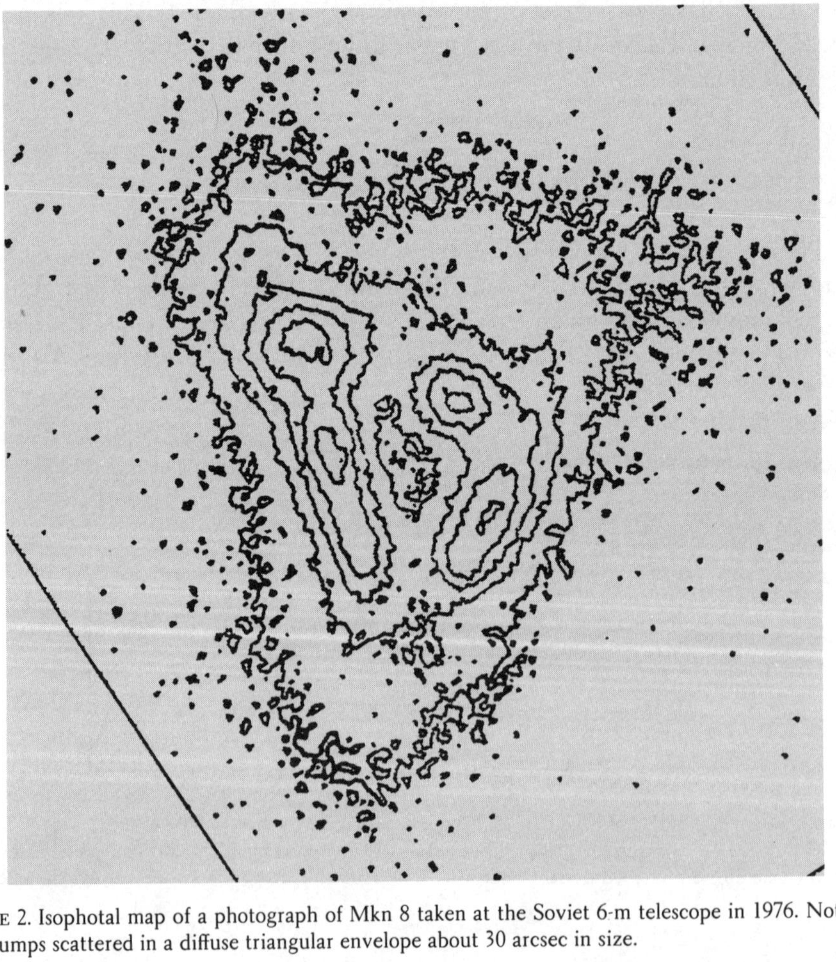

FIGURE 2. Isophotal map of a photograph of Mkn 8 taken at the Soviet 6-m telescope in 1976. Note the five clumps scattered in a diffuse triangular envelope about 30 arcsec in size.

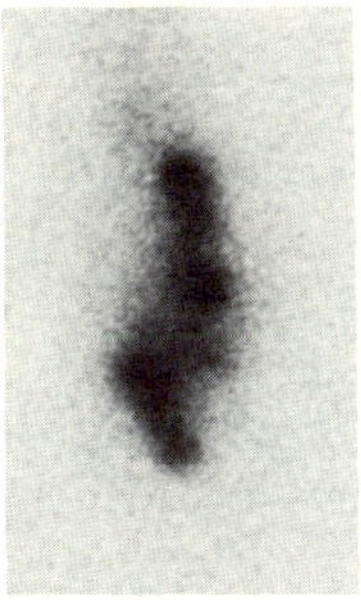

FIGURE 3. Electronograph of Mkn 296 obtained with the electronic camera at the Haute Provence Observatory 193-cm telescope. Eight clumps are distributed in an oblong envelope with a faint extension in the upward direction, about 30 arcsec long.

nosities, dimensions, and internal motions larger than those of classical irregular galaxies.

5. Optical Spectra

Meanwhile, optical spectra had been obtained for Markarian galaxies—by Sargent at Palomar, Khachikyan at Byurakan, and Duflot at Haute Provence. Very often they exhibited strong emission lines such as Hα, Hβ, [OII] λ378, [OIII] $\lambda\lambda$496, 501. These are typically found in HII regions. HII regions are vast clouds of interstellar gas, mainly hydrogen, ionized by one or two young bright hot stars of spectral type O or B. The temperature of the gas can reach 10 000 K. The Orion nebula, in our Galaxy, is such an HII region. Much larger ones are known in other galaxies: the Tarentula nebula, or 30 Doradus, in our nearby galaxy satellite, the Large Magellanic Cloud, and NGC 604 in the neighboring Triangulum spiral galaxy. These so-called giant HII regions or complexes comprise up to 100 O-B stars and up to 10^6 solar masses of ionized gas, intermixed with dust grains and molecules. They are quite exceptional, numbering only two or three per galaxy.

Sargent found several cases of spiral galaxies with a nucleus made up of a few "hot spots" which could also be giant HII complexes.

In these works spectra were obtained for some clumpy irregulars showing the same emission lines. This is a first hint of the nature of the clumps.

6. Mass of the Clumps

Other interesting information was obtained from high-dispersion spectra taken at Haute Provence Observatory, always with Caterina Casini and with Massimo Tarenghi of European Southern Observatory. Setting the slit of the spectrograph along the chain of clumps of Mkn 296, the strong Hα line was observed by us to be inclined along the dispersion. This is due to a general velocity gradient of 100 km s^{-1} along the 15 000-light-year length of the clumpy part. Assuming a simple model in solid

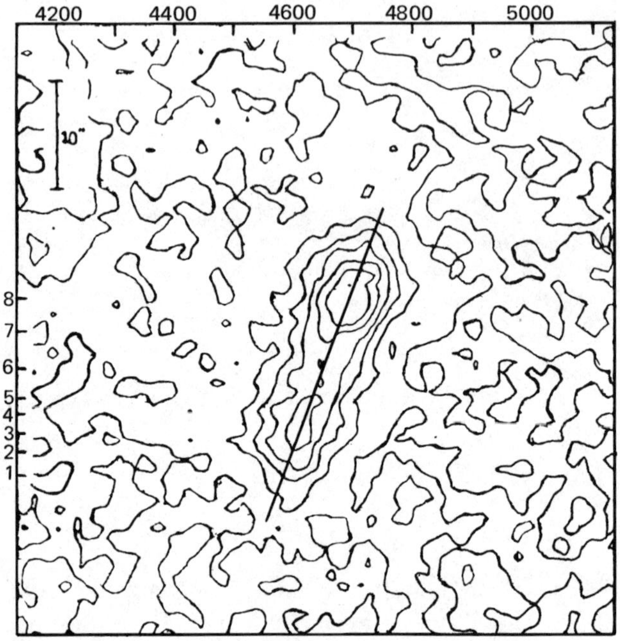

FIGURE 4. Isophotal map of the Hα line in a spectrum of Mkn 296; the slit was set along the clumps, here numbered 1 to 8, with a 10-arcsec scale; the spectral shifts are expressed in km s^{-1} with respect to the Sun. The straight line is a mean rotation curve.

rotation, dynamics give the mass of the clumpy part to be that of 10^9 suns. As there are 8 clumps, the mean share for each one is therefore 1.4×10^8 suns.

This mass is quite large! It is about 50 times larger than the ones which have been measured for the giant HII regions NGC 604 and 30 Doradus. We have here a serious hint at the gigantic masses involved in the clumps; it matches their high luminosities, around absolute magnitude -16, or 2.5×10^8 suns.

An interesting fact emerged also: The total mass of the clumpy part is three times *smaller* than the neutral hydrogen mass we found in Mkn 296 from its 21-cm line. This means that the neutral hydrogen is scattered in a large volume around the clumps, its general overall distribution having practically no effect on the dynamics of the central part. This leads us to apply for observing time with a radiotelescope which have in the 21-cm line enough angular resolving power for constructing a detailed map of the gas distribution—the Westerbork (The Netherlands) supersynthesis radiotelescope. With its 12 25-m dishes movable along a 1.5-km base line, this super-interferometer has a 20-arcsec beam.

Observations made this year in collaboration with Hugo van Woerden, of Kaptein Laboratorium, Groningen, on Mkn 7 and 8 are being reduced now. What is the shape of the gas envelope and its spatial and kinematical relationships to the clumps that will emerge from this study?

7. Ultraviolet Spectra

Galaxies with bright compact parts having near-ultraviolet excess emission like the clumpy irregulars were choice targets for the first astronomical ultraviolet satellite telescope able to reach extragalactic space. Thus Caterina Casini and I obtained two shifts on the International Ultraviolet Explorer to observe Mkn 297. Low-dispersion spectra involving about half of the clumps were obtained in collaboration with Piero Benvenuti, of the European Space Agency (Madrid). At the cost of very long 5-hour exposures, they covered the range 115–320 nm, which is unobservable from the ground.

Three interesting facts emerged from these faint records: (1) The uv continuum emission of Mkn 297 is intrinsically 400 times larger than that of 30 Doradus. This means that one clump is ~100 times more powerful than a giant HII region—fact confirming previous hints. (2) Absorption lines typical of early bright stars of types O and B were present; CII λ133, CIV λ155, FeII(8) λ162, etc. This is the first evidence for the presence of newly formed stars in the clumps. (3) The shape and intensity of

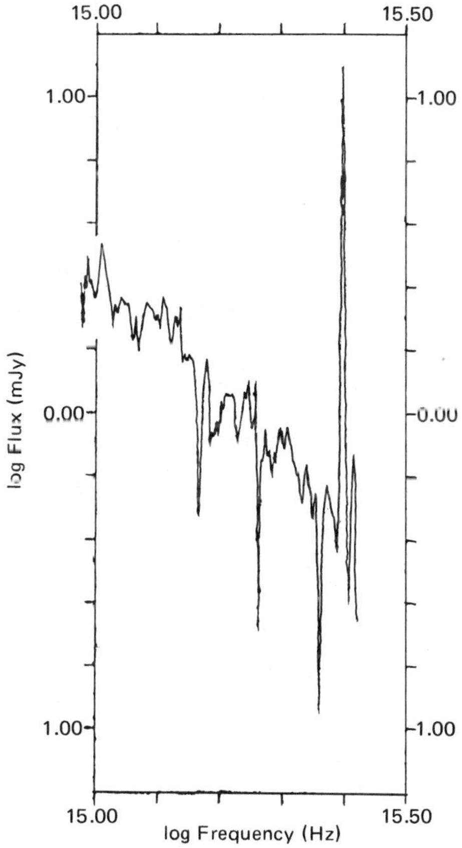

FIGURE 5. Continuum uv spectrum of Mkn 297 obtained with the IUE satellite in millijansky versus hertz. The high-frequency spike is due to geocoronal Lyman-α, and the absorptions are reseau marks.

the spectrum can be accounted for by a mixture of 4×10^5 early O stars plus 3×10^9 later type A stars. A clump therefore contains 10 000 early stars! In the second year of IUE we disposed over more shifts. We now have statistics on eleven clumps. They show that on the average one clump radiates at 155 nm, 200 times more than 30 Doradus and like 10^4 early type stars.

This provides good support for the hypothesis that clumps are what we may call *hyper*giant HII complexes and that clumpy irregulars are galaxies in which exceptional bursts of star formation take place.

8. Radio Centimeter Observations

Further support was just obtained this fall: if the foregoing hypothesis is valid, the hot gas ionized by the early stars should emit thermal radiation at radio wavelengths. Up to now, radio emission was known only for Mkn 297, at long wavelengths of 73 to 6 cm. Its spectrum is quite steep, decreasing at high frequency, which is characteristic of synchrotron radiation which is often present in galaxies at such wavelengths. By contrast, thermal radiation has a flat spectrum.

To get rid of the synchrotron radiation, one has to observe at still shorter wavelengths. In collaboration with Richard Wielebinski, of Max Planck Institute for Radioastronomy at Bonn, we applied for time on the 100-m Effelsberg radiotelescope with its new 1.3-cm receiver. During perfect weather it performed remarkably and detected 5 out of 7 clumpy irregulars at flux densities in the range 3–10 mJy.* Translated into masses of ionized hydrogen at 10 000 K, this fits the expected values around 10^8 suns. However, in order to be sure that we really deal with thermal radiation, we still have to check that at neighboring wavelengths we get the same flux densities.

9. Where Are We?

So, where are we? We found rare examples of irregular galaxies

- with 5–10 clumps of high surface brightness;
- with near-uv excess;
- with optical emission lines;
- with strong uv emission;
- large in dimensions, luminosities, neutral hydrogen contents, and total masses.

On the average a clump has diameter up to 3000 light-years, absolute magnitude -16, total mass 10^8 suns, 10^4 early bright stars, uv and 1-cm radio emission like 100 times a giant HII region.

All this points to the concept of hypergiant HII complexes.

*1 mJy (millijansky) = 10^{-26} erg cm^{-2} s^{-1} Hz^{-1}.

10. The Problems

What problems are raised by these facts?

First, about the exact nature of the clumps. Most probably they are bursts of star formation on a tremendous scale. However, in one case, Mkn 297, the long-wavelength radio spectrum has the synchrotron signature; in another case, Mkn 8, the emission line [NeIII] λ387 is visible, indicating high excitation of the gas. Thus, do some clumps have other peculiar processes going on? One could think about the synchrotron radiation emitted by supernova remnants produced by the already dead very short-lived stars.

Second, what about the age of the clumps? Multicolor photometry points to ages around 10^7 years—very young indeed; but this is not a safe determination.

There are also problems raised by the origin of such a peculiar clumpy structure: Is it the result of a disruption, as suggested by Khachikyan for Mkn 8? Is it simply the result of the tidal gravitational interaction with nearby galaxies or of the merging of two colliding galaxies? Is it due to the infall of intergalactic gas? Or is it a recent replica of what is thought to have happened to most galaxies at the beginning of the expansion of the universe? Or what else?

Problems about the evolution of such galactic structures: In case their energy is positive, would they evolve into what Vorontsov-Vel'yaminov called "nests" of galaxies, i.e., very tight small groups of a few galaxies? If their energy is negative, would they evolve into normal type galaxies through internal viscosity? Or what else?

The clue question is whether they constitute a transient state—and, if "yes," how is it produced and whence—or whether they form a different class of objects. Investigation of such galaxies can give clues on the physical processes at work. Interest also exists in

- these hypergiant HII complexes for themselves;
- the simultaneous triggering of a collection of them in a single galaxy;
- the fact that they may be the best model we could get to tackle the problem of the appearance of primeval galaxies, a subject which will come within the reach of the giant 2.4-m Space Telescope when it is launched in 1984.

11. Future Investigations

In view of these problems, we started a number of investigations in specific fields with the help of the best available instrumentation. One project is the production of detailed radio maps of "clumpies" with the Very Large Array (VLA) of the National Radio Astronomy Observatory in New Mexico, in collaboration with Dave S. Heeschen; this mainly in order to get the distribution of ionized gas. The VLA is a set of 27 25-m dishes movable along 40-km tracks; by supersynthesis it reaches a resolution of 2 arcsec at 6-cm wavelength. In the first runs we obtained very good maps which clearly show the individual clumps. We have now more telescope time to

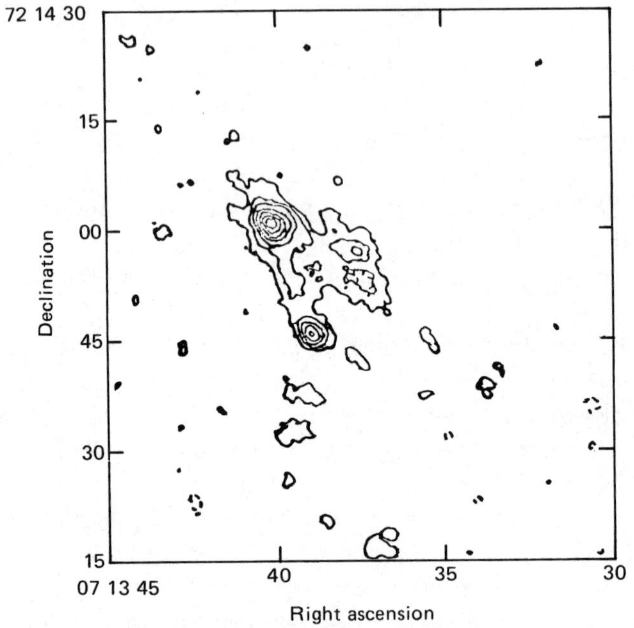

FIGURE 6. Radio map of Mkn 8 obtained with the VLA in New Mexico at 6 cm wavelength. Four clumps are well visible; note the resemblance with the optical photograph of Fig. 2. Coordinates are celestial coordinates; in particular declination is in degree, arcmin, and arcsec.

observe at 20 and 2 cm in order to get a spectral index (i.e., shape of radio spectrum) distribution. Already it appears that there is not a strict correlation between optical and radio emissivities of individual clumps. Perhaps this will allow investigation of the internal absorption of light by dust.

Another study involves the determination of the physical conditions (electron densities and temperatures) and of chemical abundances of heavy elements (mainly N and O) in the clumps—in collaboration with Ann Boesgaard of Institute for Astronomy at Honolulu. She obtained spectra of four clumpies with the Mauna Kea 2.2 m telescope which are being analyzed now. The point is that newly formed galaxies would be deficient in heavy elements, which are not yet abundantly processed from primeval gas by successive generations of stars. This is indeed an important piece of information to unravel.

Together with Wielebinski, we shall also request observations at the Effelsberg radiotelescope at 2.8 and, if feasible, at 9 mm, in order to assess the thermal or non-thermal nature of the radiation in the cm range.

A probe has been made for CO molecule emission in collaboration with Mark A. Gordon (from NRAO Millimeter 36-ft Radiotelescope at Kitt Peak) and Eugene E. Epstein (of Aerospace Corporation, Los Angeles). CO molecules emit a line at 2.6 mm, which has been observed in HII regions in our Galaxy and in a few nearby galaxies. After 40 hours of integration we only obtained an upper limit; as it is close to what should be expected from a direct comparison with normal HII regions, we applied for more time in order to get an answer to the question, "the same or less?" This should be important for the physics of what is going on in these superbursts, there being a close interplay between gas concentration, imbedded dust, stellar radiations and winds, and molecule formation. However, these are difficult observations because of the great distances of clumpy galaxies, 150 000 light-years on the average.

We also started to involve the infrared instrumentations set in the outstanding 4200-m-high site on Mauna Kea, Hawaii, in collaboration with Eric E. Becklin and Gareth C. Wynn-Williams. The 2.2-m Hawaii University and the new NASA Infrared 3-m Telescope detected Mkn 297 and 325 this summer in the near-ir range 1–5 μm. Even a preliminary map of Mkn 325 has been produced with a 7-arcsec resolution which fits the optical image. It is generally thought that the near-ir originates in supergiant cool red stars, but here astrophysics is only in its first years. In the far-ir, around 20–100 μm, preliminary tests with the NASA telescope show that several hour integration time could give detection. It is believed that far-ir radiation comes from dust grains heated at 40 K by the bright young stars. Detection would give the amount of dust, which is interesting, as it is a catalyst for molecule formation and determines the internal absorption.

Last but not least, an ode to the king of the space telescopes, Einstein or HEAO-B, which is able to produce images of far-away extragalactic bodies with a 2-arcsec resolution from x rays in the 0.15–3-kev energy range—quite a feat!

Why x rays for clumpy galaxies? If we are right, in these gigantic bursts, stars of very high mass—10 suns or more—should have been formed, too. But the higher the mass, the shorter the life, because the nuclear fuel consumption rises faster than the supply. And what is the outcome of a dying star? From experience in our Galaxy, it could be two things: a supernova remnant or an x-ray binary.

When exploding, a supernova, as we saw, ejects part of its atmosphere into the surrounding interstellar gas; shock fronts of very high temperatures (10^6 K) are created which emit bremsstrahlung radiation in the x domain.

x binaries: A large proportion of stars belong to pairs whose members revolve around one another. When one member passes through the supernova stage, its remnant may be a neutron star, with diameter around 10 km, instead of a million or so initially. If this object is inside the Roche lobe of its companion, it will swallow part of the latter's atmosphere, which, when highly compressed in the intense gravitational field around the neutron star, will reach very high temperatures and radiate in the x domain.

Thus, detection of x rays in clumpy galaxies could give interesting information, on the rate of formation of high mass stars, for example. Very rough estimates from galactic supernova remnants or x binaries, from 30 Doradus or active nuclei of galaxies, indicate that in 10,000 seconds integration time with HEAO-B a clumpy galaxy could yield 10–100 x photons, enough for a flux density determination.

Observe how in that 10-A domain photons are scarce, though very energetic—and quite expensive: they are counted one by one!—and that they appear more like particles, in contrast to the other end of the spectrum, in radioastronomy, where they appear more like waves, showering vast expanses of metallic meshes with innumerable low-energy photons.

Alas! Just as with Heeschen, we were allocated to 40,000 s observing time. But the huge instrument just failed, two of its gyros refusing to be turned on. This accident was surely an international astronomical catastrophe. Let us hope that a telerepair can be performed.

Another first class instrument is scheduled for 1984: the Space Telescope. Compared to present ground-based telescopes perturbed by the atmosphere, it will have a nearly 10 times better angular resolution, will reach 100 times fainter objects, and will be able to extend its view into the uv range.

In preparing to observe clumpies with its Faint Object Camera and its Faint Object Spectrograph, we started to use a newly constructed telescope which is probably the best in the world for high-resolution photographs, because of the excellence of its site—Pic du Midi, in the Pyrennees, at nearly 3000 m—and of its design—a completely closed dome: the 2-m French telescope. In collaboration with Gérard Coupinot and Josette Hecquet from Toulouse Observatory, we expect to obtain a series of high-quality photographs which will be scrutinized by deconvolution techniques. It is evident that structural details of the clumps will be of primary importance for their study.

Let me finish with a curious remark: Because of their remoteness, clumpy irregular galaxies require huge, sophisticated, and specialized instruments. However, a very valuable contribution to their study could also be made by amateur astronomers: to watch them for supernova apparitions. With their circa 14th magnitude brightness, the galaxies are accessible to 20-cm-diam telescopes; and it is not unreasonable to expect them to give birth to supernova at a rate of one per year, because of the large number of young stars they contain.

So, *finis* it is. Still, we may express a wish: That these *galaxies mal fichues* will significantly aid our understanding of the formation and evolution of galaxies and stars.

References and Notes

References may be found in P. Benvenuti, C. Casini, and J. Heidmann, *Nature* **282**, 272 (1979).

11

THE APORIAS OF COSMOLOGY AND THE ATTEMPTS AT OVERCOMING THEM BY NONSTANDARD MODELS

H. HÖNL*
Faculty of Physics
Universität Freiburg
Freiburg, GFR

H. DEHNEN
Faculty of Physics
Universität Konstanz
Konstanz, GFR

1. INTRODUCTION

Of the scientists of modern times Isaac Newton was the first who recognized clearly the cosmological problem and tried to solve it with the help of the general principles of mechanics and of the gravitational law developed by himself. The validity of Euclidean geometry, which includes automatically the spacelike infinity of the Universe, was evident for him as for his forerunners. Herewith, the ground elements of physics and geometry are given, which has been the basis of physics and astronomy up to the beginning of our century.

Newton argued as follows: If the total mass of the Universe is finite and has been initially homogeneously distributed within a sphere of finite radius, the mass would have been contracted in consequence of the gravitational forces. This contradicts observations. Therefore it seemed more probable to Newton that initially an infinite mass was distributed homogeneously over infinite space. Then the question arises, whether the assumption of a time-independent and homogeneous distribution of matter and consequently of a static Universe is compatible with the assumption of infinity of space. The answer is negative, so that one of the basic assumptions must be given up. Here only Einstein's general relativity leads to progress.

*Deceased March 29, 1981.

It may be interesting to discuss in this connection Kant's idea about the cosmological antinomies; it concerns exactly the question of finiteness or infinity of the Universe. Because Kant does not distinguish between infinity and boundlessness of space in consequence of the situation of mathematics of that time, his question was: bounded or unbounded Universe? Both possibilities are rejected by Kant, but for different reasons: If the Universe were bounded the question arises, what is beyond the boundary? In the opposite case of an unbounded Universe the infinity of space exceeds every possible experiences. On the other hand, there is no doubt that Kant was convinced of the infinity of space because this is a consequence of Euclidean geometry, the axioms of which were evident for him a priori in virtue of "reiner Anschauung."

More important for modern cosmology has been Kant's idea of the immutability of the natural laws with time. Especially in view of the large extrapolations in space and time, often the question arises whether the known natural laws are also valid in these very far regions. The answer for Kant is already given by the fact that the universality of natural laws is the a priori condition of exact sciences at all. And indeed, the universal validity of natural laws has been proved up to the present empirically in many cases of astrophysics.

2. The General Relativistic Cosmology

Einstein's general relativity is the unified theory of space-time and gravitation, which overcomes Newton's conception. The four-dimensional space-time continuum possesses non-Euclidean geometry and its metric contains gravitation. Geometry and gravitation are identical and are caused by the matter distribution in the Universe according to Einstein's field equations. Thus, Einstein's theory is in principle already cosmology. Any solution of the field equations of gravitation represents a cosmological model, which gives a picture of the distribution of matter in the whole space-time. In this connection besides the local geometry the topology of space is important, which means its large-scale structure. The most essential feature for cosmology is the fact that space is, in contrast to the Euclidean case, not a priori infinite; there exist infinite as well as finite spaces.

In spite of this progress the fundamental difficulties of cosmology do not vanish automatically. For this we give two examples. First we consider the Universe consisting of a large island of galaxies, outside of which space is empty. In this case, the solutions of Einstein's field equations are uniquely determined only after adding boundary conditions in the spacelike infinity. Therefore the dynamics of the system would be determined by the behavior of the metric in the spacelike infinity and the gravitational interaction of matter would be only a small perturbation of the metrical field, which plays on the whole the role of Newton's "spatium absolutum." The dynamics of motion of the galaxies would depend primarily on the boundary conditions in the spacelike infinity and not on the material interaction alone.

In order to avoid this aporia, Einstein considered in 1917 the possibility of a closed and finite Universe for which boundary conditions at infinity are superfluous. In this case the homogeneously distributed matter determines the constant curvature of the spherical space. However, one finds (G. Lemaitre, A. S. Eddington) that such a model is unstable. After a small perturbation it begins to expand or to collapse, and now timelike initial conditions for the Universe are implicated. The problem is removed only in a different direction.

Nevertheless, Einstein's paper has given rise to a search for an expansion or a contraction of the Universe, the result of which was the discovery of the flight motion of galaxies by E. Hubble in 1929. The explanation of this as a consequence of an expansion of the Universe was confirmed 15 years ago by the discovery of the electromagnetic 3K black-body background radiation. This radiation predicted by G. Gamow in 1946 in connection with the attempt at explanation of the abundance distribution of the chemical elements and detected accidentally by A. Pencias and R. Wilson in 1965 means that in the early Universe matter and radiation were in thermodynamic equilibrium at temperatures much higher than the present one and that the radiation has been cooled subsequently by an adiabatic expansion of the Universe. Remarkable are two facts: First the high isotropy of the radiation (deviation from ideal isotropy less than 1/3000), which points to an isotropic distribution of matter at the time of last emission of the radiation (recombination of the hydrogen); and secondly, the high number of photons of the background radiation, which exceeds that of the present heavy particles by the enormous factor of 10^8.

In any case the flight of the galaxies and the existence of the isotropic background radiation are empirical facts which must be taken into account in every modern cosmology.

Both facts are described usually in the framework of Friedman's standard models for the Universe. Their simplifying assumptions are (1) homogeneous and isotropic distribution of matter and (2) constancy of entropy, which means no irreversible processes, especially no spontaneous creation of matter.

On this basis it follows that the Universe would have a spacelike constant curvature, which can be positive, negative, or even zero. In case of positive curvature, the space is closed and finite (spherical or elliptical space), otherwise it is infinite. In every case, the Universe would expand; only in case of positive curvature the expansion goes over into a contraction after some time. Taking as measure for the size of the Universe the so called distance parameter $R(t)$, corresponding to the distance of two arbitrary galaxies, one finds for the change of R with time the curves given in Fig. 1.

In any case, the expansion begins at the time $t = 0$ with the value $R = 0$. At this point of time—the so-called "Big Bang"— density and temperature of matter and radiation would have been infinite, which is physically unacceptable. This result is serious all the more as S. Hawking, G. Ellis, and R. Penrose could show in 1968 that this singularity is not a consequence of the symmetries (homogeneity and isotropy) of

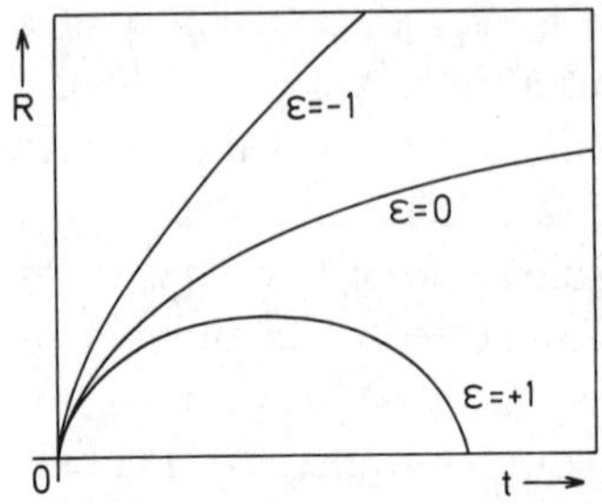

FIGURE 1. Time dependence of the distance parameter R in the case of Friedman's standard models for the different curvatures: Positive curvature ($\epsilon = +1$, spherical or elliptical space, cycloid), vanishing curvature ($\epsilon = 0$, flat space, Neil's parabola) and negative curvature ($\epsilon = -1$, hyperbolic space).

the models, but follows necessarily from the very general "energy condition" $\rho c^2 + 3p \geq 0$ (where ρ is mass density and p is the pressure of the cosmic substratum).[1]

3. The Aporias of the Standard Models

In order to overcome this unsatisfactory situation two possibilities exist. Firstly a modification of Einstein's theory or of Friedman's assumptions, which will be discussed later. Secondly a more real consideration of the behavior of matter, which at first seems to be the more plausible way. In view of the extreme conditions of superhigh densities ($\gtrsim 10^{14}$ g/cm^3) and temperatures ($\gtrsim 10^{12}$ K) near the Big Bang one has to consider that heavy elementary particles (protons and neutrons) can be produced out of radiation energy and that the strong attraction interaction of the hadrons must be taken into account. As the authors could show,[2] both phenomena lead to a high negative pressure comparable with the cohesive pressure of a van der Waals gas, in consequence of which the "energy condition" is violated. Then the initial singularity is avoided and substituted by a state of minimum size of the Universe with a maximum but finite density and temperature ($\rho \simeq 10^{14}$ g/cm^3, $T \simeq 10^{12}$ K).

Now, in case of the spherical finite space of positive curvature the possibility of a periodic continuation of the course of $R(t)$ (see Fig. 1) exists. But such a "breathing universe" with strongly periodic time behavior contradicts the law of increasing entropy, and this was to be expected in view of Friedman's second assumption. However, because of the high number density of the photons of the background radiation, the relative increase of entropy in every "aeon" is small and therefore only small deviations from the strong periodicity are to be expected. On the other hand the question remains open, why in such a Universe the entropy does not already diverge but possesses the present finite value determined first of all by the photons of the background radiation. In Section 5 we shall discuss such a breathing model with increasing entropy more in detail.

In the case of the open Friedman models with negative or vanishing curvature, the problem of the increasing entropy is not so serious, because the expansion continues forever, connected with a steady dilution of matter. But here the already discussed difficulty of the infinity of space appears. The global assumption of homogeneous and isotropic distribution of matter is an assumption ad hoc and not provable at all since it exceeds every possible experience. Because of the Big Bang the presence is influenced only by a finite region (inside of the horizon) of the infinite space.*

In any case the entropy problem remains unsolved in the framework of the standard models in consequence of Friedman's second assumption. Strongly connected with this is the fact that the standard models have no explanation for the matter content of the Universe. On the contrary, it must be given as an arbitrary initial value. This unsatisfactory situation can be solved only by a modification of Friedman's assumptions or of Einstein's theory.

4. The Steady-State Theory and the Theory of Dirac–Jordan

Attempts at overcoming the aporias of Friedman's theory have been started at first by Dirac and Jordan.[3] In their theory Einstein's field equations are modified in such a way, that (besides the gravitational coupling quantity) the mass of the Universe is not constant but increases with time connected with an increasing entropy. Later, a similar attempt was made by Hoyle, Bondi, and Gold.[4] They postulate ad hoc a spontaneous creation of matter in the Universe in such a way that the Big Bang is avoided and the dilution of matter in consequence of the expansion is compensated exactly. Such a model is homogeneous not only with respect to space, but also with time.

However, both theories contradict the Planckian distribution of the electromagnetic background radiation.[5] If these theories were realized, the number of photons per frequency unit would have to be much higher than according to Planck's formula in consequence of the permanent photon production.

There is no doubt that both theories must be given up.† On the other hand, they show clearly the difficulty of avoiding the cosmological aporias. The only remaining basic idea of these theories is the concept of increasing entropy. It seems to us that this law is so fundamental that it should be taken into account by any further cosmology.

5. Nonstandard Models and the Vacuum

Another and more plausible way consists in retaining Einstein's field equations but modifying Friedman's assumptions only. From these the assumption of constant entropy of the Universe is the most doubtful one, not only in view of the local irreversible processes of formation of galaxies and stars, but especially because of its con-

*In the absence of a Big Bang, as is the case in the steady-state theory (see Section 4), this argument against the infinite space vanishes. There remains, however, the unsatisfactory boundary condition in the spacelike infinity.

†Also the solar-relativistic effects confirm rather Einstein's than Dirac and Jordan's theory.

sequence of a constant number of particles. This is also not confirmed by experience. On the contrary, it is known that under special conditions elementary particles with finite rest mass can be produced out of the *vacuum* spontaneously. The generation of electron–positron pairs by strong electromagnetic fields is the best known example for this process.

This fact requires a reconsideration of the physical properties of the vacuum. The vacuum of modern physics is not identical with empty space—with Democritus' *μὴον*; it is to be taken for the "ground state of the Universe" and it possesses structure. That means that it contains all elementary particles virtually, in consequence of which it is polarizable and able to produce real particles and antiparticles after sufficient supply of energy. Therefore the vacuum is to be considered as its own substratum besides the classical matter and the metrical field.

From this point of view it is to be expected that the expansion of the Universe, by which the energy conservation of matter is broken in principle, has an influence on the vacuum in such a way that spontaneous particle–antiparticle generation happens. To this possibility E. Schrödinger pointed already in 1939, and later B. de Witt in 1953.[6] Strange to say, first calculations concerning this very interesting aspect were first performed only in 1967 by R. Sexl and H. Urbantke.[7] Later, several authors have investigated this question and today there exist two different directions for treating this problem.

5.1. *The Coherent Procedure*

The first one uses a *coherent* development of the (quantized) matter fields during the evolution of the Universe with time, and the number of created particles is obtained when the time evolution of the Universe is nearly stopped (S-matrix approach). This method is free from additional assumptions, but one finds that the particles are created only in the immediate neighborhood of the Big Bang with a number which is negligibly small in comparison with the number of the real particles in the Universe (factor of 10^{-22}) and therefore without cosmological importance.[8] This situation changes only if additional strong fields besides the cosmological one (e.g., electromagnetic fields) are present.

For simplicity we discuss in the following the creation of Klein–Gordon particles in an expanding 3-flat Universe, when additionally a homogeneous parallel electric and magnetic field is present. Starting form the homogeneous and isotropic 3-flat Robertson–Walker line element in the conformally flat form

$$ds^2 = R^2(\eta)\{d\eta^2 - dx^2 - dy^2 - dz^2\} \tag{1}$$

and from the conformally coupled Klein–Gordon equation*

$$[g^{\mu\nu}(i\nabla_\mu - eA_\mu)(i\nabla_\nu - eA_\nu) - \tfrac{1}{6}R_\alpha^\alpha - m_0^2]\Phi = 0 \tag{2}$$

*∇_μ means the covariant partial derivative.

for the matter field one finds with the electromagnetic 4-potential

$$A_\mu = E\eta\delta_\mu^3 - Hx\delta_\mu^2 \,(E = \text{const}, H = \text{const}) \tag{3}$$

for a homogeneous parallel electric and magnetic field the following number density of created particle-antiparticle pairs ($\hbar$, $c = 1$) [9]:

$$\begin{aligned} n &= R^{-3}\frac{(e^2E^2 + m_0^2b^2)^{5/4}}{(2\pi)^3 m_0 b} K \exp\left[-\frac{\pi m_0^2 a^2}{(e^2E^2 + m_0^2b^2)^{1/2}}\right] \\ K &= \frac{\pi e H}{(e^2E^2 + m_0^2b^2)^{1/2}}\left[\sinh\frac{\pi e H}{(e^2E^2 + m_0^2b^2)^{1/2}}\right]^{-1} \end{aligned} \tag{4}$$

Here for the distance parameter R the time dependence

$$R = (a^2 + b^2\eta^2)^{1/2} \tag{5}$$

has been chosen where the conformal time η is connected with the measured cosmological time t according to

$$t = \frac{a^2}{2b}\left\{\frac{b}{a}\eta\left[1 + \left(\frac{b}{a}\eta\right)^2\right]^{1/2} + \ln\left(\frac{b}{a}\eta + \left[1 + \left(\frac{b}{a}\eta\right)^2\right]^{1/2}\right)\right\} \tag{6}$$

Consequently, near the Big Bang ($t = 0$), where the particle pairs are created, the time dependence of R is given by

$$R = a\left[1 + \frac{1}{2}\left(\frac{b}{a^2}t\right)^2\right] \qquad (|t| \ll a^2/b) \tag{7}$$

whereas for large times ($|t| \gg a^2/2b$) the behavior

$$R = (2b|t|)^{1/2} \tag{8}$$

follows corresponding to a radiation-dominated Universe. Evidently a has the meaning of the minimum distance parameter at the Big Bang $t = 0$ and b represents its acceleration. The quantities e and m_0 are charge and rest mass of the particles and $\mathcal{E} = E/R^2$ and $\mathcal{H} = H/R^2$ the measurable (parallel) electric and magnetic field strength.

For $E \to 0$ and $H \to 0$ one gets from (4) the number of particle pairs created by the pure cosmological (gravitational) field:

$$nR^3 = \frac{(m_0 b)^{3/2}}{(2\pi)^3}\exp\left(-\frac{\pi m_0 a^2}{b}\right) \leq \frac{(m_0 b)^{3/2}}{(2\pi)^3} \tag{9}$$

Obviously a large acceleration b and a small value of the minimum distance parameter a amplify the particle creation. On the other hand, with increasing rest mass the number of created particles increases at first and decreases rapidly after crossing a maximum at

$$m_{0\,\max} = \frac{3}{2}\frac{b}{a^2\pi} \tag{9a}$$

For a quantitative comparison with observation, the value of b must be determined. As one can show,[9] b is connected with the temperature T of the background radiation [cf. (8)] according to

$$b = R^2T^2(8\pi G\sigma/3)^{1/2} \tag{10}$$

(G is the Newtonian gravitational constant, σ is the Stefan–Boltzmann constant). Insertion of (10) into (9) yields for the present stage of the Universe ($T_0 = 2.7$ K present temperature of the background radiation)

$$n_0 \lesssim \frac{m_0^{3/2}}{(2\pi)^3}\left\{\frac{8\pi G\sigma T_0^4}{3}\right\}^{3/4} \simeq 5 \times 10^{-28}\ \mathrm{cm}^{-3} \tag{11}$$

in case of protons and neutrons. Compared with the observational value $n_0 \simeq 10^{-6}$ cm^{-3} (corresponding to the present mass density of approximately 10^{-30} $\mathrm{g/cm^3}$) the result (11) is too small by the enormous factor of 10^{-22} as mentioned above.

Finally we note that the results of the coherent method depend essentially on the form of the expansion law. There exists even a strongly time-dependent expansion law *without* any particle creation.[10] This fact points to the possibility that the pure gravitational coherent procedure is an inappropriate one for the cosmological problem.

This situation is changed in the case of the existence of strong electromagnetic fields; and indeed, in view of the background radiation these were present in the early Universe. For

$$eE/m_0b \gg 1 \tag{12}$$

one finds from (4) for the created particle pairs:

$$nR^3 = \frac{(eE)^{5/2}}{(2\pi)^3 m_0 b}\frac{(\pi H)/E}{\sinh(\pi H/E)}\exp\left(-\pi\frac{m_0^2a^2}{eE}\right) \tag{13}$$

In contrast to the pure gravitational case (9), where the number of created particles increases with m_0 ($<m_{0\max}$) and the acceleration b, here their number decreases with increasing values of m_0 and b. Instead of this, the number of created particle pairs increases very rapidly for strong electric fields, whereas strong magnetic fields ($H/E \gg 1$) damp the generation strongly.

Inserting for E and H the mean electric and magnetic fields of the background radiation one finds[9] that the condition (12) is fulfilled very well and that the *present* number density n_0 of heavy particles is in accordance with that of the created particle pairs according to (13). In the limiting case $a = 0$ the expansion law (5) [cf. (8)] is exactly that of a pure radiation Universe, so that *all* heavy particles would have been created out of the vacuum.

Nevertheless, there exists also here a serious lack. The electromagnetic fields and the corresponding expansion law (5) must be given independently from the creation process. They are the same as in the standard models given as arbitrary initial values. There exists, as in the pure gravitational case (9), no back-reaction of the particle creation on the expansion law. And this is a property of the coherent procedure as such. It seems, therefore, that the coherent procedure is not appropriate for describing the cosmological situation. On the contrary, the creation process should influence the expansion law, which itself causes the particle creation in a self-consistent way.

5.2. *The Incoherent Procedure*

The aim of the second direction of investigation is an *incoherent* treatment of the matter fields. The essential step is the transition to localized particle creation *rates* per time unit, which represents, however, an additional assumption. But in this way a self-consistent treatment of the time evolution of the Universe and of the generation of its matter content is obtained. The properties of the Universe are largely not determined by initial values as above but can be deduced from provable *natural laws alone*. Because it seems to us that this is the most interesting progress in avoiding the cosmological aporias, we give a more detailed representation of the method and its results, following a paper of Schäfer and Dehnen.[11]

5.2.1. THE THEORETICAL BACKGROUND. In analogy to the calculation of the atomic transition probabilities by the time-dependent perturbation theory in the usual quantum mechanics, we determine first the particle *creation rate* at every point of time in the case of the homogeneous and isotropic time-dependent Robertson–Walker line elements:

$$ds^2 = dt^2 - R^2(t)\left\{d\chi^2 + \begin{pmatrix}\chi^2\\ \sin^2\chi\\ \sinh^2\chi\end{pmatrix}(d\theta^2 + \sin^2\theta\, d\varphi^2)\right\} \tag{14}$$

In view of the fact that the overwhelming majority of the heavy particles of the Universe, which determine also its mass, has spin 1/2, we take in contrast to (2) for describing the matter field the covariant Dirac for free spin-1/2 particles. Then the field equations for Dirac's 4-spinor ψ are given by ($c = 1$)*

$$i\gamma^\mu\psi_{\|\mu} - m_0\psi = 0 \tag{15a}$$

with the generalized Dirac matrices γ^μ

$$\gamma^\mu\gamma^\nu = g^{\mu\nu}$$

and the covariant spinor derivative

$$\psi_{\|\mu} = \psi_{|\mu} - \Gamma_\mu\psi$$

The spinorial connection coefficients are defined by

$$\Gamma_\mu = \tfrac{1}{4}h^\alpha_{(\rho)\|\mu}h^{(\rho)}_\beta\gamma^\beta\gamma_\alpha \tag{15b}$$

whereby the orthonormal tetrad field $h_{(\rho)\mu}$ fulfills the completeness relations:

* $\|_\mu$ means the covariant and $|_\mu$ the usual partial derivative.

$$h_{(\rho)\mu}h^{(\rho)}_{\nu} = g_{\mu\nu}, \qquad h_{(\alpha)\mu}h^{(\beta)\mu} = \delta^{\beta}_{\alpha}$$

The 4-flux density of probability reads

$$j^{\mu} = \bar{\psi}\gamma^{\mu}\psi \qquad (\bar{\psi} = \psi^{\dagger}\gamma^{\mu}h^{(4)}_{\mu}) \tag{15c}$$

and the energy-momentum tensor of the Dirac field is given by

$$T_{\mu\nu} = \frac{i}{2}(\bar{\psi}\gamma_{(\mu}\psi_{||\nu)} - \bar{\psi}_{||(\mu}\gamma_{\nu)}\psi) \tag{15d}$$

For m_0 we choose the mass of the *nucleons* as the most stable heavy particles.

With the particle creation rate r per *time* and *volume element*, the continuity equation of the particle density n takes the form

$$\dot{n} = -3(\dot{R}/R)n + r \tag{16}$$

whereas the field equations of gravitation are reduced in the case of the line elements (14) to the Friedman equation:

$$\dot{R}^2 = 8\pi G n m_0 R^2/(3 - \epsilon) \tag{17}$$

with

$$\epsilon = \begin{cases} 0, & \text{flat space} \\ +1, & \text{spherical space} \\ -1, & \text{hyperbolic space} \end{cases}$$

For the matter density ρ we have set for simplification

$$\rho = n m_0 \tag{18}$$

which represents a good matter model, if the kinetic and potential energy of the particles is small in comparison with their rest energy.

We note that with this matter model an effective equation of state is necessarily connected, which can be derived immediately from the continuity equation for the energy density. We shall discuss it later (see Section 7) since it is not necessary for integrating the field equations. After knowledge of the particle creation rate the differential equations (16) and (17) can be integrated immediately; the results are the functions $R(t)$ and $\rho(t)$, the latter according to (18), which will be compared with the observations.

5.2.2. PARTICLE CREATION RATE. For calculation of the particle creation rate we have to start from the equations (14) and (15). First we give the *stationary energy eigensolutions* of (15), restricting ourselves to the Robertson–Walker line element (14) with *spherical* three-dimensional space ($\epsilon = +1$) and *constant* radius of curvature R_0 (static Einstein Universe):

$$ds^2 = dt^2 - R_0^2\{d\chi^2 + \sin^2\chi(d\theta^2 + \sin^2\theta\, d\varphi^2)\} \tag{19}$$

We write the stationary energy eigensolutions which Schrödinger has found already in 1938, in the following abbreviation[12]:

$$\psi_{nkm} = \begin{cases} u_{nkm}(R_0; \chi, \theta, \varphi)e^{-iE_n t}, & E_n > 0, \quad n > 0 \\ v_{\hat{n}-k-m}(R_0; \chi, \theta, \varphi)e^{iE_{\hat{n}} t}, & E_n < 0, \hat{n} = -n > 0 \end{cases} \tag{20}$$

with the energy eigenvalues*

$$E_n = \sin(n)[(m_0^2 + n^2)/R_0^2]^{1/2} \tag{21}$$

and the quantum numbers

$$\begin{aligned} n &= \pm 3/2, \pm 5/2, \ldots \\ k &= \pm 1, \pm 2, \ldots \; |k| < |n| \\ m &= \pm 1/2, \pm 3/2, \ldots \; |m| < |k| \end{aligned} \tag{21a}$$

The positive energy eigenstates ($n > 0$, u_{nkm}) correspond later to the particle states, the negative energy eigenstates ($n < 0$, $v_{\hat{n}-k-m}$) to the antiparticle states.

The eigenfunctions u_{nkm} and v_{nkm} (we write in the following n instead of $\hat{n}$ with $n > 0$) satisfy the eigenvalue equations

$$\begin{aligned} Hu_{nkm} &= E_n u_{nkm} \\ Hv_{nkm} &= -E_n v_{nkm} \end{aligned} \tag{22}$$

with the hermitian Hamiltonian†

$$H = -i\gamma^4\gamma^a(\partial_a - \Gamma_a) + m_0\gamma^4 \tag{23}$$

Then after normalization of the probability density (15c) according to

$$\int j^\mu n_\mu \, dF_{t=\text{const}} = 1 \tag{24}$$

($n_\mu = \delta_\mu^4$) the eigenfunctions fulfill the orthonormality relations

$$\begin{aligned} \int \bar{u}_{nkm}\gamma^4 u_{n'k'm'} \, dF_{t=\text{const}} &= \delta_{nn'}\delta_{kk'}\delta_{mm'} \\ \int \bar{v}_{nkm}\gamma^4 v_{n'k'm'} \, dF_{t=\text{count}} &= \delta_{nn'}\delta_{kk'}\delta_{mm'} \\ \text{otherwise} &= 0 \end{aligned} \tag{25}$$

and represent a *complete* orthonormal function system. Accordingly the general *one-particle* state has the form ($c_{nkm'}d^*_{nkm}$ = const):

$$\psi = \sum_{n>0,k,m} [c_{nkm}u_{nkm}(R_0)e^{-iE_n t} + d^*_{nkm}v_{nkm}(R_0)e^{iE_n t}] \tag{26}$$

Now, the expansion of the Universe will be taken into account *first* by the substitution

$$R_0 \rightarrow R(t) \tag{27}$$

*In this section n is a quantum number and *not* the particle number density.
†Latin indices run from 1 to 3 only.

in $u_{nkm'}$, $v_{nkm'}$, E_n, and H. By this operation the relations (22), (24), and (25), and herewith the one-particle representation of (20), remain conserved and the energy (21) has exactly the behavior of that of a classical particle in the expanding Universe ($|\mathbf{p}| \cdot R = \text{const}$). This is in accordance with the fact that (20) after the substitution (27) represents a WKB approximation in the case of the time-dependent Robertson–Walker line element. But in spite of the substitution (27) the solutions (20) and the series (26) with *constant* expansion coefficients do not exactly fulfill the Dirac equation furthermore, because the total spinorial connection coefficients (15b) depend now on $\dot{R}$. Therefore we set *secondly* instead of (26) in the sense of a time-dependent perturbation ansatz

$$\psi = \sum_{n>0,k,m} \left[c_{nkm}(t) u_{nkm}(R) \exp\left(-i \int^t E_n \, dt'\right) \right. \\ \left. + d^*_{n\,km}(t) v_{nkm}(R) \exp\left(i \int^t E_n \, dt'\right)\right] \tag{28}$$

The differential equations for $c_{nkm}(t)$ and $d^*_{n\,km}(t)$ follow immediately from the Dirac equation (15a); one finds

$$\begin{aligned} \dot{c}_{nkm} &= -\sum_{n',k',m'} d^*_{n'k'm'} A_{nkm\; n'k'm'}(\dot{R}) \\ \dot{d}^*_{n\,km} &= \sum_{n',k',m'} c_{nkm'} A_{nkm\; n'k'm'}(\dot{R}) \end{aligned} \tag{29}$$

with the matrix elements

$$A_{nkm\; n'k'm'}(\dot{R}) = -\delta_{nn'}\delta_{-kk'}\delta_{-mm'} \frac{m_0}{2E_n^2} \frac{\dot{R}}{R^2} n \exp\left(2i \int^t E_n \, dt'\right) \tag{30}$$

Thereon we perform the transition to the *quantized* Dirac field in the usual way. We consider the spinorial field variable ψ as an operator and the field equations (15) as operator valued. Then we expand the operator ψ corresponding to (28) with respect to the one-particle energy eigenstates:

$$\begin{aligned} \psi &= \sum_{n>0,k,m} [\mathbf{c}_{nkm}(t) u_{nkm}(R) \exp\left(-i \int^t E_n \, dt'\right) + \mathbf{d}^\dagger_{nkm}(t) v_{nkm}(R) \exp\left(i \int^t E_n \, dt\right) \\ \bar{\psi} &= \sum_{n>0,k,m} [\bar{u}_{nkm}(R) \mathbf{c}^\dagger_{nkm}(t) \exp\left(i \int^t E_n \, dt'\right) + \bar{v}_{nkm}(R)\, \mathbf{d}_{nkm}(t) \exp\left(-i \int^t E_n \, dt'\right) \end{aligned} \tag{28'}$$

The operator-valued expansion coefficients $\mathbf{c}_{nkm}(t)$, $\mathbf{d}_{nkm}(t)$ fulfill the usual anticommutator relations for particle and antiparticle annihilation and creation operators

$$\begin{aligned} \{\mathbf{c},\mathbf{c}\} = \{\mathbf{d},\mathbf{d}\} = \{\mathbf{c},\mathbf{d}\} = \{\mathbf{c},\mathbf{d}^\dagger\} = \{\mathbf{c}^\dagger,\mathbf{d}\} = \{\mathbf{c}^\dagger,\mathbf{d}^\dagger\} = 0 \\ \{\mathbf{d}^\dagger_{nkm}, \mathbf{d}_{n'k'm'}\} = \{\mathbf{c}^\dagger_{nkm}, \mathbf{c}_{n'k'm'}\} = \delta_{nn'}\delta_{kk'}\delta_{mm'} \end{aligned} \tag{31}$$

and their differential equations are given by (29) in operator-valued form.

The fact that the matrix elements (30) are different from zero only in the cases

$n = n'$, $k = -k'$, $m = -m'$ means that particles and antiparticles will be created simultaneously with the same energy but with opposite momentums only; and this happens in a nonstationary universe alone because of the factor $\dot{R}$ in (30). There exists energy production in consequence of the broken time invariance of the metric, but no momentum production (no broken space symmetries).

Inserting the expansion (28′) into the operator-valued energy-momentum tensor (15d) of the Dirac field and integrating over the three-dimensional space we get the *energy operator*. With respect to (25) we find (n^μ normal of the three-dimensional hypersurface)

$$\begin{aligned} \mathbf{E} &= \int \mathbf{T}_{\mu\nu} n^\mu n^\nu \, dF_{t=\mathrm{const}} \\ &= \sum_{n>0,k,m} E_n(\mathbf{c}^\dagger_{nkm}\mathbf{c}_{nkm} - \mathbf{d}_{nkm}\mathbf{d}^\dagger_{nkm}) \end{aligned} \tag{32}$$

which is automatically diagonal; by normal ordering we get finally in view of (31)

$$:\mathbf{E}: = \sum_{n>0,k,m} E_n(\mathbf{c}^\dagger_{nkm}\mathbf{c}_{nkm} + \mathbf{d}^\dagger_{nkm}\mathbf{d}_{nkm}) \tag{33}$$

Accordingly the *particle number operator* is given in the usual form by

$$\mathbf{N} = \sum_{n>0,k,m} (\mathbf{c}^\dagger_{nkm}\mathbf{c}_{nkm} + \mathbf{d}^\dagger_{nkm}\mathbf{d}_{nkm}) \tag{34}$$

The particle creation out of the vacuum at an *arbitrary point of time* $t_0 = 0$ results as follows. One defines the vacuum state at t_0 in the usual way:

$$\mathring{\mathbf{c}}_{nkm}\,|0\rangle = \mathring{\mathbf{d}}_{nkm}|0\rangle = 0 \text{ with } \langle 0|0\rangle = 1 \tag{35}$$

whereas

$$|\mathring{p}_{nkm}\rangle = \mathring{\mathbf{c}}^\dagger_{nkm}|0\rangle, |\mathring{\bar{p}}_{nkm}\rangle = \mathring{\mathbf{d}}^\dagger_{nkm}|0\rangle \tag{36}$$

represents the one-particle and antiparticle state at t_0 normalized to unity. Then the number of particles and antiparticles at the time point $t(>t_0)$ produced out of this vacuum at t_0 is given by the following vacuum-expectation value of (34):

$$N(t) = \langle 0|\mathbf{N}(t)|0\rangle \tag{37}$$

wherein $\mathbf{N}$ is known after solving the differential equations (29) in operator-valued form. Then the *total* particle creation *rate* at the time $t_0 = 0$ is defined by the limit

$$\dot{N} = \lim_{t\to 0} N(t)/t \tag{38}$$

according to which the created particles are considered instantaneously as an *incoherent* mixture (classical matter).*

*The name "incoherent" procedure has been chosen because the correlations between the phases of the wave functions are taken into account only during the infinitesimal time interval $t(=dt)$. In contrast to this, within the "coherent" procedure the phase correlations remain conserved during the whole time evolution of the Universe.

The passage to this limit has the important advantage that the evolution of the radius of curvature $R(t)$ must be known only in the infinitesimal neighborhood of the time point t_0; therefore we can solve the differential equations (29) by an iteration method without loss of accuracy. After approximation of $R(t)$ by the tangent at the arbitrary point $t_0 = 0$

$$R(t) = R_0 + (\dot{R})_0 t \tag{39}$$

the matrix elements (30) take the form, restricting ourselves to the first order of $(\dot{R})_0$ (which is consistent within the first iteration step),

$$A_{nkm\,n'k'm'} = -\delta_{nn'}\delta_{-kk'}\delta_{-mm'} \frac{m_0}{2\overset{\circ}{E}{}_n^2} \left(\frac{\dot{R}}{R^2}\right)_0 n e^{2i\overset{\circ}{E}_n t} \tag{40}$$

Herewith and with the initial values at t_0 [see (35) and (36)] we find in the *first* integration step from the differential equations (29)

$$\begin{aligned} c_{nkm}(t) &= \overset{\circ}{c}_{nkm} + \overset{\circ}{d}{}^{\dagger}_{n-k-m} \frac{m_0}{4i\overset{\circ}{E}{}_n^3} \left(\frac{\dot{R}}{R}\right)_0 \frac{n}{R_0} (e^{2i\overset{\circ}{E}_n t} - 1), \\ d^{\dagger}_{nkm}(t) &= \overset{\circ}{d}{}^{\dagger}_{nkm} + \overset{\circ}{c}_{n-k-m} \frac{m_0}{4i\overset{\circ}{E}{}_n^3} \left(\frac{\dot{R}}{R}\right)_0 \frac{n}{R_0} (e^{-2i\overset{\circ}{E}_n t} - 1). \end{aligned} \tag{41}$$

Herewith and with the use of (34) we get finally from (37) in view of (35) and (36) the following number of particles produced after the time t out of the vacuum at $t_0 = 0$ [$\overset{\circ}{E}_n$ is inserted according to (21)]:

$$N(t) = \frac{1}{2} \sum_{n=3/2}^{\infty} m_0^2 \left(\frac{\dot{R}}{R}\right)_0^2 \frac{n^2}{R_0^2} \left(n^2 - \frac{1}{4}\right) \frac{\sin^2(t\{[(m_0^2 + n^2)/R_0^2]^{1/2}\})}{[(m_0^2 + n^2)/R_0^2]^3} \tag{42}$$

For evaluation of the sum in (42) we go over to the integral. In case of $R_0 \gg \hbar/m_0 c$ we get

$$N(t) = \frac{m_0^2}{2} \left(\frac{\dot{R}}{R}\right)_0^2 R_0^3 t \int_{m_0 t}^{\infty} x^{-5}(x^2 - m_0^2 t^2)^{3/2} \sin^2 x \, dx \tag{43}$$

and herewith we find according to (38) for the *total particle creation rate:*

$$\dot{N} = \frac{\pi}{4} m_0^2 \left(\frac{\dot{R}}{R}\right)_0^2 R_0^3 \tag{44}$$

valid at every point of time, so we set $R_0 \to R$ and $(\dot{R}/R)_0 \to \dot{R}/R$ in the following.

This result has three important features: (a) Similarly to the coherent result (9) the particle creation increases with increasing rest mass m_0 of the created particles and in the limiting case $m_0 \to 0$ the creation vanishes. Therefore no *neutrino* (or photon) *production* occurs which depends on the fact that the metrics (1) and (14) are conformally flat and (2) and (15) are conformally invariant for $m_0 \to 0$. Thus no contradiction to the observation of the background radiation is to be expected in contrast to the theories of Dirac and Jordan or Hoyle, Bondi, and Gold. But in anisotropic world models neutrino and photon production happens and leads to an isotropization of the models.[13] Perhaps this process is the real origin of the isotropy of our Universe, although anisotropies are damped also by the pure expansion of the Universe itself. Then the present isotropy would not be a de facto assumption but a necessary consequence of the physical laws. That the particle creation depends on the rest mass of the created particles in spite of the validity of the equivalence principle is a typical quantum mechanical effect. Only in classical physics do the gravitational effects not depend on the mass of the (test) particles because of this principle.

(b) The particle creation represents an *irreversible* process, because it is independent of the sign of the Hubble parameter $H = \dot{R}/R$. Therefore in the expanding as well as in a contracting universe particle creation happens proportional to the square of the Hubble parameter together with an increasing of entropy. This irreversibility is caused immediately by the use of the incoherent method.

(c) The particle creation rate is proportional to the volume $2\pi^2 R^3$ of the spherical three-dimensional space. Consequently by division through this volume we get the *particle creation per time and volume element* in the simple form

$$r = \frac{\dot{N}}{2\pi^2 R^3} = \frac{m_0^2}{8\pi}\left(\frac{\dot{R}}{R}\right)^2 \triangleq \frac{1}{8\pi c}\left(\frac{m_0 c}{\hbar}\right)^2\left(\frac{\dot{R}}{R}\right)^2 \tag{45}$$

if we go back finally to CGS units.

It is remarkable that the expression (45) allows the transition to the open universes of the line element (14) with *flat* or *hyperbolic* three-dimensional space, because in the limit of infinite space volume the quantity $\dot{R}/R$ remains (in contrast to R in the flat space) well defined and has the meaning of the Hubble parameter only. For the flat space it is proved that this heuristic argument is correct.

5.2.3. NONSTANDARD MODELS. Now we go back to the equations (16) and (17). Together with the result (45) they take the form

$$\dot{n} = -3\frac{\dot{R}}{R}n + \frac{1}{8\pi c}\left(\frac{m_0 c}{\hbar}\right)^2\left(\frac{\dot{R}}{R}\right)^2 \tag{46}$$

$$\left(\frac{\dot{R}}{c}\right)^2 = \frac{8\pi G}{3c^2} n m_0 R^2 - \epsilon \tag{47}$$

For obtaining a self-consistent cosmological solution where the metric, which means the behavior of the distance parameter R (radius of curvature in the case $\epsilon = 1$), is determined by the particles created by the time dependence of R, the coupled differential equations (46) and (47) must be solved. An exact solution has been found only in the case of vanishing curvature ($\epsilon = 0$). In both the other cases only the asymptotic behavior for small and large values of the distance parameter

$$R \underset{\gg}{\ll} 3\frac{\hbar c}{m_0^2 G}\frac{\hbar}{m_0 c} \simeq 10^{25}\text{ cm} \tag{48}$$

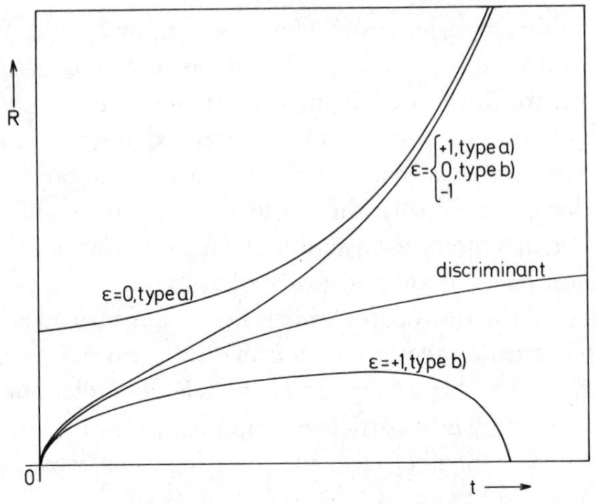

FIGURE 2. The different solution types in case of the nonstandard models.

can be calculated analytically, whereas the intermediate range must be determined by the use of a computer. The results are drawn in Fig. 2.

Evidently the behavior of $R(t)$ depends on initial values, especially on the initial value of the number of particles, as the different solution types show. But we note that the later behavior of the different solutions coincides, which means that the asymptotic behavior for large times is determined by the created particles alone and their asymptotic number does not depend on the initial values practically. Furthermore we note, that—as in the standard models—there follows from (47) in view of (18) a critical matter density $\rho_c = (3/8\pi)H^2/G = 6 \times 10^{-30}$ g/cm^3 ($H = \dot{R}/R = 55$ km/sec Mpc most probable present Hubble parameter), so that the relation

$$\rho \begin{matrix} > \\ = \\ < \end{matrix} \rho_c \leftrightarrow \epsilon = \begin{matrix} +1 \\ 0 \\ -1 \end{matrix} \tag{49}$$

is valid, according to which the geometry of the Universe can be determined by measuring ρ and H in principle. However, because of the similar asymptotic behavior of the solutions a very exact determination of ρ and H would be necessary in this region for deciding between the different geometries.

In the case of the *flat space* ($\epsilon = 0$) there exist *two* solution types: The first one (type a) starts at $t = -\infty$ from the *empty Minkowski space-time*, the second one (type b) begins with a Big Bang as the corresponding standard model; both solutions go over finally in an accelerated Universe (decelerating parameter $q = -1$) with exponentially increasing distance parameter.* In the case of type (a) the *whole matter* of

*In the case of type (a) there exist more precisely two solutions: The first one possesses an exponentially increasing distance parameter from the very beginning at $t = -\infty$ which is totally identical with the behavior of the steady-state theory. The second one has at $t = -\infty$ a *finite* distance parameter with vanishing time derivative and only *asymptotically* for "large times" an exponentially increasing distance parameter as described above. We discuss in the following the *second* case only.

the Universe *is created by the expansion* of the Universe alone. Furthermore it shows that the empty space is unstable with respect to the properties of the vacuum.

In the case of the *hyperbolic space* ($\epsilon = -1$) we find *one* solution starting from a Big Bang as in the standard model and going over asymptotically in an accelerated solution identical with that of the flat space.

In the case of the *spherical space* ($\epsilon = +1$) *two* solution types are found: The first one (type a) has a behavior similar to the solution of the hyperbolic space, the second one (type b) represents the *deformed* cycloid solution of the standard model, deformed in consequence of the irreversible particle creation. Between both types a *discriminant* solution exists with a critical initial number of particles:

$$N_c = \frac{9\pi}{4}\left(\frac{\hbar c}{m_0^2 G}\right)^2 \simeq 2 \times 10^{77} \tag{50}$$

If the initial number of particles is smaller than N_c, we get a solution of type (b); otherwise we get a solution of type (a).

We note explicitly that in the nonstandard models an *acceleration* of the Universe ($\ddot{R} > 0$) is possible and not in contradiction with any curvature. Moreover, the possibility of accelerated Universes within the framework of the nonstandard models is of essential interest, because J. Gunn and B. Tinsley have emphasized[14] that an acceleration of the Universe is best in accordance with the observations of the magnitude-red-shift relation.

Finally in the case of positive curvature ($\epsilon = +1$) the possibility is given to start with $N < N_c$ and to join one deformed cycloid (type b) to another avoiding any singularity according to the mechanism described in Section 3 till in consequence of the permanent particle creation $N \geq N_c$ is valid and the solution goes over finally into the discriminant or into type (a) (see Fig. 3). There exists no methodological difficulty in starting this breathing Universe with increasing period at the time $t = -\infty$ with vanishing volume, mass, and entropy. After an infinite number of aeons the asymptotic solution [type (a) or discriminant] is achieved with finite entropy and mass, which is created out of the vacuum alone.

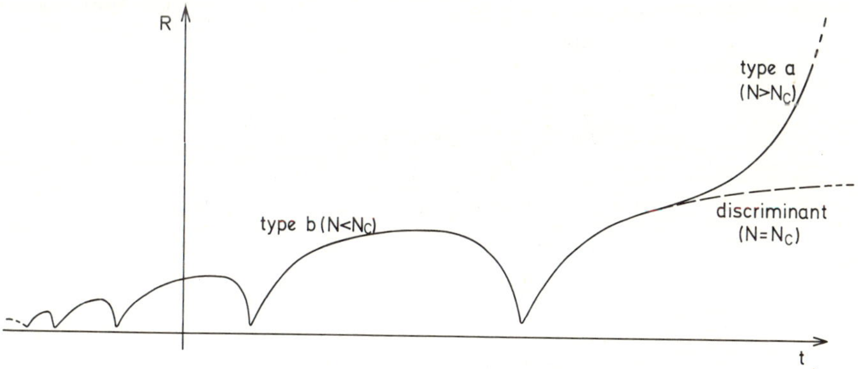

FIGURE 3. Breathing universe with increasing entropy in case of positive curvature ($\epsilon = +1$).

Therefore it may be justified to restrict ourselves to quantitative discussions of the *asymptotic* solutions valid for

$$t \gg 6 \frac{\hbar c}{m_0^2 G} \frac{\hbar}{m_0 c^2} \simeq 2 \times 10^7 \text{ yr} \mathrel{\hat{=}} R \gg 10^{25} \text{ cm} \tag{48'}$$

[cf. (48)]. According to Fig. 2 we find *for all three curvature types* the following result of an *accelerated* Universe:

$$\begin{aligned} R &\sim \exp\left(\frac{1}{9}\frac{m_0^2 G}{\hbar c}\frac{ct}{\hbar/m_0 c}\right) \\ \rho &= nm_0 = \frac{8}{\pi}\frac{m_0^2 G}{\hbar c}\left(\frac{m_0 c}{12\hbar}\right)^3 m_0 = 1.6 \times 10^{-24} \frac{\text{g}}{\text{cm}^3} \\ H &= \frac{\dot{R}}{R} = \frac{1}{9}\frac{m_0^2 G}{\hbar c}\frac{m_0 c^2}{\hbar} = 3 \times 10^4 \text{ km/sec Mpc} \\ r &= \frac{c}{648\pi}\left(\frac{m_0^2 G}{\hbar c}\right)^2\left(\frac{m_0 c}{\hbar}\right)^4 = 2 \times 10^{-15} \text{ cm}^{-3} \text{ sec}^{-1} \end{aligned} \tag{51}$$

Although the value of $r(= 10^6$ particles per year and km^3) is too small for direct observations, the values of ρ and H are too great in comparison with the observational data ($\rho \simeq 10^{-29}$ g/cm^3, $H = 55$ km/sec Mpc). But we note that in case of the flat space solution starting from the empty flat Minkowski space-time the values (51) are upper limits, so that in this case no contradiction to the observations can be concluded. On the contrary there exists an *intermediate* range, in which ρ and H take the observational values very well according to the flat space condition $\rho = \rho_c$ [cf. (49)]. But this solution without Big Bang represents a cold birth of the Universe, so that serious difficulties could appear with respect to the background radiation and to the origin of the light chemical elements.

In the *asymptotic* region mass density and Hubble parameter are constant with time: The dilution of matter in consequence of the expansion is compensated exactly by the particle creation. Obviously these asymptotic solutions are identical with that of the steady-state theory, although they have a different theoretical basis.

Finally we consider the decelerated discriminant solution of *positive* curvature ($\epsilon = +1$, $\ddot{R} < 0$), which also represents a good approximation of the solution of type (a) and (b) in the intermediate range. We find in the case of (48′)

$$\begin{aligned} R &= \left(6\frac{\hbar c}{m_0^2 G}\frac{\hbar}{m_0 c}\right)^{1/2} (ct)^{1/2} \\ \rho &= \frac{m_0}{16\pi(\hbar/m_0 c)^2}\frac{1}{ct} \\ N &= \frac{9\pi}{2\sqrt{6}}\left(\frac{\hbar c}{m_0^2 G}\right)^{3/2}\left(\frac{m_0 c}{\hbar}\right)^{1/2} (ct)^{1/2} \\ r &= \frac{1}{32\pi c}\left(\frac{m_0 c}{\hbar}\right)^2 \frac{1}{t^2} \end{aligned} \tag{52}$$

where N means the *total* number of particles of the Universe. We note that according to (52) the Mach relation between mass and radius of the Universe

$$MG/Rc^2 = \tfrac{3}{4}\pi = \text{const} \tag{52a}$$

($M = Nm_0$ mass of the Universe) is valid during the main part of the evolution of the Universe (in contrast to the situation in the standard model of positive curvature, where this relation is fulfilled only for the maximum value of the distance parameter).

Because in the case of the solution (52) the age of the Universe is given by

$$t = (2H)^{-1} = 9 \times 10^9 \text{ yr} \tag{53}$$

the present values of the quantities (52) can be expressed by the Hubble parameter and elementary natural constants as follows:

$$\begin{aligned}
R &= \left(\frac{3\hbar c}{m_0^2 G}\frac{\hbar}{m_0 c}\frac{c}{H}\right)^{1/2} = 4 \times 10^{26} \text{ cm} \\
\rho &= \frac{m_0}{8\pi}\frac{H}{c(\hbar/m_0 c)^2} = 9 \times 10^{-27} \text{ g/cm}^3 \\
N &= \frac{9\pi}{4\sqrt{3}}\left(\frac{\hbar c}{m_0^2 G}\right)^{3/2}\left(\frac{m_0 c}{\hbar}\frac{c}{H}\right)^{1/2} = 8 \times 10^{78} \\
r &= \frac{1}{8\pi c}\left(\frac{m_0 c}{\hbar}\right)^2 H^2 = 9 \times 10^{-21} \text{ cm}^{-3}\text{ sec}^{-1}
\end{aligned} \tag{54}$$

Accordingly the *present* particle creation rate amounts to 300 particles per year and km³ only and lies beyond the observational limit. The order of magnitude of the values is more or less in accordance with the observations, although the value of the matter density is greater than the observational one by at least three orders of magnitude.*

Comparing the particle number N in (54) with the initial particle number N_c for obtaining the discriminant solution, equation (50), we find

$$\frac{N}{N_c} = \left(\frac{m_0^2 G}{3\hbar c}\right)^{1/2}\left(\frac{m_0 c}{\hbar}\frac{c}{H}\right)^{1/2} \sim 40 \tag{55}$$

This means that up to 97% of the present matter in the Universe would be created by the expansion during the later stage; only 3% of the matter must be produced during foregoing aeons.

In order to perform all these self-consistent calculations in which the back-reaction of the created particles on the geometry of the Universe is included, we have introduced the assumption of a *localized* particle creation rate. But the mechanism, which causes the rate and herewith the incoherence of matter, is *unknown*. It may be that it is the same mechanism which causes the irreversibility in the thermodynamics.

*Compare, however, the remark (b) of Section VII.

6. The Most Probable Model of the Universe

Today several models for the Universe are known, but there exists only one Universe. Here the question arises, which represents the most satisfactory description of the Universe, and in connection with the foregoing considerations, what is the meaning of a "satisfactory" description?

In view of the uniqueness of the Universe, *we postulate* that its present observable properties follow necessarily from physical laws and not from contingent de facto assumptions as *boundary conditions* and *initial values*. We want to know not only *how* the Universe is constituted but also *why* it is so. Only in this way may the aporias of cosmology be avoided and a selection of the most probable model of the Universe be achieved.

Boundary conditions in the spacelike infinity can be avoided in the framework of general relativity in the case of finite three-dimensional spaces. However, the field equations of gravitation are partial differential equations and therefore initial conditions cannot be avoided in principle. Then our postulate means that the properties of later stages of the Universe should not depend noticeably on initial values. The Universe should "forget" the details of its own early history.*

Macroscopic physical systems with such a property are known within the thermodynamics for the case of increasing entropy. Therefore it seems to us that our cosmological postulate can be satisfied under the condition of increasing entropy of the Universe, whereby primordial irregularities are smoothed out additionally also by the expansion itself. And indeed, the nonstandard models support this supposition.

According to this a "satisfactory" model of the Universe should fulfill the following conditions:

1. Independence of the later stages from initial and boundary conditions. This presupposes:
   a. Finiteness of 3-dimensional space,
   b. Increasing of entropy.
2. Subsequently accordance with physical and astronomical experiences is to demand. This means especially
   a. Homogeneity and isotropy on a large scale for the present visible part of the Universe;
   b. Expansion of the Universe and an "age" of approximately 10^{10} yr;
   c. Quantitative explanation of the mean matter density of approximately 10^{-29} g/cm^3 and of its chemical composition.
   d. Quantitative explanation of the temperature of the background radiation of 2.7 K (no singularity, but an early stage of high density and temperature).
   e. Explanation of the formation of the small scale structure as galaxies or clusters of galaxies.

*We note that our assumption is incompatible with B. Carter's "Anthropic Principle." In contrast to this our standpoint is more that of C. Misner's "Chaotic Cosmology."

We are far away from a model which fulfills *all* these points satisfactorily and in detail. But we are able to exclude many models which are in conflict with one or more conditions, for instance the standard models. On the other hand, it seems to us that the nonstandard models satisfy the conditions best, especially that of Fig. 3, although some points are not clarified until now.

7. Conclusions

We point first to three interesting properties of the nonstandard models:

(a) By the continuity equation for the energy density

$$\dot{p}c^2 + 3\frac{\dot{R}}{R}(pc^2 + \gamma) = 0 \tag{56}$$

the pressure of the material substratum is defined. According to this energy creation is necessarily connected with a *negative* pressure in the *expanding* Universe. In our case we find from (56) with respect to (46) [see also (18)]

$$p = -\frac{m_0 c}{24\pi}\left(\frac{m_0 c}{\hbar}\right)^2 \frac{\dot{R}}{R} \tag{57}$$

In the case of the exponentially increasing radius of curvature of our asymptotic solutions this negative pressure is constant and responsible for the existence of this solution type within Einstein's theory. Such a *constant* negative pressure can be interpreted as a cosmological constant Λ. Together with (57) and (51) one gets

$$\Lambda = -\frac{8\pi G}{c^4}p = \frac{1}{108}\left(\frac{m_0 c}{\hbar}\right)^2\left(\frac{m_0^2 C}{\hbar c}\right)^2 = 3.5 \times 10^{-53}\ \mathrm{cm}^{-2} \tag{58}$$

in good agreement with the value assumed by Einstein and Lemaitre. This result points to the possibility that the cosmological constant, the physical meaning of which is unknown hitherto, describes in the framework of the classical theory those phenomena which become important under consideration of the effects of the vacuum described by the quantum field theory.

(b) In view of the fact, that according to Heisenberg's uncertainty relation, a particle of the mass m_0 needs the "creation time" $\hbar/m_0c^2$ (Compton time) the differential quotient (38) should be substituted more correctly by the difference quotient with respect to the Compton time. This causes a correction of the particle-creation rate (45) by a factor $\gamma \lesssim 1$, the exact value of which could not be calculated till today, because in the case of the difference quotient the higher iteration steps in Section 5.2.2 become essential. This factor, however, influences also the results (51) and (54) in such a way that the accordance with the observations is improved.

(c) Finally the nonstandard models have the property that *particle–antiparticle pairs* have been created up to today, so that the Universe would contain equal parts of matter and antimatter, which is in agreement with the well-known ideas of H.

Alfvèn and R. Omnes.[15] The high number of photons of the background radiation suggests that the early Universe has consisted of equal parts of particles and antiparticles. Only for the tiny fraction of the *heavy* particles there appears a serious difficulty: Because the overwhelming majority of matter of our Galaxy seems to consist of normal matter, a mechanism of separation of heavy particles and antiparticles should have been effective at the formation time of galaxies or even earlier. Because the particle–antiparticle creation continues during the whole evolution of the Universe, a certain amount of *intergalactic* matter and antimatter should exist.

These considerations suggest a quantitative explanation of the origin of the photons of the background radiation. Their creation is not immediately possible with the help of the generation mechanism discussed above, because in the limiting case $m_0 = 0$ the creation rate vanishes. We think that most of the primordial particle–antiparticle pairs generated at earlier times are annihilated into photons, which have been thermalized subsequently by scattering. Then also the relatively high values of the matter density in (51) and (54) would be reduced, so that the accordance with the observation would be better too. But a detailed calculation in this direction does not exist.

Nevertheless, the nonstandard models represent an essential cognitive progress which is achieved by combination of macro- and microphysics. The old idea of cosmological particle creation would be established on the usual physical laws. Because the particle creation turns out to be an irreversible process, the main objection to the standard models does not exist any longer. Furthermore, initial values and the existence of a solitary Big Bang are avoided in the case of the flat space solutions of type (a) and for the spherical space solution of Fig. 3. Here the properties of the Universe follow indeed from natural laws *alone*, and the cosmological data are given largely by elementary natural constants.

Although the nonstandard models are able to remove fundamental difficulties of the standard models, there is no doubt that herewith the aporias of cosmology are overcome not at all. One difficulty arises already from the description of the unique Universe by partial differential equations. As we have seen, contingent initial conditions cannot be avoided in principle, but only their influence on later stages can be diminished by irreversible processes.

Therefore it seems to us that aporias or antinomies are characteristic phenomena of cosmology *as such*. We show this phenomenon especially with the help of the two world-models, which we have emphasized to be the most satisfactory ones:

(1) Model $\epsilon = 0$, type (a): The initial state ($t \rightarrow -\infty$) is the vacuum of the Minkowski space-time. Out of this elementary particles (which may annihilate partially) are generated quasi without cause. This feature is similar to the unknown origin of irreversibility in the thermodynamics, whereby both phenomena may have the same root. After a short intermediate stage the Universe goes over into the final state of the steady-state model. Initial and final state are unsatisfactory in the same manner. It is also difficult to understand how a *finite* time interval of the essential evolution of the Universe is possible between two infinite quasistationary stages.

(2) Model $\epsilon = +1$, Fig. 3: This model unquestionably possesses dynamics. After an infinite number of aeons (type b) there follows finally after a *last* Big Bang the transition into an accelerated steady-state solution (type a) or into the decelerated discriminant solution. However, the initial state ($t \rightarrow -\infty$) is totally paradoxical: The evolution of the Universe starts exactly from the "naught" ($R = 0$). The Universe has "history" so long as its behavior is that of the type (b) solution (breathing Universe); finally it goes over into the "historyless" steady-state phase.

It is our opinion that cosmological aporias of this or similar kind are unavoidable within the present state of physics. It is difficult to find a reason for this situation. Of course it is justified to restrict the cosmological investigations to those parts of the Universe which are accessible by experiences. On the other hand there are no real reasons why the theory should not pass over these limits. For instance, the question of finiteness or infinity of space is probably not a mock problem. Perhaps these problems are connected immediately with the questions of stability of systems of stars or galaxies. We state that the present cosmology is unable to give a unique or at least a satisfactory answer to questions concerning the Universe as a whole. Most of these questions have philosophical aspects, and we do not try to solve these problems.

References and Notes

1. S. Hawking and G. Ellis, *Astrophys. J.* **152**, 25 (1968); R. Penrose, *Phys. Rev. Lett.* **14**, 57 (1965).
2. H. Dehnen and H. Hönl, *Astrophys. Space Sci.* **33**, 49 (1975).
3. P. A. M. Dirac, *Nature* **139**, 323 (1937); *Proc. R. Soc. London Ser. A* **165**, 199 (1938); P. Jordan, *Schwerkraft und Weltall* (Fr. Vieweg-Verlag, Braunschweig, 1945).
4. F. Hoyle, *Mon. Not. R. Astron. Soc.* **108**, 372 (1948); H. Bondi and T. Gold, *ibid.* **108**, 252 (1948)
5. H. Hönl and H. Dehnen, *Z. Astrophys.* **68**, 181 (1968); P. Jordan, *ibid.* **68**, 201 (1968).
6. E. Schrödinger, *Physica* **6**, 899 (1939); B. de Witt, *Phys. Rev.* **90**, 357 (1953).
7. H. Urbantke and R. Sexl, *Acta Phys. Austriaca* **26**, 339 (1967).
8. S. G. Mamaev, V. M. Mostepanenko and A. A. Starobinskii, *Sov. Phys.-JETP* **43**, 823 (1976).
9. G. Schäfer and H. Dehnen, *J. Phys. A, Math. Gen.* **13**, 517 (1980).
10. J. Audretsch, *Phys. Lett.* **72A**, 401 (1979).
11. G. Schäfer and H. Dehnen, *Astron. Astrophys.* **54**, 823 (1977); *ibid.* **61**, 671 (1977).
12. E. Schrödinger, *Pontifica academia scientiarum, Commentationes* **2**, 321 (1938).
13. V. N. Lukash and A. A. Starobinskii, *Sov. Phys.-JETP* **39**, 742 (1974).
14. J. E. Gunn and B. M. Tinsley, *Nature* **257**, 454 (1975).
15. H. Alfvén, *Rev. Mod. Phys.* **37**, 652 (1965); R. Omnès, *Phys. Rep.* **3C**, 1 (1972).

12

On Relativistic Quantum Theory

L. P. Horwitz*
Department of Physics and Astronomy
Tel Aviv University
Ramat Aviv 69978, Israel

1. Introduction

The construction of a formalism for the description of the dynamical properties of many-body systems which is consistent with special relativity, in the framework of classical or quantum theory, is still far from complete. Dirac's proposal,[1] in 1949, of Hamiltonian-type theories, developed[2,3] into an action-at-a-distance theory with the Lorentz group canonically represented. Currie, Jordan, and Sudarshan,[4] however, showed that when the canonical coordinates are taken as the physical coordinates of point particles and their world lines are Lorentz invariant, the only possible dynamics is that of free particles. The history of these developments, and an invaluable collection of reprinted papers, has been provided by Kerner.[4] Rohrlich and King[5,6] have developed a Hamiltonian theory in which constraint equations control the masses of individual particles and also have the role of generating the evolution of the system.[7] For N-body systems, the requirements of compatibility imply the existence of N-body forces (corresponding to the potential functions appearing in the constraint equations).

In this article, I wish to discuss a formalism for relativistic dynamics which does not contain *a priori* constraints. As we shall see, the S matrix for scattering automatically contains a delta function constraining a quantity which is essentially the sum of squared masses; the masses of individual particles are not necessarily conserved. The theory which we shall study here therefore is capable of describing collisions in which the masses of the individual particles may be different before and after the interaction. The question of the conservation of individual masses is determined by the form of the interaction and the kinematics of the collision. If, under some circumstances, individual particle masses are conserved, one might expect the collision to be describable

*Research supported in part by the U.S.–Israel Binational Science Foundation (BSF), Jerusalem, Israel.

as well by a constrained Hamiltonian formalism.[6] We shall discuss the possibility of relating the two approaches in this case.

The theory that we shall study was first proposed by Stueckelberg[8] in 1941 for a single particle in a potential or an external electromagnetic field. In order to parametrize the evolution of a system in space-time, Stueckelberg suggested using a parameter τ which would play the role of the time t as it is used in the nonrelativistic (Galilean) dynamics. Piron and I extended this theory to the many-body case by assuming that the parameter τ is *universal*, and that the evolution of an entire system is parametrized by it.[9] Feynman, in his formulation[10] of this theory by use of path integral techniques, suggested that a different τ be used for each particle. The consequences of this supposition were not worked out by Feynman,[10] but as we shall see, it leads to conservation of individual particle masses[11,12] and a theory closely related to the constrained Hamiltonian formalism.

A generalization of the formalism discussed here has been given in which particles with spin can be treated.[13,14] In order to achieve this, a timelike vector n^μ is used to define a superselection rule. The physical states of the system are therefore described by $\psi_{\tau n\mu}(x) \in \mathcal{H}_{n\mu}$; Lorentz transformations carry one superselection sector $\mathcal{H}_{n\mu}$ to another, and induce Wigner rotations in each sector which are unitarily implemented by the induced representation $U(n^\mu)$ [a $(2S + 1) \times (2S + 1)$ matrix). Reuse[15] has given a review of the work done in exploring the properties of systems with spin, along with a discussion of the work done on bound states, which we shall not treat in this article. Fanchi and Collins[16,17] have recently proposed a probabilistic foundation for a spinless theory of the type considered here.

It is perhaps appropriate to present a review of some recently developed ideas associated with relativistic quantum theory in this volume prepared as a memorial to Wolfgang Yourgrau. His deep interests in relativity and in the development of the quantum theory would certainly have evoked his encouragement to efforts in this direction. This paper is dedicated to his memory, and the memory of the light that was shed by his work on science and its history.

2. Basic Notions

We shall describe the evolution of a classical relativistic particle as a sequence of events with space-time coordinates x^μ, parametrized by a universal "time" τ which we shall call the *historical time* to distinguish it from the geometrical time t. For an N-particle system, there is an N-fold sequence of events $x_1^\mu, x_2^\mu, \cdots x_N^\mu$, parametrized by the same τ. We postulate the existence of a function K of the coordinates at these events and the canonically conjugate momenta $p_1^\mu, p_2^\mu, \cdots, p_N^\mu$; an application[9] of the usual techniques yields the canonical equations

$$\frac{dx_a^\mu}{d\tau} = \frac{\partial K}{\partial p_{a\mu}}, \qquad \frac{dp_a^\mu}{d\tau} = -\frac{\partial K}{\partial x_{a\mu}} \tag{1}$$

For the description of a system of free particles, let us take

$$K_0 = \sum_{a=1}^{N} \frac{p_{a\mu} p_a^\mu}{2M_a} \tag{2}$$

where the M_a are parameters which may not coincide with $(-p_a^\mu p_{a\mu})^{1/2}$ [we shall use metric $g^{\mu\nu}$ $(-, +, +, +)$ and take $c = 1$]. For free particles, then,

$$\frac{dx_a^\mu}{d\tau} = \frac{p_a^\mu}{M_a} \tag{3}$$

i.e.,

$$\frac{d\mathbf{x}_a}{d\tau} = \frac{\mathbf{p}_a}{M_a} \qquad \frac{dt_a}{d\tau} = \frac{E_a}{M_a} \tag{4}$$

and hence, eliminating τ,

$$\frac{d\mathbf{x}_a}{dt_a} = \frac{\mathbf{p}_a}{E_a} \tag{5}$$

the equation relating velocity to energy and momentum according to Einstein. Since

$$\left(\frac{dx_a^\mu}{d\tau}\right)\left(\frac{dx_{a\mu}}{d\tau}\right) = -\left(\frac{ds_a}{d\tau}\right)^2 = \frac{p_a^\mu p_{a\mu}}{M_a^2} \tag{6}$$

we see that the proper time coincides with τ only if $p_a^\mu p_{a\mu} = -M_a^2$.

For the case of a system of particles moving in an external electromagnetic potential $A^\mu(x)$,

$$K = \sum_a \frac{[p_a^\mu - eA^\mu(x_a)][p_{a\mu} - eA_\mu(x_a)]}{2M_a} \tag{7}$$

and similar results are obtained (the situation for which interparticle interactions are taken into account will be discussed later).

Let us define the variable

$$m_a^2 c^2 \equiv -p_a^\mu p_{a\mu} \tag{8}$$

In the nonrelativistic limit, we shall insist that the quantity[18]

$$\varepsilon_a = E_a - M_a c^2 = c(\mathbf{p}_a^2 + m_a^2 c^2) - M_a c^2 \tag{9}$$

where M_a is the limiting Galilean mass of the nonrelativistic particle, be finite as $c \to \infty$ (compared to all velocities). It follows that

$$\varepsilon_a = \frac{\mathbf{p}_a^2}{2M_a} + (m_a - M_a)c^2 + O\left(\frac{1}{c^2}\right) \tag{10}$$

and hence $m_a = M_a + O(1/c^2)$. Identifying M_a with the numerical denominator in (2), we obtain

$$K_0 = \sum_a \frac{p_a^\mu p_{a\mu}}{2M_a} = \sum_a \frac{\mathbf{p}_a^2}{2M_a} - \sum_a \varepsilon_a - \frac{Mc^2}{2} - \sum_a \frac{\varepsilon_a^2}{2M_a c^2} \tag{11}$$

If we think of an action-at-a-distance type theory for which

$$K = K_0 + V(x_1, \ldots, x_N) \tag{12}$$

where V does not depend on the momenta (and, in general, only on relative coordinates), then

$$\frac{d(ct_a)}{d\tau} = -\frac{\partial K}{\partial (E_a/c)} = -\frac{\partial K_0}{\partial (\varepsilon_a/c)}) = c + \frac{\varepsilon_a}{m_a c} \tag{13}$$

and hence $ct_a = c\tau + ct_a(0) + O(1/c)$. Since we have assumed that V is translation invariant, in the nonrelativistic limit it depends on the ct_a only up to $O(1/c)$, and hence[18]

$$K = K_0 + V \sim H - \sum_a \varepsilon_a - \frac{Mc^2}{2} \tag{14}$$

where

$$H = \sum_a \frac{\mathbf{p}_a^2}{2M_a} + V(\mathbf{x}_1, \mathbf{x}_2, \ldots, \mathbf{x}_N) \tag{15}$$

the usual nonrelativistic Hamiltonian. In this limit, the equations of motion (1) become

$$\frac{dt_a}{d\tau} = 1, \qquad \frac{d\mathbf{x}_a}{dt} = \frac{\partial H}{\partial \mathbf{p}_a}, \qquad \frac{d\mathbf{p}_a}{dt} = -\frac{\partial H}{\partial \mathbf{x}_a}, \qquad \frac{dE_a}{dt} = \frac{\partial H}{\partial t} \tag{16}$$

where we have replaced τ by $t_a \equiv t$ everywhere.

For the quantum theory, one postulates commutation relations (we shall treat here only the case of spinless particles)

$$[q^\mu, p^\nu] = i\hbar g^{\mu\nu} \tag{17}$$

for operators (in an appropriate domain) of the Hilbert space $\mathcal{H} = L^2(\mathbb{R}^4; d^4x)$ of functions $\psi(x)$, i.e., for which

$$\int |\psi(x)|^2 \, d^4x < \infty \tag{18}$$

The function $\psi(x)$ is not a function of x^2 alone; it is not Lorentz invariant, although the integral (18) is. The quantity $|\psi(x)|^2$ has the interpretation of a probability density for *events*. One might ask about the physical meaning of an amplitude which must go to zero as $t \to \pm\infty$; this does *not* imply that particles disappear in time. The wave function is defined at a particular value of τ, and we may designate it as $\psi_\tau(x)$. At a larger value of τ, the center of the distribution may move to later values of t [as for the classical free particle equations (3)], and we see that the wave packet may sweep out a track in the four-dimensional space-time corresponding to a classical world line. To see how this works explicitly for a free particle, consider the evolution equation[8,9] (we take $\hbar = 1$)

$$i \frac{\partial \psi_\tau}{\partial \tau} = K\psi_\tau \tag{19}$$

Hence, for the free particle,

$$\psi_\tau(x) = \frac{1}{2\pi} \int \exp(ip^\mu x_\mu) \exp\left(-i \frac{p^\mu p_\mu}{2M} \tau\right) \hat{\psi}_\tau(p)\, d^4p \tag{20}$$

and the stationary phase point

$$x_c^\mu \sim \frac{\bar{p}^\mu}{M} \tau \tag{21}$$

follows the classical trajectory according to a space-time version of the Ehrenfest principle.

According to (6), one could take for τ the reading of the laboratory clock (by choosing $P_L^\mu P_{L\mu}/M_L^2 = -1$ for the laboratory); one then understands the attenuation in t of the wave function for values greater or less than $(\bar{E}/M)\tau$ as the statement that the occurrence at corresponding events much in advance or behind the laboratory time (if the energy is not too large) is unlikely.

Since the wave packet has a coherent extension in t, one can consider the possibility of interference phenomena in time.[19]

Note that although the stationary phase points generate invariant world lines, the distribution of probability around these lines is not invariant, since the ψ's are not invariant functions.

In order to discuss scattering theory we must consider another essential point, that of localizability. It is well known that the question of the position operator for a relativistic particle is a rather delicate one. Following Newton and Wigner,[20] let us study briefly the mass-shell Klein–Gordon theory, for which states are defined by functions $\phi(\mathbf{p}) \subset L^2(\mathbb{R}^3, d^3p/E)$, with $E = (\mathbf{p}^2 + M^2)^{1/2}$. The scalar product is defined by

$$(\phi_1, \phi_2) = \int \frac{d^3p}{E} \phi_1(\mathbf{p})^* \phi_2(\mathbf{p}) \tag{22}$$

Suppose that $\phi_2(p) \equiv \phi_0(p)$ is the wave function defining a particle at rest precisely at the point $\mathbf{x} = 0$. Then, this function should be orthogonal to the wave function $\exp(-i\mathbf{p} \cdot \mathbf{a})\phi_0(\mathbf{p}) \equiv \phi_a(\mathbf{p})$ describing a particle at rest at the point $\mathbf{x} = a$, for $\mathbf{a} \neq 0$. This requires

$$0 = \int \frac{d^3p}{E} e^{(i\mathbf{p}\cdot\mathbf{a})} |\phi_0(p)|^2$$

which implies $|\phi_0(p)|^2/E = \text{const}$, or (some argument about phases is required[20])

$$\phi_x(\mathbf{p}) \propto \sqrt{E}\, e^{-i\mathbf{p}\cdot\mathbf{x}} \tag{23}$$

The Fourier transform of this wave function is not concentrated at the point **x**, but is distributed over a range $\sim 1/M$. Furthermore,

$$i\frac{\partial}{\partial \mathbf{p}}\phi_{\mathbf{x}}(\mathbf{p}) = \mathbf{x}\phi_{\mathbf{x}} + i\frac{\mathbf{p}}{2E^2}\phi_{\mathbf{x}} \tag{24}$$

and hence $i\partial/\partial\mathbf{p}$ does not have the (generalized) eigenvalue **x**. It is clear from (24) that the relativistic position operator is

$$\mathbf{x}_{0p} = i\frac{\partial}{\partial \mathbf{p}} - i\frac{\mathbf{p}}{2E^2} \tag{25}$$

We must therefore ask whether the operator **x**, defined by (17) to be $i\partial/\partial\mathbf{p}$ in momentum space (via Fourier transform) really corresponds to the position of a particle that might arrive in a detector. Since the wave function $\psi(p) \in L^2(\mathbb{R}^4)$ contains many momentum components, we shall study the operator[9]

$$\mathbf{x}_{0p} = \mathbf{x} - \tfrac{1}{2}\{t, \mathbf{p}/E\} \tag{26}$$

where the symmetrized product is used since t, E do not commute, corresponding to the particle's rest position even after a finite time interval. The expectation value of this operator is

$$\int d^4p\, \psi^*(p)(\mathbf{x} - \tfrac{1}{2}\{t, \mathbf{p}/E\})\psi(p) = \int m\, dm \int \frac{d^3p}{E}\psi^*(\mathbf{p}, E)\left(i\frac{\partial}{\partial\mathbf{p}} - \frac{i}{2}\frac{\mathbf{p}}{E^2}\right)\psi(\mathbf{p}, E) \tag{27}$$

where in the right-hand side of this equation, $E = (\mathbf{p}^2 + m^2)^{1/2}$ and the variable m is defined, as in (9), by $m^2 = -p^\mu p_\mu$ [we have included only the positive root in (27)]. On the left side, **x** is represented by $i\partial/\partial\mathbf{p}$, but the derivative acts only on the space part of p^μ in ψ. On the right, $i\partial/\partial\mathbf{p}$ acts both on **p**, and the mass-shell value of E. The latter must be corrected for in redefining the derivative, and this correction compensates the extra term in

$$\mathbf{x}_{0p} = \mathbf{x} + i\frac{\mathbf{p}}{E}\frac{\partial}{\partial E} - \frac{i}{2}\frac{\mathbf{p}}{E^2} \tag{28}$$

obtained by using $t \to -i\partial/\partial E$ in (26). We therefore see that the coordinate operator contained in x^μ is *the covariant relativistic generalization of the Newton–Wigner operator*. There is no corresponding Newton–Wigner time operator, since $t \to -i\partial/\partial E = -i(E/m)\partial/\partial m$, and it cannot be defined within a given mass shell.

3. Scattering and the Wave Operator

As we have pointed out in the previous section, the Ehrenfest motion of free wave packets carries the sequence of events corresponding to a free particle to $t \to \pm\infty$ for $\tau \to \pm\infty$ [this follows, for example, from (21)]. If the kinematical conditions for a collision are appropriate for the system to develop as a scattering system, then streams of events at $\tau = -\infty$, evolving as free particles, should develop to streams of events at $\tau = +\infty$, again evolving as free particles. There is, of course, the possibility that a sequence of events would move from $t = +\infty$ at $\tau = -\infty$ to $t = -\infty$ at $\tau = +\infty$; this evolution corresponds to that of an antiparticle, and we shall not treat this case in the present work (see Ref. 12 for further discussion). If ψ is a physical state, at $\tau = 0$, of the scattering system, the condition that the particles move freely at $\tau \to +\infty$ is expressed by the asymptotic condition

$$\lim_{\tau \to +\infty} \| e^{-iK\tau}\psi - e^{-iK_0\tau}\phi_{\text{out}} \| = 0 \tag{29}$$

and similarly,

$$\lim_{\tau \to -\infty} \| e^{-iK\tau}\psi - e^{-iK_0\tau}\phi_{\text{in}} \| = 0 \tag{30}$$

where ϕ_{in}, ϕ_{out} are free states evolving under the noninteracting K_0 (a theory with modified free evolution operators could be constructed as well[21]). If the set of states ϕ_{in}, ϕ_{out} are dense, (29) and (30) imply the definition of the wave operators

$$\Omega_\pm = \lim_{\tau \to \pm\infty} e^{iK\tau} e^{-iK_0\tau} \tag{31}$$

Replacing the limit by an Abel-type integral, one obtains, as in the nonrelativistic theory (for Ω_- for example)

$$\Omega_- = \lim_{\epsilon \to 0+} \epsilon \int_T^\infty d\tau\, e^{iK\tau} e^{-iK_0\tau} e^{-\epsilon\tau} \tag{32}$$

which is equivalent to (31) if the limits exist. Integrating by parts,

$$\Omega_- = \lim_{\epsilon \to 0} \int_T^\infty \frac{d}{d\tau} (e^{iK\tau} e^{-iK_0\tau}) e^{-\epsilon\tau}\, d\tau + \text{finite}$$

and hence it is sufficient for the existence of this (isometric) operator that

$$\int_T^\infty \| V e^{-iK_0\tau}\phi \|\, d\tau < \infty \tag{33}$$

for $\phi \in D$ dense in $L^2(\mathbb{R}^4)$. Since

$$\langle x | e^{-iK_0\tau} | x' \rangle = -\left(\frac{M}{2\pi i} \right)^2 e^{i(x-x')^2 M/2\tau} \tag{34}$$

the integral would exist for $\phi \in L^1$, $V \in L^2$, as for the Galilean theory.[22] However, since K is invariant, the potentials V must be invariant functions. For the two-body problems, for example,

$$K = \frac{p_1^\mu p_{1\mu}}{2M_1} + \frac{p_2^\mu p_{2\mu}}{2M_L} + V(x_1 - x_2) \tag{35}$$

$$= \frac{P^\mu P_\mu}{2M_T} + \frac{p^\mu p_\mu}{2M} + V(x) \tag{36}$$

where we may take

$$p^\mu = p_1^\mu + p_2^\mu \tag{37}$$

$$p^\mu = \frac{M_2 p_1^\mu - M_1 p_2^\mu}{M_1 + M_2} \tag{38}$$

$$M_T = M_1 + M_2, \qquad M = \frac{M_1 M_2}{M_1 + M_2} \tag{39}$$

and $x = x_1 - x_2$. Since V is invariant, it must be a function of $x^2 = x^\mu x_\mu$. Since [taking the positive branch $t = (\mathbf{x}^2 + s^2)^{1/2}$]

$$\int V(x^2)^2\, d^4x = \int V(s^2)^2\, ds \int \frac{d^3x}{(\mathbf{x}^2 + s^2)^{1/2}} = \infty \tag{40}$$

$V(x^2)$ cannot be in L^2 (the calculation given here corresponds to a study of scattering in terms of relative motion, with the total center-of-mass motion removed). The crude estimates applicable to nonrelativistic scattering are therefore inadequate for the relativistic case.

Let us consider again the requirement (33), with ϕ a wave packet that will be suitably restricted. The proof that I will give here follows that of Ref. 23: Let

$$\begin{aligned}
\|Ve^{-iK_0\tau}\phi\|^2 &= \int d^4x\, V(x)^2 \left| \int d^4y\, e^{i(x-y)^2 M/2\tau}\phi(y)\right|^2 \cdot \left(\frac{M}{2\pi\tau}\right)^4 \\
&= \int d^4z\, V\left(\frac{z\tau}{M}\right)^2 \left| \int d^4y \left[\exp\left(i\frac{M}{2\tau}y^2\right) - 1 - i\frac{M}{2\tau}y^2 + 1 + \frac{iM}{2\tau}y^2\right]\right. \\
&\quad \times \left.\frac{1}{(2\pi)^2} e^{-izy}\phi(y)\right|^2 \\
&\le [\max_x V(x)^2] \int d^4z \left| \int d^4y \frac{1}{(2\pi)^2} e^{-iz\cdot y}\phi(y)\left[\exp\left(i\frac{M}{2\tau}y^2\right) - 1 - i\frac{M}{2\tau}y^2\right]\right|^2 \\
&\quad + \int d^4z\, V\left(\frac{z\tau}{M}\right)^2 |\hat{\phi}(z)|^2 \\
&\quad + \frac{M^2}{4\tau^2}\int d^4z \left[V\left(\frac{z\tau}{M}\right)\right]^2 \left|\int d^4y\, y^2 e^{-iz\cdot y}\phi(y)\right|^2
\end{aligned} \tag{41}$$

The first term is bounded by[23]

$$[\max_x V(x)^2] \int d^4z \left| \int d^4y \frac{1}{(2\pi)^2} e^{-iz\cdot y}\phi(y)\, i\left(\frac{M}{2\tau}y^2\right)^2 \Big/ 2\right|^2$$

and hence, if $(y^2)^2\phi(y) \in L^2$, we obtain the bound

$$O\left(\frac{1}{\tau^4}\right) \max_x V(x)^2 \tag{42}$$

The second term, if $\hat{\phi}(z)$ has support in $|z^2| \ge \epsilon > 0$ (no zero mass components), is bounded by

$$\max_{|z^2|\ge\epsilon>0} V\left(\frac{z\tau}{M}\right)^2 \tag{43}$$

The third term of (41) has a similar structure to the second, but in this case we must require that the Fourier transform of $y^2\phi(y)$ have support only in $|z^2| \geq \epsilon > 0$ [i.e., that $(\partial^2/\partial z^2)\hat{\phi}(z)$ have support away from zero mass]. With this restriction, one obtains the bound

$$O\left(\frac{1}{\tau^2}\right) \max_{|z^2|\geq\epsilon>0} V\left(\frac{z\tau}{M}\right)^2 \tag{44}$$

For bounded potentials $V(x)$, and ϕ in the dense set admitted by the above conditions, (43) provides the strongest restriction on the form of V. If $V(x) \sim 1/(x^2)^\alpha$ then $[1/(z^2)^\alpha]\tau^{2\alpha}$ must be integrable. Hence

$$\alpha = \tfrac{1}{2} + \delta, \qquad \delta > 0 \tag{45}$$

We have therefore shown that wave operators exist for a nontrivial class of potentials in the framework of relativistic scattering theory.[24] With the wave operators, one can define an S-matrix in the usual way,

$$\phi_{\text{out}} = \Omega^\dagger_{\text{-ms}}\Omega_+\phi_{\text{in}} = s\phi_{\text{in}} \tag{46}$$

The proof of the existence of wave operators is not sufficient to assure a good scattering theory. It is also necessary to show that the ranges of the wave operators are dense in the space of scattering states. For application to the many-body problem, the question of cluster decomposition should be studied. Although significant progress has been made in studying these problems in the nonrelativistic case in recent years, the essential complication of the relativistic theory, i.e., the long-range behavior of invariant functions (or, alternatively, the hyperbolic character of the differential operators involved), makes them somewhat more difficult in this case.

4. Cross Sections

I shall assume, in this section, that the fundamental problems of scattering theory have been solved, in principle, and will proceed to formulate the physical problem of scattering for the (reduced) two-body case in order to obtain an expression for a measureable cross section. It will be assumed, for simplicity, that no antiparticles are involved. We shall describe the phenomena in terms of the relative coordinates, defined in (37)–(39), between the incident and target particles. The wave packets leaving the accelerator are in the asymptotic region; their centers have the free particle trajectories

$$\begin{aligned} x^*_{a\,\mu}(\tau) &= x^*_{a\,\mu}(0) + \frac{p^*_\mu}{M}\tau \\ p^*_\mu(\tau) &= p^*_\mu = \text{const} \end{aligned} \tag{47}$$

Let us choose the direction of $\mathbf{p}^*$ to be the z axis in the laboratory frame (of the target). The wave packets start at $\tau = 0$ at a point $z_a = -L$, but the other coordinates transverse to the beam, x_a, y_a, t_a, have to be averaged (the time of the emission of a packet at $\tau = 0$ is subject to uncertainty as well as the spatially transverse directions; wave packets displaced in t, relative to the target time, will sample the invariant t-dependent potential at different points). Starting with one wave packet $\psi_{\rm in}(p)$, the different constituents of the beam are given by

$$\psi^a_{\rm in}(p) = \psi^{\rho,x^0}_{\rm in}(p) = e^{-i\rho\cdot \mathbf{p}} e^{ix^0p^0}\psi_{\rm in}(p) \tag{48}$$

It is difficult to completely "cover" the range of the potential with a finite ensemble of $\psi^a_{\rm in}$'s since the invariant potential is constant along hyperbolas in the transverse direction. However, all of these hyperbolas approach the light cone asymptotically, and at large distances along the light cone, the weight of the potential per unit "area" (two space, one time dimension) becomes small. This is the same phenomenon leading to the possibility of defining wave operators in the first place. Hence, the spread of the beam can be finite in space-time, and still essentially "cover" the potential due to the target.

The probability that a particle approaching the potential region with some in-asymptote $\psi^a_{\rm in}$ will be found after the interaction in the region d^4p around the 4-momentum p is given by

$$\omega(d^4p \leftarrow \psi^a_{\rm in}) = d^4p\,|\psi^{a\ \rm scatt}_{\rm out}(p)|^2 \tag{49}$$

We shall be interested here only in the direction of the outgoing momentum and in the energy, and we will therefore integrate over the magnitude $|\mathbf{p}|$ to get the probability of the particle emerging with energy in dp^0 around p^0 (proportional to $(p^0_1 - M_1)$ if $p^0_2 \approx M_2$) and 3-momentum in $d\Omega$ around $\mathbf{p}$:

$$\omega(d\Omega\, dp^0 \leftarrow \psi^a_{\rm in}) = d\Omega\, dp^0 \int_0^\infty dp\, p^2 |\psi^{a\ \rm scatt}_{\rm out}(\mathbf{p}, p^0)|^2 \tag{50}$$

The total number of observed scatterings into $d\Omega\, dp^0$ is the sum over the probabilities for each incident packet,

$$\begin{aligned} N_{\rm scatt}(d\Omega\, dp^0) &= \sum_a \omega(d\Omega\, dp^0 \leftarrow \psi^a_{\rm in}) \\ &\cong n_{\rm inc} \int d^2\rho \int dx^0\, \omega(d\Omega\, dp^0 \leftarrow \psi^{\rho,x0}_{\rm in}) \end{aligned} \tag{51}$$

where we have assumed that $n_{\rm inc}$, the number of packets per unit area and time, is essentially constant over the range needed to cover the target. Then, the cross section is

$$\sigma(d\Omega\, dp^0 \leftarrow \psi_{\rm in}) = \frac{N_{\rm scatt}(d\Omega\, dp^0)}{n_{\rm inc}} = \int d^2\rho \int dx^0\, \omega(d\Omega\, dp^0 \leftarrow \psi^{\rho,x0}_{\rm in}) \tag{52}$$

Note that the cross section is defined as a number divided by a density and not a rate divided by a flux (defining the rate as $N_{\rm scatt}/\Delta\tau$ and the flux as $n_{\rm inc}/\Delta\tau$, where $\Delta\tau$ is the pulse length in τ, the definitions would coincide). The cross section is *three-dimensional* (similar three-dimensional cross sections have been defined by J. H. Cooke[25] and J. L. Cook[26]), including the integration over the time of interaction.

To extract the scattered part of ψ_{out}, we shall need some properties of the S operator. The intertwining relations

$$\Omega_{\pm} K_0 = K \Omega_{\pm} \tag{53}$$

imply that

$$[S, K_0] = 0 \tag{54}$$

Hence,

$$\begin{aligned} \frac{p_{\text{out}}^2}{2M} &= \langle \psi_{\text{out}} | K_0 | \psi_{\text{out}} \rangle = \langle \psi_{\text{in}} | S^+ K_0 S | \psi_{\text{in}} \rangle \\ &= \langle \psi_{\text{in}} | K_0 | \psi_{\text{in}} \rangle = p_{\text{in}}^2 / 2M \end{aligned} \tag{55}$$

Since $P^2/2M_T$ is conserved as well, by translation invariance, we can add this quantity to both sides to conclude that $m_1^2/2M_1 + m_2^2/2M_2$ is conserved, but not necessarily the individual particle masses. For $M_2 \to \infty$,

$$p^2 = \frac{M_2^2 p_1^2 + M_1^2 p_2^2 - 2M_1 M_2 p_1 p_2}{(M_1 + M_2)^2} \sim -m_1^2 - M_1^2 + 2M_1 E_1 \tag{56}$$

implying that a particle scattering on a heavy target may change its mass with a corresponding change in the energy $\Delta E = (\frac{1}{2}M)\Delta m^2$. If the particle stays on the same mass shell, its energy will also be conserved.

In the limit in which the wave packets we have used approach improper eigenfunctions of the four-momentum,

$$\psi_{\text{out}}(p) = \int d^4p' \langle p | S | p' \rangle \psi_{\text{in}}(p') \tag{57}$$

Let us define, in analogy with the nonrelativistic scattering theory, the "on-shell" T-matrix elements

$$\langle p | S | p' \rangle = \delta^4(p - p') - 2\pi i \delta \left(\frac{p^2}{2M} - \frac{p'^2}{2M} \right) T(p \leftarrow p') \tag{58}$$

The second term contributes to the scattered wave required in (50). From (50), (52), and (58), we have

$$\begin{aligned} \frac{d\sigma(d\Omega\, dp^0 \leftarrow \psi)}{d\Omega\, dp^0} &= \int d^2\rho \int dx^0 \int_0^\infty dp\, |\mathbf{p}|^2 (2\pi)^2 \int d^4p' \int d^4p'' \\ &\times \delta \left(\frac{p^2}{2M} - \frac{p'^2}{2M} \right) \delta \left(\frac{p^2}{2M} - \frac{p''^2}{2M} \right) T(p^\mu \leftarrow p'^\mu) T^*(p^\mu \leftarrow p''^\mu) \\ &\times \exp[-i\rho \cdot (\mathbf{p}' - \mathbf{p}'') + i x^0 (p'^0 - p''^0)] \psi(p') \psi^*(p'') \end{aligned} \tag{59}$$

The integrations over ρ, x^0 result in δ functions and the p'' integration can be performed immediately. For wave packets narrow enough around p^* so that $p'_{11} + p'_{11} \neq 0$, where p'_{11} is the component of p' parallel to $\mathbf{p}$ (determined by Ω), one obtains[12]

$$\frac{d\sigma(d\Omega\, dp^0 \leftarrow \psi)}{d\Omega\, dp^0} = (2\pi)^5 M^2 \int d^4p' \frac{|\mathbf{p}|}{p'_{11}} |T(p \leftarrow p')|^2 \, |\psi(p')|^2 \tag{60}$$

where

$$|\mathbf{p}| = (p^{02} + p'^2)^{1/2} \tag{61}$$

Assume now that $\psi(p)$ is appreciably different from zero only in a small region around p^*, and that in this region the T-matrix elements do not change significantly. Under these conditions, (60) becomes, independently of the shape of the wave packet,

$$\frac{d\sigma(d\Omega\, dp^0 \leftarrow p^*)}{d\Omega\, dp^0} = (2\pi)^5 M^2 \frac{|\mathbf{p}|}{|\mathbf{p}^*|} |T(p \leftarrow p^*)|^2 \bigg|_{p^2=p^{*2}} \tag{62}$$

Starting from the unitarity of S, we may also derive an optical theorem[12]

$$\text{Im}T(p \leftarrow p) = \frac{|\mathbf{p}|_{\text{lab}}}{32\pi^4 M} \sigma_{\text{tot}}(p) \tag{63}$$

In the nonrelativistic limit, as we have pointed out before in the classical case (a similar result is valid in quantum case), relative times go to zero as $O(1/c^2)$, and $V(x)$ is independent of t. This implies an additional factor $\delta(p^0 - p^{0*})$ in the T matrix. Energy conservation implies mass conservation, so then $|\mathbf{p}| = |\mathbf{p}^*|$. The expression (60) then becomes

$$\begin{aligned} \frac{d\sigma(d\Omega \leftarrow \mathbf{p}^*)}{d\Omega} &= (2\pi)^5 M^2 \int dp^0\, \delta(p^0 - p^{*0})^2\, |T_{\text{NR}}(\mathbf{p} \leftarrow \mathbf{p}^*)|^2 \\ &= [2\pi\delta(p^{*0} - p^{*0})](2\pi)^4 M^2\, |T_{\text{NR}}(\mathbf{p} \leftarrow \mathbf{p}^*)|^2_{|\mathbf{p}^*|=|\mathbf{p}|} \end{aligned} \tag{64}$$

a result identical to the usual nonrelativistic cross section, but with a factor $2\pi\,\delta(0)$, corresponding to the accumulation of events from the ensemble of wave packets redundantly distributed over time in the integral in (59).

5. Perturbation Expansion

Having obtained formulas for the calculation of differential cross sections in terms of the S matrix, it is useful to have perturbative techniques to evaluate the relevant matrix elements. There is a significant amount of physical insight to be gained as well in investigating the perturbation expansion. The procedure[12] is completely analogous to that of the nonrelativistic theory. Again, we shall confine ourselves to action-at-a-distance potential models. The perturbation expansion for particles interacting through the electromagnetic potential would be of great interest (for direct comparison with quantum electrodynamics), but it has not yet been worked out.

Let us define the interaction picture by

$$\psi_{I\tau} = e^{iK_0\tau}\psi_\tau \tag{65}$$

The evolution of interaction picture wave functions is described by

$$i\frac{\partial\psi_{I\tau}}{\partial\tau} = V(\tau)\psi_{I\tau} \tag{66}$$

where (we assume V does not depend explicitly on τ)

$$V(\tau) = e^{iK_0\tau} V e^{-iK_0\tau} \tag{67}$$

The asymptotic conditions (29), (30) for scattering imply, for appropriate potentials, that $\psi_{I\tau}$ evolves from a constant, before interaction, $\lim_{\tau\to-\infty}\psi_{I\tau} = \phi_{\text{in}}$ to a constant $\lim_{\tau\to+\infty}\psi_{I\tau} = \phi_{\text{out}}$, after the interaction. The unitary operator $U_{I(\tau,\tau 0)}$ governing the evolution of $\psi_{I\tau}$ satisfies the differential equation

$$i\frac{d}{d\tau} U_I(\tau, \tau_0) = V(\tau)U_I(\tau, \tau_0) \tag{68}$$

which, with the initial condition $U_I(\tau_0, \tau_0) = 1$, is equivalent to

$$U_I(\tau, \tau_0) = 1 - i\int_{\tau 0}^{\tau} d\tau'\, V(\tau')U_I(\tau', \tau_0) \tag{69}$$

For small V, one may hope that the iterative expansion will converge:

$$U_I(\tau, \tau_0) = 1 - i\int_{\tau 0}^{\tau} d\tau'\, V(\tau') + (-i)^2\int_{\tau 0}^{\tau} d\tau' \int_{\tau 0}^{\tau'} d\tau''\, V(\tau')V(\tau'') + \cdots \tag{70}$$

It follows from the definitions given above that the S operator is defined by

$$S = \lim_{\substack{\tau\to+\infty \\ \tau 0\to-\infty}} U_I(\tau, \tau_0) \tag{71}$$

With the help of the iterative expansion (70), the matrix elements of S can be expressed as

$$\langle\phi|S|\psi\rangle = \sum_{n=0}^{\infty} \langle\phi|S^{(n)}|\psi\rangle \tag{72}$$

where

$$\begin{aligned}\langle\psi|S^{(n)}|\psi\rangle = (-i)^n \int_{-\infty}^{\infty} d\tau_1 \langle\phi|e^{iK_0\tau 1} V e^{-iK_0\tau 1} \int_{-\infty}^{\tau 1} d\tau_2\, e^{iK_0\tau 2} V e^{-iK_0\tau 2} \\ \cdots \int_{-\infty}^{\tau n-1} d\tau_n\, e^{iK_0\tau n} V e^{-iK_0\tau n}|\psi\rangle\end{aligned} \tag{73}$$

The last integral, over τ_n, in (73) will be convergent if ψ and V satisfy (33), the condition for the existence of the wave operator. In such an absolutely convergent integral, one can insert an exponential factor $e^{\epsilon\tau n}$ with the limit $\epsilon \to 0$ implied. In this form, we can replace ψ, ϕ by improper states, and the exponential factor will control the convergence of the integrals. Inserting intermediate states, we must then evaluate

$$\begin{aligned}\langle p'|S^{(n)}|p\rangle = (-i)^n \int_{-\infty}^{\infty} d\tau_1 \int_{-\infty}^{\tau 1} d\tau_2 \cdots \int_{-\infty}^{\tau n-1} d\tau_n \int d^4q_1 \int d^4q_2 \cdots \int d^4q_{n-1} \\ \times\, e^{ik(p')\tau 1}\langle p'|V|q_1\rangle e^{-ik(q1)}\, e^{-ik(q1)\tau 1}\, e^{ik(q1)\tau 2}\langle q_1|V|q_2\rangle e^{-ik(q2)\tau 2} \\ \cdots e^{-ik(qn-1)\tau n-1}\, e^{ik(qn-1)\tau n} e^{\epsilon\tau n}\langle q_{n-1}|V|p\rangle e^{-ik(p)\tau n}\end{aligned} \tag{74}$$

where $k(q) = q^2/2M$. In carrying out the τ_n integration, one obtains

$$\int_{-\infty}^{\tau_{n-1}} d\tau_n\, e^{ik(q_{n-1})\tau_n} e^{\epsilon\tau_n} e^{-ik(p)\tau_n} = i\,\frac{\exp\{i[k(q_{n-1}) - k(p) - i\epsilon]\tau_{n-1}\}}{k(p) - k(q_{n-1}) + i\epsilon} \tag{75}$$

where the factor $e^{\epsilon\tau_{n-1}}$ will control the integral over τ_{n-1}. Carrying out the consecutive integrals and putting ϵ to zero in the first, we obtain

$$\begin{aligned} \langle p'|S^{(n)}|p\rangle &= -2\pi i\,\delta[k(p) - k(p')] \int d^4q_1 \int d^4q_2 \cdots \int d^4q_{n-1} \\ &\quad\times \langle p'|V|q_1\rangle \frac{1}{k(p) - k(q_1) + i\epsilon} \langle q_1|V|q_2\rangle \frac{1}{k(p) - k(q_2) + i\epsilon} \\ &\quad\cdots \langle q_{n-2}|V|q_{n-1}\rangle \frac{1}{k(p) - k(q_{n-1}) + i\epsilon} \langle q_{n-1}|V|p\rangle \qquad (76) \\ &= -2\pi i\,\delta[k(p) - k(p')]\langle p'|V(G_0V)^{n-1}|p\rangle \qquad (77) \end{aligned}$$

where

$$G_0 \equiv G_0[k(p)] \equiv \frac{1}{k(p) - K_0 + i\epsilon} \tag{78}$$

has the form of a Feynman propagator for a Klein–Gordon particle with "mass" $2Mk(p)$. In the form (76), we find the usual Feynman expansion for the S matrix of a particle scattering in an external potential.

In fact, let us consider, for a moment, the theory that we have worked out as applicable directly to the scattering of a particle in an external field. Taking K of the form (7), the potential is

$$V = -\frac{e}{2M}[p \cdot A + A \cdot p] + \frac{e^2}{2M} A^2 \tag{79}$$

The matrix element of V is given by

$$2M\langle p|V|q\rangle = -e(p_\mu + q_\nu)\tilde{A}^\mu(p - q) + e^2\tilde{A}^2(p - q) \tag{80}$$

The resulting expansion is essentially that of the scattering of a spinless particle in an external vector potential, as given explicitly by Rohrlich[27] and mentioned by Feynman[10] and Cooke[25] in the framework of the proper time formalism.

We recall, however, that we have developed this formalism for the relative motion in a two-body problem. The δ function appearing in (76) does not conserve the masses of the individual particles. Since [using the relations (38)]

$$\begin{aligned} p_1^2 &= \left(\frac{M_1}{M} P - p\right)^2 = \frac{M_1^2}{M^2} P^2 - \frac{2M_1}{M} P \cdot p + p^2 \\ p_2^2 &= \left(\frac{M_2}{M} P + p\right)^2 = \frac{M_2^2}{M^2} P^2 + \frac{2M_2}{M} P \cdot p + p^2 \end{aligned} \tag{81}$$

it follows from the conservation of p^2 and P that

$$p_1'^2 = p_1^2 - \frac{2M_1}{M} P \cdot q$$
$$p_2'^2 = p_2^2 + \frac{2M_2}{M} P \cdot q \tag{82}$$

where $q = p' - p = p_2' - p_2 = p_1 - p_1'$. A sufficient condition for the conservation of individual particle masses is therefore that $P \cdot q = 0$.[28] It has been suggested[29] that a potential of the form $V(x^\mu - P^\mu(x \cdot P)/P^2) \equiv V(x^\perp)$ will assure this condition, since the Fourier transform results in $\tilde{V}(q) = \delta(P \cdot q)\tilde{V}'(q)$. A potential of this type would not, however, admit of a separation of the center of mass and relative variables. In constrained Hamiltonian theory,[5,6] the two-body force is required, in fact, to have this form as a result of the condition that the constraints be first class (we shall return to this point later). Piron and Reuse[14] have postulated the existence of an irreversible process from the virtual unstable off-mass-shell state back to the equilibrium values of the masses.

Let us study the second term of the expansion (72), displaying explicitly the two-body variables. Since P^2 is conserved, we may add and subtract $P^2/2M_T$ in each denominator and in the δ function to obtain

$$\begin{aligned}\langle p_1'p_2'|S^{(2)}|p_1p_2\rangle &= -2\pi i\, \delta\left(\frac{p_1^2 - p_1'^2}{2M_1} + \frac{p_2^2 - p_2'^2}{2M_2}\right) \\ &\quad \times \int d^4q_1 \int d^4q_2 \frac{\langle p_1'p_2'|V|q_1q_2\rangle\langle q_1q_2|V|p_1p_2\rangle}{(p_1^2 - q_1^2)/2M_1 + (p_2^2 - q_2^2)/2M_2 + i\epsilon}\end{aligned} \tag{83}$$

The Feynman diagram for the intermediate state in $S^{(2)}$ would have two internal lines corresponding to the momenta q_1, q_2, but there is just a single propagator. Since we are not describing a theory of gauge interactions, such as electromagnetism, but an "action-at-a-distance" potential theory, the structure of (83) is not necessarily in disagreement with conventional field theory. However, it is interesting to ask whether there could be an effective action-at-a-distance formalism with a perturbation expansion which closely resembles that of gauge interaction quantum field theories. The resulting theory could be useful as a phenomenological potential model.

It is clear that the structure of the denominator in the propagator in (83) is due to the assignment of a single τ variable to all particles. A generalization of (74) to the following form

$$\begin{aligned}\langle p_1'p_2'|S^{(2)}|p_1p_2\rangle &= (-i)^2 \int_{-\infty}^{\infty} d\tau_1 \int_{-\infty}^{\tau_1} d\tau_2 \int_{-\infty}^{\infty} d\sigma_1 \int_{-\infty}^{\sigma_1} d\sigma_2 \int d^4q_1 \int d^4q_2 \\ &\quad \times \exp\left[i\left(\frac{p_1'^2}{2M_1}\tau_1 + \frac{p_2'^2}{2M_2}\tau_2\right)\right] \\ &\quad \times \langle p_1'p_2'|V|q_1q_2\rangle \exp\left\{-i\left[\frac{q_1^2}{2M_1}(\tau_1 - \tau_2) + \frac{q_2^2}{2M_2}(\sigma_1 - \sigma_2)\right]\right\} \\ &\quad \times \langle q_1q_2|V|p_1p_2\rangle \exp\left[-i\left(\frac{p_1^2}{2M_1}\tau_2 + \frac{p_2^2}{2M_2}\sigma_2\right)\right]\end{aligned} \tag{84}$$

in which a different τ is associated with each particle, will yield upon integration over $\tau_1, \tau_2, \sigma_1, \sigma_2$

$$4\pi^2 \delta\left(\frac{p_1^2 - p_1'^2}{2M_1}\right) \delta\left(\frac{p_2^2 - p_2'^2}{2M_2}\right) \int d^4q_1 \int d^4q_2 \langle p_1'p_2'| V | q_1q_2\rangle \times \langle q_1q_2| V | p_1p_2\rangle \left(\frac{p_1^2 - q_1^2}{2M_1} + i\epsilon\right)^{-1} \left(\frac{p_2^2 - q_2^2}{2M_2} + i\epsilon\right)^{-1} \tag{85}$$

The corresponding diagrams still have ladder structure in momentum space due to the structure of the vertices, but the generalized expansion (84) at least has the property that internal lines have their own propagators. It is also true that individual particle masses will be conserved. The interaction picture relation[11]

$$\psi_{I\tau\sigma} = \psi_{I\text{in}} + \left\{ \int_{-\infty}^{\tau} d\tau' \int_{-\infty}^{\sigma} d\sigma' + \int_{-\infty}^{\tau} d\tau' \int_{\sigma}^{\infty} d\sigma' + \int_{\tau}^{\infty} d\tau' \int_{-\infty}^{\sigma} d\sigma' \right\} V[x_1(\tau'), x_2(\sigma')]\, \psi_{I\tau'\sigma'} \tag{86}$$

has the property that for $\tau, \sigma \to -\infty$, $\psi_{I\tau\sigma} \to \psi_{I\text{in}}$ (presumably constant). Iterating (86), one obtains

$$\begin{aligned}
\psi_{I\tau\sigma} = \psi_{I\text{in}} &+ \left\{ \int_{-\infty}^{\tau} d\tau' \int_{-\infty}^{\sigma} d\sigma' + \int_{-\infty}^{\tau} d\tau' \int_{\sigma}^{\infty} d\sigma' + \int_{\tau}^{\infty} d\tau' \int_{-\infty}^{\sigma} d\sigma' \right\} \\
&\times V[x_1(\tau'), x_2(\sigma')]\psi_{I\text{in}} \\
&+ \left\{ \int_{-\infty}^{\tau} d\tau' \int_{-\infty}^{\sigma} d\sigma' \int_{-\infty}^{\tau'} d\tau'' \int_{-\infty}^{\sigma'} d\sigma'' \right. \\
&+ \int_{-\infty}^{\tau} d\tau' \int_{-\infty}^{\sigma} d\sigma' \int_{-\infty}^{\tau'} d\tau'' \int_{\sigma'}^{\infty} d\sigma'' \\
&+ \int_{-\infty}^{\tau} d\tau' \int_{-\infty}^{\sigma} d\sigma' \int_{\tau'}^{\infty} d\tau'' \int_{-\infty}^{\sigma'} d\sigma'' \\
&+ \int_{-\infty}^{\tau} d\tau' \int_{\sigma}^{\infty} d\sigma' \int_{-\infty}^{\tau'} d\tau'' \int_{-\infty}^{\sigma'} d\sigma'' \\
&\qquad + \int_{-\infty}^{\tau} d\tau' \int_{\sigma}^{\infty} d\sigma' \int_{-\infty}^{\tau'} d\tau'' \int_{\sigma'}^{\infty} d\sigma'' \\
&\qquad + \int_{-\infty}^{\tau} d\tau' \int_{\sigma}^{\infty} d\sigma' \int_{\tau'}^{\infty} d\tau'' \int_{-\infty}^{\sigma'} d\sigma'' \\
&\qquad + \int_{\tau}^{\infty} d\tau' \int_{-\infty}^{\sigma} d\sigma' \int_{-\infty}^{\tau'} d\tau'' \int_{-\infty}^{\sigma'} d\sigma'' \\
&\qquad + \int_{\tau}^{\infty} d\tau' \int_{-\infty}^{\sigma} d\sigma' \int_{-\infty}^{\tau'} d\tau'' \int_{\sigma'}^{\infty} d\sigma'' \\
&\qquad \left. + \int_{\tau}^{\infty} d\tau' \int_{-\infty}^{\sigma} d\sigma' \int_{\tau'}^{\infty} d\tau'' \int_{-\infty}^{\sigma'} d\sigma'' \right\} \\
&\qquad \times V[x_1(\tau'), x_2(\sigma')] V[x_1(\tau''), x_2(\sigma'')]\psi_{I\text{in}} \\
&\qquad + \cdots
\end{aligned} \tag{87}$$

The space-time diagrams for (87) are of the type depicted in Fig. 1, in contrast to the sum of ladder type space-time diagrams generated by the iteration of

$$\begin{aligned}\psi_{I\tau} &= \psi_{I\text{in}} + \int_{-\infty}^{\tau} d\tau'\, V[x_1(\tau'), x_2(\tau')]\psi_{I\tau'} \\ &= \psi_{I\text{in}} + \int_{-\infty}^{\tau} d\tau'\, V[x_1(\tau'), x_2(\tau')]\psi_{I\text{in}} \\ &\quad + \int_{-\infty}^{\tau} d\tau' \int_{-\infty}^{\tau'} d\tau''\, V[x_1(\tau'), x_2(\tau')]\, V[x_1(\tau''), x_2(\tau'')]\psi_{I\text{in}} + \cdots \end{aligned} \tag{88}$$

In the limit as $\tau, \sigma \to \infty$, the first three terms of the second-order part of (87) survive, and contain the full structure of the diagrams of Fig. 1. The contributions to $S^{(2)}$ contain those shown in (85), but there are additional terms with opposite signs for ϵ (in higher-order contributions, the required convergence factors do not propagate automatically, but must be put in at each stage, assuming that the dense set for the convergence of (33) is stable under multiplication by arbitrary powers of V along with the appropriate propagators).

Differentiating (86), we find the equation

$$-\frac{\partial^2 \psi_{I\tau\sigma}}{\partial\sigma\partial\tau} = V[x_1(\tau), x_2(\sigma)]\psi_{I\tau,\sigma} \tag{89}$$

and transforming away the interaction picture

$$\psi_{\tau,\sigma} = e^{-iK_{0}(1)\tau}\, e^{-iK_{0}(2)\sigma}\psi_{I\tau,\sigma} \tag{90}$$

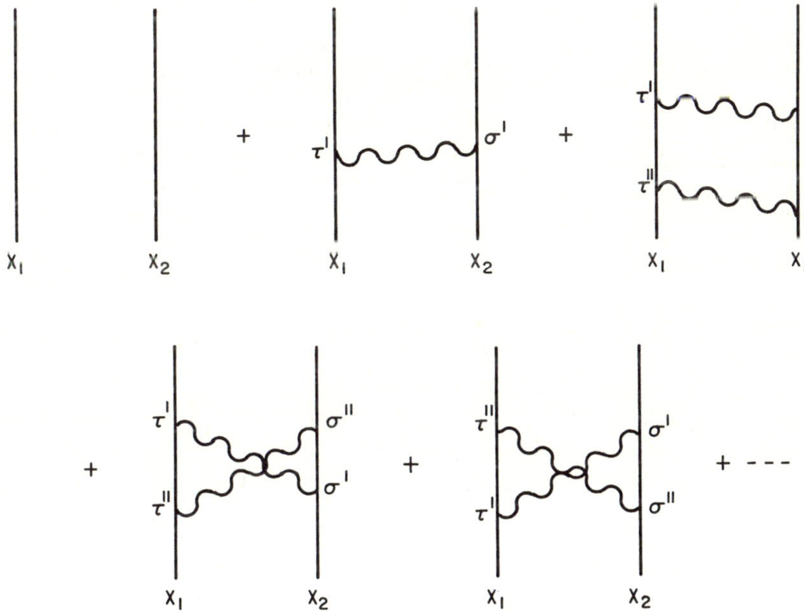

FIGURE 1. Space-time diagrams for Eq. (87).

one obtains[11]

$$\left(i\frac{\partial}{\partial\sigma} - K_0^{(2)}\right)\left(i\frac{\partial}{\partial\tau} - K_0^{(1)}\right)\psi_{\tau,\sigma} = V(x_1, x_2)\psi_{\tau,\sigma} \tag{91}$$

which does not correspond to a canonical system. An equation of the type (91) follows from writing *separate* equations for particles 1, 2 in the field of the other, i.e., as a self-consistent system*

$$\begin{aligned}\left[i\frac{\partial}{\partial\tau} - K_0^{(1)}\right]\psi_{\tau,\sigma} &= u(x_1; x_2)\psi_{\tau,\sigma}\\ \left[i\frac{\partial}{\partial\sigma} - K_0^{(2)}\right]\psi_{\tau,\sigma} &= u(x_2; x_1)\psi_{\tau,\sigma}\end{aligned} \tag{92}$$

Operating on the first of (92) with the operator $(i\,\partial/\partial\sigma - K_0^{(2)})$, we have that

$$\left(i\frac{\partial}{\partial\sigma} - K_0^{(2)}\right)\left(i\frac{\partial}{\partial\tau} - K_0^{(1)}\right)\psi_{\tau,\sigma} = \{u(x_1; x_2)u(x_2; x_1) - [K_0^{(2)}, u(x_1; x_2)]\}\psi_{\tau,\sigma} \tag{93}$$

In the reverse order, we would obtain

$$\{u(x_2; x_1)u(x_1; x_2) - [K_0^{(1)}, u(x_2; x_1)]\}\psi_{\tau,\sigma} \tag{94}$$

on the right-hand side. For the equations (92) to be equivalent to something of the form of (91), it is therefore necessary that

$$\begin{aligned}&\{u(x_1; x_2)u(x_2; x_1) - [K_0^{(2)}, u(x_1; x_2)]\}\psi_{\tau,\sigma}\\ &\qquad = \{u(x_2; x_1)u(x_1; x_2) - [K_0^{(1)}, u(x_2; x_1)]\}\psi_{\tau,\sigma}\end{aligned} \tag{95}$$

i.e., the integrability condition

$$[K_0^{(1)} + u(x_1; x_2), K_0^{(2)} + u(x_2; x_1)]\psi_{\tau,\sigma} = 0 \tag{96}$$

be satisfied. The "potentials" $u(x_1; x_2)$, $u(x_2; x_1)$ and $V(x_1, x_2)$ may, of course, depend on momenta. In the stationary case, when we can replace $i\,\partial/\partial\tau$ by $-m_1^2/2M_1$ and $i\,\partial/\partial\sigma$ by $-m_2^2/2M_2$, the equations (92) become[31]

$$\begin{aligned}[p_1^2 + m_1^2 + \phi(x_1; x_2)]\psi &= 0\\ [p_2^2 + m_2^2 + \phi(x_2; x_1)]\psi &= 0\end{aligned} \tag{97}$$

These are constraint equations which, with the conditions (96), are of the first class type.

It is possible, as we have remarked earlier, that an effective perturbation expansion for systems with electromagnetic interaction in the single τ formalism could provide a basis for the heuristic construction given above. The notion that a potential $V(x_1(\tau), x_2(\sigma))$ can act as an interaction for any τ, σ corresponds physically, in the

*With the addition of $\psi_{\tau,\infty} + \psi_{\infty,\sigma} - \psi_{-\infty,\infty} - \psi_{\infty,-\infty}$ to (86), it becomes a solution of (92) also.

framework of a world with a single τ, to the idea that some carrier of the force could be emitted at historical time σ and stored in a cloud around the world line of particle 2 for some indefinite time; it is exchanged after this time, and may be absorbed by particle 1 at the historical time τ. The electromagnetic current is an integral over τ of a vector valued density,[12,30] corresponding to the expectation value of covariant derivatives in the state ψ_τ. Hence the substitution of the A^μ field obtained from this source by means of Maxwell's equations into the evolution equation for ψ_τ can be expected, in some iterative scheme, to generate integrals over a sequence of τ variables. This program has not yet been carried through, and the question of the control of individual particle masses in scattering and a full understanding of the relation of the unconstrained single τ formalism to the constrained Hamiltonian approach remains to be resolved.

Acknowledgments

Much of the content of this article is based on joint work with Yeshiahu Lavie; the rigorous proof of the existence of the wave operator was a joint effort with Avi Soffer, and I gratefully acknowledge many discussions with them and Rafi Arshansky and Fred Rotbart about other aspects of the program as well. I also wish to thank my friend, Constantin Piron, for his continued collaboration, guidance and criticism during the years that we have been working on this subject. I wish to thank Fritz Rohrlich for many discussions on the constrained Hamiltonian formalism, for his continuing interest and encouragement, for a careful reading of this manuscript, and for his hospitality during my visit at Syracuse.

References and Notes

1. P. A. M. Dirac, *Rev. Mod. Phys.* **21**, 392 (1949).
2. B. Bakamjian and L. H. Thomas, *Phys. Rev.* **92**, 1300 (1953).
3. L. L. Foldy, *Phys. Rev.* **122**, 275 (1961).
4. E. H. Kerner, editor, *The Theory of Action-at-a-Distance in Relativistic Particle Dynamics* (Gordon and Breach, New York, 1972).
5. F. Rohrlich, *Ann. Phys.* (N.Y.) **117**, 292 (1979).
6. M. King and F. Rohrlich, *Ann. Phys* (N.Y.) **130**, 350 (1980).
7. F. Rohrlich, *Phys. Rev. D* **23**, 1305 (1981), and references therein. In particular, see I. T. Todorov, Comm. JINR F2-10125, Dubna (1976); and V. V. Molotkov and I. T. Todorov, "Gauge Dependence of World Lines and Invariance of the S-Matrix in Relativistic Classical Mechanics," Int. Rep. Ic/80/101, Trieste (1980); I. T. Todorov, "Constraint Hamiltonian Approach to Relativistic Point Particle Dynamics I," Lectures at ICTP, Trieste, 1980. For recent articles on this topic, see the *Proceedings of the Barcelona Workshop*, "Relativistic Action-at-a-Distance. Classical and Quantum Aspects," June (1981); F. Rohrlich, "Evolution and Covariance in Constraint Dynamics," SLAC-PUB-2737, Stanford (1981); L. P. Horwitz and F. Rohrlich, *Phys. Rev. D.* **24**, 1528 (1981).
8. E. C. G. Stueckelberg, *Helv. Phys. Acta* **14** 372, 588 (1941).
9. L. P. Horwitz and C. Piron, *Helv. Phys. Acta* **46**, 316 (1973).
10. R. P. Feynman, *Phys. Rev.* **80**, 440 (1950) (App. A).

11. L. P. Horwitz, Y. Lavie, and A. Soffer, in Proceedings of the VIII Colloquium on Group Theoretical Methods in Physics, *Ann. Israel Phys. Soc.* **3** (1980).
12. L. P. Horwitz and Y. Lavie, *Phys. Rev. D* (in press).
13. L. P. Horwitz, C. Piron, and F. Reuse, *Helv. Phys. Acta* **48,** 546 (1975).
14. C. Piron and F. Reuse, *Helv. Phys. Acta* **51,** 146 (1978).
15. F. Reuse, *Found. Phys.* **9,** 865 (1979).
16. J. R. Fanchi and R. E. Collins, *Found. Phys.* **8,** 851 (1978).
17. R. E. Collins and J. R. Fanchi, *Nuovo Cimento* **48A,** 314 (1978).
18. L. P. Horwitz, C. Piron, and W. C. Shieve, XIII IUPAP Conf. on Stat. Phys., ed. D. Cabib, C. Kuper and I. Riess, *Ann. Israel Phys. Soc.* **2** (1978); and to be published.
19. L. P. Horwitz and Y. Rabin, *Nuovo Cimento Lett.* **17,** 501 (1976).
20. T. D. Newton and E. P. Wigner, *Rev. Mod. Phys.* **21,** 400 (1949).
21. W. O. Amrein, Ph. A. Martin, and B. Misra, *Helv. Phys. Acta* **43,** 313 (1970); C. Dollard, *J. Math. Phys.* **5,** 792 (1964).
22. R. G. Newton, *Scattering Theory of Particles and Waves* (McGraw Hill, New York, 1966); J. P. Taylor, *Scattering Theory* (John Wiley and Sons, New York, 1972).
23. L. P. Horwitz and A. Soffer, *Helv. Phys. Acta,* to be published.
24. An alternative proof can be constructed using methods of L. Hörmander, *Math. Z.* **146,** 69 (1976) [see also M. Reed and B. Simon, *Methods of Modern Mathematical Physics III* (Academic Press, New York and London, 1979)].
25. J. H. Cooke, *Phys. Rev.* **166,** 1293 (1968).
26. J. L. Cook, *Aust. J. Phys.* **25,** 117 (1972).
27. F. Rohrlich, *Phys. Rev.* **50,** 666 (1972).
28. P. M. Pearle, *Phys. Rev. D* **1,** 979 (1970).
29. A. Komar, *Phys. Rev. D* **18,** 1881, 1887, 3617 (1978).
30. E. C. G. Stueckelberg, *Helv. Phys. Acta* **15,** 23 (1942).
31. I wish to thank Fritz Rohrlich for clarifying this connection.

13

Vacuum Tunneling in Static Space-Times

C. J. Isham
Blackett Laboratory
Imperial College of Science and Technology
London SW.7, England

1. Introduction

This article is respectfully dedicated to the memory of Wolfgang Yourgrau, philosopher and physicist. Wolfgang's energy and interests were boundless. One of his fields of study was classical general relativity and, although he never worked directly on quantum gravity, he took a keen interest in developments in that area and discussed the subject with me at great length. Wolfgang frequently expressed the view that modern mathematics had much to give to physics and I hope for this reason that he would have approved of the contents of the present paper.

There is currently much interest in the role which space-time topology plays in quantum gravity. It is also known that Minkowski space Yang–Mills theory exhibits important topological properties. It is therefore to be expected that Yang–Mills theory in a curved, topologically complex space-time $\mathcal{M}$ will yield many features of interest and possibly involve a delicate interplay between the topological structures of $\mathcal{M}$ and the internal symmetry group.

One possible approach is through the classification of gauge sectors, and hence principal fiber bundles, over the four-dimensional space-time.[1,2] Another starting point lies in the canonical decomposition of the field theory and a study of the evolution of states defined on an initial spacelike hypersurface. In the present paper the latter approach is followed and attention is concentrated on the topological properties of initial vacuum states and their subsequent quantum tunneling. The space-time is assumed to be static of the form $\Sigma \times R$ where Σ is a spatial hypersurface. This enables an analytic continuation to be performed in the time variable and the tunneling amplitudes to be described by "Riemannian" functional integrals.

The basic vacuum states are labeled by the homotopy classes $[\Sigma, G]$ of maps from Σ into the internal symmetry group G, and much of the paper is concerned with studying their properties. These classes form an Abelian group and the dual group of characters is shown to label the vacua that diagonalize the transition amplitudes.

Homotopy classes of maps between two spaces occur in various places in theoretical physics, and one of the main aims of this paper is to illustrate the general Postnikov methods for analyzing such sets. The style is deliberately expository and it is not assumed that the reader possesses more than an elementary knowledge of algebraic topology. An Appendix has been attached and contains some useful but more sophisticated results.

The analogs of $|n\rangle$ and $|\theta\rangle$ vacua are described in Section 2 and it is shown that the $|\theta\rangle$ states are gauge invariant up to a phase factor and diagonalize the tunneling amplitudes. The classification of $[\Sigma, G]$ when G is $U1$ or is simply connected is performed in Section 3 and is completed in Section 4 for the more difficult examples $G = U(n)$ or $SO(n)$ where the Postnikov machinery is required. The paper concludes in Section 5 with a short summary of the results and a few remarks concerning possible applications to the canonical quantization of gravity.

2. Vacuum Tunneling

Let us start by briefly reviewing some of the vacuum tunneling results for conventional Yang–Mills theory in Minkowski space-time.[3,4] The Yang–Mills lagrangian is

$$L = \frac{1}{4g^2} \mathbf{F}_{\mu\nu} \cdot \mathbf{F}^{\mu\nu}, \qquad F^i_{\mu\nu} = \partial_\mu A^i_\nu - \partial_\nu A^i_\mu + C^i_{jk} A^j_\mu A^k_\nu \tag{1}$$

where g is the coupling constant and $\{C^i_{jk}\}$ are the structure constants of the internal symmetry group G. It is convenient to work in the gauge $\mathbf{A}_0 = 0$ in which case the energy of a static field configuration is

$$E = \frac{1}{4g^2} \int \mathbf{F}_{ij} \cdot \mathbf{F}^{ij}\, d^3\underline{x}, \qquad i, j = 1 \ldots . 3 \tag{2}$$

A classical vacuum field is one whose energy vanishes, which from (2) means $\mathbf{F}_{ij} = 0$. This in turn implies that the potential A_i is pure gauge, i.e.,

$$A_i(\underline{x}) = \Omega^{-1}(\underline{x}) \partial_i \Omega(\underline{x}) \tag{3}$$

where $\Omega(\underline{x})$ is some function from three-space R^3 into G. In order that this field describes some local physical phenomenon it is customary (and crucial) to require that Ω tends to the unit matrix $\mathbb{1}$ as the point $\underline{x}$ tends to spatial infinity. Thus the two-sphere boundary of R^3 may be identified to a single point which is mapped to 1. However, R^3 with boundary identification is homeomorphic to a three-sphere S^3 and hence Ω may be viewed as a map from S^3 into G which takes the "point at infinity" to 1.

Any such initial configuration will of course evolve in time but since the dynamical field equations are smooth the *homotopy* class of Ω cannot change. Thus the vacuum fields are naturally labeled by the set of these classes which in fact constitute the third homotopy group of G, denoted $\pi_3(G)$. If G is any simple, non-Abelian Lie group then $\pi_3(G) \simeq Z$ (the integers) whereas if $G = U1$ (the Maxwell case) then $\pi_3(G) = 0$. Hence electromagnetism provides no interesting topological structure whereas Yang–Mills groups such as $G = SU(n)$, $SO(n)$ $(n \geq 3)$, $U(n)$ $(n \geq 2)$, G_2, etc., all have zero energy configurations labeled by a single integer (often known as the winding number). Quantum mechanical vacuum states will be wave functionals $\psi[A]$ peaked around these classical fields and are denoted $|n\rangle$.

Although there can be no classical evolution from one homotopy class to another it is possible for tunneling to occur between quantum vacuum states. As usual this may be conveniently described by analytically continuing the time variable from a real to a purely imaginary value and expressing the transition amplitude as a functional integral

$$\lim_{T\to\infty} \langle n|e^{-\hat{H}T}|m\rangle \sim \int \delta A_{(n-m)} \exp(-\int L\, d^4x) \tag{4}$$

over all field configurations $A_{(n-m)}$ which interpolate between the zero energy configurations corresponding to the initial and final winding numbers n and m.

The existence of tunneling shows that the states $|n\rangle$ are not the correct physical vacua. Indeed they are not even gauge invariant since there exist "large" gauge transformations T_m such that

$$T_m|n\rangle = |m + n\rangle \tag{5}$$

This is equivalent classically to the observation that, under a gauge transformation $w(x)$, the potential A_i of (3) becomes

$$A_i \to (\Omega w)^{-1}\partial_i(\Omega w) \tag{6}$$

whilst the homotopy class of Ωw is the sum of the classes [as elements of the group $\pi_3(G)$] of the functions Ω and w.

The physical states are expressed as Fourier sums

$$|\theta\rangle = \sum_n e^{in\theta}|n\rangle \tag{7}$$

and are hence labeled by a real number lying between 0 and 2π. These states transform with a physically irrelevant phase factor:

$$T_m|\theta\rangle = e^{-im\theta}|\theta\rangle \tag{8}$$

and diagonalize the transition amplitudes:

$$\lim_{T\to\infty} \langle\theta|e^{-\hat{H}T}|\theta'\rangle \sim \delta(\theta - \theta') \sum_p e^{-ip\theta} Z_{(p)} \tag{9}$$

where

$$Z_{(p)} = \int \delta A_{(p)} \exp(-\int L\, d^4x) \tag{10}$$

After this brief survey of the Minkowski space situation let us consider the case where the space-time $\mathcal{M}$ is curved but static with a global topology $\mathcal{M} = \Sigma \times R$ where Σ is a compact spatial three-manifold. The Yang–Mills lagrangian is (1) with the metric tensor suitably inserted and some zero-energy configurations are again of the form (3), but where Ω is now a smooth function from Σ into the internal symmetry group G. Thus the vacuum states are labeled by elements of the set $[\Sigma, G]$ of homotopy classes of such maps. Note that Σ may either be assumed *ab initio* to be compact or else may be obtained from a noncompact space which is compactified by identifying the boundary to a point, just as S^3 was obtained above from R^3. In the latter case the maps are naturally pointed (i.e., they take the point at infinity into $\mathbb{1}$) and it is convenient to assume that pointed maps are also employed in the former case after selecting some fixed but arbitrary base point in Σ.

Clearly the structure of the set $[\Sigma, G]$ will depend on the topology of both Σ and G and we will analyze this structure for a wide range of Lie groups G. This set can be made into a group H by defining $\Omega_1 \cdot \Omega_2(x) \equiv \Omega_1(x)\Omega_2(x)$ and, although it is not obvious, this group is always Abelian (see results in Sections 3 and 4). Thus the vacuum states can be written as $|h\rangle$ with h in H and quantum tunneling can occur. Since the metric is static there is no problem in analytically continuing the globally defined time parameter and, in analogy with (4), the transition amplitude may be expressed as the functional integral based on the, continued, Riemannian manifold $\tilde{\mathcal{M}}$,

$$\lim_{T\to\infty} \langle h | e^{-\hat{H}T} | h' \rangle \sim \int \delta A_{(h-h')} \exp(-\int_{\mathcal{M}} \mathcal{L}) \tag{11}$$

The states $|h\rangle$ are not gauge invariant but transform under large gauge transformations $T_{h'}$ as [cf. (5)]

$$T_{h'}|h\rangle = |h + h'\rangle, \qquad h, h' \in H \equiv [\Sigma, G] \tag{12}$$

We attempt to define physical vacua as linear sums [cf. (7)]:

$$|\chi\rangle = \sum_{h\in H} \chi(h)|h\rangle, \qquad \chi(h) \in \mathbb{C} \tag{13}$$

where the states $|h\rangle$ are normalized to one. (As shown in Sections 3 and 4, the group H is isomorphic to, or a subgroup of, a direct sum of copies of the integers and various cyclic groups and hence the states $|h\rangle$ may be orthonormalized by a Kronecker δ rather than a Dirac δ function.) Invariance under a gauge transformation is required up to a phase factor $\alpha(h')$, so that for all h' in H

$$T_{h'}|\chi\rangle = \alpha(h')|\chi\rangle, \qquad |\alpha(h')| = 1 \tag{14}$$

$$\therefore \sum_{h\in H} \chi(h)|h + h'\rangle = \alpha(h') \sum_{h\in H} \chi(h)|h\rangle = \sum_{h''\in H} \chi(h'' - h')|h''\rangle \tag{15}$$

The states $|h\rangle$ are linearly independent and hence (15) implies

$$\alpha(h')\chi(h) = \chi(h - h') \tag{16}$$

and if, without loss of generality, we set $\chi(0) = 1$, (16) gives

$$\alpha(h') = \chi(-h')$$

and hence

$$\chi(h)\chi(h') = \chi(h + h'), \qquad |\chi(h)| = 1 \tag{17}$$

Then the vacuum states $|\chi\rangle$ are labeled by the *characters* of H, namely, the homomorphisms H^* of H into the multiplicative group of complex numbers of modulus 1.

The new tunneling amplitude is

$$\lim_{T\to\infty} \langle\chi_1|e^{-\hat{H}T}|\chi_2\rangle = \sum_{h,h'} \bar{\chi}_1(h)\chi_2(h')Z_{(h-h')}$$

$$\equiv \sum_{h,h'} \bar{\chi}_1(h - h')\bar{\chi}_1(h')\chi_2(h')Z_{(h-h')} \tag{18}$$

$$\sum_{h',h''} \bar{\chi}_1(h'')\bar{\chi}_1(h')\chi_2(h')Z_{(h'')} \tag{19}$$

However, the characters are a complete set in the sense that, for some normalization constant N,

$$\sum_h \bar{\chi}_1(h)\chi_2(h) = N\,\delta_{\chi_1\chi_2} \tag{20}$$

where $\delta_{\chi_1\chi_2}$ signifies either a Kronecker δ function or a Dirac δ function according as χ is discrete or continuous. Thus the new states $|\chi\rangle$ diagonalize the transition amplitude and give

$$\lim_{T\to\infty} \langle\chi_1|e^{-\hat{H}T}|\chi_2\rangle \sim \delta_{\chi_1\chi_2} \sum_h \bar{\chi}_1(h)Z_{(h)} \tag{21}$$

Note that (9) is a special case of (21) since the character group of the integers is precisely $U1$ and is labeled by an angle taking on values between 0 and 2π.

A moment's thought shows that in certain circumstances there exist potentials which yield a vanishing F_{ij} and yet are not of the form (3). For example in the $U1$ case let Ω_1 and Ω_2 be two functions from $\Sigma \to U1$ with different winding numbers (see Section 3) and form the affine sum,

$$A_i^\lambda = \lambda\Omega_1\partial_i\Omega_1^{-1} + (1 - \lambda)\Omega_2\partial_i\Omega_2^{-1} \tag{22}$$

Because of the linear dependence of F_{ij} on A_i this also yields $F_{ij} = 0$. This means that the zero-energy solutions $\Omega_1\partial_i\Omega_1^{-1}$ and $\Omega_2\partial_i\Omega_2^{-1}$ can be connected together, even in real time, by a path which maintains zero energy. Under these circumstances it is not obvious that they should be regarded as distinguishable states. Although for the sake of completeness I have included in Section 3 a tunneling description for the pure gauge configurations; this fact should be borne in mind when considering the physical interpretation.

Note that A_i^λ cannot itself be of the form $\Omega_\lambda \partial_i \Omega_\lambda^{-1}$ as the function Ω_λ would then be a homotopy between the homotopically inequivalent functions Ω_1 and Ω_2. One wonders therefore what type of field A_i^λ is and can a similar interpolation between gauge vacua arise for other groups? The answer lies in the concept of the holonomy group. From a mathematical point of view our construction so far has been to form the trivial bundle $\Sigma \times G$ over Σ and take the canonical flat connection $A_i = 0$ which is associated with the Cartan–Maurer structure form. The different gauge fields $\Omega\partial_i\Omega^{-1}$ then arise by performing automorphisms of this bundle. This connection has the property that its holonomy group is trivial. That is, if a closed curve in the base space is lifted up to the bundle by parallel transport then the initial and final points coincide.

However, in a space Σ which is not simply connected it is possible to find flat connections (i.e., $F_{ij} = 0$) which have a nonvanishing discrete subgroup of G as holonomy group. These connections look locally like $\Omega\partial_i\Omega^{-1}$ but are patched together in a nontrivial way. Connections of this type are labeled by homomorphisms of $\pi_1(\Sigma)$ into G and automorphisms of the G bundle again generate different winding numbers. The resulting G bundle may be trivial [e.g., if $G = SU(n)$] or in certain circumstances may be nontrivial with its structure group being reducible to the holonomy group and no further [e.g., if $G = U1$ or $SO(n)$].

In general each homomorphism generates a new vacuum state with its associated homotopy transform. However, if $\pi_1(\Sigma)$ contains a Z factor then homomorphisms into G are like $e^{i\lambda\theta}$ and hence there is one-parameter family of them labeled by λ. This gives rise to the possibility of connecting together two connections via a one-parameter family in which the holonomy group changes as the parameter is varied. This is precisely what happens in the $U1$ case. For example if $\Sigma = S_1$ and $A = 1$, $B = e^{in\theta}$ then the interpolating connection is related to a function $e^{i\lambda n\theta}$ and the holonomy group for value λ is

$$\{e^{i\lambda np} \mid p \in Z\} \tag{23}$$

which may be the integers or a cyclic group depending on the value of λ.

In general the problems to be answered are the following:

1. Classify the trivial and nontrivial G bundles which are obtained from homomorphisms from $\pi_1(\Sigma) \to G$.
2. Compute the homotopy classes of automorphisms (the analog of $[\Sigma, G]$) when the G bundle is nontrivial.
3. Investigate the possibility of connecting two connections together by a one-parameter family.
4. Discuss the effects of all this on the classification of vacuum states and vacuum tunneling.

As far as (3) is concerned it seems likely that the winding numbers associated with elements of $H^3(\Sigma)$ (Sections 3 and 4) cannot be linked but it may be possible to join those connected with $H^1(\Sigma)$ [such as $U(n)$ or $SO(n)$].

This whole problem is quite sophisticated mathematically and it would be out of place to pursue it here. We will publish a second paper dealing with the subject at some later time and in the present work will concentrate on the simple pure gauge vacua and the classification of $[\Sigma, G]$. This investigation is anyway a natural prerequisite for pursuing the full problem.

3. The Structure of $[\Sigma, G]$

The main technical problem is to analyze the homotopy set $[\Sigma, G]$. The technique employed is to approximate G by a homotopically simpler space. More precisely let $G_{(3)}$ be a space such that there exists a map f from G into $G_{(3)}$ inducing isomorphisms between the first three homotopy groups of G and $G_{(3)}$ and inducing a subjective map of $\pi_4(G)$ onto $\pi_4(G_{(3)})$. (Such a map is called a *3-equivalence*; the definition of a n-equivalence is analogous. See Section A.1.). Then there is a bijection[5–7]

$$[\Sigma, G] = [\Sigma, G_{(3)}] \tag{24}$$

for any three-dimensional manifold Σ and the properties of $[\Sigma, G]$ may be found by studying $[\Sigma, G_{(3)}]$, which may be simpler.

Suitable approximations $G_{(3)}$ may be built from Eilenberg–Maclane spaces $K(\pi, n)$ which are defined by the characteristic property[5–7]

$$\begin{aligned} \pi_i(K(\pi, n)) &= 0, \qquad i \neq n \\ &= \pi, \qquad i = n \end{aligned} \tag{25}$$

where π is any Abelian group and n is any positive integer.

Now $\pi_2(G) = 0$ for all Lie groups G and therefore the product space

$$K = K(\pi_1(G), 1) \times K(\pi_3(G), 3) \tag{26}$$

has, by construction, the same first three homotopy groups as G and $\pi_4(K) = 0$. Hence if there was a map f from G into K which induced the isomorphisms of $\pi_i(G)$ with $\pi_i(K)$ $1 \leq i \leq 3$, it would necessarily map $\pi_4(G)$ onto $\pi_4(K)$ and hence K would be a suitable model for $G_{(3)}$. Unfortunately such a map may not always exist but, as discussed in Section 4, it is always possible to find a "twisted" product of $K(\pi_1(G), 1)$ with $K(\pi_3(G), 3)$ which does possess the desired properties. By a twisted product I mean a fibration over $K(\pi_1(G), 1)$ with fiber homotopic to $K(\pi_3(G), 3)$. (Recall that a fibration is rather like a fiber bundle except that the different fibers are homotopic to each other rather than homeomorphic.)

The great power of Eilenberg–Maclane spaces is that the homotopy set $[\Sigma, K(\pi, n)]$ is known. In fact

$$[\Sigma, K(\pi, n)] = H^n(\Sigma; \pi) \tag{27}$$

where $H^n(\Sigma; \pi)$ is the nth cohomology group of Σ with values in the Abelian group π. This bijection is defined as follows. Consider the cohomology group $H^n(K(\pi, n);$

π). Since $\pi_i(K(\pi, n)) = 0$ for $i < n$ we have (see Section A.2)

$$\begin{aligned} H^n(K(\pi, n); \pi) &= \text{Hom}(H_n(K(\pi, n); Z); \pi) \\ &= \text{Hom}(\pi_n(K(\pi, n); \pi)) \\ &= \text{Hom}(\pi; \pi) \end{aligned} \tag{28}$$

where Hom $(A; B)$ denotes the set of homomorphisms of A into B and the Hurewicz theorem (Section A.3) has been used, which states in this case

$$\begin{aligned} H_i(K(\pi, n); Z) &= 0, \qquad 0 < i < n \\ H_n(K(\pi, n); Z) &= \pi_n(K(\pi, n)) = \pi \end{aligned} \tag{29}$$

Now choose the element ι_n in $H^n(K(\pi, n); \pi)$ which corresponds via (28) to the identity homomorphism of π with π and define

$$T[h] = h^*(\iota_n) \tag{30}$$

where $[h]$ is the homotopy class of a map h from Σ into $K(\pi, n)$. Then T is the isomorphism from $[\Sigma, K(\pi, n)]$ onto $H^n(\Sigma; \pi)$ that is alluded to in (27).

These cohomology groups $H^n(\Sigma; \pi)$ are explicitly computable for any "named" space Σ that is likely to occur in practice, and hence expressing $[\Sigma, G]$ in terms of them is a genuine solution to the problem of analyzing the structure of this homotopy set.

Note that in the product case (26)

$$[\Sigma, G] = H^1(\Sigma; \pi_1(G)) \oplus H^3(\Sigma; \pi_3(G)) \tag{31}$$

and for a twisted product we will show in Section 4 that $[\Sigma, G]$ is a specific subgroup of this direct sum of cohomology groups.

As a simple example consider $G = SU(n)$ where

$$\pi_1(SU(n)) = 0, \qquad \pi_3(SU(n)) = Z \tag{32}$$

and furthermore (A.2):

$$\begin{aligned} H^3(SU(n); Z) &= \text{Hom}(H_3(SU(n); Z); Z) \\ &= \text{Hom}(\pi_3(SU(n); Z)) \\ &= \text{Hom}(Z; Z) = Z \end{aligned} \tag{33}$$

According to (27) and (30) there exists some map h from $SU(n)$ into $K(Z, 3)$ such that $h^*(\iota_3)$ is one of the generators of $H^3(SU(n); Z) = Z$. It is easy to show that h_* is an isomorphism:

$$h_*: \pi_3(SU(n)) \approx \pi_3(K(Z, 3)) = Z$$

and since $\pi_1(SU(n)) = 0 = \pi_1(K(Z, 3))$ it follows that $K(Z, 3)$ is a suitable model for the approximation space $SU(n)_{(3)}$. Then (24) and (27) give

$$[\Sigma, SU(n)] = [\Sigma, K(Z, 3)] = H^3(\Sigma; Z) \tag{34}$$

and we get a complete cohomological description of the homotopy set $[\Sigma, SU(n)]$.

Since Σ is a three-manifold we can apply the following general results[8]:

If Σ is compact and orientable, $H_3(\Sigma; Z) = H^3(\Sigma; Z) = Z$

If Σ is compact and nonorientable, $H_3(\Sigma; Z) = H^3(\Sigma; Z) = Z_2$

If Σ is noncompact, $H_3(\Sigma; Z) = 0$ and $H^3(\Sigma; Z) = \mathrm{Ext}\,(H_2(\Sigma; Z); Z)$ (Section A.2)

Since $H_2(\Sigma; Z)$ is torsion free this implies that, if for example $H_2(\Sigma, Z)$ is finitely generated, then $H^3(\Sigma; Z) = 0$ (35)

In particular we observe that if Σ is compact (which we are assuming) and orientable then the vacuum states $|h\rangle$ are labeled by a single integer no matter how complex the topology of Σ. Note that if we start with a noncompact Σ and do *not* identify the points at infinity (and hence do not pass to a compact space) then the vacuum state structure has no interesting topological content unless $H_2(\Sigma; Z)$ is such that $\mathrm{Ext}(H_2(\Sigma; Z); Z)$ does not vanish.

Clearly the same remarks apply to any simply connected non-Abelian symmetry group G since

$$\pi_1(G) = 0, \qquad \pi_3(G) = Z, \qquad \text{and } H^3(G; Z) = Z \tag{36}$$

and the arguments leading to (34) may be repeated. In all these cases the vacuum states may be denoted $|n\rangle$ (assuming Σ is orientable) as in the simple Minkowski space model discussed in Section 2. Thus $H = Z$ and the invariant vacuum states $|\chi\rangle$ are labeled by elements θ of the character group $Z^* = U1$. Just as in (7) we have $|\theta\rangle = \Sigma_n e^{in\theta}|n\rangle$ and the tunneling amplitude is (9) again:

$$\lim_{T\to\infty} \langle\theta|e^{-\hat{H}T}|\theta'\rangle \sim \delta(\theta - \theta') \sum_p e^{ip\theta} Z_p \tag{37}$$

As in the usual Minkowski space case the integer p can be expressed as an integral over the Yang–Mills field strengths and provides an additional topological contribution to the action

$$S_{\text{top}} = \int_{\mathfrak{M}} \mathbf{F}_{\mu\nu} \cdot {}^*\mathbf{F}^{\mu\nu} \tag{38}$$

where

$${}^*\mathbf{F}^{\mu\nu} = \tfrac{1}{2}\, \epsilon^{\mu\nu}_{\alpha\beta} \mathbf{F}^{\alpha\beta} \tag{39}$$

Such a term will lead to CP violating amplitudes once the different topological sectors have been summed. Thus the discussion above substantiates by a vacuum tunneling argument the CP violation statements made in Deser *et al.*[9]

Another example which is easily discussed is $G = U1$. Clearly

$$\begin{aligned} \pi_i(U1) &= 0, \qquad i \neq 1 \\ \pi_1(U1) &= Z \end{aligned} \tag{40}$$

and hence $U1$ is an Eilenberg–Maclane space $K(Z, 1)$. Thus we immediately obtain the cohomological classification:

$$[\Sigma; U1] = H^1(\Sigma; Z)$$

Now

$$H^1(\Sigma; Z) = \mathrm{Hom}(H_1(\Sigma; Z); Z) = \mathrm{Hom}(\pi_1(\Sigma); Z) \tag{41}$$

and $H_1(\Sigma; Z)$ is a direct sum of b_1 (the first Betti number of Σ) copies of the integers Z and assorted cyclic groups. It follows that $H^1(\Sigma; Z)$ is isomorphic to the direct sum of b_1 copies of Z and the vacuum states may be labeled by $|\mathbf{n}\rangle = |n_1, n_2 \ldots n_{b1}\rangle$. The corresponding χ-vacua are labeled by b_1 angles $\theta_1, \theta_2 \ldots \theta_{b1}$

$$|\theta_1 \ldots \theta_{b1}\rangle \equiv |\boldsymbol{\theta}\rangle = \sum_{\mathbf{n}} \exp(-i\mathbf{n}\cdot\boldsymbol{\theta})|\mathbf{n}\rangle \tag{42}$$

and the transition amplitudes are

$$\lim_{T\to\infty} \langle\boldsymbol{\theta}|e^{-\hat{H}T}|\boldsymbol{\theta}'\rangle \sim \delta^{(b1)}(\boldsymbol{\theta} - \boldsymbol{\theta}') \sum_{\mathbf{p}} \exp(-i\mathbf{p}\cdot\boldsymbol{\theta}) Z_{(\mathbf{p})} \tag{43}$$

Let $z_1 \ldots z_{b1}$ be a basis of the free subgroup of $H_1(\Sigma; Z)$. Then the integers $n_1 \ldots n_{b1}$ may be expressed in terms of the initial (or final) field configurations as line integrals over these cycles in Σ:

$$n_i = \frac{1}{2\pi} \oint_{z_i} \underline{A} \cdot d\underline{l} \tag{44}$$

and in (43)

$$p_i = n_i^{(+\infty)} - n_i^{(-\infty)} = \frac{1}{2\pi} \oint_{z_i} (\underline{A}^{(+\infty)} \cdot d\underline{l} - \underline{A}^{(-\infty)} \cdot d\underline{l}) \tag{45}$$

$$= \frac{1}{2\pi} \int_{z_i x(-\infty,\infty)} F_{\mu\nu}\, d\sigma^{\mu\nu} \tag{46}$$

where + and − superscripts refer to the final and initial vacuum field configurations. Thus (43) can be written as

$$\operatorname*{Lt}_{T\to\infty} \langle\boldsymbol{\theta}|e^{-\hat{H}T}|\boldsymbol{\theta}'\rangle \tag{47}$$

$$\sim \delta^{(b1)}(\boldsymbol{\theta} - \boldsymbol{\theta}') \sum_{p1\ldots pb1} \exp\left(-i\sum_{j=1}^{b1} \frac{p_j}{2\pi} \int_{z_j x(-\infty,\infty)} f_{\mu\nu}\, d\sigma^{\mu\nu}\right) Z_{(p1\ldots pb1)}$$

and may be contrasted with (37). Note that the action is augmented with the sum of integrals over *two*-dimensional subspaces

$$S_{\mathrm{top}} = i \sum_{j=1}^{b1} \frac{p_j}{2\pi} \int_{z_j x(-\infty,\infty)} f_{\mu\nu}\, d\sigma^{\mu\nu} \tag{48}$$

and this additional term will contribute significantly when the sum over topological sectors is performed, although the remarks made earlier concerning holonomy linking should be remembered.

It was pointed out in Ref. 9 that, in a general space-time $\mathcal{M}$, there will exist electromagnetic field configurations $f_{\mu\nu}$ such that

$$\int_{\mathcal{M}} f^*_{\mu\nu} f^{\mu\nu} \neq 0$$

provided that $H^2(\mathcal{M}; R)$ is nontrivial. This suggests adding $\int f^*_{\mu\nu} f^{\mu\nu}$ as an extra topological contribution to the action with an associated CP violating effect. We see however that this term does *not* arise from vacuum tunneling, which instead leads to (48). Indeed, in the present case

$$\int_{\Sigma x(-\infty,\infty)} f^*_{\mu\nu} f^{\mu\nu} = \int_{\Sigma x(-\infty,\infty)} f \wedge f = \int_{\Sigma x(-\infty,\infty)} d(A \wedge f)$$
$$= \int_{\Sigma^{(\infty)}} A \wedge f - \int_{\Sigma^{(-\infty)}} A \wedge f = 0 \quad (49)$$

since f vanishes on the initial and final hypersurfaces $\Sigma^{(-\infty)}$ and $\Sigma^{(\infty)}$.

The electromagnetic field energy is

$$E = \frac{1}{4} \int_{\Sigma} f_{ij} f^{ij} \sqrt{g}$$

and we have assumed above that the vacuum state corresponds to $f_{ij} = 0$. However, there is another possibility. If $H^2(\Sigma; Z) \neq 0$, the $U1$ bundle over Σ may be nontrivial and the resulting two-form f_{ij} is then closed but not exact. The energy E is now minimized, within any cohomology class, by a field strength w_{ij} that is harmonic (i.e., $dw = \delta w = 0$). There are precisely b_2 independent such two-forms where b_2 is the second Betti numbers of Σ.

Thus classical "vacuum" configurations are represented [up to elements of finite order in $H^2(\Sigma; Z)$] by harmonic two-forms and these can only locally be expressed in terms of gauge potentials. The analog of a gauge transformation is a base-point-preserving automorphism of the $U1$ bundle E onto itself, and the homotopy classes of such automorphisms will determine the analog of the $|\mathbf{n}\rangle$ vacua.

An automorphism is a base point-preserving map u from E onto E such that $u(pg) = u(p)g$ for all p in E and g in $U1$. Since u is fiber preserving and the $U1$ action is free, there exists a function $\tilde{u}$ from E into G such that

$$u(p) = p\tilde{u}(p), \qquad p \in E$$

and

$$u(pg) = u(p)g = pg\tilde{u}(pg) = p\tilde{u}(p)g \quad (50)$$

Hence $\tilde{u}(pg) = g^{-1}\tilde{u}(p)g = \tilde{u}(p)$ since $U1$ is Abelian. Thus $\tilde{u}$ defines a map $\tilde{\tilde{u}}$ on the base space-time by $\tilde{\tilde{u}}(x) = \tilde{u}(\pi^{-1}(x))$, where π is the projection of the $U1$ bundle. Hence automorphisms are equivalent to "gauge functions" from Σ into $U1$ and the classification of vacuum states is once again afforded by the cohomology group $H^1(\Sigma; Z)$.

Thus relative to a given vacuum state $|\alpha\rangle$ (where α labels a fixed $U1$ bundle) there exist a family $|\alpha; n_1 n_2 \ldots n_{b1}\rangle$. Tunneling can occur between states with the same α label but with different "winding numbers" $n_1 \ldots n_{b1}$. The idea of $|\boldsymbol{\theta}\rangle$ vacua and the associated diagonalized transition amplitudes may be developed as in the previous $U1$ case.

Note that there is no analog of this possibility when $G = SU(n)$ since $SU(n)$ bundles over a three-manifold Σ are labeled[2] by $H^4(\Sigma; Z) = 0$ and hence only the trivial bundle exists.

4. The Cases where $G = U(n)$ or $SO(n)$

Let us now discuss the more complicated situation in which both $\pi_1(G)$ and $\pi_3(G)$ are nonzero such as for example when $G = U(n)$ $(n \geq 2)$ or $G = SO(n)$ $(n \geq 3)$. Ideally the product space in (26) would serve as a model for $G_{(3)}$ but we do not know yet if this is possible. Let us try, more generally, to find a model for $G_{(3)}$ which is a $K(\pi_3(G), 3)$ fibration over $K(\pi_1(G), 1)$:

$$\begin{array}{ccc} K(\pi_3(G), 3) & \xrightarrow{i} & G_{(3)} \\ & & \downarrow P \\ & & K(\pi_1(G), 1) \end{array} \tag{51}$$

A very special fibration with fiber $K(\pi_1(G), 3)$ may be constructed as follows.[5–7,10] Let PK denote the set of paths in $K(\pi_3(G), 4)$ parametrized by $[0, 1] \subset R$, which start at the base point. Let $\pi: PK \to K(\pi_3(G), 4)$ be defined by $\pi(\omega) = \omega(1)$. Then PK is a fibration over $K(\pi_3(G), 4)$ whose typical fiber over the base point is the loop space $\Omega K(\pi_3(G), 4)$ and whose total space is contractible. The exact homotopy sequence (Section A.6) of the fibration gives

$$\pi_i(\Omega K(\pi_3(G), 4)) = \pi_{i+1}(K(\pi_3(G), 4)) \qquad \text{and hence } \Omega K(\pi_3(G), 4)$$

is an Eilenberg–Maclane space $K(\pi_3(G), 3)$ and we have the fibration ξ.

$$\begin{array}{ccc} K(\pi_3(G), 3) & \to & PK \\ & & \downarrow \pi \\ & & K(\pi_3(G), 4) \end{array} \tag{52}$$

Let f be any map from a space X into $K(\pi_3(G), 4)$. There exists an induced fibration $f^*(\xi)$ over X with total space

$$E(f^*(\xi)) = \{(x, p) \epsilon X \times PK \,|\, f(x) = \pi(p)\} \tag{53}$$

and which has the same fiber $K(\pi_3(G), 3)$ as ξ. The crucial result of Postnikov decomposition theory[6,7,10] is that there always exists a map θ from $K(\pi_1(G), 1)$ into

$K(\pi_3(G), 4)$ such that the induced fibration $\theta^*(\xi)$ has a total space which is a $G_{(3)}$. Furthermore the map p is a 1-equivalence (Section A.1). The resulting picture is

$$\begin{array}{ccc} K(\pi_3(G), 3) \xrightarrow{i} & G_{(3)} & \\ & \downarrow p & \\ & K(\pi_1(G), 1) \xrightarrow{\theta} K(\pi_3(G), 4) & \end{array} \tag{54}$$

and if θ is homotopically trivial (i.e., homotopic to a constant map) the fibration is the product (26). The main task is to analyze the map θ for various groups G of interest. Note that, by (27), θ determines and is determined by, an element of the cohomology group $H^4(K(\pi_1(G), 1); \pi_3(G)$. [Equation (27) remains true if Σ is any CW complex which an Eilenberg–Maclane space may always be assumed to be].

(a) $G = U(n)$, $n \geq 2$. In this case $\pi_1(U(n)) = \pi_3(U(n)) = Z$ and (54) becomes

$$\begin{array}{ccc} K(Z, 3) \xrightarrow{i} & U(n)_{(3)} & \\ & \downarrow p & \\ & K(Z, 1) \xrightarrow{\theta} K(Z, 4) & \end{array} \tag{55}$$

where θ determines an element of $H^4(K(Z, 1); Z)$. However, $K(Z, 1) = S^1$ and of course $H^4(S^1; Z) = 0$. Therefore θ in (55) must be homotopically trivial and the fibration is correspondingly a simple product:

$$U(n)_3 = K(Z, 1) \times K(Z, 3) \tag{56}$$

Hence we obtain a complete cohomological classification of the vacuum states:

$$[\Sigma, U(n)] = [\Sigma, U(n)_3] = [\Sigma, K(Z, 1)] \times [\Sigma, K(Z, 3)] = H^1(\Sigma; Z) \oplus H^3(\Sigma; Z) \tag{57}$$

Thus the vacuum states are labeled $|n; n_1 \ldots n_{b1}\rangle$ where n is the integer from $H^3(\Sigma; Z)$ (assuming Σ is orientable) and $n_1 \ldots n_{b1}$ are the b_1 integers from $H^1(\Sigma; Z)$. The χ-vacua are similarly easily obtained as $|\theta; \theta_1 \ldots \theta_{b1}\rangle$ and the tunneling amplitude (21) is an obvious mixture of (37) and (43).

(b) $G = SO(n)$, $n \geq 5$. The structure of $[\Sigma, SO(n)]$ depends on the value of n. In all cases ($n \geq 3$) we have $\pi_1(SO(n)) = Z_2$, $\pi_3(SO(n)) = Z$ and (54) becomes

$$\begin{array}{ccc} K(Z, 3) \xrightarrow{i} & SO(n)_{(3)} & \\ & \downarrow p & \\ & K(Z_2, 1) \xrightarrow{\theta} K(Z, 4) & \end{array} \tag{58}$$

Now $K(Z_2, 1) = RP^\infty$ and its cohomology groups are well known (Section A.4) with in particular $H^4(K(Z_2, 1); Z) = Z_2$. Thus there are two possible homotopy classes of θ giving, respectively, a trivial and nontrivial fibration over $K(Z_2, 1)$.

Consider the map f from $SO(n)$ into $SO(n)_3$ which induces the necessary relations between the first four homotopy groups. Clearly

$$f^*: H^1(SO(n)_{(3)}; Z_2) \approx H^1(SO(n); Z_2) = Z_2$$

and

$$p^*: H^1(K(Z_2, 1); Z_2) \approx H^1(SO(n)_{(3)}; Z_2)$$

since p is a 1-equivalence. Therefore

$$(p \cdot f)^*: H^1(K(Z_2, 1); Z_2) \approx H^1(SO(n); Z_2) \quad (59)$$

and if the nontrivial element of $H^1(SO(n); Z_2)$ [respectively, $H^1(K(Z_2, 1); Z_2)$] is denoted by x (respectively, l_1) we have

$$(p \cdot f)^*(l_1) = x \quad (60)$$

Now if $n \geq 5$ the cohomology of $H^4(SO(n); Z_2)$ is such that[6] $x^4(\equiv x \cup x \cup x \cup x)$ does not vanish and therefore

$$(p \cdot f)^*(l_1^4) = [(p \cdot f)^*(l_1)]^4 = x^4 \neq 0 \quad (61)$$

On the other hand the nontrivial generator of $H^4(K(Z_2, 1); Z) = Z_2$ has the property that its mod 2 reduction is just l_1^4 and from (61)

$$p^*(l_1^4) \neq 0 \quad (62)$$

Now if ι_4 denotes a generator of $H^4(K(Z, 4); Z) = \mathrm{Hom}(H_4(K(Z, 4); Z); Z) = \mathrm{Hom}(\pi_4(K(Z, 4)); Z) = \mathrm{Hom}(Z, Z) = Z$ (see Section A.3), then there are two possibilities:

$$\text{(i) } \theta \text{ is trivial and } \theta^*(\iota_4) = 0 \quad (63)$$

$$\text{(ii) } \theta \text{ is nontrivial and } \theta^*(\iota_4) \bmod 2 = l_1^4 \neq 0 \quad (64)$$

However, a general property of the Postnikov system in (54) is that the composite map $\theta \cdot p$ is homotopic to a constant[6,10] and hence $p^*\theta^*(\iota_4) = 0$. But (62) and (64) show that this is not possible if θ is nontrivial and so we deduce that θ is homotopically trivial. Thus we again have a product representation:

$$SO(n)_{(3)} = K(Z_2, 1) \times K(Z, 3), \qquad n \geq 5 \quad (65)$$

and

$$[\Sigma, SO(n)] = H^1(\Sigma; Z_2) \oplus H^3(\Sigma; Z) \quad (66)$$

Every element in $H^1(\Sigma; Z_2) = \mathrm{Hom}(\pi_1(\Sigma); Z_2)$ is clearly of order 2 and hence this cohomology group is a direct sum of copies of the cyclic group of order two (Z_2). Vacuum states are correspondingly labeled $|n; {}^{1111}_{0000} \ldots {}^{1}_{0}\rangle$, where n is the integer associated with $H^3(\Sigma; Z) = Z$ (assuming that Σ is orientable) and ${}^{1}_{0}$ represents the generator (1) or identity (0) of a Z_2 group. The dual χ states are $|\theta; \pm \pm \pm \ldots\rangle$ where $+$ and $-$ refer to the characters (the character group of Z_2 is Z_2):

$$+(1) = 1, \qquad +(0) = 1 \quad (67)$$
$$-(1) = -1, \qquad -(0) = 1$$

It is not obvious if the Z_2 characters can be represented by gauge fields, but nevertheless their presence in the general tunneling amplitude (21) will have significant effects on the physical predictions.

(c) $SO(n)$, $n = 3$ or 4. Topologically $SO4 \sim SO3 \times SU2$ and hence $[\Sigma, SO4] = [\Sigma, SO3] \times [\Sigma, SU(2)]$, which reduces the $SO4$ problem to $SO3$ since we know that $[\Sigma, SU2] = H^3(\Sigma; Z)$ [Eq. (34)]. The discussion above for $SO(n)$ $n \geq 5$ is not applicable to $SO3$ since this group is three dimensional and hence $H^4(SO3; Z_2) = 0$, which implies that x^4 vanishes identically. Suppose that θ in (58) was trivial for $n = 3$. Then

$$SO3_{(3)} = K(Z_2, 1) \times K(Z, 3) \tag{68}$$

and (Section A.5)

$$H^3(SO3_{(3)}; Z_2) = Z_2 \oplus Z_2 \tag{69}$$

However, in fact $H^3(SO3_{(3)}; Z_2) = H^3(SO3; Z_2) = Z_2$, which contradicts (69) and hence θ must correspond to the nonzero element of $H^4(K(Z_2, 1); Z) = Z_2$ so that the fibration is *not* in product form.

Given any fibration

$$\begin{array}{ccc} F & \xrightarrow{i} & E \\ & & \downarrow p \\ & & B \xrightarrow{\theta} D \end{array} \tag{70}$$

induced from the path space over D we can generate the Puppe exact sequence[2,5–7,10]

$$\to [\Sigma, \Omega E] \xrightarrow{\Omega p_*} [\Sigma, \Omega B] \xrightarrow{\Omega\theta_*} [\Sigma, F] \xrightarrow{i_*} [\Sigma, E] \xrightarrow{p_*} [\Sigma, B] \xrightarrow{\theta_*} [\Sigma, D] \tag{71}$$

where for example $\Omega p : \Omega E \to \Omega B$ is defined in the obvious way as $(\Omega p(w))(t) = p(w(t))$. "Exact" means that for any consecutive pair of maps $[\Sigma, A] \xrightarrow{\alpha} [\Sigma, B] \xrightarrow{\beta} [\Sigma, C]$, $\beta^{-1}(*) = \operatorname{Im}\alpha$, where * is the constant map from Σ into the base point of C. If these sets possess natural group structures (for example if F, E, B, D and i, p, θ are all "loops" of other spaces and maps) then the linking maps will be homomorphisms and (71) is exact in the sense of group theory.

Applying this to (58) gives the short exact sequence (assuming that Σ is connected so that $[\Sigma, Z_2] = 0$)

$$0 \to H^3(\Sigma; Z) \xrightarrow{i_*} [\Sigma, SO3] \xrightarrow{p_*} H^1(\Sigma; Z_2) \to 0 \tag{72}$$

which if it splits would reproduce (66). The topological properties of Σ determine whether or not (72) splits, and this aspect has been extensively studied by Shastri *et al.*[11] in the context of Misner–Finkelstein kinks. I will proceed in a somewhat different way in order to illustrate further the Postnikov method.

Let $r : SO3_{(3)} \to K(Z_2, 1) \times K(Z, 3)$ correspond to the generators (Section A.8)

x and a of $H^1(SO3_{(3)}; Z_2) \oplus H^3(SO3_{(3)}; Z) = Z_2 \oplus Z$. Without loss of generality we may assume that r is a fibration[5–7,10]

$$\begin{array}{ccc} F & \xrightarrow{i} & SO3_{(3)} \\ & & \downarrow r \\ & & K(Z_2, 1) \times K(Z, 3) \end{array} \tag{73}$$

and try to find the homotopy groups of F.

Applying the exact homotopy sequence (Section A.6) gives

$$\to \pi_4(K(Z_2, 1) \times K(Z, 3)) \to \pi_3(F) \xrightarrow{i_*} \pi_3(SO3_{(3)}) \xrightarrow{r_*} \pi_3(K(Z_2, 1) \times K(Z, 3)) \to \pi_2(F)$$
$$\to \pi_2(SO3_{(3)}) \to \pi_2(K(Z_2, 1) \times K(Z, 3)) \to \pi_1(F) \xrightarrow{i_*} \pi_1(SO3_{(3)})$$

$$\xrightarrow{r_*} \pi_1(K(Z_2, 1) \times K(Z, 3))$$
$$\to \pi_0(F) \to \pi_0(SO3) \to \pi_0(K(Z_2, 1) \times K(Z, 3))$$

i.e.,

$$0 \to \pi_3(F) \to Z \xrightarrow{r_*} Z \to \pi_2(F) \to 0 \tag{74}$$
$$0 \to \pi_1(F) \to Z_2 \xrightarrow{r_*} Z_2 \to \pi_0(F) \to 0 \tag{75}$$

Now, since $pr_1 \cdot r = p$ is a one-equivalence (pr_1 means project onto the first factor), r_* is an isomorphism in (75) and hence $\pi_0(F) = \pi_1(F) = 0$ but $\pi_2(F)$ and $\pi_3(F)$ are not fully determined by (74).

From (58) and (73) we have the diagram

$$\begin{array}{ccccc} K(Z, 3) & \xrightarrow{i} & SO3_{(3)} & \xrightarrow{pr2\cdot r \equiv t} & K(Z, 3) \\ & & \downarrow pr_1 \cdot r = p & & \\ & & K(Z_2, 1) & & \end{array} \tag{76}$$

Now apply the Serre short exact cohomology sequence (Section A.7) with Z coefficients to this fibration to yield

$$\to H^2(K(Z, 3)) \xrightarrow{\tau} H^3(K(Z_2, 1)) \xrightarrow{p^*} H^3(SO3_{(3)}) \xrightarrow{i^*} H^3(K(Z, 3)) \xrightarrow{\tau} H^4(K(Z_2, 1))$$
$$\xrightarrow{p^*} H^4(SO3_{(3)})$$

i.e.,

$$0 \to 0 \to Z \xrightarrow{i^*} Z \xrightarrow{\tau} Z_2 \to 0 \tag{77}$$

where $H^4(SO3_{(3)}; Z)$ is a subgroup of $H^4(SO3; Z) = 0$ and hence vanishes.

Equation (77) implies that i^* is x2 on $H^3(SO3_{(3)})$ and since t^* is by construction x1 on $H^3(K(Z, 3))$ it follows that $(t \cdot i)^*$ is a x2 map from $H^3(K(Z, 3)) = Z$ into $H^3(K(Z, 3))$. Hence $(t \cdot i)_*$ is x2 on both $H_3(K(Z, 3)) = Z$ and $\pi_3(K(Z, 3))$. But the exact homotopy sequence of (76) shows that i_* is x1 on $\pi_3(K(Z, 3))$ and hence t_* is a x2 map from $\pi_3(SO3_{(3)})$ into $\pi_3(K(Z, 3))$. This implies that r_* in (74) is also x2 and hence

$$\pi_2(F) = Z_2, \qquad \pi_3(F) = 0 \tag{78}$$

The fibration (73) can also be built by the Postnikov method and we study the diagram

$$\begin{array}{ccc} K(Z_2, 2) \xrightarrow{j} & SO3_{(3)} & \\ & \downarrow r & \\ & K(Z_2, 1) \times K(Z, 3) \xrightarrow{\omega} K(Z_2, 3) & \end{array} \tag{79}$$

Clearly ω defines an element of $H^3(K(Z_2, 1) \times K(Z, 3); Z_2)$ and this group (Section A.5) is $Z_2 \oplus Z_2$ with generators $1 x \iota_3$ mod 2 and $l_1^3 \times 1$ where ι_3 is a generator of $H^3(K(Z, 3); Z) = Z$. Thus, if l_3 denotes the nontrivial generator of $H^3(K(Z_2, 3); Z_2) = \mathrm{Hom}(\pi_3(K(Z_2, 3)); Z_2) = \mathrm{Hom}(Z_2; Z_2) = Z_2$, there are four possibilities:

$$\omega^*(l_3) = 0 \tag{80}$$

$$\omega^*(l_3) = 1 x\, \iota_3 \bmod 2 \tag{81}$$

$$\omega^*(l_3) = l_1^3 \times 1 \tag{82}$$

$$\omega^*(l_3) = l_1^3 \times 1 + 1 \times \iota_3 \bmod 2 \tag{83}$$

Choice (80) is ruled out as it would imply that the fibration in (79) was trivial and hence $SO3_{(3)} = K(Z_2, 1) \times K(Z, 3) \times K(Z_2, 2)$, which gives the false result $\pi_2(SO3_{(3)}) = Z_2$.

By the definition of r, $r^*\ (1 \times \iota_3 \bmod 2) = a \bmod 2 \neq 0$ and $r^*(l_1^3 \times 1) = x^3 \neq 0$. But $\omega \cdot r$ is homotopically trivial and hence $r^*\omega^*(l_3) = 0$. Thus neither (81) nor (82) is acceptable. Fortunately a and x are related (Section A.8) in $H^3(SO3; Z_2)$ by $a \bmod 2 = x^3$ and hence

$$r^*[(l_1^3 \times 1) + (1 \times \iota_3 \bmod 2)] = 0 \tag{84}$$

and the choice (83) must be the correct one. Thus (79) leads to the Puppe sequence:

$$H^2(\Sigma, Z) \xrightarrow{\text{mod } 2} H^2(\Sigma; Z_2) \xrightarrow{i_*} [\Sigma, SO3] \xrightarrow{r_*} H^1(\Sigma; Z_2) \oplus H^3(\Sigma; Z) \xrightarrow{\omega_*} H^3(\Sigma; Z_2) \tag{85}$$

with

$$\begin{aligned} r_*[f] &= (f^*(x), f^*(a)) \\ \omega_*(\alpha, \gamma) &= \alpha^3 + \gamma \bmod 2 \end{aligned} \tag{86}$$

Now when Σ is orientable consider the Bockstein exact sequence[7] induced by the coefficient group homomorphisms $Z \xrightarrow{x2} Z \xrightarrow{\text{mod } 2} Z_2$:

$$\begin{array}{ccccccccc} \to H^2(\Sigma, Z) & \xrightarrow{\text{mod } 2} & H^2(\Sigma; Z_2) & \xrightarrow{\beta} & H^3(\Sigma; Z) & \xrightarrow{\text{mod } 2} & H^3(\Sigma; Z) & \xrightarrow{x2} & H^3(\Sigma; Z_2) \\ & & & & \| & & \| & & \| \\ & & & & Z & & Z & & Z_2 \end{array} \tag{87}$$

It follows by exactness that the Bockstein map β is trivial and hence mod 2 is a surjective map from $H^2(\Sigma; Z)$ onto $H^2(\Sigma; Z_2)$. This implies in (85) that j_* has $H^2(\Sigma; Z_2)$ as its kernel and hence $\text{Im}j_* = \ker r_* = \{0\}$. Thus r_* is injective and $\text{Im}r_* = \ker\omega_*$, which gives the classification

$$[\Sigma, SO3] = \{(\alpha, \beta)\epsilon H^1(\Sigma; Z_2) \oplus Z | \alpha^3 + \beta \bmod 2 = 0\} \tag{88}$$

[Strictly speaking (85) may only be exact in the sense of sets since ω is not the loop $\Omega\phi$ of any ϕ. However, more sophisticated arguments[13] show that the result (88) is still true.]

It is not perhaps obvious that the right-hand side of (88) is a subgroup. However, let $\alpha_1^3 = \beta_1 \bmod 2$ and $\alpha_2^3 = \beta_2 \bmod 2$; then

$$\begin{aligned}(\alpha_1 + \alpha_2)^3 &= \alpha_1^3 + \alpha_2^3 + \alpha_1^2\alpha_2 + \alpha_2^2\alpha_1 \\ &= (\beta_1 + \beta_2) \bmod 2 + (\alpha_1^2\alpha_2 + \alpha_1\alpha_2^2)\end{aligned} \tag{89}$$

Now

$$\alpha_1^2\alpha_2 + \alpha_1\alpha_2^2 = \text{Sq}^1(\alpha_1 \cup \alpha_2) \tag{90}$$

and the Steenrod square[5–7,10] may be expressed in terms of the operators in (87) by $\text{Sq}^1 = \bmod 2 \cdot \beta$. However, as we have just seen β is trivial on $H^2(\Sigma; Z_2)$ and hence (90) vanishes and so $[\Sigma, SO3]$ really is an Abelian subgroup of $H^1(\Sigma; Z_2) \oplus Z$.

$H^1(\Sigma, Z_2)$ is a direct sum Z_2^N of, say, N, Z_2 groups and hence ω_* in (85) is a homomorphism of $Z_2^N \oplus Z$ onto $H^3(\Sigma; Z_2) = Z_2$. The homotopy set $[\Sigma, SO3]$ is isomorphic to the kernel of this map, as in (88), which is $Z_2^N \oplus 2Z$ or $Z_2^{N-1} \oplus Z$ depending on whether there exist nonvanishing α^3. This corresponds to cases where (72) respectively does and does not split as discussed by the authors of Ref. 11. This reference may also be consulted for a list of values of $[\Sigma, SO3]$ for a number of interesting examples of Σ.

The vacuum states are evidently labeled by $|n; \alpha_1, \alpha_2 \ldots \alpha_N\rangle$ where the particular element $\alpha = (\alpha_1, \alpha_2, \ldots \alpha_N) \epsilon H^1(\Sigma; Z_2)$ has to satisfy the condition $\alpha \cup \alpha \cup \alpha = n \bmod 2$ in $H^3(\Sigma; Z_2) = Z_2$. Apart from this restriction the statements on tunneling, etc., are essentially the same as before.

Note that, as in the $U1$ case, it is possible for the initial $U(n)$ or $SO(n)$ bundle to be *non*trivial over Σ and such bundles are classified by elements of $H^2(\Sigma; Z)$ and $H^2(\Sigma; Z_2)$, respectively. The computation of the bundle automorphisms is more complicated than is the $U1$ case and needs further study.

5. Conclusions

We have seen that by using Postnikov methods the homotopy classes $[\Sigma, G]$ and hence the "n vacua," may be completely classified in terms of cohomology groups of Σ with coefficients in the first and third homotopy groups of G. Thus the topological structures of Σ and G are intertwined. The results are summarized in Table I.

TABLE I. The n-Vacua for Special Choices of the Group G

Group G	$[\Sigma, G]$	$\vert n\rangle$ vacuua
$SU(n)$, G_2 Spin (n), etc.	$H^3(\Sigma; Z)$	$\vert n\rangle$
$U(1)$	$H^1(\Sigma; Z)$	$\vert n_1, n_2 \ldots n_{b1}\rangle$
$U(n)$	$H^1(\Sigma; Z) \oplus H^3(\Sigma; Z)$	$\vert n; n_1 n_2 \ldots n_{b1}\rangle$
$SO(n)$, $n \geq 5$	$H^1(\Sigma; Z_2) \oplus H^3(\Sigma; Z)$	$\vert n; {}^{111}_{000} \ldots {}^{1}_{0}\rangle$
$SO(4)$	$[\Sigma, SO3] \oplus H^3(\Sigma; Z)$	$\vert n_1; n_2; \alpha\rangle$ $n_2 \bmod 2 = \alpha^3$
$SO(3)$	$\{(\alpha, \gamma) \epsilon H^1(\Sigma; Z_2) \oplus Z \vert \alpha \cup \alpha \cup \alpha = \gamma \bmod 2\}$ (Σ orientable)	$\vert n; \alpha\rangle$, $n \bmod 2 = \alpha^3$

In all cases quantum tunneling can occur and the corresponding χ vacua diagonalize these amplitudes as

$$\lim_{T\to\infty} \langle \chi_1 | e^{-\hat{H}T} | \chi_2 \rangle \sim \delta_{\chi 1 \chi 2} \sum_{h \epsilon [\Sigma, G]} \bar{\chi}(h) Z_{(h)} \tag{91}$$

The χ are the characters of the Abelian group $[\Sigma, G]$ and their presence in (91) will have a significant effect on nonperturbative calculations involving summing over the different topological sectors. In the $SU(n)$ and $U(n)$ case, n can be expressed in terms of the Yang–Mills field strengths albeit in the $U1$ part of $U(n)$ as an integral over two-dimensional subspaces of space-time. The four-dimensional $SU(n)$ contribution will yield CP violating effects as discussed in Deser *et al.*[9]

It is possible for either $SO(n)$ or $U(n)$ bundles to be nontrivial over the initial hypersurface, in which case the different "n vacua" relative to an initial energy minimizing field configuration are generated by homotopically nontrivial automorphisms of the principal bundle. The full details of this situation require further investigation as does the significance of configurations corresponding to a nonvanishing holonomy group.

Many treatments of gauge theories employ fiber bundles on the four-dimensional space-time manifold. This may be related to the present work by considering the implications of the boundary conditions $\mathbf{F}_{\mu\nu} \to 0$ at $t = \pm\infty$ on $\Sigma \times R$. This enables the space to be conformally compactified and the end points identified to a point. The resulting space is the suspension $S\Sigma$ of Σ and principal G bundles over $S\Sigma$ are in one-to-one correspondence with homotopy classes of maps from Σ into G. If $\Omega_1, \Omega_2 : \Sigma \to G$ correspond to the initial and final vacuum states, then the associated bundle is determined by the class $\Omega_1 - \Omega_2$ in the Abelian group $[\Sigma, G]$.

A related observation is that bundles over a four-dimensional complex $\mathcal{C}$ are classified[1,2] by elements of $H^2(\mathcal{C}; \pi_1(G))$ and $H^4(\mathcal{C}; \pi_3(G))$ and that

$$H^2(S\Sigma; \pi_1(G)) = H^1(\Sigma; \pi_1(G))$$
$$H^4(S\Sigma; \pi_3(G)) = H^3(\Sigma; \pi_3(G))$$

thus establishing a link with the groups employed in this paper. The appropriate characteristic classes are the first and second Chern classes (C_1 and C_2) for $U(n)$ bundles [just C_1 for $U1$ and C_2 for $SU(n)$], the first Pontryagin class (p_1) and the second Stieffel–Whitney class (ω_2) for $SO(n)$ bundles ($n \geq 3$) plus the Euler class (e) for $SO(4)$. These classes belong to the cohomology groups:

$$C_1 \epsilon H^2(\mathcal{C}; Z), \qquad C_2, p_1, e \epsilon H^4(\mathcal{C}; Z), \qquad \text{and } \omega_2 \epsilon H^2(\mathcal{C}; Z_2)$$

For example the second Chern number of an $SU(n)$ bundle on $S\Sigma$ that interpolates between initial and final vacua $|n_1\rangle$ and $|n_2\rangle$ is $n_1 - n_2$ whilst for a $SO(n)$ bundle the Pontryagin number is $2(n_1 - n_2)$. Similarly C_1 interpolates the $n_1 \ldots n_{b_1}$ for $U(n)$ theories and ω_2 the $\binom{111}{000} \ldots \binom{1}{0}$ for $SO(n)$ theories.

Finally one may ask whether any of the ideas are appropriate for quantum gravity itself. One possibility is to study the Wheeler–de Witt quantization in which the state vector is a functional $\psi[^3g]$ of the intrinsic 3-geometry of the spatial hypersurface Σ. The natural gauge group is the group of automorphisms of the tangent bundle of Σ. If Σ is orientable and if it admits spinors then it is necessarily parallelizable and hence the tangent bundle is trivial. Triads of tangent vectors $\{e_a\}$ may be globally defined and a bundle automorphism is represented by a gauge function from Σ into $SO3$ which rotates the triads (but preserves the metric)

$$e_a \rightarrow e_b \Omega^b_a, \qquad a, b = 1 \ldots 3, \quad \Omega : \Sigma \rightarrow SO3 \tag{92}$$

$$g_{\mu\nu} = e_{\mu a} e_{\nu b}\, \delta^{ab}, \qquad \mu, \nu = 1 \ldots 3 \tag{93}$$

Chose a global triad of vectors f_a once and for all. Then any other (orientation-preserving) triad may be written as $e_a = f_b \Lambda^b_a$ for some maps Λ from Σ into $GL^+(3, R)$ (the set of 3×3 invertible matrices with positive determinant) and the corresponding metric is

$$g_{\mu\nu} = f_{\mu b} f_{\nu c} \Lambda^b_a \Lambda^c_d\, \delta^{ad} \tag{94}$$

It is natural to postulate that the wave functional ψ depends not only on $g_{\mu\nu}$ but also on the homotopy class of Λ, i.e., $\psi = \psi[^3g, [\Lambda]]$. Topologically $GL^+(3, R) = SO3 \times R^6$ and hence

$$[\Sigma, GL^+(3, R)] = [\Sigma, SO3] \times [\Sigma, R^6] = [\Sigma, SO3] \tag{95}$$

where the classification of $[\Sigma, SO3]$ was performed in Section 4. Thus with respect to the initial choice of parallelization $\{f_a\}$, the quantum state functionals may be labeled

$$\psi = \psi[^3g; \gamma, \alpha] \tag{96}$$

where $\alpha \epsilon H^1(\Sigma; Z_2)$, $\gamma \epsilon H^3(\Sigma; Z) = Z$ and $\alpha \cup \alpha \cup \alpha = \gamma \bmod 2$.

The tunneling interpretation of such functionals is problematical. First, it is not clear what is meant by a "vacuum" state—is it any solution of the Wheeler–De Witt equation? Secondly the concept of tunneling from one state (96) to another is not meaningful in this canonical formalism as there are no explicit time labels on the state vectors. Rather, the time is supposed to be extracted in some intrinsic way from the three-metric itself [e.g., by defining $t = \int_\Sigma ({}^3g)^{1/2}$] and hence in this indirect sense (96) is already a Schrödinger picture state. To develop a tunneling in time within this picture it would be necessary to first identify an implicit time variable and then study the topological effects on the remaining two degrees of freedom. This whole subject deserves further study.

Note that defining a vacuum configuration, in analogy with (3), as a spin connection $\omega_{\mu ab}$ that is pure gauge is not very satisfactory as it implies that the three-curvature vanishes on Σ. However, this is only possible[12] if the Betti numbers of Σ satisfy $b_1(\Sigma) = b_2(\Sigma) = 3$ (as for example on the torus $S^1 \times S^1 \times S^1$), which of course they may not.

Appendix: Some Useful Results in Algebraic Topology

A.1. An *n equivalence*[5–7,10] between two connected topological spaces X and Y is a map $f: X \to Y$ such that f_* maps $\pi_i(X)$ isomorphically onto $\pi_i(Y)$ for $1 \leq i \leq n$ and f_* is an epimorphism from $\pi_{n+1}(X)$ onto $\pi_{n+1}(Y)$. If Σ is a n-dimensional manifold (or more generally a n-dimensional CW complex) then $[\Sigma, X] = [\Sigma, Y]$.

A.2. If R is any ring and G is a R module the *universal coefficient theorem*[7] for cohomology states that

$$H^q(X; G) = \mathrm{Hom}(H_q(X; R); G) \oplus \mathrm{Ext}(H_{q-1}(X, R); G) \tag{A.1}$$

where $\mathrm{Ext}(A, B)$ is the set of extensions of B by A, i.e., short exact sequences

$$0 \to A \to E \to B \to 0$$

(In particular this applies for $R = Z$ and G any Abelian group.)

A.3. The *Hurewicz theorem* states that[5–7,10]

$$\pi_1(X)/(\text{commutator subgroup}) = H_1(X; Z)$$

and that in general if $\pi_1(X) = \pi_2(X) = \pi_{n-1}(X) = 0$ then

$$H_1(X; Z) = H_2(X; Z) \cdots H_{n-1}(X; Z) = 0 \quad \text{and} \quad \pi_n(X) = H_n(X; Z)$$

In this case (A.1) gives, with $R = Z$ and G an Abelian group,

$$\begin{aligned} H^n(X; Z) &= \mathrm{Hom}(H_n(X; Z); G) \\ &= \mathrm{Hom}(\pi_n(X); G) \end{aligned} \tag{A.2}$$

A.4. The Z cohomology groups of $K(Z_2, 1)$ are[7,8,10]

$$H^1 = H^3 = H^5 = \cdots H^{2n+1} = \cdots = 0, \qquad H^2 = H^4 = H^6 = \cdots H^{2n} = \cdots Z_2 \tag{A.3}$$

If α denotes the generator of $H^2(K(Z_2, 1); Z)$ then α^n is the generator of $H^{2n}(K(Z_2, 1); Z)$. The Z_2 cohomology groups of $K(Z_2, 1)$ are all copies of Z_2 with generators l_1, $l_1^2, l_1^3 \ldots$, where l_1 is the generator of $H^1(K(Z_2, 1); Z_2) \approx Z_2$ and $\alpha \bmod 2 = l_1^2$.

A.5. The Kunneth[7,8] formula for the Z_2 cohomology of the product of two spaces is

$$H^q(X \times Y; Z_2) = \bigoplus_{i=0}^{q} H^i(X; Z_2) \otimes H^{q-i}(Y; Z_2) \tag{A.4}$$

and the image of $a \otimes b \in H^i(X; Z_2) \otimes H^{q-i}(Y; Z_2)$ in $H^q(X \times Y; Z_2)$ is denoted $a \times b$

A.6. If $F \xrightarrow{i} E \xrightarrow{p} B$ is a fibration there is a long exact homotopy sequence[5–7,10]

$$\begin{aligned} &\to \pi_n(F) \xrightarrow{i_*} \pi_n(E) \xrightarrow{p_*} \pi_n(B) \xrightarrow{\partial} \pi_{n-1}(F) \to \cdots \\ &\to \pi_1(E) \xrightarrow{p_*} \pi_1(B) \xrightarrow{\partial} \pi_0(F) \xrightarrow{i_*} \pi_0(E) \xrightarrow{p_*} \pi_0(B) \end{aligned} \tag{A.5}$$

A.7. If $F \xrightarrow{i} E \xrightarrow{p} B$ is an orientable fibration in which $\pi_1(B)$ acts trivially on F, there is short exact sequence (the Serre sequence), for any coefficient ring,[6,7,10] which truncates at some point after the $H^2(B)$ term:

$$0 \to H^1(B) \xrightarrow{p^*} H^1(E) \xrightarrow{i^*} H^1(F) \xrightarrow{\tau} H^2(B) \to \cdots \tag{A.6}$$

(i) $H^{n+1}(B) \xrightarrow{p^*} H^{n+1}(E) \xrightarrow{i^*} H^{n+1}(F)$ is exact (i.e., $\ker i^* = \operatorname{Im} p^*$) if $(n \geq 1)$.

$$H^1(B; H^n(F)) = H^2(B; H^{n-1}(F)) = H^n(B; H^1(F)) = 0 \tag{A.7}$$

(ii) The dimension raising map τ is defined on $H^n(F)$, maps into $H^{n+1}(B)$, $\operatorname{Im} i^* = \ker \tau$ and $\operatorname{Im} \tau = \ker p^*$ if (A.7) and (A.8) are true $(n \geq 1)$:

$$H^2(B; H^{n-1}(F)) = H^3(B; H^{n-2}(F)) = \cdots H^n(B; H^1(F)) = 0 \tag{A.8}$$

(trivial if $n = 1$).

A.8. The Z and Z_2 cohomology groups of SO3 are (generators in brackets)

$H^1(SO3; Z) = 0,$		$H^1(SO3, Z_2) = Z_2$	(x)
$H^2(SO3; Z) = Z_2$	$(\beta(x)),$	$H^2(SO3; Z_2) = Z_2$	(x^2)
$H^3(SO3; Z) = Z$	$(a),$	$H^3(SO3, Z_2) = Z_2$	$(x^3 = a \bmod 2)$

References and Notes

1. A. Dold and H. Whitney, *Ann. Math.* **69**, 3 (1959).
2. S. J. Avis and C. J. Isham, "Quantum Field Theory and Fibre Bundles in a General Spacetime," in *Recent Developments in General Relativity*, S. Deser and M. Levy, editors (Plenum New York, 1979).
3. C. G. Callan, R. F. Dashen, and D. J. Gross, *Phys. Let.* **63B,** 334 (1976).
4. R. Jackiw and C. Rebbi, *Phys. Rev. Lett.* **37,** 172 (1976).
5. R. M. Switzer, *Algebraic Topology—Homotopy and Homology* (Springer-Verlag, New York, 1975).
6. G. W. Whitehead, *Elements of Homotopy Theory* (Springer-Verlag, New York, 1978).

7. E. H. Spanier, *Algebraic Topology* (McGraw-Hill, New York, 1966).
8. A. Dold, *Lectures in Algebraic Topology* (Springer-Verlag, New York, 1972).
9. S. Deser, M. J. Duff, and C. J. Isham, *Phys. Lett. B* **93**, 419 (1980).
10. R. E. Mosher and M. C. Tangora, *Cohomology Operations and Applications in Homotopy Theory* (Harper and Row, New York, 1968).
11. A. R. Shastri, J. G. Williams, and P. Svengrowski, *Int. J. Ther. Phys.* **19**, 1 (1980).
12. S. J. Goldberg, *Curvature and Homology* (Academic Press, New York, 1962).
13. J. W. Rutter, *Topology* **6**, 379 (1967).

14

On the Extensions of General Relativity

D. Ivanenko
Faculty of Physics
Moscow State University
Moscow, USSR

1. Successes and Difficulties of Gravitation Theory

Einstein's general relativity theory (GRT), while remaining the recognized foundation of gravitational theory, is in no way devoid of some serious difficulties, as seems to be more and more accepted by the physics community today. To declare this openly requires courage of the kind so admirably evinced by Wolfgang Yourgrau. He stimulated vivid discussions on these and other fundamental problems, which is one reason his memory will forever remain alive among his many friends and colleagues.

Without trying to present here the overall picture of difficulties plaguing GRT and list the full catalog of proposed generalizations, we should like to concentrate on the gauge theory of gravity, particularly its fiber bundle formalization, while only touching on some other, not less important, problems. For details the reader is referred to the relevant literature, which includes recent papers of our own group.[1–4]

The situation with GR difficulties reminds us of Newton's theory, whose great successes in explaining planetary motions and all terrestrial phenomena put aside grave doubts of Leibniz, Berkeley, and later Mach, about the notions of absolute space and time, and of the Cartesians about the seemingly action-at-a-distance gravitation law. But, as Einstein remarks in the preface of the valuable book of Max Jammer, *Concept of Space*, Newton achieved all that was possible at that time and, as later became clear, constructed the correct classical (nonquantum and nonrelativistic) picture of the universe. But Newtonian theory suffered a real blow in cosmology, leading to obvious infinities of gravitational forces and luminosity when a static distribution of stars in an infinite Euclidean universe was assumed. To remove both these paradoxes two interesting hypotheses were proposed:

(1) Lambert (later elaborated by Charlier *et al.*) assumed the hierarchical structure of the universe, where (in modern language, *grosso modo*) our observed metagalaxy possesses the same components, considered as "molecules," particles from the

viewpoint of a higher level, etc.—the dimensions and density assumed to guarantee the finite limit for potential and luminosity. It is very interesting that, although both paradoxes were removed by GR, Einstein even in 1922 did not protest against the hierarchical picture, and now unexpectedly it again appears in the picture of elementary particles as microuniverses (Lambertian lower level!); the dimensions of particles are of the order of the gravitational Schwarzschild-type radius with Salam's large strong gravity constant, quite analogous to the wonderful fact that the observed dimensions of the metagalaxy ($\sim 10^{28}$ cm) are of the order of its Schwarschild GR radius, so that our metagalaxy is a type of black hole, like the particle ("strong" black hole) (cf. Salam, Roman, Treder, Ivanenko, Aman, Pavsič, Recami, *et al.*).

The other proposal to remove the two paradoxes is due to Seeliger and Neuman, who introduced the short-range ("Yukawa"-type) potential $(1/r)\exp(-\kappa_0^2 r)$ instead of the Newtonian $(1/r)$, which is realized, so to speak, in modern fashion not by massless gravitons but by massive mesonlike particles of spin 2 (also Salam's strong gravity hypothesis). But these last revivals are of quite recent years. The real evolution of gravity theory has passed through special relativity (SR) and electromagnetic field theory, with the first real post-Newtonian step made by Poincaré, who applied SR to gravitation, predicting its finite velocity of propagation, and introduced four-dimensional pseudo-Euclidean flat space in the first famous papers of 1905–1906, thus establishing SR, which was also founded by Einstein in his widely known papers of 1905, etc. A further step was achieved by the Nordström–Abraham scalar relativity, but the general line led through Einstein's equivalence principle to the idea of geometrized gravity with potentials coinciding with Riemann metric components (Einstein with the participation of M. Grossmann, 1913). The final generally covariant equations were established in November 1915 in parallel papers by Hilbert (influenced by the Einstein program and not paying much attention to the covariance problem) and, some days later, by Einstein. With the Hilbert Lagrangian, the scalar of curvature R became the cornerstone of the theory, though one that was not free of difficulties, due to the not fully clarified problem of boundary conditions (cf. the recent work of Hawking and Gibbons, Obukhov in our group, etc.). (The chief point here is that, on rejecting the divergence term, one has the inconsistency with boundary conditions of the degree of the differential equations; if one uses the divergence-free Lagrangian one is confronted with its nontensor character. Let us indicate already here that the choice of the Lagrangian presents difficulties also in the Einstein–Cartan theory in relation to torsion. Many types of quadratic Lagrangians were analyzed (Buchdahl, Lanczos, Frolov-Obukhov in our group, Stelle, Hehl, Charap, Camesind, Lopez, Loos, Yang, Fairchild, etc.). The most appropriate choice seems now to be a Hilbert-type scalar, possibly constructed from generalized connections containing Cartan torsion and Weyl dilation terms and supplemented by various quadratic terms suggested by various arguments, especially the Yang–Mills field analogies. To finish this short digression about Lagrangians, let us point out that quadratic Lagrangians are necessarily introduced by quantum gravity. Quadratic Lagrangians lead in the Einstein–Cartan theory not to algebraic (as in the case of linear **R**) but to differential coupling of spin and torsion; in this manner they provide us with torsion waves and, in quantum version, with their quanta, "tordions."

Concerning the Einstein–Hilbert equations of the gravitational field one must note that both eminent authors have primarily not remarked on the necessity of the cosmological term, compulsory on formal grounds and now recognized necessary for cosmology even by its former adversaries (e.g., Zeldovich and Wheeler) and erstwhile supporters of Einstein—who introduced the first cosmological term in 1917 in his static cosmological model, as a type of antigravity force, but then erroneously removed this term after the advent of Friedmann's model of the expanding universe, even though this model does not require the absence of a cosmological term. There are, moreover, serious grounds for considering a change in the character of the cosmological term, which could have been larger at the primordial Big Bang or Planckian epoch (Linde and Ivanenko). Finally, the gravitational field equations of GRT read

$$R_{\alpha\beta} - \tfrac{1}{2}g_{\alpha\beta}R - \lambda g_{\alpha\beta} = G'T_{\alpha\beta}$$

in the standard notation; here, $G' = 8\pi G/c^4$, with G the Newtonian constant, while $T_{\alpha\beta}$ denotes the energy density tensor of any matter distribution.

The dramatic successes of GRT in removing the Mercury anomaly and in predicting the observed light deflection, time retardation, and red shift of spectral lines in the gravitational field are all well known, as are the breath-taking explanations of the galactic runaway and the nature of the isotropic background radiation as the relict radiation from the primordial state, predicted by Gamow. But now, returning to the chief point of our presentation, one should not forget the following profound difficulties besetting GRT, which we shall only briefly enumerate as at least unsolved or unclarified problems: singularities, definition of the energy of the gravitational field, and the existence of big numbers. The latter characterize the strange relation between cosmological and atomic quantities [e.g., the ratio $\sim 10^{40}$ of metagalaxy to particle dimensions, and the ratio (also $\sim 10^{40}$) of Coulombic and gravitational forces between particles], which caused Dirac to propose the hypothesis of a diminishing gravitational constant and which led Jordan, and later Brans and Dicke, to the scalar–tensor theory of the gravitation field. We must also emphasize the whole complex of interconnected problems arising from the presence or absence of symmetries in GRT, the correct definition of reference systems, etc., all connected with the problem of "boosting" special relativity to general relativity, which were so cleverly "solved" by Einstein, leaving aside the objections of Abraham and Kottler and, later, of Fock, Havas, Bondi, etc. (somehow like the situation with Newtonian theory).

One may feel that without clarifying these problems one cannot advance safely to the description of the extreme situation inside particles, for the Big Bang or Planckian regions, at energies $\sim 10^{19}$ BeV, for dimensions near $l = (hG/c^3)^{1/2} \approx 10^{-33}$ cm etc. Among many proposals to generalize Einstein's theory, we choose with the growing majority of authors the gauge theory of gravitation. As is well known, its basic idea, expressed by Weyl (1918), was to retain the invariance of Lagrangian and field equations by passing from global transformations with constant parameters to local transformations with parameters that are functions of space-time points. Then, to compensate the derivatives of parameters that appear, one is obliged to introduce the corresponding gauge fields. After the gauging of the $U(1)$ group, which led to the electromagnetic field, the second step (as remarked by A. Salam in 1973 in an instructive paper) was the gauging of the Lorentz rotation group, which resulted in gravitation field connections containing the Riemann–Christoffel term (or Fock–Ivanenko coefficients in the spinor case); the clear presentation of gravity as a gauge field was clarified later (cf. Utiyama, Brodsky–Ivanenko–Sokolik, Kibble, Frolov, and the informative 1976 review by F. Hehl *et al.*). But the very recognition of the power of gauging came when Yang and Mills compensated the internal degrees of freedom (at first isospin, afterwards other properties) in 1954. Now the gauge theory constitutes the basis of the chromodynamics of strong interaction theory, and of the Glashow–Weinberg–Salam unification of weak ("atenonic") and electromagnetic interactions. But here we concentrate only on the gravitational case with its successes and difficulties. As its first achievement one must count the natural introduction of torsion alongside the Riemann–Einstein–Grossmann curvature of space-time. Torsion can prevent collapse (Trautman *et al.*) It also leads to an interesting precession of spin in the space endowed with torsion (shown in our group by Krechet and Ponomariev, later refined by Pronin, and also developed by Trautman), which could provide a method for measuring torsion. For the measurement of torsion one can use also the autoparallel deviation of two test particles, as remarked by Obukhov. Of course, torsion, like conventional gravity, is subject to quantum-mechanical measuring limitations, as we indicated recently when developing analogous considerations for the GR fields of Bronstein, Rosenfeld, Treder, and Bergmann.

One of the most important consequences of torsion is its induction of nonlinearities, e.g., in Dirac's equation—of cubic pseudovector type, having the form proposed earlier (Ivanenko, 1938):

$$[\gamma^\alpha \partial_\alpha + m + \lambda(\bar{\psi}\gamma^\alpha\gamma_5\psi)\gamma_\alpha\gamma_5]\psi = 0$$

This equation has acquired great importance recently: It was proposed by us as the basis of prematter; this idea was developed by Heisenberg, who cited our work as the "predecessor" of his quantized theory. With Pauli, he introduced isospin in our equation. Later, with Naumov, Nguen Giac Giao, and Daniluk we generalized this by taking into account unitary "quarkian" spin and other $SU(N)$ symmetries. All this led to interesting though preliminary results for the masses and coupling constants of particles up to Ω^-, in qualitative agreement with empirical data; e.g., for the fine structure constant one gets the values $\alpha = 1/120$ to $1/115$ (vs. $1/137$!). The geometrical meaning of our cubic nonlinearity by means of torsion (Rodichev, 1961; developed by Peres, Datta, and Hehl; Krechet-Ponomariev, in our group, treated the general case) acquires a great significance, as the use of Salam's large gravity constant leads unexpectedly just to the right order of magnitude for the self-coupling constant $\lambda = \frac{3}{8} G' hc$, $G' \sim 10^{40} G$ (Sinha, Sivaram, and Ivanenko). The nonlinearities are also induced both by torsion and Weyl dilation (then of vector type) in the Maxwell–Proca as in the spinor equation and then lead to solitonic-type solutions; the equations could be transformed into the well-known sine (or hyperbolic sine)-Gordon type (Smetanin in our group, and Krechet). For a timelike vector in a space endowed with torsion (Proca–Maxwell case) we get

$$\Box u - \mu^2 \sinh u = 0,$$

etc. Now the nonlinear spinor theory (of the Ivanenko–Heisenberg type) is revived for prequarkian matter (e.g., by Terazawa), as "preons" (Salam, Pati, and Strathdee) seem to be necessary due to a multitude of quarks. We believe also (with Sardanashvily) that nonlinear spinors are the most primitive as yet known physical and mathematical (and even logical, in the sense of a binary "yes–no" scheme) objects that could be used in a first explorative, but anyhow necessary, approximation for the description of unified primordial fields. The very general Coxeter symmetry group seems suitable, even when unified with the geometrical space-time arena itself, due to violent metric and torsion fluctuations à la Wheeler and violent mutual transmutations of all fields, first considered by us, taking account of graviton $\rightleftharpoons$ particle transmutations.

Let us remark, by the way, that at this stage of extremal conditions—possibly suited to the state inside quarks at collapse near the Big Bang in the Planckian region—space-time could acquire some type of discrete structure, described by difference equations, first suggested by us (Ambartzumian and Ivanenko, 1930) and a little later by Heisenberg. The notion was further developed in various versions by Snyder, Coish, and others. Lately the idea of discreteness has been suggested by Penrose, Weizsaecker, and Finkelstein, and treated also by us in modernized form. Most popular seems to be the version involving only time discrete chronons (Finkelstein, Caldirola, and Casagrande). The possibility of passing down to real "observed" conditions of the ordinary particle mixture after some fraction of a second after the Big Bang via a sequence of symmetry breakings (including some hierarchical levels?) seems to be an attractive possibility. In this context, we mention an exact solution of Einstein's equations for a nonhomogeneous distribution of matter leading to an island-type or microuniverse "particle," as was shown by us with Krechet.

Let us return now to the main point of our considerations in this article, which is connected with the gauge theory of gravity. Here the powerful fiber bundle formalism seems quite necessary for overcoming the not clarified gauging of translations. Moreover, the introduction of a second space (internal space or fiber) alongside the conventional "base" (curved Riemann–Einstein space) seems to constitute a profoundly necessary step, possibly representing, with the whole gauge treatment of gravity, a second step of real significance after Einstein's establishment of GRT!

Let us now elaborate in more details the basic point concerning the passage from SR to GR by means of fiber bundles.

2. Fiber Bundle Approach in the Gauge Theory of Gravitation

The basic ideas of the gauge gravitation theory are still not generally accepted in spite of its more than twenty years' existence.

The principal reason for this is that the gauge character of gravity is connected with the whole complex of problems surrounding Einstein's general theory of relativity—the definition of the reference system, the (3 + 1)-splitting, the presence (or absence) of symmetries in GR, the necessity (or triviality) of general covariance, and the meaning of the equivalence principle—which led Einstein from special to general relativity. The actuality of this complex of interconnected problems is demonstrated by the well-known work of V. Fock, who saw no symmetries in general relativity, declared the equivalence principle unnecessary, and proposed even to substitute the designation "chronogeometry" for "general relativity" (see also P. Havas). Developing this line, H. Bondi quite recently also expressed doubts about the "relativity" in Einstein's theory of gravitation.

All proposed versions of the gauge gravitation theory must clarify the discrepancy between Einstein's gravitational field being a pseudo-Riemannian metric field and the gauge potentials representing connections on some fiber bundles, and the fact that there exists no group whose gauging would lead to the purely gravitational part of space-time connections (Christoffel symbols or Fock–Ivanenko–Weyl spinorial coefficients).

We shall not analyze here all attempts that have been made to overcome this difficulty. But we believe that every gauge theory, aiming to describe gravitation, must be in agreement with the basic relativity and equivalence principles of Einstein's theory. Moreover, the gauge gravitation theory is to be constructed from these principles.

For this purpose, let us compare the relativity principle (RP) and the equivalence principle (EP) of gravitational theory with the postulates of gauge theory by using the formalization of both of these theories in terms of fiber bundles.

Without entering into details and comparing here all its existing versions, one notes that RP essentially states the equivalence of all reference systems for the formulation of fundamental physical laws. So, one needs the notion of reference systems for a given theory and must define the class of their transformations. RP then requires the form conservation of physical laws with respect to the group of all transformations of the reference frames. One needs also to indicate some reference system wherein the physical laws and equations possess their well-known forms. In the case of gravitational theory, this is the reference system corresponding to the transition to special relativity which is established by EP.

In the fiber bundle approach to gauge theory, the interaction of a field multiplet $\{\varphi\}$ possessing the symmetry group G is described by the introduction of a geometrical structure (connections) on the fiber bundle λ whose sections are the matter fields $\{\varphi\}$. The basic principle of gauge theory—the local invariance—means the covariance of matter field equations under transformations of the fiber bundle λ atlases. These transformations are elements of the local group G(X), which is defined as the group of all sections of the principal fiber bundle associated with λ. In the fiber bundle formalism of the gauge theory, a reference system is established by fixing the atlas of the fiber bundle λ. This is expressed in the description of the matter fields $\{\varphi\}$ by means of a certain family of functions with values in the space of a typical fiber and with a particular gauge of the gauge fields.

One must distinguish the case when a special choice of gauging is possible, leading to the reduction of the gauge fields of the group G to the gauge fields of one of its subgroups H (the holonomy group of these fields on the fiber bundle λ is reduced to H); that is, a reference system exists where the gauge theory of the group G is effectively reduced to the gauge theory of its subgroup H. Then one has the contraction of the structure group G of the fiber bundle λ to the subgroup H. The necessary and sufficient condition for this contraction is the existence of some global section σ of the quotient bundle associated with λ possessing the typical fiber G/H. The field σ arising in such manner in gauge theory is analogous to Goldstone fields in the well-known models with spontaneous symmetry breaking.[5,6]

In the fiber bundle formalism, the Einstein gravitational field of the orientable space-time manifold X^4 is (in mathematical language) defined as a section of the fiber bundle Λ of pseudo-Euclidean bilinear forms in the tangent spaces T_x over X^4. Λ is associated with the tangent bundle $T(X^4)$, which possesses the structure group $GL^+(4,R)$. The gravitational fields and connections (Christoffel symbols), expressed by the components of these fields, provide a geometry on the tangent bundle $T(X^4)$, which is conventionally understood as the geometry of the space-time itself.

Then, in analogy to gauge theory, the choice of a reference system in gravitational theory can be defined as the fixing of an atlas $\psi = \{U_i, \psi_i\}$ of the tangent bundle $T(X^4)$, and the reference frame transformations are the local group $GL^+(4,R)(X^4)$ transformations of the atlases of this bundle.

This definition of the reference systems is close to the one used in the tetrad formulation of general relativity. The fixing of the atlas of $T(X^4)$ leads to construction at each point of a tetradic repere $\{t_x\}$, which is the image $\{t_x\} = \psi_i^{-1}(x)\{t\}$ of the bases $\{t\}$ of the typical fiber of $T(X^4)$; and changes of the atlases are followed by the tetrad transformations. The Einstein (metrical, not tetradic) formulation of general relativity corresponds to the case of purely holonomic transformations of the reference frames, when, along with a coordinate system, defined by the atlas $\psi_x = \{U_i, \varphi_i\}$ of the manifold X^4, one always chooses the reference system adjusted to it with the atlas $\psi = \{U_i, \psi_i = d\varphi_i\}$ and if this correlation is strictly retained in the passage to another coordinate frame $\gamma: \psi_x \to \psi'_x$ accompanied by the corresponding transformation $d\gamma: \psi \to \psi'$ of the reference system.

Taking all this into account, one can formulate the relativity principle in gravitational theory as the requirement of the covariance of the equations of matter fields interacting with gravitation in respect to the local group $GL^+(4,R)$, and as such it coincides with the basic postulate of local invariance in gauge theory. This means that the gravitational gauge theory can directly be constructed from RP as the gauge theory of external symmetries. Its peculiar character is manifested by the fact that external symmetry gauge fields act also on the operators of partial derivatives $\{\partial_\mu\}$ as the vectors of tangent spaces. This circumstance leads, for example, to supplementary possibilities for index contraction and permits, in contrast to the Yang–Mills case, the construction of the curvature scalar from the strength tensor of gauge fields and its choice as the Hilbert–Einstein Lagrangian of the theory. But one must emphasize that $GL^+(4,R)$ gauge theory is broader than the theory with pseudo-Riemannian metric gravitational fields, because it possesses metric fields with an undefined signature. The passage to the proper Einstein gravitation is connected with the equivalence principle.

In the conventional gauge theory, EP is as a rule not explicitly formulated. The representation of gauge fields by 1-forms of connections on fiber bundles leads already to the existence of a reference system where gauge fields are vanishing at a given point. Hence, in the gauge theory of gravitation, EP means the indication of the type of gauging only. If one accepts in symmetry language the reduction of external symmetries to the Lorentz group as the criterion for the transition to special relativity, then EP can be formulated as the requirement of the $GL^+(4,R)$ gauge theory reduction to the gauge theory of the Lorentz group. This is expressed by the condition of the restriction of the holonomy group on the tangent bundle $T(X^4)$ to the Lorentz group. Like the general case of the gauge field theory, one then has the condition for the reduction of the structure group $GL^+(4,R)$ of the tangent bundle $T(X^4)$ to the Lorentz group and hence the important fact of the existence of the global section of the quotient bundle associated with $T(X^4)$ with the typical fiber $GL^+(4,R)/SU(3,1)$. This fiber bundle is isomorphic to the fiber bundle Λ of pseudo-Euclidean bilinear forms in the spaces tangent to X^4. One sees that the fulfillment of EP leads to the existence of the metrical gravitational field globally defined on the whole given space-time manifold X^4. At the same time this permits us to interpret the gravitational field as a Goldstone-type field in the gauge theory of external symmetries.[7] The possibility for considering gravitation as a Goldstone (Higgs) field was also proposed recently by Ne'eman and Trautman in another version. But in contrast to the Goldstone fields of internal symmetries, the gravitational field cannot be made to vanish everywhere by any special choice of gauge.

The contraction of the structure group $GL^+(4,R)$ to the Lorentz $SO(3,1)$ group, and consequently to its maximal compact subgroup $SO(3)$, implies some F. Klein–Chern geometry of invariants on $T(X^4)$,[6] which enables us to interpret the geometrical aspects of gravitation in the spirit of Erlanger program. From this point of view, one may not agree with the Fock–Havas–Bondi opinion about the absence of any symmetries in general relativity.

Let us briefly remark that the realization of gravitational fields by cross sections of the quotient bundle defines their topological classification as the classification of the bundles $T(X)$ whose structure groups can contract to the Lorentz group.[6] As $SO(3,1)$ is the image of the injection of the $SL(2,C)$ group into $GL^+(4,R)$, the structure of the $SL(2,C)$ bundle on $T(X^4)$ is defined and its classification in Chern $[c_{1,2} \in H^{2,4}(\gamma,z)]$ and Pontryagin ($p_1 = c_1^2 - 2c_2$) indices proves to be possible, without passing to Riemannian metrics and $SO(4)$ bundles. For compact manifolds X^4 these indices are represented by Chern differential forms (expressed by curvature) and the integrals $c_2[X^4]$ and $\frac{1}{3}p_1[X^4]$ coincide with Euler number and the homology index (both equal zero!) of the manifold X^4, respectively.

Finally we can draw the conclusions that the gauge gravitation theory, constructed from relativity and the equivalence principles reformulated in the forms of fiber bundles, is the $GL^+(4,R)$ gauge theory of external symmetries, reduced in some reference frames to the Lorentz gauge theory, with the description of the metric gravitational field as a Goldstone-type field, analogous to Goldstone fields in models with spontaneous symmetry breaking. With all this, one returns once more also to modern Machian versions (so repeatedly advocated by Wolfgang Yourgrau) of deep connections between cosmological, gravitational, and microphysical quantum field-theoretic aspects of all of physical reality—thus taking a modest step towards fulfilling Einstein's dream of a unified geometrized theory.

References and Notes

1. D. Ivanenko, "Perennial Modernity of Einstein's Theory of Gravitation," in, *Relativity, Quanta, and Cosmology* (Einstein Centenary Jubilee Volume of the National Italian Academy of Science, Rome), M. Pantaleo, editor (Johnson Reprint Corporation, New York, 1979; Italian version published by Giunti-Barbera, Firenze, 1979).
2. Abstracts of Contributed Papers, Ninth Int. Conf. on General Relativity and Gravitation, Jena, 1980, Vol. 3 (papers by D. Ivanenko, G. Sardanashvily, Y. Obukhov, and E. Smetanin).
3. *Present Problems of Theoretical Physics* (University of Moscow, 1976) (papers by D. Ivanenko, G. Sardanashvily, F. Hehl, and V. Krechet).
4. G. Sardanashvily, *Phys. Lett.* **75A**, 257 (1980); D. Ivanenko and G. Sardanashvily, *Lett. Nuovo Cimento* **30**, 220 (1981).
5. A. Trautman, *Gen. Rel. Grav.* **11**, 287 (1980).
6. Y. Ne'eman, *Gen. Rel. Grav.* **11**, 309 (1980).
7. D. Ivanenko, "Problems of Unifying Cosmology with Microphysics," in *Physics, Logic, and History*, W. Yourgrau and A. D. Breck, editors (Plenum Press, New York, 1970).

15

Status and Perspectives of Cosmology

Toivo Jaakkola
Observatory and Astrophysics Laboratory
University of Helsinki
Helsinki, Finland

. . . *fundamentally we can know only the infinite.*
—Friedrich Engels

1. Introduction

Two mutually contradictory trends can be discerned in cosmology during the past few decades. On one hand, there has been a huge development, both in quantity and in quality, of cosmological data. On the other hand, there has been a long-lasting stagnation as regards results concerning the fundamental questions of present-day cosmology. For example, the central question posed by the standard cosmological theory, whether the Universe is open or closed, has remained unanswered.

Obviously not without connection with this paradigm, even the basic assumption of standard cosmology, viz., that the Universe is expanding, has not been firmly established. Discussion of some aspects of this problem, such as that concerning the quasar red shifts, has been lively. However, there has been very little of a systematic attempt to test the whole conception of expansion, which still remains only a hypothesis—often appearing as a mere belief—without empirical verification.

In this vague situation, the general direction should be determined from the rich already existing and continuously increasing and improving data. This means systematization of all the data of cosmological significance without too much regard of predefined theoretical frames. This task contains analysis of the test results already obtained for the familiar global and local tests. Further, a systematic search should be directed at the question of the existence, or not, of cosmological evolutionary effects—this is indeed the question which more profoundly than any other addresses the nature of the actual Universe. And, of course, the properties of the red-shift phenomenon itself—which has provided the empirical foundation stone for the standard cosmology—should be mapped carefully in order to be certain about its actual nature.

Some such systematization of data has already been made.[1-3] A brief review of this work will be presented in Section 2, which gives a personal view of the status of cosmology today. All aspects of the data lead to suspicions about the correctness of the basic assumptions of standard cosmology. This conclusion urges an alternative cosmological theory, and one such approach is outlined in Section 3.

The characteristic features of this theory are lack of expansion and other evolution on a cosmological scale. The processes which maintain large-scale equilibrium, while matter evolves locally, form the essence of the theory. Only the general conditions of these processes are here outlined, and closer study of them is a matter for the future.

Independent of the correctness of this approach, it is already obvious that understanding the diversified phenomena, occurring in stars, galaxies, and intergalactic space and governing the evolution and the state of matter in the Universe, involves deeper insight into the very foundations of science.

Wolfgang Yourgrau was a man who devoted his life to this profoundly human endeavor.

2. The Status

2.1. *Test Results of Standard Cosmology*

Figure 1 represents, as a function of the year of publication, the results on the value of the deceleration parameter q_0, as obtained by different authors with various methods (for references, see Ref. 3). One can see from the figure that, first, there has been no trend of convergence in the test results in the course of time, while the number and quality of the data have strongly increased. Second, the results appear significantly dependent on the method. The mean values of q_0 are the following: $+0.93 \pm 0.19$ from the Hubble diagrams; $+0.03 \pm 0.08$ from the local tests; -0.30 ± 0.63 (-0.85 ± 0.31, the weighted mean value) from the optical angular-diameter–red-shift tests; in the radio (θ,z)-diagrams θ fall as $\theta \propto z^{-1}$, which is steeper than predicted in any model.

Counts of galaxies have recently been made up to the 26th magnitude.[4] With the K correction used, the data fit the Friedmann model with $q_0 = 0$.

Hence, while the (m,z) tests consistently give a closed model, the local tests as well as the counts of galaxies give an open one; the (θ,z) relations point either to a positive cosmological constant or to the steady-state model, and the radio observations—which give the most trustworthy data—violate the predictions of all the relativistic expanding models. It can be seen from Fig. 1 and from the mean values given that the dependence of the cosmological test results on the method is of a systematic and stable character. Figure 2 illustrates further the inconsistency of the standard interpretation of the cosmological data.

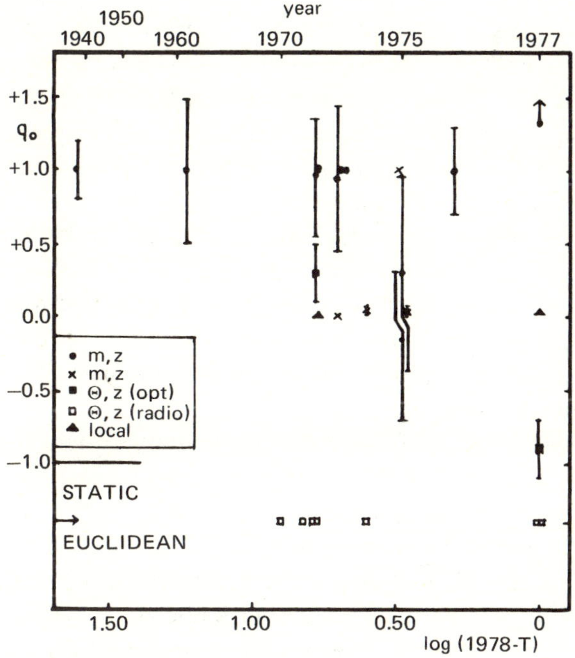

FIGURE 1. Results on the value of the deceleration parameter, represented according to the method and to the year of publication. References in Ref. 3.

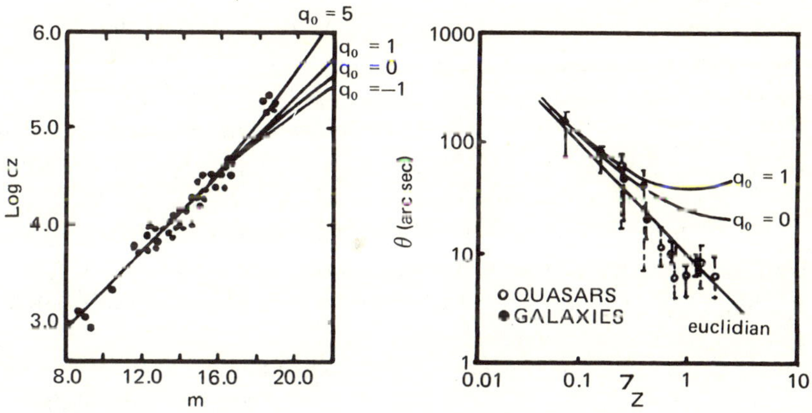

FIGURE 2. The Hubble diagram (Fig. 2a) and the angular-diameter–red-shift relation (Fig. 2b) compared with the various relativistic, expanding models. Note the inconsistency of the results. References in Ref. 3.

2.2. Are There Cosmic Evolutionary Effects?

All individual galaxies are evolving. But do the average properties of galaxies change in the course of cosmic time or do they not? The question whether the Universe as a whole is evolving or whether it does not change is perhaps the most fundamental question of cosmology. The view that the parameters of galaxies evolve indeed fits well with the big-bang theory. On the other hand, this cosmology strives for finding the global properties (geometry and dynamics) of the Universe and uses for this global empirical tests requiring similar objects at large and small distances, i.e., unevolving galaxies. To be or not to be?

If there are cosmic evolutionary effects, these would affect the global tests and might be the reason for the discrepant results discussed in Section 2.1. Evidence of possible cosmological evolution of galaxies, quasars, radio sources, and clusters of galaxies will now be briefly discussed.

There is no reliable sign of cosmological evolution of normal galaxies, i.e., the main material component of the Universe. It has been alleged that distant galaxies[5] and clusters of galaxies[6] are bluer than the nearby ones. In these considerations, the influence of the K term has been underestimated or forgotten. Because of their characteristic spectral energy distribution, the red shift dims red elliptical galaxies stronger than the bluer galaxies, and this makes the contribution of the latter apparently larger at large z than at small z. Within the galaxies there is a gradient of color, and also through this the effect of the spectrum and the red shift makes the distant galaxies apparently blue and, at the same time, apparently large. These observational effects mimic the evolution of color and size of galaxies and of clusters of galaxies. Also the contribution of active galaxies which are easier to observe tends to become large in the selection of the samples at large distances.

It has been tried to avoid the discrepancy between the q_0 values obtained from the Hubble diagram and those obtained by other methods by supposing luminosity evolution of galaxies.[7] There is no independent evidence for this evolution. The color has remained the same at least for a period corresponding to the time coverage of the Hubble diagrams implying $q_0 \sim 1$. The counts of faint galaxies definitely rule out any considerable luminosity evolution.[4] On the other hand, there are theoretical studies[8] of the capture of neighbor galaxies by the supermassive galaxies in the centers of clusters, implying that the central galaxy grows larger and brighter. This effect has been verified empirically in the *cD* clusters,[9] and requires (within the frame of the standard theory) an evolutionary correction to q_0 which makes the empirical inconsistency of the modern cosmology even worse.

It has been argued[10] that quasars present compelling evidence of cosmic evolution; specifically the counts of optically selected quasars would show that their relative space density has been much larger in the past. However, except for the brightest sample ($B < 15.7$), the counts neatly follow the homogeneous Euclidean prediction.[11] The deviation from that of bright magnitudes is too sharp and corresponds to too small red shifts to be an indication of evolution. In fact, it has been shown[11] that the deviation is mainly due to a morphological selection effect and the K term. A sample of quasars is, *ab definitio*, selected partly on a morphological basis, and it follows automatically that the nearest physically identical objects representing some definite structure (here, Sy1 galaxies) fall out of the sample. This selection effect again mimics cosmic evolution. See Fig. 3.

Perhaps the main argument in favor of cosmic evolution has been the number-flux density relation of radio sources. This was also the main reason for abandoning the steady-state cosmology about twenty years ago.

The slope of the integral counts at large flux densities is steeper than predicted in a homogeneous, static model. However, it has been pointed out[12] that the differential counts fit the homogeneous static model for a wide interval of flux density, and there is a deviation only at the highest and at the lowest flux densities to which only a few percent of the observed sources correspond. Further, similarly as found for quasars, the deviation for the bright sources happens sharply, which is not reasonable for evolution. These features are valid for the counts at 5000 MHz, but the counts at different frequencies are mutually consistent.

There are numerous possible interpretations of the steep slope for strong sources without cosmological evolution. The lack of a few tens of bright sources might be due to a statistical fluctuation[12] or to a morphological selection effect analogous to that in the quasar samples, here resulting from large angular sizes of the nearby sources. Perhaps more significant, probably there are two power classes of sources, of which the intrinsically fainter, with flat radio spectra, would be connected with dwarf satellite galaxies. The existence of satellite systems would depend on the properties of the giant galaxies; the Galaxy and the nearest galaxies obviously lack such a system. It would follow that the faint class of sources contributes to the counts only at medium and low flux densities. This would account for the steep slope for strong sources in a natural manner without any taste of a geocentric "local hole."

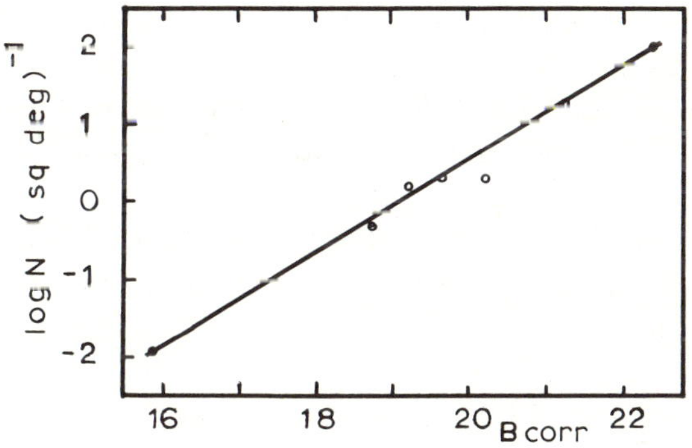

FIGURE 3. Counts of optically selected quasars after the corrections for the morphological selection effect and the K term. The former correction means inclusion of type 1 Seyfert galaxies. The line shows the prediction $N \propto 10.^{0.6m}$ Reference 11.

Faithfully to the fashion, it has been attempted to interpret the perplexing angular-diameter–red-shift relation, without any sign of the predicted minimum at $z \sim 1$, by an increase of size during cosmic time. There is no reason, however, why the change should be just such that it cancels the effects of geometry and expansion.[13] Even less one may expect such evolution connected with optical luminosity evolution to give a similar cancellation.

There is no sign of a cosmological evolution of clusters of galaxies, either. The suggested color evolution was seen to be explainable by the *K* effect. Zwicky[14] points out several structural parameters of clusters which are similar at the limit of the Mount Palomar 5-m telescope and at small distances.

The concept of an evolving intergalactic background, with increasing scales and decreasing temperature of the radiation which is thought to be the relic of the primordial fireball, belongs to the standard cosmology as an essential content. There is no direct test of the evolutionary picture of the background radiation, which will be discussed from another aspect later in this paper.

To summarize, there is no reliable evidence for the correctness of the assumption of a systematic evolution of galaxies during cosmic time. Hence this assumption does not help the situation discussed in the preceding section. As there are very many parameters to test, and even one positive result would suffice to destroy the perfect cosmological principle and give support for the big-bang theory, the failure to find any speaks in favor of cosmic evolutionary effects really being nonexistent. Some effects reported to exist are probably observational effects. If due to evolution, these would have happened unappropriately recently, at the same time as the rest of the objects show a remarkable stability of properties over the last 10^{10} years. For its part, the latter feature has been ascribed to processes like "initial bursts" with hardly any subsequent evolution. Here there is an almost biblical sound about a primary origin of the heavens. A more scientific aim would be to interpret the properties of galaxies through the processes occurring in galaxies as we see them now.

The concept of cosmic evolutionary processes in the standard cosmology appears as an entangled mixture of *ad hoc* assumptions about effects in the distant and nearby past. The entire construction of the standard cosmology can hardly pass Occam's razor. In effect hypothesis of cosmic evolution forms a place of refuge in which to escape the inconsistent results of the global tests. It is true that the inclusion of galactic evolution makes the theory more realistic. But at the same time the theory erodes to one which can hardly be tested empirically. Indeed reliable knowledge about the Universe using the global tests in the standard way seems by now improbable.

2.3. *Nature of the Red-Shift Phenomenon*

Before one studies whether the rate of expansion decelerates, remains unchanged, or even accelerates, it seems reasonable to ask whether there exists such an expansion at all!

The interpretation during the twenties and early thirties of the newly discovered

red shifts by the Doppler effect seems by now all too ready-made and premature. In fact those scientists, like Hubble,[15] who had the closest connection with the actual observations of the red shift, were skeptical in this respect. In astronomy the Doppler effect due to the motions of the planets, rotation of stars and galaxies, intrinsic motions within systems of galaxies, etc., appears always *symmetrically*: there are as many blue shifts as red shifts. Contrary to that, red shift is an *asymmetric* effect. Should not the different effects have different interpretations? This seems probable, and the search for a non-Dopplerian mechanism for the red shifts[1] in no case constitutes a scientifically uneconomic approach.

There are a few groups of scientists[16,17] that have examined the red shifts from various aspects not founded on the Dopplerian view. I summarize here results obtained in one of these groups.[17] These results are based on a systematic study[2] of the red-shift phenomenon in systems of various scales, extending from the cosmological scale, via supergalaxies, clusters, groups, and pairs of galaxies and single galaxies, to the stars, including the Sun. A red shift appears to be a universal effect in a similar sense as gravitation is universal. It is not only a cosmological effect showing itself only between the galaxies but a galactic effect as well, appearing, e.g., within our own Galaxy ten times and within the galactic nucleus 10^5 times stronger than on the average within the Metagalaxy (the latter strength is given by the Hubble constant). There is no dualism connected with the familiar concepts of "anomalous" (e.g., in quasars) and cosmological (i.e., "systematic") red shifts. These are in fact two forms of the same phenomenon with one and the same physical mechanism, and between them lies, in a systematic way, a variety of other features.

The red shift appears to be an effect not bound to the space (its expansion) but one bound to the presence of matter. The matter density (ρ) and the strength of the red shift (h) were found[2] to be correlated as

$$h(\rho) = 2.0 \times 10^{-8}(1 + 2.5\sqrt{\rho}) \quad (1)$$

where h is given symbolically in Dopplerian terms and the units are in the cgs system. See Fig. 4.

The continuity of the various red-shift effects from that appearing on the cos-

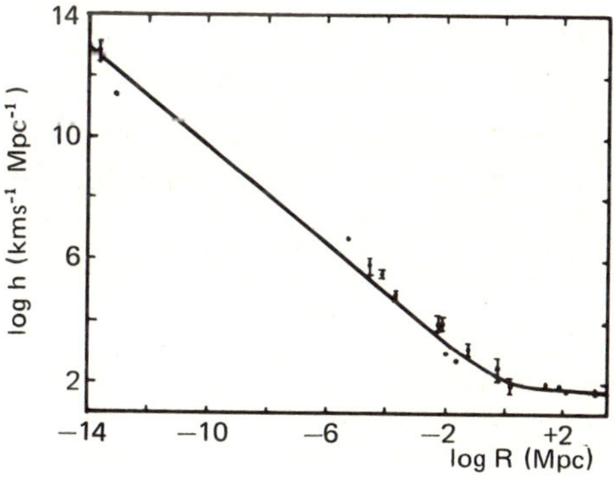

FIGURE 4. Strength of red shift in systems of various scales. Parameter h is given symbolically in km s^{-1} Mpc^{-1}. See Ref. 2.

mological scale to those appearing on the smaller scales, and relation (1) indicate that not only the latter effects but also the cosmological red shifts are non-Dopplerian phenomena. This means that there is no expansion of the Universe.

The analysis of the red-shift data which lead to such a conclusion suffers, of course, of many uncertainties due to the weakness of the data and some assumptions necessarily involved. The evidence from the analysis of red shifts should be considered only as one piece of the integral evidence which should be drawn—in one line or another—from all relevant cosmological and astrophysical data.

The rest of the present paper will be based on the hypotheses, supported by the results discussed in this and the foregoing section, that there is neither expansion nor other cosmological evolutionary effects in the Metagalaxy. A new approach to the cosmological problem is prompted also by the failure—as appears in Section 2.1 (and 2.3)—of the standard cosmology to interpret and utilize the hugely developed cosmological data.

3. Perspectives: Outlines of an Alternative Theory

3.1. *General Remarks*

Dropping the assumptions of universal expansion and other cosmological evolutionary effects immediately gives one two fundamental features of the new picture: the Universe is *static* and the Universe is *nonevolving*. As processes of evolution are encountered on any level of the structural organization of matter, *equilibrium evolutionary processes* are required in order to maintain the properties of matter unchanged on the cosmological scale. The study of the equilibrium processes will be the essence of the cosmological approach here outlined.

It will be seen later that cosmological observations point to the *Euclidean geometry of space*. Here I only point out that the minimum in the angular-diameter–red-shift relation, predicted in the big-bang models of $z \sim 1$, has never been observed.

The Universe is considered as *infinite* both in time and in space. The difficulties connected with the "creation of the Universe" at the event of the big bang are not present here. Infinity of space follows from its Euclidean nature. Also the argument of Lucretius[18] still holds good: "Since we must admit that there is nothing outside the universe, it can have no extreme point, and therefore lacks an end and a measure."

As there is neither expansion nor other cosmological evolutionary effects, *there is no characteristic cosmic time*. Further, in accordance with the available observations, the Universe is assumed to be *homogeneous and isotropic*, i.e., structureless *already on the metagalactic* scale. The two properties unite to what is known as the perfect cosmological principle.

The picture arising, represented above as it results from empirical considerations, has a long tradition in natural philosophy. One can mention here, e.g., the Greek materialist Anaximander, who understood the Universe as infinite and containing an infinite number of worlds. According to him matter is in restless motion and the reason for the motion is in matter itself, in the battle of mutually contradicting forces. Heracleitus, Virgil, and Lucretius had similar ideas.[19] The notion of the balance of the opposite forces in Nature is familiar in the old Chinese philosophy. Kant's[20] claim that the idea of time is not appropriate to the Universe itself, reflects a view about an unchanging universal entity. Engels[21] pointed out that indestructibility of matter does not mean only a quantitative permanence but a qualitative permanence as well, i.e., an unchangeableness of matter as a whole.

In the history of science the idea of the big-bang theory of a rapidly evolving Universe is indeed in a unique position.

In its proper meaning, cosmology investigates the universal: it concerns matter as a whole. If it is assumed, as in the tradition referred to, that matter as a whole is unchangeable, a study of an expanding, evolving Metagalaxy would not be cosmology in its proper meaning. It would be only a study of a particular material system in its particular phase of evolution; it would be something like an "extended geography." It should be considered as fortunate for cosmology if the Metagalaxy would prove to be homogeneous and unchanging, because in that case an examination of the Metagalaxy means an examination of the universal. In that case we see a proper sample of the Universe, and man will, in the course of the progress of science, learn to know matter as a whole. Using other words, man will learn to understand the infinite. One can say that the characterization of the Metagalaxy as a structureless, unchanging, representative part of the Universe opens the door for cosmology.

The standard definition of cosmology is: the study of the structure and the evolution of the Universe. Now, if the Universe is in the large homogeneous and isotropic (i.e., structureless) and if it does not expand or change otherwise (i.e., does not evolve), cosmology defined like this would shrink into two negative sentences. If the Metagalaxy is homogeneous, static and unchanging, cosmology will be *study of the structure and evolution of matter in the Universe*. Cosmology so defined will be far from being a "search for two parameters,"[22] H_0 and q_0. It will mean a survey of the structure and processes of matter in all its forms, using and binding together results from the various special sciences. This makes the subject very extensive and ramified, but still fixes the problem better than the former formulation, which is meaningless from a profound semantic point of view. It is also interesting that the latter formulation corresponds well to the original Greek meaning of the word: cosmology is a doctrine about the order of the world.[23]

3.2. Formulas for Global Relations

For comparison with the cosmological data available, it is worthwhile to derive first the formulas for global interrelations. A red shift means a decrease in the energy E of a photon, in a homogeneous Universe directly proportional to the distance r:

$$\int_{E0}^{E} \frac{dE}{E} = -C_1 \int_0^r dr \tag{2}$$

In order to keep up the commonly used Dopplerian terminology, which is used here purely for practical reasons, we set $C_1 = H/c$, where H is the Hubble constant and c the velocity of light. Noticing that $z = (\lambda - \lambda_0)/\lambda_0 = E/E_0 - 1$, we obtain

$$\ln(1 + z) = (H/c)r \tag{3}$$

This is the red-shift–distance relation in the static, Euclidean universe. The magnitude–red-shift relation has the form

$$m = 5 \log \ln(1 + z) + 2.5 \log(1 + z) + C_2 \tag{4}$$

where the form of the first term results from (3), the second term is Hubble's "energy effect," and C_2 is a constant equal to $C_2 = M + 5 \log c/H + 25 = M + 43.5$ if M is the absolute magnitude and $H = 60$ km s^{-1} Mpc^{-1}. The angular-diameter–red-shift relation has the form

$$\theta = (DH/c)\ln(1 + z) \tag{5}$$

where θ is in radians and the linear diameter D in Mpc. For the number–magnitude relation we obtain from Eqs. (3) and (4), finding that $r \propto N^{1/3}$, the following formula:

$$0.6m = \log N + A_2 N^{1/3} + C_3 \tag{6}$$

where N is the number of objects brighter than m, C_3 is a constant, and $A_2 = 8.1 \times 10^{-5}\, n^{-1/3}$ (n is the number density); for normal galaxies $A_2 \sim 3 \times 10^{-4}$. For the radio counts there is the relation

$$-1.5 \log S = \log N + A_2 N^{1/3} + C_4 \tag{7}$$

The second term on the right-hand side of Eqs. (6) and (7) must be taken into account in very deep counts.

3.3. *Comparison with Observations*

The formulas derived above are shown in Fig. 5. It appears that the static, Euclidean (m,z) relation corresponds very closely to the simple Hubble relation which has the same form as in the Friedmann model with $q_0 = 1$. As was seen in Section 2.1, the various results yield on the average $q_0 = 0.93 \pm 0.19$, i.e., the data fit the static, Euclidean prediction. The angular-diameter–red-shift relation for the radio sources was found to follow the inverse law. The result for clusters of galaxies,[24] symbolically $q_0 = -0.9$, deviates from the static, Euclidean relation in the other direction; this may be partly due to effects due to the K term. The summed evidence again points to the static, Euclidean solution. The optical counts[4] correspond, applying a reasonable K correction, exactly to the prediction of the static model. As for the radio counts, I refer to the discussion in Section 2.2.

Therefore, the predictions for global relations successfully fit the observations. The fit and convergence of the global tests in the present approach provide additional evidence for the static, Euclidean character of the Universe.

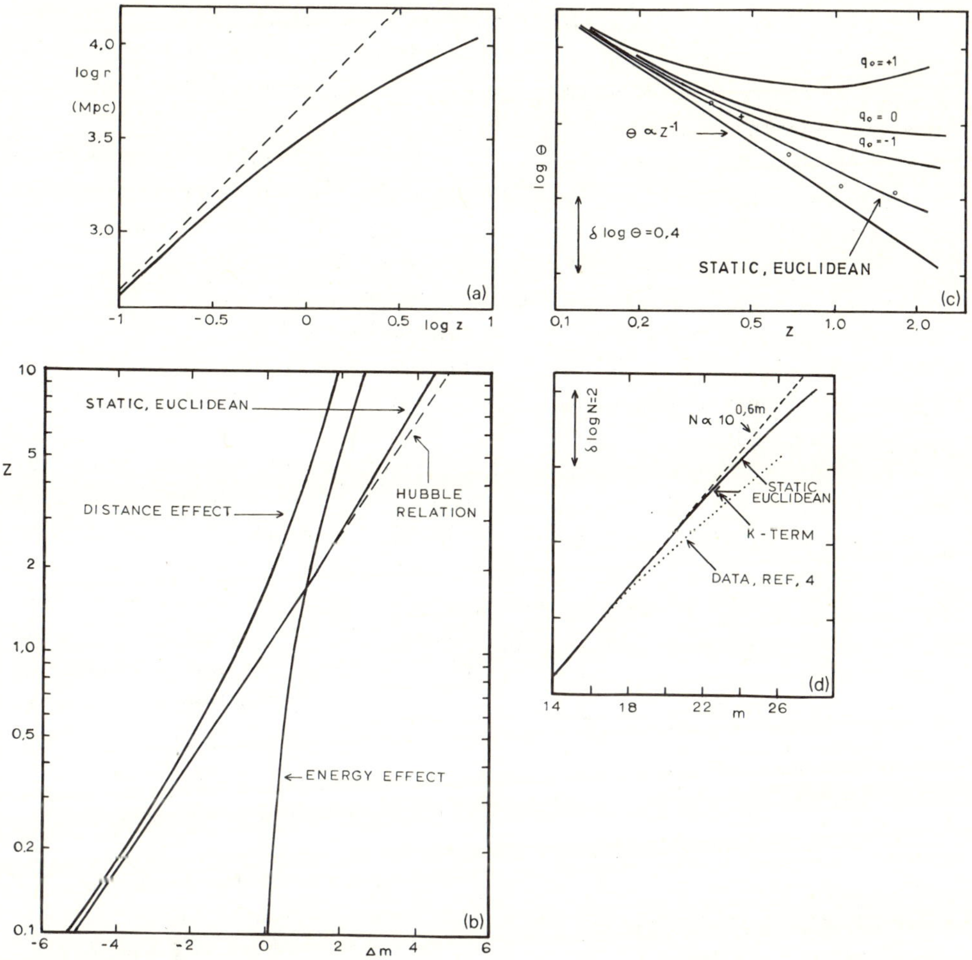

FIGURE 5. Predictions for the global relations in a static, Euclidean model. Data Fig. 5c: open circles, Ref. 13; cross, Ref. 24.

3.4. The Equilibrium of Evolutionary Processes

The equilibrium of cosmic evolutionary processes forms the very essence of the cosmological idea described here. We see any galaxy in a state of evolution; therefore we must find other evolutionary processes which maintain the galaxies as a class unchanged on the large scale. Here only some general propositions can be presented on this central topic of the theory. This is made in the hope of motivating theoretical work and observations relevant to the problem.

Fundamental cosmological parameters which must be explained are: the density distribution of matter, the distributions of morphological structure and angular momentum of galaxies, as well as the overall chemical composition and elementary particle ratios. These distribution functions define the structure of matter in the Universe. The interpretation of the value of the Hubble constant and its relation to the gravitational constant and the microphysical constants poses an even more profound problem of physics and cosmology.

Galaxies and systems of galaxies form the main components of the mass distribution. A hypothesis concerning the origin of the characteristic masses and sizes will be presented in Section 3.6. Particle systems like galaxies and their systems tend to grow dense nuclei, and a countereffect is needed to keep, in the large, the density distribution unchanged. We see such an effect in nonstationary phenomena in nuclei of galaxies.[25] The mechanism of these phenomena has been for years the theme of theoretical studies, but there is no agreement about it. A test of the hypothesis is to compare the observed outward flow of matter involved in the explosions and the theoretically calculated inward flow.

The interpretation of the morphological distribution of galaxies means giving account of the origin, of the evolution, and, necessarily in the equilibrium theory, also of the death of galaxies. The origin of galaxies may in general be attributed to the activity of the nuclei of larger galaxies. There exists much evidence of such processes. The larger galaxies, especially the cD galaxies, have obtained their identity now observed by capturing and merging other galaxies. This effect is well grounded theoretically,[8] and empirically it is supported by the structure and luminosity function of the cD clusters.[9] This effect means at the same time loss of identity, i.e., death, for many smaller galaxies.

Concerning the individual evolution of galaxies, there are reasons to believe that Hubble's morphological sequence is at the same time an evolutionary sequence. Parameters expected to change in a certain manner during evolution do indeed change in that manner along the Hubble sequence, and parameters not expected to change indeed do not; of course one must consider galaxies with a similar status in a system.

Position of the quasars in the scheme is not clear, but it seems[11,26] that a part of quasars (e.g., those with extended radio lobes) are abrupt phases after slow evolution of normal galaxies and a part of quasars (e.g., those with compact, flat spectrum radio sources) are young compact galaxies just expelled from a large galaxy.

One possible evolutionary scheme is in Fig. 6. Having two dimensions it allows inclusion of qualitatively different forms of evolution, and individual evolution on various levels of mass. The scheme is defective and possibly erroneous in details, but may illustrate the equilibrium processes of galactic structure. Of course, the system is not closed, and evolution within a larger system must be taken into account.

The observed type-distribution of galaxies gives an important test possibility for cosmology. If this is an equilibrium distribution, i.e., if the frequencies of the different types correspond to the theoretical periods of these evolutionary phases, this means strong evidence against the big-bang theory and in favor of the theory here discussed.

Angular momentum is a parameter related to morphological structure. Tidal forces of neighbor galaxies slow down the rotation. What then winds the clock up? Cosmology is here facing a task of explaining the phenomenon of the rotation of

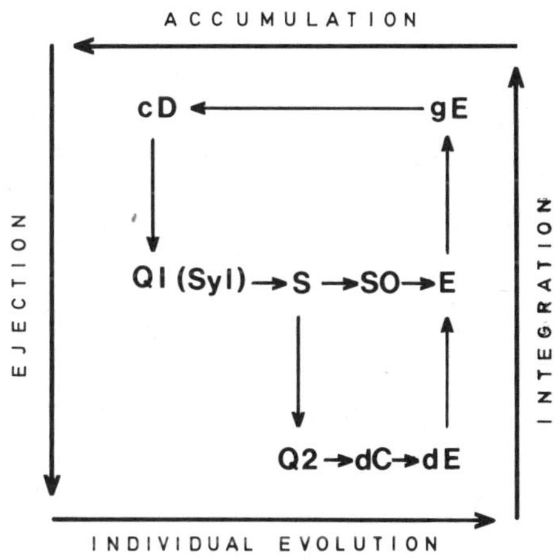

FIGURE 6. Evolutionary scheme for galaxies. Q1 and Q2 are two classes of quasars of which Q1 is thought to be identical with Seyfert 1 galaxies (Ref. 11). KEY: *d*, dwarf; g, giant (see the text).

cosmic bodies. One cannot here resort to the round the-corner solution of referring to the "primordial turbulence," as is done in the standard theory.

From the empirical point of view, rotation of galaxies is not a relict phenomenon. The spiral galaxies rotate very fast, characteristically with a constant rotation velocity out to so far as can be observed.[27] This is the primary reason for the problem of missing mass and supermassive halos in outer galaxies and in the Milky Way.[28] From that simple point one can say that galaxies are rotationally active.

One possible way of an increase of angular momentum may again be attributed to the activity of the nuclei of galaxies. When a nucleus throws out part of its mass this gives rise to outward motion also in the outer regions. It is a well-known fact in stellar dynamics that the stars which have peculiar motion in the direction of the rotation obtain most easily the velocity of escape. So these stars move outward in the equatorial plane, increasing its angular momentum. The slower stars drift, due to the encounters, out of the plane to form the slower-rotating halo. This is the same as the logs of pine in a stream always drift, without any wind, onto the shore. The total effect in this scheme is that, triggered by nuclear activity, kinetic energy in the form of random motions is transformed to one in the form of rotational motion. A similar process may play a role also in the rotation of stars and planets.

The equilibrium process for maintaining the chemical composition should then be commented on. The observed abundances are roughly uniform, but the observed differences give space for differences in evolutionary phase. It is interesting that helium is apparently deficient in the nucleus of the Galaxy[29] as well as in quasars and Seyfert galaxies.[30] There is a systematic gradient of metal content, which increases towards the central regions of galaxies. It is most important that the overall abundances of the elements form an equilibrium distribution: the abundances follow the stability of the atomic nuclei.[31]

When hydrogen burns into helium and further into metals, what turns the nuclei back into protons? Today one can present only guesses for that. Once again nuclei of galaxies may be in the focus of the problem, working as "smelting furnaces" of matter—the deviating abundance of Sagittarius A and the quasars is here to be remembered.

In the extreme temperatures of supernovae metal nuclei dissociate into helium and further into protons and neutrons,[32] and a similar process may also be connected with the activity of nuclei of galaxies. On the other hand, in a universe with an infinite time scale there is time enough for any nuclei to decay into protons even gradually. The equilibrium distribution of the elements is a general argument in favor of the hypothesis, but the details are not yet clear.

As regards the photon–baryon equilibrium, a partial solution of the problem will be discussed in the following section.

Reference to the second law of thermodynamics can be anticipated. It should be pointed out that the cosmological system here considered is not an isolated system but one connected to the entire Universe in a certain manner (see Section 3.6). Further, this system is not, unlike a gas, composed of identical particles but of several qualitatively different constituents, i.e., stars, objects in the nuclei of galaxies, interstellar and intergalactic gas and dust, as well as the gravitational, magnetic, and radiation fields. Altogether the cosmological system is too complicated for the thermodynamical theory to apply.

3.5. *The Radiation Background*

The dichotomy of galaxies and background sets for cosmology a difficult problem which will be fully solved only in the probably not very near future. In particular for the cosmology of equilibrium processes presuming large-scale equilibrium between electromagnetic radiation and other forms of matter, intergalactic space will be the tempering trough of the theory and without doubt the arena for new important discoveries.

The discovery of the microwave background radiation with roughly a blackbody spectrum has been considered, with full justification, the most important cosmological observation since the discovery of red shifts. Here the standard theory obtained, for the first time, some independent support for the idea that a big bang might have happened. But the problem is not so simple. Also the equilibrium theory predicts a *priori* an intense, isotropic background radiation—after the history of the Olbers Paradox this should not be too surprising. A blackbody distribution at some probably low temperature is also expected from the probable existence of electromagnetic–gravitational equilibrium. The spectrum should have a blackbody form only locally; the integrated spectrum should be slightly deformed by the red shift.

In principle, the energy density of the background radiation should have a value coinciding with the density due to the radiation of galaxies. The intensity of radiation due to the galaxies from within distance r is the integral

$$I_r = \frac{nL}{4\pi} \int_0^r \frac{dr}{1 + z} \tag{8}$$

where n is space density and L the luminosity of an average galaxy. Using Eq. (3), one obtains

$$I_r = \frac{cnL}{4\pi H}\left[1 - \exp\left(-\frac{H}{c}r\right)\right] \tag{9}$$

or

$$I_z = \frac{cnL}{4\pi H}\left(\frac{z}{1+z}\right) \tag{10}$$

which approaches the value

$$I = \frac{cnL}{4\pi H} \tag{11}$$

when $z \rightarrow \infty$. The function I_z and the contribution of the background intensity from the different z intervals are shown in Fig. 7. It can be seen that half of the total background light comes from within $z = 1$ and nine percent from $z > 10$.

As the integration above yielded a finite value of I, the Olbers Paradox becomes solved in a straightforward manner.

If one adopts for the parameters the values $n = 3 \times 10^{-2}$ Mpc^{-3}, $L = 3.9 \times 10^{44}$ erg s^{-1} (10^{11} stars of solar luminosity), and $H = 60$ km s^{-1} Mpc^{-1}, we obtain for the integrated radiation energy density $U_c = (4\pi/c)\, I = nL/H$ the value $U_c = 2 \times 10^{-13}$ erg cm^{-3} = 0.13 eV cm^{-3}. Within the range of uncertainties, this is the same as the observed energy density of the microwave background, U_c(obs) = 0.25 eV cm^{-3}.

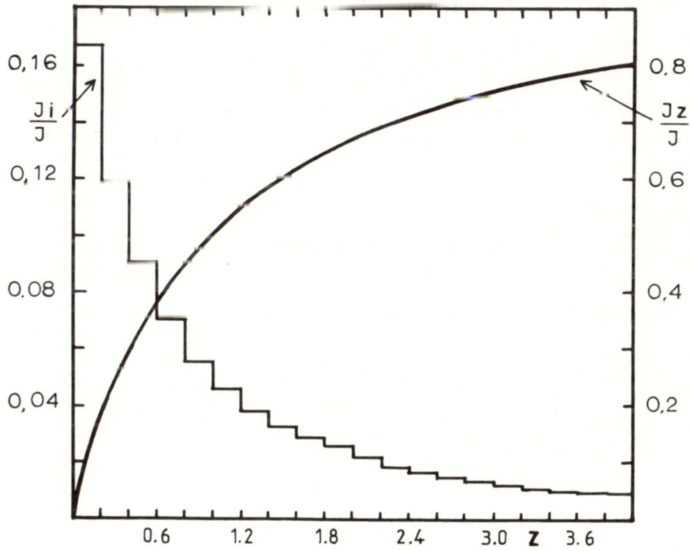

FIGURE 7. Contribution $I_z/I = z/(1 + z)$ to the background radiation from distance z, and contributions from different z intervals.

As pointed out above and as is explicitly given by Eq. (10), the red shift should, *a priori*, deform the original blackbody spectrum. A deviation from the 3°K Planck curve in this sense has indeed been observed.[33] Within the frame of the standard theory, only *ad hoc* interpretations have been suggested for this feature.

The conventional theory provides no interpretation of the observed photon–baryon number ratio $n_\gamma/n_b \sim 10^7$–10^9. In the present approach this parameter depends *a priori* on the Hubble constant determining the "half-life time" of the photons, $\bar{\tau} = \ln 2/H$, and on the mass-to-luminosity ratio of galaxies:

$$\frac{n_\gamma}{n_b} = \frac{U_c/h\nu}{\rho/m_p} = \frac{Ln_p}{Hh\nu M} \tag{12}$$

where M and L are the galactic mass and luminosity, m_p the proton mass, h Planck's constant, and ν the peak frequency of the background radiation. Therefore the problem reduces to the astrophysical problems of the emissivity of matter on the one hand and the red shift on the other hand. From Eq. (12), the predicted value is $n_\gamma/n_b \sim 7 \times 10^8$, which coincides with the observed number.

3.6. *The Problem of Stability, the Macroscopic Structure of Matter, and Mach's Principle*

The problem of the stability of infinite static models has occupied the minds of scientists in a similar way as that posed by the Olbers Paradox. Indeed, there exists an apparent gravity paradox: the gravitational force becomes indefinite in the naive static, infinite model. Philosophers like Kant, Hegel, and Engels have made reflections about a repulsive force counterbalancing the gravitational force. At the end of the nineteenth century Seeliger and Neumann suggested an absorption of gravitation over large distances by a factor $e^{-\varkappa r}$ where $\varkappa$ is a small quantity.[34] Einstein's cosmological constant is a famous variation of the idea of a repulsive force. Others are expressed in the steady-state theory and in a theory about an electric universe.[35]

The present approach considers the red shift as an interaction effect. Gravitation is one of the most probable candidates for the counterpart in that. This hypothesis would have some interesting implications. Inertia is generally believed to be caused by gravitational interaction by cosmic matter. The interaction by the masses within distance z is, with the above hypothesis,

$$a_z = -\frac{4\pi Gc\rho}{H}\left(\frac{z}{1+z}\right) \tag{13}$$

which approaches a finite value

$$a_c = -\frac{4\pi Gc\rho}{H} \tag{14}$$

when $z \rightarrow \infty$, i.e., the gravitational paradox becomes relaxed. With $\rho = 5 \times 10^{-31}$ g cm^{-3} and $H = 60$ km s^{-1} Mpc^{-1}, $a_c = 6.4 \times 10^{-9}$ cm s^{-2}. Consider a particle at an edge of a system with radius R and mass M. The acceleration of the particle due to the system is

$$a_s = -GM/R_2 \tag{15}$$

The data[36] show, as compiled in Table I, that the values of a_s are very much the same in the galaxies and in their systems, and at the same time these are equal to the cosmological value a_c. The systematic nature of these values speaks for the idea that the *cosmic masses are controlling the formation of the macroscopic structure in the Universe*. The evolved elliptical galaxies deviate from the general trend, implying that during the evolution subsequent to the formation the gravitational force of the galaxy progressively exceeds that by the cosmic masses.

It seems that we see in the large-scale structure of matter, observed as galaxies and their systems, a concrete manifestation of Mach's principle. In Eqs. (13) and (14) there is a possible explicit formulation for how the cosmic masses interfere with the local phenomena.

Parallel with the coincidence $a_c \sim a_s$ there is one between the cosmic radiation energy density U_c and the local density, U_s, due to the starlight. It is easy to deduce the similarity $U_c \sim U_s$ using the relation $a_c \sim a_s$. Hence it appears that the hypothesis made above leads to interpretation not only of the similarity of the parameter a_s for galaxies and their systems but also of the similar density of radiation from the local and from the cosmic sources.

TABLE I. Acceleration, a_s, of a Particle Lying in an Edge of a System with a Mass M and Radius R

Class of objects	log M (g)	log R (cm)	a_s (cm s^{-2})
Supergalaxies	48.7	25.5	3.4×10^{-10}
Large clusters of ellipticals	47.9	24.5	5.4×10^{-9}
Small clusters of ellipticals	47.2	24.3	2.7×10^{-9}
Small clouds of spirals	47.0	24.3	1.7×10^{-9}
Small loose groups of spirals	16.5	24.1	1.3×10^{-9}
Small dense clusters of ellipticals	46.5	23.7	8.5×10^{-9}
Compact groups of spirals	45.5	22.6	1.3×10^{-9}
Giant ellipticals	45.5	22.35	4.2×10^{-7}
M31 (Sb)	44.6	22.79[a]	7.0×10^{-9}
Milky Way (Sbc)	44.45	22.67[b]	8.8×10^{-9}
The Universe (a_c)[c]	—	—	6.4×10^{-9}

[a] R = 20 kpc adopted.
[b] R = 15 kpc adopted.
[c] The acceleration a_c due to the cosmic masses.

On the basis of Eqs. (10) and (13) we can define an "effective radius" of the Universe at $z = 1$ ($r = 3450$ Mpc for $H = 60$ km s^{-1} Mpc^{-1}), covering the volume which produces a half of the total cosmic inertia and a half of the total background light. In a sense the Universe is "effectively closed" for gravitational and electromagnetic interactions.

The equilibrium theory can hardly work without coupling between the various physical interactions. Therefore theoretical problems connected to various equilibrium processes, as well as the cosmological observations, might serve as guides in the efforts towards a unified physical theory. Such an approach has been previously attempted within the framework of the so-called large number coincidences.[37] One outcome of this hypothesis has been the supposition about the time variation of fundamental physical "constants." In the equilibrium theory not involving any cosmic time such a universal time variation of the constants is denied. Instead, a variation of the "constants" from place to place and local variation in the course of evolution are permitted. Through the finite interaction of the cosmic masses, the ratios between the global and the microphysical parameters make sense also in this case.

4. Conclusions

The cosmological test results, when interpreted within the frames of the standard theory, show an extreme dispersion. The consideration of cosmic evolutionary effects does not help the situation, as there does not seem to be any. The analysis of the redshift phenomenon further weakens the basis of the standard theory. This theory does not hence seem capable of mediating reliable knowledge about the Universe in spite of the hugely developed empirical data.

The present status in cosmology urges an alternative approach. The outlines are presented above for a theory postulating a large-scale equilibrium of all cosmic evolutionary processes, and the stability of the Universe. The predictions for the global observations confront successfully the actual data. The great lines for the equilibrium processes are clear-cut, but the specific suggestions here presented for these processes are only preliminary. There exist numerous concise empirical tests of the equilibrium theory. These tests include the global tests, the tens of tests of possible cosmological evolution and expansion, and the local distributions of numerous parameters of galaxies. The equilibrium cosmology is indeed exceptionally apt for empirical disproof or verification.

The equilibrium theory gives good reasons to expect important new findings as regards, e.g., the photon–baryon ratio, element abundances, relations between the various interactions, and the theory of rotation.

The best arena for the study of these questions and other cosmological problems is the local neighborhood, including nearby galaxies and their systems, and intergalactic space. Nowadays cosmologists strain their forces to the extreme to reach the ultimate depths of cosmic space, and become confronted with all the uncertainties following from that. Without stopping this enterprise it should be noticed that the answer may be closer, partly even in our own Galaxy. If the cosmological principle proves to be right in its perfect form, it guarantees that by studying the local phenomena, the finite, we get knowledge also about the infinite. Every local environment is at the same time cosmological. And vice versa, we encountered above tangible indications that we cannot understand the particular without regard to the general. As expressed in the phrase in the beginning, we can in the proper sense of the word become conscious only of the infinite.

Acknowledgments

The author is very grateful to K. J. Donner, P. Teerikorpi, I. Tuominen, and O. Vilhu for critical discussions.

References and Notes

1. J.-C. Pecker, in *Décalaces vers le rouge et expansion de l'Univers, IAU Coll.* 37 (1976), p. 451; T. Jaakkola, M. Moles, J.-P. Vigier, J.-C. Pecker, and W. Yourgrau, *Found. Phys.* **5**, 257 (1975); M. Moles, Thesis, Univ. Pierre et Marie Curie, Paris (1978).
2. T. Jaakkola, *Acta Cosmologica* **7**, 17 (1978).
3. T. Jaakkola, M. Moles, and J.-P. Vigier, *Astron. Nachr.* **300**, 229 (1979); additional result on $q_0 \sim 1$ for quasars: P. Teerikorpi, *Astron. Astrophys.* **98**, 309 (1981).
4. I. D. Karachentsev and A. I. Kopylov, *Pisjmo Astron. Zh.* **3**, 246 (1977); I. D. Karachentsev, *Pisjmo Astron. Zh.* **6**, 3 (1980).
5. H. Spinrad, in *The Evolution of Galaxies and Stellar Populations*, B. M. Tinsley and R. B. Larson, editors (Yale Univ. Obs., New Haven, Connecticut, 1977), p. 301; R. G. Kron, *Ibid.*, p. 334.
6. H. Butcher and A. Oemler, see Ref. 5.
7. B. M. Tinsley, *Astrophys. J.* **211**, 621 (1977).
8. J. P. Ostriker and S. Tremaine, *Astrophys. J. Lett.* **202**, L113 (1975); S. White, *Mon. Not. R. Astron. Soc.* **174**, 19 (1976).
9. A. Sandage and H. Hardy, *Astrophys. J.* **183**, 743 (1973).
10. R. F. Green and M. Schmidt, *Astrophys. J. Lett.* **220**, L1 (1978).
11. T. Jaakkola, *Astrophys. Space Sci.*, in press.
12. K. I. Kellermann, *Astron. J.* **77**, 531 (1972); I. I. K. Pauliny-Toth and K. I. Kellermann, in *Confrontation of Cosmological Theories with Observational Data, IAU Symp.* 63, M. S. Longair, editor (1974), p. 111.
13. R. D. Ekers and G. K. Miley, Indications on morphological and spectral constancy, in *Radioastronomy and Cosmology, IAU Symp.* 74, D. L. Jauncey, editor (1977), p. 109; and J. G. Bolton, *ibid.*, p. 85.
14. F. Zwicky, *Morphological Astronomy* (Springer-Verlag, Berlin, 1957), pp. 163, 164.
15. E. Hubble, *Observational Approach to Cosmology* (The Clarendon Press, Oxford, 1937).
16. H. Arp, in *Décalages vers le rouge et expansion de l'Univers, IAU Coll.* 37 (1976); G. Burbidge, *Ibid.*; A. Kipper, *Publ. Tartuskoi Astrofis. Obs.* 40 (1972) and *Astrofizika* **10**, 283 (1973); I. E. Segal, *Mathematical Cosmology and Extragalactic Astronomy* (Academic Press, New York, 1976); W. G. Tifft, *Astrophys. J.* **175**, 613 (1972).
17. References 1, 2, and 3, and references therein.
18. Lucretius, *De Rerum Natura*, transl. (Clarendon Press, Oxford), Book i, p. 959.

19. J. D. North, *The Measure of the Universe, A History of Modern Cosmology* (Clarendon Press, Oxford, 1965), p. 423.
20. See North, Ref. 19, p. 357.
21. F. Engels, *Dialectics of Nature* (Progress Publishers, Moscow, 1972), p. 37.
22. A. Sandage, *Phys. Today*, Febr. 1970, 34.
23. J. D. North, Ref. 19, p. 275.
24. P. Hickson, *Astrophys. J.* **217,** 964 (1977); Ref. 13.
25. V. A. Ambartsumian, in *Nuclei of Galaxies*, D. J. K. O'Connel, editor (North-Holland Publ. Co., Amsterdam, 1971), p. 9.
26. T. Jaakkola, K. J. Donner, and P. Teerikorpi, *Astrophys. Space Sci.* **37,** 301 (1975).
27. V. C. Rubin, *Comments Astrophys. Space Sci.* **8,** 79 (1978); V. C. Rubin, V. K. Ford, Jr., and N. Thonnard, *Astrophys. J.* **238,** 471 (1980).
28. J. P. Ostriker and P. J. E. Peebles, *Astrophys. J.* **186,** 467 (1973); J. Einasto, A. Kaasik, and E. Saar, *Nature* **250,** 309 (1974).
29. P. G. Mezger, Th. Wilson, F. F. Gardner, and D. K. Milne, *Astrophys. Lett.* **6,** 35 (1970).
30. M. Peimbert and H. Spinrad, *Astrophys. J.* **159,** 809 (1970); and *Astron. Astrophys.* 7, 311 (1970); D. E. Osterbrock and R. A. Parker, *Astrophys. J.* **143,** 268 (1966); J. N. Bahcall and B. Kozlovsky, *Astrophys. J.* **155,** 1077 (1969).
31. W. D. Harkins, *J. Am. Chem. Soc.* **39,** 856 (1917), R. C. Tolman, *J. Am. Chem. Soc.* **44,** 1902 (1922); North, Ref. 19, pp. 227, 256.
32. I. S. Shklovskii, *Stars, Their Birth, Life and Death* (W. H. Freeman and Co., San Francisco, 1978), p. 294.
33. D. P. Woody and P. L. Richards, *Phys. Rev. Lett.* **42,** 925 (1979).
34. See North, Ref. 19, p. 17.
35. R. A. Lyttleton, *Rival Theories of Cosmology* (Oxford Univ. Press, Oxford, 1960), p. 22.
36. G. de Vaucouleurs, *Publ. Astron. Soc. Pacific* **83,** 113 (1971); C. W. Allen, *Astrophysical Quantities*, 3rd ed. (University of London, The Athlone Press, 1973).
37. P. A. M. Dirac, *Proc. R. Soc. London, Ser. A* **165,** 199 (1938); A. S. Eddington, *Fundamental Theory* (Cambridge University Press, London, 1946).

16

THE CONNECTION BETWEEN THE OBJECTIVE DESCRIPTION OF MACROSYSTEMS AND QUANTUM MECHANICS OF "MANY PARTICLES"

G. LUDWIG
Fachbereich Physik
Philipps-Universität, Marburg
D 3550 Marburg, GFR

It has been shown[1–5] that the quantum mechanics of microsystems can be based on the objective description of the macroscopic apparatuses that prepare and register the microsystems. In this connection no difficulty arises with the interpretation of quantum mechanics and the interpretation of the so-called "measuring process." The problem of the measuring process is reduced to the very fundamental problem of the statistical mechanics of nonequilibrium processes.

This fundamental problem is the following. On the one hand we can develop by well-known methods (direct products of Hilbert spaces, Hamiltonians guessed by correspondence principles, etc.) a quantum mechanics of "many particles," termed briefly $\mathcal{PT}_{\text{qexp}}$. But, as the number of particles increases so do the difficulties, not only of handling but also of interpreting the theory—e.g., it is not known, even in principle, how the various "observables" of the theory are to be measured.

On the other hand, we know from experience what the experimental physicists are measuring on special macrosystems and also that these measured quantities can be looked upon as objective qualities of the systems. But we have no theory capable of telling us what those special self-adjoint operators (or perhaps other mathematical symbols?) representing these "real" measurements are. We have only some correspondence hints for guessing the desired self-adjoint operators, knowing full well that actual measurements are in no way exact or even very accurate measurements of these "idealized" observables represented by the guessed self-adjoint operators.

Is it hopeless to formulate the problem of statistical mechanics in general, and not only in a very inaccurate form, for more or less special examples? We will try to give here a general framework for establishing the connection between the objective description of macrosystems and the extrapolated quantum mechanics $\mathcal{PT}_{\text{qexp}}$. We will present the framework in three steps: First we will give a brief survey for a very simple mathematical structure for describing statistics. In the second step we will furnish a general framework (termed briefly $\mathcal{PT}_m$) for describing the macrosystems in an objective form. And, in the third step, we will explain the connection between the objective description $\mathcal{PT}_m$ and $\mathcal{PT}_{qexp}$, in the form of a mathematical embedding.

1. Statistics

To describe statistics in physics the Kolmogoroff scheme is sometimes too special. Therefore, I have proposed a quite general description in Ref. 6, Section XIII, Ref. 3, Ref. 7, Section 11, and Ref. 1, Section II, which is applicable not only in the development of quantum mechanics but also in *all* other areas of physics. We will use this description also in Section 2.

Since this description of statistics has already been published in the books and papers mentioned, we will give here only a rough draft of its structure.

1.1. *Statistical Selection Procedures*

M may be the basic set of the mathematical structure for describing statistics. The elements of M are the representative of those physical features whose statistics is to be described. The statistics is related to selection procedures by which special features can be selected as often as desired. We describe a selection procedure by a set a of M, where a is the set of those features that satisfy the postulates of an individual selection procedure. Therefore, we describe the selection procedures by a mathematical structure $\mathcal{J} \subset \mathcal{P}(M)$ which satisfies the following axioms:

AS *1.1.* If $a,b \in \mathcal{J}$, $a \subset b$ then $b \backslash a \in \mathcal{J}$.

AS *1.2.* If $a,b \in \mathcal{J}$ then $a \cap b \in \mathcal{J}$.

$\mathcal{J}$, together with the axioms AS 1.1, 1.2, is called a structure of species *selection procedure*. $\mathcal{J}$, together with a function λ, $\mathcal{T} \overset{\text{def}}{=} \{(a,b) \mid a,b \in \mathcal{J}, a \supset b, a \neq \phi\} \overset{\lambda}{\rightarrow} [0,1] \subset \mathbb{R}$ is called a structure of species *statistical selection procedure*, if, in addition to AS 1.1 and 1.2, the following axioms are valid:

AS *2.1.* If $a,b \in \mathcal{J}$, $a \cup b \in \mathcal{J}$ and $a \cap b = \phi$ then $\lambda(a \cup b,a) + \lambda(a \cup b,b) = 1$.

AS *2.2.* If $a,b,c \in \mathcal{J}$, $a \supset b \supset c$ and $b \neq \phi$, then $\lambda(a,c) = \lambda(a,b)\lambda(b,c)$.

AS *2.3.* If $a,b \in \mathcal{J}$ and $b \neq \phi$ then $\lambda(a,b) \neq 0$.

$\lambda(a,b)$ is called the probability of the finer selection procedure b relative to the coarser one a.

1.2. *Preparation and Registration*

For experiments with physical systems we will specialize the selection procedures. To perform experiments on systems and to record statistics, it is necessary to prepare these systems and to register something, what has happened in or was caused by the systems. We therefore introduce the following structures:

Let M be the basic set the elements of which are representatives of the physical systems, i.e., M is used as an index set for the systems. The preparation procedures are described by elements of a set $\mathcal{Q} \subset \mathcal{P}(M)$, with the following axiom:

APS 1. $\mathcal{Q}$ is a structure of species statistical selection procedure.

Let $\lambda_{\mathcal{Q}}$ be the probability function for $\mathcal{Q}$. The registration is described by the elements of two sets $\mathcal{R}_0 \subset \mathcal{P}(M)$ and $\mathcal{R} \subset \mathcal{P}(M)$. The elements of $\mathcal{R}_0$ are the representatives of the *registration methods,* i.e., the measuring apparatuses used. The elements of $\mathcal{R}$ are called *registration procedures* and are representatives of the digital indications of the measuring apparatuses. Therefore, we introduce for $\mathcal{R}_0$, $\mathcal{R}$ the following axioms:

APS 2. $\mathcal{R}$ is a structure of species selection procedure.

APS 3. $\mathcal{R}_0$ is a structure of species statistical selection procedure. (Let $\lambda_{\mathcal{R}0}$ be the probability function for $\mathcal{R}_0$.)

APS 4.1. $\mathcal{R}_0 \subset \mathcal{R}$.

APS 4.2. For every $b \in \mathcal{R}$ there is a $b_0 \in \mathcal{R}_0$ with $b \subset b_0$.

Let $\mathcal{Q}'$ be the set $\mathcal{Q}$ without the element ϕ and $\mathcal{R}'_0$ the set $\mathcal{R}_0$ without the element ϕ. Then we have the following definition:

Definition 2.1 We say that a and b_0 ($a \in \mathcal{Q}'$, $b_0 \in \mathcal{R}'_0$) *may be combined,* provided that if $\tilde{a} \in \mathcal{Q}'$, $\tilde{b}_0 \in \mathcal{R}'_0$, $a \subset \tilde{a}$, $\tilde{b}_0 \subset b_0$, we find that, $\tilde{a} \cap \tilde{b}_0 \neq \phi$,

$$C = \{(a,b_0) \mid a \in \mathcal{Q}', b_0 \in \mathcal{R}'_0\ (a,b_0) \text{ may be combined}\}$$

The problem of introducing axioms for the set C (i.e., introducing physical laws for the physical possibility to combine a preparation procedure a and a registration method b_0) is not trivial in general. In the case of macrosystems, this problem is not difficult to solve (see Section 2.2); but in the case of preparation and registration of microsystems this problem has a much deeper significance (see Ref. 1, Sections II and III).

In *any* case, we postulate the very plausible axioms:

APS 5.1.1 For each $a \in \mathcal{Q}'$ there exists a $b_0 \subset \mathcal{R}'_0$ such that $(a,b_0) \in C$.

APS 5.2. For each $b \in \mathcal{R}$, $b \neq \phi$ there exists at least one $a \in \mathcal{Q}'$ and a $b_0 \in \mathcal{R}'_0$ which satisfies $b_0 \supset b$, $(a,b_0) \in C$ and $a \cap b \neq \phi$.

APS 5.2 is a stronger axiom than APS 4.2.

The most general selection procedures for physical systems are introduced by the elements of $\mathcal{J}$, where

Definition 2.2. $\mathcal{J}$ is the structure of species selection procedure generated by the set

$$\{a \cap b \mid a \in \mathcal{Q}', b \in \mathcal{R}, \text{there exists a } b_0 \in \mathcal{R}'_0 \text{ which satisfies } b_0 \supset b \text{ and } (a,b_0) \in C\}$$

For $\mathcal{J}$ we postulate the following axioms:

APS 6. $\mathcal{J}$ is a structure of species statistical selection procedure ($\lambda_{\mathcal{J}}$ is the probability function for $\mathcal{J}$).

APS 7. If $a_1, a_2 \in \mathcal{Q}$, $a_2 \subset a_1$, $b_{10} \in \mathcal{R}_0$, $b_{20} \in \mathcal{R}_0$, $b_{20} \subset b_{10}$ and $(a_1,b_{10}) \in C$, then

$$7.1.\ \lambda_{\mathcal{J}}(a_1 \cap b_{10}, a_2 \cap b_{10}) = \lambda_{\mathcal{Q}}(a_1,a_2)$$
$$7.2.\ \lambda_{\mathcal{J}}(a_1 \cap b_{10}, a_1 \cap b_{20}) = \lambda_{\mathcal{R}0}(b_{10}b_{20})$$

APS 8.

$$M = \bigcup_{a \in \mathcal{Q}} a = \bigcup_{b \in \mathcal{R}} b$$

The function $\lambda_{\mathcal{J}}$ describes the most general probabilities that can be observed on physical systems. But it is sufficient to observe special values of $\lambda_{\mathcal{J}}$ which are contained in the statement of Theorem 2.1 below. To formulate this theorem we introduce the following sets:

$$\mathcal{F} \overset{\text{def}}{=} \{(b_0,b) \,|\, b_0 \in \mathcal{R}'_0,\ b \in \mathcal{R},\ b \subset b_0\}$$

the elements of $\mathcal{F}$ are called *effect processes* or *questions*.

$$\mathcal{C} \overset{\text{def}}{=} \{(a,f) \,|\, f = (b_0,b) \in \mathcal{F},\ (a,b_0) \in C\}$$

By $\mu(a,(b_0,b)) = \lambda_{\mathcal{J}}(a \cap b_0, a \cap b)$ is defined a function

$$\mathcal{C} \xrightarrow{\mu} [0,1] \subset \mathrm{R}$$

One can prove (see Ref. 1, Section II) the following theorem:

Theorem 2.1. $\lambda_{\mathcal{J}}$ is determined by $\lambda_{\mathcal{Q}}$ and μ.

$\mu(a,(b_0,b))$ is the probability of getting the "answer yes" (i.e., getting a response of b) on the "question" (b_0,b), if the systems are prepared by the procedure a.

2. The Objective Description of Macrosystems

The essential characteristic of macroscopic measurements is that these measurements can be interpreted as measurements of objective properties. We want to describe this feature by a mathematical form, and in this way we hope to clarify what is meant by objective properties and their measurements.

2.1. *State Space and Trajectory Space*

To describe these objective properties we introduce a "state space" Z. The elements z of Z are called *states*. The physical interpretation of a state space can be explained briefly by examples. In classical point mechanics the Γ space of the (p_i,q_i) can be taken as state space. In hydrodynamics the states are represented by triples of fields $(\mathbf{u}(\mathbf{r}), \mu(\mathbf{r}), T(\mathbf{r}))$, where $\mathbf{u}(\mathbf{r})$ is the velocity field, $\mu(\mathbf{r})$ the density field, and $T(\mathbf{r})$ the temperature field. For the Bolzmann description of a rarefied gas the states are given by the Bolzmann distribution function $f(\mathbf{r},\mathbf{v})$, where

$$\int_{\mathcal{V}} f(\mathbf{r},\mathbf{v})\, d^3\mathbf{r}\, d^3\mathbf{v}$$

is the approximate number of atoms in the region $\mathcal{V}$ of the μ space.

Let M_m be the set for labeling the various macrosystems. Then every special system x (system labeled by x) has at every meaningful time t a state $z \in Z$; in mathematical form: There is a function f: $M_m \times \Theta \rightarrow Z$ where Θ is a suitable time interval, in which the systems are existing. This function f is often written in the form $z_x(t)$. The function $z_x(t)$ is called the trajectory of the system x in the state space Z.

For general physical reasons (description of the inaccuracies; see Ref. 7, Section 9) it is useful to introduce in Z a uniform structure p, so that Z_p (Z with the uniform structure p) is precompact and metrizable. Then Z_p is also separable and has a compact completion $\hat{Z}_p$.

Since we want also to describe irreversible systems, we take as time scale Θ only the interval $0 \leq t < \infty$. The point $t = 0$ of the labor time scale is arranged such that the systems are prepared before $t = 0$. It is clear that the assertion $t \rightarrow \infty$ is not realistic but an idealization describing that we do not know how long the systems exist.

Every trajectory $y = z(t)$ can be identified with the graph

$$G(y) = \{(t,z) \mid t \in \Theta,\ z \in Z \text{ and } z = z(t)\}$$

as a subset of $X = \Theta \times Z$.

We want to introduce in X also a uniform structure of physical inaccuracy. Without dwelling too long on what is the best way to describe the inaccuracy in the set X, we introduce a uniform structure in X by the following fundamental systems of vicinities:

$$U_{\gamma,V} = (X_\gamma \times X_\gamma) \cup V$$

where γ is a compact subset of the open interval Θ_0: $0 < t < \infty$, $\gamma' = \Theta \backslash \gamma$, $X_\gamma = \gamma' \times Z$, and $V = \{[(t,z), (t',z')] \mid |t - t'| < \epsilon, (z,z') \in U\}$, with U a vicinity of p in Z.

X endowed with this uniform structure may be called X_p. Since the graphs are subsets of X, we want to introduce in $\mathcal{P}_0(X)$ [with $\mathcal{P}_0(X)$ the set of all subsets of X without ϕ] a uniform structure of physical inaccuracy. We do this in a "canonical" form:

Let $\mathcal{N}$ be the vicinity filter of X_p. The "canonical" filter $\tilde{\mathcal{N}}$ in $\mathcal{P}_0(X)$ is defined by the following fundamental systems of vicinities:

$$\tilde{U} = \{(A,B) \mid A \in \mathcal{P}_0(X), B \in \mathcal{P}_0(X), A \in U(B), B \in U(A), U \in \mathcal{N}\},$$

where $U(A)$ [and correspondingly $U(B)$] is the set of all $x \in X$, for which exists a$x' \in X$ with $(x,x') \in U$. $[\mathcal{P}_0(X)]_p$ may be the set $\mathcal{P}_0(X)$ with this "canonical" uniform structure.

We introduce the "trajectory space" Y as the set of all graphs $G(y)$, where $y = z(t)$ is a continuous mapping $\Theta \rightarrow Z_p$. Since Y is a subset of $\mathcal{P}_0(X)$, the uniform structure of $[\mathcal{P}_0(X)]_p$ induces a uniform structure on Y. Y with this structure may be called Y_p. The uniform structure p separates the elements of Y. Y_p is precompact, metrizable, and separable. $\hat{Y}_p$ (the completion of Y_p) is compact. We will drop the index for $\hat{Y}_p$: $\hat{Y} = \hat{Y}_p$.

We assume that the function $M_m \times \Theta \xrightarrow{f} Z$, defined above, yields a continuous trajectory for every $x \in M_m$, i.e., the function f defines another function $M_m \xrightarrow{g} Y$. This is what we mean by the objective description of macrosystems.

2.2. *Preparations and Measurements of Macrosystems*

We will not describe in mathematical detail (see Ref. 5) that we want to use only registration apparatuses that measure the trajectories of the systems. To describe the registration of trajectories we introduce a basic set $\tilde{\mathcal{R}}_m$ and a subset $\tilde{\mathcal{R}}_{0m} \subset \mathcal{R}_m$, with the physical interpretation that an element $\rho_0 \in \tilde{\mathcal{R}}_{0m}$ represents an apparatus which measures trajectories and that an element $\rho \in \mathcal{R}_m$ is a digital signal on such a measuring apparatus.

We assume that $\tilde{\mathcal{R}}_m$ is an ordered set such that for two elements $\rho_1,\rho_2 \in \tilde{\mathcal{R}}_m$ the lower bound $\rho_1 \wedge \rho_2$ exists and that the order intervals $[0,\rho]$ are Boolean algebras. We assume also that $\tilde{\mathcal{R}}_{0m}$ has (as a subset of $\tilde{\mathcal{R}}_m$) the same properties as $\tilde{\mathcal{R}}_m$ and that for every $\rho \in \tilde{\mathcal{R}}_m$ there is a $\rho_0 \in \tilde{\mathcal{R}}_{0m}$ with $\rho \leq \rho_0$.

Now we introduce a mapping $\mathcal{R}_m \xrightarrow{h} \mathcal{P}(M_m)$ with this interpretation: $x \in h(\rho)$ means that the system x is selected according to the signal ρ. We postulate that $h(\rho_1 \wedge \rho_2) = h(\rho_1) \cap h(\rho_2)$ and, for $\rho_1 \geq \rho_2$, $h(\rho_1 \backslash \rho_2) = h(\rho_1) \backslash h(\rho_2)$, where $\rho_1 \backslash \rho_2$ is the complement of ρ_2 in the Boolean algebra $[0,\rho_1]$. In addition we assume that $\tilde{\mathcal{R}}_{0m} \xrightarrow{h} \mathcal{P}(M_m)$ is injective.

We define $\mathcal{R}_m = h(\tilde{\mathcal{R}}_m)$ and $\mathcal{R}_{0m} = h(\mathcal{R}_{0m})$. Then it is possible to prove (Ref. 5) that $\mathcal{R}_{0m}$ and $\mathcal{R}_m$ are structures of species selection procedures. We postulate that $\mathcal{R}_{0m}$ is a structure of species statistical selection procedure, i.e., that there is a probability function $\lambda_{\mathcal{R}0m}$ for $\mathcal{R}_{0m}$ (see Section 1.2). We thus have the following:

APS 2. $\mathcal{R}_m$ is a structure of species selection procedure.

APS 3. $\mathcal{R}_{0m}$ is a structure of species statistical selection procedure.

APS 4.2. To every $b \in \mathcal{R}_m$ there is a $b_0 \in \mathcal{R}_0$ with $b \subset b_0$.

The elements $b \in \mathcal{R}_m$ are called registration procedures (of the trajectories) and the elements $b_0 \in \mathcal{R}_{0m}$ are called registration methods (of the trajectories).

We introduce a set $\mathcal{Q}_m \in \mathcal{P}(M_m)$ of preparation procedures in the same way as in Section 1.2, i.e., with the postulate APS 1.

Since we only register the macrosystems after time $t = 0$, i.e., after the preparation of the systems, we postulate the following:

APS 5m. $a \in \mathcal{Q}_m$, $b_0 \in \mathcal{R}_{0m}$, $a \neq \phi$, $b_0 \neq \phi$ imply $a \cap b_0 \neq \phi$.

It follows that the set C (see Definition 2.1) is of the form $\mathcal{Q}'_m \times \mathcal{R}'_{0m}$. As in Section 1.2 we introduce $\mathcal{J}_m$ and postulate APS 6, APS 7, and APS 8.

We define the sets

$$\mathcal{F}_m = \{(b_0,b) \mid b_0 \in \mathcal{R}'_{0m}, b \in \mathcal{R}_m, b \subset b_0\} \tag{1}$$

$$\mathcal{C}_m = \{a,f) \mid a \in \mathcal{Q}'_m, f \in \mathcal{F}_m\} = \mathcal{Q}'_m \times \mathcal{F}_m \tag{2}$$

and the function

$$\mu_m(a,f) = \mu_m(a, (b_0,b)) = \lambda_{\mathcal{J}m}(a \cap b_0, a \cap b) \tag{3}$$

$\lambda_{\mathcal{J}m}$ is determined by μ_m and $\lambda_{\mathcal{Q}m}$.

2.3. *The Connection between the Trajectories and the Registration and Preparation Procedures*

It is essential for any discussion of macroscopic measurements that the theory of the measuring apparatuses be known, without us possessing a theory of the dynamics of the systems undergoing measurement. To describe this situation for the measurement of the trajectories $y \in Y$, we introduce the set

$$\phi = \{(\rho_0,\rho) \mid \rho_0 \in \tilde{\mathcal{R}}_{0m}, \rho \in \tilde{\mathcal{R}}_m \text{ and } \rho_0 \geq \rho\} \tag{4}$$

and a function ψ_m

$$\phi \xrightarrow{\psi m} L(\hat{Y}) \tag{5}$$

where $L(\hat{Y})$ is a subset of the set $C(\hat{Y})$ of all real continuous functions on $\hat{Y}$:

$$L(\hat{Y}) = \{k \mid k \in C(Y) \text{ and } 0 \leq k \leq 1\} \tag{6}$$

The physical significance of $k = \psi_m(\rho_0,\rho)$ is as follows: $k(y)$ is the probability that the signal ρ on the apparatus ρ_0 gives a response if the system measured has the trajectory y; that is, $k(y)$ is the probability for $x \in h(\rho)$ if $x \in h(\rho_0)$ and $y = g(x)$, with g defined as at the end of Section 2.1.

By $h(\rho_0,\rho) = (h(\rho_0),h(\rho))$ we define in a canonical way a surjective function $\phi \xrightarrow{h} \mathcal{F}_m$. There are physical reasons for requiring a connection between ψ_m and μ_m. We introduce this connection by the postulate that, for every $a \in \mathcal{Q}'_m$, there exists a function u_a such that the diagram

$$\begin{array}{ccc} \phi & \xrightarrow{\psi m} & L(\hat{Y}) \\ h\downarrow & & \downarrow u_a \\ \mathcal{F}_m & \xrightarrow{\mu m,(a,\ldots)} & [0,1] \end{array} \tag{7}$$

is commutative and that $u_a \in C'(\hat{Y})$, with $C'(Y)$ denoting the dual Banach space to $C(\hat{Y})$, while $u_a \geq 0$ and $u_a(1) = 1$. If we define $K(\hat{Y}) = \{u \mid u \in C'(\hat{Y}), u \geq 0, u(1) = 1\}$, we have $u_a \in K(\hat{Y})$. It follows from the scheme (7), that there is a function $\mathcal{Q}'_m \xrightarrow{\varphi m} K(\hat{Y})$

$$\mu(a, (h(\rho_0), h(\rho))) = \langle \varphi_m(a), \psi_m(\rho_0,\rho) \rangle \tag{8}$$

where $\langle \ldots, \ldots \rangle$ is the canonical bilinear form for $C'(\hat{Y}),C(\hat{Y})$.

We define $K_m(\hat{Y}) = \overline{\text{co}}\varphi_m(\mathcal{Q}'_m)$, with $\overline{\text{co}}$ as the norm closed convex hull. Let $\overline{K}^\sigma_m(\hat{Y})$ be the $\sigma(C',C)$ closure of $K_m(\hat{Y})$ in $C'(\hat{Y})$. Since $K(\hat{Y})$ is $\sigma(C',C)$-compact, we have that $\overline{K}^\sigma_m(\hat{Y}) \subset K(\hat{Y})$ and that $\overline{K}^\sigma_m(\hat{Y})$ is also compact. Let $\partial_e\overline{K}^\sigma_m(\hat{Y})$ be the set of the extremal points of $\overline{K}^\sigma_m(\hat{Y})$. By the theorem of Krein-Milman we then have $\overline{K}^\sigma_m(\hat{Y}) = \overline{\text{co}}^\sigma\partial_e \overline{K}^\sigma_m(\hat{Y})$.

The elements of $\overline{K}^\sigma_m(\hat{Y})$ are called ensembles of trajectories and those of $\partial_e\overline{K}^\sigma_m(\hat{Y})$ elementary ensembles of trajectories. The elements of $\partial_e\overline{K}^\sigma_m(\hat{Y})$ describe the dynamics of the systems. But, in general, we have no laws for the dynamics without using statistical mechanics (see Section 3).

It is possible to define the support $\hat{S}_m$ of $K_m(\hat{Y})$ as a closed subset of $\hat{Y}$. The physical significance of $\hat{S}_m$ is given by the following equivalence:

$$\langle u,k \rangle = 0 \text{ for all } u \in K_m \text{ is equivalent to } k(y) = 0 \text{ for all } y \in \hat{S}_m.$$

In this sense $\hat{S}_m$ is the set of "all possible" trajectories. We define a mapping $C(\hat{Y}) \xrightarrow{s} C(\hat{S}_m)$ by $k(y) \xrightarrow{s} k(y)|_{\hat{S}_m}$. It can be proved that the diagram

$$\begin{array}{ccccc} & \xrightarrow{\psi m} & L(\hat{Y}) & \xrightarrow{s} & \\ \phi & & & & L(\hat{S}_m) \\ & \xrightarrow{h} & \mathcal{J}_m & \xrightarrow{\psi ms} & \end{array} \tag{9}$$

defines a function ψ_{ms}.

It is possible to identify $K_m(\hat{Y})$, $\overline{K}^\sigma_m(\hat{Y})$ with subsets of $K(\hat{S}_m) \subset C'(\hat{S}_m)$. Consequently, we will write $K_m(\hat{Y}) = K_m(\hat{S}_m)$, $\overline{K}^\sigma_m(\hat{Y}) = \overline{K}^\sigma_m(\hat{S}_m)$. The support of $K_m(\hat{S}_m)$ is $\hat{S}_m$. According to the representation theorem of Riesz, the set $K(\hat{S}_m)$ is the set of all σ-additive measures on the Borel field $\mathcal{L}(\hat{S}_m)$. In this way $\langle u,\eta_\sigma \rangle$ is defined, signifying with η_σ the characteristic function for the sets $\sigma \in \mathcal{L}(\hat{S}_m)$. Usually one writes $u(\sigma) = \langle u,\eta_\sigma \rangle$ and also $u(\sigma) = \langle u,\sigma \rangle$. $u(\sigma)$ is then the probability for the set σ of trajectories in the ensemble u. With the aid of $u(\sigma)$ one can write the dual form $\langle u,k \rangle$, with $k \in C(\hat{S}_m)$, as an integral

$$\langle u,k \rangle = \int_{\hat{S}_m} k(y)du(y) \tag{10}$$

Equation (10) is often called a trajectory integral or path integral. If σ is an open set of $\hat{S}_m$, then there exists an ensemble $u \in K_m(\hat{S}_m)$ with $u(\sigma) \neq 0$.

2.4. *Time Translations of the Registration and Preparation Procedures*

At first we define, for $\tau \geq 0$, the transformations T_τ for trajectories by $T_\tau y = y'$ with $y = z(t)$ and $y' = z'(t) = z(t + \tau)$. We find that T_τ is a p-uniform continuous and surjective mapping $Y \to Y$. It is therefore possible to extend T_τ as a continuous mapping $\hat{Y} \to \hat{Y}$. With $k \in C(\hat{Y})$,

$$V_\tau k(y) = k(T_\tau y) \tag{11}$$

produces a mapping V_τ of $C(\hat{Y})$ in itself. The mappings T_τ and V_τ generate for $\tau \geq 0$ a half-group.

Now we introduce, for $\tau \geq 0$, mappings $\tilde{R}_\tau$ of $\tilde{\mathcal{R}}_m$ in $\tilde{\mathcal{R}}_m$, which are order isomorphisms of $\tilde{\mathcal{R}}_m$ into $\tilde{R}_\tau\tilde{\mathcal{R}}_m$, so that $\tilde{R}_\tau$ is an isomorphism of the Boolean algebra $[0,\rho_0]$ with $\rho_0 \in \tilde{\mathcal{R}}_{0m}$ into the Boolean algebra $[0,\tilde{R}_\tau\rho_0]$ with $\tilde{R}_\tau\rho_0 \in \tilde{\mathcal{R}}_{0m}$. The physical interpretation of $\tilde{R}_\tau$ is that $\tilde{R}_\tau\rho_0$ is the same measuring apparatus represented by ρ_0, but measuring a time τ later than ρ_0. The time translation is in this sense a translation of the measuring process relative to the preparation procedures.

To formulate the consequence of this physical interpretation in a mathematical form, we introduce the following postulate. The diagram

$$\begin{array}{ccc} \phi & \xrightarrow{\psi_m} & L(\hat{Y}) \\ \tilde{R}_\tau\downarrow & & \downarrow V_\tau \\ \phi & \xrightarrow{\psi_m} & L(\hat{Y}) \end{array} \tag{12}$$

is commutative.

Similar to the time translations $\tilde{R}_\tau$ for the elements of $\tilde{\mathcal{R}}_m$ we can now introduce time translations R'_τ for the preparation procedures; but these translations can only be defined for $\tau \leq 0$, since the preparation of the systems is completed before the time $t = 0$.

The physical significance of R'_τ and $\tilde{R}_\tau$ has the consequence

$$\langle \varphi_m(R'_\tau a), \psi_m(\rho_0,\rho)\rangle - \langle \varphi_m(a), \psi_m(\tilde{R}_\tau\rho_0,\tilde{R}_\tau\rho)\rangle$$

With V'_τ denoting the dual mapping of V_τ, the diagram

$$\begin{array}{ccc} \mathcal{Q}'_m & \xrightarrow{\varphi_m} & K(\hat{Y}) \\ R_{-\tau}\downarrow & & \downarrow V'_\tau \\ \mathcal{Q}'_m & \xrightarrow{\varphi_m} & K(\hat{Y}) \end{array} \tag{13}$$

is commutative. In addition, one finds that the following diagram:

$$\begin{array}{ccc} \tilde{\mathcal{R}}_m & \xrightarrow{h} & \mathcal{R}_m \\ \tilde{R}_\tau\downarrow & & \downarrow R_\tau \\ \tilde{\mathcal{R}}_m & \xrightarrow{h} & \mathcal{R}_m \end{array} \tag{14}$$

determines a mapping R_τ. With the canonical extensions of h, $\tilde{R}_\tau$, R_τ one gets, from (14) and (12), the diagram

$$\begin{array}{ccccc} \mathcal{F}_m & \overset{h}{\leftarrow} & \phi & \overset{\psi_m}{\rightarrow} & L(\hat{Y}) \\ R_\tau \downarrow & & \downarrow \tilde{R}_\tau & & \downarrow V_\tau \\ \mathcal{F}_m & \overset{h}{\leftarrow} & \phi & \overset{\psi_m}{\rightarrow} & L(\hat{Y}) \end{array} \tag{15}$$

Employing the abbreviation $V_\tau^{(s)} = sV_\tau$ (s defined in Section 2.3), we find

$$V_{\tau 1}^{(s)} V_{\tau 2}^{(s)} = sV_{\tau 1} sV_{\tau 2} = sV_{\tau 1} V_{\tau 2} = sV_{\tau 1+\tau 2} = V_{\tau 1+\tau 2}^{(s)}$$

and the validity of the diagram

$$\begin{array}{ccc} L(\hat{Y}) & \overset{s}{\rightarrow} & L(\hat{S}_m) \\ \downarrow V_\tau & & \downarrow V_\tau^{(s)} \\ L(\hat{Y}) & \overset{s}{\rightarrow} & L(\hat{S}_m) \end{array} \tag{16}$$

From (16) and (9), there results

$$\begin{array}{ccc} \mathcal{F}_m & \overset{\psi_{ms}}{\rightarrow} & L(\hat{S}_m) \\ \downarrow R_\tau & & \downarrow V_\tau^{(s)} \\ \mathcal{F}_m & \overset{\psi_{ms}}{\rightarrow} & L(\hat{S}_m) \end{array} \tag{17}$$

3. The Embedding of the Objective Description of Macrosystems in an Extrapolated Quantum Mechanics of Many Particles

It would take too long to discuss the problem whether a quantum mechanics is a closed theory also for many particles (for details see, e.g., Ref. 5 together with Ref. 7). We shall only quote a sentence of a speech given by E. P. Wigner in 1979 at an Einstein Centennial Symposium: "The preceding arguments seem to show that ... the validity of quantum mechanics is just as clearly restricted to non-macroscopic situations as are our macroscopic theories, including the general theory of relativity, to non-microscopic situations."

Our opinion is that classical theories for macroscopic systems are not approximations to a quantum theory of many-particle systems. We consider it as an error, e.g., to view the Newtonian mechanics of our planetary system as an approximation to a "quantum theory of the planetary system" or to say that the description of tennis balls by classical mechanics is only an approximation to a "quantum theory of tennis balls." A quantum theory of tennis balls cannot be a closed theory ("g.G.-abgeschlossen" in the sense of Ref. 7, Section 10.3), since it seems to be "impossible" to make interference experiments with tennis balls as one can with electrons.

But if this is so, what then is the relation between such theories $\mathcal{PT}_m$ of macroscopic systems having the form described in Section 2 and an extrapolated quantum theory $\mathcal{PT}_{\text{qexp}}$ of many particles?

Also, a reader who does not agree that the quantum theory $\mathcal{PT}_{\text{qexp}}$ of many particles is not closed, and who accordingly believes that the macroscopic description (as given in Section 2) is an approximation to this quantum theory $\mathcal{PT}_{\text{qexp}}$, can use the following mathematical methods as a general formulation of the connection between the macroscopic description and the quantum theory of many particles.

3.1. The Extrapolated Quantum Mechanics

It is well known how one describes in ordinary quantum mechanics a composite system by direct products of Hilbert spaces and symmetrization procedures (antisymmetric or symmetric parts of the product spaces). If one wants to describe several particle numbers, superselection rules have to be taken into account. In any case, the set K of all statistical operators w (all self-adjoint operators w with $0 \leq w \leq 1$ and $\operatorname{tr} w = 1$, w commuting with the superselection rules) is a convex set. The linear span of K is a Banach space B. The dual Banach space B' can be identified with the set of all bounded self-adjoint operators (commuting with the superselection rules), and the canonical bilinear form of B, B' can be identified with $\operatorname{tr}(xy)(x \in B$ and $y \in B')$.

For our purpose it is very helpful to go over to a more comprehensive description of preparation and measurement than the foregoing, using statistical operators as ensembles (or states) and self-adjoint operators as observables. We will therefore invoke the description of preparation and registration procedures explained in Section 1. To arrive at quantum mechanics we have to add axioms concerning the possibilities of preparation and registration procedures for "microsystems" (for such axioms, see, e.g., Ref. 1 or 3). It will suffice for our purposes to introduce only a comprehensive description in the following form (see also Ref. 1 or 5).

Let $\mathcal{Q}'$ be the set of preparation procedures (without the empty set), $\mathcal{F}$ the set of effect processes or questions (see Section 1), and $L \subset B'$ the set of all self-adjoint operators g satisfying $0 \leq g \leq 1$. There are mappings $\mathcal{Q}' \xrightarrow{\varphi} K$ and $\mathcal{F} \xrightarrow{\psi} L$ with $\mu(a,f) = \operatorname{Tr}[\varphi(a)\psi(f)]$. $\mu(a,f)$ was defined in Section 1.2 as the probability that the effect process f occurs if the microsystems are prepared by the procedure a. We assume that $\varphi(\mathcal{Q}')$ is normdense in K and $\psi(\mathcal{F})$ is $\sigma(B',B)$-dense in L. The elements of K are called ensembles or states; the elements of L are called effects. For quantum theory it is essential that φ and ψ are not injective! The elements of the set $\mathcal{R}(b_0) = \{b \mid b \in \mathcal{R}, b \subset b_0\}$ and $b_0 \in \mathcal{R}_0$ are all digital signals on the measuring apparatus represented by b_0. $\mathcal{R}(b_0)$ is a Boolean set algebra. We find that $\psi_{b0}(b) = \psi(b_0,b)$ is an additive measure $\mathcal{R}(b_0) \xrightarrow{\psi_{b0}} L$ on the Boolean algebra $\mathcal{R}(b_0)$. Therefore, we employ the following definition (see also Ref. 1, Section IV):

Let Σ be a complete Boolean algebra and $\Sigma \xrightarrow{F} L$ a σ-additive measure; $\Sigma \xrightarrow{F} L$ is then called an *observable*. A subset l of L is called a set of coexistent effects if there exists an observable with $l \subset F\Sigma$.

This concept of an observable is wider than the usual one. But it is also more realistic.

The set G of the extreme points of L is the set of all projection operators. The elements of G are called *decision effects*. If for an observable $F(\Sigma) \subset G$, the observable is called a decision observable. A decision observable (with scale; see Ref. 1, Section IV) corresponds to that what is "usually" called an observable.

It has been proved (see Ref. 1, Section III) that a decision effect $e \in L$ and an effect $g \in L$ are coexistent if $eg = ge$, i.e., if e and g commute.

Now we presume the extrapolated quantum mechanics $\mathcal{PT}_{\text{qexp}}$ in the form described above. It is our opinion that this formal procedure of extrapolation to "many" particles does not guarantee that all elements of $\mathcal{Q}$ and $\mathcal{F}$ are realistic, i.e., that it is "in principle" possible to construct apparatuses appropriate for all $a \in \mathcal{Q}$ or all $f \in \mathcal{F}$.

There are many experiences and theoretical arguments that have led us to this opinion, but a comprehensive list of these will not be given here. (We already presented some comments in the introduction to Section 3.) But it is not essential at the moment to be convinced that not all imaginable preparation procedures (as elements of $\mathcal{Q}$) and registration procedures (as elements of $\mathcal{R}$) are realizable. It is only essential to be convinced that the preparation and registration procedures described in Section 2 by the sets $\mathcal{Q}_m$ and $\mathcal{R}_m$ cannot be all those gained in $\mathcal{PT}_{\text{qexp}}$ by extrapolation of the usual quantum mechanics to many particles. These considerations lead us to the following embedding of the theory $\mathcal{PT}_m$ in $\mathcal{PT}_{\text{qexp}}$.

3.2. Embedding of $\mathcal{PT}_m$ in $\mathcal{PT}_{qexp}$

It is very plausible to relate the two theories $\mathcal{PT}_m$ and $\mathcal{PT}_{\text{qexp}}$ in the following way. A preparation procedure in $\mathcal{PT}_m$ can be considered also as a preparation procedure in $\mathcal{PT}_{\text{qexp}}$ and especially as a relativistic one. A registration method in $\mathcal{PT}_m$ can be regarded also as a registration method in $\mathcal{PT}_{\text{qexp}}$, and similarly for a registration procedure. How is it possible to give a mathematical form to this idea?

According to the general considerations carried out in Ref. 7, Section 8, we introduce an embedding mapping

$$M_m \xrightarrow{i} M \tag{18}$$

with M as the set of systems in $\mathcal{PT}_{\text{qexp}}$ and M_m as the set of systems in $\mathcal{PT}_m$ (see Section 2). We postulate for this mapping that i is bijective. It would be possible to identify M_m with M by this mapping, but we will not do this.

In a canonical form, i generates a mapping $\mathcal{P}(M_m) \xrightarrow{i} \mathcal{P}(M)$. We postulate for this mapping the following relations (19) to (22):

$$\mathcal{Q}_m \xrightarrow{i} \mathcal{Q}, \quad \mathcal{R}_m \xrightarrow{i} \mathcal{R}, \quad \mathcal{R}_{0m} \xrightarrow{i} \mathcal{R}_0 \tag{19}$$

The scheme (19) describes our idea that the macroscopic preparation and registration procedures are some of the extrapolated procedures of $\mathcal{PT}_{\text{qexp}}$. The meaning of these identifications is that the corresponding probabilities in the two theories must be the same. Therefore, we postulate

$$\begin{aligned}\mu_m(a,(b_0,b)) &= \lambda_{\mathcal{J}_m}(a \cap b_0, a \cap b)\\ &= \lambda((ia) \cap (ib_0),(ia) \cap (ib))\\ &= \mu(ia,(ib_0,ib))\end{aligned} \tag{20}$$

But it is necessary to leave open the possibility that (20) is not exact but only represents such a good approximation that the differences between the left and right side cannot be tested experimentally.

(19) together with (20) represents our idea that $\mathcal{Q}_m$, $\mathcal{R}_m$ can be imagined as subsets of $\mathcal{Q}$ and $\mathcal{R}$. But there can be additional structures in $\mathcal{PT}_m$ and $\mathcal{PT}_{qexp}$ which have the same physical interpretation. We want to postulate that all those structures of $\mathcal{PT}_m$ are related to the corresponding structures in $\mathcal{PT}_{qexp}$. In this manner we have already postulated in (20) that the probabilities in $\mathcal{PT}_m$ and $\mathcal{PT}_{qexp}$ are the same.

We will only consider here as such an additional structure the "time translation" of the registration. As described in Section 2 we have postulated in $\mathcal{PT}_m$ that it makes sense to measure the state z at a time t. This postulate has as a background the assumption that we know the theory of the measuring apparatuses. There exists sometimes the opinion that it is possible also in quantum mechanics to define a measurement "at the time t." But actually there is no such possibility in quantum mechanics, since the real measuring processes involve interactions of such duration that during their performance the states of the systems under study change significantly. Only as possible idealization in $\mathcal{PT}_m$ is one able to measure the state z in such a short time that it is practically left unchanged. *Therefore it is not possible* to find a structure in $\mathcal{PT}_{qexp}$ which corresponds to the structure of a measurement "at a time t" in $\mathcal{PT}_m$.

But there is another structure, which has the same interpretation in $\mathcal{PT}_m$ and $\mathcal{PT}_{qexp}$: the time translation of a registration. We have already given a description of this time translation in Section 2.4. Now we have a corresponding time translation operator R_τ in quantum theory: a time translation $R_\tau b_0$, $R_\tau b$ of the registration method and procedures. The essential significance of the so-called Hamiltonian operator H in quantum mechanics is that it determines this time translation by the equation (see Ref. 1, Sections VII and VIII)

$$\psi(R_\tau b_0,R_\tau b) = U_\tau \psi(b_0,b)U_\tau^+ \tag{21}$$

with $U_\tau = e^{iH\tau}$.

In the following we introduce the operator $\mathcal{U}_\tau$

$$\mathcal{U}_\tau g = U_\tau g U_\tau^+ \tag{22}$$

as an operator in the Banachspace B' (see Section 3.1). Thus we can write (21) in the form

$$\psi(R_\tau b_0,R_\tau b) = \mathcal{U}_\tau \psi(b_0,b) \tag{23}$$

How does one relate R_τ in $\mathcal{PT}_m$ with R_τ in $\mathcal{PT}_{qexp}$? We postulate that these two operators have the same consequences in the probability structure, i.e.,

$$\lambda((ia) \cap (iR_\tau b_0),(ia) \cap (iR_\tau b)) = \lambda((ia) \cap (R_\tau ib_0),(ia) \cap (R_\tau ib)) \tag{24}$$

for all $a \in \mathcal{Q}_m$. Equation (20), together with (24), yields

$$\lambda_{\mathcal{J}m}(a \cap R_\tau b_0, a \cap R_\tau b) = \lambda((ia) \cap (R_\tau ib_0),(ia) \cap (R_\tau ib))$$

Introducing

$$\mu(a,(b_0,b)) = \lambda(a \cap b_0, a \cap b) = \mathrm{Tr}(\varphi(a)\psi(b_0,b))$$

we find

$$\begin{aligned}\lambda_{\mathcal{J}m}(a \cap R_\tau b_0, a \cap R_\tau b) &= \mathrm{Tr}(\varphi(ia)U_\tau\psi(ib_0,ib)U_\tau^+) \\ &= \mathrm{Tr}(\varphi(ia)[\mathcal{U}_\tau\psi(ib_0,ib)])\end{aligned} \tag{25}$$

Equation (25) is only a postulate for $\tau \geq 0$, since R_τ on the left side is defined only for $\tau \geq 0$. For $\tau = 0$ it is equivalent to (20), so that (25) is the main postulate for the embedding.

It is necessary to remember again the possibility that (25) is only valid as a very good approximation. We have postulated (25) for $\tau \geq 0$, but this equation cannot be valid for "too great" τ, since its right-hand side is in many cases an almost periodic function in τ. However, the recurrence time will exceed millions of years.

Using the function μ_m of Eq. (20), one can write the relation (25) in the form

$$\mu_m(a,R_\tau(b_0,b)) = \mathrm{Tr}(\varphi(ia)[\mathcal{U}_\tau\psi(ib_0,ib)]) \tag{26}$$

or, by (33), (9), and (47), in the form

$$\langle\varphi_m(a), V_\tau^{(s)}\psi_{mS}(b_0,b)\rangle = \mathrm{Tr}(\varphi(ia)[\mathcal{U}_\tau\psi(ib_0,ib)]) \tag{27}$$

If we have found an embedding of the form postulated above, we say that the macroscopic description $\mathcal{PT}_m$ is compatible with $\mathcal{PT}_{qexp}$. The so-called statistical mechanics of nonequilibrium processes is nothing else than the elaboration of this embedding for special cases.

3.3. *Some Consequences of the Embedding Postulate*

We have not proved mathematically the existence of such an embedding postulated in Section 3.2. If we mentioned at the end of Section 3.2 that the statistical mechanics is a procedure of elaborating the embedding postulates, we have to admit that for real systems there are no mathematical proofs of the embedding relations. Only for simple models was it possible to prove the embedding relations.

For real systems the embedding relations (mostly cast in much more specialized forms than those of Section 3.2) are taken for granted, and then one tries to deduce from these relations the dynamical laws for $\mathcal{PT}_m$ (see Section 3.4). We will therefore present only here some consequences of the embedding relations.

Connecting the mappings φ_m and ψ_{mS} of Section 2.3 with the embedding mapping i, and the mappings φ,ψ of Section 3.1, we obtain the following diagrams:

$$\begin{array}{lcl} \mathcal{Q}'_m \xrightarrow{\varphi} K(\hat{S}_m) & & \mathcal{F}_m \xrightarrow{\psi}{}_{ms} L(\hat{S}_m) \\ i\downarrow & & i\downarrow \\ \mathcal{Q}' \xrightarrow{\varphi} K & & \mathcal{F} \xrightarrow{\psi} L \end{array} \tag{28}$$

With these mappings we get, from (27) with $f = (b)_0,b)$ and $\mathcal{U}_\tau$ according to (22)

$$\langle \varphi_m(a),\psi_{mS}(R_\tau f)\rangle = \mathrm{Tr}(\varphi(ia)[\mathcal{U}_\tau\psi(if)]) \tag{29}$$

Since we assume $\psi_{mS}(\mathcal{F}_m)$ to be dense in $L(\hat{S}_m)$, the relation yields the existence of a mapping $K_m \xrightarrow{j} K_m(\hat{S}_m)$ defined by $j\varphi(ia) = \varphi_m(a)$ and with K_m denoting the norm closure of $\varphi(i\mathcal{Q}')$. The first diagram of (28) can be accomplished in the following way:

$$\begin{array}{ccccc} & \nearrow^{\varphi_m} & K_m(\hat{S}_m) & \subset & K(\hat{S}_m) \\ & & \uparrow j & & \\ \mathcal{Q}'_m & & K_m & \subset & K \\ & \searrow^{i} & \uparrow \varphi & & \uparrow \varphi \\ & & \mathcal{Q}_m & \subset & \mathcal{Q}' \end{array} \tag{30}$$

It is not so easy to accomplish the other part of (28), since $K_m(\hat{S}_m)$ must not separate the elements of $L(\hat{S}_m)$!

Let us go over from $C(\hat{S}_m),C'(\hat{S}_m)$ to another pair of dual Banach spaces. $\mathcal{J}(\hat{S}_m)$ may be the subset of $\mathcal{L}(\hat{S}_m)$ (see Section 2.3) of all elements σ of $\mathcal{L}(\hat{S}_m)$ which have a measure $u(\sigma) = 0$ for all $u \in K_m(\hat{S}_m)$. $\Sigma_m = \mathcal{L}(\hat{S}_m)/\mathcal{J}(\hat{S}_m)$ is then a complete Boolean algebra and the measures $u \in K_m(\hat{S}_m)$ are σ-additive measures on Σ_m. $B(\Sigma_m)$ may be the set of all bounded σ-additive (signed) measures on Σ_m. $B(\Sigma_m)$ is a base norm space. The base $K(\Sigma_m)$ of $B(\Sigma_m)$ is the set of all positive and normed measures. $K_m(\hat{S}_m)$ can be identified with a subset of $K(\Sigma_m)$ and, therefore, we write $K_m(\Sigma_m) = K_m(\hat{S}_m)$. $K_m(\Sigma_m)$ is norm closed in $B(\Sigma_m)$. Let $B'(\Sigma_m)$ be the dual Banach space to $B(\Sigma_m)$. $C'(\hat{S}_m)$ can be identified with a subset of $B'(\Sigma_m)$. $L(\hat{S}_m)$ is in this sense $\sigma(B'(\Sigma_m),B(\Sigma_m))$-dense in $L(\Sigma_m)$, where $L(\Sigma_m)$ is the order interval $[0,1]$ of $B'(\Sigma_m)$.

Instead of the right-hand diagram in (28) we get

$$\begin{array}{l} \mathcal{F}_m \xrightarrow{\psi_{wS}} L(\Sigma_m) \\ i\downarrow \\ \mathcal{F} \xrightarrow{\psi} L \end{array} \tag{31}$$

If we fix a special element b_0 of $\mathcal{R}_{0m}$ we can introduce the following diagram:

$$\begin{array}{l} \mathcal{R}_m(b_0) \xrightarrow{\psi_{mSb0}} L(\Sigma_m) \\ i\downarrow \\ \mathcal{R}(ib_0) \xrightarrow{\psi_{ib0}} L \end{array} \tag{32}$$

where ψ_{mSb0} and ψ_{b0} are defined by

$$\psi_{mSb0}(b) = \psi_{mS}(b_0,b), \qquad \psi_{b0}(b) = \psi(b_0,b)$$

$\mathcal{R}_m(b_0)$ and $i\mathcal{R}_m(b_0)$ are isomorphic Boolean algebras. From (32), we get

$$\begin{array}{ccc} & & L(\Sigma_m) \\ & \nearrow^{\psi_{mSb0}} & \\ \mathcal{R}_m(b_0) & & \\ & \searrow_{\psi_{ib0}i} & \\ & & L \end{array} \tag{33}$$

where ψ_{mSb0} and $\psi_{ib0}i$ are additive measures on $\mathcal{R}_m(b_0) = \{b \mid b \in \mathcal{R}_m, b \subset b_0\}$. The relation (27) takes the form

$$\langle jw, V_\tau^{(s)}\psi_{mSb0}(b)\rangle = \mathrm{Tr}(w[\mathcal{U}_\tau\psi_{ib0}i(b)]) \tag{34}$$

for all $w \in K_m$; $\langle \ldots, \ldots \rangle$ is in (34) the canonical bilinear form of the pair $B(\Sigma_m)$, $B'(\Sigma_m)$.

It is possible to complete the Boolean algebra $\mathcal{R}_m(b_0)$ relative to a metric defined by

$$d(b_1,b_2) = \langle jw_0, \psi_{mSb0}(b_1\backslash b_1 \cap b_2) + \psi_{mSb0}(b_2\backslash b_1 \cap b_2)\rangle$$

with an effective measure w_0 of $K_m(\Sigma_m)$. In this way it is possible to define the mappings thus ψ_{msb0} and $\psi_{ib0}i$ on the completion Σ_{b0} of $\mathcal{R}_m(b_0)$. These mappings are then σ-additive measures.

In the sense of classical measurements it should be possible to find a registration procedure $\tilde{b}_0 \in \mathcal{R}_{0m}$ that measures approximately all trajectories. We therefore postulate the following:

There exists an element $\tilde{b}_0 \in \mathcal{R}_{0m}$ with

$$\overline{\mathrm{co}^\sigma \psi_{mS\tilde{b}0}\Sigma_{\tilde{b}0}} = L(\Sigma_m) \tag{35}$$

According to Ref. 1, Section IV, Theorem 2.34, there exists such a subset $\Sigma_{\tilde{b}0}^{\prime}$ of $\Sigma_{\tilde{b}0}$ for which $\Sigma_{\tilde{b}0}^{\prime} \xrightarrow{\psi_{ms\tilde{b}0}} \partial_e L(\Sigma_m) = \Sigma_m$ is bijective. It follows that $\Sigma_{\tilde{b}0}^{\prime}$ is a Boolean algebra isomorphic to Σ_m and that $\psi_{mS\tilde{b}0}$ is an isomorphism. We get by the definition

$$F_{\tilde{b}0} = \psi_{i\tilde{b}0}i\psi_{mS\tilde{b}0}^{-1}$$

a σ-additive measure

$$\Sigma_m \xrightarrow{F_{\tilde{b}0}} L \tag{36}$$

i.e., an observable. This observable is called *the macroscopic observable*. We will drop the index $\tilde{b}_0$ and write F instead of $F_{\tilde{b}0}$. By an observable F is defined an affine mapping S of K in $K(\Sigma_m)$: $\mu(w,F(\sigma)) = m(\sigma)$ with $m = Sw$. This mapping can be extended as a mapping $B \xrightarrow{S} B(\Sigma_m)$. It is $S'\sigma = F(\sigma)$ if one identifies Σ_m with $\partial_{\mathfrak{R}} L(\Sigma_m)$. From (34) for $\tau = 0$ we get the following relation:

$$\langle jw,y \rangle = \mathrm{Tr}(w(S'y)) = \langle Sw,y \rangle \tag{37}$$

for all $y \in B'(\Sigma_m)$ and all $w \in K_m$. It follows that j and S are the same mappings of K_m in $K_m(\Sigma_m)$.

We therefore get from (34)

$$\mathrm{Tr}(w[S'V_\tau^{(s)}\psi_{mS\tilde{b}0}(b)]) = \mathrm{Tr}(w[\mathcal{U}_\tau \psi_{i\tilde{b}_0} ib])$$

for all $w \in K_m$. From this equation follows because of (35)

$$\mathrm{Tr}(w[S'V_\tau^{(s)}y]) = \mathrm{Tr}(w[\mathcal{U}_\tau S'y]) \tag{38}$$

for all $w \in K_m$ and all $y \in L(\Sigma_m)$. This relation is equivalent to ($\mathcal{U}'_\tau w = U'_\tau w U_\tau$):

$$\langle V_\tau^{(s)\prime} Sw,y \rangle = \langle S\mathcal{U}'_\tau w,y \rangle$$

for all $w \in K_m$ and all $y \in L(\Sigma_m)$, i.e.,

$$V_\tau^{(s)\prime} Sw = S\mathcal{U}'_\tau w \tag{39}$$

for all $w \in K_m$. From these equations follows that (38) and (39) are also valid for all $w \in \tilde{K}_m = \overline{\mathrm{co}} \bigcup_{\tau \geq 0} \mathcal{U}_\tau K_m$ and that

$$S\tilde{K}_m \subset K_m(\Sigma_m) \tag{40}$$

Since $SK_m = jK_m$ is dense in $K_m(\Sigma_m)$, also $S\tilde{K}_m$ is dense in $K_m(\Sigma_m)$. The relation (39) is equivalent to

$$V_\tau^{(s)\prime} S = S\mathcal{U}'_\tau \tag{41}$$

as mappings of $\tilde{K}_m$ in $K_m(\Sigma_m)$. Equation (41) implies that the two half-groups $\mathcal{U}_\tau$ ($\tau \geq 0$) operating on $\tilde{K}_m$ and $V_\tau^{(s)\prime}$ ($\tau \geq 0$) operating on $K_m(\Sigma_m)$ are homomorphic and that S is a homomorphism.

Equation (41) is the embedding relation we wanted to establish. Its existence guarantees that the macroscopic description of the systems in a state space is compatible with the extrapolated quantum mechanics of many particles. It is our proposal to found, on the one hand, the total statistical mechanics of irreversible processes on (41) and, on the other, to prove (41) in special cases. The latter task is very difficult, because (41) need not be an exact mathematical relation if the systems are finite systems included in finite regions. For such finite systems $\mathcal{U}_\tau$ is an almost periodic function in τ, but $V_\tau^{(s)}$ an aperiodic one. Therefore the first step of a "proof" of (41) has to be an approximation of $\mathcal{U}_\tau$ by an aperiodic half-group $\tilde{\mathcal{U}}_\tau$, i.e., $\|S\mathcal{U}_\tau w - S\tilde{\mathcal{U}}_\tau w\| \ll 1$ for $w \in K_m$ and τ not too large. One way to make such an approximation can be a limit to infinite systems (infinite particles in an infinite region; see, e.g., Ref. 8).

Another approach is to replace $\mathcal{U}_\tau$ by $\tilde{\mathcal{U}}_\tau = e^{-\tau/T}\mathcal{U}_\tau$, with T denoting a time of magnitude of millions of years.

For the second step of a proof of (41) one can use theorems, due to Nagy, concerning connections of groups and half-groups (see Ref. 9). It cannot be the task of this short article to "prove" (41) or to "found" the total statistical mechanics on (41). But to illustrate the meaning of (41), we next give a well-known example.

3.4. The Macroscopic Dynamics

We want to show by an example how the relation (41) determines the macroscopic dynamics. First we will give an extension of $F = F_{\hat{b}0}$, to the Borel algebra $\mathcal{L}(\hat{Y})$ of $\hat{Y}$. If $\sigma \in \mathcal{L}(\hat{Y})$, then $\sigma \cap \hat{S}_m \in \mathcal{L}(\hat{S}_m)$. $\sigma \cap \hat{S}_m$ defines an element of Σ_m, so that $F(\sigma \cap \hat{S}_m)$ is defined by $\Sigma_m \xrightarrow{F} L$. We define a σ-additive mapping $\mathcal{L}(\hat{Y}) \xrightarrow{\tilde{F}} L$ by

$$\tilde{F}(\sigma) = F(\sigma \cap \hat{S}_m) \tag{42}$$

The support of F is the set $\hat{S}_m$. In such a way the set $\hat{S}_m$ of the possible trajectories of the system is determined by $\tilde{F}$.

By

$$u(\sigma) = \mu(w, \tilde{F}(\sigma)) \tag{43}$$

and $u = \tilde{S}w$ we define an affine mapping

$$K \xrightarrow{\tilde{S}} K(\hat{Y}) \tag{44}$$

The support of $\tilde{S}K$ is also $\hat{S}_m$, while the norm-closure of $\tilde{S}\tilde{K}_m$ is the set $K_m(\hat{Y})$ (see Section 2.3).

The relation (31) is equivalent to

$$V'_\tau \tilde{S}w = \tilde{S}\mathcal{U}'_\tau w \tag{45}$$

for all $w \in \tilde{K}_m$.

We see that it is not necessary to know at the beginning the set $\hat{S}_m$. It is only necessary to know $\tilde{S}$. If $\tilde{S}$ is a convex mapping of total K in $K_m(\hat{Y})$ then also $\hat{S}_m$, Σ_m, and $\Sigma_m \xrightarrow{F} L$ are determined by $\tilde{S}$. But it is not possible to choose as $\tilde{S}$ any affine mapping $K \to K(\hat{Y})$; it is necessary to choose it in such a manner, that the relation (45) is satisfied.

Our example is characterized by some *additional* assumptions: The first assumption is that $\hat{Y}\backslash Y$ is of F-measure zero. We say that the dynamics of the system is continuous because Y is the set of continuous trajectories. If the dynamics is continuous, the sets

$$\sigma(\rho;\tau) = \{y \mid y \in Y, y = z(t) \text{ and } z(\tau) \in \rho\}$$

for $\rho \in \mathcal{L}(Z)$ are elements of $\mathcal{L}(Y)$ for $\tau > 0$. We will assume for simplicity that also $\sigma(\rho;0)$ is an element of $\mathcal{L}(Y)$. We abbreviate

$$\chi_\tau(\rho) = \tilde{F}(\sigma(\rho;\tau)) \tag{46}$$

Equation (46) has the consequence, that the effects $\chi_\tau(\rho)$ are coexistent (see Section 1) effects for all $\tau \geq 0$, since all the sets $\sigma(\rho;\tau)$ are elements of $\mathcal{L}(\hat{Y})$. The definition of $\sigma(\rho;\tau)$ yields

$$V_{\tau 1}\sigma(\rho;\tau_2) = \sigma(\rho;\tau_1 + \tau_2) \tag{47}$$

From (45) and (47) we find that

$$\mathrm{Tr}(w\chi_{\tau 1+\tau 2}(\rho)) = \mathrm{Tr}(w[\mathcal{U}_{\tau 1}\chi_{\tau 2}(\rho)]) \tag{48}$$

is valid for all $w \in \tilde{K}_m$.

The second assumption for our example is the following: There exists a decomposition $Z = \bigcup_\nu \rho_\nu$ of Z with $\rho_\nu \in \mathcal{L}(Z)$, $\rho_\nu \cap \rho_\mu = \phi$ for $\nu \neq \mu$ in such a way that the ρ_ν are macroscopically small but that the effects $\chi_0(\rho_\nu)$ can be taken as decision effects, i.e., as projection operators. Macroscopically small means that the states $Z \in \rho_\nu$ are practically not distinguishable by measurements.

Since the $\chi_\tau(\rho_\nu)$ are coexistent with the $\chi_0(\rho_\mu)$ and the $\chi_0(\rho_\mu)$ are projection operators, it is necessary (see Section 3.1) that the $\chi_\tau(\rho_\nu)$ commute with all $\chi_0(\rho_\mu)$, i.e.,

$$\chi_\tau(\rho_\nu) = \sum_\mu \chi_0(\rho_\mu)\chi_\tau(\rho_\nu)\chi_0(\rho_\mu) \tag{49}$$

and that

$$\tilde{F}(\sigma(\rho_\mu,0) \cap \sigma(\rho_\nu;\tau)) = \chi_0(\rho_\mu)\chi_\tau(\rho_\nu) \tag{50}$$

be valid.

Our next assumption is

$$\chi(\rho_\mu)\chi_\tau(\rho_\nu)\chi(\rho_\mu) = \lambda_\tau(\nu,\mu)\chi(\rho_\mu) \tag{51}$$

Equations (49) and (51) ensure that all $\chi_\tau(\rho_\nu)$ are coexistent. But (51) is not required for the coexistence for all $\chi_\tau(\rho_\nu)$! Now it is necessary to make an assumption for the set $\tilde{K}_m$. It seems to be plausible that the following elements

$$w_\nu = \frac{\chi_0(\rho_\nu)}{\mathrm{Tr}(\chi_0(\rho_\nu))} \tag{52}$$

are elements of $\tilde{K}_m$. Equation (52) means that it is possible to prepare the systems in such a way that the state of the systems at time $t = 0$ is a state in ρ_ν, but w_ν gives no further information.

From (48) follows that

$$\lambda_\tau(\alpha,\mu) = \frac{\mathrm{Tr}(\chi_0(\rho_\mu)U_\tau\chi_0(\rho_\alpha)U_\tau^+)}{\mathrm{Tr}(\chi_0(\rho_\mu))} \tag{53}$$

and

$$\lambda_{\tau 1+\tau 2}(\alpha,\mu) = \sum_{\nu} \lambda_{\tau 2}(\alpha,\nu)\lambda_{\tau 1}(\nu,\mu) \tag{54}$$

Equation (54) tells us that the dynamics is a Markov process.

We introduce the following projection operator of K in K:

$$Qw = \sum_{\mu} \frac{\chi_0(\rho_\mu)}{\mathrm{Tr}(\chi_0(\rho_\mu))} \mathrm{Tr}(w\chi_0(\rho_\mu)) \tag{55}$$

The dual Q' of Q is of the same form (for $g \in L$):

$$Q'g = \sum_{\mu} \chi_0(\rho_\mu) \frac{\mathrm{Tr}(\chi_0(\rho_\mu)g)}{\mathrm{Tr}(\chi_0(\rho_\mu))} \tag{56}$$

The relation (4.7) together with (4.10) is equivalent to

$$\chi_\tau(\rho_\nu) = Q'\mathcal{U}_\tau Q'\chi_0(\rho_\nu) \tag{57}$$

and

$$Q\mathcal{U}'_{\tau 1+\tau 2}Q = Q\mathcal{U}'_{\tau 1}Q\mathcal{U}'_{\tau 2}Q \tag{58}$$

Equation (58) is the famous relation that the mappings $Q\mathcal{U}'_\tau Q$ form a half-group.

The total observable $\tilde{F}$ is practically determined if the elements

$$\tilde{F}(\sigma(\rho_1;\tau_1) \cap \sigma(\rho_2;\tau_2) \cap \cdots \cap \sigma(\rho_u;\tau_u)) \tag{59}$$

are known. These elements can be calculated by recursion. At first we get, as mentioned above,

$$\tilde{F}(\sigma(\rho_{\nu 1};0) \cap \sigma(\rho_{\nu 2};\tau_2 - \tau_1)) = \chi_0(\rho_{\nu 1})\chi_{\tau 2}(\rho_{\nu 2})\chi_0(\rho_{\nu 1})$$

It follows from (45) that

$$\tilde{F}(\sigma(\rho_{\nu 1};\tau_1) \cap \sigma(\rho_{\nu 2};\tau_2)) = Q'\mathcal{U}_{\tau 1}Q'\tilde{F}(\sigma(\rho_{\nu 1};0) \cap \sigma(\rho_{\nu 2};\tau_2 - \tau_1))$$

Since $\tilde{F}(\sigma(\rho_{\nu 1},0) \cap \sigma(\rho_{\nu 2};\tau_2 - \tau_1) \partial\, \sigma(\rho_{\nu 3};\tau_3 - \tau_1))$ has to be coexistent with both $\tilde{F}(\sigma(\rho_{\nu 1};0)) = \chi_0(\rho_{\nu 1})$ and $\tilde{F}(\sigma(\rho_{\nu 2};\tau_2 - \tau_1) \cap \sigma(\rho_{\nu 3};\tau_3 - \tau_1))$ and since $\chi_0(\rho_\mu)$ are projection operators, we get

$$\tilde{F}(\sigma(\rho_{\nu 1};0) \cap \sigma(\rho_{\nu 2};\tau_2 - \tau_1) \cap \sigma(\rho_{\nu 3};\tau_3 - \tau_1))$$
$$= \chi(\rho_{\nu 1})\tilde{F}(\sigma(\rho_{\nu 2};\tau_2 - \tau_1) \cap \sigma(\rho_{\nu 3};\tau_3 - \tau_2))\chi(\rho_{\nu 1})$$

From (45) it follows again that

$$\tilde{F}(\sigma(\rho_{\nu 1};\tau_1) \cap \sigma(\rho_{\nu 2};\tau_2) \cap \sigma(\rho_{\nu 3};\tau_3))$$
$$= Q\mathcal{U}_{\tau 1}Q'\tilde{F}(\sigma(\rho_{\nu 1};0) \cap \sigma(\rho_{\nu 2};\tau_2 - \tau_1) \cap \sigma(\rho_{\nu 3};\tau_3 - \tau_1))$$

The explicit calculation shows that all the effects (59) can be expressed by the transition probabilities $\lambda_\tau(\nu,\mu)$. The result is also well known as that of a stochastic process of the *Markov* form.

4. Conclusions

The description of macrosystems in a state space is compatible with an extrapolated quantum mechanics of many particles iff the relation (41) or (45) is valid. We want to put forth the proposal to found statistical mechanics of irreversible processes on these relations. A much deeper question is to prove these relations for special systems.

References and Notes

1. G. Ludwig, *Foundations of Quantum Mechanics* (Springer-Verlag, Berlin, in press).
2. G. Ludwig, "A Theoretical Description of Single Microsystems," in W. C. Price and S. S. Chissick, editors, *The Uncertainty Principle and Foundation of Quantum Mechanics* (John Wiley and Sons, New York, 1977).
3. G. Ludwig, "An Axiomatic Basis of Quantum Mechanics," in H. Neumann, editor, *Interpretation and Foundation of Quantum Theory* (BI-Wissenschaftsverlag, 1980).
4. G. Ludwig, "The Objective Description of Preparation and Registration as a Basis of Quantum Theory," submitted to *Nuovo Cimento.*
5. G. Ludwig, *An Axiomatic Basis of Quantum Mechanics* (Springer-Verlag, Berlin, translation in preparation).
6. G. Ludwig, *Einführung in die Grundlagen der theoretischen Physik*, 4 Vols. (Vieweg, Braunschweig, 1974–1979).
7. G. Ludwig, *Die Grundstrukturen einer physikalischen Theorie* (Springer-Verlag, Berlin, 1978).
8. G. G. Emch, *Algebraic Methods in Statistical Mechanics and Quantum Field Theory* (John Wiley and Sons, New York, 1972).
9. F. Riesz and B. Sz. Nagy, *Vorlesungen über Funktionalanalysis* (Deutscher Verlag der Wissenschaften, Berlin, 1956).

17

THE CURRENT THEORY OF STRONG INTERACTIONS AND THE PROBLEM OF QUARK CONFINEMENT

STANLEY MANDELSTAM
Department of Physics
University of California
Berkeley, California 94720

Before I begin my paper, let me express the hope and, indeed, the conviction, that this volume will prove a worthy memorial to Professor Yourgrau, whose tragic death occurred so soon after the commencement of his "retirement"—a word that, in this case, would have been most misleading. While I have not seen the contents at the time of writing, I have no doubt that the volume will express the broadness, variety and depth of his interests, and his devotion to academic learning.

I should like to give what must of necessity be a very cursory outline of the theory which many physicists believe to underlie the strong interactions of elementary particles, following which I should like to say something about the most crucial problem which faces us in that theory, namely quark confinement. I shall attempt to direct my paper mainly to readers who are not specialists in particle theory or experiment. Much of what I say will be known to the specialists.

Let me begin by saying that most particle theorists are now considerably more optimistic that they may understand the theory of the strong interactions than they were, say, five years ago. Indeed, we now believe that we have a theory of the strong interactions; the theory is known as quantum chromodynamics (QCD). Thus far no numerical calculations have been performed and compared with experiment. We have yet to find a reliable approximation scheme. Nevertheless, the very general predictions of the theory fit so well with certain salient experimental facts that it is hard to believe it to be wrong. Is it possible that particle theorists are simply brainwashing one another? Some eminent theorists have denied such a possibility. I personally should not go so far, but I believe it to be unlikely.

What surprised many people is that this progress has been achieved without any conceptual breakthrough. The postulates of QCD do not differ in any radical way from quantum field theory as it was introduced in 1930 by Heisenberg and Pauli.[1] Needless to say, the postulates possess a wealth of consequences that Heisenberg and Pauli could not possibly have foreseen.

I shall divide my paper into four parts:

1. The nature of gauge theories (QCD is a gauge theory)
2. Quark models
   Scaling and the pointlike structure of hadrons
   Asymptotically free gauge theories
3. *S* matrices with spectra of linearly rising rotational sequences
   String models of hadrons
4. Hadronic states in gauge theories
   Quark confinement

Sections 2 and 3 will not appear to be related, but the work described in Section 4 will unify these two approaches. At the end of the paper I shall say a word about other elementary-particle interactions, and their relation to the strong interactions.

1. The Nature of Gauge Theories

Gauge theories are familiar to all of us, since electrodynamics is such a theory. Their distinguishing feature is that, while they satisfy the causality requirements, they *cannot* be expressed in terms of *local, physical* variables. While this statement may appear rather formal it does have direct experimental consequences, as is illustrated in a setup proposed by Aharonov and Bohm[2] (Fig. 1). Two beams of electrons travel across a field-free region but, in between the beams, there is a region with a magnetic field. As the field is turned on, the interference fringes move. This result follows from thoroughly tested hypotheses of electromagnetism and quantum mechanics, and it has also been verified directly.

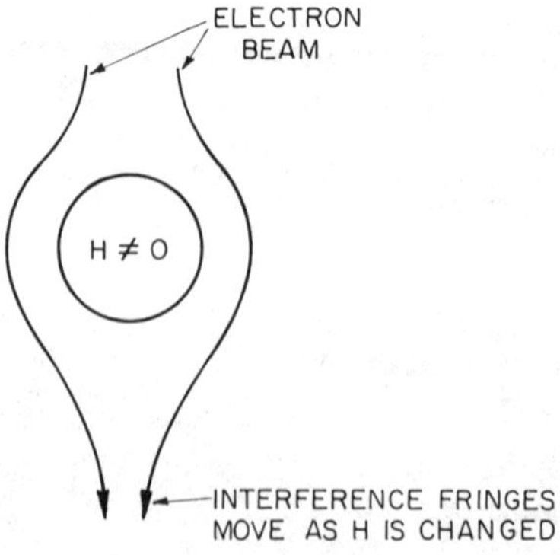

FIGURE 1. The Bohm–Aharonov experiment.

One can understand this experiment in two ways. One may say that, though the potentials themselves have no significance, the integral $\int A_i dx^i$, round a closed curve, does. This integral is nonzero for a curve surrounding the region where $H \neq 0$. Alternatively, one may say that charged particles necessarily have their Coulomb fields associated with them, and the Coulomb field penetrates into the region where $H \neq 0$.

In fact, there are two ways of formulating gauge theories. One can either introduce local variables such as the potential, but these are not physical variables. The gauge-invariant combinations thereof form the physical variables. Alternatively, one may introduce nonlocal physical variables such as a creation operator for a charged particle with its Coulomb field.

While most of this paper will be nontechnical, I do wish to write down the Lagrangian for QCD. I hope that those readers who are unfamiliar with the field-theoretic concepts should still be able to follow the rest of the paper. Let us start with the Lagrangian for quantum electrodynamics (with a scalar charged field):

$$\mathcal{L} = -\frac{1}{4}\left(\frac{\partial A^\nu}{\partial x_\mu} - \frac{\partial A^\mu}{\partial x_\nu}\right)^2 - \frac{1}{2}\left(\frac{\partial \phi^*}{\partial x_\mu} + ieA^\mu\phi^*\right) \times \left(\frac{\partial \phi}{\partial x_\mu} - ieA^\mu\phi\right) - \mu^2\phi^*\phi \tag{1}$$

The terms involving eA^μ are the interaction terms. They are given by the rule, obtained from gauge invariance, that when differentiating a charged field:

$$\frac{\partial}{\partial x_\mu} \to \frac{\partial}{\partial x_\mu} \pm ieA^\mu, \qquad \text{for } \phi, + \text{ for } \phi^* \tag{2}$$

In electrodynamics, the gauge field (the electromagnetic field with its photons) is separate from the *charged* field associated with electrons, protons, etc. Yang and Mills[3] proposed a generalization of electrodynamics in which the gauge field itself is charged. For technical reasons which do not concern us here, the Yang–Mills field is called a *non-Abelian* gauge field. The fundamental charged field, instead of only having two components, transforms according to the adjoint representation of some group. The gauge field itself transforms in a similar way. Instead of (2), we now have

$$\frac{\partial \phi^\alpha}{\partial x_\mu} \to \frac{\partial \phi^\alpha}{\partial x_\mu} + gf^{\alpha\beta\gamma}A^{\beta\mu}\phi^\gamma \tag{3}$$

where $f^{\alpha\beta\gamma}$ are the structure constants of the group. Indeed, since the gauge field is itself charged, we need introduce no other field in order to have an interaction. The Lagrangian for the pure Yang–Mills field would be

$$\mathcal{L} = -\frac{1}{4}\left(\frac{\partial A^{\alpha\nu}}{\partial x_\mu} - \frac{\partial A^{\alpha\mu}}{\partial x_\nu} + gf^{\alpha\beta\gamma}A^{\beta\mu}A^{\gamma\nu}\right)^2 \tag{4}$$

We note the similarity between the right-hand side of (3) and the terms in the bracket of (4). The group believed to be relevant to the strong interactions is $SU(3)$, whose adjoint representation is eight dimensional. Thus the indices α take eight values. In addition, we may have particles which transform according to other representations; those that transform according to the three-dimensional representation are known as quarks. The Lagrangian for a Yang–Mills field interacting with quarks would be

$$\mathcal{L} = -\frac{1}{4}\left(\frac{\partial A^{\alpha\nu}}{\partial x_\mu} - \frac{\partial A^{\alpha\mu}}{\partial x_\nu} + g f^{\alpha\beta\gamma} A^{\beta\mu} A^{\gamma\nu}\right)^2 + \frac{1}{2}\Psi_{ai}\left[\gamma^\mu\left(i\frac{\partial}{\partial x^\mu}\delta_{ab} + g\lambda^\alpha_{ab} A^\alpha_\mu\right) - m_i\right]\Psi^{bi} \tag{5}$$

where λ is the $SU(3)$ matrix analogous to the τ matrices of $SU(2)$, and i represents possible extra degrees of freedom which are not part of the dynamics of strong interaction. Later in this paper we shall discuss the physical significance of the degree of freedom associated with the $SU(3)$ group.

2. Quark Models

I assume that most readers are familiar with the fact that the gross features of hadronic structure can be understood on the assumption that the hadrons are built out of "quarks" with spin ½, and that quarks themselves have never been seen. Originally, it was thought that quarks came in three "flavors" (two values of the *I*-spin, plus a strange quark); recently further flavors have been appearing. The quark model assumes that the lightest baryons have wave functions which are symmetric in their space, spin, and flavor degree of freedom. To remove the contradiction with Fermi statistics, Greenberg[4] proposed that quarks obey a more complicated kind of statistics called parastatistics. This is equivalent to postulating a degree of freedom for the quarks in addition to the space, spin, and flavor degrees. This new degree of freedom takes three values, different values thereof being in principle indistinguishable. It was called *color* by Gell-Mann. The contradiction with Fermi statistics is now removed if the ground state of baryons is assumed to be antisymmetric in color.

The idea of quarks as actual pointlike elementary particles began to be taken seriously when experiments at SLAC indicated scaling at high energies. Scaling can be understood if it is assumed that hadrons are composed of pointlike constituents. Later experiments showed that the pointlike constituents probably had the quantum numbers of quarks.

Quantum field theory is the only mathematical formalism at present available for dealing with pointlike elementary objects. The fields themselves correspond to these objects, while the particles actually seen in nature are complicated combinations thereof. However, owing to short-distance (ultraviolet) divergences, the elementary objects should not really be considered as pointlike. In Fig. 2a, we observe that a pos-

(a) (b)

FIGURE 2. Vacuum polarization in asymptotically nonfree and asymptotically free theories.

itive charge surrounds itself with negative charges. The coupling at the center is thus stronger than the average coupling of the whole object; it is the latter that is measured experimentally. In fact, the coupling at the center becomes infinitely strong, and all short-distance (i.e., high-momentum) simplicity, including scaling, is lost.

The question may be raised: could a positive charge surround itself with positive instead of negative charges? (Figure 2b.) If so, the coupling at the center would be weak, and we would still have high-momentum simplicity. In fact, the coupling at the center is so weak that no pointlike object remains, so *again* we do not have strict scaling. Here, however, everything is much more under control, and we have scaling up to logarithms. The SLAC experiments can certainly tolerate a logarithmic deviation from scaling.

People familiar with the formalism of quantum field theory can prove that, in a theory described by local, physical fields, the second situation (Fig. 2b) cannot occur. However, in gauge theories, which are *not* so described, the proof fails. Either situation is possible. It turns out that electrodynamics is similar to nongauge theories in this respect; things become complicated at short distances. In many non-Abelian gauge theories the second situation occurs, as was discovered by Politzer[5] and by Gross and Wilczek.[6] Such theories are said to be *asymptotically free*. The experimental verification of scaling suggests that the theories underlying strong interactions possess asymptotic freedom.

One may ask why nature chooses to have a complicated non-Abelian gauge theory just to obtain asymptotic freedom. The answer may have been suggested by Landau,[7] who, in the years before his death, maintained with the utmost vigor that the short-distance complications of asymptotically nonfree theories were so severe that the theories were inconsistent. Asymptotically free theories were not known at the time, so Landau in fact rejected field theory. His arguments certainly do not apply to asymptotically free theories; if he was right, these are the only consistent theories.

We have seen that non-Abelian gauge theories involve a degree of freedom, which we have denoted by α,β,γ or a,b in (5). What is this degree of freedom? Several facts lead to the conclusion that it cannot be flavor, but perhaps it is color. The resulting theory is quantum chromodynamics, which was suggested independently by several groups (Gell-Mann, Fritzsch, and Leutwyler;[8] Gross and Wilczek;[6] Weinberg;[9] Wess[10]). The Lagrangian is (5); quarks are characterized by a color index a which takes three values, and a flavor index i. The gauge particles, known as *gluons*, have an eight-valued color index α.

We now come to a very important property of non-Abelian gauge theories, which brings us to the question of confinement. There is a kind of "conservation of difficulty" in field theories. If a theory is simple at short distances, it must necessarily be complicated at large distances—and the behavior at large distances determines the nature of the particles we actually see. We would not expect the particles to be anything like those predicted by perturbation theory. Could we explain the fact that quarks are never seen? In fact, gluons, or anything else with a color quantum number, have never been seen. Nambu suggested some time ago that confinement should be regarded as the absence of any particle which carries a color index; quark confinement is just one particular example. Before we discuss this question, I should like to say something about string models of hadrons.

3. Linearly Rising Rotational Sequences

Experimentally, it is observed that hadronic states appear to be members of approximately linear rising rotational sequences (Fig. 3), the angular momentum being roughly proportional to the square of the center-of-mass energy. The sequences cover the complete range of experimentally available energies; with baryons, where the experiments are easiest, resonances with j up to 10 have been seen.

Veneziano[11] proposed a very simple formula for the elastic two-particle S matrix. The resonances in his S matrix were infinitely narrow, but otherwise it satisfied all the required properties, and the resonances did form linearly rising rotational sequences. Starting from his formula, physicists developed the *dual-resonance model* for the complete hadronic S matrix in the narrow-resonance approximation.

It was later found that a space-time picture underlying the dual-resonance model could be obtained if it was assumed that hadrons were stringlike objects. The initial idea was due independently to Nambu,[12] Nielsen,[13] and Susskind,[14] the most comprehensive development to Goldstone, Goddard, Rebbi, and Thorn.[15] The linearly rising rotational sequences are an immediate consequence of the model. Interactions take place by successive joining and splitting of strings.

The dual-resonance model, which was constructed to explain hadronic spectra and processes involving a small number of hadrons, appeared to have little in common with QCD, which explained processes involving large number of hadrons. Could the two approaches be reconciled and, if so, could certain crucial defects in the dual-resonance model be rectified? In a very beautiful paper, Nielsen and Olesen made a suggestion for doing so, and for unifying two completely different approaches to particle theory. Their work brings me to the last part of my paper.

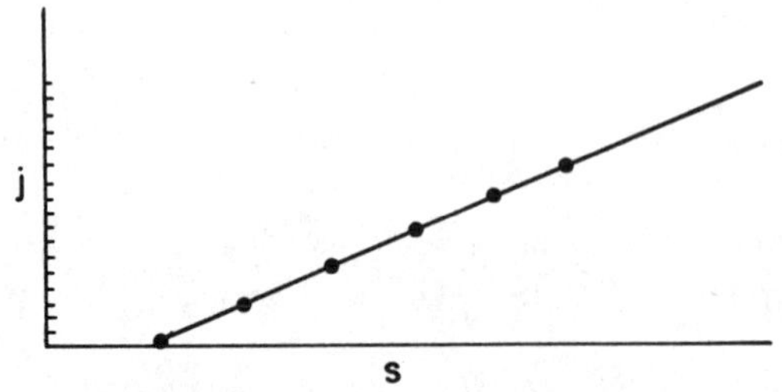

Figure 3. Linearly rising rotational sequence.

4. Hadronic States in Gauge Theories—Quark Confinement

Nielsen and Olesen[16] considered a simple model of an Abelian gauge field interacting with a charged field. Their model had the feature, which was not new to particle theory, that the vacuum was analogous to the superconducting rather than the normal state of a metal. A superconductor may be considered as a coherent quantum plasma of charged objects. In metallic superconductors, these objects are Cooper pairs. In the Nielsen–Olesen model they were the particles associated with the charged fields of the theory.

It is known that it is hard to get magnetic flux into a superconductor. The only way it can occur is in the form of quantized vortices of strength $2\pi/g$, where g is the coupling constant (e in ordinary electrodynamics). Landau and Ginzburg[17] have made a study of such vortices. The vortex itself forms a region of normal state (which has no antipathy to magnetic flux) within the superconducting state surrounding it (Fig. 4). Nielsen and Olesen made the suggestion that the strings of the dual-resonance model were representations or idealizations of Landau–Ginzburg vortices.

The original Nielsen–Olesen vortices had to be endless—either infinite in length or closed. The dual-model strings were open strings of finite length. As Nambu pointed out, finite-length Nielsen–Olesen vortices would have to end on monopoles to absorb the flux. Dirac had shown long ago that a monopole in a quantum theory would have to have strength $2\pi/g$, the same strength as that of a magnetic vortex.

The work of Nielsen and Olesen, and of Nambu, drew particle theorists' attention to the fact that, while everyone was looking for a mechanism which would explain confinement, there existed a familiar physical system with all the properties for which we were searching. A single monopole in a superconductor has infinite energy—the flux associated with it cannot spread out, but must extend from it in a tube of infinite length. However, a monopole–antimonopole pair would have finite energy, since the flux could go from one to the other, as is shown in Fig. 5, which represents a gedanken-experiment to illustrate confinement. "Elementary" monopoles are not necessary; one could use a long solenoid. The energy required to separate the pair would be proportional to the distance between them, since the flux tube carries energy. In fact, the setup in Fig. 5 (without the solenoid) may be regarded as a model of a meson, the two monopoles representing the quarks. The distance between the quarks could never exceed nuclear distances.

The quarks in QCD are the analogs of electric charges, not magnetic monopoles. We pointed out earlier in this paper that gauge theories could not be formulated in terms of local, physical variables; a charged particle had to have "electric" flux associated with it. We use quotation marks to emphasize that we are dealing with the non-Abelian analog of familiar quantities. Until recently, no one questioned the pos-

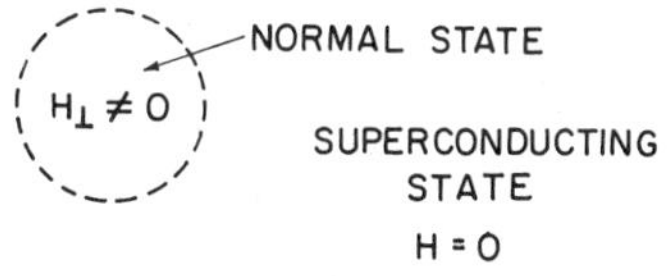

Figure 4. Cross section of a Landau–Ginsburg vortex.

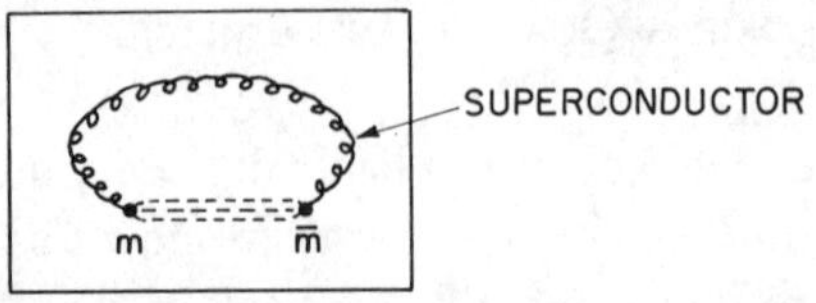

FIGURE 5. Gedankenexperiment to illustrate confinement.

sibility that "electric" flux could spread out, as it does in ordinary electrodynamics. However, if "electric" flux was unable to spread out, we should have a situation precisely analogous to monopoles in a superconductor, and quarks would be confined.

What could prevent the "electric" flux from spreading out? A very attractive possibility is that "electric" flux is confined to tubes for the same reason that magnetic flux is so confined in a superconductor. We know that the equations of electrodynamics are symmetric in electric and magnetic variables. The equations for non-Abelian gauge theories have no such *manifest* symmetry; the fundamental field is "electrically", not "magnetically" charged. Nevertheless, increasing evidence has been accumulating that QCD does possess an underlying "electric"–"magnetic" symmetry. Wu and Yang[18] showed, a number of years ago, that "monopole"-like configurations can occur. If the vacuum were a coherent plasma of such "monopoles," just as an ordinary superconductor is a coherent plasma of charges, "electric" flux would be squeezed into tubes, and we should have color confinement.

Finally, we have to ask why the vacuum is a coherent plasma of Wu–Yang "monopoles." The answer is technical, and has not yet been fully worked out. Very qualitatively, in non-Abelian theories, unlike in Abelian theories, the perturbation-theory vacuum is a very bad "trial vacuum"—it has an infinite term in its energy density which does *not* come from ultraviolet divergences. To remove this infinity, one must introduce a large number of random gauge transformations. The "monopoles" appear to do just that.

According to current ideas, hadrons consist of confined quarks linked by electric flux tubes as illustrated in Fig. 6. The diagram describes hadrons of high angular momentum, where the centrifugal force keeps the quarks apart. For low angular momentum the length of the flux tube becomes comparable with its thickness, and we may have a structure resembling the MIT bag.

5. Conclusions

This paper has been concerned with the strong interactions. As their name implies, these are the most important interactions at nuclear distances, or at distances

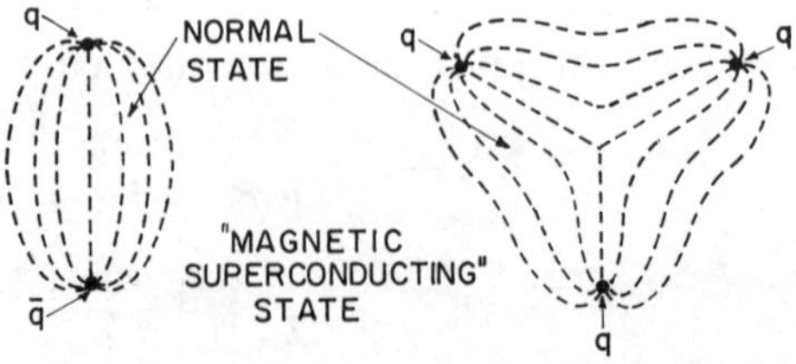

FIGURE 6. Hadrons in gauge theories.

somewhat below the nuclear scale. Such interactions are no doubt responsible for the so-called elementary strongly interacting particles, the mesons and baryons, which we actually see.

Elementary particles are also subject to other interactions. Besides electromagnetic interactions, there are the weak interactions responsible for β decay and many other decays of elementary particles. Even before QCD was formulated, Weinberg[19] and Salam[20] independently suggested that weak and electromagnetic interactions might together be described by a non-Abelian gauge theory. The Weinberg–Salam model is now the generally accepted theory of weak and electromagnetic interactions.

At nuclear distances (10^{-13} cm or thereabouts), the terms "strong" and "weak" appropriately describe the respective types of interactions. We have seen, however, that strong interactions become weaker at short distances, and eventually become as weak as "weak" interactions. This probably occurs at around 10^{-29} cm—a figure which is at present uncertain by several orders of magnitude.

At such distances, strong, electromagnetic, and weak interactions are all described by non-Abelian gauge theories of comparable interaction strengths. It is attractive to speculate that the theories can all be included in a single non-Abelian gauge theory. Such "grand-unified" theories are very much in vogue at the present time. If one of them turns out to be correct, it will be a theory of most of present physics.

However, there is one type of interaction which I have not mentioned in this paper, and that is gravitation. Central to the physics of Newton's day, this interaction has since become a subject of its own, of vital interest to astronomers and astrophysicists, but disconnected from the rest of physics. Even Einstein's beautiful theory did not really change the situation; he himself was convinced that gravitation must play an essential role in the structure of matter but, until a few years ago, only a minority of physicists agreed with him. Now elementary-particle physicists are paying more attention to gravitation. It, too, is a non-Abelian gauge theory, though of a structure somewhat different from that of other such theories. At atomic and nuclear distances, gravitation is much weaker than all other known interactions. Its strength increases at small distances, and becomes comparable to that of other interactions at the "Planck length" of 10^{-33} cm; not very different from the distance of 10^{-29} cm mentioned above. The grand unification of strong, electromagnetic, and weak interactions may itself be dependent on gravitation or it may occur at distances where gravitation is still fairly weak, but, in any case, it is tempting to conjecture that a final unification of all known interactions occurs at the Planck length.

References and Notes

1. W. Heisenberg and W. Pauli, Z. *Phys.* **56**, 1 (1929).
2. X. Aharonov and X. Bohm, *Phys. Rev.* **115**, 485 (1959).
3. C. N. Yang and R. L. Mills, *Phys. Rev.* **96**, 191 (1954).
4. O. W. Greenberg, *Phys. Rev. Lett.* **13**, 598 (1964).

5. H. D. Politzer, *Phys. Rev. Lett.* **30**, 1343 (1973).
6. D. J. Gross and F. Wilczek, *Phys Rev. D* 8, 3633 (1973).
7. L. D. Landau, A. A. Abriksov, and I. M. Khalatnikov, *Dokl. Akad. Nauk USSR* **95**, 773, 1177; **96**, 261 (1954); L. D. Landau and I. Pomeranchuk, *Dokl. Akad. Nauk USSR* **102**, 484 (1955).
8. M. Gell-Man, H. Fritzsch, and H. Leutwyler, *Phys. Lett.* **47B**, 365 (1973).
9. S. Weinberg, *Phys. Rev. D* 8, 4482 (1973).
10. J. Wess, unpublished results.
11. G. Veneziano, *Nuovo Cimento* **57A,** 190 (1968).
12. Y. Nambu, *Proc. Intern. Conf. on Symmetries and Quark Models*, Wayne State University (1969).
13. H. B. Nielsen, *Proc. 15th Intern. Conf. on High-Energy Physics*, Kiev (1970).
14. L. Susskind, *Nuovo Cimento* **69A,** 457 (1970).
15. J. Goldstone, P. Goddard, C. Rebbi, and C. B. Thorn, *Nucl. Phys.* **B56,** 109 (1973).
16. H. B. Nielsen and P. Olesen, *Nucl. Phys.* **B61,** 45 (1973).
17. L. D. Landau and B. L. Ginzburg, *Zh. Exp. The. Fiz. (USSR)* **20,** 1064 (1958).
18. T. T. Wu and C. N. Yang, in *Properties of Matter under Unusual Conditions*, H. Mark and S. Fernbach, eds. (Interscience, New York, 1969), p. 349.
19. S. Weinberg, *Phys. Rev. Lett.* **19** 1264 (1967).
20. A. Salam, in *Elementary Particle Theory*, N. Svartholm, ed. (Almquist and Forlag, Stockholm, 1968).

18

Statistical Inference by Minimal Bures Distance

Jean-Paul Marchand
Department of Mathematics
University of Denver
Denver, Colorado 80208

1. Introduction

We have developed, over the past years, a general theory of statistical inference in quantum mechanics[1–6] that has now reached a stage in which a presentation based on first principles seems appropriate.

I dedicate this presentation to the memory of my unforgettable, admirable friend, Wolfgang Yourgrau.

The general purpose of statistical inference is the determination of a state w_0 of the system S which reflects precisely the information about S that is available, no more and no less. Statistical inference is thus an information-theoretical concept, and w_0 describes the state of our knowledge of S, rather than its "real" state.

In physics, information about S may be obtained by measurement. But it is necessary to assume that "no measurement" is also information. It must, in other words, be possible to assign S on a priori state v, which reflects our knowledge of S prior to any measurement. The available information therefore consists of

(a) an a priori state v,
(b) the outcome of a partial measurement.

Let us illustrate these points.

Suppose that my colleague of the laboratory next door sends me a sample of 1000 identically prepared, completely polarized spin-½ particles, with the comment: "This ensemble of spins has been prepared in a pure state p, but I shall not tell you in which." Although the "real" state of this system is thus pure, my ignorance of its polarization compels me to assign it an a priori state v which corresponds to a homogeneous mixture of all possible polarizations p, i.e., the state whose density matrix is

$$V = \tfrac{1}{2}\sigma_0 \tag{1}$$

But there are situations in which v should be chosen differently. If my colleague had, for instance, told me that he prepared the ensemble in an equilibrium state, at temperature T, with a free spin reservoir in a constant magnetic field in 3-direction, the proper choice of v would be the Gibbs state

$$V = Z^{-1} \exp\left(- \frac{c}{kT} \sigma_3 \right) \tag{2}$$

I am now free to perform my own measurements on S. Suppose that I mcasure the expectation b for the 1-component of the spin. Although this information is, of course, not sufficient to determine the state of S uniquely, the requirement that the inferred state w_0 reflects *not less* than my information obtained from the measurement imposes on w_0 the restriction

$$w_0(\sigma_1) = b \tag{3}$$

The constraint (3) defines an equivalence class of states w, whose density matrices have the general form

$$W = \tfrac{1}{2}(\sigma_0 + b\sigma_1 + w_2\sigma_2 + w_3\sigma_3) \tag{4}$$

with w_2 and w_3 only restricted by the positivity of W. In order to chose the inferred state w_0 from this equivalence class, we must now use the postulate that w_0 reflects *not more* information than that obtained from the a priori state v and the measurement (3).

In the simple case of the homogeneous a priori state (1), the appropriate choice of w_0 is

$$W_0 = \tfrac{1}{2}(\sigma_0 + b\sigma_1) \tag{5}$$

The argument is based on the *principle of insufficient reason* and runs as follows: The state (1) is unpolarized and the partial information from the measurement is only about σ_1. Hence there is no reason to assume that w_0 shows a σ_2 or σ_3 polarization.

For more general a priori states, the problem of implementing the principle of insufficient reason is, however, less trivial, and its solution requires the formulation of a generalized principle of inference. [For the above spin system and the anisotropic a priori state (2), an explicit answer will be found in Example 5.1.]

We propose in this article an inference principle which is based on minimal Bures distance between the inferred state w_0 and the a priori state v. As our applications will show, this principle leads to a reasonable concept of inference, both in classical probability theory and in the theory of quantum-mechanical measurement.

2. Bures Distance between States

The mathematical concept of Bures distance can be discussed at various levels of generality.[7–11] We shall first formulate it in a somehow restricted context which is adapted to our needs.

Let $\mathcal{A}$ be the von Neumann algebra of bounded operators in a separable Hilbert space $\mathcal{H}$ which is generated by the physical observables (self-adjoint operators) of the system S. A *state* of S is a normal positive linear functional n on $\mathcal{A}$ such that $n(I) = 1$. Let N be the set of all states of S. (Note that N is convex but not linear.)

Let $v,w \in N$, v faithful. If $w \leq \lambda v$ $(\lambda \geq 1)$, there exists[7] a unique positive operator $T \in \mathcal{A}$, called the *Sakai operator* of w relative to v, such that $1 \leq \|T\|^2 \leq \lambda$ and

$$w(A) = v(TAT), \ \forall \ A \in \mathcal{A} \tag{6}$$

The *Bures distance* between v and w can now be defined by[8]

$$d(v,w) = [2(1 - v(T))]^{1/2} \tag{7}$$

Another useful mathematical quantity is the *Uhlmann transition probability*[9]:

$$P(v,w) = |v(T)|^2 \tag{8}$$

As mentioned above, these concepts can be generalized. A few indications may suffice for the present purpose.

(1) The Sakai theorem can be generalized to the case where the relative boundedness condition $w \leq \lambda v$ is replaced by the absolute continuity condition $w < v$. The operator T is then in general unbounded, but it still exists as an element of $L_2(\mathcal{A},v)$, the Hilbert space completion of the inner product space $\mathcal{A}$ with $\langle A_1,A_2\rangle$ — $\mathrm{Tr}(A_1A_2)$. (Note the close analogy to the classical Radon–Nikodym theorem!)

(2) In the definition of $d(v,w)$ and $P(v,w)$ the reference to the Sakai operator can be avoided altogether. Bures distance and Uhlmann transition probability can, in fact, be defined for arbitrary states[10,11] and it can be shown that $d(v,w)$ is indeed a distance function defining a metric on N.

(3) For pure states ϕ and ψ the Uhlmann transition probability reduces to the usual expression $|(\phi,\psi)|^2$. It is, however, doubtful whether it has a well defined operational significance for arbitrary mixed states v and w. We shall refer to it for mathematical convenience only.

3. The Inference Principle

In this Section we postulate an inference principle based on minimal Bures distance, and discuss some of its mathematical properties.

Let $\mathcal{A}$ again be the algebra generated by the observables of S, and $v \in N$ an *a priori* state. The partial measurement can be described by an expectation functional $w_{\mathcal{S}}$ on a subset $\mathcal{S} \subset \mathcal{A}$ of observables. Let $w|_{\mathcal{S}}$ denote the restriction of $w \in N$ to $\mathcal{S}$.

Then the equivalence class of the states on $\mathcal{A}$ which are compatible with the partial measurement of $\mathcal{S}$ is defined by

$$N(w_{\mathcal{S}}) = \{w \in N: w|_{\mathcal{S}} = w_{\mathcal{S}}\}$$

[Note that, in the case of no measurement, $\mathcal{S} = \phi$ and $N(w_{\mathcal{S}}) = N$.]

We are now prepared to formulate our inference principle:

Inference Principle: The state w_0 inferred from the partial measurement $w_{\mathcal{S}}$ is a state in $N(w_{\mathcal{S}})$ with minimal Bures distance $d(v,w)$ or, equivalently, maximal Uhlmann transition probability $P(v,w)$, from the a priori state v.

We immediately observe that $w_0 = v$, if either $\mathcal{S} = \phi$ (no measurement), or $v \in N(w_{\mathcal{S}})$ (v compatible with the measurement). This conclusion is reasonable, since in either case the measurement does not add anything to the information already contained in v. If, on the other hand, the measurement is complete and defines a unique state w, then $N(w_{\mathcal{S}}) = \{w\}$, and the inference principle implies that $w_0 = w$.

Next we consider the important case where the partial measurement defines a state $w_{\mathcal{B}}$ on a von Neumann subalgebra $\mathcal{B} \subset \mathcal{A}$.[3]

Theorem 3.1. Let $w_{\mathcal{B}}$ be relatively bounded with respect to the restriction $v|_{\mathcal{B}}$ of v (i.e., $\exists\ \lambda, \mu \in [1,\infty) \ni (1/\mu)v|_{\mathcal{B}} \leq w_{\mathcal{B}} \leq \lambda v|_{\mathcal{B}}$). Then w_0 is the unique state in $N(w_{\mathcal{B}})$ whose Sakai operator T_0 relative to v is in $\mathcal{B}$.

According to this theorem, the inferred state w_0 can now be constructed explicitly as follows:

Corollary 3.2. w_0 is the state

$$w_0(A) = v(T_0AT_0), \qquad A \in \mathcal{A} \tag{9}$$

where $T_0 \in \mathcal{B}$ is defined by

$$w_{\mathcal{B}}(B) = v|_{\mathcal{B}}(T_0BT_0), \qquad B \in \mathcal{B} \tag{10}$$

As we shall see in the next section, the Sakai operator T_0 can be viewed as a generalized Radon–Nikodym derivative. Theorem 3.1 states therefore that, in the case where $\mathcal{B}$ is a subalgebra, w_0 has the characteristic property that its Radon–Nikodym derivative is in $\mathcal{B}$.

We conclude this section by indicating the proof of the theorem.

Let w_0 be defined by (9) and (10), and let w be any other state in $N(w_{\mathcal{B}})$ of the form $w(A) = v(TAT)$. Then

$$v(TBT) = v(T_0BT_0),\ \forall\ A\ B \in \mathcal{B} \tag{11}$$

Since $w_{\mathcal{B}}$ is relatively bounded with respect to $v|_{\mathcal{B}}$, $T_0 \in \mathcal{B}$ has a bounded inverse $T_0^{-1} \in \mathcal{B}$. Inserting $B = T_0^{-1}$ in (11) we have

$$v(TT_0^{-1}T) = v(T_0)$$

and the Schwarz inequality implies

$$|v(T)|^2 = |v[T_0^{1/2}(T_0^{-1/2}T)]|^2 \leq v[T_0^{1/2}(T_0^{1/2})^+] \cdot v[(T_0^{-1/2}T)^+ (T_0^{-1/2}T)]$$
$$= v(T_0) \cdot v(TT_0T) = |v(T_0)|^2$$

According to (7) and (8), the inequality $v(T) \leq v(T_0)$ implies

$$P(v,w) \leq P(v,w_0), \qquad d(v,w_0) \leq d(v,w)$$

Hence $w_0 \in N(w_{\mathcal{B}})$ has minimal Bures distance from v. The uniqueness of w_0 follows from the criterion for the validity of the equality sign in Schwarz' inequality and the faithfulness of v.

4. Classical Systems

Let us consider the precise sense in which our definition generalizes the classical concept of inference. In order to do so, we must translate the language of classical probability into Hilbert space language.

Consider the probability space $\{\Omega,\Sigma,\nu\}$, with Ω a set, Σ a σ algebra of subsets of Ω, and ν a σ-additive probability measure on Σ. If the Hilbert space $\mathcal{H}$ is the space $L^2(\Omega,\Sigma,\nu)$ of Σ-measurable, ν-square integrable functions $f(\omega)$ on Ω and $\mathcal{A}$ the Abelian von Neumann algebra of the bounded multiplication operators A: $f(\omega) \rightarrow A(\omega)f(\omega)$, then $\mathcal{A}$ is isomorphic to $L^\infty(\Omega,\Sigma,\nu)$, i.e., the set of essentially bounded, Σ-measurable functions $A(\omega)$, and the expectation functional

$$v(A) = \int_\Omega A(\omega)\, d\nu \tag{12}$$

defines a state v on $\mathcal{A}$.

Assume now that on a σ subalgebra $\Sigma' \subset \Sigma$ a measure $\mu_{\Sigma'}$ is given such that $\mu_{\Sigma'}$ and the restriction $\nu_{\Sigma'}$ of ν are relatively bounded. Then $L^\infty(\Omega, \Sigma',\nu_{\Sigma'})$ is isomorphic to a von Neumann subalgebra $\mathcal{B} \subset \mathcal{A}$, and the expectation functional

$$w_{\mathcal{B}}(B) = \int_\Omega B(\omega)\, d\mu_{\Sigma'} \tag{13}$$

is relatively bounded with respect to $v|_{\mathcal{B}}$. Since $\mathcal{A}$ is Abelian, it follows from (9) and (10) that the inferred state is defined by

$$w_0(A) = v(T_0^2 A), \qquad w_{\mathcal{B}}(B) = v(T_0^2 B) \tag{14}$$

On the other hand, (12) and (13) imply that

$$w_{\mathcal{B}}(B) = \int_\Omega \frac{d\mu_{\Sigma'}}{d\nu_{\Sigma'}} B(\omega)\, d\nu_{\Sigma'} = v\left(\frac{d\mu_{\Sigma'}}{d\nu_{\Sigma'}}\, B\right) \tag{15}$$

where $d\mu_{\Sigma'}/d\nu_{\Sigma'}$ is the classical (bounded) Radon–Nikodym derivative. Hence

$$w_0(A) = v\left(\frac{d\mu_{\Sigma'}}{d\nu_{\Sigma'}} A\right) \tag{16}$$

If μ_0 is the measure on Σ defined by

$$w_0(A) = \int_\Omega A(\omega)\, d\mu_0, \qquad \forall\, A \in \mathcal{A} \tag{17}$$

then (16) implies that

$$\frac{d\mu_0}{d\nu} = \frac{d\mu_{\Sigma'}}{d\nu_{\Sigma'}} \in \mathcal{B} \tag{18}$$

We therefore conclude that, in the classical case, the *inferred measure* μ_0 *is characterized by the property that its Radon–Nikodym derivative* $d\mu_0/d\nu$ *is* Σ'*-measurable.*

The following example shows that this characterization of classical inference is reasonable.

Example 4.1: The Classical Position Measurement. Let Ω be the unit cube, Σ the Lebesgue measurable subsets of Ω, and Σ' the σ algebra of subsets generated by a "macroscopic" partition $\{\Delta_i\}$ of Ω. If nothing is known about the system before the measurement, we can assume that the mass density in Ω is a priori homogeneous. Hence $\nu = \lambda$ is the Lebesgue measure [note that $\lambda(\Omega) = 1$].

Suppose now that a measurement of the mass densities in Δ_i yields

$$\mu_{\Sigma'}(\Delta_i) = \mu_i$$

Then (18) implies that the inferred measure μ_0 has the density

$$\frac{d\mu_0}{d\lambda}(\omega) = \sum_i \frac{\mu_i}{\lambda(\Delta_i)} \chi_{\Delta_i}(\omega)$$

with χ_{Δ_i} the characteristic functions of Δ_i. The inferred density is thus constant in the interior of the "macrocells" Δ_i, as expected from the principle of insufficient reason.

We observe, however, that if the "box" had been known a priori to be in a gravitational field, we would have been compelled to chose a different, inhomogeneous *a priori* measure ν. In that case, the inferred measure μ_0 would have turned out to be proportional to ν in the interior of each macrocell Δ_i. We leave the details to the reader.

5. Abelian Measurement

The simplest quantum-mechanical inference problem occurs if the set $\mathcal{S}$, whose expectations can be obtained from the partial measurement, is an Abelian subalgebra $\mathcal{B} \subset \mathcal{A}$. ($\mathcal{A}$ itself is, of course, no longer Abelian.) This happens, for instance, if the experiment consists in the measurement of a complete set of commuting observables, in which case $\mathcal{B}$ is maximal Abelian. But the set need not be complete; it is sufficient that the measurement is "microscopic," i.e., can be decomposed into yes–no experiments, in which case $\mathcal{B}$ is simply the von Neumann algebra generated by the measured projections.

We first consider an algebra $\mathcal{B}$ which is generated by a discrete set $\{P_i\}$ of mutually orthogonal projections. Let

$$w_{\mathcal{B}}(P_i) = w_i$$

be the result of the measurement. From Corollary 3.2 follows that $t_0 \in \mathcal{B}$. Hence T_0 is a linear combination $\Sigma_k t_k P_k$, and (10) implies

$$w_i = v(T_0^2 P_i) = \sum_k t_k^2 v(P_k P_i) = t_i^2 v(P_i)$$

$$\Rightarrow t_i = \left[\frac{w_i}{v(P_i)}\right]^{1/2} \Rightarrow T_0 = \sum_k \left[\frac{w_k}{v(P_k)}\right]^{1/2} P_k$$

The inferred state w_0 is therefore, according to (9),

$$w_0(A) = \sum_{k,l} \left[\frac{w_k\, w_l}{v(P_k)v(P_l)}\right]^{1/2} v(P_k A P_l) \tag{19}$$

In the special case where v is proportional to the trace τ, (19) reduces to

$$w_0(A) = \sum_k \frac{w_k}{\tau(P_k)} \cdot \tau(P_k A) \tag{20}$$

This is the classical expression for the coarse-grained state relative to the partition $\{P_i\}$. We note, however, that (20) makes sense only if the projections P_i have finite range, whereas (19) is always well defined. From this we conclude, that *our notion of inference generalizes the classical concept of coarse-graining.*

Next, we consider the case where $\mathcal{B}$ is generated by a spectral measure $\{E(\Delta)\}$ which is absolutely continuous relative to the Lebesgue measure λ. With the notations $v(\lambda) = v[E(\lambda)]$ and $w(\lambda) = w_{\mathcal{B}}[E(\lambda)]$, the inferred state can be written[1]

$$w_0(A) = \int\int d\lambda\, d\mu \left[\frac{w^1(\lambda)w^1(\mu)}{v^1(\lambda)v^1(\mu)}\right]^{1/2} \frac{\partial^2}{\partial\lambda\partial\mu} v[E(\lambda)AE(\mu)] \tag{21}$$

(21) is the continuous analog of (19).

Let us consider two examples, the first of the discrete, and the second of the continuous type.

Example 5.1: The Spin. A spin ½ in a magnetic field in 3-direction (temperature β^{-1}) is partially measured by determining its 1-component $w_{\mathcal{B}}(\sigma_1) = b$. Here we set $\dim \mathcal{H} = 2$, $\mathcal{A} = B(\mathcal{H})$, $\mathcal{B} = \{\sigma_1\}''$, and define v as the Gibbs state

$$v(A) = \tau(VA), \qquad V = ½(\sigma_0 + a\sigma_3)$$

with $a = -\tanh^{-1}\beta$. A straightforward calculation of (19) yields

$$w_0(A) = \tau(W_0 A), \qquad W_0 = ½[\sigma_0 + b\sigma_1 + a(1 - b^2)^{1/2}\, \sigma_3] \tag{22}$$

(22) is the answer to the problem stated in the example of the introduction. For a "geometric" interpretation of this result cf. Ref. 6.

Example 5.2: Quantum-Mechanical Position Measurement. Here we have $\mathcal{A} = B(\mathcal{H})$, $\mathcal{B} = \{E(x)\}''$, $\{E(x)\}$ the spectral family of the position operator $Q = \int x\, dE(x)$, $w(x) = w_{\mathcal{B}}[E(x)]$ the measured distribution in configuration space. The general expression for w_0 is now easily obtained from (21).

An interesting special case has been considered by R. Benoist[12]: Let

$$\varphi(x) = [2\pi\, (\overline{\Delta x})^2]^{-1/4} \exp\left[\frac{-(x-\bar{x})^2}{4(\overline{\Delta x})^2} + \frac{i\bar{p}x}{\hbar}\right]$$

be the minimum uncertainty wave packet with expectations $\bar{x}$ and $\bar{p}$ for position and momentum, and dispersion $\overline{\Delta x}$. Let $v(A) = (\varphi, A\varphi)$ be the *a priori* state, and suppose that

$$w(x) = [2\pi(\tilde{\Delta}x)^2]^{-1/2} \int_{-\infty}^{x} dy \exp\left[\frac{-y^2}{2(\tilde{\Delta}x)^2}\right]$$

is also Gaussian with expectation $\tilde{x} = 0$ and dispersion $\tilde{\Delta}x$. It turns out that w_0, too, is a minimum uncertainty state with momentum expectation $w_0(P) = v(P)$.

6. Coupled Systems

Another quantum-mechanical inference problem consists in determining the most likely state of a coupled system of which one component has been measured.

Let $\mathcal{H} = \mathcal{H}_1 \otimes \mathcal{H}_2$ be the Hilbert space of the coupled system $S = S_1 + S_2$, $\mathcal{A} = B(\mathcal{H})$ the algebra generated by the observables of S, and v an *a priori* state on $\mathcal{A}$. If the measured subsystem is S_1, then the measurement defines a state $w_{\mathcal{B}}$ on the subalgebra $\mathcal{B} = B(\mathcal{H}_1) \otimes I$ of $\mathcal{A}$. We note that $\mathcal{B}$ is not Abelian in this case.

Let τ_1, τ_2 denote the partial traces over the spaces $\mathcal{H}_1$, $\mathcal{H}_2$. It is easy to see that $w_{\mathcal{B}}$ defines uniquely a density operator $W_1 \in B(\mathcal{H}_1)$ such that

$$w_{\mathcal{B}}(A_1 \otimes I) = \tau_1(W_1 A_1), \qquad A_1 \in B(\mathcal{H}_1)$$

If $V_1 = \tau_2(V)$ admits a (possibly unbounded) inverse V_1^{-1}, it can be proved[4] that the inferred state w_0 of S is given by

$$w_0(A) = v(T_0 A T_0), \qquad T_0 = T_1 \otimes I \tag{23}$$

with T_1 the closure of the operator

$$V_1^{-1/2}(V_1^{1/2} W_1 V_1^{1/2})^{1/2} V_1^{-1/2}$$

Moreover, the restriction of w_0 to the unknown subsystem S_2 is

$$w_0^{(2)}(A_2) \equiv w_0(I \otimes A_2) = v(T_1^2 \otimes A_2) \tag{24}$$

7. The Measuring Process

The analysis of subsystems can be applied to the fundamental question of the measuring process itself.[5,13] If $S = S_A + S_P$ now denotes the combined system of the measuring apparatus S_A and the measured probe S_P, the problem consists in inferring the state of S_P from the readings of the apparatus S_A.

Let us describe S_A. Since the apparatus is a classical system, it admits, say, n distinct positions which correspond to n orthogonal vector states $\psi_i \in \mathcal{H}_A$. The projections A_i onto ψ_i form a partition of the identity in $\mathcal{H}_A$, and the measured subalgebra is therefore generated by $\{A_i \otimes I\}$. We note that $\mathcal{B}$ is Abelian. If w_i are the probabilities of the instrument readings $i = 1,2,\ldots,n$, as obtained from a sequence of identical experiments, then the state $w_{\mathcal{B}}$ on $\mathcal{B}$ is uniquely defined by

$$w_{\mathcal{B}}(A_i \otimes I) = w_i$$

The a priori state v on $\mathcal{A}$ contains the information about $S_A + S_P$ before the measurement. It involves the knowledge of the observables of S_P which the apparatus is supposed to measure, and the accuracy with which it measures them. Mathematically, this amounts to the specification of n projections P_i in $\mathcal{H}_P$ representing the measured observables, and a correlation matrix $\epsilon_{ik} = E(A_k|P_i)$, where $E(A_k|P_i)$ is the probability that the apparatus reads "k" if S_P is in the state P_i. Since $A_k \otimes I$ and $I \otimes P_i$ commute, we have

$$\epsilon_{ik} = E(A_k|P_i) = \frac{v(A_k \otimes P_i)}{v(I \otimes P_i)}$$

Next we require that $v(I \otimes P_i) = \alpha$ is independent of $_i$. (If this were not so, v would yield some information about S_P before any measurement is performed!) From the normalization of v follows that $\alpha = 1/n$. Hence

$$v(A_k \otimes P_i) = \frac{1}{n}\,\epsilon_{ik}$$

and a simple argument shows [5] that the a priori state is

$$v(A) = \tau(VA), \qquad V = \frac{1}{n}\sum_{i,k} \epsilon_{ik}(A_k \otimes P_i) \tag{25}$$

Having defined $\mathcal{A}$, $\mathcal{B}$, v, and $w_{\mathcal{B}}$, we can now apply (23) and (24) with the following result: The inferred state of the combined system $S_A + S_P$ is[5]

$$w_0(A) = \tau(W_0A), \qquad W_0 = \sum_{i,k} \frac{\epsilon_{ik}w_k}{\Sigma_l\,\epsilon_{lk}}(A_k \otimes P_i) \tag{26}$$

and the inferred state of S_P is

$$w_0^{(P)}(P_i) = \tau_P(W_0^{(P)}P_i), \qquad W_0^{(P)} = \sum_{j,k} \frac{\epsilon_{jk}w_k}{\Sigma_l\epsilon_{lk}}\,P_i \tag{27}$$

Example 7.1: The "Perfect" Measurement. In a "perfect" measurement, the states of S_A and S_P are totally correlated, i.e., we have $\epsilon_{ik} = \delta_{ik}$. According to (27) the reduced density matrix of the inferred state of S_P is now

$$W_0^{(P)} = \sum_i w_i P_i \tag{28}$$

(28) is usually called the "reduced state". But we note that reduction occurs here as a consequence of our incomplete information about the system, rather than a "collapse" of the prepared state due to the acausal interference of the observer with the system (cf. addendum to Ref. 5).

8. Other Types of Inference

We have so far applied our inference principle to situations where the partial measurement resulted in a state $w_{\mathcal{B}}$ on a subalgebra $\mathcal{B} \subset \mathcal{A}$. The inferred state w_0 could then be constructed with the help of Corollary 3.2. There are, however, other inference problems in which $w_{\mathcal{S}}$ is given for subsets $\mathcal{S} \subset \mathcal{A}$ which are not themselves algebras. In these cases the constructive method fails, and one must proceed along the original idea of minimizing the Bures distance $d(v,w)$.

We first consider the case where the information consists of the expectation value a of a single self-adjoint operator A. The method then requires minimizing $d(v,w)$ under the constraints

$$w(I) = 1, \qquad w(A) = a \tag{29}$$

For arbitrary a priori states this leads to a difficult variational problem. But there is an important special case, for which the problem can be solved by the "commutative" method of Lagrange multipliers.

Suppose that A has a discrete spectrum $\{a_i\}$, with corresponding spectral projections $\{E_i\}$, and that the density operator V of v is a function of A. As an example, A could be the energy operator H and V the Gibbs state $Z^{-1} \exp(-\beta H)$. By an argument similar to that used for the derivation of the canonical ensemble, one can assume that the inferred state is also a function of A. Hence we have

$$A = \sum_i a_i E_i\,, \qquad V = \sum_i v_i E_i\,, \qquad W = \sum_i w_i E_i$$

From (6) follows successively that

$$W = TVT, \qquad V^{1/2} T V^{1/2} = (V^{1/2} W V^{1/2})^{1/2}$$
$$v(T) = \tau(V^{1/2} T V^{1/2}) = \tau\,[(V^{1/2} W V^{1/2})^{1/2}] = \sum_i V_i^{1/2} W_i^{1/2}$$

According to (7) and (8), the minimization of $d(v,w)$ is equivalent to the maximization of $v(T)$, and the constraints (29) read

$$\sum_i w_i = 1, \qquad \sum_i w_i a_i = a \tag{30}$$

The (straightforward) solution of this Lagrange problem is

$$w_i = \frac{v_i}{4(\lambda + \mu a_i)^2} \tag{31}$$

where the multipliers λ and μ are obtained from (30).

A more interesting class of inference problems arises if the partial measurements of the system S have been made at various times t_k, $k = 1, \ldots, n$. Let H be the Hamiltonian of S, and $U_t = \exp(-iHT)$ the group of unitary time propagators. The measurement of a subalgebra $\mathcal{B}^{(k)}$ at time t_k is equivalent to the measurement of the algebra $U_{-tk}\mathcal{B}^{(k)}U_{tk}$ at time $t = 0$. Our information at time 0 consists therefore of the expectations $w_{\mathcal{S}}$ of the set

$$\mathcal{S} = \bigcup_{k=1}^{n} \left(U_{-tk}\, \mathcal{B}^{(k)} U_{tk} \right) \tag{32}$$

We note that the union $\mathcal{S}$ of subalgebras is not itself a subalgebra of $\mathcal{A}$. We must therefore again proceed by minimization of $v(T)$.

As an illustration, we have computed a simple concrete example. (Even in this simple example, the calculations proved to be rather lengthy.)

Example 8.1: Spin Measurement at Two Times. Consider a spin ½ with Hamiltonian $H = \sigma_3$. As in Example 5.1, let v be the Gibbs state

$$V = \tfrac{1}{2}(\sigma_0 + a\sigma_3), \qquad a = -\tanh^{-1}\beta$$

We measure σ_1 at times 0 and t. According to (32), the set $\mathcal{S}$ is the union of the algebras

$$\mathcal{B}_0 = \{\sigma_1\}'', \qquad \mathcal{B}_t = \{U_{-t}\sigma_1 U_t\}''$$

Let the result of the measurement be

$$w_{\mathcal{S}}(\sigma_1) = b, \qquad w_{\mathcal{S}}(U_{-t}\sigma_1 U_t) = b_t \tag{33}$$

The problem now consists in maximizing

$$v(T) = \tau[(V^{1/2}WV^{1/2})^{1/2}]$$

under the constraints (33). The result is as follows:

$$W_0 = \frac{1}{2}\left\{\sigma_0 + b\sigma_1 + \frac{b_t - b_0 \cos 2t}{\sin 2t}\sigma_2 + \frac{a}{\sin 2t}[(\sin 2t)^2 - b_0^2 + 2b_0 b_t \cos 2t - b_t^2]^{1/2}\sigma_3\right\} \tag{34}$$

Here W_0 is the inferred state at $t = 0$. It is clear that W_0 depends on the time t at which the second measurement has been performed.

As a test of (34), we check some special cases.

(*a*) $b_0 = b_t = 0$. From (34) follows $W_0 = \tfrac{1}{2}(\sigma_0 + a\sigma_3) = V$. This is expected, since v is compatible with the constraints.

(b) $b_0 = 1$. This is a complete measurement at $t = 0$. Hence $\mathcal{B}_0$ alone is sufficient to determine the pure state $W = ½(\sigma_0 + \sigma_1)$ uniquely. The outcome of the measurement of $\mathcal{B}_t$ *must* therefore be $w(U_t\sigma_1U_{-t}) = b_t = \cos 2t$. If this is inserted in (34), we have, as expected, $W_0 = ½(\sigma_0 + \sigma_1) = W$.

9. Inference and Entropy

We have characterized the inferred state w_0 as a state in $N(w_{\mathcal{S}})$ which maximizes the functional

$$S_v(w) = v(T) \tag{35}$$

If v is proportional to the trace τ (for normalizable v this is possible only if $\mathcal{H}$ is finite-dimensional), then (6) implies that $T \sim W^{1/2}$ and hence

$$S_v(w) \sim \tau(T) = \tau(W^{1/2}) \tag{36}$$

The expression (36) is a so-called α *entropy*, with $\alpha = ½$.[14] It is therefore natural to consider the generalization (35) of (36) as a candidate for relative entropy.

We have shown in Ref. 2 that $S_v(w)$ has the following mathematical properties:

(a) Concavity [i.e., $S_v(\lambda_1 w_1 + \lambda_2 w_2) \geq \lambda_1 S_v(w_1) + \lambda_2 S_v(w_2)$, $0 \leq \lambda_i \leq 1$, $\lambda_1 + \lambda_2 = 1$].

(b) Monotonicity under Restriction [i.e., $S_{v|\mathcal{B}}(w|_{\mathcal{B}}) \geq S_v(w)$].

(c) Multiplicativity [i.e., $S_{v1\otimes v2}(w_1 \otimes w_2) = S_{v1}(w_1)\cdot S_{v2}(w_2)$].

We note that, instead of (c), additivity could simply be achieved by defining the entropy as $\log v(T)$ rather than $v(T)$. It has, however, been shown[14] that $S_v(w)$ lacks in general the property of subadditivity (for a weaker statement cf. Ref. 2).

Various relative entropies have been proposed over the past years,[15] and it is tempting to evaluate them in their information-theoretic context. In Ref. 6 we have computed the maximum entropy states for 9 relative entropies (in the simple spin model of Example 5.1) and found that they all differ. Our conclusion was that the definition (35) leads to the simplest and most satisfactory results. But it must be emphasized that our point of view is strictly limited to the problem of inference. The entropy $S_v(w)$ defined in (35) is therefore to be considered as a *statistical likelihood function for states*, and *not as a thermodynamical concept*.[16]

References and Notes

1. J.-P. Marchand, *Found. Phys.* **7**, 35 (1977).
2. S. Gudder, J.-P. Marchand, and W. Wyss, *J. Math. Phys.* **20**, 1963 (1979).
3. J.-P. Marchand and W. Wyss, *J. Stat. Phys.* **16**, 349 (1977).
4. R. Benoist and J.-P. Marchand, *Lett. Math. Phys.* **3**, 93 (1979).

5. R. Benoist, J.-P. Marchand, and W. Yourgrau, *Found. Phys.* **7**, 827 (1977).
6. R. Benoist, J.-P. Marchand, and W. Wyss, *Lett. Math. Phys.* **3**, 169 (1979).
7. S. Sakai, *Bull. AMS* **71**, 149 (1965).
8. D. Bures, *Trans. AMS* **135**, 199 (1969).
9. A. Uhlmann, *Rep. Math. Phys.* **9**, 273, (1976).
10. H. Araki, *RIMS, Kyoto Univ.* **8**, 335 (1972).
11. S. Gudder, *Intl. Journ. Theoret. Phys.* **20** 383 (1981).
12. R. Benoist, Thesis, Univ. of Denver (1978).
13. S. Gudder, *Rep. Math. Phys.* (to appear).
14. W. Ochs, *Rep. Math. Phys.* **9**, 331 (1976).
15. A. Wehrl, *Rev. Mod. Phys.* **50**, 221 (1978).
16. W. Yourgrau and A. van der Merwe, in *Cosmology, Fusion and other Matters*, E. Reines, editor (Colo. Univ. Press, Boulder, Colorado, 1972).

19

THE KNOTS OF QUANTUM THERMODYNAMICS

JAMES L. PARK AND RALPH F. SIMMONS, JR.
Department of Physics
Washington State University
Pullman, Washington 99164

> *. . . The law that entropy always increases—the second law of thermodynamics—holds, I think, the supreme position among the laws of Nature. If someone points out to you that your pet theory of the universe is in disagreement with Maxwell's equations—then so much the worse for Maxwell's equations. If it is found to be contradicted by observation—well, these experimentalists do bungle things sometimes. But if your theory is found to be against the second law of thermodynamics I can give you no hope; there is nothing for it but to collapse in deepest humiliation. . . .*
>
> —SIR ARTHUR EDDINGTON[1]

Several years ago, in the course of intense discussions with one of us concerning research problems of mutual interest, our colleague Wolfgang Yourgrau repeatedly employed the colorful term "knots"—perhaps an allusion to the story of Gordius of Phrygia—to denote challenging fundamental dilemmas in the conceptual fabric of physical theory. The galaxy of foundations problems he investigated during his remarkably productive life included, among many other interests, topics in quantal, statistical, and thermal physics. Issues both old and new which arise in efforts to unify these sciences surely qualify as excellent specimens of Professor Yourgrau's knots in natural philosophy. Hence, to honor his memory, we offer an essay on the knots of quantum thermodynamics. In order to focus mainly upon what Yourgrau used to call "tough science" as opposed to historiographic or metalinguistic analysis, we shall not dwell on the epistemological aspects but proceed immediately to the mathematical foundations of our subject.

1. SOME MATHEMATICAL NOTIONS AND NOTATIONS

To establish a mathematical framework in which to expose conflicts between quantum mechanics and thermodynamics, we begin by stating three postulates of quantum theory. The statements are essentially those given by Margenau and Park.[2]

P.1. Systems. With every physical system there is associated a complex, separable, complete inner product space, a Hilbert space $\mathcal{H}$. The Hilbert space associated with a composite system of two distinguishable systems A,B which are associated with Hilbert spaces $\mathcal{H}_A$, $\mathcal{H}_B$, respectively, is the direct product space $\mathcal{H}_A \otimes \mathcal{H}_B$.

P.2. State Preparations. To every reproducible preparation scheme **Π** for a physical system there corresponds a nonnegative definite, self-adjoint, unit trace operator ρ on $\mathcal{H}$ which contains all physical predictions regarding data gathered immediately subsequent to that preparation (cf. P.3). The trace class operator ρ, called the density operator or statistical operator, thus represents the state of the quantum system prepared in the manner **Π**.

P.3. Observables, Measurements, Data, and Ensembles. Some continuous, linear, self-adjoint operators $\{A, B, \ldots\}$ on $\mathcal{H}$ correspond to observables of the system. The arithmetic mean value A of the data yielded by measurements of observable A on an ensemble of systems prepared in the manner **Π** with associated density operator ρ is given by the continuous linear functional of A

$$\langle A\rangle = \mathrm{Tr}(\rho A) \tag{1}$$

All the density operators for the system, referred to in P.2., are elements of $T(\mathcal{H})$, the real Banach space of all self-adjoint trace class operators on $\mathcal{H}$. Within this space $T(\mathcal{H})$ there is a set of nonnegative definite ("positive") operators on $\mathcal{H}$. This set is called the positive cone $V^+(\mathcal{H})$ in $T(\mathcal{H})$. The positive cone $V^+(\mathcal{H})$ is not a subspace of $T(\mathcal{H})$ since a positive operator multiplied by a negative real scalar does not yield a positive operator.

Contained in $V^+(\mathcal{H})$ is a convex set $V_1^+(\mathcal{H})$ of positive operators with unit trace. The elements of this set are the density operators corresponding to all possible preparations of the system.

The extreme elements of $V_1^+(\mathcal{H})$ are called the pure states of the system. The pure state density operators are elementary projectors onto one-dimensional subspaces of $\mathcal{H}$. Thus the pure states correspond to vectors (rays) in $\mathcal{H}$, e.g., an eigenvector or any linear superposition of eigenvectors of observable A corresponds to a pure state. The methods used to determine the density operator associated with a preparation **Π** will be considered below.

The standard quantum mechanical observables are represented by elements of $B(\mathcal{H})$, the space of all continuous, linear, self-adjoint operators on $\mathcal{H}$. The phraseology of P.3 recognizes not only the fact that not all elements of $B(\mathcal{H})$ correspond to observables (an aspect of superselection rules) but also the converse proposition that there are observable properties of some systems which do not correspond to operators in $B(\mathcal{H})$. Included in the latter category are functions of two or more noncommuting observables[3] and, of particular interest in the present context, such thermodynamic properties as temperature and entropy. The space $B(\mathcal{H})$ is a real Banach space with the operator norm

$$\|A\|_B = \sup_{\|x\|=1} \|Ax\|, \qquad A \in B(\mathcal{H}) \tag{2}$$

The third postulate also emphasizes the conventional interpretation of $\langle A \rangle$ as the arithmetic mean of the results of measurements of A on an ensemble of identically prepared systems. When the link is made between quantum mechanics and certain features of thermodynamics, we shall find it necessary sometimes to reinterpret the linear functional $\mathrm{Tr}(\rho A)$ so that its referent shifts from an ensemble of quantal systems to a single thermodynamic system.

For a complete physical theory, the three postulates already stated must be supplemented by a causal law. In conventional quantum theory there is the standard dynamical postulate which asserts that the time evolution of a system is given by a one-parameter unitary group $\{\Lambda_t\}$ of transformations of $T(\mathcal{H})$ into $T(\mathcal{H})$. However, as we shall soon see, this elegant rule is inadequate to describe all phenomena contemplated when quantum mechanics and thermodynamics are unified.

In the canonical version of quantum statistical thermodynamics, thermal equilibrium states are represented by density operators having the familiar form

$$\hat{\rho} = \frac{\exp(-\beta H)}{\mathrm{Tr}\,\exp(-\beta H)} \tag{3}$$

where H is the energy operator. For any quantum state ρ, the statistical analog for internal energy U is given by

$$U = \mathrm{Tr}(\rho H) \tag{4}$$

and the entropy is expressed as

$$S(\rho) = -k\mathrm{Tr}(\rho \ln \rho) \tag{5}$$

When (5) is maximized subject to constraint (4), the unique result is the canonical state (3), with inverse temperature β uniquely determined by (4).

The quantal version (5) of the famous Gibbsian entropy formula was derived originally by von Neumann[4] using the admittedly obscure notion of semipermeable membranes for different pure quantum states. However, the form of $S(\rho)$ given above can be obtained in other ways[5,6]; and we shall assume, as is customary with students of this subject, that (5) is correct. The recent exhaustive study by Hatsopoulos and Gyftopoulos[7–10] of the foundations of thermodynamics and its relationship to quantum mechanics contains an especially cogent alternative derivation of (5).

The mathematical form of the quantum mechanical entropy gives rise to important consequences for the entropy of composite systems. Consider two interacting systems A and B with associated Hilbert spaces $\mathcal{H}_A$ and $\mathcal{H}_B$. The state of the composite system of two distinguishable systems is assumed to be an operator defined on $\mathcal{H}_A \otimes \mathcal{H}_B$, i.e., $\rho_{AB} \in V_1^+(\mathcal{H}_A \otimes \mathcal{H}_B)$. The entropy calculated from this state is

$$S_{AB} = -k\mathrm{Tr}(\rho_{AB} \ln \rho_{AB}) \tag{6}$$

The reduced density operator for system A is customarily defined by

$$\rho_A = \mathrm{Tr}_B(\rho_{AB}) \tag{7}$$

where Tr_B denotes the trace over $\mathcal{H}_B$. Therefore $\rho_A \in V_1^+(\mathcal{H}_A)$. The entropy calculated from ρ_A is

$$S_A = -k\mathrm{Tr}(\rho_A \ln \rho_A) \tag{8}$$

For system B there are of course expressions for ρ_B and S_B similar to (7) and (8).

The interesting inequality

$$S_{AB} \leq S_A + S_B \tag{9}$$

is referred to as the subadditivity of entropy.[11] The equality holds if and only if

$$\rho_{AB} = \rho_A \otimes \rho_B \tag{10}$$

When this is the case, systems A and B are uncorrelated.

Since the mean value of the observable A for a system described by ρ is given by the linear functional

$$\mathrm{A} = \mathrm{Tr}(\rho \mathrm{A}) \tag{11}$$

where $\rho \in V_1^+(\mathcal{H})$ and $\mathrm{A} \in B(\mathcal{H})$, it is tempting, and indeed some authors[12,13] do this, to consider entropy to be the mean value of the operator $\eta = -k \ln \rho$. However, there are difficulties encountered in making this identification. The operator η is not an element of $B(\mathcal{H})$ for all $\rho \in V_1^+(\mathcal{H})$. To see this, note via (2) that the norm of the operator η is

$$\|\eta\|_B = \sup_{\|x\|=1} \|\eta x\| = \sup_{\|x\|=1} \|-k \ln \rho x\| \tag{12}$$

Consider the case where ρ is pure, i.e., ρ is an extreme element of the convex set $V_1^+(\mathcal{H})$. Then there exists an eigenvector $x_0 \in \mathcal{H}$, $\|x_0\| = 1$ with eigenvalue $\alpha = 0$. The norm of the vector ηx_0 is unbounded:

$$\|\eta x_0\| = \langle \eta x_0, \eta x_0 \rangle^{1/2} = [k^2(\ln \alpha)^2]^{1/2} \to \infty \qquad \text{as } \alpha \to 0 \tag{13}$$

Nevertheless the entropy is well defined for the state $\rho_0 = |x_0\rangle\langle x_0|$ in the sense that

$$\lim_{\alpha \to 0} (-k\alpha \ln \alpha) \to 0 \tag{14}$$

Hence η, being unbounded, is not an element of $B(\mathcal{H})$ and does not qualify under P.3 as a quantal observable. Therefore entropy cannot rigorously be interpreted as the average value of a quantal observable.

A second peculiarity of the entropy functional involves its time evolution. In conventional quantum mechanics the time evolution is governed by a one-parameter unitary group of transformations on $T(\mathcal{H})$, a typical group element of which will be denoted by U_t. Through the correspondence $\sigma \to \mathrm{Tr}(\sigma \mathrm{A})$ we identify $B(\mathcal{H})$ with the dual of $T(\mathcal{H})$. Through this duality the transformation U_t induces a dual transformation U_t^* on $B(\mathcal{H})$. The state $\rho_t \in T(\mathcal{H})$ is related to the state $\rho_{t+s} \in T(\mathcal{H})$ by

$$\rho_{t+s} = U_s \rho_t \tag{15}$$

The average value of an observable $A \in B(\mathcal{H})$ at time $t + s$ is given by

$$\langle A\rangle_{t+s} = \mathrm{Tr}(\rho_{t+s}A) = \mathrm{Tr}(U_s\rho_t A) \tag{16}$$

This gives the time evolution of the average value in the so-called Schrödinger picture; i.e., the states change in time and the operators, unless explicit functions of time, remain fixed. The duality between $T(\mathcal{H})$ and $B(\mathcal{H})$ provides the time evolution of the average value in the Heisenberg picture; i.e.,

$$\langle A\rangle_{t+s} = \mathrm{Tr}(U_s\rho_t A) = \mathrm{Tr}(\rho_t U_s^* A) \tag{17}$$

All of this is standard and well known.

The problem with the entropy functional is that in the Schrödinger picture the state ρ depends on time but in the Heisenberg picture ρ is independent of time. Therefore the rate of change of the entropy is identically zero in the Heisenberg picture. When the time evolution is given by a unitary group, as in conventional quantum mechanics, no inconsistency arises since the entropy functional is invariant under unitary transformation, i.e.,

$$S(\rho_{t+s}) = S(\rho_t) \qquad \text{when } \rho_{t+s} = U_s\rho_t \tag{18}$$

So $dS(\rho)/dt \equiv 0$ in the Schrödinger picture also.

The reason for this invariance is that the value of $S(\rho)$ depends on the eigenvalues of ρ, which are invariant under unitary transformation. Thus we arrive at the long known but insufficiently appreciated result that *the unitary time evolution of conventional quantum mechanics is incapable of predicting a change in entropy*. This proposition leads immediately to one of the most disconcerting knots in physics.

2. The Second Law Knot

The temporal invariance of $S(\rho)$ under unitary evolution places mechanics and thermodynamics in a logical confrontation rather like the celebrated incompatibility of Galilean–Newtonian mechanics and Maxwellian electrodynamics from which emerged the theory of relativity. In particular, if the Second Law of Thermodynamics is to be retained as a basic principle of physics, then either quantum mechanics or statistical thermodynamics or both must be modified.

The Second Law mathematically characterizes thermal equilibrium in a closed system as the state which maximizes $S(\rho)$ subject to subsidiary constraints. It is generally understood also to entail that spontaneous processes of evolution toward thermal equilibrium do in fact occur, and that for closed systems the entropy never decreases.

Now, if the Second Law stands in conflict with quantum mechanics, might we reasonably consider abandonment of that law as a fundamental axiom of physics? We think not; indeed we shall proceed on the premise that the Second Law, like the great conservation laws, deserves a rather sacrosanct position in the hierarchy of physical principles. Our introductory quotation from the writings of A. S. Eddington forcefully conveys this viewpoint, but of course we cannot claim that Sir Arthur would have espoused any of the unorthodoxies advocated in the present essay.

Unitary quantum dynamics cannot be saved by denying the existence of motions in which closed systems gain entropy unless we are willing to deny the most elementary triumphs of the Second Law. Consider for example a composite system whose constituents A and B are initially uncorrelated, noninteracting, and separately in thermal equilibrium with distinct inverse temperatures, β_A, β_B. The total initial density operator is then

$$\rho_0 = \hat{\rho}_A \otimes \hat{\rho}_B \tag{19}$$

where $\hat{\rho}_A$, $\hat{\rho}_B$ are each of the canonical form (3). If internal constraints are lifted, interaction occurs, and eventually the composite system evolves to a unique state of mutual equilibrium, viz.,

$$\rho_\infty = \hat{\rho}_{AB} \tag{20}$$

where $\hat{\rho}_{AB}$ has the canonical form (3) with one final inverse temperature β_{AB}. This familiar model, which might refer to an experience as commonplace as equilibration following submergence of a piece of warm lead in a cup of cool water, is of course consonant with the Second Law since $S(\rho_\infty)$ is the maximum value the entropy functional can attain under the given circumstances. However, in general we also find in this situation that $S(\rho_\infty) > S(\rho_0)$, so that the motion $\rho_0 \to \rho_\infty$ cannot be understood within the framework of unitary mappings provided by conventional quantum theory. In other words, no conceivable Hamiltonian can generate the motion which in fact occurs as the lead cools and the water warms.

Throughout the past century the literature of physics has offered a variety of rationalizations for this paradox, so that some physicists when exposed to the problem viscerally reject it as heresy. Many others adhere to one or more of the standard explanations briefly described and criticized below; the literature citations given in each case exemplify a school of thought but are not exhaustive.

(a) The Second Law and the entropy concept are subjective or "anthropomorphic," the entropy increase representing essentially the growing obsolescence of past knowledge rather than an objective dynamical process.[14–17]

Despite our genuine appreciation of the methodology of information theory, we believe nevertheless that the thermodynamic entropy of a closed system rises independently of the informational state of the observer.

(b) Thermodynamically closed systems are mechanically open in some sense and are therefore mechanical subsystems for which $S(\rho)$ is not invariant.[18–22]

This alleged openness of thermodynamically isolated systems has been associated with various causes, including quantal uncertainties at the boundary, random external fields, and even cosmological asymmetries. In our opinion, all of these ideas share the same flaw, viz., that in any bounded mechanical system obeying a unitary law of motion the total entropy is invariant and, since the overall motion is quasiperiodic,[23] the entropy of a subsystem calculated using (7) and (8) will also be quasiperiodic and hence exhibit no Second-Law unidirectionality.

Consider for example a system A with surroundings B, prepared initially in an uncorrelated pure state

$$\rho_{AB}(0) = |\psi_A\rangle\langle\psi_A| \otimes |\psi_B\rangle\langle\psi_B| \tag{21}$$

When (21) is substituted into (6), (7), and (8), we find that

$$S_{AB}(0) = S_A(0) = S_B(0) = 0 \tag{22}$$

However, as the composite system A plus B evolves unitarily, S_{AB} remains invariant; but both S_A and S_B may increase as interactions breed correlations between A and B. This peculiar result is of course consistent with the subadditivity property (9).

Superficially, mechanical subsystem A seems to be undergoing a process of the kind contemplated by the Second Law. Unfortunately this is merely an illusion, for the motion of A plus B is ultimately periodic, so that eventually S_A and S_B decrease again to zero. Thus the Second-Law behavior of A is only temporary, and highly dependent upon the choice of an initial condition like (22) which captures S_A during an ascent phase.

(c) Irreversible phenomena should be treated by invoking special limit procedures or redefinitions of S, but without altering conventional unitary quantum dynamics.[24–27]

Included in this broad category are theories based on such notions as coarse-graining or the thermodynamic limit and the abstract theories of ergodicity and mixing, incisive analyses of which could easily fill a book. However, we do wish to record in particular our skepticism toward demonstrations that entropy may increase indefinitely but only in infinite-volume, infinite-population assemblies. Even if rigorously correct, such propositions can hardly be germane to the physical problem, since realistic systems in which entropy is observed to increase are in fact finite.

(d) Even though the Second Law is not a universal theorem of mechanics, nevertheless we tend to observe only entropy-increasing situations because by our very nature we encounter or create only systems with highly improbable initial conditions.[28,29]

A particularly dramatic statement of this view was given by Landé[30]:

> In ordinary life and in thermodynamics we deal with finite systems which deliberately are brought into a state of high entropy by branching them off from the rest of the universe. Here we have indeed, at least for a limited time span, a parallelism of time and entropy—which does not prove any such parallelism in general, ... We happen to live in a very improbable state of the world which develops into a more probable state. But if ours were not a very improbable world, there would not be any "we" to speculate about it.

This approach incorporates features of both (a) and (b), as well as a philosophical perspective which, though intriguing, is scientifically infertile, since it confers upon the Second Law knot a rather dogmatic mantle of inevitability. We prefer to believe that the knot will eventually be untied by the development of a more comprehensive dynamics in which entropy-increasing processes occur as natural motions. The question of man's relation to the universe would play no role in such a theory.

(e) *The entropy of thermodynamic systems rises due to repeated measurements, each followed by a discontinuous state change in accordance with the projection postulate of quantum mechanics.*[20,28]

This idea, a variant of (b) in which the randomizing environment is imagined in some sense to perform measurements, is based upon a popular but discredited quantum theory of measurement in which wave packet reduction is a key concept.

It was first shown by von Neumann[4] that the entropy of a system cannot decrease under the projection postulate transformation, and thereafter various authors have attempted to explain irreversibility using the projection postulatc. While it is true that the entropy can increase and the energy can be conserved under the projection postulate transformation, it is not true that the transformation drives the system to the canonical state.

According to the projection postulate, an energy measurement results in the transformation of the density operator,

$$\rho \rightarrow \rho' = \sum_n P_n \rho P_n \tag{23}$$

where P_n is the projection operator onto the eigenspace corresponding to energy eigenvalue E_n. If the energy spectrum is nondegenerate, the elementary projectors are

$$P_n = |n\rangle\langle n| \tag{24}$$

where $H|n\rangle = E_n|n\rangle$.

The energy is conserved under this transformation if

$$U \equiv \mathrm{Tr}(\rho H) = \mathrm{Tr}\left(\sum_n P_n \rho P_n H\right) \tag{25}$$

To show that (25) holds, we note that $[P_n, H] = 0$ so that

$$\mathrm{Tr}\left(\sum_n P_n \rho H P_n\right) = \mathrm{Tr}\left(\sum_n P_n^2 \rho H\right) = \mathrm{Tr}(\rho H) = U \tag{26}$$

The matrix for ρ' is diagonal in the energy representation and its diagonal elements are the same as those of ρ, i.e.,

$$\langle m|\rho'|l\rangle = \langle m\left|\left(\sum_n \left(\left|n\rangle\langle n\right|\rho\left|n\rangle\langle n\right|\right)\right|l\rangle = \langle m|\rho|m\rangle \delta_{ml} \tag{27}$$

It is possible to change the diagonal elements of ρ' if another observable A is measured which does not commute with H, but then energy would not be conserved in general.

Let $\overline{P}_n$ be the projector onto the subspace associated with the eigenvalue a_n of A, $[A,H] \neq 0$,

$$\overline{P}_n = |a_n\rangle\langle a_n| \tag{28}$$

where $A|a_n\rangle = a_n|a_n\rangle$. The projection postulate transformation is

$$\rho' \rightarrow \rho'' = \sum_n \overline{P}_n\rho'\overline{P}_n \tag{29}$$

The energy in the state ρ'' is

$$U'' = \mathrm{Tr}(\rho''H) = \mathrm{Tr}\left(\sum_n \overline{P}_n\rho'\overline{P}_n H\right) \tag{30}$$

In general $U'' \neq U$.

In the approach to equilibrium of a closed thermodynamic system, the energy is conserved as the entropy increases. The projection postulate can lead to a stationary state with the energy conserved, but unless the diagonal elements of the initial state happen to be already those of the final maximum-entropy canonical state, the projected state will not be canonical. Therefore this theory cannot explain the general validity of the Second Law of thermodynamics.

Having rejected all of the foregoing attempts to reconcile the contradiction between the Second Law and the invariance of $S(\rho)$ under unitary evolution, we now explore an alternative interpretation of that dilemma. Rather than clinging to the tradition of Hamiltonian mechanics from which the unitary dynamical postulate of orthodox quantum theory evolved, let us suppose the latter is simply incomplete. It must therefore be replaced with a new principle of motion encompassing not only the familiar unitary evolutions of conventional quantum theory but also the nonunitary evolutions required to describe thermodynamical processes.

This possibility has been suggested by several authors,[31–34] but the most persuasive arguments are to be found in the work of Hatsopoulos and Gyftopoulos,[7–10] who argue quite conclusively that some as yet unknown, nonunitary principle of quantal motion must be devised in order to achieve a satisfactory unification of thermodynamics and quantum mechanics into one uncontradictory science.

Motivated either by these thermodynamic considerations or by the desire to obtain phenomenological descriptions of the evolution of subsystems, various authors[18,19,35–38] have considered mathematical generalizations of quantum dynamics in which the laws of motion do not conserve S. These efforts have generally featured nonunitary mappings which are elements of linear semigroups, thus assuring the applicability of a substantial body of standard mathematical structures. This approach has yielded several interesting contributions to the quantum dynamics of subsystems, where entropy may rise or fall but energy need not be conserved. Moreover, Band and Park[32–34] found that for 2-level quantum systems there exist linear maps—independent of the initial density operator—which describe energy-conserving, entropy-increasing processes. The latter would be an essential feature of a unified theory of quantum thermodynamics.

Unfortunately, the present authors[39] have discovered that for N-level systems with $N > 2$, there are no linear maps describing energy-conserving, entropy-increasing motion. Consequently the search for a new principle of motion can expect no assistance from the mathematical theory of linear semigroups.

Although the correct equation of motion for ρ is not known at this point, we can give a list of necessary conditions that such an equation must satisfy. Consider a closed system with energy operator H and density operator ρ. Let $\rho(s)$ be related to $\rho(s + t)$ by a nonlinear mapping Γ_t,

$$\rho(s + t) = \Gamma_t\rho(s) \tag{31}$$

It is desirable to assume Γ_t is an element of a one-parameter semigroup G of nonlinear transformations. The usual semigroup property

$$\Gamma_s \circ \Gamma_t = \Gamma_{s+t} \in G \tag{32}$$

is required if it is assumed that time is homogeneous. Then any time s can be taken as the initial time $s = 0$, so that (31) becomes

$$\rho(t) = \Gamma_t\rho(0) \tag{33}$$

Being nonlinear, this transformation cannot be written in matrix form, but it is still mathematically possible[40] to obtain an equation of motion for $\rho(t)$ in terms of a nonlinear infinitesimal generator F of the semigroup:

$$\frac{d\rho(t)}{dt} = F\rho(t) \tag{34}$$

The conditions to be satisfied by the generalized evolution are that it (i) maintain positivity, (ii) maintain self-adjointness, (iii) preserve the trace,

$$\frac{d}{dt}[\mathrm{Tr}(\rho)] = \mathrm{Tr}\left(\frac{d\rho}{dt}\right) = \mathrm{Tr}(F\rho) = 0 \tag{35}$$

(iv) conserve the energy,

$$\frac{dU}{dt} = \mathrm{Tr}\left(\frac{d\rho}{dt}H\right) = \mathrm{Tr}(F\rho H) = 0 \tag{36}$$

and (v) increase the entropy,

$$\frac{dS}{dt} = -k\mathrm{Tr}\left(\frac{d\rho}{dt}\ln\rho\right) = -k\mathrm{Tr}(F\rho\ln\rho) \geq 0 \tag{37}$$

with equality if and only if $\rho = \hat{\rho}$, the canonical state.

These five conditions are not independent; (ii), (iii), and (v) together imply (i). To see that this is plausible, note that the entropy functional (5) may be expressed in terms of the eigenvalues $\{r_i\}$ of ρ as

$$S(\rho) = -k \sum_i r_i \ln r_i \tag{38}$$

a nonnegative definite form for all density operators, i.e., for each $\rho \in V_1^+(\mathcal{H})$. Moreover, for a self-adjoint, unit trace $\rho \notin V_1^+(\mathcal{H})$, at least one r_i would be negative and (38) would then be complex. Thus if an equation of motion satisfies (ii), (iii), and (v) for all t and the initial state $\rho(t_0)$ is an element of $V_1^+(\mathcal{H})$, then $\rho(t) \in V_1^+(\mathcal{H})$ for all t; i.e., condition (i) is automatically fulfilled. Of course, there are positive evolutions for which (37) is not satisfied so (i), (ii), and (iii) together do not imply (v).

The equation of motion

$$\frac{d\rho}{dt} = -i[H,\rho] + \frac{1}{\lambda}(\hat{\rho} - \rho) \tag{39}$$

with $\lambda > 0$ provides an example satisfying all conditions. In (39) $\hat{\rho}$, the canonical state, is determined uniquely by

$$\mathrm{Tr}(\hat{\rho}H) = \mathrm{Tr}(\rho H) \tag{40}$$

It is easy to show that (39) satisfies conditions (ii) and (iii) and (40) can be used to show that (iv) is satisfied. That (39) satisfies (v) is a bit more difficult to show, but once this is established we know from the preceding discussion that (i) is satisfied.

Substituting (39) into (37) gives

$$\begin{aligned} -k\mathrm{Tr}\left(\frac{d\rho}{dt}\ln\rho\right) &= -k\mathrm{Tr}(-i[H,\rho]\ln\rho) \\ &\quad + \frac{-k}{\lambda}\mathrm{Tr}[(\hat{\rho} - \rho)\ln\rho] \geq 0 \end{aligned} \tag{41}$$

Since the trace is invariant under cyclic permutations and since $[\ln\rho,\rho] = 0$, the first term on the right-hand side of the equality vanishes. Since $k, \lambda > 0$, the second term will be greater than zero if

$$[-\mathrm{Tr}(\rho\ln\rho)] - [-\mathrm{Tr}(\hat{\rho}\ln\rho)] \leq 0 \tag{42}$$

To show that (42) holds, we recall that for all ρ satisfying (40),

$$-\mathrm{Tr}(\hat{\rho}\ln\hat{\rho}) \geq -\mathrm{Tr}(\rho\ln\rho) \tag{43}$$

with equality if and only if $\rho = \hat{\rho}$. It follows that

$$[-\mathrm{Tr}(\rho\ln\rho)] - [-\mathrm{Tr}(\hat{\rho}\ln\rho)] \leq [-\mathrm{Tr}(\hat{\rho}\ln\hat{\rho})] - [-\mathrm{Tr}(\hat{\rho}\ln\rho)] \tag{44}$$

Using Klein's inequality in the form

$$[-\mathrm{Tr}(\sigma\ln\sigma)] - [-\mathrm{Tr}(\sigma\ln\sigma')] \leq 0 \tag{45}$$

for $\sigma, \sigma' \in V_1^+(\mathcal{H})$, with $\sigma = \hat{\rho}$ and $\sigma' = \rho$ we obtain from (44)

$$[-\mathrm{Tr}(\rho \ln \rho)] - [-\mathrm{Tr}(\hat{\rho} \ln \rho)] \leq 0 \tag{46}$$

which is precisely (42) and therefore (v) is satisfied by (39).

Although there is no physical justification for the specific equation of motion (39), it does have some attractive properties. We can express (39) in terms of the infinitesimal generator F of a semigroup of nonlinear transformations,

$$\frac{d\rho}{dt} = -i[H,\rho] + \frac{1}{\lambda}[\hat{\rho}(\rho) - \rho] = F\rho \tag{47}$$

The equivalence relation in $V_1^+(\mathcal{H})$ which partitions $V_1^+(\mathcal{H})$ into equivalence classes $[\rho U]$ of states ρ with the same energy U induces a partitioning of the semigroup G. The generator F depends on the state of the system through (40). Since F thus depends on the energy of the system there is a unique generator F for each equivalence class $[\rho U]$.

An equation of the form (47) would be applicable not only to closed systems but would also be useful for describing at least approximately an open thermodynamic system exchanging energy with some other system, in which case F would have an implicit dependence on time through $U(t)$. If the energy of the system changes slowly with respect to the characteristic relaxation time λ of the system, then the system would be very nearly in a canonical state at all times, fulfilling the classic definition of a reversible quasistatic process. The dependence of F on U is therefore desirable since it allows one to account, within the framework provided by the equation of motion, for the rapid or slow removal of constraints on the system.

All of this detail regarding possible generalized evolutions is of course conjectural; the correct principle of motion with sufficient structure to complete the theory of quantum thermodynamics and thereby to untie the Second Law knot remains to be discovered.

3. Knots in the State Concept

For a quantum mechanical system the state corresponds to a density operator $\rho \in V_1^+(\mathcal{H})$. For every reproducible preparation Π of the system there is a density operator ρ. The state of a system at any time t can be obtained by measuring the average values at time t of a quorum[41,42] of observables for the system. However, it would not be necessary to measure the average values of a quorum if one knew the state, ρ_{AB}, of the composite system consisting of system A plus preparation device B, since the state ρ_A could then be obtained from the composite state ρ_{AB} using (7).

To apply this in practice would require information about the interactions between the preparation device and the system and the states $\rho_A(0)$, $\rho_B(0)$ when A and B are initially brought together. If A and B are initially uncorrelated and the evolution of the composite system is given by the mapping Γ_t, then the state of A at time t is

$$\rho_A(t) = \mathrm{Tr}_B[\Gamma_t \rho_A(0) \otimes \rho_B(0)] \tag{48}$$

The concept of measuring a quorum of observables was introduced to avoid the above assumption and therefore circumvent the problem of knowing the entire history of the system of interest. However, the definition of the state of A as a partial trace is useful if one agrees never to ask questions concerning the *joint* observables defined on $\mathcal{H}_A \otimes \mathcal{H}_B$. The average values of joint observables depend on correlations between A and B and this information is lost in the operation of taking the partial trace.

Conventional quantum mechanics requires only that a preparation Π be reproducible to have associated with it a state ρ_A of the prepared system. For quantum thermodynamics, however, it was discovered recently that an additional restriction must be placed on the preparation. Hatsopoulos and Gyftopoulos[7–10] were the first authors to point out the need for a distinction between two types of reproducible preparations, referred to as ambiguous and unambiguous. The original definitions of these new concepts are expressed in terms of time correlations of average values. Instead of quoting those definitions, we may readily illustrate this newest quantal knot by contrasting two reproducible preparations which are both acceptable in ordinary quantum mechanics but only one of which is acceptable in quantum thermodynamics.

We consider in particular preparations $\{\Pi_i | i = 1,2, \ldots\}$ each of which when repeated generates an ensemble characterized by canonical density operator

$$\hat{\rho}_i = \frac{\exp(-\beta_i H)}{\mathrm{Tr}[\exp(-\beta_i H)]} \tag{49}$$

where β_i is determined by the mean energy U_i through

$$U_i = \mathrm{Tr}(\hat{\rho}_i H) \tag{50}$$

Since $\hat{\rho}_i$ is the thermodynamic equilibrium state, the corresponding preparation scheme Π_i may essentially entail waiting until the closed system with internal energy U_i spontaneously attains stable thermal equilibrium. An alternate procedure would be to couple the system to a thermostat with temperature $T_i = 1/k\beta_i$ until mutual equilibrium is achieved, then isolate the system.

Consider next a preparation Π_0 which combines Π_1 and Π_2 by some operational rule. For example, if the flip of a coin results in heads, a system is generated by Π_1 and if tails results, a system is prepared in the manner Π_2. We choose Π_1 and Π_2 such that the mean energy associated with Π_0 is

$$U_0 = \tfrac{1}{2}U_1 + \tfrac{1}{2}U_2 \tag{51}$$

but $\hat{\rho}_0 \neq \hat{\rho}_2$ so that $\beta_1 \neq \beta_2$.

Let the preparation Π_3 be such that

$$U_3 \equiv \mathrm{Tr}(\hat{\rho}_3 H) = U_0 \tag{52}$$

In general $\beta_3 \neq \beta_1$, $\beta_3 \neq \beta_2$.

The preparation Π_0 is ambiguous where Π_3 is unambiguous. If some thermodynamic property, e.g., temperature or adiabatic availability, is measured on systems prepared by Π_0 one will find two results, sometimes T_1, sometimes T_2 for the temperature. Any system prepared by Π_3 will, upon measurement, yield the temperature T_3. On the other hand, if the quantum mechanical energy observable H is measured on systems prepared by Π_0 and by Π_3 no distinction is possible since the result of each measurement is an eigenvalue of H and the arithmetic means of these results are the same for Π_0 and Π_3.

This distinction between ambiguous and unambiguous preparations is an attempt to illustrate the difference between quantum observables and thermodynamic properties. There are measurable properties of systems, such as temperature and entropy, which are not the mean values of some quantum mechanical operators. This difference is present even in Gibbs' original approach where the temperature is related to the modulus of the distribution and is not the average value of a function of phase.

The temperature of a single thermodynamic system is, however, related to the average value of a quantum mechanical operator H. In conventional quantum mechanics the average value of H refers to the arithmetic mean of the results of measurements of H on an ensemble of identically prepared systems. In quantum statistical thermodynamics, $\mathrm{Tr}(\rho H)$ is interpreted as the internal energy of a single system prepared in the manner Π with the associated density operator ρ. An ensemble is necessary to determine the ρ associated with preparation Π, but once ρ is known it applies to *each* system prepared by Π as far as *thermodynamic* properties are concerned. The density operator ρ will predict with certainty the entropy of a single system prepared in the manner Π, but it cannot predict with certainty the result of a measurement of the quantum mechanical energy of a single system prepared in the manner Π. It is this difference in the predictive power of ρ which necessitates the distinction between ambiguous and unambiguous preparations. Systems prepared by an unambiguous preparation all share the same values for thermodynamic quantities; thus, such an ensemble is thermodynamically "homogeneous" even though its density operator may have the form traditionally associated with inhomogeneous quantal ensembles (mixtures).

Requiring that preparations be unambiguous has important implications concerning the interpretation of the canonical state. Consider an ensemble of identical N-level systems prepared by the ambiguous preparation Π_m. We take Π_m such that each system in the ensemble is prepared in an energy eigenstate associated with energy eigenvalue E_i and such that the relative frequency of occurrence p_i of systems in state E_i is

$$p_i = \frac{\exp(-\beta E_i)}{\mathrm{Tr}[\exp(-\beta H)]} \tag{53}$$

The density operator representing the preparation Π_m is the incoherent mixture

$$\check{\rho} = \frac{\sum_{i=1}^{N} \exp(-\beta E_i)|E_i\rangle\langle E_i|}{\sum_{i=1}^{N} \exp(-\beta E_i)} = \frac{\exp(-\beta H)}{\mathrm{Tr}[\exp(-\beta H)]} \tag{54}$$

Now consider a second ensemble prepared by an unambiguous preparation Π_c of the kind described under (50) which prepares identical N-level systems in the canonical equilibrium state

$$\hat{\rho} = \frac{\exp(-\beta H)}{\mathrm{Tr}[\exp(-\beta H)]} \tag{55}$$

By construction, any system from the ensemble prepared by Π_m has come from an identifiable pure subensemble and hence the entropy is zero for each system. On the other hand, any system taken from the ensemble prepared by Π_c is in thermal equilibrium with nonzero entropy.

The two density operators $\check{\rho}$ and $\hat{\rho}$ have the same form, but the interpretation of each is fundamentally different. Requiring all preparations in quantum thermodynamics to be unambiguous eliminates preparations of the type Π_m. The state of a system prepared in thermodynamic equilibrium is not some unknown energy eigenstate but is simply the entire mixture $\hat{\rho}$, the canonical state. By the state of a system we mean all the information concerning the average values of all observable properties of the system at the present time and all the information necessary to predict the average values in the future. Since $\check{\rho}$ applies only to the ensemble it cannot be used to predict the thermodynamic behavior of one system taken from the ensemble.

Another important consequence resulting from the elimination of preparations Π_m from the theory of quantum thermodynamics is that the popular use of transition probabilities to describe the approach to equilibrium is fallacious. A transition probability is a concept taken from perturbation theory and gives the total probability that a system in one energy eigenstate of the unperturbed Hamiltonian will make a transition to another energy eigenstate. The transition is induced or caused by the perturbation. If the system is not in an energy eigenstate, as is the case when the state is $\hat{\rho}$, then there are no transition probabilities defined.

An example of the use of transition probabilities in describing the approach to equilibrium is found in the theory of paramagnetic relaxation. It is imagined that there is some number N_+ of the spin systems in the eigenstate spin up (magnetic moment parallel to the applied field) and some number $N - N_+ = N_-$ in the eigenstate spin down. The interaction of the spins with the lattice is treated as a perturbation. An equation for the rate of change of N_+ is then written as

$$\frac{dN_+}{dt} = W(- \to +)N_- - W(+ \to -)N_+ \tag{56}$$

where $W(- \to +)$ is the probability of a transition from spin down to spin up, $W(+ \to -)$ is the probability of a transition from spin up to spin down. If V_{SL} is the perturbation Hamiltonian describing the spin-lattice interaction, then to first order,

$$W(- \to +) = W(+ \to -) \propto |\langle \mathrm{up}|V_{SL}|\mathrm{down}\rangle|^2 \tag{57}$$

The rate equation above predicts the equilibrium numbers,

$$N_+ = N_- = N/2 \tag{58}$$

If the sample is at some finite temperature T, then $N_+ \neq N/2$ and the rate equation fails in its prediction. We attribute this failure to the fact that each spin system in the sample is not in some energy eigenstate but is in a mixed state.

If N_+ is interpreted as N times the average value of the projection operator onto the eigenspace of spin up, then the rate equation applies to the entire ensemble as if it were a single system, i.e., as if a single system were prepared by some unambiguous preparation. In this case the transition probabilities are not defined since the system is not in some unknown energy eigenstate.

We therefore conclude that the use of rate equations to describe relaxation to equilibrium cannot provide a fundamental understanding of the process. The fallacious character of the approach becomes clear when we realize it is necessary to reinterpret $W(- \to +)$ and $W(+ \to -)$ in order to predict the correct equilibrium state. It is customarily *assumed* that these probabilities are related by a Boltzmann factor,

$$\frac{W(- \to +)}{W(+ \to -)} = \exp(2\beta\epsilon) \tag{59}$$

where $\beta = 1/kT$ and ϵ is given by $H|\text{up}\rangle = -\epsilon|\text{up}\rangle$. This assumption is of course without justification and is in fact not compatible with the condition

$$W(+ \to -) = W(- \to +) \propto |\langle \text{up}|V_{SL}|\text{down}\rangle|^2 \tag{60}$$

4. Knots Associated with Zeroth and First Laws

The traditional aim of quantum statistical thermodynamics has been to account for the behavior of thermodynamic systems in terms of the laws of quantum mechanics; yet even the Zeroth and First Laws of thermodynamics when interpreted mechanically give rise to troublesome knots.

To demonstrate this we will briefly discuss an application of thermodynamics, i.e., a question to which thermodynamics is capable of yielding an answer, and then show the approximations necessary in mechanics to yield the same answer.

One of the fundamental problems of thermodynamics is to predict from the initial conditions the final conditions of two interacting systems. For example, if a piece of warm lead is placed in a cup of cool water, we may inquire as to the final (equilibrium) temperatures of the lead and the water. To solve the problem we rely on the Zeroth and First Laws of thermodynamics. The Zeroth Law states that two systems in equilibrium have equal temperatures, and furthermore that a third system at the same temperature is in equilibrium with the first two systems. Therefore in our example the final state is attained when the temperature of the lead is the same as the temperature of the water. We could also argue that the final equality of temperatures is a consequence of the Second Law since the total entropy of the two systems is maximum when the temperatures are equal.

The value of this final, common temperature can be predicted using the First Law, which states that the total energy of the composite system lead plus water is constant. The energy lost by the lead is gained by the water. If the energy of each as a function of temperature is known, it is then easy to calculate the final, common temperature. In order to verify that the final temperatures are indeed equal, one must separate the two systems, i.e., terminate the interaction. It is routinely assumed in thermodynamics that separating the systems affects neither the energies nor the temperatures of the two systems. Moreover, as commonly applied to composite systems, the First Law is a statement concerning the sum of the private energies and as such does not include the interaction energy. As is the case with the energy, in equilibrium the entropy is also regarded as an extensive thermodynamic quantity; thus the total entropy of the composite system is the sum of the private entropies of the constituent subsystems.

Now let us analyze the same type of problem—two systems interacting—with quantum statistical thermodynamics. Take two systems A and B with associated Hilbert spaces $\mathcal{H}_A$ and $\mathcal{H}_B$, respectively. The energy operator H and density operator ρ_{AB} for the composite system are defined on the tensor product space $\mathcal{H}_A \otimes \mathcal{H}_B$. The initial state is assumed to have the uncorrelated form

$$\rho_{AB}(0) = \rho_A(0) \otimes \rho_B(0) \tag{61}$$

We can, as discussed earlier, define the state of, say, system A at any time t as the reduced density operator ρ_A obtained by taking the trace over $\mathcal{H}_B$ of ρ_{AB}. In some cases H can be written as a sum of private energies plus an interaction term,

$$\begin{aligned} H &= H_A \otimes I_B + I_A \otimes H_B + V_{AB} \\ &= H_A + H_B + V_{AB} \end{aligned} \tag{62}$$

(The second equality gives the standard abbreviated form where identities and $\otimes$ symbols are suppressed.) Under these circumstances, we may define the private energy of A as

$$U_A(t) = \mathrm{Tr}[\rho_{AB}(H_A \otimes I_B)] = \mathrm{Tr}_A(\rho_A H_A) \tag{63}$$

Quantum statistical thermodynamics predicts that the equilibrium state $\hat{\rho}_{AB}$ for the composite system is the canonical state

$$\hat{\rho}_{AB} = \frac{\exp[-\beta(H_A + H_B + V_{AB})]}{\mathrm{Tr}\{\exp[-\beta(H_A + H_B + V_{AB})]\}} \tag{64}$$

with $\mathrm{Tr}(\hat{\rho}_{AB}H) = \mathrm{Tr}\{(\rho_A(0) \otimes \rho_B(0)H\}$.

The density operator $\hat{\rho}_{AB}$ cannot in general be factored into the tensor product of two canonical states. Further, we cannot at this point logically claim that the two systems A and B are at the same temperature because there is only a *single* (composite) system; but if the interaction energy is neglected compared to the private energies of the two systems A and B then the canonical operator can be factored into the tensor product of two canonical density operators with the same temperature parameter:

$$\hat{\rho}_{AB} = \frac{\exp(-\beta H_A)}{\text{Tr}[\exp(-\beta H_A)]} \otimes \frac{\exp(-\beta H_B)}{\text{Tr}[\exp(-\beta H_B)]} \equiv \hat{\rho}_A \otimes \hat{\rho}_B \tag{65}$$

Two independent systems can now be identified and the temperatures are indeed equal. Nevertheless, the statement of the equality of temperatures in equilibrium is only approximately true inasmuch as two temperatures can be defined and compared only if the interaction is neglected.

The First Law of Thermodynamics, viewed as conservation of the sum of private energies, is also an approximation. In thermodynamics, the entropy is maximized subject to the constraint that $U_A + U_B =$ const. In quantum statistical mechanics the constraint is that the total energy is constant, i.e., $U_A + U_B + \langle V_{AB} \rangle \equiv$ const. If $\langle V_{AB} \rangle \ll U_A + U_B$, then we recover the First Law of thermodynamics from quantum statistical thermodynamics.

That the entropy in equilibrium is additive over subsystems is also seen to be an approximation. The entropy of the composite system is calculated using (64); but from the subadditivity of entropy (9) we have

$$S_{AB} \leq S_A + S_B \tag{66}$$

with equality if and only if (65) were strictly true. Therefore the extensive property of entropy is an approximation. Evidently some familiar thermodynamical procedures are best interpreted as approximations; the art of thermodynamics is to know how to partition a system into subsystems so that these approximations are valid and the laws of thermodynamics apply.

As an example, consider two very thin sheets A and B. If we assume the interaction energy is proportional to the surface area in contact between two systems, then the thermodynamics of the composite system where the area of contact is the edges of the sheets will be different from the case where the contact is the faces of the sheets. In the first case $\langle V_{AB} \rangle \ll U_A + U_B$, but in the latter case $\langle V_{AB} \rangle$ need not be negligible compared to U_A and U_B. Simply stating that the sheets are brought into contact is not sufficient to determine if the normal methods of thermodynamics apply to the two sheets separately or only to the composite system viewed as one sheet.

Finally, we discuss an interesting knot connected with the concept of interaction as it is customarily interpreted in statistical thermodynamics. As we have seen, the interaction part of H needs to be present and large enough to provide a channel for the transfer of an appreciable amount of energy between the systems but at the same time must be, in some sense, small compared to the private energies of the interacting systems. It is not always clear what this criterion means. There are systems for which the H is not simply a sum of private energy operators onto which a small interaction term is added. If this is the case, then the interaction may not possess the properties described above. To illustrate this, consider the interaction of an N-particle gas G in a box compressed by a weight W in a gravitational field. The energy operator for W has the standard form

$$H_W = \frac{P^2}{2m} + mgZ \tag{67}$$

and the energy operator for G

$$H_G = H_G' + \sum_{i=1}^{N} V_{i\text{box}} \tag{68}$$

In this expression, H_G' contains all kinetic energies of the particles of the gas plus all interparticle potential energies V_{ij}, i.e.,

$$H_G' = \sum_{i=1}^{N} \frac{p_i^2}{2m_i} + \sum_{i,j=1}^{N} V_{ij} \tag{69}$$

The term $V_{i\text{box}}$ is the potential energy of the ith particle due to the walls of the box:

$$V_{i\text{box}}(x_i,y_i,z_i,Z) = V_0[1 - \Theta(x_0 - x_i)\Theta(x_i)\Theta(y_0 - y_i)\Theta(y_i)\Theta(Z - z_i)\Theta(z_i)] \tag{70}$$

where Θ is the Heaviside unit step function.

This potential energy contains both the position operators for the gas particle and the position operator of the weight, and is therefore an operator defined in the tensor product space $\mathcal{H}_G \otimes \mathcal{H}_W$, where $\mathcal{H}_G$, $\mathcal{H}_W$ are the Hilbert spaces associated with G, W respectively.

The total energy operator for the system is

$$H = H_G' \otimes I_W + I_G \otimes H_W + V_{GW} \tag{71}$$

where

$$V_{GW} = \sum_{i=1}^{N} V_{i\text{box}}(x_i,y_i,z_i,Z) \tag{72}$$

Clearly we cannot regard V_{GW} as being "small" in any sense. Indeed to neglect V_{GW} is to disregard the walls of the box, thereby making the system of gas plus weight fundamentally different. The point of this illustration is that we cannot give a general rule concerning the ignorability of the interaction between two systems. Whether V_{AB} can be neglected so that A and B may be regarded as separable in applying the laws of thermodynamics must be decided case by case.

References and Notes

1. A. S. Eddington, *The Nature of the Physical World* (Cambridge U.P., Cambridge, 1928), p. 74.
2. H. Margenau and J. L. Park, *Found. Phys.* **3**, 19 (1973).

3. J. L. Park and H. Margenau, Chap. 5 in *Perspectives in Quantum Theory*, W. Yourgrau and A. van der Merwe, editors (MIT Press, Cambridge, Massachusetts, 1971).
4. J. von Neumann, *Mathematical Foundations of Quantum Mechanics* (Princeton U.P., Princeton, New Jersey, 1955).
5. R. Baierlein, *Atoms and Information Theory* (W. H. Freeman, San Francisco, 1971).
6. B. K. Skagerstam, *Stat. Phys.* **12,** 449 (1975).
7. G. N. Hatsopoulos and E. P. Gyftopoulos, *Found. Phys.* **6,** 15 (1976).
8. G. N. Hatsopoulos and E. P. Gyftopoulos, *Found. Phys.* **6,** 127 (1976).
9. G. N. Hatsopoulos and E. P. Gyftopoulos, *Found. Phys.* **6,** 439 (1976).
10. G. N. Hatsopoulos and E. P. Gyftopoulos, *Found. Phys.* **6,** 561 (1976).
11. M. B. Ruskai, *Ann. Inst. Henri Poincaré* **19,** 357 (1973).
12. K. Gottfried, *Quantum Mechanics Volume I: Fundamentals* (W. A. Benjamin, Reading, Massachusetts, 1966), p. 233.
13. R. Lenk, *Brownian Motion and Spin Relaxation* (Elsevier, New York, 1977), p. 28.
14. E. T. Jaynes, *Phys. Rev.* **106,** 620 (1957).
15. E. T. Jaynes, *Phys. Rev.* **108,** 171 (1957).
16. E. T. Jaynes, *Am. J. Phys.* **33,** 391 (1965).
17. A. Katz, *Principles of Statistical Mechanics* (W. H. Freeman, San Francisco, 1967).
18. V. Gorini, A. Kossakowski, and E. C. G. Sudarshan, *J. Math. Phys.* **17,** 821 (1976).
19. G. Lindblad, *Commun. Math. Phys.* **48,** 119 (1976).
20. R. S. Ingarden and A. Kossakowski, *Ann. Phys. (N.Y.)* **89,** 451 (1975).
21. J. M. Blatt, *Prog. Theor. Phys.* **22,** 745 (1959).
22. B. Gal-Or, editor, *Modern Developments in Thermodynamics* (Wiley, New York, 1974).
23. E. C. Percival, *J. Math. Phys.* **2,** 235 (1961).
24. R. Jancel, *Foundations of Classical and Quantum Statistical Mechanics* (Pergamon, New York, 1969).
25. D. Ruelle, *Statistical Mechanics Rigorous Results* (W. A. Benjamin, New York, 1969).
26. I. Prigogine, C. George, F. Henin, and L. Rosenfeld, *Chem. Scripta* **1973,** 4.
27. I. E. Farquhar, *Ergodic Theory in Statistical Mechanics* (Interscience, London, 1964).
28. A. Landé, *New Foundations of Quantum Mechanics* (University Press, Cambridge, Massachusetts, 1965).
29. R. Peierls, *Surprises in Theoretical Physics* (Princeton U.P., Princeton, New Jersey, 1979), p. 73.
30. A. Landé, *New Foundations of Quantum Mechanics* (University Press, Cambridge, Massachusetts, 1965), p. 75.
31. J. Mehra and E. C. G. Sudarshan, *Nuovo Cimento* **11,** 215 (1972).
32. J. L. Park and W. Band, *Found. Phys.* **7,** 813 (1977).
33. W. Band and J. L. Park, *Found. Phys.* **8,** 45 (1978).
34. J. L. Park and W. Band, *Found. Phys.* **8,** 239 (1978).
35. A. Kossakowski, *Bull. Acad. Polon. Sci. Math.* **20,** 1021 (1972).
36. A. Kossakowski, *Bull. Acad. Polon. Sci. Math.* **21,** 649 (1973).
37. A. Kossakowski, *Rep. Math. Phys.* **3,** 247 (1972).
38. E. B. Davies, *Commun. Math. Phys.* **39,** 91 (1974).
39. R. F. Simmons and J. L. Park, *Found. Phys.* **11,** 297 (1981).
40. K. Yosida, *Functional Analysis*, 4th ed. (Springer-Verlag, New York, 1974), p. 246.
41. J. L. Park and W. Band, *Found. Phys.* **1,** 211 (1971).
42. W. Band and J. L. Park, *Found. Phys.* **1,** 339 (1971).

20

About the Facts of Cosmological Significance

Jean-Claude Pecker
Collège de France
Paris, France

From the beginning of this century, it may be said that cosmology has completely modified its approach of the universe; we face a new science, which rests upon a complete recognition of the importance of astronomical observations as well as upon the necessary mastering of the universal space-time geometry and the universal laws of physics.

The difficulty that appears is twofold: On one side, one is not entirely sure what the "universal" physics is. Are some laws, which are well established on the laboratory scale, valid on the scale of the Galaxy? And, if so, do they still hold for the entire Universe? A very classical example, although not quoted often in this context, is the second law of thermodynamics. Expressed in rather schematic terms, it describes and limits the laws ruling the energetic conversions. But, in practice, not all types of energies are considered on the laboratory or even on the astronomical scale. The forces of inertia—if they are due to large-scale distribution of masses in the Universe, as claimed years ago by Ernst Mach[1] [and as used, for a while, at least, by Albert Einstein,[2] who unsuccessfully invoked it to justify the equivalence principle]—those Mach forces have, so far as I know, never been taken into account in any universal thermodynamical theory. Another example can be found in the forces associated with as yet unknown particles, which may interact in many ways with the classical particles of physics; let us mention for example the W-bosons, necessary to some physical theories, but never observed, or the axions—scalar bosons—implied by considerations of a purely physical nature, both possibly of little importance at the laboratory scale, but possibly of a great importance at the cosmological scale (see, for example, Mikaelian[3]).

Hence, we can see that the theoretical approach to cosmology still lacks the needed confidence in the exhaustive character of its formulation and indeed in a fundamental way. How could we consider physics as terminated?

The situation on the side of the observational cosmology is still less satisfactory. At the beginning of the century, such a small number of facts were considered to be of a cosmological significance that one hardly took them into account. Now, they keep piling up on the desk of cosmologists. Here, some well established relation, assumed to be cosmological; there, a given measure of a given number, in principle well defined, but often badly measured; elsewhere, some qualitative facts, whose significance may be large. Hence, the cosmologists, who basically have to solve some kind of universe equations, have to satisfy a very large number of boundary conditions. It is not, in principle, an impossibility. The equations of relativity, in the twenties and still now, have to be oversimplified in order to be solved, hence their solution needs only a few boundary conditions. Adding new ones, however, makes undesirable such a large number of approximations, and, therefore, consideration of the new data may be somewhat troublesome. But it certainly does not keep one from building coherent cosmologies—at least in principle.

In practice things are quite different. A certain inertia or, to put it more gently, a certain need to still simplify the problems in order to obtain solutions that can be expressed in a sufficiently simple way, is leading cosmologists to a mental walk which can be abstracted as follows: One selects first a few facts, which one is sure then to have some cosmological signification; as for the others, either one tries to introduce them later on, as a proof of the quality of the constructed cosmology, or, if this is impossible, one rules them out, as being of no significance from the cosmological point of view.

Hence, different schools are blooming, as they select different groups of facts as being of cosmological significance, as they use different values of the "constants" that are needed to perform the integrations, and as they have to describe a given set of observations, necessarily somewhat dispersed, with a given analytical law.

Our purpose in this paper is certainly not to make an exhaustive study of the observational cosmology, which has been the basis of our lectures at the College de France in the academic years 1980–1982 (Pecker, 1981, 1982).[4] What we want to achieve here is only to give a few examples of diverging opinions on what degree of significance any given observation may have, or not have; and we shall try to show the cosmological consequences of some arbitrary choices. We shall not conclude of course, without asking a few questions.

The Olbers Paradox

There is no need to go into details of Olbers' expression of his paradox.[5] It is an observed fact that the sky appears as black, or at least dark, at night; this fact can be shown to contradict the description of the universe as an homogeneous distribution of light-sources (stars, galaxies) in a Euclidian universe. It is quite clear, indeed, that if the latter were the case, every line of sight would meet, at a large or small distance, the radiating surface of a star, of which the brightness per unit of solid angle, is independent of distance. Hence, every line of sight would meet some solarlike object, and the whole sky would have a brightness similar to that of the Sun.

The Olbers paradox has started an enormous flow of publications.[5,6,16–18,52] At first, people understood its deep cosmological signification. Either one had to abandon the Euclidian geometry, or one had to abandon the idea of a homogeneously filled universe, or else one had to introduce physical mechanisms lowering the intensity of distant sources in one way or another.

Charlier[16] and Fournier d'Albe,[17] amongst the pioneers at the beginning of the century, as well as de Vaucouleurs[18] and Mandelbrot,[52] quite recently, have given the Olbers paradox a nonhomogeneous interpretation by postulating (Charlier and Fournier d'Albe) or showing (de Vaucouleurs) that indeed the distribution of structures in the Universe is hierarchized, in such a way that the density ρ of a given volume, around the observer, is smaller the bigger its size l, but also that the rate of decrease of $\rho(l)$ with l is sufficiently high so that an integration performed from observer to infinity allows one to show that the observed brightness at a given point, selected at random, is actually negligible (Fig. 1).

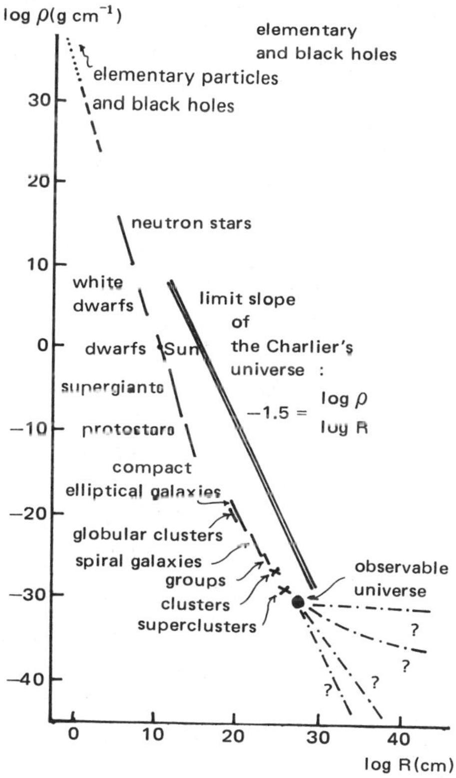

FIGURE 1. The hierarchized universe (ordinates and abscissae: logarithmic scales, cgs units).

However, how is it possible to introduce an hierarchized density in the models of the Universe? Mandelbrot, by his generalized cosmological principle,* implied more or less an average density equal to zero, a solution compatible with de Sitter's empty universe. But truly enough, the problem is observational. Is the hierarchy likely to stop, after a certain type of structure, say the superclusters of galaxies, and will the average density tend to a constant value? And what constant value then? Zero? Or what? Or will the hierarchical behavior continue for ever, unveiling, at each new penetration of observation into the deepest depths of the Universe, some new structure of the hierarchy? And how is this linked with the finite or infinite description of the universe? If the density is very small, then might the universe as well be open, and its radius infinite?

Obviously, true or not, the idea of an hierarchical universe is a disturbing one, difficult to introduce in the cosmological equations.

Naturally, as soon as Slipher's effect[19]—let us give back to Slipher the discovery of the universality of the red shifts—was known, and as soon as Hubble's[20,21] relation between distance and red shift was widely accepted, people ceased to worry about the Olbers principle, and explained it on the sole basis that very distant objects are reddened in such a large way as to become unobservable in the visible region of the spectrum. Intergalactic obscuration by dust could only add to this phenomenon.

People immediately forgot the cosmological importance of Olbers's paradox, hidden behind that of Hubble's relation; and cosmologists did not try to introduce the very concept of hierarchization in their theories. De Vaucouleurs's and Mandelbrot's papers did not alter their calm serenity either.

But is it all that simple?

There is something more to Olbers' paradox: its gravitational aspect. As a matter of fact, the gravitational force acting on any local mass results from the integration, over the total universe, of the $1/r^2$ forces due to all masses. Of course, there is a basic difference with the classical form of the Olbers paradox related to radiation. In a given point, radiation from all the universe is adding up, while, by contrast, gravitational forces may be compensating each other. However, it is easy to show that only a small amount of fluctuation about some average density is sufficient to introduce local fluctuations that are stronger than the effect of nearby masses. Hence, for example, the planets should be affected by such local fluctuations in a large way; and we know, since Kepler and Newton, and after Le Verrier and Lowell, that this is not the case. Avoiding this gravitational Olbers paradox cannot be done by resorting to the red shift (or to obscuring matter). We may even state that it could be considered a fundamental law of the Universe that "there is no antigravitation screen," despite H. G. Wells and his "cavorite"!

* *Conditional cosmographic principle*: "The condition distribution of mass is the same for all conditioned frames of reference," it being understood that the "condition" in question is such that the origin of the considered frames of reference are all *material* points. It concerns all *real* observers, and does not have the degree of unrealistic generality of the "strong (or universal) cosmological principle."

To the contrary, the hierarchy solves the gravitational problem, as it does the statement that the universe is finite and its curvature nonnegligible.

One can see that Olbers' paradox, whichever way we can turn it around, continues to have an important cosmological significance; but whether it speaks for Hubble's law, for the limitation of the universe and its finitude, or for the completely hierarchized universe, is not so clear.

Hubble's Law: The Constants of the Apparent Expansion

The Hubble relation has quite a different status. It was indeed the first group of coherent observations which very soon led to a choice amongst the physically based cosmologies. Before one knew the Hubble's relation, the purely physical considerations (such as that of stability) were the only ones to be taken into account. When the Hubble relation appeared, the cosmologists tried to fit it, assuming the red shift of spectral lines was indeed measuring the receding velocity of distant galaxies.

Hubble himself, much closer to observation, and somewhat skeptical about what one could extract from his own discoveries, and Tolman, a mathematical cosmologist, indeed had serious doubts.[21] Not only did they speak, most of the time, of "apparent receding velocities," but they suggested additional tests for the physical existence of the expansion.

But the cosmologists generally ignored this warning. In other words, they tried to find in their equations those solutions which could be described as "expanding universes."

On the one hand, admission that Hubble's relation concerns the true receding velocities implied that the Hubble ratio $v/D = H_0$ (the index zero designating the present and nearby universe) has to be considered as a basic integration constant; the same being true for the deceleration parameter q_0. On the other hand, if $R(t)$ schematizes a given solution to the (simplified) universe equations, H_0 and q_0 can be computed for any given behavior of $R(t)$:

$$H_0 = \dot{R}/R \quad \text{and} \quad q_0 = -\ddot{R}\, R/\dot{R}^2 \qquad \text{for } t = t_0$$

One difficulty is that the quantity $\tau - 1/H_0$ is only an estimation (even an overestimation, in most of the possible expanding universes) of the age of the universe. The actual value of H_0 had thus undoubtedly (when one believes in expansion) a cosmological significance. But it has to be at least as large as the older ages determined in the Galaxy, from the study of the HR diagram of globular clusters; for a long time the two figures have been racing against each other. The value of $1/H_0$ has been regularly decreasing from 530 km s^{-1} Mpc^{-1} (Hubble's first value), to 55 km s^{-1} Mpc^{-1} (Sandage and Tammann[22]), the age of the universe increasing at the same time from "at least 2 Gy" to "at least 18 Gy." It happens that the ages of the older globular clusters have ranged from 10 Gy to 20 Gy. There is not any contradiction remaining now; but in the early fifties the last figures were incompatible with the expanding universe interpretation of the Slipher red shifts.

A recent study[23] of the distance estimators has shown the biases of the usual determination of H_0 and that a value of $H_0 = 100$ km s^{-1} Mpc^{-1} (or about 8–10 Gy for the age of the universe) is much more likely than the 18 Gy accepted in 1977. Many authors (de Vaucouleurs,[18] Heidmann,[24] etc.) had, for several years, kept the same attitude. But in that case, are not the older clusters possibly even older than the overestimated age of 8–10 Gy for the universe? And then, is it reasonable to consider the expanding universe as proven beyond a doubt? We certainly do not think so. (See Fig. 2.)

Another type of difficulty, probably still more basic, is linked with the very definition of H_0. According to Segal[25] and Nicholl,[26] the statistical study of the velocity-distance relation shows it to be more likely a parabolic law than a linear one, as generally admitted. In that case, the whole construction is falling apart. Segal himself has proposed a very elaborate theory, the "chronogeometry," which accounts for the parabolic law; and the fit with observations is really remarkable. But Segal did not consider other facts of presumably cosmological significance. We cannot enter into many more details in the chronogeometry; but Segal's theories are very appealing and should be considered very seriously.

The Cosmological Background Radiation

In the triumphant years of the expanding universe, around 1950, when no one had the slightest doubt about the expansion of the cosmos or of the regular isotropy

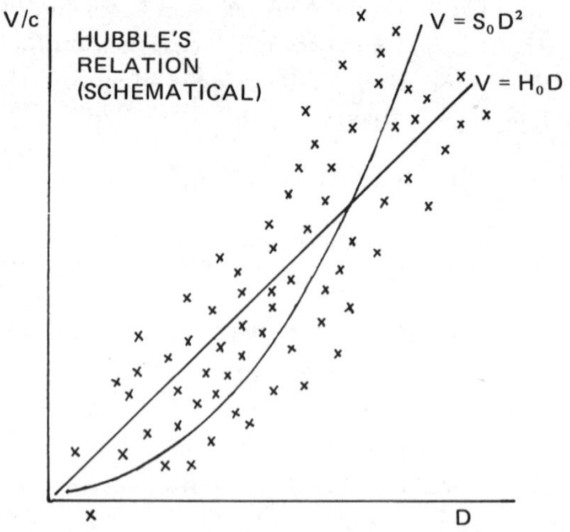

FIGURE 2. Schematizations of the velocity–distance relation (scales of abscissa and ordinates are intentionally not specified).

and uniformity of its basic nature, Gamow[27,28] was one of the most ardent propagandists of the new vistas on the expanding universe.

Quite naturally, and along the lines of, for example, Lemaitre's model of the universe,[53] the backwards extrapolation of present universe conditions leads to the "primordial" explosion of a hyperdense universe.[28] The development of Gamow's idea led him to propose that we should at present observe a remnant of this primeval physical event. The conditions near the Big Bang (this designation is, I believe, one of the most important contributions of Gamow to this question!) are those of a very hot, very dense medium, much richer in radiation than in elementary particles. In particular, Gamow established the essential relation $R(t)T(t) = R_0T_0$. To be fair, we should indeed add to his name those of Alpher and Herman[29,30]; at the same time as Gamow, or apparently earlier (in 1948) they had the same kind of ideas. Later on, they collaborated with Gamow on a few joint papers; they parted company subsequently for some unknown (to me) reasons. In this series of publications, using the above relation, these authors managed to get the value of T_0, knowing at a time near the Big Bang, some coherent values of R and T and the present value of R, i.e., R_0. I shall not enter into the details, referring the readers instead to this fascinating series of papers, in which the predictions of T_0—the present value of the temperature of the Universe, that we should observe, instead of the Olbers stellar brightness, undiluted in all directions of the universe—have indeed been ranging over a large interval up to 60 K. It would be of course very interesting, from the point of view of the history of cosmology, to make a detailed analysis of this series of papers, which were at the time highly publicized by Gamow himself, using all kind of media. These temperatures should naturally be observed in the millimetric and submillimetric spectral regions.

Years later, after unsuccessful attempts to detect this background radiation in the millimetric wavelengths, the Princeton team[31] (of which the leaders were R. H. Dicke and P. J. E. Peebles) found, really by sheer luck, that two engineers from the Bell Telephone Company did indeed discover the signal, not knowing its potential cosmological importance. These investigators published two notes, one by Dicke, Peebles, Roll, and Wilkinson,[31] the other by Penzias and Wilson.[32] In 1964, more than 15 years after the first hint about a "cosmological background remnant radiation," they had found a beautiful confirmation of the Big Bang speculation. A large amount of publicity was then given by the scientific press to this discovery, which was deemed of such paramount importance as to earn the Nobel Prize for Penzias and Wilson (although some might think that Dicke and Peebles, too, deserved it—as well as Gamow in his time).

The discovery of the 3K background radiation was of course confirmed by subsequent measurements. It gave to the Big Bang cosmology a bright confirmation and resolved beyond doubt the various hesitations that some still had vis-à-vis Lemaitre's[53] mathematical description, physically elaborated by Gamow.

The prediction, followed by the discovery, gave the 3K radiation the status of a cosmologically significant fact. But is it truly convincing?

One could note first that, in 1954, Findlay-Freundlich[33] also made the prediction of a background radiation of 1.9 to 6 K (more accurate than that of Gamow!) on the sole basis of a theory which was very different from the Big Bang. Findlay-Freundlich admitted that the red shifts were not due to any expansion (thus not measuring a true velocity) but to some aging process of the photons (a "tired light" mechanism) which were losing energy in space by collisions with other particles, other photons in the case of Findlay-Freundlich theory. The very existence of the Slipher effect and the Hubble relation implied for Findlay-Freundlich a homogeneous (or approximately so) filling of the universe with photons, the energy of which could be computed. Unfortunately, the paper by Findlay-Freundlich, published together with a note of appraisal by Max Born,[34] was counteracted immediately on the basis of astronomical and physical arguments by ter Haar,[14] and the Burbidges,[35] and not much was heard any more about it.

True enough, the Findlay-Freundlich theory was far from perfect. In particular, it created a confusion between the total number of the photons (proportional to T_0^3) and their total energy (proportional to T_0^4). The quality of the prediction appears thus *a posteriori* as a result of luck, errors compensating each other.

But, on good or bad basis, it predicted the existence of a 3K radiation. And the same can after all be said about the Alpher, Gamow, Herman computations.

At this stage, we can only conclude that, cosmologically significant or not, the 3K observations did not speak for either the Big Bang or the steady-state universe of Findlay-Freundlich. Sure enough, it was a cosmological fact, but in a sort of loose way, as it cannot be useful for testing any cosmology, since some cosmologies, very different each from the other, could indeed predict it just as well.

Later work by Vigier, Roberts, and the present author[36,37] pointed out the errors made by Findlay-Freundlich, but tried to correct them. Our first attempts to consider photon–photon collisions as a red-shifting process have been criticized in a sufficiently convincing way to lead us to consider that, after all, the photon energy might well be carried away by collisions with particles other than photons; and as a possibility (not unique but not absurd either) we later suggested the possibility of photon–boson collisions, for which the theory had been developed by Maric *et al.*,[38] in order to interpret the "abnormal" red shifts discovered in California (by Zwicky[39] and by Arp[40]), or in France (by Nottale and Karoji and Vigier[41–45], and Jaakola[8,9]). The assumed process, however, can be developed to interpret the normal Slipher shift, and the authors have shown how the 3K radiation could be produced by locally produced photons (Pecker, 1977).[46] They insisted on the fact that fluctuations of density in the universe have as a natural consequence the existence of fluctuations of the local Hubble ratio $h(r)$ [of which the actual Hubble ratio is only an integrated value $H_0 = \int_0^r h(r)dr$] and as well of fluctuations in $T(\mathbf{r})$, *both types of fluctuations being correlated. (See Fig. 3.)*

But, there exist in the picture many other cosmologies. The very elaborate G-varying cosmology of Hoyle and Narlikar[47] or that of Canuto and his associates,[48] which both found in the lunar occultation observations of van Flendern[49] some kind of observational justification, were modern versions of the classical Jordan–Dirac cosmologies and have also predicted the 3K background radiation.

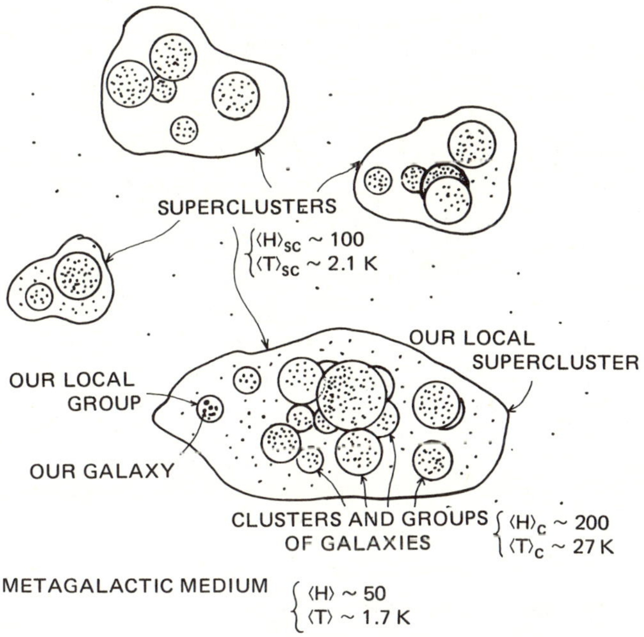

FIGURE 3. The "local" signification of Hubble's ratio and universal background radiation (after Pecker and Vigier, Ref. 37).

We are not sure either about other cosmologies advanced here and there; some authors have not been testing their ideas along the proper lines, being content instead to reproduce the Hubble relation.

Another set of observations led to some new thinking and weakened the significance of the 3K radiation as coherent with the simple-minded classical Big Bang universe: It happens that the 3K radiation is neither that of a blackbody[11] nor is it isotropic[12,50], and its nonisotropic character differs from that of the red shift; if one describes it as due to the motion of the Sun towards an apex, it is impossible to find an agreement between the solar apexes with respect to the 3K radiation and with respect to the galaxies, distant or close. (See Fig. 4.)

We should thus get interested in the 3K radiation, study it, and account for it—but we should, by all means, remove it from the pedestal where an undue confidence in the classical Big Bang has put it.

The Abundance of Light Elements

It is quite difficult to produce in the stars enough helium (^3He or ^4He) and enough deuterium (^2H) to account for their observed abundances, as measured in the vicinity of the Sun or even in a very few neighboring galaxies.

This deeply puzzled many theoretical experts on nuclear reactions, until it

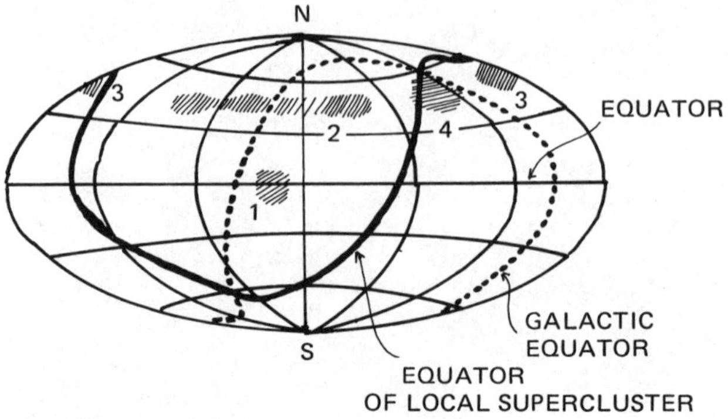

1:	Apex w.r.t. 3 K background radiation	V~ 350 km s^{-1}
2:	Apex w.r.t. nearby galaxies	V~ 300 km s^{-1}
3:	Apex w.r.t. local group	V~ 300 km s^{-1}
4:	Apex w.r.t. distant galaxies	V~ 1300 km s^{-1}

FIGURE 4. Anisotropies in Hubble's ratio and in universal background radiation (according to several authors; compilation by H. Karoji).

became clear that very high temperatures could indeed promote the build-up of these elements from hydrogen in a rather short time. This led Alpher, Bethe, Gamow, and many others later to associate the observed content of helium and deuterium with the Big Bang conditions.

Initially, when the universe is extremely hot, the progressive synthesis of elements takes place from protons and neutrons. But this process is quickly brought to a stop when the density and the temperature of the universe have fallen to sufficiently low values for the primeval composition to be considered as stable or, more precisely, capable of being modified only by local processes, notably at the dense centers of galaxies or stars, in a moderate way, affecting mostly the heavy elements.

This view is certainly compatible with what we are led to conclude about the physical conditions immediately following the Big Bang, from the knowledge of the density ρ_0 in the universe, the value of H_0 generally agreed upon, and the uncertainty in q_0; one can indeed compute instead ρ_0, assuming that ^{4}He, ^{3}He, and especially ^{2}H (the more sensitive indicator) are entirely primeval. The obtained value fits the observations reasonably well, provided one considers neutrinos as zero-mass particles—which is not obvious any more.

But, is this reasoning completely flawless? Certainly not. In the first place, as we have seen when discussing the Olbers paradox, we know little about the actual value of the density ρ_0; it might be much smaller than now contemplated. We do not know whether the concept of a mean density has any significance; for all we know, the universe, on the average, might as well be empty (as in the de Sitter models), expanding, yes, but not from an initial condition described by a hypercondensed universe and without any Big Bang, as far as we can extrapolate into the past the present universe of our observations.

In the second place, the nuclear processes are not the only ones to be active in the astrophysical media.

One knows that diffusion processes, which are able to strongly modify locally the relative abundances of various elements, are at work in the stellar interiors and atmospheres. We also know that ejection process, stellar winds, and the exchange of matter between binary components are very numerous and powerful processes; in particular, the solar wind exists and may have been much stronger still in the early stages of solar evolution. One strongly suspects the active galaxies, like the active stars, eject matter; but normal galaxies may do so during most of their life, in a moderate way. One knows actually that strong differences in metal content between halos and disks of spiral galaxies (such as ours) are observed, which are possibly due to the combination of stellar and galactic evolution or/and to diffusion processes inside the interstellar matter, as first suggested by the present author.[54] (See Fig. 5.)

The effective process could then be that, in a galaxy such as ours, helium has been kept whenever, at the time of large activity, a large amount of hydrogen has been expelled; it is suggested that this is a frequent situation, judging by the number of double radio sources symmetrical with respect to some more or less active galaxies. Hydrogen is lighted, and moreover, it is pushed by Ly α, a line which goes through the ionized region around stellar nests; the situation is in contrast to that of helium, four times heavier, which is not pushed at all, its resonance line, at 585 Å, being indeed used to maintain the ionization of hydrogen around the stellar hot young stars.[54]

In the case of deuterium, this process does not apply; but it is not either necessary to invoke the Big Bang for building deuterium atoms in the proper number. Whatever

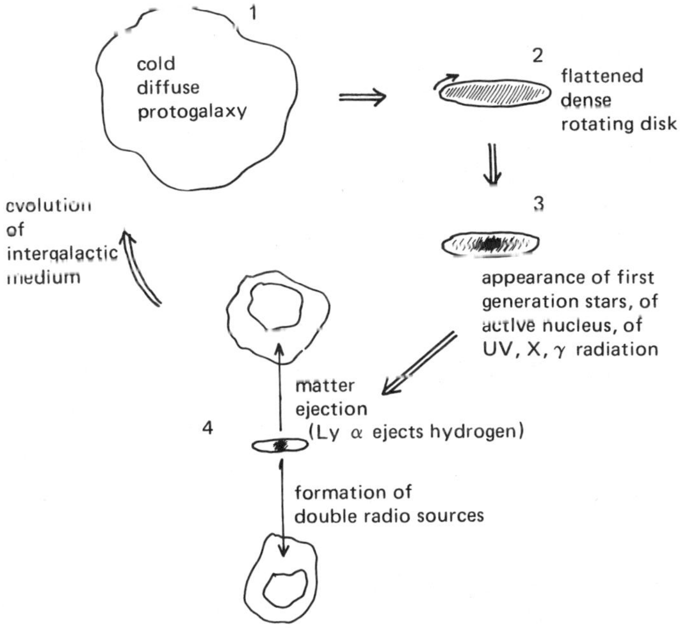

FIGURE 5. Galactic evolution and helium enrichment (schematical, after Pecker).

processes may have been building and spreading the deuterium atoms in the Galaxy, we know that there are some unusually large amounts of deuterium found in some meteorites,[10] this element might have been synthesized at the very early stages of the solar system, when the Sun was a T Tauri star, in the dense wind of energetic photons likely to be associated with those stars. In the Galaxy, the rather non isotropic distribution of deuterium seems to indicate that, whether built there or destroyed elsewhere, the measured deuterium cannot be identified with the so-called "primeval" deuterium.

In any case, the view that the Galaxy is 12–15 Gy* old is closely linked to the apriorisms of anyone believing in the Big Bang cosmologies. Although we know no object in the Galaxy older than, say 15 Gy, we are aware also that such objects would be difficult to observe; then there are, of course, white dwarfs, neutron stars, etc. (not to speak of the very hypothetical black holes). We also know (example: the pulsar of the Crab nebula) that evolution is rather rapid when a star comes to the stage of post-main-sequence presupernova; the collapse is a matter of ky, not of Gy or even My. But who knows how long they can exist after that? A white dwarf may be millions of Gy old! And we do not know enough cases similar to that of the Crab Nebula (or perhaps of the Cygnus loop) to make a correct statistical analysis of the problem. Nothing thus excludes the possibility that the Galaxy may be much older than at present accepted; hence all the time required to build or destroy any quantity of He and D is available. When one looks at humanity, one sees not any individual more than 150 years old. Yet, we know that humanity is much older; and the Earth much older still than humanity. Why not apply that kind of easy logic to the heavenly bodies?

The same reasoning may be applied to other galaxies; the only serious indications we have concern their minimum age, not their maximum.

Hence we see that the abundance of light elements, in the reference frame of the classical Big Bang theory, is undoubtedly a fact of great cosmological significance; and their actual values are constraining boundary conditions. However, if one does not adhere to this type of cosmology, one has nothing more than a simple feedback, or forced coincidence, in the agreement of the data with the theory. This shows indeed that these observations, interesting as they may be, have not the least cosmological significance!

Other Facts of Possible Cosmological Significance

The "equivalence principle," an essential element of Einstein's general relativity, and its link with the Mach principle are well known. Has anyone tried to introduce the Mach principle into a *suitable* cosmological form? Yet, one can consider through many laboratory experiments (the solar system being an "extension" of the laboratory) the equivalence principle as reasonably well established from facts.

*Obviously, one should use 1 ky = 10^3 years; 1 My = 10^6 years; 1 Gy = 10^9 years.

The occultations of many stars by the Moon allow us, when their time is measured with a sufficient accuracy, to determine the motion of the Moon, provided proper limb corrections are taken into account; van Flandern finds a significant time-variation of the gravitational constant as a necessary explanation of the observations (the radius of the universe can be regarded as constant). The Hoyle–Narlikar and Canuto cosmologies take that kind of phenomenon as a major fact of cosmological significance. Others think that the facts have not yet been established beyond doubt.

The study of the catalogues of galaxies, going as deep into space as possible, has led Peebles to demonstrate, in a rather convincing way, the existence of a cellular structure in the universe; galaxies, and clusters of galaxies, just like the wax in a bee hive, are grouped on the walls of the cells. To Souriau[13] and Fliche,[7] this observation has a high degree of cosmological significance, and they consider it one of the most interesting features of their cosmological views (essentially describing a nonhomogeneously expanding universe).

To use the same authors,[13] the space-distribution of quasars (assuming they are located at their cosmological distances), together with the distribution in the sky of the gamma-ray background radiation, provides an indication in favor of a zero-baryonic-number universe in which matter and antimatter, i.e., quasars and antiquasars, penetrate each other near the equator of the universe; this equatorial zone is then devoid of quasars and radiating in the gamma ray spectral domain.

But other authors, who do not believe in the cosmological distances of quasars, cannot take this view with confidence. For them, the red-shift distribution of quasars, the large space velocity derived from very long base interferometry which affects, one with respect to the other, the members of couples of quasars, and the everlasting mystery concerning the production of energy in quasars if they are located at their cosmological distance—all this indicates that quasars may be much closer than they appear and that a large part of their red shift is due to some phenomenon other than expansion, maybe the "tired-light mechanism" already quoted or some other mechanism, such as the one suggested by Barnothy.[51]

It is difficult to predict which new observations will, perhaps for a while or maybe forever, earn in future the title of "observational facts of cosmological significance." But we know for sure that observation will proceed and that the cosmologists will have to take it or leave it.

Any Conclusion?

It is not very easy to conclude!

Extreme points of view, which are often imposed upon the authors more by some religious or philosophical belief than by any observation, may be adopted with respect to cosmology, i.e., to the extrapolation from the observed to the not-yet-observed or even to the nonobservable.

One can swear by the steady-state universe or one can believe in the expanding universe. Or anything else. But one has to face the facts. So far, no cosmology has been able properly to describe all the facts. Accordingly, two attitudes remain possible: One is to say: Some of the facts (those we understand, with *our* cosmology) are of a cosmological significance; the others have a local origin and will need some kind of explanation not yet found in the theory of galactic evolution or in any other field of astrophysics when properly developed. The other attitude is to say: Let us wait for a moment—and at the same time develop conflicting hypotheses—until new facts will contradict some of our computations. After all, some compromises are still possible. One can think, for example, of an expanding universe without a Big Bang; or one can imagine a big bang much more ancient; or one might still deem it possible that a part of the shift (especially the normal red shifts) could be explained by mechanisms not yet proven to be operative, but still plausible, or at least not excluded.

In other words, one may dogmatize or one may not.

Our choice is made. This book is devoted to the memory of Wolfgang Yourgrau, who all his life fought dogmatism, allowing a place for all reasonable concepts, however heretical; he was a model heretic, open to new ideas, open to new science, but always keeping his own innovations within the field of rational physics. We felt that a plea against dogmatism in cosmology had here its natural place.

References and Notes

1. E. Mach, *The Science of Mechanics* (Open Court Publishers, Chicago, 1919).
2. A. Einstein, *Ann. Phys. (Leipzig)* **49** (1916).
3. K. O. Mikaelian, *Phys. Rev. Lett* **77B,** 214 (1978); *C. R. Acad. Sci. Paris, Ser. B* **288,** 93 (1979).
4. J. -C. Pecker, Annuaire du Collège de France (1981).
5. W. Olbers, *Aston. Jahrb. Spathen ed. Berlin* **150,** 110–121 (1826).
6. J. P. L. de Chéseaux, *Traité de la Comète* (M. M. Bousquet, Lausanne et Genève, 1744), pp. 223–229.
7. H. A. Fliche and J. -M. Souriau, *Astron. Astrophys* **78,** 87 (1979).
8. T. Jaakkola, H. Karoji, G. Le Denmat, M. Moles, L. Nottale, J. -C. Pecker, and J. -P. Vigier, *Mon. Not. R. Astron. Soc.* **177,** 191 (1976).
9. T. Jaakkola, M. Moles, and J. -P. Vigier, *Astr. Nachr.* **300,** 229 (1979).
10. M. Javoy, L. Merlivat, and F. Robert, *C. R. Acad. Sci. Paris, Ser. B* **288,** 375 (1979).
11. P. L. Richards and D. P. Woody, *Phys. Rev. Lett.* **42,** 925 (1979).
12. G. F. Smoot, M. V. Gorenstein, and R. A. Muller, *Phys. Rev. Lett.* **39,** 898 (1977).
13. J. -M. Souriau, personal communication (1979).
14. D. ter Haar, *Observatory* **74,** 128 (1954); *Phil. Mag.* **45,** 320 (1954); *Phil. Mag.* **45,** 1023 (1954).
15. F. Wilczek, *Phys. Rev. Lett.* **40,** 279 (1978).
16. C. V. L. Charlier, *Ark. Mat. Astron. Fys.* **4,** 1 (1908); *Ark. Mat. Astron. Fys.* **16,** 1 (1922).
17. E. E. Fournier D'Albe, *Two New Worlds: I The Infra World; II The Supra World* (Longmans, Green and Co., London, 1907).
18. G. de Vaucouleurs, *Science* **167,** 1203 (1970); *Astrophys. J.* **205,** 13 (1976).
19. V. M. Slipher, *Am. Phil. Soc. Proc.* No. 5 (1917).
20. E. Hubble, *Proc. Natl. Acad. Sci.* **15,** 168 (1929).
21. E. Hubble and R. Tolman, *Astrophys. J.* **82,** 302 (1935).
22. A. Sandage and G. A. Tammann, *Astrophys. J.* **197,** 265 (1975).
23. M. Aaronson, J. Huchra, and J. Mould, *Astrophys. J.* **229,** 1 (1979).
24. J. Heidmann, in "L'Evolution des Galaxies et ses Applications Cosmologiques," *Coll. CNRS, No. 263* (1977), p. 487.

25. I. E. Segal, *Mathematical Cosmology and Extragalactic Astronomy* (Academic Press, New York, 1976).
26. J. F. Nicoll and I. E. Segal, *Ann. Phys. (N.Y.)* **113,** 1 (1978).
27. G. Gamow, *Mr. Tomkins in Paperback*, p. 55, "Cosmic Opera" (Cambridge University Press, 1957).
28. G. Gamow, *Phys. Rev.* **74,** 505 (1948); *Rev. Mod. Phys.* **21,** 367 (1949); *Dan. Acad. Math. Phys.* **27** (10), 15 (1953).
29. R. A. Alpher, H. Bethe, and G. Gamow, *Phys. Rev.* **73,** 803 (1948).
30. R. A. Alpher and R. C. Herman, *Nature* **162,** 774 (1948); *Phys. Rev.* **75,** 1089 (1949).
31. R. H. Dicke, J. E. Peebles, P. G. Roll, and D. T. Wilkinson, *Astrophys. J. Lett.* **142,** 4141 (1965).
32. A. A. Penzias and R. W. Wilson, *Astrophys. J. Lett.* **142,** 419 (1965).
33. E. Findlay-Freundlich, *Proc. Phil. Soc.* **A67,** 192 (1954); *Phil. Mag.* **45,** 303 (1954).
34. M. Born, *Proc. Phil. Soc.* **A67,** 193 (1954).
35. G. Burbidge and M. Burbidge, *Phil. Mag.* **45,** 1019 (1954).
36. J. -C. Pecker, A. P. Roberts, and J. -P. Vigier, *C. R. Acad. Sci., Paris* **274,** 76 (1972); *C. R. Acad. Sci. Paris,* **274,** 115 (1972); *Nature* **237,** 227 (1972).
37. J. -C. Pecker and J. -P. Vigier, *Astrofizika* **12,** 315 (1976).
38. Z. Maric, M. Moles, and J. -P. Vigier, *Astron. Astrophys.* **53,** 191 (1976); *Nuovo Cimento Lett., Ser. 2* **18,** 269 (1977).
39. F. Zwicky, *Proc. Natl. Acad. Sci.* **15,** 773 (1929); *Morphological Astronomy* (Springer Verlag, Berlin, 1957).
40. H. C. Arp, *Astrophys. Lett.* **7,** 221 (1971).
41. L. Nottale, *Astrophys. J. Lett.* **208,** L103 (1976).
42. L. Nottale and J. -P. Vigier, *Nature* **268,** 608 (1977).
43. L. Nottale, Thesis, Paris VI University, June 10 1980.
44. H. Karoji, L. Nottale, and J. -P. Vigier, *Astrophys. Space Sci.* **44,** 229 (1976).
45. H. Karoji and L. Nottale, *Nature* **259,** 31 (1976).
46. J. -C. Pecker, *Coll. CNRS, No. 263,* "L'Evolution des Galaxies et ses Applications Cosmologiques," (1977), p. 451; *Coll. CNRS No. 37* on Abnormal redshifts, Balkowski, Westerlund ed., CNRS Paris (1977); Annuaire de Collège de France (1980); "Observations Mal Comprises dans le Cadre Classique," *Coll. Centenaire Einstein, College de France, 6–9 June* (1979).
47. F. Hoyle and J. V. Narlikar, *Nature* **233,** 41 (1971).
48. V. Canuto and S. H. Hsieh., *Astrophys. J.* **224,** 302 (1978).
49. T. van Flandern, *Non. Not. R. Astron. Soc.* **170,** 333 (1975).
50. B. E. Corey and D. T. Wilkinson, *Bull. Am. Astron. Soc.* **8,** 351 (1976).
51. J. M. Barnothy and M. F. Barnothy, *IAU Symp. 44,* on External galaxies and quasistellar objects (1978), p. 478.
52. B. Mandelbrot, *Fractals* (Freeman, San Francisco, 1977).
53. G. Lemaître, *J. Math. Phys. (MIT)* **4,** 188 (1976).
54. J. -C. Pecker, *Astron. Astrophys.* **18,** 253 (1972).

21

STOCHASTIC INTERPRETATION OF RELATIVISTIC QUANTUM EQUATIONS

N. CUFARO-PETRONI
Istituto di Fisica
Università di Bari
Bari 70100, Italy

J.-P. VIGIER
Institut Henri Poincaré
Paris, France

1. INTRODUCTION

Recent discussions on the Einstein–Podolsky–Rosen paradox[1] have shown that quantum mechanics implies spacelike correlations between two linear polarizers which measure the rate of coincidence between the relative orientations of pairs of photons emitted in the S state. If a forthcoming crucial experiment of Aspect[2] confirms this then the only possible "causal" (i.e., which preserves the fundamental fact that no individual particle can leave the light cone) way out of the resulting contradiction between relativity and the quantum theory of measurement seems to lie in the direction of an extension of the stochastic interpretations of quantum mechanics in terms of subquantum random fluctuations resulting from the action of a stochastic "hidden" invariant thermostat. Indeed these models (a) deduce the form of the quantum waves from the physical assumption that the stochastic jumps occur at the velocity of light; (b) interpret the preceding superluminal interaction in terms of superluminal propagation of a "quantum potential"[3] which is not carried by individual particles but results from phaselike collective motions carried by the said thermostat.

2. DIRAC'S AETHER

Since Dirac's pioneer work[4] it has been known that Einstein's relativity theory (and Michelson's experiment) are perfectly compatible with an underlying relativistic stochastic "aether" model. Inherent to this model is Einstein's idea that quantum statistics reflect a real subquantal physical vacuum alive with fluctuations and randomness. This concept of a nonempty vacuum has been recently revived not only to yield a foundation to the stochastic interpretation of quantum mechanics[3,5,6] but also to explain causally possible nonlocal superluminal interactions resulting from the Einstein–Podolsky–Rosen paradox.[7,8] Indeed, if the experiment of Aspect confirms their existence, the only way out of the resulting contradiction between relativity and the quantum theory of measurement[9] seems to lie in the direction of an extension of the causal stochastic interpretation of quantum mechanics. This assumes the existence of causal subquantal random fluctuations induced by a stochastic "hidden" thermostat proposed by Bohm, Vigier, and de Broglie.[10,11] By "causal" we imply (a) that no individual particle can leave the light cone; (b) that the sequence of causes and effects on any timelike particle motion is observer independent; (c) that one can deduce the form of de Broglie's "pilot" quantum waves (and the corresponding particle motions) from the assumption that they represent real collective drift and random motions on the top of Dirac's "aether" in which the quantum jumps occur at the velocity of light[3,12]; and (d) that possible superluminal interactions can be interpreted in terms of a quantum potential Q reflecting the particle's interaction with the thermostat which also carries their associated de Broglie waves. This potential's motion is not carried by individual particles but results from superluminal phaselike collective motion carried by the vacuum.

Dirac's "aether" rests on the idea that through any point 0 passes a flow of stochastic particles and antiparticles (described in Fig. 1 as particles moving backwards in time) whose momenta have the extremities of their four-vectors p^μ (with $p^\mu p_\mu = m^2c^2$) distributed with a uniform surface density on the two three-dimensional surfaces of the hyperboloids H^+ and H^-. They will thus remain invariant under all Lorentz transformations.

This stochastic relativistic distribution constitutes the only possible model for a physical undetectable thermostat for spin-zero particles into which we can study the relativistic analog of the classical nonrelativistic Brownian motion. Dirac has derived this from the indeterminacy principle. However, it differs from it by two new physical properties:

(a) Since the light cone behaves like an asymptotic accumulation manifold of Dirac's stochastic distribution we can assume that the corresponding stochastic jumps of a Brownian particle, submitted to its random action, occur practically at the velocity of light. Indeed, any given exchanged energy is statistically superseded by more energetic interactions.

(b) This ultrarelativistic Brownian motion includes the possibility of pair creation and/or annihilation. This is important since the mixture of particles and antiparticles has been shown to provide a realistic interpretation of possible negative probability distributions.[13]

The preceding model can evidently be extended to include spin if (following Bohm *et al.*[10] and Souriau *et al.*[14]) we endow our extended Dirac's "aether" elements,

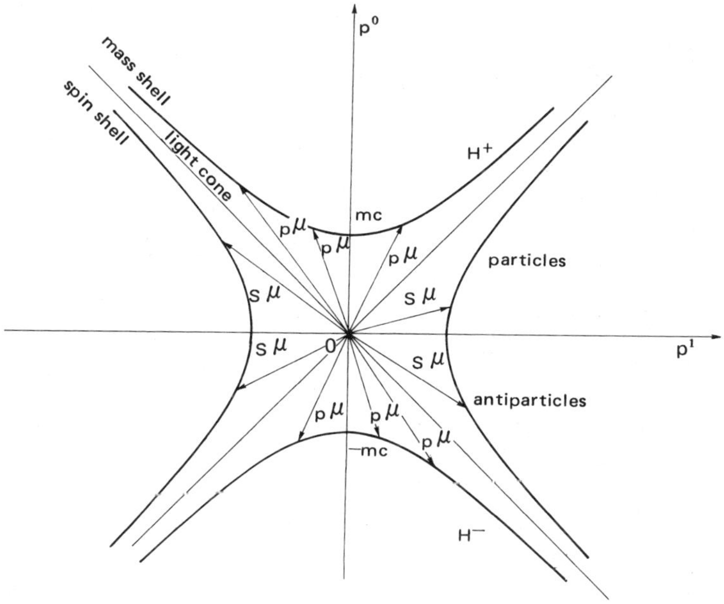

FIGURE 1. The shells correspond to the two Casimir invariants of the Poincaré group, i.e., $p_\mu p^\mu = m^2c^2$ and $S_\mu S^\mu = \text{const}$.

wave elements, and concentrated particles elements with internal rotation which represent spin's classical counterpart,[15] so that now all extended elements are characterized by an internal angular momentum $S_{\alpha\beta}$ added to their four-momentum p^μ and unitary four-velocity $\dot{x}_\mu = dx_\mu/d\tau$ (i.e., $\dot{x}_\mu\dot{x}^\mu = c^2$) which are no longer collinear. The conservation equations yield $\dot{p}_\mu = 0$ (i.e., $p_\mu p^\mu = m^2c^2$) and $p^\mu\dot{x}^\nu - p^\nu\dot{x}^\mu = S^{\mu\nu}$ with $S_{\mu\nu}S^{\mu\nu} = \sigma_0^2$.

On this basis one can introduce three classical extended models corresponding to the three spin states (i.e., $J = 0$, $J = ½$, and $J = 1$) by the three constraints $\tilde{S}_{\mu\nu}\dot{x}^\nu = 0$, $S_{\mu\nu}\dot{x}^\nu = v_0\ddot{x}_\mu$, and $S_{\mu\nu}\dot{x}^\nu = 0$. One then obtains $S_\mu S^\mu = \text{const}$ and $s_\mu s^\mu = \text{const}$ with $S_\mu = \tilde{S}_{\mu\nu}p^\nu$ and $s_\mu = S_{\mu\nu}\dot{x}^\nu$. In particles internal motions then include circular rotations of a center of matter density [denoted (c,c)] $x_\mu(\tau)$ around a center of mass [denoted (c,M)] and $y_\mu(\theta)$ moving parallel to p^μ with a constant radius R_μ (defined by $m^2c^2R_\mu = S_{\mu\nu}p^\nu$) and de Broglie's frequency $\hbar\omega = mc^2$ with respect to its proper time. This corresponds to a classical counterpart of the quantum "*Zitterbewegung*." The (c,c)'s internal Darboux–Frenet frames which now represent internal spin do not rotate ($J = 0$), rotate twice ($J = ½$), or rotate once ($J = 1$) while the (c,c) rotates once around the (c,M). Antiparticles have opposite (mirror) internal rotations. Along the (c,M)'s motion one thus recovers classically Planck's formula $A = mc\theta$ (where A represents the relativistic action and θ the path interval), which can also be written A =

$mc\theta = \hbar\alpha$: α denoting our new internal de Broglie phase and $\hbar$ its conjugate momenta. The corresponding internal rotations (oscillations) also yield $\theta = (\hbar/mc)\alpha = \lambda\alpha$, where $\lambda = (\hbar/mc)$ is Compton's classical radius, i.e., the (c,M)'s internal motion associated with one (2π) internal rotation.

On the preceding basis we can clearly introduce spin into Dirac's aether model. To every particle or antiparticle's p^μ in Fig. 1 is associated the orthogonal spacelike spin vector S_μ whose extremities will have a constant surface density on the "spin shell," i.e., timelike continuous "spin shell" surface $S_\mu S^\mu$ = const. Such a stochastic distribution complements the constant surface distribution of the p_μ's on the "mass shell" $p_\mu p^\mu = m^2c^2$ and is evidently invariant under any Lorentz transformation, so that the stochastic distribution of the (p_μ, S_μ) pairs now represent a $J = 0$, $J = ½$, $J = 1$ Dirac aether. The same result of course is obtained if we replace in Fig. 1 the p_μ's by $\dot{x}_\mu$ (since $\dot{x}_\mu\dot{x}^\mu = c^2$) and the S_μ's by s_μ.

Dirac's aether is thus now assumed to consist of a covariant mixture of $J = 0$, $J = ½$, and $J = 1$ particle–antiparticle pairs, the totality clearly defining a background sea which encompasses (as we shall later discuss) Bohm's and Vigier's subquantum level[10] de Broglie's subquantum thermostat,[11] Sudarshan *et al.*'s aether,[16] or (in the $J = 1$ limit for zero mass) the zero-point electromagnetic of stochastic electrodynamics (SED) of de la Peña[17] and Marshall.[18] As we will show later,[19–21] Dirac's aether model with spin implies the correct form of the $J = 0$, $J = ½$, and $J = 1$ wave equations if they are considered as describing the behavior of organized collective excitations (spin waves) on the top of our covariant subquantal vacuum. This results from the preceding hypothesis (a) and (b) combined with the assumption that all our stochastic motions have only short range (i.e., contact particle–particle collision-type interactions) so that random velocities and accelerations only result from direct short-range averaging over limited four-dimensional volume elements.

The wave equations can then be deduced along the two lines of demonstration utilized in nonrelativistic stochastic theory, i.e., Einstein's and Smoluchowski's Brownian motion theory[3] and Chandrasekhar's random walk (at the velocity of light) on discrete covariant four-dimensional lattices.[5]

Their corresponding de Broglie's "pilot" wavelike solutions all satisfy de Broglie's "phase correlation principle," a principle characterized by the fact that all internal (*Zitterbewegung* plus spin) rotations of the associated extended particle inhomogeneities (or nonlinear solitons), which follow the drift lines of their surrounding "pilot" wave, rotate in phase with the fluid elements which surround them. One can compare this process with a plane flying at Mach 1 within its own sound wave; the waves themselves (thus very different in nature from ordinary sound waves) can be best described as organized "spin waves" propagating on a chaos of the spinning classical $J = 0$, $J = ½$, and $J = 1$ tops which now constitute Dirac's aether with spin. As one knows,[22] all wave equations can be mapped into relativistic hydrodynamical representations where particles move (on the average) from one line of flow to another and jump (at the velocity of light) from one drift line to another. The quantum potential (which now also includes quantum torques) represents stochastic interactions (action and reaction) between our "particles" and the waves both carried by Dirac's aether. de Broglie's phase correlation clearly expresses the well-known property that oscillators can only transfer energy by contact to their immediate neighbors provided they oscillate in phase. This property is true, for example, for neighboring closed vortex tubes in classical fluids . . . or for connected spring systems. It implies that the stochastic jumps do not disturb the regular ordering of de Broglie clocks in de Broglie waves and justifies de Broglie's idea[11] that the subquantum thermostat is built with a superposition of independent $J = 0$, $J = \frac{1}{2}$, and $J = 1$ covariant thermostats.

The preceding model of Dirac's aether evidently recovers the main results of SED with one difference: The vacuum now always contains a mixture of waves *and* particles so that one cannot reduce SED's electromagnetic vacuum to a superposition of classical Maxwell waves. Indeed one sees it is now built with a distribution of plane Proca waves with Poynting vectors uniformly distributed on the photon's mass shell. Indeed if we go to a negligible mass limit of Proca's equations we see, following Marshall,[23] that Dirac's isotropic covariant distribution goes into $I(\nu) = (4\pi h/c^3)\nu^3 \cdots$ but that we cannot suppress discrete photons—a fact confirmed by experiment. Of course electron waves in Dirac's $J = \frac{1}{2}$ aether are also submitted to the Braffort Tzara zero-point e.m. vacuum as shown directly by Sudarshan *et al.*[16]

Indeed we can recover Sudarshan *et al.*'s description of the aether as a superfluid state of particle–antiparticle pairs. Indeed if we start from our covariant stochastic distribution of fermion–antifermion pairs in the $L = 0$ orbital states with zero values of the total spin $S = \frac{1}{2}(\sigma_+ + \sigma_-)$ (where σ_+ and σ_- are the spin operators of the particles in question, for example the positron and the electron, respectively), the Hamiltonian can be written in the momentum representation

$$H = \Sigma\epsilon_k c^+_{k,\sigma-} c_{k,\sigma-} + \Sigma\epsilon_k d^+_{k,\sigma+}\, d_{k,\sigma+} \quad \Sigma V_r(k,k')c^+_{k',\sigma-} d^+_{-k',-\sigma-}\, d_{-k,-\sigma-}\, c_{k,\sigma-} \qquad (1)$$

where $(c^+_{k,\sigma-}, c^+_{k,\sigma-})$ and $(d_{k,\sigma+}, d_{k,\sigma+})$ are the particle and antiparticle (creation–annihilation) operators, respectively, for the momentum state k, with $S = 0$ we have $\sigma_+ = -\sigma_-$ for the pair and the interaction.

H has the same form as that of Bardeen–Cooper and Schrieffer for superconductivity in metals except that each pair is now electrically neutral, an idea which can be generalized to any quantum number and any particle–antiparticle pair. Denoting by ϵ_k (with $V_r = V_0 =$ const for $|\epsilon_k| < \epsilon_0$, and zero otherwise) the Hartree–Fock energies of the fermion and of the antifermion we see that the ground state energy of our vacuum sea will be lowered by $\Delta E = \frac{1}{2}N(0)\Delta^2$, where Δ is the gap parameter $\Delta = \epsilon_0 \exp[-1/N(0)\, V_0]$, N_0 being the density of states on the Fermi surface. Our aether will now display long-range correlations. Particles or collective excitations (de Broglie's waves) will not exchange momentum (up to a particular momentum) and there will be negligible viscosity. However, when the energies involved exceed Δ new phenomena might arise. The energy of an elementary quasiparticle excitation E_k is given by $E_k = (\epsilon_k^2 + \Delta^2)^{1/2}$. For a pair it is $2E_k$ and if we assume that masses arise predominantly from interactions we get $|\Delta| = mc^2$. If $\Delta \gg \epsilon_k = pc$ (p representing the particles' linear momentum magnitude) the excitation of a real pair will cost $2|\Delta|$ and we see that the gap will depend on the strength of the interactions in which they take part.

This model opens many possibilities. The most remarkable is that superfluid behavior occurs when fluid velocity is less than the velocity of the elementary excitation. The critical velocity of the fluid above which the superfluidity condition will not be satisfied for the aether can be calculated from $\omega_c = |\Delta|/\hbar k$ when $\hbar k$ is the momentum of the excitation. If we choose as before k to be the inverse Compton wavelength of our excitation we get $V_c = mc^2/mc = c$ so that our vacuum implies a limiting velocity c which justifies Einstein's assumption.

3. Covariant Stochastic Equations

In the present section we will derive stochastic Nelson's equations from our relativistic generalization of the hydrodynamical model of quantum mechanics extended to spinning particles. This classical relativistic model contains three new physical features.

(I) The fluid elements (and the particles) which follow the lines of flow of the fluid with irregular fluctuations are built from extended elements in the sense discussed by Bohm[10] and Souriau.[14]

(II) The stochastic fluctuations occur at the velocity of light.

(III) The fluid is a mixture of extended particles (and antiparticles): the latter being mathematically equivalent to particles moving backward in time.[13,24]

The existence of such fluctuations (which induce in the particle a Markov type of Brownian motion) has been shown[10] to lead any initial distribution of the particles in the fluid into a limiting equilibrium distribution proportional to the fluid's average conserved drift density $\rho(x_\mu(\tau))$. This means that the fluctuations of our Madelung fluid induce on our particles stochastic jumps at the velocity of light (from one line of flow to another) and that such jumps can be decomposed into the regular drift motion V^μ plus an apparent spacelike random part U^μ with $V_\mu = dx_\mu(\tau)/d\tau$, τ representing the proper time along the drift lines: so that $V_\mu V^\mu = c^2$.

Indeed any velocity W_μ represented by a point P (with $W_\mu W^\mu = 0$) of the light cone can be decomposed into the sum of two four-velocities V^μ and U^μ, i.e., $W_\mu = V_\mu + U_\mu$ with $U_\mu U^\mu \leq 0$. Since the three independent components of W_μ determine the four components of U^μ, if one considers a particle of the preceding type it undergoes two independent types of motions: (a) regular motions along the fluid's drift lines of flow with the fluids own velocity V_μ, and (b) stochastic jumps in any direction with the velocity of light with a four-velocity W_μ satisfying $W_\mu W^\mu = 0$.

To establish (a) let us first recall that a particle or a regular fluid element (which can be compared with the stochastic particle and the thermostat's elements in the usual Brownian motion) are now represented in four-dimensional space-time by timelike hypertubes instead of timelike lines. These hypertubes can be naturally assumed to have a minimum spacelike radius $\bar{r}/2$ which yields the minimum distance $\bar{r}$ which separates two continuous particles in any spacelike section passing through their center of mass. Independently of the stochastic jumps our drifting fluid is thus comparable with a timelike set of extended fibers and the minimum time needed to pass from one of these hypertubes to the next is thus $r/c = \Delta t$ since the jumps occur at the velocity of light. This implies that the proper-time variable which corresponds to adjacent events in our stochastic model have nonzero minimum temporal separation $\overline{\Delta\tau}$.

The second step is just to generalize to our relativistic model the average velocities utilized by de la Peña and Cetto[25] to discuss the nonrelativistic theory of classical and quantum-mechanical systems. Let us start (Fig. 2) from a four-dimensional volume limited on the side by the fluid's regular lines of flow and, at both extremities, by two spacelike constant phase surfaces[26] S_1 and S_3. If the domain is small enough such surfaces are separated by an interval $2\Delta\tau$: an interval $\pm\Delta\tau$ separating S_1 and S_3 from a median section S_2. Of course $|\Delta\tau| \geq \overline{\Delta\tau}$.

As a consequence of the assumed stochastic equilibrium we can treat on the same footing the fluid behavior and an ensemble of similarly prepared particles characterized by the density $\rho(x, \tau)$ in configuration space where x_μ represents a point in four-dimensional space-time.

We can describe the average local motions of the elements of the ensemble by the selection of all particles that at proper time $\tau = \tau_2$ are contained in a small four-dimensional volume element around the point $r^\mu = r_2^\mu$. This is necessary in our model, since if one starts from a particle in its local drift rest frame (i.e., the frame in which the neighboring fluid element is practically at rest) its stochastic jumps along the light cone can bring it into any neighboring line of flow: both in the forward and backward proper time direction. As a consequence our general stochastic model implies the use of a four-dimensional stochastic space-time volume element to recover all possible stochastic jumps of each drifting particle. We have thus made the new theoretical step of introducing along with the average space positions the new concept of an average time in a four-dimensional volume element.

In order to describe the global motion of this element we select the particles that at proper time τ_2 are contained on a small section (space-volume element) of S_2 limited by the hypertubes boundary. According to our model it is possible to distinguish two

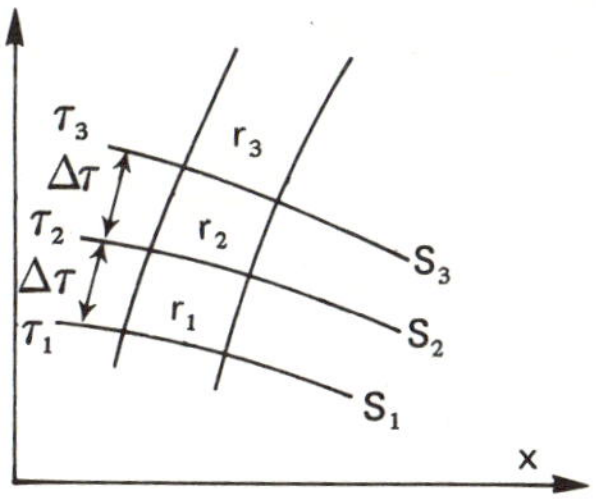

FIGURE 2. A four-dimensional volume element bounded by flow lines and by two spacelike constant-phase surfaces.

different kinds of motion of this volume element during a short interval $\Delta\tau$. Besides its motion as a whole in the hypertube (which preserves the fluid's scalar density ρ) the element will suffer variations of ρ due to the stochastic jumps which move matter from one line of flow to another and will bring fluid across the hypertubes' boundary. Generalizing de la Peña and Cetto's[25] ideas we can obtain a simplified description in terms of two quasilocal statistical velocities. If we take any one of the particles of our volume element and call r_1^μ and r_2^μ its average mean position at $\tau_1 = \tau_2 - \Delta\tau$ and $\tau_1 = \tau_2 + \Delta\tau$ we can calculate the average of $r_3^\mu - r_2^\mu$ over the subensemble defined by the particles which belong to our small volume element. We call these average values the mean and denote them with $\langle\ \rangle$. We thus write

$$r_3 - r_2 = \langle r_3 - r_2 \rangle + \delta_+ r \quad \text{and} \quad r_2 - r_1 = \langle r_2 - r_1 \rangle + \delta_- r \tag{2}$$

Since one must assume (in our model) the homogeneity, isotropy, and time independence of our stochastic mechanism, the change variable $\delta_\pm r_i$ must satisfy $\langle(\delta_+ r_i)\rangle = \langle(\delta_- r_i)\rangle$ so that we can omit the indexes from such expressions and write in general $\langle \delta r_i \rangle = 0$.

We can now derive from (2) two different velocities, i.e.,

$$b_+^\mu(2) = (r_3^\mu - r_2^\mu)/\Delta\tau \quad \text{and} \quad b_-^\mu(2) = (r_2^\mu - r_1^\mu)/\Delta\tau \tag{3}$$

whose mean values

$$V_+^\mu(2) \equiv \langle b_+^\mu(2) \rangle = \langle (r_3^\mu - r_2^\mu)/\Delta\tau \rangle \tag{4}$$

and

$$V_-^\mu(2) \equiv \langle b_-^\mu(2) \rangle = \langle (r_2^\mu - r_1^\mu)/\Delta\tau \rangle$$

are the relativistic generalization of the mean forward and backward velocities. From these one can derive the regular fluid's velocity V^μ and a stochastic velocity U^μ through the relations

$$\begin{aligned} V^\mu(2) &\equiv \langle (r_3^\mu - r_1^\mu)/2\Delta\tau \rangle = \tfrac{1}{2}(V_+^\mu + V_-^\mu) \\ U^\mu(2) &\equiv \langle [(r_3^\mu - r_2^\mu) - (r_2^\mu - r_1^\mu)]/2\Delta\tau \rangle = \tfrac{1}{2}(V_+^\mu - V_-^\mu) \end{aligned} \tag{5}$$

Now the stochastic velocity U^μ can be determined in any spacelike direction by calculating the flow between τ_1 and τ_3 of all elements which cross a drift timelike plane passing through r_2^μ and orthogonal to a spacelike direction x. Indeed let us consider (see Fig. 3) an ensemble of fluid elements (particles) which are at τ_2 in the neighborhood of x. If $\rho_{1L}(\rho_{1R})$ then represents the scalar densities in the neighborhood of $x_{1L}(x_{1R})$ at $\tau = \tau_1$ we see that these densities are related to n_+ and n_- through

$$n_+ = (x_{3R} - x_2)\rho_{3R} = (x_2 - x_{1L})\rho_{1L} \tag{6}$$

and

$$n_- = (x_{1R} - x_2)\rho_{1R} = (x_2 - x_{3L})\rho_{3L}$$

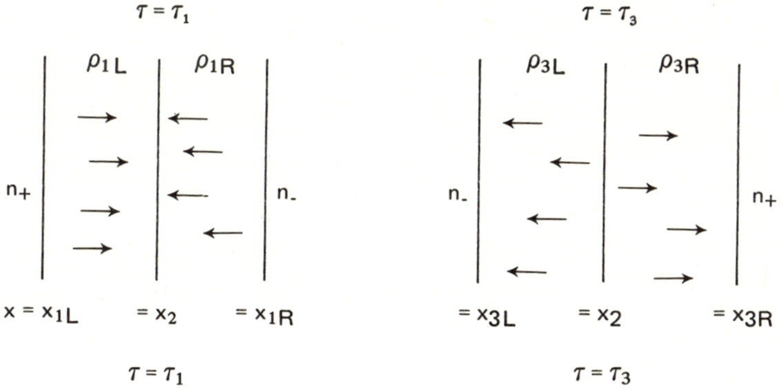

FIGURE 3. $x_{1L}(x_{1R})$ is the average position of the $n_+(n_-)$ particles at $\tau_1 = \tau_2 - \Delta\tau$ and $x_{3R}(x_{3L})$ is the average position of the same particle at $\tau_3 = \tau_2 + \Delta\tau$:$\rho_L(\rho_R)$ being the densities of particles to the left (right) of $x = x_2$.

This yields

$$x_1 + x_3 - 2x_2 = [1/(n_+ + n_-)][-\rho_{1L}(x_2 - x_{1L})^2 + \rho_{1R}(x_{1R} - x_2)^2 + \rho_{3R}(x_{3R} - x_2)^2 - \rho_{3L}(x_2 - x_{3L})^2] \tag{7}$$

which can be averaged over the ensemble. Since each of the parentheses then become $\langle(\delta x)^2\rangle$ we can write to the first approximation [with $n_+ + n_- = 2\rho(\tau_2)\Delta x$]:

$$U^\mu = \frac{\langle x_1^\mu + x_3^\mu - 2x_2^\mu\rangle}{2\Delta\tau} = \frac{\langle(\delta x)^2\rangle}{2\Delta\tau}\frac{1}{\rho}\partial^\mu\rho = D\frac{\partial^\mu\rho}{\rho} \tag{8}$$

if we define as usual the diffusion coefficient as $D = \langle(\delta r_i)^2\rangle/2\Delta\tau$ and neglect higher-order terms in $\Delta\tau$. D is always <0 since our quantum jumps are spacelike.

This is exactly the relativistic generalization of Einstein's definition[27] of the stochastic velocity in Brownian motion. We further have $V_\pm^\mu = V^\mu \pm U^\mu$ which connects the forward (particle) and backward (antiparticle) velocities with the fluid's regular drift velocity V^μ and its stochastic velocity U^μ.

The second step is to associate the two velocities needed to describe our motion to four accelerations required to describe the forward and backward changes of these velocities. To do this we require the existence of our minimum proper time interval $\Delta\tau$ which allows us to define the four accelerations

$$\begin{aligned} b_+(3) - b_+(2) &= a_+^+\,\Delta\tau + \delta_+ b_+ \\ b_-(3) - b_-(2) &= a_-^+\,\Delta\tau + \delta_+ b_- \\ b_+(2) - b_+(1) &= a_+^-\,\Delta\tau + \delta_- b_+ \\ b_-(2) - b_-(1) &= a_-^-\,\Delta\tau + \delta_- b_- \end{aligned} \tag{9}$$

which evidently lead to systematic drift and stochastic derivative operators. Indeed if we define as D_d and D_S the following operations on a general function $f(r)$ of the stochastic variable r, i.e.,

$$D_d f(r_2) = \langle [f(r_3) - f(r_1)]/2\Delta\tau \rangle$$

and

$$D_S f(r_2) = \langle [f(r_1) + f(r_3) - 2f(r_2)]/2\Delta\tau \rangle \tag{10}$$

which are evidently related with the forward (D^+) and backward (D^-) derivative operators through the relation

$$D^{\pm} = D_d \pm D_S$$

we see they thus correspond to scalar (proper time type) derivatives in timelike and spacelike directions ... and yield the drift and stochastic velocities through $V^\mu = D_d r^\mu$ and $U^\mu = D_S r^\mu$, where the dummy index 2 has been omitted. This generalizes $v_\mu = dx_\mu \, d\tau$ and lead to the preceding mean accelerations through the expressions

$$a^+_\pm = D^+ V_\pm \quad \text{and} \quad a^-_\pm = D^- V_\pm$$

Moreover a development in Taylor series yields

$$\begin{aligned} D_d f &= \frac{\partial f}{\partial \tau} + V_\mu \partial^\mu f + \cdots \\ D_S f &= U_\mu \partial^\mu f + D \partial_\mu \partial^\mu f + \cdots \end{aligned} \tag{11}$$

where the diffusion coefficient D is given as before by the relation

$$\langle (\delta r_i \cdot \delta r_j)/2\Delta\tau \rangle = D\delta_{ij} \tag{12}$$

in the drift rest frame: diffusion in time representing, as before, particle–antiparticle transition: δr_i and δr_j denoting any pair of Cartesian components of $\delta_\pm r$ .. which are assumed to be statistically independent if $i \neq j$.

The third (essential) step is to derive the covariant generalization of Nelson's equation, in our model. To do that we recall that any detailed description must start from the general equation

$$m\ddot{r}^\mu = f^\mu_d + f^\mu_S \tag{13}$$

where f^μ_d represents the drift spacelike forces and f^μ_S the purely random effects, the dot denoting proper-time derivatives. The corresponding statistical theory must, according to our model, start from the ensemble of particles which at any proper time τ_2 lie in the neighborhood of r^μ_2. The mean of the preceding relation thus becomes

$$m\langle \ddot{r}^\mu \rangle = F^\mu_d + F^\mu_S = F^\mu_d$$

where

$$F^\mu_d = \langle f^\mu_d \rangle \qquad \text{with } F^\mu_S = \langle f^\mu_S \rangle = 0 \tag{14}$$

Since the mean value of $\ddot{r}$ is taken over the same ensemble utilized to define our average velocities and accelerations in the preceding steps, it must be expressed as a linear combination of $a^{\pm}_{\pm}$. To determine these combinations, we remark that $\langle \ddot{r} \rangle$ and $\langle f_d \rangle$ can be split into two parts, i.e., a part $\langle \ddot{r} \rangle^+$ (or $\langle f_d \rangle^+$) which is invariant under proper time reversal, i.e., $\tau_3 - \tau_2 \rightarrow \tau_1 - \tau_2$, and a part $\langle \ddot{r} \rangle^-$ (or $\langle f_d \rangle^-$) that changes sign under this discrete symmetry which changes V_μ but conserves U_μ. Combining Eq. (14) with its counterpart obtained through a proper-time reversal operation we obtain the new set of equations

$$m\langle \ddot{r}^\mu \rangle^{\pm} = (F^{\pm}_{d})^\mu \tag{15}$$

We now make the final step in our demonstration of Nelson's equation by examining the implications of Eq. (15). The first implication is the importance of the proper-time relation $m\langle \ddot{r} \rangle^+ = F^+_d$ which evidently represents the stochastic generalization of Newton's law for our model. Indeed the usual four-dimensional acceleration $\ddot{X}^\mu$ of a classical point X^μ satisfies $\dot{X}_\mu \ddot{X}^\mu = 0$ (since $\dot{X}_\mu \dot{X}^\mu = C^2$) and is invariant under proper-time reversal. The same holds for our stochastic case since: (a) the drift acceleration $\dot{V}_\mu$ is orthogonal to V_μ; (b) the stochastic spacelike velocity U_μ is locally orthogonal to V_μ so that the corresponding stochastic accelerations (which vanish on the average since $\langle F_S \rangle = 0$) are thus always orthogonal to V^μ.

The second implication is that $\langle \ddot{r} \rangle^+$ must be expressed by just the linear combination of relations (9) which are proper-time inversion invariant, i.e., $(a^+_- + a^-_+)$ or $(a^+_+ + a^-_-)$ or a linear combination therefrom.

The third implication is that a mean acceleration (which corresponds mathematically to second-order proper-time derivatives) should be defined physically only by the motions of fluid elements surrounding r^μ_2, i.e., enclosed within the four-dimensional volume element limited by S_1 and S_3 utilized to define mean quantities. We deduce therefrom and from the explicit form of the a's given in Eq. (15), that the only quantity of this type invariant under $\tau \rightarrow -\tau$ is $(a^+_- + a^-_+)$. Indeed the definition of $(a^+_+ + a^-_-)$ implies knowledge of the behavior of fluid elements which lie outside our volume since it contains four-velocities of elements which are crossing S_3 and S_1 in the backward and forward directions, i.e., are leaving this volume. Moreover one sees that the combination $\langle \ddot{r} \rangle = (a^+_- + a^-_+)$ evidently represents the relativistic definition of the sum of the mean accelerations of antiparticles (a^+_-) and particles (a^-_+) passing through r^μ_2 at $\tau = \tau_2$.

As a consequence we must write relation (15) in the form

$$\tfrac{1}{2}m(a^+_- + a^-_+) = F^+ \tag{16}$$

which is exactly the relativistic generalization of the form given by de la Pena and Cetto[25] to Nelson's equation. Clearly Eq. (16) contains particle–antiparticle symmetry.

The same argument applies to the $-$ part of (15). Indeed the only combinations of $a^{\pm}_{\pm}$ that change sign under proper-time reversal are $(a^+_+ - a^-_-)$ and $(a^+_- - a^-_+)$ and the second only is exclusively defined by the motion of fluid elements between S_1 and S_3. We thus have $\tfrac{1}{2}m(a^+_- - a^-_+) = F^-$ which satisfies the continuity equation and is compatible with the introduction of the Lorentz force for charged fluid elements. Moreover these relations can be rewritten with the help of the definiers of D_d and D_S into the form

$$m(D_d V^\mu - D_S U^\mu) = (F^+)^\mu \tag{17}$$

and

$$m(D_d U^\mu + D_S V^\mu) = (F^-)^\mu \tag{18}$$

In Eq. (18), both sides tend (as they should) to zero in the nonstochastic limit. In order to clarify what is the connection between h, which is the internal frequency and/or spin, and D, which represents random jumps, we will limit the demonstration to the Klein–Gordon case since spin (as can easily be shown) complicates but does not modify substantially our demonstration.

To calculate D we now assume (a) that in our stochastic jumps D describes (on the mean) one jump only; (b) that in each individual jump the particle's (c, c) undergoes one rotation only around the (c, M). This is natural since jumps do not cancel rotation.

Assumptions (a) and (b) evidently represent the simplest possible Brownian motion scheme which preserves both the average drift motion *and* the local phase correlation *and* average ordering of the de Broglie clocks Zitterbewegung. If no interaction occurs during one stochastic jump we can consider it as the $v \to c$ limit of a free particle motion described by the Lagrangian $\delta\mathcal{L} = -mc(\dot{x}_\mu \dot{x}^\mu)^{1/2} d\tau$, where $d\tau$ represents the corresponding drift interval. As stated above during the corresponding free jump the (c, c) rotates an amount 2π so that the corresponding distance $\langle(\delta x)^2\rangle^{1/2}$ *must* be equal to the Compton wavelength $\lambda = \hbar/mc$. When going to the $v = c$ limit we thus write $\langle(\delta x)^2\rangle = c^2\Delta\tau^2$ which yields [since $D = \langle(\delta x)^2\rangle/2\Delta\tau$] the diffusion constant $D = \hbar/2m$. Moreover one can utilize the real internal rotations in our vacuum and wave elements to complete (and justify) some essential properties of de Broglie's "hidden thermostat." Indeed we first see that Dirac's aether's absolute temperature is necessarily zero. This clearly results from the Langevin equation associated with our Brownian motion, i.e.,

$$m\dot{v}^\mu = \beta v^\mu + \langle F_Q^\mu \rangle + \langle F_E^\mu \rangle \tag{19}$$

where β is the friction[28] coefficient of the Brownian particle: $\langle F_Q^\mu \rangle$ and $\langle F_E^\mu \rangle$ representing averages of the quantum (stochastic) and external force. Since we can define the vacuum's temperature T^0 by $D = (kT^0) \cdot \tau/m$, where τ represents the usual relaxation time, i.e., $\tau = m/\beta m = 1/\beta$: We get $D = \hbar/2m = k(T^0/\beta)$ so that only $T^0 = 0$ yields $\beta = 0 \ldots$, i.e., a frictionless vacuum. On this vacuum our fluid elements and/or particles undergo a Brownian Markov process at the velocity of light depicted by the relativistic Fokker–Planck equation:

$$(\partial\rho/\partial\tau) + \partial^\mu(W_\mu^\pm \rho) \pm D\Box\rho = 0 \tag{20}$$

where $W_\mu^\pm = v_\mu \pm u_\mu$.

The conservation of the average number ρ of extended oscillators within our drift tubes of flow (an essential point in our demonstration of Nelson's equations[3]) also allows an immediate definition of the temperature, entropy and heat carried by de Broglie's pilot Klein–Gordon wave. To that effect we first recall Boltzmann's famous analysis[29] of the stochastic behavior of distributions of periodic mechanical systems. Let us consider a mechanical system characterized (1) by microscopic variables q_i ($i = 1, 2 \ldots$) which oscillates rapidly with a period $\tau = 1/\nu$; (2) by macroscopic coordinates a_i which remain constant when the system is in a given stable state but changes slowly when it passes from one state to another in a reversible way. One sees immediately that this is exactly the situation in the (average) rest frame of a three-dimensional slice V_0 of a drift tube (see Fig. 2): the variables q_i correspond to the vectors R_μ: the reversible character in time resulting from the particle–antiparticle mixture. The a_i now represent the variable surface conditions on the tube's surface which describe its external conditions (surface tension etc.[22]) and their variation corresponds to the work performed on the tube from the outside.[22] Boltzmann's result can be summarized as follows.[29]

If $\mathcal{A}$ denotes the cyclic integral of the Maupertuisian action on one period of the internal motion the quantity of heat, δQ, given to V_0 (when it undergoes a reversible change of state in its rest frame) is defined by $-\delta Q_0 = V_0 \delta \mathcal{A}_0 = T_0 \delta \mathcal{S}$ where T_0 represents the temperature in its rest frame and $\mathcal{S}$ the systems entropy. Since T and Q must Lorentz transform in the same way, two equivalent possibilities can be used, viz., (a) Tolman's[30] and Eckart's[31] definitions, who assume that $\mathcal{S}$, T, and Q_0 are scalars or (b) Planck's and de Broglie's (denoted by a prime)[32] which connect them with the drift motion, i.e., $T_0 = T'_0 u_0$ and $Q_0 = Q'_0 u_0$ ($U_0 = c$ in the rest frame) so that T'_0 and Q'_0 transform like $T'_0(1 - \beta^2)^{1/2}$ and $Q'_0(1 - \beta^2)^{1/2}$ [i.e., exactly like $\nu = \nu_0(1 - \beta^2)^{1/2}$], $\mathcal{S}$ remaining a scalar. As shown by Bohm and Vigier in the stochastic model of the Klein–Gordon equation [with a wave $\psi(x,\tau) = R \exp(iS/\hbar)$] we know[10] that the rest frame's particle mowing equilibrium distribution is given by $R^2 = \psi^*\psi$. Since we also know that the action $\mathcal{A} = S$ satisfies $\dot{S} = -Mc^2$ (since $\mathcal{A} = \int_{\tau 1}^{\tau 2} Mc^2 d\tau$) we see that the maximum of the entropy $\mathcal{S}$ corresponds to a minimum of the action $\mathcal{A}$ defined from the motion of our particles, if $M^2 = m^2 - (\hbar^2/c^2)(\Box R/R)$. Any isolated particle thus minimizes the value of the action integral $S = \int -Mc^2 d\tau$, and we can write as a consequence of Boltzmann's analysis $h\nu_0 = mc^2 = kT$ so that $-\mathcal{S}/k = \mathcal{A}/h$ with $\delta Q_0 = -Mc^2$ and $\mathcal{A} = -hM/m \ldots$, where k represents Boltzmann's constant. This leads to the following consequences:

(1) The particles' internal oscillation energy $E_0 = h\nu_0 = Mc^2$ along the drift lines are in equilibrium with the wave's local temperature T.

(2) The various random paths are weighted by $\exp(i\int \mathcal{L} d\tau)$, which means that Feynman's graphs really correspond to possible random motions within Dirac's aether—the i factor resulting from the minus sign in Nelson's equation.

(3) No information can be carried by the quantum potential from one particle to another—despite the fact of the superluminal propagation of the stochastic quantum potential.

(4) The Klein–Gordon fluids satisfies the first two laws of Thermodynamics. This can be shown directly from the decomposition of the Klein–Gordon energy momentum tensor explicitly performed by Halbwachs.[22] Indeed he has deduced from the Lagrangian

$$\mathcal{L} = \partial_\mu \psi^* \partial^\mu \psi - (m^2c^2/\hbar^2)\psi^*\psi$$

the expression

$$t_{\mu\nu} = \rho v_\mu v_\nu + q_\mu v_\nu + q_\nu v_\mu + \theta_{\mu\nu} \tag{21}$$

where q_μ denotes a heat current [orthogonal to the conserved current $j_\mu = \rho v_\mu$ (with $v_\mu v^\mu = c^2$)]. From $\partial^\mu t_{\mu\nu} = 0$ one immediately obtains introducing the scalar $\omega c^4 = t_{\alpha\beta}v^\alpha v^\beta$ and writing $\omega = \rho(\epsilon + Mc^2)$ (where ϵ represents the internal energy) exactly Eckart's form[31] of the first law, i.e.,

$$MD_d\epsilon + (1/c)[\partial^\alpha q_\alpha + q^\alpha \dot{v}_\alpha] + \omega_{\alpha\beta}\partial^\alpha v^\beta = 0 \tag{22}$$

where

$$\omega_{\alpha\beta} = t^{\gamma\delta}\eta_{\gamma\alpha}\eta_{\delta\beta}$$

with

$$\eta_{\mu\nu} = \delta_{\mu\nu} + (v_\mu v_\nu/c^2)$$

As noted by de Broglie the second law is a straightforward consequence of

$$-\mathcal{S}/k = \mathcal{A}/h$$

since entropy is thus maximized along the fluid's drift lines.

(5) The heat current takes the form $q_\alpha = -G\eta_{\alpha\beta}\partial^\beta\theta$, where G represents the local thermal conductivity of the Klein–Gordon fluid and θ the local (probabilistic) temperature density $\theta = T.\, R^2 = (mc^2/k)R^2$. Indeed its calculated form[22]

$$q_\mu = \rho(\hbar^2/M^2)u^\lambda\partial_\lambda \log R\eta_{\mu\nu}\partial^\nu \log R$$

can be transformed into

$$q_\mu = [k\hbar^2/(2M \cdot mc^2)]u^\lambda\partial_\lambda \log R\eta_{\mu\nu}\partial^\nu\theta = G\eta_{\mu\nu}\partial^\nu\theta$$

4. Relativistic Quantum Equations

The deviation of relativistic quantum wave equations can now be carried along a line of demonstration which is just a relativistic generalization of the ideas introduced by Einstein and Smoluchowski into Brownian motion theory. Assuming that our particles are (1) carried along the lines of flow or a regular drift motion v_μ of extended particles associated with a collective motion on the top of Dirac's thermostat, (2) jump stochastically at the velocity of light from one average drift line of flow to another and thus (for an ensemble of identical particles with arbitrary initial positions) reach an average mean conserved distribution $\rho(x)$; one can immediately demonstrate the stochastic force law first assumed by Nelson,[33] from which one deduces (Lehr and Park,[12] Vigier,[3] Guerra and Ruggiero[21]) a stochastic wave $\psi(x) = [\rho(x)]^{1/2}\exp[iS(x)/h]$ with $v_\mu = (1/m)\,\partial_\mu S$, which satisfies the Klein–Gordon equation. Along this line we will show a derivation for all the three basic quantum mechanical equations (Klein–Gordon, Proca, and Dirac's equations). This is necessary in the stochastic interpretation of quantum mechanics (SIQM) because according to the Cartan–Pauli demonstration[15,34,35] there are only three possible basic two-dimensional spin subgroups (which preserve two-dimensional planes). Cinematically they correspond to purely Lorentz ($J = 0$), purely spatial ($J = 1$), and light cone tangent ($J = \frac{1}{2}$) subrotations in $SO(3.1)$. Mathematically[36] they correspond to $D(0)$, $D(1)$, and $D(\frac{1}{2})$ representations of the rotation group imbedded in the general representation $D(\frac{1}{2}, \frac{1}{2})$ of the Lorentz group. The usual quantization of their three classical counterparts indeed yields the usual wave equations.[15]

We come now to our derivation (we will use in this section the Stochastic notation of Guerra and Ruggiero[21]) starting from the demonstration of the Dirac equation in its Feynman and Gell-Mann[37] form. In order to do that we define

$$
\begin{aligned}
D &= \frac{\partial}{\partial\tau} + b_\nu\partial^\nu, \qquad b_\nu = Dx_\nu(\tau) \\
\delta D &= -\frac{1}{2m}\Box + \delta b_\nu\partial^\nu, \qquad \delta b_\nu = \delta D x_\nu(\tau) \qquad (\hbar = c = 1)
\end{aligned}
\tag{23}
$$

and we suppose that the drift and the stochastic velocities b_μ and δb_μ can be derived from

$$
b^\mu = \frac{A^\mu + \partial^\mu\phi}{m}, \qquad \delta b^\mu = -\frac{\partial^\mu \ln \rho^2}{2m} - \frac{\partial^\mu\rho}{m}\rho^{-1} \tag{24}
$$

where $A_\mu(x)$ is the vector potential of an external field, and $\phi(x, \tau)$ and $\rho(x)$ are real diagonal 4 x 4 matrixes. We also require that ρ is nonsingular and that ρ^2 is a trace invariant matrix which plays the role of a density matrix (but not a definite positive one because of the presence of particles and antiparticles).

The classical covariant stochastic equation and the continuity equation for our "rigid" top in an external field are now

$$
m(DD - \delta D\delta D)x^\mu(\tau) = b_\nu F^{\nu\mu} + \frac{1}{4m}\partial^\mu\sigma_{\nu\lambda}F^{\nu\lambda} \tag{25}
$$

$$
\partial^\nu(b_\nu\rho^2) = 0 \tag{26}
$$

where $\sigma^{\mu\nu} = (i/2)[\gamma^\mu, \gamma^\nu]$ are the generators of spinor transformations determined by Lorentz transformation. We will use in the following the γ-matrix notation of Bjorken and Drell.[38] Of course Eqs. (25) and (26) are 4 x 4 matrix equations. From Eqs. (23) and (24) we have for the Eq. (25)

$$\partial^\mu \left[\frac{\Box\rho}{2m} \rho^{-1} - \frac{\partial\phi}{\partial\tau} - \frac{(A_\nu + \partial_\nu\phi)(A^\nu + \partial^\nu\phi)}{2m} + \frac{1}{4m} \sigma_{\nu\lambda} F^{\nu\lambda} \right] = 0 \tag{27}$$

which shows that the term in the brackets is constant. If we pose this constant equal to zero, by rearranging the energy scale, Eq. (27) takes the following form:

$$\left(\Box - 2m \frac{\partial\phi}{\partial\tau} - \partial^\nu\phi\partial_\nu\phi - A_\nu A^\nu - 2A_\nu\partial^\nu\phi + \frac{1}{2} \sigma_{\mu\nu} F^{\mu\nu} \right)\rho = 0 \tag{28}$$

Similarly for Eq. (26) we have from Eqs. (23) and (24)

$$(2\partial_\nu\phi\partial^\nu + \Box\phi + \partial_\nu A^\nu + 2A_\nu\partial^\nu)\rho = 0 \tag{29}$$

if now we make a linear combination of Eqs. (28) and (29) with coefficients 1 and i, after multiplication by $\exp(i\phi)$ we have

$$\Big[(\Box + 2i\partial^\nu\phi\partial_\nu + i\Box\phi - \partial_\nu\phi\partial^\nu\phi - A_\nu A^\nu + i2A^\nu + 2iA^\nu\partial_\nu - 2A_\nu\partial^\nu\phi)\rho \tag{30}$$

$$+ \frac{1}{2} \sigma_{\mu\nu} F^{\mu\nu}\rho - 2m \frac{\partial\phi}{\partial\tau} \rho \Big] \exp(i\phi) = 0$$

which finally takes the following synthetic form:

$$\left[(i\partial_\nu - A_\nu)(i\partial^\nu - A^\nu) - 2im \frac{\partial}{\partial\tau} - \frac{1}{2} \sigma_{\mu\nu} F^{\mu\nu} \right] \exp(i\phi)\rho = 0 \tag{31}$$

if now we multiply this matrix equation for an arbitrary constant (x independent) spinor w and if we pose $\varphi(x, \tau) = \exp(i\phi)\rho w$ we have

$$\left[(i\partial_\nu - A_\nu)(i\partial^\nu - A^\nu) - 2im \frac{\partial}{\partial\tau} - \frac{1}{2} \sigma_{\mu\nu} F^{\mu\nu} \right] \varphi(x, \tau) = 0 \tag{32}$$

which, with the positions

$$\phi(x, \tau) = S(x) - \frac{m\tau}{2} \quad \text{and} \quad \psi(x) = \exp(iS)\rho w \tag{33}$$

gives exactly the Feynman–Gell-Mann equation

$$[(i\partial_\nu - A_\nu)(i\partial^\nu - A^\nu) - \tfrac{1}{2}\sigma^{\mu\nu}F_{\mu\nu}]\psi(x) = m^2\psi(x) \tag{34}$$

The connection between the ρ^2 matrix and the four-component Dirac spinors ψ is clarified if we consider the four spinors $\psi^r = \exp(iS)pw^r$, where w^r ($r = 1, 2, 3, 4$) are constant spinors such that (for $\overline{w}^r = w^{r+}\gamma_0$)

$$\overline{w}^r w^{r'} = \epsilon_r \delta_{rr'} \qquad \epsilon_r = \begin{cases} +1, & r = 1, 2 \\ -1, & r = 3, 4 \end{cases} \tag{35}$$

$$\sum_r \epsilon_r w^r \overline{w}^r = I$$

For example we could take

$$w_1 = \begin{pmatrix} 1 \\ 0 \\ 0 \\ 0 \end{pmatrix}, \qquad w_2 = \begin{pmatrix} 0 \\ 1 \\ 0 \\ 0 \end{pmatrix}, \qquad w_3 = \begin{pmatrix} 0 \\ 0 \\ 1 \\ 0 \end{pmatrix}, \qquad w_4 = \begin{pmatrix} 0 \\ 0 \\ 0 \\ 1 \end{pmatrix} \tag{36}$$

It is now easy to show that

$$\rho^2 = \sum_r \epsilon_r \psi^r \bar{\psi}^r, \qquad \mathrm{Tr}\, \rho^2 = \sum_r \epsilon_r \bar{\psi}^r \psi^r \tag{37}$$

are coherent with our requirement of a trace invariant ρ^2 matrix. This derivation can now be extended to spin-0 charged particles if in Eq. (24) we suppose that $\phi(x, \tau)$ and $\rho(x)$ are real functions (instead of matrixes) and that in Eq. (25) we have only the Lorentz term $b_\nu F^{\nu\mu}$ without spin term. Here, of course, $\psi(x) = \rho e^{iS}$, and Eq. (34) aduces to the Klein–Gordon equation. This spin-0 equation was, of course, the first equation deduced (Guerra and Ruggioro,[21] Lehr and Park[12]) from a covariant stochastic formalism. Finally for spin-1 uncharged (Proca) particles we can derive the wave equation with the same formalism used for spin-½ particles [here, of course, $\rho(x)$ and $\phi(x, r)$ remain 4 x 4 real diagonal matrixes, Eq. (25) becomes a free equation, and the vector field $A_\mu(x)$ of Proca particles will be $A_\mu(x) = \exp(iS)\rho c_\mu$, where c_μ is a constant four-vector]; but we can also obtain it with an "ad hoc" physical hypothesis without introducing ρ and ϕ as matrixes. Namely,[19] if ϕ and ρ are real functions, exactly as for spin-0 particles, we can use as stochastic elements the classical spinning particles of Weyssenhoff and Raabe.[39] As one knows[22] we can represent each individual particle of that type by a spacelike vector R_μ (representing the distance between the spinning particle's center of mass and the center of matter density) and thus substitute for the scalar field density ρ of the spin-zero particles the new density $\rho = a_\nu a^\nu$, where $a^\nu = \rho^{1/2} R^\nu$ is the real amplitude of a more general complex vector field whose phase factor disappears in making the scalar product for ρ. Now a straightforward calculation[21] shows that, if we choose $\partial\rho/\partial\tau = 0$ (because we are only interested, for the time being, in the time-independent wave equations) Nelson's equation and the continuity equation become, respectively,

$$\partial^\mu \left(\frac{T^{\alpha\nu\beta} T_{\alpha\nu\beta}}{2\rho^2} + \frac{a^\nu \Box a_\nu}{\rho} - \partial_\nu \Phi \partial^\nu \Phi - 2m \frac{\partial \Phi}{\partial \tau} \right) = 0 \tag{38}$$

$$a^\mu (2\partial^\nu a_\mu \cdot \partial_\nu \Phi + a_\mu \Box \Phi) = 0 \tag{39}$$

where

$$T^{\alpha\nu\beta} = a^\alpha \partial^\nu a^\beta - a^\beta \partial^\nu a^\alpha \tag{40}$$

is a tensor equal to zero in the case of a scalar field.

We can now easily derive Proca's equations. Indeed, if we recall that $a^\mu = \rho^{1/2} R^\mu$ with $R^\mu R_\mu = 1$, it is easy to show that

$$T^{\alpha\nu\beta} T_{\alpha\nu\beta} / 2\rho^2 = R^\nu \Box R_\nu \tag{41}$$

So that, if we further complete our model by the hypothesis that the constant vector R^ν satisfies an equation like

$$\Box R^\nu = c_0 R^\nu \tag{42}$$

where c_0 is a constant, we can reduce Eq. (38) to

$$\frac{a^\mu \Box a_\mu}{\rho} - \partial^\nu\Phi\partial_\nu\Phi - 2m\frac{\partial\Phi}{\partial\tau} = E \tag{43}$$

where E is a constant that we can make equal to zero by rearranging the energy scale. Equation (38) then becomes

$$a^\mu\left[\Box a_\mu - \left(\partial^\nu\Phi\partial_\nu\Phi + 2m\frac{\partial\Phi}{\partial\tau}\right)a_\mu\right] = 0 \tag{44}$$

As said in the beginning, in order to have Eq. (42) it is sufficient to adopt the classical Weyssenhoff model. As shown by Halbwachs,[22] in each particle's rest frame Eq. (42) becomes $\ddot{R}^\nu = c_0 R^\nu$, which yields exactly Weyssenhoff's equation for the free spinning particle. Moreover, in our model R^ν (the constant vector which joins the center of matter to the center of gravity of the Weyssenhoff particle) is also perpendicular to the velocity $b^\mu = \partial^\mu\Phi/m$, so that we have[22] $a_\mu\partial^\mu\Phi = 0$. Now if we consider Eqs. (39) and (44) as representing, respectively, the imaginary and real part of a total equation, it can easily be shown that (with $B_\mu = a_\mu \exp(i\Phi)$ we get

$$B^*_\mu\{[(1/2m)\Box + i\partial]/\partial\tau\}B^\mu = 0 \tag{45}$$

an equation which reduces to the following set of equations:

$$-(1/2m)\Box B_\mu = i\partial B_\mu/\partial\tau \tag{46}$$

by a Lorentz transformation which allows only one B^μ to be different from zero.

Finally, if we follow the tentative suggestion of Feynman[40] in the way proposed by Guerra and Ruggiero[22] and assume for B_μ the proper-time dependence

$$B_\mu(x, \tau) = \exp(-im\tau/2)A_\mu(x) \tag{47}$$

we find the classical set of Proca equations:

$$(\Box + m^2)A_\mu = 0 \tag{48}$$

5. Conclusion

We conclude our exposition with two remarks. The first remark is that our extended (to spin) Dirac's aether model[41] imposes different Markov processes on different classical objects which are thus described by different wave equations, i.e., the usual Klein–Gordon, Proca and Dirac equations. In that sense SIQM goes beyond the

Copenhagen interpretation of the quantum mechanics (CIQM), which introduces them *a priori* without theoretical justification.

The second remark is that the use of a real material subquantal aether implies predictions not contained in CIQM. If particles indeed behave as oscillators moving in phase with a surrounding wave of oscillations (somewhat like a plane flying at Mach-1 in its surrounding sound wave) the corresponding objective reality of de Broglie's waves opens the way to new types of experiments (proposed by Garuccio and Vigier[6] and Fitchard[42]) in which SIQM and CIQM predict theoretically different testable results. The decision on the Bohr–Einstein controversy thus now rests on the shoulders of experimental physicists.

References and Notes

1. A. Einstein, B. Podolski, and N. Rosen, *Phys. Rev.* **47,** 777, (1935).
2. A. Aspect, *Phys. Rev. Lett.* **14D,** 1944, (1976).
3. J. -P. Vigier, *Lett. Nuovo Cimento* **24,** 258,265, (1979).
4. P. A. M. Dirac, *Nature* **168,** 906, (1951).
5. N. Cufaro -Petroni and J. -P. Vigier, *Int. J. Theor. Phys.* **18,** 807, (1979).
6. A. Garuccio and J. -P. Vigier, *Found. Phys.* **10,** 797 (1980).
7. N. Cufaro -Petroni and J. -P. Vigier, *Lett. Nuovo Cimento* **25,** 151 (1979).
8. N. Cufaro -Petroni and J. -P. Vigier, *Lett. Nuovo Cimento* **26,** 149 (1979).
9. N. Cufaro -Petroni, A. Garuccio, F. Selleri, and J. -P. Vigier, *C. R. Acad. Sci. Paris* **290B,** 111 (1980).
10. D. Bohm and J. -P. Vigier, *Phys. Rev.* **96,** 208 (1954); **109,** 1882 (1958).
11. L. de Broglie, *La thermodynamique de la particule isolèe* (Gauthier-Villars, Paris, 1964).
12. W. Lehr and J. Park, *J. Math. Phys.* **18** 1235, (1977).
13. Ya. P. Terletski and J. -P. Vigier, *Sov. Phys.-JETP* **13,** 356 (1961).
14. F. Halbwachs, J. -M. Souriau, and J. P. Vigier, *J. Phys. Rad.* **22,** 26 (1961).
15. C. Fenech, M. Moles, and J. -P. Vigier, *Lett. Nuovo Cimento* **24,** 56 (1979).
16. K. P. Sinha, C. Sivaram, and E. C. G. Sudarshan, *Found. Phys.* **6,** 65, 717 (1976); K. P. Sinha and E. C. G. Sudarshan, *Found. Phys.* **8,** 823 (1978).
17. L. de la Peña-Auerbach and A. M. Cetto, *Phys. Rev. D* **3,** 795 (1971).
18. T. Marshall, *Proc. R. Soc. London* **276A,** 475 (1963).
19. N. Cufaro -Petroni and J. -P. Vigier, *Phys. Lett* **73A,** 289 (1979).
20. L. de la Peña-Auerbach, *J. Math. Phys.* **12,** 453 (1971).
21. F. Guerra and P. Ruggiero, *Lett. Nuovo Cimento* **23,** 529 (1978).
22. F. Halbwachs, *Théorie relativiste des fluides à spin* (Gauthier-Villars, Paris, 1960).
23. T. Marshall, *Proc. Cambr. Phil. Soc.* **61,** 537 (1965).
24. M. Flato, G. Rideau, and J. -P. Vigier, *Nucl. Phys.* **61,** 250 (1965).
25. L. de la Peña-Auerbach and A. M. Cetto, *Found. Phys.* **5,** 355 (1975).
26. J. -P. Vigier, *C. R. Acad. Sci. Paris* **266,** 598 (1968).
27. A. Einstein, *Investigations on the Theory of Brownian Movement* (Dover, New York, 1956).
28. V. J. Lee, *Found. Phys.* **10,** 77 (1980).
29. R. Dugas, *La thèorie physique au sens de Boltzmann* (Griffon, Neuchatel, 1959).
30. R. C. Tolman, *Theory of the Relativity of Motion* (University of California Press, Berkeley 1918); *Relativity Thermodynamics and Cosmology* (Oxford U.P., Oxford, 1935).
31. C. Eckart, *Phys. Rev.* **58,** 919 (1940).
32. L. de Broglie, *C. R. Acad. Sci. Paris* **253,** 1078 (1961).
33. E. Nelson, *Phys. Rev.* **150,** 1079 (1966).
34. E. Cartan, *The Theory of Spinors* (Hermann, Paris, 1966).
35. W. Pauli, *Theory of Relativity* (Pergamon Press, London, 1958).

36. P. Hillion and J. -P. Vigier, *Nucl. Phys.* **16,** 360 (1960).
37. R. P. Feynman and M. Gell-Mann, *Phys. Rev.* **109,** 193 (1958).
38. J. D. Bjorken and S. D. Drell, *Relativistic Quantum Mechanics* (McGraw-Hill, New York, 1964).
39. J. V. Weyssenhoff and A. Raabe, *Acta Phys. Pol.* **9,** 7 (1947).
40. R. P. Feynman, *Rev. Mod. Phys.* **20,** 367 (1948).
41. J. -P. Vigier, *Lett. Nuovo Cimento* **29,** 467 (1980).
42. E. E. Fitchard, *Found. Phys.* **9,** 525 (1979).

22

New Quantum Mechanics

C. Piron
Département de Physique Théorique
Université de Genève
1211 Genève 4, Switzerland

During the last twenty years, our understanding of quantum mechanics has made much progress and the formalism itself has been generalized. This is due to the efforts of four research groups,[1–4] which have studied different aspects of the theory. The groups of Marburg[1] and Geneva[4] have worked on the interpretation and the formalism of quantum theory itself. Randall and Foulis,[3] on the other hand, have developed generalizations of the usual statistical formalism, while Mittelstaedt and his co-workers have concentrated on the philosophical aspect of quantum mechanics. As a result, during these last years, many small modifications have been introduced into quantum theory. Put together, these small changes define a quite different theory—the new quantum theory. We want to briefly review here some of these changes. Our article is divided into four parts, treating the following subjects:

1. The realistic interpretation.
2. Continuous superselection variables, formalism, and quantization.
3. New types of Schrödinger equations.
4. Dissipative deterministic dynamics.

At the conclusion we shall evoke the celebrated Einstein–Podolsky–Rosen article[5] and furnish a proof of a generalization of Bell's inequalities which are also applicable to quantum models.

1. The Realistic Interpretation

The orthodox interpretation of quantum mechanics is associated with the Copenhagen interpretation, which is more of a framework than an interpretation in itself.

The so-called statistical interpretation is well known. It is an attempt to give a complete self-consistent interpretation of the quantum formalism. It is not in the line of the Copenhagen school; it is rather the opposite of a realistic interpretation for the very reason that here the concept of property has no meaning at all for one given individual system. Many physicists have adopted the statistical interpretation. The general feeling is that it is the only possible one. In fact, it is the one that has been most emphasized in textbooks. In spite of many apparent difficulties, Einstein's wish has been realized: A realistic interpretation, in terms of "elements of reality,"[5] has been constructed. Let us now describe this new interpretation.

By a *physical system*, we understand a part of reality conceived as existing in space and time and which is isolated or has been isolated from the rest of the universe. A theory is nothing else than a model, a mathematical model sufficiently precise to describe reality as it is, and sufficiently idealized to give intuition on its reactions to the exterior actions of the physicist. The notion of *property* is the basis of such a model. From the realistic point of view, the system *is* and it is what it is. It possesses different properties, and whether they are known or not to the physicist does not change anything of the reality of the system itself. We have to describe these properties and not at first to justify or to explain the physicist's knowledge about them. When we believe that the system actually possesses a certain property, we may decide to test this belief by performing a suitable experiment. If by performing this one we obtain what we expected, we are reinforced in our original belief, but if we obtain something else we must admit our mistake and change the theory accordingly. Thus, we can never prove a theory, we only have the possibility to falsify it. In conclusion, a physical property is always associated with some experiment and a well-defined result. If a given system is or has been prepared such that one can confirm that, in the event of this experiment, the expected result would be certain, then we will say that the corresponding property is an *actual* property (an element of reality) of the system. Moreover, other possible properties which are not "actually" actual will be said to be *potential.* The complete set of all actual properties is by definition the *state* of the system. By itself or under the influence of the exterior the system changes. This means that the state changes, some actual properties becoming potential while others, potential at the start, appear in actuality.

The main problem is to exhibit a mathematical framework capable of describing the complete family of all the actual and potential properties of the system, its possible states, and the structure inherited from these notions. To do this, we proceed in the following way. We first define the concept of *question.* A question is any experimental proposal leading to a well-defined alternative for which the terms are the "yes" corresponding to the expected result and the "no" corresponding to any of the others. Given a question α we can define an inverse, the question $\alpha^{\sim}$ obtained by exchanging the terms of the alternative, i.e., the "yes" and the "no." Given a family of questions α_i, we can define a product, the question $\pi\alpha_i$ corresponding to the following proposal. In order to perform $\pi\alpha_i$, first of all choose one of the α_i; you are free in your choice and must not always make the same choice. You then perform this α_i and attribute to $\pi_i\alpha_i$ the answer obtained for α_i. Let us now consider the set Q of all the questions which can be defined in any reasonable way for a given system. We can choose Q closed for the inverse and the product. Moreover Q contains the trivial question I, which consists in doing anything with the system and stating the answer "yes" each time; it also contains the absurd question, 0, which is just the inverse of I. With each question of Q a property of the system is associated by the following definition:

The question is "true," or the corresponding property is "actual," if the given system is such that if we would actually perform the question the answer "yes" would be certain. In the other case, the property is potential, as we have explained before.

Having thus defined the fundamental objects of the theory, let us define a structure on these objects. A question α is said to be *stronger* than a question β, i.e., $\alpha < \beta$, if whenever α is true β is true as well. Such a relation is transitive and defines a (partial) ordering on the equivalence classes. Two questions α and β are *equivalent* if and only if $\alpha < \beta$ and $\beta < \alpha$. An equivalence class is called a *proposition* and there is a one-to-one correspondence between propositions and properties.

Theorem. The set $\mathcal{L}$ of all propositions, obtained as the quotient of Q by the equivalence relation, is a complete lattice for the ordering relation.

Proof. We must prove the existence of a greatest lower bound and a least upper bound for each subset of $\mathcal{L}$. Let a_i, $i \in J$ be a subset of $\mathcal{L}$ and choose in each a_i a question α_i. Let us consider the question product $\pi\alpha_i$. Since Q is closed for the product, $\pi_J\alpha_i$ defines a proposition of $\mathcal{L}$ which is nothing else than the greatest lower bound $\wedge_J a_i$, since by definition of the product

$$\text{“}\alpha_i \text{ is true”} \ \forall i \in J \text{ iff “}\pi_{J\alpha_i} \text{ is true”}$$

On the other hand, the least upper bound $\vee_J a_i$ exists also, since this is nothing than the greatest lower bound of all upper bound of all the a_i, $i \in J$.

As one can see, the logical conjunction "and" of two properties defined by "a is actual" and "b is actual" is equivalent to a property, the greatest lower bound "$a \wedge b$ is actual." However, the logical disjunction "or," which by definition is "a is actual" or "b is actual," is not equivalent to a property, since to test the affirmation "a is actual" or "b is actual" we must in general perform two different experiments on the same system.

In addition to the lattice structure, $\mathcal{L}$ admits another one, the compatibility. A compatible complement of a proposition a is a proposition b such that

(I) a and b are complements, i.e.,

$$a \wedge b = 0 \quad \text{and} \quad a \vee b = I$$

(II) There exists a question $\alpha \in Q$ such that

$$\alpha \in a \quad \text{and} \quad \alpha^\sim \in b$$

The first axiom is the following:

Axiom C. For each proposition of $\mathcal{L}$ there exists at least one compatible complement.

For example, let us suppose that we have a beam of photons (linearly polarized). The proposal for an experiment which consists in putting a polarizer in the beam defines a question. In fact, it is possible to verify, by dispatching photons one by one, that this experiment leads to an alternative: either the photon passes through—this is the answer "yes"—or it is absorbed, and this is the answer "no." We shall define the proposition a_ϕ by specifying the orientation of the polarizer, i.e., the angle ϕ. Experience shows that to obtain a photon in such a way that "a_ϕ is true" it is sufficient to consider the photons which have traversed a first polarizer oriented at this angle ϕ. But experience also shows that it is impossible to prepare photons that are able to cross with complete certainty a polarizer oriented at the angle ϕ as well as another oriented at an angle $\phi' \neq \phi$ (modulo π). In such a case a_ϕ and $a_{\phi'}$ are complements but only if $\phi' = \phi + \pi/2$ (modulo π) are they compatible complements. As we see here, Axiom C is satisfied. For a discussion of its validity in a more general case, see B. Mielnik.[6]

We are now in a position to characterize the lattice of properties describing a classical system.

Classical Hypothesis. For each pair a,b of compatible complements we have whatever is the state

"a is actual" or "b is actual"

Such characterization is given by the following theorem:

Theroem. If the lattice of properties satisfies Axiom C and the classical hypothesis, it is isomorphic to $\mathcal{P}(\Omega)$, the lattice of subsets of Ω, the set of all states of the system.

Proof. Let us define μ as the map which applies the property $a \in \mathcal{L}$ on $\mu a \in \mathcal{P}(\Omega)$, the subset of all states such that "a is actual." We have to prove that μ is an isomorphism. According to the relation of equivalence it is trivial that μ is one-to-one. Moreover, μ preserves the greatest lower bound:

$$\mu(\bigwedge_I a_i) = \bigcap_I \mu a_i$$

From Axiom C it follows that with each property a is associated at least one compatible complement, written a', and $\mu a'$ is contained in the complement of the subset μa. If the classical hypothesis is satisfied, $\mu a'$ is exactly the complement of μa and then the compatible complement is unique. Before proving that μ is onto, let us show that each point of Ω is in the image. Given a state E, i.e., a point of Ω, the greatest lower bound p of all properties actual for E is also actual and it is the strongest property of E. If μp contains another state $S \neq E$, then q, the strongest property of S, would not be actual for E and, according to the classical hypothesis, we would have the contradiction

$$q < p < q'$$

where q' is the compatible complement of q. Let $x \in \mathcal{P}(\Omega)$; x is the image of a property—it is the greatest lower bound of the compatible complements p' associated with the strongest properties p defined by the states E not in x.

This theorem is a justification of the usual description of a classical system in terms of phase space. Here Ω plays the role of phase space and, according to the theorem, each subset of Ω represents a property which is actual if and only if the state is in the subset.

The existence of pairs of compatible complements, and of states of a system for which neither one nor the other of the properties is actual, is an experimental fact. It is not a question of logic. From a principle for the deterministic evolution of system, whatever that may be, we cannot conclude that the result of a hypothetical experiment is certain. A counterexample of the classical hypothesis is furnished by the experiment on thermal neutrons by means of a perfect crystal interferometer (see Fig. 1).[4] The pair of properties is the presence of the neutron in a half-space (region I) or in the complementary half-space (region II). It is a pair of compatible complements, and the experiment consists in applying a magnetic field in one of these regions. If the neutron is actually present in this region, its spin should turn through an angle φ according to the Larmor precession. In particular, if $\varphi = 2\pi$, nothing should change. It would evidently be the same if the neutron was actually present in the other region.

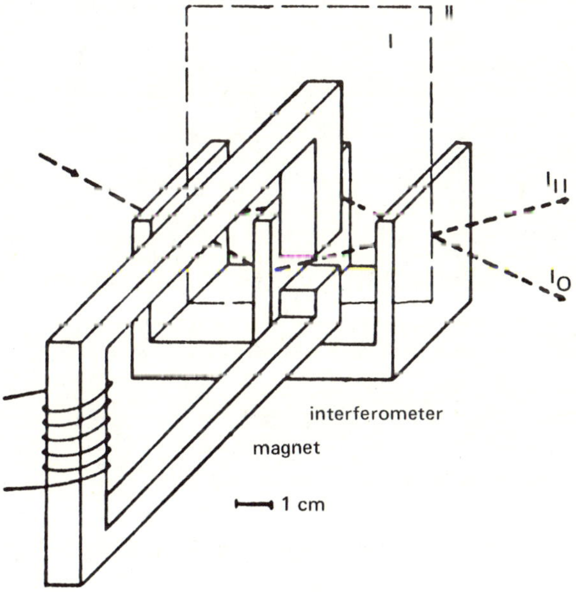

FIGURE 1. Thermal neutrons interacting with a perfect crystal interferometer.

However, such a field considerably modifies the counting rate at the exterior of the crystal interferometer. Accordingly, at the instant considered, the neutron is actually present neither in region I nor in region II, thus contradicting the classical hypothesis (Fig. 2).

Since the classical hypothesis is thus not always satisfied, it must be dropped and replaced by some weaker hypotheses. Proceeding in this way, one has constructed a formalism of general quantum theory which encompasses within the same framework and the same interpretation classical as well as quantum systems. Thus the dichotomy between classical and quantum descriptions is definitely overcome.

2. Continuous Superselection Variables—Formalism and Quantization

We want to give here just the result and not all the details of the axiomatic construction.[8] Moreover, we will not consider the most general case, but restrict our attention to a particular case which is sufficiently general to contain both the classical theory and the usual quantum theory.

Let us take a family of separable complex Hilbert spaces H_α, where $\alpha \in \Omega$ and Ω is some set of indices. We define the lattice of properties $\mathcal{L}$ by the following:

Each family of projectors $\{P_\alpha\}$, where P_α acts on H_α, represents a property.

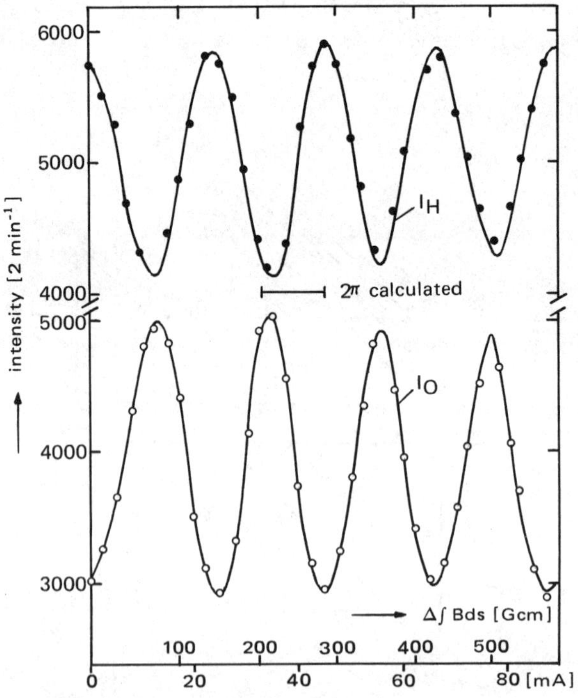

FIGURE 2. Measurement data for experimental setup of Fig. 1.

Each ray defined by a unit vector $\psi_{\alpha 0}$ in one of the Hilbert spaces represents a state.

By definition, "$\{P_\alpha\}$ is actual" if and only if $P_{\alpha 0}\psi_{\alpha 0} = \psi_{\alpha 0}$. According to this definition, the ordering relation

$$\{P_\alpha\} < \{Q_\alpha\}$$

is equivalent to

$$P_\alpha Q_\alpha = P_\alpha \qquad \forall \alpha \in \Omega$$

Finally, the compatible complement associated with $\{P_\alpha\}$ must be taken as $\{P_\alpha\}' = \{I_\alpha - P_\alpha\}$. Also the compatible complement is unique and it defines an orthocomplementation on $\mathcal{L}$, i.e.,

$$\{P_\alpha\} < \{Q_\alpha\} \Rightarrow \{Q_\alpha\}' < \{P_\alpha\}'$$

since we have

$$P_\alpha Q_\alpha = P_\alpha \Rightarrow (I_\alpha - Q_\alpha)(I_\alpha - P_\alpha) = I_\alpha - Q_\alpha$$

The pair $\{X_\alpha\}$, $\{I_\alpha - X_\alpha\}$ satisfies the classical hypothesis if and only if X_α is 0_α or I_α, $\forall \alpha \in \Omega$. Such a property $\{X_\alpha\}$ is called a *classical property*. The set of all classical properties is a sublattice of $\mathcal{L}$ which is isomorphic to $\mathcal{P}(\Omega)$. The set Ω is called the *superselection space*. If Ω contains only one point, $\mathcal{L}$ reduces to the usual quantum case, where $\mathcal{L}$ is $\mathcal{P}(H)$ the lattice of the projectors on a given Hilbert space H. On the contrary, if each H_α is of dimension 1, $\mathcal{L}$ is identical with $\mathcal{P}(\Omega)$ and describes a classical system. In general, each $\mathcal{P}(H_\alpha)$ is a sublattice of $\mathcal{L}$ and $\mathcal{L}$ itself is called $\vee \mathcal{P}(H_\alpha)$, the direct union of the $\mathcal{P}(H_\alpha)$.

Given a self-adjoint operator A in the space H there exists a spectral family, i.e., a map E from the Borel sets of R to the projectors of H, which decomposes this operator. $\mathcal{B}$, the set of equivalence classes of Borel sets, modulo an element of the kernel of E, is a complete Boolean lattice. The quotient of E by this equivalence relation is a map $\underline{A}$ which conserves the greatest lower bound and the orthocomplementation. Conversely, each map from a complete Boolean lattice to the projectors of a complex separable Hilbert space, which conserves the greatest lower bound and the orthocomplementation, can be obtained in this way.[8] Physically, such a map defines a correspondence between properties of the measuring apparatus and properties of the system. In the same way, in the general case, given a family of self-adjoint operators $\{A_\alpha\}$, we define a complete Boolean lattice $\mathcal{B}$ as $\vee \mathcal{B}_\alpha$, the set of families of elements taken in the $\mathcal{B}_\alpha$, and a map $\underline{A}$ by the relation $\underline{A}\{\Delta_\alpha\} = \{A_\alpha \Delta_\alpha\}$, where $\Delta_\alpha \in \mathcal{B}_\alpha$. According to the physical interpretation just given, we will call such a map $\underline{A}$ an observable and there will be no confusion in the following with the associated family of self-adjoint operators $\{A_\alpha\}$.

Let us consider such an observable $\underline{A}$: $\mathcal{B} \rightarrow \mathcal{L}$. The elements of $\mathcal{B}$ correspond to some properties of the apparatus, for example the different possible positions of the pointer. In a very natural way a group acts on the scale of the apparatus. This group is characteristic for the physical nature of the observable. Mathematically, this leads us to define an observable as a solution of an imprimitivity system.[9] Given a group G, a complete Boolean lattice $\mathcal{B}$ and a representation σ of the group G in the automorphisms of $\mathcal{B}$, we will say that the observable $\underline{A}$ is a solution of an imprimitivity system based on G if the following diagram is commutative:

$$\begin{array}{ccc} \mathcal{B} & \xrightarrow{A} & \mathcal{L} \\ \sigma\downarrow & & \downarrow S \\ \mathcal{B} & \xrightarrow{A} & \mathcal{L} \end{array}$$

where S is a representation of G by automorphisms of $\mathcal{L}$, i.e., according to the generalized Wigner theorem,[10] by transformations of the type

$$\{P_\alpha\} \mapsto \{Q_\alpha = U_{p\alpha^{-1}} P_{p\alpha^{-1}} U^{-1}_{p\alpha^{-1}}\}$$

where p is a permutation of the indices of Ω and U_α is a unitary or antiunitary map of the space H_α onto the space $H_{p\alpha}$.

With these concepts we are now able to explain what we will call the quantization of a physical system. We will take two examples: the Galilean and the Lorentz particles.

An elementary Galilean particle is by definition a physical system described by a lattice $\mathcal{L}$ which admits the observables momentum $\underline{\mathbf{p}}$, position $\underline{\mathbf{q}}$, time $\underline{t}$, and no other compatible and independent observables. The group G corresponds to the changes of the choice of the linear scales on the apparatuses measuring $\underline{\mathbf{p}}$, $\underline{\mathbf{q}}$, $\underline{t}$. This group is the semidirect product of R^7 by $SO(3)$, defined by the momentum translations $\boldsymbol{\pi}$, the space translations $\mathbf{a}$, the time translations τ, and the rotations R acting on $\boldsymbol{\pi}$ and $\mathbf{a}$ in the same way. There exist two and only two elementary Galilean particles which are at the same time solutions of the three imprimitivity systems $\underline{\mathbf{p}}$, $\underline{\mathbf{q}}$ and $\underline{t}$[11]:

(a) *The classical particle:* $\mathcal{L}$ is the lattice $\mathcal{P}(R^7)$ of the subsets of $R^7 = (\mathbf{p},\mathbf{q},t)$, S is the canonical representation, and

$$\begin{aligned} \underline{p}(\Delta) &= \{\mathbf{p},\mathbf{q},t) \mid \mathbf{p} \in \Delta, (\mathbf{p},\mathbf{q},t) \in R^7\} \\ \underline{q}(\Delta) &= \{\mathbf{p},\mathbf{q},t) \mid \mathbf{q} \in \Delta, (\mathbf{p},\mathbf{q},t) \in R^7\} \\ \underline{t}(\Delta) &= \{\mathbf{p},\mathbf{q},t) \mid t \in \Delta, (\mathbf{p},\mathbf{q},t) \in R^7\} \end{aligned}$$

(b) *The quantal particle:* $\mathcal{L}$ is defined via a family of complex Hilbert spaces $\{H_t\}$, $t \in R$, all identical to $L^2(R^3,dv)$. S is induced by the following unitary transformations:

$$\begin{aligned} (U(\boldsymbol{\pi})\phi)_t(\mathbf{x}) &= \exp(i\hbar^{-1}\boldsymbol{\pi}\mathbf{x})\phi_t(\mathbf{x}) \\ (U(\mathbf{a})\phi)_t(\mathbf{x}) &= \phi_t(\mathbf{x} - \mathbf{a}) \\ (U(\tau)\phi)_t(\mathbf{x}) &= \phi_{t-\tau}(\mathbf{x}) \\ (U(R)\phi)_t(\mathbf{x}) &= \phi_t(R^{-1}\mathbf{x}) \end{aligned}$$

where $\hbar$ is Planck's constant. The observables $\underline{\mathbf{p}}, \underline{\mathbf{q}}$, and $\underline{t}$ themselves are defined by the families of self-adjoint operators:

$$\{p_t^k = -i\hbar\partial_k\}, \qquad \{q_t^k = x^k\}, \qquad \text{and } \{t_t = tI\}$$

As we see from this model, due to the superselection formalism, we have been able to describe the time as an observable. Since neither $\mathbf{p}_t$ nor $\mathbf{q}_t$ explicitly depend on t, the representation given here is the Schrödinger picture. By performing an automorphism of $\mathcal{L}$ induced by a family of unitary operators $\{U_t\}$ we can pass to another representation. For example for $\{U_t = \exp[i(p_t^2/2m)t]\}$ we obtain the interaction picture. In this representation we must remember that the $\mathbf{q}_t = \mathbf{x} - (\mathbf{p}_t/m)t$ act for different values of t on different Hilbert spaces and are thus compatible, in spite of the fact that as operators in $L^2(R^3,dv)$ they do not commute.

Similarly, we can define an elementary Lorentz particle as a physical system described by a lattice $\mathcal{L}$ which admits the observables momentum-energy $\underline{p}^\mu = (\underline{\mathbf{p}},\underline{E}/c^2)$, position in space time $\underline{q}^\mu = (\underline{\mathbf{q}},\underline{t})$, and no other compatible and independent observables. The corresponding group G is now the semidirect product of R^8 by SO(3,1) defined by the momentum-energy translations π^μ, the space-time translations a^μ, and the Lorentz transformations Λ_ν^μ acting on π^μ and a^μ. Again, there exist two and only two Lorentz particles which are at the same time solutions of the imprimitivity systems $\underline{p}^\mu$ and $\underline{q}^\mu$ [11]:

(a) The classical particle: $\mathcal{L}$ is the lattice $\mathcal{P}(R^8)$ of the subsets of $R^8 = (p^\mu,q^\mu)$, S is the canonical representation, and

$$\underline{p}^\mu(\Delta) = \{(p^\mu,q^\mu) \mid p^\mu \in \Delta, (p^\mu,q^\mu) \in R^8\}$$
$$\underline{q}^\mu(\Delta) = \{(p^\mu,q^\mu) \mid q^\mu \in \Delta, (p^\mu,q^\mu) \in R^8\}$$

(b) The quantal particle: $\mathcal{L}$ is defined via the complex Hilbert space $L^2(R^4,dv\,dt)$. S is induced by the following unitary transformations:

$$[U(\pi^\mu)\phi](x^\mu) = \exp(i\hbar^{-1}g_{\mu\nu}\pi^\mu x^\nu)\phi(x^\mu)$$
$$[U(a^\mu)\phi](x^\mu) = \phi(x^\mu - a^\mu)$$
$$[U(\Lambda_\nu^\mu)\phi](x^\mu) = \phi(\Lambda_\nu^{-1\mu}x^\nu)$$

where $g_{\mu\nu} = (1,1,1,-c^2)$. The observables $\underline{p}^\mu$ and $\underline{q}^\mu$ are defined by the self-adjoint operators:

$$p^\mu = -i\hbar g^{\mu\nu}\partial_\nu \quad \text{and} \quad q^\mu = x^\mu$$

Again, the representation given here is the Schrödinger picture.

If we admit additional observables in these models, we obtain the spin-½ particles as new solutions of the same imprimitivity systems.[12]

Manifestly, the two quantal particles, the Galilean and the Lorentz, are two very different models. In the Lorentz model we have one more observable, $E = i\hbar\partial_t$, thus

t does not define a superselection variable as in the Galilean model. If we choose a frame and if we impose in the Lorentz model the corresponding coordinate t to be superselection variables, we obtain by integral decomposition a new model which turns out to be the same as the Galilean one. Obviously, such a model is not manifestly covariant for pure Lorentz transformations. This is due to the space-time breaking symmetry introduced by our choice of a time direction.

We shall see when we treat the dynamics that this symmetry breaking corresponds to a well-defined physical situation.

The advantage of our method of the so-called "quantization of a system" is that it has nothing to do with a quantization, i.e., with a procedure to add quantum corrections to an arbitrarily given classical model. We treat the classical and the quantum case on the same footing. We do not consider one model as better or more exact than the other. Of course, particles here are very idealized objects; nevertheless, both, classical and quantum systems exist in the real world. A good theory must take account of this reality.

3. New Type of Schrödinger Equations

By definition, evolution is the change of the state of the system. In the ideal case of a reversible evolution, with constant entropy, it is usually postulated that the change is described by an automorphism of the lattice $\mathcal{L}$ which induces a representation of the one-parameter group. For a physical system described by a family of complex Hilbert spaces H_α, $\alpha \in \Omega$, all identical to a given one, this condition, supplemented with some differentiability conditions, implies that the evolution is governed by the following generalized Schrödinger equations:

$$i\partial_\tau \phi_\alpha = H_\alpha \phi_\alpha \tag{1}$$

and

$$\dot{\alpha} = X(\alpha) \tag{2}$$

where $\{H_\alpha\}$ is a family of self-adjoint operators and X is a vector field on Ω.

Formally, given an observable by its family of self-adjoint operators $\{A_\alpha\}$, we can define the corresponding velocity observable by

$$\dot{A}_\alpha = i[H_\alpha A_\alpha] + \partial_\tau A_\alpha \tag{3}$$

where

$$\partial_\tau A_\alpha = \lim_{\delta\tau \to 0} \frac{1}{\delta\tau} (A_{\alpha + X(\alpha)\delta\tau} - A_\alpha)$$

In the particular case of the Galilean quantal particle we will impose Newton's Galilean-relativity principle:

(I) In any frame, $t_t = I$.

(II) There exists a constant m called the mass, and for every momentum translation $\boldsymbol{\pi}$ the velocity of the position satisfies

$$U(\boldsymbol{\pi})\dot{\mathbf{q}}_t U(\boldsymbol{\pi})^{-1} = \dot{\mathbf{q}}_t - \frac{\boldsymbol{\pi}}{m} I_t \tag{4}$$

With this principle, the Schrödinger equation reduces to[13]

$$i\partial_t\phi_t = H_t\phi_t$$

where the Hamiltonian $\{H_t\}$ is given by

$$H_t = \frac{1}{2m}[\mathbf{p} - \mathbf{A}(\mathbf{q},t)]^2 + v(\mathbf{q},t) \tag{5}$$

For the Lorentz quantal particle, since there are no superselection variables, the Schrödinger equation is

$$i\partial_\tau\phi = K\phi$$

where K is chosen to be[14]

$$K = \frac{1}{2m}\left\{[\mathbf{p} - \mathbf{A}(q^\mu)]^2 - \frac{1}{c^2}[E - V(q^\mu)]^2\right\} + \tfrac{1}{2}mc^2$$

Mathematically, this equation is perfectly consistent, but it is difficult to interpret. Since the invariant parameter τ is not a self-adjoint operator acting in $L^2(R^4, d\mathbf{v}\, dt)$, it is not an observable, it has no actual definite value, and it must disappear at the end of any calculation. In fact, during a real evolution, one always measures some variables and the final state is defined by giving the value of this variable. Such a variable which is continuously measured during the evolution is called a control parameter.[14] The control parameter is by definition one of the superselection variables of the system. Therefore, to each possible control parameter corresponds not only a different physical situation but also a different type of evolution and even a different lattice of properties.

In the classical case all these evolutions are physically equivalent, since in this particular case the control of a variable does not in principle disturb the system. Nevertheless, for comparison with the quantum case, it is useful to be able to write directly the equations relative to a given control variable. Let us therefore consider the Lorentz classical particle. As we have seen, its states are described by the points of the phase-space $\Omega = (\mathbf{p}, E, \mathbf{q}, t)$. According to the generalized Schrödinger equation, its evolution is defined by a vector field on Ω. Not all possible vector fields are realized in nature, but those which can be considered as the limit of some quantum evolution. This is the modern way to justify the symplectic structure of the classical dynamics.

In the phase space Ω we consider the restriction ω_Γ of the 2-form, $dp \wedge dq - dE \wedge dt$, to the submanifold Γ defined by $K = 0$, where K a function on Ω which can be chosen to be

$$K = \frac{1}{2m}\left\{[\mathbf{p} - \mathbf{A}(\mathbf{q},t)]^2 - \frac{1}{c^2}[E - V(\mathbf{q},t)]^2\right\} + \tfrac{1}{2}mc^2$$

—formally the same expression as in the quantum case. The evolution is given by the tangent vector field $X \in T\Gamma$ such that $i_x\omega = 0$. If t is the control parameter, we must write E as a function of the other variables:

$$E = c[(\mathbf{p} - \mathbf{A}(\mathbf{q},t))^2 + M^2c^2]^{1/2} + V(\mathbf{q},t) \equiv H(\mathbf{p},\mathbf{q},t) \tag{6}$$

and we find the canonical equations

$$\frac{dp^i}{dt} = -\frac{\partial H}{\partial q^i}, \qquad \frac{dE}{dt} = \frac{\partial H}{\partial t}, \qquad \frac{dq^i}{dt} = \frac{\partial H}{\partial p^i} \tag{7}$$

If one performs a pure Lorentz transformation, the submanifold Γ and the 2-form ω_Γ are covariant. The canonical equations for the new control parameter t' are obtained by writing the new energy E' as a function of the other new variables. This expresses the covariance of such a dynamics.

Let us come back to the case of the Lorentz quantal particle and let us suppose that the control parameter is the time coordinate t of some frame. As we have explained before, such a system is described by a family of Hilbert spaces H_t, $t \in R$, all identical to $L^2(R^3,dv)$. According to the correspondence principle, the evolution is given by the Schrödinger equation

$$i\hbar\partial_t\phi_t(\mathbf{x}) = H_t(\mathbf{p},\mathbf{q})\phi_t(\mathbf{x}) \tag{8}$$

where the family of operators $H_t(\mathbf{p},\mathbf{q})$ is formally defined by (6). With our interpretation, the covariance of such an equation can be completely understood; nevertheless, the mathematical nature of the operators which appear here is far from being clarified. Let us also remark that the same interpretation can be applied to the well-known interacting Dirac equation,[15] i.e., to

$$i\hbar\partial_t\phi_t(\mathbf{x}) = \{c[\alpha(\mathbf{p} - \mathbf{A}(\mathbf{q},t)] + mc\beta\} + v(\mathbf{q},t)\}\phi_t(\mathbf{x}) \tag{9}$$

where $\phi_t(\mathbf{x}) \in \mathbb{C}^4 \otimes L^2(R^3,dv)$ and α and β are the Dirac matrices.

Finally, as an example of applications of this kind of Schrödinger equation, let us consider a free Galilean quantal particle moving in one-dimensional space from $q = q_0$ to $q = q_1$. Looking at q_1 we measure at what time t the particle arrives. In this case the position q is the control parameter and according to the correspondence principle the "Schrödinger" equation is

$$-i\hbar\partial_q\phi_q(t) = (2mE)^{1/2}\phi_q(t) \tag{10}$$

where $\phi_q(t) \in L^2(R,dt)$ and E is the self-adjoint operator $i\hbar\partial_t$. The solution is formally given by

$$\phi_{q1}(t) = \exp[i\hbar^{-1}(2mE)^{1/2}(q_1 - q_0)]\phi_{q0}(t)$$

The expectation value of t measured at the position q_1 can be calculated. We find

$$\begin{aligned}\langle t\rangle_{q1} &= \int dt\, \phi^*_{q1}(t) t \phi_{q1}(t) \\ &= \langle t\rangle_{q0} + \left\langle \frac{m}{(2mE)^{1/2}} \right\rangle_{q0} (q_1 - q_0)\end{aligned}$$

which is what we would expect from the physical point of view.

4. Deterministic Dissipative Dynamics

The Schrödinger equation, which has been derived from considerations on reversible evolution, cannot describe relaxation phenomena as they appear in time-independent dissipative processes. Recently, relaxation phenomena have been intensely studied specially in the case of an open system.[16] In the particular case of a spin ½ system, such an equation has been proposed by Bloch. In terms of the density matrix ρ, this celebrated Bloch equation can be written as follows:

$$\begin{aligned}\dot{\rho} = &\frac{i\gamma B}{2}[\sigma_z,\rho] + \frac{M_0 - \mathrm{Tr}(\sigma_z\rho)}{2T_1}\sigma_z \\ &- \frac{\mathrm{Tr}(\sigma_x\rho)}{2T_2}\sigma_x - \frac{\mathrm{Tr}(\sigma_y\rho)}{2T_2}\sigma_y\end{aligned} \tag{11}$$

Here B denotes the magnetic field which is given, T_1 and T_2 the two relaxation times which are characteristic of the medium, and M_0 a constant which fixes the equilibrium, i.e., the assymptotic value of the magnetization ($|M_0| \leq 1$). Since the density matrix by definition is a positive operator, the Bloch equation (11) is valid in full generality only if[17]

$$2T_1 \geq T_2$$

Such an equation describes the relaxation as a nondeterministic process, since they change a state $\rho = \rho^2$ into a mixture $\rho > \rho^2$. It does not, however, describe the individual motions of the spins, since, as is well known, the density matrix does not characterize a distribution of states. For example, $\rho = \frac{1}{2}(I + \cos\theta\sigma_z)$ describes not only an ensemble of spins up and down in the ratio $\cos^2(\theta/2)$ to $\sin^2(\theta/2)$ but also an ensemble of spin directions uniformly distributed at an angle θ around the z axis. For all these reasons, the Bloch equations and other similar equations cannot explain phenomena like spin echoes.[18] Recently, Gisin has proposed an equation for the evolution of a spin ½ which rotates and at the same time relaxes to be finally in the direction of the exterior magnetic field. The Gisin equation for spin ½ [19] is

$$\partial_t \phi_t = \frac{\gamma B}{2}[i\sigma_z + k(\sigma_z - \langle\sigma_z\rangle_t)]\phi_t \tag{12}$$

where k is a positive constant which characterizes the relaxation. This is a nonlinear equation preserving the norm. Moreover, it is compatible with the interpretation of ϕ_t as the state. The solution of (12) can easily be computed; we find

$$\phi_t = \langle\exp(\gamma Bk\sigma_z t)\rangle_0^{-1} \exp\left[\frac{\gamma B}{2}(+i + k)\sigma_z t\right]\phi_0$$

Since equation (12) is linear in the magnetic field it can obviously be generalized to arbitrary magnetic fields by replacing everywhere $B\sigma_z$ by $\mathbf{B\sigma}$. The equation (12) is a particular example of a more general Gisin equation:

$$i\hbar\partial_t\phi_t = [H + ik(\langle H\rangle_t - H)]\phi_t \tag{13}$$

where H is the usual Hamiltonian of the Schrödinger equation and k is the positive constant which characterizes the dissipative process. From the equation (13) one deduces

$$\hbar d_t\langle H\rangle_t = 2k(\langle H\rangle_t^2 - \langle H^2\rangle_t) \leq 0$$

This inequality proves that the system dissipates energy if and only if the state is not a eigenstate of the Hamiltonian. The eigenstates of H are fixed states and are semistable but the ground state is stable. In general, the system relaxes to some eigenstates of the energy.

If we apply Eq. (13) to the Galilean quantal particle model with the Hamiltonian corresponding to the harmonic oscillator we find an evolution which is analogous to the classical damped oscillator. If the initial state is the coherent state defined by the point α_0 of the classical phase space, then the solution is the coherent state defined by the point α_t solution of the classical equations of the damped oscillator.[20]

Proof. We have to consider only the one-dimensional case since the variables separate in this equation. In terms of the usual creation and annihilation operators Eq. (13) is equivalent to

$$i\partial_t\phi_t = \omega[a^+a + ik(\langle a^+a\rangle_t - a^+a)]\phi_t$$

If the solution is a coherent state, it reads

$$\phi_t = \exp(-\tfrac{1}{2}|\alpha_t|^2 + \alpha_t a^+)\phi$$

where ϕ is the ground state and α_t a complex number which turns out to be $2^{-1/2}$ $(\langle q\rangle_t + i\langle p\rangle_t)$. This will be the solution if the following equation is satisfied:

$$(-\tfrac{1}{2}d_t|\alpha_t|^2 + d_t\alpha_t a^+)\phi_t = \omega[-i\alpha_t a^+ + k(|\alpha_t|^2 - \alpha_t a^+)]\phi_t$$

This imposes the two conditions

$$d_t|\alpha_t|^2 = -2\omega k|\alpha_t|^2$$
$$d_t\alpha_t = -\omega(i + k)\alpha_t$$

which are satisfied by the "classical solution":

$$\alpha_t = \exp[-\omega(i + k)t]\alpha_0$$

5. Conclusion

In conclusion of this panorama of what we call the new quantum mechanics, let us rediscuss the famous EPR paradox in the light of the new theory. The central point in a series of works on what is usually called the Einstein–Poldolsky–Rosen paradox is an ideal experiment on a correlation at distance between two identical particles. A source at the center emits two particles. In two spatially well-separated regions, one on either side of the source, one tests a given direction of the spin of the particle which appears there. One is interested in the correlation between the results in the two regions for each individual emission of the source. Let a denote the spin direction tested in one region and b the spin direction tested in the other region. In each case let us assign the values $+1$ or -1 according to whether the result of the text is positive or negative. By definition $P(a,b)$, the mean value of the product of these two random variables, is the correlation searched for. J. S. Bell is the first to have shown that, under quite general hypotheses, the function $P(a,b)$ must satisfy certain inequalities. These are the Bell inequalities[21]:

$$|P(a,b) - P(a,b')| + |P(a',b) + P(a',b')| \leq 2$$

We will proceed by proving these inequalities in the general framework of the continuous superselection variables formalism. Our proof will thus be valid in the quantum case as well as in the classical case, with or without hidden variables hypotheses. We will assume the so-called Bell's locality hypothesis, according to which the value found from the test in one of the two regions depends on the state of the two particles emitted and on the perturbation of the apparatus, but in no way on the test which is performed at the same time in the other region. Under this condition the correlation $P(a,b)$ is given by the following expression:

$$P(a,b) = \int d\mu(\phi)[2\epsilon(\phi,a) - 1][2\eta(\phi,b) - 1]$$

where ϕ is the state of the two-particle system emitted by the source, $\epsilon(\phi,a)$ the probability for obtaining $+1$ if one tests a when the state is ϕ, $\eta(\phi,b)$ the probability for obtaining $+1$ if one tests b when the state is ϕ, and $d\mu(\phi)$ the probability for having the state ϕ. Thus we have

$$|P(a,b) - P(a,b')| \leq \int d\mu(\phi)|2\eta(\phi,b) - 2\eta(\phi,b')|$$
$$|P(a',b) + P(a',b')| \leq \int d\mu(\phi)|2\eta(\phi,b) + 2\eta(\phi,b') - 2|$$

from which the Bell inequalities follows, since in any case

$$|\eta(\phi,b) - \eta(\phi,b')| + |\eta(\phi,b) + \eta(\phi,b') - 2| \leq 2$$

The assumption of the theorem is correct in any local classical hidden variables models, and this by construction. But it is also correct in many quantum models. This is always the case if the state at the moment of the measurements is the symmetrized or antisymmetrized product of a one-particle state localized in one of the regions with a one-particle state localized in the other region. However, quantum mechanics also predicts the existence of states of the kind S, i.e., antisymmetric in the spin variables. Such a state is for example

$$[f(\mathbf{x})g(\mathbf{y}) + g(\mathbf{x})f(\mathbf{y})](|+\rangle|-\rangle - |-\rangle|+\rangle)$$

where f is a function with support in one of the regions and g is another function with support in the other region. In this case, if the perturbation of the apparatus corresponds to an ideal measurement, a computation shows that

$$P(a,b) = -\cos(a,b)$$

As it is easy to see, this is an example of correlation which does not satisfy the Bell inequalities. Since the particle which has been detected in one of the regions was not actually in this region before the measurement, the corresponding apparatus perturbs the system by acting on the two regions. Such a result is not *a priori* in contradiction with the usual relativistic interactions, because no restriction has been imposed on the duration of such a process. The above interpretation has no meaning in the old quantum mechanics, where the state is only defined via instantaneous measurements, which are ideal by definition.

Such a kind of experiment has been performed but on a two-photon system.[22] The results agree with a quantum model which postulates the two photons in a state S, and, of course, they contradict the Bell inequalities.

To conclude, we would like to return to the source of the controversy, to the paper of Einstein, Podolsky, and Rosen. These authors prove (sic) the incomplete character of the quantum mechanical description in two steps. They first show that, either quantum mechanics is incomplete, or, when two operators corresponding to two physical quantities do not commute (for a given state), the two quantities cannot simultaneously possess reality. Then the authors conclude by showing (sic) on an example that the second term of the alternative is false. The example chosen is a two-particle state of the same kind as the ones contradicting the Bell inequalities. Their argument is essentially based on Bell's locality hypothesis. They suppose from the beginning that the particles are really located where they are detected, and they do not imagine that each one of the two spatially separated apparatuses can interact with each one of the two particles and perturb them. Such a point of view is in complete contradiction with the experimental results. It is the merit of the Bell inequalities to put this clearly in evidence.

References and Notes

1. G. Ludwig, *Grundstrukturen einer physikalischen Theorie* (Springer, Berlin, 1979); *Foundations of Quantum Mechanics* (Springer, Berlin, to appear); "An Axiomatic Basis of Quantum Mechanics," in *Interpretations and Foundations of Quantum Theory* (Colloque Marburg, 1979); G. Ludwig and H. Neumann, "Connections between Different Approaches to the Foundations of Quantum Mechanics" (Colloque Marburg, 1979).
2. C. Piron, *Foundations of Quantum Physics* (W. A. Benjamin Inc., Reading, Massachusetts, 1976); "La description d'un système physique et le présupposé de la théorie classique," *Ann. Fond. Louis de Broglie* **3**, 131 (1978); "Ideal Measurement and Probability in Quantum Mechanics," Symposium on Logic, Measurement and Probability in Quantum Mechanics (1979); D. Aerts and I. Danbechis, "Physical Justification for Using the Tensor Product . . . ," *Helvetica Physica Acta* **51**, 661 (1978).
3. C. H. Randall and D. J. Foulis, "Empirical Logic and Tensor Products" (Colloque Marburg 1979); "Operational Statistics and Tensor Products" (Colloque Marburg, 1979); D. J. Foulis and C. H. Randall, "Manuals, Morphisms and Quantum Mechanics," in *Mathematical Foundations of Quantum Theory* (Academic Press, New York, 1978).
4. P. Mittelstaedt and E.-W. Stachow, "Operational Foundation of Quantum Logic," *Foundations of Physics* **4**, 355 (1974); P. Mittelstaedt, "The Concepts of Truth, Possibility and Probability in the Language of Quantum Physics" (Colloque Marburg, 1979); E. W. Stachow, "The Propositional Language of Quantum Mechanics," Habilitationsschrift, Universität Köln (1980).
5. A. Einstein, B. Poldolsky, and N. Rosen, "Can Quantum Mechanical Description of Physical Reality Be Considered Complete?" *Phys. Rev.* **47**, 777–780 (1935).
6. B. Mielnik, "Quantum Logic: It is Necessarily Orthocomplemented," in *Quantum Mechanics, Determinism, Causality and Particles*. M. Flato *et al.*, editors (D. Reidel Publishing Company, Dordrecht, Holland, 1976).
7. H. Rauch, A. Zeilinger, G. Badurek, A. Wilfing, W. Bauspiess, and U. Bonse, "Verification of Coherent Spinor Rotation of Fermions," *Phys. Lett.* **54A**, 425 (1975); A. Zeilinger, G. Badurek, W. Bauspiess, U. Bonse, and H. Rauch, "Experimental Tests of some Basic Predictions of Quantum Mechanics by Neutron Interferometry," *Prog. Sci. Culture* **1** (4), 458 (1976).
8. C. Piron, *Foundations of Quantum Physics* (W. A. Benjamin Inc., Reading, Massachusetts, 1976), Chap. 3.
9. C. Piron, Ref. 8, Chap. 5.
10. G. W. Mackey, *Induced Representations of Groups and Quantum Mechanics* (Benjamin, New York, 1968).
11. N. Giovannini and C. Piron, "On the Group-Theoretical Foundations of Classical and Quantum Physics: Kinematics and State Spaces," *Helv. Phys. Acta* **52**, 518–540 (1979).
12. C. Piron, Ref. 8, Chap. 4.
13. J. M. Jauch, "Gauge Invariance as a Consequence of the Galilean Invariance for Elementary Particles," *Helv. Phys. Acta* **37**, 284 (1964).
14. C. Piron, "Un nouveau principe d'évolution réversible et une généralisation de l'équation de Schrödinger," *C. R. Acad. Sci. Paris, Ser. A* **286**, 713 (1978).
15. T. Kato, *Perturbation Theory for Linear Operators*, 2nd Ed. (Springer Verlag, Berlin, 1976), p. 305.
16. E. B. Davies, *Quantum Theory of Open Systems* (Academic Press, London, 1976).
17. A. Kossakowski, "On the general Form of the Generator of a Dynamical Semi-group for the Spin ½ System," *Bull. Acad. Polon. Sci., Sér. Sci. Math. Astron.* **21**, 649 (1973)
18. G. J. Bene, "New Spin Echo Techniques in the Earth's Magnetic Field Range," *Pure Appl. Chem.* **32**, 67 (1972).
19. N. Gisin, "Spin Relaxation and Dissipative Schrödinger-like Evolution Equations," *Helvetica Physica Acta* **54**, 457 (1981).
20. N. Gisin, "A Simple Non-linear Dissipative Quantum Evolution Equation," *J. Phys. A.* **14**, 2259 (1981).
21. J. S. Bell, *Physics* **1**, 195 (1965).
22. E. S. Fry, "An Experimental Test of Local Hidden-Variable Theory Using an Atomic Cascade in Hg^{200}," *Prog. Sci. Culture* **1** (4), 448 (1976).

23

ON NONLINEAR CLASSSICAL DIRAC FIELDS AND QUANTUM PHYSICS

ANTONIO F. RAÑADA
Departamento de Física Teórica
Universidad Complutense de Madrid
Madrid 3, Spain

1. INTRODUCTION

This paper considers some aspects and properties of the classical Dirac fields which may be relevant in order to establish some new links between classical and quantum physics and for a better understanding of some open problems in quantum theory.

But, before proceeding, it is convenient to sharpen the sense in which the term "classical" will be used. A physical theory may be called classical in two cases: (i) if it can be based on the same postulates as any macroscopic theory, such as the mechanics of Newton or the electromagnetism of Maxwell, and (ii) if it makes use of a precise space-time description of the evolution of the phenomena. Although this is not at all the case, it is normally assumed that both meanings are completely equivalent, this belief being a source of misunderstandings. For instance, it is sometimes argued that a classical theory can only be obtained by taking the limit $\hbar \rightarrow 0$. This is true if we use the first meaning, but it is absolutely false with the second one, because we may construct theories in which there is a precise space-time evolution, independently of the magnitude of the actions involved. They can be called classical although they are different in general from any macroscopic theory. In fact the classical models which will be discussed in the following cannot be obtained as a limit when $\hbar \rightarrow 0$. It is also assumed very often that the second meaning implies the nonexistence of discontinuous phenomena. This is not true as the theory of the stability of the solutions of partial differential equations shows. During the last years it has become clear that the relation between a space-time description and a deterministic theory is much more complex than it was thought to be, one of the reasons being the very unstable and even chaotic dependence on the Cauchy data which can appear in some equations.

In this review article the term "classical" will be used in the second sense. As we will see, the classical Dirac fields have some properties which allow us to say that they have a quantum-like behavior from some points of view.

2. The Classical Dirac Fields and the Pauli Principle

One of the main reasons of the extended belief that the classical Dirac fields are only of academic interest and can never be useful for understanding the natural phenomena is their supposed incompatibility with the Pauli principle. It is argued that, if electrons or the nucleons were to be described by a Dirac field following a continuous space-time evolution, in the ground state of an atom or a nucleus all the fields would be bound in the same lowest state. However, this is not the case if the space of the states is properly chosen.[1] Let us be specific and assume that the classical description of a system of n spin-½ fermions is given by n Dirac fields ψ_k and not by a single field, as in the electromagnetic case, and that the Lagrangian density is

$$L = \sum_{k=1}^{n} L_D(\psi_k) + L_2(\psi_k, A_\mu, \phi) + L_3(A_\mu, \phi) \tag{1}$$

where $L_D(\psi)$ is the usual linear expression for the Dirac field ψ, L_2 does not depend on $\partial_\mu\psi_k$ and contains the interactions between the $n\psi_k$, an electromagnetic field A_μ, and an isoscalar, scalar, or pseudoscalar field ϕ, and L_3 includes the Lagrangian densities of the free fields A_μ and ϕ. Let us further assume that

$$D_{ij}L_2 = \overline{\psi}_i \frac{\partial L_2}{\partial \overline{\psi}_j} - \frac{\partial L_2}{\partial \psi_i}\psi_j = 0 \tag{2}$$

a slightly restrictive condition which is verified by $L_2 = 0$, the minimal electromagnetic coupling, the usual $\psi - \phi$ interaction, and many nonlinear couplings between the ψ_k. It can only hold if all the Dirac fields are identical, in the sense that they have the same interactions. For instance, if ψ_i is charged and ψ_j is neutral, (2) is not verified.

It is a very simple matter to show, by operating with the equations of $\overline{\psi}_i$ and ψ_j, that if (2) is verified the currents $\overline{\psi}_i\gamma^\mu\psi_j$ are conserved and the quantities

$$N_{ij} = \int_{R^3} \psi_i^\dagger \psi_j d^3\mathbf{r} \tag{3}$$

are constants of the motion. We now define the configuration space Σ as the set of all the n components, L^2 spinorial functions $\{\psi_k(\mathbf{r})\}$, $k = 1, 2, \ldots, n$, such that

$$N_{ij} = \delta_{ij}C \tag{4}$$

where C is a constant which plays the same role as the Planck constant with which it will be naturally identified.

It is now clear that the fact that N_{ij} are constant implies that ϵ is invariant under the evolution generated by (1). The definition of the configuration space is sensible because of this invariance and because it contains all the asymptotic states in which all the fields ψ_k have the same norm C and are well-separated packets. The reader is referred to Ref. 1 for details.

If the initial state at $t = -\infty$ is given by Z Dirac fields in Σ, well-separated wave packets for instance, in the Coulomb field of a nucleus of charge Z, the extra energy can be radiated away and a bound state can be formed. However, two fields will never occupy the same state. The same applies obviously to the formation of a nucleus.

In conclusion, there is no incompatibility between classical theories and the Pauli principle. On the contrary the classical Dirac fields can obey, quite naturally, the exclusion principle.

3. The Use of Particlelike Solutions as Representation of Extended Particles

The basic concept in this section is the *solitary wave* (sometimes also called *kink*) which is a regular solution of the field equations which propagates without changing its characteristic shape and structure. We will only consider localized finite-energy solitary waves which will also be called *particlelike solutions* or even *particles* for short.

The work in this direction started in 1939, when N. Rosen[2] considered the interaction of the electromagnetic and scalar fields. Later on the self-coupled Dirac field [3,4] and the Dirac–Maxwell interaction[5] were studied. Although it does not use a Dirac field, we must mention in this connection the interesting work of Bialynicki-Birula and co-workers[6,7] on some wave equations with logarithmic nonlinearities.

In this section we will review some models which give a fair representation of a real particle, the nucleon, as a solitary wave of several classical nonlinear fields. As is known, the solitary waves in one space dimension have in some cases the very surprising property of being stable upon collisions and are then called *solitons*.[8,9] They behave as particles which scatter elastically, and in some cases they have even the right energy-momentum relation as was already pointed out in 1961 by Skyrme.[10] After the development of a new theory, the Inverse Scattering Transform Theory, to study this behavior, it has been found to be very frequent and it is now believed[11] that the partial differential equations which have solitons may form a dense set in an appropriate topology.

It is clear that, if there are solitary waves which can give a fair representation of some real elementary particles, it would be very interesting to know if they are solitons. Unfortunately, it is extremely difficult, if not impossible, to extend the present solition theory to 2 or 3 space dimensions, in spite of very strong efforts in this direction.[12] It is not even clear the correct definition of these objects in three dimensions. However, it seems that it must be more elaborate than in one dimension and that the behavior upon collisions may depend on such variables as the relative velocity or the impact parameter with the possible appearance of inelasticity, creation of new packets, and so on. In our case there is another difference to be considered. An n-solition solution consists in a single field with n bumps. However, as was suggested in Section 2, the appropriate classical representation of n spin-½ fermions might be given by n fields each with its own bump, which interact through nonlinear couplings. The statement of the problem must consider this important difference.

The interest of this question is as outstanding as its difficulty. The results of Section 2 suggest that, if the Dirac solitary waves have in some way the properties of the one-dimensional solitons, it would be possible to develop a classical theory of extended particles obeying the exclusion principle. This theory would, at least, narrow the breach between classical and quantum physics and could be useful to study some open problems such as the scattering of hadrons.

Let us now explain how a model of extended particles can be constructed by using nonlinear Dirac fields. Our starting point is an equation studied by Soler in 1970.[13] It can be derived from the Lagrangian density

$$L = \frac{i}{2}[\bar{\psi}\gamma^\mu\partial_\mu\psi - (\partial_\mu\bar{\psi})\gamma^\mu\psi] - M\bar{\psi}\psi + \lambda(\bar{\psi}\psi)^2, \qquad \lambda > 0 \tag{5}$$

It turns out that the field equation

$$i\gamma^\mu\partial_\mu\psi - M\psi + 2\lambda(\bar{\psi}\psi)\psi = 0 \tag{6}$$

admits particlelike solution of the form

$$\psi = e^{-i\Omega Mt}\left(\frac{M}{2\lambda}\right)^{1/2}\begin{pmatrix} H(\rho)\binom{1}{0} \\ iF(\rho)\binom{\cos\theta}{\exp i\varphi \sin\theta} \end{pmatrix} \tag{7}$$

where $\rho = Mr$, Ω is the frequency in units of M and the dimensionless radial functions H and F obey the equations

$$\begin{aligned} F' + \frac{2}{\rho}F + (1 - \Omega + F^2 - H^2)H &= 0 \\ F' + (1 + \Omega + F^2 - H^2)F &= 0 \end{aligned} \tag{8}$$

As we are interested in finite energy localized regular solutions we require that the solutions obey

$$F(0) = 0, \qquad \lim \rho^n F = \lim \rho^n G = 0 \qquad \text{when } \rho \to \infty$$

We refer to the quoted papers for details of the method for solving (8). Several solutions, characterized by the number of nodes, were found by numerical analysis. This suggests that there is an infinite sequence of a nodeless ground state and its radial excitations. Let us consider the nodeless solution. The energy, norm, and spin are calculated from the corresponding currents and can be written as

$$\begin{aligned} E &= \frac{2\pi}{\lambda M}\left[\Omega\int_0^\infty (F^2 + H^2)\rho^2\, d\rho + \frac{1}{2}\int_0^\infty (F^2 - H^2)^2\rho^2\, d\rho\right] \\ N &= \int_{R^3}\psi^+\psi\, d^3r = \frac{2\pi}{\lambda M^2}\int_0^\infty (F^2 + H^2)\rho^2\, d\rho \\ S &= \tfrac{1}{2}N \end{aligned} \tag{9}$$

As we see, the norm plays the same role as Planck's constant. In contrast with the linear case, there is a family of solutions depending continuously on the frequency Ω. In order to overcome the indeterminacy of Ω, Soler[13] used a minimum energy principle: The physical value of Ω is given by the minimum of $E(\Omega)$, which is found to be at $\Omega = 0.936$, the energy and the norm being

$$E = \frac{23.60}{\lambda M}, \qquad N = \frac{22.98}{\lambda M^2} \tag{10}$$

which implies $E = 1.026MN$. If the norm is equal to unity ($\hbar$ in our units) there is a mass renormalization of 2.6%. Figure 1 shows the functions F and H versus ρ for $\Omega = 0.936$. They show the typical decrease with ρ of a bound state. It is clear that all the dynamical variables are concentrated in a region whose dimensions are of the order of $1/M$.

To apply this model to real extended particles it is convenient to study the effect of the electromagnetic interaction. This can be done by adding to (5) the term [14,15]

$$L_{EM} = -\tfrac{1}{4}F'_{\alpha\beta}F^{\alpha\beta} - e\delta\bar{\psi}\gamma^{\mu}\psi A_{\mu} - k\frac{e}{4M}\bar{\psi}\sigma^{\alpha\beta}\psi F_{\alpha\beta} \tag{11}$$

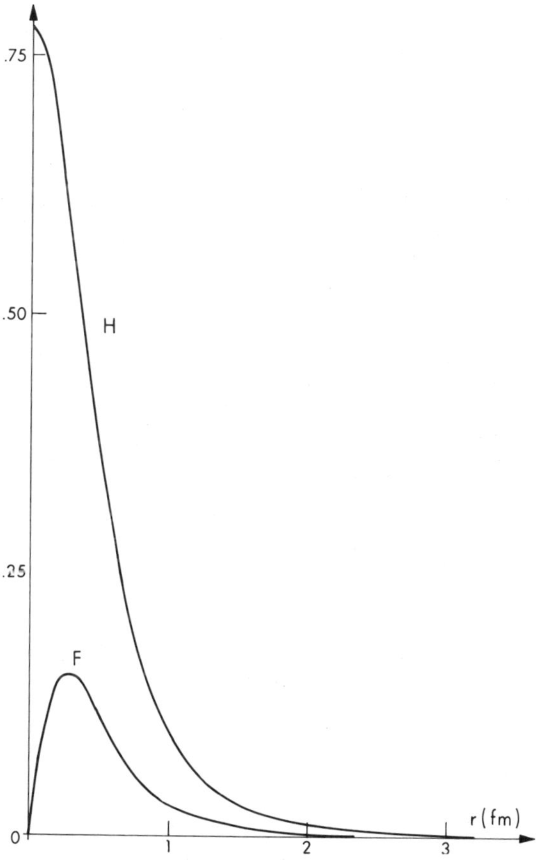

FIGURE 1. Radial functions F and H vs r in the model 1.

where δ is 0 or 1, in order to describe neutral or charged particles, and k is a coupling constant that determines the anomalous magnetic moment. The complete field equations were solved in Ref. 15 by a perturbative expansion in series of $\epsilon = e^2/2\lambda M^2$, the natural dimensionless parameter of the theory. The zero-order approximation gives just the Soler model. In it the ground state energy is equal to the nucleon mass if the mass parameter takes the value $M = 914$ MeV, the renormalization being of 2.6% as mentioned before.

To first order in ϵ, the following results were obtained: if $M = 913.7$ MeV and $\lambda = 22.98/M^2$, the solitary waves representing the proton and the neutron have the following energies and magnetic momenta:

$$E(p) = 938.26 \text{ MeV}, \qquad \mathcal{M}(p) = 2.79\mu_N$$
$$E(n) = 937.81 \text{ MeV}, \qquad \mathcal{M}(n) = 1.91\mu_N$$

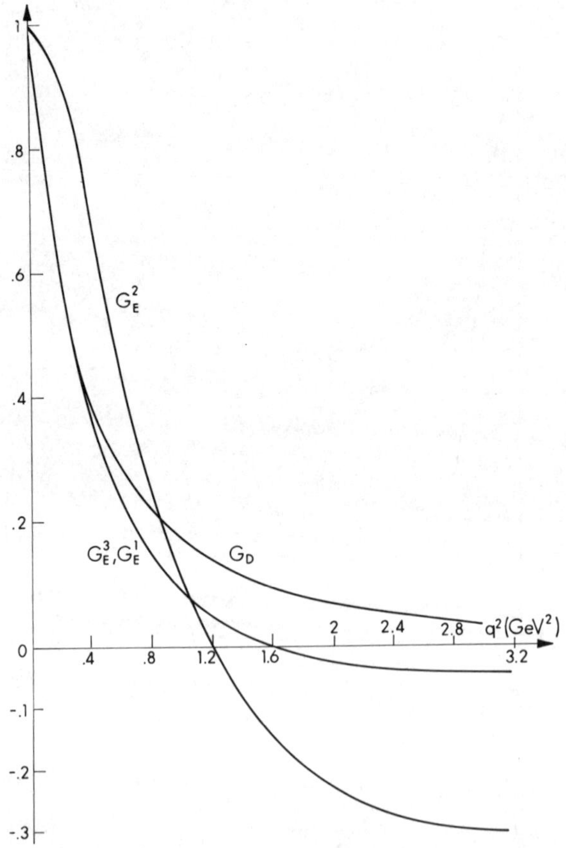

FIGURE 2. Electric form factor of the proton in the three models, compared with the dipole formula G_D. The difference between the models 1 and 3 is too small to appear in the drawing.

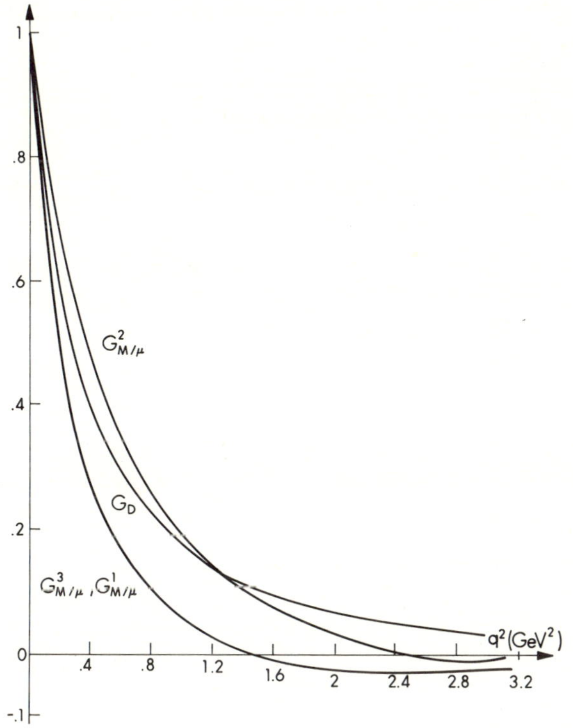

FIGURE 3. Same as Fig. 2 for the magnetic form factor of the proton.

where the magnetic momentum $\mathcal{M}$ is defined as

$$\mathcal{M} = \frac{e}{2} \int_{R^3} \mathbf{r} \times \bar{\psi}\gamma\psi \quad d^3\mathbf{r} \tag{12}$$

and μ_N denotes a nuclear magneton. The result can be considered satisfactory in spite of the wrong value of the n–p mass difference

$$E(p) - E(n) = 0.45 \text{ MeV} \tag{13}$$

However, the value of this quantity remains as one of the unsolved riddles of theoretical physics. There are two aspects of the calculations of the energy which must be emphasized: First of all, the theory is finite, in the sense that there are no infinities to be canceled, the renormalization being only finite. Moreover, it is not necessary to know in advance the electromagnetic form factors, which are an output of the calculation as well as the electric mean square radii.

In Figs. 2, 3, and 4 the electric and magnetic form factors are plotted and compared with the well-known dipole forula G_D. As we see, the results are specially good at low momentum transfer, as can also be verified on comparing the electric and mag-

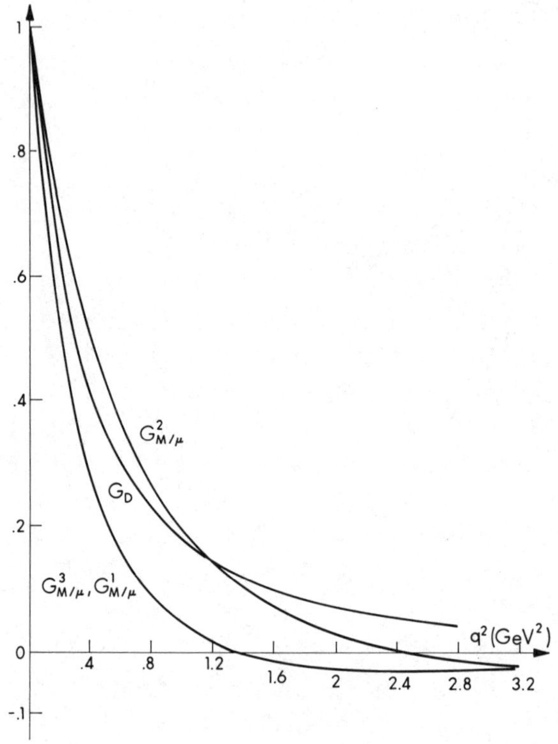

FIGURE 4. Same as Fig. 2 for the electric form factor of the neutron.

netic mean square radii of the model with the experimental values.[16,17] This is done in Table I.

As these results are encouraging, the effect of a real pseudoscalar field to represent the cloud of pions was studied in two different models,[17,18] to be called from now on models 2 and 3, the model 1 being that of Soler. They are based on the addition to the Lagrangian density (5) of the terms

$$\text{In model 2: } \tfrac{1}{2}(\partial_\mu\phi\partial^\mu\phi - m^2\phi^2) + G\bar{\psi}\gamma^5\psi\phi \tag{14}$$

$$\text{In model 3: } \tfrac{1}{2}(\partial_\mu\phi\partial^\mu\phi - m^2\phi^2) + ig\bar{\psi}\gamma^\mu\gamma^5\psi\partial_\mu\phi \tag{15}$$

TABLE I. Electric and Magnetic Mean Square Radii in the Nonlinear Model of the Nucleon[a]

	$\langle r^2\rangle_{el}$	$\langle r^2\rangle_{mg}$
Proton	$(0.789)^2$	$(0.908)^2$
Experimental value	$(0.842 \pm .015)^2$	$(0.843 \pm .028)^2$
Neutron	$-(0.373)^2$	$(0.919)^2$
Experimental value	$-(0.340 \pm .004)^2$	$(0.903 \pm .173)^2$

[a] 1 unit = 1 fermi^2.

Where G and g are coupling constants. As we see the interactions $\psi - \phi$ are the pseudoscalar coupling (model 2) and the pseudovector one (model 3).

The ground state solitary waves were obtained in both models. The spinorial field has the same form as before, while ϕ is in p wave and has the form

$$\phi = \alpha\Phi(\rho) \cos \theta \tag{16}$$

where α depends on the parameters G, g, M, and λ.

The results can be summarized as follows:

(1) The predictions of the third and first models are very similar, although it is necessary to use a different value of λ. This curious result can be stated as follows: The effect of ϕ in the pseudovector coupling can be accounted for by a change of the coefficient of $(\bar{\psi}\psi)^2$. However, it must be stressed that if $\lambda = 0$ there are no finite-energy solitary waves.

(2) The second model has a mass renormalization of about 8% and gives the right sign of the n–p mass difference. However, the radii are too small by a factor of about 1.6, which suggests that the attraction introduced by ϕ is too strong.

(3) The radii of the models 1 and 3 are almost identical and in fairly good agreement with the experimental data. This can be observed in the form factors, plotted in Figs. 2, 3, and 4.

All the details are explained in Refs. 17 and 18.

It is worth mentioning that these models are suitable for the study of the properties of the particles in their rest frames, but the problem of the scattering seems to be very difficult because of the lack of the appropriate mathematical techniques.

An important problem to consider is the origin of the self-coupling of the Dirac field. In 1950 Weyl[19] showed that they can arise as a consequence of a gravitational self-coupling produced by the spin. He considered a theory in which the metric and affine properties of the space-time are independent. The variational procedure leads to the same equations as in the standard metric theory if there is only gravitation and electromagnetism. However, if Dirac fields are present, the space-time becomes twisted and in the Minkowski limit the Dirac equation picks up a nonlinear term of the form

$$\lambda\bar{\psi}\gamma^{\mu}\gamma^{5}\psi\bar{\psi}\gamma_{\mu}\gamma^{5}\psi \tag{17}$$

It has been shown[20] that in a different model of the universe the Weyl idea does not give (17) but the term $(\bar{\psi}\psi)^2$, which is the basic self-action of the previously discussed ls. The study of the solitary waves of a Dirac field with the self-coupling (17) is therefore interesting. It was found that in this fourth model there is a family of nodeless solutions with many minima of the function E(Ω).[21] If $M = 670$ MeV and $\lambda = 6.46/M^2$, the masses of the first three minima are 939, 1123, and 1184 MeV, which are very close to the masses of the nucleon (939 Mev), the Λ (1115 MeV), and the Σ (1192 MeV). It is encouraging to have a mass spectrum of the right order of magnitude.

To sum up this section, it can be said that the classical nonlinear Dirac fields offer an interesting representation of the extended particles which seems suitable to study the nucleons.

4. The Problem of the Stability of the Solitary Waves

In order to develop the previous point of view, it is necessary that the solitary waves be somehow stable in order to maintain their identity, under small deformations at least. There exist two frequent modes of decay of solitary waves: a singular mode, in which a singularity appears at the center of the packet, and a dissipative one, in which the fields tend pointwise to zero. However, these processes are very different to the decay of the real particles, in which there are always some new particles in the final state. This suggests that the solitary waves must be stable in some sense, even if they represent unstable particles, and that a real decay could be due to coupling terms whose effect would be the production of a new wave at the expense of the old one.

In any case, the stability properties of the solitary waves are very important. However, the problem is very difficult and very little is known, the mathematical techniques being insufficient. Moreover, these properties may depend drastically on the addition of higher-order terms. For instance, Anderson and Derrick[22] showed that the solitary waves of the ϕ^4 model are unstable and that they decay by one of the above-mentioned modes. But soon afterwards, Anderson found [23] that the addition of a sixth-order term modifies the situation and that for some values of the self-coupling constants there is stability.

A very interesting result was obtained by Kumar *et al.*[24] They found that in some scalar field models the nodeless solution is conditionally stable under perturbations which preserve the charge, while the solutions with nodes do not have this property. Since the first one can be associated with the ground state and the other with its excitations, this property parallels the behavior of a stable quantum system.

The case of the Dirac field seems to be more complicated. Soler[25] considered the evolution of a spherically symmetric field with the self-coupling $(\bar{\psi}\psi)^2$ and found a very strong tendency towards the stationary solitary waves even with initial conditions very far from it. Alvarez[26] studied the cases of one, two, and three space dimensions by linearizing the equation of a small perturbation and looking for the eigenfrequencies. He found that the solitary waves are stable. Moreover, he obtained by numerical analysis a strong degree of stability in the collisions of solitary waves in one space dimension.

In order to construct models of extended particles it is necessary to consider carefully the possibility of relating the quantum transitions to the unstability of such waves as solutions of partial differential equations. It is to be expected that, although very little is known at this moment, in a short period some important new results will be obtained. We will only emphasize that it is a very difficult problem of the greatest importance.

5. The Confinement of Classical Nonlinear Dirac Fields

The confinement of the quarks is certainly the most curious and intriguing problem of the physics of extended particles. It is therefore natural to ask, in the spirit of the preceding models, whether the nonlinear Dirac fields can offer some help in the explanation of the elusiveness of the constituents of the hadrons. It turns out, if we maintain the identification between extended particles and solitary waves, that it is very easy to construct models in which the particles are bound states of several confined fields. In fact, the confinement arises quite naturally in some nonlinear models. Let us take, as an example, the case of two Dirac fields ψ_1 and ψ_2, whose Lagrangian density is

$$L = L_D(\psi_1) + L_D(\psi_2) + 2\lambda(\bar{\psi}_1\psi_1)(\bar{\psi}_2\psi_2) \tag{18}$$

where $L_D(\psi)$ is the usual Lagrangian density for the field ψ. The field equations are

$$\begin{aligned} i\gamma^\mu\partial_\mu\psi_1 - M\psi_1 + 2\lambda(\bar{\psi}_2\psi_2)\psi_1 &= 0 \\ i\gamma^\mu\partial_\mu\psi_2 - M\psi_2 + 2\lambda(\bar{\psi}_1\psi_1)\psi_2 &= 0 \end{aligned} \tag{19}$$

It is clear that these equations admit particlelike solutions, for instance $\psi_1 = \psi_2 = \psi_S$, where ψ_S is a solitary wave of Eq. (6). However, if one of the fields vanishes, the other one obeys the linear Dirac equation which does not have this kind of solution. In other words, the theory based on (18) predicts the existence of particles, all of them being bound states of ψ_1 and ψ_2, but forbids those associated only with one of the fields. Or, otherwise stated: Although the fields ψ_1 and ψ_2 manifest themselves as constituents of particles, none of them can appear as a particle. As this property parallels the assumed behavior of the quarks, we can say that ψ_1 and ψ_2 are confined fields.

It turns out that it is possible to go even further and to build a model in which the confined fields obey the curious triality and the only solitary waves which appear bind three fields or a field and its charge-conjugate one. In this paper we will only present the main ideas, the details being explained in Ref. 27.

First of all it is necessary to find the representation of the antiparticles. Soler[20] proposed to reverse the sign of the derivative term in (5). The field of an antifermion, to be denoted by the letter ϕ, obeys thus the equation

$$-i\gamma^\mu\partial_\mu\phi - M\phi + 2\lambda(\bar{\phi}\phi)\phi = 0 \tag{20}$$

To each solution $\psi(\mathbf{r},t)$ of (6) corresponds a solution $\phi(\mathbf{r},t)$ of (20) such that

$$\phi(\mathbf{r},t) = \psi(-\mathbf{r},-t) \tag{21}$$

which has the opposite frequency and in which the currents are reversed. As the additive numbers change their sign except the energy, which keeps the same value, we can interpret this solution as the antiparticle of ψ. We represent thus a particle and its antiparticle by two different fields, in clear contrast with quantum field theory.

A very simple model of a hadron can be now proposed. A particle is represented by a system of six Dirac fields ψ_1, ψ_2, ψ_3, ϕ_1, ϕ_2, ϕ_3 with the Langrangian density.

$$L = L_1 b + L_2 + L_3 \tag{22}$$

where L_1 is the linear part and contains the usual Dirac Lagrangian densities of the six fields with a change in the sign of the derivative terms $\bar{\phi}_k \gamma^\mu \partial_\mu \phi_k$:

$$L_2 = \frac{2}{3}[(\Sigma \bar{\psi}_k \psi_k)^2 + (\Sigma\, \bar{\phi}_l \phi_l)^2 + 4\,(\Sigma \bar{\psi}_k \psi_k)(\Sigma \bar{\phi}_l \phi_l)] \tag{23a}$$

$$L_3 = \lambda' \sum_{i<j} (\bar{\Psi}_{ij} \Psi_{ij})^2 \tag{23b}$$

and $\Psi_{ij} = \Psi_i - \Psi_j$; $\Psi_i = \psi_i + \gamma^5 \phi_i$. The conserved current is

$$j^\mu = \sum_k (\bar{\psi}_k \gamma^\mu \psi_k - \bar{\phi}_k \gamma^\mu \phi_k) \tag{24}$$

and can be interpreted as the baryonic current. It must be noted that the currents which are added in (24) are not conserved separately. The S wave particlelike solutions of (22) are the following bound states: (i) A baryonic solution in which $\phi_i = 0$; $\psi_1 = \psi_2 = \psi_3 = \psi_S$, with spin $\frac{3}{2}$ and positive baryonic charge. (ii) An antibaryonic solution with negative baryonic charge, charge conjugate to the preceding one. (iii) A family of mesonic solutions in which $\psi_i = \psi_S$; $\phi_i = \phi_S$; $\psi_j = \psi_k = \phi_j = \phi_k = 0$, where (i,j,k) is any permutation of (1,2,3). Their baryonic charge is zero and their spin is comprised between zero and one. The energy of these mesons is $\frac{2}{3}$ times that of the baryon. There are also solutions with two ψ's and two ϕ's or three ψ's and three ϕ's different from zero. If we identify the field ψ with a quark q and ϕ with an antiquark $\bar{q}$, we can interpret these solutions as the states (qqq), $(\bar{q}\bar{q}\bar{q})$, $(q\bar{q})$, $2(q\bar{q})$, and $3(q\bar{q})$. The fact that there are no solutions with nonzero triality as q, $2q$, $(qq\bar{q})$, and so on, is very satisfactory. The mass degeneration of the spin-one and spin-zero mesonic solutions is due to the fact that the Lagrangian (22) does not contain spin-dependent forces between the quarks. If $M \cong 400$ Mev and $\lambda \cong 23/M^2$, the masses of the baryonic (or the antibaryonic) and the mesonic solutions are, respectively, 1200 and 800 MeV, which are close to the masses of the Δ (1236) and the ρ or the ω.

In conclusion: The solutions of the model based on the Lagrangian density (22) can be put in close correspondence with the nonstrange hadrons. It seems therefore that the classical nonlinear Dirac fields can offer a very useful tool for studying the structure of these particles.

6. The Nonlinear Dirac Fields and the Point Particles

The preceding considerations point out the convenience of studying the possible existence of nonlinear self-couplings, undetected until now, in quantum physics. If this were the case, a relation between the quantum effects and the nonlinearity could be established. As quantum mechanics is a linear theory, the possibility that it is some kind of linear limit should be carefully studied. However, as its main successes have been obtained in the case of point particles without structure, the preceding ideas cannot be applied in their present form. The opposition extended hadron-point lepton appears also from our perspective.

Some results have been nevertheless obtained. For instance, the effect of a polynomic nonlinear self-coupling of the electron field in a hydrogen atom has been studied.[29-31] It was found that, even for very large values of the self-coupling constants, the nonlinear effects cannot be detected in atomic physics. Moreover, there is a very close relation between the processes of linearization and quantization, because the linearization of a nonlinear classical theory cannot be performed in a coherent way without the introduction of a constant of action. The bifurcation of the nonlinear frequencies from the linear eigenvalues was studied and a formal similarity with a truly classical system such as a rod under axial stress was found. It is perhaps convenient to mention here that some experiments are now under preparation to detect nonlinearities in quantum mechanics, although in the spirit of the previously quoted work of Bialynicki-Birula.[32] The problem of the electromagnetic self-energy was also studied as well as a classical renormalization scheme.[31]

As the Heisenberg uncertainty principle is one of the most important bases of the interpretation of quantum physics and it is closely related to the linearity of the wave equation, it is tempting to study how it should be modified by the existence of self-couplings. We will only mention that it turns out[33] that a nonlinear theory disagrees with this principle in some highly nonlinear situations which deserve to be qualified as "ultraquantum." The transition from this ultraquantum world to the quantum one can be done by a process of linearization which has some similarity with the transition from quantum to classical physics.

7. Conclusion

The general conclusion of this work is that the classical nonlinear Dirac fields are not only of academic interest but, on the contrary, they are worthy of careful consideration as they can be useful in the study of some open problems in quantum physics and of the relation between classical and quantum mechanics.

References

1. A. F. Rañada, in *Dynamical Systems and Microphysics*, edited by A. Avez, A. Blaquire, A. Marzollo, Academic Press (in press).
2. N. Rosen, *Phys. Rev.* **55,** 94 (1939).
3. R. Finkelstein, R. Lelevier, and M. Ruderman, *Phys. Rev.* **83,** 326 (1951).
4. R. Finkelstein, C. Fronsdal, and P. Kaus, *Phys. Rev.* **103,** 1571 (1956).

5. M. Wakano, *Prog. Theor. Phys. (Kyoto)* **35,** 1117 (1966).
6. I. Bialynicki-Birula and J. Mycielski, *Ann. Phys. (N.Y.)* **100,** 62 (1976).
7. J. Oficjalski and I. Bialynicki-Birula, *Acta Phys. Pol. B* **9,** 759 (1978).
8. A. C. Scott, F. Y. Y. Chu, and W. McLauglin, *Proc. IEEE* **61,** 1443 (1973).
9. A. F. Rañada, editor, *Lecture Notes in Physics* **98,** (Springer-Verlag, Berlin, 1979).
10. T. H. Skyrme, *Proc R. Soc. London Ser. A* **260,** 127; and **262,** 236 (1961).
11. V. G. Makhankov, *Phys. Rep.* **35,** 1 (1978).
12. V. G. Makhankov, G. Kummer, and A. B. Shavachka, *Phys. Scr.* **20,** 454 (1979).
13. M. Soler, *Phys. Rev. D* **1,** 2766 (1970).
14. M. Soler, *Phys. Rev. D* **8,** 3424 (1973).
15. A. F. Rañada, M. F. Rañada, M. Soler, and L. Vazquez, *Phys. Rev. D* **10,** 517 (1974).
16. L. Vázquez, *Lett. Nuovo Cimento* **19,** 37, 561 (1977).
17. A. F. Rañada and L. Vazquez, *Prog. Theor. Phys. (Tokyo)* **56,** 311 (1976).
18. L. Garcia and A. F. Rañada, *Prog. Theor. Phys. (Kyoto),* **64,** 671 (1980).
19. H. Weyl *Phys. Rev.* 77, 699 (1950).
20. A. F. Rañada and M. Soler, *J. Math. Phys.* **13,** 673 (1972).
21. A. F. Rañada, *J. Phys. A. Math. Gen.* **11,** 341 (1978).
22. D. L. T. Anderson and G. Derrick, *J. Math. Phys.* **11,** 1336 (1971).
23. D. L. T. Anderson, *J. Math. Phys.* **12,** 945 (1971).
24. A. Kumar, V. P. Nisichenko, and Yu P. Rybakov, *Int. J. Theor. Phys.* **18,** 425 (1979).
25. M. Soler, Memorias de la Academia de Ciencias, Madrid (1975).
26. A. Alvarez, *Phys. Lett.* **86A,** 327 (1980).
27. A. F. Rañada and M. F. Rañada, Preprint Universidad ComplutenseUCMFT 80/2 (1980).
28. M. Soler, Report Junta Energía Nuclear, Madrid, ftcr-98 (1970).
29. A. F. Rañada, *Int. J. Theor. Phys.* **11,** 795 (1977).
30. A. F. Rañada and J. Uson, *J. Math. Phys.* **21,** 1205 (1980).
31. A. F. Rañada and J. Uson, *J. Math Phys.* **22,** 2533 (1981).
32. A. Shimony *Phys. Rev. A* **20,** 394 (1979).
33. A. F. Rañada and L. Vazquez, *Phys. Rev. D* **19,** 493 (1979).

24

Elementary Particles as Microuniverses

Erasmo Recami*
Istituto di Fisica Teorica
Università di Catania
95129 Catania, Italy

PART I: INTRODUCTION AND DIGRESSIONS

1. Introduction

The thought that each microscopic "particle" of matter *might* be something like a whole "cosmos"—extremely reduced in size—has probably old origins. For instance, it appears in the papers (ca. 400 B.C.) by Democritus of Abdera. Namely, Democritus,[1] by reversing the analogy, spoke about giant atoms which could reach the size of our cosmos; and, in order to be clear, he added: if one of such superatoms (that constituted supercosmoses) would leave its "giant cosmos" and fall on our world, our world would be destroyed.

That thought is connected with the meditations—very common as well[2]—on the effects of *dilations* or *contractions* on the physical objects, or even on the "world."

Within the scientific arena, let us recall[3] the old idea of a "*hierarchy of cosmoses*," corresponding to very different *scale factors* and possibly organized like a series of "Russian dolls." We can say that in the microscopic analysis of matter one met, roughly speaking, a series of "Chinese boxes"; and something analogous might happen also when investigating the universe, i.e., in the direction of the *macro*, besides of the *micro*. "Hierarchical" theories were formulated,[3] e.g., by J. H. Lambert (1761) and later by Chalier (1909–1922) and Selety (1922), followed in more recent times by physicists such as O. Klein, H. Alfvén, J. E. Charon,[4] K. P. Sinha and C. Sivaram,[5] and M. A. Markov,[6] D. D. Ivanenko[3] and some other Russian authors, till

*Work supported in part by *Fondazione Somaini* (Como, Italy) and by C.N.R.

the papers by P. Caldirola, P. Castorina, G. D. Maccarrone, M. Pavšič, and the present author,[7] by A. Salam and co-workers,[8] by P. Roman and collaborators,[9] by H. J. Treder, etc.

In this article we shall essentially refer to the line followed by us (cf. Refs. 7). Our approach starts from the well-known empirical observation that the ratio $\bar{R}/\bar{r}$ between the Hubble radius $\bar{R} \simeq 10^{26}$ m of our cosmos ("*gravitational cosmos*") and the characteristic radius $\bar{r} \simeq 10^{-15}$ m of subnuclear particles equals grosso modo the ratio S/s of the strength S of the nuclear (or strong) field over the strength s of the gravitational field.* It will then be tempting to think of a *similarity* between the macrocosmos and hadrons (now conceived as "*strong microuniverses*"). We shall therefore assume cosmos and hadrons—both regarded as *finite*[7] objects†—to be *similar* systems, in a geometricophysical sense, i.e., to be systems governed by similar laws, differing only for a scale transformation which carries $\bar{R}$ into $\bar{r}$ and the gravitational field into the strong field (both fields being *a priori* tensorial, in our theory). Namely, mere dimensional evaluations will tell us that by "contracting" the cosmos by the factor $\rho^{-1} \equiv \bar{R}/\bar{r} \simeq 10^{41}$ (i.e., by transforming it into a hadronic microcosmos), the associated field strength increases[7] in the same ratio (passing from gravitational to strong).

Moreover, since the typical duration of a decay process is inversly proportional to the interaction strength, we shall analogously be able to explain why the typical lifetime of our gravitational cosmos (e.g., equal to the duration of a complete expansion/recontraction cycle, in the case of the cyclic big-bang theory $\simeq 10^{18}$ s) is multiple of the typical lifetime of hadrons ($\tau \simeq 10^{-23}$ s) by the same factor $\rho^{-1} \equiv \bar{R}/\bar{r} \equiv S/s \simeq 10^{41}$. Finally, we shall explain, always in a simple way, the fact (itself already *empirically* known, too) that the cosmos mass M equals ρ^{-2} times the typical mass m of the reference hadron considered. And so on.

Before reaching the core of our arguments, however, let us start from the distance, saying a few words about the mathematical "environment" useful to throw light on our initial motivations.‡

Moreover, since we are going to consider (besides the usual transformations) also space-time dilations and contractions, let us at this point recall a passage from the last scientific writing of Einstein[10]:

> ... From the field equations one can immediately derive what follows: if $g_{ik}(x)$ is a solution of the field equations, then also $g_{ik}(x/\alpha)$ is a solution, where α is a positive constant ("*similar solutions*"). Let us for instance suppose system g_{ik} to represent a finite-size crystal embedded in a flat space. We could then have a second "universe" with another crystal, exactly *similar* to the previous one, but dilated by the factor α. As long as we confine ourselves to a universe containing nothing but a unique crystal, we do not meet any difficulties. We realize that the size of such a crystal ("*standard* of length") is not fixed by the field equations. . . .§

*For the definitions of S, s see the following.

†For the moment, to fix our ideas, let us assume the naive model of a "Newtonian ball"—in three-dimensional *space*—for both our cosmos and hadrons. More sensible models[7] (of Friedmann type) will be considered later.

‡At the first reading, one might jump to Section 3.

§Our translation from German.

This passage is part of the "Preface" to Ref. 10, written by Einstein at Princeton on April 4, 1955 (two weeks before dying), and deals with his last unitary theory of the asymmetric field.

2. On "Projective Relativity"

Special relativity (in both its ordinary and "extended"[11] forms) refers to a pseudo-Euclidean chronotopous which is *flat* and *infinite*. One immediately realizes that such a four-dimensional background constitutes a very risky extrapolation of the *local* properties of our space and time, and does not adapt itself to the description of our cosmos. It is for instance difficult to believe that physical laws are covariant also under time translations of thousands of millions of years (in their ordinary form, at least); and so on.

An interesting step towards a space-time (s-t) that *a priori* is suitable to cosmological studies is the following one. Let us observe that the Galilei group **G** can be obtained (through a "contraction") from the Poincaré one **P** as the "*limiting case*" when $c \to \infty$. We can wonder whether the Poincaré group can be in its turn a "limiting case" of another, new group. Remaining in a four-dimensional space (only considering 10-parameter groups), in 1954 Fantappié showed that a unique new group exists, depending with continuity on a parameter R, which is reduced to Poincaré's for $R \to \infty$ and which *cannot be any more* the "limit" of any other group. Such a new group, **F**, happens to be that one of the motions into itself of a de Sitter space-time.[12] Now, the de Sitter s-t is representable as a hypersurface with equation

$$Z_0^2 + Z_1^2 + Z_2^2 + Z_3^2 + Z_4^2 = 0 \tag{1}$$

embedded in a flat five-dimensional space (here and in what follows we shall admit that some coordinates can be imaginary). From that point of view, then, the group **F** becomes the group of rotations in a flat, five-dimensional space; and this clearly shows that **F** generalizes the Poincaré group [whose homogeneous—Lorentz—part, as is well known, is isomorphic from the "complex" point of view to the group of rotations in a flat four-dimensional (1,3) space].

A useful, important *physical interpretation* of the de Sitter group **F** has been put forth in 1959 by Arcidiacono,[13] who distinguished the de Sitter s-t from the "relative" s-t *of each observer* living in it. He took into account the fact that every observer would perceive the events as though they happened in a flat s-t, any geodesics appearing to the observer as a straight line. In other words, each "relative" space-time is a geodesic representation of de Sitter s-t on a *tangent hyperplane*. Thus, the de Sitter group *becomes* represented by the projections from the center of quadric (1) and sections with the tangent hyperplane. Or, rather, **F** becomes the group of *projectivities* which transform into itself the quadric

$$X_1^2 + X_2^2 + X_3^2 + X_4^2 + R^2 = 0 \tag{2}$$

that is to say

$$R^2 = x^2 + y^2 + z^2 - c^2t^2 \tag{2'}$$

When introducing homogeneous coordinates, by setting $X_i \equiv Rx_i/x_5$, Eq. (2) becomes $x_1^2 + x_2^2 + x_3^2 + x_4^2 + x_5^2 = 0$. In conclusion, from the projective point of view, the ordinary "physical" space-time is the region *external* to the Kayley–Klein "absolute" with Eq. (2). But the projective space defined as the region external to the quadric (2) is nothing but the Castelnuovo space-time,[14] and only in this space do the mathematical expressions receive a physical interpretation. A "*(special) projective relativity*" follows,[13] which reduces to the ordinary relativity only when $R \to \infty$.

Here, let us add only the following. As is well known, Einstein built up his general relativity in such a way as to include special relativity as a particular case. Subsequently, to build up "unitary theories," others tried to enlarge the Riemannian geometry of general relativity (GR). That is to say, before enlarging the GR, they did *not* try to "bring to perfection" the special relativity. According to Ref. 13, the ordinary unitary theories turn out to be unsatisfactory also because all theories built up by enlarging the Riemannian geometry (or by passing to five- or six-dimensional manifolds) are still "based" on special relativity (SR)—set up in the Minkowski s-t—and on the Poincaré group. Such a group is not "simple"[12] and therefore splits in the chronotopical (6 parameters) rotations and (4 parameters) translations. This leads to the partition of ordinary SR in two independent parts (mechanics—of continuous media; and electromagnetism), where a sharp distinction exists between "matter" properties and "electricity" properties. In projective relativity,[12,13] on the contrary, rotations and translations merge together into the rotations of (a five-dimensional hypersphere) S^5 via the new *fundamental length* R; as a consequence, a link is found between "matter" and "electricity," while remaining, nevertheless, inside the realm of the classical theories directly founded upon groups (*"Erlangen program" for physical theories*).

If, afterwards, one wants to erect a "general relativity" starting from projective relativity (which is based on the de Sitter–Fantappié group), one expects that the new "projective general relativity" extends Einstein's gravitational theory on a cosmological scale, and therefore is particularly suited for astrophysical problems.

In order to blend the conceptions of those who prefer to rely directly upon the ds^2 and of those who on the contrary strictly appeal to the "path of groups" (step-by-step), we can take advantage of the *unifying* point of view of Cartan, who, by generalizing the idea of space, inserted the very Riemannian geometry in a group-context. In fact, following Cartan,[15] a Riemannian variety V^4 can be regarded as constituted of infinitely many—e.g., Euclidean—spaces *tangent* to it at each of its points, each one of those spaces having a *geometry* (in Klein's sense) grounded on the roto-translations group; such a geometry was called holonomous by Cartan. Those infinitely many, Euclidean space-elements are then linked together through a certain "*connection*" *law* (in this case called Euclidean by Cartan), which allows deducing *both* curvature and torsion (local properties) of V^4 by using infinitesimal closed *cycles* on the variety, *and* the "holonomy group" (global properties) of V^4 by using finite closed cycles on V^4. Vice versa, once the holonomy group is known, the connection law can be univocally determined.[12,13] Of course, what precedes can be at once extended to the cases when the tangent spaces possess a non-Euclidean geometry, based on a group $\mathbf{G}^r$ with r parameters (still in the sense of the Erlangen program). Likewise, given any holonomous (i.e., founded upon a group) geometry, *anholonomous* geometries can be constructed corresponding to it. For instance: in Minkowski s-t the holonomy group is obviously the identity, and such space-time is holonomous; on the contrary, the Riemannian s-t of general relativity is no longer holonomous: however, it admits (it being devoid of torsion) as holonomy group the Lorentz one, i.e., the group of rotations in S^4 space.

Let us summarize[13]: (i) In order to go beyond SR, Einstein jumped from a theory founded upon the rotations group $\mathbf{R}_4$ (Lorentz group) to theories which rather utilize Riemannian manifolds $V^4, V^5, V^6, \ldots$ and in this way he skipped, in a sense, the *direct* (step-by-step) *path* of groups;[15b] (ii) In the "*theory of the universes*" by Fantappié and Arcidiacono, on the contrary, the models of cosmoses (or of "universes") are built up on the basis of the rotations groups $\mathbf{R}_4$, $\mathbf{R}_5$, $\mathbf{R}_6, \ldots$, thus establishing a link among physical laws,[12] group, and geometrical *model* of the cosmos or universe. In fact, the chosen group acquires in that way the geometrical task of representing the motions into itself of the corresponding "universe model," and, from the physical point of view, of expressing in mathematical form a "*principle of relativity*." Physics, for instance, can be built up by using the "topological group," i.e., $[n(n - 1)/2]$-dimensional manifolds which possess both geometrical and group structures[12] (this is comparable[13] with what Lagrange did in his analytic mechanics, when he described a mechanical system in terms of its "Lagrangian parameters"); (iii) In order to reconcile the points of view of Einstein and of Fantappié and Arcidiacono, we can make use of the link established by Cartan between group theory and differential geometry. From this "third" point of view,[13] we can set up a series of "special relativities," based on the rotation groups $\mathbf{R}_n$, and then associate with each of them a "general relativity" by having recourse to an "anholonomous" geometry (that admits $\mathbf{R}_n$ as its holonomy group, and therefore is a Riemannian geometry).

In the Appendix we show, for example, how to build up a "general relativity" when starting from projective (special) relativity.

An Alternative Approach

If we wanted *strictly* to follow the "Erlangen program" in physics, the following alternative approaches would be available.[13,16] The investigation of de Sitter universe—projective relativity—and of the corresponding generalized Maxwell equations[13] confirms the usefulness (besides of a straightforward group-theoretical foundation of physics) of resorting to the rotations groups $\mathbf{R}_n$ of n-dimensional spaces. We saw that in such a way a succession of "universe models" is obtained, represented by the hyperspheres S^{n-1} embedded in n-dimensional spaces $E^n (n = 4,5, \ldots)$; and the problem arose of developing a "relativity" just based upon the group $\mathbf{R}_n$ of the motions into itself of the hyperspheres S^{n-1}. [Incidentally, in the groups $\mathbf{R}_n (n > 3)$, with their projective coordinates $x_i (i = 1, \ldots, n)$, there appear n-3 *universal constants*, necessary for adding square lengths to the squares of the new coordinates (which come after the first three ones, x,y,z) without violating the requirement of dimensional homogeneity[17]].

If one sets, as in projective relativity, n-4 normalization conditions, one takes back the "n-dimensional relativity" to a four-dimensional formulation (in terms of the space-time coordinates).[13] At the limit for $R \to \infty$, besides, every hypersphere S^{n-1} is reduced to a flat space E^{n-1}, and its n projective coordinates become n-1 Cartesian coordinates; consequently, the group $\mathbf{R}_n$ [with $n(n - 1)/2$ parameters] decomposes into the product of rotations $\mathbf{R}_{n-1}$ and translations $\mathbf{T}_{n-1}$ [having $(n - 1)(n - 2)/2$ and $(n - 1)$ parameters, respectively], while the normalization conditions become $n - 5$ independent equations with $n - 1$ unknowns.[13] For instance, for $n = 5$ we get the projective transformations (projective geometry); for $n = 6$ the conformal transformations *(conformal geometry)*; for $n > 6$ the Cremona-type transformations: In this way, one succeeds in applying the *algebraic geometry* to physics.

At this point, the new *alternative* approach comes in. Let us notice, in fact, that within the aforesaid group-theoretical conception of physics a particular rolc is played by the *generalized* "Maxwell equations" of the various hyperspherical universes S^{n-1}, which are covariant under the group $\mathbf{R}_n$. If we call $H_{ik} = -H_{ki}$ $(i,k = 1,2, \ldots ,n)$ the generalized "electromagnetic field," possessing $n(n - 1)/2$ distinct components, the generalized Maxwell equations then become[13]

$$\text{Curl } H_{ik} = J_{ike}, \qquad \text{Div } H_{ik} = I_k \qquad [i,k,e = 1,2, \ldots ,n] \tag{3}$$

where J_{ike} and I_k are the field "sources," and Curl, Div are understood to operate in n dimensions.

A relation has been uncovered between enlarging the basic *group* of physics and the possibility of unifying the various physical interaction fields, such a synthesis being performed by the very algebraic structure of the various rotations groups.

Particularly interesting appears to be the *extension* from the group $\mathbf{R}_5$ (projective relativity) to the group $\mathbf{R}_6$ (conformal relativity), the latter comprehending also the uniform accelerations. In Refs. 13 it has been shown, in this connection, that the corresponding generalized Maxwell equations yield a *unified theory* of matter (gravitation *plus* "hydrodynamics" of continua) *and* of electromagnetism. In particular, for $R \to \infty$, one is taken back to a flat space E^5 and the above-mentioned generalized Maxwell equations split, on one hand, into Corben's equations[16] (of the unified *gravitational-electromagnetic* field) *and*, on the other hand, into the *mechanical* equations of the generalized "hydrodynamical" field.[13] By using such a "conformal relativity" ($n = 6$), therefore, there is no need of passing—as is done on the contrary in general relativity—to "anholonomous" manifolds, but one succeeds in describing even gravitation without departing from a *strict* group-theoretical formulation of physics.

3. About "Conformal" Relativity

Historically, when due account was taken of the electromagnetic phenomena, besides of the mechanical ones, it was necessary to abandon—as is well known—Galilean relativity in favor of Einstein's. We could now wonder whether, once arrived at investigating also the nuclear and subnuclear forces, a further extension towards a new relativity should be necessary. Actually, at the beginning of Section 2, we considered—roughly speaking—the following "chain" of groups:

$$\mathbf{G}(c \to \infty; R \to \infty) \Leftarrow \mathbf{P}(c; R \to \infty) \Leftarrow \mathbf{F}(c; R) \tag{4}$$

where the *final*, de Sitter group "contains"[13] *two* universal constants (a fundamental length, R, and the speed of light in vacuum, c). But, in order to plan in a *dimensionally* correct way even only a mechanical (dynamical) theory, three universal constants are needed.[17] To lengthen the chain (4) one has, however, to leave the ten-parameter groups (i.e., the four-dimensional Minkowski space):[13] It is then easy to reach the *conformal group* **C**, with 15 parameters, which can be shown to be locally isomorphic to the rotations of a six-dimensional space. That group will allow setting up[12] the new *"conformal (special) relativity,"* a generalization of the projective one and a theory having as universe model a five-dimensional hypersphere (embedded in a flat six-dimensional space). In such a conformal relativity, now, three independent universal constants c,R,h will enter, where the third constant, h, ought *a priori* to depend on a mass (besides on a length and a time).[13,18] Let us recall at this point also what we wrote in Section 2.1.

PART II: SUBNUCLEAR PARTICLES AS MICROUNIVERSES

4. A Hierarchy of "Universes"

We can also start from a different point of view (although within a more limited framework, in a sense) for generalizing the special relativity according to the spirit of the beginning of Section 3. In the following, we shall essentially refer to work done by the present author in collaboration with P. Caldirola and M. Pavšič, as it appears from the references. Let us, namely, observe that the symmetries of the Maxwell equations have not been fully exploited by SR. In fact, the Maxwell equations are known to be covariant—besides under Poincaré transformations—even under conformal transformations[19] and, in particular, under dilations.* Moreover, we have already recalled that Einstein gravitational equations too are covariant under dilations.† Then, let us fix our attention in particular on the (space-time) *global* dilations [$\mu = 0,1,2,3$]:

$$x'_\mu = \rho x_\mu \tag{5}$$

and *postulate*[7] that physical laws are covariant also under the *global* dilations (5), where only *discrete* values of ρ are supposed to have in nature a physical counterpart.

*In the general case when charges are present, such covariance exists provided that also charges are suitably "scaled."

†When in presence of matter and of a cosmological term, such covariance is preserved, as we shall see, provided that also the gravitational charge (= mass) and the cosmological constant are suitably scaled.

(Such discrete values might be derived, e.g., by imposing suitable boundary conditions in five-dimensional spaces[20]; but we leave this problem open here.)

At this point, let us recall that natural objects seem essentially to interact via (at least) four fundamental forces: the gravitational, the "weak," the electromagnetic, and the "strong," here listed according to decreasing *strength*. For instance, the strength of the electromagnetic force is measured by the dimensionless coefficient $(4\pi\epsilon_0)^{-1}e^2/\hbar c$. In general, the strength of an interaction can be measured by the dimensionless square of the corresponding "coupling constant." Here we are interested in particular in the strength of the gravitational interaction:

$$\frac{Gm^2}{\hbar c} \simeq 1.3 \times 10^{-40} \tag{6a}$$

and of the strong one:

$$\frac{Ng^2}{\hbar c} \begin{array}{l} \nearrow \simeq 3 \\ \searrow \simeq 15 \end{array} \tag{6b}$$

where (i) G and N are the gravitational and strong *universal constants* in vacuum, respectively; (ii) quantities m and g represent the gravitational charge (= mass) and the strong charge (see the following) of one and the same hadron,[7] e.g., of a nucleon or a pion-meson. The value in Eq. (6a) is calculated for the pion mass, $m = m_\pi$; in Eq. (6b) the first number typically corresponds to the value of the $\pi\pi\rho$ coupling-constant square[21] (whilst the second number represents, more generally, the value of the $pp\pi$ coupling-constant square).

With regard to the above expression "strong charge of a hadron," let us regard the *quarks** as the actual sources of the strong field, i.e., the *real* carriers of strong charge, and let us call "color" the sign s of quark strong charges[7]; more precisely, the hadrons can be considered as endowed with *zero* total strong charge, each quark possessing a strong charge $g_i = s_i \mid g' \mid$ where $\Sigma_i s_i = 0$. The ordinary strong interactions among hadrons should, *in a sense*, originate from van der Waals-type forces.[7] In correspondence to quantity m of Eq. (6a), in Eq. (6b) the quantity $g \equiv ng_0$ will enter, quantity g_0 being the *average* magnitude of the constituent-quark charges and n being the quark number.

Let us put:

$$\rho \equiv \frac{Gm^2}{Ng^2} \begin{array}{c} \nearrow \simeq 4 \times 10^{-41} \searrow \\ \searrow \simeq 0.9 \times 10^{-41} \nearrow \end{array} \simeq 10^{-41} \tag{7}$$

*Let us recall, incidentally, that the hadron *constituents* (2 for *mesons* and 3 for *baryons*) have been named "quarks" by M. Gell-Mann. The Anglo-Saxon word quark is usually ennobled by literary citations [e.g., Gell-Mann got inspiration, as is well known, from J. Joyce's *Finnegans Wake* (1939)]. Let us here quote that Goethe properly used such a word in *Faust*, verse 292, where Mephistopheles referring to mankind exclaims: "*In jeden Quark begräbt er seine Nase.*"

and notice that, if we conventionally choose $m \equiv g$, then the "strong universal constant" N becomes

$$N = \rho^{-1}G \approx 10^{41}G \simeq \frac{hc}{m_\pi^2} \simeq 7 \times 10^{30}\ \text{m}^3\ \text{kg}^{-1}\ \text{s}^2 \tag{8}$$

Conversely, if we choose units such that $[N] = [G]$ and moreover $N = G = 1$, then we get

$$g = \frac{m}{\sqrt{\rho}} \simeq 2 \times 10^{-33}\ \text{cm} \simeq 3 \times 10^{-5}\ \text{g} \simeq \left(\frac{\hbar c}{G}\right)^{1/2} \equiv \text{Planck mass} \tag{9}$$

where we eventually chose in Eqs. (6b) and (7) the *upper* value. Equation (9) tells us, by the way, that the "Planck mass" $(\hbar c/G)^{1/2} \equiv m(\rho^{-1})^{1/2}$ is nothing but the typical *hadron* (or, perhaps, *quark*) "strong charge," in suitable units. We do *not* expect, therefore, existence of further, new small black holes—as predicted by other authors—with a mass of the order of the Planck mass, since we already met hadrons (or, perhaps, quarks) with *strong charges* of the order of Planck mass (in suitable units).

The most important observation is, however, the following one. Let us regard both hadrons ("typically" the *pions*, or the nucleons), and our cosmos as finite objects. Then, relation (7) and the fact that, when calling $R(\text{U}) \equiv R$ the Hubble radius of our cosmos[7] and $r(\text{h}) \equiv r$ the hadron *(pion)* radius in strong interactions, one gets

$$\frac{r(\text{h})}{R(\text{U})} = \frac{10^{-15}\ \text{m}}{10^{26}\ \text{m}} \simeq 10^{-41} = \rho \tag{10}$$

suggest that our cosmos and hadrons can be considered as (finite) *similar* systems* governed by similar laws that differ only in the scale factor ρ (which carries R into r and the gravitational field into the strong one). Roughly speaking, we can imagine that—by shrinking the cosmos by the factor $\rho \simeq 10^{-41}$—we can get the hadrons (see the following, and Refs. 7), that is to say that, by dilating a hadron by the factor $\rho^{-1} \simeq 10^{41}$, we can get a cosmos. In Refs. 7, indeed, after having called "*universe*" any almost isolated system, governed by *one* of the fundamental forces, we have analogously introduced a "hierarchy of universes,"[3] which can be obtained through a series of suitable *discrete* dilations (or contractions).

Drawing our inspiration, as said above, from the hypothesis that physical laws are covariant under (discrete) dilations, we are led to assume, briefly, the following[7]:

(A) *Inside our cosmos (gravitational case)* the Einstein equations, with attractive cosmological constant[5] Λ, hold $[G = 1]$:

$$R_{\mu\nu} - \tfrac{1}{2} g_{\mu\nu} R^\rho_\rho - \Lambda g_{\mu\nu} = -\frac{8\pi}{c^4} T_{\mu\nu} \tag{11}$$

*Just to fix our ideas, let us think here in terms of "Newtonian balls" both for hadrons and our cosmos. Later on, we shall adopt less naive models (namely, Friedmann models) for both our cosmos and hadrons, consistently with general relativity.

(B) *Inside hadrons* ("*strong*" *case*) the "scaled-down" Einstein equations hold [N = G = 1]:

$$\tilde{R}_{\mu\nu} - \tfrac{1}{2}\tilde{g}_{\mu\nu}\tilde{R}^{\rho}_{\rho} - H\tilde{g}_{\mu\nu} = -\frac{8\pi}{c^4}S_{\mu\nu} \tag{12}$$

Simple dimensional evaluations immediately tell us that (within our "dilation covariant" relativity)[7]

$$H \equiv \rho^{-2}\Lambda, \qquad m_G = \rho m_S \tag{13}$$

where $m_G \equiv \hbar(2\Lambda)^{1/2}/c$ and $m_S \equiv \hbar(2H)^{1/2}/c$ are the average mass—small, but finite—of the "*external*" gravitons and the average mass of the "external" strong quanta, respectively: cf. Refs. 7 and Scction 6. Moreover, the strong-charge tensor $S_{\mu\nu}$ is essentially $S_{\mu\nu} \equiv \rho^{-1}T_{\mu\nu}$, where $T_{\mu\nu}$ is *a priori* the ordinary matter tensor (containing, e.g., the Dirac spinorial functions, etc.). For example, if we require (also) the "external" gravitational interactions [see Refs. 7] to have a range of the order of $R = R(U) \simeq 10^{26}$ m, then we obtain at once[7]

$$m_G \simeq 10^{-68}\ \text{Kg}, \qquad \Lambda \simeq 10^{-56}\ \text{cm}^{-2}, \qquad m_S \simeq m_\pi \tag{13'}$$

as well as:

$$H^{-1} \equiv \rho^2\Lambda^{-1} \simeq 10^{-25}\ \text{cm}^2 \simeq 0.1\ \text{barn} \tag{14}$$

The present, elementary theory[7] allows proving in a systematic way all the empiric relations (which connect macro- with microcosmoses) heuristically discovered by Weyl, Eddington, Dirac, etc.[22]; although our own *numerology*[7] connects the *gravitational* interactions with the *strong* ones [that are, like the former, always attractive, *nonlinear*, and eventually associable with non-Abelian gauge theories: these sections propose, indeed, an *ante litteram* geometrical interpretation of those theories], and *not* with the electromagnetic ones (as suggested, on the contrary, by Dirac). For instance, it is straightforward to prove from our "dilation-covariant relativity" that the mass M of our cosmos and the mass m of the pion are linked together as follows:

$$M = \rho^{-2}m \simeq 10^{54}\ \text{kg}, \qquad m = \rho^2 M \simeq 10^{-28}\ \text{kg} \tag{15}$$

The numerology derived by our theory, moreover, does *not* predict any dependence of G on time: see Ref. 21.

Before going on, let us specify that we shall confine the models to be adopted (for both our cosmos and hadrons) within the realm of Friedmann models. In particular, we shall take advantage of the fact that the Friedmann models are compatible with the Mach principle,[23] and are embeddable in five dimensions.[24]

Moreover, always consistently with Eqs. (11), for the *spatial* part of our cosmos we can choose the simple model of the three-dimensional hypersurface of a hypersphere. Analogously we can proceed for hadrons (strong universes), so as to be able to extend, e.g., the Mach principle for them: in the sense that the *inertia* of every hadron-constituent (parton) will coincide with its strong charge (and *not* with its gravitational charge!). In such a way, we shall be able to consider an "*equivalence principle*" as *locally* valid even inside hadrons, so as to justify from the point of view of general relativity the present geometrization of the strong field [first of all inside hadrons, and then, as we shall see, even in their surroundings]. It is apparent that also the other fundamental fields could *a priori* be geometrized in the same way.

5. Inside a "Universe": Quark Confinement

Let us now find out an exact solution of Eqs. (12), inside a hadron, for a spherically symmetric distribution g' of strong charge. The geodesic equation for a (small) test constituent with strong charge g'' in the vacuum $[i,j = 1,2,3; N = 1]$,

$$\frac{d^2x^i}{dt^2} = \frac{c^2}{2} g^{ij} g_{00,j}$$

yields in the *radial* case

$$\frac{d^2r}{dt^2} = -\frac{c}{2}\left(1 - \frac{2g'}{c^2r} + \frac{Hr^2}{3}\right)\left(\frac{2g'}{cr^2} + \frac{2Hr}{3}\right) \qquad [N = 1] \tag{16}$$

The spherically symmetric distribution g' can be identified, e.g., with a quark; on the contrary, g'' must be a (small) test particle. When g'' is another quark, Eq. (16) holds only *approximately* and merely furnishes an idea about the radial behavior of g'' in the field of g'. Nevertheless, Eq. (16) yields the so-called "asymptotic freedom" of quarks (or, rather, "small constituents") for small distances r, as well as the quark (constituent) "confinement" for large values of r.

Let us examine the case of *small values* of r, when the attractive term $\propto -1/r^2$ dominates (as in the gravitational case). [Notice that the repulsive term $\propto 1/r^3$ effectively works only at *extremely* small values of r, so that the radial acceleration in the gravitational case would vanish only for $r \simeq 2Gm/c^2$!] However, if we attribute an angular momentum J with respect to g' to the considered constituent g'', i.e., if we take into account the "kinetic-energy term" in the potential, then, with the choice (8) for the measure units, we can write for small r [$r \ll r(h), r \ll 1$ fm] the total effective potential:

$$V \approx -\frac{Ng'}{r} + \frac{(J/g'')^2}{2r^2} \approx -\frac{Ng'}{r} + \frac{(J/g'')^2}{2r^2} \tag{16'}$$

We see that g'' will tend to stabilize itself dynamically at a distance r_0 from g' where the *(total)* interaction potential vanishes. This result seems to give simply a reason for the *asymptotic freedom* of hadron constituents.

In the case of *two* quarks, some approximate considerations do even allow one to write down a Regge-like relation[7] between J and the hadron mass.

For *large distances*, when $r \simeq r(h) \simeq 1$ fm, one gets[7] from Eq. (16) a radial attractive force (confining the constituents inside hadrons) which is proportional to $-r$ ($N = 1$; $[N] = [G]$):

$$F \simeq -\frac{g''}{3}(c^2Hr - 2H) \approx -\frac{g''c^2Hr}{3} \propto -r \qquad \left[r \simeq \left(\frac{6Ng}{cH}\right)^{1/3} \simeq 1\text{ fm}\right] \tag{17}$$

In other words, by applying the methods of general relativity to hadron structure, we get in a very natural way also a *confining* potential $V \propto r^2$ for large values of r. For very large values of r, attainable, e.g., when the considered hadron starts to get deformed under a high-energy collision, we would get[7] an even stronger confining force, proportional to $-r^3$:

$$F \simeq -g''c^2\left(\frac{H^2}{9}r^3 + \frac{H}{3}r + \cdots\right) \qquad [r > r(\mathrm{h})] \tag{18}$$

In conclusion, through Eq. (16), our unified (classical) approach to strong and gravitational interactions yields a completely defined (radial) potential for constituent–constituent interaction inside hadrons. Such interesting potential appears to deserve further attention. Another conclusion is that the introduction of our microuniverses (for which, essentially, the theory of general relativity can be used) can advantageously substitute models such as the "MIT bag" model.

Let us notice that all our previous results must (and can) hold also inside our cosmos, *mutatis mutandis*.

6. In the Surroundings of a Hadron (in our Space)

We may regard the *spatial* parts of our cosmos and of hadrons (time *aside*) as embedded in a four-dimensional flat *space* E^4. The problem of strong interactions *between two hadrons* (e.g., two nucleons) requires* considering what heuristically we can call the "*intersections*" of hadrons with our cosmos, such intersections being two-dimensional spherical surfaces that we just call "hadrons" *tout court*. Since (in our cosmos) two "hadrons" interact *strongly*—e.g., via van der Waals-*like* forces—we need therefore to describe the (strong) interactions between the aforesaid "intersections." To this end, when considering the motion of a hadronic test particle—possessing both strong and gravitational charges—a "*bi-scale*" theory is required (inside our cosmos) in the surroundings of hadrons and in presence of subnuclear interactions. In other words, we need to modify the gravitational Einstein equations by introducing, in the *micro*neighborhood of the above-mentioned intersections (hadrons), a strong metric deformation $s_{\mu\nu}$ affecting (only) the objects with strong charge (i.e., with scale factor $\kappa = \rho \simeq 10^{-41}$), and *not* affecting the ones with gravitational charge only (i.e., with scale factor $\kappa = 1$). Around a hadron we can assume the *gravitational* metric tensor to be $f_{\mu\nu} \cong \eta_{\mu\nu}$ (in suitable coordinates), and set

$$g_{\mu\nu} \equiv f_{\mu\nu} + s_{\mu\nu} \cong \eta_{\mu\nu} + s_{\mu\nu} \tag{19}$$

*Rather, it would require using the *fiber bundle* techniques.

where the components of the strong metric tensor $s_{\mu\nu}$ have to vanish for $r \gg 1$ fm. In Refs. 7, we proposed the following field equations (for test objects having both gravitational and strong charges, in the surroundings of a hadron, in our space):

$$R_{\mu\nu} + Hs_{\mu\nu} = -\frac{8\pi}{c^4}(S_{\mu\nu} - \tfrac{1}{2}g_{\mu\nu}S^{\rho}_{\rho}) \tag{20}$$

with $S_{\mu\nu} = NT_{\mu\nu}$; $N = G\rho^{-1}$; and where the "cosmological (strong) term" with the hadronic constant H takes care of the geometric properties of the strong field around the "source hadron."

Let us here *verify* that, at the static limit and for "weak" field ($r \gtrsim 1$ fm), we do correctly reduce ourselves to deal with a scalar field s_{00} having the required Yukawian behavior (of course, with $|s_{00}| \ll 1$ for $r \gg 1$ fm). Equation (20) in suitable coordinates writes

$$R_{\mu\nu} + H(g_{\mu\nu} - \eta_{\mu\nu}) \simeq -\frac{8\pi}{c^4}(S_{\mu\nu} - \tfrac{1}{2}g_{\mu\nu}S^{\rho}_{\rho}) \tag{20'}$$

which can also read

$$\begin{aligned} R_{\mu\nu} - \tfrac{1}{2}g_{\mu\nu}R^{\rho}_{\rho} - Hg_{\mu\nu} &= -\frac{8\pi}{c^4}S'_{\mu\nu} \\ S'_{\mu\nu} \cong S_{\mu\nu} &- \frac{c^4}{8\pi}H(\eta_{\mu\nu} - \tfrac{1}{2}g^{\alpha\beta}\eta_{\alpha\beta}g_{\mu\nu}) \end{aligned} \tag{21}$$

where the *last* term has the meaning of *interference* between the two tensorial fields. Notice that our strong-field tensor is precisely

$$\frac{\Phi_{\mu\nu}}{g'} \equiv \tfrac{1}{2}s_{\mu\nu} \cong \tfrac{1}{2}(g_{\mu\nu} - \eta_{\mu\nu}) \tag{22}$$

By *linearizing* with respect to the flat metric, from Eq. (20′) we get the linearized equations with "hadronic (cosmological) term" [$r \gtrsim 1$ fm]:

$$\begin{aligned} &\partial^{\mu}\partial_{\mu}s_{\alpha\beta} + 2Hs_{\alpha\beta} \simeq \frac{16\pi}{c^4}(S_{\alpha\beta} - \tfrac{1}{2}\eta_{\alpha\beta}S^{\nu}_{\nu}) \\ &\partial_{\mu}s^{\mu}_{\nu} = \tfrac{1}{2}\partial_{\nu}s^{\mu}_{\mu}, \qquad |s_{\mu\nu}| \ll 1 \qquad \text{for } r \gg 1 \text{ fm} \end{aligned} \tag{23}$$

Equation (23) is a (*relativistically* covariant) equation for a *massive* tensorial field.[25]

In the *static* limit ($\partial g_{\alpha\beta}/\partial t \simeq 0$), when

$$S^{\rho}_{\rho} = S_{00} = Nc^2\gamma \tag{24}$$

quantity $N\gamma$ being the density of strong-charge magnitude (and γ the ordinary mass density), one gets for the scalar field $V_{\text{ext}} \equiv \tfrac{1}{2}c^2 s_{00}$ the equation

$$\nabla s_{00} - 2Hs_{00} \simeq \frac{8\pi}{c^4}N\gamma \tag{25}$$

Finally, for a point particle with strong charge g at rest in the origin, a *spherically symmetric* solution of Eq. (25) is ($g_{00} = 1 + s_{00}$)

$$s_{00} \equiv \frac{2V_{ext}}{c^2} = -\frac{2g}{c^2 r} \exp[-r(2H)^{1/2}] \tag{26}$$

In the case of a nucleon, Eq. (6b) holds with the *second* value, $Ng^2/\hbar c \simeq 15$. It is enough to identify

$$(2H)^{1/2} \equiv m_S c/\hbar \tag{27}$$

in order to get for the (external) *field mass*[25] the value

$$m_S \equiv \hbar(2H)^{1/2}/c \simeq m_\pi \tag{27'}$$

In conclusion (for test-particle low speeds, and "weak" field) we actually obtained a scalar field with the correct *Yukawian behavior:*

$$V_{ext} \simeq -\frac{g}{r} \exp\left(-\frac{rm_\pi c}{\hbar}\right) \tag{28}$$

7. Further Remarks, and Some Speculations

If, in our space (inside our cosmos), we want to associate with ordinary hadrons—i.e., with the aforesaid "intersections" (cf. Section 6)—a *spherically symmetric* source of the strong field $\Phi_{\mu\nu}/g' \equiv \frac{1}{2}s_{\mu\nu}$, we can try to solve the "Schwarzschild-*type* problem" for the Einstein-*type* equations (20). Remember that Eqs. (20), unlike Eqs. (12), are *no longer* Einstein equations (but they are *modified* Einstein equations). In particular, we shall eventually be able to calculate the "*strong* Schwarzschild-type radii" corresponding (in our space) to hadrons. The results $r_S^{(s)}$ that we shall obtain appear to yield actually the "effective radii" of hadrons in strong interactions.

In order to perform our task, we shall write Eqs. (20) as

$$R_{\mu\nu} - \frac{1}{2}g_{\mu\nu}R^\rho_\rho - H(g_{\mu\nu} + \eta_{\mu\nu} - \frac{1}{2}g^{\alpha\beta}\eta_{\alpha\beta}g_{\mu\nu}) = -\frac{8\pi}{c^4} S_{\mu\nu} \tag{20''}$$

and then use the *trick* of transforming them into

$$\begin{aligned} R_{\mu\nu} - \frac{1}{2}g_{\mu\nu}R^\rho_\rho &\simeq -\frac{8\pi}{c^4}(S_{\mu\nu} + t^{(\text{strong})}_{\mu\nu}) \\ t^{(\text{strong})}_{\mu\nu} &\equiv -\frac{c^4 H}{8\pi}(g_{\mu\nu} + \eta_{\mu\nu} - \frac{1}{2}g^{\alpha\beta}\eta_{\alpha\beta}g_{\mu\nu}) \end{aligned} \tag{29}$$

with $S_{\mu\nu} = NT_{\mu\nu}$, $N = \rho^{-1}G$.

By setting

$$\overline{S}_{\mu\nu} \equiv S_{\mu\nu} + t^{(\text{strong})}_{\mu\nu} \tag{30a}$$

we are left with the equations

$$R_{\mu\nu} \simeq -\frac{8\pi}{c^4}(\bar{S}_{\mu\nu} - \tfrac{1}{2}g_{\mu\nu}\bar{S}^{\rho}_{\rho})$$
$$|t^{(\text{strong})}_{\mu\nu}| \ll 1 \qquad \text{for } r \gg 1 \text{ fm} \tag{30b}$$

For a spherically symmetric distribution of hadronic charge

$$t^{(\text{strong})}_{00} = \frac{\Phi_{00}}{g'} u(r) \simeq \tfrac{1}{2}(g_{00} - 1) \cdot u(r) \tag{31a}$$

the structure of Eqs. (29) suggests[26,7] writing, in analogy with what one does when in presence of an electromagnetic field

$$u(r) = \frac{1}{8\pi g'^2}(|\nabla\Phi|^2 + \tilde{\mu}^2|\Phi|^2) \simeq \frac{1}{32\pi}[|\nabla(g_{00} - 1)|^2 + \tilde{\mu}^2|g_{00} - 1|^2] \tag{31b}$$

where we put

$$\Phi \equiv \Phi_{00}, \qquad \tilde{\mu} \equiv m_S c/\hbar \simeq m_\pi c/\hbar \tag{31c}$$

Equations (31), and the following ones, are better dealt with by means of the choice (9), with N = 1.

To solve our problem, we can thus adopt an iterative procedure. For the *first iteration*, in the static limit, we can take for its zeroth-order approximation $[r \gtrsim 1 \text{ fm}]$

$$\tfrac{1}{2}(g_{00} - 1)|_I \cong \frac{\Phi^I}{g'} \simeq -\frac{g}{r}\exp(-r\tilde{\mu}) \tag{26'}$$

and get

$$u(r) \approx \frac{1}{8\pi}\frac{g^2\exp(-2\tilde{\mu}r)}{r^2}\left(\frac{1}{r^2} + \frac{2\tilde{\mu}}{r} + 2\tilde{\mu}^2\right) \tag{32}$$

We shall put as usual $\{\Phi_{00}/g' \equiv \tfrac{1}{2}s_{00} \cong \tfrac{1}{2}(g_{00} - 1) \equiv \exp[\nu(r)]\}$:

$$s_{\mu\nu} \cong g_{\mu\nu} - \eta_{\mu\nu} = 2\begin{pmatrix} \exp[\nu(r)] & & & \\ & \exp[\lambda(r)] & & \\ & & -r^2 & \\ & & & -r^2\sin^2\theta \end{pmatrix} \tag{33}$$

where ν, λ are functions still to be determined. Notice that we are looking for "*strong* black-hole"-*type* solutions, inside our cosmos and in vacuum (and considering the space outside the horizon).

By inserting Eqs. (31a) and (33) into Eqs. (29) one gets, among others, the equation[7, 26]

$$-\frac{8\pi g^2}{m'^2c^2}u(r) \approx \exp[-\lambda(r)]\left[\frac{1}{r^2} - \frac{1}{2}\frac{d\lambda(r)}{dr}\right] - \frac{1}{r^2} \tag{34}$$

where $Ng^2/\hbar c \approx 15$, and m' is the mass of the hadronic *test particle* (we can choose, e.g., $m' \approx m_q$ = average quark mass, the "test quark" being considered *a priori* as situated initially outside the horizon). The exact solution of Eq. (34) is

$$\exp[-\lambda(r)] = 1 - \frac{2l}{r} + \left(\frac{\tilde{\mu}k}{r} + \frac{k}{r^2}\right)\exp[-2\tilde{\mu}r] \tag{35}$$

where $k \equiv g^4/c^4m'^2$, and where $l = g^2m/c^2m'^2$ is an integration constant with the dimensions of a length, quantity m being the hadron mass (e.g., the nucleon mass).

Obviously, in our Schwarzschild-type geometry, the *strong* Schwarzschild-type radii $r_S^{(s)}$ will be calculated in correspondence with

$$\exp[-\lambda(r)] = 0, \qquad \exp[\nu(r)] = 0 \tag{36}$$

The first one of Eqs. (36) yields values of $r_S^{(s)}$ depending only *slightly* on $\tilde{\mu}$. Almost the same results are got, e.g., for $\tilde{\mu} \approx m_\pi c/\hbar$ or $\tilde{\mu} \approx 0$. In the simpler case $\tilde{\mu} \approx 0$, the first one of Eqs. (36) becomes

$$r_S^{(s)2} - 2lr_S^{(s)} + k = 0 \tag{37}$$

that is to say

$$r_S^{(s)} = l \pm (l^2 - k)^{1/2} \tag{37$'$}$$

For the *nucleon*, e.g., we have the two approximate solutions

$$r_S^{(s)}(\mathcal{N}) = \begin{cases} r_1 \approx 10^{-15}\ \text{cm} \\ r_2 \approx 0.8\ \text{fm} \end{cases} \tag{38}$$

While the larger value is in good agreement with the ordinary nucleon radius in strong interactions, many alternative interpretations might be suggested for the smaller value.

We have now to verify that on the Schwarzschild-type horizon also the second one of Eqs. (36) holds. But the calculation of $\nu(r)$ can be performed only at the price of further approximations. We therefore limit ourselves only to verifying that, in the present case, it is actually[26]

$$\exp[\nu(r)] \approx \{\exp[\lambda(r)]\}^{-1} \approx 1 - \frac{2g^2m}{rc^2m'^2} \tag{39}$$

It is finally worthwhile to notice that our "strong" metric

$$ds^2 = \exp[\nu(r)]c^2\,dt^2 - \exp[\lambda(r)]\,dr^2 - r^2(d\theta^2 + \sin^2\theta\,d\phi^2) \tag{33$'$}$$

together with Eqs. (35) and (39), has been shown by Mignani to be identifiable with the 't Hooft monopole metric in curved space-times.[27]

In conclusion, ordinary hadrons can be considered in our space as "strong black-hole"-*type* objects, in the sense seen above. This is a *very peculiar* sense, since, e.g., our "strong black holes" at the static limit are surrounded (for $r > 1$ fm) by a strong (scalar) field with Yukawian behavior, which has nothing to do of course with *ordinary* "strong black holes." However, if the usual physics, valid for ordinary (strong) black holes, can be extended—partially at least—to our *peculiar* "strong black holes," then many stimulating questions would arise, which have been mentioned elsewhere.[7] In particular, our "strong horizon" might play, at a classical level, a role similar to the one of the already mentioned "MIT bag." [Let us recall that strong black holes can be characterized (besides by mass, spin, electric charge, and pseudoscalar charge) also by further quantum numbers or charges, since the ordinary "short-range" fields can be regarded as "long-range" fields at their scale]. Moreover, let us add that the "classical *confinement*" here obtained for hadron constituents can be violated by quantum effects such as Hawking's. The "Hawking temperature" for a strong black hole, for example, turns out[7] to be of the order of $T \simeq 2 \times 10^{11}$ K, and corresponds *a priori* to an "evaporation time" of the order of $\Delta t \simeq 10^{-23}$ s, *unless* we do impose some stability conditions of the kind of Bohr's.[7,5] [In any quantum theory, however, quarks may be again "totally confined," if you want, by associating with their classical Schwarzschild (strong) horizon a suitable barrier of strict "superselection rules" and of "conservation superlaws."]

Everything that was previously said, of course, can be "translated" so as to hold for the case of "gravitational universes" like our cosmos.

Let us add three speculations:

First, if our cosmos is *similar* to a hadron, it might, for example, be conceived, following the calculations in this Second Part, as a *super-pion*, and therefore as constituted by one matter *half-cosmos* (or "metagalaxy"[3]) and by one antimatter *half-cosmos* (just as each pion consists of a quark and an antiquark).

Second, let us assume that for ordinary neutrons (from the point of view of our *peculiar* "strong black holes") the validity of the Second Law of black-hole thermodynamics can be extended, saying that when two black holes melt together the Schwarzschild area of the final black hole must be *larger* than the sum of the two initial Schwarzschild areas. Then, when neutrons would melt together (within the cyclic big-bang theory) at the end of an expansion–recontraction cycle, the "superneutron" born from the fusion of those 10^{80} neutrons ought eventually to possess, by *extrapolation*, a Schwarzschild horizon with area

$$S \gg 10^{80}\pi r_n^2 \qquad (r_n \simeq 10^{-13}\text{ cm}) \tag{40}$$

so that the melting process (= big-bang explosion!) should rebuild up a new object with radius

$$R \gg 10^{27}\text{ cm} \tag{40'}$$

Third, if hadrons are *similar* to our cosmos, they too could perform successive cycles of expansion and recontraction, with a period, however, of about $\Delta\tau \simeq 10^{18}/10^{41}$ s $\simeq 10^{-23}$ s. We should thus get that subnuclear particles can be regarded as *pointlike* only at certain successive, discrete positions along their trajectory (associable with a fundamental *chronon*). It is interesting that Caldirola,[28] with regard to this argument, started from a "finite difference" equation for the electron motion and ended with the conclusion that *leptons* too can be considered as pointlike objects moving on a four-dimensional de Sitter microuniverse (to which, in our ordinary space, a *spherical object* can be associated, by suitably projecting onto a tangent hyperplane, such a sphere performing successive cycles of expansions and recontractions with a period of about 10^{-23} s).

A structure of the type of the "microuniverses" could therefore be characteristic of all subnuclear particles,[7] and show a classical path to unification also of weak interactions (together with the strong and gravitational ones).

8. Conclusions

We have shown that, when applying the methods of general relativity to subnuclear particle physics, one can advantageously substitute models such as the "MIT-bag" model with our classical theory which considers hadrons to be (finite, "strong") *microuniverses*. According to our unified (classical) approach to gravitational and strong interactions, inside hadrons there is essentially a tensorial field (= "strong gravity"), and the constituents are supposed to exchange spin-2 "gluons." In other words, "scaled down" Einstein equations, with "strong" cosmological term, hold for the hadron interior, and they yield in a natural way the *confinement* and "asymptotic freedom" of the hadron constituents.

Our approach allows also writing down a (*biscale*, tensorial) field theory of hadron–hadron interactions by suggesting (modified) Einstein-type equations for strong interactions. In particular, we obtain in this way: the correct Yukawian behavior of the strong potential (for $r > 1$ fm) at the static limit; and the value of hadron radii in strong interactions. In a sense, we carry out the old idea of Riemann and later Clifford that the matter particles were merely the manifestation of a local strong curvature of space.

Our hope for the future is that the internal symmetries of the microcosmoses associated with hadrons (for instance via a projection onto a tangent hyperplane) can lead to some basic elements of the theory of quantum chromodynamics such as the "color $SU(3)$."

As a by-product, we *derived* a whole "numerology" connecting our gravitational *cosmos* to the strong microcosmoses *(hadrons)*. For further details, cf. Refs. 7.

Acknowledgments

The author is grateful to Professor A. van der Merwe for his kind interest, to Professors P. Caldirola, P. Castorina, G. D. Maccarrone, and M. Pavšič for discussions, and to Dr. L. R. Baldini and Dr. C. Tolomeo for kind cooperation.

Appendix

In this Appendix, let us, for example, hint at the *construction* of the "general relativity" by starting from projective relativity. We have then to introduce[13] an anholonomous X^4 space (in general, a variable-curvature Riemannian manifold) that admits the de Sitter group as its holonomy group. Since that group is isomorphic to the one of S^5 rotations (where S^n indicates, let us repeat, the n-dimensional hyperspherical space), we have to resort to the geometry of a Riemannian variety V^5 (which just admits as holonomy group the one of the rotations of S^5). Such a geometry of V^5 will have then to be interpreted in terms of projective differential geometry of a four-dimensional manifold X^4. It is known that the projective differential geometry of a X^n allows in fact a $(n + 1)$-dimensional interpretation in terms of the Riemannian geometry of V^{n+1}.

Following again Cartan, a space X^4 with *projective connection* is a space having the characters of a projective space in the (infinitesimal) neighborhood of each point P of it, and endowed with a projective *(homographic)* connection law between the neighborhoods of two infinitely close points of it. For such a purpose, it is necessary to provide a suitable field of quadrics Q, placed in the spaces tangent to the single points P of X^4 (cf. Ref. 13). Once the point P is fixed, the corresponding quadric $Q(P)$ constitutes the "absolute" of the local, non-Euclidean metric. The parallel transport in a Riemannian V^4 preserves the isotropic cones; *analogously*, the projective connection must yield a projective transport law that preserves the aforesaid field of quadrics Q. After having thus determined the projective connection,[13] one can build up in the usual way the "curvature projective tensor" $P_{\alpha\beta\gamma\delta}$ $(\alpha,\beta,\gamma,\delta = 1,\ldots,5)$, whose vanishing is the necessary and sufficient condition for the given space to be "projectively flat" (i.e., with constant curvature). In fact, the constant-curvature varieties are locally representable onto the Euclidean space with preservation of the geodesics.

The vanishing of the curvature Riemannian tensor leads, in GR, to getting again the Minkowski s-t; *on the contrary*, the vanishing of the curvature projective tensor $P_{\alpha\beta\gamma\delta}$ leads back, in projective general relativity, to the de Sitter s-t with constant curvature. Finally, the tensor $P_{\alpha\beta\gamma\delta}$ has the important property of *including* the torsion tensor (so that Cartan called it the "curvature-and-torsion-tensor"). Actually, at variance with what happens in the ordinary spaces endowed with affine connection, now the curvature of a projective curvature space implies a *torsion:* this is due to the fact that the de Sitter group (holonomy group of X^4) decouples, at the "relativistic" limit, in the rotations and translations of S^4, to which the "rotation curvature" and the "translation curvature" (= torsion) correspond, respectively.

References and Notes

1. A. Virieux-Reymond, "Democrito di Abdera," in *Scienziati e Tecnologi* (Mondadori; Milano, 1975).
2. See, e.g., also J. Swift, *Travels into Several Remote Nations of the World by Lemuel Gulliver* (Benj. Motte Pub., London, 1726).

3. D. D. Ivanenko, Chap. 9 in *Relativity, Quanta and Cosmology in the Development of the Scientific Thought of A. Einstein*, F. De Finis, editor (Johnson Pub. Co., New York, 1979), Vol. 1, p. 295.
4. J. E. Charon, *C. R. Acad. Sci. Paris* **A266,** 750 (1968); *Theorie de la Relativité Complexe* (A. Michel, Paris, 1977), and references therein.
5. C. Sivaram and K. P. Sinha, *Phys. Rep.* **51,** 111, (1979), and references therein.
6. M. A. Markov, *Zh. Eksp. Teor. Fiz.* **51,** 878 (1966).
7. P. Caldirola and E. Recami, *Lett. Nuovo Cimento* **24,** 565 (1979); P. Caldirola, M. Pavsic, and E. Recami, *Phys. Lett.* **A66,** 9 (1978); *Nuovo Cimento* **B48,** 205 (1978); E. Recami and P. Castorina, *Lett. Nuovo Cimento* **15,** 347 (1976); E. Recami, in *Annuario '79, Enciclopedia EST-Mondadori* (Mondadori, Milano, 1979), p. 60; E. Recami: in *Centenario di Einstein 1879–1979: Astrofisica e Cosmologia, Gravitazione, Quanti e Relatività*, M. Pantaleo, editor (Giunti-Barbera, Firenze, 1979), Chap. 18, p. 1077; E. Recami, Chap. 16 in *Relativity, Quanta and Cosmology in the Development of the Scientific Thought of A. Einstein*, F. De Finis, editor (Johnson Rep. Co., New York, 1979), Vol. 2, p. 581; P. Caldirola, G. D. Maccarrone, and E. Recami, *Lett. Nuovo Cimento* **27,** 156 (1980); See also R. Mignani, *Lett. Nuovo Cimento* **16,** 6 (1976); V. de Sabbata and P. Rizzati, *Lett. Nuovo Cimento* **20,** 525 (1977); V. de Sabbata and M. Gasperini, *Lett. Nuovo Cimento* **25,** 489 (1979).
8. A. Salam, *Ann. N.Y. Acad. Sci.* **294,** 12 (1977), and references therein; A. Salam and J. Strathdee, Phys. Rev. *D* **18,** 4596 (1978).
9. P. Roman and J. Haavisto, *Int. J. Theor. Phys.* **16,** 915 (1978).
10. A. Einstein, Preface to *Cinquant'anni di Relatività*, M. Pantaleo, editor (Editr. Universitaria, Ediz. Giuntine, Firenze, 1955).
11. See, e.g., E. Recami, editor, *Tachyons, Monopoles, and Related Topics* (North-Holland, Amsterdam, 1978); E. Recami and R. Mignani, *Riv. Nuovo Cimento* **4,** 209–290, 398 (1974).
12. L. Fantappié, *Rend. Accad. Naz. Lincei* **17,** 158 (1954); reprinted in *Opere Scelte* (U.M.I., Bologna, 1973), p. 873. See also *Rend. Accad. Naz. Lincei* **12,** 12 (1952).
13. G. Arcidiacono, *Relatività e Cosmologia* (Veschi Pub., Roma, 1979), Vol. 2; *Gen. Rel. Grav.* **7,** 885 (1976); **8,** 865 (1977); **9,** 949 (1978); **10,** 445 (1979); and **11** (1980), in press. See also E. Pessa, *Collect. Math. (Barcelona)* **24,** 151 (1973).
14. G. Castelnuovo, *Mon. Not.* **91,** 829 (1931); *Rend. Accad. Naz. Lincei* **12,** 263 (1930).
15. (a) E. Cartan, *C. R. Acad. Sci. Paris* **174,** 593 (1922); *Ouvres Completes*, partie III, Vol. 1 (Gautier-Villars, Paris, 1955); *Leçons sur la theorie des espaces à connexion projective* (Gautier-Villars, Paris, 1937).
15. (b) Cf. also, e.g., S. Weinberg: *Gravitation and Cosmology* (J. Wiley, New York, 1972), p. 92.
16. H. C. Corben, *Phys. Rev.* **69,** 225 (1946).
17. See, e.g., E. Recami and C. Spitaleri, *Scientia* **110,** 204 (1975); *G. Fis. (Bologna)* **17,** 190 (1976).
18. See, e.g., R. L. Ingraham, *Nuovo Cimento* **12,** 825 (1954); **1,** 82 (1955); **B46,** 217, 261 (1978).
19. E. Cunnigham, *Proc. London Math. Soc.* **8,** 77 (1910); H. Bateman, *ibid.* **8,** 223 (1910).
20. See, e.g., M. Pavsic, *Lett. Nuovo Cimento* **17,** 44 (1976); P. Caldirola, *ibid.* **18,** 465 (1977); *Rend. Accad. d'Italia* **1,** 19 (1939); *Nuovo Cimento* **A45,** 548 (1978); P. Caldirola, G. Casati, and A. Prosperetti, *Nuovo Cimento* **A43,** 127 (1978).
21. P. Caldirola, G. D. Maccarrone, and E. Recami, *Lett. Nuovo Cimento* **27,** 156 (1980).
22. See, e.g., H. Weyl, *Ann. Phys. (Leipzig)* **59,** 129 (1919); A. Eddington, *The Expanding Universe* (Univ. Mich. Press, Ann Arbor, Michigan, 1958); P. A. M. Dirac: *Proc. R. Soc. London, Ser. A* **165,** 199 (1938). See also P. Jordan, *Nature* **164,** 637 (1949) and Refs. therein; O. Klein, *Science* **171,** 339 (1971) and Refs. therein; G. Cocconi, *Comments Nucl. Part. Phys.* **7,** 177 (1978); J. K. Laurence and G. Szamosi, *Nature* **252,** 538 (1974).
23. D. J. Raine, *Mon. Not. R. Astron. Soc.* **171,** 507 (1975).
24. J. Rosen, *Rev. Mod. Phys.* **37,** 204 (1965); G. Szekeres, *Nuovo Cimento* **A63,** 1062 (1966).
25. C. Sivaram and K. P. Sinha, *Phys. Lett.* **B60,** 181 (1976); *Rep. Progr. Phys.* **55,** 1288 (1976).
26. Cf., e.g., D. K. Roos, *Nuovo Cimento* **A8,** 603 (1972) and references therein; E. Recami and P. Castorina, *Lett. Nuovo Cimento* **15,** 347 (1976).
27. R. Mignani, *Lett. Nuovo Cimento* **16,** 6 (1976). See also F. A. Bais and R. J. Russel, *Phys. Rev. D* **11,** 2692 (1975); P. van Nieuwenhuizen, D. Wilkinson, and M. J. Perry, *Phys. Rev. D* **13,** 778 (1976).
28. P. Caldirola, *Lett. Nuovo Cimento* **23,** 83 (1978); **27,** 225 (1980); *Riv. Nuovo Cimento* **2**(13) (1979); *Nuovo Cimento* **A49,** 497 (1979); **A45,** 548 (1978); Chap. 5 in *Albert Einstein 1879–1979: Relativity, Quanta and Cosmology*, F. De Finis, editor (Johnson Rep. Co., New York, 1979), Vol. 1, p. 187.

25

Precision Mathematics and Approximation Mathematics in Physics*

R. Rompe
GDR Physical Society
Berlin, GDR

H.-J. Treder
Akademie der Wissenschaften der DDR
Berlin, GDR

I.

The year 1908 saw the publication for the first time of Felix Klein's lectures on elementary mathematics, under the title *Elementarmathematik vom Höheren Standpunkte Aus.*[7]† In this memorable work, Klein expresses himself repeatedly on fundamental questions of mathematics and their relation to mechanics and physics. In this connection, he writes that physics does not need exact mathematics, or *Präzisionsmathematik*, in itself and can amply get along with an *Approximationsmathematik*, because the measurement data of physics that underlie any theoretical mathematical treatment are necessarily afflicted by uncertainties—called forth by the inescapable existence of limits on the precision with which measurements can be performed.

These limits or bounds are given for macroscopic bodies and structures (i.e., for aggregates of microscopic objects), on the one hand, by thermodynamic fluctuation phenomena and, on the other (even at absolute zero of temperature), by the atomic structure of the material world. Both phenomena ultimately lead us back to universal elementary constants. Thus, for example, the Nyquist formula contains the constants e, m, h, and c, in addition to the electron density u_e. For microscopic objects, the limits are generally furnished by the uncertainty relations of quantum physics and

*Translated from German by A. van der Merwe.

†The definitive, polished edition of this work appeared (in part after Klein's death in 1925) as three volumes between 1924 and 1928. The third volume treats the topic *Präzisionsmathematik und Approximationsmathematik*.

immediately express the existence of the elementary action constant h and the light velocity c. But we know that at low temperatures also "macroscopic quantum effects" can operate as measurement limits, which are likewise set by Planck's constant h.

Generally speaking, these measuring restrictions do not prevent, microscopically, the measurement with arbitrary precision of, say, position or energy; one only requires a correspondingly larger expenditure of energy (for the energy determination) or a correspondingly longer duration of time (since all measuring procedures can occur only with light velocity or less).

This view of Klein's comes very close to the thoughts of Einstein as contained in his 1921 Academy lecture, *Geometrie und Erfahrung,* in which he uttered his famous statement, "Insofar as the laws of mathematics refer to reality, they are not certain, and insofar as they are certain, they do not pertain to reality." Earlier, Einstein had posed, completely in the spirit of Klein, the question: How is it possible that mathematics, which is after all a product of human thought, independent of any experience, matches so well the objects of reality?

One must in this connection refer to Boltzmann, who frequently occupied himself with the question concerning the meaning of mathematicotheoretical formulations in physics. His views closely approach those of Klein and can throughout be regarded as determining for the development of Einstein's own insights. In his later life, Einstein developed opinions, such as those put forth in *Physik und Wirklichkeit,*[9]* according to which one can hardly claim that "mathematics is a product of human thought, independent of all experience," since it is, of course, constructed on an abstraction of the visually immediately perceivable existence of "material objects" and the relations of element and (denumerable) set that are unquestionably valid for them.

One may also ask what Einstein understood by mathematics "matching" reality. Does it mean that a definite mathematical structure corresponds uniquely to an occurrence of reality? Indubitably, this cannot be assumed; there are many real phenomena that are in no way physically connected with one another, but which can still be represented by the same mathematical structure, at least to some degree of approximation. One need only think of the Laplace equation $\nabla^2\varphi = 0$ and the diverse applications it can have in physics. It ought further to be discussed whether under "matching" one should understand that the mathematical structure fits only "globally" or also "in detail," in the sense that every particular of the mathematical structure corresponds to a particular of the real event—which is in no way self-evident beforehand.

Any physical occurrence can mathematically be treated with the aid of interpolation formulas. This, to be sure, allows us to establish, from a given law of parameters, another law of parameters, but it naturally makes no contribution to the questions regarding the contents of physics. The classic example of the treatment of a dynamical problem according to the Ptolomaic method and on the basis of physical principles is,

*Cf. Ref. 21, Part II.

of course, furnished by our planetary system viewed as an N-body problem of celestial mechanics. According to the principles of Newtonian gravitational dynamics, the motions of N celestial bodies (regarded as point masses) follow from the $3N$ Newtonian equations of motion when the gravitational forces among the masses are given by Newton's law of universal attraction. These motions are determined uniquely for all times provided, at an arbitrary time t_0, in addition to the N masses m_n ($n = 1,2, \ldots, N$), the instantaneous $3N$ position coordinates $x^i_n(t_0)$ and $3N$ velocity components $\dot{x}^i_n(t_0)$ are stipulated. Now, admittedly, in celestial mechanics only the two-body problem (sun–planet or double star) can in general be solved in closed form, viz., as a Kepler motion. But the general problem admits in principle a solution in the form of a series development whose terms and parameters can be computed in an exact manner algorithmically from the initial conditions on the basis of Newton's laws. This series expansion can be carried out to arbitrary order and determine the motions of the N bodies for all times from the initial conditions, without further reference to observations. (This is the problem solved by scholars of celestial mechanics since the time of Lagrange and Laplace.) The epicycles of Ptolemy also correspond to a series representation of the planetary orbits; there exists, however, no theoretically founded algorithm for deriving the parameters of the epicycles, starting from the initial conditions, that is, the data which fix Ptolemy's "deferents." Rather, the parameters (radius, orbital velocity, and position) of each epicycle must be drawn anew from observations. This circumstance also obtains here already for two-body problems when one does not have recourse to the laws of Kepler. (By contrast, the parameters of the epicycles belonging to the Kepler problem become theoretically determinable on the basis of Kepler's first and second laws—as Kepler, among others, showed in 1609 in his *Astronomia Nova*.)

According to Frege, mathematical structures are from a logical standpoint rigorously consistent, but otherwise empty. The connection with real events must be established by supplementary definitions, which at the same time make the assertions of the structure amenable to experimental verification. The structure itself is completely indifferent to any concrete assertion. The wave function φ, for example, in the wave equation $\Box^2\varphi = 0$ can be any one of several entities—electrical, elastomechanical, or thermal quantities, or quantities pertaining to the eigenfunctions of atoms, etc. Only when the meaning of φ in a physical sense is spelled out, will the connection be established between the mathematical structure and the verifiable and measurable phenomena of the real world.

The statement that the mathematical structure is indifferent to the physical content will sound trivial in simple cases, and nobody would break his head over what it could possibly mean that one can resolve the most diverse periodic processes into Fourier components. But this triviality—it is frequently forgotten—remains even when the mathematical structure becomes more complex and more abstract. One could say that, perhaps purely intuitively, increasing complexity and abstraction are viewed as potential sources of insights into physical circumstances; in this connection, see further below. (It is not reflected on that every advance in physics rests on the recognition of a new physical fact.) The parallel with scholasticism that here exists was pointed out by Klein, quoting particularly B. Bolzano and G. Cantor (who saw themselves in this tradition). On the other hand, it is true that increasing complexity of the mathematical structure permits a greater variety in the interpretation possibilities and the definitions establishing connections with physics. The tautological character of mathematics always naturally remains, and the connection with reality must always be additionally formulated.

Physics, which does not confine itself to qualitative assertions, requires moreover that these supplementary propositions furnish instructions for the execution of measurements. In this manner, new qualities can emerge from mathematical structures—no thanks, however, to mathematics but to the factual definitions that fall inside the scope of physics. It is in this sense that Einstein declares in his treatise, *Geometrie und Erfahrung*, that "abstract" (precision) geometry can be brought into relation with physics when one adds to its axioms also the following observation: "Solid bodies behave with respect to their orientation possibilities like bodies of Euclidean geometry of three dimensions. Thereby the precision geometry is stripped of its only formal logical character, in that experiential objects are assigned to its empty conceptional schemes."

With this supplementary proposition, physical in content, Einstein establishes the relation between precision mathematics and physics. Einstein further wrote that this view enabled him to formulate the general theory of relativity. We believe that Einstein here addressed a quite general principle, as it already became known in classical mechanics, for example, in the combination of the differentiation between the law of motion and initial and boundary conditions. Corresponding initial and boundary conditions can change the character of the equations also qualitatively. (There are more degrees of freedom with finite than with "infinitesimal" nonholonomic conditions.)

The amenability to axiomatization of geometry—which, to be sure, is not capable of being extended in like perfection to other branches of mathematics—must unquestionably be viewed in connection with directly perceivable "material objects" and their space-time relationships. Accordingly, we are able to discern already in the foundations of geometry the existence of the most general fundamental physical perceptions; and the relation between precision geometry and reality hangs together with the nature of the axioms that derive from reality, just as is the case for Newton's principles of mechanics.

We can formulate Einstein's proposal also as follows: The axioms of geometry are compatible with Newton's principles because they can be traced back to the same inventory of immediate physical experiences.

Actually, the supplementary axiom of physical content would be superfluous if the parallel axiom of geometry—on which, from the generally prevailing fundamental insights, one cannot decide—did not exist. In consideration of this one axiom, Einstein's requirement that some "quasirigid bodies" be introduced is throughout a necessity.

We may now ask whether, next to the axioms of geometry and Newton's principles of macrophysics, further equally fundamental insights can be gained. In Einstein's approach, this question has a relatively simple answer when an axiomatic representation exists of the mathematical structures belonging to the area under consideration.

If one follows Boltzmann and H. Hertz, then the Maxwell equations can be regarded as the "principle" of electrodynamics (for free fields). There are several fundamental components in statements on the material reality and on the vacuum: geometry, mechanics, electrodynamics, and special relativity (as their common ingredient). As further fundamental principles, Planck adopted Clausius' laws of thermodynamics, these determining the processes in the macroscopic material world. Finally, nonrelativistic quantum mechanics, in the axiomatic form given to it by J. von Neumann, is taken as the microscopic counterpart of Newton's principles.

As concerns the infinitesimal calculus, on whose application theoretical physics is founded, its axiomatic foundations reside in set theory. The basic rules of arithmetic and elementary logic seem to be in a situation analogous to that of elementary geometry: In the sense of D. Hilbert's "protomathematics," the axioms can be established *ab initio* from the immediately given properties of the "material objects." (But from this we should exclude axioms that lead directly outwards from elementary mathematics: the "principle of complete induction," according to Poincaré, and the "law of the excluded middle," after Brouwer.) About fifty years ago, Brouwer declared in his remarkable but also controversial paper,[4] "*Mathematik, Wissenschaft, und Sprache,*" that the axioms of logic are in fact compatible and demonstrable with our reality and the principles governing the macrophysics of physical bodies—this with the exception of what is referred to in the transfinite as the "principle of the excluded middle." Here one may, in analogy to Einstein's approach, decide to say, in addition to the "intuitive" axioms of Brouwer, that the objects of classical physics are those for which the law of the excluded middle is valid.

It is known that quantum physics has questioned the meaning precisely of this axiom. We then obtain the following remarkable situation: On the one hand, the usefulness of the geometric axioms (points, straight lines, planes, etc.) points to the existence of quantum forces, since undeniably an individualized, delimitable "material object" can be understood only on the basis of short-range exchange forces, entities which are unknown in classical physics. On the other hand, precisely these forces, which make possible the existence of denumerable "material objects," are incompatible with the axioms of arithmetic and logic, because for them, due to their indistinguishability, neither commutativity under multiplication nor the law of the excluded middle makes any sense; that is, the quantum phenomena are incompatible with the foundations of our set theory. Between the concepts of geometry and set theory there exists in a certain respect (as far as their application to physics is concerned) a complementarity in the sense of Bohr.

In view of the foregoing, one may think that—after 50 years' development of physical thought has widened our intellectual horizon in more than one respect—it is appropriate to ask to what extent the views expressed by Klein or Einstein are still valid. This is a legitimate question the more because the development of quantum physics has brought to the fore totally new and unexpected epistemological features which, according to the analysis of Bohr and Heisenberg, concern precisely the nature of measuring processes.

II.

In the third volume of his work[17] cited above, Klein treats the question: How secure actually are the experimental foundations of our natural laws? He then quotes a treatise of Newcomb from the year 1895, which posed the question as to how accurately the Newtonian law for the gravitational attraction between two masses had been tested. It was shown that the precision was not very high, under terrestrial conditions: At the astronomical distances of our planetary system, by contrast, numerical values were accurate to the fourth figure after the decimal point or better. Now, although the measurement technique has undergone an extraordinary refinement since Newcomb, one will always be left with the accuracy limits, for the reasons mentioned at the beginning.

An improvement in our knowledge of the distances belonging to the (inner) planets and the moon has resulted from a fundamental divorcing of distance determinations from celestial mechanics (i.e., from Kepler's third law). Nowadays we measure the transit times of radar signals that are reflected by the planetary surfaces (in the case of the moon, also of laser signals bounced off moon-based reflectors). With light velocity being a precisely known quantity, this method furnishes distance values that are reducible to and as reliable as the extremely accurate readings of atomic clocks. Of course, with this kind of precision, it is necessary to invoke the general theory of relativity in order to interpret the measurements. Under these circumstances, the distances that are deduced, respectively, from Kepler's law, parallax data, and radar measurements are, per definition, different quantities, which coincide only in the approximation of classical physics and when Euclidean geometry is assumed. The conversion of a distance obtained by one method into a distance found by another, requires now (in the Newtonian approximation) again the theory of gravitation, this time Einstein's general relativity theory. It should, however, be noted above all that the high internal accuracy of the measurements is no longer actually that of the planetary distances. After all, the signals are reflected by the surfaces of the planets, which objects are, of course, in no way perfect spheres. By contrast, the distances of celestial mechanics pertain to abstract points, the planetary centers of mass. The radar distances accordingly suffer from genuine inaccuracies corresponding to the profiles (and atmospheres) of the planets.

This circumstance seems to us to be of fundamental importance for the treatment of the measuring process: It should be clarified whether a refinement of measuring methods, insofar as it is associated with another measuring device, gives rise to assertions that pertain to the same object.* Obviously this situation holds equally for objects that can be assigned uniquely to classical physics and for objects of quantum physics. Analogous to what we have said about the measurement of planetary distances, one would obtain quite different atomic sizes, depending on whether the atoms are probed with slow electrons (Ramsauer effect), fast electrons, light quanta having the characteristic absorption frequency (large resonance cross section), or gamma quanta. The atomic dimensions, it is known, can differ from one another by several orders of magnitude, depending on the means of measurement. By contrast, statements about, say, the number of components of a multiplet, are obviously unaffected by our choice of measuring device. A doublet remains always a doublet and a triplet a triplet, provided, of course, that the external disturbance accompanying the measuring process does not alter the symmetry character of the atom.

What is denumerable, is otherwise not always obvious in the microscopic domain. The multiplet components, no doubt, are; but particles, for example, are not, because their number is not conserved in the high energy region of the measuring device (probing particles, e.g.), so that an uncertainty between position and time measurements, on the one hand, and the specification of the particle number, on the other, comes into play—a state of affairs to which particularly Heisenberg has drawn attention.

Our reliance on the foundations of physics acquired, however, a new quality when spectroscopic data, particularly those of astrophysics, revealed that, in the distant reaches of the universe, the basic principles of mechanics and relativity remain strictly valid in their application to the microscopic objects of quantum physics. The immutability of the constants h, c, and e (and, therefore, Sommerfeld's fine-structure constant $e^2/\hbar c = 1/137$) over a time span of more than 10 billion years ($\sim 10^{40} h/mc^2$), corresponding to distances of 10 billions of light years—and the thereby determined extreme constancy of the spectroscopically observable, and consequently also mechanical, properties of atoms—allows the conclusion that the principles of dynamics, with Newton's homogeneous flow of time and thereby the exact reversibility of time, are true propositions in the sense of a precision mathematics.

What astrophysics lacks spectroscopically are however in the first instance only algebraic relations among the lines in the spectroscopic series (and among the series themselves). These relations prove to be independent of space and time, and that is what follows from the constancy of h, c, and e and from microscopic reversibility. At the same time, the absolute positions of the lines are kinematic and dynamic variables, functions of the most diverse local effects. (Also the magnitude of line splittings is a function of the intensity of external influences, but the number of lines here depends only on their symmetries.)

*This confirms once again the correctness of Einstein's pronouncement to Heisenberg: What can be measured results from the theory.

III.

A further connection between precision mathematics and physics has its origin in arithmetic. It is known that the transcendental numbers e and π can be computed *ab initio* with arbitrary precision. "*Ab initio*" means here "from the whole numbers, by application of the basic operations of arithmetic." After the discovery of the elementary constants and their fundamental interpretations by Planck, the idea arose in physics that, just as is the case in mathematics for e and π, an *ab initio* calculation of certain physical quantities should be possible. This conception received support from the success of Lorentz' theory of the normal Zeeman effect, in which the splitting of an atomic spectral line in an external magnetic field, as it concerns the number of components, their polarization, and the magnitude of the splitting, could be computed from the size of the magnetic field strength H and the values of the elementary constants e, m, and c. When subsequently the Bohr model of the atom made it possible to compute the ionization energy of the hydrogen atom from e, m, and h, opinion gained strength that it should be worth the effort to attempt an *ab initio* calculation from the elementary constants of physical quantities appearing in all areas of physics.

In pursuit of this end, it is well known that astonishing results were achieved. The nonrelativistic one- and two-body problem, including, for example, the spectrum of the hydrogen atom under the inclusion of its nuclear motion, were found to be completely solvable, to an accuracy determined by that with which the elementary constants were known, with the aid of exact solutions of Schrödinger's equation.

The relativistic one-body problem, too, which is capable of furnishing the spin of the electron and the anomalous magnetic moment of the electron (or the g factor) can be correctly treated with an accuracy that nowadays extends to the 8th or 9th figure after the decimal point, as precisely as the elementary constants are known. If one also takes into account the "vacuum effects," through an expansion in powers of the fine-structure constant, then the agreement between theory and experiment becomes really phenomenal.

The experimental techniques perfected over the past twenty years for the measurement of combinations of elementary constants, particularly through the introduction of radiofrequency interference procedures, have actually achieved an accuracy of 1 in 10^9 or better. We here meet with a point of contact between precision mathematics and physics of which forty or fifty years ago no one had an inkling and which entitles us to the following fundamental assertion: Neither Klein, nor Boltzmann before him, was obliged to interpret the strictly dissipation-free, reversible Newtonian mechanics as an abstraction corresponding to a real behavior—which, afflicted by the known classical fluctuation phenomena, could only claim an approximate validity (cf. the angle sum π in a concrete triangle). Now it turns out that the microscopic world, the processes among atoms, electrons, and light quanta, is exactly as required by Newton's principles—absolutely reversible. The agreement that exists, on the evidence of astrophysical findings, throughout the universe in the value of the fine-structure constant, the number of multiplet components, and the absence of energy loss for light quanta traveling through vacuum are strong indications that Newton's reversible behavior is strictly obeyed in the microscopic domain. It is thus understandable that, in the fourth edition of his *Wärmestrahlung*, Planck took, from Einstein's work about light quanta, as his most natural and significant assignment the demonstration of the validity of microscopic reversibility.

IV.

It is known that Weierstrass, with the aid of the concept "the set of real numbers" was able to exactly justify the idea of a differential quotient, which was first introduced by Newton and Leibniz. On the application of the differential quotient the entire edifice of theoretical physics is constructed. Here, exact mathematics and its logically consistent interpretation of the number concept in terms of the number continuum produce, to be sure, something different from what we can directly experience in nature, where, after all, every measured curve is ascertainable only with a certain accuracy and leads to difference quotients that automatically contain certain ambiguities.

As long as one stays within the jurisdiction of classical physics, one has really no occasion for doubting the clear separation, as put forth by Klein and Einstein, between exact mathematics, on the one hand, and physics, with the "approximation mathematics" sufficient for its purposes, on the other.

But on the other side, as Weyl pointed out, the number continuum exists of clearly distinguishable, noninterchangeable individual entities, while Newton's space continuum and his "time flow," that is, the continuum of space and time, exist of mass and time points that are *not* individuals, are intrinsically indistinguishable, and therefore not capable of being characterized by their own properties. To fix and set them apart from one another, one always requires a coordinate system and the stipulation of coordinates.* This indistinguishability of the mass and time points is undoubtedly one of the most decisive anticipations of quantum mechanical findings, which has enabled Newtonian mechanics to be so extensively operative also in the sphere of all quantum phenomena that have thus far been investigated. The conception of mass points that are incapable of being individualized turns out to be a totally adequate statement about a certain aspect of reality.

We are in agreement with Weyl when he expresses the opinion that the Newtonian continuum cannot be reduced to the continuum of mathematical analysis (but rather, as he showed, to a hyperanalysis). The mathematical proof (Weierstrass) for the admissibility of the transition to the limit for the differential quotient remains therewith an internal mathematical matter. In the context of physics, more essential

*Weyl wrote: "Das Koordinatensystem ist das unvermeidliche Residuum der 'Ich-Vernichtung' in jener geometrisch-physikalischen Welt, welche die Vernunft aus dem Gegebenen unter der Norm der Objektivität herausschält."

is the physical demonstration of the adequacy for the real world of the differential quotient, as it follows (see above) from the nondissipative character of microprocesses.

Also Klein consistently did not consider it obvious that space and time should be represented by a continuum. However, it is indeed so that more profound propositions even over finite domains of numbers, for example, those occurring in probability calculus and statistics (and therefore also in atomic theory), can only be established with the use of analysis; the same holds in arithmetic, as well. Even statements concerning the distribution, etc. of prime numbers are derivable, indeed definable, only with an appeal to continuum mathematics.

Although the natural (as well as the algebraic) numbers are discrete and denumerable, the properties of number series can be comprehended in sufficient depth and generality only in terms of continuum mathematics. Also probability (and even combinatorial) calculus requires continuum analysis and therewith the transition from sums to integrals and from differences to differentials. All this holds not only for sophisticated theorems, but is also in practice unavoidable whenever a sufficiently large set of objects or events are to be examined.

The meaning of the continuity axiom for world geometry already throws light on a theorem of absolute geometry which Legendre first proved and Hilbert then made formalizable with the aid of his continuity axiom: If Euclidean geometry or, more precisely, its theorem that the angle sum of a triangle amounts to π, holds exactly (in the sense of precision geometry) even for a single finite triangle at any point of space, then Euclidean geometry will, on account of the "continuity of the continuum," hold everywhere in space.*

If, conversely, it can be shown (with the methods of measuring technique and precision mathematics) that any triangle whatsoever is not Euclidean, then nowhere in space will Euclidean geometry be accurately valid either. This holds, however, naturally only for continua that everywhere satisfy Hilbert's continuity axiom, that is, simply the axiom of Eudoxos and Archimedes, augmented by Cantor's axiom, which requires the measurability of line segments, in the sense of precision mathematics, by a laying off procedure that employs arbitrarily chosen measuring rods.

V.

In view of the foregoing, we are now confronted with the following complicated situation: In the spirit of Einstein's comments in his 1936 article "*Physik und Realität*," which in our opinion closely link up with Boltzmann's "*Grundprinzipien und*

*Euclidean (and no other) mathematics shares with classical physics an independence of absolute quantities. In classical mechanics one finds no universal constants (like h or c in quantum mechanics or special relativity). In Euclidean geometry (and only there, as Lambert already showed) there exists no (local or global) unit of length defined by "a measure of spatial curvature." In Lobachevski's "non-Euclidean" geometry and the elliptic "Riemannian" geometry, this measure of curvature is a spatial constant; in the general Riemannian manifold, by contrast, it is a continuous function of position and direction.

Grundgleichungen der Mechanik," of 1899, we consider "material objects" as given by immediate perception. These entities are discretely distinguishable, capable of being demarcated, and classifiable according to size, form, or some other physical properties. In particular, one can count and number them and arrange them by values of characteristic parameters. If by abstraction we reduce them to mass points, they lose their distinguishability property, etc., and can no longer be characterized as individuals, except through specification of their coordinates. The mass points become constituents of the physical continuum, whereby any comprehension of their behavior requires the establishment of measuring rods. The material bodies, by contrast, are discrete. They can be counted and, if necessary, listed by their properties. It is possible to construct, from material objects as well as mass points, aggregates that acquire a higher-order individuality. In particular this individuality can disappear, which represents a kind of dissipation. Just as is the case for material objects, the aggregates are capable of changing their positions and their physical properties, although they are in detail composed of mass points.

Remarkably, the individualizability of material objects, and of the aggregates composed of them, as well as the lack of individuality of mass points and the individualizability of the aggregates constructable from them, are in complete agreement with the basic laws of quantum mechanics, so that the contradiction between discreteness and continuity is seen to be deeply anchored in nature itself. Obviously, both the individualizable material object and the physical continuum composed of nonindividualizable mass points represent different sides of the same reality. If one is prepared to see in denumerable material objects the foundation of atomic theory, then one can acknowledge it as the substantiation of the approximation-mathematical aspect. The nondenumerable mass points can then be viewed as a substratum of the physical continuum, which bears a certain analogy to the continuum of real numbers, where one must heed Weyl's remark concerning the individuality of the components of the mathematical continuum. Since obviously both discrete and continuous aspects are found in reality, one possibility for the relation between exact mathematics and reality is thereby provided.

A further opinion flows from the practice of physics since Galilei: When one views matters without prejudice, then it becomes clear that the increase of knowledge in physics has resulted from a permanent interaction of theory, mathematical formulation, experimental investigations, and technical applications of fundamental physical knowledge; cf. Ref. 21, Part I. In this connection, the practical procedures are instructive. The correctness of the exact mathematical equations are routinely taken for granted and all approximation-mathematical reservations are attributed to unsharp initial and boundary conditions.

We hope to have explained above that what throughout corresponds to classical physics as the physics of objects is the entity for which the axioms of Euclidean geometry and the axioms of arithmetic and logic, i.e., set theory, possess an unlimited validity. That does not mean, though, that outside the construct of classical physics things must always be likewise. The growth of knowledge in physics since Galilei and Newton for a large part came about because investigators were not content observing nature; instead, on the basis of theoretical considerations and mathematical formulations, they experimentally, through test and measuring arrangements, sought and contrived initial and boundary conditions that were suitable for fashioning real situations out of mathematically consistent solutions. In this manner most new findings of physics have been "artificially" produced. Their connection with nature was indeed theoretically to be accepted, but it subsequently often became necessary to demonstrate, at no small expense, their validity in practice. This physics of artifacts has led to a blossoming of applied physics, which to a considerable and ever-increasing degree has also affected general technology. One need only think of the vast transfer of laboratory techniques to the area concerned with the peaceful uses of nuclear energy and to large industrial technological installations.

If, again without prejudice, one surveys the totality of our knowledge in physics, then there emerges roughly the following picture of the construction of that part of reality which falls within the competence of physics, inasmuch as it is comprehensible only in terms of the concepts that have been developed in physics. (We have briefly reported our thoughts on this matter earlier; see Ref. 21, Part II.

Physical reality simultaneously comprises (in spatial and temporal regard) the world of material particles ($m_0 > 0$), the electromagnetic (photonic) and neutrino radiation ($m_0 = 0$) produced by it, the aggregates (molecules, solids, etc.) composed of the material particles, and, finally, the vacuum. Everything we know in physics, based on observations and measurements with instruments—which themselves are aggregates of such material particles—and everything we can with their aid ascertain, pertain to this material world with its photon and possibly neutrino radiation. In this world, the grand theories of physics—geometry and relativity theory, mechanics, electrodynamics, thermodynamics, and quantum mechanics—hold sway.

The material world exists and permeates, as a component of reality, the vacuum, which is the carrier of electromagnetic phenomena traveling with the fundamental velocity c (representing the limiting transmission pace of energy and information) and the birthplace of material particles, which indeed, when passing from the vacuum into the material world, must satisfy the conservation laws characterizing the large physical theories and which, accordingly, at the moment of their creation require certain starting conditions.

In vacuum all particles enjoy a virtual existence. ("Each particle is composed of all others."—Heisenberg.) We view the events in vacuum in the framework of approximations of theories belonging to the material world. The material world and the vacuum filled with its virtual particles are already parts of physical reality. Many exact-mathematical consistent pronouncements of precision mathematics obviously appear in vacuum and are therefore also constituents of physical reality. With the realization of suitable conditions of a kind that hardly ever occur in nature, such exact-mathematical results can enter the material world as artifacts and will usually significantly broaden our physical insight. Such artifacts are obviously the exotic particles, entities characterizing our present elementary-particle physics (with $E \gg m_0c^2$ and $d \ll h/m_0c$) or the macroscopic quantum phenomena at low temperatures, special substances that are not found in nature, and substances occurring in nature but having an unnatural degree of purity or some other unusual property.

The physics of artifacts thus represents the area in which precision mathematics—naturally made concrete through additional, physically meaningful conceptual conditions in the Einsteinian sense—is most intimately associated with physics and its development.

We live in a cosmos that contains about 10^{80} baryons [Eddington's number $(hc/Gm^2)^2$]. In this vast cosmos the occurrence of extremely improbably phenomena becomes possible, such as coherent electromagnetic radiation, energy concentrations of the most unusual kind, but also processes such as the evolution of life from inanimate matter. Nevertheless, not all assertions of mathematics are realized naturally in the cosmos, because it concerns here a considerably higher order of mathematical thought. Eddington's number is, to be sure, a very large number, but it is still finite!

Thus, through materialization of virtual phenomena in vacuum, a part, but only a part, of the potential inherent in precision mathematics becomes applicable to the events of reality.

References and Notes

1. N. Bohr, *Atomphysik und Menschliche Erkenntnis*, Parts I and II (Braunschweig, 1956/1957).
2. L. Boltzmann, *Populäre Schriften* (J. A. Barth, Leipzig, 1906).
3. B. Bolzano, *Theorie der Reellen Zahlen* (Prague, 1962).
4. J. L. Brouwer, "Mathematik, Wissenschaft, und Sprache," *Monatsh. Math. Phys. (Wien)* **36**, 153 (1929).
5. G. Cantor, *Abhandlungen*, E. Zermelo, editor (Berlin, 1932).
6. A. S. Eddington, *Space, Time, and Gravitation* (Cambridge U.P., Cambridge, 1920).
7. A. S. Eddington, *The Mathematical Theory of Relativity* (Cambridge U.P., Cambridge, 1923).
8. A. Einstein, *Geometrie und Erfahrung* (Springer, Berlin, 1922).
9. A. Einstein, "Physik und Wirklichkeit," *J. Franklin Inst.* **221**, 313 (1936).
10. A. Einstein, *Aus Meinem Spätcn Jahren* (Stuttgart, 1979).
11. G. Frege, *Schriften zur Logik* (Berlin, 1973).
12. K. Fuchs, "Theorie, Wahrheit, und Wirklichkeit," in *75 Jahre Quantentheorie* (Akademie-Verlag, Berlin, 1977).
13. W. Heisenberg, *Der Teil und das Ganze* (Munich, 1969).
14. H. Hertz, *Die Prinzipien der Mechanik* (Leipzig, 1894).
15. D. Hilbert, *Die Grundlagen der Geometrie*, 7th ed. (Leipzig, 1930).
16. J. Hjelmslev, *Die Naturliche Geometrie* (Hamburg, 1923).
17. F. Klein, *Elementarmathematik vom Höheren Standpunkte Aus*, Vol. 1 (1924), Vol. II (1925), and Vol. III (1928) (Springer, Berlin).
18. F. Klein, *Vorlesungen über die Entwicklung der Mathematik im 19. Jahrhundert*, Vols. I and II (Berlin 1926/1927).
19. M. Planck, *Acht Vorlesungen über Theoretische Physik* (Leipzig, 1910).
20. H. Poincaré, *Wissenschaft und Hypothese*, 3rd ed. (Leipzig, 1914).
21. R. Rompe and H.-J. Treder, *Über Physik* (Akademie-Verlag, Berlin, 1979).
22. R. Rompe and H.-J. Treder, *Grundfragen der Physik* (Akademie-Verlag, Berlin, 1980).
23. E. Warburg, "Verhältnis der Präzisionsmessungen zu den allgemeinen Zielen der Physik," in *Physik in der Kultur der Gegenwart*, 2nd ed. (Leipzig, 1925).
24. H. Weyl, *Das Kontinuum* (Veit, Leipzig, 1918).
25. H. Weyl, *Philosophie der Mathematik und Naturwissenschaft*, 3rd ed. (R. Oldenbourg, Munich, 1966).

26

Electroweak Interaction in Polarized Electron–Positron and Electron–Electron Scattering

T. Anders, W. Jachmann, H. Salecker, and K. Wadan
Sektion Physik
Universität München
Munich, West Germany

1. Introduction

The discovery of renormalization through spontaneous symmetry breaking in non-Abelian gauge theories is one of the most important results in theoretical physics during the last decade. In this manner it has been possible to create an (almost) unified theory of electrodynamics and weak interaction. Further results along these lines are quantum chromodynamics and grand unification theories. However, up to now there is no definite experimental verification of the theory of electroweak interaction either in the standard form (Salam and Weinberg) or with a more complicated gauge group. So far, experiments on weak interaction can still be described with an old-fashioned four-fermion interaction with an additional, neutral current. There are especially no experimental indications for the predicted intermediate bosons like the $W^{\pm}$ and the Z^0. These particles cannot yet be produced because accelerator energies are still too small. But one could ask for their influence on precise experiments in quantum electrodynamics (QED) as one does quite generally in testing QED.[1] Unfortunately, accuracy is still too small to see these effects in the anomalous magnetic moments of the electron and muon or the Lamb shift in hydrogen. On the other hand such effects could be seen already in the accessible energy range in sufficiently precise electron–positron annihilation experiments. Especially annihilation with polarized $e^{\pm}$ beams into muons could show these effects clearly. Unfortunately, these experiments have considerable experimental errors, which are up to now much too large to give a definite result in this direction. Another (theoretical) uncertainty arises from the fact that muon–electron universality is needed here but not yet tested independently in this region.

In this paper we want to show that electron–positron scattering with polarization is a suitable compromise. Higher experimental accuracy can be obtained here and the Z^0 effects are still not too small. Prospects for the production of polarized electron–positron beams at high energies are good.[2] If experiments could be done with an accuracy better than 10% Z^0 effects can clearly be seen. There are already a number of different calculations of electron–positron scattering in the literature partly with special polarizations.[3] Unfortunately there is no complete agreement between the results of these works. Therefore, we will present here the results of a calculation of electron–positron and electron–electron scattering in electroweak interaction with arbitrary polarizations in the initial and final states. To avoid computational errors we have used three different methods of calculation independently. Finally we will discuss typical numerical examples for both scattering processes.

In the case of final polarizations, the effect of the z^0 contribution can be seen much better. Under these circumstances also the "natural" transverse polarizations of the initial state can be very useful. This will be discussed in a forthcoming paper.

2. Polarized Electron–Positron Scattering

In electroweak interaction (standard model) electron–positron scattering proceeds in the lowest order through the diagrams of Fig. 1. Here we have neglected Higgs-boson exchange, the contribution of which is expected to be very small in the energy region considered. From Fig. 1 we obtain the amplitude

$$M_{e^+e^-} = \frac{1}{t}[\bar{u}(p',w)\gamma_\mu u(p,x)]\,[\bar{v}(q,y)\gamma^\mu v(q',z)] - \frac{1}{s}\,[\bar{u}(p',w)\gamma_\mu v(q',z)]\,[\bar{v}(q,y)\gamma^\mu u(p,x)]$$
$$+\frac{1}{t_z}\,[\bar{u}(p',w)\Gamma_\mu u(p,x)]\,\,[\bar{v}(q,y)\Gamma^\mu v(q',z)] - \frac{1}{s_z}\,[\bar{u}(p',w)\Gamma_\mu v(q',z)]\,\,[\bar{v}(q,y)\Gamma^\mu u(p,x)] \quad (1)$$

where we have used the following notations (compare also Fig. 1):

$$\begin{aligned}(p^\mu) &\approx E(1,0,0,1), \qquad (p'^\mu) \approx E(1,\sin\vartheta\cos\varphi,\,\sin\vartheta\sin\varphi,\,\cos\vartheta)\\ (q^\mu) &\approx E(1,0,0,-1), \qquad (q'^\mu) \approx E(1,-\sin\vartheta\cos\varphi,-\sin\vartheta\sin\varphi,-\cos\vartheta)\\ s &= (p+q)^2 = 4E^2, \qquad t = (p-p')^2 \approx -4E^2\sin^2(\vartheta/2),\; u = -s-t\\ s_z &\approx s - M_{z0}^2 + iM_{z0}\Gamma_{z0}\,, \qquad t_z \approx t - M_{z0}^2 + iM_{z0}\Gamma_{z0}\end{aligned} \quad (2)$$

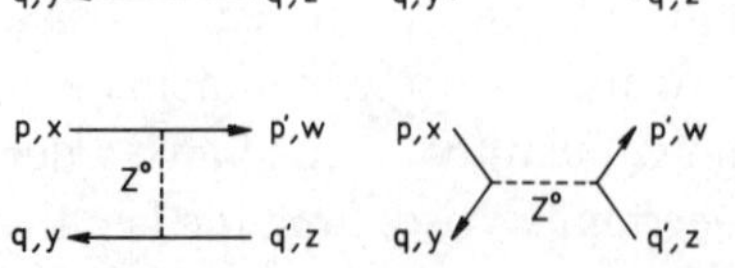

FIGURE 1. Diagrams for electron–positron scattering in the lowest order of the standard model of electroweak interaction. Higgs-boson exchange has been neglected. Spin polarization of the initial electron or positron and final positron or electron are denoted by x,y,z,w, respectively.

In the standard model we have

$$\Gamma_\mu = (g_V + g_A\gamma^5)\gamma_\mu, \qquad g_V = \frac{4\sin^2\theta_w - 1}{2\sin 2\theta_w}$$
$$g_A = -\frac{1}{2\sin 2\theta_w}, \qquad \alpha = \frac{e^2}{4\pi} \tag{3}$$

θ_w is the Weinberg angle, ϑ the scattering angle, and φ the azimuthal angle, M_{Z^0} the mass and Γ_{Z^0} the width of the Z^0.

The calculation of the differential cross section with arbitrary polarizations is not difficult but fairly complicated. We have, therefore, calculated it independently in three different ways, namely, with the algebraic manipulation systems SCHOONSCHIP and SPUC[4] and with a new method.[5] This method allows one to transform first the matrix element quite generally into a simple form from which we can directly obtain the cross section without trace calculations. We obtained in these ways (terms of order m_e/E are always neglected) in the c.m.s.

$$\begin{aligned}
\frac{16}{\alpha^2}\frac{d\sigma^{e^+e^-}}{d\Omega} = {} & |A|^2[(1 + x_\parallel)(1 + y_\parallel)(1 + z_\parallel)(1 + w_\parallel) \\
& + (1 - x_\parallel)(1 - y_\parallel)(1 - z_\parallel)(1 - w_\parallel) \\
& + 2x_\perp y_\perp z_\perp w_\perp \cos(\varphi_x + \varphi_y - \varphi_z - \varphi_w)] \\
& + |B|^2[(1 + x_\parallel)(1 - y_\parallel)(1 + z_\parallel)(1 - w_\parallel) \\
& + (1 - x_\parallel)(1 + y_\parallel)(1 - z_\parallel)(1 + w_\parallel) \\
& + 2x_\perp y_\perp z_\perp w_\perp \cos(\varphi_x - \varphi_y - \varphi_z + \varphi_w)] \\
& + |C_+|^2(1 - x_\parallel)(1 + y_\parallel)(1 + z_\parallel)(1 - w_\parallel) \\
& + |C_-|^2(1 + x_\parallel)(1 + y_\parallel)(1 - z_\parallel)(1 + w_\parallel) \\
& + 2\mathrm{Re}\{AB^*[(1 + x_\parallel)y_\perp(1 + z_\parallel)\, w_\perp e^{i(\varphi_w - \varphi_y)} \\
& + x_\perp(1 + y_\parallel)z_\perp(1 + w_\parallel)\, e^{i(\varphi_z - \varphi_x)} + x_\perp(1 - y_\parallel)z_\perp(1 - w_\parallel)\, e^{i(\varphi_x - \varphi_z)} \\
& + (1 - x_\parallel)y_\perp(1 - z_\parallel)\, w_\perp e^{i(\varphi_y - \varphi_w)}] \\
& + AC_+^*[x_\perp(1 + y_\parallel)(1 + z_\parallel)\, w_\perp e^{i(\varphi_w - \varphi_x)} \\
& + (1 - x_\parallel)\, y_\perp z_\perp(1 - w_\parallel)e^{i(\varphi_y - \varphi_z)}] \\
& + AC_-^*[x_\perp(1 - y_\parallel)(1 - z_\parallel)\, w_\perp e^{i(\varphi_x - \varphi_w)} \\
& + (1 + x_\parallel)\, y_\perp z_\perp(1 + w_\parallel)e^{i(\varphi_z - \varphi_y)}] \\
& + BC_+^*[x_\perp y_\perp(1 + z_\parallel)(1 - w_\parallel)\, e^{i(\varphi_y - \varphi_x)} + (1 - x_\parallel)(1 + y_\parallel)z_\perp w_\perp e^{i(\varphi_w - \varphi_z)}] \\
& + BC_-^*[x_\perp y_\perp(1 - z_\parallel)(1 + w_\parallel)\, e^{i(\varphi_x - \varphi_y)} + (1 + x_\parallel)(1 - y_\parallel)\, z_\perp w_\perp e^{i(\varphi_z - \varphi_w)}] \\
& + C_+C_-^* x_\perp y_\perp z_\perp w_\perp e^{i(\varphi_x - \varphi_y + \varphi_z - \varphi_w)}
\end{aligned} \tag{4}$$

Here we have used the abbreviations

$$A = \frac{s}{t} + (g_V^2 - g_A^2)\frac{s}{t_z}, \qquad B = \frac{t}{s} + (g_V^2 - g_A^2)\frac{t}{s_z}$$
$$C_\pm = -\frac{u}{t} - \frac{u}{s} - (g_V \pm g_A)^2\left(\frac{u}{t_z} + \frac{u}{s_z}\right) \tag{5}$$

The spin vectors s of the four particles (see Fig. 1; s = x,y,z or w) are decomposed into longitudinal and transversal components $s_\parallel$ and $s_\perp$, respectively, in their local coordinate systems. These coordinate systems are defined as follows: The third axis shall be the same as the direction of the momentum of the particle, the second axis shall have the same direction for all four particles, namely, perpendicular to the scattering plane, while the first axis shall lie in the scattering plane forming a right-handed coordinate system with the other axes. Since polarization in general may be incomplete, we have $|s| \leq 1$ and

$$-1 \leq s_\parallel \leq 1$$
$$0 \leq s_\perp \leq 1$$

For the azimuthal angle φ_s, we have

$$0 \leq \varphi_s \leq 2\pi$$

With $g_V = g_A = 0$ we obtain from (4) the well-known cross section for polarized $e^\pm$ scattering in QED. In case of unpolarized final states our formula (4) agrees with the cross section of Ref. 3 (Budny, 1975).

The cross section (4) shows a few interesting symmetries. These will be discussed together with the cases of final polarization elsewhere.

3. Numerical Examples for Electron–Positron Scattering

In this section we will give a few typical numerical examples. We take $E = 20$ GeV (per beam), $\sin^2\theta_w = 0.228$, and $\Gamma_{Z0} = 2.5$ GeV. This implies $M_{Z0} \approx 89$ GeV. In the initial state we consider the following polarizations: no polarization

$$1: (x_\parallel = y_\parallel = 0, x_\perp y_\perp \cos(\varphi_x - \varphi_y) = 0)$$

the two transversal polarizations

$$2: (x_\parallel = y_\parallel = 0, x_\perp y_\perp \cos(\varphi_x - \varphi_y) = +0.854, \text{i.e.}, \downarrow\uparrow)$$
$$3: (x_\parallel = y_\parallel = 0, x_\perp y_\perp \cos(\varphi_x - \varphi_y) = -0.854, \text{i.e.}, \uparrow\uparrow)$$

and the three longitudinal polarizations

$$4: (x_\parallel = y_\parallel = 0.924, x_\perp y_\perp \cos(\varphi_x - \varphi_y) = 0, \text{i.e.}, \rightarrow\leftarrow)$$
$$5: (x_\parallel = -0.924, y_\parallel = 0.924, x_\perp y_\perp \cos(\varphi_x - \varphi_y) = 0, \text{i.e.}, \leftarrow\leftarrow)$$
$$6: (x_\parallel = 0.924, y_\parallel = -0.924, x_\perp y_\perp \cos(\varphi_x - \varphi_y) = 0, \text{i.e.}, \rightarrow\rightarrow)$$

$P = 0.924$ is the maximum possible polarization degree in storage rings.[2] In the final state we consider here no polarization.

In Fig. 2 we show for these cases

$$F = (1 - \cos\vartheta)^2 \frac{d\sigma^{e+e-}}{d\Omega} \tag{6}$$

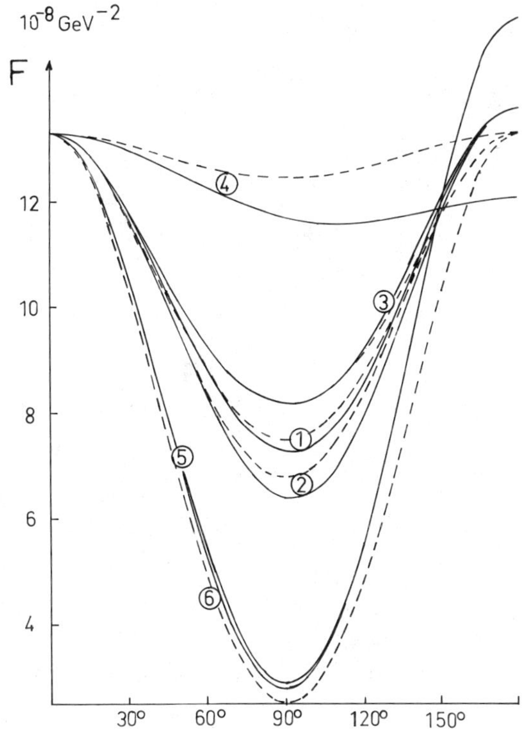

FIGURE 2. $F = (1 - \cos\vartheta)^2\, d\sigma^{e^+e^-}/d\Omega$ as a function of the scattering angle ϑ for the six initial polarization cases and no final polarization under the conditions discussed in Section 3. The QED parts of the cross section are dotted.

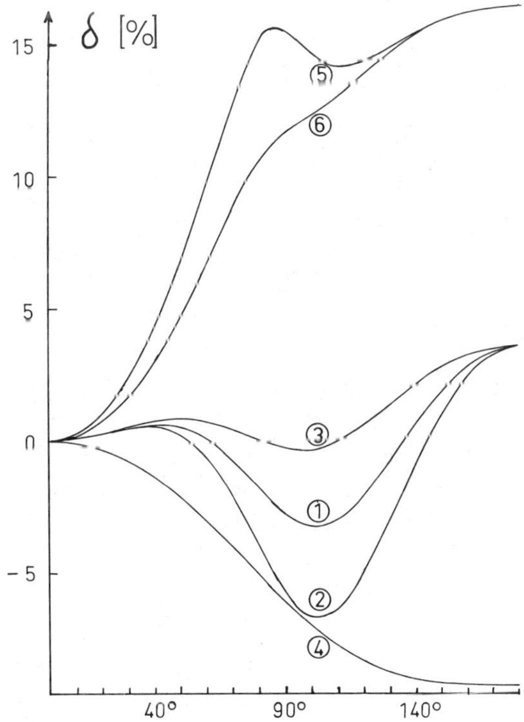

FIGURE 3. The relative correction δ (in %) for $e^{\pm}$ scattering according to Z^0 exchange (7) for the different polarizations and conditions discussed in Section 3 as a function of the scattering angle ϑ.

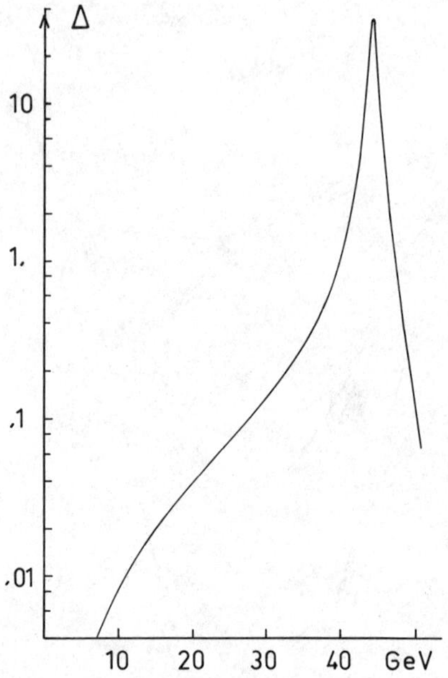

FIGURE 4. $\Delta = \delta_{\leftarrow\leftarrow} - \delta_{\rightarrow\rightarrow}$ for $e^{\pm}$ scattering as a function of the beam energy E for 90° scattering. This difference is produced by the parity nonconserving part of Z^0 interaction and, therefore, independent of radiative corrections.

as a function of the scattering angle ϑ. In this way we can avoid the trivial Coulomb singularity. The QED part shows then symmetry around 90° and the forward–backward asymmetry comes solely from the Z^0 interaction.

In Fig. 3 we have plotted the Z^0 influence as a correction (in %) to the QED contribution

$$\frac{d\sigma}{d\Omega} = \frac{d\sigma_{\text{QED}}}{d\Omega}(1 + \delta) \tag{7}$$

for the above-mentioned six initial polarization cases. We see that the two parallel configurations of longitudinal polarization (←←) and (→→) are fairly interesting. From $\vartheta \gtrsim 60°$ δ is here of the order of 10% or more. For 90° scattering we have $\delta \approx 15.5\%$ resp. 11.7%. Therefore, with an experimental error smaller than 10% one could see this deviation clearly and confirm the Z^0 contribution for the first time directly. However, radiative corrections from higher orders of QED can also give contributions to these terms of δ. Unfortunately radiative corrections for polarized beams have not yet been calculated. For this reason we are carrying out such a calculation. Only the difference between the two parallel configurations of the longitudinal polarizations will not have contributions from the QED radiative corrections since this difference is caused by the parity nonconserving terms of Z^0 exchange, which are proportional to $g_A g_V^3$ and $g_A^3 g_V$.

The first term is very small in the standard theory because of the smallness of $(g_V/g_A) = 4\sin^2\theta_w - 1 \approx -0.088$. But the second term can be appreciable. Unfortunately at $E = 20$ GeV the difference $\Delta = \delta_{\leftarrow\leftarrow} - \delta_{\rightarrow\rightarrow}$ is at most only about 3.8% but increases at higher energy. Figure 4 shows Δ as a function of the beam energy E for 90° scattering.

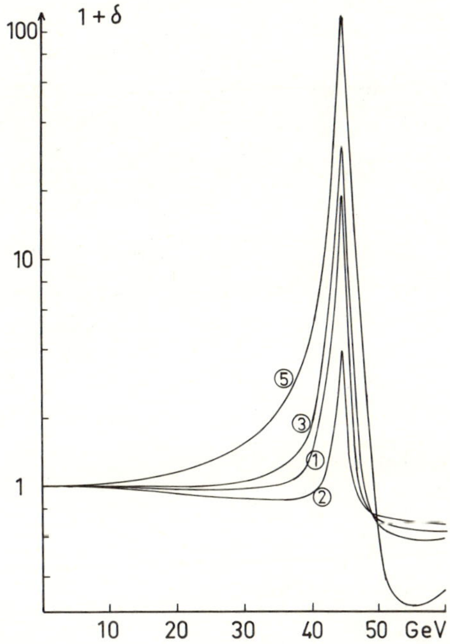

FIGURE 5. The ratio of the cross sections $d\sigma^{e^+e^-}/d\sigma_{QED} = 1 + \delta$ for $e^\pm$ scattering for several polarizations and 90° scattering as a function of the energy E also beyond the Z^0 mass.

In Fig. 5 we show the ratio of cross sections

$$\frac{d\sigma}{d\sigma_{QED}} = 1 + \delta$$

for several polarization configurations for 90° scattering as a function of energy also beyond the Z^0 mass. Together with Fig. 3 we can imagine the course of δ with energy E and scattering angle ϑ.

The cases of final polarization will be discussed elsewhere.

4. Polarized Electron–Electron Scattering

Electron–electron scattering proceeds in the standard model through the diagrams of Fig. 6. Here we have again neglected the Higgs boson exchange.

As in electron–positron scattering we have also calculated the differential cross section with arbitrary polarizations because of the complexity in the above-mentioned

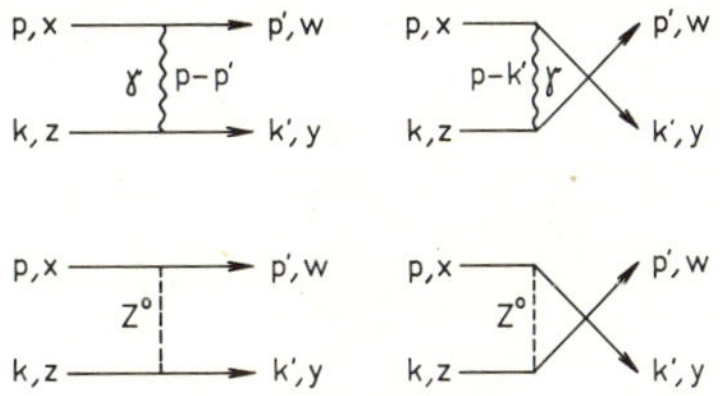

FIGURE 6. Diagrams for electron–electron scattering in the lowest order of the standard model of electroweak interaction. Higgs-boson exchange has again been neglected. Spin polarizations are denoted by x,z, in the initial, and w,y, in the final state, respectively.

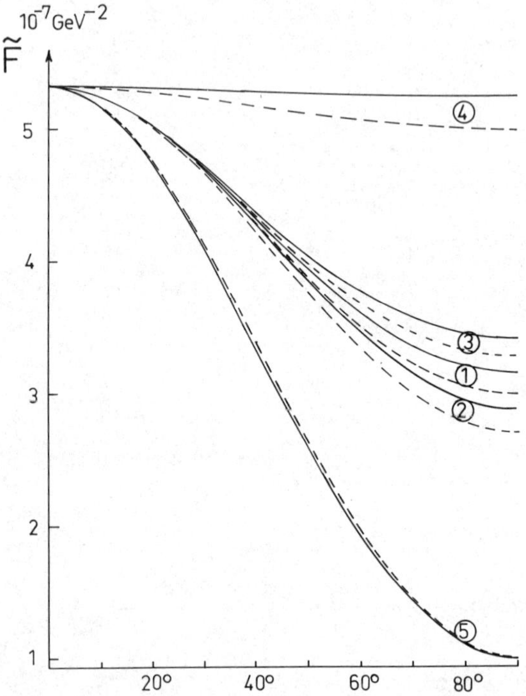

FIGURE 7. $\tilde{F} = \sin^4 \vartheta \, d\sigma^{e^-e^-}/d\Omega$ as a function of the scattering angle ϑ for the five remaining initial polarizations and no final polarization under the conditions discussed in Section 5. The QED parts of the cross section are dotted.

three different ways. Besides this we have also used the substitution law to obtain the cross section—always with the same result. Neglecting terms of order m_e/E we find in the c.m.s.

$$
\begin{aligned}
\frac{16}{\alpha^2}\frac{d\sigma^{e^-e^-}}{d\Omega} = {} & |\tilde{A}|^2[(1 + x_\parallel)(1 - y_\parallel)(1 - z_\parallel)(1 + w_\parallel) \\
& + (1 - x_\parallel)(1 + y_\parallel)(1 + z_\parallel)(1 - w_\parallel) \\
& + 2x_\perp y_\perp z_\perp w_\perp \cos(\varphi_x + \varphi_y - \varphi_z - \varphi_w)] \\
& + |\tilde{B}|^2[(1 + x_\parallel)(1 + y_\parallel)(1 - z_\parallel)(1 - w_\parallel) \\
& + (1 - x_\parallel)(1 - y_\parallel)(1 + z_\parallel)(1 + w_\parallel) \\
& + 2x_\perp y_\perp z_\perp w_\perp \cos(\varphi_x - \varphi_y - \varphi_z + \varphi_w)] \\
& + |\tilde{C}_+|^2(1 - x_\parallel)(1 - y_\parallel)(1 - z_\parallel)(1 - w_\parallel) \\
& + |\tilde{C}_-|^2(1 + x_\parallel)(1 + y_\parallel)(1 + z_\parallel)(1 + w_\parallel) \\
& + 2\mathrm{Re}\{\tilde{A}\tilde{B}^*[(1 + x_\parallel)y_\perp(1 - z_\parallel)w_\perp e^{i(\varphi_w - \varphi_y)} \\
& + x_\perp(1 - y_\parallel)z_\perp(1 + w_\parallel)e^{i(\varphi_z - \varphi_x)} + x_\perp(1 + y_\parallel)z_\perp(1 - w_\parallel)e^{i(\varphi_x - \varphi_z)} \\
& + (1 - x_\parallel)y_\perp(1 + z_\parallel)w_\perp e^{i(\varphi_y - \varphi_w)}] + \tilde{A}\tilde{C}_+^*[x_\perp(1 - y_\parallel)(1 - z_\parallel)w_\perp e^{i(\varphi_w \varphi_x)} \\
& + (1 - x_\parallel)y_\perp z_\perp(1 - w_\parallel)e^{i(\varphi_y - \varphi_z)}] + \tilde{A}\tilde{C}_-^*[x_\perp(1 + y_\parallel)(1 + z_\parallel)w_\perp e^{i(\varphi_x - \varphi_w)} \\
& + (1 + x_\parallel)y_\perp z_\perp(1 + w_\parallel)e^{i(\varphi_z - \varphi_y)}] + \tilde{B}\tilde{C}_+^*[x_\perp y_\perp(1 - z_\parallel)(1 - w_\parallel)e^{i(\varphi_y - \varphi_x)} \\
& + (1 - x_\parallel)(1 - y_\parallel)z_\perp w_\perp e^{i(\varphi_w - \varphi_z)}] + \tilde{B}\tilde{C}_-^*[x_\perp y_\perp(1 + z_\parallel)(1 + w_\parallel)e^{i(\varphi_x - \varphi_y)} \\
& + (1 + x_\parallel)(1 + y_\parallel)z_\perp w_\perp e^{i(\varphi_z - \varphi_w)}] + \tilde{C}_+\tilde{C}_-^* x_\perp y_\perp z_\perp w_\perp e^{i(\varphi_x - \varphi_y + \varphi_z - \varphi_w)}\}
\end{aligned}
\tag{8}
$$

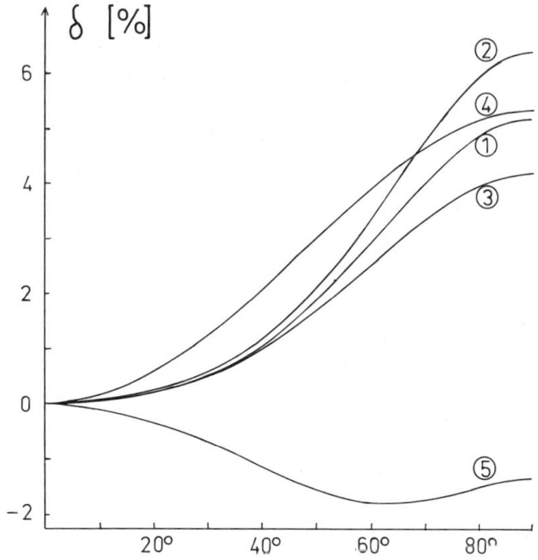

FIGURE 8. The relative correction δ (in %) as in Fig. 3 now for electron–electron scattering.

With the coefficients

$$\tilde{A} = \frac{u}{t} + (g_V^2 - g_A^2)\frac{u}{t_z}, \qquad \tilde{B} = \frac{t}{u} + (g_V^2 - g_A^2)\frac{t}{u_z}$$
$$\tilde{C}_\pm = -\frac{s}{t} - \frac{s}{u} - (g_V \pm g_A)^2\left(\frac{s}{t_z} + \frac{s}{u_z}\right) \tag{9}$$

The spin polarizations of the four electrons (see Fig. 6) are denoted here with x, z in the initial and with w, y in the final state. Other notations are equivalent to (2)

$$\begin{aligned} t &= (p - p')^2 \approx -4E^2 \sin^2(\vartheta/2) \\ u &= (p - k')^2 \approx -4E^2 \cos^2(\vartheta/2) \\ u_z &= u - M_{Z0}^2 + iM_{Z0}\Gamma_{Z0} \end{aligned} \tag{10}$$

With $g_V = g_A = 0$ we obtain again the well-known cross section for polarized electron–electron scattering in QED.

The cross section (8) shows like (4) a number of interesting symmetries, which we will discuss together with the cases of final polarization elsewhere.

5. Numerical Examples for Electron–Electron Scattering

Here we will give a few typical examples for the electroweak effects in electron–electron scattering. We again take $E = 20$ GeV, $\sin^2\theta_w = 0.228$, and $\Gamma_{Z0} = 2.5$ GeV and consider the same polarizations as in electron–positron scattering. The longitudinal cases 5 and 6, i.e. (←←) and (→→), are now identical since both particles cannot be distinguished in the final state.

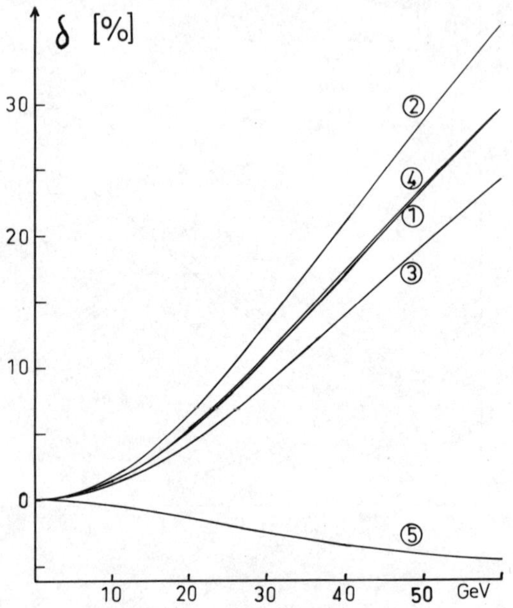

FIGURE 9. δ (in %) for the polarizations and conditions discussed in Section 5 for 90° electron–electron scattering as a function of the beam energy E.

In Fig. 7 we have now plotted

$$\tilde{F} = \sin^4 \vartheta \, \frac{d\sigma^{e^- e^-}}{d\Omega} \tag{11}$$

to show the symmetry around 90°. The pure QED parts are dotted.

In the next figure (Fig. 8) we show the correction δ to the QED contribution according to (7) for the remaining five polarization cases. We see that in electron–electron scattering the deviation is under these circumstances at most about 6%. With increasing energy E δ increases also but not in the resonancelike manner as in electron–positron scattering (Fig. 5), since in electron–electron scattering there is no annihilation contribution (see Fig. 9).

Electron–electron scattering is, therefore, not as sensitive to the special effects of electroweak interaction as electron–positron scattering.

Final polarizations will be discussed in a forthcoming paper.

REFERENCES AND NOTES

1. K. Ringhofer and H. Salecker, *Found. Phys.* **10**, 185 (1980).
2. C. Y. Prescott, SLAC-PUB-2630 (1980).
3. D. Dicus, *Phys. Rev. D* **8**, 890 (1973); R. Budny and A. McDonald, *Phys. Rev. D* **10**, 3108 (1974); C. H. Llewellyn-Smith and D. V. Nanopoulos, *Nucl. Phys. B* **78**, 205 (1974); R. Budny, *Phys. Lett.* **55B**, 227 (1975); **58B**, 338 (1975).
4. T. Anders, "SPUC Manual," Sektion Physik, Universität München (1979).
5. W. Jachmann, "Explizite Berechnung von Streuamplituden und Wirkungsquerschnitten für polarisierte Dirac-Teilchen und Photonen," Internal Report, Sektion Physik, Universität München (1980).

NOTE: Manuscript received March 30, 1981.

27

Infinity and the Natural World

Richard Schlegel*
Department of Physics
Michigan State University
East Lansing, Michigan

I.

An awareness of infinity is a major component in our contemplation of nature. We live with the possibility that the finite world of our experience is to be contrasted both with spatial domains of infinite magnitude and never-ending reaches of time. Our successes in understanding the universe must then be juxtaposed with a kind of threat that an overwhelming infinity will always forbid a comprehensive grasp.

It is one of the triumphs of the past 100 years that mathematicians, led by the work of Georg Cantor,† have achieved an explicit theory of infinite magnitudes (even granting that there are persisting problems of incompleteness). And, in the past 50 years both physics and astronomy have come to results that bear on the questions of infinite progression. On the microlevel of quantum theory, the Heisenberg uncertainty relations have told us of a limitation on indefinite decrease of magnitude: space time description cannot be physically carried to arbitrarily small intervals. There is no such positive result for indefinitely large spatial magnitudes, but in astronomy the question of infinity has at least now become a prime question in the consideration of various cosmological models.

I shall not be attempting a review in this essay of current empirical findings with respect to the possible infinities of nature. Rather, my concern will be with somewhat more general questions that are, nonetheless, directed to the relations between concepts of infinity and properties of the natural world. I will begin with a discussion of the at most denumerable infinity of the space and time extents of the universe. I shall show an application of this result in determining the validity of some cosmological

*Deceased (May 30, 1982).
†See the recent study by Dauben (Ref. 1).

models. I will then go to my major theme for this paper: a treatment of the problem of an infinite past for the universe. This question, clearly posed some two hundred years ago by Immanuel Kant as one of the antinomies of pure reason,[2] has recently been reopened in the literature,[3-6] with the presentation of allegedly definite results: on the one hand, that the universe indeed may have existed through an actual infinity of past times, and, on the other, that it is contradictory for it to have done so. I shall try to explicate both what I believe to be the correct answer and an implication from that answer for cosmology.

II.

In a formal way it is easy to pass through all the finite numerical magnitudes and reach one that is beyond any chosen finite number. For example, we consider two arcs, $r\theta$ and $(2\pi - \theta)$r, which together form the circumference of a circle of radius r. We write the ratio $z = r\theta/[(2\pi - \theta)r]$. As θ is increased from 0 to 2π, z grows from 0, and $\rightarrow \infty$ as $\theta \rightarrow 2\pi$. This means that z does go through all of the positive integers n, with $z = n$ for $\theta = 2\pi n/(n + 1)$; we might therefore want to say that we have achieved a physical expression of those integral numbers.

In fact, however, a literal physical determination of the ratio would not support that assertion. Suppose, say, that we used a micrometer to mark off the arcs $r\theta$ and $r(2\pi - \theta)$, for values approaching 2π. After a time we would come to an end point where the precision of measurement would be comparable to the magnitude of the arc $r(2\pi - \theta)$; for, as already noted, the uncertainty relations put a limit on how closely a spatial extent may be defined (and measured). Generally, the space localization can be made increasingly sharp with greater availability of energy for the interaction–measurement. Even, however, with any amount of energy there is a limit that is short of a mathematical point; for, at the size of the Planck length, $(G\hbar/c^3)^{1/2} = 10^{-33}$ cm [where G is the Newtonian gravitational constant, $\hbar = (1/2\pi)$ is Planck's constant, and c is the speed of light], a space interval has effectively formed itself within a Schwarzschild radius and no further penetration by an interacting measuring particle would be possible.

It is to be granted that the multiplicity of microlevel particles, for example, the number of atoms in a gram of matter, is impressively great by standards of ordinary human counting; and, also, that we routinely do achieve a formal infinity in mathematical procedures, as with the instanced ratio z, or, say, in integrating a physical potential, expressed as kdr/r, from R to ∞. Yet, I believe we may properly conclude that quantum-physics limitations on smallness of magnitude do prevent our physically delineating any microlevel entity about which we could say: An infinite number exist in a finite domain. Likewise, the limitations on time measurement, arising both from the quantum-theoretic uncertainty relations and from Planck-length considerations, place a limit on the minuteness of physically defined temporal intervals. Hence, for time extension, too, we cannot assert that an infinite number of physically distinct intervals exist in a finite time span.* It appears that if there are infinite quantities in the physical world we can expect to find them not in the realm of the very small, but rather only as constituents of a possibly indefinitely large macroscopic world.

III.

The infinity indicated by the symbol ∞ is not a defined number but is rather a kind of operator. Thus, it tells us that when we write $\lim_{x\to\infty} (1/x) = 0$ we are taking x to be a sufficiently large number that $1/x$ comes as close to zero as we wish to bring it. But even though symbolized as "$\to \infty$", we do not actually take x beyond the realm of finite numbers. For our considerations of physical infinity it will now be useful to introduce the first (smallest) number that has been defined as larger than any finite number. This is the number $\aleph_0$, which is the cardinal number of all the finite integral numbers (or of any set that is in one-to-one correspondence with those numbers).

We must not forget that $\aleph_0$, unlike the magnitude x of $x \to \infty$, is not within the range of finite numbers. If we write a sequence of increasing integers, $1,2,3, \ldots, n$, we always remain within those numbers, no matter how far the sequence is extended. At no point can we say that $n + 1$ carries us to the number $\aleph_0$; for, the sequence is unending, even though within the range of the finite numbers. Nonetheless, $\aleph_0$ as defined is the cardinal number of *all* the numbers n; and, of course, it follows a different set of arithmetical rules than do the finite numbers (notably, $n + \aleph_0 = \aleph_0$, $n\aleph_0 = \aleph_0$, and even $\aleph_0\aleph_0 = \aleph_0$). If all the numbers n were written down, in some kind of infinite domain, there would be $\aleph_0$ of them; to repeat, no number n could describe the multiplicity, since there is always another and different finite number $n' = n + 1$

The second transfinite number which will be of interest to us is the "power of the continuum," $\aleph$, the cardinal number of the set of mathematical points in a line segment (or higher-dimensional spatial region). The number $\aleph$ gives us a nondenumerable multiplicity, since $\aleph > \aleph_0$, and counting is customarily taken as being within the domain of integers or rational fractions (the latter also form a set with cardinal number $\aleph_0$). It can be shown that $\aleph = 2^{\aleph_0}$, where the right-hand term may be regarded as an infinite extension of ordinary exponentiation; that is, $2^{\aleph_0} = 2 \cdot 2 \cdot 2 \cdot 2 \cdot \ldots$, through an actual infinity of factors of 2.†

IV.

In any space of finite volume V, the greatest possible number of physical entities (e.g., atoms), each of effective finite volume $v > 0$, is of course the finite number n

*For discussion and references with respect both to space and time measurements see Schlegel (Ref. 7).
†See Fraenkel (Ref. 8a).

$= V/v$. (By specifying "finite volume" we exclude infinitesimal volumes; that is, we require that $1/v$ be a specific finite number.) It is readily seen that if the universe is spatially infinite the cardinal number of the set of all finite-size physical entities in the universe is at most $\aleph_0$. Thus, one could think of those matter components of the universe as being located in imaginary closed spatial shells of finite thickness, the first shell formed around some chosen point and successive contiguous shells enclosing the one of next lower number in the sequence of shells. The space of the shell structures need not be Euclidean. The physical entities could then be counted by enumerating them in each shell, starting in the shell at the origin and passing successively to the enclosing shell of next higher number. Any entity would eventually be counted, with some finite integer n and, also, the counting could be endless, since in an infinite universe there need be no limit to the number of shells. Hence, the maximum number of all the entities is $\aleph_0$. Again, we note that any entity in the infinite set, cardinal number $\aleph_0$, is counted by a finite number n, but also that there is no n at which the counting stops.

The concept of gravitational collapse of matter might suggest that physical entities of nonzero mass with zero spatial extension could be formed, in consequence of there being matter states of infinite density. Could there then be more than $\aleph_0$ of such physical entities, even with a restriction to $\aleph_0$ space units (e.g., cm) along any one cosmic spatial dimension? We readily come to a negative answer. For any transfinite cardinal number $\aleph_n$, $0 \cdot \aleph_n = 0$.[8b] Hence, even if considered to be made of matter of density $\aleph_0$, the physical entity with $v = 0$ also has zero mass.

Alternatively, a set of particles with cardinal number $\aleph > \aleph_0$, with $v > 0$ for each particle, would occupy a volume $v \cdot \aleph = \aleph$. But the total volume of space cannot exceed $\aleph_0$ (see below), so we cannot have more than $\aleph_0$ nonzero size physical particles in the universe. Our restriction applies to components of ordinary matter: neutrons, electrons, and protons. For bosons, and notably photons, there is no theoretical limit on the number which may exist in a given spatial volume. Empirically, the only known possibility for indefinitely large photon densities is in association with early high-temperature, condensed stages of the universe. At radiation temperatures above 10^{11} K many kinds of particles, including protons and neutrons, would be in thermal equilibrium with the radiation.[9] We would then have those same arguments against more than $\aleph_0$ particles that have just been presented.

If the universe were infinite, with total number of volume units (e.g., cm^3) not $\aleph_0$, but the larger number $\aleph$, we could of course have more than $\aleph_0$ physical entities of effective $v > 0$. The argument against the universe's comprising $\aleph$ volume units can be made in the same manner that denumerability was demonstrated for the set of physical entities of nonzero volumes. The successive shells around an origin point may now be regarded as constituted of hypothetical unit-volume spaces. We are then able in principle to count the number of units in any domain of space, since any unit volume can be reached from the chosen origin point (given the never-ending character of enumeration within the set of cardinal number $\aleph_0$). Hence, if any domain of space had cardinal number $\aleph$ of unit volumes, we would be able to count the members of a set of that cardinality, in violation of the definition of $\aleph$. In brief, the counting of $\aleph$ finite units in space is impossible because we progress a finite distance in counting each finite spatial unit, whereas if $\aleph$ mathematical elements are to be fitted into any finite spatial extension there can be no progression through it by an enumeration of those elements.

V.

The limitation in an infinite universe to cardinal number $\aleph_0$ for the set of physical entities or space units, each with volume $v > 0$, has some consequences with respect to cosmological theory.[10-13] Among the standard Friedmann models, with but one exception, all those with a positive cosmological constant Λ have a spatial scale factor $R(t)$ that increases exponentially with time.[13] This is true for $+$, $-$, or 0 value of the curvature index k, and for any average mass density $\rho \geq 0$. [$R(t)$ is, of course, the time-dependent metric factor that describes the observed expansion of the universe.] The exception is the oscillating Friedmann model, $k = +1$, with $\Lambda > 0$ but $\Lambda < \Lambda_E$, and R initially, $t = 0$, between $R = 0$ and R at $\dot{R} = 0$; Λ_E is the value required for Λ in the 1917 static Einstein model of the universe.

Generally, then, if we have a $\Lambda > 0$ model, for large R we can write (using gravitational units, $c = 1$ and $G = 1$)

$$R(t) = R_0 e^{\alpha t} \tag{1}$$

where $\alpha = (\Lambda/3)^{1/2}$. By a change of base we can rewrite Eq. (1):

$$\begin{aligned} R(t) &= R_0 2^{(\alpha/\log_e 2)t} \\ &= R_0 2^{\alpha' t} \end{aligned} \tag{2}$$

Now, if time t is extended forever, $t = \aleph_0$, $\alpha' t = \alpha' \aleph_0 = \aleph_0$, and Eq. (2) gives

$$R(t) = R_0 2^{\aleph_0} = \aleph \tag{3}$$

Equation (3) is contradictory; for, it requires that a finite distance R_0 will, after an infinite time, become one containing $\aleph$ space units, each greater than zero. We have seen, however, that we cannot have a nondenumberable set of such units in even an infinite line. We have a point of logic, then, against the existence of $\Lambda > 0$ Friedmann models (with the noted exception), for a universe which is to exist forever with $R(t)$ as given by Eq. (1).

We come to contradiction in Eq. (3) only if by $t = \aleph_0$ we understand an actual infinity. For the "weaker" infinity, $t \to \infty$, implying that $t = n$, where n is a finite number as large as desired, Eq. (3) yields $R(t) = R_0 2^{\alpha' n}$, and $R(t)$ is then always finite. In one sense, n as large as desired may be regarded as an acceptable infinity. However, if $t = \aleph_0$ is not admitted, the stipulation is implied that somewhere the sequence of n values is stopped (for, otherwise we would have $t = \aleph_0$). Hence, time as "forever," past or future, would seem to require $\aleph_0$ as the correct cardinal number for the set of past or future time intervals.

One nonstandard cosmological model, the steady-state cosmology[14,15] of Bondi, Gold, and Hoyle (BGH), is discredited by the appearance of a cardinality $\aleph$ after a $t = \aleph_0$. The argument can be cast as in Eq. (3), since in the BGH theory there is an exponential growth of $R(t)$, with α replaced by Hubble's constant. We can work more directly, however, with the matter-creating property that is required by the theory. The large-scale average matter density ρ is postulated in the BGH model as being always and everywhere the same, as is also the expansion parameter, Hubble's constant. New atoms, presumably of hydrogen, spontaneously come into existence at a rate such that ρ does remain constant, even with space expansion. Hence, after some time $t = \tau$ any one atom effectively becomes two atoms, and we have an exponential growth in the number of atoms N:

$$N = N_0 2^{t/\tau} \tag{4}$$

Suppose time to be measured in units of τ. Then, for $t = \aleph_0$, Eq. (4) becomes

$$N = N_0 2^{\aleph_0} = N_0 \cdot \aleph = \aleph \tag{5}$$

Each atom, however, has $v > 0$, and we have seen that for such atoms we cannot have in the universe a set with cardinal number $\aleph$. Evidently, the BGH universe could not have existed forever (as stipulated in that steady-state theory). Again, there is no problem for $t = n$ time units, where $n \to \infty$; but, if *every* n is to be allowed, the cardinal number of past units is $\aleph_0$.

VI.

The restriction to $\aleph_0$, both for the set of all physical entities of nonzero volume and for the cosmic volume of space, is perhaps rather obvious, and also, a property that seems to be accepted without controversy (even though, of course, one cannot rule out that topological possibilities allowing higher-order infinities might be proposed and found to be physically correct). Also, I know of no cogent *a priori* arguments against the *possibility* which we have assumed, that the spatial universe be infinite rather than restricted to a finite number n of nonzero space units. In contrast, the assertion of an infinite time span for the universe has seemed to some to be intrinsically untenable. In particular, as I have noted, there is recent published controversy with respect to the possible existence of the universe through an infinite past time. I will now turn to that discussion, in which implications of an infinite sequence are primarily at issue rather than points of theory or observation in contemporary cosmology.

The basic question is whether or not the universe can have come to the present by passing through the required infinite series of past time intervals (each interval assumed, now and throughout this discussion, to be of finite, nonzero magnitude).

Kant,[2] associating each interval of time with a state of the world, argued that one could never pass by "successive synthesis" through an infinity of such states since the beginning of the world, for infinity is endless. Hence the world cannot have existed forever in time.

Whitrow[3] endorses Kant's conclusion. We may write the sequence of past time intervals as

$$\ldots, -5, -4, -2, -1, 0 \tag{6}$$

The $^*\omega$ sequence (6) has no first member but has a final member, 0, symbolizing the present time. Whitrow argues that if the universe had existed through all past time, it would have reached the present by existing through $\aleph_0$ events (or intervals), since the cardinal number of elements in the $^*\omega$ sequence is $\aleph_0$. But, such an enduring involves the contradiction that one reach the present, a terminal point, even though the length of time (number of intervals) from the beginning of the universe is never-ending. Craig[6] gives further support to the Kant–Whitrow conclusion. He cites a long historical tradition of argument against the infinite past. I will center on two points: (1) a series of past intervals is completed by successive addition to the series; (2) an actual infinity cannot be completed by successive addition (since, again, there is no final addition that brings one to actual infinity). Craig therefore also concludes that "a finite time ago, the universe began to exist."

Popper[4] accepts that for a universe that has always existed the set of all past time intervals is actually infinite. He argues, however, that from any past time point there is only a finite series of intervals. Hence, the problem of forming an actual infinity by successive addition does not arise, and it is possible for the universe to have existed forever. Stating the matter in another way, Popper asserts that it is illicit to select any starting point for the universe in the $^*\omega$ sequence (which has no first point). And, *any* past time point that is considered in the sequence is only a finite number *n* of intervals in the past; the present, therefore, can be reached from that point by successive addition of intervals. Craig raises the objection against Popper that the universe which has always existed must *in fact* have come through an actual infinity of all past intervals, regardless of our being able mentally to select finite but indefinitely large subsets of those intervals.

Bell[5] concurs with the opinion expressed by Popper, that "the attempt to show by *a priori* reasoning the impossibility of time without beginning" is "doomed to failure." He (Bell) bases his argument on explicit infinite-sequence properties. He asserts (in agreement, essentially, with Popper) that Whitrow's argument would be correct if the infinite sequence of past time intervals were not $^*\omega$ but $1 + {^*\omega}$; that is, if it were necessary to start at an initial interval 1 and pass through the intervals of the $^*\omega$ sequence. Further, Bell takes issue with an assertion by Whitrow about a symmetry between infinite past and future. Future time is commonly represented by the ω sequence,

$$0, 1, 2, 3, 4, 5, \ldots \tag{7}$$

where now there is a first member, 0, again symbolizing the present, but no final member. Whitrow argues that the infinite sequence of past events "corresponds to an actually infinite (aleph-zero) sequence of future events," and that just as the latter must be rejected so must the former. But, Bell points out, even if there has been an $^*\omega$ sequence (for past events) "it does not follow that from this one can construct an infinite sequence of future events." And, indeed, we see that no subset of (6) has the property required for (7), of having a first member but no last member. It does not seem valid, therefore, to carry any arguments from possibility of an infinite past to possibility for an infinite future, or conversely.

The preceding discussion indicates that the question of an *a priori* impossibility for an infinite past is a subtle one that can apparently be reasonably argued to opposite conclusions. I shall now hope to show, however, that the correct answer is the one that does allow the infinite past existence of the world.

VII.

Let us again consider the $^*\omega$ sequence (6), taken as symbolizing all past time intervals. Popper and Bell point out that if we wish to reach the present from any past instant we need to pass through only a finite number n of intervals, and hence we will eventually come to the present from any chosen past time. The $^*\omega$ sequence does contain *only* finite numbers. This is the key point. As we noted earlier (Section 3) for an ω sequence, we do *not* include any transfinite numbers in an infinite sequence of integers. And yet, the Whitrow assertion that the cardinal number of past intervals is $\aleph_0$ is also correct; for, if we ask through how many intervals of time the world has passed, if it has existed for an infinite past time, the answer must be $\aleph_0$. An assertion that the total number of intervals through which it has existed is some finite number n would be wrong, however large n is, because there is always some yet larger number $n + 1$. We here again meet the disjunctive property of the cardinal number $\aleph_0$; it counts the set of finite numbers n, but is larger than any n, and also, one can never come to $\aleph_0$ by passing through the numbers n. (If one could, we would not have the two different rules of arithmetic: $n + 1 = n'$, also a finite number but $\neq n$, and $n + \aleph_0 = \aleph_0$).

It is helpful to draw on the distinction made by Bertrand Russell between defining the members of a set by denotation-enumeration or by a class concept. We might, for example, name the members of a number set as being 1, 2, 3, 5, 7; or, alternatively we could define the identical set as consisting of the one-digit prime numbers. But, obviously, for an infinite set definition of its members by explicit denotation is not feasible. We may, however, specify the desired set by a class concept, as when we define the ω sequence as that formed by successively adding 1 to the preceding integer, starting with 0 or, similarly, the $^*\omega$ sequence by successively subtracting 1. Russell uses the distinction in arguing against Kant's objection to an infinite past for the world. He writes[16] "'Completion by successive synthesis' seems roughly equivalent to enumeration, and it is true that enumeration of an infinite series is practically impossible. But the series may none the less be perfectly definable as the class of terms having a specified relation to a specified term." So, we are able to accept without contradiction that the world has come to the present from any past epoch in the $^*\omega$ sequence through a *finite* number n of time intervals (which conceivably could be enumerated); and yet, also, that it must therefore have existed through the defined set of all intervals, with cardinal number $\aleph_0$.

An analogy with the possible existence of an infinite spatial extension may lend elucidation. We consider an infinite straight line, whose length is taken to be $\aleph_0$ cm. Some chosen point on the line is designated as the zero point; then, points at 1-cm intervals are indicated, to the right and left, by the ω and $^*\omega$ sequences, respectively. Any point on the line may be reached by advancing from the zero point, to the right or left as is appropriate, because any point is associated with a finite integer. Yet, the line is truly infinite. Thus, we do not change its length by adding a cm length to it, because for any n, $n + 1$ is also an integer on the line; that is, the cardinality of points on the line is $\aleph_0$ and $\aleph_0 + 1 = \aleph_0$.

We might think of a messenger device, traveling at a constant and finite nonzero speed u cm/s, as coming to us, at the zero point, from any chosen distant point. In order that the device may reach us at $t = 0$ (the present time) from any point of the line the universe of line and messenger must have existed forever. For a point n cm distant it must have existed n/u s, and since all n's are on the line, the cardinality of times that the universe has existed is $(1/u)\aleph_0 = \aleph_0$. (Units may be chosen in which $1/u$ is an integer.)

The generally unobjectionable possibility for existence of an infinite line should, I believe, carry with it an acceptance of a possible infinite past time for the world. One way of defining an infinite line would be to say that it is one for which, if it is divided into unit lengths, each associated with a different integer, it is always possible to find a unit length that is at a distance different from that of any previously chosen length. Such a line must extend forever, through $\aleph_0$ units, for it would be finite if it ended at some n, however large; we found that an infinite set of past time intervals, also of cardinality $\aleph_0$, can be associated with the line. The actually infinite sequence, we must see, is formed of the ever-increasing finite.

VIII.

The universe may well not have existed forever, even though there is no contradiction in the concept of its having done so. In a companion paper published with the cited ones by Bell[5] and Craig,[6] Weingard[17] points out that present cosmological theory (essentially, the general theory of relativity) does not require a finite past, even though the galactic red shifts (and one would want to add the observed 3 K background radiation), taken with the theory, do indeed indicate a space-time singularity for the universe in the finite past. Developments still to come in cosmology may further restrict possibilities for an infinitely extended past, and perhaps some degree of closure of theory with respect to creation of the universe at a finite past time may even be achieved. A demonstration of temporal finitude would be a most notable step in natural philosophy, comparable to a firm decision with respect to spatial finitude. Rather, however, than further discussing these conjectural matters, I wish in closing to comment on requirements for physical models that might be associated with a world that has existed through all of past time.

With the discrediting in physics of Newton's absolute time we have come to associate temporal advance with progressive change in the natural world. Thus, were the universe completely static, both in manifest motion and in chemical change, we would have no grounds for considering time, or a time measure, to be a component of nature. It follows, then, that we must expect an infinite series of past cosmic change to be associated with a universe which has existed "in time" forever. I shall avoid involvement, however, with consideration of the possible past history of any particular region of space and matter–radiation. Instead I shall invoke the cosmological principle, and be concerned only with overall cosmic behavior, for a universe that is spatially finite or infinite.

Suppose that some form of progressive, one-way change defines the time advance of the universe. Then the present state has been reached after the action of that change through an infinite time (assuming, as we are, that the universe has existed forever). Simply from a qualitative consideration it seems evident that a progressive change carried through infinite time would be destructive of the world as we know it; such natural balances as those between radiation and matter, empty and maximally filled space, relatively stable and changing physical systems, would disappear. We might expect, for example, that unceasing expansion would have carried the world into virtual emptiness.

A proscription against infinitely extended progressive cosmic change was given as a postulate in MacMillan's steady-state cosmology:* "The universe does not change always in any one direction." The principle proposed by MacMillan is given firm support for certain cosmological models in the contradictions that come with existence through infinite time. Thus, we saw in Section 5 that a past infinite time for the BGH steady-state model gives us too many atoms for the available space; also, that for all but one special instance the standard positive-cosmological-constant models lead in an infinite time to a space variable that contradicts experience. Likewise, we may see Olber's paradox as arising because a "one-way" process, the emission of stellar radiation, is envisioned (quite incorrectly, we can now say) as having continued through an indefinitely long period of time.

For a cosmos which has existed forever the alternative to presence of basic one-

*See Ref. 18. W. D. MacMillan, a mathematical astronomer at the University of Chicago, formulated a pioneer steady-state cosmology in the period around 1920. He proposed that the universe is spatially infinite and has existed forever, and that stellar radiation, by some unknown mechanism "in the depths of space," forms itself into H atoms which in turn condense and give rise to stellar bodies (Olbers' paradox being thereby avoided). For further references and discussion see Refs. 12 and 19.

way action is, of course, that of eventual change of direction for any such fundamental process. The potentiality for matter–radiation conversion in either direction, given suitable circumstances, appears surely to be a natural property of prime importance in avoiding basic irreversible progression for the natural world. And, of course, we readily think of the $R(t)$ factor, which determines at a given t either the radius of the finite universe or the relative size of a proper volume in an infinite universe. Until the early 1970s an oscillating finite cosmos, with $R(t)$ as given by the zero-cosmological-constant Friedmann equation, had been widely favored for several decades; in that model we do have the possibility of an endless series of cycles of alternate contraction and expansion of the universe [although, in fact, with only a trusting assumption that a turn-around in dR/dt, required by the equations after maximum expansion, would also occur after minimum $R(t)$].

Relevant empirical evidence is still so slender that we can expect future shifts in opinion with respect to what is the realistic cosmological model. But if an infinite past is indicated, I think we cannot avoid some form of oscillating model with respect to $R(t)$. If the oscillations are strictly periodic, in time and spatial magnitude, there is the possibility of a universe that is atemporal, except for the time progression that is defined in each cycle. That is, if successive large-scale cycles in $R(t)$ are substantially identical, with no process carrying progressively from one cycle to the next, we would have no basis, other than conceptual projection, for a time measure inclusive of more than a single cycle. A past of cosmic change that is not consistently one-way and yet irregular might give a basis for time measure extended indefinitely far into the past.

One can hope that relationships between cosmic and micro properties—now beginning to be discerned in physics—will further guide our exploration of the cosmic past. Whatever the way scientific investigation carries us, I believe we can say that logically both possibilities are open: a past creation event with the beginning of time, or, existence through past time without end.

As a final word, I want to give expression to a note of caution that comes from the person we remember and honor in these essays: a man whose thought ranged so widely, and who encouraged so many others in their scientific work. In an essay on cosmology and logic, Yourgrau[20] pointed out: "There are few fundamental principles which are absolutely necessary to work in the field of cosmology, but the subject is inundated with *ad hoc* assumptions, postulates, constants. . . ." It is indeed fitting to the leadership of Wolfgang Yourgrau that we should speculate in cosmogony, but it is also in order, as he stresses in the cited paper, that we should not forget how young and tentative it still is as a science.

References and Notes

1. J. W. Dauben, *Georg Cantor: His Mathematics and Philosophy of the Infinite* (Harvard Univ. Press, Cambridge, Massachusetts, 1979).
2. Immanuel Kant, *Critique of Pure Reason*, trans. N. K. Smith (Macmillan, London, 1933), p. 397.

3. G. J. Whitrow, *Brit. J. Phil. Sci.* **29,** 39 (1978).
4. K. R. Popper, *Brit. J. Phil. Sci.* **29,** 47 (1978).
5. John Bell, *Brit. J. Phil. Sci.* **30,** 161 (1979).
6. W. L. Craig, *Brit. J. Phil. Sci.* **30,** 165 (1979).
7. R. Schlegel, *Superposition and Interaction* (Univ. of Chicago Press, Chicago, 1980), Secs. 2-4 and 3-5.
8. A. A. Fraenkel, *Abstract Set Theory* (North-Holland Pub. Co., Amsterdam, 1961): (a) pp. 110–112; (b) p. 96.
9. Steven Weinberg, *The First Three Minutes* (Basic Books, Inc., New York, 1977), Chap. VII.
10. R. Schlegel, *Nature* **193,** 665 (1962); **196,** 1015 (1962).
11. R. Schlegel, *Phil. Sci.* **32,** 21 (1965).
12. R. Schlegel, *Completeness in Science* (Appleton-Century-Crofts, New York, 1967), Chap. 8.
13. R. Schlegel, *Int. Theor. Phys.* **12,** 217 (1975).
14. H. Bondi and T. Gold, *Mon. Not. R. Astron. Soc.* **108,** 252 (1948).
15. Fred Hoyle, *Mon. Not. R. Astron. Soc.* **108,** 372 (1948).
16. Bertrand Russell, *Principles of Mathematics*, 2nd Ed. (W. W. Norton, New York, 1938), pp. 349–350.
17. Robert Weingard, *Brit. J. Phil. Sci.* **30,** 170 (1979).
18. William D. MacMillan, *Science* **62,** 63, 96, 121 (1925).
19. R. Schlegel, *Am. J. Phys.* **26,** 601 (1958).
20. Wolfgang Yourgrau, in *Essays in Memory of Imre Lakatos*, R. S. Cohen, P. K. Feyerabend, and M. W. Wartofsky, editors, Boston Studies in the Philos. of Sci. XXXIX, (Reidel, Dordrecht/Boston, 1976), p. 751.

28

Induced Superfluorescence and the Nature of the Wave–Particle Duality

F. Selleri
Istituto di Fisica
Università di Bari
Bari 70126, Italy

J.-P. Vigier
Institut Henri Poincaré
Paris, France

It is now possible in practice to observe macroscopic fluctuations in the time domain initiated by an isolated single spontaneous atomic emission of light. This is the so-called "superfluorescence effect (SF)" described in some recent papers.[1] SF produces radiation pulses with much larger amplitude than predicted for normal incoherent radiation processes. The phenomenon is similar to the so-called "laser effect," because the coherent decay of something like 10^8 atoms is triggered by a single spontaneous emission due to quantum fluctuations of the electromagnetic field in vacuum. This phenomenon, first detected in 1973,[2] has been observed in a number of near-infrared, infrared, and optical transitions.[3] Atomic cesium is used for studying SF from a pencil-shaped volume of active atoms. The SF emission from the pencil goes mainly through its end-faces into two narrow solid angles $\Delta\Omega$. The emitted light pulse reaches a peak intensity proportional to N^2 (where N is the number of excited atoms) at the so-called "delay time" $\tau_D \sim (t_R/4)(\ln N)^?$, with $\tau_R \equiv t_n/(N\Delta\Omega/4\pi)$. The emitted field behaves a classical coherent field. It has a well-defined amplitude and phase: SF emission pulses from two different samples with slightly different SF transmission frequencies produce beats. Experiment confirms that the strength of the initiating quantum noise determines the average delay time.[4] SF is set off by the first photon emitted spontaneously in the solid angle $\Delta\Omega$ and yields as predicted by theory a relative standard deviation in the delay time of 12% for $N = 10^8$ excited atoms.

The aim of the present letter is (a) to discuss the possibility of triggering the SF process from the outside (i.e., to produce "induced SF"); (b) to study its implications for the interpretation of quantum mechanics, in particular as it concerns the wave–particle duality.

The possibility to test the reality of our proposed trigger process is summarized in Fig. 1. Light from a source of atomic cesium is focused, filtered (in order to reduce its intensity to the desired level of single photons), and chopped into photon bunches (trigger pulses). The source is made of diluted atomic cesium in such a way that no SF process takes place in it. In this way one can be sure to have *incoherent* emission of single photons having the same wavelength as that which can generate the SF process within the atomic cesium pencil. These photons impinge on a semitransparent mirror M of transmission coefficient 1/2 and are either reflected toward a photomultiplier (PM1 in Fig. 1) or transmitted towards the atomic pencil in such a way as to cross it at the right angle suitable for producing SF pulses. The increased rate of SF pulses as revealed by the photomultiplier PM2 of Fig. 1 would constitute evidence of the existence of the proposed "induced superfluorescence effect." This first part of the game can also be played without the semitransparent mirror. Presence of this effect would evidently corroborate the PSV[1] theory and Vrehen's preliminary results.[5] If the existence of induced SF is confirmed, a second important result can be obtained through the reduction of the trigger pulses to the level (already reached in famous experiments of Mandel and Pfleegor[6]) where the trigger photons appear one by one. Insertion of the semitransparent mirror is also essential for the following considerations. We need a situation in which every photon reaches the pencil long before the next one goes through the filter. In this case indeed appears the theoretical and experimental possibility of separating different interpretations of quantum mechanics.

To clarify this point, we first recall that the quest for experiments capable of

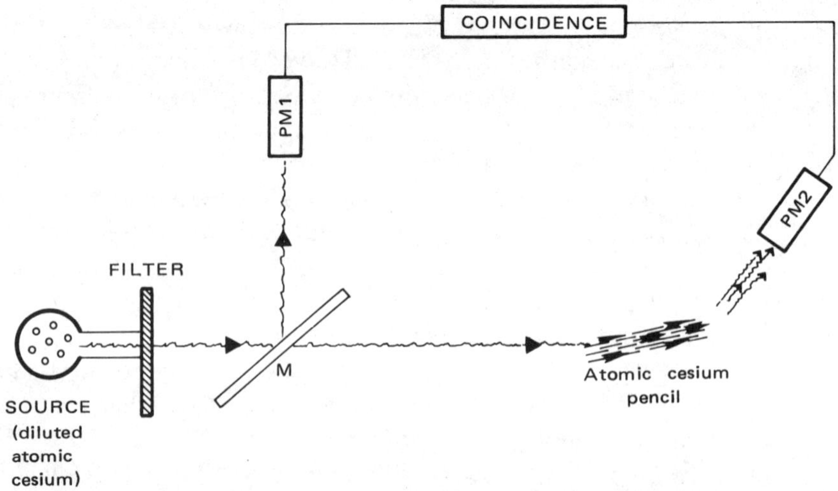

FIGURE 1. Experimental setup for testing the reality of the proposed trigger process.

distinguishing between different formulations of quantum mechanics already includes at least two examples: the paper by Fitchard[7] on different formulations of the uncertainty principle and a paper by the two of us[8] on measurements on correlated quantum systems in EPR conditions. Two extreme points of view which one can in principle distinguish experimentally are the Copenhagen interpretation of quantum mechanics (CIQM) and the statistical interpretation of quantum mechanics (SIQM). The CIQM is Bohr's original formulation of quantum theory: Quantum states are associated with individual systems, microsystems are particles and/or waves in the sense of the complementarity principle, and a measurement on a system provokes a discontinuous collapse of its wave packet. The SIQM is based on the assignment of quantum states to statistical ensembles of similarly prepared systems, and a measurement does not cause any reduction of the wave packet. The picture of microsystems seems to vary from author to author. While some seem to believe that the wave properties are only a mathematical tool for calculating the physical behavior of microsystems whose spatial nature (wave, particle, etc.) is not interesting or cannot be understood, we will stick to the interpretation of Einstein[9] and de Broglie[10] who make a stronger commitment to a precise description of micro-objects: A photon, for instance, is *at the same time* a particle and a wave, in the sense that the latter contains and guides the former in its motion. This type of SIQM has furthermore been enriched through the introduction of stochastic processes by Bohm and Vigier[11] and by de Broglie himself.[12] Recently a stochastic process at the speed of light has been suggested,[13] which allows one to deduce several interesting results.

What interests us here most are the different predictions made by CIQM and SIQM for the experiment of Fig. 1: When the semitransparent mirror *M* is present, as indicated in the figure, and if photons travel one by one, nothing can reach the atomic cesium pencil if a photon has been revealed by PM1 (at least in the case in which the M–PM1 distance is smaller than the M-pencil distance) and if CIQM is correct. In fact the observation of a γ in PM1 makes the wave packet collapse and leaves nothing propagating from M to the cesium pencil. Very different is the situation with SIQM: Here the waves are objectively real and the particle is forced to be reflected *or* transmitted by M, while the wave is reflected *and* transmitted. In this case, therefore, even when the energy-carrying particle is revealed by PM1, a wave crosses the atomic cesium pencil. This wave carries no particle but should nevertheless be able to generate induced emission (and thus SF pulses) in the atomic cesium pencil, since it has the right frequency to resonate with the transition frequency of the excited cesium atoms. Coincidences between reflected photons observed by PM1 and SF pulses observed by PM2 would therefore constitute evidence in favor of the Einstein–de Broglie SIQM. Such coincidences would disappear if M is substituted by a totally reflecting mirror. If observed this phenomenon would constitute the proof of the existence of a new level of objective reality since it is well known experimentally that single photons (in SIQM: particles) are either reflected or transmitted by a semitransparent mirror.[14] Some other proposals of observation of what Einstein called "Gespensterwellen" have already appeared in the literature.[15]

Acknowledgments

The authors are grateful to Dr. Luigi Fornari for having pointed out to them the importance of superfluorescence for the interpretation of quantum mechanics.

References and Notes

1. Q. H. F. Vrehen, M. F. H. Schuurmans, and D. Polder, *Nature* **285**, 70 (1980).
2. N. Skribanowitz *et al.*, *Phys. Rev. Lett.* **30**, 309 (1973).
3. Q. H. F. Vrehen, *Laser Spectroscopy IV* Proceedings of the Fourth Int. Conf. Rottach-Egern; Fed. Rep. Germany, June 11–15, 1979, H. Walther and K. W. Rothe, editors (Springer-Verlag, Heidelberg, 1979).
4. Q. H. F. Vrehen and M. F. H. Schuurmans, *Phys. Rev. Lett.* **42**, 224 (1979).
5. See Ref. 3
6. R. L. Pfleegor and L. Mandel, *Phys. Rev.* **159**, 1084 (1967); *J. Opt. Soc. Amer.* **58**, 946 (1968).
7. E. E. Fitchard, *Found. Phys.* **9**, 525 (1979).
8. F. Selleri and J. -P. Vigier, "Sur la description de Copenhague de la mésure de spin des systemes quantiques correlées," Bari-Poincaré preprint (1980).
9. A. Einstein, in A. *Einstein Philosopher–Scientist* P. A. Schlipp, editor (Library of Living Philosophers, 1949).
10. L. de Broglie, *La Theorie Quantique Restera-t-elle indeterministe?* (Gauthier-Villars, Paris, 1952).
11. D. Bohm and J. -P. Vigier, *Phys. Rev.* **96**, 208 (1954); W. Lehr and J. Park, *J. Math. Phys.* **18**, 1235 (1977); E. Nelson, *Phys. Rev.* 150, 1079 (1966).
12. L. de Broglie, *La Thermodynamique de la Particule Isolée* (Gauthier-Villars, Paris, 1964).
13. J. -P. Vigier, *Nuovo Cimento Lett.* **24**, 265 (1979); **24**, 258 (1979); N. Cufaro-Petroni and J. -P. Vigier, *Nuovo Cimento Lett.* **25**, 15 (1979); **29**, 149 (1979).
14. J. F. Clauser, *Phys. Rev. D* **9**, 853 (1974).
15. F. Selleri, *Nuovo Cimento Lett.* **1**, 908 (1969); A. Szczepanski, *Found. Phys.* **6**, 427 (1976); A. Garuccio and J. -P. Vigier, *Found. Phys.* **10**, 797 (1980).

29

BREAKDOWN OF PREDICTABILITY: *An Investigation into the Nature of Singularities*

K. TAHIR SHAH
International Centre for Theoretical Physics
Trieste, Italy

1. INTRODUCTION

I met the late Professor Wolfgang Yourgrau in 1977 during my summer visit to ICTP, Trieste. After some trivial conversation he asked me what kind of physics I do. My answer was sort of a lament: "I am afraid of physics" I told him. Whenever my confusion reaches a critical point I take refuge in pure mathematics. On the other hand, I want to understand nature and do not want to abandon physics. I do not recall his exact advice, but it was a pretty long discussion about science, philosophy, and mathematics. He persuaded me to devote my time to questions deeper than just computations and proofs. I am grateful for his gift of persuasion which pushed me towards this effort. I am also thankful to Professor Abdus Salam for his open invitation to all of us from developing countries to come to the Trieste Centre and make use of it in our intellectual endeavors. When Professor van der Merwe invited me to contribute to Yourgrau's memorial volume I felt greatly honored and decided to contribute an enquiry into the nature of singularities from the foundational point of view. I believe it is very close to Wolfgang's spirit; however, it is a modest enterprise of a nonexpert, as I am neither a logician nor a philosopher.

The purpose of physical theories is not only to organize the data and the concepts but to use this organization for the purpose of prediction or postdiction. In other words, extrapolation is used in one way or the other to extract more than what is given as previous knowledge. Extrapolation of relations beyond their premises of discovery leads to new discoveries. Whenever we do any operation on objects we would like to know the answer. If the answer is well defined within the logical, mathematical, and physical framework, the operation makes sense. Otherwise, we end up with an undefined object, i.e., one which cannot be identified within the given structure, and consequently the operation does not make sense. Such an undefined object is given the name *singularity*. For instance, in gravitational theory a singularity can be regarded as a place where there is breakdown of the classical concept of space-time as a manifold with a pseudo-Riemannian metric. Because all known laws of physics are formulated on a classical space-time background, they will break down at singularities. In a similar fashion, singularities show up elsewhere in physics and mathematics. Is this occurrence of singularities everywhere a unified (unique) behavior of nature or does it represent limitations on the conceptual structure of our knowledge at any given moment? It is my opinion that the occurrence of singularities is due to "incompleteness" in our knowledge. This I shall try to justify in the sequel.

In any branch of science we never get infinities (singularities) experimentally but only through theoretical manipulation. Theoretical manipulation means imposition of mathematical structure over experimental data as a sort of organization. So the origin of trouble is not nature but mathematical structures with idealized concepts. In arithmatic or algebra the relations and operations are extrapolated beyond their domain of validity where they become more or less meaningless. My viewpoint regarding infinities,* therefore, is that they are due to extrapolations from the language of finite mathematics. On the other hand, extrapolation or not, when an experience comes into conflict with a world of concepts which are already sufficiently fixed, we do not know what to do with it. We simply call it a singularity—because it has no similarity to any other mathematical object we know of, and we are unable to explain it within our existing knowledge. The only way out of this dilemma is to extend our knowledge by adding new concepts so that an interpretation, a model, or a kind of cause–effect relationship, can be established in terms of previous knowledge.

Let me explain why I am going back to logic and metamathematics to trace the origin of singularities. It appears to me that not only the operation (or rather non-operation) of division by zero is responsible for this difficulty, but there are two things involved at the epistemological level. Firstly, we tend to idealize our experience through extrapolation. For example, a point mass, a point, or a line does not exist in nature. Secondly, we expect to have well-defined results of any operation within a given theory, i.e., categoricity. In other words, we tend to believe in "completeness," which is almost never possible except for very few systems. Sometimes these extrapolations are successful in practice; for example, Cartorism has been quite a successful doctrine, but it has led to many antinomies. However, singularities are not antinomies except for the case of the infinite set that is a member of its own class, and which has led to the development of the theory of types by Russell. My philosophical attitude towards the question of interpretation and extrapolation from experience (finite!) to a purely logical—idealization, possibly infinitistic—deduction, is somewhere between that of Einstein and Kant. Let me quote them to illustrate what I mean:

> Physics is an attempt to conceptually grasp reality as it is thought independently of its being observed. In this sense one speaks of physical reality.—A. EINSTEIN

*Infinities are a particular type of singularities coming from the real number system.

> All thought acquires material content only through its relationship to sensory experiences. Pure logical thinking can give us no knowledge of the external world of experience; all knowledge about reality begins with experience and terminates in it.—E. KANT

I believe neither in "reality" independent of its being observed nor that pure logical thinking can give us no knowledge of the world. By observation I mean not the trivial, but the most sophisticated, methods one can think of that take into account all the known invariance principles. Reality is then what we observe and is liable to appear differently if methods of observation are more refined. The pencil I am using to write is made of wood, molecules, atoms, protons, neutrons, quarks, and lately of prequarks and preons with all kinds of interaction among these constituents.

The present article is divided into two parts. The first part deals with an intuitive answer on how to deal with singularities and is based on historical development of theories. The second part is a quasiformalistic approach like nonstandard models in higher-order languages and William Lawvere's axiomatic formulation of categories.

2. EXAMPLES OF SINGULARITIES

We shall analyze some of the well-known examples of singularities, most of which are related to infinity of the number system.

(a) *The Real Number System.* We shall go over details of this system to illustrate the epistemological meaning of singularities as well as the nature of the inverse of zero with respect to multiplication operation.

Consider the natural numbers $1, 2, 3 \ldots \subset \mathbb{N}$ as objects and division as the only operation known.* Let us assume that all the knowledge U, call it the universe, is just $\mathbb{N}$ and division operation. To avoid complications take $\mathbb{N}$ as a finite subset of the natural numbers. Denote $U = \{\mathbb{N}, \div\}$ such that $\div$ relates two natural numbers to a third natural number. As far as division gives elements of $\mathbb{N}$, everything is fine. What happens when division relates 2 and 5, which is not an element of $\mathbb{N}$? It cannot be interpreted in U. It is a singularity in U. The universe $U = \{\mathbb{N}, \div\}$ is incomplete. To complete it, at least partially, we must define a new concept called a fraction. A fraction by itself does not exist except as a part of the whole. Now, extend U to include fractional numbers. One can go on adding new operations and new concepts to the original universe U to give meaning to irrationals like $\sqrt{2}$, the "imaginary" $i = (-1),^{1/2}$ negative numbers, etc. One can obtain the real number system, the complex number system, and so on by the process of extension. In most of these extensions old rules remain valid and do not lead to nonunique interpretations. But the problem remains with the division of zero. The universe of reals does not have an interpretation for "inverse" of zero. The inverse of zero, or $1/0$, is an ambiguous object. The predictability in U is lost at this point. One concept common to all mathematical

*We shall use set-theoretic notions, membership, etc., unless otherwise stated.

theories is the concept of equality or identity. A statement of the form "$x = y$" always means that x and y denote the same object and each can be replaced by the other without changing the truth value of the statement. However, this concept, the most basic one, breaks down for $1/0$. For, if $1/0 = 1/0 \Rightarrow 1/0 = x$, then $1 = x \cdot 0 = 0$ or $1 = 0 \Rightarrow 0 \neq 0$ and $1 \neq 1$.

The argument is valid if, as in the case of $(-1)^{1/2}$ or $2^{1/2}$ etc., all the rules of U are valid for the undefined object, i.e., $[(-1)^{1/2}]^2 = -1$ or $(2^{1/2})^2 = 2$ remains true. Therefore, this singular object, called infinity, is a particular one. It becomes meaningless to write down such a sacred concept as equality for $1/0$. In other words, $x \neq x$ for this type of object. There are more surprises! If we take the conventional wisdom of $1/0$ to be the number 1 divided into parts "nothing" indefinitely a large number of times, we can generate wonders. For example,

$$\frac{1}{0} = \infty \Rightarrow 1 = 0 + 0 + \cdots + 0 + \cdots = \sum_{\infty} 0$$

$$\frac{2}{0} = \infty \Rightarrow 2 = 0 + 0 + \cdots + 0 + \cdots = \sum_{\infty} 0$$

$$\frac{x}{0} = \infty \Rightarrow x = 0 + 0 + \cdots + 0 + \cdots = \sum_{\infty} 0$$

(∞ = symbolic notation for indefinite).

So if you are able to sum nothing for ever you will get everything! What one would like to conclude from the above example is that some operations when extended too far may lead to wonderland. In fact, it is well known from Padoa's principle that any object having more than one interpretation is independent of the other primitive notions of the theory and, consequently, the inverse of zero (denoted 0^{-1}) is independent of the primitive notions of the number system. It is simply not possible to prove that 0^{-1} is a real or that 0^{-1} is not a real.

Thus the well-known singularity called "infinity" is neither a very small nor very big number. It is simply not a number. Conclusion: There are various possible paths to the solution of this problem. One is to enlarge the number system and go over to nonstandard models. The second is simply to avoid extrapolation—i.e., you cannot divide something into indefinitely many nothings and hence give up the continuum.

(b) A matrix A_{ij} is called singular if $\det|A_{ij}| = 0$, i.e., it is not invertible, due to the undefined status of 0^{-1}. The completeness question is obvious here. A matrix actually represents a transformation between two vector spaces. Two vector spaces are transformation T-equivalent if $\det|A_{ij}| \neq 0$. However, when A_{ij} is singular, we do not have T-equivalent vector spaces and one goes beyond this T equivalence class of vector spaces. One goes over to something else. This implies incompleteness because the existence of a noninterpretable object implies incompleteness.

(c) For C^∞ maps the situation is similar because a similar condition is imposed on the Jacobian.

(d) In the case of complex analytic functions, singularity is defined through breakdown of a property. The function $f(z) = u(x,y) + iv(x,y)$, $z = x + iy$, does not satisfy the Cauchy–Riemann equations $\partial u/\partial x = \partial v/\partial y$ and $\partial u/\partial y = -\partial v/\partial z$. Harmonicity is lost at so-called singular points.

(e) Let $g = (V,\mu)$ be a Lie algebra on vector space V with the law of multiplication μ. Consider the set of all μ having the structure of an algebraic manifold and such that the action of an algebraic group H is defined on it. With a given topology one can find the bifurcation set $\Sigma = \overline{H(\mu)} - H(\mu)$, where $H(\mu)$ is the orbit of μ. The points $\eta \in \Sigma$ do not belong to an H-equivalent class. These are singular points and form a "boundary" to the orbit. Here singularity means symmetry is lost or broken.

(f) The spontaneous breaking of symmetry can be described through breaking of harmonicity for the Lagrangian which is considered as a harmonic map. A map between two Riemannian manifolds M and N is harmonic if the divergence of its differential vanishes. Let $\phi: M \to N$, then ϕ is harmonic if $\tau(\phi) = \mathrm{div}(d\phi) = 0$. When $\tau(\phi) \neq 0$, i.e., when the tension field is present, the harmonicity is lost or broken; $\tau(\phi) = 0$ is actually the Euler–Lagrange equation.

(g) Now we shall consider singularities encountered in physical theories. In electromagnetism and gravitational theories one comes across potentials of the type $1/r$, which are obviously singular at $r = 0$. In general, potentials $V(r) = 1/r^n$ (for $n \geqq 1$) are singular at the origin and one has a problem with $1/0$. Another aspect of these potentials is that mass is considered to be concentrated at a point, which is an idealization far removed from reality. The singularity here is due to incompleteness, extrapolation, and idealization of the physical situation.

(h) *Space-Time Singularities:* The first fundamental form of space-time for a time-independent (static) spherically symmetric and matter-free field (Schwarzschild solution of Einstein's equation) is given by

$$c^2 ds^2 = c^2 \left(1 - \frac{2MG}{c^2 r}\right) dt^2 - \left(1 - \frac{2MG}{c^2 r}\right)^{-1} dr^2 - r^2\, d\Theta - r^2 \sin^2\Theta\, d\phi^2$$

It is obvious from the above quadratic form that when $2MG = c^2 r$, then $(1 - 2MG/c^2 r)^{-1} = 0^{-1}$, i.e., the coefficient of dr^2 is the inverse of zero. We shall not go into a discussion on how this singularity can be displaced to $r = 0$ by the methods of Kruskal et al. This 0^{-1} was behind the assumption that the so-called "black holes" do exist. There are some artificial means like "cosmic censorship principles" to avoid seeing what is called a "naked singularity."

A more serious description is given in terms of the incompleteness of the metric of space-time. This idea was later on extended to incompleteness of geodesics and bundles. The point where a geodesic ends abruptly is topologically different and considered as a boundary point of space-time.

(i) *Quantum Field Theory.* In quantum field theories, where perturbative methods are used, singularities show up in the calculation of S-matrix elements. These matrix elements are given in terms of a product of Green's functions and are usually of the type

$$\frac{1}{k^2 + i\epsilon} \quad \text{or} \quad \frac{m + \hat{p}}{m^2 - p^2 - i\epsilon}, \quad \text{etc.}$$

A typical self-energy divergent integral is

$$\Sigma(p) = \frac{e^2}{(2\Pi)^4 i}\sum_n g^{nn}\int dk\left(-\frac{1}{k^2+i\epsilon}\right)\gamma^n\frac{m+\hat{p}}{m^2-p^2-i\epsilon}\gamma^n$$
$$= \frac{e^2}{(2\Pi)^4 i}\sum_n g^{nn}\int\frac{dk}{k^2+i\epsilon}\gamma^n\frac{\hat{p}-\hat{k}+m}{(p-k)^2-m^2+i\epsilon}\gamma^n$$

For large values of $|k|$ the function to be integrated goes as $|k|^{-3}$ and consequently the integral diverges linearly. The point is not that the integrand becomes 0^{-1} for $|k| = 0$, but that the upper limit is given as infinity. Also, in the definition of the integral the idea of a continuous variable, a point set is used. The problem is multifold here. If we get finite answers it is pure coincidence that 0^{-1} cancels out in the process of mathematical manipulation. The renormalization is a kind of nonstandard extension, in the sense that, by saying that finite entities are observable, one eliminates infinities. In actual computations infinities are canceled out by the addition of other infinities.

3. Epistemology *Grosso Modo*

Physics is derived essentially from observation and experience; however, its universe of discourse contains concepts which are not exactly exemplified in experience, e.g., point mass, space-time continuum, and so on. As we have seen from the above examples, both formal and computational difficulties are due to idealization as well as incompleteness inherited from mathematical theories. The historical development of physical and mathematical theories leads us to the conclusion that knowledge has a kind of stratified structure. In fact, in ordinary life whenever we fail to understand, i.e., fail to exemplify in terms of our past experience or knowledge, we refer the discussion to a more profound or deeper level of knowledge. This intuition of "deeper" and deeper level of understanding and knowledge can be exploited to avoid the undefined objects. The best example of this stratification is the development of number systems from natural numbers. In this section we shall still be using the naive set theory.

Let $\Sigma\, U_p(k)$ be all knowledge at a given period, which is conveniently small, and at a given place, say on Earth.* Let $U(k)$ be a set of primitive concepts with k n-any operations ($n \geq 2$). $U_p(k)$ is a set of complex concepts by combining p primitive concepts under n-any operations. Let $\Sigma\, U_p(k)$ be all primitive and complex concepts at a given space-time period. In other words, $\Sigma\, U_p(k)$ represents all the knowledge homo sapiens have at a fixed period in space-time. All the physical data, for example, can be explained, i.e., can be given a cause–effect relationship within $\Sigma\, U_p(k)$. $\Sigma\, U_p(k)$ then represent a stratum or level of knowledge. When some physical data or a result

*It would be futile to discuss the totality of knowledge of all possible "sapiens" existing in our cosmological universe, since one can imagine only their existence.

of mathematical manipulation cannot be explained with $\Sigma\, U_p(k)$ then one is free to go over to a more refined and enlarged level of knowledge. Let us use a superscript to label the stratum as $\Sigma\, U_p^1(k)$, $\Sigma\, U_p^2(k)\ldots$. The process of going from $\Sigma\, U_p^1(k)$ to $\Sigma\, U_p^2(k)$

$$J{:}\Sigma\, U_p^1(k) \rightarrow \Sigma\, U_p^2(k)$$

is not a continuous process in the sense that just the addition of one or more concepts solves the problem. One has to modify the whole outlook. In physics this happens so often that we take it as a natural process of scientific progress.

Examples. $\Sigma\, U_p^1(k)$ = Newtonian mechanics, and $\Sigma\, U_p^2(k)$ = relativistic mechanics.

$$\Sigma\, U_p^1(k) \rightarrow \Sigma\, U_p^2(k)$$

$\Sigma\, U_p^3(k)$ — quantum mechanics $\rightarrow \Sigma\, U_p^4(k)$ = relativistic quantum mechanics.

The notation is symbolic and does not obey rules from known mathematics. $\Sigma\, U_p'(k)$ is the real number system; $\Sigma\, U_p^2(k)$ is a nonstandard extension; $\Sigma\, U_p^1(k) \rightarrow \Sigma\, U_p^2(k)$ infinitesimals are given meaning.

A singularity is then any data set/concept which represents the breakdown of knowledge—it cannot be explained via cause–effect or logical deduction. A singularity at the *ith* level is not interpretable in $\Sigma\, U_p^i(k)$, but is given meaning in $\Sigma\, U_p^{i+1}(k)$ and becomes part of the established knowledge of this next level. A detailed discussion on epistemology and laws of nature is out of place here and we shall not indulge in it. What I would like to emphasize is that a singularity is not a singularity in its own class of object or within structure defined with respect to various new operations and properties. For example, an element of enveloping algebra does not belong to algebra itself and hence has no interpretation within algebra.

4. Formalization

There are three different approaches which can be of interest in the discussion on singularities. These are as follows:

1. Nonstandard models.
2. Axiomatization of categories and Lawvere's self-adjointness.
3. Stratification scheme to avoid singularities (already described above).

Roughly speaking, nonstandard models are extensions of standard models and use higher-order formal languages, while the use of categories (and especially of variable sets) is essentially finitistic doctrine. The stratification scheme we are proposing has some similarities with the theory of types of Russell, but there are infinite or undefined objects which appear but only at different stages of knowledge. It is principally a deterministic doctrine.

(a) *Nonstandard Models*. When mathematical systems like natural numbers, elementary group theory, etc. are axiomatized within the first-order predicate calculus, one finds that this axiomatization is not categorical and that there exist models of the relevant axiomatic theory not isomorphic to the system one started with. Below we shall briefly investigate this incompleteness problem. (For notation and definitions, see the end of this chapter. An elementary "standard" system is a set $\mathcal{S}$ together with a subset $\mathcal{R}$ of the set of relations on $\mathcal{S}$ such that $\mathcal{I} \in \mathcal{R}$. The language of $\mathcal{S}$ is the subset $\mathcal{L}(\mathcal{S}) = \{p \in P(V,\mathcal{R})\ \mathrm{var}(p) \subseteq \mathcal{S}\}$ of $P(V,\mathcal{R})$. Interpreting each element of $\mathcal{S}$ and $\mathcal{R}$ as itself assigns a truth value $v(p)$ to each $p \in \mathcal{L}(\mathcal{S})$. The theory of $\mathcal{S}$ is the theory $\mathcal{T}(\mathcal{S}) = (\mathcal{R}, A, \mathcal{S})$, where $A = \{p \in \mathcal{L}(\mathcal{S}) |\ v(p) = 1$. Let $^*\mathcal{S}$ be any model $\mathcal{T}(\mathcal{S})$. We say that $\mathcal{S}^*$ is a standard model of $\mathcal{T}(\mathcal{S}) i \not\subset f$. $^*\mathcal{S}$ is isomorphic to $\mathcal{S}$, and otherwise $^*\mathcal{S}$ is called a nonstandard model of $\mathcal{T}(\mathcal{S})$. The analysis based on this nonstandard model is called nonstandard analysis and has been developed by A. Robinson.† The most important part of its is that infinitesimals do exist, although we are taught since the time of Weierstrass that infinitesimals do not exist. If $\mathbb{R}$ is the set of real numbers, then its nonstandard extension $^*\mathbb{R}$ is a non-Archimedean ordered field. Let us call elements of $^*\mathbb{R}$ hyperreal numbers. An element $a \in {}^*\mathbb{R}$ is called finite if there exists a standard real number b such that $|a| < b$. Otherwise, a is called infinite. A (finite) element a is called infinitesimal if $|a| < b$ for all standard real numbers $b > 0$. Zero is infinitesimal.

The following are the axioms of nonstandard analysis:

1. There exists a proper ordered field extension $^*\mathbb{R}$ of the field of real numbers.
2. Let f be a real-valued function defined on some subset of the set $\mathbb{R}^n$ of all n-tuples of real numbers. Then to f there corresponds a hyperreal-valued function f^* on n hyperreal variables, called the natural extension of f. The field operations of $^*\mathbb{R}$ are the natural extensions of the field operations of $\mathbb{R}$.
3. If two systems of formulas have exactly the same real solutions, then their natural extensions have exactly the same hyperreal solutions.

It is interesting to note that both infinitesimal and infinite hyperreal numbers actually exist. Sums, differences, and products of infinitesimals are infinitesimals, as is the reciprocal of an infinite number. The product of an infinitesimal and a finite number is infinitesimal (details can be found in many new textbooks).

(b) *Categorical Method*. We shall follow F. W. Lawvere, who developed a theory in which categories are axiomatically introduced without using a set theory, and from which the set theory can be derived. The importance of this development lies in the fact that one avoids the inconsistencies of set theory, especially incompleteness which is related to undefined objects or singularities. He utilizes a theory of topos, which were introduced by the Gorthendieck School as a generalization of the notion

†There is a recent formulation of nonstandard models under the title "internal set theory" by E. Nelson with an eye towards application to the divergence problem of QFT.

of topological space. With a topological space is associated a topo, namely the category of set-valued sheaves over it. Lawvere maintained that the notion of topos could be viewed as a conceptual form of the notion of higher-order language or, alternatively, as a world in which higher-order notions could be interpreted. The notion of elementary topos (ET), as defined by, e.g., Koch and Writh, frees the notion of topos from any external form of infinities. For instance, an ET is not required to have arbitrary (infinite) limits or colimits. The use of variable structures and variable sets is another important aspect of this formalism. In the first-order mathematics or in traditional set theory, the constancy of sets, i.e., constant predicate, is taken for granted. However, Robinson's nonstandard analysis and Cohen's forcing method involve passing from a system $\overline{\mathcal{S}}$ (of constant sets) to a new system $\overline{\mathcal{S}}$ that still satisfies the basic axioms for constant sets; but both methods pass through systems of variable sets. In nonstandard analysis and forcing, one starts with a model $\mathcal{S}$ of a theory of constant sets. If X (= domain) is a given constant set, then $\mathcal{S}^X : X \rightarrow \mathcal{S}$ is a system of variable sets with domain consisting of "points." If one "stops" the variation at a point of X, one gets back $\mathcal{S}$; but if onc stops the variation at an ideal point $\in [\beta(X) - X]$, one gets a new system $\overline{\mathcal{S}}$ of sets that satisfy the same elementary axioms (e.g., those expressing constancy) as $\mathcal{S}$ but will be different from $\mathcal{S}$ It will contain new "infinitesimal" elements, for example. However, in Cohen's forcing the property expressing constancy remains but, unlike the case of nonstandard analysis, deeper properties such as continuum hypothesis and axiom of constructability are destroyed in the passage from $\mathcal{S} \rightsquigarrow \overline{\mathcal{S}}$. One can speak of ideal points of topos in the following way. Any given infinitary theory can be construed as a finitary theory in which all formulas obtained by infinitary disjunction are reconsidered to be "atomic" formulas. The powerfull method of this approach has not been utilized so far either in mathematics or physics.

Remark. There are two basic aspects concerning the nature of singularities. One is the independence of singularity with respect to other primitive notions of the universe of knowledge (or what we called "stratum of knowledge"), while the second is incompleteness of the universe and consequently its enlargement. We have seen above that the axiom of choice and the existence of a continuum play a role in telling us that singularities are an unavoidable part of knowledge. Gödel's incompleteness theorem tells us that no finitely axiomatizable system is complete. The stratification of knowledge is in some way similar to the axiom of inaccessible cardinals. That is, if one performs by induction a transfinite sequence of suitable closure operations, one can still go further and presumably find inaccessible cardinals beyond. The point is not that we can go ad infinitum but that at any stratum completeness is achieved. The assumption is that we can never know "everything," but it does not forbid us to redefine singularities at each level. For a universe of knowledge, say ZF set theory, one realizes that the existence of continuum and singularities are not independent. We say this rather timidly, because we have not proved this, but it is an intuitive conclusion. We have seen that enlargement cannot completely eliminate nondefinable objects.

APPENDIX

Glossary

To avoid confusion, an explanation of words and symbols used in the text is given below.

Correctness. A proposition is correct if, within a logical system, it is deduced according to accepted rules.

Truth Content. It can be defined with respect to the totality of experience, certainty, and completeness of the system. A correct proposition borrows its "truth" from the truth content of the system to which it belongs.

Experimental Verification. An experiment can be a physical, thought world (*gedankenwelt*), or logical experiment. By "logical experiment" we mean the act of proving or verifying a proposition within the system. The principle of third exclusion is not taken for granted here.

True Relation. A relation is true if it can be verified or proved at any space-time point under identical circumstances, physically or logically.

Admissible Knowledge. This has true relations such that the truth value is always one, i.e., completely true.

Partially Admissible Knowledge. Here true relations can have a truth value between 0 and 1, i.e., relations can be partially true.

Nonadmissible Knowledge. No experiments of any sort, logical or otherwise, are possible for the purpose of verification or proof of relations. Experiment could be a "random event" in space-time, i.e., once verified but not repeatable.

Predictable. A piece of knowledge (exclude nonadmissible) is predictable if only unique interpretation can be given starting with the same set of axioms. This does not exclude more than one solution of a problem.

Interpretation. Intuitive model such that a definition can be given later on.

Notation

$A \models P$	p is a consequence of the assumptions A
$A \vdash P$	p is provable from the assumptions A
$al(n)$	at least n elements
$ar(t)$	-arity of t
$Con(A)$	set of consequences of A
$D\rho$	domain of relation ρ
$Ded(A)$	set of all deductions from A
$\mathcal{J}$	identity relation symbol
$\mathcal{L}(\mathcal{T})$	language of the theory $\mathcal{T}$
$O^k(S)$	set of kth-order objects on S
$\mathcal{P}$	Peano arithmetic

$P(\mathcal{T})$	algebra of the theory
Pred $(V,\mathcal{R})$	first-order predicate calculus on $(V,\mathcal{R})$
Prop(x)	propositional calculus on X
$\mathcal{S}'$	enlargement of $\mathcal{S}'$ with respect to σ
$^*\mathcal{S}'$	full enlargement of $\mathcal{S}'$
$\mathcal{T}(\mathcal{S}')$	theory of the system $\mathcal{S}'$
$v(p)$	truth value of p
Var(P)	set of free variables of p

Definitions

For the reader's convenience we list some important definitions.

(1) A n-ary operation on the set A is a function $t:A^n \to A$. n is called the arity of t.

(2) A type $\mathcal{T}$ is a set T together with a function ar: $T \to \mathbb{N}$ from T into nonnegative numbers. $\mathcal{T} = (T,\text{ar})$ and $T_n = \{t \in T\ \text{ar}(t) = n\}$.

(3) An algebra of type T or T-algebra is (A,t_A); $t_A:A^{\text{ar}(t)} \to A$, $t \in T$. n-ary T-algebra operations, $t \in T_n$.

(4) Let A and B be T algebras. A homomorphism $Q:A \to B$ s.t. $\forall\ t \in T$ and all $a_1, a_2 \ldots a_n \in A$ we have $\varphi(t_A(a_1, \ldots, a_n)) = t_B(\varphi(a_1) \ldots \varphi(a_n))$. φ preserves all the opertions of T. If φ^{-1}: $B \to A$ is also a homomorphism, then φ is an isomorphism and we shall write $A \sim B$.

(5) Let X be any set, F any T algebra, and $\sigma:X \to F$. F or (F, σ) is a free T-algebra on X of free generators, if for every T-algebra A and $\tau:XA$, exist a unique $\varphi:F \to A$ s.t.$\varphi\sigma = \tau$.

(6) Let $T = \{F, \Rightarrow\}$, F = 0-ary and $\Rightarrow$ = binary operations. Any T-algebra is called a proposition algebra.

(7) The propositional algebra $P(x)$ of the propositional calculus on the set X of propositional variables is the free T-algebra on X.

(8) A valuation $P(x)$ is a proposition algebra homomorphism v: $P(x) \to \mathbb{Z}_2$; $p \in P(x)$ true with respect to v if $v(p) = 1$, false if $v(p) = 0$. (Here two-valued logic is implicitly assumed.)

(9) A logic $\mathcal{L}$ is a system = {set P of propositions, $V:P \to W$ valuation set, for each $A \subset P$ a set of finite sequence of elements of P, i.e., proofs from assumption A}

(10) Propositional calculus Prop(x) = $\{P = P(x)$, $V:P(x) \to \mathbb{Z}_2$, $A \subset P(x)$ set of proofs}.

(11) Logic $\mathcal{L}$ is sound if $A \vdash p \Rightarrow A \models p$, consistent if F is not a theorem, and adequate if $A \models p \Rightarrow A \vdash p$.

(12) Set-free definition of a category: A category C is a two-sorted system, the sort being called objects A of C and morphisms f of C. The undefined terms, in context, are "A is the domain (or the codomain) of f," "k is the composite of g with f," and "t is the identity morphism of A." The first axiom asserts that every morphism f

has exactly one object A as domain and one object B as codomain, i.e., $f{:}A \to B$ (notation). Next, for all morphisms g and f, there exists a composite k of g with f if and only if domain (g) = codomain (f); i.e., $k = gf$ (notation). $[h(gf) = (hg)f]$ associativeness is also required by another axiom. Every object B has one and only one identity morphism, with domain and codomain B; $1_B{:}B \to B$ (notation). Also $1_B f = f$ for every $f{:}A \to B$ and $g1_B = g$ for every $g{:}B \to C$.

(13) A morphism of categories is called a *functor*, i.e., a functor $F{:}\,\mathbb{C} \to \mathbb{D}$ assign to each object A of $\mathbb{C}$ an object $D = F(A)$ of $\mathbb{D}$ and to each morphism $A \to B$ of $\mathbb{C}$ a morphism $d = F(f){:}F(A) \to F(B)$ of $\mathbb{D}$. Also $F(1_B) = 1_{FB}$.

(14) A morphism of functors is called a *natural transformation.*

(15) *Adjoint Functor*. Given a functor $U{:}\ \mathbb{C} \to \mathbb{X}$ and an object X of $\mathbb{X}$, a universal morphism from X under U is a morphism $m{:}\ X \to UR$, with R an object of $\mathbb{C}$, such that every morphism $f{:}\ A \to UA$ in $\mathbb{X}$ can be written as a composite $f = U(f')m$ for exactly one morphism $f'{:}\ R \to A$ of $\mathbb{C}$, i.e.,

$$\begin{array}{ccc} X & \xrightarrow{m} & UR \\ & {\scriptstyle f}\searrow & \downarrow U(f') \\ & & UA \end{array}$$

A left adjoint of the given functor $U{:}\ C \to \mathbb{X}$ is then described as a functor $F{:}\ \mathbb{X} \to \mathbb{C}$ for which there is a natural transformation $\eta{:}\ I \to UF$, $I : \mathbb{X} \to \mathbb{X}$ the identity functor, and such that each $\eta_X{:}\ X \to UFX$ is universal from X under U.

First-Order Mathematics

A mathematical theory can be defined in terms of lists of relations and axioms. A part of mathematics constructed within the first-order predicate calculus is called *first-order mathematics*.

(16) A first-order mathematical theory is a triple $\mathcal{T} = (\mathcal{R},A,C)$ where $\mathcal{J} \in \mathcal{R}$, $A \subseteq P(V,\mathcal{R})$ for some $V \supset C$ such that V-C is infinite, and var(A) = C. The set A is called the set of (mathematical) axioms of $\mathcal{T}$, the set C is called the set of (individual) constants of $\mathcal{T}$, while the language of $\mathcal{T}$ is the subset $\mathcal{L}(\mathcal{T}) = \{p \in P(V,\mathcal{R}) \mid \text{var}(p) \subseteq C\}$ of $P(V,\mathcal{R})$. A theorem of $\mathcal{T}$ is an element $p \in \mathcal{L}(\mathcal{T})$ such that $A \vdash \mathcal{J}p$.

(17) A model of the first-order theory $\mathcal{T} = (\mathcal{R},A,C,)$ is a set M together with function $\nu{:}\ C \to M$, $\psi{:}\ \mathbb{R} \to$:rel(M) such that, for some set V of variables ($V \supset C$, V-C infinite), there exists a "proper interpretation (M, φ,ψ,ν) of $P(V,\mathcal{R})$ for which $\varphi|_C = \nu$ and $\nu\,(A) \subseteq \{1\}$.

(18) A first-order theory $\mathcal{T}$ is called *consistent* if F is not a theorem of $\mathcal{T}$.

(19) The theory $\mathcal{T}$ is consistent if and only if there exists a model of $\mathcal{T}$.

(20) The theory $\mathcal{T}$ is called *complete* if, for every $p \in \mathcal{L}(\mathcal{T})$ either $\mathcal{T} \vdash p$ or $\mathcal{T} \vdash p$ ($\mathcal{T}p$ means not p).

(21) The first-order theory $\mathcal{T}$ is complete if and only if every $p \in \mathcal{L}(\mathcal{T})$ which is true in one model of $\mathcal{T}$ is true in every model of $\mathcal{T}$.

(22) Let (M_1,ν_1,ψ_1) and (M_2,ν_2,ψ_2) be models of the theory $\mathcal{T}$. Then, M_1 is isomorphic to M_2 if there exists a bijective map α: $M_1 \rightarrow M_2$ such that $\alpha\nu_1 = \nu_2$ and $(m_1,m_2, \ldots ,m_n) \in \psi_1 r$ if and only if $(\alpha(m_1), \ldots , \alpha(m_n)) \in \psi_2$ for all $r \in \mathcal{R}_n$, all $m_1, \ldots ,$ $m_n \in M_1$, and all $n \in \mathbb{N}$.

(23) The theory $\mathcal{T}$ is called *categorical* if all models of $\mathcal{T}$ are isomorphic.

(24) If the theory $\mathcal{T}$ is categorical, then $\mathcal{T}$ is complete.

References and Notes

General References on Logic, Sets, and Foundations

1. D. W. Barnes and J. M. Mack, *An Algebraic Introduction to Mathematical Logic* (Springer-Verlag, Berlin, 1975).
2. N. Bourbaki, *Theory of Sets* (Hermann, Paris, 1968).
3. P. Suppes, *Introduction to Logic* (Reinhold Van Nostrand, New York, 1957).
4. E. W. Beth, *The Foundations of Mathematics* (Harper, New York, 1966).
5. P. J. Cohen, *Set Theory and the Continuum Hypothesis* (Benjamin, Reading, Massachusetts, 1966).
6. A. A. Fraenkel and Y. Bar-Hillel, *Foundations of Set Theory* (North-Holland, Amsterdam, 1958).
7. W. Yourgrau and A. D. Breck, *Physics, Logic, and History* (Plenum Press, New York, 1970).
8. H. Margenau, *The Nature of Physical Reality* (McGraw-Hill, New York, 1950).

Physics, Singularities, and Related Ideas

9. N. N. Bogoliubov, *Introduction to Axiomatic Quantum Field Theory.* (Benjamin, Reading, Massachusetts, 1975).
10. C. Misner, K. Thorne, and J. Wheeler, *Gravitation* (Freeman and Co., San Francisco, 1973).
11. K. Tahir Shah and W. Yourgrau, ICTP, Trieste, Internal Report IC/78/105 (unpublished).
12. K. Tahir Shah, *Found. Phys.* **9**, 271 (1979).
13. F. Destefano and K. Tahir Shah, ICTP, Trieste, Internal Report IC/79/120.

Nonstandard Models and Categories

14. A. Robinson, *Non-Standard Analysis* (North-Holland, Amsterdam, 1970).
15. P. J. Kelemen and A. Robinson, *J. Math. Phys.* **13**, 1870 (1972).
16. A. Voros, *J. Math. Phys.* **14**, 292 (1973); M. O. Farrukh, *ibid.* **16**, 177 (1975).
17. E. Nelson, "Internal Set Theory", *Bull. AMS* **83**, 1165 (1977).
18. J. Tarski and Ph. Blanchard, *Acta Phys. Austrica* **49**, 129 (1978).
19. W. A. Luxemburg, *Applications of Model Theory to Algebra, Analysis and Probability* (Holt, Rinehart, and Winston, New York, 1959).
20. F. W. Lawvere, *Proc. Symp. Pure Math.* Vol. XVIII, AMS (1970).
21. A. Kuck and G. C. Wraith, Lect. Notes No. 30, Aarhus Universitet, Denmark (1971).
22. I. MacLane, *Proc. Symp. Pure Math.* XVIII, AMS (1970), p. 231.
23. F. W. Lawvere, "Variable Quantities and Variable Structures in Topology," in *Algebra, Topology and Category Theory*, A. Heller and M. Tierney, editors (Academic Press, New York, 1976).
24. D. A. Clarke, *Hierarchies and Predicates of Finite Types*, AMS, Mem. 51 (1964).
25. A. Levy, *Hierarchy of Formulas in Set Theory*, AMS, Mem. 57 (1965).
26. J. Keisler, *Actes Congr. Inst. Math.* **I**, 141 (1970).

30

Wave Functions in One-Dimensional Disordered Systems with Long-Range Hopping

Michael J. Stephen*
Physics Department
Rutgers University
Piscataway, New Jersey 08856

1. Introduction

The properties of the electronic states of disordered quantum systems are of considerable interest to solid state physicists. Most of the known exact results are for one-dimensional systems and in this case it has been established under quite general conditions that all eigenstates are localized. A review of this work has been given by Ishii.[1] Most of the discussions of this problem consider particular solutions of the stationary Schrödinger equation, i.e., solutions for given energy with boundary conditions prescribed on one edge of the system, and show that the solution grows exponentially with probability unity. It is then argued[2] that eigenstates for a long but finite system occur at those energies where an exponentially growing wave function on the left can be matched to an exponentially growing function from the right. In this way a localized state is constructed which decays exponentially away from some central region.

All the above results have been obtained for systems with short-range interactions. In this paper we investigate the properties of particular solutions of the Schrödinger equation in a one-dimensional random system with long-range interactions. The question we attempt to answer is: for what type and range of interaction do we expect only localized states (as in the short-range case) and is it possible to have extended states with long-range interaction?

*Supported in part by the National Science Foundation under grant No. DMR 78-10276.

2. The Model

The model we consider is the tight-binding Anderson[3] model with Schrödinger equation

$$Ea_i = \epsilon_i a_i + \sum_{j=1}^{n} V_j(a_{i+j} + a_{i-j}) \tag{1}$$

where E is the energy, a_i is the wave function amplitude on site i of the one-dimensional lattice, and ϵ_i is the random energy of this site. The ϵ_i are mutually independent, random variables with a common distribution. The zero of energy is chosen so that $\langle \epsilon_i \rangle = 0$ and the second moment is

$$\langle \epsilon_i^2 \rangle = \sigma \tag{2}$$

The V_j ($j = 1 \ldots n$) are the transfer energies and V_j is a matrix element for the process in which the electron hops j steps. We begin by considering the case where n is finite and later we will allow $n \to \infty$.

Equation (1) is a recursion relation involving $2n + 1$ amplitudes. The boundary conditions on the left side of the crystal are given, i.e., we know the first $2n$ amplitudes. We can then use Eq. (1) iteratively to calculate the wave function amplitudes deep in the crystal. This is most easily accomplished by writing (1) in a transfer matrix form. We define a column vector of the amplitudes by

$$\mathbf{a}_i = (a_{i+n-1}, a_{i+n-2}, \ldots, a_i, \ldots, a_{i-n})$$

Then Eq. (1) can be written

$$\mathbf{a}_{i+1} = T^{(i)}\mathbf{a}_i \tag{3}$$

where the transfer matrix $T^{(i)}$ is a $2n \times 2n$ matrix with nonzero elements

$$\begin{aligned} T_{1j}^{(i)} &= -\frac{V_{n-j}}{V_n}, \qquad j = 1 \cdots 2n, \quad j \neq n \\ T_{1n}^{(i)} &= \frac{E - \epsilon_i}{V_n} \\ T_{j+1,j}^{(i)} &= 1, \qquad j = 1 \cdots 2n - 1 \end{aligned} \tag{4}$$

By repeated application of (3) we have

$$\mathbf{a}_L = T^{(L)} \cdots T^{(1)}\mathbf{a}_1 \tag{5}$$

and the mean square amplitudes L steps from the edge are

$$\langle \mathbf{a}_L^2 \rangle = \tilde{\mathbf{a}}_i \langle \tilde{T}^{(1)} \cdots \tilde{T}^{(L)} T^{(L)} \cdots T^{(1)} \rangle \mathbf{a}_1 \tag{6}$$

where $\tilde{T}$ is the transpose of T.

The calculation of the average matrix in (6) can be reduced to the determination of the eigenvalues of a matrix recursion relation. Thus if

$$M_L = \langle \tilde{T}^{(1)} \cdots \tilde{T}^{(L)} T^{(L)} \cdots T^{(1)} \rangle \tag{7}$$

then

$$M_{L+1} = \langle \tilde{T} M_L T \rangle \tag{8}$$

We then require the eigenvalues of the matrix equation

$$\lambda M = \langle \tilde{T} M T \rangle \tag{9}$$

For eigenvalues $|\lambda| > 1$ the mean square amplitudes $\langle \mathbf{a}_L^2 \rangle$ will increase exponentially with L while if $|\lambda| < 1$ they will decrease exponentially with L. If $|\lambda| = 1$ they will not grow or decay but oscillate or remain constant.

It is convenient to write (9) as a matrix equation. Denoting the elements of the $2n \times 2n$ matrix M by $m_{ij}(i, j = 1 \cdots 2n)$ Eq. (9) is

$$\lambda m_{ij} = L_{ij,kl} m_{kl}, \qquad L_{ij,kl} = \langle \tilde{T}_{ik} \tilde{T}_{jl} \rangle \tag{10}$$

The matrix L is conveniently divided into two parts

$$L = L^{(0)} + L' \tag{11}$$

where $L^{(0)}$ is the matrix when there is no randomness ($\sigma = 0$). L' involves the randomness and only has one nonzero term

$$L'_{nn,11} = \sigma / V_n^2 \tag{12}$$

The eigenvalue problem (10) now takes the form

$$(L^{(0)} - \lambda) M = -L' M \tag{13}$$

or

$$M = (\lambda - L^{(0)})^{-1} L' M \tag{14}$$

We get the dispersion relation determining the eigenvalues λ by taking the 11 element of (14)

$$1 = (\lambda - L^{(0)})^{-1}_{11,nn} \frac{\sigma}{V_n^2} \tag{15}$$

So it is necessary to calculate the matrix $(\lambda - L^{(0)})^{-1}$ and this can be done from a knowledge of the eigenvalues and vectors of $\tilde{T}$ in the absence of any randomness.

3. Dispersion Relation

The eigenvalues of T in the absence of disorder can be determined from (1). In this case $a_j \sim e^{ikj}$ and are plane waves and the energy E is related to the wave vector k by

$$E(k) = 2 \sum_{j=1}^{n} V_j \cos kj \tag{16}$$

T is an operator which translates the system by one lattice spacing. Thus its eigenvalues are e^{ik_α}, where k_α are the values of k satisfying (16) for fixed E. It is convenient to define $V_0 = -E$ and then (16) can be written

$$\sum_{j=-n}^{n} V_j e^{ik_\alpha j} = 0, \qquad V_{-j} = V_j \tag{17}$$

We restrict $-\pi < k \leq \pi$ and the eigenvalues will occur in pairs $e^{\pm ik_\alpha}$.

We also require the left and right eigenvectors of the matrix $\tilde{T}$ when there is no disorder. The left eigenvectors are easily shown to be

$$(1/N_\alpha)(1, e^{-ik_\alpha}, e^{-2ik_\alpha}, \ldots, e^{-(2n-1)k_\alpha}) \tag{18}$$

where N_α is a normalization constant. The right eigenvectors are conveniently taken in the form

$$(1/N_\alpha)(1, b_2^{(\alpha)}, \ldots, b_{2n}^{(\alpha)}) \tag{19}$$

It is shown in the Appendix that the normalization constant is given by

$$N_\alpha^2 = -iV_n^{-1} e^{-ink_\alpha}\left(\frac{\partial E}{\partial k_\alpha}\right) \tag{20}$$

From the results (17)–(20) it is straightforward to write down $(\lambda - L^{(0)})^{-1}$ and the dispersion relation (15) takes the form

$$1 = \sigma \sum_\alpha \sum_\beta \frac{e^{i(k_\alpha + k_\beta)}}{E'(k_\alpha)E'(k_\beta)(e^{i(k_\alpha+k_\beta)} - \lambda)} \tag{21}$$

where the sums extend over all the solutions of (17) and $E'(k_\alpha) = \partial E/\partial k_\alpha$. This is the final form for the dispersion relation and we now consider some special cases for the hopping matrix elements V_j.

4. Examples

(A) Exponential Range Hopping. In this case $V_j = Ve^{-\gamma j}$ where γ is a positive constant. Equation (17) for the eigenvalues is

$$\frac{E}{V} = \frac{2e^{-\gamma}(\cos k - e^{-\gamma})}{(1 - e^{-\gamma+ik})(1 - e^{-\gamma-ik})} \tag{22}$$

For simplicity we consider the center of the band $E = 0$. Then Eq. (22) has two roots $e^{\pm ik_0}$ with $\cos k_0 = e^{-\gamma}$ and

$$E'(\pm k_0) = \pm 2V(e^{2\gamma} - 1)^{-1/2} \tag{23}$$

The dispersion relation (21) is given by

$$1 = \frac{4\sigma \sinh^2 \gamma}{V^2} \frac{\lambda(\lambda + 1)}{(\lambda - 1)[(\lambda + 1)^2 - 4\lambda e^{-2\gamma}]} \tag{24}$$

For small σ/V^2 the largest eigenvalue is

$$\lambda \simeq 1 + \frac{\sigma}{2V^2}(e^{2\gamma} - 1) \tag{25}$$

This eigenvalue leads to exponentially growing wave functions and is very similar to the case of short-range hopping. The other two eigenvalues of (24) have $|\lambda| < 1$ and lead to exponentially decreasing wave functions.

(B) Power Law Range Hopping. We investigate the case $V_j = V/j^2$. Equation (17) for the eigenvalues is now $(0 < k < 2\pi)$

$$\frac{E}{V} = \frac{k^2}{2} - \pi k + \frac{\pi^2}{3} \tag{26}$$

and again taking the case $E = 0$ there are two solutions $k_\pm = \pi(1 \pm 1/\sqrt{3})$ and

$$E'(k_\pm) = \pm\pi V/\sqrt{3} \tag{27}$$

The dispersion relation (21) is now given by

$$1 = \frac{12\sigma \sin^2 k_+}{\pi^2 V^2} \frac{\lambda(\lambda + 1)}{(\lambda - 1)(\lambda - e^{2ik_+})(\lambda - e^{-2ik_+})} \tag{28}$$

It may be shown that there is one eigenvalue $\lambda > 1$ and two with $|\lambda| < 1$ so that this is very similar to the previous case and the short-range case.

5. Conclusions

We have investigated the behavior of the mean square wave function in a random one-dimensional tight-binding model with long-range hopping. It is shown that the mean square particular solutions of the Schrödinger equation in general increase exponentially even when the hopping is long range and decreases like the inverse square of the distance. This is very similar to Borland's[2] original result for short-range hopping. The results indicate that only localized states are to be expected in one-dimensional systems even with long-range hopping.

Acknowledgments

I first met Wolfgang Yourgrau when I was an undergraduate and I attended his course in the Philosophy of Science at the University of the Witwatersrand. This was the beginning of a long and continuing friendship with him and his family. Wolfgang was a dedicated scholar, very knowledgeable in many fields, and was one of the first people to arouse my interest in science. But above all he was a wonderful human being, warm, generous, and concerned. I will always treasure my friendship and association with him.

Appendix

From the transfer matrix (4) the right eigenvectors (19) of $\tilde{T}$ satisfy

$$b_{m+1}^{(\alpha)} = e^{ik_\alpha} b_m^{(\alpha)} + \frac{V_{n-m}}{V_n} \tag{A.1}$$

for $m = 1 \cdots 2n$ with $b_{2n+1} = 0$. In order to obtain the normalization constant N_α^2 we multiply (A.1) by e^{-imk_α} $(2n - m + 1)$ and sum on m from 1 to $2n - 1$. After some rearrangement and use of (17) we find

$$N_\alpha^2 = \sum_{m=1}^{2n} b_m \, e^{-(m-1)k_\alpha} = -V_n^{-1} \sum_{j=-n}^{n} V_j j e^{-i(n+j)k_\alpha} \tag{A.2}$$

which is the same as Eq. (20).

References and Notes

1. K. Ishii, *Prog. Theor. Phys. (Suppl.)* **53,** 77 (1973).
2. R. E. Borland, *Proc. R. Soc. London* **A274,** 529 (1963).
3. P. W. Anderson, *Phys. Rev.* **109,** 1492 (1958).

31

PERCEPTION OF QUANTUM SYSTEMS

E. C. G. SUDARSHAN*
Department of Physics
University of Texas
Austin, Texas 78712

1. INTRODUCTION

Quantum mechanics presents us with a world view with which many of us have become familiar as long as it is restricted to the domain of the very small or very exceptional. It has become so much a part of the tradition of contemporary physics that a new level of intuition regarding such systems has developed for the research physicist. A high-energy physicist or an atomic physicist can often deduce in a noncomputational manner the behavior of quantum systems of interest. For example, the identification of forces with exchange of quanta or the behavior of a particle in a potential well; or even the interrelationship of the hierarchy of groups to patterns of particles and interactions are no more mysterious than the "feel" a structural engineer has for the strength of a building design or of a good driver for his automobile. The present generation of physicists have no serious preoccupation with the wave–particle dualism or the indeterminacy principle.

It turns out, however, that this mastery is achieved at a price: the language of quantum physics is different from that of the everyday world; and substantially mathematical. One uses, for example, noncommuting operators to represent the dynamical variables of a quantum system. The relationship that a quantum system bears to a classical system is often purely formal.[1] There are some approximations in calculation of physical quantities of a quantum system that can be made by making suitable use of a "corresponding" classical system: but the definition of the sense in which there is a good approximation becomes the problem. To the extent we attempt to use the classical language of definite positions and velocities for the particles we find discon-

*Research supported in part by the National Science Foundation under Contract No. PHY77-22864.

tinuities and paradoxes: very much as mapping the surface of the globe onto plane sheets of paper produces distortions and discontinuities. The "uncertainties" and "uncontrollable measurement interactions" have their origin in the use of an alien language. But language not only affects our descriptions, it affects our world view and hence our very perceptions.

This paper explores the possible modes of perception of a quantum mechanical system. It is an attempt to point out to a wider audience isolated results known to individuals. At the same time it is the intention of the author to relate it to more familiar and more immediate complexes of phenomena and perceptions. It is the fond hope of the author that he can indicate the immediate relevance of quantum models for our most precious perceptions of elegance, unity, and the dynamically creative.

Wolfgang Yourgrau was a person with a breadth of interests including questions about the perception of quantum systems. He was a rare individual who took delight in most of his activities. To his memory I dedicate this paper.

2. The Quantum System and Its States

As the prototype system let us consider a quantum particle without spin. It is defined kinematically by a triplet of canonical operators **q**, **p** of position and momentum in a three-dimensional space. They do not all commute with each other but satisfy

$$\begin{aligned} q_j p_k - p_k q_j &= i\hbar\delta_{jk} \\ q_j q_k - q_k q_j &= 0, \qquad p_k p_j - p_k p_j = 0 \end{aligned} \tag{1}$$

It is therefore meaningless to talk about a complete specification of a state in terms of values being assigned to all these variables. Instead, quantum mechanics associates a map[2]

$$F(\mathbf{q},p) \rightarrow \langle F(q,p)\rangle \tag{2}$$

which is linear, nonnegative, and normalized. This means that

$$\begin{aligned} \alpha_1 F_1 + \alpha_2 F_2 &\rightarrow \alpha_1\langle F_1\rangle + \alpha_2\langle F_2\rangle \\ F \geq 0 &\rightarrow \langle F\rangle \geq 0 \\ 1 &\rightarrow \langle 1\rangle = 1 \end{aligned} \tag{3}$$

The nonnegativity requires further explanation since this involves the ordering of operators and the notion of adjoints. Define a *-automorphism by

$$\begin{aligned} (\alpha_1 F_1 + \alpha_2 F_2)^* &= \bar{\alpha}_1 F_1^* + \bar{\alpha}_2 F_2^* \\ (F_1 F_2)^* &= F_2^* F_1^* \end{aligned} \tag{4}$$

where the bars denote ordinary complex conjugation. An element is said to be nonnegative if it is the nonnegative linear combination of elements of the form F^*F. The positivity requirement contains within it the requirement of mapping "real" elements (i.e., elements of the form $F^* = F$) into real numbers. Such a normalized nonnegative linear map is a statistical state of the quantum mechanical system.

Under suitable restrictions on the algebra of dynamical variables the statistical state can be identified in a standard way. *For example,* if we were to choose for dynamical variables only Hilbert–Schmidt operators (i.e., operators F whose "square" F^*F has a finite trace) under the scalar product

$$(F,G) = \mathrm{Tr}(F^*G) \tag{5}$$

they form an inner product space which can be completed to a Hilbert space. Since the state is a linear functional on this Hilbert space, it should be identifiable with a vector of the Hilbert space, which means that there should be a dynamical variable ρ of the Hilbert–Schmidt class which realizes the state of the quantum system. Since each one-dimensional projection is to be mapped into a real number it follows that ρ should be "real" ($\rho^* = \rho$) and nonnegative. Finally, we note that the unit operator 1 does not belong to the Hilbert–Schmidt class but it can be approximated by finite sums of orthogonal projections. So we may require that for any such sum the map should be approximating unity from below. We may therefore require ρ to be traceable and having unit trace:

$$\rho \geq 0, \qquad \mathrm{Tr}(\rho) = 1 \tag{6}$$

The quantity ρ is called the density operator (or the density matrix) realizing the state given by the mapping. Having chosen a density operator one may extend the map from Hilbert–Schmidt operators to all bounded operators including the Weyl family of operators ($\bar{\lambda} = \lambda$, $\bar{\mu} = \mu$):

$$W(\lambda,\mu) = \exp(i\lambda \cdot \mathbf{p} + i\mu \cdot \mathbf{q}) \tag{7}$$

with the map

$$\langle W(\lambda,\mu) \rangle = \chi(\lambda,\mu) \tag{8}$$

Clearly

$$|\chi(\lambda,\mu)| \leq 1 \tag{9a}$$
$$\chi(0, 0) = 1 \tag{9b}$$
$$\bar{\chi}(\lambda,\mu) = (-\lambda,-\mu) \tag{9c}$$

This function $\chi(\lambda,\mu)$ may be called the characteristic function[3] for the state. Since the Weyl family of operators form a basis for all bounded operators, we have essentially found the numerical values to be assigned to all bounded operators once the characteristic function is specified.

Since the operators $\mathbf{q},\mathbf{p}$ do not commute we could rewrite the operators in a different order: say, put all the q_j factors to the extreme right and put all the the p_j factors to the left; or vice versa. Each of these choices would assign to the *same* state the *same* density operator ρ but *different* characteristic functions χ. Thus the characteristic functions depend on the ordering convention used for the dynamical variables since $W(\boldsymbol{\lambda},\boldsymbol{\mu})$ would mean different operators (even with $\boldsymbol{\lambda},\boldsymbol{\mu}$ fixed) if the order of the $\mathbf{q},\mathbf{p}$ operators is understood differently. For simple reorderings of the kind mentioned above the change in the characteristic function is by a simple multiplication (for more details, see below).

3. The Wigner Distribution Function

For the time being let us restrict ourselves to the choices (7) and (8) and explore the ways of associating classical pictures with a quantum state. Since

$$\exp(i\boldsymbol{\lambda}\mathbf{p}) \cdot \exp(i\boldsymbol{\mu}\mathbf{q}) \neq \exp(i\boldsymbol{\lambda}\mathbf{p} + i\boldsymbol{\mu}\mathbf{q})$$

it is clear that the quantum rules (7), (8) do not define unique values for $\mathbf{q},\mathbf{p}$; we have therefore a distribution of values for $\mathbf{q},\mathbf{p}$. Is it, then, possible to view a quantum system as a statistical distribution over a classical phase space? In other words for every quantum state ρ, is it possible to define a phase space distribution function such that

$$\begin{aligned} &\int\int R(x, k) \exp(i\boldsymbol{\lambda} \cdot k + i\boldsymbol{\mu}\mathbf{x})\, d^3k\, d^3x \\ &= \chi(\boldsymbol{\lambda},\boldsymbol{\mu}) = \mathrm{Tr}[\rho \exp(i\boldsymbol{\lambda} \cdot \mathbf{p} + i\boldsymbol{\mu} \cdot \mathbf{q})] \end{aligned} \tag{10}$$

is satisfied? Further, would it be possible to choose $R(\mathbf{x},\mathbf{k})$ such that it is real, nonnegative, and normalized so that we may view it as a true probability distribution?

The answer to the first question is immediate and was given by Wigner[4] four decades ago. We simply invert (10) to obtain

$$R(\mathbf{x}, \mathbf{k}) = (2\pi)^{-3}\int d^3\lambda \int d^3\mu\, \chi(\boldsymbol{\lambda}, \boldsymbol{\mu}) \exp(-i\boldsymbol{\lambda} \cdot \mathbf{k} - i\boldsymbol{\mu} \cdot \mathbf{x}) \tag{11}$$

Moreover this answer is unique under the conditions we have framed: the solution $R(\mathbf{x}, \mathbf{k})$ so obtained is normalized and real. But unfortunately, it is in general not nonnegative. One cannot avoid the pointwise negativity of the Wigner distribution function. It is therefore idle to ask the question as to whether we could define extremal states for which the distribution collapses to a sixfold delta function signifying unique momentum and position values.

One must stress at this point that this (pointwise) negativity of the general Wigner distribution function is necessary for the quantum system to be faithfully rendered by the phase space picture; and would arise for the definition of distribution functions over any set of noncommuting random variables. In all such situations the marginal distribution (obtained by integrating out certain variables so that the remaining random variables form a commuting set) will always be nonnegative. Thus, given any set of commuting variables, we can associate nonnegative distribution functions with them.

We can now ask the question: Is it possible to find all the extremal functions from which all other distribution functions could be obtained? The answer to this is immediate: since the density operators are nonnegative and have unit trace they are *convex combinations* of one-dimensional *projections*. Any such projection may be written

$$\pi = \psi\psi^{+} \tag{12}$$

where ψ is a unit vector in the space in which $\mathbf{q},\mathbf{p},\rho$ are represented by operators. The general density operator is

$$\rho = \sum_j \alpha_j \pi_j = \sum_j \alpha_j \psi_j \psi_j^{\dagger}$$
$$\sum_j \alpha_j = 1, \qquad \alpha_j \geq 0 \tag{13}$$

where, without loss of generality we may assume that the projections are orthogonal:

$$\pi_j \pi_k = 0, \qquad j \neq k \tag{14}$$

By the convolution property of the Fourier transform it then follows

$$\int d^3x \int d^3k \; R_j(\mathbf{x}, \mathbf{k}) R_k(\mathbf{x}, \mathbf{k}) = 0, \qquad j \neq k \tag{15}$$

This clearly shows that if at least one of the distribution functions is nonvanishing everywhere in the phase space, some of the distribution functions are negative somewhere in phase space. The characteristic functions are, however, analytic in the six variables $\boldsymbol{\lambda},\boldsymbol{\mu}$ and hence they cannot vanish in a region. It follows that almost all the Wigner distribution functions are somewhere negative. It is therefore not possible to eliminate the pointwise negativity of the general Wigner distribution function. Nor can the distribution functions for projections be further decomposed.

We have therefore a curious situation: When we restrict the distribution function to any set of commuting variables we get a distribution which is nonnegative. Yet the complete multivariate distribution is not nonnegative. It is necessary that the distribution be over the entire spectrum of values in such a fashion that for extremal states which are given by one-dimensional projections the Wigner distribution function satisfies the relation

$$\iint d^3x \, d^3k \, |R(\mathbf{x}, \mathbf{p})|^2 = (2\pi)^3 \tag{16}$$

4. Alternate Phase Space Distributions

Since the Wigner distribution function is not always pointwise nonnegative, one may search for other rules of association and other methods of arriving at a suitable phase space density. Such distributions can be obtained from the concept of "normal ordering." We can rewrite the commutation relations (1) in the form

$$a_j a_k^* - a_k^* a_j = i\,\delta_{jk}$$
$$a_j a_k - a_k a_j = a_j^* a_k^* - a_k^* a_j^* = 0 \tag{17}$$

with

$$a_i = (\omega q_i + ip_i)(2\omega)^{-1/2}$$
$$a_i^* = (\omega q_i - ip_i)(2\omega)^{-1/2} \tag{18}$$

where ω is a suitably chosen angular frequency. An operator function of **q,p** is said to be normal ordered when all the **a*** operators stand to the left and all the **a** operators to the right; it is said to be antinormal ordered when the role of **a*** and **a** are interchanged. This definition of normal (and antinormal) ordering for polynomials in **q,p** can be extended to Weyl operators by the formula

$$\begin{aligned} W(\lambda,\mu) &= \exp\left\{i\left[\frac{\mu}{(2\omega)^{1/2}}\, i(2\omega)^{1/2}\lambda\right] a + i\left[\frac{\mu}{(2\omega)^{1/2}} + i(2\omega)^{1/2}\lambda\right] a^*\right\} \\ &= W_N(\lambda,\mu)\cdot \exp\left[\omega\left(\lambda^2 + \frac{\mu^2}{4\omega^2}\right)\right] \end{aligned} \tag{19}$$

where W_N stands for the normal ordered operator obtained from the Weyl operator (and dropping all the commutator terms arising from the reordering). Similarly the antinormal ordered operator is given by

$$W(\lambda,\mu) = W_A(\lambda,\mu)\exp\left[+\omega\left(\lambda^2 + \frac{\mu^2}{4\omega^2}\right)\right] \tag{20}$$

A characteristic function can be defined by

$$\chi_A(\lambda,\mu) = \langle W_A(\lambda,\mu)\rangle \tag{21}$$

It can then be shown that this characteristic function corresponds to a distribution function introduced by Kano:[5]

$$R_A(\mathbf{x},\mathbf{k}) = \mathrm{Tr}[\pi(\mathbf{x},\mathbf{k})\rho] \tag{22}$$

where $\pi(\mathbf{x},\mathbf{k})$ is the unique one-dimensional projection which satisfies

$$\mathbf{a}\pi(\mathbf{x},\mathbf{k}) = \frac{\omega\mathbf{x} + i\mathbf{k}}{(2\omega)^{1/2}}\,\pi(\mathbf{x},\mathbf{k}) \tag{23}$$

Clearly $R_A(\mathbf{x},\mathbf{k})$ is a nonnegative distribution function which can be shown to have the proper normalization.

Does this imply that we have overcome the problem of negativity in the phase space description so that we may now interpret the Kano distribution function as a true probability distribution? Not so, since the antinormal ordering can take a positive operator into a form which is formally not positive definite. The simplest is the usual Hamiltonian for the harmonic oscillator with the angular frequency

$$H = \tfrac{1}{2}(p^2 + \omega^2 q^2)$$

In the normal ordered form this leads to

$$H = H_N + ½\omega$$

while in the antinormal ordered form

$$H = H_A - ½\omega$$

We have only shifted the burden of possible negativity to the operator structure rather than the distribution function.

5. Superpositions and Surrealistic World Views

We are therefore led to conclude that a quantum state cannot be considered to be a statistical distribution over phase space in the strict sense: there is no hidden observable which leads to the quantum uncertainties. But the marginal distributions over a subset of commuting variables is always a nonnegative true probability distribution. But unlike a classical system in which probabilities may appear for an extremal state if we take a less-than-complete set of dynamical variables, for a quantum system *there is no complete set of variables* in terms of which we can see a unique phase space state; instead we have a number of possible complementary views.

This surrealistic picture in which a new facet of reality appears depending on the perspective is not restricted to the simple quantum system that we have considered, but is valid for any dynamical system admitting of noncommuting dynamical variables. If A and B are dynamical variables which do not commute:

$$AB - BA \neq 0 \tag{24}$$

we may still consider states in which A or B have definite values. (The technical problem of operators with continuous spectrum may be circumvented by constructing from them operators which have discrete spectrum.) Consider the projector $\pi(B = b)$ to a discrete value of the operator B. Without loss of essential generality we could consider it to be a one-dimensional projection. But this is not a state in which the operator A has a definite value, otherwise the commutation relation (24) would be violated. (We ignore the possibility that the commutator may vanish for this particular state.) But any decomposition into states $\pi(A = a)$ having definite values of A is done only at the price of making them not have unique values for B. The need for a distribution of values of both A and B is a direct consequence of their noncommutation.

What is the nature of the state $\pi(A = a)$ in terms of the family of projectors $\pi(B = b)$? It cannot be a probabilistic sum of the latter family since that would not lead to a one-dimensional projection. The combination involved is a "linear superposition" arrived at as follows: We realize the dynamical variables A, B as well as the states $\pi(A = a)$, $\pi(B = b)$ as linear operators on an inner product vector space. Let us denote the ray to which $\pi(A = a)$ projects by ψ_a and the one to which $\pi(B = b)$

projects by ϕ_b. There are unimodular factors which are undefined on ψ, ϕ; we choose any arbitrary but fixed choice. Then the actions of A and B are represented in the form

$$\begin{aligned} \pi(A = a')\psi_a &= \delta_{aa'}\,\psi_a \\ \pi(B = b')\phi_b &= \delta_{bb'}\,\delta_b \\ \pi(A = a')\phi_b &= \sum_c \alpha_{bc}\phi_c \\ \pi(B = b')\psi_a &= \sum_c \beta_{ac}\psi_c \end{aligned} \tag{25}$$

where α_{bc} and β_{ac} are a set of numbers (which depend on the phases chosen for the vectors). The vectors of one set are obtained by linear superpositions of vectors of the other kind. The relative phases and amplitudes of the combination are important in determining the new state. Noncommutativity *requires such superpositions.*

Conversely, in a system in which vectors are superposed (and from the superposition we derive the projectors for the extremal states) we surely have noncommutative dynamical variables, and hence the difficulty in interpreting them in terms of classical pictures. Any such attempt will lead to negativity somewhere. Such systems must be then perceived in such a manner that the coherent superpositions are essentially taken into account.

A dynamical system in which not all information can be directly perceived (but in which we can still obtain the information by yet another set of observations which obscure the initial information) is provided by elementary optics. Light is a wave phenomenon but the directly perceived quantity is the photometric intensity of light. But the intensity does not propagate; to know the propagation of light we need to know the so-called two-point correlation function given by

$$\Gamma(\mathbf{x}_1,\mathbf{x}_2) = \langle \phi^*(\mathbf{x}_1)\phi(\mathbf{x}_2)\rangle \tag{26}$$

which is the normal ordered map of the product of the (analytic signal) wave field and its *-conjugate. For quantum wave fields this map can be obtained from a distribution function discovered by Sudarshan[6] and by Glauber.[7] The specialization of Γ when $\mathbf{x}_1$ and $\mathbf{x}_2$ coincide gives the photometric intensity

$$I(\mathbf{x}) = \Gamma(\mathbf{x},\mathbf{x}) \tag{27}$$

An immediate example of the need for knowing $\Gamma(\mathbf{x}_1,\mathbf{x}_2)$ rather than $I(\mathbf{x})$ on one screen to know the propagation to the next screen is given by the two-slit interference pattern. The pattern depends not only on the intensity of illumination of the two slits. If they are in phase the central fringe is a bright fringe; if they are out of phase the central fringe is a dark fringe; and if they are incoherent there is no interference pattern at all. But in all these cases the intensity of illumination of the slits is totally unchanged. The coherence information which could not be "seen" on the first screen becomes an intensity pattern at the second screen. But at the same time the information on the intensity pattern on the first screen is lost in the interference pattern as intensity but is latent as the coherence information there. But at no time do we "see" the full informational content of the illumination.

In this case again one can attempt to bring out the latent information by using a classical format. This is done by looking for light rays in wave optics. But as expected, if we want to have an exact treatment in terms of light rays, it should include an explanation of diffraction and interference of light added to light-producing darkness. So it turns out that we end up with a ray distribution function in space which is not always nonnegative.[8] We have *bright rays* and *dark rays*. Once we agree to this we can get a quantitative realization of these typical wave patterns including their scaling properties using light rays.

6. Can We Perceive a Quantum System Directly?

Let us now ask the question: Is it then, impossible to perceive a quantum system fully and without these distortions? We would answer in the affirmative: and the garland of ideas leading to such an answer gathers many flowers from many gardens.

First of all, we recall that the problem of rendering the surface of the earth in flat maps is a problem of great interest. We have a number of ingenious map projections, but all of them include distortions and discontinuities of their own. But this does not mean that we cannot comprehend the earth's surface: we must comprehend it as the surface of a sphere. The distortions are because of our attempts to map it on an inappropriate surface. We have to elevate our perceptions from the flat sheet to the closed compact unbounded sphere.

The second example is that of the wave field of light. Once it is understood that the two-point function contains new physical information we can immediately devise an experimental arrangement like the double slit to display this in the form of intensity. The total perception of the system then involves "seeing" the two-point function and our inability to see more than one "section" of it at any time. But eventually we are able to comprehend the propagation in terms of the two-point correlation function. The ray distribution function helps us to visualize this more clearly, but it is only an aid.

Is it possible to continue this elevation of perception to truly quantum mechanical situations? In other words is it possible to perceive phase-coherent superpositions? Most of the orthodox views that have been published seem to argue against such a possibility:[9] they say this cannot be since our perception of the world is classical! Yet, as far as we have come to understand, everything under the Sun is quantum mechanical. So why do we insist that we cannot?

7. Two Modalities of Awareness

One reason is probably the structure of our language. Physics is not only conveyed by language but it molds it as well. The language of everyday use is suffused with classical physics; it therefore becomes difficult to convey the picture of phase-coherent superpositions. As our perceptions rise so would our language. But until that happens, most of our discussions about such perceptions are going to be inadequate.

Perhaps this is not unrelated to the structure of our brain and the nature of our awareness. We are all familiar with the idea of the linguistic, discriminating, legal conceptual underpinning of our "rational" thought patterns. Such mental processes could be adequately described in common language; and each step could be reasoned, assumptions spelled out, and conclusions stated clearly. This mode of functioning is essential to our articulate culture and epoch.

Yet there is another mode of awareness which is inventive, contemplative, and nonverbal. Such states of awareness are often associated with highly creative periods in one's life or with great insights; but as often it does not lead to anything describable in everyday language. In this mode of functioning, sensations, feelings, and insights are not neatly categorized into chains of distinct thoughts, nor is there a step-by-step development of a logical–legal argument-to-conclusion. Instead, patterns appear, interweave, coexist; and the sequencing is made inoperative. Conclusion, premises, feelings, and insights coexist in a manner defying temporal order.

Some neuroscientists have postulated the association of these two modes of awareness with the left and right hemispheres of the brain.[10] Different kinds of activities are then shown to be associated with one or the other of the hemispheres: arithmetic, language, and logical reasoning, are primarily left-brain functions while music and meditative insights are primarily right-brain activities.

If this is so, the total perception of a quantum mechanical system in terms of superposition could be a right-brain function: we may already be perceiving such systems but simply lack the language through which such perceptions could be conveyed. The Wigner distribution function (as well as the Kano and the Sudarshan–Glauber distribution functions) are the incomplete rendering of this perception.

For a long time physics has veered away from the subjective and the feeling level of awareness as being inexact and unsuitable for scientific study. The attempts at the total perception of quantum systems as well as the personal experience of many creative scientists suggest that this is an unnecessary restriction on physics. If all that exists and has dynamics is the domain of physics, why should modes of awareness be an exception?

Acknowledgments

The ideas discussed in this paper owe a great deal to discussions that I have had with many people whose ideas have either influenced me or pointed out the need to deviate from the orthodox view or both. These include: Yakir Aharanov, Bryce de Witt, David Bohm, Lawrence Domash, Gordon Globus, Werner Heisenberg, Brian Josephson, Ravi Kapur, Herman Mathews, Baidyanath Misra, Joseph Moyal, Narasimha Mukunda, Asher Peres, Marthanda Varma, John Wheeler, and Eugene Wigner.

REFERENCES AND NOTES

1. P. A. M. Dirac, *Principles of Quantum Mechanics* (Oxford University Press, New York, 1958), Chaps. 4 and 5; J. E. Moyal, *Proc. Cambridge Phil. Soc.* **45**, 99 (1949); T. F. Jordan and E. C. G. Sudarshan, *Rev. Mod. Phys.* **33**, 515 (1961); G. S. Agarwal and E. Wolf, *Phys. Rev. D* **2**, 2161 (1970).
2. I. E. Segal, *Ann. Math.* **48**, 930 (1949); E. C. G. Sudarshan, in *Lectures on Theoretical Physics*, Vol. 2 (W. A. Benjamin, New York, 1961).
3. J. E. Moyal, *Proc. Cambridge Phil. Soc.* **45**, 99 (1949).
4. E. P. Wigner, *Phys. Rev.* **40**, 749 (1932); *Perspectives in Quantum Theory*, W. Yourgrau and A. van der Merwe, editors (Dover, New York, 1979).
5. Y. Kano, *J. Math. Phys.* **6**, 1913 (1965).
6. E. C. G. Sudarshan, *Phys. Rev. Lett.* **10**, 277 (1963).
7. R. J. Glauber, *Phys. Rev.* **130**, 2529 (1963); C. L. Mehta and E. C. G. Sudarshan, *Phys. Rev.* **138B**, 274 (1965).
8. E. C. G. Sudarshan, *Physica* **96A**, 315 (1979).
9. N. Bohr, in *Albert Einstein: Philosopher Scientist*, P. A. Schilpp, editor (Tudor, New York, 1949); A. Einstein, B. Podolsky, and N. Rosen, *Phys. Rev.* **47**, 777 (1935).
10. Robert Ornstein, *Psychology of Consciousness* (Freeman, San Francisco, 1963); John Eccles and Karl Popper, *The Self and Its Brain*.

32

Stochastic Electrodynamics

An Overview

M. Surdin
Centre des Faibles Radioactivités
Laboratoire Mixte CNRS-CEA
91190 Gif-sur-Yvette, France

Introduction

As a sequel to the absorber theory of radiation of Wheeler and Feynman,[1] Braffort and Tzara[2] postulated the existence of a universal random electromagnetic field at the absolute zero of temperature—the zero-point field.

Applying thermodynamic considerations such as were used by Wien to establish his famous law, they derived an expression giving the energy spectral density of the zero-point field. For a one-dimensional case they found

$$\epsilon(\omega) = \frac{K\omega^3}{3\pi c^3} \tag{1}$$

where K is a constant having the dimension of action.

The application of Lorentz invariance to the spectrum of the zero-point field yields the same result. However, neither derivation furnishes the numerical value of K.

Braffort and Tzara[2] considered the behavior of a harmonic oscillator in the presence of the zero-point field and found that the energy of the oscillator, to the first approximation, is given by

$$\langle \epsilon \rangle = K\omega_0/2 \tag{2}$$

Although they did not pursue this work, it may be considered that this paper gave birth to stochastic electrodynamics (SED).

Without knowledge of the preceding paper, Marshall in two papers[3] derived independently Eqs. (1) and (2).

A succession of papers followed, where various problems in microphysics were considered.

Braffort *et al.*[4] considered the behavior of the harmonic oscillator in the presence of the zero-point field and derived the following expression for the average energy, to the second order of approximation:

$$\langle \epsilon \rangle = \frac{K\omega_0}{2}\left[1 - \frac{1}{2\pi}\left(\frac{\omega_0}{\omega_M}\right)\log\left(\frac{\omega_0}{\omega_M}\right)\right] \tag{3}$$

where $\omega_M = 3mc^3/2e^2$, m being the mass of the electron and e its electric charge. The second term in Eq. (3) represents the Lamb shift.

Surdin *et al.*[5] obtained Planck's law for black-body radiation.

Braffort and Taroni[6] considered the action of the zero-point field on a free electron moving in a uniform magnetic field H. They found that the energy of the electron is given by

$$\langle \epsilon \rangle = \frac{K\omega_H}{2}\left[1 - \frac{2}{\pi}\left(\frac{\omega_H}{\omega_M}\right)\left(1 + \log\frac{\omega_H}{\omega_M}\right)\right] \tag{4}$$

where $\omega_H = eH/mc$; the second term in Eq. (4) corresponds to the magnetic polarizability and the third is a correction which has its counterpart in quantum mechanics (QM).

Considering the zero-point field as Markovian, Surdin derived[7] Schrödinger's equation. This derivation was further elaborated by Claverie and Diner[8] and generalized by de la Peña-Auerbach and Cetto.[9]

Boyer[10] derived the London–van der Waals force between a neutral polarizable particle and a conducting wall and between two neutral polarizable particles.

Considering nuclear forces of strong interaction, Surdin obtained[11] an expression for the potential reminiscent of Yukawa's potential.

It is not the purpose of the present paper to proceed with a detailed survey of the papers published during the last two decades. The preceding examples are meant as an illustration of the results obtained in SED. All these results have their counterpart in QM if one sets $K = \hbar$.

Considering the topics of the papers cited above, it appears as though their authors' original purpose was to develop SED (a classical theory) as a means for deepening our understanding of the relation between classical and quantum physics, to alleviate the known difficulties concerning the axiomatics and interpretation of QM[12] and, eventually, to replace QM (even as they draw inspiration from it) as a physical theory.

Such an avenue of research, involving as it were a close adherence to the development of QM and full mathematical rigor, is quite legitimate.

However, bearing in mind the knowledge[13] that not all the results of SED will have their counterpart in QM and, particularly, the fact that SED has its own approach to the physical world, another avenue of research commands attention.

The main motivation for this line of research stems from the fact that the zero-point field is a *universal field*. Thus, all physical events should be considered as resulting from the action of this field. For instance, the emission or the absorption of em radiation by a hydrogen atom depend on the zero-point field, i.e., on the state of the universe. This approach should thus comprise considerations in such domains as cosmology, astrophysics, macrophysics, as well as microphysics.

In what follows both avenues of research will be explored.

As past experience with new theories has shown that one may reap numerous interesting results starting from physical considerations and using simple, albeit adequate mathematics, this is the method we pursue below. It is possible, however, that further developments will call for a more sophisticated mathematical attack.

The Zero Point Field

Consider a large cavity of volume V, bounded by perfectly reflecting walls, containing thermal radiation. The volume of this cavity may be changed adiabatically. Let $u(\nu,T)$ be the energy spectral density of radiation contained in the cavity. Here the interest is focused on the spectral density $u(\nu,T = 0)$ at $T = 0$. It is assumed that the electromagnetic nature of radiation at $T = 0$ is the same as at $T \neq 0$.

As a consequence of an adiabatic change dV of the volume V, the frequency ν of a wave packet is changed, due to Doppler effect, by $d\nu$, such that[14]

$$d\nu/\nu = -\tfrac{1}{3}dV/V \tag{5}$$

One may also consider the following argument: The radiation contained in the cavity is distributed according to a pattern of standing waves depending on the volume of the cavity. For a small relative variation of the volume of the cavity dV/V, a relative variation in its linear dimensions $\tfrac{1}{3}dV/V$ ensues. The standing wave pattern adjusts itself to the new volume conditions so that the relative variation of the wave length of a wave packet $d\lambda/\lambda$ is the same as the relative variation of the linear dimensions of the cavity; Eq. (5) for the frequency variation follows.

As the energy content of the cavity remains constant, whatever the variation of its volume, one obtains

$$du/u = -dV/V \tag{6}$$

Combining Eqs. (5) and (6), one gets

$$d\nu/\nu = \tfrac{1}{3}du/u \tag{7}$$

Integration of Eq. (7) yields

$$u(\nu,T = 0) = C^t\nu^3 \tag{8}$$

Equation (8) is the same as Eq. (1), where, for further convenience, in the three-dimensional case, one sets for the arbitrary integration constant $C^t = K/\pi c^3$.

Considering the full treatment of thermal radiation at temperature T, Fermi[15] obtains

$$3u = \nu \partial u/\partial \nu + T \partial u/\partial T \tag{9}$$

For $T = 0$ one regains Eq. (7).

The gist of the preceding is to show that the exponent 3 in Eq. (1) is due to the fact that the physical universe has three spatial dimensions. This conclusion is upheld in general relativity.[16]

Reconsidering Eq. (1), it is not conceivable that $\epsilon(\omega)$ becomes infinite as ω^3. There must exist some mechanism which prevents this tendency of $\epsilon(\omega) \to \infty$ as $\omega \to \infty$. In accordance with the absorber theory of radiation, this mechanism is the creation of a pair of harmonic oscillators. Each component of the pair behaves as an independent harmonic oscillator exchanging energy with the zero-point field.[17,18a]

To ascertain this, consider the equation of motion of a harmonic oscillator in the presence of the zero-point field:

$$-\frac{2e^2}{3mc^3}\dddot{x} + \ddot{x} + \omega_0^2 x = \frac{e}{m} E(t) \tag{10}$$

where $E(t)$ is the electric component of the zero-point field and dots indicate differentiation with respect to time. The first term corresponds to the radiation (reaction) force of the harmonic oscillator. Let

$$E_r(t) = -\frac{2e}{3c^3}\dddot{x} \tag{11}$$

Then the energy density, if one considers both electric and magnetic components, is

$$W = \frac{1}{4\pi} E_r^2(t) = \frac{1}{4\pi} \cdot \frac{4e^2}{9c^6} \dddot{x}^2 \tag{12}$$

The average density is then

$$\langle W \rangle = \frac{e^2}{9\pi c^6} \langle \dddot{x} \rangle^2 = \frac{e^2 \cdot \omega^4}{9\pi c^6} \langle \dot{x}^2 \rangle \tag{13}$$

To derive the last equation, use was made of the probability laws of distribution.[17] The spectral density is

$$w(\omega) = \frac{d}{d\omega} \langle W \rangle = \frac{4e^2\omega^3}{9\pi c^6} \langle \dot{x}^2 \rangle \tag{14}$$

But $m\langle \dot{x}^2 \rangle = K \cdot \omega_0/2$, hence

$$w(\omega) = \frac{4e^2\omega^3}{9\pi c^6} \cdot \frac{K\omega_0}{2m} \tag{15}$$

The maximum value of the spectral density of the reaction field $w(\omega)$ cannot be larger than the spectral density of the zero-point field $\epsilon(\omega)$. Equating both values gives the maximum frequency at which the harmonic oscillator exchanges energy with the zero-point field:

$$w(\omega) = \frac{4e^2K\omega_0\omega^3}{9\pi c^6 2m} = \epsilon(\omega) = \frac{K\omega^3}{3\pi c^3} \tag{16}$$

hence

$$\omega_0 = \frac{3mc^3}{2e^2} = \omega_M = 1.6 \times 10^{23}\ \mathrm{s}^{-1}$$

This means that within the framework of SED no em radiation at frequency higher than ω_M may exist.

The following comments are in order here:

- The Lamb shift in Eq. (3) and the corrective terms in Eq. (4) are the result of a cutoff at ω_M.
- The concept of materialization of em radiation at frequencies $\omega \geq \omega_M$ led to the calculation[17] of the numerical value of K and to the derivation of an expression of the potential of nuclear forces of strong interaction similar to that of Yukawa's.[11]
- The spectral density of the zero-point field Eq. (1) satisfies Lorentz invariance in empty space. The cutoff at ω_M, which would destroy the Lorentz invariance, is obtained in matter-filled space. This may be considered as an indication that Maxwell's equations might not be applicable and should be amended at such high frequencies in the presence of matter.
- Statistical considerations of hadron distribution led Hagedorn and Ranft[19] to envisage a maximum universal temperature corresponding to ω_M.
- The spectral density given by Eq. (1) does not correspond to a Markov process. However, in many cases and particularly in that of the harmonic oscillator, due to the narrowness of its bandwidth, the underlying process will be taken to be Gaussian and sometimes as white noise. The last equation in Eq. (13) is the result of a Gaussian process.
- Whereas for a process satisfying Eq. (1) an autocorrelation function does not exist, one may obtain one when a cutoff at ω_M is considered. Simple calculation yields

$$C_{zz}(\tau) = \frac{2K}{\pi c^3}\int_0^{\omega_M} \omega^3 \cos\omega\tau\, d\omega$$
$$= \frac{2K}{\pi c^3}\left(\frac{\omega_M^3 \sin\omega_M\tau}{\tau} + \frac{3\omega_M^2 \cos\omega_M\tau}{\tau^2} - \frac{6\omega_M \sin\omega_M\tau}{\tau^3} - \frac{6\cos\omega_M\tau}{\tau^4} + \frac{6}{\tau^4}\right)$$

One verifies that $C_{zz}(0)$ has a finite limit

$$C_{zz}(0) = \frac{2K}{\pi c^3}\cdot\frac{\omega_M^4}{4} = \frac{2K}{\pi c^3}\int_0^{\omega_M} \omega^3\, d\omega = \frac{K}{\pi c^3}\cdot\frac{\omega_M^4}{2}$$

as it should. For large τ, $C_{zz}(\tau)$ behaves as if it were the output of a low-pass filter with a cutoff at ω_M and the input is white noise.

It is interesting to compare the energy content of a wave packet at frequency ν_0 and that of the zero-point field centered on the same frequency. Consider a wave packet centered on frequency ν_0 of duration Δt; then the coherence length is

$$l = c\Delta t = c/\Delta\nu$$

where $\Delta\nu$ is the bandwidth of the packet. The coherence volume is

$$v = \frac{4\pi}{3 \times 8} \cdot \frac{c^3}{(\Delta\nu)^3} = \frac{\pi}{6} \cdot \frac{c^3}{(\Delta\nu)^3}$$

The energy density in the packet is then

$$\Delta_{\nu 0} = \frac{6 \times 2\pi K\nu_0(\Delta\nu)^3}{\pi C^3} = \frac{12K\nu_0(\Delta\nu)^3}{c^3}$$

The energy density of the zero-point field centered on the same frequency ν_0 and for the same bandwidth $\Delta\nu$ is

$$\Delta_z = \frac{2K(2\pi)^3\nu_0^3\,\Delta\nu}{c^3}$$

where both electric and magnetic components are accounted for. Hence

$$\frac{\Delta_z}{\Delta_{\nu 0}} = \frac{16\pi^3\nu_0^2}{12(\Delta\nu)^2} \cong 4\pi^2 \frac{\nu_0^2}{(\Delta\nu)^2}$$

This ratio depends only on the relative bandwidth of the radiation considered. The corresponding fields are in the ratio

$$\frac{E_z}{E_{\nu 0}} \cong 2\pi \frac{\nu_0}{\Delta\nu}$$

For optical frequencies, taking $\lambda_0 = 5000\text{Å}$ and $\Delta\lambda = 0.25\text{Å}$, one has

$$\frac{\nu_0}{\Delta\nu} = \frac{\lambda_0}{\Delta\lambda} = 2 \times 10^4 \quad \text{and} \quad E_z/E_{\nu 0} \cong 1.2 \times 10^5$$

Concluding this section one may state that the unique postulate of SED is the existence at the absolute zero of an all-pervading universal fluctuating em field. The properties of this field are such that it may be considered as a continuous extension of the classical thermal radiation which leads to the spectral density given by Eq. (1). A modification of this expression is obtained by considering the maximum frequency ω_M above which no em radiation exists. The intensity of the zero-point field is very large when compared to the intensities of electric fields involved in atomic phenomena. Local modifications of the spectral distribution of the zero-point field, due to events such as atomic emission or absorption, may occur. These modifications are compensated so that the spectral density remains on the average that given by Eq. (1) with a cutoff at ω_M.

Microphysics

The harmonic oscillator is a favorite of all new theories. In SED, however, as it is practically the only system where a full solution is available, it plays a specially important role, as will be seen hereafter.

As an example of the methods used in SED, in what follows the harmonic oscillator is treated in some detail.

Consider a nonrelativistic harmonic oscillator where the force $e\dot{\mathbf{x}} \wedge \mathbf{B}(t)$ may be neglected; the equation of motion is then

$$-\frac{2e^2}{3c^3}\dddot{x} + m\ddot{x} + kx = eE(t) \tag{17}$$

where $E(t)$ and $B(t)$ are the electric and magnetic components of the zero-point field. $E(t)$ is a stationary random function of zero mean value and the spectral density of which is given by Eq. (1). Let

$$\tau = \frac{1}{\omega_M} = \frac{2e^2}{3mc^3}, \qquad \omega_0^2 = \frac{k}{m}, \quad \text{and} \quad A(t) = \left(\frac{e}{m}\right) E(t)$$

then Eq. (17) becomes

$$-\tau\dddot{x}(t) + \ddot{x}(t) + \omega_0^2 x(t) = A(t) \tag{18}$$

In analogy with rational mechanics, one defines electrokinetic momentum and energy such that

$$\begin{aligned} P_{ec} &= m\dot{x} - \frac{2e^2}{3c^3}\ddot{x} + e\int E(t)\,dt \\ \epsilon_{ec} &= P_{ec}^2/2m \\ \epsilon_{tot} &= \epsilon_{pot} + \epsilon_{ec} \end{aligned} \tag{19}$$

To obtain the average values $\langle\epsilon_{pot}\rangle$ and $\langle\epsilon_{ec}\rangle$, Bernamont's method is applied: Consider Eq. (18) at time t and at time $t + \theta$, form the product of the left-hand side members and equate the result to the product of the right-hand side members; then perform an integration over an infinite range of t. A differential equation of autocorrelation functions is thus obtained, viz.,

$$\tau^2\chi^{(6)}(\theta) + \chi^{(4)}(\theta) - 2\omega_0^2\chi^{(2)}(\theta) + \omega_0^4\chi(\theta) = \langle A(t)A(t+\theta)\rangle \tag{20}$$

where $\langle x(t)x(t+\theta)\rangle = \chi(\theta)$ is the correlation function of $x(t)$. Forming the Fourier transform of both members of Eq. (20), one finds

$$\xi(\omega) = \frac{(e^2/m^2)\epsilon(\omega)}{(\omega^2 - \omega_0^2)^2 + \tau^2\omega^6} \tag{21}$$

with

$$\mathcal{F}[\langle E(t)E(t+\theta)\rangle] = \epsilon(\omega) \quad \text{and} \quad \mathcal{F}[\chi(\theta)] = \xi(\omega)$$

Hence

$$\begin{aligned}\langle\epsilon_{\text{pot}}\rangle &= \frac{1}{2}k\langle x^2\rangle = \frac{1}{2}k\int_0^\infty \xi(\omega)\,d\omega \\ \langle\epsilon_{\text{ec}}\rangle &= \frac{P_{\text{ec}}^2}{2m} = \left(\frac{k^2}{2m}\right)\left[\int_0^t x\,dt\right]^2 = \left(\frac{k^2}{2m}\right)\int_0^\infty \frac{\xi(\omega)\,d\omega}{\omega^2}\end{aligned} \tag{22}$$

The nonrelativistic case corresponds to $\omega/\omega_M \ll 1$; then, performing the integrations of Eqs. (22) and considering the first order of ω/ω_M approximation, one obtains

$$\begin{aligned}\langle\epsilon_{\text{pot}}\rangle &= \frac{K}{4}\omega_0\left(1-\frac{1}{\pi}\frac{\omega_0}{\omega_M}\log\frac{\omega_0}{\omega_M}\right) \\ \langle\epsilon_{\text{ec}}\rangle &= \frac{K}{4}\omega_0\left[1-\frac{1}{\pi}\left(\frac{\omega_0}{\omega_M}\right)^2\log\frac{\omega_0}{\omega_M}\right]\end{aligned} \tag{23}$$

Hence

$$\langle\epsilon_{\text{tot}}\rangle = \langle\epsilon_{\text{pot}}\rangle + \langle\epsilon_{\text{ec}}\rangle = \frac{K}{2}\omega_0\left(1-\frac{1}{2\Pi}\frac{\omega_0}{\omega_M}\log\frac{\omega_0}{\omega_M}\right) \tag{24}$$

An analogous treatment of the motion of an electron in a uniform external magnetic field H leads to Eq. (4).

Bearing in mind the fact that for a nonrelativistic harmonic oscillator one has $\omega/\omega_M \ll 1$ and also $\omega_0/\omega_M \ll 1$, the equation of motion Eq. (18) may be approximated by

$$\ddot{x} + \tau\omega_0^2\dot{x} + \omega_0^2 x = A(t) \tag{25}$$

Under these conditions (very narrow bandwidth) one may consider $A(t)$ as white noise. Then the probability $f(x,\dot{x})$ is found to be

$$f(x,\dot{x}) = A\exp\left[-\frac{m(\dot{x}^2+\omega_0^2x^2)}{K\omega_0\left(1-\frac{1}{\pi}\frac{\omega_0}{\omega_M}\log\frac{\omega_0}{\omega_M}\right)}\right] \tag{26}$$

where A is a normalizing constant. Within this approximation, one finds

$$\langle\epsilon_{\text{tot}}\rangle = \frac{K}{2}\omega_0\left(1-\frac{1}{\pi}\frac{\omega_0}{\omega_M}\log\frac{\omega_0}{\omega_M}\right)$$

If one considers, as does Heisenberg,[21] that the uncertainty in the position δq is given by the value x_1 of x at which $f(x)$ is e^{-1} of its maximal value and that the uncertainty in the momentum is δp given by the value $m\dot{x}_1$ of $m\dot{x}$ for which $f(\dot{x})$ is e^{-1} of its maximal value, then one has, neglecting the corrective term in Eq. (26),

$$x_1^2 = K/m\omega_0, \qquad m^2\dot{x}_1^2 = mK\omega_0$$

hence

$$m\dot{x}_1 x_1 = K, \qquad \text{i.e.} \quad \delta p \cdot \delta q = K \tag{27}$$

Considering now an ensemble of harmonic oscillators in equilibrium with radiation at temperature T, one has to add to the classical $\langle \epsilon_{cin} \rangle = kT/2$ and $\langle \epsilon_{pot} \rangle = kT/2$ the mean energy of interaction between oscillators due to the zero-point field. Such considerations[5] lead to Planck's law for black body radiation, viz., the average energy of the harmonic oscillator becomes

$$\langle \epsilon_{tot}(\omega,T) \rangle = \frac{K\omega}{2}\left(\frac{2}{e^{K\omega/kT} - 1} + 1\right) \tag{28}$$

For purposes of comparison of the distribution of Eq. (28) with the Gibbs distribution of the ensemble of harmonic oscillators, viz.,

$$\langle \epsilon(\omega,T) \rangle = \frac{\Sigma_n E_n e^{-E_n/kT}}{\Sigma_n e^{-E_n/kT}} \tag{29}$$

one sets $K\omega/kT = x$, then the expansion of $1/(e^x - 1)$ in powers of e^x has a well-known unique solution[22]

$$\frac{1}{e^x - 1} = \frac{e^{-x} + 2e^{-2x} + 3e^{-3x} + \cdots}{1 + e^{-x} + e^{-2x} + e^{-3x} \cdots}$$

using this expression and taking into account the last term in the right-hand side of Eq. (28) and rearranging, we find

$$\langle \epsilon(\omega,T) \rangle = \frac{\Sigma_{n=0}(n + 1/2)K\omega e^{-(n+1/2)K\omega/kT}}{\Sigma_{n=0} e^{-(n+1/2)K\omega/kT}}, \qquad n = 0, 1, 2, 3 \ldots \tag{30}$$

comparing Eqs. (29) and (30), one obtains for the higher energy levels of the harmonic oscillator

$$\epsilon_n = (n + 1/2)K\omega, \qquad n = 0, 1, 2, 3, \ldots \tag{31}$$

The above result may be used to derive the expression of ϵ_n for the excited states of the hydrogen atom.

In classical mechanics the orbit of a body moving in a central force field is planar. Consider in the plane of the orbit two orthogonal axes 0x, 0y. The motion of an electron in the Coulomb field $-e^2/r$ in the presence of the zero-point field can be considered as the resultant of the motion of two one-dimensional harmonic oscillators, having the same resonant frequency, oscillating independently along the two orthogonal axes. Let $\omega_0^2 = e^2/mr_m^3$ be the resonant frequency; r_m is the "average radius" of the orbit. As was noticed above, owing to the fact that the resonance curve of the oscillator is very narrow, the average kinetic and potential energies are the same, so that

$$\begin{aligned} (n_1 + 1/2)K\omega_0 &= m\omega_0^2\langle x^2 \rangle \\ (n_2 + 1/2)K\omega_0^2 &= m\omega_0^2\langle y^2 \rangle \end{aligned} \tag{32}$$

Hence

$$(n_1 + n_2 + 1)K/m\omega_0 = \langle r_n^2 \rangle \tag{33}$$

Since the two harmonic oscillators are independent, one may write

$$n_1 + n_2 = n, \qquad n = 0, 1, 2, 3 \ldots \tag{34}$$

Replacing (34) in (33) and ω_0 by its value, one obtains

$$r_n = (n + 1)^2 K^2/e^2 m, \qquad n = 0, 1, 2, 3 \ldots \tag{35}$$

which is the radial quantification relation in QM.

The concept of the harmonic oscillator is paramount in two further cases, viz., the nuclear forces of strong interaction, where it is shown that this force is of electromagnetic origin, and the de Broglie wave.

For nuclear forces of strong interaction two cases are considered:

(a) A nucleon in the vicinity of a conducting wall. The zero-point field induces on the nucleon an electric dipole which oscillates randomly. The oscillating dipole interacts with its image in the conducting wall; the resulting interaction modifies the interaction of a nucleon with the nucleus (the wall).

(b) The interaction of two nucleons. The induced dipole on nucleon A induces a dipole on nucleon B which interacts with the dipole induced on nucleon A. This modifies the interaction between two nucleons.

Performing the calculations, where the dipoles are represented as harmonic oscillators, one obtains[11] relations where the interactions (a) and (b) are represented by integrals over the whole domain of frequencies ω. If one uses the cutoff at ω_M, the result is

$$U_W \cong -\frac{3.5}{4\pi} K\omega_M \cdot \frac{1}{(2R\omega_M/c)} e^{-2R\omega_M/c} \left[1 + \frac{2}{(2R\omega_M/c)} + \frac{2}{(2R\omega_M/c)^2} \right] \tag{36}$$

for the nucleon–wall configuration, where R is the distance of the center of the nucleon to the wall.

Let $V_0 = (3.5/4\pi)K\omega_M$, then $V_0 \cong 29.5$ MeV; and $\beta = c/2\omega_M$, then $\beta \cong 0.9 \times 10^{-13}$ cm. For the nucleon–nucleon configuration one has

$$U_N \cong \frac{10.6}{8\pi} K \cdot \omega_M \frac{1}{(D\omega_M/c)^2} e^{-2D\omega_M/c} \times \left[1 + \frac{2}{(D\omega_M/c)} + \frac{5}{(D\omega_M/c)^2} + \frac{6}{(D\omega_M/c)^3} + \frac{3}{(D\omega_M/c)^4} \right] \tag{37}$$

where D is the distance between centers of the nucleons.

It is of interest to compare U_W to U_N for the minimal distance of approach for the nucleon–wall and nucleon–nucleon configurations. If one considers the minimal distance as the "radius" of the nucleon $R_0 = 2\beta$ and $D_0 = 4\beta$, one finds $U_W/U_N \cong 6$. This fact is interpreted as follows: If two nucleons are positioned in the proximity of a nucleus then their maximum mutual energy of interaction is negligible, in the first approximation, as compared to their energy of interaction with the nucleus. This property is equivalent to the "saturation of the binding energy."

The preceding is valid whether the nucleons are electrically charged or not; there is no distinction between protons and neutrons. The potential of Eq. (36) becomes that of Yukawa's for $R \gg 2c/\omega_M$.

It is of interest to show that the de Broglie wave derives its origin from the zero-point field.

Consider the motion of a particle of mass m and let its velocity, called the longitudinal velocity, at instant t be v. The zero-point field acts on the particle which behaves as a harmonic oscillator. Consider first a charged particle; under the action of the zero-point field the particle is displaced, thus polarizing the medium in which it moves. The medium reacts on the particle in a more or less complicated manner depending on the boundary conditions of the medium. However, for small displacements, the force acting on the particle is proportional in the first approximation to its displacement x, and tends to restore its dynamical equilibrium. The same mechanism is valid for a neutral particle, since the zero-point field induces on the particle an electric dipole, which is then displaced due to differential forces acting on the separated charges of the dipole.

Let v_0 be the component along an axis of the maximum velocity acquired by the particle under the action of the zero-point field. As the width of the resonance is very narrow, one may equate the average kinetic and potential energies:

$$\tfrac{1}{2}mv_0^2 = \omega_0^2\langle x^2\rangle \tag{38}$$

It is supposed that to account for the magnetic component of the zero-point field energy contribution equal to that of the electric component is considered. Thus

$$\frac{1}{2}\,m\omega_0^2\left(\frac{1}{2}\frac{v_0^2}{\omega_0^2}\right) = \frac{1}{2}\,K\omega_0 \tag{39}$$

The relative width of the resonant line is given by

$$\Gamma/\omega_0 = \frac{2}{3}\,\frac{e^2}{Kc}\,\frac{v_0^2}{c^2} \tag{40}$$

Consider three axes moving with the particle and having their origin at the center of the particle. The longitudinal axis coincides with the direction of the longitudinal velocity v at instant t, the other two axes, mutually perpendicular, are in a plane passing through the particle and perpendicular to the longitudinal axis. One prescribes the condition that v_0 is such that the average kinetic energy due to the zero-point field is equal to the kinetic energy of the longitudinal motion. Hence

$$\tfrac{1}{2}\langle\tfrac{1}{2}mv_0^2\rangle = \tfrac{1}{2}mv^2 \tag{41}$$

Combining Eqs. (39) and (41) and setting $\lambda = 2\pi v/\omega_0$, one obtains

$$K = \frac{v^2m}{\omega_0} = \frac{\lambda vm}{2\pi} \quad \text{or} \quad \lambda = \frac{2\pi K}{mv} = \frac{2\pi K}{p} \tag{42}$$

Since $2\pi K = h$, Eq. (42) is de Broglie's equation. Equation (42) has the following physical interpretation: The harmonic oscillator, the "center" of which moves with velocity v, emits and receives an em wave of wavelength λ. This wave accompanies the particle with velocity v.

It is interesting to note that the electric charge does not enter Eq. (42), which means, as stated above, that the moving particle may be electrically charged or may be neutral.

Interesting consequences, such as the calculation of the linewidth and the explanation of Young's interferences, were considered.[13]

To derive Schrödinger's equation, a somewhat different mathematical technique is used.

Consider the one-dimensional generalized Langevin equation

$$\ddot{x} + \beta\dot{x} - \frac{1}{m} \cdot F(x,t) = A(t) \tag{43}$$

where $F(x,t)$ is an external field of force and $A(t) = (e/m)E(t)$, $E(t)$ being the electric component of the zero-point field. The next step of the derivation is to write the Fokker–Planck equation (FPE) giving the probability $f(x,p,t)$ with $p = m\dot{x}$. The FPE for a process which is not Markovian, as is the underlying process of $A(t)$, is somewhat involved and necessitates a sophisticated mathematical treatment. However, when using such treatment and in order to obtain finite results, one has at some state of the development to resort to approximations.[9] Here, while obtaining the same useful result, one may simplify the derivation by noting, as has been done above, that, due to the smallness of $\tau = 2e^2/3mc^3$, the resonant curve of any system describable by Eq. (43) is very narrow. One can then consider $A(t)$ as Markovian; the corresponding FPE is then[23]

$$\frac{\partial f}{\partial t} + \frac{p}{m} \cdot \frac{\partial f}{\partial x} + F(x,t)\frac{\partial f}{\partial p} = \beta\frac{\partial}{\partial p}(fp) + \frac{Sm^2}{2} \cdot \frac{\partial^2 f}{\partial p^2} \tag{44}$$

where

$$S = \lim_{\Delta t \to \infty} \frac{1}{\Delta t} \left\langle \left\{ \int_t^{t+\Delta t} A(t)\, dt \right\}^2 \right\rangle$$

For the harmonic oscillator, where β and S are calculable, one finds that these quantities are proportional to τ; so that to the first approximation, $\tau = 0$, which is the case considered in QM for Schrödinger's equation, the FPE becomes

$$\frac{\partial f}{\partial t} + \frac{p}{m} \cdot \frac{\partial f}{\partial x} + F(x,t)\frac{\partial f}{\partial p} = 0 \tag{45}$$

Of course, writing $\tau = 0$, one neglects the energy dissipation effects in the system. Full consideration for $\tau \neq 0$ was undertaken.[7]

Consider now the Fourier transform of the distribution function $f(x,p,t)$ given by

$$\rho(x,\xi,t) = \int_{-\infty}^{+\infty} f(x,p,t) \exp(2i\xi p/K)\, dp \tag{46}$$

which, in view if Eq. (27), is a legitimate operation. On using Eq. (46), Eq. (45) becomes

$$\frac{\partial\rho}{\partial t} - \frac{iK}{2m} \cdot \frac{\partial^2\rho}{\partial\xi\partial x} - \frac{2i\xi}{K} F(x,t)\rho = 0 \tag{47}$$

In terms of a new set of variables, namely, $r = x + \xi$ and $r' = x - \xi$, Eq. (47) assumes the form

$$\frac{\partial\rho}{\partial t} - \frac{iK}{2m}\left(\frac{\partial^2}{\partial r^2} - \frac{\partial^2}{\partial r'^2}\right)\rho - \frac{i}{K}(r - r')F(x,t)\rho = 0 \tag{48}$$

Using the mean value theorem, one has

$$(r - r')F\left(\frac{r + r'}{2}, t\right) = \int_{r'}^{r} F(u,t)\, du = -[V(r,t) - V(r',t)]$$

where V is the potential function of the external field force. Equation (48) thus becomes

$$\frac{\partial\rho}{\partial t} - \frac{iK}{2m}\left(\frac{\partial^2}{\partial r^2} - \frac{\partial^2}{\partial r'^2}\right)\rho + \frac{i}{K}[V(r,t) - V(r',t)]\rho = 0 \tag{49}$$

One now considers a function ψ such as

$$\rho(r,r',t) = \psi(r,t)\psi^*(r',t)$$

Equation (49) may then be decomposed into

$$\frac{\partial\psi}{\partial t} - \frac{iK}{2m} \cdot \frac{\partial^2\psi}{\partial r^2} + \frac{i}{K} V(r,t)\psi = 0 \tag{50}$$

and a similar equation for ψ^*, when $r = r'$.

Equation (51) is the QM Schrödinger equation if one sets $K = \hbar$.

For the harmonic oscillator Eq. (50) reads

$$iK\frac{\partial\psi}{\partial t} + \frac{K^2}{2m}\frac{\partial^2\psi}{\partial r^2} - \frac{m\omega_0^2 r^2}{2}\psi = 0 \tag{51}$$

If, instead of the Fourier transform of $f(x,u,t)$ of Eq. (46) where $u = m\dot{x}$, one considers the alternate possibility

$$\Pi(u,\eta,t) = \int_{-\infty}^{+\infty} f(x,u,t) \exp(2ix\eta/K)\, dx \tag{52}$$

with a new set of variables $p = u + \eta$ and $p' = u - \eta$, one obtains the following differential equation for the harmonic oscillator:

$$iK\frac{\partial \Pi}{\partial t} - \frac{p^2 - p'^2}{2m}\Pi + \frac{K^2\omega_0^2 m}{2}\left(\frac{\partial^2}{\partial p^2} - \frac{\partial^2}{\partial p'^2}\right)\Pi = 0 \tag{53}$$

Let $\Pi(p,p',t) = \Phi(p,t)\Phi^*(p',t)$; then Eq. (53) yields

$$iK\frac{\partial \Phi}{\partial t} - \frac{p^2}{2m}\Phi + \frac{K^2\omega_0^2 m}{2}\frac{\partial^2 \Phi}{\partial p^2} = 0 \tag{54}$$

and a similar equation for $\Phi^*(p',t)$.

In QM, Eq. (54) results from an application of the Fourier transform

$$\Phi(p,t) = (2\pi\hbar)^{-1/2}\int_{-\infty}^{+\infty} \psi(r,t)\exp[(-ip\cdot r)/\hbar]\,dr \tag{55}$$

to the Schrödinger equation.

The *derivation* of Schrödinger's equation (50) was obtained by a procedure that starts from the FPE, Eq. (45), which is a classical equation whereby the physical reality is completely described. $f(x,p,t)$ is a probability distribution function of the kind used in classical statistical mechanics; it describes in phase space an ensemble of similarly prepared systems. To obtain Schrödinger's equation in configuration space, one applies the Fourier transform of Eq. (46), which averages over all momenta p, thus suppressing them. The conjugate transformation equation (52) averages over all space variables x, and a conjugate equation to Schrödinger's in momentum space is obtained.

It appears as if in SED probabilities are to be interpreted according to Einstein[24] and, of course, there are no "hidden variables."

The zero-point field concept may be used to impart a new luster to the vexing problem of "hidden variables." Consider[25–27] the group of experiments which were designed to compare the predictions of theories of hidden variables with those of QM. The experimental results appear to be in agreement with QM predictions.

The various experiments use essentially the same setup: An optical source emits in cascade two em radiations of two different wavelengths. Using appropriate filters, the two radiations are separated. On its path each radiation traverses a polarizer and is received by a photomultiplier. One observes the variation of the number of coincidences, at the output of the two photomultipliers, as a function of the angle between the planes of polarization of the two polarizers.

The outstanding effect of these experiments is that the two em radiations acquire on emission a polarization dependent on the setting of the polarizers. This result is in conformity with QM.[28]

In SED, although the theory of atomic radiation and absorption has not been fully formulated, it was shown above how atomic higher-energy levels stem from the zero-point field. Considering the above experiments, the emitting source is "informed" through the zero-point field of the boundary conditions, so that the emission of optical radiations is effected in accordance with the boundary conditions, i.e., the setting of the polarizers. This consideration led Surdin[29] to propose an experiment whereby the boundary conditions, i.e., the setting of the polarizers, may be varied after the emission at the source was effected and before its reception by the photomultipliers.

An interesting problem is to consider the effect of the zero-point field on the entropy of a system. Here, once more, one resorts to the harmonic oscillator, since it is the best-known system.

Consider the sum-over-states

$$Z = \sum_{n=0}^{\infty} e^{-\epsilon_n \beta} \tag{56}$$

where $\beta = 1/kT$. For the harmonic oscillator, according to Eq. (31), $\epsilon_n = (n + \frac{1}{2})K\omega$; then

$$Z = \sum_{n=0}^{\infty} e^{-(n+1/2)K\omega\beta} = e^{-K\omega\beta/2} \frac{e^{K\omega\beta}}{e^{K\omega\beta} - 1} \tag{57}$$

The entropy S is given by

$$\frac{S}{k} = \ln Z - \beta \frac{d}{d\beta} \ln Z \tag{58}$$

For the harmonic oscillator one finds

$$\frac{S}{k} = -\ln\left(1 - e^{-K\omega/kT}\right) + \frac{K\omega/kT}{e^{K\omega/kT} - 1} \tag{59}$$

the same relation that obtains in QM.[30]

For large temperatures T, one has

$$S = k \ln T + k \ln \frac{ek}{K\omega} \tag{60}$$

The first term in the right-hand side of Eq. (60) is the classical term; it is the second term which is the contribution of the zero-point field. One notes that, whereas the entropy increases, i.e., the disorder of the system increases, for increasing T, the entropy decreases when ω increases. The effect of the zero-point field is to increase the order of the system.

At absolute zero, $T = 0$, one has $S = 0$, and the Nernst theorem, the third law of thermodynamics, is satisfied. This would not be the case if the zero-point field were not present.

To conclude this section, a discussion of whether the photon concept is necessary is given. It will be noted that in the preceding the notion of the photon was not evoked, since the theory could proceed without it. Can this attitude be held throughout? In particular, could one in the framework of SED explain the results of experiments specially designed[31] to prove the necessity of the photon concept?

In 1969 Lamb and Scully,[32] on the one hand, and Franken[33] on the other, have shown that the photoelectric effect could be explained by using QM to account for the properties of matter and classical theory for the radiation field, without the use of the photon concept.

Several special experiments were designed (see Ref. 31) to settle the question of the necessity of the photon concept. The most advanced and sophisticated experiment, and also the most convincing, is that of Clauser.[31] Briefly, the setup is the following: An optical source emits in cascade two em radiations—first 1, then 2, on two different wavelengths. Using appropriate filters, the two radiations are separated into two beams. On its path, each beam encounters a beam splitter, and each split beam is then received by a photomultiplier. For beam No. 1 one thus has 1D, for the direct beam, photomultiplier, and 1R, for the reflected beam, photomultiplier. Similar nomenclature is used for the photomultipliers of beam No. 2. One then performs delayed coincidences at the outputs of the photomultipliers. If the number of delayed coincidences is plotted as a function of the delay time, one finds that this number N(1D-2R), N(1R-2D) decrease exponentially with increasing delay time, corresponding to the mean life time of the atomic intermediate state (between 1 and 2). However, N(1D-1R) and N(2D-2R) are constant, independent of the delay time, and correspond to accidental coincidences.

This result is accounted for if one considers that the emitted *photon*, because of the beam splitter, is found either in 1D (2D) or in 1R (2R), but not in both at the same time. This outcome is in conformity with QM and with the photon concept; it is contrary to classical theory, where a beam consisting of a wave packet, when split into two halves by the beam splitter, should be found in 1D and in 1R at the same time.

In this experiment and in similar, albeit simpler, ones, such as the interference experiment with a plane parallel plate, the essential fact is the behavior of a wave packet at the surface of separation of two media—the metallic film in the case of the beam splitter.

The complete theory of such effects in the framework of SED has not yet been fully elaborated. However, the conjecture is that at the surface of separation the behavior—either the passage through or the reflection—of the wave packet depends on the conditions of the electric component of the zero-point field on the separation surface. This conjecture is justified by the order-of-magnitude calculation of the preceding section for the ratio of the zero-point electric field to the electric field of the wave packet. As the phase of the electric component of the zero-point field is random, the probabilities of reflection and of the passage through of the wave packet are equal. Such considerations account for the result of Clauser's experiment without the necessity to call upon the photon concept.

MACROPHYSICS

A model, derived from the zero-point field notion, of the magnetic field of rotating bodies is presented next.

Consider an electrically neutral rotating body. The zero-point field induces electric dipoles in the atoms of the body. The appearance and the disappearance of these dipoles is a random phenomenon. It is considered in what follows that, due to the rotation of the body and for reasons of symmetry, the dipoles align themselves radially in planes parallel to the equator, directed either towards the axis of rotation or in the opposite direction. The resultant magnetic field is a random dipolar field either parallel or antiparallel to the direction of rotation, when the electric dipoles are aligned in the inverse direction or in the direction of the axis of rotation, respectively.

Consider all those electric dipoles, the electric charge center of which are situated, at a given instant, on a cylinder of radius r centered on the axis of rotation. The magnetic field created by these dipoles, rotating around the axis of rotation, is the resultant of two contrary fields—one created by the positive charges, the other by the negative charges of the electric dipoles.

These considerations and the fact that one has to deal with random phenomena, led to the calculation of the correlation function of the magnetic field. One may thus derive formulas giving the mean square of the amplitude of the magnetic field created in the center of the rotating body for various geometrical configurations, such as a thin disk, a sphere, etc.[17]

For a sphere of uniform density rotating around one of its diameters at angular velocity Ω, one obtains

$$\langle H^2 \rangle = \frac{1}{8} \frac{G}{c^2} \frac{M^2 \Omega^2}{R^2} \tag{61}$$

where G is Newton's constant of gravitation, M the mass of the rotating sphere, and R its radius. The value thus calculated, or rather $\pm \langle H^2 \rangle^{1/2}$, corresponds to the amplitude of the magnetic field at the center of the sphere, parallel to the direction of rotation or in the opposite direction.

The analysis of the variation in time of the magnetic field is based on the following remarks: Eq. (61) is obtained when *all* the induced dipoles are aligned in the *same* direction, either towards the axis of rotation or in the opposite direction. Superposed in this equilibrium state are fluctuations wherein dipoles change their direction randomly. These fluctuations occur with a time constant of the order of the em relaxation time T_L of the rotating body. For a sphere of radius R, electrical conductivity σ, and permeability μ, one has

$$T_L = \frac{4}{\pi} \sigma \mu R^2 \tag{62}$$

If, due to fluctuations, a domain of the body has its dipoles directed in the direction opposite to that of the rest of the body and if this domain has sufficient extension, considering the "cooperative" nature of the phenomenon—the tendency to reverse direction is larger the larger the actual number of reversed dipoles—then the ensemble of the dipoles of the body will reverse their direction. This corresponds to the reversal of the direction of the magnetic field. The reversal occurs, as confirmed experimentally, within a very short time interval τ, much shorter than T_L, so that $1/T_L \ll 1/\tau$.

One has here the ingredients of a fluctuating system with two modes subject to cooperative effects. Numerous such systems, subject to cooperative effects, have been considered. Most of them respond to the same mathematical treatment which will be sketched below.

Our conclusions may be summarized as follows:

- The magnetic field of a rotating body is due to a random process, resulting in a dipolar magnetic field pointing either in the direction of rotation or in the opposite direction.
- The absolute values of the mean amplitudes of the magnetic fields in both directions are equal.
- The average value of the field, calculated over a time interval large compared to the em relaxation time T_L, is zero.
- The zero crossings of the magnetic field occur at instants distributed at random, in fact, according to a stationary Poisson distribution.

These considerations are borne out by experimental results on the earth's paleomagnetism.[17]

Consider now the cooperative aspect of the phenomenon. Let $\rho_1 = n_1/n_0$ and $\rho_2 = n_2/n_0$ be the fractions of the total number n_0 of the dipoles directed, respectively, towards the axis of rotation and in the reverse direction. The differential equation of the population evolution relative to this type of problem is well known; it is an equation similar to the "competition equations" of Volterra. For ρ_1 one has

$$\frac{d\rho_1}{dt} = -\frac{1}{T_L}\rho_1 + \frac{1}{\tau}\rho_1\rho_2 \tag{63}$$

the second of Volterra's equations is replaced here by the condition $\rho_1 + \rho_2 = 1$. The first term on the right-hand side of Eq. (63) corresponds to the population evolution at equilibrium, whereas the second term corresponds to the reversal phase and expresses the cooperative nature of the process. The study of Eq. (63) is effected in two stages: First the reversal phase is considered, then the equilibrium phase.

Considering the fact that $1/T_L \ll 1/\tau$, one may replace Eq. (63) by

$$\frac{d\rho_1}{dt} = \frac{1}{\tau}\rho_1\rho_2 \tag{64}$$

using the relation $\rho_1 + \rho_2 = 1$; one gets

$$\frac{d\rho_1}{dt} = \frac{\rho_1}{\tau} - \frac{\rho_1^2}{\tau} \tag{65}$$

For solving this type of problem one usually[34] introduces an order parameter

$$\psi = (\rho_1)^{1/2}\, e^{i\varphi} \tag{66}$$

where Ψ is a function of the time t and, for extended bodies, also of the space coordinates. φ is a phase, a function eventually of the same variables. Here φ takes on the value 0 or π. Equation (65) then becomes

$$\frac{d\Psi}{dt} = \frac{1}{2\tau}\Psi - \frac{1}{2\tau}|\Psi|^2\Psi \tag{67}$$

The equilibrium is obtained for $d\Psi/dt = 0$. The three equilibrium positions correspond to $\Psi = 0$, an unstable equilibrium, and to $|\Psi|^2 = 1$, i.e., $\Psi = \pm 1$. These two solutions correspond to stable equilibrium positions, where the dipoles are directed either toward the axis of rotation or in the opposite direction.

However, Eq. (67) should be completed by the adjunction of a fluctuating term $F(t)$, resulting from the action of the zero-point field, thus

$$\frac{d\Psi}{dt} = \frac{1}{2\tau}\Psi - \frac{1}{2\tau}|\Psi|^2\Psi + F(t) \tag{68}$$

If the random term $F(t)$ did not exist, the system once at a stable equilibrium position will remain there indefinitely. It is due to fluctuations, i.e., to the existence of $F(t)$, that the passage from one stable equilibrium position to the other is made possible.

At equilibrium, the system fluctuates, its evolution equation is Eq. (63), where the second term in the right-hand side is zero; the system responds to fluctuating oscillations with a time constant T_L.

The above considerations were used[35] to interpret the results of an experiment on the magnetic field of rotating bodies. This experiment will here be only briefly described; for a fuller description see Ref. 36.

For evident reasons, the configuration of the rotating body should have a symmetry of revolution. To rotate a sphere presents technological difficulties; a cylindrical form was therefore chosen. It was considered desirable to experiment with two cylinders of widely different materials; thus one of the cylinders (T) was made of a tungsten alloy of density $\delta = 17.6$; the other cylinder (B) was made of bronze of density $\delta = 8.8$; both cylinders had the same geometry. Simple considerations, involving mechanical constraints and the requirement that the peripheral velocity should be lower than the velocity of sound in air at the target angular velocity $\Omega = 2\pi \times 1000$ s^{-1}, led to the choice $R = 5$ cm for the radius of the cylinder. To avoid flexural oscillations of the body, the height of the cylinder was chosen to be $h = 15$ cm.

To rotate such cylinders at the angular velocity given above, air bearings and an air turbine were used (electric motors were banned because of possible magnetic perturbations).

Preliminary considerations, supported by order-of-magnitude calculations, have shown that the direct measurement of the magnetic field presented insuperable difficulties. Another method, simpler and more economical, was used. The existence of the magnetic field was deduced from measurements of its effects. The magnetic field of a rotating body is a dipolar random field, either parallel or antiparallel to the direction of rotation. During a reversal the field decreases from $+H_0$, become zero, and then increases in the opposite direction to become $-H_0$. Barring small fluctuations during a time interval of the order of T_L, the average amplitude of the field remains $-H_0$, then the field goes through the reverse succession of events to become $+H_0$.

In a cylindrical coil, the axis of which coincides with the axis of the rotating cylinder, the reversals of the magnetic field induce an emf in a form of a sequence of short pulses $E(t)$, all of the same amplitude, alternately positive and negative, occurring at instants distributed according to a stationary Poisson distribution.

The method of detection is based on the specificity of the signal $E(t)$ autocorrelation function $C_{EE}(\tau)$. The measuring procedure is the following: When the cylinder is rotating at a constant angular velocity Ω, one plots the total autocorrelation function $C_T(\tau)$ as a function of τ. Then, the cylinder being at rest, one plots the noise autocorrelation function $C_{NN}(\tau)$. The independence of signal and noise leads to

$$C_{EE}(\tau) = C_T(\tau) - C_{NN}(\tau) \tag{69}$$

Formulas were derived which allowed the calculation of the magnetic field from the knowledge of $C_{EE}(\tau)$.

Various possible parasitic effects were considered and taken care of.

As the experimental results are in good agreement with those predicted by the theory and from the internal consistency of results obtained for each cylinder and between cylinders, one may conclude that the existence of the magnetic field of a rotating body was demonstrated by detecting its effects.

ASTROPHYSICS

As an application of the results of the preceding section the magnetic field of celestial bodies will be briefly reviewed.[37]

In the first part of Table I the theoretical and the corresponding observed values of the magnetic fields of some celestial bodies are given, as well as those relative to the laboratory experiment. The second part of the table is devoted to the corresponding values of the planets.

The following observations are of interest:

- The formulas used to calculate $\langle H^2 \rangle$ are derived for bodies of uniform density and uniform rotation.
- The formula for the sphere giving the mean square field at the center is Eq. (61).
- As most of the observed values of the magnetic field are referred to the equator and as Eq. (61) gives the field at the center, it appeared that the theoretical expression of H_{eq} should be of interest. Expressions of the magnetic field created by a rotating sphere of uniform volume distribution of charges are known.[38] It is shown that $H_{eq} = \frac{1}{5}H_0$, where H_0 is the field intensity in the center of the sphere.

TABLE I. Magnetic Fields of Some Rotating Bodies

Body	M (g)	R (cm)	Ω (s^{-1})	$H_0 = \langle H^2 \rangle^{1/2}$ (G)	$H_{eq} = \frac{1}{5}H_0$ (G)	H_{exp} (G)
Neutron star	1.99×10^{33}	3×10^6	12.5	2.5×10^{13}	5×10^{12}	10^{12}–10^{13}
Sun	1.99×10^{33}	6.96×10^{10}	2.8×10^{-6}	2.4×10^2	48	3–50
Galaxy	2.3×10^{44}	3.8×10^{22}	8×10^{-15}	0.6×10^{-6}		10^{-6}–10^{-5}
Experiment cylinder B	1.1×10^4	5.0	2826	10^{-8}		10^{-8}
Mercury	3.3×10^{26}	2.44×10^8	2.04×10^{-6}	8×10^{-3}	1.6×10^{-3}	$(2\text{–}3.5) \times 10^{-3}$
Venus	4.87×10^{27}	6.05×10^8	2.4×10^{-8}	5.9×10^{-4}	1.18×10^{-4}	$\leq 3 \times 10^{-4}$
Earth	5.98×10^{27}	6.378×10^8	7.2×10^{-5}	2.0	0.4	0.31
Mars	6.46×10^{26}	3.4×10^8	7.0×10^{-5}	0.4	0.08	$(6\text{–}60) \times 10^{-5}$
Jupiter	1.9×10^{30}	7.14×10^9	1.8×10^{-4}	1.48×10^2	29.6	11–14
Saturn	5.69×10^{29}	6.0×10^9	1.73×10^{-4}	50	10	0.2–1
Uranus	8.73×10^{28}	2.6×10^9	1.6×10^{-4}	16	3.2	
Neptune	1.03×10^{29}	2.48×10^9	1.1×10^{-4}	13.9	2.8	
Pluto	5.98×10^{26}	3×10^8	1.13×10^{-5}	6.9×10^{-2}	1.38×10^{-2}	
Moon	7.35×10^{25}	1.738×10^8	2.6×10^{-6}	3.3×10^{-3}	6.6×10^{-4}	$<5 \times 10^{-5}$

- The formula for a thin disk (Galaxy) is

$$\langle H^2 \rangle = \frac{2G}{c^2}\frac{M^2\Omega^2}{R^2}$$

- Except for the value H_{exp} for the earth and the experimental cylinder, all the other values are not directly measured.
- The conditions of uniform density and uniform rotation are approximately satisfied in the case of the earth and may be in that of the neutron star; they are satisfied, of course, in the case of the experimental cylinder. Formulas giving $\langle H^2 \rangle$ for nonuniform rotation and density can be obtained. However, since no reliable information is available about the variation of the density and of the angular velocity within the celestial bodies, such calculations were not made. Thus, the agreement between theoretical and experimental results within a factor of 3 is considered to be adequate.
- The formulas giving $\langle H^2 \rangle$ have three parameters, viz., M, R, and Ω; considering the above table, the ranges of the parameters are $M \to 1\text{–}10^{40}$, $R \to 1\text{–}10^{22}$, and $\Omega \to 1\text{–}10^{17}$. There are only a few laws in physics which apply over such extended ranges.
- The agreement between the calculated and the observed values of Table I may be considered as good except for Mars, the Moon, and Saturn. In the case of Mars it was suggested[37] that the magnetic field is reversing or has reversed recently. The discrepancy between calculated and observed values of the magnetic field of the Moon led to interesting considerations concerning the distribution and the dimensions of magcons.[37] As for Saturn, the discrepancy

between the observed and the calculated values has not been completely accounted for and may be attributed to any of the following factors: a nonuniform distribution of mass or of rotation or else the existence of slightly ionized rings, modifying the distribution of the magnetic field as compared to that of a uniform sphere.

Another outcome of the magnetic field of rotating bodies is the theory of the origin of the solar system. The contour of the theory is the following: The nebula has contracted into a central body (the sun) which rotates fast enough to become unstable and eject matter, forming a ring, a sort of parking ring, around the central body. Because of its rotation, the central body acquires a magnetic field. Considering Hoyle's magnetic mechanism,[39] the enhanced magnetic intensity impedes outward ejection of matter from the parking ring. It is only during the reversal of the magnetic field, when the absolute value of its amplitude is low, that matter is ejected to form the first ring around the parking ring. During the period when the magnetic field is constant the radius of the ring increases. The angular momentum it gains by Hoyle's mechanism is at the detriment of the angular momentum of the central body and the parking ring. When the next magnetic field reversal occurs, matter is ejected anew, forming the second ring, which in turn gains angular momentum at the expense of the angular momentum of the central body and the parking ring. The process continues until the last ring (Mercury) is formed. The ejected matter is thus found in several rings. Considering energy fluctuations between rings at equilibrium, the Titus–Bode law is obtained. The computed numerical values for the amplitude of the magnetic field, the average time interval between field reversals, the angular momentum of the different rings and their radii, and the angular momentum of the central body are all in good agreement with observations.

Cosmology

The sole condition required by SED is that once the zero-point field was postulated its spectral density should be given by Eq. (1). This condition is so general that almost any cosmology is eligible.[16] However, from the point of view of SED, the steady-state universe and the bright universe cosmologies are of particular interest.

The zero-point field is inherent in the steady-state universe. Let L be the luminosity of sources of red shift between z and $z + dz$, $z = \nu_e/\nu - 1$, where ν_e is the frequency at the emitter and ν is the frequency at the receiver. Let n be the number density of these sources; then in the steady state model one has[40]

$$\epsilon(\omega)\, d\omega = \frac{2Ln}{H\omega_e^4}\, \omega^3\, d\omega \tag{70}$$

where H is Hubble's constant.

It appears, thus, that the spectral density of all sources in this universe has the same frequency dependence as the zero-point field. Moreover, stipulating some particular conditions to be satisfied by the sources, one may identify[18a] the spectral density given in Eq. (70) with that of Eq. (1). By the way, one notes that K, the constant that gives the scale of the spectral density of the zero-point field, is not a "fundamental" constant, as is $\hbar$, its homologue in QM.

The steady-state model was "abandoned by most investigators today" (1973) not least because it gives no satisfactory account of the 2.7 °K background radiation.[16] However, in 1978–1980 it was shown[18a,b] that one should not reject altogether this cosmology.

To derive Eq. (1), a large cavity, bounded by perfectly reflecting walls, containing radiation was considered. A universe delimited by a perfectly reflecting boundary may be termed a "bright universe," since no radiation escapes through this boundary; viewed from the "outside" this universe appears as a "black hole." If the limiting surface is the Schwarzschild surface Σ, then its radius is

$$R_s = 2GM/c^2 \tag{71}$$

Considering the values of the "radius" and the mass of our known universe, Eq. (71) is approximately satisfied. The bright universe cosmology was considered[18a,41] with interesting results.

Comments

The "maturity" of a theory might be appraised by its unifying powers and by the fewness of the number of adjustable parameters used. SED, the theory presented here, is founded on the sole, quite plausible postulate, viz., the existence of a fluctuating electromagnetic field at the absolute zero of temperature—the zero-point field. Not all the fundamental problems, of which the outstanding one is the full treatment of the hydrogen atom, have as yet found their solution in the framework of SED. The examples presented above show, however, the possibilities of the theory in different domains of physics and its unifying powers. Thus, the gravitational forces are shown[17] to have their origin in the "residue" of the zero-point field; whereas the nuclear forces of strong interaction are derived from the zero-point field where the cutoff frequency ω_M was taken into account. The nuclear forces of strong interaction were considered as the interaction of an oscillating dipole, induced by the zero-point field on a nucleon, with its image in a conducting wall or in another nucleon. For nuclear forces of weak interaction it is conjectured that one has to consider the interaction of two dipoles (harmonic oscillators), oscillating in antiphase, with their image.

The origins of all physical phenomena appear, thus, to stem from the zero-point field.

An em emission of a wave packet by an atom may be considered as the formation of a "coherent" wave packet above the "sea" of em radiation of the zero-point field. However, the formation of a coherent wave packet entails the formation of a complementary coherent packet; for lack of better designation call it an "anti-wave-packet."

The wave packet and its anti-wave-packet, when combined, leave the "sea" undisturbed. Interesting considerations based on such a model give an indication of its potentialities.

SED does not require the use of adjustable parameters. However, in some cases, in order to be able to proceed with calculations, adjustable parameters as well as some secondary hypotheses were used. One such hypothesis is the assumption that all induced dipoles in a rotating body are aligned in planes parallel to the equator, pointing towards the axis of rotation or in the opposite direction. It is conjectured that such hypotheses will not be needed when a more sophisticated approach is used.

Most of the results referred to in this paper were obtained using simple mathematics. But the use of more sophisticated mathematics is not excluded; moreover, it is hoped that mathematical techniques will be developed specially for use in SED.

The development of novel theories, such as SED, present their own peculiar difficulties, which may or may not be solved by their originators. Their dissemination, however, involves editors and journals. It is thanks to editors of the caliber of Wolfgang Yourgrau who combined a spirit of adventure, so indispensable in science, with a vision as progressive as his cultural background was diverse, that such theories reach their proper audience.

References and Notes

1. J. A. Wheeler and R. P. Feynman, *Rev. Mod. Phys.* **17,** 157 (1945).
2. P. Braffort and C. Tzara, *C. R. Acad. Sci. Paris* **239,** 1779 (1954).
3. T. W. Marshall, *Proc. R. Soc. London* **A276,** 475 (1963); *Proc. Cambridge Phil. Soc.* **61,** 537 (1965).
4. P. Braffort, M. Surdin, and A. Taroni, *C. R. Acad. Sci. Paris* **261,** 4339 (1965).
5. M. Surdin, P. Braffort, and A. Taroni, *Nature* **210,** 405 (1966).
6. P. Braffort and A. Taroni, *C. R. Acad. Sci. Paris* **264B,** 1537 (1967).
7. M. Surdin, *Int. J. Theor. Phys.* **4,** 117 (1971).
8. P. Claverie and S. Diner, *Ann. Fond. Louis de Broglie* **1**(2), 73 (1976).
9. L. de la Pena-Auerbach and A. M. Cetto, *Found. Phys.* 8(3/4), 191 (1978); *J. Math. Phys.* **20,** 469 (1979).
10. T. H. Boyer, *Phys. Rev. A* **7,** 1832 (1973).
11. M. Surdin, *C. R. Acad. Sci. Paris* **280B,** 337 (1975).
12. T. Bastin, editor, *Quantum Theory and Beyond* (Cambridge Univ. Press, Cambridge, 1971); W. Yourgrau and A. van der Merwe, editors, *Perspectives in Quantum Theory* (MIT Press, Cambridge, Massachusetts, 1971; and Dover Publ. Inc., New York, 1979).
13. M. Surdin, *Ann. Fond. Louis de Broglie* **4**(2), 139 (1979).
14. E. Fermi, *Thermodynamics and Statistics* (The University of Chicago Press, Chicago, 1966), p. 94.
15. E. Fermi, Ref. 14, p. 95.
16. C. W. Misner, K. S. Thorne, and J. A. Wheeler, *Gravitation* (Freeman, San Francisco, 1973).
17. M. Surdin, *Ann. Inst. Henri-Poincaré* **15**(3), 203 (1971).
18(a). M. Surdin, *Found. Phys.* 8(5/6), 341 (1978).
18(b). M. Surdin, *Found. Phys.* **10**(1/2), 175 (1980).
19. R. Hagedorn and J. Ranft, *Nucl. Phys.* **B48,** 157 (1972).
20. J. Bernamont, *Ann. Phys. (Paris)* **7,** 71 (1937).
21. W. Heisenberg, *Z. Phys.* **43,** 172 (1927); M. Jammer, *The Conceptual Development of Quantum Mechanics* (McGraw Hill, New York, 1966).
22. R. C. Tolman, *Principles of Statistical Mechanics* (The Clarendon Press, Oxford, 1938).

23. S. Chandrasekhar, *Rev. Mod. Phys.* **15,** 1 (1943).
24. A. Einstein, *The Born-Einstein Letters* (McMillan, New York, 1971), letter No. 106; L. E. Ballentine, *Am. J. Phys.* **40,** 1763 (1972).
25. S. J. Freedman and J. F. Clauser, *Phys. Rev. Lett.* **28,** 938 (1972).
26. J. F. Clauser, *Phys. Rev. Lett.* **36,** 1223 (1976).
27. E. S. Fry and R. C. Thomson, *Phys. Rev. Lett.* **37,** 465 (1976).
28. O. Costa de Beauregard, *Ann. Fond. Louis de Broglie* **2**(4), 231 (1977).
29. M. Surdin, *Phys. Lett.* **66A,** 261 (1978).
30. E. Fermi, *Thermodynamics and Statistics* (The University of Chicago Press, Chicago, 1966), p. 87.
31. J. F. Clauser, *Phys. Rev. D* **9,** 853 (1974).
32. W. E. Lamb and M. O. Scully, *Polarisation, Matière et Rayonnement,* Edited by Société Française de Physique, Presses Universitaires de France, Paris, 1968 (Plenum Press, New York, 1969).
33. P. A. Franken, *Atomic Physics* (Proc. Conf. Atomic Phys., June 3–7, 1968) (Plenum Press, New York, 1969).
34. H. Haken, *Rev. Mod. Phys.* **47**(1) (1975).
35. M. Surdin, *Ann. Fond. Louis de Broglie* **5**(2), 127 (1980).
36. M. Surdin, *J. Franklin Inst.* **303,** 495 (1977).
37. M. Surdin, *Nuovo Cimento* **2C,** 527 (1979).
38. E. Durand, *Magnétostatique* (Masson, Paris, 1968), p. 158.
39. F. Hoyle, *Q.J.R. Astron. Soc.* **15,** 28 (1960).
40. S. Weinberg, *Gravitation and Cosmology* (J. Wiley & Sons, New York, 1973), p. 459; R. Hobard, *Found. Phys.* **6,** 473 (1976).
41. M. Surdin, *Spec. Sci. Techn.* **3**(3), 327 (1980).

33

On the Increase of Entropy in the Carleman Model II*

A. Wehrl
Institut für Theoretische Physik
Universität Wien
Vienna, Austria

Statistical mechanics covers a large area within the scientific ouevre of Wolfgang Yourgrau. And perhaps the field that fascinated him most was the problem of *increase of entropy* and the *second law of thermodynamics*. Many papers as well as a well-known textbook[1] are the eloquent testimony to his preoccupation with this subject.

One of the first important steps towards the understanding of the phenomenon of irreversibility was taken by Boltzmann in 1872 when he derived his celebrated transport equation. His equation has become an essential part of statistical mechanics. Its successes "aroused the admiration *and* the suspicion of Boltzmann's contemporaries." And even today, in spite of the efforts of many generations of physicists, there are still many problems left open, and progress is made little by little only.

Yourgrau was always aware of the necessity of investigating the fundamental problems connected with the Boltzmann equation on a sound mathematical basis. Roughly three classes of problems may be distinguished: First, the *derivation* of the equation itself and precise statements concerning the limit in which one may obtain the equations of motion exhibiting irreversible behavior from reversible Hamiltonian dynamics.[4–7] Second, the *mathematical structure*. Up to now, the status of our knowledge about the existence and the behavior of solutions is far from satisfactory. In fact, there are no powerful mathematical tools yet available to handle nonlinear integro-differential equations such as the Boltzmann equation.[8] The third class of problems

*Work supported in part by "Fonds zur Förderung der wissenschaftlichen Forschung in Österreich," Project number 3569.

deals with macroscopic consequences, for instance the *hydrodynamic limit*. Again, there is still much work to do.

In order to get a profound insight into these problems, Yourgrau, in the last period of his life, became interested in a very simple model which was proposed by Carleman in the 1930s but was published only in 1957, eight years after his death.[9] This model is an extremely simplified analog of the Boltzmann equation. It is one-dimensional and the velocities of the molecules are allowed to take two discrete values, $\pm c$, only. Thus the "distribution function" in the Boltzmann equation, $f(\mathbf{x},\mathbf{v},t)$, is replaced by two functions $u_+(x,t)$ and $u_-(x,t)$, denoting the density of particles with velocity $+c$ or $-c$, respectively, at point x and time t. The "gas" is not confined, but x varies over all points of the real line $(-\infty,\infty)$. It is further assumed that there are only two types of interactions, viz., two "+" particles go over into two "−" particles, and *vice versa*, the probability for both processes to occur within one unit of time being the same number σ. Then the Boltzmann equation, which in absence of external forces reads

$$\frac{\partial f(\mathbf{x},\mathbf{v},t)}{\partial t} + \mathbf{v}\nabla f(\mathbf{x},\mathbf{v},t) = \int d^3\mathbf{v}_1\, d\Omega\, |\mathbf{v}-\mathbf{v}_1|\sigma(|\mathbf{v}-\mathbf{v}_1|,\theta)[f'f'_1 - ff_1]$$

(for the notation, cf., e.g., Ref. 11; in particular, f' is the final distribution, i.e., the distribution after a collision) "translates" into a system of two equations:

$$\frac{\partial u_+}{\partial t} + c\frac{\partial u_+}{\partial x} = \sigma(u_-^2 - u_+^2)$$
$$\frac{\partial u_-}{\partial t} - c\frac{\partial u_-}{\partial x} = \sigma(u_+^2 - u_-^2)$$

Of course, the Carleman model is rather unphysical. But the remarkable thing about it is that with its aid one can prove almost all those results which one would like to obtain for the Boltzmann equation itself—as, for instance, the existence of solutions for a wide class of initial conditions or a rigorous treatment of the hydrodynamic limit. (A survey of results was given by Kaper and Leaf.[10]) There is some reason to believe that it should be possible to extend the methods for the Carleman equations to the case of the Boltzmann equation.

Just as for the Boltzmann equation, the H theorem holds for the Carleman model: The quantity

$$-\int (u_+ \ln u_+ + u_- \ln u_-)\, dx$$

never decreases in time.[9] In a short note,[11] published shortly before his death, Yourgrau and the present author have shown that there exists a large class of generalizations of the H theorem: Let f be a concave function, defined on the half-line $(0,\infty)$, which is once continuously differentiable; then, for

$$S_f \equiv \int [f(u_+) + f(u_-)]\, dx$$

one gets

$$\frac{d}{dt} S_f(u) \geq 0$$

Thus, not only does entropy never decrease, but the same is true for all "quasientropies"[12] (or Csiszar's measures of information[13]). As a consequence, all Renyi entropies[14,15]

$$S_\alpha \equiv (1 - \alpha)^{-1} \ln \int (u_+^\alpha + u_-^\alpha)\, dx$$

never decrease. In information-theoretical language, this means that all sensible measures of the lack of information are nondecreasing; or, in other words, information is lost, or chaos is approached, in the strongest possible way.[16]

This phenomenon was first studied, in the quantum-mechanical case, by Uhlmann,[17] later on by the present author,[18] and, in the classical framework, by Ruch and collaborators.[19,20] It is nowadays called "*chaos enhancement*" (an expression due to M. Ingleby and E. Lieb, unpublished). It is known that several types of linear equations (master equations[21,22] and Fokker–Planck[23] equation) exhibit such a behavior. Besides the Carleman model, another example of a *nonlinear* equation showing chaos enhancement was given later on by Crell and Uhlmann.[24]

Some remarks apply to the "quasientropies" $S_f(u)$. First, it is easy to see that

$$\|u\|_\infty \equiv \sup_x \{u_+(x), u_-(x)\}$$

is not increasing. [Let $\alpha \to \infty$ in our previous generalized H theorems; then $S_\infty(u) = -\ln \|u\|_\infty$.] Tartar,[25] by a different method has even shown that $\|u\|_\infty \to 0$. Hence,

$$S_f(u) \to f'(0)$$

(which may of course be $+\infty$).

Second, one has to encounter some existence problems.[12] In order to avoid ugly discontinuities, it is necessary that $f(0) = 0$. Then the set of possible values of $S_f(u)$ for density distributions [i.e., mathematically speaking, for functions $u_\pm(x)$ that are ≥ 0 such that $\int(u_+ + u_-)\, dx = 1$] is the (possibly extended) interval

$$[\lim_{x\to\infty} f(x)/x,\ f'(0)]$$

If both ends of this interval are infinite, then $S_f(u)$ need not exist (and in fact, this is the most frequent case). Therefore, to be careful, let us suppose that $\lim f(x)/x > -\infty$. We shall, for the rest of this paper, call the class of all concave functions, vanishing at 0, with this property and which, in addition, are continuous at 0, "good functions" and denote them by G. If $f \in G$, then $S_f(u)$ exists for all density distributions u but clearly it may still be $+\infty$.

It should be remarked that, in order to prove theorems about G, it is sufficient to prove theorems about the nonnegative "good" functions, which we will denote by

G^+. This is clear from the following lemma and the identity $S_f(u) = S_g(u) + \lambda$, where $g(x) = f(x) - \lambda x$.

Let $f \in G$, and let $\lambda \equiv \lim f(x)/x$. Then $g(x) \equiv f(x) - \lambda x \in G^+$.

The proof is quite simple: g is concave and $g(0) = 0$. Therefore, for $x < y$,

$$g(x)/x \geq g(y)/y \geq \lim_{y \to \infty} g(y)/y = \lambda - \lambda = 0$$

We now want to proceed to a more detailed investigation of chaos enhancement. Our following results do not refer solely to the Carleman model, but are of more general consequence. This means that we no longer want to be restricted to two-component distributions (u_+, u_-), but will rather consider arbitrary finite or infinite multicomponent distributions $u = (u_1, u_2, \ldots, u_n)$ or $u = (u_1, u_2, \ldots)$. To fix the notations: $u \geq 0$ means $u_i \geq 0$ for all i;

$$\|u\|_1 \equiv \sum_i \int |u_i| \, dx, \qquad \|u\|_\infty \equiv \sup_{i,x} |u_i(x)|$$

Our previous definition of chaos-enhancement generalizes to:

A density distribution u (i.e., $u \geq 0$, $\|u\|_1 = 1$) is more chaotic than another one, v, if and only if, for all $f \in G \cap C^1$,

$$S_f(u) = \Sigma \int f(u_i) \, dx \geq \Sigma \int f(v_i) \, dx = S_f(v)$$

(C^1 denotes the class of all once continuously differentiable functions.)

We shall use the symbol $u \succ v$. Of course we could also have written $u \succ v$ if and only if $S_f(u) = \|f(u)\|_1 \geq \|f(v)\|_1 = S_f(v)$ for all $f \in G \cap C^1$. By virtue of our previous lemma, we can replace G by G^+ in the definition.

The restriction to differentiable concave functions is not very convenient, and we would like to get rid of it. As a first step in this direction, let us consider broken-linear "good" functions. Such functions are determined by two sequences of numbers, $0 = \xi_0 < \xi_1 < \xi_2 < \ldots$ and $0 = \eta_0 < \eta_1 < \eta_2 < \ldots$ such that $\xi_n \to \infty$ and $\eta_1/\xi_1 \geq \eta_2/\xi_2 \geq \ldots$. If $\xi_i \leq x \leq \xi_{i+1}$, $f(x)$ is given by

$$f(x) = \frac{\xi_{i+1} - x}{\xi_{i+1} - \xi_i} \eta_i + \frac{x - \xi_i}{\xi_{i+1} - \xi_i} \eta_{i+1}$$

Now it is easy to construct an increasing sequence f_k of "good" nonnegative C^∞ functions even such that $f_k \uparrow f$. Therefore, by monotone convergence, $\|f_k(u_i)\| \to \|f(u_i)\|$, etc., and we obtain that $u \succ v$, or even $\|f(u)\|_1 \geq \|f(v)\|_1$ for all $f \in G^+ \cap C^\infty$, implies $\|f(u)\|_1 \geq \|f(v)\|_1$ for all "good" broken-linear functions.

On the other hand, every good nonnegative function, differentiable or not, can be approximated below by an increasing sequence of broken-linear good functions [for some given sequence ξ_i, simply take $\eta_i = f(\xi_i)$ and refine that sequence properly]. Thus it turns out that the following definitions are equivalent: $u \succ v$ if and only if $\|f(u)\|_1 \geq \|f(v)\|_1$ for all f (i) $\in G \cap C^1$, (ii) $\in G \cap C^\infty$, (iii) $\in G$. In the previous definitions, G can be replaced by G^+.

Let, in particular, $f(x) = f_s(x)$, where $f_s(x) = 0$ for $x \leq s$ and $f_s(x) = s - x$ for $x \geq s$. Then $u \Upsilon v$ implies

$$\sum_{\{x:u_i(x)\geq s\}} \int (u_i - s)\, dx \leq \sum_{\{x:v_i(x)\geq s\}} \int (v_i - s)\, dx$$

This can be interpreted in the sense that the portion of u exceeding an arbitrary number s is smaller than the one of v; in other words, the distributions u_i appear to be flatter than the v_i on the whole.

In case of *one-component* distributions, the above relation exhibits very appealing features if one considers rearrangements.[26] The (one-sided) rearrangement $\bar{u}$ of the density distribution u is defined as follows: Let $M(y)$ be the measure of the set where $u(x) \geq y$. The rearrangement $\bar{u}$ is the inverse of $M(y)$, i.e., $\bar{u}(M(y)) = y$. Since $M(y)$ is decreasing, $\bar{u}$ is, with the possible exception of an at most countable set of values, well defined and decreasing too. One may complete the definition by demanding that $\bar{u}$ be right continuous at those exceptional points.

Then u and $\bar{u}$ are said to be "equimeasurable," which in particular means that, for every measurable function f,

$$\int f(u)\, dx = \int f(\bar{u})\, dx$$

Hence $u \leftrightarrow v$ if and only if $\bar{u} \leftrightarrow \bar{v}$. [There is a minor point since $\bar{u}, \bar{v}$ are defined on $(0, \infty)$ only; this is quite immaterial, however.] Inserting f_s for f, we get

$$\int_{\{x:u(x)\geq s\}} [\bar{u}(x) - s]\, dx \leq \int_{\{x:v(x)\geq s\}} [\bar{v}(x) - s]\, dx$$

Define

$$U(x) = \int_0^x \bar{u}(x')\, dx', \qquad V(x) = \int_0^x \bar{v}(x')\, dx'$$

(Note that these functions are continuous, increasing, and concave.) Then our last relation reads, if we let $s = u(x)$ and take into account that $\bar{u}$ and $\bar{v}$ are decreasing,

$$U(x) - x\bar{u}(x) \leq V(x) - x\bar{v}(x)$$

for all $x \geq 0$.

To exploit this relation, we first remark that, for small x, $\bar{u}(x) \leq \bar{v}(x)$; otherwise we could choose s sufficiently large to obtain

$$\int_{\{\cdots\}} [\bar{u}(x) - s]\, dx > \int_{\{\cdots\}} [\bar{v}(x) - s]\, dx$$

Consequently, $U(x) \leq V(x)$. Now let $x_0 = \inf \{x: U(x) > V(x)\}$ provided that this exists at all. We have just shown that $x_0 > 0$. By continuity of U and V, $U(x_0) = V(x_0)$. Therefore, $u(x_0) \geq v(x_0)$ and, for x exceeding but being sufficiently close to x_0, $u(x) > v(x)$. Again we arrive at a contradiction by choosing s properly. Hence, $U(x) \leq V(x)$ for all x.

The function $U(x)$ admits another characterization, viz.,

$$U(x) = \sup \left\{ \int_{-\infty}^{\infty} u(x')\phi(x')\,dx' : 0 \leq \phi(x') \leq 1, \int \phi(x')\,dx' \leq x \right\}$$

This is a consequence of the rearrangement inequality

$$\int u\phi \leq \int \overline{u\phi}$$

$\bar{\phi}$ being easily determined to be the characteristic function of the interval [0,x].

Let us now investigate the converse: whether $U(x) \leq V(x)$ implies $u \succ v$. That means that we want to know whether, for "good" f,

$$\int [f(u) - f(v)]\,dx \geq 0$$

In order to avoid bothersome infinities, let us suppose $f'(0) < \infty$; by a monotone convergence argument this turns out to be sufficient for $u \succ v$. Also, we may suppose f to be sufficiently differentiable.

Clearly,

$$\int [f(u) - f(v)]\,dx = \int [f(\bar{u}) - f(\bar{v})]\,dx$$

A concave function stays always below its tangents. Therefore,

$$f(\bar{u}) - f(\bar{v}) \geq f'(\bar{u})(\bar{u} - \bar{v})$$

By partial integration,

$$\int f'(\bar{u})(\bar{u} - \bar{v})\,dx = f'(\bar{u})(U - V)\Big|_0^{\infty} - \int \frac{df'(\bar{u})}{dx}(U - V)\,dx$$

The first term vanishes. As concerns the second one, let us assume for the moment that $df'(\bar{u})/dx$ exists, and thus is ≥ 0 (because f is concave and $\bar{u}$ is decreasing). Hence, by virtue of $U \leq V$

$$\int f'(\bar{u})(\bar{u} - \bar{v})\,dx \geq 0$$

It is justified, however, to forget about existence questions of $df'(u)/dx$, because we can consider the integral in question as a Stieltjes integral

$$\int f'(\bar{u})\,d(U - V)$$

and use the formula for partial integration of Stieltjes integrals.[27]

Having reduced the discussion of the relation $\succ$ to a simple question of majorizing one function by another, our findings tell us that the relation $\succ$ is transitive, i.e., $u \succ v$, $v \succ w$ implies that $u \succ w$. This is clear since $U \geq V$, $V \geq W$ implies $U \geq W$. Every transitive relation generates an equivalence relation by defining $u \sim v$ if and only if $u \succ v$ and $v \succ u$. In our case this means just that $u \sim v$ if and only if $U = V$, or, since U and $\bar{u}$, or V and $\bar{v}$, respectively, are in a one-to-one correspondence, if and only if $\bar{u} = \bar{v}$.

To mention just one simple consequence: if $u \succ v$, $u_1 \succ v$, then, for any convex combination $w \equiv \lambda u + (1 - \lambda)u_1$ $(0 \leq \lambda \leq 1)$, $w \succ v$. Suppose this combination is a genuine one, i.e., $0 < \lambda < 1$. What then does $w \sim v$ imply? At first, clearly $W(x) \leq \lambda U(x) + (1 - \lambda)U_1(x) \leq V(x) = W(x)$; thus, $u \sim u_1 \sim v$. Choose ϕ with $0 \leq \phi \leq 1, \int \phi \, dx = x$, such that $W(x) = \int w\phi \, dx$. This way we obtain $W(x) = \int u\phi \, dx = \int u_1\phi \, dx$, too. In other words, for any $s \geq 0$, the sets $\{x : w(x) \geq s\}$, $\{x : u(x) \geq s\}$, and $\{x : u_1(x) \geq s\}$ are the same. Therefore, $w = u = u_1 = v$ almost everywhere; i.e., the distributions $w \sim v$ are extremal in the convex set $\{u : u \succ v\}$.

So much for a short, incomplete outlook at a field of research which has begun to open up only quite recently. The author, who still had the privilege of collaborating with Wolfgang Yourgrau on this subject, thinks that the example of the Carleman model demonstrates how intricate and subtle the problems of the second law and of the approach to equilibrium or chaos really are. There are still many surprises in store.

References and Notes

1. W. Yourgrau, A. van der Merwe, and G. Raw, *Irreversible and Statistical Thermophysics* (Macmillan, New York, 1966; revised Dover edition, 1982).
2. L. Boltzmann, *Wiener Berichte* **66**, 275 (1872).
3. G. E. Uhlenbeck, in *The Boltzmann Equation*, E. G. D. Cohen and W. Thirring, editors (Springer-Verlag, Berlin, 1972).
4. H. Grad, in *Handbuch der Physik*, Vol. 12, S. Flügge, editor (Springer-Verlag, Berlin, 1958).
5. O. E. Landford III, in *Dynamical Systems, Theory and Applications*, J. Moser, editor (Springer-Verlag, Berlin, 1975).
6. O. E. Landford III, in *International Conference on Dynamical Systems in Mathematical Physics*, Soc. Mathèmatique de France, Astèrisque **40**, 1975.
7. H. Spohn, *Rev. Mod. Phys.*, **52**, 569 (1980).
8. C. Cercignani, *Theory and Application of the Boltzmann Equation* (Scottish Acad. Pr., Edinburgh, 1975).
9. T. Carleman, *Problémes Mathèmatiques dans la Théorie Cinétique des Gaz* (Almqvist and Wiksell, Uppsala, 1957).
10. H. Kaper and G. K. Leaf, "Initial Value Problems for the Carleman Equation," preprint, Argonne, 1979.
11. A. Wehrl and W. Yourgrau, *Phys. Lett.* **72A**, 13 (1979).
12. A. Wehrl, *Found. Phys.* **9**, 939 (1979).
13. I. Csiszar, *Stud. Sci. Math. Hung.* **2**, 299 (1967).
14. A. Renyi, *Wahrscheinlichkeitsrechnung* (Deutscher Verlag der Wissenschaften, Berlin, 1961).
15. W. Thirring, *Lehrbuch der mathematischen Physik*, Vol. 4 (Springer Verlag, Vienna, to appear).
16. A. Wehrl, *Rev. Mod. Phys.* **50**, 221 (1978).
17. A. Uhlmann, *Wiss. Z. Karl-Marx-Univ. Leipzig, Math.-Naturw. R.* **20**, 633 (1971); **21**, 427 (1972); **22**, 139 (1973).
18. A. Wehrl, *Rep. Math. Phys.* **6**, 15 (1974).
19. E. Ruch, *Theor. Chim. Acta* **38**, 167 (1975).
20. E. Ruch and A. Mead, *Theor. Chim. Acta* **41**, 95 (1976).
21. A. Uhlmann, *Rostocker Physikalische Manuskripte* **2**, 45 (1977).
22. A. Wehrl, in *Recent Advances in Statistical Mechanics*, A. Corciovei, editor (Central Institute of Physics, Bucharest, 1980).
23. B. Crell, T. de Paly, and A. Uhlmann, *Wiss. Z. Karl-Marx-Univ. Leipzig, Math.-Naturw. R.* **27**, 229 (1978).
24. B. Crell and A. Uhlmann, preprint KMU-QFT-3, Leipzig, 1979.
25. L. Tartar, in *Dynamical Systems*, Vol. 1, L. Cesari *et al.*, editors (Academic Press, New York, 1976).
26. G. H. Hardy, J. E. Littlewood, and G. Polya, *Inequalities* (Cambridge Univ. Press, Cambridge, 1934).
27. F. Riesz and B. Sz.-Nagy, *Vorlesungen über Funktionalanalysis* (Deutscher Verlag der Wissenschaften, Berlin, 1956).

34

The Klein Paradox Revisited

Harald Wergeland
Seminar of Theoretical Physics
University of Trondheim
7034-NTH Trondheim, Norway

1. History

Dirac's equation for the electron was—and perhaps still is—the greatest common achievement of quantum theory and special relativity. Its established domain, ranging from spectroscopy to the later developments of quantum electrodynamics, seems incontestable.

However, already amidst the overwhelming initial triumphs—fine structure, anomalous Zeeman effect, Klein–Nishina formula, etc.—Oskar Klein[1] (1928) pointed out a glaring paradox implied by that beautiful equation:

Electrons may penetrate an electrostatic potential barrier

$$e\varphi(x) = \begin{cases} V, & 0 < x \\ 0, & x < 0 \end{cases} \tag{1}$$

even when their kinetic energy, $E - mc^2$ is lower than the barrier.

Since in Klein's example the barrier was infinitely broad this could not be associated with the wave mechanical tunnel effect. It is truly a paradox: Electrons, too slow to surpass the potential, may still only be partially reflected,

$$\text{reflection coefficient: } R < 1$$
$$\text{transmission coefficient: } T = 1 - R > 0!$$

Actually there are three different cases to be distinguished:

(a)	$V + mc^2 < E$,	$R < 1$	normal penetration
(b)	$V - mc^2 < E < V + mc^2$,	$R = 1$	total reflection
(c)	$mc^2 < E < V - mc^2$,	$R < 1$	Klein's paradox

The relativistic connection of energy and momentum

$$p^2c^2 = (E - e\varphi)^2 - m^2c^2 \tag{2}$$

shows that p is imaginary in case (b), real in cases (a) and (c).

In the latter case, which is the one of interest here, Klein[1] found the reflection coefficient,

$$R = \left(\frac{1 - r_1r_2}{1 + r_1r_2}\right)^2$$

where

$$r_1 = \left(\frac{E - mc^2}{E + mc^2}\right)^{1/2}, \qquad r_2 = \left(\frac{V - E - mc^2}{V - E + mc^2}\right)^{1/2} \tag{3}$$

Even for an infinitely high barrier, i.e., $r_2 = 1$ and energies ~1 MeV, R is less than 75%!

From (2) and (3) it thus appears that as soon as the barrier is sufficiently high: $V > 2mc^2$, electrons may transgress the repulsive wall—seemingly defying conservation of energy.

Likewise a square well potential of any height $V > 2mc^2$, cannot bind electrons in closed stationary states, except at some discrete energy levels within the band of total reflection $V \pm mc^2$ around the rim.

It should perhaps be noted that this need not affect the current hypotheses of "quark confinement," since in these theories the barrier is associated with the scalar term mc^2 in Dirac's equation, and *not* with the time component of a four-potential $\{0,0,0,\varphi\}$. In that case the Klein paradox disappears, as seems first to have been noted by Dosch, Jenssen, and Müller[2] (1971).

At an early stage, Klein's paradox was rendered somewhat less acute by calculations of Sauter[3] (1931) which show that a smoothly rounded barrier may be practically impenetrable. Consequently, the discontinuous potential (1) in Klein's example came to be regarded as problematic inasmuch as it involves an infinity of the electric field strength:

$$-\text{grad}\ \varphi(x = 0) = \infty$$

In those days (and maybe even now?) a certain critical field strength $\mathcal{E}_c \sim m^2c^3/e\hbar \sim 10^{13}$ V/cm, was surmised to play a role.

Later on, however, Hylleraas[4] (1943) showed that the transmission coefficient can be just as efficiently depressed by means of the staircase potential

$$e\varphi(x) = \begin{cases} V, & a < x \\ V_1, & 0 < x < a \\ 0, & x < 0 \end{cases} \tag{4}$$

provided the intermediate step V_i,a is suitably chosen.

Since this is no less discontinuous than Klein's potential (1) it appears from then on not so obvious that the paradox is connected in any simple way with the polarization of the vacuum and its eventual "dielectric breakdown."

An even more decisive turn in the whole problem ensued from the study of Jenssen and collaborators,[2] namely, their result on the penetration of potential hills, i.e., barriers of *finite* breadth. They discovered that for potential hills, the transmission will attain the value $T = 1$ for certain energies! ($E < V - mc^2$).

And such "windows" are present whether the potential is continuous or not. In Jenssen's words: "A potential hill of sufficient height acts as a Fabry–Perot etalon for electrons, being completely transparent for some wave lengths, partly or completely reflecting for others."

This result enhances Klein's paradox and even prefigures an experiment to test the penetration of electrons through macroscopic barriers of electrostatic potential, exceeding their kinetic energy.

A natural objection to the paradox was that Klein had treated Dirac's ψ function as a probability amplitude instead of working with the appropriate matrix elements $\langle \text{vacuum} | \psi | 1 \text{ electron} \rangle$.

And it was perhaps not fortuitous that Oskar Klein—himself one of the originators of the "second quantization"—should hint at the paradoxical features of the negative energy states, which were later to be disposed of by Dirac in the electron–positron theory.

However, even if this observation is important, it does not immediately justify the enunciation of some textbooks that the paradox is resolved as soon as field quantization is taken into account.

Nor is it correct, as has been propounded in the same context, that the classical wave aspect of the ψ function can lead to reflection coefficients exceeding unity (!). This belief originates in an erroneous reproduction of Klein's solution to the boundary-value problem.

As was pointed out by Jenssen *et al.*,[4] the conditions

$$R + T = 1 \tag{5}$$

R, T nonnegative—which are indispensable for any meaningful description in terms of reflection and transmission—are implied by the Dirac equation, whether ψ be conceived as a probability amplitude or more generally as field operator.

Actually the invariance of the integral

$$\int_\sigma \bar{\psi} \gamma_\mu \psi \, d\sigma^\mu$$

where σ is any spacelike surface, not only secures the conditions (5) as soon as the boundary-value problem of the Dirac equation is correctly solved. It admits also a meaningful definition of a one-particle probability density even when the states of negative energy are ignored.

Certainly these latter must sometimes be taken into account. However, in a Cauchy problem of the Dirac equation, such as the construction of wave packets—it turns out that the spectrum of positive energies alone, is sufficient to cover a very wide class, e.g., one in which the expectation value

$$\langle 1|\psi^\dagger(t_0)\psi(t_0)|1\rangle = \langle \text{vac}|\psi(t_0)\psi^\dagger(t_0)|\text{vac}\rangle = \delta(x - x_0) \tag{6}$$

condenses to a delta function.

Hence sharp localization of one electron at one instant t_0 is thus definable.

Not possible in this way however, are wave packets of compact support: $\psi = 0$ everywhere outside a finite interval. To construct such an object (wave packet?) one needs also the negative energy states. But associating these with the positrons we step out of the Grassmann algebra of the electron field.

Nor is it possible by way of the positive energy spectrum of the free electron to achieve complete Einstein causality. A positive energy wave packet is slightly *acausal* in the following sense:

Just as in the nonrelativistic case its diffusion can be described by

$$\Delta x^2(t) = \Delta x^2(t_0) + U^2 \cdot (t - t_0)^2$$

However, in contradistinction to ordinary nonrelativistic wave mechanics $[U_{\text{n.r.}} = \hbar/m\Delta x(t_0)]$—which is of course patently noncausal—the diffusion velocity U will here be bounded as it should be in a relativistic theory by the velocity of light:

$$\frac{U(\Delta^2 \to 0)}{c} \sim 1 - \frac{\Delta^2/\lambda^2}{|\log(\Delta^2/\lambda^2)|}$$

where $\Delta^2 = \Delta x^2(t_0)$ and $\lambda = \hbar/mc$, the Compton wavelength.

Hence, one has a sort of "causality in the mean," but not a pointwise causality, since the amplitude: $\psi(t > t_0)$, strictly speaking will be finite everywhere although it vanishes exponentially within a Compton wavelength beyond the light cone: $x - x_0 = \pm c(t - t_0)$.

It is also to be admitted that Dirac's position operator x is nonlocal to the extent of the "Schrödinger Zitterbewegung" or its characteristic property of mixing positive and negative energy states; this, however, does not preclude an absolutely sharp localization by way of positive energy states in the sense of Eq. (6).

In general the indeterminacy of position can be computed in two equivalent ways:

$$\Delta x^2 = \langle x\rangle - x_0^2 = \begin{cases} \text{either } \langle\psi|x^2|\psi\rangle - x_0^2 & (7) \\ \text{or } \quad \langle x\psi|x\psi\rangle - x_0^2 & (7') \end{cases}$$

Here ψ is a superposition of positive energy states only: $E > mc^2$. The ket $|x\psi\rangle$, on the other hand, can only be expanded in the complete set: $E \gtrless \pm mc^2$, of free electron states. The contributions to $x\psi$ from the negative energies are of the order of $\hbar/mc$ and oscillate with the Schrödinger Zitter frequency $2mc^2/\hbar$. Only when these terms are included does the result of (7′) agree with (7). The latter, however, can be evaluated without any regard at all to the negative energy states.

This may illustrate the rather modest part played by the negative energy states as long as pairs can occur in virtual processes only. In Klein's example this can be secured by taking the energy of the incident electron to be below the threshold: $3mc^2 \approx 1.5$ MeV, for pair production.

The paradox has also been countered on the ground that an electron cannot be localized with greater precision than one Compton wavelength: $\mathrm{Min}(\Delta x) \sim \lambda$. In view of Eq. (6), however, this latter statement is perhaps best understood to mean that the number operators N_1 and N_2 for electrons inside volumes v_1 and v_2, respectively, do not commute:

$$[N_1(t), N_2(t)] \neq 0, \qquad |x(1) - x(2)| < \hbar/mc$$

unless the distance from any point in v_1 to any point in v_2, well exceeds the Compton wavelength. Hence it is not directly applicable to bona fide one-particle states.

And so far, no connection of the Klein paradox with any one of the nonlocal features of the electron has been demonstrated.

Anyway, the nonlocalizability of the electron needs some qualifications as indicated. And it seems premature to announce it as the resolution of the Klein paradox before consensus is reached on some very elementary details of the problem, which we shall now review.

2. The Boundary-Value Problem

As in Klein's example we shall restrict the discussion to one dimension, although that is not essential, and has ultimately to be relinquished in a calculation fully including the inelastic processes.

We may then write the solutions of the Dirac equation in the plane wave form

$$u(\pm p,E)\exp[i(\pm px - Et)/\hbar]$$

and fix the longitudinal component of the spin at $s_x' = \hbar/2$ say, since σ_x commutes with the Hamiltonian

$$\mathcal{H} = \mathcal{H}^0 + e\varphi(x)$$

Thereby the spinors u are for the present purposes reduced to two components, and the Hamiltonian to a 2×2 matrix:

$$\mathcal{H}^0 = -i\hbar c\rho_1\partial_x + \rho_3 mc^2$$

with

$$\rho_1 = \begin{pmatrix} 0 & 1 \\ 1 & 0 \end{pmatrix}, \qquad \rho_3 = \begin{pmatrix} 1 & 0 \\ 0 & -1 \end{pmatrix}$$

Approaching our one-dimensional scattering problem by way of the stationary states, we have to solve

$$\mathcal{H}\psi = E\psi \tag{8}$$

with E = const. > 0. That is to say: The energy of our one-particle system is positive, but

$$\mathcal{H} = \begin{cases} \mathcal{H}^0 + V & \text{inside the barrier} \\ \mathcal{H}^0 & \text{outside the barrier} \end{cases} \tag{9}$$

Outside the barrier, the solution of Eq. (8) can thus be written

$$\psi_1 = [a_1 u(p,E)\, e^{ipx/\hbar} + b_1 u(-p,E) e^{-ipx/\hbar}] \cdot e^{-iEt/\hbar} \tag{10}$$

where the u's are positive energy free particle spinors:

$$u(\pm p,E) = \begin{pmatrix} 1 \\ \pm r_1 \end{pmatrix}, \qquad r_1 = \frac{pc}{E + mc^2} = \left(\frac{E - mc^2}{E + mc^2}\right)^{1/2}$$

Inside the barrier, on the other hand, the spinors must satisfy

$$(\pm p_2 c \rho_1 + \rho_3 mc^2 + V - E) u(\pm p_2, E - V) = 0 \tag{10'}$$

where $p_2c^2 + mc^2 = (E - V)^2$. Since per supposition $E < V$, this is formally the equation satisfied by the "negative energy" spinors:

$$u(\pm p_2, -E_2) = \begin{pmatrix} \mp r_2 \\ 1 \end{pmatrix}, \qquad E_2 = V - E$$

$$r_2 = \frac{p_2 c}{E_2 + mc^2} = \left(\frac{V - E - mc^2}{V - E + mc^2}\right)^{1/2}$$

But Eq. (10) does not justify the off-hand conclusion that in this region we have to invoke holes or positrons. The stationary state (8) under consideration is one of positive energy, as reflected by the positive frequency time factor in the solution

$$\psi_2 = [a_2 u(p_2, -E_2) e^{ip_2x/\hbar} + b_2 u(-p_2, -E_2) e^{-ip_2x/\hbar}] e^{-iEt/\hbar} \tag{11}$$

The phases of the two partial waves are $(\pm p_2 x - Et)/\hbar$, and not at all $(\pm p_2 x + E_2 t)/\hbar$, as would befit a negative energy state: $\mathcal{H}\psi = -E_2\psi$, corresponding to the local momentum p_2 inside the barrier.

The constants a,b are determined up to a common normalization factor by the boundary conditions. Before using them, however, it is instructive first to look at the transport of matter outside and inside the barrier.

Under the simplifications chosen, the Dirac current density is proportional to

$$\mathcal{J} = \psi^\dagger \rho_1 \psi$$

Outside the barrier we have by Eq. (9)

$$\rho_1\psi = \begin{pmatrix} 0 & 1 \\ 1 & 0 \end{pmatrix} \psi_1 = \left[a_1 \begin{pmatrix} r_1 \\ 1 \end{pmatrix} e^{ipx/\hbar} + b_1 \begin{pmatrix} -r_1 \\ 1 \end{pmatrix} e^{-ipx/\hbar} \right] e^{-iEt/\hbar}$$

$$\psi^\dagger = [a_1^*(1,r_1)e^{-ipx/\hbar} + b_1^*(1,-r_1)e^{+ipx/\hbar}]\, e^{+iEt/\hbar}$$

which gives

$$\mathcal{J}_1 = 2r_1(|a_1|^2 - |b_1|^2) \tag{12a}$$

the interference terms cancelling out by orthogonality.

In this region the term $b_1 u(-p,E) \exp[-i(px + Et)/\hbar]$ thus represents the current from right to left, i.e., the reflection.

Inside the barrier, however, one finds by Eqs. (10) and (11) mutatis mutandi

$$\mathcal{J}_2 = 2r_2(|b_2|^2 - |a_2|^2) \tag{12b}$$

In this region therefore, the roles of the partial waves are interchanged: $b_2 u(-p_2, -E_2) \exp[-i(p_2 x + Et)/\hbar]$ here represents transport of matter from left to right, i.e., transmission of the incident flux, despite the phase velocity $-E/p_2$ of this wave is from right to left. Its group velocity, however, is positive:

By the definition (10) of p_2 the group velocity is

$$\frac{\partial E}{\partial p_2} = -\frac{p_2}{V - E}$$

opposite to the direction of the local momentum inside the barrier, and thus positive for the b_2 wave.

3. Examples

3.1. *The Infinitely Broad Barrier Eq. (1)*

The condition that ψ be continuous gives two equations between the amplitude constants $a_1 b_1 a_2 b_2$ in Eqs. (9) and (10).

Further, since in Klein's example, electrons are impinging from the left, there can be no influx of matter from the right. That is, we have to set

$$a_2 = 0 \tag{13}$$

in view of Eq. (12b).

Herewith the system is determinate: $\psi_1(x = 0) = \psi_2(x = 0)$ gives, by Eqs. (10) and (11)

$$\left.\begin{aligned} a_1 + b_1 &= r_2 b_2 \\ r_1(a_1 - b_1) &= b_2 \end{aligned}\right\} \rightarrow \left\{\begin{aligned} \frac{b_1}{a_1} &= \frac{r_1 r_2 - 1}{1 + r_1 r_2} \\ \frac{b_2}{a_1} &= \frac{2r_1}{1 + r_1 r_2} \end{aligned}\right.$$

Thus by Eqs. (12a) and (12b) the reflection and transmission coefficients come out in agreement with Klein's formulas Eq. (3)

$$R = \left|\frac{b_1}{a_1}\right|^2 = \left(\frac{1 - r_1 r_2}{1 + r_1 r_2}\right)^2$$

$$T = \frac{r_2}{r_1}\left|\frac{b_2}{a_1}\right|^2 = \frac{4 r_1 r_2}{(1 + r_1 r_2)^2}$$

which obviously comply with the conservation of matter, Eq. (5).

3.2. *Barrier of Finite Breadth*

$$e\varphi(x) = \begin{cases} 0, & l < x \\ V, & 0 < x < l \\ 0, & x < 0 \end{cases}$$

In this case we have three pairs of amplitudes $a_1 b_1$ $(x < 0)$ $a_2 b_2$ $(0 < x < l)$ and $a_3 b_3$ $(x > l)$. Continuity of ψ at the edges then according to Eqs. (9) and (10) gives the four equations

$$(x = 0):\ a_1\begin{pmatrix} 1 \\ r_1 \end{pmatrix} + b_1\begin{pmatrix} 1 \\ -r_1 \end{pmatrix} = a_2\begin{pmatrix} -r_2 \\ 1 \end{pmatrix} + b_2\begin{pmatrix} r_2 \\ 1 \end{pmatrix}$$

$$(x = l):\ a_2\begin{pmatrix} -r_2 \\ 1 \end{pmatrix} e^{i\eta} + b_2\begin{pmatrix} r_2 \\ 1 \end{pmatrix} e^{-i\eta} = a_3\begin{pmatrix} 1 \\ r_1 \end{pmatrix} e^{i\delta} + b_3\begin{pmatrix} 1 \\ -r_1 \end{pmatrix} e^{-i\delta}$$

where $\eta = p_2 l/\hbar$.

Corresponding to the "Ausstrahlungsbedingung" (12) in Klein's example we have now to set $b_3 = 0$, meaning: no incident current from the right.

Eliminating $a_2 b_2$, the linear relations between the amplitudes can be written in the form

$$\begin{pmatrix} a_1 \text{ incident} \\ b_1 \text{ reflected} \end{pmatrix} = \begin{pmatrix} \alpha, & \beta^* \\ \beta, & \alpha^* \end{pmatrix}\begin{pmatrix} a_3 \text{ transmitted} \\ b_3 \text{ incident} = 0 \end{pmatrix}$$

where

$$\alpha = \cos\eta + \frac{i}{2}\left(r + \frac{1}{r}\right)\sin\eta\, e^{i\eta}$$

$$\beta = \frac{i}{2}\left(r - \frac{1}{r}\right)\sin\eta\, e^{-i\eta}$$

$$r = r_1 r_2$$

Rewriting it as a scattering relation:

$$\underset{\text{outgoing}}{\begin{pmatrix} a_3 \\ b_1 \end{pmatrix}} = S \underset{\text{ingoing}}{\begin{pmatrix} a_1 \\ b_3 = 0 \end{pmatrix}}$$

one finds

$$S_{a_3a_1} = \frac{2r\, e^{-2i\eta}}{2r\cos\eta + i(1 + r^2)\sin\eta}, \qquad S_{b_1a_1} = \frac{i(r^2 - 1)e^{-2i\eta}}{2r\cos\eta + i(1 + r^2)\sin\eta}, \ldots$$

and so forth.

In this way one can associate transmission and reflection coefficients with squares of matrix elements:

$$\begin{aligned} T &= |S_{a_3a_1}|^2 = \frac{4r^2}{4r^2 + (1 - r^2)^2 \sin^2(p_2 l/\hbar)} \\ R &= |S_{b_1a_1}|^2 = \frac{(1 - r^2)^2 \sin^2(p_2 l/\hbar)}{4r^2 + (1 - r^2)^2 \sin^2(p_2 l/\hbar)} \end{aligned} \tag{14}$$

and the conservation equation (5) with the unitarity $S^{-1} = S^\dagger$, of a scattering operator.

Equations (14) show that in general the transmission coefficient is larger than for the infinitely broad barrier. It will even reach the value $T = 1$, whenever the breadth l is a half-integral multiple of the de Broglie wave length h/p_2, inside the barrier. As shown by Jenssen *et al.* this peculiar type of "Ramsauer effect" occurs for continuous potential hills as well. In any case it seems to exacerbate the paradox.

3.3. *Continuous Barrier*

A somewhat undesirable step in the preceding examples is the appearance of the negative energy spinors Eq. (10), especially since these play such a part in some attempts to resolve Klein's paradox along the lines of the hole theory.

It is to be noted, however, that negative energy spinors do not show up at this stage when working with continuous potentials.

We shall here consider a very simple one:

$$e\varphi(x) = V_0 e^{2x}, \qquad x\epsilon[-\infty,+\infty] \tag{15}$$

Solving the appertaining Dirac equation

$$(-i\hbar c\rho_1\partial_x + \rho_3 mc^2 + V_0 e^{ax} - E)\psi = 0 \tag{16}$$

in the form

$$\psi = af_a + bf_b \tag{17}$$

where the two fundamental solutions f_a, f_b are so chosen as to coincide asymptotically with free particle spinors,

$$f^a_b(x \to -\infty) \sim \begin{pmatrix} E + mc^2 \\ \pm pc \end{pmatrix} e^{\pm ipx/\hbar} \tag{17'}$$

at $x = -\infty$ where the potential vanishes, one finds:

$$f^a_b = N_s z^s e^{z/2} \begin{pmatrix} s(\epsilon^+ + is)F + i(\epsilon\epsilon^+ - s^2 + sz)G - isz\dfrac{\alpha + 1}{2s + 1}H \\ \\ -s(\epsilon^- + is)F + [s(\epsilon + \epsilon^-) - \epsilon z]G + \epsilon z\dfrac{\alpha + 1}{2s + 1}H \end{pmatrix} \tag{18}$$

Here

$$z = 2iV_0/\hbar ca, \qquad s^a_b = \pm ip/\hbar a$$

$$\epsilon = E/\hbar ca, \qquad \alpha = s - i\epsilon$$

$$\epsilon^+ = (E \pm mc^2)/\hbar ca, \qquad N_s = \frac{pc}{E \pm pc} \cdot e^{i\pi s/2}$$

and F, G, H are the confluent H.g. functions:

$$F = {}_1F_1(\alpha; 2s; z), \qquad G = {}_1F_1(\alpha + 1; 2s + 1; z), \qquad H = {}_1F_1(\alpha + 2; 2s + 2; z)$$

The amplitudes a, b are now completely determined by the Ausstrahlungsbedingung at $x = +\infty$. For large positive x the asymptotic expressions

$$f^a_b(x \to +\infty) \sim \Phi^a_b + \Gamma^a_b \tag{19}$$

approach a sum of two (mutually orthogonal) spinors:

$$\Phi^a_b = \phi^a_b \begin{pmatrix} 1 \\ 1 \end{pmatrix} \exp[i(Eax - V_0 e^{ax})/\hbar ca]$$

$$\Gamma^a_b = \gamma^a_b \begin{pmatrix} 1 \\ -1 \end{pmatrix} \exp[-i(Eax - V_0 e^{ax})/\hbar ca]$$

where

$$\phi^a_b = \pm q\epsilon^\dagger_\pm \cdot \frac{e^{\pi(\epsilon\mp} \pm q/2)}{i\epsilon_\pm} \cdot \frac{\Pi(\pm 2iq - 1)}{\Pi(i\epsilon)}$$

$$\gamma_b^a = \frac{1}{2}\mu\epsilon_\pm^+ \cdot \frac{e^{\pm \Pi q/2}}{i\epsilon_\pm} \cdot \frac{\Pi(\pm 2iq)}{\Pi(-i\epsilon_\mp)}$$

$$q = p/\hbar ca, \qquad \epsilon_\pm = \epsilon \pm q, \qquad \mu = mc/\hbar a$$

Taking into account the time-dependence $\exp(-iEt/\hbar)$, the phase of the Γ waves according to Eq. (19) will be

$$\left(\frac{V_0}{ac} e^{ax} - \frac{E}{c} x - Et\right) \Big/ \hbar$$

and thus their phase velocity

$$\left(\frac{\delta x}{\delta t}\right)_{ph=\text{const}} = \frac{Ec}{V_0 e^{ax} - E}$$

will be in the same direction as the local momentum in this region:

$$-i\hbar\partial \log \Gamma/\partial x = (V_0 e^{ax} - E)/c, \qquad ax \gg \log(E/V_0)$$

while the group velocity

$$\frac{\partial E}{\partial p} = -\frac{pc^2}{V_0 e^{ax} - E} (\sim -c!)$$

has the opposite direction, i.e., from right to left.

Hence in the superposition (16), representing the stationary scattering state, a and b must be chosen in such a way that the Γ terms vanish:

$$a\Gamma_a + b\Gamma_b = 0 \tag{20}$$

This can also be confirmed by checking the asymptotic form of the current $\mathcal{J}(x \to +\infty)$: Writing in the superposition (16)

$$a\Phi_a + b\Phi_b = \Phi, \qquad a\Gamma_a + b\Gamma_b = \Gamma$$

one sees that

$$\mathcal{J} - \psi|\rho_1\psi(x \to \infty) \sim |\Phi|^2 - |\Gamma|^2$$

By the choice (17′) of the fundamental solutions, the reflection coefficient is evidently $|b/a|^2$ or by Eq. (20)

$$R = \left|\frac{\gamma_a}{\gamma_b}\right|^2 = \left(\frac{\epsilon_+^+}{\epsilon_-^+}\right)^2 \cdot \left(\frac{\epsilon_-}{\epsilon_+}\right)^2 \cdot \left|\frac{\Pi(i\epsilon_+)}{\Pi(i\epsilon_-)}\right|^2 \cdot e^{2\pi q} \tag{21}$$

Using the identity

$$\left(\frac{\epsilon_+^+}{\epsilon_-^+}\right)^2 \cdot \frac{\epsilon_-}{\epsilon_+} = \left(\frac{E + mc^2 + pc}{E + mc^2 - pc}\right)^2 \cdot \frac{E - pc}{E + pc} = 1$$

and the mirror formula for the factorial function [$\Pi(z) = \Gamma(z + 1)$], this can be written

$$R = \frac{1 - \exp[-2\pi(E - pc)/\hbar ca]}{1 - \exp[-2\pi(E + pc)/\hbar ca]} < 1 \tag{21'}$$

Hence we have the Klein paradox—as ever.

It is remarkable that the final result (21) is completely independent of the constant V_0 in the potential (15). In the limit $a \to \infty$, $V_0 > 0$; the potential (15) tends to a "wall":

$$e\varphi(x) = \begin{cases} \infty, & x > 0 \\ 0, & x < 0 \end{cases}$$

and the reflection coefficient (21) to

$$R = \frac{E - pc}{E + pc}$$

which is precisely Klein's result [Eq. (3) with $r_2 = 1$] for an infinitely high barrier.

3.4. *Status Praesens*

These examples may show that a critical field strength is not conspicuous on the level of pure potential scattering.

Whether the calculations be phrased in the particle $\leftrightarrow$ field, or in their historical wave mechanical form, makes no difference. The relevant elements of the scattering matrix: $\langle 1_{p'} \text{ out} \mid 1_p \text{ in}\rangle$, $\langle p' \text{ out} \mid p \text{ in}\rangle$, respectively, are *numerically* equal. Hence, second quantization does not affect the final result in this approximation.

Since the paradox is a gross effect—not a question of the utmost quantitative accuracy—it seems unlikely that the whole refinement of invariant perturbation theory could be essential to this problem. At any rate, quite a laborious calculation is required to obtain the S matrix more completely by the rules of quantum electrodynamics.

Since the electrostatic potential $\varphi(x)$ may be large, the conventional expansion in powers of e is not directly applicable. Yet, the knowledge of exact solutions to the equation of motion for the Dirac field ψ coupled to that (c-number) field, furnish the starting point for a Furry representation[5] from which a perturbation expansion can proceed.

In this picture the external field $\{0,0,0,\varphi\}$ is accounted for in the motion of the Dirac field operators ψ. These in turn induce the quantized em field $\{A_1A_2A_3 0\}$, while the time dependence of the state functional $\Psi[\psi, A]$ is determined by

$$i\hbar \frac{\delta\Psi}{\delta\sigma} = -\frac{1}{c}(\mathbf{J} \cdot \mathbf{A})\Psi$$

that is, by the radiative interaction alone.

In the conventional, charge symmetric form of quantum electrodynamics the electric current is defined as $J = e[\overline{\psi},\gamma\psi]/2$; its vacuum expectation value can differ from zero in this picture.

The practical feasibility of such an approach will clearly depend very much upon the form of the commutation relations in the Furry picture. As can be seen from Eqs. (18) these will not be simple for the potential (15). But other choices of external field could possibly make them more tractable.

Again one may speculate on the merits of Green's functions constructed in the standard way, as compared to those available by restriction to the positive energy spectrum, at the price of a slight acausality (and also charge asymmetry) of the theory.

For energies below the pair treshold ($3mc^2$) the most obvious correction to the pure potential scattering is the radiation loss, i.e., the Bremsstrahlung. This latter may of course be considerable, and accordingly the conservation equation (5) will have to be understood as a sum total, including all the inelastic reflections and transmissions through the barrier. Or in scattering terminology: $SS^\dagger = 1$ can only be displayed when none of the outgoing channels are ignored.

Anyway, some explicit computations by this or another method would be desirable in order to substantiate—or eventually to disprove—the current answers to the Klein paradox. As long as no program of this sort is implemented it must remain a challenge.

References and Notes

1. O. Klein, Z. *Phys.* **53,** 157 (1929).
2. F. Sauter, Z. *Phys.* 69, 742 (1931).
3. E. A. Hylleraas, *Vidensk. Akad. Oslo Math. Naturv. Kl.* (1943) No. 2.
4. H. G. Dosch, J. H. D. Jenssen, and F. Müller. *Phys. Norvegica* 5, 151 (1971).
5. W. Furry, *Phys. Rev.* **81,** 115 (1951).

35

Einstein Nonlocality, Space-Time Structure, and Thermodynamics

H. D. Zeh
Universität Heidelberg
Heidelberg, GFR

The problem of a possible connection between the different arrows of time observed in physical phenomena has recently gained interest,[1–3] but has still to be considered as open. Moreover, the arrow most important for our everyday life—the thermodynamic one—is plagued by ambiguities in the statistical definition of entropy, which apparently has to be based on some observer-related concept of "coarse-graining" or "relevance".[3,4] A feature common to all useful relevance concepts (formally expressed by idempotent operators in the space of phase-space densities or density matrices) is some kind of locality required if entropy is to be approximately additive,

$$S = \int d^3r s(r)$$

Additivity would be lacking in a pure ensemble definition of entropy

$$S_e = -\int dp\, dq\, \rho(p,q) \ln \rho(p,q)$$

since $\rho(p,q)$ necessarily contains correlations between different particles even over astronomical distances. They are usually explained as caused by interactions in the past. In particular, the additivity (or locality) of entropy neglects macroscopic correlations which are very *relevant* though of little numerical contribution.

The problem of locality has assumed particular importance in quantum theory since Bell[5] demonstrated that the nonlocality of quantum correlations cannot be explained by an ensemble interpretation of the wave function (or by incomplete information), but is instead a property of "reality"—a concept that can be maintained at least in Everett's interpretation of quantum theory.[5–7] These quantum correlations—now often referred to as Einstein nonlocality—follow from the superposition principle, as the general state of a composed system is a superposition

$$\Sigma c_{mn} \varphi_m^{(1)} \varphi_n^{(2)}$$

of its subsystem states, not simply a product state $\varphi^{(1)}\varphi^{(2)}$. They cannot be considered as small corrections for macroscopic systems for dynamical reasons.[8] In other words: *Local* entropy is large in quantum theory even for a complete description of the universe, and without any *further* coarse-graining (in addition to locality).

The correlated pure states can be mapped into ensembles of uncorrelated states—each ensemble given by the diagonal form of a density matrix—by applying the idempotent operation of omitting the nondiagonal matrix elements connecting the subsystems. The usefulness of this procedure for making predictions follows from the assumption of a retarded nature of the correlations, that is, of their having local causes but no local effects—in particular no effects on measurement apparatuses or observers. This assumption introduces an arrow of time, but it is most remarkable at this point that the classical concept of *local* causes and effects has to be introduced into a manifestly nonlocal quantum world. The origin of this locality can be traced back to the psychophysical parallelism at the fundamental level of a theory of observation.[3,6]

The nonlocal nature of "reality" raises new problems connected with the global space-time structure, as with horizons and singularities, especially when this structure is itself subject to quantization. In order to be able to disregard the quantization of the global space-time structure, assume that our reduction (or Everett) branch of the state vector contains only superpositions of space-time structures not too different from some background structure—an assumption which may readily be justified dynamically, since neutral boson fields tend to collapse towards coherent states.[9] Notice that conventional entropy is always defined with respect to this branch (as a property of the "relative state," in Everett's terminology).

Given the global space-time structure, a future horizon as, for example, defined by the Schwarzschild radius is a natural cutoff for correlations with respect to an observer outside the horizon. Hence, the outside world is equivalent to an *ensemble* of states, such as that described by the density matrix for black body radiation.[10] This equivalence refers to the potential reductions or branchings of the universal state inferred by (local) observations.[8] Hence, the CPT violation predicted for phenomena of strong quantum gravity by conversion of pure states into mixtures[20] simply reflects that of conventional measurementlike processes.[18] The fact that the horizon (and therefore the black body temperature) is observer dependent,[11] as is the concept of entropy,[3] strengthens the remarkable analogy of black hole dynamics and thermodynamics,[12] and the case for their relation to the "relative state" concept.[3,5] The horizons obscure from the outside observer the space-time singularities which limit the existence of Cauchy surfaces required for a complete set of "initial" data in classical (conceptually local) physics, and for the definition of a general pure state in quantum physics.

The postulate of "cosmic censorship"[13] introduces an arrow of time in the space-time structure. It requires that there be only future singularities into which something may fall but no past singularities from which something can emanate. If a future singularity is created by the gravitational collapse of matter, and if it disappears later on by black hole evaporation, the ensemble entropy of the universe has increased according to the statistical treatment of this process.[10] This statistical treatment contains the thermodynamical arrow (by assuming that advanced correlations may be disregarded in this universe)—it does *not derive* a connection between the censorship postulate and the retarded nature of correlations.

Another arrow in the global structure is defined by the cosmic expansion and the big bang singularity in the past. Plausible arguments have been presented[14] for its connection with the thermodynamical arrow; these have in turn been severely challenged.[2] A particular argument in favor of a connection between these arrows may be obtained from the rapidity of the expansion, which is, in a sense, faster than light[15]: The size of "zones of causality," that is, fractions of the universe possessing a nonvanishing common intersection of the past light cones of all their space points, increases with time and approaches zero with vanishing age of the universe. (The widespread opinion that it would require an extremely improbable accident for the different zones of causality to simultaneously come into existence "without knowing about one another" is based on a too-conventional notion of causality.) If it is assumed that there were no nonlocal quantum correlations initially (just after the singularity), the growing size of zones of causality gives rise to an enormous information sink, just as is needed for a justification of Boltzmann's *Stosszahlansatz*. In this way local entropy may increase without violating the law of conservation of ensemble entropy, which in fact remains zero for a pure state of the universe.

This argument is of further interest for a refinement of an experimental test of quantum nonlocality by means of Bell's inequality[16]: In its presently most advanced form the two polarization filters used in this experiment are given a certain direction by some last-minute decision. Even if this experiment can be arranged in such a way that a light signal of the decision cannot reach the other counter before this has registered the second particle, the decision itself could in principle have an unknown cause in the backward light cone of the second counter, and hence be correlated to the latter. If the experiment could be refined in such a way that the orientation of the first filter were triggered by a photon of the cosmic background radiation arriving from the direction opposite to the second counter, any common cause were excluded and a correlation would have had to be preestablished in the "creation" at the big bang ("conspiracy").

The nonexistence of quantum correlations connecting different zones of causality would be equivalent to an initial local state of the universe (a state of type $\Pi_r|\alpha_r\rangle$, with $|\alpha_r\rangle$ being the state at point r). This could not be the relativistic vacuum.

The assumption of nonexistence of such initial quantum correlations appears plausible to our prejudiced minds. But problems might arise if the same properties were assumed to hold with respect to future singularities such as the possible final singularity in a recollapsing universe, or singularities behind future horizons. It is even an open question whether both conditions are consistent with each other. These problems were the reasons for Penrose' objections[2] against a relation between the arrows of thermodynamics and space-time structure. They led him instead to propose a special *initial* space-time structure (the vanishing of the Weyl tensor). Just as information may flow into correlations between the growing zones of causality and thereby give rise to a thermodynamical arrow of time, so may information hopefully flow into the increasing generality of the space-time structure according to Penrose' model.

However, the possible problems of consistency when applying such assumptions to past as well as to future singularities are not obviously relaxed by the *change* of specific model assumptions for special initial *and* final conditions. They are instead avoided in both of these models by applying either one or the other assumption to initial singularities only, thereby introducing a fundamental time-inversion asymmetry of our factual world.

Is this really necessary? Special conditions at the final singularity in a recollapsing universe are not disturbing yet as the recollapse is only a possibility. It may be sufficiently far away if it exists at all. It is possible that macroscopic correlations (those being bound to have local effects) will gain importance with time and ensure consistency with the second half of our universe' life time. Otherwise a rigorous proof of inconsistencies could possibly exclude a recollapsing universe.

But what should happen at black hole singularities? We would not be surprised to see an astronaut falling into a black hole horizon. We would be surprised to see an astronaut coming out of a white hole of comparable mass if this existed as a lagging core of the big bang, because evolution inside the white hole did not have enough time for him to develop from statistical initial conditions. Our expectation violates the symmetry between white and black holes. However, there could be phenomenological laws valid near black holes and to be discovered as "antithermodynamics" (conspiracy), that prevent complex objects from falling into black holes (or at least from sending messages before they do)!

This problem obtains quite novel aspects in quantum theory: The description of the world in classical terms or pictures (as we have just done) corresponds to the "relative states" of certain Everett branches. The branching itself (not only the selection of a branch as it results from this branching!) seems to be observer dependent.[3] Two observers inside and outside the horizon (as defined from the spatial infinity of the outside observer) would be in different situations, thus experiencing very different Everett branchings. As mentioned above, conventional entropy is calculated with respect to the relative state. Hence, even the thermodynamical arrow may be observer-dependent. What an outside observer (with respect to his Everett branch) describes as the ensemble of states inside the horizon (measured by Bekenstein–Hawking's black hole entropy), may combine for a potential "insider" as a complete Everett universe by means of certain phase relations which facilitate the vanishing of quantum correlations between inverse zones of causality. He would calculate the entropy of his inside world from his Everett branch, and possibly obtain a much smaller result.

The situation appears related to Hawking's conjecture[17] that black and white holes are indistinguishable to an outside *observer* in an *equilibrium world*—probably a self-contradictory construction. The character of a hole with respect to an outside observer seems to depend on whether this outside world is expanding or collapsing.

So probably does the thermodynamical and therefore the psychological arrow as it appears to this observer, in this way possibly restoring the overall time-reflection symmetry. The most symmetric quantum state is a superposition of an expanding and collapsing world. It is separated into its components by the symmetry-breaking Everett decomposition[18] if this is appropriately defined.[3] This may explain cosmic censorship.

To sum up: The possibility of an essential observer dependence of what we are used to call our universe (Wheeler's "observer-participancy"[19]) now seems to be indicated from two sides: by the recent developments of black hole dynamics and by the problem of a fundamental interpretation of quantum theory. This may be encouragement to consider such unconventional ideas a bit more seriously. (NOTE: Since this paper was submitted, several related papers have appeared in the literature.[21–24])

REFERENCES AND NOTES

1. P. C. W. Davies, *The Physics of Time Asymmetry* (University of California Press, 1977).
2. R. Penrose, in *General Relativity*, S. W. Hawking and W. Israel, eds. (Cambridge University Press, Cambridge, 1979), p. 581.
3. H. D. Zeh, *Found. Phys.* **9**, 803 (1979).
4. R. Zwanzig, in *Boulder Lectures in Theoretical Physics*, Vol. 3 (1960), p. 106.
5. H. Everett III, *Rev. Mod. Phys.* **29**, 454 (1957).
6. H. D. Zeh, *Epist. Lett.* 49.0, **26**, 21 (1980).
7. Ref. 2, p. 585.
8. H. D. Zeh, *Found. Phys.* **1**, 69 (1970).
9. O. Kübler and H. D. Zeh, *Ann. Phys. (N.Y.)* **76**, 405 (1973); H. D. Zeh, *Found. Phys.* **3**, 109 (1973).
10. S. W. Hawking, *Commun. Math. Phys.* **43**, 199 (1975).
11. P. C. W. Davies, *J. Phys. A* **8**, 609 (1975); W. Unruh, *Phys. Rev. D* **14**, 870 (1976).
12. J. D. Bekenstein, *Phys. Rev. D* **7**, 2333 (1973).
13. R. Penrose, in *Confrontation of Cosmological Theories with Observational Data*, M. S. Lougain, editor (Reidel, Boston, 1974).
14. T. Gold, editor, *The Nature of Time* (Cornell U.P., Ithaca, New York, 1967).
15. W. Rindler, *Mon. Not. R. Astron. Soc.* **116**, 663 (1956); S. Weinberg, *Phys. Rev. Lett.* **36**, 294 (1976).
16. J. F. Clauser and A. Shimony, *Rep. Progr. Phys.* **41**, 1881 (1978).
17. S. W. Hawking, *Phys. Rev. D* **13**, 191 (1976).
18. H. D. Zeh, *Found. Phys.* **5**, 371 (1975).
19. J. A. Wheeler, *Frontiers of Time*, Enrico Fermi School of Physics (Varenna, 1977), course 72.
20. R. M. Wald, *Phys. Rev. D* **21**, 2742 (1980).
21. D. N. Page, *Phys. Rev. Lett.* **44**, 301 (1980).
22. P. C. W. Davies, *Quantum Gravity 2*, C. J. Isham *et al.*, eds. (Claredon Press, Oxford, 1981), p. 183.
23. R. M. Wald, *Quantum Gravity 2*, C. J. Isham *et al.*, eds. (Clarendon Press, Oxford, 1981), p. 224.
24. R. Penrose, *Quantum Gravity 2*, C. J. Isham *et al.*, eds. (Clarendon Press, Oxford, 1981), p.245.

PART B

ISSUES IN BIOLOGY, PHILOSOPHY, AND THE HISTORY OF SCIENCE

1

Biology and Physics: *Reflections on Reductionism*

Francisco J. Ayala
Department of Genetics
University of California
Davis, California 95616

The Questions

The relationships between biology and the physicochemical sciences, and between organisms and inorganic matter, are of considerable interest to the philosophy of science. Many papers, books, and symposia have been dedicated to these relationships in recent years (e.g., Refs. 1 and 2). The issues at stake are sometimes called "the problem of reductionism" or "the question of reduction." Few philosophical issues have been more actively debated in recent years, particularly among scientists, than the question of reduction. These debates, however, often involve several different issues, not always properly distinguished. Issues about the relationship between biology and the physical sciences, or between organisms and their physical components, arise in at least three domains, which may be called "ontological," "methodological," and "epistemological." I shall identify the issues raised in each domain, and then consider each domain in turn.

Reductionistic questions arise, first, in what may be called the ontological, the structural, or the constitutive domain. The issue here is whether or not physicochemical entities and processes underlie all living phenomena. Are organisms constituted of the same components as those making up inorganic matter? Or, is it the case that organisms consist of other entities besides molecules and atoms? Other questions related to the previous ones are the following: Are organisms nothing else than aggregations of atoms and molecules? Do organisms exhibit properties other than those of their constituent atoms and molecules?

Secondly, there are reductionist questions that might be called methodological, procedural, or strategical. These questions concern the strategy of research and the acquisition of knowledge, the approaches to be followed in the investigation of living beings. The general question is whether a particular biological problem should always be investigated by studying the underlying (ultimately, physical) processes, or whether it should also be studied at higher levels of organization, such as the cell, the organism, the population, and the community.

The third type of reductionistic questions concern issues that may be called epistemological, theoretical, or explanatory. The fundamental issue here is whether or not the theories and laws of biology can be derived from the laws and theories of physics and chemistry. Epistemological reductionism is concerned with the question whether biology may be dispensed with as a separate science because it represents simply a special case of physics and chemistry.

Distinction of the various kinds of questions being asked in debates about reductionism is the first step towards solving the issues. Much argumentation and confusion have resulted from failure to identify the question being argued in particular instances. It is not untypical, for example, to see a reductionist concerned primarily with the ontological question, accusing a self-proclaimed antireductionist of having vitalistic ideas, while the latter may be an epistemological antireductionist but also in fact an ontological reductionist.

Atoms and Organisms

In the ontological or constitutive domain, the reductionist–antireductionist controversy in its extreme form resolves into the mechanism versus vitalism issue. The mechanist position is that organisms are ultimately made up of the same atoms that make up inorganic matter, and of nothing else. Vitalists argue that organisms are made up not only of material components (atoms, molecules, and aggregations of them) but also of some nonmaterial entity, variously called by different authors, entelechy, vital force, *élan vital,* radial energy, and the like. Aristotle (384–322 B.C.), the great Greek philosopher who was also the best biologist of his time, is sometimes said to have been the first systematic proponent of vitalism. The modern mechanism–vitalism controversy dates from the seventeenth century when René Descartes (1596–1650) proposed that animals are nothing else than complex machines. Early in the twentieth century, vitalism was defended by philosophers such as Henri Bergson (1859–1941), and by some biologists, notably Hans Driesch (1867–1941). At present this extreme form of vitalism has no distinguished proponents among biologists, and few if any among philosophers.

Vitalism has been excluded from science primarily because it does not meet the requirements of a scientific hypothesis; because it is not subject to the possibility of empirical falsification and therefore leads to no fruitful observations or experiments. Moreover, all available evidence indicates that organisms and life processes can be explained without recourse to any substantive nonmaterial entity.

Ontological reductionism claims that organisms are exhaustively composed of nonliving parts; no substance or other residue remains after all atoms making up an organism are taken into account. Ontological reductionism also implies that the laws of physics and chemistry fully apply to all biological processes at the level of atoms and molecules.

Ontological reductionists do not necessarily claim, however, that organisms are nothing but atoms and molecules. The inference that because something consists only of something else, it is nothing but this "something else" is an erroneous inference, called by philosophers the "nothing but" fallacy. Organisms consist exhaustively of atoms and molecules, but it does not follow that they are nothing but heaps of atoms and molecules. A steam engine may consist only of iron and other materials, but it is something else than iron and the other components. Similarly an electronic computer is not only a pile of semiconductors, wires, plastic, and other materials. Organisms are made up of atoms and molecules, but they are highly complex patterns, and patterns of patterns, of these atoms and molecules. Living processes are highly complex, highly special, and highly improbable patterns of physical and chemical processes.

A much debated reductionist question that belongs in the ontological domain is whether organisms exhibit "emergent" *properties*, or whether their properties are simply those of their physical components. For example, are the functional properties of the kidney simply the properties of the chemical constituents of that organ? It must be pointed out, first, that the question of emergent properties is not exclusive to biology, but arises for all complex systems with respect to their parts. The general formulation of this question is whether the properties of a certain kind of object are simply the properties of other kinds of objects, namely, their component parts, organized in certain ways.

Whether complex systems exhibit emergent properties is largely a spurious issue that can be solved as a matter of definition. Consider the following question: are the properties of common salt, sodium chloride, simply the properties of sodium and chlorine when they are associated according to the formula NaCl? If among the properties of sodium and chlorine we include their association into table salt and the properties of the latter, the answer is "yes." In general, if among the properties of an object we include the properties that the object has when associated with other objects, it follows that the properties of complex systems, including organisms, are also the properties of their component parts. This is simply a definitional maneuver that contributes little to understanding the relationships between complex systems and their parts.

In common practice we do not include among the properties of an object all the properties of the systems resulting from its association with any other objects. There is good reason for that. No matter how exhaustively an object is studied in isolation, there is usually no way to ascertain all the properties that it may have in association with any other object. Among the properties of hydrogen we do not usually include the properties of water, of ethyl alcohol, of proteins, and of human beings. Nor do we include among the properties of iron those of the steam engine.

The question of emergent properties may also be formulated in a somewhat different manner. Can the properties of complex systems be *inferred* from knowledge of the properties that their component parts have in isolation? For example, can the properties of benzene be predicted from knowledge about oxygen, hydrogen, and carbon? or, at a higher level of complexity, can the behavior of a cheetah chasing a deer be predicted from knowledge about the atoms and molecules making up these animals? Formulated in this manner, the issue of emergent properties is an epistemological question, not an ontological one; we are now asking whether the laws and theories accounting for the behavior of complex systems can be derived as logical consequences from the laws and theories that explain the behavior of their component parts. Epistemological questions of reductionisms will be discussed later.

Research Strategies

One outstanding characteristic of living beings is a complexity of organization recognized in their common name, "organisms." There is a hierarchy of levels of complexity that runs from atoms, through molecules, macromolecules, organelles, cells, tissues, multicellular organisms, populations, and communities. Some biological disciplines focus on one or a few of these levels of complexity of organization. Cytology is the study of cells, histology is the study of tissues, ecology studies populations and communities. Yet the biological disciplines are distinguished more by the kinds of questions asked and the kinds of answers sought than by the levels of organization that are investigated.

Methodological reductionism is the claim that the best strategy of research is to study living phenomena at increasingly lower levels of complexity, and ultimately at the level of atoms and molecules. For example, genetics should seek to understand heredity ultimately in terms of the behavior and structure of DNA, RNA, enzymes, and other macromolecules, rather than in terms of whole organims, which is the level at which the Mendelian laws of inheritance are formulated. Methodological reductionism has its counterpart in what may be called methodological compositionism.[3] This is the claim that in order to understand organisms we must explain their organization—how organisms and groups of organisms come to be organized and what functions the organization serves. According to methodological compositionism, organisms and groups of organisms must be studied as wholes, not only in their component parts.

Methodological reductionism in its extreme form would be the claim that biological research should be conducted only at the level of the physicochemical component parts and processes. Research at other levels is, allegedly, not worth pursuing or is at best of only provisional value, since biological phenomena must ultimately be understood at the molecular and atomic levels. Methodological compositionism in its extreme form makes the opposite claim; namely, that the only biological research worth pursuing is that at the level of whole organisms, populations, and communities. Research at lower levels of organization may, alledgedly, be good physics or good chemistry but it has no biological significance.

It is unlikely that any scientist would thoughtfully sponsor the extreme forms of either compositionism or reductionism advanced in the previous paragraph. Extreme methodological reductionism would imply the unreasonable claim that genetic investigations should not have been undertaken until the discovery of DNA as the hereditary material, or that a moratorium should be declared in ecology until we can investigate the physicochemical processes underlying ecological interactions. Similarly, extreme methodological compositionism would imply that understanding the structure of DNA or the enzymatic processes involved in replication is of no significance to the study of heredity, or that the investigation of physicochemical reactions in the transmission of nerve impulses is of no interest to the understanding of animal behavior.

A moderate position of methodological reductionism points to the success of the analytical method in science, and to the obvious fact that the understanding of living processes at any level of organization is much advanced by knowledge of the underlying processes. Moderate methodological reductionists would claim that the best strategy of research is to investigate any given biological phenomenon at increasingly lower levels of organization as this becomes possible, and ultimately at the level of atoms and molecules.

The positive claims of moderate methodological reductionists are legitimate. The analytical method is of great heuristic value; much is often learned about a phenomenon through the investigation of its component elements and processes. In biology, the most impressive achievements of the last few decades are those of molecular biology. But there is little justification for any exclusionist claim that research should always proceed by investigation of lower levels of integration. The only criterion of validity of a research strategy is its success. Compositionist as well as reductionist approaches, synthetic as well as analytic methods of investigation, are justified if they further our understanding of a phenomenon, if they increase knowledge. Reductionist and compositionist approaches to the study of a biological problem are complementary; often the best strategy of research is an alternation between analysis and synthesis.

Investigation of a biological phenomenon in terms of its significance at higher levels of complexity often contributes to the understanding of the phenomenon itself; compositionist investigations are also heuristic. For example, it is doubtful that the structure and functions of DNA would have been known as readily as they were if there had been no previous knowledge of Mendelian genetics. The problem of the specificity of the immune response of antibodies proved refractory to a satisfactory solution as long as antibodies, antigens, and their structure alone were taken into consideration. The natural selection theory of antibody formation emerged only when antibodies were considered in their organismic milieu. Although the idea of clonal selection was first logically inadequate and quite vague, it had an enormous heuristic value in helping to understand how the specificity of antibodies comes about.[4]

Methodological reductionism and compositionism are sometimes based on convictions about the possibility of epistemological reduction. A methodological reductionist might claim that research should be pursued at increasingly lower levels of organization because he is convinced that ultimately all biological phenomena will be explained by the laws and theories of the physical sciences and thus that the only knowledge of lasting value is that acquired at the level of atoms and molecules. A methodological compositionist might claim that epistemological reductionism is impossible, and therefore that full understanding of a biological phenomenon requires knowledge of its significance for higher levels of integration. Methodological and epistemological reductionism are nevertheless separate issues. Methodological reductionism is concerned with the strategies of research and the acquisition of knowledge. Epistemological reductionism deals with the organization of knowledge and the logical connections between theories. An epistemological reductionist, for example, might nevertheless accept compositionist approaches to research because of their heuristic value. Similarly, epistemological antireductionists often claim that biological research should be pursued at all levels of integration of the living systems, including the atomic and molecular levels.

Epistemological Reductionism

When philosophers of science speak of reductionism they generally refer neither to ontological nor to methodological issues, but to epistemological reduction. In biology the question of epistemological (theoretical, explanatory) reduction is whether the laws and theories of biology can be shown to be derived as special cases from the laws and theories of the physical sciences.

Science seeks to discover patterns of relations among vast kinds of phenomena in such a way that a small number of principles explain a large number of propositions concerning those phenomena. Science advances by developing gradually more comprehensive theories, i.e., by showing that theories and laws that had hitherto appeared as unrelated can in fact be integrated in a single theory of greater generality. For example the theory of heredity proposed by Mendel can explain about many kinds of organisms diverse observations such as the proportions in which traits are transmitted from parents to offspring, why progenies exhibit some traits inherited from one parent and some from the other parent, and why the offspring may exhibit traits not present in their parents. The discovery that the behavior of chromosomes during meiosis is connected with the Mendelian principles made possible the explanation of many additional observations concerning heredity; for example, why certain traits are inherited independently from each other, while other traits are transmitted together more often than not. Further discoveries made possible the development of a unified theory of inheritance of great generality, which explains many diverse observations, including the distinctness of individuals, the adaptive nature of organisms and their traits, and the discreteness of species.

The connection among theories has sometimes been established by showing that the tenets of a theory or branch of science can be explained by the tenets of another theory or branch of science of greater generality. The less general theory (or branch of science), called the secondary theory, is then said to have been reduced to the more general or primary theory. Epistemological reduction of one branch of science to another takes place when the theories or experimental laws of a branch of science are shown to be special cases of the theories and laws formulated in some other branch of science. The integration of diverse scientific theories and laws into more comprehensive ones simplifies science and extends the explanatory power of scientific principles, and thus conforms to the goals of science. Epistemological reductions are of great value to science because, as special cases of the integration of theories, they greatly contribute to the advance of scientific knowledge.

The reduction of a theory or even a whole branch of science to another has been repeatedly accomplished in the history of science.[5,6] One of the most impressive examples is the reduction of thermodynamics to statistical mechanics made possible by the discovery that the temperature of a gas reflects the mean kinetic energy of its molecules. Several branches of physics and astronomy have been to a large extent unified by their reduction to a few theories of great generality such as quantum mechanics and relativity. A large sector of chemistry was reduced to physics after it was shown that the valence of an element bears a simple relation to the number of electrons in the outer orbit of the atom. Parts of genetics have been to some extent reduced to chemistry after discovery of the structure and mode of replication and action of the hereditary material, DNA.

The impressive successes of these reductions, and in particular the spectacular achievements of molecular biology, have led some authors to claim that the ideal of science is to reduce all natural sciences, including biology, to a comprehensive physical theory that would provide a common set of principles of maximum generality capable of explaining all observations about natural phenomena. Some authors have gone so far as to claim that the only biological research worth pursuing is that contributing to the explanation of biological phenomena in physicochemical terms.

Conditions for Reduction

Nagel[5] has formulated the two conditions that are necessary and jointly sufficient to effect the reduction of one theory or branch of science to another. These are the condition of *derivability* and the condition of *connectability*. Epistemological reduction takes place when the experimental laws and theories of a branch of science are shown to be special cases of the experimental laws or theories of some other branch of science. The condition of derivability simply states that in order to reduce a branch of science to another it is necessary to show that the laws and theories of the secondary science can be derived as logical consequences from the laws and theories of the primary science.

No terms can appear in the conclusion of a demonstrative argument that do not appear in the premises. The reduction of one theory to another takes the form of a deductive argument in which one of the premises is the primary theory, and the conclusion is the secondary theory. For the deduction to be logically valid there must be another premise that establishes the connection between the terms of the primary theory and the terms of the secondary theory. This is the condition of connectability. Generally the experimental laws and theories of a branch of science contain distinctive terms that do not appear in other branches of science. To accomplish an epistemological reduction it is necessary that suitable connections be established between the terms of the secondary science and those used in the primary science. This may be accomplished by redefining the terms of the secondary science using terms of the primary science. The reduction of thermodynamics to statistical mechanics required the definition of terms such as "temperature" by means of terms such as "kinetic energy." The reduction of theories or experimental laws of genetics to physicochemistry requires that terms such as "gene" and "chromosome" be defined by means of terms such as "hydrogen bond," "nucleotide," "deoxyribonucleic acid," "histone protein," and the like. Whenever the conditions of connectability and derivability are satisfied, the epistemological reduction of a theory to another becomes logically feasible. If all the experimental laws and theories of a branch of science can be reduced to experimental laws and theories of another branch of science, the former science will have been completely reduced to the latter.

It bears repetition that epistemological reduction is not a question whether the *properties* of a certain kind of objects, such as organisms, result from the properties of other kind of objects, such as their component parts. This problem is an issue of ontological reductionism, which can be solved by a convention as to what is to be included among the properties of component parts. The reduction of one science to another is rather a matter of deriving a set of *propositions* from another such set. Scientific laws and theories consist of propositions about the natural world. The question of epistemological reduction can only be settled by concrete investigation of the logical consequences of propositions, and not by discussions about the nature of things or their properties. It is of course a legitimate epistemological question to ask whether *statements* concerning the properties of organisms can be deduced logically from statements concerning the properties of their physical components.

It follows from the previous comments that questions of epistemological reduction can only be properly answered by concrete reference to the actual state of development of the disciplines involved. Certain parts of chemistry were reduced to physics after the modern theory of atomic structure was advanced half a century ago. That reduction could not have been accomplished before such a development. If the reduction of one science to another is not possible at the present stage of development of two given disciplines, claims that such a reduction will be possible in the future carry little weight since such claims depend on the hypothetical development of as yet nonexisting theories.

From Biology to Physics?

There are some extreme positions about the question of epistemological reductionism that can be quickly discounted. Some substantive vitalists have claimed that biology is irreducible in principle to the physical sciences because living phenomena are the manifestation of nonmaterial principles, such as vital forces, entelechies, and so on. Epistemological antireductionism is then predicated on ontological antireductionism. Vitalism is not an empirical hypothesis because it does not lend itself to the possibility of empirical falsification. Moreover, the origin, structure, and functions of organims can be explained without recourse to nonmaterial components or principles.

At the other end of the spectrum would be the claim that the epistemological reduction of biology to the physical sciences is possible, and indeed the most important task of biologists at present. The impressive successes of molecular biology during recent decades have moved some people to claim that the only worthy and truly scientific biological investigations are those leading to the explanation of biological phenomena in terms of the underlying physicochemical components and processes. Nevertheless, the epistemological reduction of biology to the physical sciences is not possible at present. In the current stage of scientific development, a great many biological terms such as organ, species, consciousness, mating propensity, fitness, competition, predator, and many others, cannot adequately be defined in physicochemical terms. Nor is there any class of statements and hypotheses in physics and chemistry from which every biological law could logically be derived. Neither the condition of connectability nor the condition of derivability—the two necessary conditions for epistemological reduction—is satisfied.

A moderate reductionist position is probably not uncommon among biologists. It is claimed that, although the reduction of biology to physicochemistry cannot be effected at present, it is possible in principle and is indeed a goal to be actively sought. The factual reduction of biology to the physical sciences is made contingent upon further progress in the biological or physical sciences or both. This moderate form of epistemological reductionism is often based on convictions about ontological reductionism. It is generally accepted by biologists that living beings are exhaustively made up of physical components. It does not follow, however, that organisms are *nothing but* physical systems. Ontological reductionism does not entail epistemological reductionism. From the fact that organisms are exhaustively composed of atoms and molecules, it does not logically follow that the behavior of organisms can be exhaustively explained by the laws advanced to explain the behavior of atoms and molecules.

The claim that the reduction of biology to physicochemistry will eventually be possible is contingent upon unspecified, and at present unspecifiable, scientific advances. It is, therefore, a position that cannot be convincingly argued. Moreover, there are reasons to believe that complete reduction of biology to physics will never be possible. Popper has shown that no major case of epistemological reduction (including such model cases as the reduction of thermodynamics to statistical mechanics) "has ever been *completely* successful: there is almost always an unresolved residue left by even the most successful attempts at reduction" (Ref. 6, p. 260; see also Ref. 7). It does not follow, however, that biologists should not attempt to reduce their theories to those of the physical sciences, whenever such an undertaking seems likely to be successful. On the contrary, epistemological reductions are most successful forms of scientific explanation. A great deal is learned from epistemological reductions even when they are unsuccessful or incomplete, because much understanding is gained by the partial success, and valuable problems arise from the partial failure.[6] In biology, the reduction of Mendelian genetics to molecular genetics has been far from completely successful.[7] Yet there can be little doubt that much has been learned from what has been accomplished up to the present.

Some authors have claimed that the reduction of biology to physics and chemistry is impossible in principle because biological disciplines employ patterns of explanation that do not occur in the physical sciences.[3,8] Historical explanations are sometimes mentioned as distinctive of the biological and social sciences. Historical explanations play a role in evolutionary theory, although this theory is primarily concerned with causal explanations of evolutionary processes. Historical patterns of explanation, however, occur in some physical sciences such as astronomy and geology. On the other hand, as I have proposed elsewhere,[9] teleological patterns of explanation are appropriate in biology but appear to be neither necessary nor appropriate in the explanation of natural physical phenomena.

References and Notes

1. A. Koestler and J. R. Smythies, *Beyond Reductionism* (Hutchinson, London, 1969).
2. F. J. Ayala and Th. Dobzhansky, editors, *Studies in the Philosophy of Biology* (Macmillan, London, and University of California Press, 1974).
3. G. G. Simpson, *This View of Life* (Harcourt, Brace, and World, New York, 1964).
4. G. M. Edelman, "The Problem of Molecular Recognition by a Selective System," in F. J. Ayala and Th. Dobzhansky, editors, *Studies in the Philosophy of Biology* (Macmillan, London, 1974), pp. 45–56.
5. E. Nagel, *The Structure of Science* (Harcourt, Brace, and World, New York, 1961).
6. K. R. Popper, "Scientific Reduction and the Essential Incompleteness of All Science, in F. J. Ayala and Th. Dobzhansky, editors, *Studies in the Philosophy of Biology* (Macmillan, London, 1974).
7. D. Hull, *Philosophy of Biological Science* (Prentice-Hall, Englewood Cliffs, New Jersey, 1974).
8. F. J. Ayala, "Biology as an Autonomous Science," *Amer. Sci.* 56, 207–221 (1968).
9. F. J. Ayala, "Teleological Explanations in Evolutionary Biology," *Phil. Sci.* 37, 1–15 (1970).

2

The Age of Analysis Reassessed

Heinz Beran*
Austrian Radio Corporation
Vienna, Austria
British Broadcasting Corporation
London, United Kingdom

Perhaps I should first of all explain that the term "Age of Analysis" is a fairly cheap label which is used to simplify or to underline one aspect, because it is clear that there has been philosophical analysis ever since the Greeks started to philosphize. Likewise, the 17th century could be called the Age of Reason, when of course reason has existed and will exist regardless of any age; and the 5th to the 15th centuries have been called the Age of Belief, while there has been and there will be belief as long as people hope for help from a higher being or seek the blessing of a church.

On the other hand, I have myself chosen this label, because it seems the best "shortspeak" for something that began at the turn of the century. Of course, I am concentrating on Vienna and two kinds of analysis which exploded there at that time. It has to do in the first place with the changed position of philosophy—the fact that philosophy lost out against science.

I remember a man on the Isle of Man—we were both interned there—who told me he knew the answer to the question "Wie ist Philosophie als Wissenschaft möglich?" (How can philosophy be a science?) It is quite a serious question; it only sounds so funny in its Kantian formulation, expecially to English ears; by the way, he did not give me any answer in the seven or so months that we were together there. Originally, *philosophia* was love of wisdom, and comprised everything the Greeks speculated on. Karl Popper says: "Historically all Western science is an offspring of Greek philosophical speculation about the cosmos."

As the offspring grew up and became independent, it turned against its father

*Originally from the Institute of Psychology, University of Vienna, now freelance author in London, mainly for the Austrian Radio Corporation and the BBC; a pupil of Bühler, Carnap, and Schlick and a friend of Karl Popper since their student days.

and said "You must be like me, or go." This was done either by shedding all problems that could not be dealt with in a scientific way—metaphysical questions, for instance, were declared to be meaningless, ethics was declared to be not statements of fact—or, eventually, by dealing with philosophy even more radically: In 1895 a professor of physics from Prague was called to Vienna and given the chair of philosophy—Ernst Mach, whose name now is best known to those who are interested in the relation of the speed of an airplane to that of sound. This was quite revolutionary. His successor, Boltzmann, kept the chair of physics when he became professor of philosophy and later (in the early twenties) Moritz Schlick, a German, came to Vienna to take the same chair. Schlick wrote: "Philosophy, which never talks sense but utters meaningless words, will soon disappear, because philosophers will find that their audience, tired of empty tirades, have gone away."

I do not want to bring Wittgenstein in here—we shall have to deal with him later in more detail—only so much: Wittgenstein took this antiphilosophical development to absurd lengths and, looking back, a modern Cambridge philosopher, Bernard Williams, says with a certain regret and possibly disbelief: "When you had analytically taken the problems apart, you would find that many of the traditional questions of philosophy had not just been solved, but had disappeared. There were really people around who were saying the whole of philosophy would be over in 50 years."

To put your minds at rest, it is not over after more than 50 years—philosophy had a new role perhaps, but it is not over; more of that later.

But why did this happen? Not only because science took such tremendous steps forward, but also because religion had been undermined—first the Catholic Church and then religious faith itself—in the course of a tremendous social change which led to the open society. The result was that Man questioned not only his social position but also his position in the world; old values had disappeared, and there was nothing to hold on to any more. In this situation, philosophers like Bertrand Russell were concerned with a new certainty. Russell wrote in *Reflections on My Eightieth Birthday:* "I wanted certainty in the kind of way in which people want religious faith."

This desire for certainty, the ruthless search for absolute certainty, now of course achieved with scientific methods, led to the discarding of all those problems that human beings have been thinking about ever since they started thinking, led to the shedding of great parts of what formerly was called philosophy and to the eventual reduction of philsophy to the study of everyday language, without even attempting to build a new system—if only of language, not to speak of a philosophic system.

And in order to achieve certainty, everything has to be analyzed: Lies have to be uncovered, untruths discovered. This certainty reminds me of Karl Popper explaining to me the background of the *Communist Manifesto* fifty years ago. He said: "In those days it was very dangerous to be a socialist. Marx had to give his followers confidence by telling them that it was certain that the proletariat would win in the end, that there were laws in history that could be scientifically proved." The first Marxist interpretation of Marx in fact.

It is worthwhile looking at this. Popper does not here argue that it was wrong of Marx to assume that there were laws in history like laws in science; he does not here enter into the argument—he has meanwhile done so at great length when he pointed out the wrongs of historicism, but then, when he talked to me 50 years ago, he used the then fashionable analytical method quite naively. His interpretation of Marx's motives could also be called psychoanalytical—or, if you like, since it was power and fear of persecution that was implied, an Adlerian interpretation.

This kind of analysis must be distinguished from the one I have hinted at till now. If I may recapitulate with Bertrand Russell what this logical analysis was: Russell writes, "Modern analytical empiricism differs from that of Locke, Berkeley and Hume by the incorporation of mathematics and its development of a powerful logical technique. It is thus able in regard to certain problems to achieve definite answers, which have the quality of science rather than of philosophy."

But now we come to this other kind of analysis; it does not take anything at its face value, but wants to dig deeper. Marx speaks of an intellectual superstructure obviously built on a material foundation. Your theories therefore depend on what society you come from and what method of production this society has arrived at. Freud analyzes your dreams by going into the unconscious, and finding out why you dream what you dream and why you say what you say. Not *what* you say, but *why* you say it. To the great majority of laymen the term "analysis" must have been conscious first in the term "psychoanalysis."

But psychoanalysis is only one part of the great analytical explosion that took place in Vienna in the first third of our century; the logical positivism of the Vienna Circle and linguistic philosophy equally belong to this era which I have here called the Age of Analysis, as does twelve-tone music. Why did it all start in Vienna, and why at that time?

Partly it has to do with the state of the Austro-Hungarian monarchy. It was prosperous and had at long last begun to become industrialized to become a modern state, but at the same time it was obvious that the seeds of its destruction had been sown: Nationalism was rife in the multinational monarchy, and in Vienna a strong social democratic movement had begun to make its voice heard. Then there was the world outside with its new ideas about philosophy; not only Schlick but Carnap, too, came from Germany, and Wittgenstein brought what he had learned from Russell and Moore back to Vienna. Freud had studied in Paris. But of course Vienna was the very best bed into which to plant all these ideas, and it would bear fruit most readily, because Vienna is an extraordinary place, a melting pot of nationalities, races, and ideas. The Viennese furthermore possess in their original makeup what could be called an analytical strain.

Vienna, this cemetery of the peoples' migration, on the northern lines of the Roman Empire, where Celts and Slavs and Germanic tribes mixed, where there was the threat of Asian invasion since the Huns, the Avars, and the Magyars came through the Pannonian plains, Vienna, which was besieged by the Turks as late as 1683, made its inhabitants both aware of the threat of the foreigner and the ability to live with him. The Viennese live with people who speak other languages, and understand that language has different layers, that one can play with words. Perhaps the most amusing example of this is what the Viennese call *Schmäh*—telling fibs, but in such a way that if the partner of a conversation shows signs of being aware that this story is a *Schmäh*, a lie or a fabricated story, you can turn the thing into a joke at once, no offense meant and none taken. This applies also to conversation with a young or not so young lady who is not really expected to take everything that is said as pure truth, but who nevertheless enters into the spirit of the game by appearing greatly flattered. A language game then this *Schmäh*, if not quite Wittgenstein's idea of such a game.

The advent of liberal ideas, the growth of industry and banking, greater prosperity, and the chance for Jews to go to universities, even if they were hardly ever allowed to get chairs at universities, meant that the Jewish contribution to this analytic explosion was considerable. They had come from the various parts of the Monarchy—Bohemia, Moravia, Hungary, or Galicia—and filled the ranks of doctors, lawyers, and journalists; no wonder that these newcomers, newcomers in great numbers—there had been Jews living in the ghetto or in small family groups throughout the thousand years of Austrian history—at long last liberated, were drawn into the whirlpool of new ideas, the new fashions in philosophy, psychology, and art.

There is a marvellous story in Freud's *Jokes and their Relation to the Unconscious* which shows a special tendency not to take anything at its face value; a special trend towards an analytic attitude: "Two Jews meet in a railway carriage in a Galician station. 'Where are you going?' asks one. 'To Krakau,' answers number two. Number one gets very angry. 'When you say you are going to Krakau, you want me to think you are going to Lemberg. But I happen to know that you are in fact going to Krakau; so why are you lying?'"

More seriously, there is a beautiful word in the Germanic language of the European Jews who went from the Rhineland to Eastern Europe; in Yiddish, the word is *klären*, which means to think, but etymologically it comes from *klar*, which is clear, as in *klarmachen*—to make clear, to analyze.

Hebrew scholars through the centuries had to interpret the Holy Scriptures and the Law, much in the way that medieval monks had to interpret the Holy Scriptures, ie., explain obscure passages and retranslate texts to make them clearer and therefore better understandable. The text was revealed from God and therefore sacrosanct; all one could do was to interpret and analyze, again and again.

Enough said about why the analytical explosion happened at that time and place. Now, as I approach my central topic, the reassessment of the Age of Analysis, I hesitate because I feel that it might all sound a little too facile if I say that psychoanalysis, logical positivism, and linguistic philosophy have so to speak run their course, and it is time to say what has remained and what we can dispense with in these systems. I cannot speak with the authority of a Professor Hayek, who the other day at the LSE in his inimitable manner finished off Marx and Freud in one fell swoop, saying that our century could be called the age of superstition, because people think they know something when in fact they don't, and he said this especially in connection with Marx and Freud.

I understand and agree with the argument that Marx and Freud claim to have constructed scientific systems when in fact the scientific character of these theories can be queried. Professor Gellner has recently called Marxism "a human social metaphysic, a specification of the general features of either the human or the social–historical situation." This is, I feel, a more scientific assessment. The term "metaphysics" is now being used in a much broader sense than in the days of logical positivism and, in fact, considered part of the scientific endeavor. But if one accepts the older version when metaphysics was practically the same as religion, one still does not have to call anything superstition simply because it cannot be falsified by scientific methods.

With Freud, things are even more complicated. There is a great amount of scientific discovery in his work; observation of neurotic behavior and of clinical symptoms, there are new models: The fundamental one of the Id, the sum total of our animal drives, our life-force or libido, where the pleasure principle rules supreme; then on the borders of reality the Ego is formed, and this Ego is led by the sum total of fathers' or other admired figures' shining example. Pardon my simplification! The Oedipus complex is nowadays accepted and absorbed in everyday language, but of course here we come up against the problem of Freud's symbolism and method of interpretation. I agree with Professor Wollheim that Freud, "by the power of his writings and by the breadth and audacity of his speculations, revolutionized the thought, the lives, and the imagination of an age." Wollheim then writes:

> He contradicted, and in some cases he reversed, the prevailing opinions of the learned as well as of common people, on many of the issues of human existence and culture. He led people to think about their appetites and their intellectual powers, about self-knowledge and self-deceit, about the ends of life, and about man's profoundest passions and about his most intimate failings, in ways that would have seemed to earlier generations at once scandalous and silly.

But, I should add, his own and following generations, too, have thought of much that he taught as scandalous. And the resistance against the disclosures of Freudian psychoanalysis has been one of the constant phenomena of its history. Let me explain this resistance. If I tell somebody that he wants to kill his father and go to bed with his mother, he will obviously be shocked. If he were willing to admit to such wishes, he would be very discreet about them. He would not tell anyone. Not even the family. Least of all the family. So, it must be repressed. It must be put in some deep store from which it can only be brought out by the association method. Let me explain the difficulty of accepting Freud's method of asking his followers to accept unpleasant suppressed truths by quoting something that is less controversial. I don't think nowadays many people really care whether Moses was an Egyptian or not. But, in *Moses and Monotheism* (1939), Freud first suggests that the reader should accept the possibility that Moses was the illegitimate son of an Egyptian princess as a kind of hypothesis, which may be unpleasant to consider but nevertheless is offered to the reader. And then, a few pages further, quite suddenly, Moses *is* an Egyptian; his monotheism is that of Akhnaton: here, in this way of presenting a new model, I find the danger of Freud's methods.

Then there is another problem: Therapeutic success is no proof of truth and of scientific value. Quite simply, the therapy depends on these stages: First you must establish a relation of trust between the analyst and the analyzed; the patient must be willing to listen to the analyst, must trust him. Secondly, you must have a lever by which you tear him out of his neurosis. Sex is a very good lever because it goes very deep into the personality of the patient, but one can use other levers like power and so on. Thirdly, when one has torn him out of his neurosis one must give him a new theory which can be used as a kind of point of crystallization, something to hold onto, and that can differ very widely. Jung, for instance, said some patients are better treated by Freudian, others by Jungian, others by Adlerian explanations. And finally, one must be able to get rid of the patient. The history of Analysis is full of cases where the patient is well when the analyst is there and ill when the analyst isn't. That what finally helps a neurotic patient out of his neurosis is not all that relevant, I think can be seen from the story of Freud's son, Martin, who, as a boy, one day went on an ice rink and the boys made a long chain in such a way that the boy who was furthest out from the center was subject to large centrifugal forces. Martin Freud was at the end of the chain and was thrown against a gentleman who was quietly skating on this same ice-rink in Vienna. The man was angry and boxed his ears. Little Martin was very upset, and when he went home for lunch had still obviously not got over this experience. Father Sigmund said to his son after lunch, "Come to my office; we must talk." Martin came out after a little while and was obviously very relieved and very happy. But—he didn't know what his father had said to him.

On the other hand, terms like "repression," "sublimation," "abreaction," "Freudian slip," etc., are absorbed into everyday language and figure in the practice of all sorts of psychotherapists and psychologists, so one must say absorption can be considered to be the second pole of two dialectical opposites: Psychoanalysis would then lie between resistance and absorption. As far as the status of psychoanalysis in the world today is concerned, one must note a certain decline; a few years ago I looked through 65 pages of an official publication listing scientific research currently being undertaken in British universities within the fields of psychology and psychiatry, to discover only one project which mentions the term "psychoanalytic" at all, and that in a manner which suggests its findings may well prove unfavorable.

And that brings me back to the main theme of this essay: to the problem of certainty and the search for scientific truth. Quantitative and behavioristic research have now beaten analytical studies, because they are considered more scientific. And in psychiatry, especially in America, they have turned to behavior therapy, sensitivity training, nude marathons, Zen, and to drugs. This latter development is especially regrettable, because it short-circuits all attempts at solution which would teach us more psychology. But, in a way, this so-called progress of medical science cannot be stopped—men are so impatient. And these drug people even get popular support when they define the "Superego" as that part of the Ego which is soluble in alcohol. This does not, however, mean that the breadth and audacity of Freud's thought has disappeared without a trace; former pupils and friends with their new schools, the whole post-Freudian development goes on. It is noteworthy that amongst the recent post-Freudians there are some who have adopted Adlerian ideas, for instance Heinz Hartmann's ego psychology and Ludwig Binswangers existential analysis. More importance is attached to the ego; the person is viewed as a whole being, the agent of its actions. Generally the emphasis on sexuality is being rejected, while the importance of interpersonal relations and of cultural background are stressed. But there is no need to decide between sex and power. Freud himself, when told that Alfred Adler had been invited to the United States, said: "Presumably the object is to save the world from sexuality and base it on aggression."

One final remark on Freud's relations with Vienna. I think Professor Wollheim was a little too severe in his judgment when he wrote about Freud: "For Vienna he had always expressed intense dislike, and he found it hard to bear the facile judgment of the world that psychoanalysis owed everything to a city from which he, in point of fact, had never experienced anything except indifference, hostility, and rejection."

As against that I should like to quote the master himself; in the Preparatory Note I to *Moses, His People, and Monotheist Religion* he writes: "But psychoanalysis, which in the course of my long life has gone everywhere, still possesses no home that could be more valuable for it than the city in which it was born and grew up." Vienna, by the way, has now after all these years acknowledged Freud's greatness; the old flat in the Berggasse has been turned into a museum and Anna, his daughter, has been honored and has lectured at Vienna University.

Let us now look at relations between Vienna and those men who were the protagonists of the other kind, the second type of analytical explosion, logical analysis, later called logical positivism, i.e., the Vienna Circle, and linguistic philosophy. Two authors, Janik and Toulmin, have written a book called *Wittgenstein's Vienna*, an account of the life, the times, and the culture of Habsburg Vienna before World War I, the Vienna of Sigmund Freud, Arnold Schönberg, Adolph Loos, Oskar Kokoschka, modernism, Mayerling, and Ludwig Wittgenstein, the great philosopher whose epochal work was formed during the decline and fall of the Austro-Hungarian Empire.

This panopticum of Vienna at the beginning of the century also deals with Fritz Mauthner and Karl Kraus, Weininger and Schnitzler, Hoffmannsthal and Richard Strauss, and others—the quantity of information is overwhelming, but I think the idea cannot be upheld that this Vienna was Wittgenstein's, except in the sense that he happened to live there part of the time and had a certain influence on some intellectuals, and that his special kind of analytical genius fitted into the place and the era. Many of you know the beautiful story Mrs. Mickey brought back from Vienna: This old lady in Vienna said, "Oh yes, I knew the Wittgenstein family. There was the gifted brother, Paul. He was a pianist, and he lost one arm in the World War, but he became a very great virtuoso, playing the piano with one hand. But the other, Ludwig, he was a funny boy. You know, in winter he used to skip with a rope on the wet grass without any shoes on. And then, one day, he went to England . . . and was never heard of again."

However, we know that when Wittgenstein came back from England he brought Russell's and Moore's ideas with him, and the *Tractatus Logico-Philosophicus* (1921) he wrote during the first world war became the text most influential on the Vienna

Circle. In his *Philosophy and Logical Syntax*, Rudolf Carnap writes

> I, as well as my friends in the Vienna Circle, owe much to Wittgenstein, especially as to the analysis of metaphysics [you remember metaphysics is meaningless according to Wittgenstein and the logical positivists, even if in a different way] and on the point just mentioned I cannot agree with him. In the first place he seems to me to be inconsistent in what he does. He tells us that one cannot state philosophical propositions and that whereof one cannot speak thereof one must be silent; and then instead of keeping silent, he writes a whole philosophical book. Secondly, I do not agree with his statement that all his propositions are quite as much without sense as metaphysical propositions. My opinion is that a great number of his propositions (unfortunately not all of them) have in fact sense; and that the same is true for all propositions of logical analysis.

Wittgenstein had talked about his famous ladder, that is to say, after you have climbed up on the steps of his kind of philosophical analysis you should throw the ladder away. Wittgenstein says toward the end of the *Tractatus*, "Anyone who understands my propositions will recognise that they are senseless." A difficult type I should say. In general, I find Wittgenstein to be what is called a philosopher's philosopher. We, that is, we ordinary people, are not at all surprised if we don't understand when two bridge players talk shop, always assuming that we are not bridge players ourselves; we accept that we cannot follow the debate between two atomic physicists or experts in stochastic processes. But somehow, when Wittgensteinians start talking about the wallpaper behind the radiator and why we assume that the wallpaper behind the radiator, there, where we cannot see it, continues or perhaps does not continue to be like it is above the radiator, we think: Ah, this is easy; we can learn something. But somehow very soon we get lost, and then we get angry, but really quite unfairly, because here philosophers talk shop. We are not philosophers, so why should we understand?

The logical positivists on the other hand are easier to understand, very clear in their language: Science is all, the rest is nonsense. There is an emphasis on clarity, against wooliness, against Victorian hypocrisy. But there were difficulties about the problem of verification, a very central point for all concerned with science—it appears that they could not formulate the verification principle satisfactorily—and about induction, difficulties which, I believe, only Karl Popper managed to overcome; and he, as you know, did not belong to the *Wiener Kreis*, in fact he was their Majesties' loyal opposition in those days.

Since the logical positivists allowed besides the propositions that are empirically verifiable only those that are demonstrable, like the propositions of mathematics and logic, there were many mathematicians in the Vienna Circle: Hans Hahn, Karl Menger, and Kurt Gödel. When I was a young student in Vienna, I heard a story which is full of significance about three mathematicians who were discussing a highly metaphysical problem, namely, what paradise was like. One of them said, "Paradise is the place where the squaring of the circle is possible." But the other two were horrified about this idea; the three agreed finally on the formula: Paradise is the place where the squaring of the circle is more possible than here. Squaring of the circle is one of three geometrical problems which have puzzled mathematicians for 2000 years since Greek times; it asks how can one construct a square equal in area to that of a given circle. This story, by the way, must be apocryphal because sometime in the 19th century it had been proved that such a construction is logically impossible.

But they were nice people, the Vienna Circle; there was Otto Neurath with his pictorial statistics and Edgar Zilsel the philosophical popularizer who taught philosophy to us students of the Institute of Education of the City of Vienna, including Karl Popper, who constantly argued with him. The optimistic belief in the power of scientific endeavor was certainly an influence on the Viennese school reform, as of course was Alfred Adler's Individual Psychology. On certain points, however, the logical positivists disagreed, namely, whether the basic propositions in which philosophical analysis terminates refer to immediate experience—the so-called "sense-data school"—or to material objects—the physicalist school, which eventually leads to the theory that mind and body are one. Neurath was one of the great protagonists of this orthodox Monism. Linked with this, there was a difference of opinion amongst the logical positivists as to whether Truth is a relation of correspondence between propositions and extralinguistic reality, or one of coherence between propositions. This is important because it leads us on to Wittgenstein Two, that is the Wittgenstein of the *Philosophical Investigations*(1953). Bertrand Russell, speaking about the logical positivists, says in *My Philosophical Development:*

> There has been two views about empirical statements: one that they were justified by some relations to facts; the other that they were justified by conformity to syntactical rules. But the adherents of WII [Wittgenstein Two] do not bother with any kind of justification and thus secure for language an untrammelled freedom which it has never hitherto enjoyed. The desire to understand the world is they think an outdated folly. This is my most fundamental disagreement with them.

Russell, who had helped to establish Wittgenstein in this country (England) by getting the *Tractatus* published, could not understand Wittgenstein in his second phase, nor did and does Popper understand. Philosophers like arguments and disputes. Personally I believe this is inherent in their job. Professor Ayer in a discussion with Brian Mages said

> I mean science is not by any means always a bed of roses, but still in science you do put up a hypothesis and you get a test and the test is positive or negative. And then if it's positive you are pleased and if it's negative you try again. In philosophy one never knows where one is, one never knows when one has got a problem solved, whether one has a problem properly posed. And, of course, some people now think there aren't any problems.

This is a typical *cri de coeur*. It comes from the fact that our computer obviously has a limited capacity. Like a wild animal in a cage, the philosophers seem to run from one end of the cage to the other as near the fencing as possible, eternally to and fro. One philosopher is for determinism, then the other must be for indeterminism; one is for everything being matter, the other for everything being mind; one lot wants to reduce everything to one original cause, (they are monists), the others see an eternal duality or plurality; one lot believes that there is nothing in our mind that does not come through our senses, the other lot insists that we see what we want to see, there are preestablished patterns on our mind which control our picture of the world; and so on. Mind you, the Viennese philosophers are especially argumentative and critical of one another. This is Coffeehouse tradition. Karl Kraus, who wasn't a philosopher, but a poet and critic, had three heroes: Shakespeare, Nestroy, and Jacques Offenbach. But the list of his dislikes or hates is much longer: He was against Freud and Wagner, against Hoffmannsthal and Max Reinhardt, against Richard Strauss and Werfel, Bahr and Lehar, and of course most of all against the Neue Freie Presse.

So we are not greatly surprised when we read the amazing story of the Brief Encounter at Trinity College Cambridge between Wittgenstein and Popper. I have two versions of the story, one from Popper's autobiography, the other one from the mouth of Professor Braithwaite, next to whom I had the honor of sitting at the dinner during the Wittgenstein Symposium at Cambridge, organized by the Austrian Institute and the Faculty of Philosophy in Cambridge.

Popper was invited by the Moral Sciences Club at Cambridge, as he says, to read a paper about some philosophical puzzle. He did not like this Wittgensteinian formulation and said he would speak about, "Are there philosophical problems?" Popper got into trouble straight away by criticizing the secretary who had invited him to read a paper on puzzles. Wittgenstein jumped up and said the secretary had acted on his own instruction. When Popper insisted that if he thought there were no genuine philosophical problems he would not be a philosopher, Wittgenstein again jumped up and spoke about puzzles and the nonexistence of philosophical problems. Popper interrupted him and read a list of philosophical problems, "Do we know things through our senses? Do we obtain our knowledge by induction?" These Wittgenstein dismissed as being logical rather than philosophical questions. Popper then referred to the problem whether potential or perhaps even actual infinities exist, a problem which Wittgenstein dismissed as mathematical. Popper then mentioned moral problems and the problem of the validity of moral rules. Popper writes: "At that point Wittgenstein, who was sitting near the fire and had been nervously playing with the poker, which he sometimes used like a conductor's baton to emphasize his assertions, challenged me: 'Give an example of a moral rule.' I replied: 'Not to threaten visiting lecturers with pokers.' Whereupon Wittgenstein in a rage threw the poker down and stormed out of the room, banging the door behind him."

Version Braithwaite: Wittgenstein was sitting opposite Popper and was slowly moving his chair towards him, picking up the poker on the way, and when Braithwaite was frightened that something might happen he casually took the poker out of Wittgenstein's hand in order to stir up the fire—the one in the fireplace of course. In the discussion, reports Braithwaite, Popper had been asked to define rationality, which he said he could not, adding: "But I know what irrational is, namely to threaten someone with a poker."

Popper himself says that after Wittgenstein's departure there was a very pleasant discussion in which Bertrand Russell was one of the main speakers. What I would give to hear Russell's version of the incident! Braithwaite, so says Popper, paid him a compliment by saying that he, Popper, was the only man who had managed to interrupt Wittgenstein in the way Wittgenstein interrupted everyone else.

What has happened to the men who emerged with the great analytical explosion in Vienna? Freud, as I have indicated, is known all the world over; even in Austria everybody knows his name. Oddly, because he came to England when Hitler entered Austria, his greatest successes were in the United States. But even there by now Freudian psychoanalysis has become, as Freud himself feared it would, a medical speciality, a kind of bloodless surgery, rather than a psychological science. It has locked itself away in its own institutes set apart from other academic disciplines and other therapeutic ideologies. There is certainly little ecumenical spirit. On the other hand, so much of Freud's ideas have been absorbed in other psychological and psychotherapeutic schools, watered down, almost perverted, one might say—in systems like the Californian humanist psychology, that hotchpotch of deep massage, transcendental meditation, and Gestalt therapy. But they also have Victor Frankl, the existentialist logotherapy man in San Diego. He, too, comes from Vienna via Auschwitz, Dachau, and Theresienstadt.

In Vienna there seems to be much more of an ecumenical movement amongst psychiatrists, psychotherapists, and psychologist, Freudians and Adlerians work together in clinics, and even speak to their knife-and-drug colleagues.

Logical positivism has more or less died: Carnap died in America; Schlick was killed by a maniac in the thirties in Vienna; here in England, Ayer seems to hold the fort with little inner conviction. However, in the United States, a new school, the Logical Empiricists have taken over from the logical positivists and carry on the tradition. There are always new developments in the States: Now we hear of Roy Schafer, who wants to replace Freud's causal-deterministic "why" language by a nondeterministic "what" language.

Wittgenstein, whose Phase Two was based on his new testament, *Philosophical Investigations* (his first testament was the *Tractatus Logico-Philosophicus*), in Phase Two has been compared to Freud because he looks at philosophy as a kind of therapy; cf. his images of the fly in the jar—and he wants to help the fly to escape.

In this country he is firmly entrenched in Cambridge and Oxford, a little like the Freudians in America, locked away in their colleges, talking to each other in complicated puzzles. In the States, Wittgenstein seems to be on the up and up—especially in California where they stress the mystical aspect. In Vienna they hardly know him; there are professors of philosophy in the provinces, in Salzburg and in Graz, who include him in their teaching, but in Vienna at the famous old University they teach Catholic Existentialism (Gabriel and Mader).

As you know, there are quite a few people (and not only arch reactionaries and fascist "pigs"), who consider all these exponents of the analytical explosion, both logical and psychological, as rather dangerous characters. They say that all moral values were destroyed by either calling them meaningless or pointing out that they all are the result of repression and that religion is an illusion. And they look at the analysts in the most general sense of the word, like people have looked at the atomic and then the hydrogen bombs—like some of these scientists looked at themselves after Hiroshima and Nagasaki. I daren't decide a question like the following: Is Marcuse, who taught a combination of Marx and Freud, in some way responsible for at least part of the ideology of the student terrorists in Germany? Now, we are not really concerned with Marcuse, because he for once isn't Viennese. But, on the other hand, do we want to put them all into prison and ask them to drink a cup of hemlock? Do we want to burn the books of all analysts? I am sure you would say no, and fortunately there seems to be a safety valve attached to all these systems, a self-correcting rudder, because after a time the more extreme propositions are found untenable and rendered harmless, in fact the comparison with the atom bomb does not really apply. One must, however, say that the front-rank analysts have all ignored one very important aspect of the human mind and that is the social aspect. Alfred Adler, on the psychology side, and Otto Neurath, on the logical analysis side, were the only ones who could be thought to have considered (because they were Social Democrats) the social aspect; add Allers, the Catholic Adlerian, inasmuch as he took notice of the importance of Christianity. All the others lived in a world of their own, unaware of the great socially relevant movements of their day.

In this respect, as in others, Vienna has provided an antidote in the person of Karl Popper, who became known in this country and in America first through his *The Open Society and its Enemies*, which he wrote in New Zealand during the war. It was published in 1945. Popper is here reestablishing the rights of the individual, warning against all totalitarian solutions suggesting piecemeal social engineering as the best method of social progress, cementing his arguments against historicism, that is against the belief that there are laws in history as there are thought to be laws in science, especially when such historical laws are employed as justification for interfering with the freedom of individuals and for using violence.

Popper comes from the same stable as the analysts I have talked about. Indeed, his *Logik der Forschung*, his first and maybe most important epistemological work, came out in Vienna in 1934, a year after all analytical endeavors were stopped by Hitler in Germany, and four years before Hitler came to Vienna to complete his task in this respect. *Logik der Forschung* did not appear in English until 1959, when it was published in London and New York as *The Logic of Scientific Discovery*. Popper argues against using induction as the method of acquiring new knowledge, and against verification as the method of making it valid. Deduction, which used to mean inferring from general laws to special cases, becomes with Popper the method of constructing hypotheses, which through experiment and observation can be falsified. Excuse the terrible simplification, please. If he brought back moral problems into philosophy in his *Open Society*, he reintroduces metaphysics, even if in a slightly new form, in his other works. In *The Self and its Brain* he writes: "Metaphysical hypotheses are important for science in at least two ways: First of all, in order to have a general picture of the world, we need metaphysical hypotheses. Secondly, in the actual preparation of our research we are guided by what I have called 'metaphysical research programmes.'" As Popper points out, Darwinism is metaphysical because it is not testable, and yet it is invaluable, because without it our knowledge could not have grown as it has done since Darwin. Popper is a pluralist where the logical positivists were monists. But, most important, he is an indeterminist. And that brings us back to the problem of philosophy versus science. Since the beginning of this century, the great philosophical discoveries have all been made by science. Einstein found that separate concepts of absolute time and absolute space must be replaced by the four-dimensional continuum of space-time. Heisenberg established the indeterminacy principle, also called the uncertainty principle, which states that it is impossible to measure the position and the momentum of a particle simultaneously with more than strictly limited precision. Jacques Monod, the French biologist, in his *Le hasard et la nécessité* (chance and necessity) has suggested that life emerged from inanimate matter by an extremely improbable combination of chance circumstances, and that this may not merely have been an event of low probability, but of zero probability—in fact, a unique event. This is a quote from Popper's paper on "Scientific Reduction and the Essential Incompleteness of all Science." There he says later, "The origin of life (like the origin of the universe) become an impenetrable barrier to science ..." or, "the undreamt-of breakthrough of molecular biology has made the problem of the origin of life a greater riddle than it was before: we have acquired new and deeper problems." Finally, we have to mention here Gödel's theorem, which says that in every arithmetical system there is a formula which, if the system is consistent, can neither be proved nor disproved. You can make an axiom of this formula for a higher system, but in this hierarchy of systems there will always be one formula the truth of which cannot be demonstrated. This means that Frege's and Russell's dream of a deductive system in which all mathematical truths could be deduced from a handful of axioms cannot be realized.

Popper says: "Since all physical science uses arithmetic, Gödel's incompleteness theorem renders all physical science incomplete." We now find ourselves in an interesting situation. Science itself has proved the arguments for its incompleteness. It can therefore surely not keep up its arrogant posture vis-à-vis philosophy on the grounds that philosophy is not an exact science. The other day I found in Bernard Williams' book on Descartes the following passage:

> The Reformation had questioned the traditional sources of religious knowledge, but had not produced any consensus, and was not going to produce any, on what others might be. The controversies surrounding these issues helped to generate movements of scepticism, not only with respect to religion itself, but with respect to other forms of supposed knowledge. If it were objected that religious belief had no true foundation, defenders of Christianity could reply that things were not better with secular beliefs.

Still this is a rather negative argument. There are, I think, stronger arguments once we have accepted that the only certainty that science can offer is that there is uncertainty in principle, which at the moment at least appears to be insurmountable.

Firstly there is that deeply rooted desire for regularity which Popper too has noticed. In *Objective Knowledge* he writes:

> It was first in animals and children but later in adults that I observed the immensely powerful need for regularity—the need which makes them seek for

> regularities; which makes them sometimes experience regularities when there are none; which makes them cling to their expectations dogmatically and which makes them unhappy and may drive them to despair and to the verge of madness.

I think this observation must lead to greater tolerance towards metaphysical and even to religious beliefs.

Secondly, also from *Objective Knowledge:*

> A pragmatic belief in the results of science is not irrational because there is nothing more rational than the method of critical discussion, which is the method of science. And although it would be irrational to accept any of its results as certain, there is nothing better when it comes to practical action; there is no alternative method which might be said to be more rational.

That means that, for practical purposes, we assume a great measure of certainty, for instance, when we send two men to the moon and count on getting them back safely to earth. So, maybe, this talk about uncertainty is a little exaggerated, a bit neurotic.

I found it interesting that Popper sometimes finds formulations which transcend his normally so rationalistic attitude. He writes, also in *Objective Knowledge,* "If there was not a real world, as rich and even much richer than the world we know so superficially from our daily life, and if the study of this world was not the main task of philosophy, then I would not be interested in philosophy." Not interested? What an argument! And I find it a sign of new tolerance that the latest book by Sir Karl Popper is written together with Sir John Eccles, the physiologist: *The Self and its Brain*—by Popper, the agnostic, and Eccles, the Catholic, who say:

> I think that science has gone too far in breaking down man's belief in his spiritual greatness and in giving him the idea that he is merely an insignificant material being in the frigid cosmic immensity. Now, this strong dualistic-interactionist hypothesis we are here putting forward certainly implies that man is much more than is given by this purely materialistic explanation. I think there is mystery in man.

I know that in Austria today there is also a great deal of such mutual tolerance amongst philosophers and psychologists. One is pleased to rediscover the formerly rather neglected great men of analysis, even if one is not ready to accept all their conclusions. Theologians and philosophers talk to each other, as do Freudians and Adlerians. It will take some time until all this is digested and until a new generation of philosophers and psychologists will emerge. Some people demand too much of Vienna; I think it is enough to have had a musical explosion in the 18th century, when Haydn, Mozart and Beethoven lived in Vienna, and the analytical explosion at the beginning of the 20th century. This analytical explosion really shook the world.

As has become quite clear, I am *impressed* by the analytical revolution, but my *sympathy* belongs to Popper who, whilst being a man of science, leads us back to the mainstream of philosophy, to Socrates' "I know that I know nothing," and to Immanuel Kant's moral attitude. If my prejudices have led me to do less than justice to the analytical revolution, I herewith apologize.

3

Transphyletic Animal Similarities and Predictive Toxicology

Jay Boyd Best
Colorado State University
Fort Collins, Colorado

Several years ago, I published an article entitled "The Evolution and Organization of Sentient Biological Behavior Systems" in one of the books edited by Wolfgang Yourgrau.[1] In that article, I presented the rudiments of a theory to account for the "higher behaviors" that I had observed in the primitive flatworm, planaria. Although a variety of subsequent evidence, in addition to that which was available to me at the time, has lent further support for it and I am even more convinced now than then of its essential correctness, it was, I realize with hindsight, a pretty wild theory relative to the prevailing views of the time. Wolfgang exhibited a great deal of courage in acting as editorial midwife for its delivery. In his other role as "pediatrician" of ideas, I hope that he would not be displeased with this sequel describing the further growth of "the child."

Planarians are probably similar to the first animals to evolve such features as bilateral symmetry, dorso-ventral asymmetry, encephalization, and a true brain in their head. Although planarians such as *Dugesia dorotocephala* (cf. Fig. 1) are only about 1/5,000,000th the mass of an adult human, their ratio of brain to body weight approximates that of a rat. Their brains, as well as those of their turbellarian relatives, the marine polyclad flatworms, are probably similar to those first evolved approximately a billion years ago.

When I first set out to train planarians in learning tasks over twenty years ago, with preconceptions shaped in terms of notions such as "tropisms," "taxes," "kineses" proposed by noted earlier investigators,[2,3] I was surprised to find that planarians exhibited behaviors more like those expected from higher animals,[4,5] i.e., more like those of a miniature psychological behavior system than the rigidly reflexive organism depicted in classical descriptions. I termed these behaviors "protopsychological"[6] to

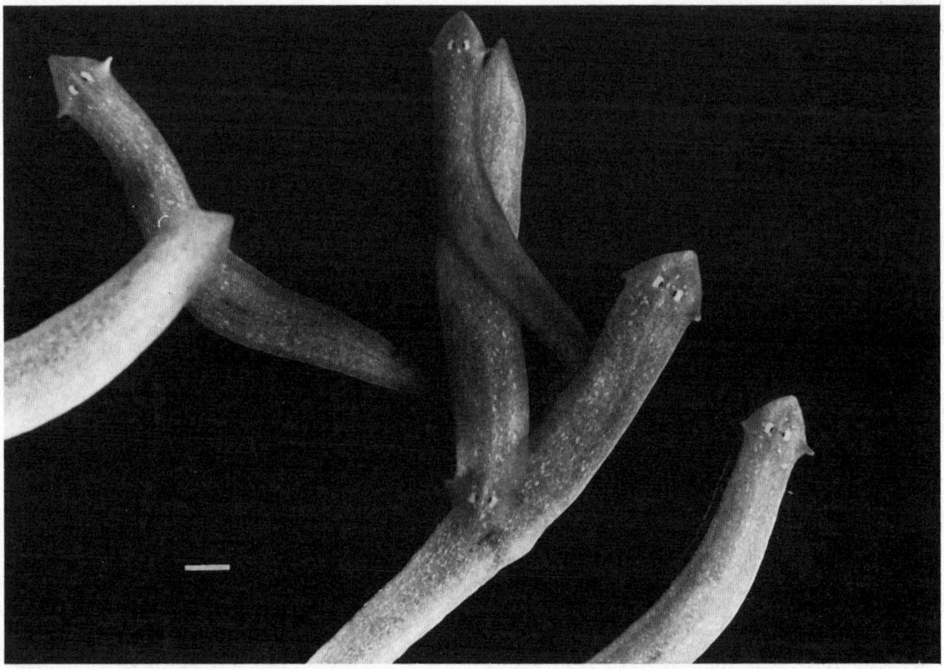

FIGURE 1. Photograph of normal asexual planarians of the species *Dugesia dorotocephala*. The small bar indicates 1 mm. The two eyelike structures in the head really are eyes.

emphasize this similarity to higher animal behavior and the inadequacy of the oversimplified mechanistic descriptions, terminology, and interpretations of these early investigators. Such complex behaviors have also been noted independently by other investigators studying marine polyclad flatworms.[7]

Various optical, electronmicroscopic, chemical, and physiological studies indicated corresponding similarities in the anatomical, neurochemical, and functional organization of the planarian nervous system, as well as a number of its other physiological systems, to those of higher animals. I suggested that the reason planarians appeared to manifest the behaviors of a miniature psychological behavior system was because their nervous system, in simplified miniature, shared many important features of organization with those of higher animals, that the brains of higher animals had, in their evolution, preserved, and merely enlarged upon, a basic foundation plan of organization that had been laid down in the primordial flatworm brain.

This idea further suggested that there was a basic plan of central nervous system organization, common to all bilaterally symmetric encephalized animals with brains,[8,9] on the main sequence of evolution. I added the qualifier "main sequence" to exclude those species which have evolved into such specialized modes of existence, e.g., parasitism, involving degeneracy, loss, or such drastic specialization of these common features, that they represent evolutionary *culs de sac*.

These similarities in the organization of the central nervous system and the behavioral patterns it gives rise to are, I suggested, a special instance of a broader pattern of basic similarity in biochemical, physiological, and cellular organization in bilaterally symmetric encephalized animals of the main evolutionary sequence. Transphyletic similarities in metabolic pathways, e.g., the Embden–Meyerhoff glycolytic pathway for anaerobic catabolism of hexoses, the Krebs cycle, fat catabolism, synthesis of steroids, etc., have long been accepted. Similar consensus prevails in regard to similarity of the functional roles of various subcellular organelles across phyla, as well as individual cell types, e.g., Golgi apparatus "packages" materials for export, mitochondria carry out oxidative phosphorylation, rough endoplasmic reticulum makes proteins for export, nerve fibers conduct impulse signals, muscles contract, whether these are in flatworms, humans, nematodes, or crayfish. They are good "tricks" for carrying out their required functions. They are efficient and they work. It probably took nature a long time playing its Monte Carlo game of anagrams with DNA nucleotide sequences to "learn" how to write the molecular computer program for each of these useful subroutines. Once such a necessarily complex message sequence was evolved, it would behoove its heirs to preserve it, to minimize the chances of further alterations in such message sequences. The organism might build upon such message sequences, redeploy them for various purposes, but the selection pressures to preserve such successful sequences in their genome would be great. Some of the ways of doing this, e.g., gene duplication, were discussed by Stebbins.[10] Perhaps such sequences are more vigorously patrolled by repair enzymes or translocated away from mutational "hot spots," but it appears likely that, by whatever means, extraordinary measures are taken to conserve them.

The idea that the supracellular organization, i.e., organization of physiological systems, of metazoans tends to be preserved is more or less an extension of the above considerations.

This article will expand upon this idea of transphyletic similarity in the physiological organization of animals, then discuss practical problems in assaying toxicological hazards to humans and the ideas of similarity that underly the choices of animals as "miner's canaries" for such purposes, and how these, together with certain economic and logistical considerations, give rise to an "uncertainty principle" in toxicological prediction that is more or less a counterpart of Heisenberg's uncertainty principle in physics. It will develop the idea that the various transphyletic similarities noted between flatworms and vertebrates in regard to their physiological and biochemical organization suggest corresponding similarities in their pharmacology and toxicology. Experimental results indicating such pharmacological and toxicological similarities and the potential value of planarians as alternative "miner's canaries" will be summarized.

1. Transphyletic Similarites

Since this is not intended as a review article, I will make no attempt to be exhaustive or complete in surveying a few of the many important features of physiological organization that appear to be analogous across phyla.

Planarians model, in simplified form, many of the physiological and biochemical systems of higher animals, e.g., muscles, glands, phagocytic reticular cells, goblet cells for mucus secretion, neurons, neurosecretory cells, an integument of columnar epidermal epithelial cells that undergoes continuous renewal from differentiating stem cells, eyes, neuroglial cells, chromosomes of conventional metaphase appearance.[11–22]

As mentioned previously, they have a bonafide brain in their head. In both planarians and marine polyclad flatworms, this is a compact bilobed structure, which exerts rostral control over locomotor patterns and other simpler subunits of behavior, e.g., feeding responses, fissioning, organized in a segmental nervous system.[7,23–25] As in the case of higher animals, orchestration of more complex behavioral sequences, e.g., learning, pursuit of prey, perseverance in following through such a pursuit and capture in the presence of distracting stimuli, location of food or prey, depend upon the integrity of the brain.[7] Comparable features of neural organization have been noted in the lobster swimmeret system.[26]

As in higher animals, the planarian central nervous system appears to utilize acetylcholine, dopamine, norepinephrine, and serotonin, packaged into light and dense core synaptic vesicles, as neurotransmitters.[27–33] Their eyes possess some interesting features anticipatory of those found in higher animal sensory systems: they are suspended in a cradle of fine muscle fibers suggestive of some degree of directional control; there is a pupillary opening through which light enters and which, with higher light intensities, is shielded by movement of melanophores that reduce the amount of light entering the photoreceptive region; there are, in addition to conventional sensory conduction fibers projecting from photoreceptors to brain, efferent nerve fibers that project from the brain to synapse onto the photoreceptors, with the evident function of modulating the sensitivity of these photoreceptors.[19,20] Such control, by the brain, of incoming stimuli by peripheral receptors, has received general acceptance in the case of the higher vertebrates but, I believe, has not been considered to be a feature of primitive nervous systems. These photoreceptors are, however, as in the case of other invertebrates, rhabdomeric rather than ciliary in their evolutionary origin, i.e., the photoreceptive portions of the photoreceptor cells are evolved from specialization of microvilli rather than neural cillia.[19,20,34–38]

Functional organization of neuroendocrine control of reproduction and other somatic functions reveals interesting analogies between planarians, other invertebrates, and the higher vertebrates. There are abundant experimental demonstrations that the rates of release of the protein, polypeptide, and steroid hormones of the mammalian or avian hypothalamic–hypophyseal–adrenal cortical (or gonadal) axis can be drastically modified by stimulus cues of a social or psychological character (e.g., cf. 39–41). Chemoreceptive cues appear to be the most commonly used in the subprimate mammals. One interesting such system, that contains several similarities to a control system that I and my co-workers studied in planarians,[23–25,42] is involved in the olfactory mediated pregnancy block phenomenon observed in female mice.[43–52] There is an important component of stimulus novelty to the effect: strange males, especially of different genotypic strain, placed in the presence of a female previously inseminated by a stud with which she is familiar, produce a drastic reduction of her incidence of conception. As in the case of the planarian system investigated by us, this pregnancy block effect produces an inhibition of reproduction when the mice are crowded into groups.[46] The pregnancy block appears to be mediated through the hypothalamic–pituitary axis.[49,53] Since the stimulus cues are olfactory, the sensory information is initially sent to, and processed through, the rhinencephalon and, in some manner, entails the release, and alteration of levels, of biogenic amines such as serotonin, noradrenaline and/or dopamine. There is a considerable body of evidence that such biogenic amines are involved in controlling the release of polypeptide hormones from neurosecretory cells of the hypothalamus and, through these, the release of protein hormones from the pituitary.[41,54-57]

Since far less manpower and funds have been invested in research on invertebrates, the story is obviously much more incomplete in their case. Yet, insofar as the evidence is available, it would appear as though there are analogous systems operative in these phyla. It is my view, shared with others, that such similarities in the organization of these neurohormonal control systems are a special instance of a much broader pattern of similarity in the physiological organization of metazoan animals.[8,9,31,32,58] Some of these correspondences are worth noting.

In insects there are neurosecretory cells in the *pars intercerebralis:* secretions (most likely polypeptide or protein) from these control the release of thoracotrophic hormone from the *corpora cardiaca* and juvenile hormone from the *corpora allata;* thoracotropic hormone regulates the release of ecdysone (a steroid) from the thoracic glands which appear to be its primary target tissue.[59] At least in the desert locust, and probably in a variety of other arthropods, the neurosecretory cells of the brain are under the control of the chemoreceptive cells of the antennae which respond to pheromones produced from the epidermis of other individuals.[60] This system thus appears to correspond to those involved in the release of adrenal or gonadal steroids in birds and mammals.

It is well known that water balance of mammals is under control of the neurosecretory system of the hypothalamic–posterior pituitary complex. Water balance of the desert locust also seems to be under the control of a neurosecretory system as is that of the mollusc, *Lymnae stagnalis.*

In *Crustacea,* five different neurohormones have been extracted from the neurosecretory cells clustered at the base of the eyestalks. Four of these are polypeptides and one, the diabetogenic hormone (DH) is a protein.[61,62] In some respects, DH, which has also been called "hyperglycemic hormone" (HGH), is reminiscent of glucagon in mammals: injection of DH causes a rise in blood glucose levels in the crustacean; extirpation of the eyestalks and associated neurosecretory glands that produce DH causes a drop in blood glucose levels and a concommitant increase in the hypodermal glycogen stores.[63]

In both vertebrates and arthropods, these neurosecretory cells are anatomically located in the brain close to its juncture with the segmental nervous system and the region of entry of the optic and chemoreceptive tracts into the brain.

Planarians, such as *Dugesia dorotocephala* and *Dugesia tigrina*, use two different alternative methods of reproduction, sexual and asexual. In sexual reproduction, they differentiate testes and ovaries (they are hermaphrodites), copulate with other sexually mature members of their own species, and deposit fertilized eggs which then subsequently develop into little planarians. Asexual planarians have no such gonads. However, with adequate food intake they grow longer and, when they are sufficiently long, undergo fissioning in which the anterior portion of the animal tears away from the posterior portion. The caudal fragment subsequently develops a new head and the rostral fragment a new tail to yield two complete but smaller planarians, which are genetic clones of each other. When asexual individuals differentiate gonads, i.e., become sexually mature, they forego fissioning. This sexual maturation appears to be under control of their brain. In the laboratory, some strains remain asexual indefinitely, continuing reproduction by fissioning for many years. However, grafting of the anterior part of a sexual planarian onto the posterior of such asexual planarians induces gonad formation in the latter.[64] Feeding extracts of sexually mature planarians to asexual ones also induces gonad formation.[65]

As mentioned above for arthropods and vertebrates, planarians also have systems of neurosecretory cells, anatomically situated in the region of the brain close to the junctures with the segmental nervous system and the optic tracts.[12,30,32,66,67] Changes in the number and histological appearance of these have been observed during sexual maturation and regneration of a new caudal portion.[66,67] The anatomical locations of such neurosecretory cell clusters in flatworms, as well as in earthworms, snails, slugs, and arthropods, are comparable to the hypothalamic location of such cells in vertebrates. Such neurosecretory cells, recognizable by their distinctive morphology, are the most common device for transduction of neural events into hormonal release[68] and trophic control of somatic vegetative functions by the brain.

The planarian brain also mediates social control of asexual reproduction. Isolation or reduced population densities, as well as decapitation, causes asexual *Dugesia dorotocephala* to fission.[23–25] Increased group size, or population density, suppresses fissioning. Soluble pheromones or metabolites discharged into the water, or onto the surfaces, of their habitat play no role in this effect. Contact chemoreception of chemical groups on the body surfaces of cohorts appears to be involved; the social sensors responsible for such detection are clumps of neural cilia distributed in the cephalic margins.[42] Similar clumps of neural cilia are also used in the chemoreceptive organs involved in olfactory sensing of the pheromonal compounds that appear to be the most important modality of interanimal communication in vertebrates and arthropods: such structures are utilized in the antennae of insects, chemoreceptive organs of lampreys and catfish, and the olfactory endings of mammals;[69–71] all of which, it should be noted, are located in anatomically comparable regions; they extend from the most rostral surfaces of these various animals. These detect the particular chemical messages of interest and translate the news into nerve impulses for transmission to the brain via sensory nerve fiber connections.

Photoperiod modulation of reproductive processes is important in higher animals: reproductive events are synchronized to environmental photoperiod, provided it is circadian, i.e., approximately 24 hours. Although details are still unclear, photoperiod modulated synthesis of melatonin from serotonin appears involved in a primary pacemaker function.[72–74] Experimental evidence, still incomplete, suggests an analogous system in planarians.[75] Intact *Dugesia dorotocephala* fission only during the dark phase of a 24-hour photoperiod, following reduction in population density. However, decapitated planarians fission indiscriminately during both light and dark phases. Placement of such decapitates in solutions of melatonin inhibits fissioning, which is resumed following transfer of these to ordinary habitat water. Effects of day length, i.e., illuminated phase duration, on sexual reproduction of *Dugesia tigrina* have been reported.[76]

As indicated earlier, the monoaminergic neuronal systems of the limbic system, utilizing dopamine, serotonin, and noradrenaline, appear, in vertebrates, to be importantly involved in the physiological linkage and integration of photoperiodic and olfactory stimulus inputs to the brain with the neurosecretory cell systems of the hypothalamus. Such monoaminergic neurons have also been demonstrated in the nervous systems of the earthworm, *Lumbricus terristris*[77,78] and molluscs,[79,80] as well as in the nervous system of planarians.[27–29,31] In view of the various analogies mentioned above and the roles such neurons seem to play in higher animals, it seemed not unlikely that they might play an analogous role in planarians and be influenced by the same neuropharmacological agents that are effective in mammals.

Prompted by such expectations, we tested planarians with methylergonovine, a nonhallucinogenic lysergic acid derivative that is a structural analog to, and blocks the action of, serotonin in mammalian systems. We also tested diethyltryptamine, another serotonin analog, that is hallucinogenic in humans. Both inhibited planarian fissioning at concentrations comparable to their effective dosage levels in mammals, but the hallucinogenic drug produced other behavioral effects that were quite different from the nonhallucinogen. Planarians normally alternate periods of activity with periods of quiescent inactivity that we refer to as "sleep." In such "sleep" states they adopt a characteristic posture in which they are slightly contracted with their ventral surface attached to the walls or bottom of the habitat container. Planarians in methylergonovine exhibited a curious sleep posture: they did not attach to the habitat surfaces and just drifted with the water currents in a slightly curved bow-shaped posture. However, when aroused, their movements appeared quite normal. Planarians in the hallucinogen, on the other hand, never slept, but maintained an incessant motion with exaggerated locomotor waves that were ineffectual in forward propulsion.

The drug, α-methyltyrosine is known to deplete the brains of mammals of serotonin and norepinephrine because it inhibits the enzymes tryptophane hydroxylase and tyrosine hydroxylase necessary for the metabolic synthesis, in brains of higher animals, of these two neurotransmitters from dietary amino acid precursors. Pretreatment of planarians with this drug reduced the effect of population density on fissioning incidence and, at higher pretreatment concentrations, produced subsequent suppression of fissioning.[81]

The segmental nervous system of planarians is facilatory to fissioning[25] and seems to contain most of the serotonergic neurons and relatively higher levels of serotonin than the brain which is inhibitory to fissioning and predominately catecholaminergic and dopaminergic.[27,28,31] It is thus interesting that reserpine, which is known to deplete serotonin in the brains of higher animals and was used earlier as a tranquilizer and for hypertension therapy in humans, suppresses fissioning of surgically decapitated planarians.[75]

It is of great theoretical importance to such comparative considerations that leu-enkephalin and β-endorphin, which appear related to the mechanisms of pain control and the action of opiates in mammalian brains, have recently been demonstrated, using immunological techniques, in neurons of the earthworm nervous system.[82]

Planarians were not found to be sensitive to curarelike compounds, e.g., tubocurarine, flaxedil, succinylcholine.[83] This is consistent with the specific selective action of these drugs on the acetylcholine receptors of the myoneural endplates of striated skeletal muscle. The muscle fibers of flatworms much more resemble smooth muscle of higher animals than straited skeletal muscle; myoneural junctions of flatworms more nearly resemble the finely arborized innervation of vertebrate smooth muscle than the myoneural endplate characteristic of vertebrate skeletal muscle.[84,85]

However, sensitivity of planarians to nicotine, i.e., significant reductions in fissioning and locomotor activity at 4 to 5 ppm and 50% lethality at 60 ppm suggests that "nicotinic-type" acetylcholine receptors, similar to those in the mammalian autonomic nervous system, play an important role in the planarian nervous system. The very close quantitative correspondence in nicotine toxicity for both planarians and mammals should also be noted.

Recently reported studies of Keenan and Koopowitz[86] are of importance to the ideas of this essay. Although it is almost axiomatic that tetrodotoxin blocks fast sodium channels in axonal membranes,[87,88] sodium components of action potential in coelenterates[89] and sodium currents in protozoan membranes[90] are not blocked by tetrodotoxin (TTX). That such tetrodotoxin sensitive sodium channels first evolved in the flatworm nervous system is strongly suggested by the demonstration of such channels in brain cells of a polyclad marine flatworm.[86] There is another important aspect to their findings: both TTX sensitive and TTX insensitive neurons were found in such flatworm brains; however, the TTX sensitive neurons were multipolar neurons. Such multipolar neurons tend to predominate in vertebrate brains, while monopolar neurons are more common in invertebrate brains, further suggesting that such free living flatworms were the common ancestral progenitor for both vertebrate and invertebrate lines of brain evolution.

One of the most striking features of planarians is the capability of a fragment of the animal to reform or regenerate a complete planarian. Thus, if one surgically transects a planarian, e.g., *Dugesia dorotocephala*, the tail regrows a new head and the head forms a new tail to yield two complete, but smaller, planarians. A fragment as little as 1/20th of the original planarian can remodel itself into a complete individual.

This capability for regeneration in planarians arises from primitive cells, called neoblasts.[14–18,91–95] The neoblasts, which are the stem cells of the planarian, are totipotent, i.e., they can differentiate into any of the specialized cell types, including neurons and muscle. This differentiation has been studied with both light and electron microscopy.[14,17,18] Mitosis is restricted to neoblasts[96] and increases following the initiation of the regeneration process by surgical transection.[21] Neoblasts are selectively destroyed, and the capability for regeneration eliminated, by x-irradiation.[14–16] The dose–response kinetics of such neoblast killing and its relationship to lethality of the organism have been studied.[97–102]

Although rates of neoblast mitosis increase during regeneration, it is appreciable even in the nonregenerating planarian. Various differentiated cell types become senescent and are lost; these are replaced by differentiating neoblasts.[17] Neoblasts sequestered for such differentiation are replaced by mitosis of the remaining neoblasts. Well-fed asexual *D. dorotocephala* grow long, fission, and the fragments regenerate into complete but smaller planarians which grow long, fission, etc. All the new specific cell types that must be formed in such growth are supplied by differentiation from this "fountain" of neoblasts.

This process, despite superficial differences, bears important similarities to those involved in mammalian and avian embryogenesis and the stem cell systems providing for renewal of intestinal mucosa, skin, and blood cells in adults.

2. Chemical Specification of Anatomical Form

If one sculpts or molds the shape of an object, e.g., a polyhedron, sphere, or gargoyle, then the shape implies little about the fine structure or molecular structure of the material comprising the object. However, if the shape of an object is the result of a molecular process such as crystallization, then the shape may imply a great deal about the molecular or fine structure of the object.

In conventional crystallization processes, the thermal motion of the elementary atoms, ions, or molecules comprising the aggregate provides a randomizing process through which various structural arrangements of the crystal can be dissolved and reformed. This permits transition of the ensemble into various aggregation states. Aggregation arrangements are thermodynamically favored that have the lowest free energy compatible with the packing constraints imposed by the volume of the elementary units. The lowest free energy arrangement is that which corresponds to perfection of the crystal and the growth of larger crystals at the expense of smaller ones.

The complex three-dimensional form of a protein, i.e., secondary and tertiary structure, is essential to its functional role. Yet, the specification for only its primary structure, i.e., the linear sequence of amino acids comprising it, is contained in the nucleotide sequences of the DNA. It is now commonly accepted that primary structure determines secondary structure in a manner not too dissimilar from those processes involved in crystal formation, i.e., the string of amino acids coils and kinks about, subject to various electrostatic, van der Waal, and hydrophobic interactions of the amino acids comprising the string; the native functional form of the protein is that configuration which minimizes the free energy of the ensemble.

Dissassociated cells of metazoan embryos in a hospitable medium tend to reaggregate into structures resembling those from which they were originally derived.[103–107] The process by which this occurs appears to be analogous to the mechanisms involved in crystal formation or in the determination of secondary and tertiary structure of proteins by primary structure.[8] Such dissociation and reaggregation experiments probably model the essential mechanisms involved in the embryological formation of organs and tissues from their component cell types.

Steinberg[107] has given a useful formulation of important aspects of this process, which will be described briefly. Such disassociated cells tend to engage in amoeboid movements.[103] These amoeboid movements, analogous to thermal movements in the case of molecules, enable transition from one arrangement to another. There are interfacial adhesion forces which are dependent upon the types of cells with surfaces in contact with each other. These interfacial adhesion forces stabilize cell aggregations by contact inhibition of amoeboid movement and by actually sticking the cells together.* The tendency to adhere, i.e., "stickiness," of the ith type of cell to one of the jth type can be measured by the free energy of adhesion, F_{ij}; $-F_{ij}$ would be the work required to separate two such adhering cells. Thus, with some fixed number of cells of different types arranged into some arbitrary aggregation, there will be some number, N_{ij}, of contacts between cells of the ith type with cells of the jth type. The resulting free energy of the aggregate will be

$$F_a = \tfrac{1}{2} \sum_{i,j} N_{ij} F_{ij}, \qquad i, j = 1, \ldots, k$$

The cells will tend to arrange themselves into that aggregate which minimizes F_a. Thus, for k different cell types and N_i cells of each type (where $i = 1, \ldots, k$), the geometry of the aggregate that the cells will form can be determined by specifying the $k \times k$ symmetric matrix of the F_{ij} values. Steinberg demonstrated experimentally and theoretically that for three embryonic cell types, endoderm, mesoderm, and ectoderm, certain inequality relationships among the six independent values determining the matrix suffice to cause reaggregation into a normal hollow spherical configuration with endoderm inside, ectoderm forming the outer layer, and mesoderm sandwiched in between. As the number of cell types becomes larger and the matrix increases in size, it becomes difficult to deduce the required relationships among the F_{ij} elements of the matrix that are necessary to yield any particular anatomical aggregation. Never-

*Contact inhibition also tends to shut down mitotic activity of dividing cells and there is some ambiguity in the meaning of the term in this respect, but Steinberg does not address this aspect.

theless, the formulation permits one some insight into the manner in which the kind of messages carried in the DNA determine anatomical form.

By specifying the surface proteins on the various cell types the DNA determines the values of the matrix elements, which in turn determine the aggregation shape yielding minimum F_a. It should be noted, however, that such a way of determining anatomical structure would be generally intolerant of mutations in the messages specifying the cell surface proteins responsible for determination of basic anatomical foundation plan. For example, a single mutation altering the surface protein in the third cell type would usually alter the values of all matrix elements in the third row and third column of the matrix. Such alterations can give drastic, and in most cases nonviable, modifications in anatomical arrangement of the cell aggregates. For example, Steinberg[107] found that the value for F_a associated with the normal configuration of reassociated cells was only slightly less than one in which endoderm formed the outer layer of the sphere and ectoderm the inner layer; correspondingly, a minor fraction of the time the cells would reassociate into such an everted embryonic structure which, had further development been allowed, would surely have not been viable.

Viewing morphogenesis in terms of such crystallizationlike processes, determined by cell surface characteristics affecting intercellular adhesiveness, permits explanation of several important phenomena.*

First, it explains why basic plans of anatomical and physiological organization tend to be preserved in evolution, i.e., the basis of genomic conservatism. Once a viable anatomical arrangement of basic functional cell types has been evolved, the viability of subsequent genomic modifications is contingent on that progenitor arrangement. Any subsequent drastic modification in that progenitor arrangement has a very high probability of rendering nonviable one or more of the various later evolutionary developments predicated on it. There probably are other viable progenitor arrangements and F_{ij} matrices which specify only a slightly modified progenitor arrangement, or even a whole different set of cell surface proteins which yield the same F_{ij} matrix, but the transition is not attainable by any sequence of single mutagenic steps that bequeath a viable genome to the heirs after each such step. Most mutations involve a single protein and, as noted above, alteration in a single surface adherency protein will alter an entire row and column of the F_{ij} matrix. There is thus a high premium, i.e., selection pressure, to minimize alterations in those DNA messages that specify such organizational ground plans. There is probably a corollary to this: those parts of the genome that are evolutionarily old messages, or translated in early ontogenic or differentiation stages, are less mutable than those which are evolutionarily new or invoked in later stages of ontogenesis or differentiation.

*There is probably a final stage in organogenesis in which some cells are subsequently laced together by intercellular filaments or by outright fusion as in the formation of muscle fibers from myoblasts or giant axons from several neuroblasts. But these processes are simply extensions and not the primary determiners of organogenesis; they are perhaps analogous to the stabilization of protein tertiary structure by formation of –S–S– linkages.

It is this genomic conservatism and way in which such DNA messages are translated into anatomical and physiological organization that are responsible for the transphyletic patterns of similarity in such organization.

Second, it provides an explanatory rationale for relating the teratogenic, benign tumorigenic, and malignant tumorigenic effects produced by mutagens and putting them on a continuum. More will be said of this later.

Though philosophically interesting, neither I nor others perceived much practical import to the above ideas. It has subsequently become apparent that this was an oversight. Such theoretical concepts provide the basis of an idea that may be useful in the solution of several pressing practical problems. To explain requires a necessary digression.

3. Practical Problems

We are currently threatened by impermeation into our environment of thousands of different toxic chemicals, associated with industry, mining, energy production, agriculture, and "military preparedness." We additionally expose ourselves to hundreds of other chemical compounds in the form of pharmaceuticals, food additives, contraceptives, cosmetics, fire retardants, etc. We are plagued with choices of alternative risks, such as, for example, whether the use of nitrites as a food preservative poses greater risks than the botulism poisoning it guards against, or whether new and promising drugs are going to produce unforseen side effects such as a generation of thalidomide babies, or high risks of cancer in our teenage daughters 18 years hence, or whether the benefits of saccharin as a sugar substitute for diabetics and the obese outweigh its carcinogenic risks.

The large variety and chemically low (but biologically effective) concentrations of environmental toxicants make it prohibitively expensive and impractical to use chemical analysis on an item-by-item basis for routine monitoring of our foods, soil, water and air for any significant portion of them. Furthermore, there must be some basis for inferring a link between such chemical agents and the undesirable biological effects imputed to them. Although some paranoia may be justified, it is not, by itself, adequate grounds for linking a chemical agent with pathological occurrences.

There is also another problem in addition to that of deleterious effects on humans. This concerns effects on other valued species, e.g., fish kills from aquatic pollutants, near extinction of raptors such as the Peregrine falcon from the egg shell fragility produced by the insecticide DDT.

Toxic responses to chemical agents can be established by several different techniques: (1) epidemiological studies; (2) studies on model organisms or systems; (3) studies on the species one wishes to protect; (4) chemical structure–function relations, in which one utilizes relationships that have been established in the past between the chemical structure of compounds and the type of biological effect that is apt to be produced by a compound of a given structure that has not yet been tested.

Epidemiological approaches seek, by studying the correlations between accidental or occupational exposure of humans to various of these agents and the incidence of various diseases in those exposed, to uncover the source and nature of the agent responsible. Although such approaches are necessary, critical decisions are frequently delayed until a generation of miscarriages, malformed babies, leukemia cases, irreversible neuropathy cases, etc. have convinced us of the need for remedial action. This is obviously unsatisfactory. Humans should only be "guinea pigs" of last resort. We need better ways of detecting such chemical hazards than exposing human beings to them. But there are also other problems of a scientific nature associated with such methods. Correlation does not, in itself, demonstrate causation. It is frequently difficult to distinguish those factors which are causally correlated from those that are only incidentally correlated. There are also problems of an instrinsically statistical character: the greater the number of suspect contributory circumstances, the greater the number of classification categories and the more stringent must be the significance level to distinguish the incidence of disease in some subpopulation from that expected by chance sampling fluctuations.

4. Model Organisms and "Miner's Canaries"

Nineteenth century coal miners knew no chemistry, but they did know there were odorless and invisible gases down in the mines that could kill them. So they would take a canary bird along with them. If the canary looked sick or passed out, the miners beat a hasty retreat from the mine. In their simple and intuitive way, they correctly perceived that what was bad for the canary was bad for them. Or at least this was so for the set of gaseous toxicants that threatened them. The canary was being used as a model organism bioassay system to provide the miners with an early warning of hazardous levels of environmental toxicants.

Recently I saw a news photo of an inspector in a storage bunker for "nerve gas" bomb canisters at the Rocky Mountain Arsenal. In addition to wearing protective clothing and a gas mask, he carried a small portable cage containing a white rabbit. It was obvious that he was using the rabbit as a "miner's canary" for detection of the dangerous nerve gases.

A number of variants of this idea has been deployed. Mice, rats, rabbits, hamsters, guinea pigs, dogs, and monkeys have been used as surrogate humans to assess the toxicities of a variety of chemical and radiological agents, and judge the potential hazards they pose to humans.

Such usages presuppose that the response of the assay animal to the toxicants in question is similar to that which would be exhibited by a human, except that the "miner's canary" should be expendable and, when used for detection, more sensitive and respond sooner. Conversely, such similarity in toxic responses provides evidence for similar toxicologically vulnerable physiological and biochemical systems in the test organisms and humans. For example, we now know that methane and carbon monoxide were the two gases that posed the greatest hazard to coal miners. Since the canary bird has a high rate of oxidative metabolism and concomitant requirement for oxygen, it is sensitive, and responds quickly to, replacement of atmospheric oxygen by methane. Also, since the canary uses blood hemoglobin to transport oxygen from its lungs to tissues and cytochrome oxidase to utilize that oxygen, and since the reactions of both are competitively blocked by carbon monoxide, it is also sensitive to this toxic gas. Similarly, the rabbit, used by the Rocky Mountain Arsenal inspector, is sensitive to military nerve gas because it, like humans, uses acetylcholine as a neurotransmitter and acetylcholinesterase to inactivate it; the organophosphate compounds of the nerve gas poison both rabbits and humans by combining with acetylcholinesterase and preventing it from inactivating acetylcholine.

Cyanide would also poison all of the species mentioned above because all depend upon cytochrome oxidase for the utilization of oxygen and cyanide inhibits cytochrome oxidase.

Recitation of such examples, well known to anyone who has taken an introductory biochemistry course, illustrates a point alluded to previously. The reason "miner's canaries" work as toxicological surrogates for humans is because they use the same toxicologically vulnerable physiological and biochemical mechanisms. Conversely, if the toxicologically vulnerable mechanisms of an organism are different from those of humans for some category of toxic agents, then that organism will not be a good miner's canary for those toxicants. For example, some strains of wild rabbits are very resistant to atropine because they possess an enzyme, atropinase, which rapidly inactivates it; poisonous plants such as Jimson weed, which contain lethal quantities of atropine, probably provided the selection pressure for evolution of this useful trait. Humans would be ill advised to rely on such rabbits as miner's canaries to ascertain which plants are safe to eat.

Genomic conservatism and the physiological similarities which derive from it are thus important to the range of species options from which miner's canaries with similar toxicological response mechanisms and sensitivities can be chosen.

The practical importance of this will be made evident in the sections to follow.

5. Miner's Canaries and Decision Paradigms

One can distinguish three types of toxicological decision paradigms for which we need appropriate "miner's canaries":

1. "*Screening procedures*" in which we know the chemical identification of a particular compound that we wish to use to produce certain desired results, e.g., a herbicide, but we need to know whether it is safe or, under the conditions of anticipated usage, apt to produce undesirable toxic side effects in humans or other valued species.
2. *Early warning detection* for poisonous concentrations of any of hundreds or thousands of different possible chemicals in our environmental milieu, where the circumstances do not permit restriction of the scope of our surveillance to some small subset of these at some particular time.
3. *Situations in which we need overall toxicity estimates of complex mixtures*, e.g., shale oil retort waters and their treated effluents, for which detailed chemical analyses are too expensive and impractical for routine application, or in which the resultant toxicities may not be readily deducible from such chemical analyses because of complex interactive effects.

6. The Injection Well Problem

An interesting practical example of the second decision paradigm listed above is provided by the use of "injection wells" for the disposal of unwanted toxic wastes. These wells are drilled down into permeable strata of the same type that provide subsurface aquifers for water wells. The toxic wastes are then forced down the well into the permeable stratum, usually without regard to whether this stratum is shared with water wells. The Environmental Protection Agency estimates that there are hundreds of thousands of such injection wells in operation in the United States. There is thus a reasonably high probability that a considerable number of water wells are at risk from contamination by these toxic chemical wastes. But which ones? And from what chemical compounds? Eager to avoid legal liability or adverse publicity, the various private companies, military installations, or governmental agencies involved maintain a clandestine veil of self-protective secrecy about such operations and, even if they know, are less than cooperative in disclosing the kinds and amounts of such chemicals disposed of in this manner. Conventional analyses for "water quality" will not reveal the parts per million, or billion, that constitute hazardous levels for many of these chemicals in drinking water. Court orders to force disclosure require "probable cause." Detailed chemical analyses for the thousands of potentially possible pollutants, for any substantial fraction of water wells, would be prohibitively expensive. So how does one ascertain which wells are contaminated without waiting for some epidemiologist to notice a statistically significant cluster of persons sickening in a given locale?

One practical method would be to utilize suitably sensitive and inexpensive aquatic animals, e.g., *Daphnia*, minnows, planarians, to test the water from some adequate sampling of water wells in a region. Such tests can be conducted to yield numerical values of the toxicity for the water from each of the wells sampled. From such values and the locations of the wells, one could construct, in much the same manner that one constructs a weather map, a map showing the isotoxicity contours in a region and use this map to obtain the approximate location of the source of such contamination, and identify the injection well and its owner responsible for the contamination.

7. Testing Statistics, Economics, Morality, and the Law

Since similarity is important for use of an organism as a miner's canary, why, one might ask, not use the most similar subhuman species, i.e., primates such as monkeys or chimpanzees? Sometimes it cannot be avoided, but there are good reasons why not. These are part of a more general problem in toxicological testing and inference.

Few disagree concerning levels or dosages of agents that quickly produce severe, distinctive toxic reactions in a large portion of the human population. Areas of controversy involve toxicant levels which produce such response symptoms in only a small portion of the population after long delays, or in which there is a "normally" high incidence of the malady in those not exposed to the particular toxicant in question.*

This is illustrated by problems involved in the testing of carcinogenic agents. Consider an agent that, under realistic conditions of exposure, causes cancer in 0.1% of the population exposed, but which is so ubiquitously distributed that most of the population is placed at risk. This means that one out of a thousand of the exposed population will develop cancer as a consequence of being so exposed. This may not seem like such a high risk, but such an agent would produce 200,000 additional cancer cases among the approximately 200,000,000 population of the United States. Viewed

*Apropos of this point, it has been suggested that man-made chemicals do not account for most cancers because there is no great difference in age-corrected cancer incidence between the populations of underdeveloped and technologically advanced countries (Ref. 108); the underlying assumption behind this inference being that the populations of the more advanced countries have greater exposure to man-made carcinogenic agents. Handler reasons from this that the great preponderance of cancers probably arise from natural, yet unidentified, carcinogenic agents. This may be overly simplistic and fallacious reasoning. In the first place, residues of the more persistent of some of these compounds have been found in the tissues of animals sampled from the most remote areas of the globe. Secondly, it is known, but not widely publicized, that some of the largest producers of chemicals that have been banned in the United States by the Environmental Protection Agency have been exporting these to less developed countries with more tolerant regulatory policies. Thirdly, workers in underdeveloped countries, who are employed in mining or industrial facilities usually are less protected from hazardous substances by the laws regarding worker safety that have been established in the more developed countries. So, through the operation of these and other factors, the peoples of underdeveloped countries may not be as removed from exposure to carcinogenic chemicals as Handler assumes. However, it should be admitted that there indeed are naturally occurring carcinogens, e.g., the plant flavenoid, Quercetin (Ref. 109).

in this way, the burden of human suffering and death from the agent would be large and unacceptable. The medical costs for treating (frequently unsuccessfully) and caring for such patients, at the somewhat conservative figure of $40,000 per patient, would be eight billion dollars.

Yet, unless the cancers are of an unusual type or clustered among some distinguishable set of people, such an incidence would not be easy to detect epidemiologically, since approximately a quarter of the population develops cancer anyway.

Such statistics also bedevil laboratory studies using model organisms. To reliably detect an incidence of 0.1% in a group of experimentally exposed animals, the test group should contain the order of 10,000 subjects, or even more if there is an appreciable incidence in the untreated controls. To reduce the necessary number of animal subjects to tractable size, e.g., a few hundred per test group, investigators resort to a controversial practice that has been challenged in legal proceedings. They increase the dosage of the agent in the experiment to higher levels in order to increase the response incidence into a detectable or measurable range, e.g., large enough dosages that several percent or more of the test animals develop cancers. However, the dosages or concentrations required to achieve this may be several orders of magnitude greater than those encountered in real-life conditions outside the laboratory. The assumption underlying this practice is that the incidence of the toxic response, e.g., cancer induction, is more or less linearly related to the dosage, which may or may not be true. It also assumes that high dose toxicities are mediated by the same mechanisms as low dose toxicities which, in a number of instances that could be recited, is clearly not true. This has led to considerable scientific discussion and debate regarding what is sometimes referred to as "the low dose problem," i.e., what is the true shape of the dose–response curve at low dosages, concentrations, or dose rates?

This question is not merely another of those bits of arcane minutia that academic scientists like to ponder. It is of great practical importance. In an attempt to provide some clear legal criterion for purposes of enforcement of the Food, Drug and Cosmetics Act, passed by the United States Congress in 1958, the "Delaney clause" was included, which authorizes the Secretary of Health to ban the inclusion, into foods, drugs, or cosmetics, of any chemical that has been demonstrated to cause cancer in man or in any animal test species.[110]* Dosage levels for such cancer induction are not mentioned. This, in effect, means that if, for example, some investigator finds that administration of some chemical to a treatment group of say 400 mice, at a dosage of 100,000 fold the human exposure to be realistically expected from that chemical, yields 20 cancers in this group and none in an equal sized control group, then this chemical could, and probably would, be banned under the Delaney clause. This is the type of situation that was at issue in the controversy regarding the use of saccharin as a noncaloric sweetener.

*The clause was introduced into the original 1958 act by Representative Jim Delaney (New York) and retained unchanged in the amended act of 1972.

Consider the hypothetical, but realistic, situation depicted in Fig. 2. The vertical axis depicts the number of cases to be expected per 100,000 individuals at risk as a function of the dosage shown on the horizontal axis. To simplify the example, it has been assumed that mice are to be used in testing and that they have approximately the same cancer proneness as humans. The "maximum tolerable" human incidence from the agent has been arbitrarily chosen as 0.15 cases per 100,000 or 300 cases out of the 200,000,000 population of the United States. However, with resources available, the minimum detectable incidence in laboratory tests with mice would be about 100 cases per 100,000, i.e., 0.1%. Even this would represent an abnormally large study involving at least 10,000 subjects placed at risk to obtain an average of 10 cases with

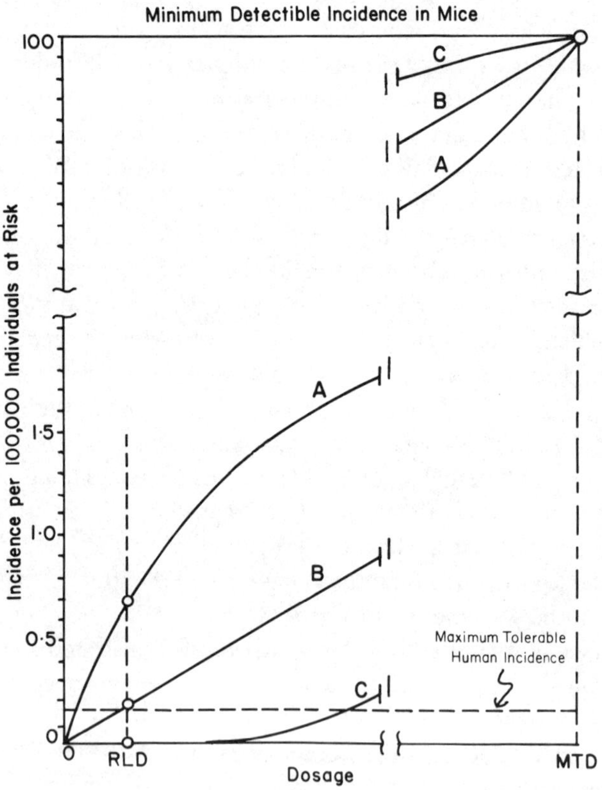

FIGURE 2. Graphical depiction of the relationship between the toxic response incidence and dosage for a test population of mice and the human population at risk, to show the importance of shape of the dose–response curve in the low-dose domain. Values chosen are hypothetical, but realistic. MTD indicates the "mouse test dose" that must be used to obtain an observable incidence (10) with the largest feasible test population of mice (10,000). RLD indicates the "real life dose" to which the human population would be exposed. For simplicity, it has been assumed that the sensitivities of mice and men are similar. Three possible low-dose response curves *A*, *B*, and *C* are shown, with the incidences to be anticipated in the human population exposed to RLD.

the mouse test dose (MTD). RLD denotes the average "real life dose" to which the human population would be exposed under conditions of anticipated usage. Three possible, although by no means exhaustive, low-dose–response curves *A*, *B*, and *C* are shown. If the chemical were not banned, and all the 200,000,000 persons in the U.S. were placed at risk from it, curves *A*, *B*, and *C* predict 1360, 360, and none as the respective number of cancer cases to be expected. Thus, if *A* or *B* were correct, the Delaney criterion would rightly reject the compound as unacceptable; however, if *C* is correct, it would wrongly result in the rejection of a useful compound and the loss of the millions of dollars that probably went into its development.

It is clear from this example that ignorance of the shape of the low-dose–response curve, together with the Delaney criterion, may conspire to wrongly eliminate needed chemicals from the market place and nullify large investments in them.

These considerations also show another important aspect of the problem: the smaller the number of animals in the test population, the greater must be the ratio of the test dosage to the real-life dosage to be assured of observing any cases of cancer in the test group, thus increasing the dose range which must be spanned by extrapolation with an unknown low-dose–response function and increasing the uncertainty in estimates of the number of cancer cases which would be induced under real-life use of the chemical. It should be emphasized that this uncertainty is over and beyond the conventional "confidence limits" problem encountered in the fitting of a function of specified form to experimental points. As will be indicated presently, the closer the phylogenetic relationship of the test animal species to man, the more restricted is the size of the group that it is practical to use.

There is another related problem in the logistics and costs of such experimental testing: cancer does not appear until some time later, which means that the test population has to be maintained, along with the control population, for extended periods of time. As a general "rule of thumb," in the case of mammals, the duration of this necessary waiting period is correlated with the average life span of the species. Also, within standard mammalian test species, both life span and size are correlated with the closeness of the phylogenetic relationship to man. Thus, for mice, with a life span of about two to three years, one may typically wait five months for such cancers to become manifest; for rats, with life spans of three to four years, this time increases to the order of seven to nine months; for Beagle dogs, with life spans of ca. 12 to 14 years, this waiting time increases to several years. Comparable problems occur in regard to other delayed responses, e.g., birth defects.

A corresponding economics goes along with the above. As one progresses from mice to primates, not only does the length of the experiment increase, but the cost per day per animal also increases. Thus, for example, costs for testing the carcinogenicity of a compound on small rodents, e.g., rats and mice, in a minimally adequate manner have been variously estimated at $150,000 to $300,000,[111,112] although current inflation rates have probably increased these costs by this date of writing. Such time and costs increase markedly as one goes to using higher animals such as dogs. To study the effects of chronic, "low-level," radiation, e.g., 2 to 10 rads per day,* on a thousand or so dogs, the costs rise to something in the neighborhood of $15,000,000 to $20,000,000 and the time frame extends to ca. 15 years. Comparable studies with primates would be yet another order of magnitude more expensive and require even longer periods of time.

It should be noted, however, that these cost estimates given for mice, rats, dogs, and primates would not be adequate to finance studies with enough animals to reliably detect cancer incidences as low as 0.1%, which would, as indicated previously, still be unacceptably high for the human population of the United States.

8. Toxicology's Uncertainty Principle

The examples discussed above serve to demonstrate what I believe to be toxicology's version of the uncertainty principle.† As in Heisenberg's theorem, it has to do with the operational methods one has for obtaining information and certain kinds of "trade-offs" one is forced to make so that certainty in one variable is obtained at the cost of a loss in certainty in another. The following might be offered as a rough preliminary statement of it.

8.1. *Definitions*

Let PK be some measure of phyletic kinship of the miner's canary with humans, D_T denote the test dosage (or dose rate) at which one could obtain a detectable response in the test group of miner's canaries, D_R denote the dosage (or dose rate) that, in the real human population at risk, would cause the maximum number of cases that could be tolerated, t_M denote the length of time required for manifestation of the toxic response in the test, and t_T the time required to complete the experimental tests on the miner's canaries. Let P_{TM} denote the probability that a chemical compound, arbitrarily chosen from the set of those of interest, will act through similar component mechanisms, in both the miner's canary and humans, to produce comparable toxic reactions in both.

8.2. *Statements*

Under conditions of limited resources (of animals, facility space, personnel, and funds) and optimum deployment of these:

*Note that this dose rate is still orders of magnitude higher than those considered acceptable for humans.
†Heisenberg's uncertainty principle states that the uncertainty in the momentum multiplied by the uncertainty in the position of a particle will always be greater than or equal to Planck's constant.

1. D_T/D_R is positively correlated with PK and, for high PK, appreciably greater than unity.
2. The uncertainties in estimates of D_R are positively correlated with PK.
3. The value of t_T is greater than t_M and both are positively correlated with PK.
4. The number of chemicals that can be tested is inversely correlated with PK.
5. The "backlog" of untested chemicals (which need to be tested) will be positively correlated with PK.*
6. P_{TM} is positively correlated with PK and approaches, but does not attain, unity for high PK.

Relaxation of the requirements for a high PK thus involves a "trade-off": one gains in the first five considerations and loses certainty in the sixth. Conversely, if one insists on a high degree of phylogenetic kinship (PK), one sacrifices in the first five considerations and gains increased certainty of correspondence in the toxicological response mechanisms (P_{TM}).

Certainty of similarity in toxicological response mechanisms between the model test organism and humans is gained at the expense of greater uncertainty regarding the quantitative dose–response relations in the low dosage domain of primary interest, as well as longer delays in obtaining the required information and increasing the backlog of untested chemicals.

In view of the trade-off costs, in information, involved in insisting on a high degree of phylogenetic kinship between the model test organism and humans, it behooves one to ask how much phylogenetic kinship is necessary? If one exercises discretion in choice of model organisms, how much does P_{TM} necessarily decrease as one relaxes the requirement for high PK?

9. When Is an Animal an Animal?

The Delaney clause, mentioned previously, does not specify any restriction on the type of animal that is to be used for demonstration of the carcinogenic potential of a chemical. It merely states " . . . animal" period.

In the biological classification of living creatures, animals include protozoans, coelenterates, flatworms, nematodes, annelids, molluscs, arthropods, fish, amphibia, reptiles and birds, as well as mammals. Although possible, it seems doubtful that either popular or legal conceptions are this broad. For example, the "Animal Research Committee," that is administratively responsible for overseeing compliance with the Federal "Humane Animal Practices Act" among the researchers of my university, has repeatedly ruled that my experiments with planarians lie outside of their domain of responsibility. Neither have accounts of my work in the popular media evoked any

*It has recently been estimated that the backlog is approximately 70,000 chemicals (Ref. 113).

negative reaction from antivivisectionist groups. Does this mean that popular and legal conceptions extend only to encompass those animals which are adequately large and hairy, or feathered, or with scales? I'm not sure. Why, the reader might ask, frivolously belabor this point?

The truth of the matter is that most toxicological testing is currently conducted for legal rather than scientific purposes. Thus most such testing programs of any magnitude, whether performed by private companies, Federal regulatory agencies, or independent testing firms, are colored by the real possibility that the results will be used as evidence in the adversary situation of a hearing or trial in a court of law.

Science and the law have never been very comfortable with one another. Large sums of money, the closing of major manufacturing facilities, the public welfare, recall and withdrawal from the market of a product representing millions of dollars of investment in the expectation of more millions of profit, hundreds of worker's jobs, etc. are commonly at stake in such hearings and trials. Judges and juries usually have minimal scientific training. It can be assumed that attorneys for the opposition will attempt to cast test results of expert witnesses in the worst possible light, emphasizing flaws and uncertainties, casting suspicion on deviations from standard protocol.

In such trial or hearing situations, novelty and unconventionality, departures from traditional practice, new procedures, rationales, and philosophies of testing, whatever their merits, tend to be viewed with suspicion. Law is as much a matter of tradition and precedent as explicit statement in congressional acts, or wording of statutes. I am unaware of any legal decisions up to the present time that set any limit, in regard to the Delaney criterion, on the requisite degree of phylogenetic relationship that an animal species must have to man in order for it to be an acceptable toxicological test species; however, I must admit that I am not an attorney or legal scholar. Aware of such potential future usage of their results, toxicologists generally tend to utilize test species and procedures which they feel would fall comfortably within the accepted lay preconceptions of an "animal," e.g., mice, white rats, rabbits, guinea pigs, dogs, monkeys, and, sometimes but more rarely, birds and fish.

10. A Possible Solution

If, as it appears, lower invertebrates such as planarians and earthworms are much more similar in their physiological and biochemical organization than previously recognized then one might anticipate corresponding similarity in their responses to toxic agents. Some of the preceding discussion suggests that P_{TM} may not decrease too much as the requirement for high PK is relaxed, especially if one restricts the use of the invertebrate test species to exclude certain categories of drugs or compounds; for example, excluding the use of planarians for tests involving curarelike compounds.

Planarians are inexpensive to use and maintain. One can use much larger numbers of animals than would be feasible with mammals. We currently maintain approximately 40,000 *Dugesia dorotocephala* on about 25 cents worth of raw beef liver a week and a few hours of workstudy student time. Such a colony will yield 5,000 to 10,000 planarians per month without depletion. Experiments involving 1,000 individuals under 50 different treatment conditions can be set up on an ordinary table top. At relatively modest costs, one could easily expand the scale of such tests to use 100,000 animals per treatment group. This, together with the fact that they generally respond sooner and appear more sensitive to most mammalian toxicants (cf. the following section), could yield information about low dose range effects that are currently unavailable with mammals as test animals. Because the experiments can be conducted more rapidly, and more different compounds tested concurrently, the backlog of untested compounds could be reduced.

11. Acute Toxicity Studies on Planarians

Toxicity studies restricted to two-week exposure–observation periods were conducted on planarians, using various representatives of major chemical categories of common toxic pollutants. These compounds included those of heavy metals such as mercury, cadmium, copper, zinc, and lead, chlorinated hydrocarbons such as chlordane, toxaphene, pentachlorophenol, and polychlorinated biphenyls, anticholinesterase pesticides such as malathione, carbaryl, and 2,2 dichlorovinyldimethylphosphate (the active component in "Pest Strips"), compounds of arsenic, phenolic compounds, cyanide, ammonia, nicotine, and benzene. Lethality, frequency, and type of visible lesions and ulcerations, as well as fissioning frequencies were observed as a function of concentration and duration of exposure.

Figure 3 shows examples of some of the sublethal visible lesions and ulcerations obtained in such tests (compare with the normal appearance shown in Fig. 1). Such head resorptions or dissolutions are the most common of these.

Figure 4 shows an example of the way that incidence of lethality depends upon concentration and duration of exposure to a toxicant, e.g., cadmium.

Some of these results are summarized in Table I along with available toxicity data on other common test species shown for comparisons.

We have, in addition, tested planarians with a limited number of complex pollutant mixtures, e.g., shale oil process water and cigarette smokes derived from cigarettes with various known tar/nicotine contents.

Except in the case of arsenate, the sensitivities of planarians to these various toxic agents correspond well with those of mammals except that planarians are usually more sensitive and respond sooner. There is considerable variability in the sensitivities of mammals to the various toxicants, depending upon the species, mode of administration, and toxic response index used. Such intramammalian variability imposes a limit on the extent to which one can expect them to correlate with any model system.

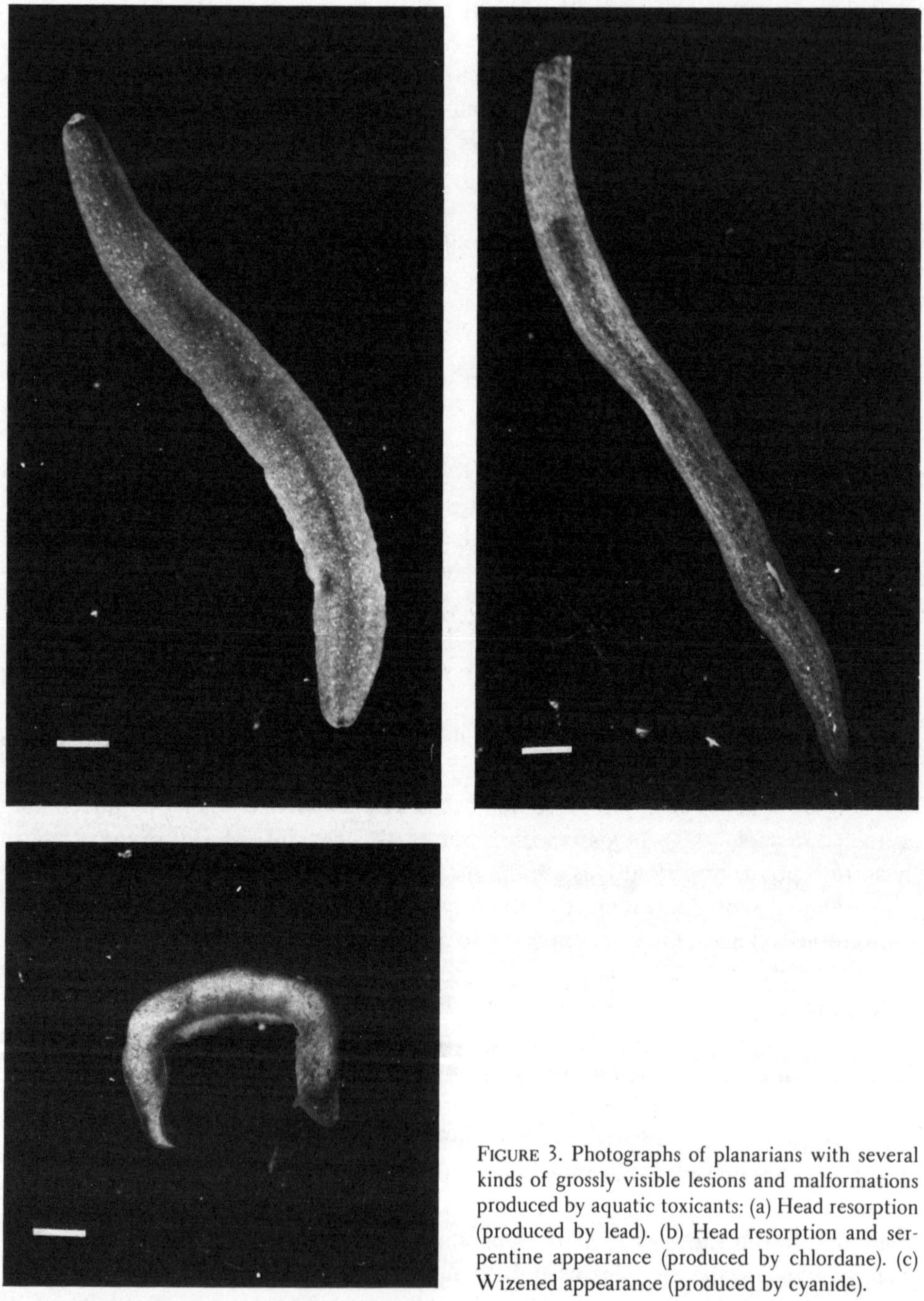

FIGURE 3. Photographs of planarians with several kinds of grossly visible lesions and malformations produced by aquatic toxicants: (a) Head resorption (produced by lead). (b) Head resorption and serpentine appearance (produced by chlordane). (c) Wizened appearance (produced by cyanide).

TABLE 1. Comparison of Planarian *(D. dorotocephala)* with Mammalian Toxicities for Some Representative Compounds[a]

Compound	Toxic concentrations[b]			Route	Species
Sodium arsenite	10 day LC50[c]	=	25–50 ppm	Water conc.	Plan.
	Fiss. inhib.[c]	=	1–10 ppm	Water conc.	Plan.
	LD50	=	19 ppm	I.P.	Mouse
	LD50	=	41 ppm	Oral	Rat
	TDLo (Terat.)	=	10 ppm	I.P.	Mouse
Sodium cacodylate	6 day LC50[c]		82 ppm	Water conc.	Plan.
	LDLo		500 ppm	Subcut.	Rabbit
Cadmium sulfate ($3CdSO_1\ 8H_2O$)	10 day LC50[c]	=	1–5 ppm	Water conc.	Plan.
	Ulcer and malf.[c]	=	0.2–1 ppm	Water conc.	Plan.
	Fiss. inhib.[c]	=	0.2 ppm	Water conc.	Plan.
	LD50	=	69 ppm	ipr.	Mouse
	TDLo Teratog.	=	2 ppm	I.V.	Hamster
	LD50	=	27 ppm	Subcut.	Dog
Caffeine	Ceph. lesions[c]			Water conc.	Plan.
	TXC50	=	100–200 ppm	Water conc.	Plan.
	LC50[c]		350 ppm	Oral	Mouse
	LD50	=	620 ppm	I.P.	Mouse
	Terat. TDLo	=	200 ppm	Oral	Rabbit
	LDLo	=	355 ppm	I.V.	Dog
	LD50	=	175 ppm	Oral	Human
	LDLo	=	192 ppm		
Chlordane	10 day LC50[c]	=	1.4–3.5 ppm	Water conc.	Plan.
	LDLo	=	40 ppm	Oral	Human
	LD50	=	283 ppm	Oral	Rat
	LD50	=	100 ppm	Oral	Rabbit
	LDLo	=	20 ppm	I.V.	Rabbit
	LD50	=	343 ppm	I.P.	Rat
$CuSO_4 \cdot 5H_2O$	1 wk. LC50[c]		2–5 ppm	Water conc.	Plan.
	LDLo	=	4 ppm	I.V.	Rabbit
	LDLo	=	50 ppm	I.V.	Mouse
	LD50	=	7 ppm	I.P.	Mouse
Sodium cyanide	5 day LC50[c]	=	1–5 ppm	Water conc.	Plan.
	Fiss. inhib.[c]	=	1–5 ppm	Water conc.	Plan.
	LD50	=	6.4 ppm	Oral	Rat
	LDLo	=	2.8 ppm	Oral	Human
	LD50	=	5 ppm	I.P.	Mouse
	LD50	=	5.8 ppm	Subcut.	Guinea pig
Ethanol	10 day LC50[c]	=	0.7–1.0%	Habitat water	Plan.
	Fiss. inhib.[c]	=	0.5%	Habitat water	Plan.
	LDLo	=	0.6%	Oral	Human
	LD50	=	0.55%	Oral	Dog

continued overleaf

TABLE 1 (continued)

Compound	Toxic concentrations[b]			Route	Species
Lead acetate $3H_2O$	1 week LC50[c]	=	20–50 ppm	Water conc.	Plan.
	Fiss. inhib.[c]	=	1 ppm	Water conc.	Plan.
	Ulcer and malf.[c]	=	5–10 ppm	Water conc.	Plan.
	TDLo	=	82 ppm	Oral	Rat
	LDLo	=	204 ppm	I.P.	Rat
	LD50	=	120 ppm	I.V.	Rat
	LD50	=	120 ppm	I.P.	Mouse
Malathione	1 day LC50[c]		1.5–5 ppm	Water conc.	Plan.
	Fiss. inhib.[c]	<	0.5 ppm	Water conc.	Plan.
	TDLo	=	857 ppm	Oral	Human
	LD50	=	1401 ppm	Oral	Rat
	LD50	=	340 ppm	I.P.	Rat
	LD50	=	50 ppm	I.V.	Rat
	TLm96	<	1–10 ppm	Aquatic	
Mercuric acetate	Fiss. inhib.[c]	<	.10 ppm	Water conc.	Plan.
	Vis. lesions[c]		.3 ppm	Water conc.	Plan
	4 day LC50	<	1.0 ppm	Water conc.	Plan.
	LD50	=	76 ppm	Oral	Rat
	LD50	=	62 ppm	Oral	Mouse
	LD50	=	4.4 ppm	I.V.	Mouse
	TLm96	<	1–10 ppm	Aquatic	
Methyl mercury chloride	Fiss. inhib.[c]	=	.03 ppb	Water conc.	Plan.
	Teratogenesis[c]	=	.1–.2 ppm	Water conc.	Plan.
	Behav. neurotox[c]	=	20–80 ppb	Water conc.	Plan.
	5 day LC50[c]	=	.3–.4 ppm	Water conc.	Plan.
	LD50	=	16 ppm	I.P.	Mouse
	LD50	=	11 ppm	I.P.	Rat
	LDLo	=	15 ppm	I.V.	Rabbit
	LD50	=	21 ppm	Oral	Guinea pig
	LD50	=	7 ppm	I.P.	Guinea pig
NH_4OH	2 day LC50		4–90 ppm	Water conc.	Plan.
	LDLo	=	43 ppm	Oral	Human
	LD50	=	350 ppm	Oral	Rat
	TCLo (eye)	=	2 ppm	—	Human
	TLm96		10–100 ppm	Aquatic	—
Pentachlorophenol	10 day LC50	=	.06 ppm	Water conc.	Plan.
	Fiss. inhib.		.03 ppm	Water conc.	Plan.
	LDLo	=	29 ppm	Oral	Human
	LD50	=	50 ppm	Oral	Rat
	LD50	=	56 ppm	I.P.	Rat
	LD50	=	100 ppm	Subcut.	Rat
	TDLo	=	46 ppm	Subcut.	Mouse
	LDLo	=	70 ppm	Oral	Rabbit
Phenol	4 day LC50		22–72 ppm	Water conc.	Plan.
	LDLo	=	180 ppm	I.V.	Rabbit
	LDLo	=	420 ppm	Oral	Rabbit
	TDLo	=	14 ppm	Oral	Human
	LDLo	=	140 ppm	Oral	Human
	LD50	=	250 ppm	I.P.	Rat
	TLm96		10–100 ppm	Aquatic	—

TABLE 1 *(continued)*

Compound	Toxic concentrations[b]			Route	Species
PCB (Aroclor 1254)	10 day LC50	=	2–4 ppm	Water conc.	Plan.
	Ulcer. and H.R.	=	.2–.5 ppm	Water conc.	Plan.
	LD50		1300–4500 ppm	Oral	Rat
	LDLo		2000–3200 ppm	Skin	Rabbit
	TDLo		1200 ppm	Oral	Rat
	LD50		358 ppm	I.V.	Rat
Methylergonovine	Fiss. inhib.	=	.5 ppm	Water conc.	Plan.
	EC contrax.	=	.5 ppm	*In vitro*	Rabbit uterus
	LD50	=	23 ppm	I.V.	Rat
Shale oil process water (Omega-9)	4 day TL50	=	0.4%	Flow through	Rainbow trout 11–18 cm
	Loss of egg hatchability and fry survival		0.2%	Habitat water	Rainbow trout
	5 day LC50	=	0.4%	Habitat water	Plan
	Hd. resorption and ulceration		0.1%	Habitat water	Plan
	LD50	=	3.3% of body wt.	Oral	Rat
Shale oil process water (Omega-9)	4 day TL50	=	0.5%	Habitat water	Rainbow trout 11–18 cm
Inorganic components	5 day LC50[c]	=	0.5%	Habitat water	Plan Plan
	Hd. resorption and ulceration[c]	=	0.1–0.2%	Habitat water	
Toxaphene	10 day[c] TC50 (Vis. lesions)	=	0.3 ppm	Water conc.	Plan.
	LDLo	=	40 by volume. IN	Oral	Human
	LD50	=	60 ppm	Oral	Rat
	LDLo	=	50 ppm	Oral	Dog
	LD50		1 ppb	Water conc.	Fish

[a]The values shown for planarians were determined by the authors, nearly all the values for mammals and aquatic animals were taken from the National Institute of Occupational Safety and Health's *1976 Registry of Toxic Effects of Chemical Substances*, edited by Christensen and Fairchild.

[b]In all cases where the substance was administered to a mammal by oral, intraperitoneal (I.P.), or intravenous (I.V.) routes the dose in mg/kg has been expressed as parts per million (ppm) of body weight, which has the same numerical value. When the substance was administered by inhalation, the concentration in the inhaled air is expressed in parts per million (ppm), by volume. In the case of the planarians and aquatic toxicities, the ppm refers to the concentration of the substance, on a weight basis, in the aqueous habitat. Abbreviations: LD50: The dose killing 50% of the test animals; LC50: The concentration killing 50% of the test animals; LDLo: The lowest dose reported to have caused death; TCLo: The lowest concentration reported to have produced a toxic reaction; Fiss. inhib.: The concentration in the habitat water of the planarians that will produce a clearly discernible inhibition or delay in their incidence of fissioning; TLm96: The generalized aquatic toxicity rating based upon the death or toxicity of the compound to various aquatic life over a 96-hour exposure period; EC Contrax: The effective concentration for obtaining the contractile response from rabbit uterine muscle in oxygenated physiological saline *in vitro*.

[c]From our own experiments.

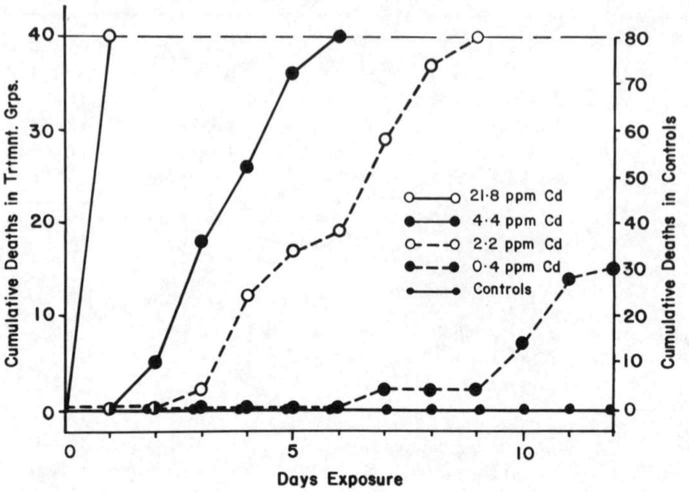

FIGURE 4. Cumulative lethality of asexual planarians of the species *D. dorotocephala*, as a function of the duration of exposure to various habitat water concentrations of cadmium. Planarians were allocated to specimen bowls, each containing 10 planarians and 50 ml of water or cadmium sulfate solution, at zero time. A total of 40 planarians were used for each of the various cadmium concentrations shown, 80 planarians were used for the water controls. The bowls of planarians were observed, cleaned, and the solution or water changed each day at the same time. No deaths occurred in the controls. All planarians exposed to 21.8 ppm of cadmic ion died within the first day. The 5-day LC50 estimate is greater than 2.2 and less than 4.4 ppm cadmium. Lag time decreases and rate of death increases with increasing cadmium concentration.

Where such data were available for comparison, there appeared to be a high degree of correspondence in sensitivities of planarians and other aquatic test species, e.g., trout, minnows, *Daphnia*, and freshwater glass shrimp, to the various toxicants tested, except in the case of toxaphene. Although planarians are a couple of orders of magnitude more sensitive than mammals to toxaphene, they are not nearly as sensitive as fish. For unclear reasons, the tissues of fish concentrate toxaphene several-thousand-fold above habitat water concentrations; this concentrating effect probably explains the extreme vulnerability of fish to even very low aquatic concentrations of toxaphene.

Mercuric compounds, especially organomercuric compounds such as methylmercuric chloride, phenylmercuric chloride, etc., are serious environmental toxicants that can cause irreversible neurotoxic symptoms, renal damage, and death at higher concentrations in adult humans and, at lower concentrations, neurological birth defects. The syndrome resulting from chronic low-level exposure to such compounds has been referred to as "Minamata disease" after the large human epidemic of such poisonings that occurred in the Minamata Sea area of Japan. Exposure to water concentrations of methylmercuric chloride greater than 0.5 ppm caused lethality as the primary response in the planarian, *Dugesia dorotocephala*. Surgically decapitated planarians, regenerating new heads in 0.1 and 0.2 ppm methylmercuric chloride, showed grossly visible teratogenic defects in new head development. Fissioning inhibition was, however, the most sensitive indicator. As indicated previously, social control of fissioning is mediated through the brain of asexual *Dugesia dorotocephala*. Figure 5 shows the effects of increased group size on such fissioning in normal habitat water, and the effect of 0.1 ppb concentration of methylmercuric chloride in their habitat water on fissioning of similar sized groupings. Thus it would appear that, in planarians as in humans, the primary effects of low concentrations of methyl mercury are neurotoxic, except that planarians are considerably more sensitive than humans or standard aquatic test species.

These neurotoxic effects probably derive from the high degree of selective affinity of such mercury compounds for sulfhydryl groups and their essential role in nerve membrane depolarization: reactions of such mercuric compounds with the sulfhydryl groups of nerve membrane have been demonstrated to block the ability of the membrane to depolarize.[114]

Cadmic compounds also produced fissioning suppression in planarians at low concentrations, i.e., at about 50 ppb or greater.

As indicated previously, one of the needs for "miner's canaries" involves the assessment of toxicities of complex mixtures such as oil shale "process waters." Poten-

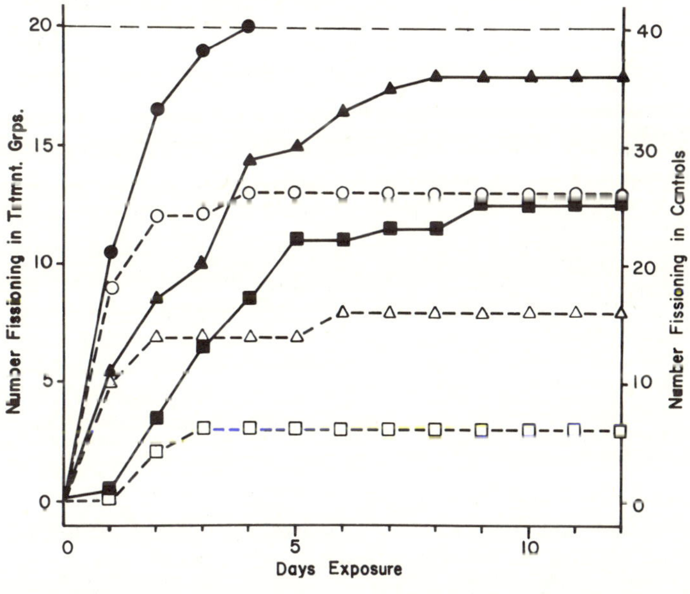

FIGURE 5. Joint effects of group size and aquatic habitat methyl mercury chloride (MMC) on the cumulative number of fissionings of asexual planarians of the species *D. dorotocephala* as a function of time. Planarians were allocated to specimen bowls, each containing 50 ml of water of MMC solution, at time zero, in groupings of 2, 5, and 10 planarians per bowl. A total of 40 planarians were tested under each control grouping, 20 were tested for each of the group sizes in 0.1 ppb of MMC. The number of fissionings were observed, tail fragments removed, and the water or solution changed at the same time each day. Increased group size and MMC both depress the rate of fissioning.
Key: Controls: ●—●, 2/b; ▲—▲, 5/b; ■—■, 10/b; 1.0 ppb: ○--○, 2/b; △--△, 5/b; □--□, 10/b.

tial threats to sources of oil from third world countries have given impetus to develop oil shale extraction as a major alternative energy source. Such oil shale process waters are one of the major environmental problems associated with such an industry.

Shale oil process water is coproduced in equal quantities or greater with *in situ* thermal extraction of shale oil.[115–118] Chemically complex, variable in composition, difficult and expensive to analyze by standard methods of water analysis, toxic, such waters are, without adequate treatment, unsuitable for consumption, irrigation, or other usages. If national goals of production of a million barrels of shale oil per day from the Green River Formation of Utah, Wyoming, and Colorado are to be met, discharge of untreated process water effluents into native waters would render 300,000,000 barrels of such waters per day too toxic for fish survival. This poses a serious environmental problem for the semiarid western states.

Planarians were tested with such waters to see how they compared with the much more expensive rainbow trout currently used for bioassay monitoring of such water toxicities. Results with the one sample for which comparative data were available revealed an LC50 value between 0.3% and 0.4%, which corresponds to the values of 0.4–0.5% reported for rainbow trout with this same sample of process water.[119] Both planarians and fish were an order of magnitude more sensitive than the toxicities reported for oral ingestion of such waters by mammals.[117]

Cigarette smokes from a variety of brands were quantitatively captured in water traps and planarians used to assay the toxicities of such trap waters. This was done to ascertain whether they might be used as an inexpensive model system for assessment of the hazard posed by atmospheric pollutants of complex chemical composition. These trap water toxicities for *Dugesia dorotocephala* correlated well with Federal Trade Commission values for the tar and nicotine content of these cigarette brands.

Discussion of the potential of planarians and other such inexpensive lower invertebrate systems for toxicological modeling of carcinogenic and teratogenic responses requires some brief description of the probable mechanisms involved in such longer-term responses.

12. Carcinogenesis—A Brief Overview

Viruses, ultraviolet and ionizing radiations, and various categories of chemical agents have all been demonstrated to be carcinogenic. Genetic factors also play a role. However, since it is now believed that environmental factors, i.e., radiation and carcinogenic chemicals, may account for the majority of human cancers, I will focus on these.

There is a high correlation between the capability of a chemical compound to produce mutations in bacteria and its carcinogenic potency. This gave support to the idea that the primary initiating event in carcinogenesis involves somatic mutation[120–123] and also provided the basis for a quick and inexpensive method for testing compounds for their carcinogenic potential, i.e., by using bacteria to assess their mutagenicity.[112,124,125] Largely due to the energetic advocacy of Bruce Ames, the most widely used of such tests is the so-called "Ames test" in which the number of reverse mutations of a nutritionally deficient strain of *Salmonella typhemurium* exposed to the compound are used as an indicator of the mutagenicity of the compound. Other systems, e.g., mammalian cell cultures, yeast, and *Drosophila*, are also used to a lesser extent to estimate the mutagenicity of compounds.[112]

It became evident, however, that various compounds, e.g., polyaromatic hydrocarbons such as benzpyrene and dimethylbenzanthracene, which occur in coal tar, were much more carcinogenic in mouse and rat tests than their mutagenicity in the Ames test would indicate.

Metabolic activation of such compounds, called "Procarcinogens," into the actively mutagenic epoxide derivative is required;[123,126] the microsomal oxidase enzymes required for such activation are present in mouse and rat tissues, but not in *Salmonella*.[112,124,125] But only certain chemical structural configurations of polyaromatic hydrocarbons are carcinogenic, or, more properly, undergo activation into a carcinogenic derivative.[120,127–129] In addition to the entire molecule being coplanar, quantum mechanical calculations reveal that they possess high electron density regions called K-regions.[127,128] Such K-regions permit molecules to be metabolically converted, i.e., "activated," into reactive epoxides that react covalently with DNA to disturb its message and produce mutational changes.[123,129–131] However, alkylating agents, e.g., aliphatic epoxides, are both mutagenic and carcinogenic without need for prior metabolic activation; alkylating agents react covalently with DNA and attach an alkyl group to it.

Such activating enzymes, i.e., microsomal oxidases, appear to be adaptively formed for the detoxification of xenobiotic substances, e.g., chlorinated hydrocarbons. NADH-cytochrome-b_5-reductase is an important component of this enzyme system and is generally considered diagnostic for it.[132] Carcinogenic activation is an unfortunate perversion of this enzyme system that works to the detriment of the organism.

Initiation involves the production of an irreversible cellular change, an unrepaired mutation of a particular kind, that is a necessary, but not sufficient, condition for cancer development (p. 670 of Ref. 120; Ref. 133).

Promotion refers to the still poorly understood sequence of events required to fully transform such mutant somatic cells into neoplastic cells.[133,134] "Promoters" facilitate promotion and hasten the emergence to tumors from initiated cells; this capability does not derive from mutagenicity. The pesticide, DDT, and the sedative, phenobarbitol, have been demonstrated to be promoters for tumorigenesis in the livers of rats previously fed an initiator.[135] Initiation is believed to occur rapidly if not almost instantaneously. Promotion, however, may normally require 5 to 9 months in small rodents and years in humans and other larger mammals.

13. Germinal Tissues as Targets for Carcinogens

Some number of cell divisions, or cell growth and DNA replication appears necessary for promotion. In adult metazooans, cell types which still retain the capability for division are probably much more susceptible to neoplastic transformation than those that have lost this capability in the course of differentiation. One might thus anticipate that stem cell systems, e.g., skin, intestinal mucosa, haemopoietic systems, as well as liver, might be likely substrates for tumorigenesis because of their capability to undergo the events necessary for promotion.

An abbreviated description of the stem cell system that produces the white blood cells called granulocytes serves to illustrate some important features of the normal dynamics of all such systems and the way these go awry in neoplastic disease such as granulocytic leukemia. The average lifetime of a mature circulating granulocyte is only about 6 to 9 hours, so they have to be continually replaced by new ones produced in bone marrow. This is done as follows: Committed granulocytic stem cells in the marrow are capable of either undergoing mitosis to produce two such stem cell daughters, or they may undergo differentiation to become myeloblasts, which require several days to mature into promyelocytes and these several days more to develop into mature granulocytes which are then released into circulation; as they mature, myeloblasts undergo further divisions, which means that a number of myeloblasts later mature into promyelocytes (at which state division ceases) for every stem cell that undergoes differentiation into a nascent myeloblast; this process, called "cell amplification," is normally limited by the loss of such mitotic capabilities as the myeloblasts mature into promyelocytes. Normally, except for stresses such as hemorrhage, infections, etc., various homeostatic regulatory mechanisms maintain the populations of such stem cells, blast cells, etc. relatively constant, even though each is continually gaining and losing members by processes of the type described.

In myelocytic, e.g., granulocytic, leukemia there occurs some defect or impediment in the differentiation of the myeloblasts derived from a malignant stem cell progenitor. Such myeloblasts do not divide faster than normal myeloblasts, but they continue to remain as primitive blast cells instead of terminating this divisional phase by maturing. This results in a great expansion in the number of myeloblasts and substantially reduced numbers of mature blood cell types.

Primitive cells, e.g., haemopoietic stem cells, myeloblasts, lymphoblasts, leukemic cells, and planarian neoblasts, look very much alike, under both optical and electron microscopy. A description of one would suffice for all. They usually have a relatively small volume of cytoplasm in relation to nuclear volume, many free or polysomal ribosomes, and endoplasmic reticulum is scanty or absent. Occasionally, as in the case of the adult form of chronic myelogenous leukemia with its "Philadelphia chromosome" anomaly, or planarian neoblasts with their "nuclear satellite bodies"[17,18] one may be fortunate enough to have a distinguishing cytoplasmic marker, but these are the exception rather than the rule. One cannot, for example, usually look at any particular "blast cell" of a patient with acute myelogenous leukemia and distinguish with certainty between a stem cell, a normal immature myeloblast, and a leukemic cell, except perhaps in more advanced stages of the disease when some of the leukemic cell population may become aneuploid.

The point is that cancer cells are usually not distinguishable from normal primitive cells, except that there are too many of them and their arrangements in relation to the normal cellular arrangements of the afflicted organs are abnormal.

14. Chemical Induction of Tumorigenesis and Teratogenesis in Planarians

At least two species of planarians of the genus *Dugesia* have been found to develop spontaneous tumors[136]; however, the cell types composing them did not appear abnormal, suggesting that such tumors are more analogous to those that would be classified as "benign" in vertebrates. But the incidence of such spontaneous tumors in the species *Dugesia dorotocephala* is either zero or very small.

However, Foster[137,138] has induced tumors, with a high yield, in *D. dorotocephala*, by treatment with several polyaromatic mammalian carcinogens; such tumors were progressive and ultimately led to the death of the animal; they were primarily composed of primitive cells. Thus, in their essential characteristics, they appear to resemble tumors that would be classified as "malignant" neoplasia in vertebrates. Foster also described (with drawings) interesting teratogenic effects in the case of some of these tumors, e.g., supernumery heads growing out of the tumor mass. Although she did not report the incidence of appearance of such tumors as a quantitative function of time, her reports suggest a latency of the order of approximately three months for their development.

We also have induced tumors and teratogenic effects in *D. dorotocephala* by exposure to the mammalian carcinogens, dimethylbenzanthracene, benzpyrene, and benzanthracene (cf. Figs. 6 and 7). Some of these tumors appear analogous to benign tumors, e.g., skin papillomas of mammals, in that their histological composition of cell types reveals no significant abnormality. Others show the concentrated predominance of primitive cells, i.e., "granularity" or "cellularity," characteristic of malignant tumors. As anticipated (cf. previous discussion), electronmicrographs of these primitive cells reveal the ultrastructural features of neoblasts.

Yet other of these planarians show teratogenic effects, e.g., three eyes instead of two, or a second head growing from the caudal end of the animal, or, occasionally, a supernumary head growing out of a tumor mass (Fig. 7).

All of these responses are comparable to ones noted in exposures of mammalian systems to mutagenic or carcinogenic agents.[139,140] In teratomas, i.e., malignant neoplasias associated with embryonic development of mammals or birds, it is not uncommon for such teratogenic development to be associated with the tumor, e.g., supernumary teeth growing out of skin.

Benign tumor development may presage malignant tumor development. For example, daily application of a carcinogenic chemical to an area of mouse skin frequently gives rise initially to benign papillomas in the area which subsequently becomes the locus of development for skin cancer.

The Steinberg formulation[107] discussed previously permits one to consider these three kinds of responses of planarians and mammals to carcinogens on a continuum.

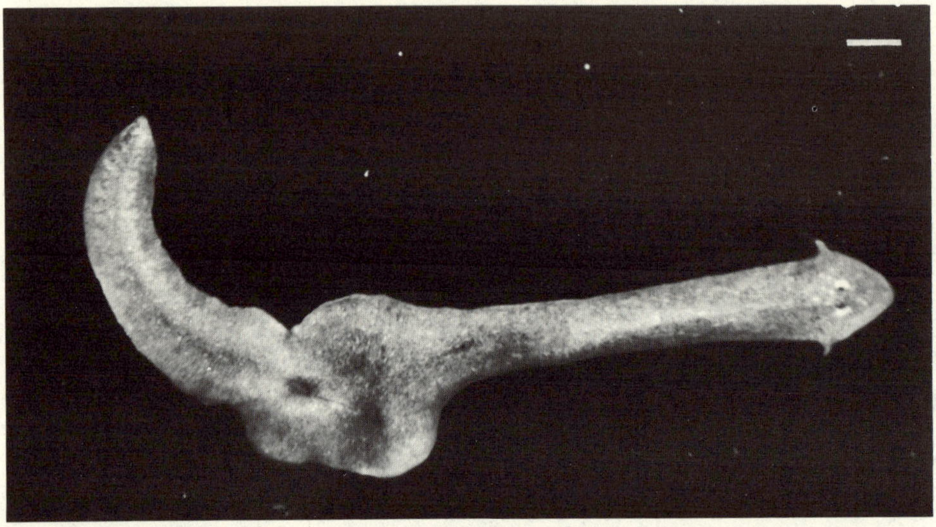

FIGURE 6. Photograph of an asexual planarian *(D. dorotocephala)* that has developed a tumorlike growth following feeding of a polyaromatic mammalian carcinogen in an uncooked egg yolk vehicle. Compare the appearance of this planarian with the normal planarians shown in Fig. 1. The horizontal bar indicates 1 mm.

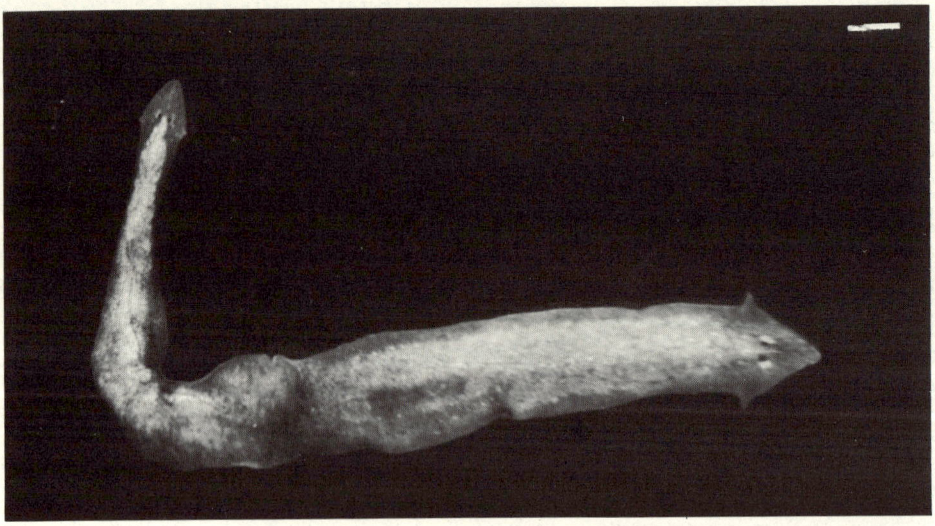

FIGURE 7. Photograph of an asexual planarian *(D. dorotocephala)* that has developed a tumorlike growth and a supernumerary head growing out of the region of the tumor following feeding of a polyaromatic mammalian carcinogen in an uncooked egg yolk vehicle. Compare the appearance of this planarian with that of the normal planarians shown in Fig. 1. The horizontal bar indicates 1 mm.

It has been demonstrated that mutagenic alterations of the DNA, comparable to those involved in carcinogenic initiation, result in molecular alterations of the cell surface, which alters its tendency to adhere to adjacent cells and exhibit contact inhibition. Such alterations are probably reflected in the F_{ij} values of the Steinberg matrix. Small alterations that do not disturb too greatly the inequality relations among the F_{ij} values determining the normal anatomical structure would give rise to the still relatively normally organized supernumery eyes, heads, etc. If one did not examine too large an area of histological section, the chances are that the cellular types and arrangements in that section area would appear fairly normal. The imperfection of the "crystal" becomes evident only in the macroscopic domain. More severe disturbances in these F_{ij} relationships would give rise to the benign tumors. In these, histological examination of smaller domains would reveal normal differentiated cell types cohering together with more or less normal appearing local order, but imperfection is manifest in examination of larger domains; although not malignant neoplasia, there is obviously something a little wrong in the way the cells "crystallize," i.e., associate, together, which is probably indicative of some, still mild, alterations in the cell surfaces so they do not associate together quite properly.

Malignant tumors would represent an even more severe disturbance of these F_{ij} relationships as well as a failure of the primitive blast cells to differentiate into nonmitotic stages of maturity. Once this state of affairs occurs, blast cell proliferation is no longer self-limited by maturation of the blast cells into nonproliferative specialized cell types. The reason such primitive blast cells predominate in such tumors is not that they divide more rapidly than normal stem or blast cells but that they just do not quit.*

A plausible mechanism can be offered to explain how mutation, i.e., "initiation," in a single planarian neoblast could give this spectrum of graded morphogenetic responses:

As an initiated, i.e., mutated, stem cell undergoes the series of mitotic divisions that appears necessary to complete promotion, i.e., transformation into malignancy, phenotypic expression of altered adhesion proteins on the cell surface occurs in graduated stages. Since there is a stochastic component to the neoblast division necessary to replace those randomly sequestered for differentiation into specialized cell types, there will, after some adequate waiting period, be the potential for such sequestration of clonal descendents of the mutant in all stages of promotion. With the turnover of the various cell types that is continually occurring in asexual *D. dorotocephala*, together with the potential for graduated disturbances of the inequality relationships between the F_{ij} elements, one has, by the mechanisms discussed above, the potential for the three types of morphogenetic responses described.

Since, in postembryonic mammals, the stem cells have only a limited range of

*The popular use of transplantable tumor strains gave a false impression of tumor cell kinetics. After serial passage through "generations" of mouse hosts, fast growing variants of the tumor cell population are selected for. However, *in vivo* cell kinetic measurements on primary neoplasias indicate that the tumor cells usually divide no faster, or even slower, than normal blast or stem cells.

differentiation capabilities, their morphogenetic responses to mutagens are restricted to benign and malignant tumor formation. The pluripotentiality of mammalian embryonic cells permits them to manifest teratogenic as well as tumorigenic responses. Planarians model the responses of both adult mammalian stem cells and embryonic developmental stages to such mutagens.

The three procarcinogens (benzpyrene, dimethylbenzanthracene, and benzanthracene) used in our experiments require metabolic activation into the actively mutagenic form. Several observations suggest that planarians do, in fact, possess enzyme systems analogous to those microsomal oxidases involved in mammalian carcinogen activation.

First is that the planarians exhibited tumorigenic responses to these procarcinogens. Second is that planarians have been demonstrated to metabolize DDT to DDD and DDE, and that the enzymes for doing this involve the inductive formation of NADH-cytochrome-b_5-reductase.[141,142] As indicated earlier in this essay, such systems appear to be responsible for carcinogen activation in mammals.

15. Summary and Conclusions

Further ramifications, of conceptions formulated previously to account for the "higher behaviors" exhibited by planarians[8] are explored.

Free-living flatworms such as planarians and marine polyclads are probably representative of the first animals to develop bilateral symmetry, encephalization, brains, and other important features of anatomical and physiological organization characteristic of vertebrates. Most of the major functional categories of cell types, e.g., neurons, neuroglia, chemoreceptors, photoreceptors, epithelium, endothelium, reticular cells, goblet cells, neurosecretory cells, and muscle, are present in such flatworms. The variety of detailed similarities in metabolic pathways, functional roles of subcellular organelles, organization and chemistry of the nervous systems, functional organization and roles of neurohormonal control systems, basic histological and functional anatomical arrangements, from flatworms to arthropods and vertebrates, indicates a higher degree of basic similarity among such animals than has generally been appreciated. Such similarities, together with certain other considerations of a theoretical nature, suggest that large portions of the genome have been conserved, with minimal modification, in the evolutionary heirs of their flatworm progenitors. This has important practical as well as theoretical implications.

"Miner's canaries," i.e. nonhuman organisms used as toxicological surrogates for humans, provide the primary method of choice for assaying environmental toxicants and new chemical compounds, intended for drug, food, or cosmetic usage, for their human hazard potential.

The fact that most toxicological testing is done for legal purposes with results apt to be used as evidence in formal court hearing procedures conducted by nonscientists in an adversary relationship imposes extrascientific considerations on such experiments. One consequence is a reluctance to utilize unconventional test procedures and test systems. Another arises from the "Delaney clause" of the U.S. Food, Drug and Cosmetics Act that provides legal authority for exclusion of any chemical that has been demonstrated to cause cancer in humans or animals, from foods or cosmetics intended for human consumption or application. Lack of dosage specification by the Delaney criterion permits investigators to use much larger dosages than those encountered in real-life conditions of usage in order to detect otherwise small cancer incidences in test animal groups of limited size. Neither the Delaney criterion nor court decisions place restriction on the range of animals that might potentially be used for such tests. However, popular conception and administrative rulings regarding kinship of the test species to man do play an important, though vaguely defined, role in choices of test species.

Economic and logistical considerations, together with considerations of phylogenetic kinship and similarity of toxicologically vulnerable response mechanisms, give rise to an uncertainty principle in toxicological testing and prediction that is analogous to the Heisenberg principle in physics. Insistence on close phyletic kinship of the test organisms to humans results in loss of information about low dose effects, longer delays in obtaining answers, and test information on a smaller fraction of those chemicals for which such test information is needed. Conversely, relaxation of the close phyletic kinship requirement permits more information about low dose effects, shorter delays in obtaining answers, and test information on a larger fraction of those chemicals for which such test information is needed. However, relaxation of the requirement for close phyletic kinship entails a greater chance that the test organism may not have the same toxicologically vulnerable response mechanism as humans for a given chemical.

Because of these considerations and the need to (1) obtain toxicological information faster and more inexpensively for an increasing backlog of untested compounds, (2) detect hazardous levels in the environment of an increasing variety of toxic chemicals, there has been increasing need for and interest in alternative organisms and test systems for use as "miner's canaries."

Transphyletic similarities in the physiological and biochemical organization of animals, and the genomic conservatism underlying them, are important to the question of toxicological similarities. Conversely, toxicological and pharmcological similarities provide evidence for physiological and biochemical similarity. There is thus an important connection between genomic conservatism, comparative toxicology, and the phyletic range from which appropriate miner's canaries can be chosen.

Acute, i.e., limited to two weeks, pharmacological and toxicological studies on planarians using representatives of a variety of chemical categories of toxic agents, e.g., heavy metals, serotonin and monoaminergic analogs, agents which interfere with cholinergic transmission, chlorinated hydrocarbons, pesticides, as well as complex toxicant mixtures, e.g., shale oil process water and cigarette smokes indicate a good correspondence between toxic sensitivities of planarians and standard vertebrate test animals, except that planarians are usually more sensitive and respond sooner. Since planarians do not utilize myoneural endplates of the type associated with striated skeletal muscle of vertebrates, they are, as might be anticipated, insensitive to drugs with a curarelike action.

Planarian regeneration and remodeling may provide a useful short-term system for assessment of teratogenic toxicants.

Longer-term studies, two to four months, involving exposures of planarians to carcinogens, produced benign tumors, malignant tumors, and teratogenic effects. Planarians appear to model major features involved in induction of tumors and teratogenic responses by carcinogenic agents. These various responses are all viewed as part of a continuum of the effects of somatic mutations on the differentiation and surface adhesion properties of primitive germinal cells, e.g., vertebrate embryonic cells, adult vertebrate stem cells, planarian neoblasts: there is an increasing perturbation of the relationships between the adhesional free energy terms, F_{ij}, of the Steinberg matrix. Asexual planarians model features of both mammalian stem cell systems and mammalian embryogenesis.

Because of persistent and pervasive similarities in the physiological and biochemical organization of bilaterally symmetric encephalized animals, and the genomic conservatism that gives rise to them, we may have a rational basis for comparative toxicology and the diverse practical toxicological problems that involve it.

Flatworms may be valuable adjuncts to the mammals, birds, fish, bacterial, and mammalian cell culture systems currently used in toxicological testing.

References and Notes

1. A. D. Breck and W. Yourgrau, *Biology, History, and Natural Philosophy* (Plenum Press, New York, 1972).
2. G. S. Fraenkel and D. L. Gunn, *The Orientation of Animals* (Reprinting of 1940 edition by Dover Publ., Inc., New York, 1961).
3. J. Loeb, *Forced Movements, Tropisms and Animal Conduct* (Reprinting of 1918 edition by Dover Publ. Inc., New York, 1973).
4. J. B. Best and I. Rubenstein, "Maze Learning and Associated Behavior in Planaria," *J. Comp. Physiol. Psychol.* **55**, 560–566 (1962).
5. J. B. Best and I. Rubenstein, "Environmental Familiarity and Feeding in a Planarian," *Science* **135**, 916–918 (1962).
6. J. B. Best, "Protopsychology," *Scientific American* 54–62 (1963).
7. H. Koopowitz, "Feeding Behavior and the Role of the Brain in the Polyclad Flatworm, *Planocera gilchristi*," *Anim. Behavior* **18**, 31–35 (1970).
8. J. B. Best, "The Organization and Evolution of Sentient Biological Behavior Systems," in *Biology, History and Natural Philosophy*, A. D. Breck and W. Yourgrau, editors (Plenum Press, New York, 1972), pp. 37–78.
9. J. B. Best, "Learning and Neurophysiology of Planarians: Transphyletic Similarities in the Organization of Brains and Psychological Behavior Systems," in *Behavioral Genetics: Simple Systems*, James Wilson, editor (Colorado University Press, Boulder, Colorado, 1973), pp. 55–84.
10. G. L. Stebbins, "The Evolutionary Significance of Biological Templates," in *Biology, History and Natural Philosophy*, eds. A. D. Breck and W. Yourgrau, editors (Plenum Press, New York, 1972).
11. L. Hyman, *The Invertebrates: Platyhelminths and Rhynchocoela*, Vol. 2 (McGraw-Hill, New York, 1951).

12. M. Morita and J. B. Best, "Electron Microscopic Studies of Planaria. II. Fine Structure of the Neurosecretory System in the Planarian *Dugesia dorotocephala*," *J. Ultrastructure Res.* **13**, 396–408 (1965).
13. K. J. Pedersen, "Slime-Secreting Cells of Planarians," *Ann. N. Y. Acad. Sci.* **106**, 424–442 (1963).
14. E. Wolff, "Recent Researches on Regeneration of Planaria," in *Regeneration*, D. Rudnick, editor (The Ronald Press Co., New York, 1962), pp. 53–84.
15. E. Wolff and F. Dubois, "Sur une method d'irradiation localisee permittant de mettre en evidence la migration des cellules de regeneration chez les planaires," *C. R. Soc. Biol.* **141**, 903–906 (1947).
16. E. Wolff and F. Dubois, "Sur la migration des cellules de regeneration chez les planaires," *Rev. Suisse Zool.* **55**, 218–227 (1948).
17. M. Morita and J. B. Best, "Electron Microscopic Studies of Planarian Regeneration. II. Changes in Epidermis During Regeneration," *J. Exp. Zool.* **187**, 345–374 (1974).
18. M. Morita, J. B. Best, and J. Noel, "Electron Microscopic Studies of Planarian Regeneration. I. Fine Structure of Neoblasts in *Dugesia dorotocephala*. *J. Ultrastructure Res.* **27**, 7–23 (1969).
19. K. Carpenter, M. Morita, and J. B. Best, "Ultrastructure of the Photoreceptor of the Planarian *Dugesia dorotocephala*. I. Normal Eye," *Cell Tissue Res.* **148**, 143–158 (1974a).
20. K. Carpenter, M. Morita, and J. B. Best, "Ultrastructure of the Photoreceptor of the Planarian *Dugesia dorotocephala*. II. Changes Induced by Darkness and Light," *Cytobiologie* **8**, 320–338 (1974).
21. J. B. Best, S. Hand, and R. Rosenvold, "Mitosis in Normal and Regenerating Planarians," *J. Exp. Zool.* **168**, 157–168 (1968).
22. J. B. Best, M. Morita, and J. Noel, "Fine Structure and Function of Planarian Goblet Cells," *J. Ultrastructure Res.* **24**, 385–397 (1968).
23. J. B. Best, A. B. Goodman, and A. Pigon, "Fissioning in Planaria: Control by the Brain," *Science* **164**, 565 (1969).
24. J. B. Best, W. Howell, V. Riegel, and M. Abelein, "Cephalic Mechanism for the Social Control of Fissioning in Planarians. I. Nature of the Feedback Cue and Switching Characteristics," *J. Neurobiology* **5**, 421–442 (1974).
25. J. B. Best, M. Abelein, E. Kreutzer, and A. Pigon, "Cephalic Mechanism for the Social Control of Fissioning in Planarians. III. CNS Centers of Facilitation and Inhibition," *J. Comp. Physiol. Psychol.* **89**, 923–932 (1975).
26. W. J. Davis, "Neuronal Organization and Ontogeny in the Lobster Swimmeret System," in *Advances in Behavioral Biology, Vol. 7: Control of Posture and Locomotion*, R. B. Smith, K. B. Pearson, R. S. Smith, and J. B. Redford, editors (Plenum Press, New York, 1974), pp. 437–455.
27. J. H. Welsh and L. D. Williams, "Monoamine Containing Neurons in a Planarian," *J. Comp. Neurology* **138**, 103–116 (1969).
28. J. H. Welsh and M. Moorhead, "The Quantitative Distribution of 5-hydroxytryptamine in the Invertebrates, Especially in their Nervous Systems," *J. Neurochemistry* **6**, 146–169 (1960).
29. J. B. Best and J. Noel, "Complex Synaptic Configurations in Planarian Brain," *Science* **164**, 1070–1071 (1969).
30. M. Morita and J. B. Best, "Electron Microscopic Studies of Planaria. III. Some Observations on the Fine Structure of Planarian Nervous Tissue," *J. Exp. Zool.* **161**, 391–411 (1966).
31. J. H. Welsh and E. C. King, "Catecholamines in Planarians," *Comp. Biochem. Physiol.* **36**, 683–688 (1970).
32. T. L. Lentz, *Primitive Nervous Systems* (Yale Univ. Press, New Haven, 1968).
33. T. H. Bullock and D. Nachmansohn, "Cholinesterase in Primitive Nervous Systems," *J. Cell Comp. Physiol.* **20**, 239–242 (1942).
34. R. M. Eakin, "Lines of Evolution of Photoreceptors," in *General Physiology of Cell Specialization*, D. Mazia and A. Tyler, editors (McGraw-Hill Book Co., New York, 1963).
35. R. M. Eakin, "Evolution of Photoreceptors," in *Evolutionary Biology II*, T. Dobzhansky, M. K. Hecht, and W. C. Stene, editors (Appleton-Century-Crofts, New York, 1968).
36. E. K. MacRae, "Observations on the Fine Structure of the Photoreceptor Cells in the Planarian *Dugesia tigrina*," *J. Ultrastruc. Res.* **10**, 334–349 (1964).
37. E. K. MacRae, "Fine Structure of Photoreceptors in a Marine Flatworm," *Z. Zellforsch.* **75**, 469–484 (1966).

38. Y. Kishida, "Electron Microscopy Studies on the Planarian Eye. I, II." *Sci. Rep. Kanazawa Univ.* **12,** 75–142 (1967).
39. F. H. Bronson, B. F. Eleftheriou, and H. E. Dezell, "Strange Male Pregnancy Block in Deermice: Prolactin and Adrenocortical Hormones," *Biol. Reprod.* **1,** 302–306 (1969).
40. K. E. Dewhurst, D. J. El Kabir, G. W. Harris, and B. M. Mandelbrote, "A Review of the Effect of Stress on the Activity of the Central Nervous–Pituitary–Thyroid Axis in Animals and Man," *Confinia Neurologiea Separatum* **30,** 161–196 (1968).
41. G. Giuliani, "Studies on Gonadotropin Release during Stressful Situations and the Role of the Central Nervous System," in *Physiology and Pathology of Adaptation Mechanisms*, E. Bajusz, editor (Pergamon Press, New York, 1969), pp. 293–309.
42. A. Pigon, M. Morita, and J. B. Best, "Cephalic Mechanism for the Social Control of Fissioning in Planarians. II. Localization and Identification of the Receptors by Electronmicrographic and Ablation Studies," *J. Neurobiol.* **5,** 443–462 (1974).
43. H. M. Bruce, "An Exteroreceptive Block to Pregnancy in the Mouse," *Nature* **184,** 105 (1959).
44. H. M. Bruce, "A Block to Pregnancy in the Mouse Caused by the Proximity of Strange Males," *J. Reprod. Fertil.* **1,** 96 (1960).
45. H. M. Bruce, "Further Observations on Pregnancy Block in Mice Caused by the Proximity of Strange Males," *J. Reprod. Fertil.* **1,** 311 (1960).
46. H. M. Bruce, "Olfactory Block to Pregnancy among Grouped Mice," *J. Reprod. Fertil.* **6,** 451 (1963).
47. H. M. Bruce, "Smell as an Exteroreceptive Factor," *J. Anim. Sci.* **25** (Suppl.), 83 (1966).
48. H. M. Bruce, "Absence of Pregnancy Block in Mice when Stud and Test Male Belong to an Inbred Strain," *J. Reprod. Fertil.* **17,** 407 (1968).
49. H. M. Bruce, "Continued Suppression of Luteotrophic Activity and Fertility in the Female Mouse," *J. Reprod. Fertil.* **4,** 313 (1962).
50. H. M. Bruce and A. S. Parkes, "Hormonal Factors in Exteroreceptive Block to Pregnancy in Mice," *J. Endocrinol.* **20,** 29 (1960).
51. A. S. Parks, "The Role of Odorous Substances in Mammalian Reproduction," *J. Reprod. Fert.* **1,** 312 (1960).
52. A. S. Parks and H. M. Bruce, "Olfactory Stimuli in Mammalian Reproduction," *Science* **134,** 1049 (1961).
53. T. L. Avery, "Phermone-Induced Changes in the Acidophil Concentration of Mouse Pituitary Glands," *Science* **164,** 423–424 (1969).
54. J. A. Coppola, "Brain Catecholamines and Gonadotropin Secretion," in *Frontiers in Neuroendocrinology*, L. Martini and W. F. Ganong, editors (Oxford Univ. Press, London, 1971), Chap. 5.
55. W. F. Ganong, "Brain Amines and ACTH Secretion," in *Hormonal Steroids*, Proc. 3rd Inter. Congr. Hormonal Steroids, Hamburg, 7–12, Sept. 1970, V. H. T. James and L. Martini, editors (Excerpta Medica, Amsterdam, 1971), pp. 814–821.
56. K. Fuxe, T. Hokfeldt, and G. Jonsson, "The Effect of Gonadal Steroids on the Tubero-infundibular Dopamine Neurons," in *Hormonal Steroids*, V. H. T. James and L. Martini, editors (Excerpta Medica, Amsterdam, 1971), pp. 806–813.
57. F. Anton-Tay and R. J. Wurtman, "Brain Monoamines and Endocrine Function," in *Frontiers in Neuroendocrinology*, L. Martini and W. F. Ganong, editors (Oxford Univ. Press, New York, 1971), pp. 45–66.
58. V. Dethier, "Microscopic Brains," *Science* **10,** 1138–1145 (1964).
59. K. C. Highnam and L. Hill, *The Comparative Endocrinology of the Invertebrates* (American Elsevier Publ. Co., New York, 1969).
60. W. Loher, "The Chemical Acceleration of the Maturation Process and Its Hormonal Control in the Male and the Desert Locust," *Proc. R. Soc. London Ser. B* **153,** 380–397 (1960).
61. L. H. Kleinholz, "Separation and Purification of Crustacean Eyestalk Hormones," *Am. Zool.* **6,** 161–167 (1966).
62. L. H. Kleinholz, F. Kimball, and M. McGarvey, "Initial Characterization and Separation of Hyperglycemic (Diabetogenic) Hormone from the Crustacean Eyestalk," *Gen. Comp. Endocrinol.* **8,** 75–81 (1967).
63. A. A. Abramowitz, F. L. Hisaw, and D. N. Papandrea, "The Occurrence of a Diabetogenic Factor in the Eyestalks of Crustaceans," *Biol. Bull.* **86,** 1–5 (1944).

64. R. Kenk, "Induction of Sexuality in the Asexual Form of *Dugesia tigrina*," *J. Exp. Zool.* 87, 55–69 (1941).
65. M. Grasso, "Esperimenti sul Controllo Study of Sexual Induction in the Asexual Agami di Planarie," *Boll. Zoologia* **38**, 532 (1971).
66. Th. Lender, "The Role of Neurosecretion in Freshwater Planarians," in *Biology of the Turbellaria*, N. W. Riser and M. P. Morse, editors (McGraw-Hill, New York, 1974), pp. 460–475.
67. Th. Lender, "Endocrinologie des Planaires," *Bull. Soc. Zool. Fr.* **105**, 173–191 (1980).
68. B. Scharrer, "The Nuerosecretory Neuron in Neuroendocrine Regulatory Mechanisms," *Amer. Zool.* 7, 161–169 (1967).
69. A. A. Bronstein and V. P. Ivanov, *Zh. Evolyut. Biokhim. Fizol.* **1**, 251–261 (1965). Cited by Ya. A. Vinnikov in G. H. Bourne, editor, *Structure and Function of Nervous Tissue*, Vol. II (Academic Press, New York, 1969).
70. Ya. A. Vinnikov, "The Ultrastructure and Cytochemical Bases of Function of the Sense Organ Receptors," in *The Structure and Function of Nervous Tissue*, G. H. Bourne, editor, Vol. II (Academic Press, New York, 1969).
71. A. Y. deLorenzo, "Electron Microscopic Observations of the Olfactory Mucosa and Olfactory Nerve," *J. Biophys. Biochem. Cytol.* **3**, 839–848 (1957).
72. J. Axelrod, "The Pineal Gland: A Neurochemical Transducer," *Science* **184**, 1341–1348 (1974).
73. S. Brinkley, S. E. MacBride, D. C. Klein, and C. L. Ralph, "Pineal Enzymes: Regulation of Avian Melatonin Synthesis," *Science* **181**, 273–275 (1973).
74. G. Brown, L. Grota, G. Bubenik, L. Niles, and H. Tsui, "Physiologic Regulation of Melatonin," in *Melatonin: Current Status and Perspectives*, N. Birau and W. Schloat, editors (Pergamon Press, New York, pp. 95–112 (1981).
75. M. Morita (personal communication).
76. C. Vowinckel and J. R. Marsden, "Reproduction of *Dugesia tigrina* under Short-Day and Long-Day Conditions at Different Temperatures," *J. Embryol. Exp. Morph.* **26**, 587–609 (1971).
77. S. Rude, "Catecholamines in the Ventral Nerve Cord of *Lumbricus terrestris*," *Comp. Biochem. Physiol.* **28**, 747–752 (1969).
78. S. Rude, "Monoamine-Containing Neurons in the Nerve Cord and Body Wall of *Lumbricus terrestris*," *J. Comp. Neurol.* **128**, 397–412 (1966).
79. D. C. Sweeney, "Dopamine: Its Occurrence in Molluscan Ganglia," *Science* **139**, 1051 (1963).
80. D. C. Sweeney, "The Anatomical Distribution of Monoamines in a Fresh-Water Luvalve Mollusc, *Sphaerium sulcatum* (L)," *Comp. Biochem. Physiol.* **25**, 601–614 (1968).
81. J. B. Best (unpublished results).
82. J. Alumets, R. Hakanson, F. Sundler, and J. Thorell, "Neuronal localization of Immunoreactive Enkephalin and β-endorphin in the Earthworm," *Nature* **279**, 805–806 (1979).
83. J. B. Best and M. Morita (unpublished results).
84. M. Morita, "Electron Microscopic Studies of Planaria. I. Fine Structure of the Muscle Fiber in the Head of the Planarian, *Dugesia dorotocephala*," *J. Ultrastructure Res.* **13**, 383–395 (1965).
85. H. Koopowitz and D. W. Ewer, "Observations on the Myo-neural Physiology of a Polyclad Flatworm: Inhibitory Systems," *J. Exp. Biol.* **53**, 1–8 (1970).
86. L. Keenan and H. Koopowitz, "Tetrodotoxin Sensitive Action Potentials from the Brain of the Polyclad Flatworm, *Notoplana actiola*," *J. Exp. Zool.* **215**, 209–213 (1981).
87. T. Narahashi, J. W. Moore, and W. R. Scott, "Tetrodotoxin Blockage of Sodium and Conductance Increase in Lobster Giant Axons," *J. Gen. Physiol.* **47**, 965–974 (1964).
88. T. Narahashi, "Chemicals as Tools in the Study of Excitable Membranes," *Physiol. Rev.* **54**, 813–889 (1974).
89. P. A. V. Anderson, "Ionic Basis of Action Potentials and Bursting Activity in the Hydromedusan Jellyfish, *Polyorchis penicillatus*," *J. Exp. Biol.* **78**, 299–302 (1979).
90. Y. Saimi and C. Kung, personal communication cited by L. Keenan and H. Koopowitz: Tetrodotoxin sensitive action potentials from the brain of the polyclad flatworm, *Notoplana actiolo*. *J. Exp. Zool.* **215**, 209–213 (1981).
91. F. Dubois and E. Wolff, "Sur une methode dirradiation localisee (45) permettant de mettre en evidence la migration des cellules de regeneration chez les planaires," *Compt. Rend. Soc. Biol.* **141**, 903–906 (1947).

92. F. Dubois, "Demonstration de la migration des cellules de regeneration des Planaires par la methode des greffes et des irradiations combinees," *C. R. Acad. Sci.* **226,** 1316 (1948).
93. F. Dubois, "Contribution a letude de la migration des cellules de regeneration chez les Planaires dulcicoles," *Bull. Biol. France Belg.* **83,** 213 (1949).
94. T. Lender, "Factors in Morphogenesis of Regenerating Freshwater Planaria," in M. Abercrombie and J. Bracket, editors, *Advances in Morphogenesis,* Vol. 2, p. 305 (Academic Press, New York, 1962).
95. T. Betchaku, "Isolation of Planarian Neoblasts and their Behavior *in vitro* with Some Aspects of the Mechanism of Formation of the Regeneration Blastema," *J. Exp. Zool.* **164,** 407 (1967).
96. M. Morita and J. B. Best (unpublished results).
97. C. S. Lange, "An Outline of Studies on the Cellular Basis of Planarian Radiation Lethality," *J. Physiol. (London)* **197,** 54–55 (1968).
98. C. S. Lange, "Studies on the Cellular Basis of Radiation Lethality. I. The Pattern of Mortality in the Whole-Body Irradiated Planarian (Tricladia paludicola)," *Int. J. Radiat. Biol.* **13,** 511–530 (1968).
99. C. S. Lange, "Studies on the Cellular Basis of Radiation Lethality. II. Survival-Curve Parameters for Standardized Planarian Populations," *Int. J. Radiat. Biol.* **14,** 119–132 (1968).
100. C. S. Lange, "Studies on the Cellular Basis of Radiation Lethality. IV. Confirmation of the Validity of the Model and the Effects of Dose Fractionation," *Int. J. Radiat. Biol.* **14,** 539–551 (1969).
101. C. S. Lange, "Studies on the Cellular Basis of Planarian Radiation Lethality. V. A Survival Curve for the Reproductive Integrity of the Planarian Neoblast and the Effect of Polyploidy on the Radiation Response," *Int. J. Radiat. Biol.* **15,** 51–64 (1969).
102. C. S. Lange and C. W. Gilbert, "Studies on the Cellular Basis of Radiation Lethality. III. The Measurement of Stem Cell Repopulation Probability," *Int. J. Radiat. Biol.* **14,** 373–388 (1968).
103. J. P. Trinkhaus, *Cells into Organs* (Prentice Hall, Englewood Cliffs, New Jersey, 1969).
104. J. P. Trinkhaus and T. L. Lentz, "Direct Observation of Type Specific Segregation in Mixed Cell Aggregates," *Develop. Biol.* **9,** 11–136 (1964).
105. A. A. Moscona, "Analysis of Cell Combinations in Experimental Synthesis of Tissues *in vitro,*" *J. Cell. Comp. Physiol.* **60** (Suppl. 1), 65–80 (1962).
106. A. A. Moscona, "Development *in vitro* of Chimaeric Aggregates of Dissociated Embryonic Chick and Mouse Cells," *Proc. Natl. Acad. Sci. U.S.* **43,** 184–144 (1957).
107. M. S. Steinberg, "Reconstruction of Tissues by Dissociated Cells," *Science* **141,** 401–408 (1963).
108. P. Handler, Dedication address, Northwestern Univ. Cancer Center, May 18 (1979).
109. J. T. MacGregor (personal communication).
110. U. S. Congress, *Federal Food, Drug and Cosmetic Act as Amended* (U. S. Government Printing Office, Washington, D.C., August, 1972.
111. *Harvard Medical School Health Newsletter* (1979).
112. M. Holstein, J. McCann, F. A. Angelosanto, and W. W. Nichols, "Short Term Tests for Carcinogens and Mutagens," *Mutation Res.* **65,** 133–226 (1979).
113. *Wall Street Journal,* page 1, September 4, 1980.
114. F. Huneeus-Cox, H. L. Fernandez, and B. H. Smith, "Effects of Redox and Sulfhydryl Reagents on the Bioelectric Properties of the Giant Axon of the Squid," *Biophys. J.* **6,** 675–689 (1966).
115. D. S. Farrier, R. E. Poulson, Q. D. Skinner, J. C. Adams, and J. Bower, "Aquisition, Processing and Storage for Environmental Research of Aqueous Effluents Derived from in situ Oil Shale Processing," in *Water Problems Associated with Coal Conversion and Shale Oil Production,* Second Pacific Chemical Engineering Congress, Vol. II, American Institute of Chemical Engineers, New York (1977).
116. D. S. Farrier, L. W. Harrington, and R. E. Poulson, "Integrated Compliance and Control Technology Research Activities for *in situ* Fossil Fuel Processing Experiments," LETC/RI-79/6. Tech. Info Center, U.S. Department of Energy, September (1979).
117. D. S. Farrier, J. E. Virgona, T. E. Phillips, and R. E. Poulson, "Environmental Research for in situ Oil Shale Processing," *Eleventh Oil Shale Symp. Proc.,* James H. Gary, editor, April 12–14 (Colo. School of Mines Press, Golden, Colorado, 1978).
118. J. P. Fox, D. S. Farrier, and R. E. Poulson, "Chemical Characterization and Analytical Considerations for an in situ Oil Shale Process Water," LETC/RI-78/7. Tech. Info Center, U.S. Dept. Energy, November (1978).

119. A. D. Anderson, M. E. Lebsack, G. M. DeGraeve, D. S. Farrier, and H. L. Bergman, "Toxicity of an *in situ* Oil Shale Process Water to Rainbow Trout and Fathead Minnows," *Arch. Environ. Contam. Toxicol.* **9**, 171–179 (1980).
120. A. Goldstein, L. Aronow, and S. M. Kalman, *Principles of Drug Action* (John Wiley and Sons, New York, 1974).
121. M. Meselson and K. Russell, "Comparisons of Carcinogenic and Mutagenic Patency," in H. Hiatt, J. Watson, and J. Winsten, editors, *Origins of Human Cancer*, Cold Spring Harbor Symp., pp. 1473–1481 (1977).
122. P. Brookes, "On the Interaction of Carcinogens with DNA," *Biochem. Pharm.* **20**, 999 (1971).
123. M. J. Cookson, P. Sims, and P. L. Grover, "Mutagenicity of Epoxides of Polycyclic Hydrocarbons Correlates with Carcinogenicity of Parent Hydrocarbons," *Nature New Biol.* **234**, 186 (1971).
124. B. N. Ames, "Identifying Environmental Chemicals Causing Mutations and Cancer," *Science* **204**, 587–593 (1979).
125. J. McCann and B. N. Ames, "The Salmonella/Microsome Mutagenicity Test: Predictive Value for Animal Carcinogenicity," in H. Hiatt, J. Watson, and J. A. Winsten, editors, *Origins of Human Cancer*, Cold Spring Harbor Symp. Cold Spring Harbor Laboratory, Cold Spring Harbor, New York, pp. 1431–1450, New York (1977).
126. J. A. Miller and E. C. Miller, "The Metabolic Activation of Carcinogenic Aromatic Amines and Amides," *Prog. Exp. Tumor Res.* **11**, 273 (1969).
127. A. Pullman and B. Pullman, "Electronic Structure and Carcinogenic Activity of Aromatic Molecules. New Development," *Adv. Cancer Res.* **3**, 117 (1955).
128. A. Pullman, "The Theory of Chemical Carcinogenesis and the Problem of Hydrocarbon–Protein Interactions," in *Biopolymers*, Symposia No. 1, John Wiley & Sons, Inc., New York, (1964), pp. 47–65.
129. E. Boyland, "Polycyclic Hydrocarbons," *Brit. Med. Bull.* **20**, 121 (1964).
130. V. T. Oliverio and C. Heidelberger, "The Interaction of Carcinogenic Hydrocarbons with Tissues. V. Some Structural Requirements for Binding of 1,2,5,6-dibenzanthracene," *Cancer Res.* **18**, 1094 (1958).
131. E. Cavalieri and M. Calvin, "Molecular Characteristics of Some Carcinogenic Hydrocarbons," *Proc. Natl. Acad. Sci. USA* **68**, 1251 (1971).
132. H. Remmer, H. Greim, and J. B. S. Chenkman, in *Methods in Enzymology*, R. W. Estabrook and M. E. Pullman, editors, Vol. X (Academic Press, New York, 1969).
133. I. Berenblum and P. Shubik, "A New Quantitative Approach to the Study of the Stages of Chemical Carcinogenesis in the Mouse's Skin," *Brit. J. Cancer* **12**, 383 (1948).
134. W. R. Bryan and M. B. Shimkin, "Quantitative Analysis of Dose-Response Data Obtained with Three Carcinogenic Hydrocarbons in Strain C3H Male Mice," *J. Nat. Cancer Inst.* **3**, 503 (1943).
135. C. Peraino, R. J. M. Fry, E. Staffeldt, and J. P. Christopher, "Comparative Enhancing Effects of Phenobarbital, Amobarbital, Diphenylhydantoin and Dichlorodiphenyltrichloroethane on 2-Acetylaminofluorene-Induced Hepatic Tumorigenesis in the Rat," *Cancer Res.* **35**, 2884–2890 (1975)
136. G. S. Lange, "Observations on Some Tumors Found in Two Species of Planaria—*Dugesia etrusca* and *D. ilvana*," *J. Embryol. Exp. Morphol.* **15**, 125–130 (1966).
137. J. Foster, "Induction of Neoplasms in Planarians with Carcinogens," *Cancer Res.* **23**, 300–303 (1963).
138. J. Foster, "Malformations and Lethal Growths in Planarians Treated with Carcinogens," *Natl. Cancer Inst. Monograph* **31**, 683–691 (1969).
139. H. Kalter, "Correlation between Teratogenic and Mutagenic Effects of Chemicals in Mammals," in *Chemical Mutagens: Principles and Methods for their Detection*, A. Hollaender, editor (Plenum Press, New York, 1971).
140. J. A. DiPaolo and P. Kotin, "Teratogenesis Oncogenesis: A Study of Possible Relationships," *Arch. Pathol.* **81**, 3 (1966).
141. R. L. Baldwin and M. R. Wells, "Effect of DDT on NADH-Cytochrome b_5 Reductase Activity in the Freshwater Planarian, *Phagocata velata*," *Bull. Environ. Contam. Toxicol.* 428–430 (1978).
142. J. Phillips, M. Wells, and C. Chandler, "Metabolism of DDT by the Freshwater Planarian, *Phagocata velata*," *Bull. Environ. Contam. Toxicol.* **12**, 355–358 (1974).

4

The Wronging of Wright

Stanley L. Jaki
Seton Hall University
South Orange, New Jersey

The name of Abraham Gotthelf Kaestner is today familiar only to a few scholars, except perhaps in Göttingen where he spent the last 46 years of his life (1719–1800) as professor of mathematics and where his memory was honored, immediately after his death, with a marble bust.[1] In his lifetime his name was on the lips of everybody in German academic circles, professors as well as students. The latter flocked to him from afar and for good reasons. It was Kaestner who raised in Germany the teaching of mathematics to a level far above to the elementary niveau which is the staple feature of Christian Wolff's once famed textbooks. Kaestner, an avid student of Gottsched, who in the 1730s began to reshape German language, presented even the most abstract topics with elegance and clarity. In addition to being an excellent teacher, he had an eye on occasion for real depths, as shown by his postscript to the dissertation (1763) of his student, G. S. Klügel, who surveyed 30 purported proofs of Euclid's parallel postulate. Kaestner himself tried, in vain, to construct one such proof.[2] His interest in the topic did not, however, fail to produce fruits indirectly. Bolyai's father was a student of Kaestner and so was J. M. C. Bartels, who had later, at the University of Kazan, Lobatchevsky as one of his students. Gauss, the third great personality in this connection, provided a less flattering reflection on Kaestner. While a student at Göttingen, Gauss avoided Kaestner's lectures as being too elementary. In Gauss' eyes, Kaestner was "the greatest mathematician among poets of his times and the greatest poet among mathematicians of the same times."[3]

Kaestner's penchant for stylistic beauty was one reason for his refusal to read the works of Kant, whom later Heine rightly charged with having established the belief among German academics that profundity of thought requires obscure style as its vehicle.[4] "I am proficient in twelve languages," the elderly Kaestner was wont to quip with an eye on Kant's *Critique of Pure Reason* and subsequent works, "I will not learn a thirteenth."[5] Through his poetical works Kaestner earned wide though ephemeral fame. He poetized at times even as a scientist. His observations with a primitive telescope of the comet of 1742 made him so elated as to present his findings in verse. Owing to the generosity of an optician, whose sister later became Kaestner's first wife, Kaestner observed the famous six-tailed comet of 1743–1744 with a better instrument. A year later Kaestner and his optician friend began the observation of sunspots with a telescope that had a focal length of twenty-six feet. Kaestner's interest and expertise in astronomy was indeed such as to obtain for him, in 1763, a year after the death of Tobias Mayer, the post of director of the Observatory of Göttingen. His work there was characterized by traits that were the hallmark of Mayer's work: quest for accuracy in observations and distrust of theorizing.[6] Neither Mayer nor Kaestner would have subscribed to the attitude of Herschel, who years later declared that if he were to choose between two errors—speculating too little or too much—he would prefer to be guilty of the latter.[7]

By 1763 Kaestner had been for nine years in Göttingen as professor ordinarius of mathematics. He went there in 1754, after having served for eight years as professor extraordinarius of mathematics in Leipzig. Kaestner's salary of a hundred thalers a year was so low that he was forced to supplement it by doing translations for various publishers interested in scientific works. Among such works translated by Kaestner was Robert Smith's *Compleat System of Opticks* (1738), from which later Herschel instructed himself in the optical part of astronomy. Kaestner competed, and successfully so, for the Prize set for 1751 by the Academy in Berlin with a philosophical essay, and in all likelihood he received monetary reward for at least some of his contributions to journals and magazines. The *Hamburgisches Magazin*,[8] a monthly collection of essays relating to the various fields of learning, was certainly a commercial venture which paid its authors well. Had Kaestner not looked for some monetary reward, it is not likely that he would have sent there in mid-1752 a thirty-page-long review.[9] It was but the elaborated form of its Latin version which he sent simultaneously to a recent venture of the Göttingen faculty, the *Recensiones de libris novis*.[10] The review was about a book, *An Original Theory or New Hypothesis of the Universe*,[11] published in 1750 in London, whose author, Thomas Wright of Durham, was not entirely unknown by 1752 in Germany, and certainly not in Hamburg.

Wright's book had already been reviewed—and enthusiastically—in the first weeks of January 1751 in another Hamburg periodical, the *Freye Urtheile*,[12] much less prestigious than the *Magazin*, although with a large circulation in northern Germany. That review, by an anonymous author who did not seem to be an expert in matters astronomical, has been many times recalled owing mainly to the fact that young Kant acknowledged his indebtedness to it in writing his cosmogonical work, the *Allgemeine Naturgeschichte und Theorie des Himmels* (1755). The two reviews of Kaestner remained unnoticed until very recently[13] in spite of considerable interest in Wright's cosmology during the past two or three decades. A detailed account of Kaestner's reviews seems to be of interest for reasons that go beyond the, at times pedantic, preoccupation with the completeness of the historical record.

Quite possibly a copy of Wright's book, which was published through subscription, was secured by the University of Leipzig on the instigation of Kaestner, who may have read its review in the *Freye Urtheile*. Kaestner certainly read the book thoroughly and seized more than one opportunity to make evident, in critical footnotes, his familiarity with past astronomical discoveries and literature. Most importantly, he read the book with his own mind's eyes. To be sure, there were some similarities between the minds of Wright and Kaestner to enable the latter to notice more than one positive aspect in it. Not surprisingly, Kaestner, as would any reader of Wright's book, was much impressed by the engravings on 32 Plates with which Wright decorated his book. They showed no small talent. In his first remark on the book, Kaestner expressed his regret that in his account—thirty pages long—of the book it was not possible to give a sample of its magnificent printing and of the beauty of the engravings. To compensate for this Kaestner wanted to present its contents as extensively as possible. In this endeavor Kaestner failed conspicuously because his empiricism predisposed him from the start against appreciating the speculative and hypothetical elements in the book. In fact, Kaestner, who dealt in a mere seven lines with the contents (various views of the cosmos from the Greeks through Bruno and Newton to Derham) of the first of nine Letters comprising Wright's book, attuned his reader from the start against Wright's main scientific thesis, by giving an almost two-page-long account of the Second Letter. That Letter, which dealt with the various degrees of scientific certainty, and above all of hypotheses, was aimed, as Kaestner pointedly noted, at conditioning favorably the reader's mind to some very hypothetical propositions in the book. To be sure, Kaestner was impressed by Wright's ingenious example of the mind's advance well beyond what was perceived by the eyes. The example, presented by Wright also in an engraving, was a small fragment of a disk showing a mere fraction of a set of concentric circles with an additional and minute set of concentric circles on the outer circle. While the fragmentary figure could remain a mystery to the untrained mind, it conjured up immediately to the mind of an astronomer the system of planets and satellites. On such a basis, Kaestner remarked, the author elaborated on the role of analogy in scientific reasoning and in the formation of ever better concepts of the universe in particular. Kaestner summed up briefly the Third Letter dealing with the size, motion, and coordination of planets and comets around the sun. He did so also with the Fourth Letter in which Wright offered considerations on behalf of the existence of a planetary system around each and every sun.

On the contrary, seven pages were devoted by Kaestner to the contents of the Fifth Letter, which dealt with the order, distances, and number of stars and with the extent of the starry realm. Kaestner agreed with Wright, and here again the esthetic sensibilities of the two seemed to match, that the three most beautiful sights in nature were the sunrise over the sea, a green landscape crowned with the rainbow, and the Milky Way. In coming to Wright's remark that Ovid's description of the Milky Way was an indication of the poet's belief that the Milky Way was a congeries of stars, Kaestner could not refrain from taking aim, in a note, at some contemporary Christian poets who still preferred the pagan myths to solid science in presenting the truths of revealed religion. Kaestner's next barb was aimed at Wright himself. In another note he deplored Wright's failure to establish with a micrometer the exact position of the many stars he spotted in the Milky Way at thc foot of Antinous. Thus Wright's illustration, Kaestner claimed, was no real progress over that in Galileo's *Sidereus nuncius* and was a regress in comparison with an engraving which appeared in the *Mémoires de l'Académie des Sciences* in Paris for 1692–1693. There the most conspicuous among the many stars in the Crib had their positions determined by micrometer.[14] Several historical notes by Kaestner accompanied his full reproduction of Wright's list of double stars, novae, and nebulous stars. Somewhat less detailed was Kaestner's account of Wright's reasoning aimed at estimating from the orbits of comets the minimum distance to the nearest stars. Underlying the reasoning was the assumption that the gravitational field of the nearest star had no influence on the longest cometary orbit.

Kaestner devoted eight pages to the Sixth Letter, and over half a dozen notes, none of them favorable. Kaestner did not fail to add a touch of sarcasm to his report that Wright believed himself to have provided the ground "for an elucidation of the hitherto unexplained, so states the author, phenomenon of the Milky Way."[15] Kaestner, as will be seen, failed to say much about that explanation and what he said was unenthusiastic. He did not suggest at all that the explanation was truly novel and much less did he wonder why it had failed to be proposed beforehand.[16] Nor was he openly critical of it. He offered his criticisms on patently secondary and irrelevant points. Thus, he seized on Wright's recall of the Mosaic account of creation: "It is difficult to see why this reference, which had been elaborated by Burnet, Whiston and others, is necessary."[17] To Wright's insistance that all stars must be of the same nature and all of them ought to be active, that is, moving, or else the whole universe would be disorderly, Kaestner added a long note which gives a perfect taste of the reluctance of an empiricist to appreciate hypotheses and of the insensitivity of a mathematician bent on logical rigor to consider analogies:

> Too many words and few rigorous conclusions. To what extent can we assert that all stars are of one and the same nature? We can do this at most insofar as they all have their own light, for we know nothing else of them. Who will however accept it without proof as an incontrovertible truth? All bodies which have it in common that they radiate their own light, must, however, either be at rest or in motion. Cannot one say by the same token that all planets, being dark bodies, are of one and the same nature, and yet one would even less be justified in concluding that none of them can orbit around the other, but that all of them being of the same nature, must orbit directly around the sun[?] Yet, it cannot be denied that one is inclined to ascribe to all self-luminous bodies either motion or rest although one has no ground to ascribe to some of them this, to some others that property but then one has to take such inclination for a conjecture and not for a secure truth. What a semblance of conclusive evidence is offered in these latter words of the author! Who can from the notion of excellence and importance of a celestial body decide whether it is more fitting for it to be at rest or to be in orbit? Would the suns be inactive if they stayed at rest in the focus of their planetary orbits, although they would, through their attraction, keep the planets in their

> tracks and perhaps would, through their axial revolutions drag them around? It seems as if for the author effectiveness and cruising would be the same.[18]

Wright's claim that, if no other things were in existence than the ones we see, we would already be in the best state we could ever achieve for ourselves, provoked a similarly sharp rebuke. Kaestner seemed to be intent on discrediting Wright on all counts that really mattered scientifically by attacking him relentlessly on incidentals. This was particularly in evidence in Kaestner's comment on Wright's remark that the Milky Way, certainly an orderly work of the Great Artificer, looked irregular because we did not view it from its very center. Not content with unnecessary criticism, Kaestner poked fun at the English and demeaned Wright with a pun:

> Who in England is so peculiar as to be bothered by the apparent irregularity of the Milky Way! And what is the usefulness of such a preachment in showing that something, whose orderliness is so much above our perception, is not [necessarily] disorderly? If a swallow looked upon a princely palace as disorderly, because it would not find it constructed as its own nest, would it be better to resort to a swallowwright [Schwalbenwright] to enlighten that [other] swallow that the reason for this is that it looked at the palace eccentrically; would it not be simpler to remind that swallow of its being but a swallow?[19]

Such was hardly a dignified tone. Nor did it reflect a basic sympathy with the book. In its absence, Kaestner could only wrong Wright even where Wright was right and farsighted.[20] To be sure, there was some rubbish in Wright's book but few books on science were entirely free of it around 1750 and even later. Too much sensitivity to its presence could only deprive for good the reader of something precious and damage the reputation of the book, though not permanently. Bent on destroying the book Kaestner could not refrain from an outburst of indignation even when it was not at all justified. Thus on finding Wright speculate whether the axial rotation of the sun could and would be the cause of the planet's orbital motion, Kaestner exclaimed:

> About the "could" there is no doubt. The chief point is the "would." Who can find there more reasons for asserting than for denying it?[21]

Kaestner of course acknowledged that in support of the foregoing effect Wright offered more than rhetoric. Wright thought that the different estimates of the inclination of the ecliptic were suggestive of such an effect; Kaestner was all too eager to survey the evidences and find fault with Wright's dicta. Obviously, Wright failed to take into account the inaccuracies of values quoted by ancient authors. Kaestner also took issue with Wright's familiarity with modern authors. From Kaestner's belaboring this point over three pages his reader could but gain the impression that the effect in question was central in Wright's train of thought. Wright was once more wronged.

The same happened again when Kaestner took up the contents of the Seventh Letter. Although the crowning piece of the book from the scientific viewpoint, this Letter received only three pages in Kaestner's review.[22] Worse, Kaestner largely missed one of the two reasons why Wright proposed two models for the Milky Way. Kaestner nowhere mentioned the whitish nebulous hue into which, according to

Wright the light of stars would coalesce if many of them were behind one another in endless rows, within two imaginary parallel planes. Wright also argued, though wrongly,[23] a point not perceived by Kaestner, namely, that if the stars were within a spherical shell their light would also form, to someone looking along the tangent of the shell, an appearance exactly like the Milky Way. In connection with these two models Kaestner spoke only about the possibility of concentric motion of many stars in either of them.[24] He concluded with obvious irony:

> One can just as well set forth in other hypotheses various spherical surfaces which are within one another and orbit around a common center with perhaps different laws and directions and one must grant that at least the drawings of the author look very beautiful in this respect.[25]

Such was a classic example of damning with faint praise.

The Eighth Letter containing conjectures on the size of the stellar universe, and analogies to intimate its enormity, did not fare any better under Kaestner's acid pen. At the very outset, Kaestner took Wright to task on an utterly incidental detail. To forestall the objection that in speaking of the starry realm he forgot the earth, Wright recalled the story according to which Aristotle told Alexander the Great, who complained for not finding the city of Macedonia on a map of Greece, that obviously it was too small to be noted. Kaestner was now at his pedantic best:

> Hardly is the story true about Aristotle and Alexander. If I am not mistaken, it is told by Socrates who invited a young man, proud of his landed properties, to find them on a map of Greece.[26]

Equally pedantic, ironic and what is worse, prolific, was Kaestner in respect to another secondary matter, namely, Wright's method of trying to intimate the great celestial distances by intimating the measure of the latter through a comparison with the size of England:

> An Englishman, who found it too difficult to understand when one first spoke to him of the earth's diameter in terms of English feet, and of the celestial distance in terms of the earth's diameter, would in my opinion form just as little a notion of the whole of England. He cannot form a notion of a thing which he can not see in its entirety and he can never see the whole of England in such a way. Still there is an advantage in all this because he hears the well-known name of the Fatherland pronounced and since he does not know what exactly belongs to it in order to represent to himself its size exactly, he still thinks to have a notion of it although he really has none.[27]

Clearly, it was of not much use to praise, as Kaestner did in the remaining part of the note, Wright's beautiful drawings.

Kaestner displayed almost malice in the next breath as he took issue with Wright's further elaboration of the point that the nearest star was at least three times as far from the sun as the long axis of the great comet of 1650 whose aphelion he believed to be the farthest among all cometary aphelia.[28] In this case, Wright argued, there would be no disturbance in the celestial realms. Kaestner was ready with another long note which began with a reference to the mutual perturbations of Jupiter and Saturn. Then Kaestner waxed analytical as he went for the big kill, flogging a presumably dead horse. He had no eyes for its vitality. The farce was all the more complete because Kaestner had a point, though not one touching on the real issue:

> What is after all the meaning of disorder? Deviations from the easiest laws of nature we have been able to grasp. These laws make up only a small part of the overall orderliness of the universe which can turn out to be such as not to be governed by the very limited orderliness which we comprehend.[29]

To be sure, Kaestner was willing to admit that other stars do not seem to influence the motion of the sun's planets. But he had strong reservations concerning comets. Was not Wright one of those who traveling through a country felt as if it was their own? Perhaps so, but by hunting his prey down relentlessly Kaestner made a big blunder. Carrying logic to the point where it becomes the art of going wrong with confidence, Kaestner recalled Fontenelle's opinion that our planets may be emissaries from other stellar domains.[30] He was completely oblivious to the fact that whereas through an extraordinarily fortuitous coordination of a great number of parameters this may have been the case with one planet, the same possibility for six other planets defied all likelihood. To spare himself such a blunder, Kaestner should have only reminded himself of Newton's advance warnings against hypotheses (espoused by Whiston, Buffon, and others) that the actual planetary orbits could be the result of capture by the sun of foreign bodies intruding into its domain or of a near-collision between such bodies and the sun.[31] Kaestner now stretched logic to its utmost where it becomes absurdity. What could be stated of planets could also be said of comets, Kaestner argued, as he handed down his self-defeating indictment of Wright's reasoning: "What is proposed by Mr. Wright is a conjecture, not a certainty."[32] Did Wright offer more than conjectures worth considering?

Kaestner, who passed over lightly some of Wright's weightiest "conjectures," found place for such incidentals as the reduction by Wright, for the sake of illustration, of the solar system to a scale corresponding to the size of London:

> Mr. Wright tries to give an idea to Englishmen about the relation of our planetary system to the nearest stars by reference to various edifices in London. If the golden sphere on the spire of St Paul's Cathedral in London represented the sun, then the [orbit of the] earth would go through Marylebone, Mercury through the Tower, Saturn through Cliefden or Chelmsworth, and the nearest fixed star would be where St. Peter's in Rome is.[33]

With St. Peter's and St. Paul's having been thrown in for good measure, discourse could readily turn to by far the weakest and most repugnant pages of the book. By speculating about the starry realm, Wright ultimately wanted to serve the spirit and not the mind. As one could expect, Wright ran thereby the risk of exposing himself to well-deserved strictures. Only strictures if not contempt could be earned by Wright's views that since even a body which traveled 20,000 yards in an hour would reach the center of the Milky Way, where he placed one abode of the blessed, only in 300 million years, the soul destined to that place had to be incorporeal! To reproduce here the long note which Kaestner added in this connection would be anticlimactic.

About the Ninth Letter, which dealt with the immense number of Milky Ways demanded as eternal abodes by the immense varieties of the blessed and the damned, Kaestner's critical remark was brief. Its first part sounded as if spoken with a tongue in cheek. The second part was devastating though well merited:

> This last letter should be translated in full to convey its loftiness and inspired character to the German reader. As to the one who merely wants an extension of his scientific knowledge, it contains nothing new.[34]

Kaestner's final summary of the entire book betrayed again his failure to perceive a huge nugget all too well exposed in at times all too familiar, at other times repulsive discourse:

> The entire work of Mr. Wright shows, in addition to a lively and inventive mind, good knowledge of astronomy, at least in its easier and more concrete parts. How remotely is one obligated to give assent to his new ideas, can easily be judged from the foregoing. The engravings are of extraordinary beauty and perhaps some readers would wish that this craftsmanship had been expended on real matters and [not] all too often on representations that owe their existence only to the author's imaginative powers.[35]

The author, Thomas Wright, is of course incomparably better remembered today than is the once famous professor of Göttingen and for good reasons. Thus Kaestner's review deserves more than the comment: reviewers . . . revilers. Book reviews played a part not only in literary history, but in the history of science as well. When Wright's book was published, the history of science was still largely the history of great books, an epoch that saw its last great moment in the publication of the *Origin of Species*. It would of course be an exaggeration to suggest that Darwin's masterpiece owed its immediate success entirely to Huxley's brilliant review of it in the *Times*.[36] Again, it would be wrong to think that Newtonian physics would have quickly triumphed in Cartesian France had the first French review of the *Principia* not praised the mathematics in that book with the proviso that physics was entirely absent from it.[37] Still, had the "big" review of Darwin's book been entrusted, say, to a Samuel Butler, something would have been sensed immediately of the fallacy of several decades of effort devoted implicitly to the purpose of proving that there was no purpose.[38] Again, had Huygens been asked to write that first French review of the *Principia*, Cartesians in France may have much sooner realized that Descartes' physics offered nothing for the purposes of mathematically accurate experimentation and observation.[39]

Such considerations are particularly applicable to the impact which Thomas Wright's *An Original Theory or New Hypothesis of the Universe*, published in 1750, may have had if properly reviewed. To be sure, there were many wrongs in Wright's book. Indeed its ultimate thrust could easily appear monumentally and incurably wrong. As the Age of Enlightenment was coming into its own, what could appear more revolting than the very last of Wright's exquisitely drawn plates illustrating that book? It showed globes upon globes representing so many stellar universes, from the center of each of which a divine eye was staring at the reader reminding him of the ultimate purpose assigned by Wright to his bold cosmological vision of immense varieties of universes, as so many abodes for the infinite classes of the saved and the damned.

Such was hardly the light called for by some scientifically most portentous pages in the book. In those pages the stellar universe first appeared as it actually is, a system of nebulae of which our Milky Way is one. The visual appearance of the Milky Way was explained by Wright on the basis of the assumption that a very large number of stars was confined within a space resembling a thin disk. He also argued that within such a space the stars must be in orbital motion around the center. If indeed all science is cosmology,[40] Wright's bold grasp of the structure of the Milky Way should seem to be one of the greatest advances ever made in the history of science. It was actually a most overdue advance and also an advance which went largely unrecognized for a long time. The story is too well known to call here for more than the briefest outline. The least remembered aspect of the story is that in addition to the foregoing, correct model of the Milky Way Wright also proposed the spherical shell model which he seemed to prefer. Far better remembered is the fact that a year before Wright published his book, Johann Heinrich Lambert, a self-educated tutor of the children of Count von Salis in Chur, had the insight that the visual appearance of the Milky Way was caused by the fact that an unusually large number of stars were confined within a lentil-shaped space.[41] It was, however, only twelve years later that Lambert put this idea into print in his *Cosmologische Briefe* (1761) together with an idea also voiced by Wright that the stars were in orbital motion within such a space. Shortly after Lambert published his *Briefe* he heard of an "Englishman" who had already put in writing similar ideas and whose book was even considered for translation into German.[42] But the thrust of the hearsay report about Wright's book may have been so emphatic about its ultimate "spiritual" perspective that the otherwise devout Lambert felt no encouragement to search for a copy of Wright's book, soon a rarity even in England.

Lambert's reference to the "Englishman" is in his reply to an inquiry from Kant who wanted to clear up questions of priority. Kant could certainly satisfy himself from Lambert's reply that the latter had no knowledge of his own discussion of the Milky Way in the *Allgemeine Naturgeschichte*. That Lambert expressed no interest in that reply in obtaining a copy from Kant of the *Allgemeine Naturgeschichte* could hardly be satisfying to its author, but this is another matter. In writing his *Allgemeine Naturgeschichte* Kant was much indebted, as was already noted, to a review of Wright's book which appeared in early 1751 in the *Freye Urtheile*. The review greatly differed from the only review which appeared in English of Wright's book. The English review was short, although not to the extent of containing no reference to the most valuable detail in Wright's book, namely, his explanation of the Milky Way as a flat disk of stars. The reference lacked, however, even a touch of surprise, let alone enthusiasm. Yet by emphasizing the incorrectness of the shell model the reviewer, who identified himself as S. L., cast indirect doubt on the crucial usefulness of the disk model. Being published in the pages of the *Gentleman's Magazine*,[43] it was in all likelihood read by all notable cultivators of astronomy in England around 1752. Among them were Bradley, Maskelyne, Michell, and others, but none of them took notice. More than a generation later English astronomical literature was still so void of meaningful references to the Milky Way that Herschel could present his own views of it as an original insight.[44]

Neither details, nor enthusiasm, were lacking in the review in the *Freye Urtheile* although, unlike that still unidentified S. L.,[45] its author did not give the impression of being an accomplished astronomer. The review deeply impressed young Kant, still in his twenties, though well beyond his university years, during which he had failed to obtain any systematic training in mathematics or the physical sciences, including astronomy. Without ever claiming categorically that he had conceived the correct idea of the Milky Way independently of his exposure to that review, he built around that idea a cosmogony in which, as it was aptly noted about a hundred years ago, the "primitive reign of chaos was little likely to terminate."[46] In fairness to Kant, he was original in one respect, namely, his identification of other nebulae as so many Milky Ways or stellar systems.

In view of the scientifically infelicitous character of every aspect of Kant's cosmogony which takes up much of the *Allgemeine Naturgeschichte*, a question may arise: Had the review of Wright's book in the *Freye Urtheile* been critical of Wright's book and had in it Wright's explanation of the Milky Way been slighted, would Kant have ever made his foray into cosmology and cosmogony? The same question can be formulated in another way: Would Kant have ever ventured to the waters of cosmology and cosmogony had he been exposed to an actually unfavorable review of Wright's book? The question is not purely speculative for at least two reasons. For one, Kant passed up many signal opportunities to call attention to his explanation of the Milky Way until news reached Germany in the late 1780s about Herschel's discoveries, an indication that he may have indeed had nagging doubts about its correctness. Prior to 1791 he made no communication about it to any astronomer in Germany, let alone in France or in England. One would expect the opposite had he been truly convinced of the originality of that explanation and of its correctness, and last but not least, of his competence in matters scientific. Much more tangible is the second reason which relates to the two reviews of Wright's book by Kaestner. Both of them were "wronging" Wright by distracting from what was truly valuable in his book. Since neither Lambert nor Kant had ever referred to Kaestner's two reviews of Wright's book, it is possible to assume that those reviews may not have come to their notice. As to Lambert, he lived through the better half of the 1750s in a somewhat isolated situation as the family tutor of Count von Salis in Chur. Although the Count was interested in books, it is unlikely that he had subscribed to the short-lived *Recensiones de libris novis*. Had he done so, Lambert, who read voraciously, would have undoubtedly spotted the short Latin review in it of Wright's book toward the end of 1752. It is even safer to assume that the much longer German review in the *Hamburgisches Magazin* would not have escaped Lambert's attention had that periodical been available in the Count's library. Of course, Lambert, while traveling with his charges through Göttingen in 1757, may have learned of the reviews by Kaestner, with whom in the subsequent years he exchanged many letters, which, however, never touch on this subject.

While one can be fairly certain about Lambert's unawareness of Kaestner's reviews, the same can hardly be said with the same confidence about Kant. Since Göttingen was the closest major German university to Königsberg, the *Recensiones* were most likely available in the University Library there. As to the *Magazin*, it was the best that came, in the way of culture, out of Hamburg, a town of commerce and shipping. Its availability in several copies in Königsberg can be assumed with the greatest probability. Kant may have seen it also in the private library of M. Knutzen, professor extraordinarius of mathematics, and a friend of Kant, and all the more so as Knutzen himself contributed to the *Magazin*.[47]

There remains the further possibility that following his perusal of the review in the *Freye Urtheile*, Kant learned of Kaestner's review in the *Magazin*, but convinced of its low quality, he deliberately kept silent about it lest he would unnecessarily dispose Kaestner, a rising star in the German academic scene, against him. Had Kant however seen *only* Kaestner's review[s] of Wright's book, the outcome might have been tantalizing enough to allow for a guess or two. Even from Kaestner's longer review Kant would have learned but a skeleton of what Wright proposed about his two models of the Milky Way. (Happily for Kant, the review in the *Freye Urtheile* had almost nothing about the spherical model, which may have sidetracked Kant, a scientific amateur.) Assuming then that Kant did not perceive the Milky Way in its true form independently of what he read in the *Freye Urtheile*, and should he have read only Kaestner's review(s), he may not have had enough information and encouragement to write the *Allgemeine Naturgeschichte*. Its cosmogonical part about the solar system was, of course, conceiveable without references to the Milky Way. Laplace and others wrote cosmogony with no reference to the Milky Way and nebulae. But in Kant's book the truth of his theory of the evolution of the solar system was closely tied to his explanation of the Milky Way. From Kaestner's review(s) alone he could receive neither solid information on this point, nor inspiration and encouragement. His book might not have been written—with no particular loss for scientific history.[48] What is true of books—*habent fata sua libelli*—is true of book reviews as well. They have a fate, at times most unintended, even when they wrong what is right.

References and Notes

1. Concerning biographical details about Kaestner, see *Allgemeine Deutsche Biographie*, Vol. XV, Neudruck (Duncker and Humblot, Berlin, 1969), pp. 439–451.

2. See "Kaestner, Abraham Gotthelf," by G. Goe in *Dictionary of Scientific Biography*, Vol. VII (Charles Scribner's Sons, New York, 1973), pp. 206–207.
3. *Allgemeine Deutsche Biographie*, Vol. XV, p. 446. Gauss was, of course too much of a genius to profit of Kaestner's lectures, however above the average. In fairness to Kaestner, the mathematician, it should be recalled that the physician-astronomer Wilhelm Olbers credited his no small expertise in celestial dynamics to his having attended Kaestner's lectures in the late 1770s.
4. See H. Heine, *Germany*, translated by C. G. Leland (John W. Lovell, New York, 1892), Vol. I, p. 140.
5. Quoted in "Kaestner" by C. M. P. in *Biographie universelle, ancienne et moderne*, Vol. 21 (chez Madame C. Desplaces, Paris, [1860]), p. 387.
6. As may be gathered from E. G. Forbes' Introduction to *Tobias Mayer's Opera inedita: The First English Translation of the Lichtenberg Edition of 1775* (Macmillan, London, 1971).
7. *The Scientific Papers of Sir William Herschel*, J. L. E. Dreyer, editor (Royal Society, London, 1912), Vol. I, p. 223. Significantly this remark of Herschel is in one of his epoch-making papers, "On the Construction of the Heavens."
8. With the subtitle, *oder Gesammlete Schriften zum Unterricht und Vergnügen aus der Naturforschung und der angenehmen Wissenschaften*. Publication began in 1747. The contents of the first 26 volumes (1747–1763) are tabulated in a massively meticulous *Dreyfaches Universal Register und Repertorium*.
9. Band X, Stück 2, pp. 151–80. The review, signed by the letters A. G. K., is explicitly attributed to Kaestner in the *Register* and will be referred to as Longer Review.
10. *Fasciculus quintus, seu Anni MDCCLIII* [1753] *qui relationum secundus est primus* (Gottingae Vid. Abr. Vandenhoekii Bibliog. Acad.), pp. 49–60; quoted as Shorter Review. About Kaestner's authorship of the unsigned review the prime evidence is the listing of nebulae, double stars, and novae, practically identical with the list in the longer review. Another, more circumstantial piece of evidence will be noted later.
11. The work is available in a facsimile reprint which includes the first publication of Wright's *A Theory of the Universe* (1734) and an introduction and notes by M. A. Hoskin (Macdonald, London, 1971).
12. For a full text of that review, see F. Krafft, *Immanuel Kant: Allgemeine Naturgeschichte und Theorie des Himmels*, mit einem wissenschafthistorischen Nachwort (Kindler Verlag, Munich, 1971), pp. 200–211.
13. In my translation, with introduction and notes, of Kant's cosmogonical work, *Universal Natural History and Theory of the Heavens* (Scottish Academic Press, Edinburgh, 1981), p. 222.
14. Kaestner's reference is not exact. He must have had in mind the report of Cassini and de la Hire which they read before the Académie des Sciences on June 30, 1692, on their observations of the passage of Mars across the "nebulous" star in Cancer, the previous May. The text of their report, printed in *Mémoires de l'Académie Royale des Sciences. Depuis 1666 jusqu'à 1699. Tome X* (par la Compagnie des Libraires, Paris, 1730), pp. 115–118, is followed by a foldout engraving of the position of some 40 stars of the Crib set against a scale determined by micrometer. Cassini and de la Hire also made the remark (p. 118) that Galileo's drawing was so inexact as to make impossible its identification with the actual grouping of stars forming the Crib.
15. Longer Review, p. 160.
16. See on this, Chaps. 4 and 5, "Galilean Myopia" and "Newtonian Distraction," in my *The Milky Way: An Elusive Road for Science* (New York: Science History Publications, 1972).
17. Longer Review, p. 161.
18. Longer Review, pp. 161–62.
19. Longer Review, p. 163.
20. Wright was, of course, wrong on several points, of which the chief was his belief in the correctness of the spherical shell model, a main reason why the chapter dealing with Wright in my *The Milky Way* was given the title, "Wright's Wrong." The other principal reason for this lies in the fact that Wright, a decade or so after the publication of his *Original Theory*, reverted to a quasi-Aristotelian explanation of the Milky Way. The evidence for this latter point became public in 1968 with the publication, with an introduction and notes, by M. A. Hoskin of Wright's manuscript, *Second or Singular Thoughts upon the Theory of the Universe* (The Dawsons of Pall Mall, London).
21. Longer Review, p. 164.
22. Longer Review, pp. 168–171.

23. Concerning this point, see Illustration V in my *The Milky Way*.
24. Actually, instead of speaking of stars within a spherical shell, Kaestner rendered Wright's idea as stars fixed on a rotating sphere! See Longer Review, p. 170.
25. Longer Review, p. 171.
26. Longer Review, p. 171.
27. Longer Review, p. 172. It seems that Kaestner looked somewhat askance at England and Englishmen. In the Shorter Review he took to task Wright for not giving appropriate justice to the German (!) Kepler's contributions and added: "What measure of truth would be contained in the science of astronomy if of the two Germans (because the Prussian [Copernicus] may rightly be counted among the Germans) one had not drawn up the true system of the world, and the other [Kepler] had not explained, with a genius not inferior to Newton, the planetary orbits" (p. 50).
28. This is obviously a printing error in Kaestner's review. Wright speaks of the comet of 1680.
29. Longer Review, p. 174.
30. Actually what Fontenelle stated was that, in accordance with his advocacy of Descartes's system, comets were planets transferred from one planetary system into another. See the Fifth Evening in his *Entretiens sur la pluralité des mondes* (1686).
31. For a critical review of these theories, see Chap. 4, "Convenient Collisions," in my *Planets and Planetarians: A History of Theories of the Origin of Planetary Systems* (Scottish Academic Press, Edinburgh, 1978).
32. Longer Review, p. 175.
33. Longer Review, p. 176.
34. Longer Review, pp. 179–180.
35. Longer Review, p. 180.
36. Moreover, it was the unexpected reluctance of a staff writer of the *Times* to do the review that provided Huxley with the opportunity. See G. Himmelfarb, *Darwin and the Darwinian Revolution* (W. W. Norton, New York, 1962), p. 264.
37. See E. J. Aiton, *The Vortex Theory of Planetary Motions* (Macdonald, London, 1972), p. 114.
38. The most forceful intimation of the relevance of Butler's criticism of Darwinian theory was provided, incidentally, in a review by James Gray, professor of zoology at Cambridge, of J. Huxley's *Evolution in Action:* "No amount of argument, or clever epigram, can disguise the inherent improbability of orthodox [Darwinian] theory; but most biologists feel it is better to think in terms of improbable events than not to think at all; there will always be a few who feel in their bones a sneaking sympathy with Samuel Butler's scepticism." *Nature* **174**, (1954), p. 279.
39. The realization came only with d'Alembert and Maupertuis, who in fact blamed in the 1740s French infatuation with Descartes for the backwardness of physics in France.
40. An often quoted Popperian phrase which owed its apparent originality to the temporary discredit which Comtean, Machist, and logical positivists succeeded in grafting on the science of cosmology. For details, see my Gifford Lectures, *The Road of Science and the Ways to God* (University of Chicago Press, Chicago, 1978).
41. For details, see my translation, with introduction and notes, of J. H. Lambert, *Cosmological Letters on the Arrangement of the World-Edifice* (Science History Publications, New York, 1976), pp. 7–8.
42. See Ref. 41, p. 25, and *The Milky Way*, pp. 199–200 for a translation of the passage in question.
43. In the July 1751 issue, Vol. 21, pp. 315–317. For a discussion of its contents, see *The Milky Way*, pp. 195–196.
44. A circumstance all the more interesting because Herschel owed a copy of Wright's *Original Theory*. For details, see M. A. Hoskin, *William Herschel and the Construction of the Heavens* (W. W. Norton, New York, 1964), pp. 115–116.
45. I submit here that S. L. stood for Smart Lethieullier, a claim which will be supported by documentation elsewhere.
46. In his "Kant as a Natural Philosopher" [*American Journal of Science* **5**, (1898), p. 97] G. F. Becker quotes this statement as having been made in 1882 by "a well known historian of astronomy."
47. He did so in 1749. For details, see my translation, *Universal Natural History and Theory of the Heavens*, p. 224.
48. The coming into popularity of Kant's cosmogonical book from the 1850s on was due to the rise of German nationalism, to the popularity of Darwinian evolutionism, of Haeckelian monism, and last but not least, to Marxist scientism as formulated by Engels. For details, see Ref. 13, pp. 51–60.

5

On the Historical Development of the Indistinguishability Concept for Microparticles

Alfred Kastler
Université Paris
Paris VI, France

The kinetic theory of gases, founded in the 18th century by Daniel Bernoulli, was further developed during the 19th century by Clausius and Maxwell, and crowned by the achievements of Ludwig Boltzmann's statistical mechanics. At the time Boltzmann was developing his method, the hypothesis of the discontinuous structure of matter, i.e., of the existence of atoms and molecules, was becoming increasingly a matter of certainty for physicists. Boltzmann's colleague at the University of Vienna, Joseph Loschmidt, indicated in the year 1865 a method for determining the number of molecules in a mole (the so-called Avogadro's number N_A) and the order of magnitude of atomic sizes (1 Å $= 10^{-8}$ cm). We know today that Avogadro's number—determined by a dozen independent methods—has the numerical value $N_A = 6.025 \times 10^{23}$, an enormously large number.

If we wanted to count the number of molecules contained in one cubic centimeter of the ambient air, we would, counting at a rate of one molecule per second, require 100 billions of years, a time equal to about ten times the age of the universe!

Boltzmann started from the idea that the internal energy of an ideal monatomic gas is equal to the sum of the energies of the translational motion of its molecules:

$$E = \sum_{i=1}^{N} \tfrac{1}{2}m(u_i^2 + v_i^2 + w_i^2)$$

He further advanced the postulate that, for a given energy E, the equilibrium state of a gas was given by the most probable distribution of its molecules over "velocity space," i.e., the space relating to the molecular velocity components u, v, and w.* Boltzmann himself expressed the latter by the following statement (Boltzmann, *Abhandlungen* II, p. 193):

> Man sieht also, dass man in der Tat berechtigt ist, zu sagen: diejenige Zustandsverteilung, welche unter allen die wahrscheinlichste ist, estspreche auch dem Zustande des Wärmegleichgewichtes ... Ich glaube aber nicht, dass man berechtigt ist, dies ohne weiteres als etwas Selbstverständliches hinzustellen, mindestens nicht, ohne vorher sehr genau definiert zu haben, was eigentlich unter der wahrscheinlichsten Zustandsverteilung zu verstehen ist.
>
> [We therefore see that one is indeed justified in saying: The distribution of states which is among all those given the most probable, corresponds to the state of thermal equilibrium.... But I do not believe that one is justified in setting this forth without further ado as something which is self-evident, at least not before having exactly defined what is to be understood by the most probable distribution.]

Now, how can one define and calculate the probability W of a distribution of states? On this point Boltzmann encountered a great difficulty: The "contents" to be distributed are the molecules, possessing a discontinuous structure. By contrast, the "container," into which these molecules are to be distributed, is the velocity space, a continuous structure. To be able to define a "distribution" and to calculate a number of different distributions of molecules over velocity space, Boltzmann was obliged to divide this space into a definite number of "elementary domains," which he called "cells." Let him express himself again, through his fundamental paper published in 1877, entitled "Beziehung zwischen dem zweiten Hauptsatz der mechanischen Wärmetheorie und der Wahrscheinlichkeitsrechnung":

> Die Zahl der lebendigen Kräfte (der Moleküle) ist eine diskrete ... Es entspricht diese Fiktion freilich keinem realisierbaren mechanischen Probleme, wohl aber einem Probleme, welches mathematisch viel leichter zu behandeln ist.
>
> [The number of states of molecular kinetic energy is a discrete one ... This fiction corresponds, of course, to no realizable mechanical problem, but no doubt to a problem which mathematically can be treated much more easily.]

To realize this program, Boltzmann divides velocity space into elementary domains between the limits u and $u + du$, v and $v + dv$, w and $w + dw$, and thus each with the extension $du\, dv\, dw$. He continues

> Es mag für den ersten Anblick befremden, da die Zahl der Moleküle, wenn auch gross, so doch endlich ist, während du, dv, dw mathematische Differentiale sind. Doch muss diese Annahme bei näherer Überlegung als selbstverständlich bezeichnet werden. Denn alle Anwendungen der Differentialrechnung auf die Gastheorie beruhen auf derselben Annahme ... Die obige Annahme besagt weiter

*The concept of the six-dimensional phase space, or the so-called μ space, which is the product of the three-dimensional x, y, z position space and the three-dimensional u, v, w velocity space, was introduced by J. Willard Gibbs towards the end of the 19th century.

nichts, als dass man die Grenzen für *u, v, w* so weit nehmen kann, dass bereits sehr viele Moleküle dazwischenliegende Geschwindigkeitskomponenten haben, und trotzdem noch alle diese Moleküle so ansehen kann, als ob sie mit denselben Geschwindigkeitskomponenten begabt wären.

[At first sight this may seem strange, since the number of molecules, though large, is finite, whereas *du, dv, dw* are mathematical differentials. But, on closer examination, this hypothesis must be regarded as obvious, seeing that all applications of differential calculus to the theory of gases are based on this same assumption. ... The foregoing assumption asserts nothing more than that one can take the limits for *u, v, w* so large that indeed a large number of molecules will have velocities inside these limits, while nevertheless we may regard all these molecules as having the same velocity components.]

We have to come back to this definition of elementary domains. In Boltzmann's time a "molecule" was viewed as an extrapolation of the concept of a macroscopic "object" to submicroscopic dimensions. It was quite natural to consider these molecules as particles which could be distinguished from each other and individually identified. A "distribution" was to be characterized by the individually labeled molecules in each cell of velocity space. Such a microdistribution Boltzmann called a "complexion." The probability W for obtaining a state of given energy of the gas, a so-called "macrostate," was then defined by the number of complexions corresponding to such a state. Boltzmann tacitly assumed that each complexion had the same elementary probability, equal to unity.

Adopting this convention, the probability W of a macrostate is given by the expression

$$W = N!/N_1!N_2! \ldots N_i! \ldots N_g!$$

This expression represents the number of possible permutations of molecules between different cells. The remarkable thing about this formula of Boltzmann is not its numerator but its denominator. The factors in this denominator imply that permutations of molecules inside a given cell do not create a new complexion and, therefore, should be excluded; according to Boltzmann "we may regard all these molecules as having the same velocity components."

This property of the velocity space cells, viz., that the velocity coordinates of molecules inside the same cell cannot be distinguished from one another, is no classical concept. It is a concept which could be justified only much later by Heisenberg's uncertainty relations. Did Boltzmann have in some way a presentiment of these relations?

Let us now try to analyze more thoroughly Boltzmann's concept of the cells of velocity space. The word "cell" was used by Boltzmann for the first time in his book published in 1896, entitled *Vorlesungen über Gastheorie (Lectures on the Theory of Gases)*. In this book we read on page 40

Teilen wir den ganzen (Geschwindigkeits-) Raum in sehr viele (g) gleichgrosse Volumenelemente *w* (Zellen), so ist die Anwesenheit des Geschwindigkeitspunktes eines Moleküls in jedem solchen Volumenelement mit der Anwesenheit in jedem anderen Volumenelement als ein gleichmöglicher Fall zu betrachten. . . . Der Fall, dass alle Moleküle gleiche und gleichgerichtete Geschwindigkeit haben, würde dem Fall entsprechen, dass alle Geschwindigkeitspunkte in derselben Zelle liegen. . . . Wir fragen nach der relativen Wahrscheinlichkeit der Einteilung der verschiedenen Geschwindigkeitspunkte unserer Moleküle in unsere Zellen *w*, wobei selbstverständlich die Zelleneinteilung, daher auch die Grösse einer Zelle *w*, die Anzahl der Zellen g und die Gesamtzahl *N* der Moleküle und deren gesamte lebendige Kraft als unveränderlich gegeben betrachtet werden müssen.

[If we divide the whole (velocity) space into very many (g) equally large volume elements *w* (cells), then the presence of the velocity point of a molecule inside any such volume element and its presence in any other such volume element are to be considered as cases of equal probability . . . The situation where all molecules would have the same velocity, in the same direction, would correspond to the case where the velocity points of all molecules lie inside the same cell . . . We inquire after the relative probability for the distribution of the various molecular velocity points over our cells, where obviously we must regard as given the division into cells and, therefore, also the cell size *w*, the number g of cells, and the number *N* of molecules and their total kinetic energy.]

In his earlier original papers Boltzmann uses the symbol *du dv dw* for the elementary domains of velocity space, and, as we have seen above, he considers *du*, *dv*, and *dw* as differentials. For him these mathematical symbols denote infinitely small domains, which can be extrapolated to zero. (We have seen that he tried to justify this.) But on this point Boltzmann harbors an illusion, for, on making such an extrapolation, one arrives necessarily at a situation where each cell contains only one molecule or none, and at this stage all factors of the denominator, would equal 1! or 0!, i.e., unity. *W* would thus reduce to *N*!, a constant independent of the temperature and the pressure of the gas, and this constant would be useless and without interest.

We know today that the cells in phase space must have a finite extension, otherwise they would be meaningless. Involuntarily, Boltzmann, when postulating his cells of velocity space, has introduced a "quantization" in the container of the molecules. He thus advanced the first "quantum" assertion. As his text shows, he attributed to the cells two properties which we know today are irreconcilable. First, they should be so small that variations of the molecular coordinates inside a cell remain inobservably small; second, they should be so large that each cell contains a large number of molecules (or, more precisely, velocity points). We know today that both these properties cannot belong to cells of the same size. We may call *Kleinzellen* (small cells) the cells possessing the first property. They are small, but finite, their size being determined by h^3, the cube of Planck's action constant, in the six-dimensional molecular phase space (μ space).

Under normal conditions for an ideal gas, most of such cells are empty, a few contain a single molecule, and only a very few more than one. Within a narrow energy domain a great number g_i of such cells may be considered together as forming a *Grosszelle* (large cell in velocity space); the number N_i of molecules inside such a cell is required to be a large number.* It is assumed that the extension of the energy domains in question remains small enough for us to regard the molecular energy as being essentially a constant E_i within each domain. We do not know by whom and when this distinction between small cells and large cells (or domains) was first introduced after Boltzmann. Einstein[1] uses it in his fundamental paper written in 1925.

On making the foregoing distinction between cells and domains, the number of complexions corresponding to a particular macrostate of given total energy is furnished by

$$W = N! \prod_i (g_i^{N_i}/N_i!)$$

The new factors $g_i^{N_i}$ measure the number of possibilities for distributing N_i molecules among the g_i cells of the energy domain E_i.

If we denote by $n_i = N_i/g_i$ the mean occupation number of a cell inside such a domain, then $n_i \ll 1$ for an ideal gas under normal conditions. But the N_i (as well as the g_i) are in general large numbers, so that it is legitimate to use the Stirling approximation in any calculation of $N_i!$ Thus,

$$\ln N_i! = N_i \ln N_i - N_i$$

According to Boltzmann, the entropy S of a state of the gas is proportional to the logarithm of the probability W. We can therefore write

$$S = k \ln W$$

where k is "Boltzmann's constant" introduced by Planck:

$$k = R/N_A$$

R being the ideal gas constant and N_A Avogadro's number. One accordingly gets

$$S/k = N \ln N - \sum_i N_i \ln (N_i/g_i)$$

In phenomenological thermodynamics the entropy of a homogeneous gas is a so-called extensive quantity, which has the property of additivity. This means that S should be proportional to the number N of molecules of a gas sample. But in our formula the term $N \ln N$ prevents S from exhibiting the property of additivity. Statistical physicists were uncomfortable with this fact, which they referred to as the "Gibbs paradox." Planck, in his book on heat-radiation,[3] asserts that this term is "*eine physikalisch bedeutungslose Konstante*" ("a constant without physical interest") and proposes to drop it, and Einstein, in his 1925 paper calls it a "*belanglose Konstante*," which means the same. But this term is in fact not without interest: It has a physical meaning, which results necessarily from Boltzmann's hypothesis that the molecules of a one-component gas are distinguishable, i.e., that they can be labeled, and that a permutation of

* The name *Grosszelle* is used by A. Sommerfeld in Ref. 2.

two molecules creates a new complexion. However, if this is so, then the molecules must be considered as different in some way from one another, which means that each Boltzmann molecule is a special "isotopic species" of the gas, and that the gas is composed of as many isotopes as it contains molecules. Thus the mixing of two samples of the same gas must be considered equivalent to the mixing of isotopes. In this case an "entropy of mixing" enters into play; this leads to the term $N \ln N$, which is thus shown not to be "without interest."

It follows from these considerations that Einstein's statement (in his 1925 paper) that the Boltzmann gas does not obey Nernst's theorem, is incorrect. Einstein's conclusion results from the dropping of the "term without interest." Let us keep this term in the expression of S, and calculate the entropy for the temperature $T = 0°K$. At this temperature all molecules are assembled in the lowest energy state E, which involves only one cell: $g_i = 1$. In this case, $N_i = 0$, $g_i^{N_i} = 1$, and $N_i! = 0! = 1$, for all states $i \neq 1$, while $N_1 = N$ and $N_1! = N!$ for the lowest state. We thus see that in this case $W = 1$ and $S = 0$, so that Nernst's theorem is fulfilled.

The first scientist to have considered the indistinguishability of microscopic particles and its effect on their statistical behavior is J. Williard Gibbs.

In his book, *Elementary Principles in Statistical Mechanics*, which he published in 1901, shortly before his death, he states in the last chapter, on "systems composed of molecules," the following:

> Especial interest attaches to the dynamics of systems composed of a great number of entirely similar particles, or for a great number of particles of several kinds, all of each kind being entirely similar to each other. . . .

He then asks "how the term phase is to be understood in such cases. If two phases differ only in that certain entirely similar particles have changed places with one another, are they to be regarded as identical or different phases?" His answer is: "If the particles are regarded as indistinguishable, it seems in accordance with the spirit of the statistical method to regard the phases as identical." Since such "phases are not altered by the exchange of places between similar particles," he continues: "We shall call them phases determined by generic definitions, or briefly, generic phases. But we shall also be obliged to discuss phases defined by the narrower definition (so that exchange of position between similar particles is regarded as changing the phase) which will be called phases determined by specific definitions, or briefly, specific phases." He concludes that: "the number of specific phases embraced in one generic phase is represented by the continuous product $n_1!\, n_2! \ldots n_h!$," where n_h is the number of identical particles of a given species. We know that this product is correct only in the case where the number of particles in an elementary cell of phase space (μ space) is much smaller than unity: $n_i \ll 1$. It is not true in the general case where quantum statistics have to be applied (cf. the citation of Dirac at the end of this chapter).

Let us now jump to the work of Max Planck. Towards the end of the nineteenth century he tried for several years to find an expression for the entropy of blackbody radiation, which means radiation inside a closed cavity at temperature equilibrium.

He knew that this radiation is independent of the size and the shape of the cavity and of the nature of its walls (Kirchhoff's law). According to Stefan's law, the total energy density u is proportional to the fourth power of the absolute temperature T and the spectral energy density, defined as $u_\nu = du/d\nu$, which goes to zero for $\nu = 0$ and for $\nu = \infty$, must be a universal function of the temperature and of the frequency ν of the radiation. In 1893 Wien showed that u_ν obeys the displacement law

$$u_\nu = \alpha\nu^3 f(\nu/T)$$

A little later he specified the function f and proposed

$$u_\nu = \alpha\nu^3 \exp(-\beta\nu/T)$$

a formula known as Wien's law.

The measurements in the visible and the ultraviolet region of the spectrum were in agreement with this law, but the experimental data in the far infrared, which became known in the spring of 1900, showed that for small values of ν/T Wien's law was not obeyed, that instead in this region u_ν at constant frequency ν is proportional to the absolute temperature T. If this result is combined with the displacement law, one gets, as Planck showed, for small values of ν/T the asymptotic result

$$u_\nu = \gamma\nu^3 \frac{T}{\nu} = \gamma\nu^2 T$$

Planck also studied the relation which exists at temperature equilibrium between the spectral density u_ν of blackbody radiation and the mean energy u_1 of a material oscillator of frequency ν. Such a "resonator" (which can be represented schematically by an electron elastically bound to a positive center) exchanges energy with the surrounding electromagnetic radiation, absorbing and emitting it. After a long and elaborate calculation, Planck established the relation

$$u_\nu = g_\nu u_1 \qquad \text{with } g_\nu = 8\pi\nu^2/c^3$$

c being the velocity of light in vacuum.

Planck showed also that the mean entropy s_1 of such a resonator is related to its mean energy u_1 by the relation

$$s_1 = \phi(u_1/\nu)$$

This relation can be considered as equivalent to Wien's displacement law. If we consider the two aforementioned asymptotic laws, the one valid for large values of ν/T and the other for small values of ν/T, and, if we write them for u_1 instead of u_ν, they take the form

$$u_1 = \alpha' \exp(-\beta\nu/T) \qquad \text{for large } \nu/T \qquad \text{(a)}$$

$$u_1 = kT \qquad \text{for small } \nu/T \qquad \text{(b)}$$

where $\alpha' = \alpha c^3/8\pi$ and $k = \gamma c^3/8\pi$.

Planck now tried to find a general formula for u_l which for large values of ν/T reduces to Eq. (a) and for small values of ν/T to Eq. (b). By "a happy guess" *(glück-liches Erraten)* he arrived at

$$u_l = \frac{\alpha'\nu}{\exp(\beta\nu/T) - 1} \qquad \text{with } k = \frac{\alpha'}{\beta}$$

If reciprocally we replace β by α'/k and if we write, with Planck, h for α', then we obtain

$$u_l = \frac{h\nu}{\exp(h\nu/kT) - 1}$$

from which it follows that

$$u_\nu = \frac{8\pi h\nu^3}{c^3}\frac{1}{\exp(h\nu/kT) - 1}$$

This is Planck's famous law of blackbody radiation, which was presented by him to the Berlin Physical Society on 19th October 1900, and has been confirmed experimentally by all subsequent measurements.[4]

The foregoing formula contains two natural constants, whose numerical values can be found from measurements on blackbody radiation. Planck was able to calculate them from the results of already existing measurements. The constant h has the physical dimensions of an action (energy multiplied by time or better energy per frequency interval). The constant k is identical with Boltzmann's constant $k = R/N_A$. The relation $u_l = kT$ is the expression of the "equipartition theorem." As a consequence of Boltzmann's statistics, the mean energy corresponding to one degree of freedom of a system is $\frac{1}{2}kT$. A linear oscillator possesses two degrees of freedom (potential energy + kinetic energy), and so its mean energy is kT. The equipartition theorem is valid in classical physics, but in quantum theory it is valid only in the asymptotic case where the quantum jumps of energy are small compared to kT.* Planck strangely makes no mention of the equipartition theorem in his paper of 1900.

The equipartition theorem was applied by Lord Rayleigh[5] to blackbody radiation in a paper he published in 1900 (Planck took notice of it only later on), which we now consider. In this paper, published in the *Philosophical Magazine*, Lord Rayleigh, starting from the analogy with his own work on acoustics, shows that the electromagnetic radiation inside a hollow cavity is composed of a discontinuous finite set of standing waves such that the number of waves in the frequency interval between ν and $\nu + \Delta\nu$ is given by

$$\Delta g = \frac{8\pi\nu^2}{c^3} V\, d\nu$$

* The relation $u_l = kT$ corresponds to $u_\nu = (8\pi\nu^2/c^3)kT$, a relation often called the Rayleigh–Jeans law, valid in the far infrared.

in a cavity of volume V. The energy of radiation in this frequency interval is equal to

$$\Delta E = u_\nu V \Delta\nu$$

so that the mean energy of a standing wave becomes

$$u_1 = \frac{\Delta E}{\Delta g} = \frac{c^3}{8\pi\nu^2} u_\nu$$

There consequently exists between u_ν and u_1 the relation

$$u_\nu = g_\nu u_1 = \frac{8\pi\nu^3}{c^3} u_1$$

and we see that the mean energy of a standing wave of the cavity is the same as the mean energy of a Planck resonator.

As already mentioned, Planck took no notice of this work of Lord Rayleigh. It was only in 1910 that Peter Debye applied Rayleigh's method to blackbody radiation. We will return to this point later on.

Planck, after having found his formula by "a happy guess," tried to establish a theoretical foundation for it. At this point he turned his attention to Boltzmann's statistical considerations, which before he had considered rather with skepticism. Boltzmann himself had encouraged him to do so. In his paper of 1897 on irreversible radiation processes[6] he wrote at the beginning

> Es ist sicher möglich und wäre jedenfalls dankenswert, einen dem Entropiesatz analogen auch für die Strahlungserscheinungen aus den allgemeinen Gesetzen derselben nach den gleichen Prinzipien wie in der Gastheorie abzuleiten. Es würde mich daher freuen, wenn sich einmal zu diesem Zwecke die Ausführungen Hrn Plancks über die Gesetze der Zerstreuung elektrischer Planwellen an sehr kleinen Resonatoren als nützlich erweisen würden.

> [It is certainly possible, and it would in any case be rewarding, to deduce for the radiation phenomena from their general laws, an analog of the entropy law, using the same principles as in the theory of gases. It would therefore gladden me if Mr. Planck's considerations about the scattering of electric plane waves by very small resonators could be proved useful for this purpose.]

Planck took up Boltzmann's suggestion and tried to define and calculate, in the way Boltzmann had done it for a gas, a distribution law for the macrostates of radiation. Identifying the most probable distribution with the state of equilibrium, he attempted to deduce an expression for the entropy of blackbody radiation. In Boltzmann's problem, molecules had to be distributed in velocity space. Boltzmann had been obliged to "quantize" this space, to divide it into a discrete number of cells. For Planck the problem was to distribute radiation energy inside a cavity among resonators (or among modes of standing waves). In this case, the container (the resonators or the modes) had a discrete structure, but its contents, the energy to be distributed, exhibited a continuous structure. What Boltzman had done for the container, Planck had to accomplish for its contents, viz., give it a discontinuous structure. He accordingly divided the radiation into equal elements of energy ϵ. Let N_i be the number of these elements in the frequency interval $\Delta\nu_i = 1$ for a volume $V = 1$. They have to be distributed among $g_i = 8\pi\nu^2/c^3$ resonators (or standing waves).

We can write

$$u_\nu = g_i u_1 = N_i \epsilon$$

where u_1 is the mean energy of a resonator. Planck's own words are the following[7]:

> Es ist einleuchtend, dass die Verteilung der N_i Energieelemente auf die g_i Resonatoren nur auf eine endliche ganz bestimmte Anzahl von Arten erfolgen kann. Jede solche Art der Verteilung nennen wir nach einem von Hrn Boltzmann *für einen ähnlichen Begriff* gebrauchten Ausdruck eine "*Complexion.*"
>
> [It is obvious that the distribution of the N_i elements of energy among the g_i resonators can be done only in a finite and definite number of ways. Each such mode of distribution we call a "*complexion,*" after the expression used by Mr. Boltzmann *for a similar concept.*]

For the number of possible complexions corresponding to a macrostate of given energy, Planck introduced a value borrowed from the theory of combinations:

$$W_i = (N_i + g_i - 1)!/N_i!(g_i - 1)!$$

He asserts:

> Die Wahrscheinlichkeit eines jeden Zustandes ist proportional der Anzahl der ihm entsprechenden Complexionen, oder mit andern Worten: irgendeine bestimmte Complexion ist ebenso wahrscheinlich als irgend eine andere bestimmte Complexion. . . . Dieser Satz bildet den Kernpunkt der ganzen vorliegenden Theorie; sein Beweis kann in letzter Linie nur durch die Erfahrung geliefert werden.
>
> [The probability of each (macro-) state is proportional to the number of complexions corresponding to it, or, in other words, any particular complexion has the same probability as any other complexion . . . This statement is the central point of the whole theory under discussion; it can be proved in the final analysis only by experience.]

With this given value of W Planck succeeds in fact in establishing his radiation formula, with the condition that for each frequency $\epsilon = h\nu$. This was the way quantum theory was born.

But why did Planck not use Boltzmann's expression for W in its (enlarged) form:

$$W = N! \prod_i g_i^{N_i}/N_i! \qquad \text{(Boltzmann)?}$$

Why, instead, did he borrow from the theory of combinations another expression

$$W = \prod_i (N_i + g_i - 1)!/N_i!(g_i - 1)! \qquad \text{(Planck)?}$$

Planck remains silent on this point. It is possible that he first tried Boltzmann's expression for W. But this would result in Wien's radiation law, not Planck's. To escape from this deadlock, he would then have tried the other formula, which does lead to the desired goal.

What is the essential difference between Boltzmann's complexions and Planck's complexions, and between Boltzmann's method of counting and Planck's method of counting?*

In both cases fundamentally different concepts are involved. But by using the same name "complexion" for two (entirely different) things, and by mentioning that his method was "similar" to that of Boltzmann, Planck has obscured the differences in question. In this way a confusion was created in the mind of physicists, which took a quarter of a century to be cleared up. For, despite a very clear analysis published by Natanson in 1911[9]—in a paper which unfortunately remained unnoticed and unmentioned—the question became entirely clarified only in the years 1924–1926, and indeed through the efforts of Einstein and of Dirac. Einstein's criticism, written in 1909,[10] reads: "Weder Herr Boltzmann noch Herr Planck haben eine Definition von W gegeben. Sie setzen rein formal W = Anzahl der Komplexionen des betrachteten Zustandes." ["Neither Mr. Boltzmann nor Mr. Planck have given a definition of W. They set purely formally W = number of complexions of the state under consideration."] Einstein's criticism is not entirely justified, for, Planck says quite explicitly (the citation above): "Der Beweis kann in letzter Linie nur durch die Erfahrung geliefert werden." ["The proof can be given in the final analysis only by experience."] Thus there is no implication of a logical deduction justifying the method of counting. But what one can rightfully reproach Planck for is that he did not insist on the difference between his definition of a complexion and that of Boltzmann.

Wherein does this difference exist?

As mentioned above, Boltzmann considered the molecules as objects distinguishable from each other, which could therefore be individually identified and labeled by a number of distinctive marks. For him a complexion is defined not only by the number of molecules in a given cell, but by the distinctive label of each molecule. (This corresponds to what Gibbs calls a "specific phase.") Planck, by contrast, treated the energy elements ϵ as indistinguishable, completely equal entities which cannot be labeled. In this case an elementary microstate can be defined only by the number of entities in each cell. (This corresponds to what Gibbs calls a "generic phase.")

To avoid confusion further on, we may agree to replace Planck's "complexion" by the word "microstate," and reserve the word "complexion" for Boltzmann's concept. For Boltzmann, each complexion has the elementary probability 1. The probability of a macrostate is then equal to the number of complexions contained in it. For Planck, on the other hand, it is the microstate which has probability 1, and Planck's W is thus given by the number of microstates corresponding to a macrostate.

* Boltzmann's method is discussed by Planck in Ref. 8.

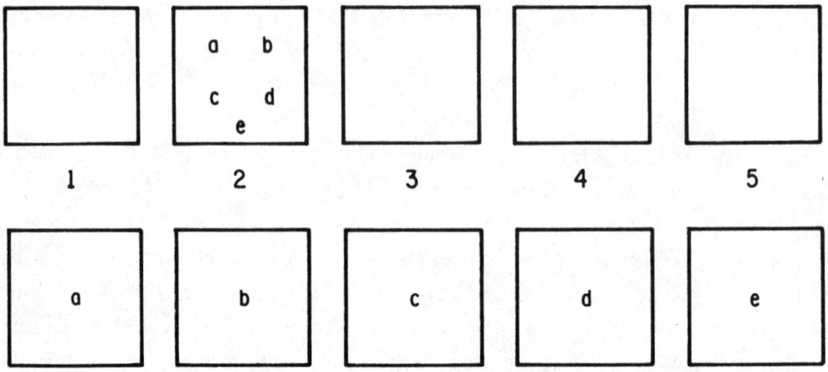

FIGURE 1. The distribution of distinguishable particles among cells.

The statistical behavior of indistinguishable particles is entirely different from the statistical behavior of distinguishable objects, as Gibbs had already foreseen.

We may illustrate this for a simple example: the distribution of $N = 5$ particles among $g = 5$ cells. We do not consider all possible distributions, but only those for which the particles either all occupy one particular cell or are uniformly distributed over the cells, one particle to a cell. We label the cells by the numbers 1 to 5 and distinguishable particles by the letters *a*, *b*, *c*, *d*, *e*, while indistinguishable particles are represented by small circles, the same for all the particles.

In the case of distinguishable particles (Fig. 1), we see that the situation where all particles are sharing the same cell (here cell No. 2) can be realized in only one way. This represents a complexion. The number of complexions where all particles are together in one of the five cells is $W = 5$. But in the situation where the particles are distributed uniformly over the cells, the number of complexions equals the number of permutations of a group of five letters, i.e., $W = 5! = 120$.

Let us consider now the case of indistinguishable particles (Fig. 2). If the particles share the same given cell, this corresponds to a microstate. Thus there are five microstates where the five particles are together in one of the cells: $W = 5$. But if the five particles are uniformly distributed, one to a cell, this can be realized by only one microstate: $W = 1$.

To summarize, in the case of distinguishable particles, each complexion has the elementary probability $W = 1$, and the probability of a macrostate equals the corresponding number of (Boltzmann's) complexions. In the case of indistinguishable particles, each microstate has the elementary probability $W = 1$, and the probability of a macrostate equals the number of (Planck's) microstates contained in it.

The statistical behavior of both types of particles is *a priori* completely different. In the case of n distinguishable particles, their uniform distribution over n cells is $(n - 1)!$ times more probable than their concentration in one cell. In the case of n indistinguishable particles, their concentration in one cell is n times more probable than their uniform distribution over the cells. If we do not mind an anthropomorphic

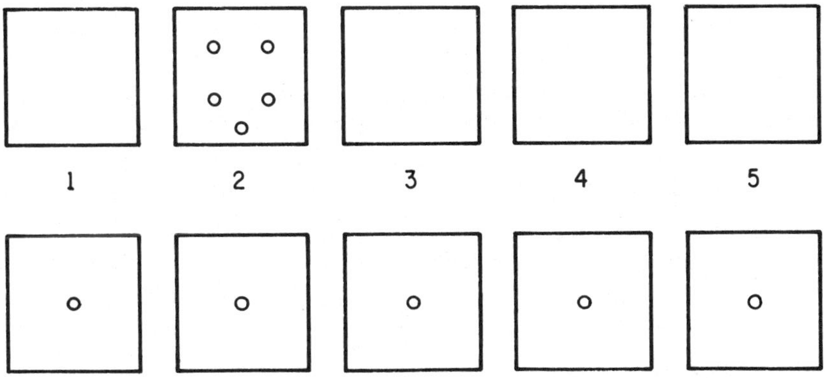

FIGURE 2. The distribution of indistinguishable particles among cells.

analogy, we may say: Distinguishable particles show a tendency to spread uniformly and indistinguishable particles to cluster together.

From everyday experiences we are familiar with distinguishable objects and their behavior. It is a natural tendency to extrapolate, with Boltzmann, this behavior to microparticles (atoms and molecules). The behavior of indistinguishable particles has for us a strange character. Even Einstein, whose mind was attuned to a high degree of abstraction, had the impression that the behavior in question is "von ganz rätselhafter Art" ("a very mysterious one") (Einstein, 1925, loc. cit.). So, it is no wonder that it took such a long time before Planck's method of counting was understood (even by himself) and finally identified as Bose–Einstein statistics. But we must remember that, in this method, the "cells" are not domains of arbitrary size, but "small cells," whose size h^3 in phase space is determined by Planck's quantum of action, which can be understood as an expression of the uncertainty relations.

We know that Planck himself adopted his own creation, the discrete structure of energy, only with hesitation. He wanted to restrict discreteness to the exchange of energy between radiation and matter only.

It was Einstein who put forward a more radical view:

1905, the interpretation of the photoelectric effect[11];
1907, the quantization of material oscillators[12];
1909, the calculation of the fluctuations of radiation energy[13];
1917, the analysis of energy exchange between radiation and atoms[14]; and so on.

He concluded that the radiation energy itself is composed of energy quanta and that, moreover, these quanta possess momentum, a hypothesis which was so beautifully confirmed by the discovery and interpretation of the Compton effect in 1922–1923. On the other hand, Debye,[15] in 1910, combined Rayleigh's structure of cavity radiation (the discontinuous modes of standing waves) with Planck's method of counting and demonstrated in this way that it was possible to avoid the introduction of Planck's artificial resonators. As each standing wave occupies the whole cavity, this also shows that the radiation quanta inside a cavity cannot be considered as localized particles.

We may now turn our attention to Ladislas Natanson's paper,[9] published in 1911. He defines the different ways for distributing discrete particles among discrete cells and demonstrates the possibility of three modes of distribution, depending on whether the particles, on one hand, and the cells, on the other, are indistinguishable or distinguishable entities. If both the particles and the cells are indistinguishable, the distribution can be characterized only by the number of cells containing a given number of particles. Natanson calls this mode "*Verteilungsart*" ("distribution mode"). The second mode considered by Natanson deals with the case where indistinguishable particles or elements are to be distributed over distinguishable cells. This mode he calls "*Anordnungsart*" ("collocation mode"). In this case the distribution is defined by the number of particles in each given cell. It corresponds to Planck's microstate. Finally, Natanson defines a third mode where both the cells and the particles are distinguishable and allow individual labeling. This mode, which Natanson calls "*Zuordnungsart*" ("association mode"), corresponds to Boltzmann's complexion, where the individuality of the molecules in each cell is recognized.

Natanson shows clearly that Planck's method of counting the probability of a macrostate corresponds to the collocation mode. Let us cite Natanson[9]:

> Es mag . . . daran erinnert werden, dass bei allen bekannten Beweisen für die Plancksche Formel dasselbe Prinzip (der Anordnung) angewandt wird. . . . Es genügt offenbar nicht, die Anzahl der Anordnungen zu zählen; man muss auch dartun, dass verschiedene Anordnungen die gleichen Chancen haben, vorzukommen. Und dass dies wirklich der Fall ist, können wir nur annehmen, wenn wir voraussetzen, dass bei einem Prozess der Wahrscheinlichkeitsberechnung die Energiehalter als identifizierbar behandelt werden können, und dass die Energieeinheiten, die in jeder Hinsicht vollkommen gleich sind, nicht als identifizierbar behandelt werden können. Da unser Verfahren in letzter Linie auf dieser Annahme beruht, so erscheint es naturgemäss, es ohne weiteres als die Grundlage der Theorie anzusprechen. Anscheinend hat man dem Umstande nicht genügende Bedeutung beigelegt, dass wir tatsächlich keine andere Möglichkeit haben, die Berechtigung des Planckschen Verfahrens zur Wahrscheinlichkeitsberechnung nachzuweisen, ausser der Berufung auf den experimentellen Beweis, durch den wir die Schlussfolgerungen der Rechnung gestützt finden.

> [We may recall that for all known proofs of Planck's formula the same principle (of collocation) is applied . . . But obviously it is not sufficient to count the number of collocations; it is also necessary to show that different collocations have the same chance to be realized. And that this is really so we may only assume if we presuppose that, in a process of probability calculus, the containers of energy can be treated as identifiable, and that the energy elements, which are completely identical in every respect, cannot be treated as identifiable. As our method is founded in the final analysis on this hypothesis, it seems natural to adopt it, without further ado, as the foundation of the theory. Apparently one has not bestowed sufficient significance on the circumstance that we have in fact no other possibility for justifying Planck's method of calculating a probability, except for the appeal to the experimental proof, by which the conclusions of the calculation are found to be supported.]

And Natanson adds:

> Die Schwierigkeit ist von der Art, dass sie naturgemäss im Zusammenhang mit dem Gegenstande auftritt. Planck nimmt als Ausgangspunkt eine vollkommen bestimmte Regel zur Bewertung der Chancen an, und durch eine besondere Berufung auf die Erfahrung finden wir *a posteriori*, dass die Voraussetzung richtig ist. Aber niemand hat anscheinend jemals versucht, ihre Richtigkeit auf Grund allgemeiner Prinzipien zu erweisen, und soweit ich es zu übersehen vermag, kann dies auch nicht geschehen.
>
> [The difficulty is such that it naturally appears in connection with this subject. Planck takes as starting point a very definite rule for evaluating the chances, and, by a particular appeal to experience, we find *a posteriori* that the presupposition is correct. But nobody seems ever to have tried to deduce the validity of the rule from first principles, and, as far as I can see, in our opinion this cannot be done either.]

Clearer than Natanson did, Planck's method cannot be characterized and defended. He answers Einstein's objection. Unfortunately, neither Planck nor Einstein took public notice of Natanson's analysis; nowhere have we found it mentioned.* This is rather surprising, since Einstein had relations, and correspondence, with Natanson.[16]†

And so it happened that it took another 13 years before Einstein, inspired by Bose's paper,[17] worked out the statistics of indistinguishable particles and transposed it from photons to atoms and molecules. Curiously, Bose does not mention the property of the indistinguishability of light quanta. In our opinion, Bose's principal merit is to have shown that one can identify the standing modes of a cavity with the cells of extension h^3 in phase space (Ehrenfest's μ space).

In his three papers, published in 1924–1925, in which Einstein[18] develops Bose-Einstein statistics and discusses its difference from Boltzmann statistics, he also does not mention the indistinguishability of particles. Replying to a criticism of Ehrenfest, he speaks of "Quanten bzw. Molekülen, die als nicht von einander statistisch unabhängige Gebilde behandelt werden" ["Light quanta or molecules which cannot be treated as independent elements"] and "eine gegenseitige Beeinflussung der Moleküle von vorläufig ganz rätselhafter Art" ["a mutual interaction between the molecules which provisionally is of a quite mysterious nature"]. Obviously, at this moment he has not reached the conclusion that indistinguishable entities have essentially a different behavior from distinguishable ones.

That the fundamental property of microparticles is their indistinguishability was established only in the following years through analysis of the symmetry properties of microparticles, especially by Dirac.[19] In his fundamental work published in 1926, in the section entitled "Systems Containing Several Similar Particles," he examines a

*Erratum added in proof: In fact, Planck mentions Natanson's paper in his report on the first Solvay Conference of 1911 (Reprinted in M. Planck, *Physikalische Abhandlungen und Vorträge*, Vol. II, p. 277).

†Probably at this time (1911) Einstein's mind was fully occupied by the puzzle of gravitation. Only when he had solved this problem (1916) did he turn his attention again to quantum theory.

system of two electrons and asserts the following for the case that one of the particles is in state m, the other in state n:

> The question arises whether the two states $(m\,n)$ and $(n\,m)$, which are physically *indistinguishable* as they differ only by the interchange of the two electrons, are to be counted as two different states or as only one state . . . The two transitions $mn \rightarrow m'n'$ and $mn \rightarrow n'm'$ are physically *indistinguishable* . . . Hence to keep the essential characteristic of the theory that it shall enable us to calculate only observable quantities, one must adopt the second alternative that (mn) and (nm) are only one state.*

Dirac shows that it is possible to satisfy this condition by "symmetrization" of the eigenfunction of the two-particle system. This can be done in two ways: Either

$$\psi_{mn} = \psi_m(1)\psi_n(2) + \psi_n(1)\psi_m(2)$$

(symmetrical combination), where 1 and 2 designates the two particles, or

$$\psi_{mn} = \psi_m(1)\psi_n(2) - \psi_n(1)\psi_m(2)$$

(antisymmetric combination). In the first combination the exchange of the two particles does not change the value of ψ_{mn}, in the second combination the exchange of the two particles changes the sign of ψ_{mn}. As any observable quantity depends only on the square of ψ_{mn}, this has no experimental consequences. So both solutions are possible. Dirac adds, "The theory at present is incapable of deciding which solution is the correct one." But, he remarks, "an antisymmetric eigenfunction vanishes identically when two electrons are in the same orbit . . . which is just Pauli's exclusion principle." Dirac concludes, therefore, that electrons have to be treated according to the antisymmetric combination. We know today that the symmetric total eigenfunction corresponds to Bose–Einstein statistics, which is obeyed by particles with integer spin, while the antisymmetric total eigenfunction corresponds to Fermi–Dirac statistics, which is obeyed by particles with half-integer spin.

Thus, thanks to Dirac, the indistinguishability of microparticles has gained definitely a legal status in quantum physics. This indistinguishability of the particles is intimately related to their wave nature.

When Einstein—in 1925—calculated the fluctuations in the number of bosons, he found the same two terms as he had found earlier for the energy of radiation. This would have led him necessarily to the wave nature of matter, even if he had ignored the ingenious work of Louis de Broglie. The waves of particles that occupy different cells of phase space are incoherent. But waves of particles in the same cell of phase space correspond to coherent vibrations, so that such particles obey the same wave function.

Finally, let us mention that the well-known formulas for boson fluctuations, viz.,

$$\overline{\Delta n^2} = n + n^2 \qquad \text{(in a small cell)}$$

* Compare this to the question raised by Gibbs a quarter of century earlier (loc. cit.).

and

$$\overline{\Delta N^2} = N + N(N/g) \qquad \text{(in a large cell or domain)}$$

where $n = N/g$, can only be applied in the case where $g \gg 1$, which means that an energy domain must contain a large number of cells. To establish this formula, the Stirling approximation is applied to N and to g. But there are experimental cases where g is small. This is especially the case for Einstein condensation of bosons, for in this case all N particles condense in the lowest-energy domain, which contains only one cell. The statement that Einstein condensation is accompanied by large fluctuations is not correct.[20] We have shown elsewhere[21] that in these cases the fluctuation formula has to be corrected:

$$\overline{\Delta n^2} = (n + n^2) \cdot \frac{g - 1}{g + 1}$$

This has especially to be considered for the mono-mode operation of a laser, where fortunately the fluctuations remain small.

References and Notes

1. A. Einstein, *Sitzungsber. Preuss. Akad. Wiss.* (1925), p. 3.
2. A. Sommerfield, *Vorlesungen über Theoretische Physik,* Vol. 5 (Akademische Verlagsgesellschaft, Leipzig, 1952), Sec. 29c, p. 209.
3. M. Planck, *Vorlesungen über die Theorie der Wärmestrahlung,* (Barth, Leipzig, 1906), Sec. 149, p. 154.
4. M. Planck, *Physikalische Abhandlungen und Vorträge,* Vol. I (Vieweg, Braunschweig, 1958), p. 154.
5. Lord Rayleigh, *Phil. Mag.* **49,** 539 (1900); reproduced in *Scientific Papers,* Vol. IV (Cambridge University Press, Cambridge, 1892–1901), p. 483. A calculation error in this paper was corrected by Rayleigh in *Nature* **71,** 559 (1905); and **72,** 54, 243 (1905); *Scientific Papers,* Vol. V (Cambridge University Press, Cambridge, 1902–1910), p. 248.
6. L. Boltzmann, *Wissenschaftliche Abhandlungen,* Vol. III (Chelsea Publishing Co., New York, 1968), p. 618.
7. M. Planck, *Ann. Phys. (Leipzig)* **4,** 553 (1901); *Physikalische Abhandlungen und Vorträge,* Vol. I (Vieweg, Braunschweig, 1958), p. 698.
8. M. Planck, *Ann. Phys. (Leipzig)* **9,** 629 (1902); *Physikalische Abhandlungen und Vorträge,* Vol. I (Vieweg, Braunschweig, 1958), p. 731.
9. L. Natanson, *Phys. Z.* **12,** 659 (1911).
10. A. Einstein, *Phys. Z.* **10,** 185 (1901).
11. A. Einstein, *Ann. Phys. (Leipzig)* **17,** 132 (1905).
12. A. Einstein, *Ann. Phys. (Leipzig)* **22,** 180 (1907).
13. A. Einstein, *Phys. Z.* **10,** 185, 817 (1909).
14. A. Einstein, *Z. Phys.* **19,** 301 (1917).
15. P. Debye, *Ann. Phys. (Leipzig)* **33,** 1427 (1910).
16. *Einstein: A Centenary Volume,* A. P. French, ed. (Heinemann, London, 1979), p. 94.
17. S. W. Bose, *Z. Phys.* **26,** 178 (1924).
18. A. Einstein, *Sitzungsber. Preuss. Akad. Wiss.* (1924), p. 281; and (1925), pp. 8 and 18.
19. P. Dirac, *Proc. R. Soc. London* **112,** 661 (1926).
20. A. Sommerfield, *Vorlesungen über Theoretische Physik,* Vol. 5 (Akademische Verlagsgesellschaft, Leipzig, 1952), pp. 273–274.
21. B. Decomps and A. Kastler, *C. R. Acad. Sci.* **257,** 1087 (1963).

6

Philosophical Semantics

Morris Lazerowitz
Smith College
Northampton, Massachusetts

It has always been thought that technical, reasoned philosophy is the comprehensive science of reality and brings under its special scrutiny such fundamental concepts as *space, time, contingency, mind, causation, language*. It has been said that the task of philosophy is to delineate "the generic traits" of reality; and even if its scope were less than this, the value placed on philosophy would fade away if it were thought incapable of yielding ontological information. The process of value-deflation has in fact set in, for the classical notion of philosophy has begun to wear thin. The philosophical tree of knowledge has so far borne only phantom fruit and created intractable disputation. A closer look at philosophy can no longer be dismissed or thought unnecessary. In the last fifty years, the work of G. E. Moore, the logical positivists, and Ludwig Wittgenstein has made it uncertain *what* a philosopher is talking about when he says that motion does not exist or that no two terms are really synonymous, although it is still natural to think that the first statement is about the occurrence of motion and the second about the existence of synonymy in a language.

Philosophy is Not Empirical

Wittgenstein has said, with the air of giving utterance to a commonly recognized fact: "Philosophical problems are, of course, not empirical."[1] To my knowledge, he made hardly any concrete attempt to show the correctness of this metaphilosophical thesis; but if correct, it would help us understand why a time-honored investigation which professedly wishes to discover truth about what there is employs none of the methods of the natural sciences. It is hardly necessary to point out the fact that there are no philosophical laboratories or the fact that philosophy has no recourse to special observations of its own, as against those made in the sciences and in ordinary life. The implication would seem to be that philosophy is neither observational nor experimental. His remark underscores the inappropriateness of saying that observation or an experiment shows that motion does not occur or that synonomy does not exist.

It is not necessary to look long in the literature to find accounts of philosophical investigations which on a normal reading can only be understood as describing empirical procedures. Three such procedures are well known and appear to be taken at face value by philosophers. One is a procedure which seems to have recourse to experiment. Another appears to make use of observation. And a third seems to resort to a search. In each case the investigation, whether by experiment, or careful scrutiny, or a search, is conducted within the mind. These procedures, if they are as described by philosophers, go against Wittgenstein's contention that philosophical problems are not empirical; and at the very least it will have to be *shown* that philosophers misdescribe the kind of work they do.

To take up first the notion that philosophers obtain results by conducting experiments. C. H. Langford, for example, describes what appears to be an experiment in the laboratory of his mind: "When we try to envisage the unitary meaning of a statement of this kind ["Men exist and men do not exist"], we find that this is quite impossible, and that therefore it has no single meaning, but rather one meaning corresponding to one part of the verbal expression, and another to another."[2] Similarly, Wittgenstein, in an attempt to refute Platonic realism, invites us to make the mental experiment of trying to think of the meaning of an expression without thinking of the expression. In his own words:

> Make the following experiment: say and mean a sentence, e.g., "It will probably rain tomorrow". Now think the same thought again, mean what you just meant, without saying anything (either aloud or to yourself). If thinking that it will rain tomorrow accompanied saying that it will rain tomorrow, then just do the first activity and leave out the second.—If thinking and speaking stood in the relation of the words and the melody of a song, we could leave out the speaking and do the thinking just as we can sing the tune without the words.[3]

The well-known outcome of his experiment is that the meaning of an expression is not something which is an "accompaniment" of it, but is rather the use the expression has in the language.

If we can put to one side for a moment the idea suggested by the words of these philosophers, and reflect on the words themselves, it can easily be seen that they are not actually being used to describe an empirical procedure, the experimental attempt to conceive or think or envisage an idea or proposition. It is necessary to keep in mind the difference between logical, or conceptual, impossibility and what might be called "natural" impossibility, i.e., between impossibility built into concepts and impossibility built into phenomena, psychological as well as physical. It is important to be clear about the difference, for example, between the impossibility of physically being in a certain place while being in another place a mile away and the impossibility of dashing from the one place to the other in a minute, and about the difference between the impossibility of carrying out in the mind the decimal expansion of π to the thousandth place and the impossibility of conceiving a six-sided pentagon. What is logically impossible to conceive is not something that transcends anyone's powers of conception; instead, it is by its very nature not subject to being conceived. This means that an expression whose meaning is a logically impossible concept or proposition has no descriptive function in the language. The expression does not thereby lack meaning, but the meaning it has is such as to prevent it from functioning descriptively. It is readily seen that the likeness between the following sentences tends to conceal an important difference between them:

(α) It is impossible to carry out in one's head the decimal expansion of π to the thousandth place.
(β) It is impossible to conceive of the completed expansion of π.

For (α) tells us what it is which is declared impossible: the phrase "carries out in one's head the decimal expansion of π to the thousandth place" describes a mental feat which is beyond the powers of ordinary minds. In this respect (β) differs radically from (α). The phrase "conceives the completed expansion of π" does not refer to a theoretical act of conceiving something, the reason for this being that "completed expansion of π" has no use to refer to an expansion. It makes *no descriptive sense* to say, "Mt. Everest has been climbed, and to match this feat the completed expansion of π has been conceived."

Paralleling this difference is the difference between experimentally attempting to carry out in one's head the expansion of π to the thousandth place and experimentally attempting to grasp in thought the completed expansion. The implication of the fact that the completed expansion of π is *logically* inconceivable, and not that it is psychologically impossible to conceive, is that it is logically impossible to *try* to conceive it. To put this more explicitly in terms of the language involved, the fact that it makes no sense to speak of thinking of the completed expansion of π implies that it makes no sense to speak of trying to think of the completed expansion—any more than it makes sense to speak of trying to bring to a stop a stationary object. This comes down to the linguistic fact that the expression "conceives the completed expansion of π" has no descriptive use in the language; and this implies that the expression "tries to conceive the completed expansion" has none either. There is no more trying to think of a completed expansion of which you have no idea than there is trying to milk a goat that does not exist or trying to remember a face which has never been seen. A sentence of the form "experimentally tried to think so-and-so" refers to no experiment if "so-and-so" describes nothing.

These remarks apply directly to the philosophical experiments made by Langford and Wittgenstein. And this fact confirms Wittgenstein's general observation about the nature of philosophical problems: a philosophical experiment is not an empirical experiment, whatever else it may be. The general maxim may be stated in the following way: if the outcome of an investigation is that something is logically impossible, the investigation is not empirical.

To see this, consider the putative *Gedankenexperiment* reported by Langford.

His attempt, and failure, to envisage the unitary meaning of the sentence "Men exist and men do not exist" is an attempt and a failure in semantic appearance only. Parenthetically, it is worth noting that there is no attempting anything without the possibility, in principle, of succeeding, and there is no failing to achieve something without the possibility, in principle, of achieving it. When Langford tells us that he found it "quite impossible" to envisage the proposition expressed by "Men exist and men do not exist," as against the propositions expressed by the two parts of the conjunctive sentence, he implies that the impossibility is one of logic, that it is conceptual, and not an impossibility in nature. It is a correct use of descriptive English to say, "The King's men tried to put Humpty Dumpty together again, but failed"; but it is only a semantic joke to say, "Langford tried to combine the two propositions *Men exist* and *Men do not exist* into a single proposition and found that his mind was not up to the task." It would not be a correct use of language, as he employs it, for him to go on and say, imitating Hume: "If anyone upon serious and unprejudiced reflection thinks he has envisaged the unitary meaning which I have failed to envisage, I must allow him that he may be in the right as well as I." The reason why the language he used to refer to his putative experiment describes no empirical procedure is that since the sentence "Men exist and men do not exist" expresses what is logically impossible, in his use of the word "proposition" it does not express a single proposition. That is, Langford is so using the word "proposition" that it does not apply to what is normally referred to as a logically impossible proposition: an indicative sentence which has a self-contradictory meaning no more expresses a proposition than does "greatest prime number" describe a number. The consequence of his contracted use of the word "proposition" is that the phrase "unitary meaning of a self-contradictory sentence"[4] could not have been used by him to refer to the subject of a thought-experiment, something the achievement of which he found "quite impossible."

Consider now the reason Wittgenstein adduces against the Platonic theory that the literal meaning of an expression is an entity which accompanies the expression and can exist apart from it. It is presented, verbally, in the form of an experimental attempt to think of the meaning by itself, an attempt comparable to trying to think of a person's mirror image actively continuing by itself, in the absence of the person. The supposed outcome, unlike that of the mirror image, is that it is impossible to think of the meaning apart from its expression. As in the preceding case the impossibility has to be taken as logical. For Wittgenstein did not mean to leave it an open possibility that Platonic philosophers might have succeeded where he failed. This possibility could be excluded with complete assurance only by *making* the word "impossible" in the sentence "It is impossible to think of the meaning of an expression apart from the expression" denote the impossibility of logic.

Philosophical investigations are in some cases presented as falling under an empirical procedure which employs observation and induction. To all appearances, instances of a certain class are inspected and generalizations are constructed on the basis of the findings. But if Wittgenstein's dictum is correct, the appearance that natural science is being practiced is a bogus facade behind which a different kind of work is being done. It needs to be said immediately that if it turns out that philosophers misrepresent their investigations, as in the preceding cases, there is no hint of conscious deception to be found in their writings. They are more like a magician who, because of deep inner needs, believes in his own tricks rather than like the rascally tailors in the fairy tale who were aware of their hoax. Berkeley observed that "could men but forbear to amuse themselves with words"[5] they would soon agree with him that matter does not exist. And if Wittgenstein's dictum turns out to be correct, it will lend plausibility to the suspicion that philosophers prefer to amuse themselves with words to working with things. Indeed, a number of Wittgenstein's later thoughts have already given substance to the uneasy notion that the dwelling philosophers have built for themselves with so much pride and in which they find their pleasure and security is an insubstantial shadow cast by the invisible manipulation of words.

Berkeley is an example of an important philosopher who seeks knowledge of things by resorting to observation. He undertakes to determine whether causation exists amongst phenomena, and he appears to do this by carefully scrutinizing them. The reader of the following passage might well note the imagery it conjures up in his mind:

> All our ideas, sensations, notions, or the things which we perceive, by whatsoever names they may be distinguished, are visibly inactive: there is nothing of power or agency included in them. So that one idea or object of thought cannot produce or make any alteration in another. To be satisfied of the truth of this, there is nothing else requisite but a bare observation of our ideas. For, since they and every part of them exist only in the mind, it follows that there is nothing in them but what is perceived; but whoever shall attend to his ideas, whether of sense or reflexion, will not perceive in them any power or activity; there, is, therefore, no such thing contained in them.[6]

Berkeley's assertion that *all* ideas are causally inactive goes beyond the statement that his ideas are without exception causally inactive, and it is to be supposed that the generalization is arrived at inductively from his experience.

When Berkeley reports that inspection discloses no agency in his ideas and that they are all visibly inactive, the picture we get is that of someone who looks at the people in a room and reports that all of them have upturned noses. In advance of looking at the people there is no telling what the shapes of their noses are, and in advance of attending to his ideas Berkeley could only speculate about their inactivity. In the case of the people's noses observation shows a rather surprising thing, and it is clear that observation *might* show a different and less surprising group of people: looking *could* reveal the opposite of what it in fact happens to reveal. In Berkeley's case also, if looking is relevant to and required for the determination of the facts, looking could, possibly, have shown the opposite of what Berkeley reported. The mark of an empirical investigation is that the opposite of what it discovers remains conceivable and could, therefore, have been discovered instead. The outcome of an investigation which employs observation cannot imply the *logical* impossibility of the opposite.

A second reading of Berkeley's statement will bring to our attention an omission which it is natural to overlook. It contains no hint of what it would be like for an idea to be active. It turns out that the term "inactive idea," which at first glance appears to have descriptive content, i.e., to function the way "indolent adolescent" functions, is not linked to "active idea," and this omission is not an oversight. It tells us that the term "active idea," in Berkeley's way of speaking, has been given no descriptive assignment. The following sentence, which continues the above-quoted passage, makes this evident: "A little attention will discover to us that the very being of an idea implies passiveness and inertness." To say that the *very being* of an idea implies inertness is just another way of saying that being an idea *entails* being inert. Hence, in Berkeley's use of the word "idea" the phrase "idea that is not inert," or the term "active idea," denotes a logically impossible concept. The term does not refer to anything which observation could, in principle, reveal. Looking is not relevant to determining whether ideas are inactive. For if observation could discover for us that ideas are inactive, it *could*, in principle, discover the opposite, which is ruled out as being logically impossible. The sentence "Ideas are visibly inactive" has been given no use to refer to anything that is or could become visible to an observer. If Berkeley had given it such a use, the phrase "visibly active idea" would make descriptive sense. But this he prevents by his declaration that "the very being of an idea" implies inertness.

There remains to be considered the third kind of case in which a seemingly empirical investigation is conducted to determine the truth-value of a philosophical claim. Essential to this kind of investigation would seem to be a search, and is so represented by Hume. The search for agency, or natural necessity, is described by Hume in the following well-known passage:

> . . . I consider in what objects necessity is commonly supposed to lie; and, finding that it is always ascribed to causes and effects, I turn my eyes to two objects supposed to be placed in the relation, and examine them in all the situations of which they are susceptible. I immediately perceive that they are *contiguous* in time and place, and that the object we call cause *precedes* the other we call effect. In no one instance can I go farther, nor is it possible for me to discover any third relation betwixt these objects.[7]

Hume seems to be reporting the negative outcome of an empirical search, and to the assertion that agency does not exist he joins a challenge: "If anyone think proper to refute this assertion, he need not put himself to the trouble of inventing any long chain of reasoning, but may at once show us an instance of cause where we discover the power or operating principle."[8] He stands ready to be convinced if others can exhibit to him what he himself has not been able to find; and this seems a scientifically correct attitude to take.

If we reexamine Hume's account of his quest and its outcome we see that it issues in two negative results, not just one; and they are not compatible with each other. One is that agency does not exist in the physical world. The other is that we have no *idea* of agency: "All ideas derive from and represent impressions. We never have any impression that contains any power or efficacy. We never, therefore, have any idea of power."[9] The idea of agency is both "impossible and imaginary."[10] And it cannot be both. For if, like the idea of a winged horse, it is imaginary and fictitious, there could be in principle a corresponding reality; and if, like the idea of a cube with eleven edges, it is impossible, there could be no corresponding reality: if it were both imaginary and impossible, it would be such that there both could be and could not be something answering to it. It is now easy to see that Hume's report of his procedure, instead of detailing an empirical search, describes one in verbal appearance only. An inquiry cannot even be begun, much less carried on, if its outcome is that there is no idea of what is being sought. There is no more looking for something answering to an idea that does not exist than there is opening a window that does not exist. There is no such thing as finding something that answers to a nonexistent idea, and neither is there such a thing as looking for something that answers to a nonexistent idea.

The cases considered, all of which appeared to involve either an experiment or careful observation or a search, have turned out, in agreement with Wittgenstein's claim about the nature of philosophical problems, not to be as represented. We may be puzzled to know what Hume's philosophical investigation actually is and what Berkeley's examination of ideas is, but it is certain that they are not the empirical inquiries it is natural to take them to be. Hume's demand to be shown an instance of causation over and above what are normally regarded as instances of causation, such as the action of one billiard ball on another, may be a semantic deception of a piece with the method he described. But on the surface at least it appears to be a legitimate request.

Wittgenstein's claim about the logical character of philosophical problems implies not only that philosophical investigations are not empirical but also that philosophical theories are not empirical. Without arguing the matter, it would seem clear that a factual proposition cannot be a proper answer to a nonfactual question. If, for example, the question raised by Zeno regarding the existence of motion is not empirical, the statement which Diogenes supported by standing up and walking in view of everyone would not be a relevant answer to the question. And we could understand why the problem was not thought solved, and also why we tend not to treat Diogenes' demonstration seriously, and even to take it as a joke.

A relevant answer to a philosophical question will have to share its logical character with that of the question; and if a philosophical question is not a request for empirical information, a philosophical theory does not supply empirical information. Kant argued for the view, still held by many philosophers, that a proposition could be both *a priori* and about things, which implies that it could be informative of what there is and also be nonempirical. There is no question that philosophical theories appear to be about the world, and it is widely taken for granted that they are. The possibility that suggests itself is that philosophical questions are requests for *a priori* information about things, and philosophical theories are logically necessary statements which are about things but do not require empirical testing.

PHILOSOPHY IS NOT A PRIORI

Wittgenstein once stated, "The great problem around which everything I write turns is: 'Is there an order in the world *a priori*, and if so what does it consist in?'"[11] His answer in the *Tractatus*, 5.634, was that "There is no *a priori* order of things." Both the philosophical question and the philosophical answer appear to be about the world; and if they are, the one is a request for *a priori* information which the other provides. In a number of places in his later writings he asserts, either explicitly or by implication, that the words "can," "must," and "cannot" when they occur in philosophical sentences are used in their logical sense: logical possibility, logical necessity, and logical impossibility.[12] A. J. Ayer's sentence, "It is logically impossible for a sense-experience to belong to the sense-history of more than a single self,"[13] is an illustration: what it says is equivalently and perhaps more naturally said by the sentence, "A sense-experience *cannot* belong to the sense-history of more than one self."

The importance of determining whether philosophical propositions are *a priori* in character, and make declarations about what is logically possible, impossible, or necessary, goes without saying. Before taking up this matter, the possible relationships between empirical propositions and logically necessary ones must be investigated. The question as to whether an *a priori* proposition can, in principle, tell us anything about the world also requires examination. And it is important to do this without making use of *philosophical* theories about the nature of necessitation, such as conventionalism and Platonism.

A logically necessary proposition, one which has its truth-value by inner necessity,[14] is such that its actual truth-value is its only theoretically possible truth-value. Using C. I. Lewis' symbol for logical possibility, "p is necessarily true" becomes "$\sim \Diamond \sim p$", and the point about the truth-value of an *a priori* proposition is made perspicuous in the following identities: $\sim\Diamond\sim p = \sim\Diamond(\sim p)$, $\sim\Diamond\sim(\sim p) = \sim\Diamond p$. By contrast, an empirical proposition is such that the truth-value it happens to have does not crowd out the possibility of its having the opposite truth-value. The fact that p is true does not eliminate $\Diamond\sim p$, and the fact that p is false does not eliminate $\Diamond p$; that is, $p.\Diamond\sim p$ and $\sim p.\Diamond p$ are consistent conjunctions. An empirical statement is one whose actual truth-value is not its only theoretically possible truth-value, and this is also the case with regard to contingent statements. This fact makes it natural to identify them. However, there is a special class of propositions whose actual truth-values are not their only possible truth-values, but to which some philosophers prefer not to apply the term "empirical." These are so-called basic propositions, such as *I am in pain* and *There looks to me to be a lake in the distance*, which although they could be false are not open to verification procedures by the persons asserting them. I can say with sense, "Perhaps my pain is caused by an abscessed tooth," but I cannot with sense say, "Perhaps I am in pain." Allowing the difference between basic propositions (which count as contingent) and nonbasic empirical propositions, the important thing for the present investigation is the property of possibly having a truth-value other than the actual one. In order to avoid a monotonous vocabulary, the words "empirical" and "contingent" will be used interchangeably.

The problem of determining how an *a priori* proposition might be related to an empirical proposition comes down, for the present purpose, to finding the answers to the following two questions:

1. Can an *a priori* proposition entail an empirical proposition? That is, can a proposition which has only one possible truth-value entail a proposition which has two possible truth-values?
2. Can an *a priori* proposition be entailed by, or be deducible from, an empirical proposition? That is, can a proposition which has two possible truth-values entail a proposition which has only one possible truth-value?

And there is a connected question, concerning the relation of an *a priori* proposition to reality:

3. Can a proposition which has only one possible truth-value present a claim about things?

To make a parenthetical remark about (3), if philosophical propositions are *a priori*, such that their truth-values are discoverable "without leaving concepts,"[15] and if logically necessary propositions are capable of conveying information about the world, then we can understand why philosophers have not found it necessary to make use of laboratories or scientific aids to the senses, such as microscopes or sonar devices.

The first of the three questions would seem hardly to stand in need of investigation. Hume stated that nothing contingent can follow from what is necessary. This would seem obvious, and it could pass without discussion except for the frequently asserted statement that a proposition that is necessarily true is also true as a matter of fact. The normal reading of the phrase "true as a matter of fact" is the reading it has in a sentence like "It is as a matter of fact true that Brutus stabbed Caesar." What is in fact true could in fact be false: it could, in principle, be true that Brutus did not stab Caesar. The implication of the claim that what is necessarily true is also in fact true is that logically necessary propositions have empirical consequences. This thesis underlies much of the reasoning of theologically oriented metaphysicians like Descartes, Leibniz, and Spinoza, who profess to demonstrate truths about the world from the necessary existence of a supreme being.

Looked at by itself, uncolored by philosophical theories in which it might be embedded, the claim that what is necessarily true is also factually true can readily be seen to be false and Hume's thesis to be correct. A proposition with only one possible truth-value cannot imply a proposition with more than one possible truth-value, for the reason that a necessarily true antecedent precludes any other than the implied truth-value from being a possible truth-value of the consequent. In other words, the truth-value imposed on what is implied by a necessary statement will be its only possible truth-value, which means that the statement is not empirical.

The second question, regarding whether it is possible for an empirical proposition to have *a priori* consequences, or whether a proposition with two theoretically possible truth-values can entail a proposition which has only one possible truth-value, has been answered in the affirmative by philosophers and logicians. Wittgenstein, for example, has said, "A tautology follows from all propositions,"[16] his reason perhaps being that a tautology "says nothing." Putting aside his special view about what a tautology says, a tautology has only one possible truth-value, and if it follows from all propositions, it follows from empirical propositions. An apparently plain case of a tautology being a consequence of a contingent proposition is a statement such as, *If poltergeists exist, then either poltergeists exist or poltergeists do not exist.* To many logicians the statement that a tautology, and in general any logically necessary truth, is deducible from a contingent proposition, presents itself as intuitively evident. C. I. Lewis has attempted to fortify this intuition with the following well-known demonstration (given here in somewhat abbreviated form):

If p is true, then either p is true and q false or p and q are both true:

$$p. \prec :p.{\sim}q. \vee .p.q \qquad (1)$$

The consequent of (1) is equivalent to and replaceable by

$$p.{\sim}q \vee q \qquad (2)$$

Since we have

$$p. \prec .p.{\sim}q \vee q \qquad (3)$$

we may infer

$$p. \prec .q \vee {\sim}q$$

Lewis goes on to say, "Thus tautologies in general are deducible from any premise we please: the theorem

$${\sim}\Diamond{\sim}q. \prec .p \prec q$$

states a fact about deducibility."[17]

Despite protestations to the contrary, maintaining that a necessary truth is deducible from any proposition, for example, that *A mouse is an animal* is deducible from, or can be "figured out,"[18] or calculated from, *There is a mouse in Westminster Abbey*, would be to introduce a paradox as violent as any encountered in philosophy. A logician who announces it is able to do so with no show of embarrassment only because it is tacitly understood to be a piece of *Alice Through the Looking Glass* logic, a game played with the term "deducible from," and is understood as such by those who debate it. There is no question that *A mouse is an animal* is strictly implied by *There is a mouse in Westminster Abbey* and that $(a + b)^2 = a^2 + 2ab + b^2$ is strictly implied by *Brutus stabbed Caesar*. But these implications are not paradoxical. They only produce an air of paradox when they are described as cases of deducibility, or entailment. How a logician can with complete assurance say, ". . strict implication, $p \prec q$, coincides in its properties with the relation 'q is deducible from p'"[19] very likely has the following explanation: to his explicit definition of "$p \prec q$" he has added an invisible Pickwickian redefinition of "deducible," which permits him to replace "$p \prec q$" by "q is deducible from p." Hidden redefinition is the recipe which makes the properties of strict implication coincide with those of deducibility.

Without attempting to give a full account of deducibility, which would require seeing how it works in various contexts, one general condition which a pair of propositions must satisfy in order for one of them to be deducible from the other needs to be highlighted. This is that the denial of the putative implication between them, i.e., asserting the antecedent and denying the consequent, gives rise to an inconsistency *between* them. Letting "ϕ" denote inconsistency, a conditional statement, If p, then q, will be a case of q following from, or being deducible from, p only if $p.{\sim}q$ implies $p\phi \sim q$. If there is no inconsistency between p and ${\sim}q$, regardless of any inconsistency within ${\sim}q$ itself, there is no entailment between p and q, and there is no deducing q from p. There is no inference-bridge *from* p to q unless p is inconsistent with ${\sim}q$. Where the inconsistency is absent, q will be logically idle, and plays no role in the deduction of q.

Consider again Lewis' demonstration of the theorem, ${\sim}\Diamond{\sim}q. \prec .p \prec q$, which, according to him, shows that strict implication is the same as entailment. It is readily seen that his proof is reversible. Beginning with $p \prec p{:}p. \prec .q \vee {\sim}q$, from which $p. \prec .q \vee {\sim}q$ is detached, we arrive at the first line, $p. \prec {:}p.q. \vee .p.{\sim}q$:

(a) $p \prec p{:}p. \prec .q \vee {\sim}q$
(b) $p. \prec .p.q \vee {\sim}q$
(c) $p. \prec {:}p.q. \vee .p.{\sim}q$

Read in reverse order, we can see what has happened. An implication which does not rest on an inconsistency between antecedent and denied consequent, viz., $p. \prec .q \vee {\sim}q$, rides in on an implication, i.e., $p \prec p$, which does rest on an inconsistency between antecedent and denied consequent. An implication which holds simply because the negation of its consequent embodies an inconsistency is carried to the conclusion, $p. \prec .q \vee {\sim}q$, by an implication-statement in the negation of whose consequent the antecedent is involved. A hitch-hiker is represented as a paying passenger by the use of a distributive rule of logic.

Whatever else may be involved, if there is a relation of deducibility *between* two propositions, there will be an inconsistency between the antecedent and the negated consequent and not only an inconsistency in the negated consequent. When both antecedent and consequent of a statement of the form *if p, then q* are empirical, the impossibility of its denial, ${\sim}\Diamond(p.{\sim}q)$, will rest on p's being inconsistent with ${\sim}q$; and the inconsistency, and also the logical impossibility, will vanish with the suppression of p. When p is contingent and q necessary, suppressing p clearly will not remove the inconsistency in ${\sim}q$. There is a special class of cases in which a necessary proposition appears to be entailed by a contingent one, for example, *if p, then* $p \vee \sim p$. Here the antecedent is a component of the consequent. The denial of this implication gives rise to an inconsistency that obtains between antecedent and negated consequent, between p and $\sim p.p$; and it needs no second glance to see that the inconsistency between them simply iterates the inconsistency within the negated consequent. We have $\sim\Diamond(p.\sim p.p)$, which may be rewritten $\sim\Diamond(\sim p.p.p)$, which is nothing more than $\sim\Diamond(\sim p.p)$. There are not *two* inconsistencies, one inconsistency between $\sim p$ and p and a second within $\sim p.p$; there is only one. The claim that $p \vee \sim p$ (or *if p, then p*) is deducible from p is only an apparent instance of a logically necessary proposition being entailed by a contingent proposition; for *if p, then* $p \vee \sim p$ is only an artificially expanded version of $p \vee \sim p$, and deflates into it. As for the general proposition, *if p, then* $q \vee \sim q$, where p is not a component of the consequent, there is no inconsistency between antecedent and negated consequent, and the implication is not an entailment.

An implication-claim the denial of which results in an inconsistency between antecedent and negated consequent which merely reduplicates an inconsistency within the negated consequent fails to satisfy a condition for one proposition to be deducible from another, and is not a correct entailment-claim. Thus, the statement,

If there is a mouse in Westminster, then a mouse is an animal

or the statement,

If there are four goats in the barn, then 2 goats + 2 goats = 4 goats

cannot correctly be restated as an entailment, such that its consequent follows from its antecedent, although it is correctly restatable as a strict implication. To make an immediate application of the principle that an *a priori* proposition cannot be deduced from one that is contingent, or that a proposition that has only one possible truth-value cannot be calculated from a proposition which has two possible truth-values, the metaphysical claim that the contingent is grounded in the necessary can be seen to be in obvious violation of this logical principle. All arguments which profess to demonstrate the existence of a necessary being from the existence of a contingent thing can be known to be invalid without examining them.

The third question which requires investigating is whether it is possible for an *a priori* truth, e.g., the proposition that a material thing can be in one place only at any given time, is capable of providing ontological information, that it can inform us of what there is in the world. Bertrand Russell has said that 2 + 2 equals 4 even far out in space, and the suggestion of his words is that the arithmetical truth states a fact about things in at least some regions of space in addition to our own region. It is not to be questioned that many philosophers have thought that at least some logically necessary propositions are about things and supply us with knowledge of them. Leibniz, for instance, distinguished between truths of fact and truths of reason, and held that the latter were true for all possible worlds. He also held that there was no real difference between the two kinds of truths, that a truth of fact was an infinitely complex analytic proposition—which would seem to imply that there is only one possible world. This idea perhaps underlies all philosophical thinking, empiricist as well as rationalistic thinking, and any other kind of thought that counts as philosophical. For the dream of the philosopher is to be able to learn what is in the Book of Knowledge, not by studying its pages, but by consulting his mind. Kant gave expression to this dream with his conception of synthetic *a priori* propositions, which unlike analytic propositions, may be called truths of reason which augment our knowledge of the things referred to by their subject-terms, and thus provide onotological information.

It would seem clear that at least some propositions which are true by logical necessity, namely, tautologies, make no declaration about the contents of any possible world. For instance, Wittgenstein maintained that tautologies "say nothing," and hence say nothing about the world. If we hold that they are true for all possible worlds, then we shall have to go on to hold that this is so because they are not falsifiable by any possible world; and they are not falsifiable because they say nothing about any possible world. It requires neither subtle reasoning nor acute perception to see that saying unicorns either exist or do not exist, or that if there are no unicorns, then there are no unicorns, is to say nothing whatever about the existence of unicorns. It does not serve as a possible answer to the question, "Do unicorns exist?" The same thing is seen to hold for tautologies of any degree of complexity. Consider for a moment the slightly more complex tautology, *If q, then either p or q*. To say, for example, that if pelicans exist, then either storks or pelicans exist, is to say nothing about the existence of either pelicans or storks, and hence is to say nothing about what there is in any possible world. It is the same with tautologies the components of which refer to properties of things, i.e., to the nature of what there is. The statement that Jones' necktie is lavender or not lavender says nothing about the color of Jones' necktie, and the statement that what Jones is wearing around his neck is a scarf or is not a scarf says nothing about the nature of what Jones is wearing around his neck.

Wittgenstein has said that tautologies are not pictures of reality,[20] that "They do not represent any possible situations,"[21] and this fits in with the idea that tautological propositions say nothing about what there is. The question now is whether any proposition that is characterized by logical necessity can be a "picture," or description, of things, i.e., whether a proposition whose actual truth-value is its only possible truth-value is prevented from saying anything about things by the way it possesses its truth-value. To put the question in another way: Do propositions which share with tautologies the attribute of being necessarily true, and therefore like them have no truth-conditions, also share with them the property of standing in no "representational relation to reality"?[21]

One reason for thinking that a proposition which is unconditionally true makes no declaration about any possible world is that it cannot, in principle, be falsified by any conceivable reality, nor can it be made true by any conceivable reality. For being unconditionally true, no condition of things and no state of affairs could *make* it true, and being necessarily true, no conceivable reality could upset it. The truth-value of a logically necessary proposition is not subject to change; hence, no imaginable change in the world can affect it. The conclusion to draw with regard to *a priori* true propositions would seem to be that they are ontologically mute. Their truth-value is not determined by what the world is like, and, therefore, they say nothing about what the world is like: they do not "stand in any representational relation to reality." The consequence of this line of reasoning is that such an assertion as "A shilling cannot be in two different pockets at the same time" says nothing about where a shilling cannot be; it says nothing about shillings. And the assertion that a mouse is an animal says nothing about what a mouse is; it says nothing about mice. Russell's remark that 2 plus 2 equals 4 even far out in space makes no factual claim about the relationship between the arithmetical equation $2 + 2 = 4$ and its range of operation in space.

Some philosophers contend that a distinction is to be made amongst the properties of things, which if correct implies that a proposition can be both about things and *a priori* true. It is easily seen that Kant's conception of synthetic *a priori* propositions implies the Aristotelian view that some amongst a thing's properties are logically essential to it, without which the thing cannot be. Consider the following statements:

Every effect has a cause.
Every change has a cause.
Every change is abrupt.

The first statement is analytic and only iterates a component of the subject-term. The second, according to Kant, makes an attribution; it ascribes a logically necessary property to what is referred to by the subject-term. The third makes a contingent attribution, ascribes a property to occurrences which they could fail to have. The first is unfalsifiable in every possible world, because it, like the tautology that if there are goats, then there are goats, says nothing about any possible world. The second says what holds good for all possible worlds: it gives expression to an informative cosmic invariant. The third is a truth for only some possible worlds.

C. I. Lewis takes the proposition, *A cat is an animal,* to be analytic, but he apparently construes it to be a proposition which attributes a property to a thing. *A cat is an animal* is represented as being a subject–predicate proposition on a footing with *A cat is a mouser,* the difference between them being that the first ascribes an essential property and the second a nonessential, or "accidental," property to the subject. Lewis' idea is that the claim made by the proposition, *A cat is an animal,* "can be assured by reference to the meanings of 'cat' and 'animal' without recourse to further empirical evidence. But it also might be established—as well established as most laws of science—by generalization from observed instances of cats."[22] It is not easy to see how the truth-value of a proposition could be determined *both* by an examination of the meanings of the words used to express the proposition and also by an examination of things to which the subject term refers. Putting this aside for the moment, what is clear is that some philosophers have the idea that certain properties of a thing are entailed by it, as against other properties which characterize it but are not logically essential to it.

A feature that is entailed by a thing will have to be essential to the *nature* of the thing, such that to suppose something is of that nature but lacks the feature is to entertain an inconsistent concept. It is like supposing that there is a cat which yet is not an animal or that there is a cat which is in more than one place at a time. The expression "the nature of a thing" carries with it the possibility of endless philosophical obfuscation. Here it means nothing more esoteric than the concept under which a thing falls or to which it answers. Talk about concepts under which things fall is a nonverbal, substitutive way of talking about words which apply to the things. It is a linguistically oblique way of talking about words which are general names of things, words like "cat" and "shoe." And a feature or a property that is essential to a thing, or is entailed by the concept to which the thing answers, is a criterion for the application of a word that is the generic name of the thing. In general, the rules for the use of a term determine the *meaning* of the term, which is nothing in addition to them. It is this linguistic fact which made it possible for Wittgenstein to give us the formula for avoiding a part of metaphysics: Don't ask for the meaning of a word, ask for its use.

It will be clear that an essential or an entailed property of a thing of a given kind is a criterion for the application of a general word to things of that kind, which is to say that its presence makes it correct to apply the word to them. A term which denotes an entailed property, e.g., "animal," unlike a term which denotes a nonessential property, e.g., "mouser," does not have a use to describe or characterize what answers to a general name such as "cat." Talk about essential properties is an ontologically formulated way of talking about verbal matters, a way which makes it look as though attributes of things are being referred to. The fact that the sentence, "A cat is an animal," expresses a logically necessary proposition is equivalent to the fact that the sentence, "'Animal' applies to whatever 'cat' applies to" expresses a true verbal proposition. The verbal sentence makes it plain that the nonverbal equivalent does not attribute a property to a thing. The fact that talk about essential properties comes down to talk about the application of terms makes it an impropriety of language to say "*This* cat is an animal and *that* one is an animal, too," as against saying "*This* cat is a mouser and *that* one is a mouser, too."

We can now see what is at work behind Lewis' contention that the proposition, *A cat is an animal*, is open to verification by observing cats. Whether by mistake or by unconscious intent, Lewis takes a word denoting a feature whose presence determines the correctness of applying a term to a thing to be a word for a characterizing property of what is referred to by the term. Lewis' conventionalist claim that the truth-value of the proposition expressed by the sentence "A cat is an animal" can be ascertained by reference to the meanings of "cat" and "animal" comes nearer the truth: "animal" gives expression to a feature whose presence is required for the correct application of "cat." This is the reason why the phrase "cat but not an animal," unlike "cat but not a mouser," has no descriptive function in the language.

The view that amongst the properties of an object, such as a cat or a tree, there are some that are logically necessary to it is either the result of a semantic misrepresentation or of a grammatical alteration of some sort. The philosophical view about the essential nature of things fades away when words denoting essential properties are seen in a correct linguistic light, i.e., when their actual use in the language is made perspicuous. What we are left with, after the philosophical impediment has been removed, is the idea that a logically necessary proposition has no ontological import and says nothing, either true or false, about things. Regardless of appearances, a proposition whose actual truth-value is its sole possible truth-value makes no reality-claim, either directly or by implication. It neither entails nor is entailed by any proposition which makes a factual assertion about the world. The outcome of this for philosophical statements, construed as *a priori*, is as disconcerting as it is perplexing: they say nothing whatever about the reality to which they appear to refer. With respect to things, actual or possible, existent or nonexistent, all necessary statements share with tautologies the feature of being ontologically uninformative.

It is hard not to think that the goal of philosophy is knowledge of reality. But if that is its true goal, philosophers can only gain knowledge of reality by recourse to empirical investigations of things. And if neither the methods nor the theories of philosophy are empirical, then philosophers live in a dream. Spinoza's conception of a systematic *a priori* science of what there is will have to be discarded as an intellectual chimera. If, however, philosophy is an empirical discipline, it is strange and mystifying that it has developed no experimental nor observational techniques, especially since in our own time the natural sciences have made such remarkable advances in techniques of investigation.

Two Faces of Philosophy

Philosophy appears to present two entirely different faces to philosophers themselves: an empirical face and an *a priori* face. But neither of the Janus faces gives a correct picture of philosophy. The empirical pictures the theories bring up in our minds do not portray what actually takes place in the doing of philosophy. The *a priori* picture equally misrepresents, or better still, conceals the nature of philosophical activity. One piece of evidence against the view that represents philosophical activity as being like that in mathematics is that philosophical theories and their supporting arguments have remained in debate for so long. Considered as an *a priori* discipline, one thing that stands out in stereoptican relief is that it is a riot of entailment-claims. This by itself is a sufficient reason for rejecting the view that philosophical utterances are *a priori*. The utterly strange fact that philosophers show no discernible anxiety about the chaotic condition of their subject and, moreover, have no realistic expectation that it will ever be anything more than an expanding conglomerate of *sic et nons*, fortifies the evidence against the *a priori* view.

We can now say what a philosophical theory is not, but we are not yet in a position to say what it is. One thing is certain. This is that a philosophical utterance is somehow able to create the appearance of stating a fact about the world, an appearance so convincing that it has the power permanently to bewitch the mind. What stands in need of explanation is the mechanics of this beguiling appearance, i.e., the semantic work which goes into its production. A further thing which requires an explanation is what it is in the *nature* of philosophical problems which makes it possible for endlessly unresolved disputation to revolve around them.

The later Wittgenstein has expressed two views about the nature of philosophical statements. One is that a philosophical statement describes or misdescribes the actual use of language. Thus, the statements that motion is impossible and that we do not really see things, misdescribe the use of terms in everyday speech. The other view, which has not been investigated by philosophers, represents a philosophical theory as introducing semantically gerrymandered usage. On this account the two statements that motion does not exist and that things are not really seen present revised pieces of terminology: the one declares the suppression of the word "motion" and the other presents a contracted use of "sees" which prevents its application to things.

One explanatory advantage which the second view has over the first is immediately evident: it enables us to explain the permanent elusiveness of *all* of its problems, however minor. Endless disputation is made possible because no fact, ontological or linguistic, is in contest. The question debated by William James' learned colleagues as to whether the man who goes around the tree also goes around the squirrel which goes around the tree while constantly facing him is capable of endless disputation because the question is neither about what occurred nor about the actual use the term "goes around" has in the language. The question is a request for an academic decision as to whether a known expression should be applied to an unusual case, one which differs in certain respects from a normal case. Depending on whether a person feels the difference to be important or unimportant, he will decide one way or the other. It is taste, to use Hume's word, not fact, which determines one's answer to the question. It is the same with a philosophical answer. A philosopher who asserts "No one really sees things" has made the decision to banish from the language, in what Moore calls a "philosophic moment," the phrase "sees things." His view gives expression to a semantic preference, and opposing views give expression to opposing preferences, all this being part of a language game. It is understandable that such a game can go on endlessly. And it is understandable also that the semantic game could, because of the ontological mode of speech in which it is formulated, be taken as scientific disagreement about fact. Wittgenstein has observed that philosophers suffer from the confusion of thinking themselves to be upsetting a fact-claiming proposition when what they are actually doing is "rejecting a form of symbolism."[23] They imagine that a philosophical problem "concerns a fact of the world instead of a matter of expression."[23]

Consider now Russell's assertion that there are no illusions of sense. He wrote: "The first thing to realize is that there are no such things as 'illusions of sense.' Objects of sense, even when they occur in dreams, are the most indubitably real objects known to us."[24] If we read these words literally, and take the expression "illusion of sense" to have its familiar use, their implication is that nothing in fact occurs to which the expression "illusion of sense" applies. Plainly, it *could*, as a theoretical possibility, be the case that sense illusions do not occur, that no one dreams, that sticks partly immersed in water never look bent, and that magicians actually, and not just in appearance, saw ladies in half. But it is not to be thought for a moment that Russell wished to imply anything of the sort. We cannot construe his words, "The first thing to realize ..." as being intended to call our attention to a fact we might have overlooked and which we might wish to check. He is not using the phrase "illusion of sense" in the normal way and claiming that we are mistaken about the nature of the cases to which we apply it. When he tells us that the sense objects that occur in our dreams are "real" we cannot rightly understand him to be saying that the people and things we dream about are actually present in our dreams, that images in our dreams are the things they represent. Russell has written that only a philosopher could fail to distinguish a post from the idea of a post,[25] and although Russell himself was a philosopher we cannot think that on this point he suffered confusion. He is not using the expression "illusion of sense" in the accepted way, which he knows as well as anyone else, nor is he misdescribing the use it has. Instead, he is keeping ordinary language and putting another beside it. He academically redistricts the phrase "illusion of sense," reduces its range of application to zero; and he does this in the form of speech in which nomenclature is not mentioned. Keeping ordinary language and adding his contrived expression to it are sufficient to fashion a semantic trap[26] which has held philosophers captive from Protagoras on.

With the term "object of sense" Russell does a comparable thing: he subjects it to semantic manipulation. The expression normally is used to refer to objects like shoes, trees, and automobiles, in other words, to things we are said to perceive. Russell stretches its use to cover mental images, dream contents, afterimages, etc. Doing this amounts to stretching the application of the word "object" or "thing" to cover what in everyday language do not count as objects or things. When Hume says that "every perception is a substance, and every distinct part of a perception a distinct substance,[27] he makes the same maneuver with the word "substance." Instead of advancing a theory about the nature of images, tastes, smells, and the like, he presents a rigged use of the word "substance." He places his semantic contrivance alongside of everyday speech, and in this way creates an erroneous but powerful idea in the minds of his audience and in his own mind as well. By means of unwitting linguistic duplicity he generates the intellectual illusion that he is revealing a discovery about the inner nature of substance.

Consider, next, Russell's statement, "We cannot point to a time itself, but only to some event occurring at that time," from which he infers that " ... there is no reason in experience to suppose that there are times as opposed to events ..."[28] It is not clear from these remarks whether Russell wishes to say that we are unable to point to "a time itself" because there are no "times as opposed to events," or urges this for another reason. One cannot point to a goat that does not exist and one cannot point to a goat that is hidden from view. The first *cannot* is the *cannot* of logic: it is logically impossible to point to a nonexistent thing. The second *cannot* is the physical *cannot:* it is physically impossible to point to something if one's hands are tied or if a high wall hides it from view. But what we are physically prevented from doing, we *could* in principle do.

To leave this point for a moment, it will be evident that the nonexistence of "a time itself" is not a correct inference from our inability to point to it, any more than the nonexistence of a goat can be a correct inference from the fact that we are prevented from pointing to it by its being hidden. In general, if it is argued that it *follows* from our not being able to point to something that it does not exist, then the reason for our not being able to point to it is that it does not exist. Russell has turned the argument around and put the cart before the horse: what he presents as the premise of his argument should be the conclusion and what he presents as the conclusion should be the premise. His reason for maintaining that we cannot point to "a time itself," as opposed to an occurrence or an object, is that it does not exist. But now what could be meant by saying that "a time itself" does not exist and that for this reason we cannot point to it? He would not say that it might exist, and that if it did exist we could point to it, just as we could point to a centaur if it did exist or to an avalanche if it occurred. We have to suppose that what he wishes to convey is that a time, as opposed to something which exists or occurs at that time, is such that by its *nature* it cannot be pointed to. Some metaphysicians have argued that time is self-contradictory, which would seem to imply that time, or a particular time, can no more be pointed to than can a seven-faced cube. But Russell has something else in mind, which is indicated by his assertion that we can point to an "event occurring at that time." The implication is plain: we can point to things and events, but not to time. And this is so not because time, or a time interval, does not exist, but because time cannot by its very nature be pointed to.

The explanation of why we cannot point to time, or to a time interval, is not hard to discover. Time is not a thing nor is it an occurrence or a process, which is to say that the word "time" is not the name of a thing or of an occurrence. Wittgenstein has observed that when we come upon a substantive we tend to look for a substance,[29] and poets and mystics attest to the substantial correctness of this observation when they speak in various ways of time as an occult cosmic river. It is logically impossible to point to what is denoted by the word "time" and by related temporal nouns, because temporal terms are not object-denoting terms like "brick" or "explosion." The reason for stating that we cannot point to "a time itself" as opposed to an occurrence at that time is to highlight this fact. As temporal nouns are used in the language it makes no literal sense to speak of pointing to what they refer to. One philosophical way of marking the difference between what might be called substantival nouns and grammatical nouns is to suppress grammatical nouns, academically exorcise them from the language. Maintaining that time or that a time by itself does not exist is an ontological way of expressing the decision to eliminate temporal nouns. Again, to express it in Wittgenstein's way, Russell puts his linguistic innovation alongside of unaltered everyday language, in which temporal terminology functions in the usual ways; and doing this creates the idea that the nonexistence of time is being declared.

To return for a moment to the metaphysical proposition that the contingent is grounded in the necessary. It was stated earlier that the metaphysical proposition stands in violation of the principle that an empirical statement cannot entail a logically necessary one. But this is not enlightening. Some able philosophers have the idea that unlike "exists," which has no use to describe what it refers to, the term "necessarily exists" does have a use to characterize some things, a use to say what they are in addition to saying that they are. Kant wrote:

> . . . if I take the concept of anything, *no matter what*, I find that the existence of this thing can never be represented to me as absolutely necessary, and that, whatever it may be that exists, nothing prevents me from thinking its nonexistence . . . If I am constrained to think something necessary as a condition of existing things, but am unable to think any particular thing as in itself necessary, it inevitably follows that necessity and contingency do not concern the things themselves; otherwise there would be a contradiction.[30]

It is not clear what Kant means by the words "contingency does not concern the things themselves," but he has made plain what construction is to be placed on the words "necessity does not concern things." This is that "necessarily existing thing" is an expression which has been given no application to anything, "no matter what." The noun term "necessary being" has been given no descriptive or characterizing use, and its function in the sentence "the contingent is grounded in the necessary" is wholly grammatical. It functions as a grammatical, not as a semantic, noun term.

Kant's remark that contingency does not concern things is puzzling since he states that ". . whatever it may be that exists, nothing prevents me from thinking its nonexistence." The implication is that being a thing *entails* being such that it could fail to exist. The further implication is that the concept of a necessarily existing thing implies an impossibility. The implied verbal fact is that the phrase "necessarily exists" has no application to *things*. Now then, if one of a pair of antithetical terms is denied a use in the language, the other loses its use. Kant appears to give indirect recognition to this linguistic fact by maintaining that neither "necessary" nor "contingent" has a use to characterize things: the claim that "necessary" has no such use implies the claim that "contingent" has none.

We are now able to understand how terminology is being used in the sentence "The contingent is grounded in, or implies, the necessary." What has happened is that metaphysicians have stretched the ordinary use of "contingent" (which normally implies being dependent on or determined by something) in such a way that it applies to things which entail the *logical* possibility of their own nonexistence. Now this stretched application of the word "contingent" is used to justify the introduction of the antithetical expression "noncontingent thing" or "necessary being." All this is done without assigning a new meaning to these terms or giving them actual applications. A familiar word in everyday use is by academic fiat reedited, and it is used to introduce a manufactured antithetical term, one which is descriptively idle. This is done in a form of language which creates the impression that the nature of things is under investigation by a higher-order science. The philosopher may truly be described as practicing sleight-of-hand semantics.

References and Notes

1. L. Wittgenstein, *Philosophical Investigations* (Basil Blackwell, Oxford, 1953), p. 47.
2. C. I. Lewis and C. H. Langford, *Symbolic Logic* (The Century Co., New York and London, 1932), p. 447.
3. L. Wittgenstein, *The Blue Book* (Basil Blackwell, Oxford, 1958), p. 42.
4. Reference 2, p. 477. In Langford's way of speaking, "unitary meaning of sentence *S*" = "proposition expressed by *S*".
5. G. Berkeley, *Principles of Human Knowledge*, A. C. Fraser, editor (Oxford, Clarendon Press, 1871), Part First, Sec. 24.
6. Reference 5, Sec. 25.
7. D. Hume, *A Treatise of Human Nature*, L. A. Selby-Bigge, editor (Oxford, Clarendon Press, 1888), Bk. I, Pt. III, Sec. XIV, p. 155.
8. Reference 7, p. 159.
9. Reference 7, p. 161.
10. Reference 7, p. 158.
11. L. Wittgenstein, *Notebooks 1914–1916*, transl. G. E. M. Anscombe (Basil Blackwell, Oxford, 1961), p. 53e.
12. Reference 3, pp. 55–6.
13. A. J. Ayer, *Language, Truth and Logic* (Victor Gollancz, London, 1951), second edition, p. 125.
14. Kant's phrase.
15. Kant's phrase.
16. L. Wittgenstein, *Tractatus Logico-Philosophicus* (The Humanities Press, New York, 1961), 5.142.
17. Reference 2, p. 251.
18. One child's expression for obtaining a conclusion from premises.
19. Reference 2, p. 251.
20. Reference 16, 4.462.
21. Reference 16, 4.462.
22. C. I. Lewis, *An Analysis of Knowledge and Valuation* (Open Court Publ. Co., LaSalle, Illinois, 1946), p. 91.
23. L. Wittgenstein, *Wittgenstein's Lectures, Cambridge, 1932–1935*, Alice Ambrose, editor (Rowman and Littlefield, Totowa, New Jersey, 1979), p. 69.
24. B. Russell, *Our Knowledge of the External World* (Open Court Publ. Co., Chicago, Illinois, 1914), p. 85.
25. B. Russell, *The Principles of Mathematics* (W. W. Norton and Co., New York, 1938), second edition, p. 451.
26. Wittgenstein's "fly bottle."
27. Reference 7, Bk. I, Pt. IV, Sec. V, p. 244.
28. B. Russell, *Our Knowledge of the External World* (Open Court Publ. Co., Chicago, Illinois, 1914), p. 117.
29. Reference 3, p. 1.
30. I. Kant, *The Critique of Pure Reason*, transl. Norman Kemp Smith (Toronto, Macmillan, 1929), p. 515. Italics mine.

7

Social Exemplifications of Physical Principles

R. Bruce Lindsay
Brown University
Providence, Rhode Island

Many years ago in the brashness of youth the writer prepared an article on the possible use of physical principles and concepts in the description and understanding of social phenomena.[1] He called attention to the earlier efforts of social scientists like August Comte, Herbert Spencer, and Lester F. Ward to apply physical concepts and laws more or less directly to sociological explanation. He emphasized the difficulties encountered in the use of such analogies, e.g., the attempt to introduce a "social force" analogous to "force" in mechanics, overlooking the highly specific meaning attributed to the term in physics, not always clearly grasped by the nonphysicist and indeed for a long time not even too precisely clear to many physicists and engineers. It was only later that his attention was drawn to the ideas and criticisms of Vilfredo Pareto (1848–1923), who in his monumental work *Trattato di Sociologia Generale* (1916)[2] stresses in great detail the same difficulties.

In the writer's 1927 article above referred to, he proceeded somewhat blithely to ignore the doubtful value of looking for social analogies of physical principles and sought to find in social phenomena illustrations of the Le Chatelier–Braun principle in physics, usually called the principle of mobile equilibrium or the generalized law of inertia. This states that if any one of the factors determining the state of a physical system in equilibrium is altered by external action, the other factors will then change in such a way as to oppose the change in the first. The writer sought social examples of the application of this principle in the proverbial inertia exhibited by human institutions, particularly with respect to attempts at social reform. The idea then seemed plausible and persuasive and received a little favorable attention from some sociologists. As a suggestive way of describing important aspects of social phenomena, it seemed to carry a kind of common sense tag. As a way of providing a deeper understanding of social processes it seemed to possess some interest, but its prescriptive value in the actual handling of social problems appeared dubious. The writer ultimately came to the opinion that progress in sociological theory is best achieved not by trying to carry over the concepts and laws of physics into sociology but by the development of concepts and theories based on the direct observation of social phenomena plus some philosophical imagination on the part of the sociologist.

But over the years there has recurred from time to time the thought: do not social exemplifications of physical principles have some value for somebody? And is not this value to be found in their heuristic and pedagogical suggestions? In other words, can such social exemplifications of physical principles illuminate our understanding of the principles themselves and lead us to ask more intelligent questions about them? Can they possess pedagogical value for those striving to understand physics without the intention of becoming physicists or even scientists of any sort? Affirmative answers to these questions will justify attention to the social examples even if their actual use in the development of social theories is questionable. It is the purpose of the present essay to suggest that such social exemplifications of physical principles do indeed often have heuristic and pedagogical value. We begin with the principles of thermodynamics, a fruitful source of social examples.

The Principles of Thermodynamics

Consider at the outset the "first law" of thermodynamics, more appropriately called the first principle, namely, that of the conservation of energy. This states that given a closed system of particles of any physically conceivable kind, charged or uncharged, and capable of interacting with each other in any fashion, as long as the system is completely isolated from its surroundings, its total energy remains constant. Alterations in the internal state of the system can take place, leading to transformations of energy from one form to another, as for example, in the production of heat or in various kinds of chemical reactions, both nuclear and extranuclear, but the total energy remains unchanged. The German physicist Rudolf Clausius (1822–1888) felt it important so to generalize the principle as to make it applicable to the universe as a whole and so stated it as follows: "Die Energie der Welt ist konstant," a statement placed by J. Willard Gibbs (1839–1903) at the beginning of his famous memoir "On the Equilibrium of Heterogeneous Substances" (1876). It must be confessed that not all professional thermodynamicists have been hospitable to Clausius' extrapolation. In any case it must also be emphasized that in neither its restricted or extended forms has the principle ever been established as an experimentally verified law, except for simple, ideal classical dynamical systems subject to certain specific kinds of forces. In its general content it has proved to be a remarkably useful hypothesis, analytical deductions from which have never failed experimental verification in the cases in which experimental test has been possible. Though the principle is often referred to as the "first law of thermodynamics," it is not a law in the usually accepted sense in the methodology of physical theory. We must indeed recognize that the principle is in fact a tautology: it is really a *definition* of energy as an entity representing a kind of constancy or invariance in the midst of changes. In the historical development of physics much attention has been paid to the search for invariants in physical phenomena. Of all such invariants that have been turned up, energy has proved to be the most fruitful. It would appear that a principle displaying the generality of energy conservation should enjoy many modes of expression; this has turned out to be the case. Many of these modes have direct relevance to the behavior and fate of man in society. Of all these, one of the most suggestive is the so-called "principle of compensation," made famous by the Sage of Concord, Ralph Waldo Emerson (1803–1882).[3] Though the essay in which the principle is enshrined displays much of the author's pleasingly cryptic style, what the principle says in effect is: for every plus in human experience there is a minus; associated with every good there is a compensatory bad; every encountered advantage is accompanied by a corresponding disadvantage. This is an exemplification in human experience of the principle that in the physical universe every appearance of energy in one form is compensated by the disappearance of energy in another form. We can find many other examples of this compensatory effect in human experience, often indeed referred to in common parlance as the inevitable "ups and downs" of human existence. We are continuously advised that it is wise to "take the bitter with the sweet as the sweet and bitter come." Consider the fact that no one has ever been able to invent a gambling game in which everyone concerned wins: if someone wins, someone else has to lose. A somewhat more flippant way of expressing the social content of the first principle of thermodynamics is: "there is no such thing as a free lunch." (*Someone* always pays!) The current humorous literature is full of such cliches as: "You cannot win," which also exemplify the first principle. Of course, the realization of this inexorable limitation imposed on all human activity does not keep people from gambling. Indeed, one may confidently state that an all-pervasive theme of all accounts of human activity in newspapers and novels is the attempt of the individual to somehow circumvent the application of the first principle of thermodynamics to himself or herself. The same point crops up in one of the most emphatic versions of the first principle, namely, that found in the Bible in the third chapter of the Book of Genesis (19th verse). Here we find the Lord admonishing the sinners of the garden, who have eaten the forbidden fruit, in the famous words. "In the sweat of thy brow shall thou eat bread." What better statement of the first principle can you find? Here again we encounter the everlasting human effort to get around the laws of nature: how many there are who struggle everlastingly to ensure that they eat their bread in the sweat of someone else's face! In any case, we may be sure of one thing! There must be sweat (actual or metaphorical) in someone's face if our so-called civilization is to endure.

Before we take up the question of the heuristic and pedagogical value of such human experience expressions of the first principle of thermodynamics let us take a look at the second principle, commonly called the second law, the law that has put entropy on the map, or at any rate on the backs of motor cars, where we read "Save

America From Entropy." Involved here is a rather more subtle and not too generally understood situation. It will be recalled that Clausius sought to apply this principle also to the universe as a whole, phrasing it as follows: "The entropy of the universe strives toward a maximum value." Accepting this universal applicability (recognizing, of course, that many if not most professional thermodynamicists do not; why should they?—they are not interested in the universe, but in applying thermodynamics to engines and chemical reactions), and ignoring for the moment some very important purely local exceptions, we can state the principle roughly in common terms as follows: "Whenever a transformation of energy takes place it is accompanied by a certain loss in the ability to repeat that precise transformation." This loss, which someone has epitomized as the lost opportunity to do work, when quantized corresponds to the associated increase in entropy. From the standpoint of human beings the most important exception to the principle is found in the local behavior of living things that can "consume" entropy or undergo biological changes in the process of their production and maintenance in which there is an actual *reduction* in entropy locally. But in the interactions of human beings with their natural environment and with each other the second principle predicts an inevitable and irrevocable increase in entropy. This energy degradation accompanying all natural energy transformations leads, of course, to the eventual "solution" of the energy problem, namely, the heat death of the universe envisioned, perhaps with some relish, by Clausius. So far as our planet earth and its fundamental source of energy, the sun, are concerned, even in the relative short term, this increase in entropy is not ignored, for the total amount of energy transformation available to this system is finite. We are, of course, here leaving out of account the possibility that earth's inhabitants will ultimately migrate to another planet in their endeavor to escape momentarily from the inexorable effects of the second principle.

It is scarcely surprising that the significance of the principles of thermodynamics for human beings has found its way into the vernacular. Thus, it has been suggested that if the first principle can be epitomized in the warning "You cannot win," the second principle may equally well take the form: "You cannot even break even!" This emphasizes that no matter what the social milieu the inescapable degradation in the possible transformation of energy in our experience clamps a boundary condition on all social progress. From the long point of view this may be looked upon as an exemplification of the enormous odds faced by human beings in the attempt to establish a satisfactory way of life for all people dwelling on our planet.

The present writer has taken a rather broad view of the significance of the second principle of thermodynamics in the life of men.[4] He prefers to consider the whole development of civilization as an attempt by mankind to counter the effect of thermodynamics. For the very creation of what we call civilization may be considered as a process of moving from more probable, primitive states of existence to less probable, sophisticated states. The latter could only have been brought about by man's interference with his environment—they could not be expected to happen just by themselves! The second principle implies that on a statistical basis states of existence inevitably move from low probability to higher probability of occurrence. These considerations led the writer to suggest a new ethical imperative for individual action, to be called the "thermodynamic imperative": *act in all things so as to consume as much entropy as possible by introducing as much order as possible in your life and eschewing disorderly and random conduct calculated to break down the regulated behavior on which our whole civilization rests.* Pains have been taken to emphasize that this imperative, like the Golden Rule or Kant's categorical imperative, is urged on human beings as individual persons, who obviously have a right to treat it as they would any other ethical imperative, i.e., endeavor to follow it or ignore it. It is patent to all that ignoring it is indeed the life style of countless numbers of people. It should be stressed that in offering this imperative the writer is suffering under no particular illusion of piety. He feels merely that in promoting an imperative suggested by one of the most fundamental principles of natural science he may be helping to restore some of the humanism which many have felt has been sorely lacking in scientific theories and principles.

We now pause to consider the question: Do the exemplifications of the principles of thermodynamics in human experience and social relations possess any heuristic and educational value? From the standpoint of the physics, chemistry, and engineering student thermodynamics is a difficult discipline. Anything that can enable the student to grasp more readily the fundamental ideas involved is certainly worthy of attention. As a human being the student is well equipped to feel at home in the field of social relations since they are a vital part of his or her everyday experience. Hence a study of the examples of thermodynamics in social phenomena can reinforce the student's grasp of the subject. At the same time, these exemplifications can encourage the raising of valid questions about social problems (e.g., the value of the thermodynamic imperative) which might otherwise never be considered. Hence, the heuristic value of the process! It is scarcely necessary to emphasize the need for the use of great care in order to avoid misconceptions. But that is the function of a good teacher.

Let us now go on to consider other illustrations of our theme.

The Ergodic Principle

Classical statistical mechanics deals with the behavior of dynamical systems possessing a very large number of degrees of freedom or with ensembles of many systems.[5] In the so-called microcanonical ensemble of Gibbs, for example, its phases are confined to a shell in phase space characterized by a range of energy values from E to $E + \Delta E$. The phase of each system in the ensemble is given by the total number of configurational coordinates and component momenta associated with the degrees of freedom of the system. As time passes each system follows a path in the phase space to which the ensemble is limited. The principal task of statistical mechanics is to evaluate the time average values of quantities characterizing the systems in the ensemble. It does this by means of the fundamental postulate that this time average in the case of any particular quantity is equal to the average of the same quantity over the phase space. In order to make this postulate plausible Maxwell and Boltzmann felt it necessary to assume that in general the phase curve of every system of constant energy must eventually pass through every point of the phase surface corresponding to this energy. Boltzmann called this surface an *ergoden* or energy surface. In English this has been termed an ergodic surface, and the assumption made by the two founders of statistical mechanics became known as the ergodic hypothesis.[6] From the beginning doubts arose about its plausibility. Even Maxwell realized that there are special cases of motions of systems for which it clearly will not hold, but he felt that for a confined system of particles colliding not only with each other but also with the walls of the confining vessel, sufficient randomness is inherent to make the hypothesis plausible.

With the development of more precise mathematical concepts of the density of points in a continuum it could be proved that the strict ergodic hypothesis of Maxwell and Boltzmann is false. In treatises on statistical mechanics it was then replaced by the so-called "quasi-ergodic hypothesis," according to which there exist systems of constant energy of many degrees of freedom for which the representative path in phase space passes eventually infinitely close to every point on the ergodic surface. During the past 50 years there has been much discussion and some controversy about the state of the quasi-ergodic hypothesis. It is improbable that the precise mathematical problem connected with this hypothesis will be solved in a hurry.[7] However, the very existence of statistical mechanics as a viable and fruitful physical theory with a host of successful applications in the calculation of physical quantities of importance in thermodynamics provides a heuristic reason for accepting its basic principles at any rate as acceptable approximations. Physicists have for a long time put up with the situation that practically nothing in the foundations of physical theories can meet the meticulous demands of fundamental logical analysis as envisaged by pure mathematics. Yet from a pragmatic standpoint statistical physics "works" as well as other successful physical theories. In the light of this it seems that we are justified in erecting the quasi-ergodic hypothesis to the rank of a principle, with due regard to its obvious limitations, and hence talk about the "ergodic principle."

We certainly do not have far to go to find a social exemplification of the ergodic principle. It takes the following form: "Whatever can happen will happen if you wait long enough." Everyone is familiar with the corollary known in the vernacular as Murphy's law: "If anything can go wrong, it will, if you wait long enough." The genuine pessimist is inclined to supplement this statement with the clause: "and you usually don't have to wait very long!" In order to bring these statements more in line with the quasiergodic hypothesis we ought perhaps to say: "*Almost* anything that can happen will happen if you wait long enough." The polar ice caps can certainly melt under appropriate terrestrial conditions as brought about by the behavior of the sun; but our human experience has simply not been long enough to see it happen. Examples of the principle abound in our experience in the shape of phenomena commonly called "accidents" because they do not occur as part of the regularly accepted pattern of events in our experience like sunrise and sunset. They happen when two "world lines" come sufficiently close together. Of course, there are plenty of "near misses," as when, for example, we narrowly escape seriously cutting ourselves when shaving. These can be looked upon as corresponding to the *quasi* aspect of the ergodic hypothesis. When Mr. Micawber in Dickens' famous novel was forever waiting for "something to turn up," to take care of his financial difficulties, he was exemplifying his confidence in the plausibility of the ergodic hypothesis. Almost everything in our daily life experience can provide an illustration of this famous principle and there are, of course, many vernacular expressions of it. The more random our behavior, or as we say, the less care we display in going about our business, the more likely we are to wind up as horrible examples of the principle in action. Our eventual death provides the crowning illustration of its significance. It is true that while we live, if we are cautious enough, we have the chance to prolong the time intervals between our human "phases" so as to minimize encountering the unfortunate phase points of life we call "accidents." in this sense we can, as it were, fight against the results of the ergodic principle in somewhat the same way in which we can endeavor to combat the impact of the second principle of thermodynamics. In neither case, of course, can we ever win, but the fight itself should constitute a sense of satisfaction.

No close inspection is necessary to make one aware of the fact that physics provides many more principles whose social exemplifications are worthy of consideration. They are not all indeed equally persuasive. Let us next consider the principle of complementarity.

Complementarity

Niels Bohr (1885–1962) was led by reflection on the principle of indeterminacy of Heisenberg to develop his so-called principle of complementarity.[8] It will be recalled that according to the Heisenberg principle, given two canonically conjugate variables like the position and momentum of a particle or the energy of the particle and the time at which it is observed, there exists an inequality relation connecting the average error in the simultaneous measurement of each, such that the product of the two errors is always greater than Planck's constant of action, h. This result is usually interpreted to mean that the attempt to secure maximum possible precision in the assignment of a value to the one observable (e.g., position) can only lead to a corresponding increase in the lack of precision in the assignment of a value to the other (e.g., momentum).[9] Bohr came to look at this in the following way: in the case of two observables represented in quantum mechanics by canonically conjugate variables any method of assigning a value to the one variable is complementary to assigning a value to the other. Both methods are essential to produce complete precision (i.e., zero average error) for both variables; neither by itself can accomplish this. If the one method succeeds in assigning a completely precise value to the observable for which it is appropriate the other method fails completely to assign a value to its observable. *Both* methods, which may appropriately be called complementary, are needed for a complete simultaneous specification of both observables. Bohr called the general content of this assertion the principle of complementarity and was so impressed with what he felt to be its importance in quantum mechanics that he sought to apply it not only to other physical problems but also to other domains of human experience. One obvious physical extension is to the wave–particle problem in atomic physics. An electron, for example, behaves like a particle under certain experimental arrangements but behaves like a wave in other circumstances, a realization of which promoted the early development of wave mechanics by L. de Broglie (1892–) though its significance had been actually appreicated by the famous Irish mathematician, physicist and astronomer William Rowan Hamilton a hundred years earlier. Bohr seized on this as another example of complementarity. The experimental arrangement in which the electron behaves like a particle (e.g., when it passes through a single slit in an otherwise opaque screen) is complementary to the setup in which it behaves like a wave (e.g., when it is given the chance to pass through either of two parallel slits in the screen, or when it is scattered by the atoms in a single crystal, as in the early experiments of Davisson and Germer and G. P. Thomson, and produces the well-known optical interference effects). Each arrangement automatically excludes the result of the other, but both are necessary to provide a complete picture of the nature of the electron.

Bohr, who was always interested in assigning a deeper philosophical meaning to physical principles, sought to extend the complementarity idea to human situations outside the field of physics. In the case of the social interaction of human beings he called attention to the complementary aspects of *love* and *justice*. Thus in dealing with people, either individually or *en masse*, the more we exhibit love for them in spite of their obvious misbehavior the less justice we do to their shortcomings and vice versa. The idea is persuasive and helps to reinforce one's conception of the complementarity principle. A little thought serves to turn up other examples which, with a little stretch of the imagination, seem equally reasonable. Consider the complementary character of the statistical study of a large group of individuals with respect to their possible mortality, compared with the specific study of each individual from the same point of view. From the study of mortality tables based on large groups of people of varying ages a life insurance company can predict with considerable accuracy how many people in a given population and in a given age range will die in a given period of time. But in so doing the company foregoes any ability to tell *which* individual people in the given sample will die. On the other hand, a competent physician examining a given individual may be able (within a certain experimental error) to provide a rather accurate assessment of the latter's chance of surviving for the period in question. But he thereby foregoes ability to predict anything about the group of which his examinee is a part. The more or less precise examination of the individual is complementary to the study of the group.

Complementarity is not confined to mortality in the domain of general human experience. The study of any property connected with human beings when conducted group-statistically is complementary to the study of such a property in each individual making up the group. Language as a means of communication between individuals of a group is a social device of enormous importance. It provides another illustration of the complementarity idea. If, for example, in reading a paragraph of text we concentrate attention on each word in a given sentence to study its various meanings and its syntactical significance in the sentence we lose control of the meaning of the sentence as a whole. On the other hand, if as we commonly do in reading, we concentrate on the sentence as a whole in order to extract meaning from it, we abandon the effort to assign a specific meaning to each individual word. The two procedures are complementary in character. Both are really necessary for the complete understanding of the written material.

As another example of the complementarity idea, consider an observer looking at a picture. He can contemplate the picture as a whole and thus gain a general impression of its character and significance. In so doing he ignores and foregoes a detailed examination of each part of the picture. On the other hand, meticulous concentration on each individual part enables the observer to tell how the picture is put together in detail but excludes an understanding of the picture as a whole. The two procedures are complementary. But here I think we encounter a possible legitimate criticism of the significance of the complementarity idea. The two approaches to the study of a picture just mentioned by no means exhaust the experience involved in the observation of the picture. Instead of merely looking at it in normal light with ordinary eyes, we may illuminate it with light of various frequencies, visible and invisible, e.g., ultraviolet light or even x rays. One can find so many complementary aspects, indeed, that one is apt thereby to reduce the process to triviality. Or in any given aspect of experience, we may easily find ourselves interested in many ways of looking at it which are not at all complementary but which are nevertheless deemed essential to a complete understanding of the experience. To what extent this criticism reduces the value of the human experience exemplification of complementarity as an important principle in quantum mechanics must be considered a debatable matter.

We close our brief examination of the social exemplifications of physical principles with a consideration of teleological ideas.

Teleological Principles

The teleological idea or the concept of purpose is involved in several physical principles, notably Hamilton's principle and the related but not identical principle of least action, Hertz's principle of the straightest path, Gauss' principle of least constraint, and Fermat's principle of least time.[10] They all effectively state that things take place in the physical world, e.g., the motions of systems of particles, in such a way as to make a certain function assume a stationary value under certain boundary conditions, usually a minimum as compared with all possible values satisfying the given conditions. For example, Hamilton's principle says that for a conservative dynamical system the motion between any two instants of time is such that the time integral of the difference between the kinetic and potential energies taken between these two time instants has a stationary value. It is as if the system had a certain purpose to satisfy. This means that the actual motion between the initial and final states at the initial and final times, respectively, takes place in such a way that the integral in question is either greater than it would be for any other possible motion between these states or less than any other possible motion between these states, it being understood that the initial and final states are the same for all the possible motions being compared. In the majority of cases to which the principle has been applied in classical mechanics, the stationary value is a minimum. With respect to the possible idea of purpose involved, philosophers have argued that since Hamilton's principle is *not* necessary for the deduction of the actual motion of dynamical systems, i.e., the Newtonian or Lagrangian equations are fully adequate, we have no real logical ground for insisting that nature imposes a teleological requirement on motions in our experience. Nevertheless the fact that one can exhibit the principles of mechanics in a teleological guise is persuasive, since it serves to tie physical principles with an idea basic to the interpretation that human beings give too much of their ordinary experience, particularly in their relations with other human beings. A rational individual is said to arrange his actions so as to be sure of achieving his fundamental desires, whether it be to accumulate wealth or gain power over his fellow men. In particular the aim here is almost always to try to attain the given desired end at *minimum* cost in human effort. This strongly suggests a heuristic connection with the minimum principles of physics.

Let us consider an example of a social idea directly suggested by such a minimum principle. This is the so-called principle of least effort, much promoted earlier in this century by the sociologist George Kingsley Zipf (1902–1950). In his book *Human Behavior and the Principle of Least Effort* Zipf sought to set up social laws exhibiting the result of minimum expenditure of human effort.[11] One of the more famous of these relates to the statistics of language communication, which we have already mentioned as one of the most important manifestations of human social behavior. We illustrate by examining Zipf's law connecting the frequency of occurrence of a given word token (e.g., the word "large") in a long book like a Victorian novel or a month's file of a major newspaper and the rank order of that word taken with respect to other word tokens, i.e., 1, 2, 3, 4, . . . , where 1 indicates the word token occurring most frequently, and so on. Zipf found that no matter what the nature of the written material is, so long as it is large enough in scope, the following relation holds

$$P_n = An^B \tag{1}$$

where n is the rank order of the word token whose frequency of occurrence is P_n. The latter is measured by the actual number of occurrences of the word token in question. The quantities A and B are constants depending on the nature and quantity of the material being studied. Zipf further found that for the cases which he examined the value of B was very close to minus unity, making the plot of P_n against n a rectangular hyperbola, reminiscent of the curve of pressure versus volume in the idea case of Boyle's law of gases. This result appears at first to be almost trivial, since the word tokens of greatest frequency of occurrence obviously correspond to highest rank order, that is, the smallest value of *n* and vice versa. But the significant thing here is the rather precise mathematical form of the relation, now known as Zipf's law of frequency—rank order of words in a language. His result has, of course, been found for a great variety of material in many languages.

Zipf saw in his law an illustration of the principle of least effort. His argument may be summarized as follows: To him least effort is exemplified by a balance between the desire of the person communicating in a given language to make do with the smallest number of word tokens consistent with the ability of the recipient of the communication to identify the words correctly, along with the concomitant desire of the recipient to receive as many words as possible in order to avoid having the same word token stand for too many different meanings, leading to possible lack of understanding of this message. Zipf provided no strictly mathematical derivation of his law and his considerations must be considered as qualitative in character.

An analytical derivation of Zipf's law has indeed been provided by the French mathematician Benoit Mandelbrot.[12] His analysis need not be repeated here, but it is worth emphasizing that he developed it by assigning to each word token in a large body of language a *cost* depending in general on the length of the word, the longer words being obviously assumed to be more costly. Mandelbrot then proceeded to set up an expression for the average cost per word in the assemblage of words being examined, in terms of the frequency of occurrence. He then *minimized* this average cost subject to the condition that the average information rate per word shall be constant. Using the information theory of Shannon he then found an expression for the frequency of occurrence of each word token in terms of the cost of that word. Finally, he obtained the relation between the cost per word and the rank order. Substitution into the expression resulting from the minimization procedure led to the relation between word frequency and rank order or the generalized form of Zipf's law. In this derivation the idea of least effort which serves as the initial basis for Zipf's considerations enters in the minimization of the average cost per word. It may be considered a very plausible exemplification of the teleological concept embedded in the minimum principles of physics.

Summary and Conclusions

It has been the aim of this article to examine certain social and other human experience exemplifications of physical principles. The principle of mobile equilibrium, the principles of thermodynamics, the ergodic hypothesis, complementarity, and the stationary principles of mechanics have been scrutinized and examples provided relating to the ordinary experiences of human beings. It is suggested that such exemplifications possess heuristic value in the sense that they suggest further questions illuminating the meaning of the principles involved and casting further light as well on the human experiences. They also possess pedagogical value by encouraging students of physics to relate physical ideas more closely with ordinary human social experience.

References and Notes

1. R. B. Lindsay, "Physical Laws and Social Phenomena," *Sci. Mon.* **25**, 127–132 (1927).
2. Vilfredo Pareto, *Trattato di Sociologia Generale* (1916); English translation with the title *The Mind and Society* (Harcourt-Brace, New York, 1935).
3. Ralph Waldo Emerson, "Compensation," in *Essays by Ralph Waldo Emerson* (Thomas Nelson and Sons, New Century Library, New York, n.d.), pp. 73ff.
4. R. B. Lindsay, "Entropy Consumption and Values in Physical Science," *Am. Sci* **47**, 376–382 (1959); see also R. B. Lindsay, *The Role of Science in Civilization* (Harper and Row, New York, 1963; Greenwood Press, Westport, Connecticut, 1973), pp. 290–298.
5. R. B. Lindsay, *Introduction to Physical Statistics* (John Wiley and Sons, New York, 1941; Dover Publications, New York, 1968), Chap. 5.
6. Paul S. Epstein, "Critical Appreciation of Gibbs' Statistical Mechanics," in *A Commentary on the Scientific Writings of J. Willard Gibbs*, Volume Two (Yale University Press, New Haven, Connecticut, 1936), pp. 465ff.
7. A. I. Khinchin, *Mathematical Foundations of Statistical Mechanics* (Dover Publications, New York, 1949).
8. Niels Bohr, "Kausalität und Komplementarität," *Erkenntis* **6**, 293 (1937); see also Niels Bohr, *Atomic Physics and Human Knowledge* (John Wiley and Sons, New York, 1958).
9. R. B. Lindsay and Henry Margenau, *Foundations of Physics* (John Wiley and Sons, New York, 1936; Dover Publications, New York, 1957), pp. 418ff. Also Ox Bow Press, Woodbridge, Connecticut, 1981.
10. R. B. Lindsay and Henry Margenau, Ref. 9., pp. 112ff.
11. George Kingsley Zipf, *Human Behavior and the Principle of Least Effort* (Addison-Wesley Press, Cambridge, Massachusetts, 1949); see also Colin Cherry, *On Human Communication* (Technology Press, MIT, Cambridge, Massachusetts; and John Wiley and Sons, New York, 1957).
12. B. Mandelbrot, "An Informational Theory of the Structure of Language Based upon the Theory of the Statistical Matching of Messages and Coding," in *Proceedings of London Symposium on Applications of Information Theory*, 1952 (Butterworth Scientific Publications, London, 1953); see also Colin Cherry, Ref. 11, pp. 105ff and 209ff.

8

Seneca's Ethics Viewed from a Modern Standpoint

Henry Margenau
Yale University
New Haven, Connecticut

Wolfgang Yourgrau's interest comprised many fields. It was strong in physics, philosophy of science, and history of science and included the urge to understand nonscientific disciplines in general scientific terms. This was clear in numerous stimulating conversations between him and the present author. In his article published in *Vistas of Physical Reality* (E. Laszlo and E. B. Sellon, editors, Plenum Press, New York, 1976), entitled "Some Reflections on Philosophy of Science," he offers and advances a pluralistic approach to philosophy of science, suggesting that no single epistemology can support all subjects, not even all sciences. Nevertheless he tended to dismiss purported disciplines which reject the scientific spirit. Recalling this, and aware of his general interest in history, I offer for Yourgrau's Memorial Volume a paper which interested him and which might exemplify a quasiscientific approach to a subject which appears rudderless without some relevance to science. It was presented in Cordoba, Spain, on the nineteen hundredth anniversary of Seneca's birth and published in a volume in his honor entitled *Actas Del Congreso Internacional De Filsofia* in 1967.

Ethics and Science

After nineteen centuries it is difficult to say something new about Seneca's ethics. Our philosopher has been read and reread, translated, interpreted, praised, and criticized. The present essay is hardly a new interpretation; it abstains from praise and criticism. But it attempts to look at Seneca from an unusual point of view and calls attention to those aspects of his writings which seem to have a special relevance for our time.

The point of view is that of a philosopher of science who sees ethics as an endeavor to control human actions in a manner not wholly unlike the way in which science predicts events. It embodies a conviction that ethics, when fully developed, exhibits a radical parallelism with the structure of science, although it is not itself a science. I am not asserting that ethics receives its norms and its impetus from science. The parallelism under consideration is a formal one which suggests that ethics solves the basic problem of generating norms out of incidental facts of human behavior in the same way in which science rises from particular observations to universal laws.

Scientific analysis moves between two extremes. At one end are the observations, the facts, the sensory data which are to be explained. They are "given" in a simple sense. We are exposed to them as we are to fate, and just like fate, the facts of observation seem devoid of intrinsic meaning. It is therefore the task of science to make them meaningful, to show forth relations between them and to establish a framework of concepts by means of which facts can be understood and predicted. This process engages the faculties of reason, introduces concepts or constructs which are not immediately given in sensation and appeals to fundamental propositions, so-called postulates or axioms, which cannot themselves be demonstrated by reference to observable facts.

Specifically, what happens is this. At one terminus of theoretical science stand the postulates, at the other the empirical facts it wishes to explain or predict. As examples I mention the postulates of Euclid, which are designed to explain the spatial structure of the actual world; Newton's laws of universal gravitation, which ultimately rationalize what happens to falling bodies, planets, satellites, and rockets; or the law of supply and demand, which pretends to make sense of economic affairs. The postulates are not generalizations of empirical facts as Bacon and the empiricists would have us believe; they are "free constructions of the human mind"—to quote a phrase of Einstein's. Their validity lies in the agreement of their consequences with observed facts. True, they are often suggested by the facts, but their fertility and richness reside in their universal claim which cannot be established through any finite set observations. I distinguish carefully here between universality and truth. A postulate can be universal but wrong, i.e., not in conformity with fact. In that case science abandons the postulate and replaces it by another. This movement accounts for the dynamic success of science, which does not harbor absolute truth.

From this point of view, then, I would like to characterize science by its three major phases:

1. *postulation,*
2. *the working out of the implications* of the postulates,
3. *the testing of the postulates* in the realm of observation.

The scientist rarely worries about the origin of the postulates; they are injected into the stream of thought by some blessed individual and are then recorded by the historian and used by the scientist. Nor does he concern himself greatly with the logic of testing, or verification. Analysis shows, however, that the latter process, too, involves purely postulational elements (stipulations as to closeness of agreement that must exist between prediction and observation if the observation is to be regarded as confirming the prediction). We see, therefore, that phases 1 and 3 are intrinsically tied to human commitments, whereas the middle one, which constitutes mainly the work of the mathematician and the theorist, is a search for the analytic consequences of 1 and an endeavor to match them against 3.

In ethics, too, one can distinguish these three phases but in a vastly different form. Science tries to explain and state facts, hence its language is in the indicative voice. Ethics raises the "ought," hence it must use the imperative. To the postulates of science there correspond, therefore, the basic *commandments* of the ethical code. Verification of predictions in science relates to observational facts and involves postulational commitments in respect to correspondence between theory and fact. In ethics, there are no predictions to be verified, but there is *behavior* to be *validated*. This validation is possible through the postulation of *primary values* which, when attained, allow the predicate "good" to be assigned to the action which realized them. The fact that these primary values, which are principles of validation, are freely chosen does not detract from their effectiveness, any more than the free choice of postulates invalidates science.

Examples of ethical postulates are the *Ten Commandments*, the *Golden Rule*, the *Analects of Confucius*; examples of primary values are survival, human happiness, self-fulfillment, and "the peace that passeth understanding." But what is the counterpart of item 2, the explication of the basic postulates? It is the process of the spinning out of the consequences of the moral imperatives in terms of human living, their application to cases, in short, *casuistry*. That word has acquired an offensive meaning of pedantry and uselessness, its employment seems to detract from the ethic that tolerates it. I believe that this attitude is indicative of the weakness of ethical thinking in our time, and that Seneca is a magnificent antidote to this paralyzing debility.

In sum, then, the three ethical elements which stand in parallelism with the three major phases of science are

1. commandments,
2. casuistry,
3. choice of principles of validation in terms of which moral behavior can be validated.

I shall now attempt to locate and formulate these elements in Seneca's writings.

Primary Values

Seneca identifies his primary values with the *summum bonum* of the Stoics, imperturbability, compliance with fate, harmony of the soul, as he calls it. Nowhere is this more beautifully expressed than in the hymn of Cleanthes, which contains the lines (in Cicero's translation)

> Duc, O parens, celsique dominator poli
> Quocumque placuit: nulla parendi mora est,
> . . .
> Ducunt volentem fata, nolentem trahunt.

In the philosopher's own words: "What *must* be *shall* be; and that which is a necessity to him that struggles is little more than choice to him that is willing."[1]

This primary value, Seneca holds, leads to happiness. Yet he escapes the error of the hedonists who take happiness to be pleasure, or who take it to be in itself a well-defined state of mind. His wisdom never commits him to an irrational and impossible calculus of happiness. Nor does he use the hairsplitting arguments of earlier Stoics to demonstrate *why* happiness is the primary value: he simply defines it to be "the peace that passeth understanding" and leaves it there, having given ample descriptive substance to make clear that his readers understand the meaning of the phrase.

He also affirms frequently that the primary value is virtue, that virtue engenders peace of mind, which is identical with happiness. So far as I can see, injection of the term virtue is a concession to the philosophic jargon of his day; it is a linguistic artifact and a useful one, and his adoption of it avoids the Socratean pitfall of equating virtue to rational understanding, or to conformity with nature in an earlier sense. To Seneca, virtue means "living conformably." He refuses to say, with Cleanthes, that it is "to live conformably to nature," or with Chrysippus that it is "living according to your experience in the course of nature," and he disagrees rather explicitly with Diogenes Babylonius who speaks of virtue as "prudence and reason" which select "things according to nature." Seneca's "living conformably" is nothing more than a reemphasis of the last line of the hymn of Cleanthes: living willingly in accordance with fate. Virtue, wisdon, and *summum bonum* are synonyms for this attitude of mind. There is no deeper logic, no Aristotelian argumentation behind this simple posit, and I believe that this unassuming simplicity, this frank and unpretentious avowal of the primacy of the Stoic goal, set in the context of his magnificent prose is one of the chief causes of Seneca's appeal to moralists through nineteen centuries.

His disposal of Aristotle's weaseling concept of virtue as moderation is too impressive to be left unmentioned. He says: "The *Stoics* will not allow a wise man to have any passions at all. The *peripatetics* temper them; but that mediocrity is altogether false and unprofitable. And it is all one as if they said that we may be a little mad, or a little sick."[2]

Having said this about Virtue, which is by definition the Stoic summum bonum and therefore more or less identical with tranquillity of mind, it behooves us to comment briefly on the way in which Seneca uses the plural, virtues. Of these he lists many, and each of them, while vaguely conducive to the summum bonum, designates a singular trait of its own. Taken collectively, these virtues do not compose Virtue in any logical sense. There is the suggestion that a person possessed of all virtues will in fact attain the primary value, but there is no proof of this, nor can there be a relation of logical entailment of Virtue by virtues. What Seneca is in effect saying is that, *in the empirical process of living,* the virtues engender the primary goal. We see here that the virtues play the role of imperatives in the ethical system, and Seneca's moral genius assures us that obedience to them will result in a realization of the ultimate good. What these imperatives are and in which manner they function in human experience will form the content of Section III.

Let us return to Virtue, the primary value. In the first part of my discourse I drew a parallel between ethics and science, asserting that the principles of validation in either field, primary values in one and verification in the other, are postulational and therefore incapable of derivation from prior truths. I believe Seneca is the first moral philosopher who saw this point. The *summum bonum* is what it is and requires no demonstration; Virtue is the end of life. Were it derivable from something else it would not occupy this unique place in his ethical system. In *De vita beata* we find this admonition: "You mistake when you ask, 'What is that for the sake of which I seek Virtue?' for you then ask for something higher than the highest. Do you ask what I seek from Virtue? Herself! For she has nothing better, she is her own reward."

If Virtue is a primary value, independent and *a priori* relative to all other moral concepts, ethics is an open system displaying an affinity with metaphysics and religion, for these latter are ordinarily relied upon in Western philosophies to supply human motivation for the acceptance of the primary values. This is true in Seneca. There is, for him, the suggestion of a reward for virtue in the beyond. The soul of the good man is elevated after death to a direct contemplation of the eternal verities which, as in Plato's philosophy, are imperfectly experienced in life; whereas souls unworthy of salvation are returned, after death, into the material processes of nature, facing again rebirth and death. His view differs from Christian doctrine and approaches Oriental thought inasmuch as it denies eternal bliss even to the soul of the virtuous. For it remains as an individual in its spiritual "heaven" only until the end of this world. After that it loses its identity and merges with Jupiter, the universal spirit. Apparently, when a new cosmic cycle begins, a new fate is enacted and new souls appear.

But to return to ethics. Seneca's system sets forth its primary values with clarity and in bold strokes; it avoids the fallacy of reductionism which afflicted its predecessors; it defines Virtue with factual precision and takes it to be the ultimate principle of validation which confers goodness on human acts. We have called it tranquillity of the soul. Let it be understood, however, that it is not to be equated to apathy. Seneca is sensitive to beauty of style. No man who can write the Consolations to Marcia can be accused of indifference to the fate of friends.

Imperatives

The methodology of ethics in terms of which I have chosen to analyze Seneca's moral writings is nowhere acknowledged explicitly in the body of his works. For our philosopher was an ethical innovator, a creative moralist whose spirit soars above the pedantic details of method. Nevertheless one senses in many places that he is aware of the need and the simplicity of the "precept–casuistry–primary value" structure, and he regrets the proliferation and sophistication which even in his day the discipline of ethics had undergone. Listen to this phrase:

> The Wisdom of the Ancients, as to the government of life, was no more than certain precepts what to do, and what not; and men were much better in that simplicity; for as they came to be more learned, they grew less careful of being good. That plain and open virtue is now turned into a dark and intricate science, and we are taught to dispute, rather than to live.[3]

He feels that, because of the greater complexity of life, a stronger attempt must be made to set forth the precepts, i.e., the imperatives, in a manner palatable to an educated taste, and this leads him to exhibit them under the smoother phraseology of virtues. But it is not difficult to restore the Senecan commandments from his exquisite book *De beneficiis*.

There he propounds that man's actions, to be ethical, must be beneficent, and then he proceeds to decompose the one supreme command, be beneficent, into a spectrum of virtues. Each virtue is equivalent to a commandment in Judeo-Christian terminology, and the book is a magnificent exegesis of his table of virtue-precepts. They are not presented in systematic form in his own writings; indeed every attempt is made to loosen the rigor of the discourse by creating the appearance of informality for the sake of pleasing a reluctant reader. Still, so far as I can see, the skeleton holding the flesh is there; the catalogue of precepts is visible beneath the rhetoric and it is neither deficient nor redundant. The sequential items even bespeak a hierarchy of virtues, from the most desirable to those which are more commonplace.

In modern philosophy Seneca's virutes would probably be called values, moral values. There is supposed to be a branch of learning called value theory, and ethics is part of it. This tendency, which I deplore, springs from the mistaken belief that there exists a dichotomy between facts and values, the one being the province of science, the other a domain of the humanities and the liberal arts. Untold damage has been done to the fruits of Western thinking by this divisive insistence, but having called attention to it elsewhere[4] I shall not belabor the point here. Relevant, however, is the fact that Seneca never speaks of values, nor of a theory of values relying upon its own peculiar justification. This linguistic matter by itself is not especially significant because our word "value," in its unspeakable indiscriminacy, would be hard to render. In Seneca's Latin the word *valor* had a wholly different connotation. It might seem, therefore, that the issue of facts versus values, science versus ethics, had no occasion to arise. This, however, is a false conclusion, for Seneca gives every evidence of having been aware of the relation between ethics and science, as all his writings show. To him, physics and ethics are complementary aspects of nature, which includes human nature. He states on numerous occasions that there is no conflict and often uses arguments drawn from physics to enhance his ethical dissertations. True, science has moved on (while in my opinion ethics has not proceeded one step beyond Seneca except in useless proliferation), but when carefully inspected the harmony proclaimed by him is still discernible at the base of all complexities. This perennial insight, to me, has always added greatly to the charm of Seneca's writings.

What, then, are his imperatives? They are all found in his *De beneficiis,* as well as scattered through all other moral essays. He says, in essence, "Let all your acts be distinguished by beneficence, i.e. (and here is the detailed list of virtues), by honesty, obedience, integrity, honor, generosity, humanity, friendship, piety, gratitude, prudence, justice, gentleness, courage, and frugality," This is the skeleton, but it is not delivered as a set of bare bones. His prose furnishes the flesh and blood, for each is exemplified and made unforgettable by some striking illustration that gives it meaning and vital force.

Most of his examples are well known to the classical scholar and hardly need recounting here. They lift his works from the level of didactic philosophy to the plane of edifying literature. Yet I cannot forego reminding you of the way in which he illustrates integrity and honor by reciting the story of Rutilius, who was sent into exile by a tyrannical government. His friends inform him that they are planning to incite a civil war in the hope of terminating the tyranny and bring him home. This elicits Rutilius' reply: "God forbid, for I had rather my country should blush for my banishment than mourn for my return."

Whence does our author derive these precepts? This question, significantly, remains unanswered. True, one finds motivation. There is often an indication that persons who fail in exercising the virtues are condemned by their countrymen, a plausibility consideration without demonstrative cogency, introduced clearly as a moral advertisement. Closest to a justification comes perhaps this passage from *De beneficiis*:

> The very meditation of (the foregoing virtues) breeds good blood and generous thoughts ... In short, the art and skill of conferring benefits is, of all human duties, the most absolutely necessary to the well-being both of reasonable nature and of every individual; as the very cement of all communities and the blessing of particulars. He that does good to another man, does good also to himself; not only in the consequence but in the very act of doing it: for the conscience of well-doing is an ample reward.

When subjected to logical scrutiny these explanations, and the tenor of all of Seneca's writings, reveal that precepts have no prior causes in ethics, that they are postulational as are all the basic truths in every other human discipline. Man must commit himself to them without asking why, for they are causally self-sufficient. Nor do they contain by analytic implication the *summum bonum* for which man ought to strive. Neither our philosopher nor anyone else can prove that an obedient person or a generous one will necessarily attain tranquillity of mind and its attendant happiness. But we are told, on the authority of a wise and persuasive man, that people who have in fact practiced the virtues did promote for themselves and others the *summum bonum.* Entailment of goal by precepts is not through rational implication; it is through the pragmatic course of human living. Demonstration, deduction from postulates in science is here replaced by the ethical laboratory called history, experience of mankind.

CASUISTRY

That phase of ethics which corresponds to the derivation of specific laws and theorems from the postulates of science is the explication of the basic imperatives or virtues through living examples, through actual cases. It is properly called casuistry and forms, as noted in the beginning, an indispensable part of every successful moral system. For a statement of general precepts or commands is never sufficient without elaborate qualifications that make it relevant to the vital needs of a given culture at a given time. Commands without explication are as empty as the law of universal gravitation without specification of the meaning of mass and distance or, for that matter, as useless as Kant's categorical imperative without stating the criteria (happiness, survival of society, or whatever) for the validity of a universal law.

There is, I think, a noteworthy historical correlation between ethical systems that worked successfully and the practice of casuistry. Current philosophical disputations in ethics manifest a disdain for it, and as a result of such deprecation moral guidance has almost totally slipped from the agency of ethics to that of religion. The morals of Western civilizations would collapse if religion were taken away—Confucianism survives in its absence. And Confucianism is full of casuistic admonition. Communism, too, seems to get along without religion, and there, likewise, behavior is taught to the young and uninitiated by assiduous attention to the specific application of the basic commands. The fact that we call this activity propaganda does not make it less effective.

The needfulness of casuistry in this systematic sense is perfectly clear in our own culture. We are dedicated to a basic code expressed in essence by the ten commandments, the Golden Rule, and the love of neighbor. But what is the meaning of a commandment like "thou shall not kill" in a civilization addicted to wars? Evidently, what we mean is "kill some people but not others"; and when we speak of love we tolerate the attitude that holds personal affection of a person to be compatible with his murder. Whether we condone our ethics or not, it is simply not correctly defined by its norms. The fifth commandment is meaningless in a world that admires heroism in war. It needs qualification in terms of what we have called casuistry, possibly a painful addition to the lofty moral philosophy which fills our books because it will reveal embarrassing difficulties and inconsistencies in our moral thinking.

Seneca is not guilty of this glaring deficiency which marks our time; indeed his glory rests in his superb ability to make his virtues live by a selective appeal to cases. In *Happy Life* he waves aside the high-flown expressions of the Stoics and proceeds to "discourse the point at the rate of ordinary men." Much of his fourth epistle is devoted to ways of striking a proper balance between the "decrees of philosophy" and the particulars of admonition. Both are necessary, he affirms, the one in order to be "well informed in the whole," the other

> . . .because they are very useful to us in conjunction with other helps. First, as they refresh the memory, and secondly, as they bring us a more distinct view of the parts, which we saw but confusedly in the whole . . . I say that the very authority of the adviser goes a great way in the credit of the advice. . . . Let a man be never so right in his opinion, he may be yet more confirmed in it by admonition.

And his prescription covers the minutest detail, even to the point where ethics becomes mere civility. He feels, again in contrast with most moral writers of our day, that courtesy is conducive to good behavior, and since courtesy and civility can be learned, an effort to teach and explain them is worthwhile. Listen how he instructs the giver of favors:

> There should be no delay in a benefit but the modesty of the receiver. If we cannot foresee the request, let us however immediately grant it, and by no means suffer the repeating of it. It is so grievous a thing to say "I beg"; the very word puts a man out of countenance; and it is a double kindness to do the thing and save an honest man the confusion of a blush. It comes too late that comes for the asking. . . .

At a later place:

> The manner of saying or doing anything goes a great way in the value of the thing itself. It was well said of him that called a good office that was done harshly and with an ill will, a *stoney piece of bread*. It is necessary for one that is hungry to accept it, but it almost chokes a man in the going down. There must be no pride, arrogance of looks, or tumor of words in the bestowing of benefits; no insolence of behavior but a modesty of mind, and a diligent care to catch at occasions, and prevent necessities. A pause, an unkind tone, word, look or action, destroys the grace of courtesy. It corrupts a bounty when it is accompanied with state, haughtiness, and elation of mind in the giving of it.[5]

Critics who approach Seneca from a point of view oriented toward systematic moral philosophy have justly lamented the excessive volume of his casuistic advice. To appreciate him fully one must recognize that he is a systematic philosopher of a derivative kind, and not a very original one even in his avowal of stoicism. But he shares this trait with most creative moralists; his writings have far greater similarity, even as regards verbose discussion of cases, with the *Old Testament*, the *Upanishads*, the *Analect*, and the *Talmud*.

Today it is generally regarded as improper to assign moral value to unreflective, habitually good behavior; to be counted as good an action must be deliberate and intended. An ethical robot is a contradiction in terms. Notice how this conficts with certain Oriental views according to which the truly good individual, for instance the buddha who has achieved nirvana, is merged in God and thereby *incapable* of doing evil. Seneca's ethic stands between these extremes. In true wisdom he sets forth that custom and habit can be conducive to ethical behavior, indeed can itself merit the attribute *good*—in complete contradiction to the Kantian belief according to which only consciously willed, premeditated acts have ethical significance.

Casuistry, as we have seen, is the elaborate effort to adjust the moral imperatives to concrete situations, to explain them in terms of examples and to give them power over men's minds. The last phase of this effort, enforcement, often engages that mysterious monitor called conscience. In my cursory study of a variety of ethical systems I have found this concept interesting and rather crucial, for its treatment reveals, more than any other single feature, the most fundamental metaphysical commitments of each system. Impractical theorists either dismiss conscience as unreliable* or make too much of it ("conscience is God's voice in the human soul"). At any rate, most philosophers who spurn casuistry neglect or discredit the human conscience. Seneca is no theorist, and he neither discredits conscience nor does he maximize its role. He takes it for granted and uses it in his persuasive appeal for good behavior. "The greatest punishment of an injury," he says, "is the conscience of having done it; and no man suffers more than he who is turned over to the pain of repentance."

Seneca is sometimes considered as a Christian moralist. I wish to conclude this discussion of his exegesis of the meaning of Virtue by showing in what essential respect he differs from men like St. Paul. The apostle lays great stress upon love, charity, and compassion. Here is what Seneca has to say after defining clemency!

> We must be careful not to confound clemency with pity; for as religion worships God, and superstition profanes that worship, so should we distinguish between clemency and pity, practicing the one and avoiding the other. For pity proceeds from a narrowness of mind that respects rather the fortune than the cause. It is a kind of moral sickness, contracted form other people's misfortunes . . . I'll give a shipwrecked man a plank, a lodging to a stranger, or a piece of money to him that wants it: I will dry up the tears of my friend, yet I will not weep with him, but treat him with constancy and humanity, as one man ought to treat another.

Conclusion

In this account I have imposed upon the moral philosophy of Seneca a structure which he himself nowhere suggests. This structure was initially derived from an analysis of science and involves three related phases of a moral system:

1. primary values whose achievement validates the ethical precepts,
2. imperatives, and
3. the explication of the imperatives which leads from them as postulates to all the minor precepts for practical living.

By choosing this structure as the skeleton of Seneca's moral philosophy I may encounter the criticism of having done violence to his thought. To this indictment I plead guilty in one respect: His thought is amorphous, sometimes untidy, and it may

*Cf. for example Schopenhauer and Freud.

be said to defy the insertion of all bones of reason. Hence the use of the skeleton introduces foreign matter.

However, I cannot concede that my treatment has falsified the quality of his thought. On the contrary, when reviewing his writings, which I had first read with fascination as a Latin student about 40 years ago, from this point of view they suddenly became alive to me, to make not only moral but even logical sense, and they disclosed so many messages which seem important today that the preparation of this paper acquired a measure of evangelism and took on a significance I did not initially foresee.

References and Notes

1. Seneca, *Epist.*, 9.
2. Seneca, *Epist.*, 10.
3. Seneca, *De vita beata* (transl. Roger L'Estrange).
4. H. Margenau, *Ethics and Science* (Princeton U. P., Princeton, New Jersey, 1964).
5. Seneca, *De beneficiis* 7.

9

Physics and Philosophy

André Mercier
Universität Bern
Berne, Switzerland

Wolfgang Yourgrau* loved to speak to mixed audiences composed of colleagues, younger staff members, and students (and even "ladies") on topics which allowed him to make big excursions from physics, his original professional field, to various philosophical domains where he found reasons to show that physics is at the same time a glorious enterprise and a discipline which has to show modesty in view of its own limitations. He was not afraid to say simple things, which more or less everybody thought they had known already for ages, but the way he said them was the originality of his talks. Though there is no question of either imitating or paraphrasing him, this paper will also attempt to present things which may be well known; however, the relationships which are expected to follow from the manner they are presented may have escaped the awareness of many a scholar.

1. The Relationship of Physics and Philosophy

The first remark is that it makes a difference, whether we speak about "Physics and Philosophy" or about "Philosophy and Physics." Not only does it acoustically sound differently (the first phrase sounds the more definite of the two), it also indicates a different intention of the understanding of the relationship between the two disciplines. By choosing the first one, the autonomy of physics is insisted upon and the particular contribution it yields to the relationship is enhanced.

*Wolfgang Yourgrau was once visiting professor at the University of Berne, Switzerland, which gave us the opportunity to deepen matters of common interest in the field of the philosophy of science, especially of physics, and the relationship of physics to philosophy.

2. Physics of Old

Now, physics has detached itself from philosophy of old during centuries and millenaries. One hears it said that the Presocratics were physicists; later Aristotle wrote a treatise on physics. But the ancient Greek had a totally different conception of Nature (φυσις) from ours. They never intended to have a physics which would allow them to build a dam, install turbines at the end of the waterfall streaming from the dam, and produce electricity to be brought into the houses and industrial areas. Much rather, they had an awe for the secrets of nature, and what the Presocratics did is scarcely worthy of being called physics, for even though they proposed a few elementary "principles" of the constitution of all kinds of matter, like air, fire and so on, and even predicted an eclipse or admitted the possibility that the Earth be round, they did not build a physics in the sense in which we understand it today to be a "science." The only such proposed "element" which is of great concern is the απειρον, but it is not a *physical* element, it is a mathematical concept.* As far as Aristotle's physics is concerned, it has simply proved to be wrong, and what we owe to him is not a physics proper, but the method of analysis as the correct idea to approach the problems of science. We also read that the Arab philosophers helped overcome the shortcomings of Aristotle's physics. But not even that is really correct; for, the man who did so was J. Philopon, a Christian who acted as secretary to the academic society of his town, Alexandria, but whose heterodoxy in religious matters was the cause of his flight from the town, so that he went away up to Baghdad and wrote not only *Contra Aristotelem* but laid down in Syriac language his ideas about the impetus which was to be learnt from him by the Arabs and brought down to Galilei (who by the way read him in spite of his works being forbidden by the Church).

Aristotle's physics was wrong. It could not explain the motions of bodies (celestial or terrestrial) and other phenomena satisfactorily. In spite of being the initiator of the analytic kind of thought which is still at the basis of contemporary scientific methodology, he did not succeed in founding physics. Of course, he did not dispose of sufficient mathematical means, but this is not the only,† nor the main, reason for his failure. The main reason was that he attributed to bodies the tendency to move towards their natural final place instead of looking for an efficient cause from outside.

At the dawn of modern philosophy, Descartes invented the powerful analytical geometry, but even that was not enough to express physics in its "correct" form, and the physics he imagined by filling space with vortices proved also to be totally unacceptable, even though the idea of relating physics to geometric properties of space was not so very absurd, since Einstein would show three hundred years later how this can be worked out satisfactorily. However, this requires a totally new conceptualization of space and time.

*Pythagoras might be excepted. He somehow "anticipated modern theoretical physics." But he was in the Indo-Druidic tradition, which is different both from the pre-Socratic and the peripatetic ones (cf. Ref. 1).

†Pythagoras and Archimedes were able to use the mathematics of their times quite correctly for purposes of physics.

It was Newton who, thanks to the invention of calculus and owing to the correct idea of an external cause due to forces producing accelerations of bodies, founded the new mechanics which since bears his name and has served as the paradigm of all subsequent physics. At the same time, physics was no longer a part of philosophy, even though it was called—and remained for a long time called—*philosophia naturalis:* natural philosophy. In Newtonian mechanics, nature is an object, of which a model is given by a system of postulates and laws. There are no secrets left unlifted owing to an awe; on the contrary, physics is intended to disclose all of nature's secrets by discovering the laws of interaction between the various constituent elements of the world "according to the theories put forward by the human mind."

We can therefore say that nearly two thousand years of speculation starting with an awe for a nature full of secrets and ending in a construction of the mind, have led to the detachment of physics from philosophy, giving physics—and consequently science—the necessary autonomy to allow it to stand, and to develop, in its own right.

3. Critique of Positivism and of the Rejection of Metaphysics

From this time on, mechanics was characterized by such a success in its experimental verification and accompanied by such efficiency in its applications, that all other developing chapters of physics were modeled upon it, viz. hydrodynamics, electrodynamics, thermodynamics . . . , yielding a "physics" in a general sense which in its turn was crowned with a comparable success, so much so, that all sciences attempted at doing the same along analogous lines. Physics has been ever since considered the paradigm of science, so much so that—and as an ideological consequence of the *engoûment* connected with this—owing to the immense difficulty of conceiving metaphysics, in a way similar to physics metaphysics lost its capacity to also be accepted as a "science"; and, further, Auguste Comte and positivism arising from his thinking felt authorized to conceive of a (so-called positivistic) classification of the sciences in which successively number, space, time, mass, matter, life, soul, and society constitute by mere addition the objects of artithmetic, geometry, kinematics, mechanics, physics, biology, psychology, and sociology. Originally, Comte had forged the word *sociophysics*, which, however, encountered little success, and he replaced it later by the word *sociology*, which quickly became a popular idea, and this classification of the "sciences" came to appear during the second half of the 19th century totally satisfactory; especially, methaphysics was excluded from it, since there is no object in the list which could by adding it to the preceding ones give it a sense and content; therefore, metaphysics was soon declared dead.

However, the positivistic classification of the sciences does not stand a severe critique in the perspective of our days. First of all, mathematics (composed, in the positivistic classification, of arithmetic and geometry) is not a "science." Indeed, the various objects of mathematics are not real objects (called beings of fact: "*êtres de fait*" in the Cartesian epistemology), they are entities and constructions of the mind (i.e., beings of reason: "*êtres de raison*" according to Descartes), which already makes the various mathematical disciplines quite different from the disciplines arising from the field of "positive" science or science proper. Therefore, it does not make good epistemological sense to list number and space along with the other items considered by the sciences proper. Moreover, the phrase "mathematics is not a science" means that mathematics is not an enterprise leading to knowledge—in Latin: it is not a *scientia;* much rather, it is of methodological nature, it is a means to reach conclusions, a very powerful means indeed; it is, to put it in Latin, a *potentia.* The fact that mathematics with its various branches is not a scientia, but a potentia, not a science but a power, by no means reduces its importance; on the contrary, owing to the purity of its intellectual origin, it is the only power which is not usurped (e.g., all political power is usurped), and from this very fact, every science—which endeavors to use mathematics following the example given by physics—is powerful too; but since science, especially physics, is a grasp of nature, this means that science has a power over nature, and from this comes the danger which has been recognized in our days by ecologists and the like.

4. Rehabilitation of Metaphysics

Another difficulty towards the other end of the positivistic classification concerns the soul. Psychology, as the word says, seems to be a "science of soul" just as biology is assumed to be the science of life. But while looking at the problems put by psychology as it stands today, one is astonished that none of them deals with "soul." Much rather, they all deal with behavior. It appears doubtful that soul is an object of consideration: for, if we have a soul, it does not serve as an object, it serves exclusively as a "sense of the divine." Therefore soul is more comparable with the physical senses than with the physical "body." It is not the spiritual body or "mind," it is the sensorial capacity, i.e., the capacity which that spiritual body has "to grasp," and hence psychology should not be called "psychology." But if soul is a spiritual sense—since physics develops on the experimental plane as the use of the physical senses and of their prolongations the physical apparatuses, and of mathematics on the theoretical plane, there should exist a "science" parallel to physics using the soul on the experiential plane and some sort of dialectics parallel to mathematics on the theoretical plane: And indeed that parallel science exists: it is metaphysics.

Now, metaphysics seemed dead and in any case never followed the application of mathematics* to scientific purposes of research about the laws of interaction governing the behavior of nature. Mathematics results from the specialization of ordinary language along the logical analysis of this language. However, this specialization

*Attempts at submitting metaphysics to the systematic and exclusive use of mathematics, as are, e.g., proposed by Bunge or others, do not allow one to encompass the actual scope of metaphysics, which arises from the radical incommensurability between the finitude of plural objects and the infiniteness of the One Being—an incommensurability which mathematics does not master!

allows one only to grasp the relationship between finite entities. Thanks to methods of complete induction and the definition of continuity yielding the continuous ensemble, it can at most grasp asymptotically or by contrast with discontinuity the notion of the Infinite and conceptualize it mathematically. But it cannot make sense of the radical incommensurability which reigns between finitude and infinity, or—as a corollary—between multitude or plurality and oneness or uniqueness. But precisely that incommensurability is the source of metaphysical speculation. For, it is it which the soul senses. Therefore, if there is a science to be called metaphysics, one should not look for it along the positivistic line, even though the mathematical disciplines, unable to truly grasp the radical incommensurability, have been for reasons just given taken out of the list; for metaphysics must make use of a discursive method which does not result from a specialization of ordinary language along the logical line towards mathematics; and, indeed, metaphysics has followed another line, which is that of dialectics, beginning in a very remote antiquity, being elaborated by Plato, leading then to Plotinus, restored, amplified, and formalized by Hegel (which does not mean that Hegel's dialectics is its last form, but which is wrongly continued along the Marxist proposal, for dialectical materialism makes the mistake of claiming that dialectics is anchored in a matter conscious of its own historical destiny, an error analogous to that made by Aristotle when he had his physics rest on the claim that the motion of bodies resulted from the tendency of these bodies to follow and reach their own ends). Dialectics is capable of further development. Seen in that perspective, metaphysics is far from being dead, it is a science with its own kind of objects, but a science *sui generis* for which the experiential aspect does not follow from interactions between finite beings (like gravitation, electromagnetism, etc.) but follows from the suffering of the world owing to the incommensurability just mentioned; whence the problematics of metaphysics: in relationship with God (and—dialectical—proofs of God), the soul (and the proof of the soul), freedom, love etc. Consequently, especially the concept of soul as is envisaged in ordinary scientific (sometimes called experimental—not to be confused with metaphysically experiential) psychology is not the concept of soul as is envisaged in metaphysics. That psychology is a psychophysics in the Comtean sense analogous to sociophysics.

We can illustrate this difference between physics and metaphysics by the following simplified scheme of their respective objectives:

(i) There are various forms of energy, like kinetic energy, electromagnetic energy, heat, energy developed by the absorption of food by animals, even money is a form of energy when it is exchanged for the acquisition of goods. (Besides energy, there are other magnitudes to be similarly dealt with in exchanges.) Now, we can—and we must—say about the exchanged magnitudes, especially about energy: The more we give, the less we have. Physics deals with that kind of exchange and so do the sciences in some sort of extension.

(ii) But there is something different from energy and the other exchangeable quantities. Of it, we can and must say: The more we give, the more we have. What is that second kind, for which no "physical" law of conservation applies? It is love (and what goes along with love). Precisely metaphysics deals with these kinds.* They make it possible for metaphysics to be meaningful because the absence of conservation of exchange is only possible in relation with the incommensurability of the metaphysical. So, metaphysics is alive.

There is a further reason, why the positivistic classification has to be rejected. It arises from time being of a nature totally different from the nature of the other items listed in the classification. For if, on the one hand, the conceptualization of the notion of time in physics seems to make it similar to, or of the same nature as the nature of, space on the one hand, mass, matter, and so forth on the other hand, actually time keeps a nature of its own which nothing else possess, viz., the fact that first every physical (generally speaking material) thing depends upon time, and that second every mathematical "thing" has no reference to time. Now, there again space has to be understood correctly: if one means space as it is determined by the axioms of one particular geometry (Euclidean, Riemannian, . . . ; one-, two-, three- . . . pluridimensional), then it is as a geometrical (or mathematical) entity totally outside the temporal changeability of material things; if, however, the position of things as referred to such and such space of coordinates is meant, then the coordinates defining this position are dependent upon time, even in the case of rest, which is a peculiar form of time dependence, viz., (momentary) time independence. (This argument can be generalized to any property of matter or field.) Time in physics (and in all the sciences *per extensionem*) is *the independent variable* upon which everything else of scientific relevance is (in principle) dependent; and if there are within a "theory" such entities as are found to be necessarily independent of time, then they are invariable or invariants of time and as such very important magnitudes. If, moreover, they are also independent of the way they are measured (coordinates independent), i.e., universal, then they are most important. Such are the universal constants, be they of theoretical nature like $\hbar$, c, e, k as building stones of the theories, be they of empirical nature as constants of interaction to be measured and numerically determined in terms of the theoretical constants within the frame of each theory like G, ϵ_0, and the constants of strong and weak interactions.

But then, if Time is different from both what stands left and right of the list of entities considered in the positivistic classification, this whole positivistic classification falls into pieces and must be—if one insists on making a classification—replaced by another one. A classification which is compatible with all the above critique would require (i) that a distinction be made between mathematics and the (positive) sciences, (ii) that a separate classification of the mathematical disciplines be founded on suitable categories, (iii) that another, separate classification of the sciences be founded on categories which will have to reveal the various dependences upon time—call them if you please temporal categories, like reversibility for the science of mechanisms, i.e., physics proper (planetary motion, electromagnetism without radiation, etc.), irrevers-

*Of course, psychology claims to deal also with love, but the way it does so is totally inadequate and unable to cope with case (ii) mentioned above.

ibility for the science of machines or engines, i.e., thermodynamics (referring especially to the increase of entropy for isolated systems), reproducibility for the science of geno-systems—as the first consideration of biology (where generations issue new generations in spite of each system being due to die as a machine on account of thermodynamics and of its relative isolatedness), etc., (iv) that metaphysics be given a suitable place in the complete scheme, especially as the science which deals with the transcendence beyond the temporal.*

5. The Factual, the True, and the Correct

All the argumentation given so far amounts to a critical view on science made from outside science, and physics in particular. This is a philosophy of science (of physics). Philosophy has among its tasks to reflect upon the sciences, especially upon physics. This is one aspect of the relationship "physics and philosophy." It has led us to the view that mathematics is not a science, that metaphysics is a science (*sui generis*). Metaphysics especially is found to fulfill the double criterion for a discipline to be a science, viz., to be theoretical and experiential (sometimes called experimental). However, its theoreticality goes along the dialectical instead of the logico-mathematical as for the "ordinary" sciences, and its experientiality goes along the suffering of the world towards the grasp of the Absolute (the Infinite and One, if you please) instead of experiment on finite beings towards the grasp of their interactions. A particular argument which shows that mathematics is not a science consists in saying that if, on the one hand, scientific propositions about laws of nature can and must be either verifiable or falsifiable (cf. Popper), that is if generally speaking assumed scientific truth has to be checked by the facts, mathematics on the other hand is never checked by any fact. Here, we must clearly distinguish between three orders of epistemological nature: factuality (which is never "true," hence propositions like "Napoleon had a son" is not true or false, it is factual or unfactual), truth (which remains always assumed as long as no facts falsify it; hence, truth has to be verified by facts; e.g. Maxwell's equations combined with Lorentz' force), correctness (of mathematical reasoning, the main criterion of which is noncontradiction at any step of the application of the accepted rules of reasoning); but the correctness of mathematics is never submitted to (factual) verification, it is itself—besides the factual verification—a second criterion for the disclosure of truth, viz., that the (theoretical) establishment of would-be truths (as deduced from assumptions or scientific postulates and laws) be made (mathematically) correctly.

*Physical reductionism, i.e., the view that thermodynamics is reducible to mechanics by means of statistical mechanics, that furthermore biology will also be reducible to physics, etc., is wrong, for it is not true that thermodynamics is reduced to mechanics, etc. (This has been commented upon elsewhere at length.) Reductionalism is an elementary form of scientism, which is itself a wrong doctrine; it is an ideology.

At this place, and as an intermediary conclusion illustrated by the argumentation developed especially in Sections 3 and 4, we may say that if, on the one hand, physics has detached itself from philosophy of old, it has not lost all relationship to philosophy, for a philosophy of physics (and generally speaking a philosophy of science) has arisen out of the process, i.e., a critical examination of the way physics is built up and works.

6. Four Main Problems for the Philosophy of Physics

It is therefore of some importance to sketch here what are some of the main problems of this philosophy of physics. They have been at least partly already touched upon in the above, but let us name them now one after the other. We shall choose four of them: one of a more metaphysical concern, physical time; one of both ontological and epistemological concern, determinisms; one of epistemological and methodological concern, interaction; and one of a classificatory concern.

(i) Physical Time. Any science proceeds by conceptualization. That is, notions originally understood by some kind of intuition are made into a concept, the character of which is to receive its proper place and to play its proper role in a conceptual frame relating the various concepts among one another by formalized ways of discourse which require that each concept always keeps the same significance (or changes its significance if the formal relationships are changed). Time is one of the most fundamental of all concepts, for it arises from one of the deeply innate ideas (in the Cartesian sense) or notions of the human mind. Its conceptualization depends upon the discipline which is tackling its problematics: theology, metaphysics, the various sciences, etc. It not only yields one of the most difficult questions of metaphysics; it is also found to be the most fundamental concept of physics, because all the other concepts are assumed time dependent according to the various physical theories at hand. It makes a difference whether the conceptualization is made, e.g., on the basis of a Newtonian, or of an Einsteinian gravitodynamics. For, in the Newtonian picture, time is a one-dimensional continuum endowed with the same structure as that of the natural numbers, whereas in the Einsteinian picture of general relativity, space-time, which is the new concept of time,[2] is a four-dimensional continuum into which the light-cone structure is built. These are examples, so the "problem of time" in physics divides into various argumentations according to the physical theory at hand.

By being the most fundamental question of a physical conceptualization and, at the same time, owing to the deep concern metaphysics has for it, time and its possible physical or metaphysical conceptualization are at the very border which distinguishes physics from metaphysics. And since the question persists of whether a metaphysics can subsist in spite of the invading success and pervading expansion of science, especially of physics, the problem of the compatibility of physics and metaphysics is found to be most acute precisely around the concept of time elaborated in one and the other case. We shall leave it at that, for this matter has been dealt with elsewhere.[2,3]

(ii) Determinisms. The problem of a determinism is connected with the problem of time as is posed in each theory. Therefore, one has to consider several forms of determinism, or determinisms. This problematic has been dealt with in detail elsewhere[4]; we shall not go into details here. The general conclusion to draw from the considerations developed there, however, is important: First, determinism is not to be interpreted as a fixed mechanism footed in the reality of physical things, for, indeed, there are various determinisms according to the various theoretical approaches of the physical; hence, determinism is, if you please, not of ontic nature, i.e., it is not rooted in reality. To make determinism an ontological problem and to conclude, e.g., that an antinomy arises between determinism and free will, is therefore a fallacy. We "believe" we see a determinism in physical reality because we approach that reality by means of such and such theoretical construction of our mind and "believe" the construction to be final and identical with reality. But the belief is an unjustified projection. Actually, determinism is also a conceptual business. And since there are various conceptions (or conceptualizations) of time according to the various physical theories, there are various determinisms. By the way, none of them is a radical indeterminism; especially quantum theoretical determinism is not, for it is a very respectable determinism since it makes very good sense to predict quantum theoretical magnitudes such as cross sections, which are excellently confirmed by measurements. However, quantum theoretical determinism is different from (but weaker than) Laplacian determinism, which itself is different from (and weaker than) Einsteinian determinism (see ref. 4), etc.

(iii) Interactions. Whereas in a tradition going back to Aristotle's physics, especially mechanics was originally conceived as the study of the motion of bodies, it is now conceived as the understanding of interactions among "bodies." However, these bodies are no longer the bodies of old. Even the antique idea of atoms (including its prolongation in the so-called "elementary" particles) does not appear any more to yield the last constituents of matter; much rather, "atoms," i.e., in modern language, particles, are nothing but the result of the application of the more fundamental idea of quantization upon the primary manifestation of matter as a field or complex of fields. For these reasons, the concept of a force is ill adapted to express the various interactions, because it refers to only one body (susceptible of acceleration), whereas interaction necessarily assumes partners. Meanwhile, what interaction is—or interactions are (for there may be at least two categories of such)—is not yet well understood today. It can philosophically—i.e., ontologically, if you please—though very vaguely be said, that it is the fate of matter to be in time, the general cause for all properties of matter to be time dependent, and the like. The epistemological problem posed by the conceptualization of interaction has scarcely been tackled at all and remains a problem for deeper consideration.

(iv) Reversibility, Irreversibility, Reproducibility, "Life," etc. Material systems can be classified according to such categories of behavior in temporal respect.

There are, first, systems which can be called mechanisms and which behave in such a way that if the flow of time were reversed ($dt \rightarrow -dt$, e.g., in Hamilton's equations) together with either space symmetry ($dq \to -dq$) or momentum symmetry ($dp \to -dp$) and, in certain cases, charge conjugation, nothing is changed in their integral physical behavior. This is the contents of the PTC theorem (where only space symmetry is envisaged in P together with time reversal in T for relativistic reasons, but where it could be possible, though not intuitive, to replace P and T by reversal of all the canonically conjugate momenta including that which is conjugate to t, however not identical with energy!). Examples of such systems are the planetary system, the atom, etc. Such systems are at the same time in principle periodic, as, e.g., the planetary motion is. It is true that beside planets, there are comets, which do not seem to necessarily fulfill a condition of periodicity; but this is due to the fact that ellipses may degenerate into parabolas or even be replaced by hyperbolas owing to changes of sign of parameters. Thus, "mechanisms" are such systems as behave reversibly, and are in principle periodic.

Secondly, there are systems which, when isolated, are subject to an increase of their entropy. They may be called machines (or engines). A machine contains at least one mechanism (whereas a mechanism does never contain a machine and is *qua* mechanism not equivalent to a machine). Contrarily to a view held by physical reductionists, a machine cannot be reduced to a mechanism. Therefore the physcis of mechanisms can at best be considered as a special case of thermodynamics where no engines are present, whereas the converse is wrong. But it remains clear that the mechanisms included in a machine behave according to the physics which is proper to them.

There are systems capable of begetting generations, i.e., of reproducing "life." They may be called vitalisms, or genetisms, for it is the main feature of living systems to be able to reproduce themselves before death. A vitalism includes at least one machine. All machines included in a vitalism are submitted to the principles of thermodynamics; this is the reason why in the sequence of generations death happens again and again to members of the sequence. But the sequence goes on (or stops sometimes owing to exterior factors or to loss of reproducibility) by virtue of the principles governing vitalisms, which are capable of reproduction.

We might go on in establishing the list of these categories. It appears in our days as a method of classification preferable to the positivistic one.

We shall not try to increase the enumeration of "problems" for a philosophy of physics (and of science). Especially, we shall not refer to problems issued by the history of science. The ones mentioned above were meant as important illustrations and also as examples which have not (yet) been dealt with by an epistemology which, in recent times, has become purely formal and has very much neglected items of really positive scientific concern.

7. Conclusion

What can we conclude?

First of all, in spite of the detachment from philosophy undergone by physics, philosophy can be said to remain very close to physics in every respect of the consideration of objective knowledge.

Then, physics and metaphysics are not incompatible. If, however, some teaching of physics is found contradictory to some teaching of metaphysics, we must accept that either is, eventually both are, wrong and need revision. Especially, since metaphysics claims to deal with that which transcends finitude, it attains distances beyond the reach of physics and may forget the latter and not conversely; hence it is more likely that physics may act as a corrective of (erroneous) methaphysics than conversely, but it is not impossible that metaphysics may redress a clumsy physics. The relationship of the two, which applies in the field of objectivity since both physics and metaphysics are sciences, has its analog in the field of subjectivity, i.e., within the arts, when one considers the relationship between poetry and music. Indeed, poetry is the (artistic, or subjective) counterpart of metaphysics; it deals with the same objects and intends the same overcoming or transcendence of finitude, whereas music is very much like physics since it deals with the dynamics of finitude in forms within non-animate matter. It is well known that music is one excellent touchstone of poetry in songs and the like.

This compatibility can be generalized. The view, held by some positivists and materialists especially at the end of the 19th century, that physics in particular and science (metaphysics excluded) in general have made religion unacceptable, is devoid of ground. Science, physics especially, and religion are compatible. For both their origin and their scope differ and should not contradict each other. Religion does not consist in believing "stories" (of genesis and the like) which are shown by science to be wrong; religion relies on surrender to a call from a divine voice which tells man not to be afraid of the world in which he is. (For example, the Arabic word "Islam" means surrender.) Philosophy on the other hand consists in cleaning at each step the environment of a thoughtful situation arrived at from which or through which the next intellectual step may be and shall be taken to reach a new thoughtful situation. The sciences are instrumental in that action of cleaning. The call and the cleaning are not incompatible, they are complementary, if you please, and they may even often lead to the same conviction, e.g., about God or the self or what have you. But they may also act as corrective to each other.

One of the great teachings of 20th century physics as compared with the boast of 19th century science is that it has acknowledged the limitations of science. We do not say merely "limits," but limitations, meaning that physics is certainly progressing, but precisely because it progresses, it is never all-encompassing; but especially and moreover, it is not made for the grasp of a nonphysical problematics. The same must be said of science in general. We owe that teaching to both Einstein and to Bohr more than to any one else. If physics serves as the paradigm of the sciences, it does so owing to historical reasons, but still more so, it seems, because it is the most exact of the sciences by virtue of both its use of the most advanced mathematics and its application of the most elaborate experimental technique. But even then, it should not be confused with the alpha and omega of knowledge and wisdom.

References and Notes

1. Ward Rutherford, *The Druids and their Heritage* (Gordon & Cremonesi Publishing, London & New York, 1978), pp. 46 and 47.
2. Cf. A. Mercier, "Physical and Metaphysical Time," *Epistemologia* **I**, 337–352, (1978).
3. A. Mercier, "Tiempo físico y tiempo metafísico," in Proc. First World Congress of Christian Philosophy, Cordoba-Embalse, Argentine, 1979. In press (Sociedad Catolica Argentine de Filosofia, Vol. III, Cordoba, 1981), p. 1469.
4. A. Mercier, "Forms of Determinism, Objectivity, and the Classification of the Sciences," in *Perspectives in Quantum Theory. Essays in Honor of Alfred Landé,* W. Yourgrau and A. van der Merwe, editors (MIT Press, Cambridge, Massachusetts, 1971).

10

Einstein, Spinoza, and God

Arne Naess
University of Oslo
Oslo, Norway

"Einstein and Spinoza" is a tempting theme for a Spinoza scholar and Einstein admirer, as well as for a Spinoza admirer and relativity specialist. In what follows I shall pick out a small part of the many subthemes which today deserve renewed consideration: God, Nature, determinism, and timelessness.

When Einstein was invited to contribute to the *Spinoza-Festschrift 1632–1932* he declined: "Unhappily love for Spinoza is not enough to justify the writing of a dissertation about him." Unlike Sigmund Freud, who declined for a similar reason, he added some substantial words about Spinoza's philosophy. He was the first, according to Einstein, who has "with real consistency applied determinism to human thinking, feeling and action." He adds that this not only requires consistency in thinking, but also "an unusual purity *(Lauterkeit)*, greatness of mind *(Seelengrösse)*, and—modesty." These requirements prevent determinism from being universally accepted among those "who fight for clearness and consistency."

These words attest to the deepness of feeling with which Einstein defended his deterministic outlook. There are, however, many open questions as to exactly what this outlook implies. The word "determinism" has never had a single, fairly definite meaning, and there is little reason to believe that it ever will. In what follows its relation to concepts of God and essence will be mentioned.

"I have often felt and occasionally also stated that Einstein stands in particularly intimate relation to the God of Spinoza."[1] These words by his old friend and collaborator Arnold Sommerfeld deserve to be taken seriously. "Many a time," says Sommerfeld, "when a new theory appeared to him arbitrary or forced, he remarked: "'God doesn't do anything like that.'"[2] Taken in isolation, the remark might be used to support the belief that Einstein's God was a transcendent God, but what he has said and written makes it clear that his God was immanent in nature. This is a cardinal idea for Spinoza. Humans are genuine parts of nature, therefore they partake in God's determining power—however modestly and fragmentarily.

The expression *Deus sive Natura*, "God or Nature," Nature *with capital N*, occurs four times in the writings of Spinoza. It is clear that one must have the distinction between *natura naturans* and *natura naturata* in mind when interpreting this famous expression. It points to Spinoza's conceptual distinction between an active and a passive aspect of a supreme totality for which he does not have a particular term. God is the name for the active aspect, not for the totality.

It seems that Einstein's feelings concerning harmony and lawfulness *in* nature make a similar distinction relevant. The God of Einstein is not *identical* with nature or the universe. God is immanent. The immanence is a different kind of relation, a relation difficult to articulate conceptually. Einstein uses the verb "to reveal" in the famous cable which Einstein felt he had to send as a reply to a cable "Do you believe in God?": "I believe in Spinoza's God, who reveals himself in the harmony of all being,[3] not in a God who concerns himself with the fate and actions of men." God is revealed *in* something, namely, all being, with which he is *not* conceptually identical. On the other hand, God is not *apart* from all being.

Spinoza does not use the term "revealed in," but prefers "expressed through." God is expressed through the modes, the particular beings. God is not something *apart* from these expressions. (This may sound heretical to some Spinoza scholars, but seem to me to be essential to his system if it is to be understood today.)

The characterization of Spinoza as a "pantheist" is often interpreted in the direction that he identifies God with the universe or the world. It is therefore a better terminology to call his view "panentheist" ("God in all"). The difference is essential in discussions of the "problem of evil." That is a still more formidable problem for pantheists than for panentheists. As a panentheist Einstein may have less difficulty in defending his theory of harmony and his maxim that God is sophisticated, but not malicious. (Incidentally, interpreting the maxims of Einstein, one should take into account his particular sense of humor. Uttered by a Spinoza, they would carry a heavier philosophical burden.)

Both Spinoza and Einstein leave questions open when talking about harmony, order, and simplicity. Clearly they do not refer to the actual state of affairs, for instance the state of international politics. Einstein says in a letter to C. Lanczos "that the problem of gravitation made him a 'believing rationalist': the physically true is logically simple, that is, it has unity at the foundation."[4] This feeling or intuition leads us again to the question of "God or Nature," *Deus sive Natura*.

That "God" for Spinoza is not another name for "natura" is seen most convincingly if we put the term *natura* wherever Spinoza uses the term *Deus*. In some places this results in nonsense or at least in statements which Spinoza would not admit as valid. Even when inserting *Natura* with capital N, there are difficulties. Perhaps Spinoza's own development from being a young Jewish theologician to a consistent philosopher was never quite completed. Einstein did not undergo a similar dramatic development and might have more easily identified God with Nature.

Does Spinoza's God not "concern himself with the fate and actions of men"?

Human beings share in the double aspect of *natura naturans* and *natura naturata*, they are expressions of God or Nature, and as such, are free *(homo liber)* and *to an extent* determine their fate and actions thenselves. There is no God somehow *outside* human beings that might concern himself with men. The power used by a human being who acts, not compelled from the outside, but in harmony with his or her essence, is part of the power of God (*Ethics*, Part 4, proof of theorem 4).

The question of power leads us to a second great and controversial area of Einstein's thought: that of determinism.

As for Spinoza, determinism for Einstein was primarily a belief in order and harmony, as opposed to chance and chaotic antagonisms, not as opposed to causelessness.

The determinism of Spinoza has, in contrast to that of Einstein, often been labeled "rigid," "absolute," "merciless." There are quotations from Spinoza which support such a view. But if we take Part Five, and the last sections of Part Four of his *Ethics* seriously, and not as expressions of inconsistency within his system, these labels must be rejected as misleading. As expressions of God and as part-taker in God's power, men are themselves determining their actions. They are creators of something genuinely new. Each thing and therefore each human being "expresses in a certain definite way the Power of God [Nature] by which He [Nature] is and acts". (*Ethics*, Part 3, proof of Theorem 6). Humans are part of *natura naturata*, but also a genuine part of the creative *natura naturans*. In so far their actions are determined by their nature or essence, they are free. Freedom is self-determination. Freedom as determination from one's own nature and essence, rather than from something external, foreign, or even antagonistic to it, is the only freedom rationally conceivable. Lack of determination, that is, *chance or arbitrariness*, does *not* provide freedom. Our freedom is not secured by a dice-playing God.

Harmony among people is for Spinoza a consequence of acting freely and not compelled from something external such as dominating passions. Einstein's unceasing fight against cruelty and oppression seem to imply a belief in the *possibility* of harmony of the kind Spinoza contemplates. The fight was one fought against irrationality. But Einstein explicitly rejects narrow or pure rationalism in favor of religiosity. For Spinoza there was no problem here: Love of God *(amor Dei intellectualis)* is the highest good "we can seek according to the dictate of reason *(ratio)*." Since the time of Spinoza the distance of the meaning of the term "reason" from that of *ratio* has increased. It is now mostly used for what Spinoza would call "calculating reason" *(ratiocinatio)*. Spinoza would agree with Einstein that rationalism in today's sense is too narrow as a basis for fighting cruelty and injustice.

In the terminology of Spinoza freedom and rationality imply harmony with the nature or essence of the actor. The free actions are performed within the framework of laws of nature, not outside. The term "order of nature" is pertinent because as expressions of "God or Nature," the laws are not antagonistic or arbitrary from the standpoint of men, who are themselves expressions of "God or Nature."

Spinoza is, however, eager to eradicate anthropocentric ideas of order and purpose. Men may understand more and more of the laws, as laws of Nature, but it implies a gradual dismissal of notions of law due to his passive emotions, his original slavery under his passions. The genuine laws do not compel or coerce but make freedom possible.

Sir Karl Popper tried to save Einstein from his determinism, which "amounted to the view that the world was a four-dimensional Parmenidean block universe in which change was a human illusion, or very nearly so."[5] Popper wanted him to fully acknowledge "the reality of time and change," and an "open" universe, "one in which the future was in no sense contained in the past and the present, even though they do impose severe restrictions on it."[6]

It is not quite clear what Einstein answered in these discussions, but he was scarcely moved in his basic views. The limitation of the approach of Popper is due to his comparison of Einstein's view to that of Parmenides rather than to that of Spinoza. The latter comparison would hardly have made the discussion simpler, but certainly more to the point.

Insofar as laws of nature strictly deserve that name, they are in Spinoza's view eternal. But this does not mean the same as that they are permanent in time. Differences in time are *irrelevant*, because these laws are "outside of time." Consider the formula for π. Even if true, it is not to the point to characterize its validity by saying that it certainly is, was, and will amount to the same, 3.14159. . . . Its validity is timeless rather than permanent.

The timeless laws of Nature are like the law of *logos* of Heraclitus: They are completely consistent with universal and pervasive physical change. "You cannot step twice into the same river," but there is an eternal *logos in* the changing river. Change and time are real, but so are the timeless laws. Without the particular things with limited life-span in time, there would be no eternal aspect of reality and no "God or Nature."

This could be said, so far I can see, by both Einstein and Spinoza without the slightest difference in basic meaning.

Einstein acknowledged that his proposals in the form of physical theories are free creations of the mind, and that his laws of course are hypothetical, but he tended to hold that insofar as they agree with the basic laws of nature, the time dimension does not enter. Neither do intrinsic statistical factors.

According to Bohr, he and Einstein had a good-humored dispute about whether Spinoza, if he lived today, would have agreed with him or with Einstein concerning the basically satisfactory character of statistical quantum theories as expressions of physical reality.

If we limit ourselves to the most careful and precise formulations of Einstein's negative conclusion, it is difficult to see how Bohr could have argued with success that Spinoza would have agreed with his view rather than Einstein's. But perhaps Bohr at that time was not confronted with Einstein's most careful formulations. The latter stresses that there is much truth in quantum theories, but as long as there is a basic statistical factor, they are incomplete: a complete theory would have to be nonstatistical, deterministic in Spinoza's sense.

Some Spinoza scholars will interpret Spinoza as Popper interprets Einstein: that the future somehow is contained and determined by the past. But then one misses the emergence of the future through the actions of *natura naturans*, including the actions of the more or less living beings. One need not disbelieve in the *genuinely new* in what happens. There is nothing in the *Ethics* corresponding to the past-present-future mechanical determinism of Laplace. What happens is *not* arbitrary, *not* by chance. But from that it does not follow that what happens is contained and (fully) determined by the past.

Einstein sometimes talked as if he believed that humans are determined to act by forces over which we have no control whatsoever. In an interview he is reported to have said the following—intended to counteract the overestimation of his own personal achievements:

> "I claim credit for nothing. Everything is determined, the beginning as well as the end, by forces over which we have no control. It is determined for the insect as well as for the star. Human beings, vegetables, or cosmic dust—we all dance to a mysterious tune, intoned in the distance by an invisible piper. [Quoted in R. W. Clark, *Einstein*, p.422]

So far as I can see, it is only the complete determination from something distant which here contradicts Spinoza. He stresses our power in shaping events as part of God's or Nature's creative power. Einstein danced to the particular tune of Einstein when he acted rationally *(ex ratione)*, that is, acted from "the necessity of our own nature" *(ex necessitate nostrae naturae)*. Einstein often seemed to act that way, for instance in the discussion about the completeness of quantum theory!

Mustering a power extremely small compared to the total power of Nature, determination from far away is considerable, but not overwhelming: No part of God's power is of zero magnitude, and the specific power of Einstein was and is part of the power of God. "The power of man, as far as it is expressive of his own actual essence, is part of the infinite power, that is, essence, of God or Nature." (*Ethics*, Part 4, proof of Theorem 4.)

Do these lines *solve* any problems? Scarcely. The notions of time, timelessness, and determination are difficult to grasp and belong to the ultimate concepts of philosophy. It may be argued that if a philosopher does not grasp what his colleague says in these matters—despite repeated serious attempts at communication—not much can be done except noting this interesting fact. There may be irreducible differences among the "freely invented" (Einstein) conceptualizations through which reason tries to grasp reality.

References and Notes

1. P. A. Schilpp, editor, *Albert Einstein: Philosopher-Scientist* (Open Court, La Salle, Illinois, 1949), p. 103.
2. Arnold Sommerfeld, in Ref. 1, p. 103.
3. Translation by Sommerfeld. Quotation from *New York Times*, 25 April 1929, p. 60, col. 4.
4. Quoted by G. J. Holton, in *Boston Stud. Phil. Sci.* 6, 186 (1970).
5. *Unended Quest*, revised ed. (Fontana, 1976), p. 129.
6. References p. 130.

11

The Satellites of Jupiter, from Galileo to Bradley

John North
University of Oxford
Oxford, England

The satellites of Jupiter are hardly to the forefront of the typical astronomer's consciousness, despite the remarkable findings of the Voyager mission. Even historians of astronomy tend to pass over the satellites in silence, perhaps thinking of them as nothing more than a trivial extension of the solar system—and paralleled by roughly comparable systems of moons around Mars, Saturn, Uranus, and Neptune. In the seventeenth century, however, they carried a cosmological message of great importance, for they were first seen at a time when the old and the new world systems were contending for the favor, not merely of astronomers, but of a significant fraction of the educated world. They were seen by Galileo in 1610. In 1676 Ole Rømer made use of them to show that light takes time to travel. My account runs for roughly half a century beyond this date, stopping at Bradley because, as I hope to explain, he marks the end of the first, largely empirical, phase of investigation of the satellites. I do not mean by this that no further empirical work was done—on the contrary, the most dedicated work of this sort was still in the future. Bradley's proof of the aberration of light nevertheless clinched the argument for the finite velocity of light, at least in the eyes of reasonable men. He was one of the first to allow for the velocity of light in tables of the four satellites then recognized, and he it was who first saw that the inequalities in their motions are interconnected—and thus possibly a consequence of gravitational interactions. In a sense, therefore, he opened the way to the theoretical studies of this problem by Euler, Bailly, Lagrange, Laplace, and others.

[Author's Note: For the sake of those who are not familiar with the Jovian system, so far as it is known at present, it should be pointed out here that the satellites are numbered in the order of their discovery. They fall into three groups. The inner satellites (V,I,II,III,IV) revolve in much the same plane, with a direct motion. The next three (VI,VII,X) revolve in the same sense, but in three different planes; but then we have VIII, IX,XI, and XII, all in different planes but with retrograde motion. I am mostly concerned with I,II,III, and IV.]

The Discovery of the Satellites

The effect of the discovery of the satellites by Galileo on his own intellectual development was itself dramatic. The story is well known: by the beginning of 1610 he had made telescopes in the Dutch manner, the most powerful of them giving a magnification of about thirty. This he turned to the skies in January 1610, to discover that the Moon was clearly mountainous, that the Milky Way is a collection of individual stars, and that Jupiter is accompanied by lesser planets. On 7 January he saw three, but took them to be merely bright stars, close to Jupiter, and in a straight line.* The following day they were, to his surprise, so arranged that it seemed to him that Jupiter must have a *direct* motion, whereas all the standard ephemerides, the "computations of the astronomers," in Galileo's words, suggested a *retrograde* motion for the day in question. On 9 January the sky was cloudy, but on the 10 January one of the "stars" had disappeared while the other two again seemed to have changed sides. (Similar phenomena had shown themselves a few days earlier to Simon Mayr, if we are to believe his story. His findings however, which were not to be published until 1614,† had a much slighter cosmological impact than Galileo's.) Galileo now decided that Jupiter's motion could not be responsible for this strange behavior, and by 11 January he had decided that Jupiter must be accompanied by three wandering stars, comparable, as he says in the *Sidereus nuncius,* to Venus and Mercury who revolve around the Sun.‡ In this book, the *Sidereal Messenger,* he published his new telescopic discoveries. The book appeared at Venice in March 1610, and a day-by-day account of Galileo's observations of the satellites was there published, ending only ten days before the publication.§ The satellites proved to follow Jupiter in his retrograde and his direct motions, said Galileo, adding that they settled the doubts of those who, while accepting the Copernican system as a whole, were disturbed at the thought that our own Moon was unusual in revolving around the Earth, which in turn moved round the Sun. (One consequence of this was that the Moon indulged in a double motion of a sort thought contrary to the canons of Aristotelianism.) But our own eyes show us, he said, four stars which wander round Jupiter as does the Moon around the Earth,

*For the page of the ms. describing the following sequence of observations, see Galileo[1] [**3** (2), p. 427, and **3** (1), pp. 35–37, 80–81], and Drake.[2]

†References to his works are given below. Mayr (Mayer, or Marius) is said to have made a partial announcement of his discovery in an astrological calendar of 1612, but an even earlier printed reference is in the introduction to Kepler.[3]

‡Galileo.[1]

§Galileo[1] [**3** (1), p. 81].

the five bodies together circuiting the Sun in about twelve years. His parting words in the book amount to a hastily formed hypothesis to explain the appreciable fluctuation in the brightness of the satellites. This fluctuation he put down to the interposition of Jupiter's atmosphere.* This hypothesis, which seems to us so innocuous, must have been profoundly irritating to the Aristotelians, who would have seen here yet another threat to their system in the rival "elmental" sphere of Jupiter's "envelope of vapors."

The *Sidereus nuncius* was Galileo's first printed testimony to his belief in the Copernican hypothesis.† The first printing of 500 copies sold out immediately, and he achieved instant fame. A second edition was published in 1610 in Frankfurt. By the summer of 1610 he had resigned his chair at Padua and returned to Florence as mathematician and philosopher to the Grand Duke of Tuscany, in honor of whose family, the Medici, Galileo had, with great foresight, named the new satellites of Jupiter.‡ He was also made chief mathematician at the University of Pisa, without teaching duties. Galileo was a maestro at the art of orchestrating his own fame, but the excitement engendered by his discovery of the satellites was as spontaneous as such things can be. Within months, Kepler had published two short works on the subject, the *Dissertatio cum Nuncio Sidereo* (Prague, 1610),§ written before he had seen the new planets himself, and the *Narratio de observatis a se quatuor Jovis satellitibus* (Frankfurt, 1611), after he had seen them (between 30 August and 9 September).|| The second book, or rather pamphlet, was at once reprinted in Florence. It was, after all, by the Imperial Mathematician, and was useful ammunition in the battle for the respect of the Jesuits of the Collegio Romano, a battle which Galileo won on his visit to Rome early in 1611.¶

For two years, Galileo steered a reasonably safe theological course. His troubles began in 1613, with the publication of his letters on sunspots.** It would take me too

*Galileo[1] [3 (1), pp. 95–96].

†See, however, the earlier *Lettera a Jacopo Mazzani* (30 May 1597), Galileo[1] (2, pp. 197–202). This received a small circulation, hence my word "printed" rather than "published."

‡Following his example, Cassini named the satellites of Saturn the "Ludovician" planets, in honor of his own patron, *Le Roi Soleil*. The individual names still used for the first four of Jupiter's satellites were provided by Simon Mayr. See P. Humberd.[4] See also¶ note on p. 695 below, and Galileo[1] [3 (1), pp. 94, 136].

§Why, asked Kepler, should Jupiter be circled by four moons, if there is no one on the planet to admire the spectacle? He concluded that both our own moon and Jupiter may be inhabited, and that we may some day fly to them. See Kepler[5] (2, pp. 497, 502), and Kepler.[4] Antonio Maria Schyrlaeus de Rheita[7] indulged in a similar conjecture. If Jupiter has inhabitants, he said, they must be bigger and more handsome than we, and in the shape of a pair of spheres. He considered whether they might still be in a state of moral innocence. Huygens in his *Cosmotheoros*[8] describes Jupiter and his satellites by analogy with the Earth and Moon, in ways reminiscent of de Rheita. Galileo[1] (5, p. 220), on the other hand, violently rejected the notion that there are living creatures on the planets.

||Kepler[5] (2 pp. 510–513).

¶Christoforo Clavius to Galileo in Florence, 17 December, 1610, Galileo[1] (10, pp. 484–485). Clavius says that the Medicean planets had been observed at Rome, congratulates Galileo, and provides some data. There are various letters concerning Galileo's triumphant visit to Rome in Galileo[1] (11, pp. 78–84).

**Galileo, 10 reprinted in Galileo[1] (5, pp. 71–245).

far afield to consider either this controversy, or Galileo's personal fortunes, but I must say more about his work on the satellites of Jupiter, if only as a corrective to the view that they represented to him no more than a qualitative cosmological spectacle, a Copernican system in miniature, as it were.

Already, whilst he was in Rome to show off his astronomical findings to the cardinals, to the Jesuits, and to the members of the Academia dei Lincei, Galileo wrote to Belisario Vinta that just as God had let him discover them, so He would let Galileo determine the laws of motion of the new planets, even though this required nightly observation.* His observations of them were indeed carried on for many years, but his tables of their motion, which were to have been published by a follower (the Olivetan and Professor at Pisa, Vincenzo Renieri), were lost to the sight of his contemporaries. Renieri actually completed the tables after Galileo's death in 1642, but was unable to publish them before he himself died in 1647, and not until the nineteenth century were the tables found—in a private library in Rome.† Accurate as even Galileo's early tables were, for their time, all were thus without influence, although many of Galileo's observations were used by later astronomers to assist in the derivation of the periodic times of the satellites.‡

Galileo's observations were made from the time of the discovery until the end of 1619—first at the Villa delle Selve, near Florence, then in Rome, and finally at the Villa Segni, in Bellosguardo, from whence he moved to Arcetri.§ He followed the movements of the Medicean planets, therefore, for somewhat less than one cycle of Jupiter.‖ His first technique of angular measurement is one which I have discussed previously, in connection with Thomas Harriot's work on sunspots.¶ It involved gauging angular distance in terms of the apparent field of view. Quite apart from the fact that it is unsuitable when the field is dark, it is unreliable to the extent that the field of view is a function of the size of the eye pupil. In 1612 Galileo made himself a crude micrometer.** A scale was placed *outside* the telescope, and was viewed with one eye, whilst the other was applied to the telescope. This crude device worked moderately well in expert hands.††

One of the principal attractions of the Jupiter system was that it presented a

*Galileo[1] (**11**, pp. 79–80).

†Galileo[1] [**3** (2), pp. 403–424; **18**, passim].

‡Some early data were published in the "Discourse Concerning Things Which Float on Water," for example; cf. Galileo.[1] Galileo communicated more accurate information than that in the *Sidereus nuncius* to G. B. Agucchi. See Galileo[1] (**11**, p. 205, 9 September 1611, pp. 214–215, 7 October, 1611, pp. 225–227, 29, October 1611).

§Galileo[1] [**3** (2): *I Pianeti Medicei*]; S. Drake.[11]

‖One condition for satisfactory value for the periodic times is that averages are taken over long intervals between moments when Jupiter is at an identifiable point of its orbit.

¶(pp. 145–146).

Galileo[1] [3** (2), p. 446].

††See letters, 548 and 561 of G. L. Ramponi to Galileo in Galileo[1] (**11**, pp. 133–136 and 159–162) where such measurements are discussed. R. M. McKeon[13] argues that such a scale was used by Galileo. See also A. Borelli.[14]

challenge to astronomers. Here was a restricted system with fast-moving planets for which the earlier history of astronomy might be regarded as a sort of rehearsal. Over the next two centuries the truth was slowly appreciated: the satellites presented all the problems of conventional planetary astronomy, and some others besides. I will consider the complications as they were appreciated historically, beginning with the most obvious, namely, that we observe the system from a moving Earth.

Making the approximation that the orbits of the Earth, Jupiter, and those of his satellites that were then known all lie in the same plane, a reasonable idea of what can be expected of a single satellite can be had from a drawing in correct proportion; and if you graduate this with scales (of angle or of time) of the sort used on medieval planetary equatoria, you will have a useful nomogram for evaluating actual and expected phenomena. Galileo designed such a device, called in Italian "giovilabio." More precisely, I should say that there are several related nomograms in the mss., reproduced in facsimile in the National Edition of Galileo's writings.* The principles implicit in the nomograms were incorporated into a brass *giovilabio* which is now in the Museum of the History of Science at Florence, and which has been described by Maria Luisa Righini Bonelli.† It is certainly Galilean in general conception, although it is probably by a late contemporary. I shall have more to say of its design and its engraved tables later.

The Satellites and the Longitude Problem

The brass *giovilabio* was meant not only to allow the expert astronomer to arrive rapidly at the probable disposition of the satellites, but to assist the mariner who wished to determine longitude at sea. This was a particularly vital subject for most of those who investigated the satellite orbits during the seventeenth and early eighteenth centuries. There is even reason for supposing that Galileo was driven to develop the pendulum-controlled clock by his ambition to solve the longitude problem through the use of Jupiter's satellites.‡ This method was first mentioned in a letter dated 11 November 1635, from Galileo to Jean de Beaugrand.§ Nine months later Galileo outlined the method, and offered it *gratis* to the States General of the United Provinces of the Netherlands.|| The States General had followed the example of Philip III of Spain, who in 1598 offered large sums of money to the first who should solve the problem of longitude. Huygens speaks of the "long promised reward," but the details

*Galileo[1] [**3** (2), pp. 486–487; cf. partial diagrams at pp. 477, 479, 481, 483].

†Bonelli.[15]

‡The first use of the pendulum for timekeeping was apparently a medical one, namely, for taking the patient's pulse (published in 1602 by Sanctorius, generally attributed to Galileo). See especially Ariotti.[16]

§Galileo[1] (**16**, p. 344).

||Galileo[1] (**16**, pp. 463–468, letter 3337).

are obscure. The governments of Venice (?),* Britain (1714), and France (1716) all offered similar rewards, although the British were the only government ever to pay, when a satisfactory solution was forthcoming.†

By the mid-eighteenth century Jupiter's satellites were out of favor for this purpose, chiefly because it was so difficult to make precise telescopic observations from a rolling and pitching ship. Throughout the previous century, however, astronomers had lived in the hope that they would solve the practical problems entailed, especially that of making a compact and powerful enough telescope. In a letter to the States General, dated 13 August 1636,‡ Galileo had given a masterly summary of the problem as a whole, together with an indication that he knew very little about the practical realities of seafaring. His clock was for keeping time only from midday to the sightings—but of course it would not have been accurate without a stable platform, any more than could the mariner have viewed the planets without some leveling device. (Remember that good telescopes were around two meters or more in length!) One could not risk losing sight of Jupiter for any length of time, since eclipses of the satellites lasted for only a minute or so. It was these eclipses, of course, which were to be used as the phenomena to fix a universal time, so to speak, and it was the times of their occurrences that Galileo now felt he was in a position to tabulate with reasonable accuracy.

The Dutch admiral Laurens Reael showed interest, but asked about the stabilizing of the telescope.§ In March 1637 Galileo begged Elia Diodati to assure Constantijn Huygens that he had a solution to this problem, and he himself sent it to Reael in June.‖ It involved a massive universal joint, with one hemispherical component moving inside a second, fixed to the ship. The two were to be separated by water or oil, the gap between the hemispheres being preserved by some eight or ten springs.¶ The arrangement simply could not have been made to work effectively. Perhaps I should add here that, despite numerous attempts to solve this technolgically difficult problem, a solution became feasible only in our own century, with the possibility of large, high-speed, gyroscopes.** In a final letter to Galileo, Reael tells him that his method is implausible, and impracticable for his sailors, "who are rude people, men only superficially acquainted with mathematics and astronomy . . . and who still find insuperable the problem of using your discovery on a moving ship, continually tossed about."†† For his work, the States General nevertheless offered Galileo the gift of a gold necklace, but he declined the offer.‡‡ According to Gould, the method of finding longitude

*I have included Venice, following Gould,[17] but have so far been unable to confirm the claim elsewhere.

†Gould.[17] The British Board of Longitude disbursed more than £100,000 before 1828, when the Board was finally abolished.

‡Galileo[1] (**16**, pp. 464–467).

§Galileo[1] (**17**, pp. 39–41).

‖Galileo[1] (**17**, pp. 46–49, 96–105).

¶Ariotti[16] (p. 368); Galileo[1] (**16**, pp. 96–99).

**Gould[17] (p. 7) lists eight later attempts to design a stabilized observing point, between 1719 and 1858.

††Galileo[1] (**17**, pp. 116–117, 22 June 1647); Ariotti[16] (p. 369).

‡‡Galileo[1] (**17**, p. 371).

from observations of Jupiter's satellites has never been successfully applied at sea.* This should not obscure the fact, however, that the method was frequently used in the seventeenth and eighteenth centuries on *land*; nor should it obscure the fact that the intense study of the problem of Jupiter's satellites had motives which were largely utilitarian, that is, had to do with the longitude problem. For this, moreover, they were by many thought to offer the *best* solution.†

Priority Disputes

As Galileo and others found, the motions of the satellites were more complex than had at first been suspected. Galileo's figures for the mean motions are much more accurate than is generally recognized, although it is a matter of taste, whether or not they justify the energy he threw into the project. Simon Mayr's tables of the satellite motions have occasionally been claimed as more accurate, which as a whole they were not, although it would be quite wrong to go to the extremes of Robert Grant, who wrote that Mayr

> . . . who contended for the independent discovery of the satellites, resolved to strengthen his claims by the construction of tables of their motions. The crude labours of this impudent pretender were, however, no sooner given to the world than they fell into deserved oblivion.‡

Grant was here falling into line with propaganda of Galileo's making. Galileo and Mayr were old antagonists, but not until 1623, in *Il Saggiatore*, did Galileo go into print maintaining that Mayr's *Mundus Jovialis* of 1614§ was a complete plagiarism.‖ Galileo was of course stung by Mayr's claim to priority, but Mayr had also tried to name the satellites the "Brandenburg stars,"¶ after his own patrons, thus jeopardizing Galileo's standing vis-à-vis the Medici. On Mayr's side, one might point out that he claimed to have begun to observe the satellites one day after Galileo,** and that the *tables* in his book antedate anything comparable by Galileo; they incorporated more accurate motions than Galileo's first published motions, as will be seen from my table of comparison. I shall raise other points of comparison as the opportunity arises.

*Gould[17] (p. 7)
†For this note see the account in Halley[18] of his reduction of Cassini's tables, discussed again here below.
‡Grant[19] (p. 79).
§Mayr.[20]
‖Galileo[1] (6, pp. 197–372).
¶He also called them for the first time by their customary names, Io (I), Europea (II), Ganymede (III), and Callisto (IV), as well as "the Mercury," "the Venus," "the Jupiter," and "the Saturn" of Jupiter. He acknowledges that Kepler first suggested the now customary names to him. Hevelius used the names "Saturn" etc.
**Mayr[20]; see the *Observatory* reference at p. 372.

The Galileo/Mayr controversy is one capable of arousing strong emotions, or so it seems, for a prize essay on the subject by Josef Klug (the only entrant), written at the end of the last century,* was so hostile to Mayr that one of the judges appointed by the Dutch Academy of Sciences, namely J. A. C. Oudemans, joined with Johannes Bosscha to defend Mayr.† The controversy continued for some time, and more recent reference to it may be found in a summary article by J. H. Johnson, and an excellent article, introducing a number of contemporary documents for the first time, by Pietro Pagnini.‡

The controversy raises three separate questions: Was Mayr capable of observing Jupiter's satellites early in December 1609(O.S.), the date at which he claims to have seen three of them? Second, did he draw up his tables from Galileo's published data? Third, did he falsify the dates assigned to his own observations, assuming that we suppose them to be genuine? Priority disputes may be tedious, but the noise created by them is a part of history. Unfortunately, this example cannot be settled except in terms of the characters of the disputants and the potential interdependence of Mayr's data and Galileo's relevant publications.§ Klug argued that Mayr could have relied on measurements taken from the plates of the third *Solar letter* of 1613, but Bosscha showed that had he done so he would have found results much less accurate than those he published.‖ It appears that the chief inaccuracy in Mayr's tables was due to badly chosen *radices* for the motions. His estimate of the diameters of the satellites, of the radii of their orbits, and of the mean motions, are good and on the whole better than any Galileo had then made public. (A cut-and-dried comparison is impossible, since Galileo changed his parameters from time to time, and there is little to be gained by chronicling the changes here.)

Mayr's book does not read like an exercise in studied deception, and a reasonable conclusion might be that he it was who first saw three of the satellites, and who first realized that they were moving, but that Galileo first saw four satellites, and did so independently. That Mayr is telling an honest tale in the *Mundus Jovialis* seems all the more probable since a letter of his, reproduced in the preface of Kepler's *Dioptrice* of 1611, tells, in outline, much the same story.¶ (Why did this not irritate Galileo? Did he not see it until after 1614?) At this level of naive discovery, their findings were virtually simultaneous. What is not at all obvious—and here I follow Pagnini**—is that Mayr was able to *explain* the observed phenomena before he had read the *Sidereus nuncius* in 1610. Mayr probably reached a full understanding of what he saw

*Klug.[21]

†Oudemans and Bosscha[22] and Bosscha.[23]

‡Johnson[24] and Pagnini.[25]

§These are: *Sidereus nuncius* (1610), *Discorso interno alle cose che stanno in su l'acqua* . . . (1612), the plates of the third *solar letter* (1613)—see Galileo[1] (5, pp. 241–245). Of less significance is the earlier correspondence between Agucchi and Galileo, for which see Klug.[21]

‖See Oudemans and Bosscha[22] and Bosscha.[23]

¶Kepler[3] (pp. 27–28).

**The crux of the argument is the absence of Mayr's name in certain key letters dating from 1610.

only towards the end of 1610. This is compatible with his good faith as regards the wording of the preface to the *Mundus Jovialis,* but it means that he is to be placed at a lower level of originality than Galileo, however accurate an observer he might have been.

In addition to Galileo's and Mayr's, another important series of observations of Jupiter's satellites was Thomas Harriot's. Harriot was in possession of a copy of the *Sidereus nuncius* by June 1610, and he began his systematic observation of the satellites on 17 October 1610(O.S.).* At first he could see only one, and not until December 14(O.S.) could he see all four of the satellites known to Galileo.† It is necessary here to remind ourselves that in 1610 there were telescopes and telescopes. William Lower observed with a friend, and could see not one of the satellites, as he tells Harriot in a letter of 4 March 1611.‡ Peiresc similarly found it impossible to obtain a telescope equal to the task before November 1610.§ Harriot has left us records of his own observations of the satellites made between October 1610 and February 1612, and calculations of their motions. Dr. John Roche has made an analysis of these unpublished ms. records. He has found that in 1611‖ Harriot made numerous calculations of the periods of the satellites, sometimes using only Galileo's published data, sometimes only his own, and sometimes comparing his own with Galileo's data. Making such a comparison he found the sidereal period of the first satellite to be 42.4353^h (correct value: 42.4582^h). In preferring to establish sidereal rather than syndoic periods, Harriot was wiser than Galileo and Mayr. For some reason he failed to use the same methods with the other satellites, and indeed did not persist in his satellite observations after February 1612—perhaps because he was disappointed with his observing conditions.¶ Much of what Rigaud has to say about Harriot's Jupiter observations is unfortunately incorrect.** He was wrong, for example, to say that Harriot gave no indication that the satellites could be off the ecliptic.†† Harriot called attention to the fact on several occasions,‡‡ although he made no attempt to lay down a theory of latitude.

There is ample testimony to the excitement caused by Galileo's announcement of his discovery of the moons of Jupiter,§§ and it is hardly surprising that others—including Mayr, Harriot, the Jesuits at the Collegio Romano,‖‖ and Agucchi—imme-

*Harriot[26] (ff. 1–14); article by J. D. North in Shirley[12] (pp. 136–137). On 1 June 1610 (O.S.) Jupiter was in conjunction with the Sun, and the satellites would have been difficult to see then and for several weeks thereafter.

†Harriot[26] (f. 2).

‡Halliwell.[27]

§Rigaud.[28] See the "Supplement . . . with an Account of Harriot's astronomical Papers," p. 28.

‖Roche.[29]

¶Roche.[29]

**North in Shirley[12] (pp. 146–147).

††Rigaud[28] (p. 25, supplement).

‡‡For example Harriot[26] (f. 4).

§§See, for example, Kepler[6] (pp. 9–10).

‖‖Galileo[1] [3(2), pp. 861–864]; Christoph Scheiner;[30] de Rheita[7] (p. 282); P. Gassendi;[31] Delambre;[32] Riccioli;[33] Hérigone (Cyriaque de Mangin);[34] Jacques Godefroy, see Brown;[35] Fontana;[36] for Hevelius see Delambre[32] (2, p. 437) and Hevelius.[37]

diately followed him in recording similar observations. I could refer also to the announcement of observations by such men as Giuseppe Bianchini, Christoph Scheiner, Schyrlaeus de Rheita, Pierre Gassendi, Pierre Hérigone, Jacques Godefroy, or Francesco Fontana, but this would only be to pile the obvious on the obvious. It seems to me that the study of the satellites took a more serious turn with the writings of Gioanbatista Odierna* and Giovanni Alfonso Borelli.

Odierna's Three Inequalities and Borelli's Celestial Mechanics

Odierna's *Medicaeorum ephemerides* . . . was published at Palermo in his native Sicily in 1656. It was dedicated to the Grand Duke Ferdinand II of Tuscany, and indeed Odierna makes an abortive attempt to rename the satellites after members of the Medici house—Ferdnipharus (IV), Cosmipharus (III, the brightest), Victripharus (II, after Victoria, wife of Ferdinand), and Principharus (I, after the heir-apparent). "Florence" was to be the face of Jupiter, and "the Arno" the bands across the face.† What makes Odierna's work noteworthy, however, was neither such window-dressing as this, nor his data for the mean motions of the satellites,‡ but rather the introduction of new parameters into the theory of the satellite motions. He enlarged upon a theme first published by Mayr, namely, that of motion in latitude. Galileo had expressed the opinion (in his polemic against Mayr in the introduction to *Il Saggiatore*)§ that the satellites are always in a plane parallel to the ecliptic. When Jupiter is in the ecliptic they will therefore appear to us to be in a straight line, on this view; and when, for example, *Jupiter is north of the ecliptic*, the satellites will seem to be *north* of the main planet when in the *inferior* position. They will be *south* when in the *superior*. When Jupiter has a southerly latitude, the appearances will be reversed. Mayr took a different view, namely, that the satellites were in orbits permanently inclined to the ecliptic. He gave tables for the satellites' latitudes based on estimated inclinations of the orbits (10, 10, 12, and 15 seconds of arc for I–IV, respectively), but in his explanation of the tables failed to take into account the movement of the parent planet in latitude. Neither Galileo nor Mayr therefore gave an adequate account of movements in latitude. (And for the record, I should here note that Galileo's attempt to discredit Mayr fails when he argues that he could not have seen the movements in latitude he claimed to have seen in the first half of 1611.) As a matter of fact, the satellites are in orbits virtually coincident with Jupiter's equator, and the latter is inclined at about 3°04′ to Jupiter's orbit. Unless we know the *direction* of Jupiter's axis, or alternatively, the nodes of the orbits, we cannot make any progress with the determination of lati-

*"Hodierna" in most early printed sources.

†He used alternative names for the satellites: Alphipharus (I), Betipharus (II), Cappipharus (III), and Deltipharus (IV). These are a Graeco-Latin alphabetical compromise.

‡His figures were inferior to those of the best of his predecessors.

§Galileo.[38]

tudinal movements. Odierna was a careful observer, capable of dismissing on empirical grounds the opinions of Galileo and Mayr, if not of reaching a completely satisfactory theoretical interpretation himself. Against Galileo he noted that conjunctions may take place more than half a diameter from Jupiter's center. Against Mayr and Galileo he found from observation that the satellites are *always north* of Jupiter when *superior*, and *south* when *inferior*.

Odierna set forth three sorts of periodic inequality. The first was a consequence of changes in satellite latitude. Delambre seems to believe that he began by taking the plane of the satellites to be inclined at 45° to that of Jupiter's orbit, and that he then expressed the maximum latitudes in tenths of a disk and found for the four satellites 1°59′, 3°7′, 5°6′, 8°29′—which he later found inexact.* What Delambre does not seem to have realized is that these figures fit together with Odierna's maximum distances for the satellites in such a way as to suggest that he believed them all to lie in a plane inclined (on my calculation) at about 3°21′ to the ecliptic. This is by no means a despicable result.

Odierna's second inequality, according to Delambre, was the annual parallax, and the third was due to Jupiter's movement round the Sun.† From Delambre's account it is hard to decide exactly how these were calculated, but it is doubtful whether together they gave more than that correction which is a function of the Jupiter–Sun–Earth angle, and which will be discussed later. Odierna was perhaps the first to note that the outermost of the four satellites then known is (unlike I–III) not invariably eclipsed at superior conjunction with the shadow cone.‡ He was also aware that with I, and usually also with II, only immersion or emersion is observed. (The unobserved phenomenon is hidden by Jupiter.)§ He listed past eclipses, some of which were spurious, but whether because they were imagined, whilst he observed under poor conditions, or whether because erroneously retrodicted, it is impossible to say. Odierna is the first to stress the importance of eclipse observations, and his tables bring eclipse calculations to the fore. He failed to account accurately for eclipse durations, but those who know the corresponding difficulties in connection with eclipses of our own Moon will forgive him his shortcomings. His ephemerides of the satellites, running from 1656 to 1676, are not as accurate or as thorough as the few who possessed telescopes might have liked, but they did help to prepare the ground for later ephemerides.

The ephemerides of 1656, with a memorable terminal date of 1676, were just half-expended, so to speak, when Borelli published his work *Theoricae mediceorum planetarum ex causis physicis deductae* (Florence, 1666). Borelli's strategy was to describe the inequalitites in the motion of our own Moon and then to transfer them to the

*Delambre[32] (2, p. 329). The original is rare, and I have not seen a copy.

†Delambre[32] (2, p. 329).

‡We know that the eclipse fails to occur when the orbital plane (of IV) is inclined at more than about 2° to the Sun–Jupiter line.

§For III and IV, both phases are usually visible. In general, eclipses are seen on the western side of Jupiter before opposition and on eastern side after opposition. Near opposition the shadows may be behind the disk, as seen from the Earth, so that eclipses are invisible.

theory of the motion of Jupiter's satellites. His book is perhaps best known for its celestial mechanics, that is, for its explanation of the elliptical orbits of planetary bodies* in terms of three types of forces: (1) a central force; (2) rays of force acting like spokes from the central body, i.e., rays which, with the rotation of that central body on its axis, move the planet round; (3) a centrifugal force. The satellites of Jupiter were supposed to be simultaneously acted upon by Jupiter's rays [as under (2)], and the rays of the Sun, and these influences were thought to combine to give the satellites their complex motions.† Borelli's planetary mechanics is not a subject for a brief aside, and I shall leave it well and truly alone, merely pointing out that it follows upon a general description of the satellite motions.‡ In his preface, Borelli tells the reader the circumstances under which the book was written. Campani sent a fine telescope to the Grand Duke, who then commanded Borelli to use it to study Saturn and Jupiter and to investigate the accuracy of Galileo's tables of the satellites' motions. Accordingly, in the summer of 1665 Borelli began his work, in an observatory in the castle of San Miniato, near Florence.

Borelli was one of those who finds it easier to lay down the lines of a bold program than to fill in the details. Working by analogy with Keplerian cosmology, he postulated ellipitical orbits for the satellites, with moving lines of apsides, inclined orbits with moving nodes (retrograde, of course), evection and variation—all as in the case of our own Moon. There were the three inequalities reckoned by Odierna, and in principle many others besides. But when it came to the extraction of new empirical parameters, all of Borelli's grandiose theory came to nothing. He made age and ill-health his excuse, but it is quite clear that the analysis was as far beyond his talents as were the necessary observations beyond the reach of seventeenth-century instruments.

It is difficult to decide how important were good telescopes to progress in this part of astronomy. If we consider the satellite observations of Hevelius, as recorded in *Selenographia* (Danzig, 1647),§ we find that the derived motions were decidedly inferior to those by Galileo, Mayr, and Harriot well over thirty years previously. Hevelius complains of his telescopes, but it is hard to believe that they were no better than those of the 1610 era. Nevertheless, whilst Hevelius could see the great spot on Jupiter, he could not see the bands which Odierna had named "the Arno." Hevelius gave almost the same figures for the maximum elongations of the satellites from Jupiter as his predecessors had done. All were estimates made in terms of the diameter of Jupiter's visible disk, usually loosely set at one minute of arc. Needless to say, the crudity of such an estimate made nonsense of Borelli's ambitious scheme, although it must be granted that he had recognized the virtue of a form of micrometer (here essentially a

*Delambre[32] (p. 333) says incorrectly that Borelli indicates no physical cause for the motions of the satellites.
†Borelli[14] (pp. 45–81).
‡Koyré[39] and other works cited there.
§See §§ note, p. 697 above.

scale in the focal plane of the eyepiece), which he says was communicated to him by Buono.*

Gian Domenico Cassini

If one may talk of a "heroic" period in the history of Jupiter's satellites, it must be taken to begin in or around 1664, when Gian Domenico Cassini,† still in Italy, obtained fine telescopes of long focal length from his acquaintances in Rome, the great telescope makers Giuseppe Campani and Eustachio Divini. I shall consider Cassini's observations as a natural prelude to Rømer's, and those which followed in their wake in England.

Within weeks of obtaining the new telescopes, and whilst using one of Campani's with a length of about 5 meters, Cassini was observing the shadows of satellites II and III on Jupiter's disk when he noticed an uncharted spot. Some days later he saw two or three movable dark spots which he assumed to be clouds, and bright marks, which he took to be volcanoes. He explained what he saw, when in 1665 he published *Tabulae quotidianae revolutionis macularum Jovis* . . . ‡ He was a lover of analogy, and his analogies with known phenomena did not stop with clouds and volcanoes, but extended to the theory of the satellite motions. He observed a notably large spot on the face of the planet, and was at length able to use it to settle the planet's period of rotation at 9^h56^m—say 10 hours.§ Now the Sun rotates once in about 28 days, and it provides Mercury—assuming the solar-ray theory of celestial motion—with a period of revolution of about 88 days. Should the innermost satellite not therefore have a periodic time of about 31.4 hours? In fact satellite I has a period of about 42.5 hours, a third as much again. (Cassini knew nothing, of course, of satellite V, the nearest now known to Jupiter, with a period of nearly 12 hours.) Cassini, who is here following Kepler's own reasoning, but who lacked Kepler's analytical skills, decided that Jupiter was less efficacious than the Sun, each judged vis-à-vis his planets. As others before him had done, Cassini alluded to the possible habitation of Jupiter.

The tables of Jupiter's rotation were published on 12 June 1665. On 22 July of the same year there appeared from Cassini's pen *Lettere astronomiche al signor ab. Falconieri spora l'ombre de pianetini di Giove*. This work reveals that Cassini, through a study especially of shadow transits, was aware of inequalities of only a few minutes of time. He studied many shadow phenomena, and made a restricted attempt to correlate them with variations in Jupiter's ecliptic latitude. He also made a half-

*Hevelius[37] (p. 145).
†Cassini[40] and Taton.[41]
‡Cassini.[42]
§A modern figure for the rotation period is $9^h\ 50^m\ 30^s$ (equatorial), increasing to the figure quoted by Cassini, in the temperate zones.

hearted attempt to find algebraic relationships between central distances, periodic times, and transmit times, for the satellites.

Dissatisfied with existing tables of the motions of the satellites, Cassini observed them carefully for more than two years, before publishing his *Ephemerides Bononienses mediceorum syderum* . . . in 1668. Apart from further descriptive material concerning Jupiter's disk, the work was not unusual in character, and yet the sheer accuracy of predictions made from the tables was impressive, to contemporary eyes. Picard, in particular, lavished praise on them, and almost simultaneously with their publication Cassini received Colbert's invitation to come to the then recently founded Académie Royale des Sciences, in Paris. Cassini's reputation throughout Europe was considerable, and although in retrospect he appears to have had a knack of making wrong conjectures, his observational work was already both extensive and accurate.

Cassini regarded his visit to Paris as a temporary affair, but in fact he was never to return permanently to Italy, so generous were the terms of his stay. His work on the satellites of Jupiter continued, although naturally enough he had—as always—many other irons in the fire. He apologized later for the inaccuracies in his 1668 tables, due, he maintained, to the haste with which he was pressed to publish them. The apology is in the introduction to *Les Hypothèses et les tables des satellites de Jupiter*, published in Paris in 1693, the culmination of many years of careful work. In the same year, Cassini published his *De l'Origine et du progrès de l'astronomie*, in which some pertinent remarks are made about the satellite problem. Seventeen years had passed since Rømer's greatest achievement, and still Cassini was reluctant to accept the idea that light travels with a finite velocity. Edmond Halley, who reduced Cassini's new tables to the meridian of London and the Julian calendar,* could not understand Cassini's reluctance to utilize the "equation of light."† If we look at *De l'Origine*, however, we find evidence that Cassini had himself considered the possibility of such an equation, and had rejected the idea. This evidence supplements a number of indirect statements by Montucla,[43] Du Hamel,[44] the *Mémoires* of the Academy,[45] and others, quoted by I. B. Cohen in his fine memoir on Rømer.‡ Here, writing in 1693, Cassini gives us the reasons for his decision, reasons which we shall find far from cogent. The passage is interesting enough to be quoted at length:§

> The observations which the Academy made of the satellites of Jupiter provided an occasion to examine one of the most beautiful problems of physics, which is, to know whether the movement of light is successive, or whether it is effected in an instant. The times of two successive emersions of the first satellite during one of Jupiter's quadratures were compared with the times of two successive immer-

*Halley.[18]

†Halley[18] (p. 239).

‡Cohen.[46] Note a small error in regard to the date given in the *Histoire*[45] for Cassini's publication of his conjecture, August 1675, not 1674.

§Cassini.[47] The methods which Galileo proposed, and which are called "useless," involve signaling with widely separated lanterns. See Galileo[1] (**8**, p. 88).

> sions of the same satellite during the opposite quadrature; and although the light of the satellite during the first quadrature covers less distance to reach the Earth when it approaches Jupiter than in the second quadrature when Jupiter is separating from the Earth, and although this difference amounts to 60,000 leagues at the very least as between the two occasions, nevertheless, scarcely any appreciable difference was found between the two intervals of time. This gave occasion to believe that the observations one can make at the surface of the Earth, or even throughout space as far as the Moon, do not suffice to form a definite conclusion on the problem, and that, as a result, the methods which Galileo proposed for this effect in his mechanics, are useless. It is not that the Academy, in the series of observations, failed to notice that the time of a considerable number of immersions of the same satellite is appreciably shorter than a similar number of emersions, which may be explained by the hypothesis of the successive movement of light; but that this did not seem enough to convince that the movement is really successive in fact, for one is not certain that this inequality of times may not be produced either by the eccentricity of the satellite, or by the irregularity of its movement, or by some other cause at present unknown, which in time one might be able to clarify.

Delambre was perhaps the first to draw attention to the fallacy in Cassini's procedure.* I will rework the problem and assume the modern value of c, the velocity of light. We may reasonably take the Earth's approach at quadrature to be directly towards or away from Jupiter throughout the interval of time between successive observations of the satellite's reaching a particular recognizable position (e.g., maximum elongation). The technique is to measure this interval when the Earth is at opposite quadratures. If T is the true period of satellite I, and t the time for light to travel between the two points on the Earth's orbit from which the satellite is successfully observed, then the periods measured at opposite quadratures will be $T - t$ and $T + t$. The difference, $2t$, was what Cassini wished to establish. To a reasonable approximation, taking the Earth to move uniformly round a circular orbit, $2t = 4\pi aT/cY$, where a is the radius of the orbit and Y is the length of the year. We find that $2t$ is a little more than 30 sec, when $T = 42.5$ hr. For the record, the corresponding times for the other three known satellites are 61 (II), 123 (III), and 287 (IV) sec (all to the nearest second, in keeping with the above formula, although the implied accuracy is somewhat spurious, in fact, since the approximations made are invalid for large intervals).† Had he observed the outer satellites, or taken, as did Rømer, the time for a large number of revolutions of the satellites, he would have found the appearances much harder to explain. To be fair to Cassini, he did believe that Rømer's hypothesis was inconsistent with what was known of the outer satellites, and this was ostensibly the reason why, after Rømer had proved his point to the satisfaction of most of the

*Delambre[32] (2, p. 736).

†Grant[19] (p. 80) repeats a remark by Maraldi, without correcting it, to the effect that if errors in the eclipse times depend on a finite velocity of light then "they should be equal for all the satellites when the earth was in the same part of her orbit." My rough figures show clearly that if this means what it seems to mean, then it is mistaken.

best astronomers in Europe, Cassini nevertheless continued to ascribe the observed inequality to an unknown cause. A few followed his example—notably Giacomo Maraldi, his nephew, who was congratulated by Fontenelle for doing so.*

Since it seems that Cassini thought of the equation of light before Rømer, it is worth asking whether he found the idea elsewhere. The astronomical test for a finite velocity of light which Descartes proposed, in terms of the concomitant delay in the perceived eclipse of the Moon, had been mentioned at intervals ever since Descartes explained it in a letter to Beeckman in 1634.† It is possible that Descartes' proposal put the idea into Cassini's mind, but it seems to me not improbable that Huygens, whom Colbert had also brought to Paris, had considered the test carefully. It is well known that he discussed the Cartesian test at length, together with Rømer's "ingenious proof," in the first chapter of his *Traité de la lumière*.‡ This was first published in 1690, but we have evidence that the lunar eclipse test was "in the air" at least in 1665, when Hooke published *Micrographia*, for he there refers to it.§ Descartes' ideas on the velocity of light were constantly under discussion in Paris, especially in connection with the "proof" he gave of the law of sines, and I cannot help feeling that the general principles of eclipse tests must have been actively discussed publicly in the period terminating in Rømer's discovery.

How severely should we judge Cassini for the stand he took against Rømer's interpretation of the satellite observations? One has to remember that Cassini was the first to bring tables of satellite motions to such a state of perfection that they were reasonably reliable for years on end. For the first satellite, this was especially true; and since the same techniques and care had been applied to the others, was it unreasonable for Cassini to postulate unknown causes for the inequalities? We know that the irregularities in the outer motions are due to mutual perturbations, and these should indeed be identified with a part of Cassini's "unknown causes." But of course they do not cover the equation of light, and Cassini never really faced up to Halley's simple remark, made when he reduced Cassini's tables to the London meridian. It is hard, said Halley, "to imagine how the Earth's Position in respect of *Jupiter* should any way affect the Motion of the *Satellites*."‖

I will return shortly to the English treatment of Cassini's work, but first I should note some of his findings in regard to the satellite orbits. In his tables of 1693 they were all placed in a plane inclined at 2°55′ to the ecliptic, with nodes fixed at within two degrees or so of 10^s 14°30′ longitude.¶ His nephew Maraldi claimed to find from a study of eclipses that the orbital inclination of satellite II was changing. (He made

*For reference to the often-quoted letter see Grant[19] (p. 81), and Cohen[46] (pp. 348–349).

†Descartes.[48]

‡Huygens.[49]

§Hooke,[50] quoted by Cohen[46] (p. 337). Strangely enough, Hooke does not appear to have been convinced by Rømer. See Cohen[46] (p. 354, note 84). Hooke's argument can seemingly be paraphrased: "If light is as fast as *that*, then I don't see why its travel should not as easily be considered instantaneous"!

‖Halley[18] (p. 239).

¶Cassini[66] (pp. 43, 52, 99). For some of Cassini's predictions, which would disprove the claims of Galileo, Mayr, and Odierna concerning satellite latitudes, see Cassini.[51]

it 3°33′ in 1707.) Both men were often working beyond the limits of meaningful observations, and the wonder is that they came so near to the truth. (Their only way of finding the nodes, for example, was to search for the eclipses of longest duration, and there were many obvious uncertainties even so. As already explained, one never sees both phases of an eclipse of satellite I, and rarely both of II.)

To give an idea of the accuracy attainable with Cassini's tables of 1693, he listed four Paris observations (three immersions and one emersion) of the first satellite, all timed to the nearest second. The errors involved in calculating the phenomena were +2 sec, +58 sec, +53 sec, −9 sec.* It will later be shown that the errors are much greater with the other satellites; but in the *first* the astronomer was seemingly provided with a celestial timepiece accurate to better than a minute, and the longitude problem for places on land was solved, in the sense that a single observation of an eclipse of Jupiter's first satellite, combined with a star transit (or its equivalent), should give the terrestrial longitude within 15′(arc).† A whole series of observations should obviously give much better results, and especially if they were made simultaneously from some central observatory. The tables of 1668 had already led the French to launch a program of longitude determination the world over—along the French coast, in Cayenne, Egypt, the Cape Verde Islands, and the Antilles, for example.‡ Cassini's Paris observations served to control and coordinate the rest. Well into the eighteenth century the method was a standard one for the longitude of land bases—witness Lacaille's determination of the longitude of the Cape in 1750.§ The Cayenne expedition led by Richer in 1672–1673 had, as is well known, a more serious object than that of merely determining the longitude of the place. One of its prime purposes was to determine the parallax of Mars during the opposition of 1672,‖ when the planet was simultaneously observed by Richer at Cayenne, and Cassini and Picard in Paris. (The measured solar parallax of 9″.5 was 8% high, and inevitably affected estimates of the velocity of light made on the basis of Rømer's method, but most writers were wise enough to state only that light took such and such a time to transverse the diameter of the Earth's orbit.)¶ Needless to say, one of the most significant of all the French longitude expeditions was that which brought Picard, Bartholin, and Rømer

*Flamsteed put the maximum error at about 4 min. He explained at some length the method of finding longitude on land.

†Note Halley's enthusiasm for the technique: "[Cassini's] Account of the *Longitudes* observed, has put it past doubt that this is the very best way, could portable telescopes suffice for the work." Halley[18] (p. 237). He looked forward to the invention of shorter telescopes, manageable on board ship. Note that Newton observed Jupiter's satellites with his first reflector—a short telescope. See Newton.[52]

‡Cassini[47] (pp. 40–42) and passim.

§Lacaille.[53,54] Note Bradley's determination of the longitudes of Lisbon and New York by the same method—from observations made by others and compared with his own at Wanstead. In his mss. (e.g., Bradley 35, Bodleian Library) he lists longitudes for places the world over—from Peking to Paraguay.

‖"Observations astronomiques et physiques faites en l'Isle de Cayenne," in Cassini[47] (p. 2).

¶Rømer (according to Huygens) 22^m; Horrebow 28^m20^s; Halley 17^m; Wargentin 16^m26^s; Delambre $16^m 26^s.4$. The last is only about 10^s too low. Cohen[46] (pp. 351–353) gives some calculations which show that Rømer's data could have been used to give a more accurate value than that actually found.

to the island of Hven, to measure the longitude of Tycho's observatory, Uraniborg.* Again, Cassini's observations, made in Paris, served as a control. Picard returned to Paris with the information he had sought, with the original records of Tycho's observations, and with Rømer, whom he had persuaded to work at the Academy.

Rømer and the Finite Velocity of Light

Rømer's discovery—or "hypothesis", if you accept Cassini's attitude to it—is not well documented,† and we have no certain way of knowing how he calculated the time for the travel of light, although what we might call "Cassini's method" (already sketched), using eclipses of satellite I, was undoubtedly the framework of his calculation. In September 1676 he announced to the Academy that the eclipse predicted for 9 November would be ten minutes later than "naive" calculations based on previous eclipses would seem to suggest. When observations confirmed his prediction, he could state with some confidence that the cause of the delay was the finite velocity of light, and that its value was such that light would take 22 minutes to cross the diameter of the Earth's annual orbit. The published accounts are relatively brief. A translation of that in the *Journal des Scavans* appeared in the *Philosophical Transactions*,‡ and indeed it was this version that Huygens first saw. As explained, there is nothing there to augment Cassini's analysis of the satellite motions from a theoretical point of view. What Rømer managed to do was to dismiss the doubts that had prevented Cassini from reaching the same conclusion, and that Maraldi used as ammunition in a polemic against Rømer—doubts, that is to say, as to the simplicity of the problem. Was there not potentially a *multiplicity* of interconnected sources of irregularity in the motions? Rømer answered Maraldi's criticisms in letters to Huygens,§ showing that he regarded the first satellite as a simple case (with the light equation alone significant), and the outer satellites as presenting all the problems that Cassini had suggested. He never published anything to suggest that he had made a systematic investigation of the satellites II, III, and IV. It is possible that he failed to obtain inconsistent results, and that his intuition told him to abandon the subject with right on his side.

After Rømer, astronomers were left in a curious situation. The best available "calculus" of the Jupiter satellites was drawn up according to principles involving a mysterious "second equation" which no one appeared to be in any hurry to rationalize by the application of Rømer's principle. I have mentioned Halley's skepticism in response to Cassini's 1693 tables. Halley was content to make some small emendations, and merely raise his eyebrows at Cassini's techniques. The first complete systems to make consistent use of the equations of light were by Pound[58] and Bradley. I

*"Voyage d'Uraniborg . . ." with Cassini.[47]

†Most of his mss. were lost in the great fire that destroyed much of Copenhagen in 1728. See Eibe and Meyer,[55] containing a hitherto lost work.

‡See Rømer.[56]

§Huygens.[57]

will try to illustrate the main differences between the early systems I have mentioned, and this by reference to that class of instruments which goes under the generic name of *Jovilabe.**

SOME INEQUALITIES—THE JOVILABE

The first and greatest inequality in the apparent motions of the satellites about Jupiter as viewed from the Earth is that produced by the annual motion of the Earth. In Fig. 1 (taken for convenience from Lalande's *Astronomie*, Vol. III), angle *SIT* is the annual parallax of Jupiter, which may be as much as 12° or so, and the interval between conjunction with the Sun and conjunction with Earth, i.e., the time the satellite takes to cover 12°, is appreciable [1^h25^m (I), 2^h50^m (II), 5^h44^m (III) and 13^h24^m (IV)]. This inequality was, however, of slight interest to those concerned with the prediction and use of eclipses.

Eclipse phenomena would not recur at equal intervals of time, even if light traveled instantaneously and the angular motion of the satellite about the planet were constant, since the planet itself moves with varying velocity in its orbit around the Sun. (One might explain this simply by saying that the "starting line", i.e., the shadow, is moving with variable velocity.) The satellite must perform its own (sidereal) revolution, so to speak, and then move further through the angle traversed by the shadow, before it is eclipsed. Cassini,[51] in 1676, was critical of Galileo and other astronomers who had ignored this inequality. As he said, they took as simple a movement com-

*This word, as Delambre points out with a trace of disdain, joins a Latin word with a Greek. Delambre[32] (2, p. 331). We have met it already in Galileo's Italian "*Giovilabio.*" The Latin form is "*Jovilabium.*"

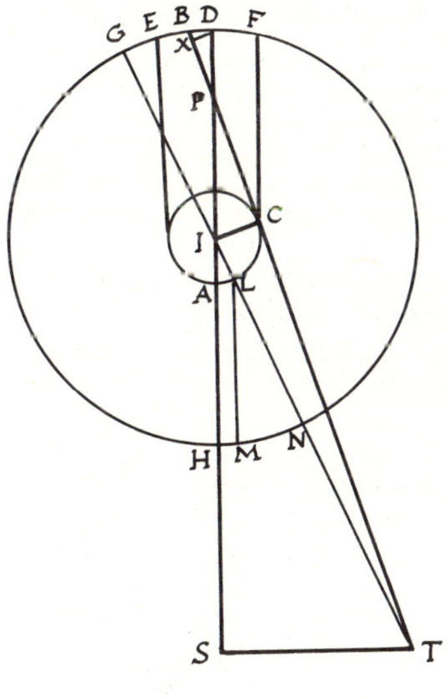

FIGURE 1. The effect of the annual motion of the Earth (T) around the Sun (S) in producing the first inequality in the apparent motion of a satellite (moving on the larger circle about Jupiter, the smaller circle—all on a very distorted scale). SIT is the annual parallax of Jupiter. The figure is redrawn (in its entirety, for completeness) from Lalande's *Astronomie*, Vol. 3, Fig. 244.

pounded from a uniform and a nonuniform motion. If this factor is not taken into account, the only way of obtaining reasonably accurate data for the mean motions is by observing Jupiter at the beginning and end of complete revolutions round the Sun, i.e., at intervals of about 12 years. This is more than most early observers were prepared to do, before publishing their findings. Galileo's manuscripts nevertheless show that he was aware of the problem. On one page, for example, he had drawn an eccentric orbit for Jupiter and marked part of it with what seem to be Jupiter's positions at ten-day intervals. Whereas the observations of 15 June 1610 to 17 March 1611 are given without correction for Jupiter's prosthaphairesis, those of 17 March to 16 July 1612 do have this equation incorporated.* As mentioned earlier, Harriot in his best calculations also took this inequality into account.

The simpler "Galilean" scheme (adopted by his contemporaries, of course, by reason of its simplicity) is well illustrated in the *Jovilabe* now in Florence (Fig. 2). The lower circle represents the Earth's orbit, with the Sun at its center, and the long (slotted) arm the line of sight to Jupiter. The four circles within the large upper scale represent the orbits of the satellites, and the transversals their separation from Jupiter, to a unit of the radius of Jupiter's disk. (The advantage of this unit is that it changes—as Jupiter's distance changes—in the same ratio as the apparent sizes of the orbits.) The parallactic angle (±12°) I mentioned a moment ago is measured on the small

*Galileo[1] [**3** (2), pp. 489–542].

FIGURE 2. *Jovilabe* attributed to Galileo. (Courtesy of Museo di Storia della Scienza, Florence.)

TABLE 1. Table of Mean Motions of Satellites I–IV

	I				II				III				IV			
Source	d	h	m	s	d	h	m	s	d	h	m	s	d	h	m	s
Berberich, calculated for 1612–14	1	18	28	34	3	13	17	42	7	3	58	48	16	18	0	0
Galileo, published 1612	1	18½			3	13⅛			7	4			16	18		
Galileo, Bellosguardo tabs	1	18	28	28	3	13	17	28	7	3	59	50	16	17	51	38
Mayr, published 1614	1	18	28	30	3	13	18		7	3	56	34	16	18	9	15
Harriot MS	1	18	28	48	3	12	33	36	7	7	33	36	16	16	0	0
Odierna	1	18	28	44	3	13	18	15	7	4	1	26	16	18	14	33
Florence giovilabio, deduced	1	18	28	31	3	13	17	40	7	3	58	49	16	17	59	57
Cassini	1	18	28	36	3	13	17	54 3	7	3	59	39 22	16	18	56	50
Hevelius	1	18	28		3	13	18		7	3	57		16	18	9	
Rømer's mechanism (Derham)	1	18	28	58	3	13	20	0	7	4	0	0	16	18	0	0
Newton, *Principia*	1	18	27	34	3	13	13	42	7	3	42	36	16	16	32	9
Bradley	1	18	28	36	3	13	17	54	7	3	59	36	16	18	5	8
Delambre (seconds quoted to nine places of decimals)	1	18	28	35.9	3	13	17	53.7	7	3	59	35.8	16	18	5	7.0

scale at the top. The sizes of the circles for the satellite orbits are to scale among themselves, but there is no need for them to be drawn to the same scale as the *Earth's* orbit, although the distance between Jupiter and the Sun must be to scale with this.

The significant failing of the instrument is not that the distance between Jupiter and the Sun is constant (in fact it is exactly 5 times the Sun–Earth distance), but that there is no means of compensating the satellite motions (in the tables engraved on the plate) for the varying velocity of the shadow-line. One other notable deficiency is less interesting: there are no radices for eclipse times marked on the plate, and auxiliary tables would have been needed. There are also no tables for yearly motions. The instrument does, nevertheless, have one unexpected virtue. When we come to calculate the parameters on which the existing tables are based (see Table 1), they seem to be of extraordinary accuracy, even better than Cassini's. This is something which can hardly be properly discussed, since we do not know the date of the instrument, but some comment should be made on the remark by Maria Luisa Righini Bonelli to the effect that the tables bear comparison with Galileo's "Bellosguardo" tables* At first sight the latter are given only to minutes of arc, whereas the *Jovilabe's* tables are to seconds. If we average over periods of 3000 days in the Bellosguardo tables, however, we find that the Galilean mean motions are of the same accuracy as those on the instrument. A comparison can be made from Table 1. Cassini's accuracy can obviously be easily exaggerated.

According to Gassendi's biography of him,† Peiresc (d. 1637) had the idea of

*Bonelli[15] (p. 419).

†Paris (1641), translated by W. Rand as Gassendi.[59] See Book 2, p. 145: "Also he caused a mechanicall Theorie or Instrument to be made like the Vulgar one of *Peurbachius*; that the Roots of the Motions being praesupposed, the Places of the *Medicean* Stars might be calculated for years, moneths, daies, and hours." Gassendi goes on to say that Peiresc gave satellites the names of members of the family of the Medici: Catharine (IV), Mary (III), this being the brightest, Cosimus Major (II), Cosimus Minor (I).

mechanical or graphical representation of the satellite positions. His journal, kept between 24 November 1610 and 21 June 1612, contains a record of numerous observations of the positions of the satellites. Gassendi tells us that he refrained from printing tables of the satellites' motions, lest he detract from Galileo's honor. They were made with the help of Jean Lombard and other assistants, who traveled to Marseilles, Malta, Cyprus, and Tripoli (Lebanon), where the positions of the satellites were recorded (in local time, of course), later to be compared with observations made by Peiresc in Aix-en-Provence.* The first accurate longitude differences by this method were thus obtained. Peiresc sustained his interest in this subject to the end of his life, and some idea of the importance to be attached to his work can be had from the value he assigned to the span of the Mediterranean in longitude, 41°30′ as against the very inaccurate 60° commonly assumed.

A third Jovilabe is described in the *Medicaeorum ephemerides* of Odierna (1656). The work, according to Delambre,† is very rare, but might well have transmitted the idea to later writers. Yet another Jovilabe, but this time one to reach a large audience through its publication in the *Philosophical Transactions* of the Royal Society, was described by Flamsteed[60] in 1685. This is easily understood from what has gone before, and so similar is it to the instrument now in Florence that one must suspect some sort of influence (Fig. 3). There is now no circle for the Earth's orbit, but a small parallactic arc is placed, as before, above the circles representing the orbits of the satellites. Flamsteed seems to take it that the user will have no difficulty in evaluating the annual parallax of Jupiter. I will not rehearse the procedure for using the instrument, which is meant only to supplement tables of eclipses of the satellites: their disposition is to be found with reference to the last eclipse. There are scales graduated in days and hours giving the angular movement of the satellites directly. There is again, however, no allowance for the varying angular velocity of the axis of Jupiter's shadow.

Another Jovilabe was described by William Whiston‡ in a work of 1738: *The Longitude Discovered by the Eclipses, Occultations and Conjunctions of Jupiter's Planets*, and no doubt there have been others.§ More to the point, however, is that Jovilabe which Cassini himself designed and used, made of sheets of card (Fig. 4). Weidler[64] gives a description of it, from details provided by Maraldi, whom he met on a visit to Paris. It comprises a graduated circle crossed by a transparent alidade (made of horn, and of a width such that it can represent Jupiter's shadow). There are four centrally pivoted movable disks in due proportion, representing the orbits of the satellites, and graduated around the edge at the appropriate mean motions with days

*Brown[35] (p. 490).

†Delambre[32] (2, p. 331). Note Odierna's use of a pendulum and meridian transits of stars for timing satellite phenomena—a very Galilean procedure. Note that there is a copy of this rare work in the British Library: 531.I.12(3).

‡Whiston[61] had more than 20 years previously published a proposal for finding longitude by the regular firing of explosive shells from vessels along the traffic lanes of the world's oceans. In 1721 he made a proposal to use the dip-circle. See Whiston.[62]

§Thus Jerome le Francais (Lalande)[63] mentions Cassini's instrument and also extends the basic principle in a graphical device of his own.

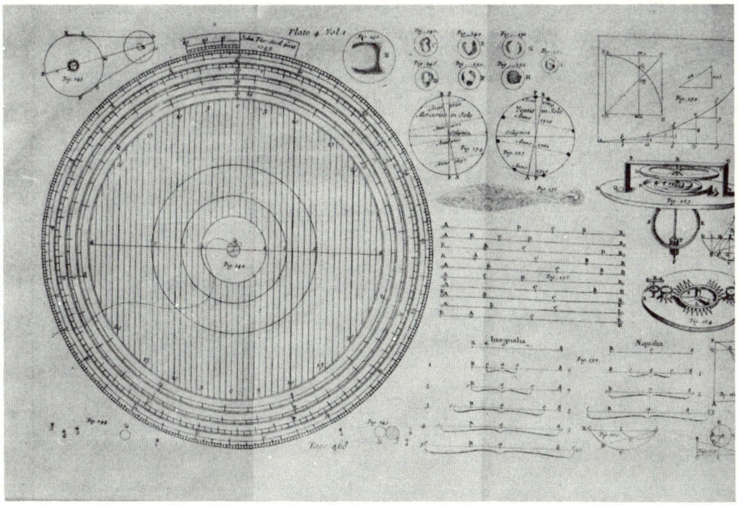

FIGURE 3. The *Jovilabe* designed by John Flamsteed, 1685.

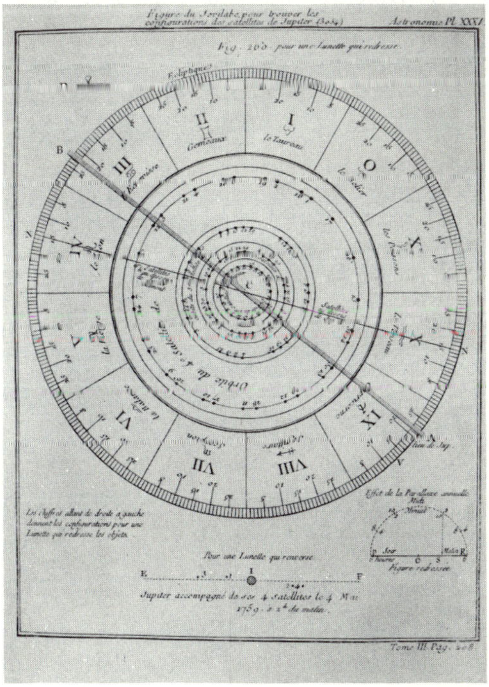

FIGURE 4. Cassini's *Jovilabe*, as printed in Lalande's *Astronomie*, Vol. 3, Pl. xxxvi.

of the month. Each of these circles is to be set with the graduation "1" opposite the satellite mean motion on the first day of the month in question (this motion being taken from tables). The diameter through the origin of the outer scale represents the shadow axis. The positions of the satellites as seen from the Earth are their orthogonal projections on a line at right angles to the alidade, once this has been set to the geocentric longitude of Jupiter. In other words, the alidade is the line of sight from the Earth to Jupiter.

The advantages of Cassini's *Jovilabe* over its precursors lies in its use of the *geocentric* longitude of Jupiter. The angle between the shadow axis and the line of sight may therefore in principle be correctly set. On the Italian instrument this is not so, simply because Jupiter's position, whether heliocentric or geocentric, is left out of the reckoning.

Cassini's second inequality, which depends on the (heliocentric) position angle of Jupiter, is taken into account in none of these *Jovilabes*. In his 1693 tables he reckoned the inequality to be a maximum when Jupiter and the Sun are in opposition, and a minimum for conjunction, the difference being 14^m10^s for the first satellite, i.e., the time for it to travel 2°; which is to say that eclipses occur 14^m10^s sooner in the former case than the latter. This, as already explained, is Rømer's equation of light, but Cassini (1) made it depend on the same parameter (2°) for all satellites;* and (2) made it independent of the eccentricity of Jupiter. Edmond Halley, in his London version of Cassini's tables, drew attention to the fact that they were not reliable, especially for the third and fourth satellite, for which he found errors of about 15 and 62 minutes, respectively.† These errors could not have been taken care of completely, however, in terms of the equation of light.

James Pound and James Bradley, and Satellite Interactions

It seems that James Pound and his nephew James Bradley were the first to introduce the equation of light into tables of the satellite motions. Bradley was born in the year of Cassini's tables, 1693, and as the discoverer of the aberration of light he was more than a little concerned with its velocity. Bradley had begun his astronomical work at his uncle's house in Wanstead, and from the very first he kept careful records of Jupiter's satellites, comparing his observations with predictions based on Cassini's tables. It was Pound who introduced him to Halley, and it is worth remembering that had Pound been prepared to give up his ecclesiastical living he would in all probability have been given the Savilian Chair of Astronomy at Oxford which went to Bradley in 1721. Halley had been Savilian Professor of Geometry until a year before, when he was created Astronomer Royal—a post which went to Bradley at Halley's death in 1742.

*Cassini.[65,66] The ideal equations in time units for the four satellites following Cassini's mean motions, on my calculation, are $14^m\ 02^s$ (IV). Cassini implicitly gives $14^m\ 10^s$ (I), $28^m\ 27^s$ (II), $57^m\ 22^s$ (III), $2^h\ 14^m\ 07^s$ (IV). (It is possible to deduce other parameters close to these.)

†Halley[18] (p. 255).

The satellite tables of the three men are almost as tangled as their careers. Pound's tables owed something of their style to Cassini's. They were sent to Halley, who printed them in the *Philosophical Transactions*,[67] and who also included them in his own collected tables, where they are added as an appendix to other tables of the satellites, sent to Halley by Bradley. (Halley's tables, printed in 1719, were not published until 1749, seven years after his death.)[68] Pound's tables (of the first satellite, like Halley's) are remarkably clear and well designed. He quotes his auxiliary parameters in millesimals, makes all his equations additive, and of course improves upon Cassini's mean motions. But his most memorable statement comes at the end of his explanations, where he tells the reader that many years of observation have convinced him "that the second inequality of this Satellite proceeds from the Progressive Propagation of Light, *and is common to all the rest of the Satellites*" (my italics).[58] He goes on to quote 7 minutes as the time for light to travel the mean radius of the Earth's orbit—a figure soon to be improved upon by his nephew's measurements of aberration. And he explains that he has added a *third* equation, to account for the varying distance of Jupiter from the Earth as a result of the eccentricity of the planet. (The second equation continues to take care of that part of the distance variation which is a function of the Jupiter–Sun–Earth angle. This might not seem to be the best of independent variables, but it was chosen for the historical reasons I have explained.)

In a sense, Pound's tables represent the end of the first great period in the history of Jupiter's satellites, a period of just over a century. With the complete incorporation of Rømer's light equations into the scheme for effecting a geometrical interpretation of the observations, the way was now open to the next phase, which might be said to have culminated in the celestial mechanics of Laplace. This new phase had begun with Newton—and I must at least mention in passing his attempted treatment of the motions of the nodes and apsides of the satellites by analogy with his theory of the motion of our own Moon.[69] Historical periods are not as neatly defined as those of the planets, however, and although I shall leave Newton aside, I would like to close by mentioning the work of Bradley, which in another sense marks the transition to the new style of research on the satellites of Jupiter.

Bradley's observations of the satellite eclipses, as listed in a Bodleian ms. (Bradley 35), run from 1712 to 1732. They are supplemented by lists of other astronomer's observations and many sheets of computations. There are also drafts (c. 1733?) of canons for the use of satellite tables. Bradley's relevant manuscripts (and I should mention here also Nos. 16 and 28) have the property of being assembled from documents of many different periods of his life. Although the greater part of the material is printed, Rigaud's edition will need to be supplemented by the manuscripts by anyone who wishes to form a clear picture of the development of Bradley's thought on the subject as a whole. This is more than I can do here, but I will indicate briefly the essence of his achievement.

Bradley's tables tend to follow the Cassini–Pound style. The published tables use the low figure of 7 minutes* (for the light time for the solar distance) rather than the

*Halley[68] (sig. Ffff 2v).

8^m13^s from aberration measurements, but only because those measurements were in the future when the tables were sent to Halley.* In Bradley's observations on his tables he shows that he is not a little perplexed by the satellite motions, which show inequalities possibly arising, he says, from orbital eccentricities of apsidal motions; "But by what we can collect from the motions of the second satellite, it is probable that they may be occasioned by the mutual action of the satellites on each other."† This passage is occasionally quoted in isolation as one of history's splendid conjectures, but the thinking behind it was far from casual. Bradley ruled out eccentricity of the orbit; it would have to be large to account for some of the rapid changes of velocity, and then it would be too large to be compatible with the remaining observations. He then asked about the *period of the errors*, and found that

> It nearly answers to the time the three inferior satellites take in returning to the same situation with respect to each other, and to the axis of the shadow of Jupiter, which is 437 days, or after 123 revolutions of the second satellite. After this period, the like errors return, nearly in the same order; but in the intermediate time, that is, after sixty revolutions, this satellite will deviate ten, twenty, thirty, and even sometimes forty minutes of time from its rate of motion during the seven preceding, or the seven following months. Now because the satellites are not found in the same place in the heavens after the aforesaid period is completed, it is possible these errors may vary somewhat on that account. And if the orbit of this satellite be likewise eccentric, as the late observations seem to make it, the inequalities arising from both causes must be very intricate and not easily to be separated by observation alone.‡

Bradley was mistaken to suppose an appreciable eccentricity, but he was right about the intricacy of the problem, and he was the first man to hit upon the *interrelations* between the satellites in their deviations from uniformity. He went on to say how the errors of satellites I and III are not so great as those of II, but seem to arise from the same causes. He expressed the hope that some geometer, "in imitation of the great Newton," would explain the irregularities on the basis of the theory of gravitation. The commensurability of the motions to which he had drawn attention was in fact to be one of the keys of the solution, arrived at by degrees in the later eighteenth century. Bradley had seen that in approximately $437^d.15$ the satellites make exactly 247 (I), 123 (II), and 61 (III) revolutions with respect to the shadow axis. The Swedish astronomer Pehr Wilhelm Wargentin, the last of the great purely empirical astronomers to work—in his case almost exclusively—with the problem of Jupiter's satellites, found the same result independently, and published it in 1741, before the Bradley–Halley publication.

It was left to Jean-Sylvain Bailly to show, in the 1760s, how the mutual attractions of the first three satellites gave rise to this periodicity in the inequalities.§ Bailly made much use of Clairaut's theory of the Moon. In 1766 the Académie Royale des Sciences offered a prize for a theoretical investigation of the satellite problem, and Lagrange's

*His ms. tables seem, even so, to preserve the maximum equation of $14^m\,00^s$.
†Rigaud[28] (p. 81) there translated from Halley's Latin original.
‡Rigaud[28] (pp. 81–82).
§Bailly.[70] His initial findings were given to the Académie, and published by that body in 1762 and 1763.

successful memoir was so comprehensive in its analysis that even Delambre found it somewhat frightening.* Lagrange considered not only the attraction of Jupiter, but also the simultaneous effects of the Sun (which Newton had done),[69] of the three satellites (following Bailly, to some extent), and of the oblate mass of Jupiter (which Halley, in 1694, had conjectured might be of significance).† Laplace later added another factor: the plane of Jupiter's orbit does not coincide with its equator. The motion of each satellite was, as he showed, determined by three differential equations of the second order, involving six arbitrary constants—24 for all four satellites. There are seven other constants needed. He, Delambre, and Lalande spent much labor on the computation of the 31 elements, and when the tables which resulted were published in the third edition of Lalande's *Astronomie*, in 1792, they were far more accurate than anything produced by the dedicated Wargentin and Cassini, and all their empirical forebears.

The moral was clear: beyond a certain point of complexity, the astronomer can only observe what his theoretical scheme tells him is there. Knowledge of the inequalities in the satellite motions remained in a state of relative confusion until the theory of gravitation and mathematical analysis came to the rescue. Puzzles would even then have remained without Rømer. But without the help of Newtonian gravitation, the satellites had made a reputation for Galileo; they had provided a spectacle, for those fortunate enough to own a telescope; they had inspired thoughts which hovered between natural theology and space fiction; they had led to the invention of the pendulum clock, and to a limited solution to the problem of longitude; and with Ole Rømer they provided what was historically one of the two most important proofs of the finiteness of the velocity of light. And after Newton, and within two centuries of their discovery, they were to stimulate perhaps more intensive mathematical activity than any other object of astronomical study.

ACKNOWLEDGMENT

This chapter is based on a lecture given at the University of Aarhus, Denmark, to celebrate the tercentary of Ole Rømer's discovery of the finite velocity of light. Bearing in mind Wolfgang Yourgrau's interest in the mode of propagation of light, I hope it is an appropriate memorial to him. Its publication was much helped by Dr. J. J. Roche of Linacre College, Oxford.

REFERENCES AND NOTES

1. Galileo Galilei, *Le opere*, Antonio Favaro, editor, 20 Vols., (Barbera, Florence, 1890–1909; repr. with additions, Barbera, Florence, 1929–1939 and 1965).
2. S. Drake, *Discoveries and Opinions of Galileo* (Doubleday, New York, 1957).
3. J. Kepler, *Dioptrice* (Augustae Vindelicorum, 1611).

*Delambre[32] (**2**, p. 398).
†Halley[18] (p. 253).

4. P. Humberd, "Le baptême des satellites de Jupiter," *Rev. Questions Sci.* **117,** 171, 175 (1940).
5. J. Kepler, *Astronomi opera omnia*, Christian Frisch, editor, 8 Vols. (Frankfurt-Erlangen, 1858–1871; repr., Olms, Hildesheim, 1971-).
6. J. Kepler, *Conversation with Galileo's Sidereal Messenger*, Edward Rosen, editor (Johnson Reprint, New York & London, 1965).
7. A. M. Schyrlaeus de Rheita, *Oculus Enoch et Eliae* (Antwerp, 1645).
8. C. Huygens, *Cosmotheoros* (The Hague, 1698).
9. J. Kepler, *Narratio de observationibus* . . . (Frankfurt, 1611).
10. Galileo Galilei, *Istoria e dimonstrazioni intorno alle macchie solari e loro accidenti* . . . (Rome, 1613).
11. Stillman Drake, "Galileo and Satellite Predictions," *J. Hist. Astron.* **10,** 75–95 (1979).
12. J. W. Shirley, editor, *Thomas Harriot, Renaissance Scientist* (Clarendon Press, Oxford, 1974).
13. R. M. McKeon, "Les débuts de l'astronomie de précision," *Physis* **13,** 225–230 (1971).
14. A Borelli, *Theorica Mediceorum planetarum* (Florence, 1666), pp. 142–144.
15. M. L. Righini Bonelli, "Galileo, l'orologio, il giovilabio," *Physis* **13,** 412–420 (1971).
16. P. E. Ariotti, "Aspects of the Conception and Development of the Pendulum in the 17th Century," *Arch. Hist. Exact Sci.* **8,** 329–410 (1972).
17. R. T. Gould, *The Marine Chronometer, its History and Development* (J. D. Potter, London, 1923), pp. 11–17.
18. E. Halley, in *Phil. Trans. R. Soc.* **18** (214), 237–256 (1694).
19. R. Grant, *History of Physical Astronomy*(Society for the Diffusion of Useful Knowledge, London, 1852).
20. S. Mayr, *Mundus Jovialis* (Nuremberg, 1614). An English translation by A. O. Prickard is in *Observatory* **39,** 367–381, 403–412, 443–452, 498–503 (1916).
21. J. Klug, "Simon Marius aus Gunzenhausen und Galileo Galilei . . . ," *Abhandl. math.-phys. Kl. Königlich Bayerischen Akad. Wiss.* **22,** 385–526 (1906).
22. J. A. C. Oudemans and J. Bosscha, "Galilée et Marius," *Arch. Néerl. Sci. Exactes Nat.*, 2nd Ser., **8,** 115–189 (1903).
23. J. Bosscha, "Réhabilitation d'un astronome calumnié," *Arch. Néerl. Sci. Exactes Nat.*, 2nd Ser. **12,** 258–307, 490–528 (1907).
24. J. H. Johnson, "The Discovery of the First Four Satellites of Jupiter," *J. Brit. Astron. Assoc.* **41,** 164–171 (1930-1931).
25. P. Pagnini, "Galileo and Simon Mayr," *J. Brit. Astron. Assoc.* **41,** 415–422 (1930–1931).
26. Thomas Harriot, MS Petworth House HMC 241/IV.2.
27. J. O. Halliwell, ed., *Letters Illustrative of the Progress of Science* (Historical Society of Science, London, 1841), pp. 38–40.
28. S. P. Rigaud, *Miscellaneous Works and Correspondence of the Rev. James Bradley* (Clarendon Press, Oxford, 1832).
29. J. Roche, "Harriot, Galileo, and Jupiter's Satellites," *Archeion* **108,** (1982).
30. Christoph Scheiner, *De maculis solaribus* (Augsburg, 1612), pp. 27–31.
31. P. Gassendi, *Opera omnia*, Vol. 4 (Lyons, 1658).
32. J.-B. Delambre, *Historie de l'astronomie moderne*, 2 Vols. (V. Courcier, Paris, 1821).
33. G. Riccioli, *Almagestum novum* . . . (Bologna, 1651, 1653), p. 489.
34. P. Hérigone, *Cursus mathematicus*, 6 Vols. (Paris, 1634–1642), in particular Vol. 5.
35. H. Brown, "Nicolas Claude Fabri De Peiresc," *Dictionary of Scientific Biography*, Vol. 10, pp. 488–492 (Scribners, New York, 1974).
36. F. Fontana, *Observationes* (Naples, 1646), Tract 6, Cap. 2.
37. J. Hevelius, *Selenographia* (Dantzig, 1647), p. 526.
38. Galileo Galilei, *Il Saggiatore* (Rome, 1623), pp. 4–5.
39. A. Koyré, *La Révolution astronomique* (Vrin, Paris, 1961), Part 3.
40. J. D. Cassini, *Mémoires pour servir a l'histoire des sciences . . . suivi de la vie de J.-D. Cassini* (Bleuck, Paris, 1810.) The author is the great grandson of the subject Gian Domenico.
41. R. Taton, "Gian-Domenico Cassini," *Dictionary of Scientific Biography*, Vol. 3, pp. 100–104 (Scribners, New York, 1971).
42. G. D. Cassini, *Tabulae quotidianae revolutionis macularum Jovis* . . . (Rome, 1665), p. 328.
43. J. E. Montucla, *Histoire des mathématiques* (Paris, 1758), Vol. 2, p. 516.
44. J.-B. Du Hamel, *Regiae scientiarum academia historia* (Paris, 1698), p. 145.

45. Anon., *Histoire de l'Académie Royale des Sciences* (Paris, 1707), p. 78.
46. I. B. Cohen, "Roemer and the First Determination of the Velocity of Light", *Isis* **32,** 327–79 (1940).
47. J. D. Cassini, "De l'origine et du progrès de l'astronomie," in *Recueil d'observations faites en plusieurs voyages. . .pour perfectionner l'astronomie et la géographie*, Messieurs de l'Académie Royale des Sciences, pp.38–39 (Paris, 1693).
48. R. Descartes, *Oeuvres*, C. Adam and P. Tannery, editors (Vrin, Paris, 1969), Vol. 1, letter 57, pp. 307–312.
49. C. Huygens, *Oeuvres completes* (La Société Hollandaise des Sciences, Amsterdam, 1967), pp. 463–469.
50. R. Hooke, *Micrographia* (London, 1665), p. 56.
51. J. D. Cassini, extract from a letter written to the *Journal des Sçavans* printed in *Phil. Trans. R. Soc.*, No. 128, 681–683 (September, 1676).
52. Isaac Newton, *Correspondence*, H. W. Turnbull, editor, p. 3 (Cambridge University Press, Cambridge, 1959).
53. N. L. De Lacaille, "Observations astronomiques faites à l'Isle de France pendant l'année 1753," *Mém. Acad. R. Sci.*, 44–54, especially 45 (1753).
54. N. L. De Lacaille, "Divers observations . . . etc.," *Mém. Acad. R. Sci.*, 94–130, especially 105, 120, 129, 130 (1754).
55. T. Eibe and K. Meyer, *Ole Rømers Adversaria* (Copenhagen, 1910).
56. O. Roemer, "Démonstration touchant le mouvement de la lumière.," *J. Sçavans* **4,** 233–236 (1676); translated in *Phil. Trans. R. Soc*, **12,** 803–804 (1677).
57. C. Huygens, *Oeuvres complètes* (La Société Hollandaise des Sciences, The Hague, 1899), Vol,8, pp. 30–35.
58. J. Pound, "New and Accurate Tables for the Ready Computing of the Eclipses of the First Satellite of Jupiter, by Addition Only," *Phil. Trans. R. S. 30, No. 361, 1021–1034, especially 1034 (1719).*
59. P. Gassendi (tr. W. Rand), *The Mirrour of True Nobility and Gentility* (London, 1657).
60. J. Flamsteed, in *Phil. Trans. R. Soc.* 15, No. 178, 1262–1265, and Table 2, Fig. 2, facing 1251 (1685).
61. W. Whiston, *A New Method for Discovering the Longitude* (London, 1715).
62. W. Whiston, *The Latitude and Longitude found by the. . . dipping needle*, (London, 1721).
63. J. Lalande, *Astronomie* (Paris, 1764), Vol. 3, pp. 197–202; second edition (Paris, 1771), Vol. 3, pp. 292–298.
64. J. F. Weidler, *Explicatio Jovilabii Cassiniani* (Wittenberg, 1727).
65. J. D. Cassini, "Tabulae motuum . . ," in *Recueil d'observations faites en plusieurs voyages*, Messieurs de l'Académie Royale des Sciences, pp. 9, 40, 103 (Paris, 1693).
66. J. D. Cassini, "Les Hypothèses . . ," in *Recueil d'observations faites en plusieurs voyages*, Messieurs de l'Académie Royale des Sciences, p. 52 (Paris, 1693).
67. E. Halley, in *Phil. Trans. R. Soc.* 30 (361), 1021–1034 (1717–1719).
68. E. Halley, *Tabulae astronomicae*, John Bevis, editor (London, 1749).
69. Isaac Newton, *Principia mathematica* (London, 1687), Book III, Prop. 23, Problem 5.
70. J.-S. Bailly, *Essai sur la theorie des satellites de Jupiter* (Paris, 1766).

12

EMERGENT PROPERTIES

D. J. O'CONNOR
University of Exeter
Exeter, England

I first met Wolfgang Yourgrau in the spring of 1950 at a philosophical congress held at the University of Cape Town. The congress, which was said to be the first of its kind ever held in the country, attracted philosophers from all over the Union of South Africa, as it then was. I had not expected to enjoy an occasion on which I felt likely to be swamped by alien outlooks, Afrikaans-speaking existentialists, and Calvinist theologians from the high veldt. But I need not have worried. Wolfgang's enthusiasm, wit, friendliness, and general ebullience made the occasion a worthwhile one for me and, indeed, for most of the conference members.

I was later a colleague of his at the University of the Witwatersrand at Johannesburg for a short time before returning to England. We had a great many discussions on philosophical topics, not only in the philosophy of science but on all the problems of philosophy which lay within his wide field of interest. One of these was the question of emergent properties.

In choosing to write about emergent properties, I was influenced by a number of considerations. Firstly, it is a topic of fairly central interest in the philosophy of science. It is, for example, closely allied to that of reduction, the notion of explaining the findings of one science by the principles of another. This has always been a major concern of philosophers of science of the positivistic school or of those who are fascinated by the goal of the unity of science. Secondly, it is a concept of a rather pervasive kind in the philosophy of science. We find it in the controversies about mechanism and vitalism and in other parts of the philosophy of biology; in discussions about consciousness in the philosophy of mind, about primary and secondary qualities in the theory of knowledge, and about meaning in the philosophy of language. Thirdly, it is not a topic that has been overdiscussed in recent years. Yet, lastly, it is not one that is so far out of fashion to sound dull or bizarre—indeed, some of the recent work of Karl Popper has given it a new interest. What I have to say falls into two parts: first, I shall look at the general question: what criteria would a property have to satisfy before we can usefully, if only tentatively, classify it as emergent? I shall then discuss one of the more interesting treatments of the subject that has been published, that of Professor Walter Elsasser.

I will start with some historical remarks. Philosophical discussion of emergent properties seems to have arisen from considering the supposed implications of the theory of evolution. Darwin, in *The Origin of Species*, had insisted on continuity and gradualism in the development of the world. "Natural selection," he wrote "will banish the belief in the continued creation of new organic beings or in any great and sudden modifications of their structure." But the evidence produced in *The Origin of Species*, massive and well-ordered as it was, failed to convince even friendly critics on several points. Objections to Darwin's version of the theory were published which he acknowledged with typical candor but which he was unable to resolve. We now know that it was lack of knowledge of the mechanisms of heredity that created these difficulties.

At the turn of the twentieth century, when Mendel's discoveries came to be appreciated, experimental biologists had turned their attention to just those large and sudden variations that Darwin had considered unimportant for evolution. De Vries, in particular, in his work on the evening primrose *(Oenothera lamarckiana)* discovered that numbers of striking variations of single characters, many of them novel, occurred in the progress from one generation of plants to the next. He called these changes "mutations" and suggested, as had Bateson before him, that these discontinuities might be important for the process of evolution. Thus early in the present century, the interest of biologists had been excited in the study of the emergence of novel and apparently discontinuous variations of the properties of plant and animal species. It was soon appreciated that a convergence between the study of Mendelian heredity and mutations offered a key to the mechanism of evolution that had not been available to the nineteenth century. (And it also showed why metaphorical extensions of the doctrine of evolution to the study of society or of language or of any other system without a genetic basis were no more than empty and misleading figures of speech.)

These historical remarks may provide some explanatory background to the interest in so-called emergent properties taken by some philosophers in the early part of the twentieth century. The well-known names here are those of Lloyd Morgan and Samuel Alexander. On their view, nature shows clear evidence of discontinuities which introduce novel features into the world. Moreover these novelties emerge at new levels of development. In general, the so-called "emergent evolutionists" take an over generous view of the process of evolution. For example, Lloyd Morgan in his Gifford Lectures talks of evolution as "the name we give to the comprehensive plan of sequence in all natural events." He thus includes progress from the inorganic physicochemical world to the biological and then to the sphere of consciousness as levels of evolution. But this is too wide a use of the term in the light of present knowledge of the causes and process of evolution. Only those changes which can be traced to alterations in the genetic basis of the changing individual can be called evolutionary, unless we are to stretch the meaning of the term "evolution" in a purely metaphorical and unexplanatory way. He should perhaps have used some such phrase as "cosmic process" or "world development" instead.

However that may be, Morgan and Alexander develop the concept of emergence in terms of three basic notions, levels of development, novelty, and unpredictability. There are difficulties in understanding the meanings of these terms, especially as different exponents of the doctrine of emergence give different lists of "levels." Lloyd Morgan distinguished a physicochemical, a biological, and a conscious level, with God as the final emergent. Alexander added space-time as a basic level.

The concept of level in their writings is intuitive and imprecise. They seem to have had in mind something of this sort: if we suppose a class L_1 of events in which properties $p_1, p_2, \ldots, p_n$ are manifested, a particular constellation of these properties in a particular relation will be a sufficient condition for the emergence of a new class L_2 of events with properties $q_1, q_2, \ldots, q_n$, none of which are manifested at any lower level. In general, a level will be the domain of a particular science or group of sciences, as physics and chemistry for the material level, biology for that of living matter, psychology for that of behavior and consciousness.

However, it is not possible to predict, on the basis of knowledge of properties $p_1, \ldots, p_n$ by themselves that the properties of L_2 will emerge. So, according to Lloyd Morgan, although a knowledge of chemistry will enable us to predict the molecular weight of carbon disulfide from a knowledge of the atomic weights of sulfur and carbon, its other properties such as its smell or its flammability could not be so predicted. This allegation of unpredictability is an important feature of early theories about emergence. Alexander indeed went so far as to suggest that the phenomena of emergence must be accepted with what he called "natural piety." He seems to have meant by this that requests for explantion of emergent properties are neither useful nor even seemly. An emergent property, he said, in a flourish of metaphysical obscurantism, "admits no explanation."[1]

This is an extreme attitude to emergent properties, and more hard-headed critics were quick to point to its defects. Alexander's statement "A being who knew only mechanical and chemical action could not predict life" invited the reply that prediction is never possible anyway to anyone who simply observes phenomena. He needs at least, in addition to his observations, a background of relevant theory and a deductive system capable of deriving the prediction from the observations and the theory. But the notion of predictability does require a little more attention.

Prediction is not a well-defined concept. If we ask what conditions have to be satisfied before we allow that a given statement P is a genuine prediction, we can give a rough description: (i) at time T_1, prior to T_2, I describe a certain event E and say that it will occur at T_2; (ii) I possess evidence at T_1 that can justify the prediction; (iii) I deduce from this evidence, by a valid process of reasoning, the statement embodying the prediction. These conditions are loosely stated. The description may be more or less accurate, the evidence more or less complete, and the process of inference more or less sketchy. Indeed, the most spectacular and important scientific predictions are precisely those where description, evidence, and techniques of inference are far from complete. Compare a present-day team of space scientists predicting the orbit of a new satellite with Mendeleev's forecast of the discovery of element 32. He predicted that it would have an atomic weight of 72 and a specific gravity of 5.5, estimates which were closely approximate to the real values. But the theoretical background available to him was tenuous in the extreme and he relied on the significance of regular fluctuations in the valencies of the known elements once they had been arranged in order of their molecular weights. The space scientists have, by contrast, a complete theoretical background and deductive technique which yield results of high accuracy. But Mendeleev made an important scientific advance while the space scientists are just plodding journeymen.

Moreover, predictions may be difficult or impossible for a variety of reasons. The necessary mathematical techniques may not be available, the complexity of the calculations may be beyond human powers, available methods of observation or measurement may be inadequate; and so on. So the notions of predictability and its opposite are too poorly defined to offer a clear basis for the concept of emergence. So too with the third of the ideas basic to emergence—novelty. If the novelty is merely psychological, in that it has not been experienced before, whether it is predictable will depend on whether or not it is a determinate form of some determinable property whose other forms are familiar. An example of this would be the first calculation of the escape velocity of a projectile from the earth's gravitational field. And this would hardly rank as a case of emergence. But if the property in question lies outside all previous experience, it clearly could not appear in the conclusion of a meaninfgul prediction. But its unpredictability would simply be a matter of logic and it would not represent one of Alexander's metaphysical emergents.

Hempel[2] has elucidated the notion of emergence by specifying three conditions:

(1) We have first to agree what are to be taken as the parts of a given whole before we can say what counts as an emergent property. We could not infer the volume of a brick wall from the size and number of its molecules, although we could make the inference if we knew the size and number of the bricks. So before we make any predictions about resultant properties in a compound we have to define the parts we are concerned with and the relations in which they stand to each other. For example, the difference between diamond and graphite lies in the relations which the carbon atoms have to each other, that is, in the nature of the crystal structure.

(2) If we understand a property to be emergent simply on the grounds that we cannot infer its occurrence from a knowledge of *all* properties of its parts, then no whole could have any emergent properties at all. The property of forming white cubic crystals with a salty taste is a compound emergent property of the combination of chlorine and sodium. And, while it is true that we could not predict the occurrence of this property from a knowledge, however detailed, of sodium and chlorine taken in isolation, if we count relational properties as properties, occurrence of the emergent

quality in question follows trivially. Thus we can only prevent the notion of emergence from being vacuous by specifying, in each case, just those attributes and parts of the complex which we are taking into account in ranking the new supervenient quality as emergent.

This requirement has the consequence, surely unintended by Hempel, that any candidate for the status of emergence will either be vacuously disqualified or admitted as emergent relatively to a restricted selection of evidence. And, on such grounds, *any* property at all might qualify—but only at the cost of its emergent status being a function of our ignorance and so totally uninteresting.

(3) Hempel's third qualification is that admission of a given property to the status of emergence must be relative to the theoretical background available at the time. There are, for example, many chemical properties which would have ranked as emergent in the middle of the nineteenth century simply because the theoretical background available then could not have enabled the best chemist in the world to make a rational prediction about them. For example, so much of present knowledge about chemical reactions depends on knowledge of the electron structure of the elements which gives a rational basis for a systematic theory of chemical reactions.

The outcome of Hempel's account of emergence (and his general approach is followed by many other writers) seems to be that if we knew enough about the workings of nature, no properties would be ranked as emergent. If indeed "sufficient knowledge" means knowledge sufficient to establish this conclusion, the statement is empty. There would then be nothing more or interesting to say about the subject. But if it is an open question whether nature is such that this sufficient knowledge is available, there may be something more to investigate.

The two attitudes to emergent properties that we have looked at so far, the position of Lloyd Morgan and Alexander on the one hand and that of Hempel, Grelling, and Arthur Pap on the other, represent two extremes: either emergent properties are a metaphysical mystery or they represent remediable gaps in our scientific knowledge. Pap indeed added what he called a "semantic" proviso that properties that were unpredictable in that they lay outside the conceptual range of the science in question would have to be defined ostensively. But otherwise his position (represented in Chapter 19 of his *Introduction to the Philosophy of Science*) is not markedly different from that of Hempel.

If indeed this was all that could be said on the subject, it would be a pretty empty topic. Let us see if it can be made more interesting.

If we are to do this, it will be necessary to talk of natural processes, events, casual lines, and the like instead of talking in terms of statements and theories. Nagel has protested against this way of talking:[3]

> The concept of reduction as the deduction of properties from other properties is potentially misleading and generates spurious problems, . . . because it suggests that the question of whether one science is reducible to another is to be settled by inspecting the "properties" or alleged "natures" of things rather than by investigating the logical consequences of certain explicitly formulated *theories* (that

is, systems of statements). For the conception ignores the crucial point that the "natures" of things, and in particular of the "elementary constituents" of things are not accessible to direct inspection and that we cannot read off by simple inspection what it is that they do or do not imply. Such "natures" must be stated as a theory.

If we understand the point of Nagel's protest, it will be safe to disregard it. And I shall disregard it for two reasons. First, that if we talk simply in terms of theories, the whole question of emergence becomes trivial in the way I have already mentioned. "Emergent" if it has any application at all will mean "emerging in the gaps of our scientific ignorance." But, more importantly, if a theory is to be worth anything, it must describe, explain, and relate the facts and processes that we are studying. It is these that are prior to the theories about them.

The growth of knowledge of the natural world has divided up the sciences according to their subject matter. Physics deals with the various forms which energy can take and the laws according to which energy is transformed from one form into another. Chemistry takes the different fundamental types of matter and traces the ways in which they combine; biology deals with the structure and working of living matter; and psychology with the regularities of human experience and human and animal behavior. If we look at this division, it is clear that it establishes a certain hierarchy and we can represent the relationship between the chief sciences as a partial ordering of sets or, more mundanely, as a series of Chinese boxes. For simplicity, let us take physics as the outermost box enclosing chemistry, which encloses biology, which, in its turn, encloses psychology. Assuming this model, we can say that the domain of physics is the largest and that it encloses the rest in that the laws of physics operate also in the realm of chemistry, biology, and psychology. Everything, including human behavior, is in accordance with the laws of physics in the sense that nothing that happens in the world is exempt from those laws.

At first sight, however, each of the enclosed sciences has its own characteristic phenomena and its own laws governing these phenomena. Learning all about physics does not tell us all, or indeed, much about chemistry. Chemistry tells us something, but again not much, about biology; and so on. Human beings, for example, are physical objects and we are vulnerable to gravitation, electricity, and other natural forces. Our physiological processes proceed according to physical and chemical laws; our heredity is traced by genetics; and so on. But within all this there still seems to be room for specifically psychological phenomena and laws which operate within a sort of reserved logical space which is not completely occupied by physical, chemical, and biological happenings.

Proponents of the doctrine of the unity of science would wish to say that this Chinese box model is a superficial one; and that if we had a more adequate knowledge of physics we would see that it would take over chemistry in that the facts of chemistry would be explicable in terms of physical laws. They would claim further that the facts of biology are ultimately explicable in terms of chemical laws; and that psychology is not just a science in a biological context in that psychological subjects are living organisms but that it is actually a special case of biology. And were this claim correct, chemistry would be shown to be a special case of physics, biology of chemistry, and psychology of biology, meaning that the facts and laws of the narrower science will ultimately be explicable in terms of those of the wider one. And so, since "being explicable in terms of" is presumably a transitive relation, every event in the world is ultimately explicable in terms of the laws of physics.

This claim has, of course, an important bearing on the status of emergent properties. The most characteristically emergent of all properties are those of living matter and of consciousness. To preanalytic common sense, these seem to emerge at the levels appropriate to biological and psychological investigation. But if the claims for the unity of science can be upheld, they must be understood as no more than phenomena of residual ignorance.

If we pursue the Chinese boxes analogy, we are presented with two alternatives. The view which accepts emergent properties as genuine features of the world which are not simply temporarily unexplained discontinuities in the seamless web of science must look at them in the following way. The causal processes (or causal lines, to use Russell's convenient phrase) which originate in the realm of physics run right through the whole of nature, physical, chemical, and biological levels alike. But other causal lines originate at the chemical level, if we are to suppose that some chemical events are emergent; others originate at the level of biology; and others still at the conscious level. To say that they *originate* is not, of course, to say that they emerge mysteriously out of no previous causal line at a lower level. For example, sensation, a form of consciousness, is known to be at least the partial outcome of processes for which physical transmissions of energy are necessary conditions. But these necessary conditions do not completely account for them. Indeed, it is logically absurd to suppose that any conjunction of necessary conditions could be a sufficient condition. (On the standard interpretation of the phrases "necessary and sufficient conditions" such a supposition would be the elementary fallacy of denying the antecedent.)

The alternative view embodied in the hypothesis of the unity of science is that physical processes are sufficient conditions for chemical processes, chemical for biological, biological for psychological. The apparent discontinuities associated with the "emergence" of a new level of events will be shown to be illusory by the advance of the sciences. Indeed, the history of physics and chemistry during the past hundred years has given good grounds for supposing that the two sciences are merging into one. As Popper, who is no believer in the unity of science, has recently said: "It would not be very surprising if the hopes, held for a long time, that chemistry can be reduced to physics were to come true, as indeed, they seem to be doing."[4] Since we cannot forecast the future of science, we have to be content with looking at the consequences of the unity of science hypothesis.

(1) The first obvious consequence is that the world, on this hypothesis, is a deterministic one. Determinism is a nice simplying theory which disposes at one fell swoop of a number of difficult philosophical issues. But it has, as we all know, serious disadvantages and weaknesses. Nor does the fact that, at the submicroscopic level, there

are (or may be) breaches in the deterministic picture really make determinism more acceptable. For if the world is a determined system at the macroscopic level at which we all live, then for all human purposes, we are part of a determined world.

(2) But there is an odd logical point about this hypothesis of causal continuity that is worth noting. If physical processes are really a sufficient condition for chemical, and chemical for biological, and biological for psychological, then, since "being a sufficient condition of" is a transitive relation, physical processes are a sufficient condition of psychological processes, that is, of consciousness. But if X is a sufficient condition of Y, then Y is a necessary condition of X. So that, on this account, consciousness would be a necessary condition of the processes of physics, of those particular processes, that is to say, which are the sufficient conditions of any conscious states. Thus we would have an easy proof of a form of idealism which I offer to any devotees of Hegel and Bradley who may still be around.

This conclusion is clearly unacceptable. And since there seems nothing wrong with the form of the argument, it must be that the premise is at fault. That is, physical processes are not a sufficient condition for chemical, biological, and conscious events. But if they are not, it looks as if we must accept the emergence of physically inexplicable events at some or all of the chemical, the biological, or the psychological levels.

Clearly, to make the rejection of emergent properties a tenable hypothesis we would have to find some way of expressing it that would not make physical processes the sufficient conditions of those of the sciences lying within physics, so to speak. It is not obvious how this could be done. We could not say, for example, that physical processes are *necessary* conditions, by the same stages, of consciousness. For this would have the equally unacceptable consequence that mind was a sufficient condition of physical events. So the hypothesis will have somehow to be expressed without reference to necessary and sufficient conditions or any covertly equivalent notions.

For simplicity, I shall concentrate on the alleged emergence of biological processes from the physical and chemical. A good deal of attention has been given to this alleged emergence by some of the most distinguished physicists and chemists of the present century—Schrödinger, Bohr, Delbrück, Wigner, Polanyi, and Elsasser, to mention a few names. Schrödinger's *What is Life?*, a very prescient essay, written in 1943, some ten years before the discovery of the structure and working of DNA, adopts a slightly ambiguous attitude to the question: are the processes of living matter emergent upon or continuous with those of the physicochemical world? On the one hand, he says, "The obvious inability of present day physics and chemistry to account for such events is no reason at all for doubting that they can be accounted for by those sciences."[5] And on the other hand, he is worried by the contrast between the statistical mechanism of physics which produces order from disorder and the extraordinarily stable order of the gene which resists so surprisingly the molecular disturbances of its environment, and produces so reliably replicas of itself over many generations in temperatures of up to 37°C or more.

A further apparent discontinuity between the physical and the biological levels is the capacity of living organisms to drink orderliness (in Schrödinger's phrase) from their environment, that is, to feed on and increase negative entropy. Such activities are, of course, in statistical terms, immensely improbable; and any satisfactory philosophy of biology has to be able to account for them.

These contrasts have seemed to some other physicists to resist any explanation in terms of what we know or can reasonably conjecture. Indeed, some of them have argued plausibly that on the basis of established physical and chemical knowledge the behavior of living matter is inexplicable. E. P. Wigner, for example, argues that if we assume the standard theory of quantum mechanics together with an assumption that has proved useful in other physical derivations, it can be shown that the probability of self-reproducing states in a material environment is zero.[6]

The physicist who has devoted most attention to the difficulties of connecting biology with physics and chemistry is Walter Elsasser. In two books and a number of papers he has drawn attention to differences between the physical and the biological levels that he regards as decisively settling the impossibility of explaining all the phenomena of living matter in physicochemical terms.

Elsasser's treatment of the relation between physicochemical and biological phenomena is interesting in that, although he is dealing with questions in which other writers have invoked the concept of emergence, he does not use the word. Instead, he talks of "biotonic" or "organismic" phenomena and laws. By this he means those phenomena and laws of living matter which cannot be fully explained in mechanistic terms. But he does not, as some supporters of emergence have done, simply assume that these phenomena represent a novel and unpredictable level of reality. He tries to explain just why biological ways of happening have the properties that they do have while admitting that the laws of physics do hold throughout the workings of organisms. To deny this is, he believes, a characteristic of vitalism; and this approach he rejects. It is useful to contrast Schrödinger's conclusion, which would probably be shared by most scientists at the present time, with Elsasser's.

Schrödinger said, at the conclusion of his essay that "living matter, while not eluding the 'laws of physics' as established up to date [that is, up to 1943] is likely to involve other 'laws of physics' hitherto unknown, which, however, once they have been revealed, will form just as integral a part of this science as the former."[7] Elsasser, on the contrary, writing over twenty years later, says: "general laws of biology which are not deducible from physics will have a logical structure quite different from what we are accustomed to in physical science."[8] But how could such nondeducibility be established?

Elsasser's first difficulty is over information storage. In the animal this is preserved in three forms: first, the genetic information represented in the gene structure; secondly, the structure of the adult organism, and thirdly, memory. The nature of the chemical storage in the gene is now known, although we still have no more than conflicting hypotheses about memory storage. Presumably, however, it must consist in some physicochemical structure or process. But we are now faced with the question that bothered Schrödinger but which he did not completely resolve. How is all this information coded at the molecular level to be protected from the continuous and deleterious environment of molecular thermal disorder in which the code-bearing structures are immersed? Acquired memories are indeed fading and fallible though far less so than they ought to be on the hypothesis of their subjection to continuous molecular noise. But genetic storage is phenomenally well preserved through individual life spans and through generations. How can this be understood in terms of known physical processes?

Information storage at the genetic level can be understood in terms of the ordering of nucleotides on the DNA molecule. Here we know the *mechanism* of information storage. What Elsasser is puzzled about is the stability of this information in the face of a hostile environment of thermal disorder. Schrödinger's answer to the question: How does information stored in the genes survive the disordering tendency of heat motion? Was it that the DNA molecule (in contemporary terms) formed what he called an "aperiodic crystal." Such crystals, he said, "represent the highest degree of well-ordered atomic association that we know of."[9] These are kept in shape by the Heitler–London forces "strong enough to elude the disorderly tendency of heat motion at ordinary temperature."[10] Elsasser, without explicitly contesting this point, clearly does not accept it. But the issue is a straightforward one in quantum chemistry. Are the Heitler–London forces which bond the DNA molecule stronger than those of molecular thermal disorder at ordinary body temperatures? I do not know if the answer to this question is available; but presumably it should not be too difficult to find out. If Schrödinger is right, Elsasser's doubts on this particular issue give no support to his belief in "biotonic laws."

His second difficulty is with information storage at the somatic level. This occurs in two ways: (1) the endogenous information which originates in genetic structure is manifested in the adult organism in the enormous complexities of the individual's structure, function, and chemical individuality. This is spectacularly well preserved throughout the life of the individual. (2) The information that originates outside the organism is preserved (in higher organisms at least) in cerebral memory.

These two types of information differ from the first in that not only do we not know how this information is stored but we have absolutely no evidence for the existence of any mechanisms capable of storing it, or, in the case of memory, of retrieving it. Moreover, as Elsasser remarks, all this information is stored, contrary to all sound engineering practice, in the soft tissues of the animal. So the storage of information in the individual organism is not related either to mechanical or to chemical stability. Indeed, metabolic activity is very high in those tissues that store information; but the reliability of the storage appears to be quite unaffected by this.

These considerations are unpersuasive insofar as they appear to rely on arguing from what we do not know to be the case to what we know not to be the case. Arguments of this sort which rely on present gaps in scientific knowledge have been refuted so often in the past by new discoveries that it is surely imprudent to rely on them.

Doubtless there is no evidence from anatomy for any storage organ in the brain or for any scanning device. But knowledge of neural anatomy and physiology is still in a very defective state. And what we must do if we are to establish the reality of emergent or biotonic laws or properties in the working and behavior of organisms is to show that such laws and properties are continuous with (though not fully explainable by) physical and chemical laws and events. Elsewhere in his writings Elsasser fully recognizes this.

He adds one further point based on information theory. The information content of adult organisms is greater than that of the germ cell, even if we exclude the accretion of information from external sources which is stored as memory. It is not only, he claims, that environmental noise can only degrade information and not add to it. But the actual structural and functional complexity of the phenotype is much greater than that of the genotype. Although biological "information" is transformable from one form into another, storage with adequate redundancy complete enough to represent the adult organism "does not exist in the germ cell."[11] Indeed, he concludes that the only discernable difference between organism and machine is in lack of mechanical information storage. "We postulate," he says, "that the maintenance of information in the organism is not, in general, effected by mechanistic means."[12]

In taking his case further than this, Elsasser relies on two premises. First, he accepts Bohr's principle of generalized complementarity. Bohr's own statement of this principle is as follows: "the strict application of those concepts which are adapted to our description of inanimate nature may stand in a relationship of exclusion to the consideration of the laws of the phenomena of life."[13] This is because, in addition to the limits set to observation by the quantum theory of the atom, the interference in the workings of the organism which would be required by observations minute and exact enough to determine those workings would destroy the organism. Secondly, Elsasser adds that the "immense" numbers of molecular states and activities involved in biological processes cannot be measured by ordinary physical means, as can, for example, the "structured" or large-scale properties of the individual. (The term "immense" here is a technical term in Elsasser's writings.[14] A given number is "immense" if its logarithm is a very large number.) Elsasser gives a vivid idea of the numbers involved here in the following way: We can set an upper limit to the membership of a class of organisms by making an estimate of the total number of cells on the earth. Suppose that all organic matter is evenly distributed over the earth's surface with 10^9 cells per square centimeter. The total number would be less than 10^{28}. Now take into account the fact that cells alter their molecular configurations very rapidly owing to their metabolism. Count each cell as a new unit once a minute—a very modest assumption. If life exists on earth from an early geological age until the death of the sun (say, 15×10^9 years or 10^{16} minutes) the total number of cell configurations would be 10^{44}. This is a large number but not an immense one, since its logarithm is only 44.

Elsasser next introduces the concept of "homeogeneous" states of atoms. Atoms in the same quantum state are "homogeneous"; this means that for practical purposes they can be regarded as identical. And although in a piece of inorganic material, a lump of quartz, say, there may be large numbers of atoms in different quantum states, so far as the macroscopic physical properties of the material are concerned, "only their averages are physically significant".[15] But in living matter the situation is very different. Here we have what Elsasser calls "radical inhomogeneity." The varying quantum states of the atoms of the molecules in the cells of the organism are essential for the working of the cells. Moreover, the number of physiologically significant microscopic states *is* an immense number. (Suppose that there are 10^{12} atoms in a typical cell and 4 quantum states available to each atom. Even on this modest supposition, the number of physiologically significant states is certainly, in Elsasser's terms, "immense.") The point has been put forcibly by Dr. Kenneth Denbigh in a recent book[16]: on Elsasser's calculations, the number of quantum states available to a single cell greatly exceeds the number of cells that could be placed in the whole universe.

These considerations introduce a radical unpredictability into the workings of biological material. Although we can forecast reliably the macroscopic "structured" properties and behavior of an organism (as, for example, when an embryologist forecasts the development of a fertilized egg) the combined consequences of Bohr's generalized complementarity and the immensity of the numbers of reacting molecules in a metabolizing cell make the detailed development of such a cell unpredictable. The first type of problem is in Elsasser's term "biotonic" because such predictions, although common and reliable, are not based on regularities deduced from the laws of physics. But as we move from gross structured properties, like the superficial anatomy of a frog, to the smaller components we get progressively involved with finer microscopic detail. And here says Elsasser, "the statistical scatter of the structured* properties will become larger and their definition becomes progressively more vague; ultimately we arrive at the immense variety of molecular configurations which must differ from one organism to the next."[17] Elsasser's argument concludes with an assertion of the compatibility of physical laws with biotonic laws: "biotonic modes of behavior always cluster round a nucleus of mechanisms which they extend and supplement."[18] I have omitted, for brevity, a good deal of the detail of his thesis: but this omission does no injustice to his case because that rests basically on the concept of *unpredictability*. Now there are, as we have seen, several different reasons why a given outcome of a process might be taken to be unpredictable. But whatever the cause, the question whether a particular event is predictable or not is an *epistemological* question which proves nothing about the nature of the events which are the subject of the prediction. The detailed progress of certain processes may well be unknowable by human beings; but this fact tells us nothing about the nature of the process which we are unable to know. And mere lack of knowledge, for whatever reason it arises, cannot serve as a basis for conjectures about the nature of whatever it is that we cannot know. Elsasser's claim that biotonic processes are consistent with and fill out physical processes has, as a necessary condition, a certain type of unpredictability.

*The "structured" properties of an organism are those which are common to all members of a class.

And the fact that we are unable to make certain types of prediction gives us no grounds for the kind of positive conclusion that he wants to draw. For what we are unable to know may well turn out, if our ignorance could *per impossibile* be removed, to have just the kind of nature that would wreck Elsasser's conclusion.

Thus Elsasser's elaborate and ingenious attempt to demonstrate that there are properties of biological events and processes that cannot be deduced from, although they are consistent with, the laws of physics and chemistry must be held to fail. It rests, in two different ways, on gaps in our knowledge. First, we have no evidence for mechanical devices and processes for information storage in the organism. Secondly, there are essential features of organic development and behavior which are unpredictable. But both of these conclusions represent arguments from ignorance. If it were true, for all time, that biotonic processes cannot be shown to be completely explicable in physical and chemical terms, that would be no more than an interesting fact about human knowledge. It would not justify any conclusion about the nature of biological processes.

From what I have said, it may seem that there is very little that can usefully be conjectured about emergent properties. Either we treat them, with Lloyd Morgan and Alexander, as metaphysical mysteries, or, with Hempel and the positivists, as unfinished business—gaps in a pattern of knowledge that we may still hope to see filled provided it be filled in such a way as to preserve the unity of science. But if these are the only alternatives, a nonreductive account of emergent properties is impossible; and if they are not, it is still unclear what such a nonreductive account would amount to.

It is a virtue of Elsasser's attempt to deal with what he called the "biotonic" phenomena of the organic world that, if he is right, we can assume without contradiction that both the laws of physics and chemistry and biotonic laws as well operate compatibly in the organic world. But his strategy is too weak to yield the kind of positive information that we require. In arguing that it is an exclusive property of organisms that they do not store information by mechanistic means, he is arguing from the imperfections of present knowledge. Moreover, in trying to show that there is no contradiction between physical and biotonic laws, he is still arguing from human ignorance. He does establish that any classes of organisms must be immensely small in numbers compared with the number of microscopic configurations that may be assumed by members of these classes. He proves, in consequence, that the membership of a class must be exhausted before any contradiction between physical and biotonic can be operationally established. But what may not be available to human knowledge may, for all that, really be the case.

In any event, to have shown that *if* biotonic laws exist, they are compatible with those of physics, would still be too weak a conclusion. We need to know and to be shown that the hypothesis of biotonic laws is required to explain the characteristic workings of living beings—their metabolism, their resistance to entropy, that they are open systems, that they preserve homeostasis, and the rest. Until we know that physics cannot account for these properties, we do not have to assume emergent or biotonic ways of working. And when we do have to assume them, we still have to explain how they are related to the physicochemical levels from which they are said, unhelpfully, to "emerge."

References and Notes

1. S. Alexander, *Space, Time and Deity* (Macmillan, London, 1920), Vol. 2, p. 46.
2. Hempel, *Aspects of Scientific Explanation* (Free Press, New York, 1965), pp. 259 ff.
3. E. Nagel, *The Structure of Science* (Routledge, London, 1961), p. 364.
4. K. Popper, *Objective Knowledge* (Oxford University Press, Oxford, 1972), p. 290.
5. E. Schrödinger, *What is Life?* (Cambridge University Press, Cambridge, 1945), p. 2.
6. E. P. Wigner, "The Probability of a Self-Reproducing Unit," in *The Logic of Personal Knowledge: Essays Presented to Michael Polanyi* (Routledge, London, 1961), pp. 231–238.
7. Reference 5, pp. 68–69.
8. W. Elsasser, *Atom and Organism* (Princeton University Press, Princeton, 1966), p. 4.
9. Reference 5, p. 77.
10. Reference 5, p. 85.
11. W. Elsasser, *The Physical Foundation of Biology* (Pergamon Press, London, 1958), p. 121.
12. Reference 11, p. 122.
13. *Atomic Theory and the Description of Nature* (Cambridge University Press, Cambridge, 1934), pp. 22–23.
14. "We define a number A as immense when log A is a large number, it being purely a matter of convention when a number is called 'large'." "Semiformal Representation of Organismic Concepts" in *Proc. Natl. Acad. Sci. U.S.A.* **54**, 1432 (1965).
15. Reference 11, p. 154.
16. K. Denbigh, *The Inventive Universe* (Hutchinson, London, 1975).
17. Reference 11, p. 156.
18. Reference 11, p. 169.

13

The Theory of Relativity and Our World View*

A. R. Peacocke
Clare College, University of Cambridge
Cambridge, England

1. What the Theory of Relativity Is Not—And What It Is

In a much admired and widely used textbook of modern physics, based on lectures given to his classes at the California Institute of Technology, Richard Feynman turns aside in his exposition of relativity in physics to the views of what he calls "cocktail-party philosophers".[1] He describes these, while slyly hinting that the category may well include some professionals, as those who say, "Oh, it is very simple: Einstein's theory says all is relative!" or "That all is relative is a consequence of Einstein, and it has profound influences on our ideas," or "It has been demonstrated in physics that phenomena depend upon your frame of reference." As he rightly goes on to point out, if these remarks genuinely indicate all there is to Einstein's theory of relativity, then one can legitimately wonder why so simple an idea underwent such a long period of gestation and needed the intellectual acumen and power of an Einstein to uncover it. The "cocktail-party philosophers" have, of course, failed to perceive the depth of the theory and how it allows us to make definite predictions about the physical world. It is consistency with experiment that led, rather rapidly, to acceptance of the strange ideas contained in the special theory of relativity; and it was indeed as a result of reflection on certain experimental phenomena that Einstein first proposed the special theory.

The "cocktail-party philosphers" have too readily and superficially transferred the scientific awareness of the relativity of different frames of reference from physical observations (of signals, length, and times) to the cultural relativities which are to be discerned in ethics, social judgements, and political and religious attitudes—relativi-

*Previously published in *Einstein Lived Here*, M. Goldsmith, A. MacKay, and J. Woodhuysen, editors (Pergmon Press, Oxford, 1980).

ties based on quite other evidence and considerations. Einstein's theory of relativity is strictly restricted to the cognition and interpretation of physical phenomena and any analogies to other kinds of "relativism" are superficial and can be highly misleading. Einstein's theory, in itself, warrants no such *general* relativizing in these areas of human concern.

Indeed, those who make such extensions may have been misled by the word "relativity," the usual shorthand title of Einstein's great intellectual creation. For the theory, far from saying that everything is relative, is concerned above all to exclude what is relative in the results of observations, made from different moving systems, and to arrive at statements of laws which are the same whether the phenomena are described by one observer or another. In brief, the theory's aim is to arrive at what is *invariant* in the physical situation and at what is independent of a particular observer. We shall discuss this more fully below.[2]

Nevertheless, setting aside the observations of the "cocktail-party philosophers," it remains true that Einstein's theory of relativity has profound consequences, as Bertrand Russell said, for "our habits of thought"[3] and for our conception of the world in which we live—and also for the relation of our perceptions to it. That the theory has not already had these repercussions, now a hundred years since Einstein's birth, is testimony to how little education in scientific ideas there is, even now, among both intellectuals and the generality of the population as a whole. Here our intention is to describe some of the implications of Einstein's theory for the thinking of anyone concerned with the world and man's relation to it—implications which cannot be set aside in any philosophical or theological interpretation of the world.

The principal features of Einstein's theory of relativity may, following Wenzl,[4] be identified as follows: (1) For all systems moving uniformly in straight lines, the equal validity of not only the Newtonian laws of mechanics but also Maxwell's equations for the electromagnetic field; and the constancy of the velocity of light (c) with respect to *all* moving systems. Hence, by means of the Lorentz transformations, there follows the relativizing of spatial and temporal measurements and, furthermore, their coordination into a four-dimensional space-time continuum. (2) The equivalence of mass (m) and energy (E), as expressed in Einstein's equation $E = mc^2$, which thereby connects those previously differentiated concepts of classical physics. (3) The curvature of the space-time continuum by mass and energy in such a way that the *metric* (i.e., its mathematical characteristics purely as space-time) becomes the expression for what appears to us as matter-energy and as gravitational fields. (4) The world is a spatially limited, finite continuum, conforming to a non-Euclidean geometry, and of increasing radius (though not necessarily existing for an "infinite time").

2. Some Negative Consequences of Einstein's Theory

By the "negative" consequences we mean those effects of the theory of relativity that have iconoclastically demoted the status of certain concepts previously held to be given features of human thought, that is, given through the very nature of the intellect in the manner of Kant's synthetic *a priori*. These negative consequences are such that M. Capek, in his important work on *The philosophical impact of contemporary physics*, asserts:

> The present transformation of physics is far more radical than the famous "Copernican revolution" of the sixteenth century ... the transition from the closed world to the infinite universe was not excessively difficult for human imagination: the earth merely exchanged its position with that of the sun, while the celestial spheres were swept away. The effort of imagination required for such steps was relatively small ... the new Newtonian view of the universe was as pictorial as the old Aristotelian one; indeed, it was even more so. After elimination of secondary nonvisual qualities ... its visual character became even more pronounced....
>
> Today we are in the midst of a far more radical transformation of our view of nature. The most revolutionary aspect of this transformation consists in the fact that the words "picture" and "view" lose entirely their etymological meaning. As the so-called primary qualitites of matter now join the secondary qualities in their exit from the objective physical world, it is clear that the future conception of matter ought to be devoid of *all* sensory qualities, including even those which are subtly and implicitly present in seemingly abstract mathematical notions.[5]

There has been, as Karl Heim has said,[6] a "twilight of the gods" of classical physics—of absolute space, time, object, and determinism. There has been an almost Wagnerian *Götterdämmerung* in which the golden prize of absolute intellectual ascendancy over nature's laws, which these gods had tried to grasp, has been finally sunk in the inscrutable depths of nature's intrinsic reality from which it had been so roughly alienated. The Valhalla of the world those gods had constructed was consumed by the fires, first of the relativity, then of the quantum theory. Perhaps one of the first consequences of the theory of relativity is to induce a new scepticism within and towards science itself. For, as Feynman has put it,[7] the first discovery resulting from Einstein's principle of relativity is that "even those ideas which have been held for a very long time and which have been very accurately verified might be wrong ... we now have a much more humble point of view of our physical laws—everything *can* be wrong!"

Let us now consider some of the absolute "gods" of the classical, Newtonian physics that Einstein demonstrated to have feet of clay. Firstly, *space* was held to be homogeneous, three-dimensional in character, and Euclidean in geometry. It was logically prior to any objects existing "in" it. It was physically inert, infinite in extent, and continuous, that is, infinitely divisible. It had an absoluteness which for Newton seemed to be one of the God-given features of creation: "Absolute space, in its own nature, without regard to anything external, remains always similar and immovable."[8] All these premises were overthrown by Einstein (except, perhaps, its infinity, though not in the sense of limitlessness; and except, possibly, its infinite divisibility). Most important of all, the relation of juxtaposition, which seems so basic to any commonsense, Euclidean concept of space, cannot be understood coherently, for the denial of absolute simultaneity—as we shall see, an essential feature of Einstein's theory—removes any possibility of there being absolute juxtapositions.

Time was a second fundamental concept in the classical physical picture of the world—the single dimension in which instants followed each other. Like space it too was homogeneous, independent of any objects of events "in" it, inert, infinite, continuous. As Newton put it: "Absolute true and mathematical time, of itself and from its own nature, flows equably without relation to anything external, and by another name is called duration."[9] In the special theory of relativity this conception is modified but not completely subverted. Although absolute simultaneity is lost, it exists for events taking place at the same location. Further, the succession of events at such isotropic locations and of events which form causal chains is independent of the choice of the frame of reference. The only type of succession whose order depends on the frame of reference is that of causally unrelated events. Indeed the concept of *causality* is affected by Einstein's theory of relativity only to the extent that we now have to recognize that causal influences can never be transmitted through the universe at a speed greater than c.

Although the metrical scale of time intervals (the "dilatation of time") depends on the frame of reference, this is a distortion only from the perspective of the relative velocity of the observer. Even this could, however, be disturbing to certain affective assessments of the "progress" of a time regarded as all-embracing and the same for all observers. Thus, the second line of the poet's "One far-off divine event / To which the whole creation moves" needs to be rewritten, according to Bertrand Russell,[10] as "To which some parts of the creation move, while others move away from it". As he says further:

> . . . the question whether, on the whole, there is progress in the universe, may depend upon our choice of a measure of time. If we choose one out of a number of equally good clocks, we may find the universe is progressing as fast as the most optimistic American thinks it is; if we choose another equally good clock, we may find that the universe is going from bad to worse as fast as the most melacholy Slav could imagine.[10]

In the general theory of relativity, the concept of time is more fundamentally transformed. Here time loses its classical features of: homogeneity and uniformity (its dilatation in a gravitational field is not merely related to a particular frame of reference), independence of physical content, and causal inertness. Moreover, certain scientific developments, subsequent to Einstein's general theory, lead to the idea that time may have a beginning (e.g., the "hot big bang") and so it may not be infinitely divisible.

Even before the full blossoming of quantum mechanics, Einstein's special theory of relativity had rendered impossible the classical notion of *matter* as something impenetrable that fills space. For the interconvertibility of matter (as mass, m) and energy ($E = mc^2$) blurred the distinction between space (the square of whose differential with respect to time was proportional to kinetic energy) and mass (supposedly calculable by summing a property of each component of a material body; actually dependent also on the binding energy of those components).

Likewise the classical idea of *motion* was also profoundly altered. If a "particle" of matter is but a certain local configuration ("curvature") of space-time, as it is according to the general theory of relativity, how can it be said to "move in" a "space" that itself constitutes its own nature? "Motion of a body" is, in the general theory, simply a series of changes in the local curvature of space-time; only the continuity of these changes can provide any grounds for identifying a "body" that can be said to be "in motion." Thus Einstein's theory completely undermines the natural distinction between motion and that which moves, and the obliteration of this distinction was completed by the subsequent representation of particles as waves in quantum mechanics. For this reason, many authors prefer to regard "events" as constituting reality rather than particles.

The triumphal progress of science from the 17th to the 19th centuries was founded on the conviction that the human *observer* could impartially make observations and thereby derive an objective account of the external world on which all observers could agree—as in Newton's claim, whatever his practice, that he did not invent hypotheses but deduced them from appearances or phenomena.[11] This corporate intersubjectivity of its cognitive assertions is still the basis for the intellectual prestige of natural science. However, the status of this classical "observer" has been profoundly modified by Einstein with respect to all observations made between frames of reference moving relative to each other, in a way which only becomes experimentally manifest when there are involved large velocities (v such that $v^2/c^2 \not\ll 1$). Developments in atomic physics, such as the formulation of Heisenberg's Uncertainty Principle, finally transformed the classical, impartial, objective observer into what Richard Schlegel has called "The Impossible Spectator."[12]

This process of destroying the image of the classical observer clearly began with Einstein's realization that the apparently objective "primary" qualities of mass, length, time, and velocity, thought to characterize external objects in themselves, were not in fact independent of the knower but were dependent on the relation of the object to the observer, in particular on their relative velocities and accelerations. This kind of observer-dependence of observations of "external" objects that Einstein expounded must not be thought to license a widespread and universal exaltation of human subjectivity, as perhaps Feynman's "cocktail-party philosophers" might be tempted to adduce. For the limitations that Einstein quantified for each human observer are purely *physical* limitations that would be equally applicable to a set of instruments located in the same frame of reference. Moreover the "dilation of time" and the "contraction of space," together with the constancy of c, are themselves intersubjectively formulable and quantifiable. So Einstein's ideas provide no support for any general subjectivism in science, let alone in other realms of thought. What they do emphasize is that science arises from the interaction between the world of nature and ourselves; there is no access to any "objects-in-themselves" apart from the acts of observation. His theories throw "into higher relief the limitations of his [man's] own personal experiences."[13] As Sir James Jeans argued, modern physical science does not suggest

> . . . that we must abandon the intuitive concepts of space and time which we derive from individual experience. These may mean nothing to nature, but they still mean a good deal to us. Whatever conclusion the mathematicians may reach, it is certain that our newspapers, our historians and story-tellers will still place their truths and fictions in a framework of space and time; they will continue to say—this event happened at such an instant in the course of the everflowing stream of time, this other event at another instance lower down on the stream, and so on.
>
> Such a scheme is perfectly satisfactory for any single individual, or for any group of individuals whose experiences keep them fairly close together in space and time—and, compared with the vast ranges of nature, all the inhabitants of the earth form such a group. The theory of relativity merely suggests that such a scheme is private to single individuals or to small colonies of individuals; it is a parochial method of measuring, and so is not suited for nature as a whole. It can represent all the facts and phenomena of nature, but only by attaching a subjective taint to them all; it does not represent nature so much as what the inhabitants of one rocket, or one planet, or better still an individual pair of human eyes, see of nature. Nothing in our experiences or experiments justifies us in extending either this or any other parochial scheme to the whole of nature, on the supposition that it represent any sort of objective reality.[14]

Though Jeans fails to emphasize in this passage the impetus of the search for invariance in Einstein's thought, he does show how far we have come from the view of the scientist as the objective spectator of classical physics—or rather *should* have come, for the latter image still prevails in the popular image of the scientist. This is another example of how a century after his birth Einstein's ideas have yet to make a truly general impact.

The negative consequences of Einstein's theories are well summarized by Bertrand Russell's curt remark: "one thing which emerges is that physics tells us much less about the physical world than we thought it did."[15] Indeed many of what were taken to be true statements about the world turn out, in the theory of relativity, to be merely definitions.[16] This is the logical significance of the theory[16] and this is what gives it an iconoclastic aura. But the absolute gods of classical physics were not just dethroned by Einstein, something positive was put in their place. To this we now turn.

3. The Implications of the Theory of Relativity

3.1. What We Know—Our Picture of the World

In the theory of relativity, the relations of positions and times as measured in one frame of reference (coordinate system) and another are not what we would have intuitively expected from our ordinary ideas about *space and time*. Take the positions and times (x, y, z, t) of an event as measured by one observer with respect to his frame of reference (which we will, for purposes of exposition, call "stationary"). Now take the positions and times (x', y', z', t') of the same event measured by another observer, whose frame of reference is moving at a velocity u relative to our first frame of reference along the positive direction of the x axis. For our present purposes, we assume that this x is in the same direction as the x' axis and that the events under consideration are localized upon it. Then the relations between positions and times as measured in the two coordinate systems are given by

$$x' = \gamma(x - ut)$$

and

$$t' = \gamma\left(t - \frac{ux}{c^2}\right)$$

where $\gamma = (1 - u^2/c^2)^{-1/2}$ (additionally, $y' = y$, and $z' = z$). It is noticeable in the first of these Lorentz transformations that a difference between a space (x) and a time measurement (t), with appropriate multiplying factors, produces a new space measurement (x'); similarly, in the second such transformation this difference, again with the appropriate factors, produces a new time measurement (t'). So in the *space* measurements as seen by one observer there is mixed in not only the space but also the *time* measurements of the other. For this reason the separation between events is measured in terms an interval (s) such that

$$s^2 = c^2t^2 - x^2 - y^2 - z^2 = c^2t'^2 - x'^2 - y'^2 - x'^2$$

This quantity (s^2) is invariant under the Lorentz transformation. The intervals (s) are said to be "spacelike" when s^2 is negative and "timelike" when it is positive. An event is a given point (x, y, z, t) in this four-coordinate ("space-time") system.

Because of the loss of absolute simultaneity in Einstein's relativity theory the concept of a great three-dimensional "now," of worldwide instants, loses all physical significance. It is no longer possible, as in classical physics, to take an "instantaneous" three-dimensional cuts across the four-dimensional process of the world. They do not correspond to anything objective in nature and can be only conventional. Through its incorporation of space into time, the theory of relativity leads to the *dynamization of space*. This is Capek's view.[17] He argues against the more static interpretation of relativity denoted by the phase "the spatialization of time" on the grounds that causal links ("world lines" in a four-dimensional representation) remain "incurably successive, thus conferring the *dynamic* character upon the world they constitute."[18] While there are no absolute juxtapositions in the world, there are absolute successions; and the dynamization of space is an inherent feature of the general theory of relativity in which space-time (or time-space, as he prefers to call it) is merged with its changing physical content.

Thus the theory of relativity

> ... shows that space and time are neither ideal objects nor forms of order necessary for the human mind. They constitute a relational system expressing certain general features of physical objects and thus are descriptive of the physical world.... These conceptual systems [of space and time in the theory of relativity] describe relations holding between physical objects, namely, solid bodies, light-rays, and watches. In addition these relations formulate physical laws of great generality, determining some fundamental features of the physical world. Space and time have as much reality as, say, the relation "father" or the Newtonian forces of attraction.[19]

One danger of "spatializing time," as distinct from "dynamizing space," is that it is tempting to make deceptive analogies between movement in space and the progression of time, so that time has a direction, which can change, just as direction of motion in space may change. This generates the so-called problem of the "direction of time" which some authors, such as Capek,[20] regard as largely illusory, being (he would urge) based on "false kinematic analogies," while others such as Reichenbach,[21] still regard as an unsolved real problem. Both agree that time is an asymmetrical relation, for it establishes a genuine serial order, that of causal chains—a concept which

> ... can be shown to be the basic concept in terms of which the structure of space and time is built up. The spatio-temporal order thus must be regarded as the expression of the causal order of the physical world.... Time order, the order of *earlier* and *later*, is reducible to causal order; the cause is always earlier than the effect, a relation which cannot be reversed.... [22]

But, Reichenbach goes on to argue:

> For the theory of relativity, time is certainly an asymmetrical relation, since otherwise the time relation would not establish a serial order; but it is not unidirectional. In other words, the irreversibility of time does not find an expression in the theory of relativity.[23]

The directionality of time has to be based on much-debated considerations, related to thermodynamics and even to the indeterminacy principle. On these grounds, other than the theory of relativity, it might well be claimed, with Capek, that there has been "a reinstatement of *becoming* into the physical world"[24]—a conclusion consonant with what the cosmologist and biologist observe concerning the emergence of new forms of complexity of inorganic and living matter, respectively, as time "proceeds." The notion of the universality of this evolutionary process has, according to the physicist H. K. Schilling, given time a new meaning as "that of the carrier or locus of innovative change."[25] Indeed, we live in an "inventive universe."[26] Strictly speaking, such inferences go beyond what may be deduced from Einstein's ideas, but who can doubt that his ideas have initiated not only a profound reconsideration of the whole concept of time in relation to gravitational and electromagnetic phenomena, but have also made us much more open to rethinking the nature of time in the light of the cosmic and evolutionary processes we observe?

3.2. *How We Know—Epistemology*

Science is the mode of human enquiry that attempts *par excellence* to attain objectivity. It tries to make formulations and formulizations that give a view of reality independent of the observer, a view that is impartial and capable of universal agreement (the requirement of intersubjectivity). This never meant that anyone's sense perception of that reality should be the same for all, but at least Newton's space and time seemed to be objective because everyone thought they could order their sense data satisfactorily by means of these categories of interpretation. But when different spaces and times came to be seen as mathematically possible, a cleavage appeared to have been made between the subjective time and space of each particular observer and possible public, formalized spaces and times. For a theory to be valid in prediction, it came to be realized,[27] it must also have a criterion of objectivity that resides within some formal property of the theoretical scheme purporting to correspond to reality. Einstein looked for patterns or operations in which the fundamental laws are not changed; that is for laws of mechanics that, unlike Newton's, would remain unchanged by the Lorentz transformation—a transformation which, together with the constancy of c, was already required if the immutability of Maxwell's laws of electrodynamics was to be maintained. For Einstein's theory of relativity, objectivity becomes equivalent to *invariance* of physical laws, in the form of differential equations: it resides in the basic froms of theoretical statements rather than in observations and perception. The significance of this approach has been well expressed by Margenau:

> The idea of invariance is the nucleus of the theory of relativity. To the layman, and sometimes to the philosopher, this theory represents quite the contrary, a set of laws which allow for variability from one observer to another. This one-sided conception is linguistically implied by the word relativity which does not characterize the theory as centrally as it should. The true state of affairs can be seen when attention is directed to the aforementioned postulate of objectivity, which requires that the basic laws (the differential equations of highest order used in the description of reality) shall be invariant with respect to certain transformations. From this the variability, or relativity, of detailed observation may be shown to follow as a logical consequence. To give a simple example: the basic laws of electrodynamics involve the speed of light, c. If these laws are to be invariant, c must be constant. But the constancy of c in different inertial systems requires that moving objects contract, that moving clocks be retarded, that there can be no universal simultaneity, and so forth. To acheive *objectivity* of basic description, the theory must confer *relativity* upon the domain of immediate observations. In philosophic discussions, too much emphasis has been placed upon the incidental consequences, doubtless because the spectacular tests for theory involve this consequence.[28]

Thus the theory of relativity appeals to an important general principle, more powerful than anybody might have supposed, namely, that the laws of phenomena should be the same regardless of the relative motion of the frames of reference of the observers of the phenomena. Ever since the 16th century physics had allowed for the individual spectator's point of view, but before Einstein it had never so radically or so profoundly built this awareness into its fundamental methodology, and so into its epistemology. So, in making invariance of the laws its aim, Einstein's theory of relativity at the same time gave an incisive answer to the problem of objectivity.[29] The theologian T. F. Torrance attributes an even wider significance to Einstein's effect on the concept of objectivity:

> . . . both observational objectivity and methodological objectivity fall away and are superseded in the face of the new conception of the real world. The causal absoluteness of space and time erected by modern [classical] science is now seen to be essentially an artificial device designed to cope with the relativity of appearance, but that kind of corrective is no longer relevant. Instead, there emerges a new and profounder conception of objectivity grounded in the invariant relatedness inherent in the universe, . . . in which many fields of relations and forces cohere together through the natural correlation of their different levels . . . in such a way that ordinary experiences and even so-called appearances figure as real on their own level in correlation with other levels of reality, and are thus treated as relative aspects of the totality of existence.[30]

No doubt Einstein's drive towards the discovery of invariance in the multiplicity of observations available from different frames of reference was motivated by his conviction that *simplicity* of theories is vital to the theoretical physicist's enquiry:

> The aim of science is, on the one hand, a comprehension, as *complete* as possible, of the connection between the sense experiences in their totality, and on the other hand, the accomplishment of this aim *by the use of a minimum of primary concepts and relations*. (Seeking, as far as possible, logical unity in the world picture, i.e., paucity in logical elements.)[31]

and

> Our experience . . . justifies us in believing that nature is the realisation of the simplest conceivable mathematical ideas.[32]

"Simplicity" here can scarcely mean lack of sophistication; Einstein is upholding an intuition held by many physical scientists since Copernicus. It was the root of his refusal to accept the Copenhagen interpretation of the Uncertainty Principle of quantum mechanics which accepted that there *could* be a basic indeterminism in physical events. So he began his search for "hidden variables" not previously discerned in atomic physics. He could not believe, as he remarked in a famous letter to Max Born, in "the dice-playing god" but only "in the perfect rule of law in a world of something objectively existing."[33]

Einstein's quest for invariance and simplicity in the mathematical interpretation of physical reality was not simply a deduction from observation—in the way it was for Newton. As Lenzen says, in his comprehensive study of Einstein's theory of knowledge, for Einstein "Experience may guide us in our choice of mathematical concepts; experience also remains the sole criterion of the serviceability of a mathematical construction for physics; but it cannot possibly be the source from which a theory is derived."[34] For Einstein, as later stressed by M. Polanyi,[35] mathematical theories of physical reality are free creations of imaginative and integrative human thought. It may well be surmised[36] that Einstein's own emphasis on the creative role of the theoretician, together with his success in putting it into practice, gave 20th century scientists a new sense of the personal, creative role of the human mind in developing their concepts and theories. This ability of the human mind never failed to astonish Einstein:

> The very fact that the totality of our sense experiences is such that by means of thinking (operations with concepts, and the creation and use of definite functional relations between them, and the coordination of sense experiences to these concepts) it can be put in order, this fact is one which leaves us in awe, but which we shall never understand. One may say "the eternal mystery of the world is its comprehensibility".... In speaking here concerning "comprehensibility", the expression is used in its most modest sense. It implies: the production of some sort of order among sense impressions, this order being produced by the creation of general concepts, relations between these concepts, and by relations between the concepts and sense experience, these relations being determined in any possible manner. It is in this sense that the world of our sense experiences is comprehensible. The fact that it is comprehensible is a miracle.[37]

Torrance believes that this distinctive element in Einstein's throught now characterizes the whole of science; he regards it as but a rigorous extension of the basic way of organizing our ordinary experiences and knowledge, through which we know things in accordance with their natures and what they are in themselves.[36,38] He goes on to make this the basis of his own "theological science."[39] Whether or not we follow him this far, Wenzl's remarks cannot be denied:

> The theory of relativity has simultaneously made us freer and richer by showing in the realm of physics, not merely by way of abstract advice but in concrete performance, that our intellectual capacity of knowledge reaches farther than our sensory capacity of perception. It is the same step in the realm of ideas as that in geometry from Euclidean to non-Euclidean geometry. Our perceptual capacity is limited to three dimensions of a homogeneous continuum of the curvature zero. Our thought-capacity reaches farther. We do not feel ourselves imprisoned in the world of appearances and of the workaday world. But neither do we need to resign ourselves so much, as did Kant, who could say nothing aside from his forms of perception and the forms of thought. We do not assert that with the theory of relativity we have reached reality as such: concerning its inner essence we shall never be able to speak otherwise than by analogy and supposition. But we have obviously come closer to objective reality than by way of our perceptual capacity. Philosophically this means at the same time that we have moved farther away from materialism even through the development of physics. For now we assert of matter only that it is something which is expressed and can be expressed differential-geometrically, by means of differential equations for a many-dimensioned continuum.[40]

Clearly Russell's assertion that Einstein's work has resulted in "changes in our habits of thought" and will "have great importance in the long-run"[3] was something of an understatement.

3.3. Whether We Know: Realism

The quotation above from Wenzl goes beyond methodological and epistemological matters to refer to the question of realism. Einstein, himself, by reality meant *physical* reality:

> The belief in an external world independent of the perceiving subject is the basis of all natural science. Since, however, sense perception only gives information of this external world or of "physical reality" indirectly, we can only grasp the latter by speculative means. It follows from this that our notions of physical reality can never be final. We must always be ready to change these notions—that is to say, the axiomatic structure of physics—in order to do justice to perceived facts in the most logically perfect way.[41]

As Margenau points out,[42] this passage makes a threefold distinction between an *external world*, the observer's *perception* (said to be "indirect") of that external world, and our *notions* of it—a hornet's nest of philosophical problems. Subsequent to Einstein, later interpreters of the theory of relativity have come to very different conclusions. Compare that of Russell:

> It throws very little light on time-honoured controversies, such as that between realism and idealism . . . The "subjectivity" concerned in the theory of relativity is a *physical* subjectivity, which would exist equally if there were no such things as minds or senses in the world.[3]

with that of Wenzl:

> To the question, what both: matter and energy, are, the general theory of relativity gives us the [following] answer: they are that which creates the metric. The metric of the space-time-continuum is the expression of the intensity of the stresses of their essence . . . The most noteworthy fact is this: that it is possible to comprehend and represent the physical magnitudes (mass, impulse, force, energy) as producers of geometrical characteristics as they occur in the theory of planes, and to be able to treat the field-forces also [in terms of] differential geometry. For this is by no means self-evident nor simply the invention of a great mathematico-artistic ingenuity; yet *in this possibility lies the decisive objective epistemological value of the general theory of relativity*.[43]
>
> What is it possible to say concerning it [reality] today at all? Nothing, except that it is a [type of] reality which effectively confronts us and which is so well mathematically describable that it can be regarded as the realization of mathematical structures and forms. . . . Being effectual for each other and towards us, being the expression and the expressibleness of the orderly relations in mathematically couchable forms: these are today the attributes of "material reality." Is this peculiar? Mathematically expressibility, we already said, is marvellous. But what lies at the base of material appearance is a being-related-to and a becoming-effectual.[44]

Torrance goes even further:

> For the new [Einsteinian] science, however, these [Newtonian] notions of "reality" wither away, as soon as the dichotomies (between the apparent and the real, subject and object, form and being, etc.) that lie at their root are eliminated,

together with the artificial correctives they forced upon science but which are now seen to be incompatible with the actual world of empirical fact. Here the real world is that which forces itself upon our inquiries in the imperious light of its own intrinsic order, is one in which intelligible structure and material content exist in mutual interaction and interdetermination. This is a world in which relations between bodies are just as real as the bodies themselves, for it is in their interrelations that things are found to be what and as and when they are. Thus, as relativity theory has brought to light, space and time are not only fused with one another but fused with the matter and energy of the universe in such a way as to constitute an unresolvable four-dimensional continuum with rationally and physically objective properties. . . . The real world confronts us, then, as a continuous integrated manifold, in which structure and substance, form and being, are inseparably conjoined in the immanent relatedness of the universe. . . . Since the real world exhibits itself to scientific inquiry in this profound unity of form and being, or structure and substance, it not only gives evidence of its existence as an intelligible system independent of our perceiving and conceiving of it, but thereby discriminates itself perpetually from our scientific constructs about it, and remains in its independence the final judge of their truth or falsity. All this implies a powerful restoration of *ontology*, not only for the philosopher or the theologian, but also for the scientist, who precisely as scientist finds that it is "existence and reality," that he wishes to comprehend.[45]

These latter two expositions (of Wenzl and Torrance) of the nature of the "reality" implied by the theory of relativity are more positive and definite than Einstein himself might have supported, for he himself seemed to retain a certain scepticism about the meaning of the term:

> Behind the tireless efforts of the investigator there lurks a stronger, more mysterious drive: it is existence and reality that one wishes to comprehend. But one shrinks from the use of such words, for one soon gets into difficulties when one has to explain what is really meant by "the reality" and by "comprehend" in such a general statement.[46]

In light of the intense philosophical debates of the last decade[47] on realism in relation to the concept of "truth" in science *(inter alia)* and its applicability (for example, to propositions, sentences, and statements) we should be wise to share Einstein's scepticism, if for somewhat different reasons, and be cautious in making any inferences from his theories about the nature of reality.

4. Time in Relativistic Physics and Theology[48]

Theological models of the relation of God to the world and the cosmological models of relativistic physics for the most part use entirely different resources for their construction. Nevertheless each has to face the question of how *time* is incorporated into its models. We shall first turn to an examination of certain aspects of the understanding of time in theology.

The principal stress in the Judeo-Christian doctrine of creation is on the dependence on God of all entities and events: it is about a perennial relationship of all-that-is to God, and not about the beginning of the Earth, or of the whole universe at a point in time. The phrase "the whole universe" is of course ambiguous since it seems to imply a boundary or limit "beyond" or outside which in some sense, God exists—and this will not do at all, for God would then be in a "beyond" or "outside" which would be but an extension of the same framework of reference as that in which the "whole universe" is conceived to exist. To avoid this conceptual impasse, the principal stress in the doctrine of creation has been an affirmation that any particular event or entity would not happen or would not be at all were it not for the sustaining creative will and activity of God.

One of the most influential attempts to understand the relation of God, time, and creation was that made by St. Augustine in the eleventh chapter of his *Confessions*, in which he addresses himself to those who ask "What was God doing before he made heaven and earth?" He describes the paradox as follows:

> "If he was at rest," they say, "and doing nothing, why did he not continue to do nothing for ever more, just as he had always done in the past? If the will to create something which he had never created before was new in him . . . how can we say that his is true eternity, when a new will, which had never been there before, could arise in it? . . . The will of God, then, is part of his substance. Yet if something began to be in God's substance, something which had not existed beforehand, we could not rightly say that his substance was eternal. But if God's will that there should be a creation was there from all eternity, why is it that what he has created is not also eternal?"[49]

This provokes St. Augustine to undertake a profound analysis of our experience of time. From it he concludes that the world was created along with and not in time. Time itself is a feature of the created cosmos and therefore no "act of creation" can be located at a point within created time itself. There is no time without events and God's eternity is not just endless *temporal* duration but a mode of existence which is qualitatively different from that of successive temporal experience. Augustine addresses God thus:

> How could those countless ages have elapsed when you, the Creator, in whom all ages have their origin, had not yet created them? What time could there have been that was not created by you? How could time elapse if it never was? You are the Maker of all time. If, then, there was any time before you made heaven and earth, how can anyone say that you were idle? You must have made that time, for time could not elapse before you made it. But if there was no time before heaven and earth were created, how can anyone ask what you were doing "then"? If there was no time, there was no "then."[50]
>
> It is therefore true to say that when you had not made anything there was no time, because time itself was of your making. And no time is co-eternal with you, because you never change; whereas, if time never changed, it would not be time.[51]
>
> Grant them [those who ask the questions "what was God doing before he made heaven and earth?" or "How did it occur to God to create something, when he had never created anything before?"], O Lord, to think well what they say and to recognise that "never" has no meaning when there is no time. If a man is said

> never to have made anything, it can only mean that he made nothing at any time. Let them see, then, that there cannot possibly be time without creation . . . Let them understand that before all time began you are the eternal Creator of all time, and that no time and no created thing is co-eternal with you, even if any created thing is outside time.[52]

These questions formed part of the later famous debate between St. Thomas Aquinas and St. Bonaventure.[53] Although both accepted that the world had a beginning in time, on the basis of the revelation in the Scriptures, the former thought that the world could have existed from eternity (that is, for an infinite time)—though, in fact, the biblical revelation shows it does not so exist—whereas the latter argued that, if the world is created, then time necessarily had a beginning. On this latter view, it is more accurate to say the world was created *with* time, which seems to be Augustine's position in the *Confessions*.

The bringing togther of time and "things" in such close connection, so that both are part of the created order, is entirely in accord with the outlook to which Einstein has accustomed us. For, unlike Newton, we can no longer regard time as a flowing river of endless duration, as a mode of extension into which events are inserted as if time had an entirely discrete and unrelated kind of existence from the events themselves and from the matter participating in them. For it is certain distinctive metric features of space and time, or rather space-time, which Einstein has shown us constitute the gravitational forces of matter, which is now known to be interchangeable with energy, a concept which itself involves an intimate relation to the concept of time. Time, in post-Einsteinian physics, is an integral and basic constituent of nature, an aspect of space-time. Hence, on any theistic view, it has to be regarded, like the rest of the created world, as owing its existence to God, as St. Augustine perceived. It is this "owing its existence to God" which is the essential core of the Judeo-Christian doctrine of Creation.

In investigating and making theoretical deductions about the remote history of our universe, scientific cosmology cannot, in principle, be doing anything which can contradict such a concept of creation. Aided by our radio-telescopes and other instruments we may, or may not, discover if there was a point in space-time when the universe, as we can observe it, began. But whatever we may be able to infer about what may have happened on the other side of the "big bang," any grounds for believing that the cosmos has derived and contingent being, and that God alone is Being in himself, would not be perturbed. The doctrine of creation itself would be unaffected, since it concerns the relationship of all the created order, including time itself, to their Creator, their Sustainer and Preserver.

Nevertheless there is an important feature which the scientific perspective inevitably reintroduces into the idea of creation in the form just described. This is the realization that the cosmos which is sustained and held in being by God is a cosmos which has always been in process of producing new emergent forms of matter. It is a world which is still being made. On the surface of the Earth, for example, man has emerged from prehistoric forms of life and his history is still developing. Any static conception of the way in which God sustains and holds the cosmos in being is therefore precluded, for the cosmos is in a dynamic state and has evolved conscious and self-conscious beings, who shape their environment and have the power of choice.

That the world was in flux and change, with all its corollaries for the destiny of the individual man, has been reflected upon since the ancient Greeks. But that the matter of the world developed in a particular direction to more complex and ultimately thinking forms was not established knowledge. The people of Israel, and following them, the Christian church, have always believed in the providential hand of God in human history, with the nonhuman world usually being regarded simply as the stage for that drama. Science now sees man as part of "nature" and both together as subject to continuous development. Any static conception of the relation of God and the world is therefore precluded, for if the emergence of new forms of matter in the world is in some way an activity of God, then it must be regarded as his perennial activity and not something already completed. The scientific perspective of a cosmos in development introduces a dynamic element into our understanding of God's relation to the cosmos which was previously only implicit in the Hebrew conception of a "living God" dynamic in action, and in the formulations of (for example) the Nicene Creed.[54]

5. A Musical Epilogue

This new awareness of "becoming" in the natural processes of the world represents a convergence of many strands of scientific thought, including the biological and cosmological, as well as those stemming from Einstein's theory of relativity. But that theory, especially in its general form, has formulated a conception of space-time which goes far beyond the power of our imagination to picture. However, our experience of time as being characterized by succession and the varying "metrics" of the passage of conscious time which we observe according to our state of mind do give some clues to how we might now model conceptually the dynamic nature of physical reality by *imageless dynamical patterns*.[55] That reality is characterized[56] by: the incompleteness of becoming and its pulsational character; the compatibility of the emergence of novelty with the causal influence of the past; the individuality of events within the continuity of the flux; the fictitious character of instantaneous acts and consequently, the impossibility of instantaneous space; the replacement of the relation of juxtaposition or co-instantaneity of points by that of co-becoming (of contemporaneity of the "causal tubes"). As Capek says, "All these features defy any attempt at visualization".[56] A source of models of, or analogies for, such imageless, dynamic patterns that Capek and a number of authors have discovered is to be found in the musical experience. A. N. Whitehead long ago noted that in his organic theory (of the nature of the physical world):

> . . . a pattern need not endure in undifferentiated sameness through time. The pattern may be essentially one of aesthetic [i.e., qualitative] contrasts requiring a lapse of time for its unfolding. *A tune is an example of such a pattern.* Thus the endurance of the pattern now means the reiteration of its successions of contrasts.[57]

More recently, Capek has developed this idea—a little cautiously, afraid that he may be misunderstood to be attempting to reinstate auditory, or other secondary qualities into the physical world, which is certainly not his intention.[58] What he hopes to offer is a "key to the understanding of the nature of the type of 'extensive becoming' that seems to constitute the nature of physical reality."[58] He wants to find in the auditory experiences of music an imageless, dynamic pattern for time, the "locus of innovative change."[25] Let Capek speak for himself:

> Let us consider a piece of music—for instance, a melody or better a polyphonic musical phrase. It is hardly necessary to underscore its successive character. As long as its movement is going on, it remains incomplete and in its successive unfolding we grasp in the most vivid and concrete way the incompleteness of every becoming. At each particular moment a new tone is added to the previous ones; more accurately, each new moment is constituted by the addition of a new musical quality. . . . The relation of. . . . arithmetical units to their sum total is the same as the relation of *the parts to the whole in space.*
>
> In the musical experience of melody or polyphony the situation is considerably different. The quality of a new tone, in spite of its irreducible individuality, is tinged by the whole antecedent musical context which, in turn, is retroactively changed by the emergence of a new musical quality. The individual tones are not externally related units of which the melody is additively built; neither is their individuality absorbed or dissolved in the undifferentiated unity of the musical whole. The musical phrase is a *successive differentiated whole* which remains a whole in spite of its successive character and which remains differentiated in spite of its dynamic wholeness. Like every dynamic whole it exhibits a synthesis of unity and multiplicity, of continuity and discontinuity: . . . it is neither continuity in the mathematical sense of infinite divisibility nor is it the discontinuity of rigid atomic blocs.[59]

> Every musical structure is by its own nature unfolding and incomplete; so is cosmic becoming, the time-space of modern physics. The musical structures, in virtue of their essentially temporal nature, cannot be subdivided *ad infinitum* without being destroyed . . . As Whitehead says, "a note of music is nothing at an instant, but also requires its whole period to manifest itself" [*Science and the Modern World,* Ref. 57, p. 54]. For this reason musical wholes—like physical processes—are not infinitely divisible; in either case durationally instants are mere ideal limits, arbitrary cuts in the dynamic continuity of becoming.
>
> . . . [In] concrete temporal experience the emergence of novelty is possible, so to speak, only on the contrasting background of its immediate past; in a similar way a new musical quality of the (provisionally) last tone acquires its individuality in contrast to, as well as in connection with, its antecedent musical context.[60]

Capek even finds in his "auditory model" a way of understanding the new meaning of spatiality with which Einstein has replaced the concept of static Newtonian space, even though nothing in a bare succession of tunes corresponds to anything even remotely analogous to the relation of space. However,

> The situation . . . is different when we turn our attention to the dynamic structure of polyphony. The dynamic pattern in this case is more complex. In a contrapuntal composition two or several melodically independent movements, whether harmonious or dissonant, are going on. The component melodic movements, besides

> each being unfolded successively, are also in a certain sense *beside* or *alongside* each other, and this relation "beside" is analogous to the relation "beside" in space.[61]

Music, it has often been noted, is the only means of human communication that cuts across all barriers of language, race, class and creed and unites in an experience of reality at once both personal and corporate. So it is not surprising that Whitehead and, more extensively, Capek have resorted to an art form independent of the visual (and so Euclidean) imagery to which even language ties us. To take a quite different context: the search for a "model" which would represent how God might be conceived as involved in a dynamic creation that continually manifests new potentialities, has led the author to musical resources for expression. I have suggested[62] that we could perhaps regard God the Creator as somewhat like a bellringer, ringing all the possible changes, all the possible permutations and combinations he can out of a given set of harmonious bells—though it is God who creates the pattern of the "bells" too. Or, perhaps better, he is more like a composer who, beginning with an arrangement of notes in an apparently simple tune, elaborates and expands it into a fugue by a variety of devices of fragmentation and reassociation; by turning it upside down and back to front; by overlapping these and other variations of it in a range of tonalities; by a profusion of time-sequence patterns, in which the interplay of sounds flows in an orderly fashion from the chosen initiating ploy. Thus does a J. S. Bach create a complex and harmonious fusion of his seminal material, both through time and at any particular instant; beautiful in its elaboration, it only reaches its consummation in the last few bars when all the threads have been drawn into the return to the home key—the key of the initial melody whose potential elaboration was conceived from the moment it was first expounded. In this kind of way might the Creator be imagined to unfold the potentialities of the universe which he himself has given it. He appears to do this, I have argued, by a process in which the creative possibilities inherent within the fundamental entities of the universe and in their interrelations become actualized within a temporal development which is itself shaped and determined by those self-same inherent potentialities that he conceived from the very first note.

Such recourse to music for an image of the temporal dynamic process that the theory of relativity opened up would not, I suspect, have been repudiated by Einstein himself. For, so his eldest son has recalled, "Whenever he felt that he had come to the end of the road or into a difficult situation in his work, he would take refuge in music, and that would usually resolve all his difficulties."[63]

References and Notes

1. R. P. Feynman, R. B. Leighton, and M. Sands, *The Feynman Lectures on Physics* (Addison-Wesley, Reading, Massachusetts, 1963), Vol. I, p. 16-1.
2. Section 3.2.
3. B. Russell, *ABC of Relativity*, 3rd revised ed., F. Pirani, editor (Allen & Unwin, London, 1969), p. 133.

4. A. Wenzl, "Einstein's Theory of Relativity Viewed from the Standpoint of Critical Realism, and its Significance for Philosophy," in *Albert Einstein: Philosopher-Scientist*, The Library of Living Philosophers, Vol. VII, P. A. Schilpp, editor (Open Court, La Salle, Illinois, 1970), pp. 584–585.
5. M. Capek, *The Philosophical Impact of Contemporary Physics* (van Nostrand, Princeton, 1961), pp. 378, 379.
6. K. Heim, *The Transformation of The Scientific World View* (S.C.M. Press, London, 1953), p. 24.
7. R. P. Feynman, *et al.*, Ref. 1, pp. 16–2, 16–3.
8. Isaac Newton, *Philosophiae Naturalis Principia Mathematica* (*Mathematical Principles of Natural Philosophy*) transl. A. Motte, revised F. Cajori (Univ. of California Press, Berkeley, 1934), Scholium II to Definition VIII, p. 6.
9. Reference 8, Scholium I to Definition VIII, p. 6.
10. Reference 3, p. 136.
11. Reference 8, General Scholium, p. 547.
12. R. Schlegel, "The Impossible Spectator," 6th Centennial Review Lecture at Michigan State University, 12 May 1975, *The Centennial Review*, pp. 218, 230.
13. R. W. Clark, *Einstein—The Life and Times* (Hodder & Stoughton, London, 1973), p. 111.
14. Sir James Jeans, *The New Background of Science* (Cambridge Unviersity Press, Cambridge, 1933), pp. 95, 96.
15. Reference 3, p. 134.
16. *Vide* H. Reichenbach, "The Philosophical Significance of the Theory of Relativity," in Ref. 4.
17. Reference 5, p. 383ff.
18. Reference 5, p. 385 (italics added).
19. Reichenbach, Ref. 4, p. 302.
20. Reference 5, p. 366, 7.
21. Reichenbach, Ref. 4, pp. 303–307.
22. Reichenbach, Ref. 4, p. 303.
23. Reichenbach, Ref. 4, p. 305.
24. Reference 5, p. 395 (italics added).
25. H. K. Schilling, *The New Consciousness in Science and Religion* (S. C. M. Press, London, 1973), p. 126.
26. K. Denbigh, *An Inventive Universe* (Hutchinson, London, 1975).
27. H. Margenau, "Einstein's Conception of Reality," in Ref. 4, p. 253.
28. Margenau, Ref. 4, p. 254.
29. Margenau, Ref. 4, p. 252.
30. T. F. Torrance, "The Integration of Form in Natural and in Theological Science," in *Science, Medicine and Man* **1**, 152 (1973).
31. A. Einstein, *Out of My Later Years* (Thames & Hudson, London, 1950), p. 63.
32. A. Einstein, "On the Method of Theoretical Physics," in *The World As I See It* (Covici-Friede, New York, 1934), p. 36.
33. Quoted by Max Born in *Natural Philosophy of Cause and Chance* (Clarendon Press, Oxford, 1949), p. 122.
34. V. F. Lenzen, "Einstein's Theory of Knowledge," in Ref. 4, pp. 361, 362.
35. M. Polanyi, *The Tacit Dimension* (Routledge & Kegan Paul, London, 1967); *Personal Knowledge* (Routledge & Kegan Paul, London, 1958).
36. Reference 30, p. 154.
37. Reference 31, p. 61.
38. T. F. Torrance, Address to Templeton Foundation, 1978.
39. T. F. Torrance, *Theological Science* (Oxford University Press, London, 1969).
40. Reference 4, p. 605, 606.
41. Reference 32, p. 60.
42. Reference 27, p. 249.
43. Reference 4, p. 592, 593.
44. Reference 4, p. 603, 604.
45. Reference 30, p. 150, 151.
46. Einstein Address at Columbia University, 1934, Reference 32, p. 137f.

47. See, for example, H. W. Putnman, *Philosophical Papers*, Vols. 1 and 2 (Cambridge University Press, Cambridge, 1975), "What is Realism?," *Proc. Arist. Soc.* 76, 177–194 (1975/6); *Meaning and the Moral Sciences* (Routledge & Kegan Paul, London, 1978); M. Dummett, "Can Truth be Defined?" in *Frege: Philosophy of Language* (Duckworth, London, 1973), Chap. 13; "Truth" *Proc. Arist. Soc.* **59,** 141–167 (1978/9); G. Evans and J. Mc.Dowell, editors, *Truth and Meaning* (Oxford University Press, Oxford, 1975).
48. The following section incorporates material (p. 376ff.) from an article by the author on "Cosmos and Creation" in *Cosmology, History and Theology*, W. Yourgrau and A. D. Breck, editors (Plenum Press, New York & London, 1977), pp. 365–381.
49. St. Augustine, *Confessions* [XI], 10, transl. R. S. Pine-Coffin (Penguin Classics Edition, Hammondsworth, 1961), p. 261.
50. Reference 49, para. 13, pp 262–263.
51. Reference 49, para. 14, p. 263.
52. Reference 49, para. 30, p. 279.
53. *Vide* F. Coplestone, *A History of Philosophy*, Vol. II (Burns & Oates, London), pp. 262–265 and p. 363ff.
54. God the Holy Spirit as "The Lord and Giver of Life."
55. Reference 5, p. 379.
56. Following Capek, Ref. 5, p. 378.
57. A. N. Whitehead, *Science and the Modern World* (Macmillan, New York, 1926), p. 193 (my italics).
58. Reference 5, p. 399.
59. Reference 5, pp. 371, 372.
60. Reference 5, pp. 372, 373.
61. Reference 5, p. 377.
62. A. R. Peacocke, Third 1978 Bampton Lecture in *Creation and the World of Science* (Clarendon Press, Oxford, 1979).
63. Quoted by R. W. Clark, Ref. 13, p. 115.

14

Science Beyond Science

Herbert Pietschmann
Institut für Theoretische Physik
Universität Wien
Vienna, Austria

1. Scientific Method

When Galileo Galilei—more than 300 years ago—opened the new era of science, he defined "nuova scienza" (new science) by its goal: quantitification by measurements. And at the celebration of the 400th anniversary of Galilei's birth, R. P. Feynman expressed so clearly the continuation of this idea till today:[1]

> . . . the sciences have developed in a very good way directly and continuously from his original, in the same spirit as he developed. And as a result there are no more witches and ghosts.
>
> That is in fact almost a definition of science today; the sciences that Galileo was worried about, the physics, mechanics and such things, have of course developed, but the same techniques worked in biology, in history, geology, anthropology and so on. We know a great deal about the past history of man, the past history of animals, and of the earth, through very similar techniques. With somewhat similar success, but not quite as complete because of the difficulties, the same systems work in economics. But I would be ashamed to tell Mr. Galileo, that it does not really work very well. For example, there is an awful lot of studying of education, particularly in teaching arithmetic—but if you try to find out what is really known about what is the better way to teach arithmetic you will discover that there is an enormous number of studies and a great deal of statistics, but they are all disconnected from one another and they are mixtures of anecdotes, uncontrolled experiments, or very poorly controlled experiments, so that there is very little information as a result.

Feynman points out two distinct facts about Galileo's method, the method of science: Firstly, it has proven to be enormously powerful in finding, ordering, and explaining facts; not only in physics and mechanics, but also in a multitude of other fields. Secondly, it seems to run into difficulties in some areas like education or teaching. If we accept this point of view, we can ask a very crucial question: Is it not demanding too much to apply this very same method without any adaptation to *all* fields of human activities? If so, can we adapt the method in such a way that it *becomes* applicable to a wider domain, keeping its tremendous critical power by predictions and at the same time becoming more flexible, more open to include also less rigid areas such as teaching or education?

Before we even try to approach this question, let us remember the essentials of scientific method.

No doubt most scientists would agree that the outstanding success of scientific method is its predictive power. We even judge theories, hypotheses, or models according to their predictive power. How does science achieve this? Unlike mathematics, scientific laws and theories cannot be proven in the strict sense. Thus, they do not form positive certainties. We can never say without any doubt that *all* charged bodies follow Coulomb's law, that *no* magnetic monopole exists, not even that no *perpetuum mobile* can be constructed. But we can be pretty certain about some of these statements.

How can we be certain?

Although we cannot prove a theory to be correct, we can refute (or modify or amend) a theory if it is incorrect. In this way—as Karl Popper[2] has so clearly shown—we approach better and better understanding of our surroundings by eliminating more and more false ideas. Trial and elimination of error is thus the main ingredient of scientific method. I have shown elsewhere[3] how this method is used in the actual day-to-day life of the working physicist. Here, I would like to put more emphasis on the mechanism of predictions and refutations or corroborations of theories.

In order for a theory to be able to make predictions, it must be free from internal contradictions and its notions must be clearly defined. Within the realm of Galilean (or "classical") science, test experiments must be suggested in an unambiguous way and its results must be uniquely given. Let me give an example of present-day physics. The theory of unified electroweak (electromagnetic plus weak) interactions of Glashow, Salam, Weinberg[4] and others predicted a number of phenomena which no other theory knew. Most prominent was the prediction of the weak neutral current, which was found by experiments. From it, a second prediction could be derived, parity violation in the atomic shell. At first, this was not found, but more and better experiments gave different answers, so that at least the theory is no longer refuted by these early findings. Another precise prediction, the existence of neutral and charged intermediate bosons at well-defined masses, is awaiting experimental test as soon as higher energies become available.

Before the weak neutral current was found experimentally, an alternative theory was formulated[5] which predicted new heavy leptons instead of the neutral current. Apart from this difference, it had all other appealing facts as the surviving theory (such as economy of particles, renormalizability, and the like). Thus we had two alternatives: One predicting a neutral current, the other one predicting no neutral current (but heavy leptons instead). Experiments refuted the latter theory by establishing the existence of the neutral current.

What I want to emphasize is that it is quite obviously possible at a given time to have contradicting predictions such as: "There is a neutral weak current" and "there is no neutral weak current." But within classical science, these predictions must *stem from different theories!* Experiments then decide once and for all which of the theories is right and which is wrong. It is impossible within classical science that one and the same theory makes predictions such as "phenomenon A exists or does not exist."

I have emphasized the word "prediction." It is quite clear that any theory can be indifferent towards some phenomena. For instance, a singly produced Λ particle may decay into a proton and a negative pion or it may decay into a neutron and a neutral pion. Thus it is correct to say that the theory allows a Λ particle to decay either through this or through that channel (or through still some others). But we do not call this a prediction! What the theory does predict is the probability, or the statistics for a large number of Λ decays. It does not predict the result of an individual Λ decay.

Let me reiterate, that within classical science, contradicting predictions must stem from different theories, of which we believe one to be right, the other one wrong. Experiments decide which one survives.

Could this be an indication why classical scientific method runs into difficulties when it is confronted with areas like education or teaching? Because emotions cannot be neglected in these fields and emotions may well be ambivalent?

Let us try a *Gedankenexperiment.* Suppose the problem is to learn a certain poem by heart. We want to find out the best attitude of the teacher to succeed in convincing the pupils to do the job. He can be persuasive, authoritarian, fatherly, or many other ways. Within classical science, we want to define all these attitudes as objectively as possible. Suppose we succeed. We then want to measure the result for a given attitude by noting the percentage of pupils who have done the job.

The underlying assumption in applying classical science is that a given attitude—if clearly defined—leads to a unique result. But our pupils may have ambivalent feelings towards the teacher. In one case, the ambivalence may be balanced towards a positive reaction (*all* of them learning the poem) by very subtle and irrational processes. In another case, the balance may go towards a negative reaction (*none* learning the poem) by small unobservable differences in behavior.

Suppose we have two theories, one of them predicting the positive reaction, the other one the negative. Both of them are then refuted *and* corroborated by experiments and classical scientific method has failed to do its job. The reason is the ambivalence, which in fact could best be taken into account by a theory which predicts both results; but that is not allowed within the realm of classical science!

As long as we do have two different theories with opposite predictions, they can go on fighting each other endlessly and both can rest their argument on empirical results. This is the situation bemoaned by Feynman as one in which "there is very little information as a result."

2. On the Development of Knowledge

In Section I, I have said that we approach better and better understanding of our surroundings by eliminating more and more false ideas. Let us analyze this process in closer detail.

At each given time in history, we have something like a canonical scientific view of the world. By this, I mean the totality of accepted theories, models, and hypotheses. We assume, of course, that this canonical view is free from internal contradictions, though it is not always shown in each case. The work of the scientist has two complementary aspects. On the one hand he has to order, clarify, and archivate this canonical knowledge. On the other hand he has to search for new emipirical facts, which do not fit into the order, which are not clear within this knowledge; in short, he has to question the validity of canonical knowledge. To some extent we can identify the first aspect with "science of the lecture hall," the second aspect with "science of the laboratory" which—to my deep regret—differ from each other sometimes to an uncomfortable degree.

Let us now concentrate on the second aspect, the search for new facts which do not fit into the canonical scheme. To be precise, suppose we have a direct contradiction of some finding with predictions of existing theories. Such a contradiction is always the germ for a lively process, which changes and improves our canonical view. During this process, we can distinguish four phases.

2.1. *Phase 1: The Critical Attitude*

One of the most important elements of scientific method is its critical attitude. It requests that we should not simply believe anything before we have investigated it thoroughly. In other words, whenever a contradiction of canonical knowledge with a new finding comes up, the first reaction *must be* that the finding is erroneous. Experiments must be repeated, discoveries thoroughly questioned even to the point of assuming possible (conscious or subconscious) fraud. (Remember the story of the Piltdown skull.) It would be dangerous to be too little critical; but it is equally difficult not to be overcritical. The sad story of the meteorites,[6] whose existence had been denied until 1803, is an example of being overcritical. The N-rays,[7] whose properties had been investigated in detail until they were shown not to exist, form an example of too little criticality. Unfortunately, I do not know of any way to establish the right measure of criticality other than learning from history—and we all know how impossible that is.

2.2. *Phase 2: Independent Investigation of the New Finding*

After the existence of a new finding has been established beyond doubt by critical analysis, we enter a new phase. The contradiction with canonical knowledge still exists, but it is—for some time—laid aside, so that the properties, measurable quantities (and so on) of the finding can be established in peace. In physics, this phase is sometimes called "phenomenological analysis" and it precedes the incorporation of the new phenomenon within a more complete theory. It is the phase in which the proper terminology, the best parametrization or description are worked out. In this phase, it is very important that the discrepancy with canonical knowledge is played down. We all know the influence of a powerful theory on the result of measurements; new finding has to be kept *independent* from canonical knowledge in order to avoid bias in its analysis. Once the finding is securely rooted in its proper terminology or set of measurements, we enter the third phase.

2.3. *Phase 3: Connection to Canonical Knowledge*

Now, the original contradiction gets into focus again. But after phase 2 it is no longer possible to question the validity of the new finding. In order to eliminate the contradiction we have two choices. We can either restrict the domain of applicability of (part of) canonical theories or we can invent an additional hypothesis. (Details of this mechanism have been explained in Refs. 2 and 3.) Typical examples for additional hypothesis are the Pauli exclusion principle or even Adams and Leverrier's postulate of a new planet (Neptune) to eliminate the discrepancy between Uranus' observed orbit and its prediction from Newtonian mechanics. An historical example for a limitation of the domain of applicability is Bohr's postulate of 1913 that classical electrodynamics does not apply to the electrons in atomic orbits.

I want to stress the often provisional character of phase 3. Although the contradiction is eliminated (the connection to canonical knowledge has been made), it is hardly satisfactory in every case. Additional hypothesis can be the final answer (as in the case of Neptune), but they can also require deeper understanding (as in the case of the exclusion principle). A limitation of the domain of applicability is almost always nothing but a first step towards deeper understanding.

This is only achieved when we enter the fourth phase.

2.4. *Phase 4: Modification of Canonical Knowledge*

If this step is necessary to eliminate a persistent contradiction, it may cause a scientific revolution in the sense of Thomas Kuhn.[8] I want to emphasize that it is not always necessary to go that far. We have already mentioned the case of the Uranus discrepancy, which was solved to complete satisfaction by the prediction and subsequent discovery of Neptune. This was not so with the discrepancy of Mercury's orbit. Leverrier tried the same method and predicted a new planet between Mercury and the sun. But it was not found. A bigger measure was asked for: Einstein's general theory of relatively, which formed a true scientific revolution, a genuine modification of canonical knowledge.

Let us look back to the origin of this path: A contradiction is observed—say, the discrepancy of Mercury's orbit with predictions from classical mechanics. It is critically questioned, phenomenologically investigated, a trial to eliminate it with an additional hypothesis fails and finally it causes a modification of canonical knowledge. In the new picture, it is no longer a contradiction, it can be predicted.

We see that a contradiction is not always simply eliminated, it can be the fruitful origin of a great improvement of canonical knowledge. Thus a true scientist must be quite ambivalent towards contradictions. On the one hand, he hates them because they disturb the picture which he holds dearly. (They go against his prejudice that canonical knowledge must be free from contradictions.) On the other hand, he loves them because they may be the root for enrichment of this same knowledge through a scientific revolution. Just as in phase 1 where we could not find a criterion for the right measure of criticality, there is no way to distinguish between "bad" contradictions (which just disturb our picture and are not even worth our investigation) and "good" contradictions (which eventually enrich our knowledge). Again, history must decide. Who could have told in advance that N-rays are not worth investigating but meteorites would have been? And what about the "Monster of Loch Ness" or "Unidentified Flying Objects" (to name but two contemporary examples)? The overwhelming majority of scientists would agree that they are *not* worth any further investigation. Yet, just how sure should we be?

But let us return to more solid grounds. It may seem that the ultimate step within classical science is the incorporation of a contradiction by means of a modification of canonical knowledge. In this way the contradiction ceases to exist as such and we again have a new canonical knowledge, free from contradictions. But I would like to single out one lonely case, which—to me—seems to transcend classical science: quantum mechanics. I think we can correctly place it beyond phase 4 in its own right. In this case, we have opened phase 5.

2.5. Phase 5: Transcending Classical Science by Incorporation of a Dichotomy

In the case of quantum mechanics we have obtained a wealth of predictive power and an unheard of unification of canonical knowledge by relaxing once and only once the principle of classical science that one and the same theory must not make contradicting predictions. Indeed, within quantum mechanics, we predict the microscopic states (protons, neutrons, electrons, or the like) to have continuous properties (waves with interference) as well as discrete properties (particles with localized collisions). We call it a dichotomy, particle–wave duality, or complementarity. It has become part of canonical knowledge in the form of the so-called "Copenhagen interpretation" of quantum mechanics. Its founding father is Niels Bohr. The working physicist has no problem in uniquely deriving predictions and in understanding experimental results within this framework. But it seems that in reflecting upon this step, we are not always sure whether it is a welcome improvement of classical science or a provisionally acceptable compromise. Many alternatives have been and still are suggested but none could so far live up to seriously challange all the achievements of canonical quantum mechanics. It is interesting to observe that other fields of science often look upon quantum mechanics as a paradise which they also hope to reach. In their reply to Jacques Monod, Eigen and Winkler[9] recall the dichotomy of quantum mechanics and claim similarities to the situation of biology today. And in his discourse with Skinner's behaviorism, Carl Rogers[10] writes

> If we choose to utilize our scientific knowledge for free men, then it will demand that we live openly and frankly with the great paradox of the behavioral sciences. We will recognize that behavior, when examined scientifically, is surely best understood as determined by prior causation. This is the great fact of science. But responsible personal choice, which is the most essential element in being a person, which is the core experience in psychotherapy, which exists prior to any scientific endeavor, is an equally prominent fact in our lives. We will have to live with the realization that to deny the reality of the experience of responsible personal choice is as stultifying, as closed-minded, as to deny the possibility of a behavioral science. That these two important elements of our experience appear to be in contradiction has perhaps the same significance as the contradiction between the wave theory and the corpuscular theory of light, both of which can be shown to be true, even though incompatible. We cannot profitably deny our subjective life, any more than we can deny the objective description of that life.

Suppose that quantum mechanics really is a step beyond classical science, a new phase 5 in the development of our knowledge. Does it have to remain a singular achievement? Or can we learn from it how to extend the method of science to systems which are inherently dichotomous?

3. A Fresh Look at Quantum Mechanics

During the "heroic" years of physics in which quantum mechanics was developed, many objections were raised even by those who helped to create it. Schrödinger's cat[11] is a well-known example, but the most persistent cause for dissatisfaction is perhaps the Einstein–Podolsky–Rosen paradox.[12] Let us recall its essential features. Suppose a spinless particle decays into two photons. In its rest frame the photons have opposite momenta. As long as we do not measure their spin vectors, they are undetermined. But as soon as we do measure the spin of one of the photons, its direction becomes fixed, and the spin of the (far away) other one "collapses" into the opposite direction to preserve angular momentum. Einstein could not accept this, because he thought it implies "telepathy." Schrödinger called it "magic" and de Broglie "upsetting our accepted ideas pertaining to space and time."

In our days, the paradox has been quantified. Bell has derived an inequality[13] which can be tested experimentally. Clauser and Shimony, being personally involved in test experiments, write about Bell's theorem,[14] that it satisfies two criteria (except for "pathological instances of local realistic theories"):

(i) The hypotheses seem to be inescapable for anyone who is committed to physical realism and to the nonexistence of action at a distance.

(ii) Discrepancies with the predictions by quantum mechanics occur in at least one situation which is experimentally realizable.

Hence, experiments can decide and they favor quantum mechanics. Thus, Clauser and Shimony write:

> The conclusions are philosophically startling: either one must totally abandon the realistic philosophy of most working scientists, or dramatically revise out concept of space-time . . .
>
> Because of the evidence in favour of quantum mechanics from the experiments based upon Bell's theorem, we are forced either to abandon the strong version of Einstein, Podolsky, Rosen's criterion of reality—which is tantamount to abandoning a realistic view of the physical world (perhaps an unheard tree falling in the forest makes no sound at all)—or else to accept some kind of action-at-a-distance. Either option is radical, and a comprehensive study of their philosophical consequences remains to be made.

I agree with this conclusion, but being a working physicist myself, I would first like to look at possible consequences for methodology rather than philosophy. Suppose we simply accept the result as evidential fact. Can we learn from it to extend the method of science?

Maybe we should abandon the *totally* realistic philosophy of most working scientists rather than totally abandon any realistic philosophy. Maybe we can—within the good tradition of science—proceed step by step.

What should be the first step beyond classical science?

Let us recall that within classical science it is excluded for a theory to make contradicting predictions. We do not say that the spin of one of the photons is either up or down (with respect to a certain measuring direction) but we say it is undetermined, it is not predicted. What bothers us so tremendously is obviously the fact that as soon as one of the spins is measured, the other one is exactly predicted. In other words, the *correlation* is predicted although we do not know how it can possibly be produced.

Instead of bemoaning this fact, can we not learn from it for a new step into unknown territory? It is indeed worthwhile to contemplate this possibility!

4. Transcending Classical Science

Why can classical science not allow for a situation in which a theory makes contradicting predictions? The answer is simply the necessity for experimental tests. A statement "Experiment *A* leads to the result *a or* non-*a*" is no prediction because it cannot be tested! Thus even the phrasing "contradicting predictions" is nonsensical. As long as we think of scientific endeavor, predictions and refutations or corroborations must be our basis.

But what if we do not even ask for a prediction of a certain experimental result (*a* or non-*a*) but request the prediction of correlations? Suppose we allow a theory to *state:*

Experiment A leads to the result *a* or non-*a*

(This is not yet a prediction.)

The *prediction* of the theory may be

If A leads to *a*, experiment *B* leads to *b*
If A leads to non-*a*, experiment B′ leads to *c*

B and *B′* may be (but not necessarily be) the same experiment.

If we accept this possibility, we have clearly transcended classical science by a new scientific method, which contains classical science as a special case. For classical science predicts "correlations of order 0":

Experiment A leads to result *a*

The first step beyond it is the one just mentioned, "correlation of order 1":

Experiment A leads to a; if so experiment *B* leads to *b*
Experiment A leads to a or non-*a*; if so experiment *B′* leads to c

But we can develop a full series of new scientific domains, each new one containing all the previous ones. Let me just indicate a possible "correlation of order 2":

Experiment A →	a ⇒ *B* →	*b* ⇒ C → *d*
		or non-*b* ⇒ C → e
	or non-a ⇒ *B* →	c ⇒ C → f
		or non-c ⇒ C → g

Evidently, the number of predictions in each case rises in general with 2^n where *n* is the order of correlation. Let me therefore restrict my considerations to the first scientific domain beyond classical science.

The essential new step is that we no longer ask for any *reason* of the correlation. We just test a prediction. (Thus we do abandon totally realistic philosophy, but only up to a point, not totally!) In much the same way as we had to give up the notion of particle orbits but still could test predicted differential cross sections in quantum mechanics.

If we can accept this step into a new domain of science, the Einstein–Podolsky–Rosen paradox ceases to be a problem. Measuring the spin of one of the photons with respect to a given axis is experiment *A*. The result is either *a* (up) or non-*a* (down). Measuring the spin of the other photon is experiment *B*. If *A* gives "up" (*a*), *B* gives "down" (*b*); if *A* gives "down" (non-a), *B* gives "up" (c). (In this case, *B* = *B′*.) Any question as to how the information was transmitted is as unphysical in the new domain as the question of precise location and momentum of a particle was in the classical domain.

So far we have not achieved too much. Instead of answering the pressing problem posed by the Einstein–Podolsky–Rosen paradox, we have simply requested not to consider it a problem by defining a new domain of science (where these questions should not be asked). But as soon as we can live with this new concept, we can try to extend it. And this brings me back to Feynman's worry described in the beginning. Within the new domain, ambivalence can be treated scientifically without doing too much violence to it. Let me take up the (admittedly simplified) *Gedankenexperiment* about teaching again. We have seen in Section I that we cannot sensibly predict the result even in a well-defined situation. But in our new domain of science, this is not asked for. What we do want to find out (to incorporate as predictions of a theory) is whether we can find correlations of the result (be it positive or negative) with other observations.

Suppose we can establish such correlations. We should not ask for their reasons but rather try to think in terms of these correlations, describe their connections and interdependences. (We should not ask for *causal* reasons, we may, however, ask for *historical* reasons! What was the essential fact in the developing life of the class which *led* to this particular result, and so on!)

The realistic philosophy of classical science is based on the possibility of complete separation of objects and observers. Quantum mechanics has taught us that there are limits to this assumption. Thus it led us beyond classical science into a domain where realism in its naive form can no longer be the basis for our models. Can we possibly expect that phenomena such as teaching, in which human beings with their feelings and emotions play an important role, are not even more complex than quantum mechanics? Yet we are tempted to approach them with methods of classical science, methods which are identical to those applicable to and most useful in celestial mechanics for example.

Personally, I am convinced that in extending science, we should not only apply the same methods to an ever widening field of phenomena; but we should also adapt our methods whenever we face a situation in which our classical approach seems to violate rather than embrace the facts and observations. I am very optimistic, that fields like psychology, group dynamics, economy, and even medicine may benefit tremendously by relaxing the assumption of naive realism which worked so well in classical physics.

References and Notes

1. R. P. Feynman, *Nuovo Cimento Suppl.* **4,** 492 (1966).
2. K. Popper, *The Logic of Scientific Discovery* (Harper & Bros., New York, 1961).
3. H. Pietschmann, *Found. Phys.* **8,** 905 (1978).
4. A. Salam, *Elementary Particle Theory* (N. Svartholm, Stockholm, 1968); S. Weinberg, *Phys. Rev. Lett.* **19,** 1264 (1967); S. Glashow, J. Iliopoulos, and C. Maiani, *Phys. Rev.* **D 2,** 1285 (1970).
5. H. Georgi and S. L. Glashow, *Phys. Rev. Lett.* **28,** 1494 (1972).
6. W. Gentner, *Naturwissenschaften* **50,** 6 (1963).

7. I. M. Klotz, *Sci. Am.* **242,** 122 (1980).
8. Thomas Kuhn, *The Nature of Scientific Revolutions* (Univ. of Chicago Press, Chicago, Illinois, 1955).
9. M. Eigen and R. Winkler, *Das Spiel* (Piper, München, 1975).
10. C. R. Rogers, *On Becoming a Person* (Constable and Co., London, 1961).
11. For example, E. Schrödinger, *Naturwissenschaften* **23,** 844 (1935).
12. A. Einstein, B. Podolsky, and N. Rosen, *Phys. Rev.* **47,** 777 (1935).
13. J. S. Bell, *Physics* **1,** 195 (1965).
14. J. F. Clauser and A. Shimony, *Rep. Prog. Phys.* **41,** 1891 (1978).

15

God as a Cosmic Guest-Worker

Roman U. Sexl
Institut für Theoretische Physik
Universität Wien
Vienna, Austria

In 1687 Isaac Newton laid the foundations for the new world-view of physics in his famous *Philosophiae Naturalis Principia Mathematica*. The elliptical orbits of the planets, discovered by Kepler, were explained convincingly in terms of the law of universal gravitation. Newton was even able to calculate the small deviations of the planetary orbits from a perfect elliptical form. Their origin was the mutual perturbation of the planets due to their gravitational attraction. This effect was most important for Jupiter and Saturn, the largest planets of the solar system. The observation showed that these planets deviated more and more from the original orbits.

This was a disturbing result. The solar system seemed to be unstable. Did this imply that the new physics had been built on shaky foundations? In the letters exchanged between Newton's pupil Clarke and the philosopher, mathematician, and physicist Leibniz this point was discussed at great length. Newton and Clarke believed that God would intervene every once in a while as a "Cosmic Guest-Worker" in order to rearrange the planetary system:

> Tis not a diminution, but the true glory of his workmanship, that nothing is done without his continual government and inspection. The notion of the world's being a great machine, going on without interposition of God, as a clock continues to go without the assistance of a clockmaker; is the notion of materialism and fate, and tends, (under pretence of making God a *supra-mundane intelligence*,) to exclude providence and God's government in reality out of the world. . . . The present arrangement of the solar system will be disordered in the course of time according to prevailing laws of motion and will then perhaps be improved.[1]

Leibniz did not agree with this point of view. A planetary system where God had to intervene frequently in order to reshuffle the planets seemed an imperfect creation indeed:

> Sir Isaac Newton, and his followers, have also a very odd opinion concerning the work of God. According to their doctrine, God Almighty wants to wind up his watch from time to time: otherwise it would cease to move. He had not, it seems, sufficient foresight to make it a perpetual motion. Nay, the machine of God's making, is so imperfect, according to these gentlemen; that he is obliged to clean it now and then by an extraordinary concourse, and even to mend it, as a clock-maker mends his work; who must consequently be so much the more unskilful a workman, as he is oftener obliged to mend his work and to set it right.[2]

Clarke replied to this

> The improvement of the solar system by God is to be understood only in a relative way. The present structure, the later disorder and the following reconstruction are in the same way part of God's plan which he has made in his original, perfect idea.[3]

This controversy shows the important role which theological arguments played at that time in the very foundation of natural science. Newton himself had studied theology extensively and has published a book on the prophet Daniel and the Apocalypse.

The problem of the mutual perturbations of Jupiter and Saturn occupied the minds of most famous physicists and astronomers for over a century. It was only in 1799 that Laplace found a physical solution of the problem of the "great anomaly of Jupiter and Saturn." During two orbits of Saturn around the sun Jupiter completes its celestial path five times. The two planets come close to one another again and again at the same position and the perturbations build up slowly. This effect lasts, however, only for roughly 500 years. Since the ratio of the orbital periods is not exactly 2:5 the point of greatest approximation shifts slowly in space and after 900 years the planets return to their original paths.

This explanation finished God's brief tenure as a "Cosmic Guest-Worker" in the universe. To Napoleon's question concerning God, Laplace answered at the beginning of the 19th century: "Sire, I do not need this hypothesis." Within a century the mutual relations between physics and theology had changed completely.

This episode in the history of physics does not stand alone. Again and again God was called as a guest-worker into the cosmos whenever an equation did not lead to the desired result or an effect remained unexplained. Whenever the former riddle was solved by a better theory the guest-worker lost his position again.

Even in our present cosmology there might be a similar situation. The red shift of the galaxies observed for the first time by Edwin Hubble in 1923 is usually explained by a universal expansion of the universe which originated in a big bang about 13 billion years ago. The cosmic black body radiation discovered in 1965 by Penzias and Wilson provides a further support for this hypothesis, since the radiation can be seen as a relic of the high temperatures which were prevalent in the early universe.

Who banged big? It is only too convenient to call upon theology as an auxiliary science and to confer on it the responsibility for the first fraction of a second of the universe. In this brief interval an obliging God should supply the physicists with the desired amount of matter. Thereafter God is requested to retire again to his heavenly sanctuary and to leave the further course of events confidently to the physicists. Even this tiny role which physics provides for God will be abolished if theories prove to be successful which try to explain the origin of the universe and of matter in a self-consistent way.

Not only in physics, but also in philosophy God has frequently been called upon in his role as a guest-worker whenever a philosophical system suffered from internal conflicts. Soon after Descartes starts with his methodological doubts and answers them by his famous "Cogito ergo sum" an obliging God will not deceive him and thereby makes further doubts superfluous. Bishop Berkeley follows a similar path. "Esse est percipie" is his first principle (to be is to be seen—the present motto of high society). Who guarantees the stability of matter, if everything exists only in our observation? Again the *Deus ex machina* is called in, a God who observes everything everywhere and thereby guarantees the continuity of the world and the matter it contains.

The use of God for such auxiliary functions needed for the maintainance of a philosophical or physical theory does not seem to be the best connection between science, philosophy, and theology. To me Einstein's God seems to be much more consistent, a God who reveals himself in the beauty and harmony of the laws of nature. Even more consistent is a God who permits men a rational, intellectual reconstruction of this world in the form of scientific or philosophical ideas.

The considerations presented above have drawn a very negative picture of the interconnections between science, philosophy, and theology. This relation is actually seen very often only in this negative way and the trial of Galilei serves as the most famous example to prove this point.

A historically important and positive connection between these central aspects of human intellectual life remains, however, often undiscussed. Twenty-five centuries of philosophical and theological tradition have led to a critical discussion of many concepts which are indispensable for present science. Many questions have first been discussed and answered in a philosophical or theological setting and have reappeared only later in the formalism of mathematical physics. Let me just mention one example. In volume 11 of his "Confessions," St. Augustine poses the question

> "What did God before He *made heaven and earth?*" I answer not as one is said to have done, merrily (eluding the pressure of the question), "He was preparing hell for pryers into mysteries."

Then the serious answer follows:

> For whence could innumerable ages pass by, which Thou madest not, Thou the Author and Creator of all ages? or what times should there be, which were not made by Thee? or how should they pass by, if they never were?

Here one question is posed in two different theological settings: What has God done before he created time? Why did not God create everything one year earlier? The answer is in both cases that time and matter form an inseparable union. In our time general relativity has led to a similar answer. It is meaningful to speak about the time coordinate only after the big bang. Time exists only after matter and space have been created.

Similar connections between theological, philosophical, and scientific questions can be found also in other fields. In ancient Greece, philosophers have asked what matter might be. They found the alternative of atomism versus continuum theories which dominated the ideas about matter for many centuries. Only with the advent of quantum field theory were new aspects found. According to this theory, each micro-object has both atomic (particlelike) properties and continuum properties, since it is part of a quantum field. The continuum problem which led Zeno to a series of famous paradoxes is thus resolved in the theories of physics by the particle–wave dualism.

The critical discussion of many concepts which became relevant for science later is only one of the contributions of the philosophical and theological traditions to the ideas of modern physics. Of similar importance was the idea of a universe and matter created by God, obeying the laws imposed by Him. The discovery of these laws thereby became one of the most important tasks for men. In these ideas about the laws of nature I see (following Carl Friedrich von Weizsäcker) the most relevant contribution of theological ideas to the origin of present science.

References and Notes

1. *The Leibniz–Clarke Correspondence*, H. G. Alexander, editor (Manchester University Press, Manchester, 1956), Clarke's first reply.
2. Mr. Leibniz's first paper, Ref. 1.
3. Clarke's second reply, Ref. 1.
4. *The Confessions of St. Augustine* (Boston, 1843), [XI], 14.
5. Reference 4, [XI], 15.

16

A Model of Reality as Mind

Henryk Skolimowski
University of Michigan
Ann Arbor, Michigan

I. The Pigeon Methodology vs. Creative Evolution

The 20th century philosophy has mummified our understanding of the mind. Instead of exploring the creative and extraordinary aspects of mind, it continually attempted to reduce it to the scope of activities characteristic of pigeons. If you use the pigeon methodology, you are bound to arrive at a pigeonlike understanding. Let me try to trace the historical circumstances that led to the elevation of the pigeon methodology as the tool of universal understanding. One terminological note: when I say "the 20th century philosophy," I mainly mean empiricist-bound, analytically oriented philosophy of the Anglo-Saxon persuasion that dominates our present universities and that has exerted a considerable influence on our thinking all over the globe in the second part of the 20th century.

Karl Popper is right: The road to understanding of the philosophical scene often leads via the understanding of the background knowledge situation. But Popper exaggerates when he maintains that to understand the knowledge situation is to understand the vicissitudes of science.

Now, there were some serious problems with the scientific understanding of the world already in the second half of the 19th century. With the discovery of non Euclidean geometries, space in the Newtonian sense started to totter. Since the absoluteness of space is one of the basic assumptions of the Newtonian system, to learn that space of the universe does not have to comply to Euclidean geometry was implicitly to admit that the foundations of Newtonian physics were cracking; or at least made uncertain. As the result of absorbing the shock (of the realization that many different geometries are possible, within which we can describe the physical cosmos) *conventionalism* was born—an ingenous doctrine of Henri Poincaré and Pierre Duhem, who claimed that systems of knowledge we develop do not necessarily describe reality faithfully, in a one-to-one way; rather the system of knowledge one develops much depends on the system of axioms one accepts to begin with. We have a great deal of liberty which system of axiom to choose, for instance, while developing geometry. This was an ingenious way of resolving the problem of non-Euclidean geometries.

Conventionalism solved one problem, but it opened up a Pandora's box to many other problems. In particular, it profoundly undermined the very notion of truth as expressed through science; that is to say, it undermined the classical or correspondence notion of truth according to which truth consists of a correspondence between reality *R* and our description of it *D*, so that we can claim that science aims at descriptions of reality which are true. Once we admit that the choice of basic concepts and of the conceptual framework is up to us, we, so to speak, slightly unhinge the classical notion of truth. I say "slightly unhinge" because at the time it appeared that the classical edifice of knowledge, as presented by science, could be saved by artful modifications of the framework.

Then problems started to appear in physics, such as radioactivity and a host of other phenomena which were clearly beyond the domain of the Newtonian paradigm. Those problems were by and large solved by the 20th century physics: Einstein's theory of relativity, Bohr's quantum theory, Heinsenberg's principle of uncertainty. While we welcomed those specific extensions of physics, we have not fully realized, let alone absorbed and digested, their consequences as they pertain to our theories of knowledge and of mind. True enough, quite a number of philosophical theories were created as the result of the emergence of quantum theory. However, the main problems, that is, of the relation of scientific truth to reality and concerning the role of the mind in those ever new extensions of physics, have been by-passed, neglected, almost ignored.

Let me be emphatic: since conventionalism emerged (in the second half of the 19th century) we have not come to terms with the notion of truth, and with the notion of reality that science purportedly describes, faithfully, adequately, unequivocally. I discount the enormous *logical* literature on the subject of truth, including Tarski's epoch-making formulation of the classical notion of truth (of 1933), for all these works are but elegant *formulations* and *reformulations*, showing the prowess and the subtlety of our logical apparatus, but not the new comprehension of reality that science explores and renders in *true* statements.

While putting forth the proposition that philosophy did not absorb in any depth the consequences of the changes in science we must not forget Karl Popper, who indeed was so struck with the fact that even most entrenched scientific theories (such as Newton's) finally fall and are falsified that he decided to build a new epistemology on the grounds of this finding. Popper's distinctive philosophy of science takes the clue from Einstein (as the overthrower of Newton) and claims that all knowledge is tentative. But at the same time it attempts to salvage and justify the superiority of scientific knowledge over all other forms of knowledge. Although Popper's epistemology admirably meets the challenge of Einstein, it is ill at ease with regard to quantum theory. Popper's specific works in quantum theory are not quite congruent with the basic thrust of his epistemology based on conjectures and refutations; "refutation" in quantum theory must be quite different in nature than in classical physics due to the difference in the nature of phenomena. Popper's philosophy of science has still less to offer as far as present particle physics is concerned and the wonderful fluid universe it unveils, which Fritjof Capra has called the eternal dance of Shiva. A most startling aspect of this fluid universe is that the observer and the observed are so intimately connected that they are inseparable from each other. Thus the very notion of objectivity is undermined, (see especially Fritjof Capra, *The Tao of Physics*). When objectivity is undermined, we are in an altogether new cognitive situation; and we have not yet quite comprehended this situation. Let me emphasize, Popper's philosophy has only attempted to meet the challenge of Einstein vis-à-vis Newton; it did not attempt to make sense of the most recent stages of particle physics. My theory of cocreative mind attempts to meet the challenge of our present understanding as posed by recent extensions in physics and in other domains of knowledge. Thus Popper's main problem was the challenge of Einstein. My main problem is the challenge of the new physics and the new cosmos it has unveiled to us. Popper's response was within the theory of knowledge, or epistemology. My response is within the theory of mind, which to me appears to be the clue to our new understanding.

Before I leave Popper let me suggest that his epistemology was already strained within the framework he has chosen for himself, particularly as he tried to accommodate the classical notion of truth within the epistemological framework based on the explicit premise that all knowledge is tentative. If all knowledge is tentative, then it would *seem* to follow that all truth is relative. Relativism of truth was the last thing Popper wanted to admit. Consequently, he invented a new concept, that of approximation to truth: newer theories which refuted older ones are better and preferable because they *closer* approximate the truth. The notion of *approximation* to truth has been ridden with difficulties which Popper never resolved, in spite of some very nice semiformal *formulations* which he found for the concept. How can we know that we "approximate" the truth if we do not know the truth; this something we approximate? Popper has never explicitly given up the classical notion of truth; for this would mean that scientific knowledge is not superior to other forms of knowledge. Why? Becasue the cornerstone of scientific theories is empirical refutability, in other words, the assumption that there is an empirical reality out there which science faithfully describes; and because it does so, we can compare our theories with nature herself. The meaning of refutability rests on the assumption that science does describe reality in an unequivocal and faithful way. Although Popper did not give up the classical notion of truth explicitly, he did give it up implicitly. As the result of Thomas Kuhn's reconstruction of science, Popper seemed to have agreed with Kuhn that theories are not refuted in the actual scientific practice. Rather like old soldiers, they fade away. Even if Popper did not agree with this specific formulation of Kuhn, he conceded the fact that theories are never *ultimately* refuted. Since the publication of Kuhn's: *The Structure of Scientific Revolutions* (1963), Popper has been rather mute about the importance of falsifiability. Now, I do not wish to go into minutia of present debates in philosophy of science. Every shop talk is ultimately a bore. One must not engage in it particularly when one contributes to the Yourgrau Memorial Volume for Wolfgang Yourgrau was a man who was after real understanding and not minutia of technical virtuosity. My main point is that since the rise of conventionalism we have lost our grasp of the classical notion of truth and also of the notion of reality that science purportedly *describes,* that is, in the classical sense of the term "describes." Even the best of the 20th century thinkers, such as Popper, have been unable to resolve the dilemmas that 20th century knowledge has posed to our comprehension.

When Newtonian physics began to totter and could no longer be seen as expressing the unshakable Laws of Nature, this gave rise to many ad hoc theories in science and in philosophy. With Ernst Mach we observe the shift from the correspondence theory of truth to the coherence theory of truth. Since science could not claim to be the guardian of truth, understood as a faithful description of reality out there, scientists and philosophers decided that perhaps we should consider statements and theories as true insofar as they are coherent with the rest of accepted knowledge.

Out of the many attempts to make sense of the post-Newtonian physics, perhaps the most radical, at least in its conceptual implications, was that of Percy Bridgeman, who conceived the doctrine of *operationalism.* Operationalism was a radical attempt to avoid any metaphysics; and indeed to avoid the troublesome concept of "reality." According to Bridgeman physical concepts did not need to have their ontological equivalents in the reality outside physics. The meaning of a concept, Bridgeman insisted, is a set of operations we perform with it. "Meaning is to be sought in operations," wrote Bridgeman in 1934. Then the definitions become increasingly diluted. In 1938 a more liberal definition was provided: "Operations are a 'necessary' but not a 'sufficient' condition for the determination of meanings." This formulation was weakened still further in 1952: "The operational aspect is not by any means the only aspect of meaning." There is a considerable discrepancy between the first and the last formulation. Although the scientific concepts were to be characterized by means of operational definitions, the concept of operation has never been clearly defined by Bridgeman. It is obvious that Bridgeman himself has given up the idea that meanings are to be sought in operations.

This story is well known and one may wonder why I rehearse it here again. For an important reason: to make it quite clear that it was a pervading crisis in the foundation of Western knowledge that brought about those pseudosolutions like operationalism and the empiricist criterion of meaning. These semantic strategies were taken on face value and gave rise to a host of new theories, including theories of mind. *The Concept of Mind* by Gilbert Ryle, originally published in 1949, is a crowning achievement of the whole epoch bent on attempting to find salvation through semantics. I will return to Ryle shortly.

Operationalism was not an isolated obscure doctrine but in many ways epitomized the spirit of the times. Quite independently of Bridgeman, a group of thinkers, who did their philosophy in the coffee houses of Vienna, arrived at the celebrated empiricist criterion of meaning, according to which only empirically verifiable statements are meaningful—all others are meaningless, save tautologies. I refer here to the Vienna Circle and the spectacular rise of their philosophy which managed to eclipse nearly all other philosophies in the midcentury.

Also making its career in the early decades of the 20th century was behaviorism: first in its crude formulation by J. B. Watson, then in its more "sophisticated" version of B. F. Skinner. Behaviorism actually was a doctrine quite apart from the other two. But it had the same purpose: to eliminate everything complex, subtle, and human and reduce it to the stuff of pigeons. This is by no means an exaggeration, for the methodology which was generated by behaviorism was one whose purpose it was to study pigeons. And yet, in all seriousness it was extended to the study of human beings. The heyday of behaviorism and its methodology is now over. It now appears quite strange that we could have taken such a crude doctrine so seriously. Yet it was taken seriously.

Behaviorism, operationalism, logical empiricism, and other forms of positivism were all developed within a larger philosophical envelope of the time and this was ontological materialism, often coupled with atheism. The purpose of nearly all the new "isms" (that somehow signified the Brave New World) was the same: to reduce all other layers and aspects of human existence to inanimate matter.

This is therefore the context for viewing prominent theories of mind of the Anglo-Saxon philosophy of the second half of the 20th century. Ryle's *The Concept of Mind* is a magnificent and scintillating book. Yet, as brilliant as his achievement is, it is but an offering on the alter of the reductionist ethos which so impressed itself on our thinking that it renders the mind as a curiously lifeless, uncreative, and unimaginative faculty.

Ryle is admirably lucid about his intentions, as he writes

> This book offers what may with reservations be described as a theory of mind. But it does not give new information about minds. We possess already a wealth of information about minds, information which is neither derived from, nor upset by, the arguments of philosophers. The philosophical arguments which constitute this book are intended not to increase what we know about minds, but to rectify the logical geography of the knowledge which we already possess.

The rectification of the logical geography of our knowledge about minds becomes a very labored process, and it finally leads to Ryle's theory of logical types (of mind's activities). The semantic footwork is inventive and brilliant. But the whole venture is simply reductionistic: the idea is not to understand MIND as it is, and as it works, but to reduce it to its observable by-products. Ryle's is a materialist theory of mind. It is also a behaviorist and operationalist theory, as it tries to avoid the problem of mind by studying its outwardly observable behavior. Thus in Ryle we see the synthesis of materialism, operationalism, logical empiricism, and behaviorism combined. The result is the virtuosity in applying the pigeon methodology which obscures real understanding of what mind is.

Ryle is a pivotal point of the materialist–reductionist–operationalist tradition. His work set the tone for the next decades of endless epicycles on the theme of the semantic–materialist theory. This now has become a tradition and quite an industry, sometimes called inquiry into mental concepts: pretending to be in the domain of the theory of mind and purportedly explaining the life of mind, but actually as detached from it as a dry leaf is detached from a healthy, growing tree.

The tradition here outlined was born of the crisis in the foundations of western knowledge which is still with us. This tradition, perhaps inadvertently, has created a monumental body of distinctions and semantic refinements which contribute but little to our understanding of the world at large. Now if a body of knowledge which through its verbal virtuosity obscures rather than illumines the purposes of our understanding can be called scholasticism, then the semantic empiricist tradition of the 20th century deserves the name of the New Scholasticism.

The tradition with which I wish to identify myself is one which stems from new physics: the post-Newtonian and post-Einsteinian physics. Philosophically it is the tradition of Whitehead and Teilhard de Chardin. In short, it is the tradition which takes the notion of creative evolution seriously and which insists that the understanding of the meaning and the vicissitudes of evolution, and our lives within it, is of importance second to none. Within this tradition the study of mind becomes an attempt to understand life unfolding, an attempt to understand ever new emergent qualities of evolution leading to the human species as one of the emergent qualities of evolution; leading further to the rise of self-consciousness and the extraordinary role of mind in this process; leading still further to the realization that there is a most intricate feedback between reality and mind; that each codefine the other, and indeed reality can be conceived as a form of mind—which is the conclusion I wish to arrive at at the end of this paper. Before I sketch what I call the ecological theory of mind, let me briefly survey some historical theories of mind.

2. Theories of Mind

The history of Western philosophy is a marvelous tale of human inventiveness and of human follies. Theories of mind are as numerous as theories of reality. But with some singular exceptions these two—mind and reality—are separated from each other, and viewed independently of each other. Some pre-Socratic philosophers knew better than that. One of them, Parmenides, said "No mind, no world." In these four words a whole magnificent insight is contained. During earlier centuries and millenia we did not have sufficient knowledge, and perhaps also sufficient courage, to translate this insight into a complete model of Mind/Reality.

Existential and social traumas of Western man have often been caused by the way in which we split and atomized the world around us. Cartesian dualism, which radically separates the mind from the body and mind from nature, has been at the root of many of our misconceptions and quite a few of our dilemmas. In order to overcome this dualism we must create a unified theory within which mind and reality can be treated as aspects of each other. The evolutionary-transcendental theory of mind, which I shall here present (and which I call for short, the *ecological* conception of mind) treats mind as coextensive with reality; and treats reality as a form of mind. This theory draws from the insight of Parmenides: "No mind, no world"; but more importantly: it builds on the conception of man as the custodian and beneficiary of ever new emerging sensitivities.

Before we go into the details of the ecological conception of mind, let us take a brief look at some of the theories of mind which have been most influential in the West and which still hold sway over our minds.

Empiricism is both a theory of knowledge and a theory of mind. It holds that there is nothing in the intellect that has previously not been in the senses. Mind is conceived here essentially as a *tabula rasa*, a white sheet on which experience writes its designs. The only active role of the mind is that it *allows* experience to write on it. Out of rudimentary experiences first impressions are formed. These impressions are transformed into forms of knowledge. How these transformations of raw experience and of impressions occur has never been explained by empiricism.

Empiricism as a theory of mind is a gross caricature of what is going on in our mind, and what we know about the marvels of human comprehension. One wonders *why* it has ever been taken so seriously and, indeed, propagated by intelligent people with zeal and commitment. The main reason for the acceptance of this clearly defective theory, in my opinion, has been ideological: the empiricist theory of the mind removes human knowledge from the authority of the church and particularly from the church's creeds and dogmas. Empiricism maintains that *everything* is acquired from the physical universe via the senses, thus instructs us that no authority of any sort needs to be obeyed. The only authority is that of our senses.

In the long run the physical universe becomes the only reality which we not only explore but also worship as all other gods are dethroned; and our senses become not only our authorities but also our deities. A theory of knowledge which sought to over come the overbearing dogmatism of religious orthodoxies has itself become a sterile dogma: not only detrimental to the quest for knowledge which goes beyond the mere physical surface, but also detrimental to the deeper quests for human meaning which are frustrated by the extensions of empiricism that insistently claims that there is nothing beyond the reality of the senses. In spite of its obvious defects, empiricism is still present around us and manifests itself in those oversimplified theories of natural and social science which are still favored and perpetuated in our academia.

Predating empiricism, but also running parallel to it, we find theories which do not consider mind as a tabula rasa but which on the contrary attribute an active role to it. Many of these theories claim that mind is endowed with capacities and propensities that are inborn, therefore *a priori*. These are rationalist theories of mind.

Among the rationalist theories of mind at least three should be mentioned: Plato's, Berkeley's, and Kant's. Plato envisaged mind as active but only so far as it recognizes, really "remembers" the Forms: ideal, incorruptible, unchangeable blueprints to which all objects of our knowledge and indeed all objects of existence must comply. Objects are what they are because ideal Forms predating their existence are embodied in them.

Berkeley's conception of mind is active in the extreme. Bishop Berkeley maintained that "esse percipi": it exists only insofar as it is perceived. All existence is, in a sense, a figment of our imagination. Things are "brought" into existence through acts of our perception, and exist only insofar as we perceive them. The consequence is subjective idealism: there is no reality independent of our perception. An ingenious and startling doctrine, particularly when defended by the scintillating mind of Bishop Berkeley, but making nonsense of all that we know: and also making nonsense of evolution itself, especially evolution as striving towards greater and greater complexity, attainment, perfection.

The third type of rationalist theory of mind is that of Immanuel Kant, which comes closest to the ecological conception of mind and yet is still far apart. Kant reversed the whole process as envisaged by empiricists. Instead of objects impressing themselves on the mind, Kant claimed that it is the other way round—it is the mind, its specific structure and its specific categories that are imposed on the objects outside which are shaped according to the categories of the mind. There is the conformity between objects and the mind, but this conformity comes about by the mind imposing its order on things. We perceive a certain order in the outside world and we structure it according to certain regularities because we cannot do otherwise. The structure of our mind continuously impresses itself on the order of reality. The order of reality is really the order of the mind.

Kant claimed that we cannot know the ultimate reality, "things in themselves." We only know the appearances of things. For we structure according to the categories of the mind, but only the appearances of reality. The whole conception of reality is opaque in Kant and indeed one big question mark. There is, as a matter of fact, no room for evolution in the Kantian system, and this was a major reason for the collapse of Kantian philosophy. With the growth of science in the 19th century, when we discovered non-Euclidian geometries and non-Newtonian physics, and accepted the notion that time is not absolute (as both Newton and Kant imagined), the unalterable structure of the mind was undermined; according to some—invalidated.

Kant's major shortcoming was to envisage the mind's categories and mind's structure as fixed and absolute. In postulating the active role for the mind, and in claiming that the world is shaped and determined by the categories of our knowledge, his influence has been great and lasting. *His* active mind was second to none.

We should add parenthetically that Kant's heritage has been continued (though in a changed idiom) throughout the 19th century and throughout the present century. What Kant attributed to the mind—the shaping and determining of reality—various other thinkers attributed to language. First came Poincaré with his ingenious conception of conventionalism; then came Ajdukiewicz, Benjamin Lee Whorf, and Quine, who radicalized Poincaré's conventionalism; and, more recently, Chomsky has crusaded for the recognition of man as language animal. Each of them has recognized language as a codefiner of knowledge and of reality. Each was close to seeing that reality and mind codefine each other.

From these theories of mind both empiricist and rationalist, I wish to distinguish the evolutionary transcendental theory of mind, which I call the ecological conception of mind, in order to distinguish it from the other would-be evolutionary theories. This ecological conception of mind is much closer to rationalist theories than to empiricist ones. The hallmark of the ecological theory is that it recognizes mind not only as active, but as cocreative: as made not only of thinking grey cells but of the *sensitivities*.

The rationalists were on the right track. But they lacked the evolutionary dimension. Hence their conception of mind was constrained and ultimately ridden with irresolvable dilemmas, which neither Plato, nor Berkeley, nor Kant could escape.

The ecological theory of mind wants to make sense of Parmenides' insight: "No mind, no world"; wants to make sense of the mind as a product of evolution unfolding; and of this part of our intellectual endowment which has to do with the acquisition of new sensitivities.

3. Man as the Self-Sensitizing Animal

One of the distinguishing features of the ecological theory of mind is that it considers man as made of fields of sensitivities. For we are bundles of quivering sensitivities. Man has been defined as a rational being; as a political being; as Homo Faber. Yet what is most important about man is that he is a *sensitive* being; that he is endowed with sensitivites; that via sensitivities he is making his evolutionary ascent. Man is, in short, a self-sensitizing, that is, self-transcending and a self-perfecting being. New sensitivities are new windows which enlarge the horizons of our world; are also the vehicles by which we carry on the evolutionary journey; and through which we make ourselves into more human and more spiritual beings.

When the first amoebas emerged from the primordial organic soup they were victorious because they acquired a new sensitivity enabling them to react to the environment in a semiconscious manner, which was the beginning of all learning. For learning is a capacity, a sensitivity, to react in a feedback way, to the environment and its conditions. The glory of evolution starts when organisms begin to use their capacities, thus their sensitivities, in a conscious and deliberate manner to further their well-being.

From the organic soup via the amoeba to the fish; from the fish via reptiles to primates; from the primates via the chimpanzee to man—this has been a continuous and enthralling story of the acquisition and refinement of ever-new sensitivities.

When matter started to sense and then evolved the eye (as the organ of its new sensitivity) this was an occasion of great importance, for reality could be now seen, could be articulated according to the power of the seeing eye. No eye to see, no reality to be seen. It is the eye that brought to reality its visual aspect. The existence of the eye and the existence of the visual reality are aspects of each other. One cannot exist without the other. For what is the seeing eye that has *nothing* to see? And what is the *visual* reality that has never been seen?

The seeing of the eye is a form of sensitivity through which we articulate reality around us. Seeing is one of many sensitivities. They are all products of the articulation of evolution. But they are not just passive repositories of the evolutionary process. Through them we slice, apprehend, and articulate what we call reality. There is no more to reality (for us) than our sensitivities can render to us. Sensitivities are articulators of reality. The emergence of every new form of sensitivity is a new window on the world.

With new sensitivities we articulate the world in new ways; we elicit from the world new aspects. The power of sensitivities is the power of cocreation. No aspect of reality imposes itself on us with an irresistable force; we take it in and assimilate it only when we acquire a way of seizing and comprehending it; when we come to possess an appropriate sensitivity that is able to process it for us.

The power of creation is the power of articulation. This is how nature goes about its business, by endlessly articulating. And such is the story within the human universe: by acquiring new sensitivities we acquire new powers of articulation, thus we acquire new powers of creation. Sensitivity, therefore, holds the key, not only to our understanding of evolution, but to the understanding of ourselves.

Let us probe this question a bit further. Are we not confusing thinking (and knowledge) with this vague stuff called sensitivity? No. Thinking is a form of sensitivity. It is a form of seeing with an immediate recall of past experiences as stored in our evolutionary layers. Paradoxically: thinking is not the kind of faculty we often think it is. It was not inserted into us, at a certain stage of our evolution, as a gift from somebody who said: "now cerebrate." Thinking nearly always occurs within a larger framework of our experience, and of the experience of the species, and this experience makes thinking much more than mere cerebration. Thinking is one of the many threads of which the tapestry of our sensitivities is woven, is only one aspect of our evolutionary endowment. And this endowment is a mine which is often mind boggling.

In our evolutionary journey, elementary perceptions (of the amoeba sensing its physical environment) give way to illuminations—when human beings created art and religion. Philosophy, art, and religion—as well as knowledge, including science—are forms of human illumination.

Through sensitivites evolution is articulated. Through sensitivities the mind of the human being is created. Through sensitivities the scope of our humanity is delineated. Through sensitivities matter is transformed into spirit.

All thinking is light we shed on the objects of our understanding. This light, when it illumines life, becomes reverence for life. Reverence for life is a form of human sensitivity towards it; and at the same time is a form of thinking about it. Thinking so conceived can be seen throughout all traditional cultures. Plato's fusion of truth, goodness, and beauty is a manifestation of it.

In the making of symbols we have found another way of augmenting ourselves. For symbols have facilitated a new important stage of our evolutionary articulation: by developing symbolic codes we have brought art, religion, and philosophy to fruition; in the process we have articulated ourselves as social, cultural and spiritual beings.

Truth, goodness, love, and beauty are vehicles of our sensitivity. They are an outgrowth of our early "natural" sensitivities which, through symbolic transformations, became instruments of illumination and, furthermore, of spiritual articulation. What we call "spirituality," "the religious feeling," "the sacred," and "the divine" are all expressions signifying an enhanced sensitivity, an enhanced capacity of the individual to react to the world and to transcend the limitations of matter.

To define the human being as a sensitive animal, as one who forms himself through the acquisition and enlargement of his sensitivities, is to pay homage to the openness of man's future and also pay homage to the attainment of evolution. Only this concept of man is right which makes sense of man's future while making sense of his past.

4. Sensitivity, Consciousness, Mind, Reality

Let us return to our earlier story of the evolutionary ascent. When the first amoebas started to articulate themselves from the original sea of the primordial, organic soup, this was at once a triumph of life ascending and a triumph (still muted at the time) of consciousness arising. For amoebas started to *react* to the environment in a deliberate and semiconscious manner.

From this point on, the evolutionary tale is one of augmentation of consciousness and the continuous acquisition of new *sensitivities* through which organisms react to the environment in ever more knowing and purposeful ways. As their sensitivities multiply organisms elicit from the environment more and more. They draw from reality in proportion to their ability to receive it and transform it. At this point we can see that their reality was outlined by the nature and scope of their consciousness and their sensitivities; reality was outlined by the nature of their mind.

There is, therefore, an intimate relationship between our total evolutionary endowment in terms of consciousness, and all the knowing powers we possess, and the nature of reality we construct, receive, and recognize. We simply cannot find, see, or envisage in reality more than our senses, our intellect, our sensitivities, our intuition (and whatever other evolutionary endowments we possess) allow us to find and see. The more sensitive and knowing we become, the richer and larger becomes our reality. When we say "our," we do not mean in the sense of the idiosyncratic, subjective perception. But we mean in terms of the capacity of the species. What is beyond the species and the mind of the species may be reality *in potentio* but not reality as we know it; our concept of reality is reality as we know it.

While receiving reality or any aspect of it, the mind (according to the ecological theory) always processes it. In processing it, the mind actively transforms reality. Let us reflect on the meaning of the two expressions: "processing reality" and "transforming reality." They are both fundamentally inadequate. For they suggest that there is such a thing as an autonomous reality "out there" to which the mind applies itself and on which it works. Such a picture is fundamentally misconceived. There is no such thing as *reality as it is*, which the mind visits and on which it works. Reality is always given together with the mind which comprehends it: in the act of comprehension, which is at the same time the act of transformation. We have no idea whatsoever what reality could be like *as it is*, because always when we think of it, when we behold it (in whatever manner) *reality is invariably presented to us as it has been transformed by our cognitive faculties*.

The organism's interaction with reality is a dual process of being in it and articulating it: by grasping onto any aspect of reality the organism invariably articulates it. Reality is never given to the organism (human or otherwise) except in forms of interactions; thus, in the form of continuous articulations and transformations specific to a given organism. We never *just* receive reality.

Mind is part of the real. It is a fragment of evolution unfolding itself. But rather a special fragment: once it has emerged, it acts as a refracting instrument. It "bends" reality according to its peculiar laws, propensities, and faculties. Mind is this particular part of reality which is both a part of reality and also apart from reality. Hence it acts as the mirror of reality. This double nature of mind makes it difficult to talk of mind as the stuff of reality. By saying that it is "of reality" and "within reality" we expect to find a slot whereby the mind can be shown as existing "objectively" among other things. But mind is not this kind of thing. Any situating of it in reality is really situating reality within it. Hence the ecological theory is at the same time an interactive theory of mind. Reality and mind constantly interact with each other. There is no other way of grasping reality but through the mind. For mind is continually shaping reality. We have not said creating reality, but only shaping it; creatively shaping, we should add.

The rise and development of the mind is essentially the story of dim light reflecting upon itself and becoming brighter light. In its evolutionary development the mind has not only been continually transformed but continually transforming. The mind, as I have argued throughout, is not to be limited to its one layer embodied in our abstract logical capacities, but must be seen as the total capacity of the organism to react intelligently and purposefully.

The reality for the amoeba has been something less than for the fish, and still less than for the human being. The richness and multifariousness of the experience of reality is in proportion to the organism's capacity to receive and decipher it, to emphasize the point. The more primitive the capacities, or to put it in another way, the more primitive the mind, the more primitive the furniture and experience of reality. The more versatile and exquisite the mind, the more versatile the reality, and the richer the experience of it. Paradoxically: the organism receives from reality as much as it puts into it. It "enriches" and transforms reality through its capacity of articulation, through its versatile mind. And the notion of God (conceived as a being capable of everything, thus omnipotent) makes perfect sense in this evolutionary scheme. If there were a being whose mind is infinitely creative and thus capable of transforming the world (reality) at will, this being would be God. But reality for God would be a strange thing: not a permanent realm but something that is constantly created and recreated; a rather uncomfortable notion for us, earthlings, used to being rooted in something permanent. One general conclusion follows however; the more creative the person the more impermanent his environment (reality).

The wonder and mystery of the mind in evolution is its capacity to enlarge reality as it grows and transforms itself. The mind is but a product of evolution unfolding, but it is the most extraordinary product. It is the light which penetrates and illumines. The mind, being part of the real, sends a beam across to other parts of the real and by illuminating them brings them back to the source of the light: the mind itself. By illuminating the darkness and bringing it to light it elevates nonbeing into being. The vision and the seeing cannot be separated from the eye. What the eye is to the act of seeing, the mind is to the act of comprehending reality.

There are those who tell us that our mind, at any rate brain, is terribly underused, as we employ perhaps only 5% to 10% of its total capacity. This is a view which we would like to believe but one which is far from the truth. If anything, our brains seem to be overloaded nowadays. Hence those repeated breakdowns.

We are an aspect of the universal mind or the total mind, and our limitations reflect the limitations of this larger mind. We can become much brighter but only if the *total mind* as developed by the whole humanity (thus evolution) becomes much larger and brighter.

The conception of mind as coextensive with reality explains why we cannot suddenly increase our mind tenfold, fivefold, or even twofold. Our mind is limited by the boundaries of reality cocreated by it; is limited by the scope of human knowledge, by the nature of human understanding. If there are individuals who have radically transcended these boundaries, more likely than not we have locked them up in lunatic asylums

He who radically transcends the scope of his mind radically transcends the scope of his reality. This is something with which we are familiar: an experience when suddenly our reason is ablaze and reality around us dramatically changes. But there are biological and psychological limits to this process of transformation: by broadening and changing the nature of reality around us, we uproot ourselves, we deracinate ourselves from the reality we have been rooted in; we lose our security, our sense of belonging; we become schizophrenic. There are therefore biological and psychological limits to what we call a radical expansion of the mind.

How does the expansion of the mind occur within the scope of the universal mind or the mind of the species? It is like climbing unconquered mountains. Before the North Face of the Eiger was conquered, it was deemed unconquerable. Once it was scaled it has become yet *another* route. The psychological curtain of impossibility has been removed. The mind has learned that this is now a possibility, and within the sphere of "reality."

Before Mount Everest was conquered, it was the ultimate human frontier. Once it has been climbed, it has become yet another mountain; to the point that Reinhold Messner, of Italy, wants to take on Everest alone, only with his rucksac, intending to spend ten days all by himself in the final stages of the ascent.

The mind is not the slayer of the real. It is the illuminator of the real. Without the mind there is darkness. And there is then no sense whatsoever with which we can talk about reality; even in our silence. There is no way of grasping what is out there except through the net of our mind. "Out there" is always an aspect of the "within" (the mind); that is, as far as our comprehension goes; and what is beyond our comprehension *is* beyond our comprehension.

The ecological theory of mind is not an expression of old-fashioned idealism which denies or mystifies reality. It is rather an expression of *suprarealism.* For it accounts for all the stages of the real in its evolutionary unfolding: it makes sense of the reality of the amoeba, of the reality of primitive cultures, of the reality of the scientific culture, of the reality of esoteric traditions.

The objectivized mind is one specific crystallization of our evolutionary journey. The scientific view of reality is one specific way of its reception and its transformation. Yes, the reception of reality in the scientific key is its transformation: through numerous filters which science places between us and the real out there. Some of these filters are sophisticated indeed, such as Schrödinger's equations and other mathematical symbols through which we filter reality.

5. From Plato via Kant to the New Cosmological Unity

One of the great questions of Western philosophy has been how to secure simultaneously the objectivity of knowledge and the reliability of morals, so that our mind moves in the universe of order and our heart dwells in the universe of compassion, the latter universe not being merely subjective and whimsical but one that is grounded in a larger design of the cosmos.

Plato's answer to this dilemma, which was made acute by two different schools of thought: the Eleatics, on the one hand, and the Sophists, on the other, was by inventing the Forms: absolute, everlasting, imperishable, which underlie the existence of all objects and of all virtue. All learning is but the recognition by the soul of that knowledge which it possessed before it was embodied. When the soul is incarnate "it is contaminated with imperfection of our body"; it then partially forgets the knowledge it originally possessed and is reminded of it by the earthly copies of the Form. Thus: all knowledge is "remembering" (anamnesis). Everything that we learn is the unveiling of what we have known, is the lifting up of the clouds of obscurity brought about by our senses and our bodily entrapments, so that the soul can see again. Right knowledge means the recognition of the whole plan of creation, including our place in it: our duties and responsibilities. By recognizing the true nature of the world, the soul recognizes the path of the "right life" and the place of virtue in it.

The objectivity of morals and reliability of man's knowledge is achieved through the same means: the soul's apprehension of the underlying Forms—which are the guarantors of *being* external to man and also guarantors of the moral order which is the inner order of being.

Twenty-two centuries later Immanuel Kant addressed himself to the same cardinal problem: how to secure objectivity of morals and reliability of knowledge. But he found that there are different grounds for securing each: the moral law was to be grounded in the sovereignty of man, which is unique and independent of the physical universe. The objective grounds for reliable knowledge Kant found in the fixed structure of the mind, which, through its unalterable and permanent categories (imposed on the world) makes the world comply to the order of the mind. Kant moved anchorage from the Forms, existing independently of the mind, to the mind itself in which these forms, under the guise of the categories, reside. Kant's overall design could be summarized as follows: the starry heaven above you (as structured by the mind and grasped by science) and the moral law within you (as guided by the moral imperative, which requires that we treat every human being as an end and never as a means only).

Two centuries after Kant we must address ourselves to the same dilemma. For as the result of the reign of empiricism we have inherited a shattered world in which everything is in pieces, "all coherence gone." And we must bring again some coherence to the world. To attempt to bring a new coherence is not only a requirement of the mind which cannot cope with the universe of incoherence and chaos; more importantly, it is an existential necessity of man who must *belong*, otherwise he feels alienated and paranoid. The needs of coherence and belonging are among the basic needs of man; their satisfaction is as important as the satisfaction of the need for shelter and food; on the spiritual level—belonging is our shelter, coherence is our food.

Bringing about a new coherence does not mean bringing about a new rigid order. Our new coherence must be seen in the dynamic flow of evolutionary forms. Evolutionary becoming is our new guarantor of being. *Becoming* properly conceived serves as the basis of a new moral order: in the process of becoming, being acquires *all* its attributes, including moral ones. Morality is the becoming of man as a responsible moral agent; simply as the *human* being.

Plato attempted to establish the unity of knowledge and morals by going *downward*; through connecting things with the underlying Forms. Within the evolutionary model we are establishing this unity by going *upward:* by observing how the process of becoming molds the variety of being in ever new ontological, moral, and spiritual forms.

Being is always being of something. Being which is not being of something does not make sense. The attributes through which being articulates itself punctuate and enliven the process of becoming. Particular forms of being are, like flowers, specific cystallizations of the process of becoming. Form (in Plato's sense) could have become

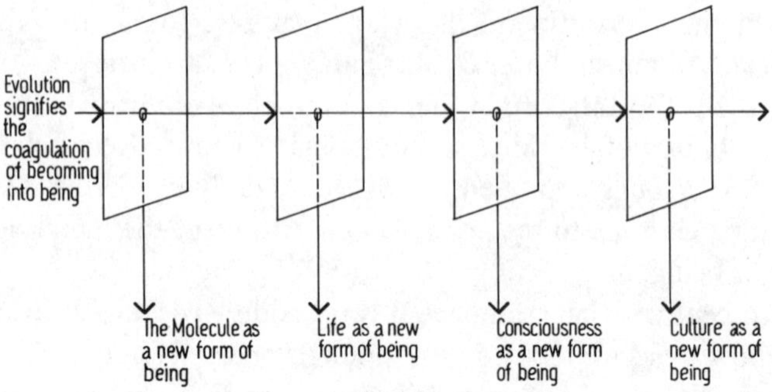

FIGURE 1

the Form only after we have acquired a new conceptual sensitivity: the power of self-reflection, the ability to conceptualize in the marvelous way Plato did, the ability to *make* the Form the underlying reality of all being. In the absence of any evolutionary model, which would explain the nature of becoming, Plato had to resort to the authoritativeness of Being, really the authoritarianism of the Form—which is just one aspect of becoming.

Becoming is the passage from nonbeing into being. Evolution conceived as the articulation of becoming into being can be presented by a series of screens which symbolically show the process of coagulation or crystallization of latent potentialities into new states of being (Fig. 1).

Evolution is the entire ground for this form of becoming cum being which is called knowledge, and for this Form of knowledge cum being which is called morals. Morals and knowledge are the off-shoots of evolution which, like tender but powerful tentacles, now attempt to give a new shape and form to evolution itself; that is on the human level.

The essence of being is a time-induced articulation of becoming. *Evolution is becoming which continually implodes into being.* In the very idea of being which comes to existence through the *articulation* of becoming we find the underlying objectivity for both morals and knowledge.

There is thus a unity between the real and the moral as Plato postulated. But it is on a different level and in a different framework than Plato had thought. The process of articulation of evolution is the guarantor of the validity of the forms of life; it is also the guarantor of objectivity of morals and the reliability of knowledge. In this process we must not forget the role of the interactive mind.

The interactive mind is the expression of the sum total of the sensitivities which went into its making. These sensitivities are filters through which mind views and receives reality. Mind is *within* reality and, at the same time, *of* the reality that surrounds it. Like the light within, it illumines reality; by illuminating, it articulates; by articulating, it cocreates.

Let us represent:

Mind I—mind in the narrow sense; Mind II—the subtotal of all sensitivities of which Mind I is made; and Mind III—reality, within which both Mind I and Mind II reside—by three concentric spheres (sphere$_{I}$; sphere$_{II}$; sphere$_{III}$; of which each is merging with the next, and of which each is an aspect of the other two. (see Fig. 2).

To say that *reality* is a form of Mind may strike some as an unjustified semantic liberty. But it is not. As mind is part of the real so reality is part of the mind.

Sphere$_{I}$ and sphere$_{III}$ are aspects of each other. Consider the statement: "The symmetry of this pale-pink carnation is a beautiful phenomenon." Insofar as our carnation is a *phenomenon*, it belongs to the sphere of reality (sphere$_{III}$). Insofar as it is "symmetrical," "pale-pink," and "beautiful," it belongs to the sphere of mind (spheres$_{I,II}$). Insofar as we make a verbal utterance about it, it belongs to the sphere of the discoursive mind (Mind I). Insofar as we appreciate its total beauty, it belongs to the sphere of sensitivities (Mind II).

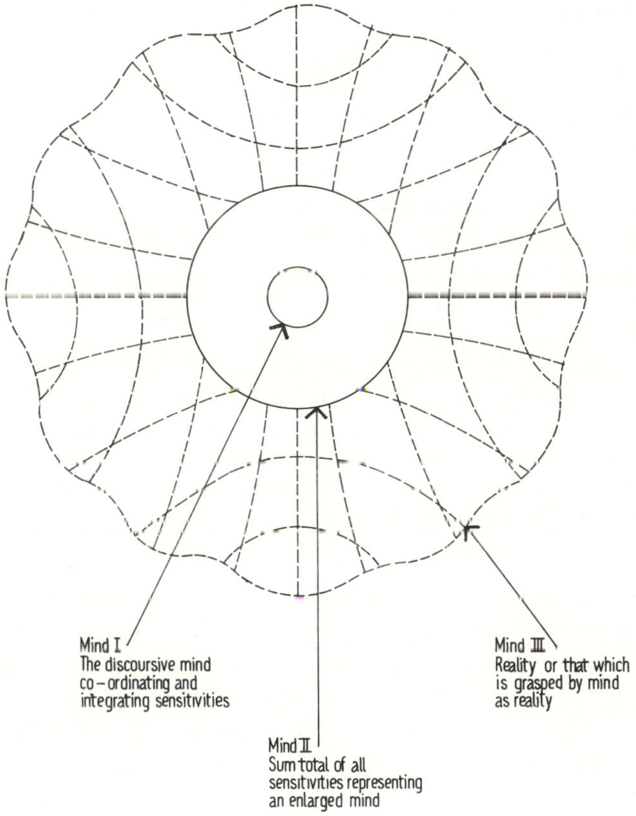

FIGURE 2

Sphere$_{III}$ (reality) is "mind," but not in the same sense in which sphere$_{I}$ is. True. However, *within the sphere of reality itself we can hardly find two different things which are real in the same sense.*

The diagram, showing the three spheres merging into each other, means to convey the essential *evolutionary* unity of our cosmology. We are emphasizing here evolutionary and cosmological unity but not homogeneity. Mind I and Mind III are aspects of each other but at the same time exhibit different stages and different modes of the articulation of evolution. Mind III, being of more recent origin, incorporates Mind I, which has provided the material for the making of Mind III, which in turn makes *human* sense of Mind I.

Our language, dominated by the precepts of empiricism, is recalcitrant to allow us to express the new cosmological unity which clearly goes against its grain. I shall therefore call on poets. Thus William Blake says:

> But to the eye of the man of Imagination,
> Nature is Imagination itself.
> AS MAN IS, SO HE SEES.

The 13th century Persian poet, Mahmud Shabistari, expresses similar insights:

> The world has become a man, and man a world.
> There is no clearer explanation than this.
> When you look well into the root of the matter,
> He is at once seen, seeing eye, and things seen.
> . . .
> In every atom there are a hundred blazing suns.
> If you cleave the heart of one drop of water,
> A hundred pure oceans emerge from it.
> If you examine closely each grain of sand,
> A thousand atoms may be seen in it.

And from another poem:

> You who wander in the desert away from your own consciousness,
> Come back to yourself to find all expertise summed up in you,
> You are the way and the reality of perfection,
> One in whom the great consciousness of God dwells.
>
> From *Gulistaneh Raz*

The poet Cummings says: "Always the more beautiful answer who asks the more difficult question." Instead of asking ourselves petty questions in an outmoded frame of reference—for empiricism *is* an outmoded frame of reference—we must have the courage to ask ourselves new bold questions, for in their significance lies our salvation. The question which I have been asking myself in recent years has been: Can we hope to develop a new unitary cosmology which would be so conceived that it renders the universe as home for man; which would provide the basis for our reconciliation with nature, and which would provide an outline of a new theory of mind? In my book *Eco-Philospehy—Designing New Tactics for Living* (Marion Boyars, 1981) I have

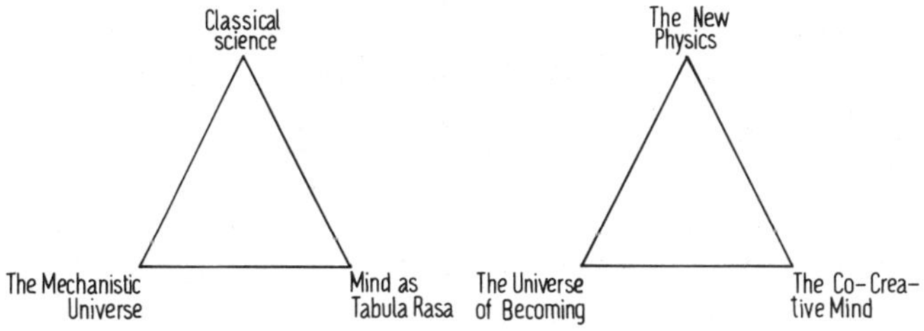

FIGURE 3

attempted to outline the rudiments of this new cosmology, particularly with regard to the universe as being home for man.

In this essay I addressed myself to the role of the mind in the universe of becoming. The becoming of the universe is inseparable from the becoming of the mind. This insight is in perfect harmony with the new ontological vistas unveiled to us by present particle physics which has abandoned the rigid, deterministic Newtonian framework and which came to recognize that—on the ultimate level of analysis—the observer and the observed merged inseparably.

Let me put the matter in general terms: there is no *objective* reality in the absolute sense, as there is no such thing as objectivity independent of our cognitive faculties. We do not photograph this (purportedly objective) reality in our scientific theories. The whole process, which we break into three component parts—the reality out there, the process of photographing it, and the process of developing the negatives, that is to say the process of formulation of theories and of knowledge—are aspects of each other, cannot be torn from each other as they codefine and coconstitute each other. The nature of our mind is the nature of our knowledge is the nature of our reality. (Try to *be* an amoeba and "think" what your "knowledge" and "reality" would be like then.).

I am aware that I am proposing some rather uncomfortable notions. But the history of modern science has been but the story of uncomfortable *and* improbable notions that have become subsequently built into the floor of our understanding of

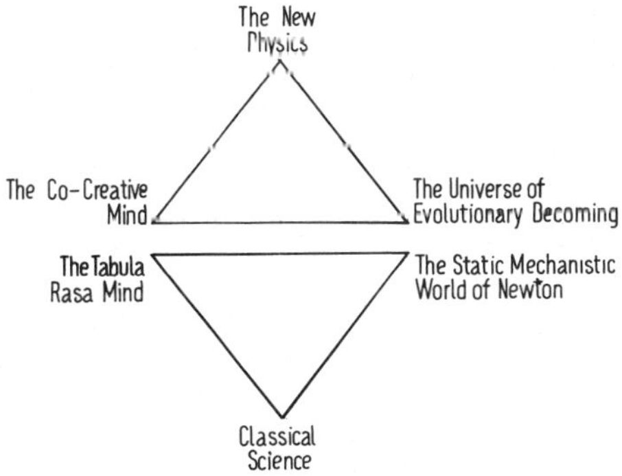

FIGURE 4

"reality." Ours is the universe of becoming: the universe of emergent qualities, the universe of new forms of understanding, which, although they at first appear as conceptual shocks, after a time resolve themselves as new illuminating insights into the nature of things.

In every culture there is an intimate unity between the concept of reality, the concept of knowledge, and the concept of the mind. The three are bound together in a triangular codefining relationship. This can also be observed in Western culture of the past three centuries. If we contrast classical physics with present physics we can present them by two triangles (Fig. 3).

Within the empiricist tradition the mind is conceived as a tabula rasa—a clean sheet of paper on which experience does all the work. There is a congruence between the static (dead) universe empiricism postulates and its concept of static, entirely passive mind. With the rediscovery of the universe of becoming, which the new physics supports in a variety of ways, the role of the mind of necessity must be redefined.

We should be aware that in our day and age the two triangles coexist with each other (Fig. 4), but in a rather confused way, which is causing—through the discrepancies they generate—a great deal of conceptual stress and quite a bit of intellectual paranoia; to say nothing of the forsaken sense of belonging.

The ecological theory of mind wants to remove this paranoia as it encourages us to take possession of the new freedom in which, unanchored from the shores of determinism, we shall be floating through the new unfolding universe. And aided by new Imagination we shall be discovering new forms of understanding of which philosophers and scientists have never dreamt in their narrowly conceived rational ivory towers of academia.

The emergence of the mind is one of the lasting mysteries. To conceive of mind as a mystery is less mysterious and less mystifying than to fathom that the mind is a mere brain working according to physiological, mechanistic laws alone; for it is to admit that mystery is part of the natural order of things. Mind and Imagination are bound together. The nature of Imagination gives us a far better clue to the understanding of the mind than a hundred neurophysiological studies. The nature of Imagination is wonderfully mysterious. And wonderfully mysterious is the quality of the universe that Mind/Imagination render to us. The unfolding universe is at the mercy of our Mind/Imagination for it is our Mind/Imagination.

17

A Matter of Chance

J. L. Synge
Dublin Institute for Advanced Studies
Dublin, Ireland

"But," cried Alice angrily, "the cards haven't been shuffled at all!"

To explain how all this came about, we must go back to tell how one morning Alice met the Queen of Hearts accidentally in the street.

"Whither away so bright and early?" asked the Queen merrily.

Alice blushed. "Well, your Majesty," she said with some hesitation, "I'm going to buy a pack of cards."

"Not a bad idea," said the Queen, "But why are you blushing?"

"It's like this," said Alice. "My father and mother are very old-fashioned, and they think all games of chance are wicked. And since card-play is a game of chance—or is it a game of chance? And what is chance anyway?"

"One question at a time, my dear," said the Queen smiling. "Yes, I agree that card-play is a game of chance; but then I really don't know what chance is! Suppose you go and buy your pack of cards, and then we shall pay a visit to the Knave of Spades. He knows nearly everything, or at least he seems to, more or less that is, because after all how could he possibly know *everything*? It makes me quite dizzy even to *think* about everything, let alone *knowing* everything. But hurry away now and get those cards, and we shall be in time to catch the Knave of Spades at his morning coffee. He is always at his best just then—he says that his brain needs lots of caffeine, whatever caffeine is, but I suppose it's what makes coffee coffee."

Alice was just about to ask how coffee could not be coffee, but the Queen of Hearts waved her in the direction of the shop where cards were sold. So Alice went into the shop and very shyly bought a pack of cards, feeling quite wicked as she did so. But she pulled herself together, reflecting how much she wanted to know what chance means. "How does that poem go?" she asked herself. "Longfellow wasn't it?

> There's many a shaft at random sent,
> Finds mark the archer never meant.

The shaft is an arrow of course, shot from a bow. But how could the archer shoot at random? He might shoot at an apple, like William Tell, but how could he shoot at *random?* Perhaps after all it might have been better to buy a bow and arrows, and then go along to someone skilled in archery—perhaps the King of Clubs"—and she imagined herself curtseying to that rather fierce-looking King, and she started to make up a little speech: "Please your gracious Majesty, I am only a little girl, but I want to learn to use the bow and arrows, and in particular please teach me to shoot at random." But of course he might be in a bad temper and just send her away. Perhaps after all it was better to start with the cards; the Queen of Hearts was a very kind person, and she would never let the Knave of Spades be rude to her if she asked him to tell her what chance is.

The Queen of Hearts was waiting for her. "Come along quick," she said, "or maybe the Knave of Spades may have gone out before we get there." So they hurried, and in due course arrived at the palace of the Spades and were shown up to the study in which the Knave of Spades was sitting at a large table covered with books and papers. He hardly glanced up when they came in. "Excuse me," he said in an excited voice. "It is a stochastic process, but I think I have got it. Just a moment, please, and do not interrupt me."

The Queen of Hearts winked at Alice, and waved her hand to indicate that she should glance at the walls. They were covered from floor to ceiling with shelves full of books, some laid in sideways above the others because there was no room left. "Can he possibly have read all those books?" Alice asked herself, and then, as the Knave of Spades went on with his work in silence, she dozed off and dreamed that he had at last attended to his visitors and was saying something like this: "It is highly probable that probability is statistical and that statistics is probabilistic, but, never mind, everything will in the end be as clear as mud, or at least as clear as something or other."

Alice woke up with a start, because the Knave of Spades was actually striding up and down the room, muttering to himself something very like what Alice had thought he was saying. Meanwhile the Queen of Hearts had gone to sleep too.

Alice was frightened to find herself alone with this strange man, and she pinched the Queen of Hearts gently to wake her up. "I was just having such a nice dream," she said yawning, "Oh, yes, I remember, this little girl has bought a pack of cards, and has come to ask you about them. Please stop walking up and down like a lion in a cage. Sit down and tell her what she wants to know."

The Knave of Hearts frowned but sat down at his table. "What do you want to know?" he asked Alice.

"Please," she said timidly, "I want to know how to play cards."

"You have come to the wrong place," he said with a kind smile. "I never play cards myself, or hardly ever. There is too much chance in it, and when people play for money . . ."

Alice interrupted him. "You said chance, didn't you? Now that is just what I want to know about. Please tell me what chance is. And how you can shoot an arrow at random!" she added breathlessly.

"Archery is not in my line at all," retorted the Knave impatiently, "but if you want to know what chance is, well, you could hardly have come to a better person. However, it would be silly for me to give you a sort of formal lecture on the subject. Words, words, words. No, since you have there in your hand a pack of cards, it should not be hard to give a practical demonstration. Come up here to the table, and let us have a look at the cards."

But at this moment the door opened and the Three of Spades came in with a tray. "Ha!" cried the Knave, "you come in the nick of time. But I have guests. So take this tray away, and bring back a larger pot of coffee, with cups and saucers and sugar and cream for three. And be quick about it."

"But please," said Alice, "my father and mother don't let me drink coffee. I am too young, they say."

"Fiddlesticks!" said the Knave. "Did your father and mother give you permission to find out what chance is? Compared to coffee, that is in the ratio of an elephant to a mouse, or of the sun to the moon, or of a thingumabob to a whatdoyoucallit. But of course you need not drink coffee. Provided you don't stop me having mine. And as for her Majesty here, it will help her to keep awake. After all, she is the chaperone, and I might be a bad man, not to be trusted in the instruction of a young girl. What," said he, addressing the Queen of Hearts, "is, in your estimation, the probability that I am a bad man?"

The Queen gave him a quizzical glance. "I think," she said after a little thought, "I would put it at zero decimal nine—or possibly nine recurring. I have no statistics on the matter, or at least only biased statistics. But if I put the probability of my falling asleep again during this interview at, let us say, one in a thousand, I think we may safely proceed, provided you are not too boring!"

When the large pot of coffee arrived, the Knave of Spades poured cups for the Queen and himself, but Alice shook her head when he offered to pour one for her. She sat down beside him and laid the pack of cards on the table. "Open it," said he, "and be sure you don't leave any cards in the packet. Spread them out face-upward, but keep them in order. I see a joker. Throw him away. Fifty-three is an unlucky number. Are all the rest there? Yes, arranged in order, one suit after another, and each suit in its proper order. The manufacturers leave nothing to chance."

"Please," said Alice timidly, "I wish you wouldn't use that word. If you are going to show me what chance is, I get muddled when you use it as if I understood it already."

"A good point!" said the Knave. "I see I shall have to watch my step with this young lady. It was a random remark."

Alice pouted but let it pass. "What do we do next?" she asked.

"Shuffle the pack," said the Knave of Spades.

"How do you do that?" asked Alice, glancing up at him to see whether he was pulling her leg.

"Experienced card-players," said he, "do it automatically. It is really wonderful to watch, quite magical. I am clumsy at it. But I will show you a way to do it. We start with order and seek disorder. Incidentally, that is in accordance with the second law of thermodynamics, but if you want to know about that, you had better go to the Ace of Diamonds. He knows all about statistical mechanics, or rather he knows all that is known, and he is obviously the man to instruct you in that, that is, if you want to be instructed."

"I want to shuffle this pack of cards," said Alice, a bit impatiently. "You remind me of the Red Herring." But she was sorry she said that, because the Knave looked a little displeased at being compared to the Red Herring whom the Court regarded as a low-class politician.

"Do you know how to cut the pack?" he asked.

Alice hesitated. She did not like the idea of cutting these nice cards. She shook her head.

"Well," said he, "it is very simple. You arrange the cards in a pile with their faces down. Then you grasp the pack from above, remove some from the top and place them at the bottom. Like this." And he did it.

"Are they now in a state of disorder?" asked Alice.

"No," said the Knave, "be patient. We must do this many times. And do it in a random way. I mean, if you merely took off the top card and placed it at the bottom, it is clear to the meanest intelligence that after fifty-two such operations, the pack would be precisely as it was at the beginning—in perfect order, as it came from the manufacturers."

Alice frowned at the word "random," but did not protest, because such protests were clearly useless. Instead, she merely asked how many times it was necessary to cut the pack in order to obtain a state of disorder, and then blushed to think that the Knave might be thinking that she was trying to make a silly pun.

"That is a hard question to answer," he replied. "I suppose you might say that a state of complete disorder would never be reached, no matter how often you cut the pack. But perhaps we might settle for a thousand times."

"That would take a long time," said Alice pensively, "If card-players really take all that time to shuffle the pack, I don't see how they have any time left to play a game. And besides you would have to tell me just where I should stick in my nails in order to make the cut. Should I stick them in at the middle of the pack?"

"Of course not," said the Knave impatiently. "If you cut the pack precisely at the middle, with twenty-six cards above the cut and twenty-six below, why, when you had done it twice, the cards would be in the same order as they were before you started. And no matter how long you went on doing it, it would go on recurring every second time. No, you have to cut the pack at random. Otherwise you will never reach a really random state."

This made Alice angry. "How can I cut the pack at random, if I don't know what that word means?" she cried.

"Well," said he gently, trying to appease her, "you must do it without thinking. You must switch off your brain, if you know what I mean."

"I don't," she retorted. "I just can't switch off my brain. Even when I am asleep, my brain keeps on working in dreams."

"I see your difficulty," he admitted. "But I have just thought of something. The Two of Clubs is not very bright. In fact, he is almost an imbecile. His brain, if he has any, works at a very low level. Suppose I get him to produce a perfect state of disorder in this pack. He is rather slow in his motions, but he might cut the pack once every ten seconds. He would do it six times in a minute and in an hour he would do it . . ."

"Six hundred and sixty times," interrupted Alice, proud of her knowledge of the multiplication table. "Then how long would he take to do it a thousand times? You have to divide a thousand by six hundred and sixty, and I can't do long division in my head." Then she realised with a shock that she had done the multiplication wrong. But she was ashamed to admit it, and anyway it probably wouldn't matter.

"Neither can I," said the Knave of Spades, "and I seem to have mislaid my pocket calculator. But it is obvious that if he does it six hundred and sixty times in an hour, if we give him two hours he will cut the pack over a thousand times and that will be all to the good—a better state of chaos. So I shall send for him now and let him get to work in the next room. I wouldn't like to leave him here, as he might reduce my books and papers to a state of perfect disorder."

"It seems a good idea," said Alice. "Perhaps, after all, an imbecile could shuffle cards better than me." But that sounded a bit sarcastic. So to make amends, she went on. "I didn't mean to be rude. But, if the Two of Clubs is a bit of an imbecile, how are you going to tell him what to do with the cards?"

"Very simple," replied the Knave. "I shall put the pack of cards on the table in front of him. Then I shall cut the pack a few times to show what that means, but I shall be careful to restore the original order of the cards before I leave him. And, talking of order, it is hard to remember the conventional order of the suits, so I think I shall write numbers, from one to fifty-two, on the faces of the cards; but he won't see these numbers, because I shall tell him not to turn them over. And I shall promise him a bag of sweets as a reward for a job well done. Any questions?"

"Will you stay with him while he is doing this?" asked Alice.

"No," said the Knave. "I don't think that would be wise. He might be influenced by my presence. If I happened to smile, ever so slightly—and perhaps about something that had nothing to do with the cards—if I smiled ever so gently when he made a cut near the top of the pack, he might go on doing that, thinking that it would earn him a bigger bag of sweets. No, I shall leave him alone."

"But will you tell him to cut the pack at *random?*" asked Alice, with emphasis on that word.

"No," replied the Knave. "That would only confuse him. His mind is already in such a state of chaos, that it will come naturally to him. And now I hope you ladies will be my guests to lunch at the Five Aces round the corner. If you would be so kind as to go and secure a table, I shall follow as soon as I have fixed up the Two of Clubs at his task. We have lots of time—two hours. Then we shall come back, and Alice will see how order is turned into disorder."

Over a bottle of wine at the Five Aces, the Queen of Hearts and the Knave of Spades talked of many things while Alice listened. When the Queen gaily offered her a sip of wine—"It would help you to understand what disorder means!"—Alice shook her head firmly. Having prolonged the meal to allow the Two of Clubs time to finish his job, they returned to the Knave's study and sat down until the full period of two hours had expired.

Then the Knave led them into the adjoining room; the Two of Clubs had gone, and the cards stood in a neat pile on the table with their faces down. Alice's heart beat rapidly at the anticipation of finding out the meaning of a lot of puzzling words—chance, random, disorder, and so on, and even in the back of her mind harder words like probability and statistics. But this last word made her blush slightly because she had heard some older girls talking about their statistics and it seemed there was something immodest about it.

The Knave too was excited, but in a different way. He wondered how many runs of consecutive numbers there would be, and he tried to make a rough estimate based on the number of cuts the Two of Clubs would have had time to make, and he had to take into consideration the temptation to cut the pack in the middle, balanced by a puckish tendency to make a cut close to the top or close to the bottom just for the hell of it. Obviously it would require more time than he had to make preliminary calculations before examining the pack, and he got out a notebook to record the findings in this actual event. He would examine them later.

The Queen, somewhat elated after a very successful lunch, regarded the whole affair as a rather delightful piece of nonsense, and was prepared to break into genial laughter whatever the outcome of the shuffling might prove to be.

With his notebook on the table beside him, and a pencil in his right hand, the Knave of Spades proceeded to lift off the cards one by one with his left hand and lay them face up on the table, beside the pack. He had, as promised, numbered the cards from one to fifty-two in the order in which they had come from the manufacturer.

The first card he lifted off was numbered 1, and the Knave solemnly wrote down 1 in his notebook. The next was 2, the third was 3, the fourth was 4, and the Knave looked a little surprised. But he went on. There followed 5, 6, 7, . . . right up to 52.

It was then that Alice cried angrily: "But the cards haven't been shuffled at all!"

"I admit that that is a possible explanation of the present phenomenon," said the Knave, looking a little flustered. "But it is not the only one. It is possible that the Two of Clubs did what Alice suggested earlier—namely, that he cut the pack precisely at the middle every time, and then the only deduction is that he did it an even number of times. Now the probability of even versus odd is one half, and so what we now witness is nothing strange, for anything that occurs when there is a probability of one half is not unexpected."

"But," said Alice, "you said you would tell him to cut the pack at random. Does that mean cutting it at the middle? And if the Two of Clubs is not very bright, he might not have been able to find the middle. And if he made a little mistake about this, every time he cut the pack, wouldn't some of the cards near the middle get a bit mixed up?"

"Yes, indeed they would," said the Knave. "But they haven't got mixed up. The order is perfect, as you have seen. But indeed he need not have cut the pack at the middle. He might have cut it one fourth from the top each time, and then the original order would have been restored after four cuts. Or indeed," he mused, "if he had chosen any rational number, a proper fraction of course, and one with 52 as denominator (because otherwise he would have had to split a card, and that he did not do), and used that fraction as basis for each cut, then in the course of time the original order would have been restored. Or am I making a mistake?"

"I don't know," said Alice. "Perhaps you are and perhaps you aren't. But is that what you call random? I mean, just repeating the same thing over and over again?"

"That depends on what you mean by 'the same thing,'" replied the Knave. "If you toss a coin or throw dice, you may think you are doing the same thing every time, but you are not. If you did, then if the coin showed heads the first time, it would show heads the second time and the third time, and always. And if the die came up with six the first time, it would come up six every time."

"My! My! My!" cried the Queen of Hearts laughing gaily. "What a fuss you are making about nothing! When we bought a new car recently, the Registration Office gave it the number AZ 9999. People went round gossiping about undue royal influence—*they* couldn't get such an interesting number for their cars. But that was all nonsense. We simply took what was coming, like everybody else. It just happened. People said that, even if you left out the alphabetical part and left out 0000 (that would be a queer number, too vacuous, if you know what I mean), why the probability of four nines was . . . well, what was it?" she asked the Knave.

"Ten to the minus four," he answered, "or one over ten thousand. I think what you are getting at is this: if you have cards numbered 1 to 52, then any order of these numbers is equally probable, and so we should not be surprised if those cards turn up in their natural order. Is that it?"

"Quite so," said the Queen. Alice was about to say "Quite so" too, but she thought that would be repetitive, so she said, "Yes, indeed" instead.

"But," said the Knave, "it is not as simple as that. To illustrate, suppose you told your chauffeur that you wanted to go for a drive in the country, and he was completely crazy with no sense of direction, what would be the probability that he would bring you home? He would pursue a random walk, as they call it (in this case a random drive), and the probability of a return to the palace would seem to be zero—well, not quite, because the number of roads is finite, and even the craziest driver would arrive home sometime, though it mightn't be for years. But actually the probability is one—you would certainly get home again."

"I don't see what a drive in the country has to do with this pack of cards," said the Queen with a trace of irritation.

"I was merely trying to illustrate . . ." said the Knave pathetically. But the Queen cut him short. "Before I married the King of Hearts, I was a barmaid, and Sir Ernest Rutherford used to drop in from time to time to explain about atoms. But I gave him a flea in the ear when people told me that he only did this because he thought that a theory was no good unless it could be explained to a barmaid. Well, I suppose there are barmaids and barmaids."

"I was merely trying to illustrate," went on the Knave. "The pack in its natural order—I mean with the numbers running from 1 to 52 in order—that is like the palace. Shuffling is like taking a drive in the country, and getting home again is like getting back to the natural order of the cards. The crazy driver is the good shuffler—the person who turns order into disorder . . ."

But here Alice interrupted. "What is disorder? If you started off with 52 and went down to 1, would that be disorder? When I leave my room at home untidy, my mother tells me I must not be disorderly, and when I ask her why she says that then I won't be able to find things I want. And of course that is true. Is there disorder when you can't find what you are looking for? But then my teacher says that long ago there were no people to look for things, and does that mean that everything was in disorder, or was everything in order like those cards?" She stopped a bit out of breath because she was not used to saying so many things one after another.

The Knave of Spades thought this over for a few minutes, and nobody spoke. At last he said "The question you ask is a very deep one. It raises the whole question of objective reality. And I don't know . . ."

"I wish you wouldn't use such long words," interjected Alice.

"Yes, indeed," said the Queen of Hearts, "when Rutherford was sipping his beer he would lean over the counter to show me a little green ball he always carried in his pocket. 'That's an electron;' he would say; 'but of course electrons are much smaller,' and he would point up to the electric bulb. 'Millions and millions of those little fellows being forced along—no wonder they get hot!' He never talked to me about objective reality. What do you mean by that anyway?"

"Rutherford was a great physicist," said the Knave, "but I don't think he ever wrestled with the really deep questions. It is widely believed nowadays that it is improper to talk about things that cannot be observed. I mean, of course, observed in principle, but I must confess I don't really know what 'in principle' means."

At this moment the door burst open and in came the Ace of Diamonds. When he saw Alice and the Queen of Hearts, he was taken aback. "Am I intruding?" he asked.

"Not at all," said the Knave of Spades. "You have come in the nick of time, when I was getting a bit out of my depth. I mean, about observations and all that. Of course I know about real observations—you look through a microscope or you take a photograph and examine it after it has been developed. But I mean observation 'in principle,' if you know what I mean."

"Indeed I do," replied the Ace of Diamonds. "Take statistical mechanics, for example—classical statistical mechanics, to keep things relatively simple." At this Alice pouted, for he glanced at her as he said this, and she was very sensitive about being patronized. "Well," he went on, "in classical statistical mechanics you may start by thinking of the molecules of a gas as little elastic spheres. I feel sure that is the way Maxwell thought about them. Now if two elastic spheres strike one another, they rebound in a certain manner which is easily worked out."

"You are getting away from the point," said the Knave of Spades. "You cannot actually observe the collision of two molecules. But if you imagine yourself to be a tiny being, much as you are, with your senses as they are, particularly the sense of sight, then you would be able to see two molecules colliding and bouncing off one another. And that is what you mean by observing the collision 'in principle,' or is it?"

"I am speaking," said the Ace of Diamonds, "about what we must now regard as a rather primitive state of science. We now know that molecules are not little elastic spheres. I am just trying to make things easy to understand."

"But," retorted the Knave, "is it permissible to observe in principle things that are false? One of the things that everybody learns, if he tries to understand the rudiments of quantum theory, is that the position and momentum of an electron cannot both be observed. And if not, why not?"

"As I said before," said the Ace a little wearily, "I am trying to keep things on an easy level."

"Please go on about the little elastic spheres," said Alice. "I would like to hear more about them. I have no trouble at all in imagining them colliding and bouncing back. But what happens next?"

"Yes, indeed," said Ace, relieved to find that he had at least one sympathetic listener. "From that simple mental picture of bouncing spheres it is possible to say how fast the molecules in the air in this room are traveling. Of course they are not all traveling at the same speed. At any moment we know how many are going slow and how many are going fast, to express things in simple terms. This is one of the simplest problems in statistical mechanics."

"Statistics indeed," cried the Queen of Hearts, "They say you can prove anything by statistics."

"Nonsense," said the Knave of Spades drily. "Statistics are facts, and you can't prove anything by facts. Facts are the end product, so to speak. The word 'statistics' originally meant facts which a statesman should know if he wants to be able to run a country. A census, for example—that is an important statistic in any country."

"But," interjected Alice, who by this time was beginning to feel a bit confused ("Perhaps this is disorder," she said to herself), "but that cannot be what the Ace of Diamonds was talking about. I think he mentioned statistical mechanics, and that must be something different—at least I would think it was."

"Of course it is," said the Ace of Diamonds. "Of course it is something different altogether, although the behavior of a crowd in the street might interest a statesman and also a physicist. I mean a crowd wandering about rather aimlessly—not a crowd marching with banners and clenched fists and shouting slogans no one can make out. I remember . . ."

But what he remembered was not stated, for the Knave of Spades butted in. "It is rather a pity," he said, "that the great Gibbs called his work 'statistical mechanics.' It has nothing to do with the sort of statistics a statesman might be interested in. He should have called it 'probabilistic mechanics,' except that the word 'probabilistic' is rather long. But what he was concerned with was probability."

"Oh," cried Alice, "that is just what I want to know. What does the word 'probability' mean?"

"That is for you, friend," said the Ace of Diamonds, turning to the Knave of

Spades. "I am using probabilities all the time, but it is now so much a part of me—of my way of thinking—that I would be hard put to define it."

The Knave was silent for a minute, while the others waited for his words with great expectation. Then he said: "What good are definitions? Just words about words, and then you will ask for definitions of the other words. In the case of statistics, I took you back to the origin of the word, somewhat facetiously, because I know as well as you that the word has wandered far out of the domain of statesmanship. But it might be no harm to start by doing the same for the word 'probability.'"

"Yes," said Alice, feeling that at last they might be getting somewhere, "what did the word mean long ago?"

"As the meaning of words becomes more obscure," said the Knave, "people add on syllables. Any word of seven syllables is suspect. The Romans were satisifed with three—*probare,* meaning to prove (much like our word, isn't it?) or to test (like probing). We have gone a long way from there. No one today would confuse what is probable with what is provable. Or would they?"

"Certainly not," said Alice quickly, "I can *prove* that the angles at the base of an isosceles triangle are equal. But if I said they are *probably* equal, my teacher would laugh at me, and I don't like being laughed at!"

"Your comment is duly noticed," said the Knave. "My question was merely rhetorical. We have moved a long way from ancient Rome. At least we think we have. In the beginning—or, well, not quite the beginning—there were monkeys, then there were men, and finally, as the final evolution, there were cards like us. We were the playthings of men, and they found themselves compelled to bet, sometimes winning money with good luck or losing it with bad luck. And sometimes they cheated, creating good luck for themselves until they were found out and dealt with appropriately."

"I think," said the Queen of Hearts, "you are spinning this out to avoid defining probability. Like Alice, I want to know what the word means."

"Yes," said the Knave. "I am spinning it out, because I feel that whatever I say is bound to be misunderstood. What is a pun?"

"I think," said Alice, "that a pun is when you use the same word twice but with different meanings. I can't just think of one at present. It is, my teacher says, a low form of humor."

"As far as probability goes," said the Knave, "the pun isn't even funny. It gives wise men headaches, and then they say rather silly things. And yet one can't help feeling that there is something in what they say after all."

"You are still spinning it out," interjected the Queen of Hearts impatiently. "For heaven's sake, say something understandable—something concrete, something we can really think about, not just a maze of words."

At this point the Ace of Diamonds felt called upon to break the deadlock. "It was Max Born," he said, "who put forward the idea that Schrödinger's famous equation must be interpreted probabilistically, and . . ."

But here Alice put her hands over her ears, rather dramatically, and shouted (with her ears covered she did not realize she was shouting) "Please, please, these long words give me a headache. Prob-ab-il-ist-ic-al-ly, that makes seven syllables, and surely we can talk to each other with shorter words."

"Well," said the Ace, smiling, "I sympathize with you. But a person doesn't really *think* in words. You can say 'prob' if you like, and once you get used to it, it should work all right. Born said that the square of the psi-function—or rather the square of its modulus—is a measure of the probability of finding the electron in a little cell in space."

"Oh, now I see!" cried Alice. "I have two pockets in my pinafore, and when I want to blow my nose in a hurry I can never remember which pocket I put my handkerchief in. I wonder if you could help me," said she to the Ace of Diamonds, "for it is very embarrassing when you feel a sneeze coming on not to be able to find your handkerchief."

"I am very glad you asked that question," he replied, "because it brings up a matter of great importance. Your handkerchief may be a very dainty little handkerchief, but it is enormous—really enormous—in comparison with an electron, or a molecule, or the objects discussed in what is called microphysics. When it is a matter of handkerchiefs, you should turn to our friend here, the Knave of Spades. Explain to Alice, please, the statistics of pocket handkerchiefs!"

"I admire your modesty," said the Knave. "It is good to hear that you do not claim to know everything. I accept the challenge. Let me adopt the attitude of a linguistic purist. That is, I separate statistics from probability. At least for a start. Well then, I have at present only the vaguest ideas about the statistics of pocket handkerchiefs. Of course I could make a guess. I could make the hypothesis that each person carried one handkerchief, and, since most people are right-handed, the handkerchief would be in their right-hand pocket. But then I would reflect that in a man's suit there is, or used to be, a pocket on the left breast intended for a handkerchief, and indeed it is easy to see that it is easier for a right-handed man to get a handkerchief from such a pocket than from, say, the right-hand pocket of his trousers. But—"

"For heaven's sake," cried the Queen of Hearts, "get to the point! What about the female sex? You are a male chauvinist—whatever that means—but please do *not* give us a lecture on that."

"Yes, indeed," said the Knave of Spades. "I confess I was drawing it out too long. My point was in brief—I was only leading up to it—that statistics (in the purist sense of the word) is concerned with *facts*. So, if we want statistics of handkerchiefs, the first step is to prepare a questionnaire—a form with a number of questions on it—and train clerks to go from door to door with authorization to demand answers from all citizens. When the forms have been filled, they would be brought to the office of the Minister of Statistics for analysis."

"And what then?" asked Alice.

"Well," said the Knave, "you could apply to the Ministry of Statistics and ask how many little girls in your age-group carried their handkerchiefs in the right-hand pocket of their pinafores and how many carried their handkerchiefs in their left-hand pockets. Of course you would have to pay a fee for this information, because the collection of information costs money, quite a lot of money if it is done thoroughly."

Alice looked a bit disappointed, and the Ace of Diamonds was bored. "Why not take a random sample?" he growled. "It would cost much less."

"Yes," interjected the Queen of Hearts, "like an opinion-poll. They are usually wrong, but they are lots of fun."

"This is not a subject for joking," said the Knave of Spades sternly. "Random sample indeed! Now what pray do you mean by that?"

The others all started to speak at once so that nobody heard anything but their own voices. Then the Ace of Diamonds said with emphasis: "You are being rather absurd. In every process of mass production it is common practice to test a few of the products, a random sample, in order to make sure that the standard of the product is maintained. Are you telling us that this cannot be done? Or what are you getting at?"

"Merely clarification," retorted the Knave unabashed. "Or are you opposed to clarification?"

"Is it necessary to be silly in order to be clear?" asked the Ace satirically.

"Please don't quarrel about this," said Alice. "It is really all my fault because I asked a question about handkerchiefs in pinafores, and by now I have quite forgotten what that question was. My father says that it is silly to be rude and rude to be silly. So I think I will try to change the subject by asking another question."

The Knave of Spades was a little annoyed by Alice's intervention because he had expected the Ace of Diamonds to get angry and had his countermeasures prepared. But now he thought he had better let Alice ask her question and see whether he could not twist things his own way. So he nodded approval, and Alice stated her question: "My father plays Scrabble with me every Sunday afternoon. I suppose all of you know the rules. There are a hundred tiles and each player picks up seven of them at random."

The Knave of Spades pursed his lips quizzically at that word. Alice blushed, but continued with some irritation: "Oh, yes, I know I must not use that word, but the tiles are lying face down and their backs look all the same. Some people put the tiles in a bag and draw them out without even seeing their backs. You don't think we cheat, do you?" she added angrily.

"Of course not," said the Knave. "Let the word pass; you seem to have justified it."

"Well," continued Alice, "a few Sundays ago the first two tiles my father drew were both blank tiles, and these can be used for any letters—so he won fifty points for making a seven-letter word, and got quite excited. He said it would not happen again for ten years. As I was busy trying to make up a word for myself, I forgot to ask him what he meant, so now I am asking you."

Although her question was addressed to the Knave of Spades, the Ace of Diamonds could not resist answering. "There are one hundred tiles," he said, "and there are two blank tiles. So the probability of drawing a blank at the first go is 1/50. Having drawn that blank, one remains among 99 tiles, and the probability of drawing it is 1/

99, or roughly 1/100. So the probability of doing what your father did is about 1/5000. Since you play once a week, and there are roughly 50 weeks in the year, it will not happen again for about a hundred years. Your father, in his excitement, made a mistake—not ten years but a whole century—but his idea was right, and I hope you now see what he had in mind."

The Ace smiled blandly, pleased at his own speed. But the Knave of Spades had planned his strategy.

"It would be very interesting," he said, "if Alice and her father reported to the Ministry of Statistics the results, in full detail, of all their games of Scrabble. Each game could be filmed as it proceeded, and the procedure analyzed statistically. But actually the whole game would be irrelevant, if it is only a question of the first two tiles drawn—whether they are both blank or not. Unfortunately Alice's father is not likely to last for a century, nor even Alice herself, alas, and so they cannot hope to find out by actual observation whether the Ace's prophecy is true or false. But then, after all, both Alice and her father are irrelevant. What is involved is merely a simple question in elementary probability."

"Yes, indeed," said the Ace of Diamonds, "you can certainly abstract Alice and her father from the game, but surely you will agree about the century?"

"We have come to the cross-roads," replied the Knave solemnly. "You'll take the high road and I'll take the low road, or vice versa, and which gets where first is not really what counts. I know I bored you all by talking too much about statistics, but I wanted to drive home that statistics is concerned with facts, or should be, if we are to appreciate by contrast the mysterious nature of probability."

"Oh, come now," said the Ace, "aren't you being a bit—what shall I say—well, mysterious yourself? Physicists use probability on all levels and don't seem to find much of a mystery in it. What about the probable error which every conscientious experimenter records?"

Now the Knave of Spades had forgotten the rules used by experimenters to calculate probable errors, and so he was relieved when Alice broke into the discussion. She said that only last Sunday, in their game of Scrabble, her father had drawn two blanks in the first seven tiles, and she found it hard to believe that he would have to wait a whole century before he repeated his earlier success.

"Ah," said the Ace of Diamonds, "I'm afraid you've got the wrong idea about probability. Your father might do it again next Sunday. When I mentioned a century, I was only giving an 'expectation.' Probability only works when there is a large number involved somewhere. Now if your father went on playing that game every Sunday for a million years and then . . ."

"Oh come off it!" cried the Queen of Hearts. "What rubbish you are talking to the poor child! I'm sure her father carries life insurance, and they have his life expectation written down somewhere. I would be very much surprised if it says a million years."

"Does he have to die when his life expectation runs out?" asked Alice, opening her eyes very wide.

"Of course not," said the Ace reassuringly, "of course not. But if you had ten thousand fathers, it would be probable . . ."

"But she hasn't got ten thousand fathers," broke in the Knave of Spades angrily. "She has only one. The so-called laws of probability do not apply to a single event, which in this case is the death of Alice's father."

"This conversation seems to have taken rather a morbid turn," said the Queen of Hearts satirically, "and I suggest that the Knave should tell us just what probability is."

The Knave of Spades cleared his throat, massaged the lobe of his left ear, cleared his throat again, massaged the lobe of his right ear, and spoke as follows: "It is a question I have often asked myself, and lots of other people have done the same, coming up with a variety of answers. On the whole I am inclined to agree with Feller, or at least to agree with what I think Feller meant. The theory of probability is a game carried out according to certain rules, the object being to calculate one probability from another. For example, if you tell me that, in tossing a coin, the probability of heads is one-half, then the probability of three heads in succession is one-eighth. But if in eight triple throws you don't find three heads appearing together, or if they appear together more than once, you don't get your money back, so to speak. The word 'probability' itself remains undefined—it is a branch of mathematics. However, Feller realized that the word was in common use with various rather vague meanings, and suggested that his probability might be called 'statistical probability.' Personally, I think that was a bad choice—it is almost impossible to clean the dirt of ages off a word, and, as I said earlier, I believe that it is a mistake to mix up statistics with this purely mathematical theory."

"Then you deny that probability is a physical theory?" asked the Ace of Diamonds.

"Would you say that the theory of differential equations is a physical theory?" countered the Knave. "I have seen a book in which the differential calculus was presented in physical terms, and I expect that this made it easier for students to understand. But if they went far enough in physics, they would come to realize that even grandfather clocks are not precise harmonic oscillators—that precise harmonic oscillators are intellectual creations, not real things, and then they would be put into intellectual confusion as to the relationship between mathematics and physics."

The Queen of Hearts yawned. "My intellect is getting confused, such as it is. Let's call it a day. Why not get that little ruffian in here and ask him about shuffling those cards?"

"Oh, the Two of Clubs! Yes indeed," said the Knave of Spades, "I had quite forgotten about him."

After some telephoning the Two of Clubs was located, and he came in with a slight swagger, accompanied by his cat.

"I want my bag of sweets," he said in a determined voice.

"Yes, certainly," said the Knave, "but first tell us how you shuffled this pack of cards."

"Well," said the Two of Clubs, "it was like this. You told me how to shuffle. Okay. But just when I was going to start the damned cat came scratching at the door and meowing, so I had to let him in. 'Now, Mike,' said I to him, 'you just lie down in that corner and let me get on with this job!' But not a bit of it. Up he jumps on the table and goes for those cards like they were a lot of mice. Scatters them all over the floor, higgledy-piggledy. So what was I to do? Why, if I was to shuffle the cards the way you said, I had to start by putting them back the way they were, and how was I to do that? Well, I found they had numbers on them. Now I don't know multiplication tables or any of that tizzy, but I do know my numbers. I mean one comes after another, if you know what I mean, but I'm kind of slow at it, hard to remember whether six comes before seven or after it, and so on. So it took me the full two hours to get the cards back the way they were, if that was the way they were, and there was no time to do the shuffling, as you called it. Please, may I have that bag of sweets?"

"Give him two bags!" cried the Ace of Diamonds in high glee. "A shrinker of entropy! A violator of the second law of thermodynamics!"

But Alice crept silently away full of sadness. If only she hadn't made that silly mistake in multiplication, the Two of Clubs would have had time to shuffle the cards and she would have learned what disorder meant, and chaos, and all that. However, she pulled herself together and decided to save up for a bow and arrows; she would learn to shoot at random, whatever that was, and she hoped that her instructors would not talk so much.

18

ON THE NATURE AND ORIGIN OF LIFE

ALBERT SZENT-GYÖRGYI
National Foundation for Cancer Research
at the Marine Biological Laboratory
Woods Hole, Massachusetts

Biology, hitherto, did not contribute considerably to the foundations of physics. On the other hand it expects its basic problems to be solved by physics.[1] Biology, being essentially the study of life, *βιοσ*, its main problem is: what is life? The question must have been asked innumerable times without finding a satisfactory answer. This is an unusual situation, experimental science being based on the experience that Nature answers intelligent questions intelligently. If she is silent, something may be wrong about the question. The question is wrong, because science rarely asks what things are, and is mostly contented with working on how they behave. Newton never asked what gravitation was.

We can formulate our question about the nature of life in a different way, which is more in the language of science. We can ask, for instance, what is the difference between "animate" (living) and "inanimate" (non living). We can try to compare the two bit by bit noting the difference in terms of exact science. The difference between the two is so great that it is but very rarely that the doubt arises into which group a certain system belongs. The difference is so great that we are justified in calling them two different physical states. What distinguishes animate from inanimate is, in the first place, the wonderful subtle reactivity of the living which is mostly in sharp contrast to the rigidity and lack of reactivity of the inanimate.

Since the living must have originated from the nonliving, we are justified to ask what kind of reactions can lend reactivity to the inanimate. Before trying to answer this question we may do well to ask: at which level of organization do we expect to find the difference? As is generally known, most material systems, animate or inanimate, can be decomposed into molecules, while molecules can be decomposed into atoms and atoms into subatomic particles, electrons, protons, and atomic nuclei. These subatomic particles can be decomposed into still smaller units, but this decomposition demands magnitudes of energy which are not available in living systems. As biologists, we have thus reached the "bottom of Nature" with subatomic particles and need not think about the lower dimensions. This renders our problem very much simpler. When trying to understand life we have to think only about three kinds of elementary particles—electrons, protons, and atomic nuclei. All the three have a mass, a charge, and all the electrons have what is called a "spin" and a coordinated magnetic moment. In the atom these forces are carefully compensated, the positive charge of the nucleus by the negative charge of the electrons, while the magnetic forces of the spinning electrons are compensated to a great extent by their pairing up. During the billions of years which elapsed since the "big bang" these particles had time to find their energy minimum, which is their most stable state. In this state they need external energy to be lifted from this state and undergo reactions.

This leads to the question, is there a method by which reactivity can be lent to stable atoms of molecules? There is one, a relatively simple one, which consists of transferring an electron from one atom or molecule to another. By taking an electron out of the system, it has to upset the whole balance: upset the electric and magnetic balance and leave a "hole," a half-occupied orbital behind. So the question arises how single electrons can be transferred from an atom or molecule to another, making both units contain an unpaired electron. An atom or molecule containing an unpaired electron is no longer called an atom or molecule. It is a "radical." The great reactivity of radicals, as compared to atoms of molecules, is common knowledge.

The basic level for the biologist is the electronic level with its three subunits. There is a wide gap between this and the level of molecules which has been little touched upon by biology or fundamental physics and which is an almost virgin field of research.

The question is how single electrons can transfer between atoms or molecules. They will transfer spontaneously if the two atoms or molecules in question collide and orbital overlap is established whereby the two units form a common electronic system linked by the overlapping orbital. If the two units have a different electron affinity, then we can expect that the particle with a greater electron affinity will retain an electron which belonged earlier to the other. The technical term for such an electron transfer is "charge transfer." If a whole electron is transferred, it transforms both particles into radicals, one with a negative and the other with a positive charge.

This leads to the question how the first biradical was created. When life originated, our globe was dark, was not in touch with free oxygen, and had to have a reducing atmosphere consisting mainly of "electron donors," that is, particles which were more prone to give up than to take up an electron.

From this point on we have to be led by speculation, but the end of our speculation can be checked by comparing its result with actuality.

The early global atmosphere was dominated by electron donors. This excluded the existence of a permanent electron acceptor. All the same, these electron acceptors had to contain single small unsaturated atomic groups, which, theoretically, still had place for an additional electron, though they may not be strong enough to actually attract electrons from other particles. Such atomic groups are the carbonyl, C=O and the C=C, C and O, C and C atoms linked by a double bond. When two of these groups are put together side by side, their electronic system may fuse and they can activate one another, forming a fairly strong electron acceptor. Two C=O groups can be strengthened still further by adding a methyl group which introduces an orbital of low energy by making a ketone out of an aldehyde which can easily take up an electron. The resulting substance is methylglyoxal (Fig. 1), a fairly strong acceptor.

At this point we can make our first attempt to test our thoughts and see whether reality supports them. It was almost seventy years ago that two British investigators and a German investigator (H. D. Dakin and H. V. Dudley[2] and C. Neuberg,[3] 1913) discovered simultaneously a most active and apparently ubiquitous enzymic system, the "glyoxalase" which can inactivate methylglyoxal, converting it into *d*-lactic acid. Nature does not indulge in luxuries, and if there is such a very powerful and apparently ubiquitous enzyme as the glyoxalase, then its origin cannot lie very far behind the beginnings of life.

It may have the appearance of free radicals which opened the way to differentiation, the building of more complex structures and functions. If this is true, then it is logical also to assume that higher degrees of development and differentiation demanded a greater difference in electron affinity between donors and acceptors. There must have been an enormous increase in this difference when owing to the cooling of our globe its water vapor envelope collapsed and the sunlight could reach the surface of the globe. Life has used the energy of this light to separate the elements of water, hydrogen, and oxygen. Oxygen is a very powerful electron acceptor which opened the possibility of a higher level of differentiation, the end result of which is us. Accordingly, evolution had to consist of two periods, the first period being an anaerobic and dark one in which the low level electron affinity between acceptors and donors made only the development of very primitive forms of life possible which did not leave many traces behind, not exceeding the bacterial or algal level. The appearance of oxygen opened the way to a higher differentiation and development which may be still far from its end point.

There are reasons to believe that under specific conditions this development can take a reverse course and under influence of some damage the process reverses and differentiation turns into dedifferentiation. In this case the end result may be cancer, which can be shown to have a poor oxygen activation and, accordingly, has to have a lower level of differentiation. This may be pure speculation, but it may show the way towards a correction by the support of the acceptor power.

Another line of thought, which also supports the correctness of the ideas pre-

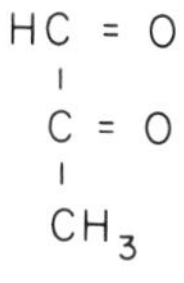

FIGURE 1. Methylglyoxal.

sented, has led P. Gascoyne and myself[4] to experiments which indicated that ascorbic acid (vitamin C) may be the catalyst of charge transfer which explains why the megadoses of the vitamin can actually prolong life of terminal cancer patients.

It was almost forty years ago that I suggested that electronic orbitals in living matter may form band patterns which promote electronic mobility. My proposition was unanimously rejected, the main reason being that at that time already many proteins had been isolated, purified, and crystallized and studied, but none showed a striking conductivity. This contradiction has solved itself, since purified, crystallized, water-free proteins which were the material of most of the measurements are inanimate matter. Studies[5,6] of my associates have shown that the electric conductivity of proteins can be increased by orders of magnitude by treatment with methylglyoxal.

References and Notes

1. A. Szent-Györgyi, *Proc. Natl. Acad. Sci. USA* **74**, 2844–2847 (1977).
2. H. D. Dakin and H. V. Dudley, *J. Biol. Chem.* **14**, 155–157 (1913).
3. C. Neuberg, *Biochem. Z.* **49**, 502–506 (1913).
4. P. R. C. Gascoyne and A. Szent-Györgyi, *Biol. Bull* **159**, 474 (1980).
5. R. Pethig and A. Szent-Györgyi, *Proc. Natl. Acad. Sci. USA* **74**, 226–228 (1977).
6. S. Bone and R. Pethig, *Submolecular Biology and Cancer* (Ciba Foundation Symposium 67, new series 1979), pp. 83–105.

19

Science, History, Philosophy, and World Views

Herman Tennessen
University of Alberta
Edmonton, Canada

1. The History of Science Reintroducing the Need for World Views—and the Futility of the Attempts at Accomplishing Such a Goal

1.1. The Integrated vs. the Disintegrated View of Science

In one of my contributions to the XVth World Congress of Philosophy (17–22 September 1973, Varna, Bulgaria) I take my point of departure in the following rather trite observation:

> There is something peculiarly redundant in the very idea of *humanistic psychology*. Whenever a psychologist, *qua* psychologist, involves himself in whichever conceivable psychological enterprise, that enterprise is not only, needless to say, bound to be a human enterprise but, in point of fact, an enterprise more or less precisely and definitely conceptualized as a direct or indirect contribution to man's understanding of man, viz., the all-human endeavor to figure out how things within man and around man operate—and what, after all, man's *lot* within the widest conceptual scheme may be. By the same token, "nonhumanistic psychology" is bound to designate nothing but a *contradictio in adjecto*. Its nonexistence is a notational necessity. How anyone could ever come to believe that it might make sense to argue as though there actually were such a thing, has only psycho-patho-semantic interest. And the beguiling key concepts at work here, I shall suggest, are: "Science," "scientific," plus possibly "objectivity," "theory-independent observations" or "percepts," "refutation," "autonomy," "autarchi," and related notions implicitly suggesting, *inter alia*, an image, a model of a theory/observation asymmetry, alleged to inexorably permeate all science—and metascience—theoretical activity.[1]

There is no denying the historical importance of such an image or model. It constitutes an integral part of what I have elsewhere referred to as *"the old, disintegrated view of the scientific enterprise."*[2] One may roughly distinguish six characteristics of the activity once considered "scientific," but which may presently *at the best* kindle a mild paleontological interest:

(1) The sciences were, according to this view, totally autonomous, completely emancipated from anything that is related to man, society, philosophy, conceptual schemes, frames of reference, total world views, let alone "metaphysics," or philosophical systems."

(2) The process of justification, testing and validation of scientific theories, hypotheses, laws, and observational journals was considered *a rational process* to be carefully distinguished from the more erratic pragmatics and heuristics dealing with favorable conditions for discovery and invention of theories.

(3) The sciences were in a literal sense, *accumulative*, facts being added to facts, new, more comprehensive and accurate theories being substituted for old ones. Theories tended to grow less cognitively different, less mutually inconsistent, and more generally comprehensive. This led in the long run, one hazarded to speculate, to a *unification* of science.

(4) One envisioned a parallel process of step-by-step *approximation to the truth*.

(5) There were claims made to alleged "discoveries" that certain phenomena were physically, biologically, etc. "*impossible*," adding up to an *a priori* part of science.

(6) On the other hand: "philosophical" reflections by man were viewed as a kind of reflection *sui generis*, using logical analysis and leaving empirical investigations to the so-called "sciences."

The presurmised humanistic/scientific contraposition is fundamentally underbraced by those very same blunderous misconceptions of all scientific undertakings as have merely enjoyed historical curiousness, at least during the last couple of decades.[3] What I have in mind here is not just the steadily increasing number of well-argued apotropaic remonstrances against the incessant perpetration of the historically influential Vienna Circle's thought model, according to which it is relatively simple to distinguish such entities as, e.g., "perception"/"interpretation," "fact"/"hypothesis," "observation (protocol) sentences"/"theoretical sentences" . . . etc.[4] Nor am I simply concerned with that well-known area within the field of theories of theories where the last generations of metatheoreticians have reluctantly resigned vis-à-vis an overwhelming amount of readily available, pertinent testimonies—from theories of perception and cognition to *Grundlagenforschung*, history of science, and general science methodology and general epistemology, all amounting to an unconditional renunciation of the extravagant fatuity of any attempt to retain *in vacuo*, as it were, such notions as "percepts," "perceptions," "sense," "observations," "facts," ("that which is the case") *per se*, i.e., as autonomous, self-contained entities, "things," "objects," resting somewhere "out there patiently awaiting to be discovered by some brisk and assiduously enterprising scientist, or in turn to be employed as more or less crucial evidence for his hypothesis."[5] I am much more seriously concerned with the whole myth of *scientism* as a belief in some sort of *scientific world view* miraculously emanating from the main bulk of so-called "scientific results." I see absolutely no reason why anything merely announced as "scientific results" should ever necessitate a change or even the slightest modification of man's orientation in life, his personal philosophy or general world view. The quondam so popular perpetrators of scientism should not have agreed, I suppose. They, it appears, seemed to have looked upon "*science*" as a mighty, monstrous, autonomous machinery that produced correct views and opinions, *in abstracto*, as it were—completely independent of its integral part of a human endeavor, its setting in society, of the individual and human *interests*, the motives, purposes, and intentions of those who attended to it, oiled it, served it, improved it, spoiled it, or neglected it.[6] This vision was, needless to say, a disastrous delusion. It simply presented itself as nonsensically silly—for human beings and their *Weltanschaung* purposes—to attempt the pursuit of science totally independent of the philosophic and general system-building enterprise. It soon transpired that only insofar as a human being is autonomous and articulate enough to have value priorities, action priorities, ontological priorities, can scientific results have a *rational* power of influencing, let alone drastically changing his basic attitudes, his "philosophy," his "world." Only under such circumstances do sentences expressing "results" of scientific research become sufficiently trenchant to touch, in a rational manner, upon personal situations. We are forced to realize the necessity of a precise and consistent framework in order to make differentiative judgements—either appreciative or disclaiming—with regard to any phenomenon that appears to the consciousness. Let us, for instance, say that one of my learned friends alleges to have actually made one of these earth-shaking, thrilling "scientific" discoveries. He has "found" some startling "scientific fact" or what not. *I can only engage in what would deserve to be called a "rational rencontre" with my friend, provided I can meet his claim with a clear and distinct cognitive latticework, a grid, with a fabric, a texture fine enough to fix-in his so called "fact" and trace its possible impact upon the whole structurer's putative foundations, as well as the effect it—according to the system—ought to have on man's currently most pressing priorities of any kind:* action, intention, volition, cognition, perception, emotion, evaluation. . . . Otherwise, the effect would be no more than an "irrational" or "mysterious" one in the sense that it would merely induce in me an uncontrolled disposition, a mood, like rainy or sunny weather would; or the faces of the moon[7]; or some mind-twisting drugs. I might adore his "discovery," his "fact," detest it or neglect it; it does not matter: my reaction would have only *psycho(patho)logical* interest. And so would any hindsight rationalization I would be prepared to doctor up.

Thus I have demonstrated the imperious necessity of human "worlds."

In what I take to be the all-embracing sense of "world," which I strive to employ here, my "world" would necessarily embrace *everything, comprising inter alia the only conceptual framework from which I can possibly draw the standards for evaluating anything.* And this not only includes all and every appraisal of the validity or veridicality of my total view, or the "reality" of my "world," but also the simplest assessment of the meaningfulness or absurdity of this very question. I see, at the present, no way for me to arrive at an assessment of my world without presupposing a frame of reference entailed *in* and *by* that world of mine. Were I to employ a *different* frame of reference (acquired from where?) and apply it to my world, in what sense of "*my world*" would it then be my world to which it was applied? Certainly not in the sense of my total view, my "*system*" or "synthesis"—which is, after all, the traditional "syncretic" philosophical word for a consistent, principally all-embracing total view, with its logic, ontology, epistemology, value system, etc.

Only one thing, I repeat, can here be assumed with some degree of certainty: World views—whether alleged to be "scientific" or not—never automatically emerge from an ever-so-abundant assemblage of so-called "scientific facts." There cannot, as it turned out, be any such thing as "a scientific world view," no established scientific knowledge in any sense of "knowledge" that can be found among the last generation students of the scientific enterrpise.[8] Efforts towards a creative world orientation should consequently not be hampered by that narrow, disintegrated view of "science" nor by the "scientism" tentatively sketched above. On the contrary, anyone involved in such efforts should rather be encouraged and enabled to utilize and participate in the humanly integrated science enterprise, but only provided they, as it were, "know who they are" to the extent of having at least provisionally articulated their basic priorities of evaluation and action. Because, at a given time, in a definite human society, only one or a small number of views on a certain topic were considered scientifically "respectable," the "scientific" coercion would tend to induce conformity, spurious agreement, and other-directedness. In short: "the old, disintegrated image of science" *did* in fact become *impersonalizing and dehumanizing*.

On the other hand, the inability of any scientific enterprise *per se* to provide a world view should not make it less attractive to the so-called "humanists," let alone "humanistic psychologists." Rather to the contrary. Many of those who turn their backs on scientific activity do so because of misunderstandings and misconceptions like those mentioned above, as to what is really going on, and lose thereby an invaluable opportunity for participating in a most central, human undertaking, indeed one that in the long run may prove to be the sole, collective, human enterprise for this remarkable species that man after all is. It is in this connection worth taking note of the fact that in the midtwentieth century a revolutionary movement was started, which completely changed the image of science. With some simplification, one could, as I do, speak of this new image as "*an integrated image of science, stressing the human side of this thoroughly human enterprise.*" And we find the rapidly expanding fields of *history of science* forming the main bridge between "humanistic" and "scientific" undertakings. These fields provide a particularly strong appeal to researchers who see themselves as representing a "humanistic" orientation. Even to professional "belletristic" writers, such as Arthur Koestler, the development of science has revealed itself as an exciting, exhilarating, and colorful drama. *The sciences, then, according to this new integrated image, are, when articulated with care, not autono-*

mous; they constitute the most exactly formulated and interpersonally testable parts of existing, or more often potential, general views about man and the world.

Efforts to isolate what might properly have been called: "autonomous" or "pure science" have failed, and will continue to fail. Scientific revolutions, or more generally, any substantial, noteworthy change in scientific tradition, makes science—represented by the small elite of creative scientists—shift loyalties within the area of total or near-total, world views: "global conceptual systems."

History of philosophy, history of ideas, and contemporary philosophical movements furnish the most comprehensive examples and studies of such general views. Philosophers who have been trained as scientists are able to make broader frameworks of contemporary science more explicit and help to work out *typologies of general world views* with the ultimate aim of covering any possible, coherent, total view of man and his world.

1.2. The "Ins" and "Outs" of World Views in Science

Among the common characteristics of dominant trends in the early part of the twentieth century was a tendency to deride all philosophical attempts at total world views. Thus, a contribution by this author to a series of studies dedicated to Rudolf Carnap on his 70th birthday[10] starts out with the following observations:

> In the olden days, when foreheads were wrinkled and thoughts profound, philosophers were rarely in doubt about their proper mission: To synthesize all available—or even all conceivable—significant knowledge into one universal theory or system. Poly-historic system-builders became, as we all know, gradually obsolete and finally extinct when human knowledge increased so explosively during the enormous expansion and differentiation of the natural ("hard") sciences in the 17th and following centuries. He drowned helplessly in an Amazon-flood of unsurveyable, isolated information. And his philosophical descendants, totally incapable as they were, of coping with the new situation, made the fatal choice of seeking themselves a little still part of the stormy river where they, undisturbed by the onrushing sciences could devote themselves to their encapsulated problem and theory formulations: "does matter (really) exist?" "does man (really) have a free will?" "time (or space) is unreal"; "what is the essence of truth?" "everything is really mind (or matter)"; "only I exist"; "there are no physical objects", . . .

Around the turn of the century (in fact from 1893 on) Swedish philosophers in Uppsala, particularly Hägerström and, later on, Phalén,[11] became increasingly convinced that most of the words and word-meanings on which traditional philosophical discussion seemed to turn, were enveloped in ambiguities and confusions, giving perpetually rise to all kinds of pseudoproblems and pseudosolutions. They were especially concerned with words like: "good," "evil," "freedom," "knowing," "time," etc. And so great was their concern, that they in fact recommended as the *sole legitimate* enterprise for philosophers: to seek to arrive at *meanings* of such philosophical key words, claiming that some sort of what Kant had called "*conceptual analyses*" were a requisite—a *sine qua non*—for the stating and solution of whatever was considered "philosophical problems." It was, so one said: "a turning point in philosophy" when one "discovered" that these alleged problems and theories were, as one declared, in fact "nonsensical," "a meaningless play with words," "a systematic abuse (or misuse) of language" . . . "*eine sinnlose Aneinanderreihung von Worten*". . . . By this the philosophers were brought out of the intellectual deadwater of traditional, academic philosophy and given a more limited but respectable assignment: "conceptual analysis," "linguistic clarification and elucidation," "pursuit of meanings," "close examination of the way words are ordinarily (naturally? usually? most frequently?) used," "dissolution of dilemmas arising from imperfect understanding of the conceptual apparatus" . . . etc. This avenue of escape from intellectual stagnation was in many ways a pleasant one. For once, it permitted the philosopher to perform what he could persuade himself to conceive as philosophically relevant activity without demonstrating his scientific illiteracy, and without trespassing on the cultivated fields of well-established scientific disciplines. He was operating in scientific no-man's-land. Unfortunately, this is no longer the fact. After the more recent rise and sudden growth of the behavioral sciences: psychology (including psycholinguistics and psychosemantics), social sciences, anthropology, and, especially, systematic lexicography and empirical linguistics, semantics, and pragmatics, this wasteland of the nomadic philosophers had gradually been inhabited and systematically cultivated by scientifically more reliable homesteaders. In this way some so-called "ordinary" "language" "philosophers" of the 1940s and 1950s, with a study of ("ordinary," "everyday," or "natural," "stock," "standard," etc.) English as their proclaimed major interest, were facing a predicament not altogether different from the one which our ancestors, the polyhistoric system builders and other ("syncretic") academic philosophers equally unsuccessfully tried to evade. By disclaiming the connection with relevant empirical sciences, this branch of "philosophy" fossilized into a new sterile system of encapsulated problem and theory formulations.

A particularly interesting situation arose when *p.t.* philosophically interesting problems of language had not yet been tackled by the scientists within any ramiculated branch of existing science disciplines (e.g., linguistics). The philosopher, then had to choose between keeping his hands clean, at the cost of ignorance or relevant matters, or engaging in empirical research himself. It turned out that, confronted with this choice-situation, the so-called "ordinary" language oriented, analytical philosophers again made a fatal choice of opting for ignorance as the lesser of the two evils. This is roughly how some of the most inveterate ("analytic"), academic armchair philosophers—first in Cambridge and Oxford, then in Hong Kong, Australia, U.S.A., and Canada—came to turn away from philosophy and turn themselves into highly speculative, do-it-yourself, green-horn "scientists" of lexicography and of a hopelessly outdated linguistic dilettantism; a move which, ironically enough, was characterized as "*The Revolution in Philosophy*"!

No wonder then that already during the slow decomposition of this sorcerous semantic soothsaying often travestied under the guileful cover of "linguistic philosophy," the minds of many Continental European and Scandinavian philosophers were turned towards the works of some of the more consistent and sophisticated of the old syncretic philosophers and system-builders, such as, say, Spinoza. And more recently, there is even a quite influential trend towards speculatory romanticists, like Fichte, Schelling, the young Hegel, the young Karl Marx. . . . However, the whole explanation for this new trend lies not entirely in the fact that philosophers long have been fed up with the carping, captious philosophical haberdashery as perpetrated by a few slow-dying survivors of the old "conceptual analysts," and the late John Austin's and the later Wittgenstein's opiating echoists.

Still more significant, I trust, are some of the reasons that I, in what follows, shall endeavor to illuminate. They are probably most usefully viewed in connection with a rather prevailing current in contemporary philosophy of science and general research methodology. Vaguely and preliminarily intimated, this may well be seen as predominantly characterized (1) by a tendency to relinquish the hitherto unmitigated abandoning of any implicitly or tacitly implied, assumed, or presupposed, total world view, and (2) by the deliberate initiation of efforts that *might* lead to a proliferation of existing scientific theories, *inter alia* by replacing fundamental premises and trusted procedures by radical innovations. This, in turn, has been seen by some as an attempt at preparing *man* as a sender or reciever, as a producer or consumer of scientific research, for the third millennium. It appears rather important, so it is said, that man does not all of a sudden find himself there as a bewildered Mr. Square in Wonderland. *If* it is true that we cannot utter a sentence without inadvertently evincing an underlying world view, and *if* it is true that this total world view (of which we are, as it were, *more* or *less* ignorant) determines what we are bound to consider true/false, right/wrong, meaningful/meaningless, reasonable/unreasonable, rational/irrational, good/bad, desirable/undesirable . . . —*then* it does indeed seem obvious: *Man must seek to make his total world view explicit, articulate it so as to permit him to evaluate it, if possible, compare it to other total views, and thus make it psychologically possible to choose the one best suited to meet the unimaginable vicissitudes of the approaching, kaleidoscopically capricious millenium.*

I, for one, happen to hold that such an attempt is futile; it cannot be done in the sense of being, in fact, notationally impossible. Although I admit that I do not see how I can seriously make any pronouncement, any *nonsuperficial* assertion whatsoever without ultimately presupposing a total world view, and although I agree to the desirability of unshrouding it, I find all attempts at an explicitly articulated world view ridden with paradoxes and absurdities.

The objectives of the present chapter are bound to appear rather baroque. I shall first argue in favor of the necessity for attempting a total, uniform, consistent, and explicit world view. Next I shall argue that, given any conceptual system of which I can possibly conceive, this task is not only an onerous one, but in principle impossible. I shall furthermore attempt to expose the silly fatuity of what may be called the "survival-test approach" for determining which world is the real one. And finally I bring in arguments in defense of my general position. In point of fact, I shall argue that what *prima facie* appears to be an exercise in futility is actually a vitally important activity for anyone attempting fully to realize the predicament of modern metascientists.

1.3. "Scientific" World Views?

There is, as I have earlier argued, no reason a modern metascientist should ever tend to suppose that so-called "scientific results" would ever necessitate a change or even a modification of his personal philosophy or general world view. There is, I repeat, no such thing as a scientific world view, no established scientific knowledge in any sense of "knowledge" that can be found among those students of scientific enterprise with whom we shall be concerned in the present chapter.

I am not particularly confident that I shall succeed in convincing others of the tenability of the views I here propose. It should be stressed, however, that only very special aspects of the total scientific enterprise are to be considered, viz., those concerned with its capacity for surprise, for unlimited novelty and unlimited diversity of interpretation. Science offers a variety of other aspects, some of which may give the impression of inescapable, lawful development, necessity, inevitability, and all that sort of thing. It is the aim of history of science, at least according to one of the many *traditional* conceptions of historical research, to make us "understand" that science was *bound to* develop exactly the way it did, given social and whatever other forces may have been operating at any given time. But it so happens that this rigid, monastic, and totalitarian aspect, is *not* the one with which we shall here concern ourselves.

Perhaps in part as a result of the obvious influence of scientific research on the main development of warfare, one could discern in the postwar years a craving for a deeper understanding of science in its relation to general history and to modern industrial civilization. In particular, the intensive study of the history of science by professional historians or scientists with *genuine historical training*, brought forth a wealth of astonishing details and also of new points of view. A revolutionary movement was started, which still seems to be gathering momentum. With some simplification, one could speak of a *new integrated image* of science, stressing the *human* side of this human enterprise. *The sciences, then, I repeat are, according to this new integrated image, at least when articulated with care, not autonomous; they constitute the most exactly formulated and interpersonally testable parts of existing, or more often potential general views about man and the world.*

Efforts to isolate what might properly have been called "pure science" have failed; and I am confident they will continue to fail. Scientific revolutions, or more generally, any substantial, noteworthy change in scientific tradition makes, as I said, science, "—represented by the small elite of creative scientists,—shift loyalties within the area of general, near-total, world views."

History of philosophy, history of ideas, and contemporary philosophical movements furnish the most comprehensive examples and studies of such general views. Syncretic oriented philosophers, who have also been trained as scientists, are able to make broader frameworks of contemporary science more explicit and help to work out *general world views* with the ultimate aim of covering any possible, coherent, total view, irrespective of whether it happens to fit the views dominant among prominent scientists of the contemporary scene.

This new and *integrated view* of "hard" as well as "soft" (e.g., social and behavioral) sciences may, therefore, readily be seen to evoke a revival of the once long dreaded "metaphysical" concern with so-called "total views" offering a global framework and, hence *in-principle solutions* to all conceivable problems which are not taken to be mere questions of details. If the sciences are seen as attempts at establishing intersubjective and fairly precisely formulated general world views, the question I earlier mentioned seems inevitable: How can we ever get to know which of these total or near-total views is right, (the most) veridical, tenable, reliable, confidence inspiring . . . ? Or a gain, if you wish: which wolrd *is* the *real* one? And how on earth would we know for certain . . . ?

An initial, but rather fundamental obstacle seems embarrassingly obvious: In the sense of "world" which I shall strive to employ here, my "world" necessarily embraces *everything,* comprising *inter alia* the only conceptual framework from which I can possibly draw the standards for evaluation of anything. And this does not only include any appraisals of the veridicality of my total view, or the "reality" of my "world," but the assessment of the meaningfulness or absurdity of this very question. For an erstwhile very chic and à la mode choice of conceptual framework any questions as to the reality of our world were to be deemed a misuse of language, "*eine sinnlose Aneinanderreihung von Worten*" or words to that effect. *I* see at any rate no way for *me* to arrive at an assessment of *my* world without presupposing a frame of reference entailed *in* and *by* that world of mine. Were I, as I said, to employ a *different* frame of reference and apply it to my world, in what sense of "*my world*" would it then be my world to which it was applied?

Jimmy Gibson, who is quite undiscouragable in his endeavor to elevate me to his position, once supplied me with a reprint of his, with some additional arguments for adopting *realism* as a total system: The real world is the realist one! I did not have to look at the content of the article: How *could* there conceivably be anything recognizable as "reason (or arguments) for realism"? *Either*: The argumentation takes place *within* a realist world employing a realist frame of reference—in which case the reasons or arguments may be reasons or arguments solely to someone already well within a realist world. *Or*: The argumentation transcends the realist frame—in which case the validity of the argumentation is obviously dependent upon the acceptance of a *non-* or *a-* or *un-* or *sur-* realistic world with a corresponding conceptual frame.

Thus: If the explicit, cognitive integration of scientific knowledge is *the* most important dimension of knowledge perfectioning, it is also the one marred by the most spectacular paradoxes. It seems that as scientists as well as philosophers we must either choose a thoughtless, scatterbrained superficiality or, if we try to be thorough, complete and consistent, we end up in a soup à la Gibson.

1.4. The Survival-Test Approach

It has been said—and it is undoubtedly going to be said again: Man has been doing pretty well as a species with the world that he, unarticulated, and maybe unknowningly, has taken to be the real one. It has, you might say, stood up to the *survival test.* The history of philosophy knows quite a few biologically, genetically, or evolutionary oriented approaches to such questions—if questions are what they are—as what there is, what we can know, or what fundamental conceptual frame do we (choose to) have for our evaluations, etc. For example: Xenophanes, Plato, Democritus, Epicurus, Hume, Simmel, Nietzsche, Wittgenstein, Austin, Popper, Max Black, Don Campbell—just to mention a few.

Nietzsche, for instance, was astounded by what he saw as a general philosophical disregard for, or omission of, this approach: The philosophers have thought of Man (and his "Nature") and the basis for our evaluations as being changeless, immutable. "Lack of evolutionary sense is the philosopher's most fatal hereditary defect." They take their point of departure in the *presently* existing man and attempt to reach their goal: A veridical philosophy (value theory included), by means of conceptual analysis. They seem to have a vague notion of Man as an *aeterna veritas,* as something which remains the same amidst all the world's violent turmoil. The philosophers, Nietzsche continues, refuse to recognize that man's faculty of cognition—like all other human faculties—has developed from something inferior, and is bound to continue its explosionlike evolutionary process into something entirely different and, it often seems, according to Nietzsche, something vastly superior. "There is not a single sentence among what we today should look upon as adequate transmitters of our most important, surest and most indisputably, significant assertions, which will not, at another stage of insights perfection, become an object of ridicule and painful shame."

But philosophers are completely undisturbed by this evolutionary face, and regard the human cognition faculty as complete, immutable, and incorrigible. This is an attack on the whole incredible optimism that permits man to rely on the obviously untenable conjecture that the tracks of thought which mankind has haphazardly chanced to follow for half a dozen millennia or so, are bound to yield infallible insights into the human conditions, the world, and man's lot in it. Particularly ludicrous, incidentally, are the notions of a moral world order, offering "meaning," "purpose," etc. to LIFE in general, as well as to the life of every single individual. It is constantly forgotten that the tracks along which our thoughts are running were laid down long before man reached even the present evolutionary stage of his cognitive faculties. They represent the logic or, if you want, the *lack* of logic, of long bygone millennia. Philosophy must adopt a genetic, evolutionary orientation, not the least with regard to ontology and epistemology, the theory of knowledge[13] and philosophy of science.

The schism, the hitherto unbridgeable abyss, between philosophy on one side, and on the other exact, rigorous, painstakingly meticulous and methodical science, originated when philosophers first asked the question: "Which understanding of the world and of man's fate therein, is most likely to permit man to lead a life in happiness, contentment and give him quietude and a long-lasting peace of mind?" The philosophers, unknowingly or surreptitiously, confused truth and happiness. They surrendered the high ideals of intellectual honesty for undisturbed "*suification*." They chose a world view to satisfy their feelings rather than their intellect. Thus spake Nietzsche.

This is one of the main themes in Nietzsche's *Menschliches, Allzumenschilches; ein Buch für freie Geister* (1878), viz.: "Man's wishful thinking, our Pollyanna-like credulity when it comes to believing that what is useful and joyful in life, or rather, any basic *theoretical* position which permits man to sustain and to go on with his existence, will, *eo ipso*, turn out to be the most probable or tenable position." On the contrary, argues Nietzsche, the more attractive (in this respect) a world view ("*Weltanschauung*") off hand appears to us, the more likely it is that we have deceived ourselves, or surreptitiously left the narrow path of veracity—and hence, the more important it is that we are on our guard against such seductive temptations, and that we constantly reexamine and reappraise our position in light of any such unmasking self-exposures.

Similar ideas were advanced by Georg Simmel,[14] particularly with regard to what we can *know* of what there is. And his conclusion is, roughly, this: There is no *reality*, no *real world*, that man has to learn to *know* in order to adjust to it so he can survive. The model is misleading. It is the other way around: The only world man can possibly know is determined by the knowing capacities man has developed for his *survival*. It is not the awareness of or alertness to some objectively valid reality or truth that has permitted man to react rationally, effectively, purposefully. . . . Quite to the contrary: Man has come to call such ideas, representations, images . . . "valid," "true," "real," etc., which have happened in the long run to mediate life-promoting actions. There are no reasons to entertain any mysterious, or as Newton said, "occult," correspondence or isomorphy with some transcendental "*Real World*."

To slightly more contemporary philosophers such as the later Wittgenstein, the late Austin, and Max Black and others, ordinary, everyday English in its stock, standard employment, may be usefully thought of as crystallizations into linguistic rules or general modes of response to the *real world* (such as it is!) which our ancestors, through endless generations, have on the whole found advantageous to survival. This sets the limits of our language, limits which cannot be transgressed without generating nonsense. Thus the limits of language set the limits for what man can possibly hope to know.

It is saddeningly salient, however, that this fascinating line of thought rests upon rather arbitrary and unconvincing assumptions. For one obvious thing: It does not touch upon the problem of what it is that is so good about survival. Why multiply these mean allowances of breath that life is? Why perpetuate such drippingly sentimental soap opera, the *samsara* unmerry-go-round of life and death? Why prolong this wild, banal, grotesque, and loathsome carnival in the world's graveyard?

And might it not be, as Nietzsche and Schopenhauer undoubtedly would have argued, that such a silly obsession with survival is bound to have colored our cognitive spectacles? Perhaps Bergson is right: That only if man can divorce his consciousness from its action-directedness can he ever grasp "what there is." Man has been, as it were, too egoinvolved in the world—too anxious to get ahead. A real task-involvement in the world would presuppose a disinterested contemplation. This might well be biologically harmful or even fatal to man as a species. But in a dying glimpse man would behold the real world and know by himself that that is what there is and what can be known.

It follows, morevoer, from the Bergsonian position that a language with a conceptual frame, a vocabulary, syntax, semantics, and pragmatics fashioned to fit a *survival* project or other practical or quotidian objectives, is bound to fail when it comes to impractical, antibiological, transquotidian contemplations of what there is and what we can know. *A transquotidian enterprise requires a transquotidian language.* That a linguistic locution has, as Austin once put it: "stood up to the survival test," would be a strong argument *against* employing it in an essay concerned with articulating a view on which world is the real one and how we would know.

We could even make a much milder claim. We could choose to accept the survival preference, accept the desirability of planning for the perpetuation of the human species to the best of our abilities. The problem, however, remains: What guarantee do we have that a total view with its specific conceptual frame, its cognitive structure, and whatever, which has been retained *up to now* (possibly by the most natural of all natural selection processes), which has been "*put to the survival test,*" if you want—what guarantee do we have that we are not just perpetuating a cognitive dinosaur, doomed to extinction if it does not show a degree of flexibility which would permit, if necessary, a total abandonment of all the characteristics that had secured its survival *so far*? That a certain way of perceiving, conceiving, knowing, understanding, thinking, feeling, evaluating, or speaking has been biologically successful from time t_1 to t_n does *not* in itself represent a promise of successfulness at time t_{n+1}. Unless, of course, we have already presupposed a total view, a "world" where the future is bound to resemble the past. But the vicious circle would be painfully apparent if we were to argue that the future has *always* resembled the past—so far! "So far" is right. If the future is not bound to resemble the past—in the future, it seems that anyone anxious for biological survival had better be prepared for a complete and radical reevaluation of all assessments of how we can know anything.

Against this background it becomes again an obvious imperative that we concern ourselves with such seemingly self-destructing paradox-ridden questions as: Which world is the real one and how would we know?

By the way of a preliminary summary then:

(1) It may well be that from an *evolutionary* point of view, limitless flexibility is the important way to survival.

(2) Limitless flexibility may require a radical and complete metamorphosis of man.

(3) The attainment of a radical and complete metamorphosis may force man to reconsider *all* that he has so far taken for granted.

(4) A total reevaluation presupposes the *possibility* of a paradigm shift in man's

most fundamental parameters, his conceptual scheme, his widest possible frame of reference.

(5) Such a shift in a global frame of reference is virtually cognitively identical to a shift in total view, a shift of "worlds," as it were.

(6) Given all presently known, available parameters, conceptual schemes, frames of reference . . . , an attempt at viewing one's own total view, comparing it to other "worlds," let alone argue so as to strengthen or weaken one's conviction that a certain "world" is *the real one*—that is all nothing but a fatuous exercise in futility. Except that it is vitally important.

1.5. Additional Difficulties Concerning Global World Views in Science

In the history of philosophy the so called "syncretic" system-builders proposed, as we have seen, solutions to *all* problems which were not taken to be questions of detail. The conventional classifications of the problems comprised logic, ontology, epistemology, methodology, ethics, and aesthetics.

Suppose now a philosopher who devotes himself to disclaiming knowledge within such "higher-order" categories.

Applied to a set of fundamental ontological beliefs or basic premises, whether they are ture or false, tenable or untenable, professed ignorance involves a claim to know something that is true of the whole world, or of whatever is most fundamental within a certain area of it. The vastness of the pretension is of the same order, whether it is in the beilef in knowing or in the belief in ignorance. If I maintain it to be true that I do not know, for instance, whether the ultimate source of knowledge is experience, I must regard myself as already acquiescing in the use of a conceptual framework that includes such terms as "knowledge," "experience," and "ultimate source." If I know that I ignore the question of whether or not Aristotle's ontology is tenable, I must, *inter alia*, know what it is that I ignore.

If I claim not to know whether a definite comprehensive ontology—Spinoza's, say—or set of basic principles—Aristotle's, for instance—is tenable or untenable, my explicit or implicit frame of reference is itself total, comprising all there is in *the world*, whatever *that* is, encompassing even the widest horizons of great philosophers.

The effect of assuming one global framework rather than another makes itself felt, or is shown, over the whole area of reasoning or action covered by the categories in the framework. A conscious modification of a formerly implicit frame of reference may result in modifications not only of what we conceive to be true or false, but even of what we conceive to constitute a *search* for truth. This happens, for instance, when our attention is directed to *research itself as an object of research* and to forms of agnostic or sceptical conclusions as a particular instance of a possible result of research about research.

Suppose we develop a very strong form of epistemological scepticism, avoiding the classical inconsistency attributed to extreme scepticism by only a hair's breadth.

Once having developed this scepticism we might start to reflect upon our own developments and try to make explicit some of the implicit assumptions, rules, and premises which come to play in *it*. Thus we might succeed in grasping a fragment of the frame within which we were working as we developed our sceptical epistemology. Our explicit scepticism brings to light the "animal faith" we had in the conceptual structure and the fundamental assumptions underlying the arguments for our scepticism and the statements expressing it.

Thus we may grow to doubt our sceptical methodology once its principles have come to light. To what extent, then, is it possible for us successively to shun the enjoyment (or assumption) of truth with regard to total views of which we come to get a few glimpses? How far back can we consistently go towards "knowing nothing?"

It is usually assumed that a total viewpoint taken up in the neighborhood of "I know nothing" is at least in part due to the choice of an "I know" concept in the direction of "I have incorrigibly true knowledge." This is due to a frame of reference which allows for the possibility that once knowledge is corrigible, any belief backed by standard evidence in the field is automatically knowledge; that all we need is enough evidence. But at least some strong, sceptical attitudes are characterized, not by a localized doubt concerning certain mathematical or perceptual statements usually classed as *absolutely certain*, but by an all-pervading, diffuse feeling of questionableness relating to *all* conceptualizations, and indeed to any position whatsoever. Here the very act of saying or thinking that anything *is* so-and-so, rather than such-and-such, is not intimately accompanied by clear and articulate questioning, but is done in a mode of general questionableness. Reflections in this sceptical attitude may find *expressions*, but not directly as *assertions* or statements, propositions, etc.

Here, of course, an important point arises: If the sceptic does not *assert* or state anything, if he fails even to pose an articulate question (based on the acceptance of an alleged definite conceptual framework), then his scepticism hardly amounts to anything that philosophers can dispute. Perhaps they may be influenced by his behavior, but only in the way their moods and tempers may be affected by the weather. This may be conceded, but it should be borne in mind that important parts of philosophical literature are writings thrown into our discussion which, taken as *wholes*, may not be expressive of any definite position. Consider, for example, Plato's early dialogues and Kierkegaard's *Concluding Unscientific Postscript.* Within such writings we can recognize strings of reflections and limited conclusions. There are, however, trends of reflections obviously undermining the limited conclusions. Therefore no position or conclusion seems consistent with all reflections. As regards conclusions, they are self-destructive. Nothing is left. The intention of the author may not have been to impart an opinion on any subject at all except within subordinate *strings* of reflections.

There are certainly many advantages in thinking of philosophy as an articulated search for something, giving rise, at its best, to the establishment of conclusions concerning some sorts of clearly formulated questions. It might, on the other hand, not be too inappropriate to view it, taken as a whole, as an essentially aimless activity, its reflections occurring in strings, that now and then, and for certain stretches of time, condense into definite sets of opinions, but which mostly go beyond such products, releasing the thinker from his past work.

Naturally, if the published works of philosophers are taken to reveal the essence of philosophy, one inevitably gets the impression that a philosophical achievement consists mostly in arriving at conclusions in an orderly fashion. But to look at it in this way involves ruling out those philosophers who are capable of *epoché* or abstention from commitment to any position in spite of being clearly aware of the arguments for and against.

But in conclusion of this note about a general *epoché*, I would say that the articulate sceptic need *not* be thought of as choosing a set of strong requirements for "knowing," such as "incorrigibility." He may choose any among a variety of requirements, including weak ones. But even then "knowing" would be a kind of achievement, the end of a process of search. That knowledge has been achieved is something that must be believed before the unbeliever is to be refuted. And that is the important point.

1.6. *Does Explication of One Global Frame of Reference Always Involve the Introduction of Another?*

If we have an inclination to study the foundation rather than the superstructure, the presuppositions and assumptions rather than the conclusions of a system of beliefs, our interest is inevitably turned from the explicit to the implicit. If we study an argument and end up with a strongly sceptical conclusion in terms of its requirements, this indicates that our attention is turned towards those stringent requirements for "knowledge" which are implicitly assumed in arriving safely at the conclusion. Then our attention is apt to shift towards the question of how one is able to find out what concepts a sceptic implicitly uses. What are the methodological beliefs in this case? And once explicated, are we prepared to accept these formerly implicit assumptions? Quite possibly a shift to a wider or deeper frame *(n)* will make us sceptically uneasy about any former sceptical uneasiness—or perhaps just disinterested in what those (*n*-1) methodological beliefs led us to conclude.

The writers who, in a few sentences, prove that "scepticism" is incoherent, usually presume a very weak concept of knowledge, a concept such that we practically speaking are incapable of opening our mouths without claiming to know something.[15] Thus B. Lonegan writes

> Am I a knower? The answer, Yes, is coherent, for if I am a knower, I can know that fact. But the answer, No, is incoherent, for if I am not a knower, how could the question be raised and answered by me? No less, the hedging answer, I do not know, is incoherent. For if I know that I do not know, then I am a knower; and if I do not know that I do not know, then I should not answer.

Here, the answer "no" is construed as equivalent to "I know that I am not a knower"—and the answer "I do not know" as equivalent to "I know that I do not know." Such an equivalence would only hold, however, provided nothing or next to nothing is added in cognitive content by adding "I know that—" to a sentence. There is no good reason why a person inclined towards scepticism also should lean towards an unusually weak concept of "knowledge."

The shifting of attention from one frame, *A*, to another frame, *B*, in which *A* has been conceived, proceeds within a new implicit framework in which both *A* and *B* can be conceived and explicated. As we move back we step immediately and inevitably from one frame in a regressive series to another. In this predicament it is tempting to try to do without this talk of "frames." Can we not dispense with it and thus also with the predicament? Do we really *use* or *have* an implicit frame of reference? How did we manage to get this idea involved in our reflections? What presuppositions did we use here? This kind of reflection about our frame-of-reference-thinking is also a way of exceeding our own explicit assumptions. As an assumption it is itself a proper object for study. And since it is only an assumption about the nature of our methodological beliefs in general, it would seem to be dispensable even if they are not. Nevertheless, dropping this terminology does not seem to help us to answer the questions that gave rise to it.

It seems that we are caught in a trap, unable to free ourselves and make a "fresh start"; we can inspect all the beliefs we have had until the moment of critical inspection, but we never reach the critical inspection itself. It is tempting here to quote some relevant reflections of Sören Kierkegaard[16] concerning "The System," that is, in our context, any attempt to explicate a fundamental global frame of reference:

> The System begins with the immediate, and hence without any presuppositions, and hence absolutely; that is, the beginning of the System is the absolute beginning. This is quite correct, and has also been sufficiently admired. But before one started with the System, why is it that one did not raise the second, equally, aye, precisely equally important question: *How does the System begin with the immediate? That is to say, does it begin with it immediately?* The answer to this question must be an unconditional negative.

We have referred above to the *intended* global character of a framework explicated as a set of fundamental beliefs or assumptions. The unavoidable slip into higher-order frames and the resulting infinite regresses suggest that it is in principle impossible to formulate a set which has the intended character. The character of wholeness refuses to reveal itself in what we grasp and formulate in discursive thinking. The impossibility of formulating a set of fundamental principles that is global, suggested by the inspection of beliefs itself involving uninspected methodological principles, is analogous to that of blowing up a balloon from the inside. To inspect the set one must do something analogous to blowing new air in from outside. Then, to inspect that process you have to regard the outside air as itself enveloped by another layer, and so on. This illustrates the regressive character of explicating global frames of reference.

This reasoning and the conclusion itself can, however, be made the object of critical inspection. Pieces of its frame of reference can be brought to light, tacit assumptions can be made explicit. This activity, however, even if highly successful, does not necessarily furnish a basis for a rejection of that reasoning or its conclusions. It may be left "suspended in the air," because the rejection is itself only an example of the kind of things rejected.

But there is another aspect of the impossibility of settling on a fundamental set of principles, and that emerges not in the question of attempts at increasing our ignorance but, rather, in the closely related attempts to increase and multiply our doubts, with a view to establishing a platform, however small, of truths which no "sane," reasonable man could possibly dispute.

It will be seen that differences of procedure determine extreme differences in the resulting platforms and that differences in fundamental orientation or vision seem to determine the procedures.

Descartes *doubted*, using his maxim *de omnibus dubitandum est*. But he conceived himself, René, born 1496, etc., to be inspecting beliefs he, René, had cherished. This egocentric frame of reference was never touched by his doubt. There is nothing very strange in this situation as long as we take "all" *(omnibus)* as it is often used in everyday situations, as an expression of less than strict universality. If "all" is successively widened in scope, there will still be some frame of reference which is left untouched by doubt. For Descartes it was natural to persist in retaining himself, or his thinking, or his doubting, as something that breasts the waves of critical inspection. This fundamental orientation gave a frame which was inconsistent with a doubt that *he* was there all the time, *doubting*. Thus, as a consequence of his particular orientation, he grasped a first indubitable truth.

What I should like to stress here is the intimate relation between the conscious, explicit conviction that he, René, was doubting and therefore, of course, existing, and the conception of the task that he, René, should doubt. As long as René's conception of his task is not challenged, the result follows. All the conceptualizations, points of view, and beliefs inherent in the *egocentrically* conceived task and its implementation are then left untouched by his doubt. The consequences of the conceptualizations are experienced by René as absolute truths and therefore as a starting-point for system building.

If we now turn to Baruch de Spinoza who, presumably, as an admirer of Descartes in his youth, had the same idea of radical doubt and egocentric introspection, the situation is slightly altered. Baruch seems to have had a personality and background somewhat different from that of René's. Baruch is led to a critical inspection of concepts of "the ego," especially in its relation to "God." Baruch could not be Baruch and doubt without "God." And critical inspection of the task of doubting was quite possible within the God-centered frame. Philosophical mysticism was so deeply seated in Baruch that everyday pursuits were strange except within a definite framework; but not one in which "God" was sought from somewhere outside "God." Rather, the framework was one that comprised "God" and contained the possibility of explaining Spinoza himself and his capacity to doubt. His starting-point is *deocentric*, not ego*centric* like Descartes', and consequently such formulations as "I doubt therefore I exist" must have been foreign to him as a basic conceptualization of self-evident truth.

The highest kind of knowledge, the "third," is conceptualized by Baruch in terms of "eternal things" and it would be rash to identify the relation here of the subject knowing to the object known with the relation between a concrete person or an ego and an empirical or, more generally, natural object of some sort.

In some sense it is justifiable to say that the third kind of knowledge *can be* without *being held* by any person at all. "Baruch" cannot "have" the third kind of knowledge in any ordinary, naturalistic sense of having knowledge.

Thus, the task of doubting everything will have to undergo renewed conceptualization if it is to be understandable at all from the point of view of a philosophical mystic or a combined mystic and rationalist like Spinoza. Therefore, the assertion, "I, Baruch, born in 1632, doubt," or, more generally, an assertion of the kind, "I, So-and-so, doubt," is not found to be a constant, definite presupposition; rather it is modified and changed with the renewed conceptualization of the task. More specifically, it changes with changing conceptualizations of *the I, the ego, the self,* the personality and its relation to what is *not I, not the self,* etc. Thus, well within the framework of the third kind of knowledge, both the task to doubt all, and the result that it cannot be doubted that Baruch doubts, are absurd.

The easiest way *into* Spinozistic thinking may be through the epistemological section of *De intellectus emendatione*. These sections are written within a less regorous framework than the propositions, definitions, axioms, and proofs of the first part of the *Ethica*. Moving from the exoteric and didactic to the esoteric point of view, bridges or ladders are thrown away, and an exposition is retained which is colored by the system itself, from the first section to the last. As an example, take the frequent paralleled epistemological and ontological phrasings in the *Ethica* (*in se est et per se concipitur*, etc.). It seems that Spinoza takes it for granted from the start of the exposition in the *Ethica*, that if certain epistemological propositions are true then certain ontological ones are also true, because their object is *ultimately* the same.

The comparison between Descartes and Spinoza is carried out within the framework of a (roughly suggested) theory of *ultimate orientations*. I doubt the adequacy of this framework, however. When one goes deep into the philosophy of Descartes, it seems that one's acceptance of the theory of ultimate orientation is itself affected. It acquires a Cartesian quality and is no longer suitable as a neutral frame within which both systems, Descartes' and Spinoza's, can be understood. On the other hand, going deep into the philosophy of Spinoza, the same tends to happen, Descartes gradually getting absurd or incomprehensible. *My conclusion is, therefore, that the framework accommodates both the views only at a very superficial level, and that in fact the views are incomparable in so far as they are conceived as global.*

The hunt for any "natural" resting point is as unrealistic as to reach out for the horizon. Maybe what we have tried to do is to make explicit what, by its very nature, cannot be made so, and perhaps all the explicit frames of reference from the first to that of the *n*th order are on the same level in relation to a kind of personal, implicit, total view or global frame of reference. Maybe we "have" a kind of preconscious total view or frame—so total that all the higher-level reflections are in some way placed within it.

1.7. *Our Preconscious Total View: More Paradoxes of Total Views, or Global Conceptual Schemes in Science*

If the preconscious view is a kind of matrix within which all attempts to unravel or bring to light the concepts and categories of any implicit frame of reference are to proceed, it cannot be identified with a philosophy like "naturalism," "realism," naive or sophisticated, or any other intended systematizations of "common sense." These are explicit and verbalized and are defenseless when attacked by means of frame-of-reference dialectics. But is it then "a view?" The term suggests something that can be grasped and inspected. This in turn implies the possibility of making it explicit.

Conceptions of explicit total views, as found in the history of philosophy, are ridden with paradoxes. Either a global frame is explicit but fragmentary, or it is total, but implicit. An analogous conclusion can be reached concerning the ordinary use of the term "view." "Views" are *of something from somewhere.* This somewhere is not part of that something—so we cannot have a total view in this sense, comprising what is viewed and the viewpoint. Admittedly, it would be misleading in ordinary situations to call a view fragmentary because it did not include its own viewpoint, but then reflections on total views are themselves rather extraordinary.

It is perhaps only after studying attempts by philosophers and others to elaborate global or near-global systems that we are led to ask: What makes syncretic-oriented systematizers, with "totalitarian" aspirations, believe in the possibility of reaching their goal? And it is only after considering what seem to be the unavoidable paradoxes, or contradictions, that we are led to talk about and imagine a kind of view, or, rather, disposition, that we have *before* making philosophical enquiries—a "totalitarian" disposition which makes it appear reasonable and even important to elaborate an explicit total view. We may refer to such an initial view as "preconscious" in the sense that parts of it, perhaps any part whatever, can be made the object of our concentrated attention and will then appear to us as fresh, verbal expressions of something we had expressed already in indirect or nonverbal ways.

The study of the paradoxes inherent in conceptions of explicit total views makes it important to enquire whether any of the great philosophers really intended to elaborate such views. Aristotle's doctrine of absolute and final principles (for instance, the Principle of Contradiction), is an attempt to base all his thinking and the resulting all-embracing system upon an intuitively certain foundation. It testifies to his greatness that he knows he cannot *argue* in favor of his own first principles because this would imply the existence of a still deeper layer of principles than the "first principles." That is, it would involve a contradiction, first principles being secondary, ultimates being penultimates or antepenultimates. Recently neo-Thomists have tried to elaborate a total view consistent with basic Aristotelian views and modern scientific knowledge. But perhaps the most celebrated undertaking since medieval times in the field of all-embracing, supreme syntheses is that of Spinoza in his *Ethica.* This book, together with his *Treatise on the Improvement of the Understanding,* in which some of the methodology of the *Ethica* is stated, represents a system which plainly intends to answer all *main* questions. And this includes questions of framework.

The supreme systematizers have always omitted details, relegating the work to the formal and empirical sciences working within a fixed ultimate framework.

In the conception of the *Encyclopedia of Unified Science* (by Otto Neurath and others), the vision of a scientific total view and that of an encyclopedia are merged into one image. Adopting the "onion-structure," that is: starting with two volumes of basic fields and adding two sets of volumes, treating the same fields in greater and greater detail, the *Encyclopedia of Unified Science* was conceived as resulting in a total system. The difference between a total view and an ever so enormous encyclopedia, as here conceived, is that in the encyclopedia there must be details, whereas no single detail is necessary in order to make a total view complete. In a total view there is just enough said to cover the essentials or principles.

Bentham tried to formulate all-embracing general views, and also to elaborate some of them in the greatest detail. His total view would not have been less complete, however, if he had not, say, prescribed a definite color for ballot boxes. Whatever our views regarding the "depth" of particular philosophers, one may safely assume that most of the great philosophical systematizers intended to form explicit total views. The question of their genesis is therefore a very real one.

1.8. The Genesis of the Belief in the Possibility of Total Views

How might it be explained, from a psychological or social science point of view, that man started to talk—or started to believe he has been talking or can talk—intelligibly, even intelligently, about his total view—his logic, ontology, epistemology, value-system, etc.—in general? How is it that he has come to conceive of the possibility of regarding his own total view as one explicatable view among others?

Perhaps the belief has developed in this way: It may strike us that a person always thinks and believes in some definite way, that is, in only one of the many possible ways we can envisage, the range of possibilities being implicitly determined by our own (observer's) frame of reference. In order to articulate clearly and succinctly what distinguishes our own view from that of the other, we proceed to make explicit not only his view, but also our own. And we set about this latter job as though it were of the same kind as the first, that is, as the job of describing the other's view, something already accommodated well within our own framework. However, we ourselves are not accommodated within our own framework, and this makes the second job totally different from the first. To carry it through seems to me to be of the same order of difficulty as to eat not only part of oneself, but all of oneself. The analogy which generates the belief in the possibility of such an explication is spurious. A view which comprehends other views does so by pinning the various views it comprehends to something else, the accomodating viewpoint. But this latter can in turn be explicated only by being accommodated in the same way, by being pinned down onto something else, and so on.

Since the views about the other person's views are conceptualized well within one's own frame of reference, our own views as contrasted with his are also thus placed. Being primarily interested in contrasts, the part is taken for the whole. That is, one's own ultimate frame is overlooked.

The same goes for another possible way in which this belief is fostered, namely, when we look back to examine earlier phases of our own lives.

"My particular education has equipped my mind with such elaborate shock-absorbing buffers and elastic defenses that everything seen and heard became automatically transformed to fit a preconceived pattern" (Koestler).[17] From the belief in an all-embracing knowledge of one's own mind as it was at a previous stage of development, only one small but errant step is required in order to conceive oneself as having at the present moment, and knowing that, a definite general outlook which can be made verbally explicit as one outlook among others.

In psychology, as in other sciences of man, concepts of total view have been introduced, and it has apparently been taken for granted that they can be used in research to make neutral, adequate classifications of individuals or groups. Let me quote some words from a strong believer in total views:

> Our revised, more dynamic and concrete conception of an ideology may not be defined as the complete system or cognitive assumptions and affective identifications which manifest themselves in, or underlie, the thought, speech, aims, interests, ideals, ethical standards, actions—in short, in the behavior—of an individual human being.[18]

Walsby's belief in an underlying ideology has striking similarities with the belief in a god that manifests itself in all that happens in the world. It seems that there is a strong Hegelian trend in research about ideology, much influenced by the sweeping concepts of Karl Mannheim. In his *Ideology and Utopia*, Mannheim writes: "Here we refer to the ideology of an age or of a concrete historico social group, e.g., of a class, when we are concerned with the characteristics and composition of the total structure of the mind of this epoch or of this group."[19]

The ideology of a person observed by Walsby would have to be described and classified in relation to a framework that has to be so comprehensive as to *completely* embrace that of the observed person. The ideology of an age or of a historicosocial group observed by Mannheim must be transcended in every respect by that of the group or age of Mannheim. The "total structure of the mind" of Professor Mannheim himself must in the way of divine intellects furnish a frame of reference and conceptual structure of a most comprehensive or value-neutral kind.

Concepts such as these are potent factors in building up images of "the fascist," "the communist man," and of other creatures which cannot be reached by ordinary channels of communication because of their fundamentally different total views from that of the observer. One cannot argue with them, but they understand "the language of power," etc., etc.

Imre Hermann, Levy Brüehl, Karl Mannheim, Walsby, and many others do not seem to doubt for a moment that the fundamental beliefs and attitudes of others, for instance their logic, can be described and compared with each other—irrespective of how different they are from that of the investigator. In cases where the victims of a total description are so-called primitives, the observer is rarely confronted with a reversal of roles—the victim describing the total view of the scientist. But if the social scientist is faced with appreciative, either critical or indignant, verbally nonprimitive proponents of the systems, he is apt to become acutely aware of at least *some* of his own assumptions, and is led to talk—or believe he could talk—about his own general frame of reference. He may thus be led to believe that he has a total view (capable of being verbalized) during the process of rationalizing his implicit assumption, i.e., that he is able to discover and adequately describe the total views of others. What he more or less uncritically has imputed to others, he now feels compelled to impute to himself as well. He insists that he has a total view and is willing to verbalize it, using words such as "the world," "man," "society," "liberty," "progress," etc., etc. And as for "absurdity:"

> As scientists as well as philosophers, we must either choose a thoughtless, scatterbrained superficiality, or, if we try to be thorough, complete and consistent, we end up in an amorphous incongruity. In either case: There is no way to bar, in some common or common-sensical way the intrusion in explanatory systems of "absurdities." There exists no logical possibility of *ruling out* any notion or locution as "absurd," independent of a frame of reference as furnished by some articulated or unarticulated global conceptual scheme. This rules out any way of establishing cross-field dimensional interrelationships between objects of research, except on a mere terminological level in a quotidian vernacular, where the depth of semantic intention is approaching zero, *i.e.*, the amorphous "everyday speech." Consequently, were our struggle towards the greatest possible enhancement of research and metaresearch conditions to be crowned by success, the necessity would seem to arise for explicitly articulating consistent, global conceptual schemes.[20]

2. The Refutation of Final "Refutations": Theories' Survival of "Death"

2.1. *Historiography and "Refutation"*

As an indication of the direction in which recent historiography moves, let us discuss the problem of refutation and refutability.

Describing scientific developments, the historiographer cannot avoid interpreting events referred to by scientists in terms such as "theory *T* has now been *refuted*," "this law, *L*, must now be considered *established*," "my hypothesis, *H*, is now *verified*." Let us now look into questions raised in relation to "refutation." Various historical questions apply to all descriptions of "refutations" of theories. The question here is: exactly *which* theory was intended to be tested? Exactly what was the relation between the theory in question and the "fad," *x*, which, by the scientists or the historian or both, was said to have falsified, refuted, weakened (etc.) *the* theory?

As regards this "fact," *x*, we find reference not only to "observations," but to "data," "phenomena," "laws," and to other theories which were conceived as being verified and now were found to be inconsistent with that theory which was put to a test. The image of refutation as one of "*confronting theories with observation*" is clearly presupposing a belief in certain specific varieties of empirical philosophy.

The disagreements among scientists as to the exact status of "refutation" (or "nonrefutation") at any definite time makes it inappropriate for the historiographer to adopt the rule that *a theory is refuted merely because it is said to be refuted by some scientist or other*. This rule may be considered part of a comprehensive set of rules covering a vast number of happenings considered normal features within scientific development itself. The rules may be roughly formulated and summarized as follows: "When a presumably competent scientist says *x* has now taken place and *x* is a normal part of the scientific enterprise proper, then *x has* taken place and the historian's job is simply to report this in his historical account of the actuality of this happening." Considering the scope of this rule, it might be called "the mirror principle" or the principle of echolalia, i.e., historiography seen as echoes of some authoritative pronouncements. If two scientists disagree about what has actually taken place, the historian may permit himself to survey what is considered tenable among his contemporaries and, on the basis of that, decide that one of the scientists was to be considered more competent than the other. In any case the historian acts in this respect as a mere servant.

It is an important feature of the new historiography that the historian of science is considered to be just as free in assessment of truth value and importance of scientists' utterances as the historian of politics is in relation to utterances of politicians. The usual methodology of critical reading of texts or listening to tape recordings is mercilessly applied to utterances even of the most honored geniuses in science.

Applied to questions of refutation, this means that if the historian has documents referring to 10 scientists' opinions (within a certain time interval) about the status of refutation or nonrefutation of a theory, his primary job will be historically to understand and explain the utterances, but he may also offer his judgment as to the truth value, tenability, etc. of the utterances in the light of later development and in the light of a philosophy of science. In so doing he cannot avoid adopting a terminology and conceptual framework that differs from one or even from all the 10 scientists studied, and even is, at least in part, considered controversial among his own contemporaries. Acting in this way, the historian of science does no more than assert himself as an historian belonging to dominant contemporary schools of historical research and writing.

2.2. *Refutation Conceived as Notational Falsification*

Let us inspect various views about happenings referred to as "refutations." Karl Popper seems to tend to accept the existence of catastrophies of such a kind that a theory, because of that kind of happening, is *conclusively and permanently* refuted, and that it therefore is, and must be, unconditionally discarded. Popper does not recognize instances in which a "refuted" theory crops up again and is accepted for a second time.[21] "There is no survival after death."

The strength of belief in the decisiveness of refutation by observation is perhaps due to the brevity of scientists' existence or, say, to the dominance of the logical point of view in "science of science" and a conception of refutations which presents them as instances of the formula

$$(H_1 \,\&\, S \cdot \supset \cdot O) \,\&\, O \qquad \text{confirmation}$$
$$(H_1 \,\&\, S \supset \cdot O) \,\&\, \neg O \qquad \text{disconfirmation}$$

Thus in the logic of propositions: "if a theory *p* implies a possible observation *q*, and non-*q* is observed, the theory is false."

Popper mentions with approval the Einsteinian theory of gravitation as a specimen of a refutable (and therefore scientific) theory. The total solar eclipse on March 29, 1919 could have yielded observations that might have *refuted* the theory. Einstein predicted that the position of certain stars would be different from their usual position if those stars were seen very close to the sun. " ... the impressive thing about this case is the risk involved in a prediction of this kind. If observation shows that the predicted effect is definitely absent, then the theory is simply refuted. The theory is *incompatible with certain possible results of observation*. . . ."[22]

Is there any historical justification for a belief that theories—apart from some extreme and exceptional cases—have been conclusively refuted in this way? If the photos taken on March 29, 1919 had been *interpreted* to show the stars in their usual position, the theory would scarcely have been considered conclusively refuted. But even if taken to be refuted, it would almost certainly not have been discarded on grounds of one single disconfirmatory *instance*.

Here, also, a methodological and normative or optative question is relevant: *Ought* the theory to have been discarded if observation had "shown" that the predicted effect did not occur? And also a metaquestion: To what extent does it make sense to have a methodology of science demanding that scientists *should* do so and so, quite contrary to their own opinions?

This history of wave and particle theories shows what effect disconfirmations may have on the status of *competing* theories: one theory is taken to account "better" for certain subclasses of relevant phenomena than certain others. Both are considered to have their strong and weak sides. There is dissatisfaction with one or both. But it would be misleading to say they are "considered refuted" as long as scientists still use them and trust them within certain limits. The same kind of competitive situation—on an even greater scale—was realized during the long competitive history of Ptolemaian and Copernican theories. Both faced a tremendous *number* of disconfirming instances, and even a considerable number of *kinds* of disconfirmations.[23] But were they ever *refuted* by observation? (Parallaxes first established 1838 and 1839.[24])

Speaking from the point of view of faithful, but critical, reading of historical sources, there are no sets of sufficient and necessary conditions (of the kind envisaged in modern methodology of science) under which a scientist has used the phrase "*this* refutes the theory" (or similar phrases). There have been situations in which we—as historians today—feel inclined to say that "the observations certainly were completely damaging to the theory"; but we shall have to add: "nevertheless no scientist conceived of them as refuting." And there have been situations such that we might plead: "How could scientists possibly have called *this* a refutation?"

Any adoption, on the part of historians, of a definite concept of refutation, for instance by listing sets of necessary and sufficient conditions of "refutation by observation," will surely go against some occurrences of the terms as used by eminent scientists. And, if adopted, it will represent a methodological *stand* that is controversial today and perhaps unknown as recently as, say, 100 years ago.

By way of summary, then, it is an open question what we *as historians* might mean by saying that a theory was or was not *refuted* by a certain set of *observations*. I shall not attempt to pose as a historian, but rather as a philosopher, and thereby feel free to take up a terminology that suits my ultimate positions. It shall, hopefully, soon be fairly clear as to which options I shall choose.

2.3. The End to Terminal Refutations

The soundness of *ad hoc* hypotheses invented to save a theory clearly depends in part upon what is *subsequently* done. Galileo applies *ad hoc* hypotheses to Copernicus, but, in the words of Feyerabend, Galileo subjected Copernicus' hypotheses "to a relentless examination that produces new and apparently independent evidence." In two other cases, discussed by Feyerabend, nothing like this is done. The talk about the observer that preserves the second law of thermodynamics "does not increase our knowledge, it just describes the existence of a clash between theory and fact in a manner that makes the clash invisible."[24] Are we safe in assuming then that all cases of continued acceptance in face of series of disconfirming observational instances conform to the description by Popper of the "conventionalist strategem,"[25] or is it rather an "evasion of falsification?"[26]

To introduce *ad hoc* auxiliary hypotheses or *ad hoc* assumptions to change a definition seem for Popper to be techniques by which the scientist "evades falsification."[27] But if such introduction is not peculiar to situations of disconfirmation, but rather, as it turns out, pervasive and fundamental in scientific practice, then the falsification itself would seem to be based on *previous* "evasions" of some other kinds.

If we are to believe the new historiographers of science, *ad hoc* hypotheses have been introduced in great numbers, and definitions are perpetually changed—without loss to what seems to be scientific progress. New theories have little chance of survival without large doses of such "protections." Sometimes, or even normally, they cannot even be stated without *ad hoc* assumptions. Hence, in one concluding sentence: *Expanding research requires an indefinite number of auxiliary hypotheses, some purely ad hoc.*

They are "purely" *ad hoc* in the sense that at the time of introduction, the scientist has no idea as to exactly how to supply independent evidence for the hypotheses. He may not even have any clear intention to look for such evidence. And the hypotheses may well lack all mitigating properties, such as connecting the hypotheses in interesting ways with other hypotheses.

This liberality or charity is understandable, considering the overwhelming number of auxiliary hypotheses, untested assumptions, and the doubtful postulates that are necessary in order to ensure a relation of strict derivability between more or less abstract sentences of a physical theory and a concrete prediction that clearly and unambiguously can be tested by an observation (or a finite set of observations) in the technical psychophysical sense of observation. The liberality is thus necessarily required even from the statics point of view of theorizing, viz., the clarification of what is already won by (theoretical) science.

If what is derived and predicted does not seem to actually come about, this must of course be duly noted and remembered. But why discard the whole theory if there are not alternative theories at hand which are *in all important respects* superior?

2.4. *The Importance of Presupposing*

In the face of an instance of disconfirmation of a theory there is little reason to avoid substituting some of the untested auxiliary propositions with some new ones, such that the false prediction either is classed as irrelevant or even in accordance with the theory. This does not imply that we try "to save the lives of untenable systems."[28] *Falsity does not imply untenability*. Tenable is what is worth holding, keeping. "Untenable" is what cannot be held on to, given a certain set of scientific norms and norms of intellectual honesty. But historically considered, theories of some importance were always, or mostly, in conflict with at least some observations; there were always *prima facie* disconfirming instances. Conflicts are treated in a "spirit of compromise." This is due both to the complexity of relations between theory and observation and to the limitation of the human existence, intellect, and imagination. Few are capable of modifying old theories in a fruitful way. And still fewer are capable of finding suitable new ones. It is easy to throw a theory overboard, but what do we do then? "What next?" is the question of expanding, creative, active research.

An example of a fairly recent, successful *ad hoc* change in assumptions in order to save a hypothesis is offered by the history of the neutrino. By postulating an almost massless and chargeless particle, the existence of which to physicists "seemed outstandingly impossible to verify," an important hypothesis about the conservation of momentum and energy could be saved. So called "data" or "facts" which were considered strongly disconfirmatory could be seen as miraculously changed into wonderfully confirmatory instances by the change in hypotheses "elsewhere." Later on, physicists were able to give independent, empirical confirmation of the existence of such a particle, thus "all turned out happily in the end."[29]

In contrast to this example, one might study the defense of the modified phlogiston theory after 1785. Conant concludes[30] his study of this "rearguard action" as he calls it, with these words:

> By the end of the century Priestley was almost alone in defending the doctrine of phlogiston. The story of the last days of the phlogiston theory is of interest, however, in illustrating a recurrent pattern in the history of science. It is often possible by adding a number of new special auxiliary postulates to a conceptual scheme to save the theory—at least temporarily. Sometimes, so modified, the conceptual scheme has a long life and is very fruitful; sometimes, as in the case of the phlogiston theory after 1785, so many new assumptions had to be added year by year, that the structure collapsed. [*Vide*: Ptolemaic cosmology!]

The moral here is this: Heuristically, there is no definite limit to what can be done with regard to hypotheses or theories in order to save what for certain purposes is thought of as important to save. The only thing is that it is wise to be clear about the price one must pay—and the dichotomy "accept as true versus reject as false" is rarely relevant. There were reasons *to work with* the neutrino hypotheses—or the more recently suggested "tachyons"—whether one "believed" in their existence or not. In expanding, active, open research there is little place for resting in *belief* or *disbelief*. The question is rather: What appears to be the best research strategy right *now*? "Saving *x*" most often means something like "finding out how to continue to work with *x*," whereas "accepting *x*" usually means "working with *x*." It is a question of "accepting as *tenable*," not of "accepting as *true*." These moralizing comments are designed to reveal some of my own bias: the perspective of expanding, active, open research; not, for example, that of scientific knowledge. I am, at this point, not interested in knowledge, but in research.

Confronted with an observational or any other kind of instance of disconfirmation, scientists sometimes resort to an auxiliary assumption, postulate, or presupposition of a kind that is untestable, or is likely to remain untested because of the difficulties involved in carrying out a decisive test. When *ad hoc* propositions are intended to "save" a theory in this way, and the theory is treated as if no cases of disconfirmation had ever been noted, we have cases of methodologically objectionable *ad hoc* assumptions. If "*ad hoc* hypotheses" are to retain a derogatory use, and that is, I suppose, still likely, some such cases have to be adopted as paradigmatic.

2.5. *History of Science and "Refutation"*

All that observations can furnish, however numerous and diversified, are *instances* of disconfirmation or *instances* of confirmation, or corroboration or strengthening support. To each instance there are normally attached specific initial conditions or, more generally, relevant observational traits of the test situation. Each instance necessitates careful evaluations. They are rarely reported even in detailed observational reports, because properly trained researchers—according to the standards within a certain community at a certain epoch—are assumed to make reasonable evaluations. But *their* significance for the conclusions of a certain scientist is on par with premises explicitly mentioned by him. In the light of this I prefer in the following to speak of disconfirmatory "instances of a theory" and of "confirmatory instances" rather than of "disconfirmation (confirmation) of a theory." Among other things, this way of speaking may hamper the common inclination to identify disconfirmation with falsification or refutation. There are scarcely *instances* of falsification or refutation of a theory. A so-called "instance" of falsification would probably make it a false theory, and the "instance" of refutation would likewise refute it. There is no place for, let us say, three falsifications (or refutations) and five nonfalsifications (or nonrefutations). But there may well be three disconfirmatory and five confirmatory instances. The "evidence" may be in part negative, in part positive.

The conclusion is *not* as follows: "Theories are never falsified or refuted"; but rather it is more like this: "Theories which have long, dramatic histories may be said at any given time to have a certain—possibly low—status in relation to efforts of verification, refutation, confirmation, etc." Viewed in this, the wide perspective of the historian of science, the changes of status inevitably are much less conspicuous than it appears to the practicing scientist. His judgments are far from *sub specie aeternitatis,* but part of his particular research strategy. What, to the scientist who is personally involved, seems to be a perfect instance of an *experimentum crucis* is to the historian only one interesting test in a long series. The result, if negative, may tend to lower the status of the theory within the scientific community for possibly as long a duration as a generation or two; but a revival is never out of the question. What is *decisive* when changing a research strategy in a given situation is, by the scientist, often *spoken of as if* representing a clear proof or disproof, "verification" or "falsification" in a strict cognitive sense of these terms. It is justifiable for a modern historian to apply a more "voluntaristic" terminology, perhaps one of which the scientist would strongly disapprove.

Considering situations of refutation, the perspective of individual scientists may, however, show great variation. The modern historian does, therefore, try to reconstruct *all* that "was in the air" at the moment—whether expressed in documents or not. Within the relevant scientific community—usually a fairly small group—there may be many things in the air, such as: a more or less generally accepted theory formulated in a fairly definite way at a time, *t*; a report, maybe somewhat differently interpreted, of a case of observational disconfirmation: several possible modifications of the theory which seem to change the chains of derivations such as to prevent the most unfortunate predictions from being made, and several possible modifications of auxiliary hypotheses or assumptions "disconnecting" the (unaltered) theory with the unfavorable observational report etc. Whatever the momentary decision of each member of the group, the many factors will *continue* to, as it were, "be in the air." Decisions are easily revoked—a long sequence of things may take place before anything is published and matters are stabilized. Sometimes all quiets down. The disconfirmations have been duly recorded and will hopefully be remembered, but little can be done about it. Scientists just continue applying and working with the theory since there is no plausible alternative at the time.

2.6. *Degrees of Veridicality or Falsidicality*

In the light of the above reflections one ought to distinguish as clearly as necessary between (a) "discovering cases or instances of disconfirmation or negative outcome of predictions" and (b) "refutation by observation." The latter expression is well adapted when assessing the status of simple generalizations expressed in everyday vocabulary. "All ravens are black. No, there on the rock right in front of me is a white one. Conclusion: refuted by observation!" Questions of how to assess the status of tenability of a *theory* involve questions of how to judge instances of reported disconfirmations, but only as a part. The assessment *might* (*within* a definite general tradition) be summed up in a sort of index. The United Nations and other institutions use indexes of overpopulation, standard of housing, eating etc., listing a great number of factors which are given more or less arbitrary weight. The index would certainly contain an assessment of what Popper calls the *severity* of the test. If the theory states something about matter in complete generality (e.g., waves associated with matter in any form whatsoever suffer refraction due to changes in the velocity of propagation of light), and the observation relates to a very strange, peculiar, newly discovered state of matter, the test is to be classed as "very severe." "How could it be expected that any theory at that moment could stand up to such a test?" it would be asked, if the theory failed.

Considering the complexity of any adequate report, however concise, assessing the status of tenability (acceptability) of a theory, terms such as "refutation" and "falsification" can best serve as rough indication of prevalent attitudes of the scientist involved. That is, these terms are shorthand for "considered refuted," "taken to be falsified," etc.

3. By Way of Summary

Two theories considered in fierce disagreement with each other, or so fundamentally different as to be conceptually incomparable, may nevertheless be "completely confirmed" in the (dubious) sense that all "established" laws within their intended field of valid reference can be derived from them.

A distinction is here relevant, viz., between (a) empirical and (b) theoretical laws. Let us tentatively adopt a terminology as suggested by Carnap: The empirical laws contain only terms referring to observables, theoretical laws contain at least one term referring to an unobservable. Boyle's laws are empirical, the corresponding laws in molecular physics (describing microevents) are theoretical.[31] We may then say: theoretical laws are at the complete mercy of theories in so far as the terms used in their formulations are given physical meaning through the theories and the theories are expressly made or constructed so as to make those laws derivable. Or, we may adopt a somewhat different distinction: An empirical law is a law the terms of which are neutral in relation to the competing theories of that epoch, whereas a theoretical law is one that contains at least one term the meaning or import of which changes with change in theories. Kirchoff's law may serve as illustration. It may be thus formulated: "The ratio between the powers of emission and the powers of absorption for rays of the same wavelength is constant for all bodies at the same temperature." With shifts in theories concerning rays, the key terms "emission," "absorption," and "temperature" shift their meaning or at least their theoretical import. The cognitive content of the law therefore changes with changing theories. Kirchoff's law is best conceived as theoretical.[32]

The agreement of a theory with established laws is mainly or wholly (according to Duhem) a question of derivability. This means that it depends mainly on inventions in mathematics and logic. Progress in physics can therefore not be expected to come from disagreement between a dominant theory and a physical law. New laws, or heavy disconfirmations of some of the old, may necessitate a modification of the theory, but scarcely a change of theoretical ideas. The basic *conceptual framework* of theory cannot be hit since both confirmations *and disconfirmations* are *in terms of* that framework. With appropriate modifications of the theory, disconfirmatory instances are changed into confirmatory or irrelevant data—retaining the conceptual framework. If, therefore, two theories are mutually inconsistent or conceptually incompatible or incommensurable, they may nevertheless both be confirmed or both disconfirmed by the same theoretical laws.

Considering empirical laws, the same conclusion holds true, but necessitates perhaps a few more words. "Copper expands when heated" has three key terms used for observables of daily life, and none that are specific of any competing theories. Two mutually inconsistent theories may both permit the derivation of "Copper expands when heated," but from "rules of correspondence" or "*Zuordnungsdefinitionen*" (Reichenbach) which connect the observables with distinctly different theoretical, conceptual frameworks. Because of the one-many relations between the terms of the empirical laws and the theoretical terms, the confirmation of one theory does not imply a disconfirmation of theories contradicting the first one or theories with an incommensurable conceptual framework.

Suppose that in the time interval $(t_2 - t_1)$ a number of natural laws are said to be established and that an accepted theory T_1 (at time t_2) is found to be inconsistent with one or more of the laws. Some would take T_1 to be refuted, others not. The room for genuine disagreements is seen to be considerable since, (a) it is not to be found out *a priori* whether T_1 needs to be in disharmony with any single observation, and, (b) even if in disagreement, it is not specified whether there are other theories, T_2, T_3, . . ., T_{n-1}, which could take over the functions of T_1 in a better way. Taking refutation to justify abandonment, both conditions must apply in order to class the theory as "refuted."

Let us use a simple illustration from the calculus of propositions: That q implies r and p is equivalent to non-q, does not exclude that p implies r. The conjunction

$$p \supset r \;.\&.\; (p \;\&\; q) \;.\&.\; p \supset r$$

has the relative strength $s = .500$, that is, the strength of a conjunction of two variables out of three, not the relative strength, $s = 1.00$, of a contradiction. From the laws—corresponding to a part of q, the rest being so-called "*Randbedingung*"-statements—the observations (in our illustration: r) are derivable; *but this does not exclude that they also are derivable from the theory* T_1 (in the illustration: p) in spite of the inconsistency of T_1 with the laws (in our illustration: the equivalence of p with non-q).

Thus, any reported refutation of a theory by natural laws, or any other theoretical statement "established" by observation, is relative to the outcome of attempts to link the theory to observations *through other channels than the refuting laws.* Now, to relate theories to observations is a crucial task for researchers, and imperfect solutions mean imperfect decidability of the weight of the "refutation."

In the light of these considerations, it might be found to be rather surprising that *new* theories often are in conspicuous conflict with some well-established laws.[33] But if it was expected from a new theory that, even in its first version, it should not contradict *any* established law, there would be so few proposals for new theories as to practically stop all theoretical developments of the sciences. Furthermore, an established law can be drastically modified without losing its admirable relation to observations. Such modification may subsequently bring the law into harmony with the theory.

This may justify a short discussion of a point made by Duhem. He concludes his treatment of testing theories with the claim that the "only experimental check on a physical theory which is not illogical consists in comparing the *entire system of the physical theory* with the *whole group of experimental laws,* and in judging whether the latter is represented by the former in a satisfactory manner."[34] This might be true if we substitute "alogical" for "illogical." If laws already formulated in terms of a definite theory are badly represented, badly "accounted for" or badly "covered" by the theory, then it certainly argues against the theory; but of more decisive importance are comparisons of competing theories in relation to carefully selected experiments. The comparison is alogical in the sense that the conceptual framework by means of which the experiment gets its theoretical import is different for the different theories. When concepts are in the way different and mutually intranslatable in practice, the comparison cannot be "logical" insofar as the theories seem to speak of different things. They are not confirmed with the same experiment as judged from the sentences expressing its theoretical import.

The way Duhem is putting things, the work with theories is narrowed down in an unnatural way: To design and test a theory gets to be a work in logic and mathematics. What he calls an "experimental" check is very little if at all different from a logicmathematical check: The relation between law and theory is one of derivability. If testing a theory is testing relation it means a test of derivability. Comparison of competing theories, and assessment of tests applied to both, necessitate an *assessment* of the relation of the theories *to observations,* not only to laws. A theory may be victorious because of its relation to certain crucially important observations—in spite of conflicts with many so-called "established laws."

Rutherford once said retrospectively: "I was perfectly aware when I put forward the theory of the nuclear atom that according to classical theory the electron ought to fall into the nucleus. . . ."[35] The conflict was not only with the only competing theory then at hand, but with laws derivable from the theory and based on an immense number of confirmatory instances. Nevertheless, Rutherford's theory did not conflict with any observations! Nobody had observed whether electrons tend to fall into the nucleus. If laws, as proposed by Mach should be what he calls "*denkökonomische*" condensations of observational data, conflict with laws would be very much more serious than it is. But the relation between the number of observationally established instances of a law and the total number of instances covered by it is as that between a finite and an infinite number.

By way of summary of the summary: our main concern has not been to decide for or against particular conceptions of refutation and refutability, but to point to the room for a *diversity of conceptions*. The procurement of this room is securely effected by systematic development of the *differences of descriptions by scientists and by historians of cases of refutation*, of tests, of concrete instances of confirmation and disconfirmation, and of the "frontier of research."

The construction of global, or rather *near*-global systems or syntheses is possible (with certain restrictions). This involves taking a stand with regard to questions on the foundations of science. A system sufficiently global to render a "scientific fact" rational can legitimately claim truth and intersubjective validity in its own "system-laden" language. All consistent, comprehensive views can claim an equal validity status greater than zero. The expression "validity" is used such that the first links of the argument chains within a system are provisionally taken to be true or correct (acceptable), until a sufficient number of embarrassing inconsistencies appear within the system. They may then be modified. A near-global system does not necessarily have to confine its claim to a *relative* truth for the system (i.e., truth only in relation to its own premises or postulates). Two or more mutually incompatible systems can both, or all, claim to "correspond with reality." The conception of the relation of near-total systems to reality does not *need* to be the particular kind which rules out the possibility of there being an *in*definite number of mutually incompatible systems which all account "equally well" for the same reality or part of such reality. The rules for the use of "reality" make the notion of "reality" a kind of *regulative* or *directive* idea. It is neither a word nor a notion which stands for any definite or indefinite substructure.

The deplorable state of affairs that "facts" (produced by scientific enterprise) are acognitive, arational, etc., if not determinable within systems (conceptual frameworks, philosophical systems, cognitive grits, etc.) is here presupposed. The claim is shown to be untenable, that some solution or salvation may be found in an adaptation of a language which Stephen Körner employs when communicating with his lab assistant, or the late Otto Neurath when prattling with his cabby . . ., and assumed to furnish a common ground for philosophers as well as (Kuhn–Feyerabend-oriented) scientists. The choice is again between settling for unambitious, superficial chatter or precisely the quotidian language so as to transgress the boundaries into available systems. This chapter is a polemic against linguistic restrictionism and a plea for linguistic transgressionalism—within science and philosophy.

In other words: Were our struggle towards the greatest possible enhancement of research and metaresearch conditions to be crowned by success, the necessity would seem to arise for explicitly articulating consistant, total world views. Secondly, the world views ought, ideally, to be *comparable*, viz., in order to establish which world is likely to provide the most fertile soil for future research. According to the present probe into problems in this connection, neither task seems to make any sense at all. But if this, then, is something of which it is worth being aware, then that would suffice to vindicate any ever-so-futile-looking ventures into the question as to which world is the real one, and how on earth we would know. On the other hand: This enquiry has undoubtably its own unclarified, uncodified, assumptions and whatever. Its result, one might well suspect, can only be established as valid within *one* of an undetermined multitude of field dimensions, and are hence *at the most* tenable within one specific world, to wit: mine. That is: If I can comprehensibly be said to "exist *in*," or "*have*" if you want, one specific world. Which is in itself a highly dubious and curious proposition.[36]

References and Notes

1. Further on this paradox *vide*, e.g., H. Tennessen, "Metaphysics and the Beguilement of Everyday Speech," *Proceedings of the XVth World Congress of Philosophy*, Sofia, Bulgaria (1975), pp. 597–601; "Quandaries of Quotidianism," *Methodol. Sci.* **12**(2), 81–93 (1979); "The Very Idea of a Humanistic Psychology," pp. 63–108, *Humanistic Psychology. Concepts and Criticism*, L. Mos and J. R. Royce, editors (Plenum Press, New York, 1980).
2. H. Tennessen, "Science of History and Notions of Personality," in W. Yourgrau, editor, *Physics, Logic, and History* (Plenum Press, New York, 1970); A. Naess, *The Pluralist and Possibilist Aspect of the Scientific Enterprise* (George Allen & Unwin, Ltd., London, 1972).
3. Reference 1.
4. Reference 1.
5. H. Tennessen, "On What There Was," in R. E. Olson and A. M. Paul, editors, *Contemporary Philosophy in Scandinavia* (Johns Hopkins Press, Baltimore, 1972).
6. H. Tennessen, "Exit Absurdity, Enter Conceptual Schemes," *Methodol. Sci.* **11**(2), 114–123 (1978).
7. Reference 6.
8. Consider, for example, the writings of: J. Agassi, "Toward a Historiography of Science," *History and Theory*, Beiheft 2 (Mouton, S'Gravenhage, 1963); "Corroboration versus Induction," *Brit. J. Phil. Sci.* 9, 311–317 (1958–1959); "Empiricism and Inductivism," *Phil. Stud.* **14**, 85–86 (1963); and "The Mystery of the Ravens," *Phil. Sci.* **33**, 394–402 (1966); Paul Feyerabend, "Problems of Microphysics," in R. G. Colodny, editor, *Frontiers of Science and Philosophy*, Vol. 1, *University of Pittsburgh Series in the Philosophy of Science* (University of Pittsburgh Press, Pittsburgh, 1962); "How to be a Good Empiricist—A Plea for Tolerance in Matters Epistemological," in B. Baumrin, editor, *Philosophy of Science*, Vol. 2, The Delaware Seminar (1962–1963) New York, London, Sydney: (Interscience Publishers, New York, 1963); and "Problems of Empiricism," in R. G. Colodny, editor, *Beyond the Edge of Certainty*, Vol. 2, *University of Pittsburgh Series in the Philosophy of Science* (Prentice Hall, Englewood Cliffs, New Jersey, 1965); Ingemund Gullvag: *Absolutisme og relativisme i erkjennelsesteorien*, Filosofiske Problemer nr. 20 (Universitersforlaget, Oslo, 1955); *Knowledge, Truth, and Reality*, Vols. I–IV (Institute for Social Research, Oslo, 1961); *Truth, Belief, and Certainty* (Det Kgl. Norske Videnskabers Selskabs Skrifter, Trondheim, 1964); "Scepticism and Absurdity," *Inquiry* 7(2), 151–163 (1964); and *Referanse, Mening og Eksistens*, Filosofiske Problemer nr. 34 (Universitetsforlaget, Oslo, 1967); Jürgen Habermas, "Knowledge and Interest," *Inquiry* 9(4), 285–300 (1966); *Theorie und Praxis* (Luchterland Verlag, Berlin, 1963 and 1967); "Toward a Theory of Communicative Com-

petence," *Inquiry* **13**(4), 360–375 (1970); and *Technik und Wissenschaft als "Ideologie"* (Suhrkamp Verlag, Frankfurt am Main, 1970); W. Kneale, *Probability and Induction* (Clarendon Press, Oxford, 1949); and "Probability and Induction," *Mind* 60, 310–317 (1951); Stephen Körner, *Conceptual Thinking* (Dover Publications, New York, 1959); Tom Kuhn, "The Structure of Scientific Revolutions," *International Encyclopedia of Unified Science, Foundations of the Unity of Science*, Vol. II, No. 2 (The University of Chicago Press, Chicago, 1962); H. E. Kyburg, Jr., "The Justification of Induction," *J. Phil.* **53**, 392–400 (1956); and *Probability and Inductive Logic* (The Macmillan Company, Collier-Macmillan, Ltd., London, 1970); Imre Lakatos, "Changes in the Problem of Inductive Logic." in I. Lakatos, editor, *The Problem of Inductive Logic* (North-Holland Publishing Co., Amsterdam, 1968); and "Criticism and the Methodology of Scientific Research Programmes," paper read before a meeting of the Aristotelian Society, London, October 1968; Arne Naess, "Physics and the Variety of World Pictures," in P. Weingarten, editor, *Grundfragen der Wissenschaften und ihre Wurzeln in der Metaphysik* (Julius Springer Verlag, Berlin, 1967); *Scepticism* (Routledge & Kegan Paul, London; Humanities Press, New York, 1968—Chapter IV in particular); "Language of Creative Research and Language of Science: A Contrast," *Estratto de Linguaggi nella societa e nella technica, Edizioni di commita-Milano* (1970); *Vittenskapsfilosofi* (Universitetsforlaget, Oslo, 1971); *A Dialogue on Induction* (University of Chicago Press, Chicago, in press); also Reference 2; Gerard Radnitzky, *Contemporary Schools of Metascience (2nd ed): The Psychology of Knowing* (Gordon & Breach, New York, 1972). I am particularly indebted to Arne Naess, Reference 2. More often than not my positions as stated above are mere metaphrases of some points already made by Arne Naess.

9. H. Tennessen, "Which World is the Real One, and How Would We Know?" Invited address to the American Psychological Association Conference, Miami Beach, September 3rd, 1970.
10. H. Tennessen, "Permissible and Impermissible Locutions. Carnap's Principle of Tolerance vs 'Ordinary Language Philosphy,'" *Logic and Language. Studies Dedicated to Professor Rudolf Carnap on the Occasion of His Seventieth Birthday*, B. H. Kazemier and D. Vuysje, editors (D. Reidel Publishing Co., Dordrecht, 1962), p. 220f.
11. Anders Wedberg, *Fran Bolzano till Wittgenstein, Filosofins historia III* (Bonniers, Stockholm, 1966).
12. F. Nietzsche, *Also Sprach Zarathustra* (Alfred Kröner Verlag, Leipzig, 1883, I Teil), p. 8.
13. F. Nietzsche, *Menschliches Allzumenschliches* (Alfred Kröner Verlag, Leipzig, 1878).
14. *Vide*, H. Tennessen, "Simmel's tillemping av seleksjonslaeren pa erkjennelsesteorier:[1] Tenkningens objektivitet," Oslo University, 1937, pp. 23–31: *Integrasjon og erkjendelse*, Filosofiske Problemer nr. 18 (Universitetsforlaget, Oslo, 1955 and 1960); "Whereof One Has Been Silent, Thereof One May Have to Speak," *J. Phil.* **LVIII**(10), 263–274 (1961); on George Simmel's "Ueber eine Beziehung der Selectionslehre zur Erkenntnistheorie, *Arch. systematische Phil.* **I** (1895).
15. B. Lonegan, *Insight—A Study in Human Understanding* (Prentice-Hall, Englewood Cliffs, New Jersey), p. 329.
16. Søren Kierkegaard, *Afsluttende uvidenskabeligt Efterskrift til de philophiske Smuler* (1846) Soren Kierkegaards Samlede Vaerker, Bd. 7 (Ejnar Munksgaard, Copenhagen, 1901–1906), p. 101.
17. A. Koestler, *The God that Failed*, here quoted after M. Polanyi, "The Stability of Beliefs," *Brit. J. Phil. Sci.* (1952)
18. H. Walsby, *The Domain of Ideologies* (Tait and Son, Ltd., Glasgow, 1947), p. 201.
19. Karl Mannheim, *Ideologie und Utopie* (Suhrkampf, Bonn, 1929), p. 49.
20. H. Tennessen, "On the Conceptual Absurdity of 'Conceptual Absurdity,'" *Phil. Forum* **XII**, 584–592 (1976).
21. It is unclear how Popper's alleged antisurvivalist position (vis-à-vis theories) is to be most reasonably interpreted. The position is definitely untenable if one allows for the usual evolution of a theory, retaining the same *name* for the theory. *Vide*, e.g., H. Tennessen, *Problems of Knowledge* (van Gorcum, Assen, 1980), pp. 33–35.
22. K. R. Popper, *Conjectures and Refutations* (Routledge & Kegan Paul, London, 1963; Basic Books, New York, 1962).
23. *Vide*, e.g., P. Feyerabend, 1970 and 1970b (and *Pro-Paramenides* a.o. . . .)
24. Friedrich Wilhelm Bessel measured the parallaxis of Gl Cygni in 1838 and Thomas Henderson that of Centauri in 1839. T. Ferris, 1979, pp. 7, 8, and 180; and H. Tennessen, 1976b, p. 236.
25. Feyerabend's *Pro-Paramenides*, Ref. 23, p. 4.
26. K. Popper, Ref. 22, p. 31.

27. K. Popper, Ref. 22, p. 31.
28. K. Popper, Ref. 22, p. 42.
29. The exposition here is Robert Cohen's, i.e., his comments to Adolf Grünbaum's paper in M. Wartofsky 1963.
30. *Harvard Case, Historians in Experimental Science*, Conant and Nash, editors (Harvard University Press, Cambridge, Massachusetts, 1957), Vol. 1, p. 111.
31. R. Carnap, *Philosophical Foundations of Physics* (Basic Books, New York, 1966), pp. 266ff.
32. H. Tennessen, Ref. 5, p. 116f and H. Tennessen, 1978, p. 14f.
33. Concerning this conflict *vide*, e.g., P. Feyerabend, "Problems of Empiricism," Ref. 8, and "Problems of Empiricism, in *The Nature and Function of Scientific Theories*, R. G. Colodny, editor (U. Pittsburgh Press, Pittsburgh, 1970), p. 20.
34. P. Duhem, *The Aim and Structure of Physical Theory* (Atheneum, New York, 1962), p. 200.
35. Rutherford is here quoted from G. Holton and Roller, *Foundation of Modern Physical Science* (Addison-Wesley, London, 1962), p. 624.
36. *Vide*, H. Tennessen, 1970, p. 54.

Bibliography

Austin, John L., on H. Tennessen's "What Should We Say?" In H. Tennessen, *Language Analysis and Empirical Semantics* (University of Alberta Bookstore, Edmonton, 1964), Sec. 10, pp. 143–150.

Bondi, H., *Cosmology* (Cambridge University Press, 1969).

Bunge, M., *Method, Model and Matter* (D. Reidel Publishing Co., Dordrecht-Holland, 1973).

Butterfield, H., *The Origin of Modern Science* (Bell, London, 1951).

Campbell, D. T., "Evolutionary Epistemology," in P. A. Schilpp, editor, *The Philosophy of Karl Popper*, Vol. 14, I & II (Open Court Publishing Co., LaSalle, 1974), pp. 413–463.

Carnap, R., *Philosophical Foundations of Physics* (Basic Books, New York; 1966).

Cohen, R., Comments (to Grünbaum's paper) in Wartofsky, editor, *Boston Studies in the Philosophy of Science*, 1961/62 (D. Reidel Publishing Co., Dordrecht-Holland, 1963).

Colodny, R. G., editor, *Beyond the Edge of Certainty* (Prentice-Hall, New York, 1965).

Colodny, R. G., editor, *The Nature and Function of Scientific Theories* (University of Pittsburgh Press, Pittsburgh, 1970).

Conant, J. B. and Nash, editors, *Harvard Case Histories in Experimental Science*, Vol. 1. (Harvard University Press, 1957).

Courant, R., "Bernhard Riesmann und die Mathematik der Letzten Hundert Jahre," *Naturwissenschaft*, **XIV** (1926).

Gregory, R., *Eye and Brain*, (McGraw Hill, New York, 1966).

Gregory, R., *The Intelligent Eye* (Weidenfeld & Nicholson, London, 1970).

Gullvaag, I., *Referanse, mening og ehsistens* (Oslo University Press, Oslo, 1967).

Habermas, J., "On Systematically Distorted Communication," *Inquiry* **13,** 205–219 (1970).

Habermas, J., *Knowledge and Human Interest* (Beacon Press, Boston, 1971).

Habermas, J. *Review Symposium* by N. Lobkowicz, F. R. Dallmayr, C. K. Lenhardt, M. A. Hill, and C. Nichols *Phil. Soc. Sci.* **2,** 193–210 (1972).

Horkheimer, M., *Kritische Theori*, Vol. II (S. Fischer, Frankfurt a. M.; 1968).

Horkheimer, M., and T. W. Adorno, *Dialektik der Aufklärung* (S. Fischer, Frankfurt a. M.; 1969).

Kyburg, H. E., "Recent Work in Inductive Logic," *Am. Phil. Quart.* **1,** 1–39 (1964).

Kyburg, H. E., "Comments on Salmon's *Inductive Evidence*," *Am. Phil. Quart.* **2,** 10–12 (1965).

Kyburg, H. E., "Salmon's Paper," *Phil. Sci.* **33,** 147–151 (1966).

Kyburg, H.E., Jr., "Review of Patrick Suppes and Jaakko Hintikka, editors, *Aspects of Inductive Logic*," *J. Phil.* **LXV**(10), 323–328 (1968).

Lenz, J. W., "Carnap on Defining *Degrees of Confirmation*," *Phil. Sci.* **23,** 230–236 (1956).

Mates, B., "Philosophical Scepticism and the Logical Antinomies" *Akten des XIV Internationalen Kongresses für Philosophie, III. Logik, Erkenntnis- und Wissenschafts-theorie, Sprachphilosophie, Ontologie und Metaphysik* (Herder & Co., Wien, 1969).

Naess, A., *Erkenntnis und wissenschaftliches Verhalten* (Jacob Dybwad, Oslo, 1936).
Naess, A., *Filosofiske Problemer* (H. Aschehoug & Co., Oslo, 1941).
Naess, A., *Usaklighetsanalyse* (Universitetsforlaget, Oslo, 1947).
Naess, A., *An Empirical Study of the Expressions "True," "Perfectly Certain" and "Extremely Probable"* (Jacob Dybwad, Oslo, 1953).
Naess, A., *Interpretation and Preciseness, a Contribution to the Theory of Communication*, Skrifter ut gitt av Det Norske Videnskaps—Akademi i Oslo, II Hist. Filos. Klasse 1953. No. 1 (Johan Dybwad, Oslo, 1953).
Naess, A., *Wie Fördert Man Heute Die Empirische Bewegung?* (Universitetsforlaget, Oslo, 1956).
Naess, A., *Notes on the Foundation of Psychology as a Science* (2nd ed.) (Universitetsforlaget, Oslo, 1960).
Naess, A., *Empirisk Semantik* (Svenska Bokförlaget, Stockholm, 1961).
Naess, A., "Can Knowledge Be Reached?" *Inquiry* **4**(4), 219–228 (1961).
Naess, A., *Symbolsk Logikk* (3rd ed. by Jens-Erik Fenstad) (Universitetsforlaget, Oslo-Bergen, 1962).
Naess, A., "The Inquiring Mind," *Phil. Today* **5**(3/4), 184–205 (1961).
Naess, A., *Communication and Argument* (Bedminster Press, Totowa, New Jersey 1965).
Popper, K. R., "Degree of Confirmation," *Mind* **47**, 143–149 (1954–55).
Popper, K. R., "'Content' and 'Degree of Confirmation': A Reply to Dr. Bar-Hillel," *Brit. J. Phil. Sci.* **6**, 157–163 (1955–56).
Popper, K. R., "A Second Note on Degree of Confirmation," *Brit. J. Phil. Sci.* **7**, 350–353 (1956–57).
Popper, K. R., "A Third Note on Degree of Corroboration or Confirmation," *Brit. J. Phil. Sci.* **8**, 294–302 (1957–58).
Popper, K. R., *Of Clouds and Clocks* (Washington University, St. Louis, Missouri, 1966).
Quine, W. O., *From a Logical Point of View* (Harvard University Press, Cambridge, Massachusetts, 1953 and 1961).
Quine, W. O., *Word and Object* (M.I.T. Press, Cambridge, Massachusetts, 1960).
Quine, W. O., *Selected Logic Papers* (Random House, New York, 1966).
Quine, W. O., "Notes on Existence and Necessity," *J. Phil.* **40**, (1943).
Quine, W. O., *The Ways of Paradox* (Random House, New York, 1966).
Radnitzky, G., *Continental Schools of Metascience I & II* (Akademiförlaget, Göteborg, 1968).
Salmon, W. C., "Rejoinder to Barker and Kyburg," *Am. Phil. Quart.* **2**, 13–16 (1965).
Salmon, W. C., "Rejoinder to Kyburg," *Phil. Sci.* **32**, 152–154 (1965).
Salmon, W. C., *The Foundation of Scientific Inference* (University of Pittsburgh Press, Pittsburgh, 1966).
Tennessen, H., *Theorien für Stereoskopie, mit besonderer Berücksichtigung der Ausnützung der Querdisparation für binokulare Tiefensehen* (Universitetsforlaget, Oslo, 1939).
Tennessen, H., "The Fight against Revelation in Semantical Studies," *Synthese* **VIII**, 231–267 (1950).
Tennessen, H., and Gullvaag, I., *Logical Analysis and Definiteness of Intention* (Universitetsforlaget, Oslo, 1955 and 1960).
Tennessen, H., "Note on the Confusion of Evidence and Illustration" Vol. **LVI**(18), 733–736 (1959).
Tennessen, H., "On Worthwhile Hypotheses," *Inquiry* **2**(3), 183–198 (1959).
Tennessen, H., "Evidence and Illustration," *Synthese* **XI**(3), 274–276 (1959).
Tennessen, H., "What Should We Say?" *Inquiry* **2**(4), 263–269 (1959).
Tennessen, H., "Logical Oddities and Locutional Scarcities," *Synthese* **XI**(4), 376–378 (1959).
Tennessen, H., "Vindication of the Humpty Dumpty Attitude towards Language," *Inquiry* **3**(2), 185–198 (1960).
Tennessen, H., "Empirical Semantics and the Soft Sciences," In *International Congress for Logic, Methodology and Philosophy of Science.* (Abstract of contributed papers) (Stanford University Press, Stanford, California, 1960).
Tennessen, H., "On Making Sense," *J. Phil.* **LVII**(24), 764–765 (1960).
Tennessen, H., "Knowledge versus Survival," *Inquiry* **16**, 407–414 and "Rejoinder to Naess," *ibid.*, 417–418 (1973).
Tennessen, H., *Introduction to Empirical Semantics* (University of Alberta Bookstore, Edmonton, Alberta, 1962).
Tennessen, H., "Thirty-Six Lectures on Topics in Philosophy of Science" [University of Alberta Bookstore (Mimeo), Edmonton, Alberta, 1963].

Tennessen, H., *Language Analysis and Empirical Semantics. Eighteen Essays* (University of Alberta Bookstore, Edmonton, Alberta, 1964).

Tennessen, H., "Ordinary Language *in Memoriam*" *Inquiry* **8**, 225–249 (1965).

Tennessen, H., "History is Science: Preliminary Remarks towards an Empirical, Experimentally Oriented Behavioural Science of History," *Monist*, **55**(1), 116–134 (1969).

Tennessen, H., "Om det finnes absurde setninger," *Norsk Filosofisk Tidsskrift* **4**(2), 55–88 (1969).

Tennessen, H., "Five Futile Attempts to Show that the Problem of Induction Is No Problem," Paper read before *The Boston Colloquium of Philosophy of Science*, May 1963, forthcoming.

Tennessen, H., "Beyond Competence. A Plea for Linguistic-Communicative Transgressionalism," *Proceedings from the VIth European Congress of Linguistics*, 1976, pp. 374–389.

Tennessen, H., "The Petrified Skeleton in the Closet of Philosophy of Science," *Methodol. Sci.* 8(2–3), 125–135 (1975).

Tennessen, H., "Scientists in Vain Wants of World Views," *Methodol. Sci.* 9(3), 120–128 (1976).

Tennessen, H., "Against Linguistic Restrictionism in Science Methodology," *Methodol. Sci.* **11**(1), 34–46 (1978).

Tennessen, H., "Permissible and Impermissible Locutions. Carnap's Principle of Tolerance vs. 'Ordinary Language Philosophy,' *Logic and Language. Studies Dedicated to Professor Rudolf Carnap on the Occasion of His Seventieth Brithday*, B. H. Kazemier and D. Vuysje, editors (D. Reidel Publishing Co., Dordrecht, 1962), p. 220f.

Tennessen, H., *Problems of Knowledge* (van Gorcum, Assen, 1980).

Tennessen, H., "From the Testimonies of the Senses to the Paradoxes of World Views," in *Annals of Theoretical Psychology*, Vol. II (Plenum Press, New York, 1982), pp. 1–18.

Tennessen, H., "Science of History and Notions of Personality," in *Physics, Logic, and History*, W. Yourgrau, editor (Plenum Press, New York, 1970).

Yourgrau, W., "Herman Bondi: Assumptions and Myth in Physical Theory," *Inquiry* **13**(3), 332–334 (1970).

20

Queries about Inquiries

Håkan Törnebohm
University of Gothenburg
Gothenburg, Sweden

Introduction

The main task I set myself in this essay is to raise a number of queries about scientific investigations. Deliberations pertaining to some of these questions lead to positions which give rise to new batches of questions.

The first query is this one: How should a scientific inquiry be modeled?

A model of inquiries will be constructed called "the running research program model." This model will be used in dealing with this question.

What kind of inquiries are there?

A systematics based on the notion of running research program will be presented. This systematics gives rise to several further questions. One of them is this one:

How is theory related to experience?

Having dealt with this question in a summary way, I will proceed to the further question:

In what contexts are inquiries imbedded?

This query provides an opportunity for introducing the notions of inquiring system and field of research. Every inquiry takes place in an inquiring system, which in its turn is imbedded in a field of research. Fields of research are ingredients in contexts of higher and higher orders. Moving upwards on a contextual ladder we arrive at the world of science, which in its turn is imbedded in the world.

How should one picture the world as seen from the vantage point of someone living in a sociocultural world with very strong scientific traditions?

How is the world of science related to the whole world? How is it constituted? What are its most prominent features?

These questions will be dealt with in the latter part of this essay, where I will be concerned with a number of issues in contemporary theory of science, with paradigms, with growth patterns in fields of research, with philosophies inside fields of research, and so on.

In particular I will be concerned with hermeneutical aspects of positivistic research. This question will be taken up: What items in a field of research are subject to attempted interpretations? This question provides an opportunity for introducing the notion of mentality, which seems to me to be a very important one. This essay contains far more open questions than answers.

A Model of Inquiries

(1) Every inquiry in the factual sciences has a territory = a part of the world to be investigated. Let X stand for a territory.

(2) Every inquiry is imbedded in a context which may be referred to as an inquiring system. Let $\hat{X}$ stand for an inquiring system with the territory X.

More about inquiring systems in a later section.

(3) Every inquiry involves *transformations of results*. We may represent this aspect of an inquiry in this way:

$$\rightsquigarrow (R, \ldots,) \rightsquigarrow$$

where R stands for a result.

(4) The forthcoming results are *controlled* by means of critical assessments.

(5) Combining (3) and (4) we may provisionally represent an inquiry as follows:

$$\rightsquigarrow (R, \mathrm{Cr}, \ldots,) \rightsquigarrow$$

Where Cr stands for a critical assessment.

(6) Every inquiry is *steered* or *governed* step by step by means of tactical plans made up on the basis of critical assessments of current results.

(7) Combining (5) and (6) we may represent an inquiry as follows:

$$\rightsquigarrow (R, \mathrm{Cr}, \mathrm{Pl}) \rightsquigarrow$$

(8) It remains to give a structure to the first component R, in the triad (R, Cr, Pl).

A result R may be conceived of as a triad, composed of an assembly H of hypotheses, an assembly I of instruments, and an assembly P of problems or tasks.

In sum $R = (H, I, P)$.

Comments

(1) Untested hypotheses in H are always accompanied by items in P, i.e., tasks of testing them.

(2) Confirmed hypotheses in H *may* also but need not be accompanied by tasks. Let h be a confirmed hypothesis. Let h refer to a state of affairs s in the territory X. Then h may be accompanied by a task to explain s.

(3) The assembly of *I* includes "software" instruments such as schematisms, i.e., modes of representing items in X. I may also include "hardware" instruments such as artifacts in a laboratory.

(4) Instruments in *I* are employed in carrying out tasks in *P*. They may also be products formed within an inquiry, which means that an assembly of tasks *P* may include tasks to improve on or to construct new "software" or "hardware" instruments.

I will now give a name to the pentads (*H*, *I*, *P*, Cr, Pl).

(9) Such pentads will be called "*research programs*" and a sequence of such pentads will be called "*a running research program.*"

(10) A running research program $\rightsquigarrow$ (*H*, *I*, *P*, Cr, Pl) $\rightsquigarrow$ where the hypotheses *H* refer to states of affairs in a territory *X* represents an inquiry about this territory.

I will now spell out this conception: An inquiry is conceived as follows.

(a) A result (*H*, *I*, *P*), composed of an assembly of hypotheses *H*, an assembly of "software" or "hardware" instruments *I*, and an assembly of tasks or problems *P*, is submitted to a critical assessment Cr.

(b) On the basis of a verdict after such an assessment of a result at hand, a researcher makes up a plan Pl, which will steer or govern acts of transforming a result (*H*, *I*, *P*) into a subsequent result (*H*′, *I*′, *P*′), which

(c) again will be submitted to a critical assessment Cr′ accompanied by a new tactical plan Pl′ and so on, until the inquiry is terminated and

(d) the final result is reported.

Tasks Suggested by the Model

The model $\rightsquigarrow$ (*H*, *I*, *P*, Cr, Pl) $\rightsquigarrow$ gives rise to a number of questions for further deliberations. I will set them out, but will refrain from discussing all of them.

(1) What criteria are employed in critical assessments of results?

(2) How are such and such criteria to be justified?

(3) What is the structure of tactical plans?

Tactical Plans

A tactical plan Pl is composed of decisions

(a) on what tasks to select from the assembly *P*,

(b) on what instruments from the assembly *I* to be used in carrying out selected tasks. It frequently happens that the resources at hand are insufficient to carry out selected tasks. In such cases a tactical plan will also include

(c) specified requests for additional resources in the form of new instruments to be imported or to be constructed or in the form of new items of the kind *H* to be employed.

Examples

If a researcher wants to test an hypothesis h, then he will make up a testing plan Pl, which includes requests for evidence bearing on h. In cases of this sort the requested resources are of the H-kind.

When Newton faced the task of deducing Kepler's laws from an assembly of explanatory assumptions, including his gravitational hypothesis, he needed a new mathematical technique, i.e., a resource of the kind I, in order to carry out this task.

A plan Pl includes in addition to the items already mentioned also

(d) a working plan, which specifies how a researcher is to go about the task or tasks which he has selected.

I have given a list of three questions which are suggested by the running research program model. This list will now be extended.

Preparations of Inquiries

(4) How are inquiries prepared? The running research program model suggests this outline of an answer:

To prepare an inquiry is to build a structure which may be represented as an *initial research program,*

$$(H_0, I_0, P_0, \mathrm{Cr}_0, \mathrm{Pl}_0)$$

This conception suggests that a preparation of an inquiry should include these operations:

(a) To select a part of the world X to be submitted to an inquiry (selection of a territory).

(b) To focus attention on certain aspects of X (focusing). Let X_0 stand for those aspects of X which are selected. [Newton selected to pay attention to motions in the planetary system in his gravitational research.]

(c) To assemble hypotheses about X_0 from various sources from previous research about X_0, from research about some other territory Y, which in the opinion of the researcher(s) is sufficiently similar to X_0 for confirmed hypotheses about Y to be regarded as possibly valid also for X_0, and from the speculative imagination of the researcher. Let H_0 be the initial assembly of hypotheses thus obtained.

(d) To assemble tasks or problems to be dealt with. Untested hypotheses in H_0 are accompanied by tasks to test them. Confirmed hypotheses in H_0 *may* be accompanied by explanatory tasks etc. Let P_0 stand for the tasks assembled in the preparation of an inquiry.

(e) To assemble appropriate "software" and possibly also "hardware" instruments to be used in dealing with tasks belonging to P_0. Let J_0 stand for such an instrument.

(f) The triad (H_0, I_0, P_0) may be submitted to a critical assessment Cr_0 which may give rise to a modification of it. Some hypotheses in H_0 may be eliminated as an outcome of a verdict that they are incredible and may be replaced by others which are regarded as more promising and so on.

The final step in the preparation of an inquiry is

(g) To make up an initial tactical plan Pl_0 which will govern the first steps in the inquiry.

Yet another question suggested by the running research program is this one

(5) What kinds of inquiries are there?

This question will be dealt with in the next section.

A Systematics of Inquiries

There are different kinds of inquiries in the factual sciences. Here I will only be concerned with inquiries which may be characterized as positivistic ones, i.e., they are concerned with no other attributes than neutral ones (see below).

(1) Inquiries of this kind may be partitioned into

(*E*) *Empirical* inquiries, where the term "empirical" is used in a restricted sense and

(*S*) *Synthesizing* inquiries, which of course are empirical also, but in a wider sense.

Theoretical inquiries in physics aiming at the construction of explanatory patterns or at the construction of theories which include more than one explanatory pattern exemplify inquiries of the kind S. Historical narratives are other examples of inquiries of the kind S. They correspond to theoretical research producing explanatory patterns and theories.

(*E*) Inquiries of the kind E may be partitioned into two kinds:

(0) Inquiries producing *primary* information. These are inquiries of order (0). Primary information or data consists of items such as results of measurements; such as the results obtained when historical documents have been submitted to source criticism; such as descriptions of traces from the past in the form of fossils, archeological artifacts and so on.

(1) Inquiries producing single confirmed hypotheses such as Boyle's or Ohm's laws. These are inquiries of the order (1).

Within inquiries of order (1) these hypotheses are synthesized neither into explanatory patterns nor into cognitive structures of a higher order than explanatory patterns. It belongs to the tasks within inquiries of the kind S to carry out such syntheses.

(*S*) Inquiries of the kind S may be partitioned into three kinds, viz.,

(2) Inquiries producing explanatory patterns. These are inquiries of order (2).

(3) Inquiries producing special theories or syntheses corresponding to special theories within fields of research such as historical ones. These are inquiries of order (3).

And finally

(4) Inquiries producing fundamental theories or syntheses corresponding to such theories in for example historical research. These are inquiries of order (4).

Comments on Inquiries of the Kind S

(1) A special theory about a territory *X* is a structure which includes several explanatory patterns. Newton's theory of gravitation includes explanations of the motions of planets, the motions of falling bodies, the motions of projectiles, the motions of double stars, and so on.

An historical narrative of order (3), for example of Rommel's campaigns in Africa during World War II, includes several explanations of what happened during those campaigns.

(2) A fundamental theory refers to patterns which are supposed to be common to several territories X, Y, Z, ... It includes assumptions about such patterns.

Newton's theories about physical space and physical time and his theory about causality are examples of syntheses of this kind. These three theories linked together constitute a fundamental theory on which classical mechanics is based.

The hypotheses in a fundamental theory may be referred to as world-picture assumptions.

(3) Special theories in physics are not isolated, or if they are, their being so is regarded as a defect of them.

Special theories are linked together into families in this way: Every theory in a family includes world-picture assumptions which belong to the same fundamental theory.

Examples

Classical mechanics is a family based on Newton's theories of space, time, and causality.

The special and the general theories of relativity are other examples of families. The special theory is based on a fundamental theory which is incompatible with the fundamental theory on which classical mechanics is based. The same is true for the general theory of relativity.

Do historians produce syntheses corresponding to fundamental theories in physics? Do they produce syntheses corresponding to families of theories in physics? I leave these questions open.

I will now sum up the systematics of inquiries presented so far in a figure (Fig. 1) which also presents the running research program model.

The systematics presented in Fig. 1 suggests several questions:

(1) How are inquiries of one order related to inquiries of another order?

(2) In particular how is theoretical work related to empirical work?

(3) How may world-picture assumptions acquire the status of knowledge?

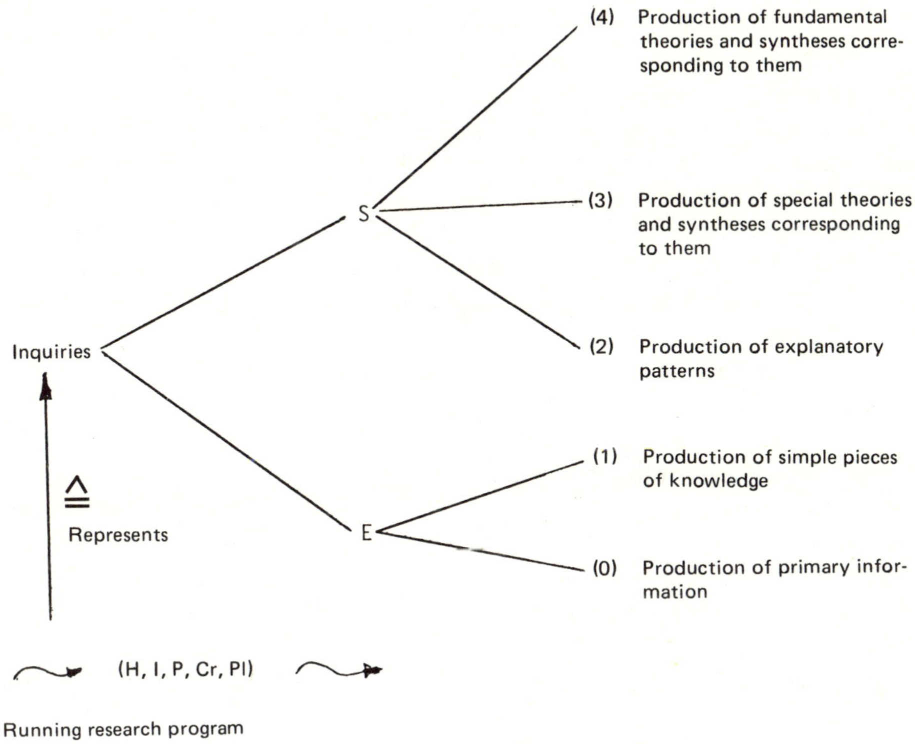

FIGURE 1. The running research program.

(4) How are explanatory patterns included in special theories?

(5) How are special and fundamental theories related to each other?

(6) What is the structure of syntheses of order (2), of order (3), and of order (4)?

(7) What criteria are employed in critical assessements of the products of inquiries of order (0), of order (1), of order (2), of order (3), and of order (4)?

(8) How may fundamental theories be refuted and replaced by other and hopefully better ones?

To deal with these questions and others which are also suggested by Fig. 1 would require a lengthy treatment, which ought to include several case studies of inquiries of the various orders (0), (1), . . . , (4). I have dealt with some of them elsewhere (see the references). In this essay I will restrict my attention to one question alone, the classical question about the relation between theory and experience.

THEORY AND EXPERIENCE

How is theory related to experience?

A tactical plan includes requests for resources which a researcher needs in addition to those at hand in order to carry out tasks which he has selected.

This item of tactical plans is a crucial one if we want to understand how theory is related to experience.

(1) When a physicist wants to explain a state of affairs s in a territory X, which is described by a *new* confirmed hypothesis h, then he has to frame at least *one* other new hypothesis h^*, to be inserted into an explanatory pattern. The hypotheses in an explanatory pattern should form a deductive argument according to the criteria employed by physicists when they assess the results of their inquiries of order (2).

Thus h^* is one of the premises and a description h' of s is a conclusion. h' may be different from h. This is frequently the case, as shown by many examples from the history of science.

The hypothesis h^* will be accompanied by a task to test h^*. A testing plan for h^* includes a *request* for evidence including primary evidence.

If h' is different and incompatible with the confirmed hypothesis h describing s, then a researcher faces the task of discriminating between h and h'. In order to be able to do so a discriminatory testing plan will be made. This includes a *request* for primary evidence of a kind which is favorable to one of the competing hypotheses h and h' and unfavorable to the other one. In order to obtain such evidence it is often necessary to refine techniques of measurements, a typical task for experimental physicists.

We see thus that explanatory work is bound to give rise to tasks belonging to inquiries of the kind E.

The results of inquiries of the E-kind may in return give rise to tasks in inquiries of the kind S. Suppose that an hypothesis h^* framed by a theoretician in an inquiry of order (2) is disconfirmed as the result of experimental work, then a theoretician has to frame another hypothesis in place of h^*.

We note thus that inquiries of the kind E will give rise to tasks within inquiries of the kind S.

There is thus an *exchange relation* between empirical and theoretical work.

Results in one kind of research give rise to tasks in inquiries of the other kind.

In the account so far offered of the relation between theoretical and empirical research requests are made by theoreticians and executed by experimentalists. The reverse relation that requests are made by experimentalists to theoreticians is also quite frequent.

Experimental physicists need the assistance of theoreticians when they construct new instruments of measurement. Theories are used by them as "software" instruments in building "hardware" instruments.

I have now given arguments in favor of the following thesis.

Thesis. The relation between theory and experience may be construed as an *exchange relation*, a give-and-take relation, between theoretical and experimental work or more generally between inquiries of the kind S and inquiries of the kind E.

These exchange relations involve requests by one party, theoreticians or experimentalists, giving rise to tasks undertaken by the other party.

Contexts in which Inquiries Take Place

In what contexts do inquiries take place? How are contexts of such and such a kind to be conceived? In order to prepare my deliberations I will introduce the notion of contextual trees.

Contextual Trees

A person belongs to several *contexts:*

1. He belongs to a family.
2. He is an employee in a firm.
3. He is a member of a hobby club, etc.

Some contexts to which a person belongs are comparatively permanent. Others are more occasional and short-lived. An example of a short-lived context is the audience of a theatrical performance.

Now each one of the three first-mentioned contexts belongs to one or more contexts of a higher order.

A family belongs to a group of relatives.

A firm belongs to a branch. It belongs also to the enterprises in a particular region, etc.

A hobby club belongs to the institutions for recreational activities in a particular region, etc.

We may proceed in the same fashion with the last-mentioned contexts and relate them to contexts of yet higher orders.

A branch, for example, belongs to an industry, etc.

On the relation between an entity, conceived to be a context or not, and a context of which it is an ingredient, we may erect a tree structure as shown in Fig. 2.

Contextual trees are very important patterns in the world, because they are present everywhere, in nature and in the sociocultural world and within that part of the sociocultural world where research is carried out.

Inquiring Systems

I will now apply the notion of contextual tree to inquiries. To simplify the discourse as much as possible I will disregard all branchings, that is I will erect a *contextual ladder* rather than a tree.

Inquiries will be placed on the lowest rung.

On the next rung I will place *inquiring systems*. They are contexts of order (1) to which inquiries to which I assign the order (0) belong.

Remark. The order (0) is assigned to every inquiry thus also to inquiries of higher order than order (0) according to the systematics displayed in Fig. 1.

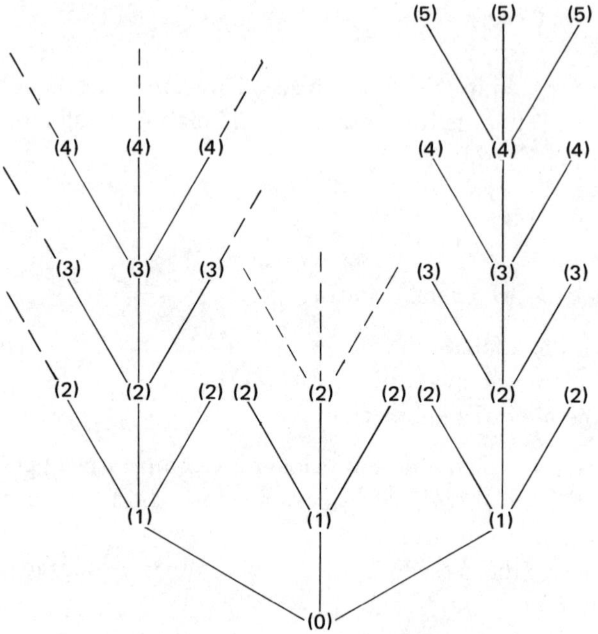

FIGURE 2. A contextual tree.

What is an inquiring system?
Let X be the territory of a particular inquiry.
An inquiring system $\hat{X}$ with this territory is an open system.
It contains these components:

1. At least one researcher;
2. *One* inquiry, no more, which is represented by the running research program model; and
3. Other activities performed by the researcher(s) in $\hat{X}$ besides those which belong directly to the inquiry, such as gathering secondary information from other inquiring systems, *compilation*, and such as *reporting* the final results of the inquiry to outsiders.

The environment of the inquiring system $\hat{X}$ includes the following:

(1) The territory X. A researcher in $\hat{X}$ gathers *primary* information from and about X. He produces knowledge about X.

(2) Other inquiring systems $\hat{Y}$, $\hat{Z}$, ...

Researchers in $\hat{X}$ gather *secondary* information from some of them and use it as resources in the inquiry that belongs to $\hat{X}$. They issue secondary information, which will be received by some of the inquiring systems in the environment of $\hat{X}$.

Figure 3 shows a picture of an inquiring system $\hat{X}$ and its relations to its environment.

Comments on Figure 3

The box $\hat{Y}$ may represent more than one inquiring system, likewise the box $\hat{Z}$. Some of the inquiring systems represented by the box $\hat{Y}$ may be identical with some of the inquiring systems represented by the box $\hat{Z}$.

The arrow represents possible actions of $\hat{X}$ on X.

If X is an organization existing now, then an inquiry about X in $\hat{X}$ may very well give rise to changes in X, intended or not.

The notion of inquiring system enables us to relate an inquiry in a natural way to other inquiries to which it is linked via flows of *secondary* information, to its territory to which it is linked via flows of *primary* information, and last but not least to one or more actors which carry out the inquiry.

If we want to study these important links, then we need the notion of inquiring system.

Remarks. The notion of inquiring system is *not* a model, as a running research program is a model of an inquiry. If we want to model an inquiring system, we need to model

1. all its ingredients including the actor or actors,
2. the couplings of these ingredients, and
3. its environment and its links to it.

The notion of inquiring system gives thus rise to further tasks within inquiries about inquiries.

I will not engage in any one of these tasks at this place, but will later on take up this question: What attributes of researchers determine their inquiries?

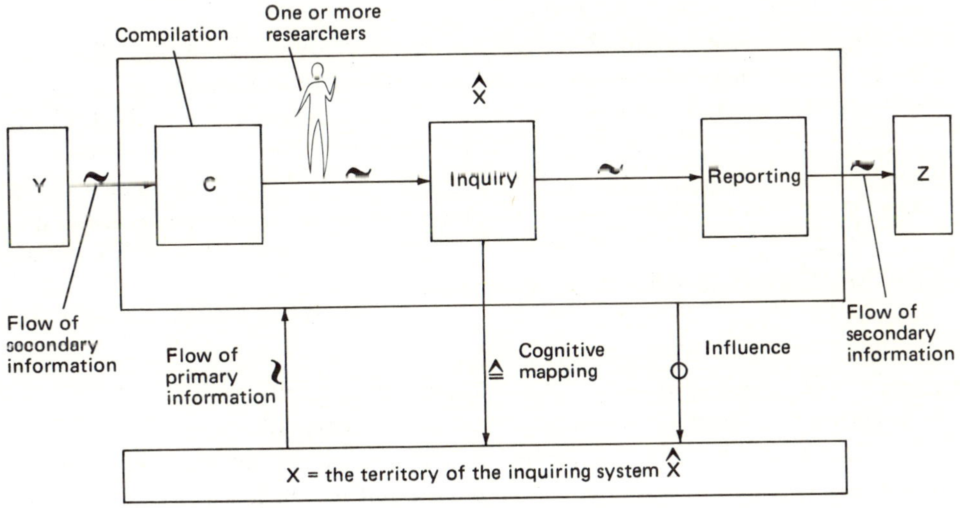

FIGURE 3. An inquiring system $\hat{X}$ and its relations to its environment.

Fields of Research

I proceed to move up yet another step on the contextual ladder from inquiring systems to *fields of research*. Fields of research are contexts of order (2).

A field of research is a complex composed of several (two or more) inquiring systems.

A field of research F has a territory that is composed of the territories X, Y, Z, . . . of those inquiring systems $\hat{X}$, $\hat{Y}$, $\hat{Z}$, . . . which are included in F.

Not every assembly of inquiring systems forms a field of research. There should be some restrictions on membership.

One such restriction is that the territories X, Y, Z, . . . of its members should be related to each other by common characteristics. They should belong to a natural part of the world.

For example, they should all belong to the nonliving part of nature or to the biosphere or to the sociocultural world.

I am painfully aware of how vague this specification is, but this vagueness seems to be unavoidable. It is rooted in the nature of things. Fortunately there exists a species of fields of research where it is possible to be a bit more precise. This species will be dealt with in the next section.

Aggregated Inquiring Systems

(1) X, Y, Z, . . . are the territories of the inquiring systems $\hat{X}$, $\hat{Y}$, $\hat{Z}$, ...

(2) The researchers in these inquiring systems have found out (or at least they assume so) that there are common features in all these territories. Let U stand for a common pattern in the territories X, Y, Z, ...

(3) Assumptions about U are included in the inquiries belonging to the systems $\hat{X}$, $\hat{Y}$, $\hat{Z}$, ... These assumptions are world-picture hypotheses.

(4) The inquiring systems $\hat{X}$, $\hat{Y}$, $\hat{Z}$, . . . produce special theories (H, I, P) about X, (H', I', P') about Y, (H'', I'', P'') about Z, and so on.

(5) The "text parts" of these theories H, H', H'', . . . include the same or at least equivalent (they may be expressed in different languages) world-picture assumptions about U. Let W be the assembly of these assumptions about U. We may therefore say that W overlaps with H, with H', and with H'' etc.

(6) W is the "text part" of a fundamental theory (W, I_W, P_W) about the common pattern U in all the territories X, Y, Z, ...

(7) The special theories produced in the inquiring systems form a *family* under the above stated conditions based on the same fundamental theory (W, I_W, P_W).

I am now prepared to introduce the notion of *aggregated inquiring system*. An *aggregated inquiring system* is a field of research in which a family of theories is produced.

A Corollary

Fundamental theories are produced in aggregated inquiring systems rather than in their individual members.

Examples of Aggregated Inquiring Systems

Classical mechanics as a field of research belongs to this species.

Terminological Remark

I am here using the term "classical mechanics" to refer to the assembly of inquiring systems which has produced this family.

This term is frequently used in another sense. It is used to refer to the family of theories *produced* by physical research rather than to the research that has produced it.

The fundamental theory (W, I_W, P_W) in classical mechanics contains Newton's theory of physical space, his theory of physical time, his theory of causality, and a general theory of substances which includes the assumption among others that every piece of matter possesses mass.

Other examples of aggregated inquiring systems in physics are those which have produced classical electrodynamics, the special theory of relativity, and quantum mechanics. The families of theories in physics are based on fundamental theories which sometimes are mutually inconsistent. Such inconsistencies are very important. They give rise to changes of fundamental theories, changes which historians of physics frequently refer to as revolutions.

The notion of aggregated inquiring system and the correlative notions of family of special theories and of fundamental theory gives rise to a number of questions:

(1) What kind of problems or tasks are included in the component P_W of a fundamental theory?

P_W includes the task of testing components of the "text part" W, which brings us to the next question.

(2) How are world-picture assumptions tested?

If two fundamental theories (W, I_W, P_W) and (W', I'_W, P''_W) are inconsistent, then either P_W or P'_W or both include tasks to reform at least one of them.

We are thus led to the following questions:

(3) How are fundamental theories reformed?

(4) How is the result of an attempt to reform a fundamental theory to be assessed in comparison with the original theory? This question is closely related to a theme which is much debated in contemporary theory of science. I am referring to Kuhn's and Feyerabend's contention that paradigm shifts are like changes in fashion rather than progressive moves.

When a fundamental theory changes then the family of theories that are based on it has also to be changed. This brings us to the following question:

(5) What *regimes* characterize the developments of aggregated inquiring systems?

We are here reminded of Kuhn's well-known views on developments in science, according to which *regimes* called normal science are interrupted by more or less severe crises, revolutionary periods. Every such tumultuous period is followed by a new and different regime.

Remark. I use the term "regime" rather than Kuhn's own term "paradigm" in order to stress a rather obvious analogy between how Kuhn conceives a scientific change and what happens on the political arena where political regimes are interrupted by crises sometimes amounting to revolutions, subsequently followed by new and different political regimes.

This list of questions may be extended to include quite a large number of questions, but I will neither do so nor do I intend to discuss any of the queries above.

Higher-Order Contexts

What is the next batch of contexts above fields of research?

This question gives rise to some embarassment.

Should we place a science like physics or like psychology or like history on the next rung, the rung of order (3), or what other contexts should be placed there? I really don't know.

The notion of field of research is so vague that an established discipline such as oceanography may very well qualify as an example of a field of research.

There are on the other hand established sciences such as biology or sociology which are rather aggregates of several fields of research than single fields of research.

In face of my embarrassment regarding what kinds of contexts ought to be placed on the next rung above that one on which fields of research are placed, I refrain from assigning any particular order to those contexts of which fields of research are ingredients. I simply mention some such contexts in a list. Some of them may very well be regarded as species of fields of research.

(1) *Monodisciplinary* fields of research;

(2) *Multidisciplinary* fields of research.

These are complexes composed of inquiring systems with territories belonging to different traditional disciplines.

Multidisciplinary fields of research may be classified into

(a) Short-lived ones. They exist only in the form of multidisciplinary projects.

(b) Long-lived, possibly permanent ones. Oceanography is a good example. Human ecology is a more recent example.

Multidisciplinary fields of research of the kind (b) may be

(b1) unestablished, or

(b2) established ones.

The question of establishment is a question about how research is institutionalized. The disciplines that are cultivated and taught in most modern universities are established, because they are institutionalized.

I continue the list of contexts with this species.

(3) Research–practice complexes.

Such a complex is composed of inquiring systems belonging to one or maybe to several established disciplines on the one hand and one or possibly several fields of practical activities on the other hand.

The technical disciplines are examples of contexts of this kind. I am referring to technical disciplines in a wide sense:

(a) to disciplines that are based on the physical sciences, technological disciplines in a restricted sense;

(b) to technical disciplines that are based on the biosciences, e.g., medical and agricultural sciences, etc.; and

(c) to technical disciplines that are based on the social sciences, and psychology, e. g., business economics and political economics.

I will not outline the climbing on "the contextual ladder" upwards step by step any further. Instead I will jump to the highest rung and look at the context we find there. Then I will move downwards from that rung until I come back to fields of research and inquiring systems again.

A World View

C. P. Snow talks about two cultures, the culture of the natural scientists and technologists, culture I, and the culture of literary men and humanists, culture II.

A person belonging to one of these cultures views the world in a different way than someone belonging to the other culture.

C. P. Snow's two cultures belong to the modern Western world. The culture of the church fathers and the culture of the prophet Mohammed and his followers are very different from C. P. Snow's two cultures. How a person looks at the world depends thus very strongly on his niche in the sociocultural world. There is no such thing as *the one and only correct* world view, but one world view may be more correct than another one.

Someone who belongs to C. P. Snow's culture I is most likely to be strongly fascinated by Mario Bunge's world view, as presented in his very important treatise *Basic Philosophy*, Volume 4.

Bunge's World View in a Very Brief Outline

The world is a system composed of systems, viz.,

(a) chemical systems,
(b) biochemical systems,
(c) living systems,
(d) mental systems,
(e) social systems.

Bunge presents his world view in a rigorous manner making use of mathematical tools. He exploits heavily recent results of modern science in his construction work.

His world picture is probably more true than most other world views and I feel certain that it must appeal strongly to persons living in C. P. Snow's culture I. But I am not so sure that people living in Snow's culture II are enamored with this picture to the same extent. They may feel that he misses some very significant features of the world, such as the significance of religious thought, literature, art, and music. Anyhow, I will try to offer a somewhat different picture of the world, which to be sure is far less rigorous and sophisticated than Bunge's world view.

Outline of an Alternative World Picture

(1) There are two compartments of the world:

(a) the sociocultural sphere SC and
(b) nature N.

I use the term "sociocultural" to indicate that it may be impossible to separate culture from society. It is very hard to imagine a society, however primitive, without a culture, or a culture without a society.

(2) Nature is divided into

(a) nonliving nature, which might be called the "physissphere" Ph, and
(b) living nature, the "biosphere" B.

There is also an intermediate zone between these two spheres

(c) the world of biomolecules BM.

(3) The sociocultural sphere overlaps nature. This area of overlap SCN may be described as nature colonized by man which is true in our time or as nature encroaching upon the sociocultural world, which was true long before our time. This part of the world SCN contains agricultural plots, roads, junk heaps, and polluted lakes. It includes also all material artifacts which man has manufactured, such as books, butter, and guns. This picture of the world is portrayed in Fig. 4.

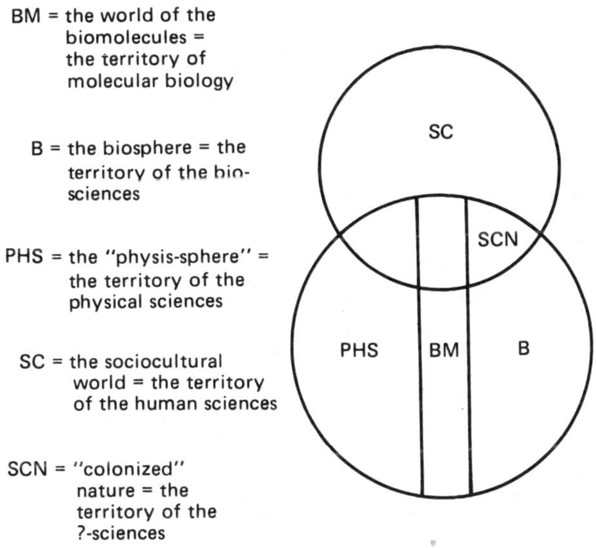

FIGURE 4. A picture of the world.

Comments

The region SCN of the world is examined within a very heterogenous assortment of fields of research.

Historians of art study some artifacts but not all.

Archeologists study some artifacts but not all.

Agricultural scientists study cultivated plots but not the whole realm SCN. Historians of technology study some features of SCN but do not examine SCN as a whole and so on.

The new discipline of human ecology has possibly the most integrative approach to this area of the world, but it is not all-inclusive either.

I have indicated the heterogenous character of the assortment of fields of research which have territories in SCN by means of the phrase "the territory of the ? sciences."

I will now offer a few broad characterizations of the various compartments of the world.

(4) There are developments in every compartment.

(a) There are lines of development from the physissphere through the world of biomolecules into the biosphere.

(b) There are lines of development from the simplest to the most complex forms of life.

(c) Ecosystems rise and decline.

(5) There are processes of development in the areas SC and SCN.

(6) The developments in nature are very different from those in the sociocultural

world at large including the area SCN. The times scales and the mechanisms are different. More about this later on.

(7) The inhabitants of the world carry attributes, i.e., properties and relations of three main kinds. To present them as simply as possible let us take a book as an example.

(a) The color of its cover, the number of pages, and the number of lines of a particular page are attributes of the same kind as we find everywhere in the physissphere. They may be referred to as *neutral* attributes.

(b) The meanings of the words, the sentences, and the chapters are attributes of quite a different kind from the neutral ones. It is appropriate to call them *meanings*.

The term "meaning" will be used in the widest possible sense in this chapter, so as to subsume other meanings as well as those that are attached to linguistic phenomena. Gestures and facial expressions may have meanings. A national flag and a cross have meanings. Musical compositions have meanings. Actions have meanings. Intentions of actions are their meanings. Artifacts have meanings. A possible use of a tool for instance is a meaning it has.

It would be extremely difficult if not impossible to define the concept of meaning in the wide sense in which it is used in this essay. It is no different from the concept of matter in this respect. The concept of meaning is similar to that of game. Both of them are family concepts in Wittgenstein's sense.

(c) A book has also a third kind of attributes. It has this third kind of attributes in so far as it is worth reading. An attribute of this kind may be referred to as significance or value. The notion of significance or value is another example of a family concept in Wittgenstein's sense.

Critical assessments of a literary work or a sculpture are concerned with attributes of the third kind.

Critical assessments within inquiries are concerned with the significance or value of results: It is a significant property of an hypothesis how true it is. It is a significant relation between a problem and a method how well the method is fitted to the problem. It is a significant relation between an alleged piece of evidence e and an hypothesis *h*, how relevant e is with respect to *h*, etc.

It is convenient to refer to these three kinds of attributes as

1. attributes of the kind *N* or *N*-attributes,
2. attributes of the kind *M* or *M*-attributes, and
3. attributes of the kind *S* or *S*-attributes, respectively.

(8) All attributes of the physissphere belong to the kind *N*.

This thesis is backed up by the physical sciences. Our ancestors and people living in parts of the sociocultural world, into which the natural sciences have not yet penetrated, reject thesis (8). Nature is full of meaning and significance to them.

(9) There may exist attributes of the kinds *M* and *S* in the biosphere at least among animals with advanced neural systems living together.

(10) The sociocultural world and the area SCN is full of meanings and values. They play major roles in the processes of development in those parts of the world. The actors in developments are constantly involved with meanings and values. Meanings and values are interpreted and assessed, destroyed, modified, and created all the time and everywhere, specially so in our time and in our part of the sociocultural world.

(9) The three kinds of attributes are related to each other in this way:

(a) Carriers of *S*-attributes always carry *M*-attributes as well. A book worth reading must also be intelligible.

(b) Carriers of meanings always carry neutral attributes as well. A readable book must have pages with lines printed on them.

The Place of Science in the World

In this section I will move one step downwards from the highest-order rung in the contextual ladder above inquiries.

There we meet the world of science. The word "science" will be used here in a wider sense than the normal one. Humanities will be included in the world of science.

Where is science situated in the world?

It is a part of the sociocultural world and it extends into the area SCN. All "hardware" instruments used in inquiries belong to SCN. So do scientific writings of every kind. So do buildings of various kinds in which research is carried out, etc.

The world of science includes the following:

(a) *Formal sciences*, branches of mathematics and formal logic. These sciences lack territories or rather the researchers in these fields create conceptual territories such as groups, lattices, rings, fields, and topological spaces.

(b) *Factual sciences*. These sciences have territories either in the sociocultural world, the human sciences, or in the physissphere, the physical sciences, or in the biosphere, the biological sciences, or in the world of biomolecules, molecular biology, or finally in the world SCN, human ecology, and an allotment of other sciences such as archeology, history of arts, and history of technology, etc.

(c) *Technical sciences*. These sciences are rooted in factual sciences in physics, in chemistry, in biology, in psychology, or in social sciences. Technical sciences have territories just as the factual sciences do, but researchers in these fields deal with their territories in a different manner from the way in which scientists in the factual sciences deal with theirs. They are not merely concerned with producing knowledge, they help practicians to transform the world and to build new structures into it. This conception of the world of science is portrayed in Fig. 5. The world of science transforms these parts of the world in which it is included and is in its turn strongly influenced by its environments in the areas SC and SCN.

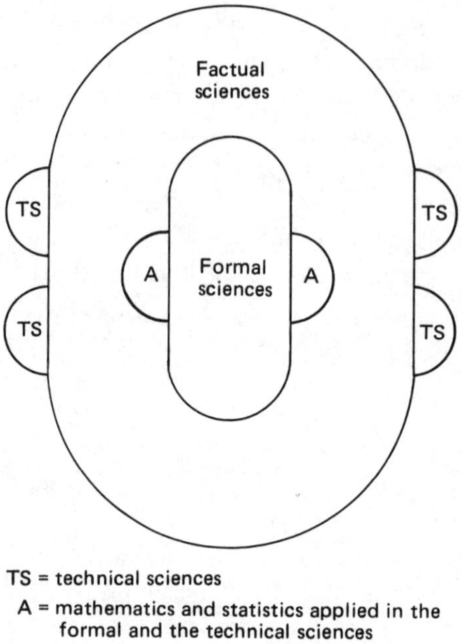

FIGURE 5. A conception of the world of science.

Different Kinds of Factual Inquiries

The picture of the world presented above suggests a simple classification of inquiring systems into

(a) positivistic,
(b) hermeneutical,
(c) critical, and
(d) hermeneutical–critical inquiring systems.

(See Fig. 6.)

(a′) An inquiring system is *positivistic,* if the inquiries in it are concerned only with neutral attributes. As there are no other attributes than neutral ones in the physissphere, research carried out in the physical sciences is and must be positivistic.

(b′) An inquiring system is *hermeneutical,* if the main work in it aims at interpretations of meanings.

(c′) An inquiring system is *critical,* if the main work in it aims at critical assessments of value attributes in its territory.

(c″) An inquiring system is hermeneutic–critical if interpretations are combined with critical assessments. The nature of the sociocultural world as portrayed above suggests that the human sciences should include inquiries of all the four kinds mentioned above. I do not intend to discuss hermeneutical and critical inquiries. They are so different in many respects from positivistic research that the ideal of a unified science of a positivistic kind seems to be an idle dream (perhaps a nightmare). The realities of the world forbid its realization.

Hermeneutic and Critical Features of Inquiring Systems

The location of science in a world full of meanings and values strongly suggests this thesis:

Thesis. There are hermeneutic and critical ingredients in all fields of research, not excluding positivistic ones (see Fig. 6). Critical ingredients in positivistic inquiries are represented by the running research program model of positivistic inquiries. I refer to the fourth component in a research program (*H*, *I*, *P*, Cr, Pl).

No more will be said about this ingredient of inquiring systems. The hermeneutic ingredients of inquiring systems call for some reflections.

The Place of Man in Science

How are researchers situated in fields of research? This question is subsumed under the more general question:

How is man situated in the sociocultural world?

There are always horizons wherever a man is situated in this world.

There is a horizon for what a man can imagine.

There is a horizon for what a man can know and another one inside this horizon for what he knows.

There is a horizon for what a man can understand and another one inside this horizon for what he understands.

A scientist engaged in research within a particular field is always located in a *research situation.* He changes his research situation by means of his mental and physical actions.

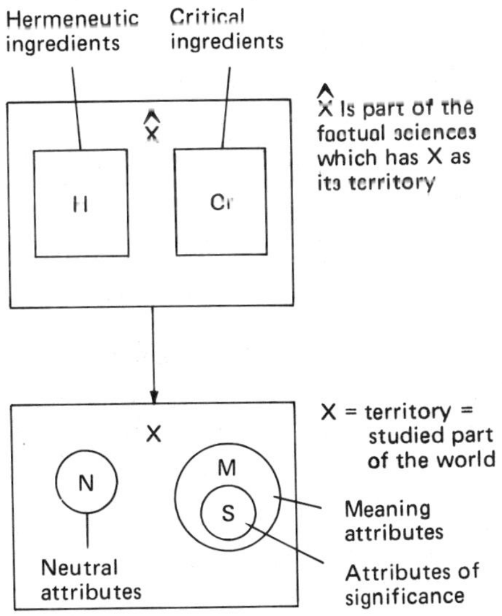

FIGURE 6. A simple classification of inquiring systems.

One research situation RS is transformed into another one RS′, which in its turn is changed into yet another one RS″, and so on.

A research situation is the situation of a scholar in a field of research.

How such a person transforms a research situation depends to a large extent on his horizon of understanding, on how he interprets what he finds in a research situation.

What kind of items does a scholar attempt to interpret and sometimes succeeds in interpreting in a research situation? This question is closely related to the question: What are the hermeneutic ingredients in a field of research?

Here follows a list of items:

(1) *persons*, the scholar himself, his collaborators, other colleagues, suppliers, and receivers of secondary information etc.;
(2) pieces of primary and secondary *information;*
(3) *problems or tasks;*
(4) software as well as hardware *instruments;*
(5) acts of *reasoning;* and
(6) *physical actions* such as operations on and with the help of artifacts.

I will not deal with all these kinds of carriers of *M*- and *S*-attributes in fields of research but will confine my attention to items (1) and (5).

MENTALITY

Developments within a field of research depend very much on the mentalities of individuals and of groups of scholars.

The notion of mentality is therefore a very important one. I will attempt to come to grips with it in this section.

First a list of statements will be presented to serve as partial specifications of this notion. Afterwards I will raise a complex question, such that answers to it contribute to the construction of a notion of mentality.

(1) To understand a person is to understand his mentality.

(2) A group of persons such as a team of researchers may have a common mentality.

(3) The notion of mentality is different from that of personality. Only individuals have personalities. There is no such thing as a collective personality but there are collective mentalities.

(4) What happens in a field of research depends to a large extent upon the mentalities of the active researchers.

Now to the question: What attributes of a researcher determine his work?

What is done by a person in a field of research depends

(a) on his *interests*, what he wants to do;
(b) on his *competence*, what he can do and what he permits himself to do; a competence is composed of

(1) a *cognitive* component, such as a capacity to raise and solve problems,
(2) an *ethical* component, a sense of responsibility for what he is doing and virtues such as impartiality and critical rationality, and
(3) an *esthetic* component, a style and a taste; and finally

(c) on his *paradigm*.

The paradigm of a researcher is a complex of opinions he holds composed of

(1) general opinions on the territory of his field of research; these opinions constitute his *world picture*;
(2) general opinions on his field of research, on how it is related to other fields, on dominant traditions in the field, on what constitutes good research, etc.; these opinions constitute his *view on science*;
(3) general opinions on kinds of tasks which seem to him important and which he would like to work on;
(4) general opinions on ways of going about these tasks; these opinions constitute his methodology; and finally
(5) general opinions on desirable results; these opinions may take the form of an *ideal of science*.

The complexes of opinions mentioned under (3), (4), and (5) may profitably be brought together into a complex which may be characterized as a *research strategy*.

I have now brought together all pieces I need as building stones to construct the notion of mentality and in combination with it two other notions, those of paradigm and strategy of research.

I. A *mentality* of a scholar = (his interests, his competence, his paradigm).
II. A *paradigm* of a scholar = (his world picture, his view on science, his research strategy).
III. A *research strategy* = (opinions on desirable tasks or problems, methodological ideas, opinions on desirable results).

Comments

(1) Paradigms serve as regulators in inquiries.

(2) No one can understand anybody's mentality completely, one's own not excluded.

(3) It is not possible to articulate all opinions belonging to a world picture or a view of science.

(4) It is possible for a scholar to fool himself and others about what opinions he really holds about the world and about his field of research and about those opinions that constitute a research strategy.

(5) No research situation is completely transparent to a researcher situated in it, or otherwise expressed.

(6) The horizon of understanding of a researcher divides every research situation in which he finds himself into a transparent and an obscure part.

(7) Faulty paradigms are likely to give rise to crises in a field of research.

(8) Misunderstandings of and failures to understand the mentalities of researchers in a field of research are obstacles of inquiries and may also in other ways bring about crises in research.

What Kuhn calls a paradigm is a regime in a field of research characterized by a number of scholars having the same mentality. A tradition of research is characterized by its followers having similar paradigms.

A school of research is an institution for transmitting knowledge, skills, and paradigms from one generation to the next.

Philosophies in Science

Kuhn's well-known scheme of the development of particular fields of research is probably not applicable to every field of research. Kuhn has not claimed that it is, so I am not now criticizing him. I believe that fields of research in many human sciences exhibit other growth patterns. What growth patterns? I leave this question open. Instead I will consider a sample of related questions:

(1) What is the structure of scientific revolutions as distinct from normal science?

(2) Why is a normal science regime followed by a period of "revolutionary science"?

(3) What do scientists do in such a tumultuous period?

Question (1) will be left open. Now to question (2).

A normal science regime is not free from crises, but scholars tend to neglect them so long as the regime lasts.

What are the causes of these crises?

A main cause of crises is faults in the paradigms of researchers. A world picture may be faulty. A methodology may not be suited to problems regarded to be important ones. An ideal of science may lead astray, etc.

Paradigms are never completely transparent. This means that researchers may very well fail to understand why they run into critical research situations.

If they take a crisis seriously as they have to do if it is serious enough, they have to deal with diagnostic and therapeutical tasks.

This is precisely what happens in those periods which Kuhn describes as revolutionary science.

They deal with a number of queries about paradigms, about issues pertaining to their worlds pictures, to their views on science, and to their strategies of research.

In their dealings with these queries scholars try to articulate opinions belonging to paradigms. They criticize some of their previous opinions, modify some of them, and they may add new ones.

They engage in discussions on various controversial issues.

These queries are so similar to traditional philosophical problems and the dealings with them are so similar to philosophical thinking and philosophical discussions, that it is reasonable to pose this thesis:

Thesis. There are philosophies *in* fields of research. There are philosophies *within* physics, *within* biology, *within* history, and so on.

These philosophies are extremely important parts of fields of research. They deserve to be studied by theoreticians of science at least to the same extent as inquiries and their results deserve to be studied, especially so if a theoretician of science has the ambition to be of service to scholars in various branches of science.

Philosophers *of* science should be familiar with philosophies *in* science, if they want to take a useful part in discussions of philosophical questions *in* physics or *in* biology or *in* psychology or in. . . . They should also know the contexts in which these philosophies *in* science are imbedded. They should know how they are related to inquiries. Here philosophers of science need the services of theoreticians of science, who in cooperation with historians of science try to model various facets of the world of science in its sociocultural setting.

BIBLIOGRAPHY

L. von Bertalanffy, *General Systems Theory* (George Braziller, Inc., New York, 1968).
M. Bunge, *Basic Philosophy, Vol. 4,* (Reidel Publishing Co., Dordrecht, 1979).
M. Bunge, *Method, Model, and Matter* (Reidel Publishing Co., Dordrecht, 1973).
M. Bunge, *Scientific Research , Vols. I and II* (Springer Verlag, New York, 1970).
E. Cassirer, *An Essay on Man* (Yale University Press, New Haven, 1944).
C. W. Churchman, *The Design of Inquiring Systems* (Free Press, Cambridge, Massachusetts, 1968).
L. Fleck, *Enstehung und Entwicklung einer Wissenschaftlichen Tatsache,* (B. Schwabe and Co., Basel, 1935).
N. R. Hanson, *Patterns of Discovery* (Cambridge University Press, Cambridge, 1958).
G. Hempel, *Aspects of Scientific Explanations* (Free Press, New York, 1965).
M. Hesse, *Forces and Fields* (Thomas Nelson and Sons, Ltd., London, 1961).
G. Holton, *Thematic Origins of Scientific Thought: Kepler to Einstein* (Harvard University Press, Cambridge, Massachusetts, 1973).
C. Houson, *Method and Appraisal of the Physical Sciences* (Cambridge University Press, Cambridge, 1976).
W. C. Humphreys, *Anomalies and Scientific Theories* (Freeman, Cooper, and Co., San Francisco, 1968).
K. Hübner, *Kritik der Wissenschaftlichen Vernunft* (Verlag Karl Alber, Freiburg/München, 1978).
F. Händle and S. Jensen, editors, *System-Theorie und Systemmechanik* (Nymphenburger Verlagshandlung GmGH, München, 1974).
T. S. Kuhn, *The Structure of Scientific Revolutions,* Second Edition (University of Chicago Press, Chicago, 1970).
I. Lakatos and A. Musgrave, eidtors, *Criticism and the Growth of Knowledge* (Cambridge University Press, Cambridge, 1970).
E. Lazlo, *Introduction to Systems Philosophy* (Harper and Row, New York, 1973).
S. Lindholm, *Vetenskap, Verklighet och Paradigm* (Science, Reality, and Paradigms) (Gebers, Stockholm, 1979).
B. E. Notting, *The Human Element in Research Management* (Amsterdam, 1959).
J. Piaget, *Main Trends in Interdisciplinary Research* (Allan and Unwin, London, 1973).

K. Popper, *Unended Quest* (Fontana/Collins, Glasgow, 1976).

H. F. Spinner, *Begründung, Kritik und Rationalitat, Bd I* (Vieweg und Sohn, Braunschweig, 1977).

H. F. Spinner, *Theorien und Metatheorien* (mimeo) (Suhrkampf Franfurtam, Main, 1972).

C. P. Snow, *The Two Cultures and a Second Look* (Cambridge University Press, Cambridge, 1964).

F. Suppe, *The Structure of Scientific Theories* (University of Illinois Press, Urbana, 1970).

W. Steegmüller, *Theoriendynamik* (Springer-Verlag, Berlin, 1973).

S. Toulmin, *Human Understanding*, Vol. 1 (Clarendon Press, Oxford, 1972).

H. Törnebohm, "An Essay on Knowledge-Formation," *Z. Allg. Wissenschaftstheorie*, **VI/I** (1975).

H. Törnebohm, *A Paradigm Shift in Theories of Gravitation* (Akademiförlaget, Stockholm, 1977).

H. Törnebohm, "Forskningsstrategier" ("Research Strategies"), No. 113 in *Reports from the Department of Theory of Science*, No. 1, Göteborg (1979).

H. Törnebohm, "Förklaringar, Teorier och Modeller" ("Explanations, Theories, and Models"), No. 110 in *Reports from the Department of Theory of Sciences*, No. 1, Göteborg (1979).

H. Törnebohm, "Inquiring Systems and Paradigms," in *Essays in Memory of Imre Lakatos*, R. S. Cohen, P. Feyerabend, and M. W. Wartofsky, editors (Reidel Publishing Co., Dordrecht, 1976).

H. Törnebohm, "Om Forskningsfält II. Tre Fallstudier från Fysikens Äldre Historia. Arkimedes, Torricelli och Newton" ("Fields of Research II. Three Case Studies from the History of Physics. Archimedes, Torricelli, and Newton"), No. 82 in *Reports from the Department of Theory of Science*, No. 1, Göteborg (1976).

H. Törnebohm, "Paradigm i Vetenskapsteorin. Del I" ("Paradigms in the Theory of Science. Part I"), No. 91 in *Reports from the Department of Theory of Science*, No. 1, Göteborg (1977).

H. Törnebohm, "Paradigm i Vetenskapsteorin. Del II" ("Paradigms in the Theory of Science. Part II"), No. 100 in *Reports from the Department of Theory of Science*, No. 1, Göteborg (1978).

H. Törnebohm, "Paradigms in Fields of Research," in *The Logic and Epistemology of Scientific Change*, I. Niiniluoto and R. Tuomela, editors (Amsterdam, 1979).

H. Törnebohm, "Vetenskapskritik" ("Science Critique"), No. 102 in *Reports from the Department of Theory of Science*, No. 1, Göteborg (1978).

H. Törnebohm, "Vetenskapsteori och Kontrollerande Vetenskapkritik" ("Theory of Science and Controlling Science Critique"), No. 105 in *Reports from the Department of Theory of Science*, No. 1, Göteborg (1978).

21

The Nature of the Physical World

J. O. Wisdom
Wilmont House
Castlebridge
County Wexford, Ireland

Wolfgang Yourgrau was both a scientist (physicist) and a philosopher of science. I have chosen a subject which involves, I think, both areas. I am primarily concerned with discussing the nature of matter, and along with that naturally go the concepts of space and time.

The nature of matter has been discussed by two very different classes of thinkers. One was philosophers and the other was scientists. Philosophers, sad to say, scarcely added to our knowledge of the subject. Scientists, by which I mean only physicists, were not given to discoursing on the concept explicitly on its own; it is by implication that we find scientific theories of the nature of matter implied in the physical theories they produced. If philosophers added nothing worthwhile, physicists, though they have contributed very greatly indeed, have left chaos. If we are to get anywhere with the subject, it is from the physicists that we must start. Nonetheless, physicists imply at least that they must have had some ordinary commonsense notion of matter to start them off; though this is more evident among the philosophers. So I would ask the reader's indulgence if I begin with the philosophers; but if he does not find them much to his taste he can skip them and go on to what the scientists have to say.

Philosophical Conceptions of Matter

Perhaps the best conception of matter that the ordinary man has is to be found in Broad. A piece of matter is something that can be picked up and put down, has some size, and color, smell, and taste, and the like; and matter is the common underlying basis of electric light bulbs, shoes, horses, and so on.

For many centuries philosophers have, I think, started from some such position and have tried to give some sophistication to the position just described. If we begin with Aristotle, as one of the most influential thinkers who considered the subject, matter was undoubtedly some such entity as was common to all material things—a highly abstract conception—in fact, he regarded it as a "substance." And "substance" in English comes from the Latin "substantia," which means something standing underneath, an underpropping, or something that supports qualities, or something in which qualities inhere or stick. So one develops the notion here of an abstract entity with qualities like color, sound, shape, and size hooked on to it. One notes in passing that matter is conceived of as a *substance that itself lacks qualities*; it possesses qualites attached to it, but does not possess any in its own right. This conception was rife throughout many centuries. It was rife throughout the period of scholasticism, which lasted for a millenium, and reappeared in the Renaissance, being made prominent by Locke. I think it is fair to say that the conception held by Aristotle, the medievalists, and Locke was the same in all. And I think it is fair to say that it is a representation in slightly sophisticated terms of the conception of the ordinary man. Now the attempt at sophistication ran into philosophical difficulties, because the idea of substance as having *no* characteristics in its own right led Locke to make this explicit in a form that was almost a joke. This came out most clearly in the topic of abstract general ideas, of which matter was a particularly important example. Locke illustrated the idea of the abstract general idea of triangle in the following way. He said it was a triangle that was neither equilateral, nor equiangular, nor right-angled, nor isosceles, nor scalene, but was *all and none of these at once*. It must be admitted that Locke very neatly expressed the broad if strange idea that has animated philosophy from Aristotle onwards and reflects something in the thinking of the ordinary man. However that may be, it called forth howls of ridicule from Berkeley. Moreover, one should remark that it is one of the very few self-contradictions to be found in the whole history of philosophy.

But I think we should also bear in mind that, idiotic as it may seem and indeed is, it reflects something of mankind's thinking about the world around him. And if mankind ends up in a self-contradiction, well that may mean only that the problem is very difficult, that is to say, that it is no easy task to come at the nature of the world around us.

About the same time came Descartes, who really added very little. He regarded matter as a substance—of that there is no question—and he specified that the substance, matter, consisted of "extension," by which, following Broad, I think he meant shape and size. In other words he was more or less Lockean though not quite, for he attributed the properties of shape and size to matter itself.

From Locke's absurd characterization I turn to the not unusual philosophical reaction to a difficult problem, i.e., when philosophers cannot satisfy themselves that a standard, normal, commonsensical view makes sense, they go to the most extraordinary extremes to provide an alternative; and in doing so they, in my view, waste the time of philosophers for a few decades, sometimes for centuries. Berkeley proposed the most outlandish theory of the nature of the natural world. This has been explained in many places with various interpretations, which I do not propose to go into now, beyond remarking that the classical interpretation of Berkeley was that matter was a mental structure in the mind of God. This is undoubtedly a wrong interpretation of Berkeley, but he did regard matter as a manifestation, like a language, of the behavior of nature, as a manifestation of the activity of God. To make this strange conception more explicit: The perceptual observations we make combine together, like words in a sentence, and convey a meaning which is a mental product of its author. Hume, following, adopted much the same attitude, but did not produce the same theory to explain material phenomena; he simply thought that the sensations with which we are endowed were associated in our minds and were not hooked on to anything such as matter—or indeed to anything else. And Humeans, a couple of centuries later, followed much the same line. Oddly enough, Schopenhauer, in the early part of the nineteenth century, before the modern Humeans, took the Berkeleian line, though the atheistic underlay—as in Hume—was different. This Berkeley–Hume–Mill–Schopenhauer–Ryle line may be described as positivistic or nominalistic; but the main point about it is that it is an extreme reaction against the notion of matter (and with some of them against other notions as well), which these philosophers could not stomach.

Going back to the Cartesian period, around the time of Berkeley and Hume, came Leibniz. Now Leibniz was a great scientist, but he also played a part in the classical line of philosophy, and he considered matter from a philosophical point of view—he was one of the few scientists to discuss this question—he regarded it as an extreme form of mental structure, and was therefore not altogether unlike Berkeley: the natural world consisted in "monads," which are mental entities but fade into materiality at the edges. It is not very easy to make much sense of this, and it had very little philosophical influence, but I think, as we shall see, it had an interesting scientific influence perhaps on Boscovich, even if it did not in itself provide us with any real knowledge of the nature of matter.

After Leibniz and most of the others mentioned, except for Schopenhauer, came the great work of Kant. Now he was a scientist (astronomer) in his spare time, and a very good one, and he was very much concerned with the nature of matter and of the natural world. For him, matter was a totally unknowable entity. But for scientific and practical purposes he held a doctrine of phenomena which, in effect, he turned into a phenomenalistic interpretation of sensory experiences. So to a certain extent he said nothing new here. He was not, however, an out-and-out positivist or phenomenalist, for he held that Newtonian science was true absolutely, and as Newtonian science presupposed the notion of matter, Kant clearly believed that, in some way or other, matter did exist. Since he regarded it as unknowable he gave us very little help. Although Hegel did not discuss matter overtly, a word should be said about him. He had a very peculiar view of the world, which I will attempt to express like this: He held that the intellectual world of ideas, the most abstract and general ideas of all, culminated in a unified fashion in which all the major intellectual concepts were united in one, called the Absolute Idea. Now matter was not one of these. One could invoke something like a complementarity principle and say that the complement to the Absolute Idea of thought was the world of nature or matter. To introduce a little jargon, one could say, drawing on Hegel's notion of dialectic, that matter is "the Other of the Absolute." Whether after some further centuries this will prove to have been an extraordinarily valuable insight, I do not know; but at the moment it does not seem to offer us anything helpful.

Before leaving the philosophers we should jump forward a century to Russell and Moore in the twentieth century. Russell thought of matter as a bundle of sensations, in this following in the classical tradition of Berkeley, Hume, Mill, and so on. He gave no really coherent account of this, so he did not advance our knowledge of the subject with the aplomb that he did in other fields. Moore perpetrated a most bizarre proof of the existence of matter, which at the end of a medium length paper, comes out as follows: I hold up two hands; you see them; a hand is a material thing; therefore, two material things exist. Moore was such an astute thinker, it is amazing that he should have permitted himself such sophistry. Of course there are material things. But the problem is not so much to prove that fact as to interpret their nature.

I have left out a school of philosophers before Aristotle, consisting mainly of Leucippus and Democritus; they held that matter consisted of atoms, meaning very small entities that could not be divided into parts. And together with a theory of "swerve," which accounted for their motion, they thought of matter not only as a material substance but also as something inherently coupled with laws of motion. I suppose it is fair to say that in a sense they fathered the modern chemical notion which came later in Dalton in the last century; and the notion of swerve could even be said to be found in Einstein.

How, then, shall we view the contribution of philosophers from Aristotle or the pre-Socratics to the present day? A pretty poor performance maybe, but what did it amount to? It amounted, on the one hand, to the version of the commonsense view of the world held by the ordinary man, which it attempted to express in a slightly sophisticated way with at least a small amount of precision; and this cannot but have affected the thinking of scientists later on in some degree. In the second place, there was a reaction against this in the bizarre kind of thinking that attempted to be very hardheaded, but was in effect, simply opposed to everything connected with common sense. This possibly absurd reaction has found an echo in some modern physical thinking, as we shall see. So it is not entirely without its influence and interest. And thirdly, there is the pre-Socratic invention of the atom. We are left with the impression that, in philosophy, the earlier the better. The very early attempts had something to them, and further attempts led only to deterioration.

One more general comment on the philosophical scene: It seems highly probable that what philosophers were basically concerned with was the problem of explaining the seen world, largely by the unseen. They may have taken different views about how much could be included in the seen, and what aspects of it should be excluded from what is seen without any doubt. I think the notion of matter as an abstract entity lying behind the seen world is presupposed in the thinking of nearly all scientists. If we consider the many classes of negativists, called "positivists" since they first arose, phenomenalists, and so on, even they were trying to give an account of the seen world, though they did so by denying that there was a problem, or more simply put, by claiming that no explanation was needed. The atomists, who had a wider conception of the physical world, not only wished to explain the seen world in the sense of what is quite still and static in front of us, to be observed, but also wished to explain movement; not only had an unseen explanation for it, but even endowed it with a fairly specific form.

To all this I would add a further point: that mankind's conception of matter is usually restricted to solids. We do not find much in the history of philosophy about liquids, though there is a mention in Thales, who held that a liquid, namely water, was the ultimate reality behind all seen things, and we also find a mention in Heraclitus—once again the early Greeks score heavily. But almost no one included the conception of a gas. It is true that the immediate successors of Thales did attribute the ultimate reality to air; though both with Thales and his successors explanations in terms of water or air were explanatory; these were also part of the general conception of matter. So on this score we must end with the rather uninspiring conclusion that mankind has had only the most rudimentary conceptions of matter.

With this, I conclude my remarks on the philosophy of the subject, the most positive point, perhaps, being that the purpose of the philosophizing through the millenia was to explain the observations we have of the seen world.

Scientific Conceptions of Matter

The Greek conception of an atom already referred to and put under the heading of philosophy should also be included under the heading of science, but here I am mainly concerned with the post-Renaissance era, from Newton to the present day.

Newton was unquestionably an atomist; he regarded matter as very small corpuscles bound together by some force or other; he was not very specific about it. But his conception of matter did include more than that, because his gravitational theory involved a force of attraction between lumps of matter acting at a distance. Moreover, it would seem that he regarded this power of attraction as exerting itself at infinite speed, or, otherwise expressed, that it exerted its power instantaneously—he did not like the notion of "action at a distance" but he seemed to be forced to adopt it.

Since physics has been so largely Newtonian ever since, one might get the impression that there was nothing else offered as an alternative, but this is not so, and in fact, very important alternatives were around, even from his time. In Newton's own day there was the wave theory of Huyghens. Now although this was, of course, a theory of light, and a wave theory, not a particle theory, not an atomic theory, nonetheless light was regarded as a material in some sense or other by both Newton and Huyghens, so here we have a dichotomy straight away from the seventeenth century between an atomic theory and a wave theory—and the battle between these two has been going on ever since.

But at the moment I wish to stray not into wave theory but into an alternative point of view that arose not so very long after Newton. Boscovich, who was a Yugoslav mathematician and physicist in the eighteenth century (in fact he covered most of it)—and it is possible that he has not even yet received his due as a physicist—Boscovich invented the notion of a point-atom. The notion may, of course, appear bizarre and unintelligible, but is none the worse for that, in fact that feature is characteristic of many fundamental concepts of physics, and physics flourishes very well on such concepts. This particular notion is analogous to the notion of a Euclidean point, which has been notorious as a contradictory geometrical conception, for the Euclidean point meant a particle of no size. The Boscovich notion was, in effect, the same thing. The difference was that the Euclidean point was a position, while Boscovich's point-atom was a piece of matter.

Now the question might be raised, why should any physicist take such a notion seriously? To this a very simple answer may be given, namely, that Faraday, at the turn of the nineteenth century, put forward a not altogether dissimilar notion, but much wider, and just as unintelligible to his peers at the time; this was the notion of "tubes of force." At first sight, of course, it might seem that there is no conceptual difficulty about this notion of Faraday's, but when one looks into it one sees that he meant something totally nonmaterial. Tubes of force were nonmaterial tubes but they existed in space. Since a point-atom for Boscovich would enter into a gravitational pull, there is not after all such a vast distance between his notion and Faraday's.

This nonmaterial kind of entity that was supposed to populate the material world reached its zenith with Maxwell in the middle of the nineteenth century (unless we prefer to reserve that credit for Einstein) with his conception of the ether. Now it is not to be forgotten that Maxwell had great difficulty in understanding Faraday for some years, but took him immensely seriously, studied him very closely, and in the end, mastered Faraday's ideas and developed them. He was indeed the first to do so. Maxwell, whose magnificent electromagnetic theory began to make its appearance in 1862 and came to full fruition in 1864, could not manage to handle the phenomena of electricity, magnetism, and light without the notion of an ether as an all-pervasive medium existing throughout all space in which these forces or entities existed and operated. Maxwell's functional theory of electromagnetics is with us still, but the ontological underlay, the ether, has had to be given up. The reason is that certain experimental evidence to do with the velocity and behavior of light turn out to be incompatible with the ether. Still we see to what lengths abstract conceptions can govern the thinking of the greatest physicists from Boscovich to Faraday and Maxwell, and here it is worthy of remark that Leibniz, though concerned with totally different matters, really was father to the same tradition; for his monad was, in effect, the same sort of thing as the Boscovich point-atom, and Leibniz, one feels, would have experienced no difficulty at all in following Faraday's notion of tubes of force, nor indeed in following Maxwell's notion of the ether (and of "displacement current"). A peculiarity of the conception of the ether is that it is never quite clear whether the ether was material or nonmaterial. Conceptually, it seems to have been nonmaterial, but something in which material things could operate, and something that could exert influences on material things in the same sort of way as other material things could. In particular, the ether was supposed to be able to slow up the passage of light much as water can. What we gather from all this is that physicists of this cast of mind are willing to embrace a conception that is somehow material and somehow nonmaterial.

We may notice this peculiar paradoxical quality in the classical controversy that has developed between atomism and wave theory, for atoms—apart from Boscovich—have always been regarded as material things, whereas waves have a peculiarly nonmaterial character. It might be going too far to say that the wave approach of Schrödinger is the heir to the nonmaterial approach of Boscovich, Faraday, and Maxwell; still, even if this is an exaggeration, it is an exaggeration worth making, perhaps.

We now enter the really modern period which began with the notion of a "quantum" due to Planck towards the end of the last century, perhaps more accurately dated at 1901. Now Planck's concept of a "quantum" hinges fundamentally on energy (though the notion of a quantum is still more abstract than that of energy). Energy was, of course, the dominant concept all the way from Newton onwards but so far we have not paid it much attention and perhaps it is time we did. Energy of the kind in question is energy in virtue of the motion of a piece of matter. If the piece of matter is at rest, then it does not have this kind of energy. Still, when it has it, energy is material in the sense that it is rooted in the mass of the piece of matter—whatever that may be. Planck was not attempting to alter the physical notions of matter at all. His fundamental discovery concerned a peculiarity of the way matter behaves: this was his discovery that, when a body is radiating heat, it emits the heat in little puffs, so to speak—one quantum may be emitted, or two quanta, or a hundred-and-one quanta but not ninety-nine and a half. Now this conception is certainly atomic. It does not add very much to our conception of matter, but it does add something, and that something is really very significant or at any rate very different from anything that had ever been thought of before: The quantum theory asserted that there was some feature of matter that could exist only in discrete amounts, whereas all down the ages it had been supposed that matter could be divided, and divided again, and redivided continuously, at least in principle. It is true that the notion of the atom deriving from the Greek atomists was such that the atom was not further divisible, but no one had entertained the idea that there was in fact a definite limit below which one could not go.

Following on Planck, it is appropriate to mention the important contribution made by Rutherford and Bohr. According to the celebrated Rutherford–Bohr theory of the atom, the core of the atom consisted of protons, and around it circulated, somewhat as in the solar system, varying numbers of electrons. This, at least in its initial form, has the look of being an atomic theory. It connects with the quantum theory in that the behavior of the electrons had quantum properties.

In connection with certain anomalies in the behavior of atoms and waves, some of Bohr's successors deny that there are anomalies once we give up the notion of material particles or waves underlying the observed phenomena—here they are exactly like the positivists.

But in the twentieth century came the astonishing conceptions of Einstein. I think one should class him as an atomist—mass, at least, was for him very material—but what made Einstein's conceptions so very peculiar was his concern with the concepts of space and time. Space-time (if we hyphenate them as his theory required) had the peculiar property of being *dynamic,* for if a material object moved in a special way, such as a curve, this was ascribed to a peculiarity in the nature of the space-time in which it moved, and one even finds the notion that matter itself is a peculiar entity created by a distortion in the space-time. At least matter, space, and time became inextricably intermingled with Einstein. We note that Einstein was following more in the tradition of Faraday and Maxwell than in the tradition of Planck, Rutherford, and Bohr, insofar as he introduced two highly abstract conceptions, as they did, and endowed them with dynamism. Further, the concept of a "field" played a huge part in Einstein's physics, and this was a notion that he derived from Faraday and Maxwell. In fact, the concept of a field adds something very peculiar and hitherto unthought of to the notion of matter.

But we are not yet quite at the end of the strange conceptions that arose in the early part of the century. Schrödinger developed wave mechanics partly based on the earlier work by de Broglie. Now Schrödinger's work was in the tradition of Planck, Rutherford, and Bohr, not in the tradition of Faraday and Maxwell. Superficially, there might appear to be nothing conceptually very new here, but in fact there is, because Schrödinger's concept of a wave was firstly more general than any kind of wave that ever is experienced in water, air, and so on, and secondly, it is a wave that is entirely abstract for it undulates on nothing at all. We cannot but notice how far we have departed from philosophical common sense here in that for the ordinary man or for philosophers down the ages, waves would have been properties *of* something—waves *of* water, waves *of* air, or whatever, but no one had conceived of there being just bare waves, and I think it is fair to say that an abstract general wave of this sort is entirely nonmaterial—even though in Schrödinger's physics waves do have effects upon material things.

How, now, are we to try to summarize this small group of very disparate conceptions?

We may note in the first place that it is not a simple dichotomy between particle or atom and wave, for an atom was never simply just a little bit of matter; it had dynamic properties. The earlier forms of atomic theory undoubtedly made a dichotomy between matter and space and time; for the atom was a material entity dropped like a lump of sugar into a container consisting of space and time. That, as we would now regard it, is a crude notion which has disappeared in the Einstein revolution. So the conception of an atom now is at least the much more sophisticated one of a piece of matter intimately bound up with the structure of space and of time.

We may note also that the notion, which Newton himself disliked, that of gravitational effects and the motion of light taking place at infinite speed, has also disappeared in the Einstein development. So one could say that atomic theory had improved very greatly even though at the risk of being outside the realm of ordinary intelligibility. One notes also the connection between the atom and dynamism, and the notion of a field.

Now although I have said that the tradition instigated by Faraday and Maxwell leads up to Einstein, this is an overstatement because the wave theory of light and the wave theory of electromagnetic waves in Maxwell also leads up to Schrödinger. For in Schrödinger we find that matter consists in nonmaterial waves, but that these waves can affect the behavior of ordinary material kinds of matter.

The difference between the Einstein universe and the Schrödinger universe is quite stark, and brings to a head the dichotomy that has been in physics for centuries. A former student of mine, now Professor Cliff Hooker, once expressed the difference between the two in a very neat form; he said that in relativity, Einstein introduced a revolutionary conception of space and time, but stuck to a classical notion of matter, whereas in quantum mechanics Schrödinger relied on a classical notion of space and time but introduced or reutilized a totally revolutionary conception of matter. I think Hooker is entirely right (even if slightly too strongly put). What his remark brings to the surface is that there is no way of fitting relativity and quantum mechanics together because they have entirely disparate conceptions of the three fundamental entities, space, time, and matter. A glorious opportunity, therefore, opens up before the philosopher to think out a new conception altogether of space, of time, and of matter, which shall be serviceable both for relativity and for quantum mechanics. This is a project that could be tackled either by a philosopher or by a physicist—though it might be remarked that a physicist contemplating such an enterprise would be taking up a piece of philosophy; and if physicists think that philosophy is no part of their job, the position here indicated shows them their mistake, for the philosophical development of a new conception of space, time, and matter is surely one of the top priorities of the day.

Concluding Remarks

What are we to make of this plethora of heterogeneous basics which rather resembles a stew into which the cook has piled every kind of herb (from thyme to hemlock)? I think it would be a mistake to seek a solution to a problem that has been growing steadily for a very long time, and which has defied solution for two thousand-six-hundred years. All I propose to do is to highlight the problem and the main factors that appear relevant to inventing a solution, but I shall add very little in the way of guesswork.

The problem seems to me to be to explain sensory appearances and motion in the physical world.

Two postulates would seem to have been made by all thinkers who take the problem seriously, that is everyone except positivists in general and quantum mechanics positivists in particular. All others maintain that there is some sort of substratum underlying sensory appearances which possesses dynamic properties.

The central ideas that have made their mark constitute a fairly short list: point centers of force; transmission of force; the hypothesis that centers of something (perhaps centers of force, perhaps centers of matter) set up a field; transmission of quanta; matter regarded as a function of space-time.

But this is not quite all because certain quantum mechanical properties of matter, space, and time, must be recognized. Thus, from the uncertainty principle, which we have to accept since it is a logical mathematical consequence of Schrödinger's wave equation (which is here assumed), it easily follows that mass is not zero but must be quantized. From the same principle it follows, also easily, that length and time must be quantized. From the equations of de Broglie it is easy to derive that frequency and wavelength must be quantized also. Moreover, since velocity is obviously quantized, we have the important result that the velocity of light is quantized. It is an obvious consequence that force must be quantized also. Apparently, therefore, so must a field.

Alas, we are, as yet, a very long way from understanding how to introduce the quantized concepts that are basic to both quantum mechanics and relativity into relativity theory itself. The only progress that I suppose one could claim over a century is that utterly abstract notions like these are indispensible to whatever theory we develop, and there is no need any longer to be suspicious of them. We are long familiar with how to work with them, even though they have no ordinary understandable meaning.

Although, when a state of science is in such a state of chaos, and is so backward, the best course to pursue is that of the angels, that is to fear to tread where fools rush in.

The justification for casting a rapid eye over the history of the subject is not because the historical approach is a sound one or necessarily leads anywhere, but simply because some of the historical ideas may contain a glimmering of something important which may easily become overlooked.

I will now rush in with one very small comment, so brief that it will not waste anybody's time very much.

One present state of the inventory of fundamental physical forces is that there are four of these. In addition to the classical electromagnetic and gravitational forces, there are two newly discovered ones: a "weak interaction" and a "strong interaction," about which little is known. These may be regarded as internal with respect to the nucleus, insofar as they are interactions between minute entities only, as contrasted with the classical forces which can operate on macroscopic entities. The two new interactions operate on such tiny entities that they are hard to detect. Since we may take the two classical ones as quantized, there may be short intervals in which they are not operating. What I mean by this, i.e., by quantization, is that not only do various quantities exist only in discrete and not in continuous amounts, e.g., that action less than h cannot exist, but that in consequence there will be minute quantum gaps in interactions, like the chugging of a railway steam-engine. Now it might be conceivable that these two internal forces could rear their heads to exercise some influence further afield during intervals when the surface forces are inactive through a quantum coincidence. If the internal ones are quantized also, while they might sometimes be capable of operating while the surface ones were out of action; at other times all four might be resting. On the other hand there is the faint possibility that the internal ones might be continuous and not quantized forces at all. If it were the case that there are times when the internal forces get through, and other times when they do not—due to their being quantized—then we might be on the track of explanation of some very special phenomena.

If I have put this in a way that is confusing to the physicist, a simple model will bring out what I mean. A conjurer sometimes produces four rings which he can hang down interlocked as a chain. But, invisible to the audience, there is a small gap in each enabling him to separate them all. The idea suggested above is that, through quantization, all four might on occasion simultaneously separate and produce an atypical effect.

I do not know whether Wolfgang Yourgrau would have agreed with any of this, but I hope at least he would have been interested. I have had considerable doubt as to whether a paper such as this was really worth writing, or rather publishing; so I shall be content if I have succeeded in underlining the appalling inconsistencies underlying modern physics, greatest of all our sciences, and if I have succeeded in showing that the problems involved are both problems of physics and problems of philosophy (of physics)—indeed the boundary between the two ceases to be of importance.

22

EARLY ATTEMPTS AT A UNITARY UNDERSTANDING OF NATURE

JAMES F. WOODWARD
Department of History
California State University
Fullerton, California

INTRODUCTION

It is well known that since the turn of the century, and particularly after the development of general relativity theory by Einstein, many attempts have been made to construct a unitary theory of matter and all of its interactions. This work, for the most part, was carried on by a handful of physicists (including Einstein) and was not considered to be in the mainstream of the advancement of physical understanding. In part this was due to a demonstrable lack of success in accomplishing this goal by those so engaged. Particularly intractable was (and is) the problem of interrelating the gravitational and electromagnetic fields in a physically nontrival way. During the past decade or so enthusiasm among physicists for the possible unification of our theoretical understanding of fundamental phenomena has been growing. The work that has inspired this shift in attitude is the ostensibly successful unification of the weak and electromagnetic interactions in the non-Abelian gauge theory of Weinberg and Salam, and the experimental confirmation of key predictions of this theory. The strong and gravitational interactions remain to be incorporated into a fully articulated theory, but many seem confident that this may be accomplished within the foreseeable future. For example, Freedman and van Nieuwenhuizen have recently remarked

> In the past 50 years remarkable progress has been made in identifying the elementary particles of matter and in understanding the interactions between them. Of course, many problems remain to be solved; two of the most fundamental ones concern gravitation. First, it is not understood how gravitation is related to the other fundamental forces. Second, there is no workable theory of gravitation that is consistent with the principles of quantum mechanics. Recently a new theory

> of gravitation called supergravity has led to new ideas on both these problems. It may represent a step toward solving them.[1]

It is not so well known, however, that the present optimism with regard to unitary theories of matter and its interactions is not the first time in the modern period (historically speaking) that such enthusiasm has been felt. Almost a century and a half ago similar views were current. Compare, for instance, with the quote above, Charles Babbage's remark that

> Ever since the period when Newton established the great law of gravity, philosophers have occasionally speculated on the existence of some more comprehensive law, of which gravity itself is a consequence. Although some have considered it vain to search for a more general law, the great philosopher himself left encouragement to future inquirers; and the time, perhaps, has even now arrived, when such a discovery may be near its maturity. It would occupy too much space to introduce many illustrations of this opinion; there is, however, one [theory] which deserves attention, because it is not merely a happy conjecture, but the hypothesis on which it rests has been carried out by its author, through the aid of profound mathematical reasoning, to many of its remote consequences.[2]

The theory to which Babbage was referring was that devised in the mid-1830s by Ottaviano Fabrizio Mossotti.[3] In view of the renewed interest in unitary theories, it is perhaps of interest to reexamine Mossotti's theory and the reaction it engendered, particularly Faraday's reaction to the theory.

Mossotti's Theory

The era in which Mossotti constructed his theory was one in which a number of now resolved issues had yet to be fully elucidated.[4] While Dalton's *chemical* atomic theory had become widely accepted by the 1830s, the *physical* atomic theory was far from well received. Although electricity was known to be associated with matter, neither the one-fluid theory, where positive electricity was viewed as a property of matter and negative electricity was treated as a fluid, nor the two-fluid theory, in which both electricities were treated as fluids distinct from matter, was generally accepted. The caloric theory of Lavoisier and Carnot, however, was widely accepted, and the basic constituents of matter, particles of some sort, were viewed as being cloaked in atmospheres of caloric. The interrelationship of electricity and magnetism, and electromagnetic induction had only recently been discovered by Oersted and Faraday, respectively. The transverse wave theory of light was only then slowly gaining acceptance. And while morphological studies in crystallography had been underway for some time, the short-range forces believed responsible for crystal structure and cohesion were an as yet unopened book.

Even in metaphysical assumptions as regards the way in which nature should be understood, there was no universal consensus. On the continent thought was heavily influenced by the legacy of Cartesian and Leibnizian metaphysics and the formal/mechanistic approach, with its attendant successes, employed by Lagrange and Laplace. In England, and to some extent in Germany, on the other hand, the dominant metaphysical concept, mixed and sometimes at odds with the vestiges of naive Newtonianism, was the doctrine of the unity of nature.[5] From this sprang the doctrines of the conservation of force (not energy) and the interconvertibility of the forces of nature. The principle of the conservation of energy later emerged from these doctrines.[6] And they also influenced Faraday's work, which we shall consider presently.

Mossotti, having been educated at Milan and having spent several years seeking a position in England, was exposed to both the British and continental views on physical theory. Thus his work, published in 1836, is both formal, and its object was to reduce all interactions of matter—including gravitation—to electric interactions. As such, Mossotti's theory probably has the distinction of being the first formal unitary theory of matter and force.

Mossotti's primary object in his memoir of 1836,[7] contrary to what one might expect, was not to demonstrate the electrical origins of gravitation. It was to show explicitly, in the first approximation, that the one fluid theory of electricity of Franklin and Aepinus, with one additional assumption, is not only compatible with Newtonian gravitation, but, more importantly, could serve to explain the detailed structure of gross matter. In particular, Mossotti, sought to elucidate the nature of what he called "intermolecular" forces and to show with mathematical rigor from first principles that stable configurations of "molecules" should occur were his additional hypothesis granted. The additional supposition he made is that the electrical forces of attraction and repulsion differ slightly.

Mossotti began his *memoir* of 1836 with a discussion of the relative merits of the one- and two-fluid theories of electricity, pointedly noting that Poisson's formalism—then widely believed to be compatible solely with the two-fluid theory—is equally applicable to both. Mossotti then suggested that there are reasonable grounds for rejecting the two-fluid theory—unless one of the fluids is supposed to be condensed into and indissociable from matter, reducing, in effect, the two-fluid theory to the one-fluid theory. Mossotti's reason for rejecting the two-fluid theory was relatively simple: the equilibrium distribution of two freely moving electrical fluids in a body made up of particles of matter, irrespective of their relative proportions, will be homogeneous. Thus, if merely attractive gravitational powers are associated with the ponderable particles of matter of the body viewed as distributed through the electrical fluids, the body will undergo gravitational collapse—an idea considered manifestly at variance with experience and patently ridiculous until quite recently. As a consequence, Mossotti argued one of the electrical fluids must be associated inseparably with ponderable matter (the atoms of molecules in his usage) to preclude collapse. Ignoring, for the moment, the question of gravitation, this yields immediately the one-fluid theory of electricity wherein matter (nucleons) *repels* other matter and attracts electricity (in the form of a negatively charged ether).

Now, according to Mossotti, two explanations of gravitation are possible. One is the orthodox view that *in addition* to repulsive electrical forces, inertial matter is endowed with attractive powers of a Newtonian nature. The other is that already alluded to—that the coupling constants for electrical attraction and repulsion differ slightly. Mossotti felt that the ascription of both attractive and repulsive powers to inertial matter was redundant if gravitation could be accounted for as a residual electric effect. He, therefore, opted for different coupling constants and set out to show that both gravitation *and* short-range intermolecular forces were a consequence of this hypothesis.

First, Mossotti assumed space to be filled homogeneously with a negatively charged electrical ether, the atoms of which were so small that it could be treated as a continuous fluid. Next, he imbedded atoms of positively charged inertial matter in this ether-filled space. Initially, the atoms of matter would repel each other, and attract ether atmospheres to themselves. But once they had acquired ether atmospheres, they should form molecules and undergo long-range gravitational interactions. Mossotti then noted that

> To apply the formulae [that stipulate the inverse square interactions of matter and ether] which we have found, for the purpose of representing [inter]molecular action, to the phenomena of the interior constitution of bodies [atoms and molecules], requires methods of calculation which are not yet developed, and which must become still more complicated when the arrangement of the molecules, their form and their density, are taken into consideration.[8]

Ultimately he used this as a justification to introduce the simplifying assumption that molecules may be taken to be spherical distributions of matter of uniform density with ether atmospheres ranged around them.

Mossotti first considered the general problem of the stationary, or equilibrium, distribution of the ether in the presence of arbitrary distributions of matter. One can write down immediately the scalar potential functions for the interaction of the ether with itself, F, and the matter with ether, G. They are

$$F = \int_v \frac{f\rho}{|\mathbf{r}' - \mathbf{r}|} dv$$

where f is the coupling constant for the interaction between like charges, ρ is the ether density at $\mathbf{r}'$, F is computed for $\mathbf{r}$, and the integration is to extend over all space except those regions occupied by matter; and

$$G = \int_v \frac{g\rho'}{|\mathbf{r}' - \mathbf{r}|} dv$$

where g is the coupling constant for the interaction between unlike charge, ρ' is the density of matter (positive charge), and the integration is to extend over all matter in space. Now, the force, F, on the ether of density ρ_0 located at $\mathbf{r}$ will be

$$\mathbf{F} = -\rho_0 \nabla F + \rho_0 \nabla G$$

F is likewise equal to the negative gradient of the pressure in the ether at **r**, or

$$-\nabla P = -\rho_0 \nabla F + \rho_0 \nabla G$$

Mossotti next claims, by analogy with an argument of Laplace,[9] that

$$p = \tfrac{1}{2} k \rho_0^2 \tag{1}$$

where k is a constant. For the moment we will assume the validity of this relation. Taking the gradient of both sides, we have

$$\nabla p = k \rho_0 \nabla \rho_0$$

which, when substituted into the equation for F, yields

$$-k \nabla \rho_0 = -\nabla F + \nabla G \tag{2}$$

which may be integrated term by term to give

$$-k \rho_0 = -F + G + C \tag{3}$$

where C is a constant of integration.

Equation (3) does not uniquely stipulate ρ_0 since F is a function of ρ and ρ_0 is merely the value of ρ at some point. It does, however, prove that ρ_0 calculated for an ensemble of particles of matter is merely the algebraic sum of the contribution for each particle considered separately. That is, the superposition principle applies in calculating ρ_0 in the presence of several particles of matter. To find ρ_0 we take the divergence of both sides of Eq. (2) obtaining

$$-k \nabla^2 \rho_0 = -\nabla^2 F + \nabla^2 G$$

At r we will have

$$\nabla^2 F = 4\pi f \rho_0$$
$$\nabla^2 G = 0$$

So

$$\nabla^2 \rho_0 = \frac{4\pi f}{k} \rho_0 \tag{4}$$

With Mossotti's simplifying assumption, mentioned above, the solution of Eq. (4) is straightforward. Consider a spherical particle of matter of radius δ centered on the coordinate origin with a radically isotropic charge distribution. In this case, for $r \geq \delta$, Eq. (4) becomes

$$\frac{\partial^2}{\partial r^2}(r\rho_0) = \frac{4\pi f}{k}(r\rho_0) \tag{5}$$

by virtue of the fact that

$$\nabla^2 \rho_0 = \frac{1}{r}\frac{\partial^2}{\partial r^2}(r\rho_0)$$

Taking $\alpha \equiv (4\pi f/k)^{1/2}$, Eq. (5) has the well-known solution

$$r\rho_0 = ce^{\pm\alpha r}$$

The boundary condition that $\rho_0 \rightarrow 0$ at infinity eliminates the positive exponent solution, so

$$\rho_0 = ce^{-\alpha r}/r \tag{6}$$

This quantity, ρ_0, must be added to the uniform ether background density to obtain the total ether density in the environs of a spherical particle of matter.

In order to obtain Mossotti's force law, consider the mutual force, F, acting between two identical particles of vanishingly small radius δ. Invoking Gauss' law, we have,

$$F = \frac{fq^2}{R^2} - \frac{gqQ}{R^2} \tag{7}$$

where R is the distance separating the particles, and q the source strength of matter. The first term on the right-hand side of Eq. (7) is the mutual force due to the action of the particles on each other; the second term is due to the action of the ether atmosphere of one particle on the other. Thus Q is the source strength of the ether atmosphere of each particle interior to a sphere of radius R (centered on either molecule). Since the uniform ether background does not contribute, Q is

$$Q = 4\pi C \int_\delta^R re^{-\alpha r}dr, \qquad \delta \simeq 0$$

Carrying out the integration and substituting in Eq. (7) we get,

$$F = \frac{fq^2}{R^2} - \frac{gq}{R^2}\frac{4\pi C}{\alpha^2}[1 - e^{-\alpha R}(1 + \alpha R)]$$

With the obvious normalization $4\pi C/\alpha^2 = q$, this equation becomes

$$F = \frac{q^2}{R^2}(f - g) + \frac{gq^2e^{-\alpha R}}{R^2}(1 + \alpha R) \tag{8}$$

For very large R, the exponential term is essentially zero, and since g is greater than f by hypothesis, an inverse square attraction obtains. The short-range behavior of this force, however, is obviously non-Newtonian. It is depicted in Fig. 1. In Mossotti's eyes the most interesting feature of this force law was that $F = 0$ if

$$e^{-\alpha R}(1 + \alpha R) = 1 - (f/g)$$

Therefore, if the molecules are separated by a distance $R' \simeq 1/2\,(f/g)^{1/2}$, they will experience no mutual force. Moreover, the equilibrium at this point is stable since displacement in either the positive or negative R direction gives rise to a force that tends to restore the condition of equilibrium. This theory, at once explaining (at least

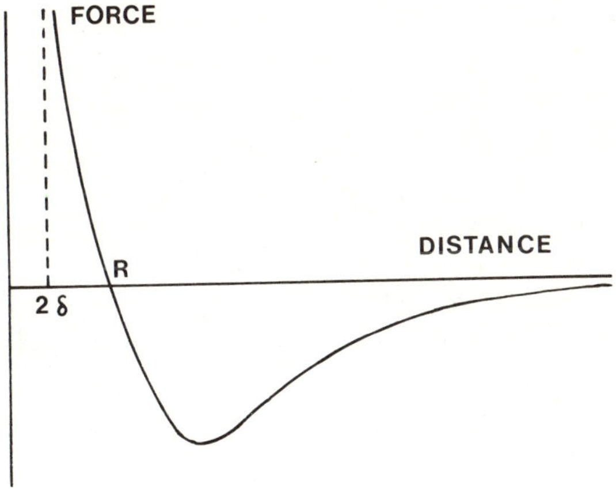

FIGURE 1. Mossotti's force law for two identical interacting particles. The ordinate is the force (positive values indicating repulsion, negative values attraction), and the abscissa the distance of separation of the centers of the particles. R' is the point of stable equilibrium separation.

in principle) intermolecular and gravitational forces from simple first principles, must be reckoned one of the most ingenious, if ultimately unsuccessful, theories devised in the nineteenth century.[10]

THE REACTION TO MOSSOTTI'S THEORY

When Mossotti's paper was first published, partly because it was quickly translated into English, it caused quite a stir. Most of the reaction that led to publication occurred in Britain, and very likely much of the British response was stimulated by the warm endorsement of Michael Faraday. Besides, England was fertile ground for a theory like Mossotti's because of the previously mentioned metaphysical predisposition of English scientists toward a unitary view of nature.

Faraday, in particular, although he had not then fully published his ideas, had decided on the basis of his experiments that electrical action was communicated by inductive polarization of the successive "contiguous" particles of matter in space.[11] Mossotti's theory is obviously compatible with this view. The competing theories of the structure of matter then extant—Dalton's billiard ball atoms and Boscovich's force singularities—are not. Thus, it is not difficult to imagine Faraday's delight with Mossotti's theory.

Faraday was unable, however, to judge the accuracy of the mathematical part of Mossotti's paper. Accordingly, he wrote to William Whewell to get his opinion on this aspect of Mossotti's work.[12] In his letter to Whewell, Faraday, who was normally very cautious when expressing his opinions on the theories of others, displayed uncharacteristic enthusiasm for Mossotti's theory; in his words, he was, "exceedingly struck with it. . . ." One might expect Faraday's favorable reaction to have been a consequence of the obvious compatibility of Mossotti's theory of "molecular" structure with his molecular polarization theory of electric induction. But this seems to have been of secondary importance to him. It was Mossotti's explanation of gravitation that had arrested his attention. As he wrote to Whewell, Mossotti's theory, "relates essentially to electricity and deduces all the phenomena of gravitation from it *and it is this which makes the interest to me* for his view jumps in with my notion which I think I mentioned to you that Universal Gravitation is a mere residual phenomenon of Electrical attraction and repulsion." [Emphasis added.]

Faraday was so pleased with Mossotti's theory that shortly after learning of it he made it the subject of his Friday evening lecture at the Royal Institution on 20 January 1837. Before giving the lecture, he evidently solicited the opinion of Charles Babbage on the suitability of the topic. Babbage responded with a note written at two in the morning on the day of Faraday's lecture.[13] Babbage, to say the least, was not nearly so struck as Faraday by Mossotti's theory. His criticism of Mossotti's theory, as expressed in this note, was that beyond a certain distance from the nucleus of a molecule the repulsion of adjacent ether atoms on one another would exceed the attraction of the nucleus, and the radiation of ether atoms away from the nucleus beyond this critical distance would ensue. Further, the action of ether atoms impinging on the molecule from other molecules undergoing similar "radiation" (which would agitate the ether atoms at the critical distance) would abet this process.

Babbage then suggested that if this were the case, gravitation should be the consequence of a mechanism similar to that of Lesage,[14], and, presumably, not that of Mossotti. Babbage's argument is, of course, specious, for it rests on the fallacious assumption that attraction by the nucleus of a molecule on ether atoms is exceeded by repulsion in the atmosphere at some finite distance. This is simply not true. Lesage's theory, furthermore, has absolutely nothing to do with Mossotti's theory. Faraday's reaction to Babbage's note, however, must remain conjectural, but he cannot have been overly influenced by Babbage as he did give a favorable account of Mossotti's theory in his lecture.[15]

Shortly thereafter, Babbage published his thoughts on Mossotti's theory in a note appended to his essay in the *Bridgewater Treatise* series.[16] By then he had favorably revised his opinion of Mossotti's theory, as can be seen from the quotation at the outset of this essay. In this note he described (inaccurately) Mossotti's theory and discussed some of its possible ramifications, suggesting ultimately that,

> . . . the whole of chemistry, and with it crystallography, would become a branch of mathematical analysis, which, like astronomy, taking its constants from observation, would enable us to predict the character of any new compound, and possibly indicate the source from which its formation might be anticipated.[17]

Babbage had clearly warmed to Mossotti's theory. Gone is any misleading reference to Lesage's theory. But the arguments on the finite radius of the ether atmosphere, its agitation and "radiation" are all still present.[18] And Babbage also advanced the suggestion that three substances, as opposed to the two in Mossotti's theory, might better explain observations—without, however, working out any formal consequences of his hypothesis.

Babbage's comments stimulated aspersive remarks from Thomas Exley.[19] Exley purported to prove that Mossotti's theory rendered the predictions: If an atom exerts an attractive (or repulsive) force on another atom at a given distance, the force will be in the same sense (attraction or repulsion) at all distances; and, molecules exert no force on each other. These conclusions are, of course, simply wrong, and doubtless were quickly recognized as such.

All of this discussion led to what, from one perspective, is unquestionably the most interesting reaction to Mossotti's theory: Samuel Earnshaw's paper, "On the Nature of the Molecular Forces which Regulate the Constitution of the Luminiferous Ether."[20] Although Earnshaw nowhere explicitly cites the work of previous authors, it is clear that his investigation was stimulated by Babbage's remarks on the stability of Mossotti's ether.[21]

Earnshaw's main points (using now current terminology and notation) can be briefly recapitulated as follows. First, he introduced three assumptions:

1. The ether consists of discrete particles which, when displaced in any direction from their equilibrium positions, should *vibrate*. That is, the ether should be stable against perturbation of its parts.
2. The forces of the ether that act on other parts of the ether should decrease in strength with increasing distance from the source.
3. The ether, if it has a mutual repulsion for itself, should be held together by attractive forces to preclude its unchecked expansion.

It is by no means obvious that Mossotti's theory requires proposition 3. But this ahistorical cavil notwithstanding, it is beyond doubt that propositions 1 and 2 must obtain if the ether is to serve as the medium of transmission of arbitrarily oriented periodic disturbances.

Earnshaw first considered the case where the force acting between the particles of the ether obeys a $1/r$ potential, V, where r is the distance separating any two particles under consideration. He calculated that the ith component of the force, F_i, acting on a particle of the ether displaced a distance Δx_i from its equilibrium position is

$$F_i = \frac{\partial^2 V}{\partial x_i^2} \Delta x_i \tag{9}$$

Now Laplace's equation also applies, so

$$\frac{\partial^2 V}{\partial x_1^2} + \frac{\partial^2 V}{\partial x_2^2} + \frac{\partial^2 V}{\partial x_3^2} = 0 \tag{10}$$

at the equilibrium point from which the ether particle is infinitesimally displaced.[22] Earnshaw then noted that either each term of Eq. (10) is identically zero, or at least one term must be positive and one negative. If each term is zero, by Equation (9), no forces act on the ether particle when it is displaced. If $\partial^2 V/\partial x_i^2$ is negative, Eq. (9) integrates to

$$\Delta x_i = A \cos(\alpha t + B) \tag{11}$$

where A, B, and α are constants. Thus for $\partial^2 V/\partial x_i^2$ negative, the forces produced by the displacement tend to restore the initial configuration. If, however, $\partial^2 V/\partial x_i^2$ is positive, Eq. (9) integrates to

$$\Delta x_i = Ae^{\alpha t} + Be^{-\alpha t} \tag{12}$$

and the resulting force does not tend to restore the initial configuration. Since one of the $\partial^2 V/\partial x_i^2$ must be positive for Eq. (10) to obtain, there is always at least one direction in which displacement of an ether particle is unstable. Moreover, it matters not whether the force with the $1/r$ potential is attractive or repulsive. (This result is now known as Earnshaw's theorem.)

To evade this result, each of the $\partial^2 V/\partial x_i^2$ must be negative, which means their sum must be negative definite. This can only occur for force laws proportional to $1/r^n$ where n is not equal to 2 since, in general, Eq. (10) is

$$\nabla^2 V = (n - 2) \int \frac{q}{r^{n+1}}\, dx^3 \tag{13}$$

q being the charge density and the integration extends over all space except the immediate environs of $r = 0$. Earnshaw shows that for a repulsive force, if $n > 2$, proposition 1 is satisfied, and that an inverse square attraction can also be added that will balance the equilibrium state repulsive force, thus meeting the requirement of proposition 3.

Earnshaw's argument seems to vitiate the physical plausibility of Mosotti's ether, and, evidently, to have quelled interest in Mossotti's theory. For the next several decades, all but Faraday ignored Mossotti's theory. Faraday repeatedly brought up Mossotti's theory in his "speculative" writings.[23] But this should not be taken to mean that Faraday was a *dévoté* of Mossotti's theory for the remaining years of his life. It was not Faraday's style to become the follower of any particular formal theory. Theories were merely guides to experiments that might falsify the theory for Faraday. And Faraday's thought was more subtle than that found in Mossotti's theory, as can be seen in his electrogravitational induction experiments of the late 1840s.

Faraday's Electrogravity Experiments

Of the many, many series of experiments that Faraday undertook in his long and fruitful career, unquestionably the boldest was that where he hoped to show that the gravitational and electromagnetic fields are inductively coupled.[24] Gravitation, with its seemingly irreducible Newtonian action-at-a-distance character, Faraday had long recognized as the one force that was least likely to yield to his developing field-theoretic approach to nature. But for Faraday, the interconvertibility principle, mentioned above, was almost beyond question. For example, in 1845, Faraday opened the nineteenth series of his *Experimental Researches in Electricity* with the remark that,

> I have long held an opinion, almost amounting to conviction in common with many other lovers of natural knowledge, that the various forms under which the forces of matter are made manifest have one common origin, or, in other words, are so directly related and mutually dependent, that they are convertible, as it were, one into another, and possess equivalents of power in their action.[25]

If gravitation conformed to this principle as Faraday believed it must, then gravitation should be convertible into other species of force. It is important to note here that Faraday was not referring to the transformation of gravitational energy into some other form (heat for example) by the use of mechanical intermediaries. Rather he thought that it was necessary to show that by suitable manipulation, the gravitational field itself could be made to produce other fields—specifically an electromagnetic field.

The likelihood that the conversion of gravitational force into electric force could be achieved with the means at his disposal, Faraday realized, was very remote. It was with some hesitancy, then, that he began to formulate ideas for feasible experiments to test the convertibility of gravity into electricity. The first of these thoughts are entered in his *Diary* on the date 19 March 1849.[26] Here, after indicating his desire to show the convertibility of gravity into electricity, he first considered the problem of delineating the gravitational analog of polarity. One need only recall that Faraday viewed the propagation of electromagnetic induction as a process of progressive polarization, and that he had already recognized the fundamental differences between mono and bipolar forces, to appreciate his desire to construct the analog of polarity for a monopolar field (gravitation).[27] He suggested that perhaps, "*to*, and *fro*, that is, the ceding to the force of approach of Gravitating Bodies, and the effectual reversion of the force or separation of the bodies, quiescence being the neutral condition. . . ." might be the appropriate analog.[28] Granting this possibility, he then reasoned that two bodies undergoing approach or recession might give rise to currents in themselves and/or surrounding bodies—particularly in coils of copper wire. Restating Faraday's ideas in contemporary terminology, he supposed that a configuration and motion of matter that will give rise to quadrupole gravitational radiation should give rise to an induced conjugate electromagnetic radiation field. This radiation field would then induce currents in surrounding circuits. Such an effect was presumably the result, in Faraday's view, of circuital electric fields that were supposedly induced around the gravitational lines of force coupling the bodies as they underwent changes in tension caused by the bodies changing relative position. It is, of course, not difficult to conjecture the analogy upon which Faraday based this speculation: it is the induction of a circuital magnetic field about the electric field of moving charges.

After speculating on experimental and empirical possibilities, suggesting that the weight of a body falling through a current-carrying helix might be affected, and hypothesizing that the parallel and perpendicular (relative to the gravitational field of the earth) masses of a body might differ, he then hazarded the guess that

> If [convertibility] exists, perhaps even then, it is the effort of gravity which at the moment of relaxation only produces the effect. Then an accelerated fall or rise [acceleration added to or subtracted from that of gravitation that is] would be required to produce a *continuous* effect of current or what ever else is produced.[29]

Translating once again into current terminology, Faraday suggested here that an electromagnetic field may only be induced when gravitational radiation produced by mechanical (nongravitational) accelerations occurs.[30] Faraday seems to have been casting about for any plausible effect that might be detectable without worrying about the fetters of strict fidelity to any particular theory. On one point, however, he seems to have been unwilling to compromise; namely, "The unchangeability (FINAL) of Gravity [is] not affected by any of these forerunning suppositions[s], for the final effect when [the two] bodies [are] at rest in relation to each other would be ever the same."[31]

Faraday realized that his ideas, clearly bold and daring, were mere speculation, for he closed his *Diary* entry for 19 March with the comment that

> ALL THIS IS A DREAM. Still, examine it by a few experiments. Nothing is too wonderful to be true, if it be consistent with the laws of nature, and in such things as these, experiment is the best test of such consistency.[32]

It was not long after, that Faraday made his preliminary experiments. These consisted of dropping cylindrical samples of copper, glass, bismuth, and shellac about one inch in diameter and several inches long through a small helix which was attached to a galvanometer. The results were negative. No detectable currents were produced in the coil by the falling bodies. Faraday thought the distance of fall, a few feet, might be too short to achieve high enough velocities to produce a detectable effect.

After this initial failure, Faraday abandoned this work until the following August. When he started experimenting again, his seriousness of purpose and the scale of the experiments had increased significantly. Writing of these experiments, he commented that,

> It was almost with a feeling of awe that I went to work, for if the hope should prove well founded, how great and mighty and sublime in its hitherto unchangeable character is the force I am trying to deal with, and how large may be the new domain of knowledge that may be opened up to the mind of man ...[33]

Clearly, Faraday had no illusions about the potential enormity of the work he was undertaking, and thus inspired, he proceeded with the first phase of his new experiments. These continued into early September of 1849, and involved dropping and raising a new helix and various core cylinders, both together and separately, through a distance of 36 feet. Once again, after debugging the apparatus, the results were negative.

Faraday, who was persistent almost to the point of obstinacy at times, was not yet ready to give up. In fact, anticipating that a null result in the experiments just described might be obtained because the effect might be too small to be observable in a "one-shot" experiment, he had considered various machines that might be constructed which would produce a cumulative, and hopefully measurable, current as a consequence of his presumed effect. The first possibility he considered, described in late March of 1849, consisted of a rotating ring (the exciting mass) located within stationary coils wound as depicted in Fig. 2. He conjectured that equal and opposite currents might be excited in the coils at the points *a* and *b*, but because no effect was supposed to result from motion of matter orthogonal to the gravitational field of the Earth, no opposing currents would arise at points c and *d*. Given Faraday's assumption, this would seem a simple and appealing device. But it was never built, for he had hit upon a more versatile device.

The machine that was built for Faraday, delivered to him on 3 September 1849, was constructed so as to cause a cylindrical core, surrounded by a stationary helical coil, to vibrate back and forth through a distance of a few inches. The arrangement of this machine is shown schematically in Fig. 3. The leads from the coil were attached to a mechanically actuated commutator and this was connected to a galvanometer. This made it possible to reverse the circuit synchronously with the frequency of vibration of the core. The commutator was also designed so that the relative phase of circuit reversal and core position in the coil could be adjusted. Thus, Faraday was able to see if currents were generated in the coil either when the core was at maximum velocity, or at maximum acceleration in its motion. Trials and modifications occupied the remainder of the month of September and continued sporadically into October, but irrespective of the disposition of the apparatus, no genuine effect could be found. Though he was not to conduct further experiments on gravitation for a decade (and

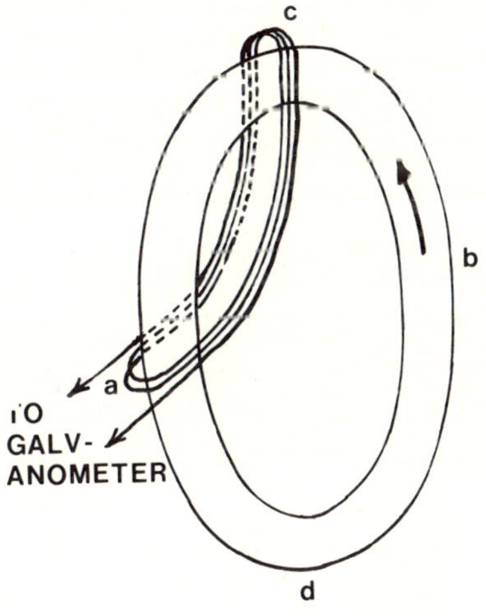

FIGURE 2. A schematic diagram of the rotating ring apparatus considered by Faraday to test the electrogravitational induction hypothesis. By spinning the ring in a vertical plane, Faraday conjectured that electrogravitational effects would manifest themselves at points a and *b*, but would be absent at points c and *d* (because no change in the tension of the gravitational lines of force linking the ring to the Earth occurs at points c and *d*). The hypothetical effect was to be detected by four coils wound like the one shown in the diagram.

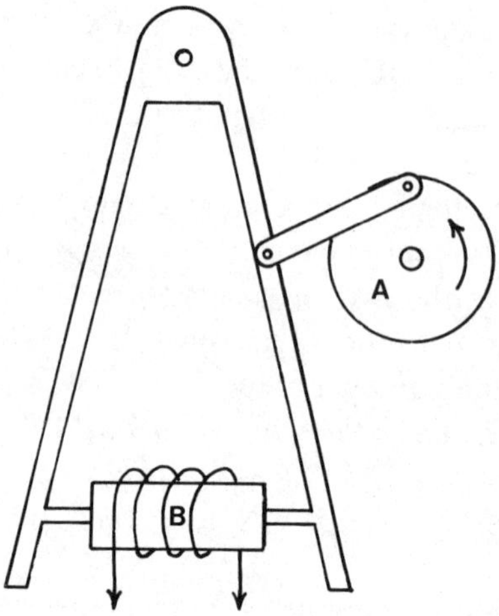

FIGURE 3. A schematic diagram of the apparatus actually employed by Faraday in his attempt to detect electrogravitational induction. By turning the wheel *A*, the sample *B* is set into lateral oscillation with an amplitude of a few centimeters. When operated with the sample oscillating vertically, the ensuing variation in the tension of the lines of gravitational force coupling the sample to the Earth was supposed to produce a periodic circuital electric field that would be detected by the coil coaxially mounted about the sample.

these two would yield null results), Faraday did not let the null results of his experiments shake his, "strong feeling of the existence of a relation between gravity and electricity, though they give no proof that such a relation exists."[34]

EPILOGUE

Subsequent developments along the lines pursued by Mossotti and Faraday can be briefly sketched. Aside from Faraday's abiding interest in Mossotti's theory, by the 1840s it was largely ignored and remained so until it was resuscitated by J. C. F. Zöllner, professor of astrophysics at Leipzig, in the late 1870s.[35] Zöllner succeeded in interesting W. Weber in Mossotti's theory. And for a few years it was again in the limelight. The theory was again briefly resurrected at the turn of the century by H. A. Lorentz in his attempt to construct a field-based theory of gravitation.[36] Although there have been several theories of gravitation constructed on the basis of electrodynamic analogs since the work of Lorentz, to my knowledge, Mossotti's different coupling constants hypothesis has not been employed in the past 80 years.

The fate of Faraday's ideas is a little more complicated. Faraday himself returned to work on his electrogravity concept and performed more experiments thereon in his last major series of researches in 1859–60.[37] But he failed to stimulate anyone else to pursue this research in his own era. It was not until the 1890s that any work on Faradayan lines took place, and when it took place, no connection with Faraday's earlier work was made.

The work of the 1890s and later that is *implicitly* based on Faradayan notions of the direct coupling of electromagnetism and gravitation was stimulated by the observation during a total solar eclipse that the corona was disposed in such a way that it suggested the presence of a solar magnetic dipole field (which was obviously not ferromagnetic in origin). This suggested to A. Schuster that planetary and stellar magnetic fields might be induced by the rotation of those bodies.[38] This idea was endorsed by Kelvin, who found it plausible that shearing of the "circumambient aether" of a celestial object might induce a magnetic field.[39] Schuster sporadically followed up on this work until the outbreak of World War I. Shortly after the War, H. A. Wilson explored the Schuster hypothesis by trying to detect an apparent "charge" on the Earth when it was "observed" in translational motion,[40] and W. F. G. Swann attempted to generate a magnetic field by spinning a copper sphere.[41] As in the case of Schuster's investigations, nothing of interest was found. Interest in magnetogravitational coupling then languished for two decades, until P. M. S. Blackett revived the idea in the mid-1940s.[42] Considerable discussion and the measurement of the Earth's magnetic field in deep mines (as a test of the hypothesis) ensued. Blackett himself executed an experiment in the early 1950s where he tried to detect a hypothetical magnetic field of a gold cylinder corotating with the Earth using a sensitive astatic magnetometer.[43] As with his predecessors, Blackett's results were negative and work on the idea was abandoned.

Is, then, this litany of negative results and unrealized hopes extending from Faraday's efforts in the 1840s to the recent past purely of historical interest? Perhaps not. Several years ago, when I became interested in Faraday's work on electrogravitational induction chiefly from a historical point of view, it occurred to me that if the hypothesis were slightly restructured and clarified, and modern techniques with a different experimental configuration were employed, the admittedly almost naive electrogravitational induction hypothesis could be tested to orders-of-magnitude higher accuracy than had hitherto been done. I, of course, anticipated a null result.

The restructuring of the hypothesis is very simple and straightforward. Wilson's result clearly requires that electrogravitational induction only occur (if at all) when matter is accelerated. Since, in general relativity theory, the local state of inertial motion is gravitational free-fall, electrogravitational induction should only appear when matter is *nongravitationally* accelerated. Consider now the case of a small mass in gravitational free-fall near the surface of the Earth from a Faradayan point of view. The mass can be viewed as gravitationally coupled to the rest of the matter in the universe by "lines" of gravitational "force." Since the distant matter in the universe dominates this coupling, distribution of these "lines" is essentially radially isotropic. When the mass strikes the surface of the Earth, it is very sharply given a nongravitational acceleration and, in analogy with radiation from an accelerated charge in electrodynamics, a "kink" in the gravitational "lines of force" is propagated outward at the velocity of light. Now, the electrogravitational induction hypothesis requires that an inductively coupled electromagnetic wave accompany the disturbance in the gravitational field produced at impact. In other words, the hypothesis requires that electromagnetic radiation be coupled to gravitational radiation.

We need not inquire into the details of the gravitational wave or the hypothetically coupled electromagnetic radiation to devise an experiment to test this hypothesis. We need merely note that if the hypothesis is correct, at the instant of impact, the accelerated mass *must* appear to *transiently* acquire an electric charge since the source of the electromagnetic field is, by definition, electric charge. Small, transient electric charges are easily detected with commonly available apparatus.

This experiment has been executed; and after an exhaustive series of tests, a small positive effect that appears not to be ascribable to conventional causes remains.[44] The charge induced at impact, presumably of electrogravitational origin, obeys the equation

$$Q^* = \beta ma \tag{14}$$

where Q^* is the induced charge, m the accelerated mass, a the acceleration, and β the electrogravitational induction coupling constant. For the substances tested, $\beta \simeq 3 \times 10^{-11}$ statcoulombs/dyn. Note that the *magnitude* of β is very nearly equal to the inverse of the speed of light.

If we tentatively accept the results of this experiment as evidence for the electrogravitationally inductive coupling of electromagnetic waves to gravitational waves, we may pose a further, testable question: Is Eq. (14) universally valid? In particular, does electrogravitational induction occur when matter is accelerated, but no gravitational radiation is produced? Consider, for example, the case of a cylinder of matter spinning about its axis of symmetry. Since the time derivative of the mass quadrupole moment is zero, no gravitational radiation is generated by the spinning cylinder. From the naive Faradayan viewpoint, however, the gravitational "lines of force" coupling the cylinder to the rest of the matter in the universe are presumably being sheared at a constant rate, and one might expect to find a coupled electromagnetic field.[45] Equivalently, looking at the sources of this hypothetical field instead of the fields themselves, Eq. (14) implies

$$\rho_c^* = \beta \rho_m a \tag{15}$$

where ρ_c^* is the electrogravitationally induced charge density generated in matter of mass density ρ_m undergoing an acceleration a with respect to a local inertial frame of reference. Therefore, if we spin a cylinder of inner and outer radii R_1 and R_2, respectively, and length L at an angular frequency ω about its symmetry axis, it is easy to show that if Eqs. (14) and (15) are universally valid, they require that a charge Q^* be induced in the cylinder given by

$$Q^* = \pm \tfrac{2}{3} \pi \beta \rho_m L \omega^2 (R_2^3 - R_1^3) \tag{16}$$

where the $\pm$ sign has been introduced because the hypothesis does not permit one to stipulate before the fact whether the induced charge will be positive or negative. If electrogravitational induction only occurs when gravitational waves are produced, no charge will be generated in the rotating cylinder.

The prediction of Eq. (16) may be checked with apparatus like that depicted

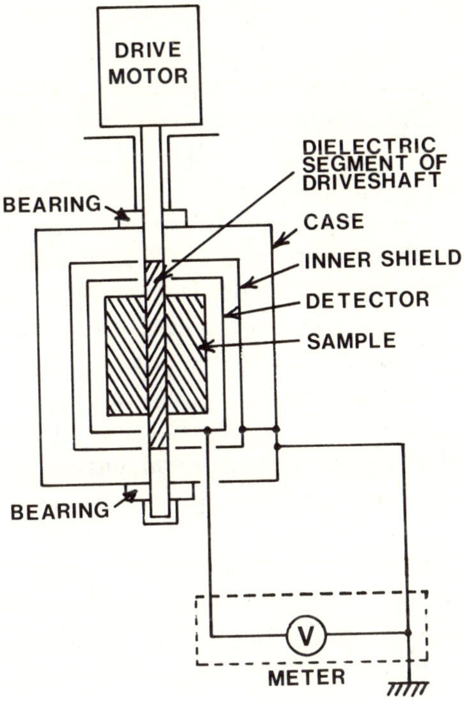

FIGURE 4. A schematic diagram of the apparatus used to test the hypothesis that electrogravitational induction occurs when a cylinder is spun about its axis of symmetry. The charge hypothetically induced in the sample by the centripital acceleration of its parts capacitatively induces the same charge of the exterior of the detector. This charge produces a potential difference between the detector and the inner shield which is measured with an electrometer. The helium atmosphere in the case is continuously drawn off at the guide bearings, dessicated, cooled, and recirculated to the central part of the case.

schematically in Fig. 4. A copper cylinder with $R_1 = 0.95$ cm, $R_2 = 3.15$ cm, $L = 5.9$ cm, and $\rho_m = 8.9$ g/cm^3 is spun on a dielectric driveshaft (so that any charge generated will not be shorted to ground). Taking $\beta = 3.3 \times 10^{-11}$ statcoulombs/dyn, as suggested by the impact experiment, this gives

$$Q^* = 1.1 \times 10^{-7} \omega^2 \text{ statcoulombs}$$

or

$$Q^* = 3.7 \times 10^{-17} \omega^2 \text{ coulombs} \tag{17}$$

Any charge generated in the sample when it is spun is capacitatively detected by an aluminum can which surrounds the sample and is mounted with nylon screws in the inner shield of a Faraday cage which shields the sample and detector from external electromagnetic fields. The potential of the detector relative to the grounded inner shield is measured with a Kiethly (model 616) digital electrometer operated as a high input impedance voltmeter. With the measured capacitance of the detection circuit, the charge induced in the sample as it is spun is calculated.

Although this experiment is simple in principle, in practice it turns out to be moderately complicated. For example, trace amounts of contaminants or moisture in the case can produce leakage paths that lead to erratic results. And thermoelectric effects, unless stabilized, can produce spurious effects. Constant recirculation, dessication, and cooling in a liquid-nitrogen cold trap of the gas (helium) in the case and thermal isolation of the inner shield and detector obviate these problems. Likewise, coupling of assymetric AC signals to the detection circuit can lead to results that look deceptively like those sought. To eliminate this problem, the outer case housing the shield and detector is fabricated of thick aluminum, the experiment is conducted in an electromagnetically shielded room, the instrumentation and its power supplies are shielded, and, most importantly, a pneumatic motor is employed to spin the sample.

When these and other straightforward precautions are taken, one obtains results like those displayed in Fig. 5. Inspection of Fig. 5 reveals that within the experimental accuracy (about 10% of the predicted values for $\upsilon > 100$ Hz), both the predicted magnitude and ω^2 dependence of the hypothetical effect are corroborated by this experiment. Thus, if the observed effect is electrogravitational in origin, the results of this experiment suggest that Eqs. (14) and (15) are indeed universally valid, and that $\beta = 3.3 \times 10^{-11}$ statcoulombs/dyn. The obvious question at this juncture is: Can the

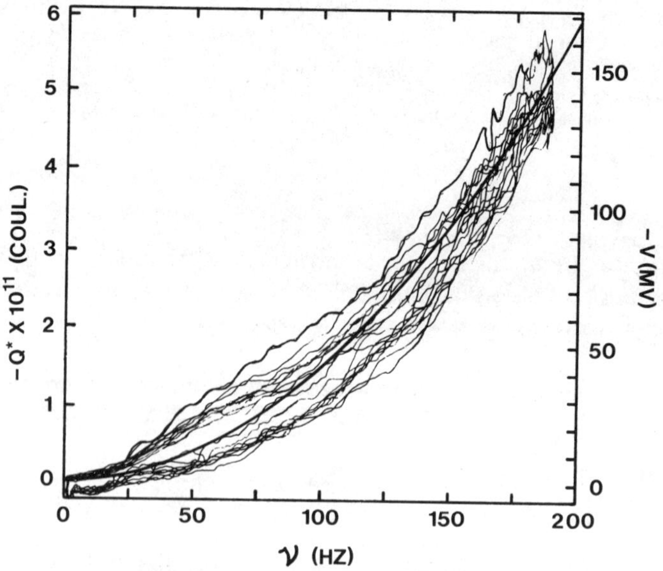

FIGURE 5. Induced charge versus spin frequency for 10 spin-up/spin-down cycles recorded on an X, Y recorder. The heavy solid curve is the theoretical prediction of Eq. (17). The voltage scale on the right is the voltage recorded by the electrometer. The Q^* scale is derived from the voltage scale by taking into account the capacitance of the detector (350 ± 35 pF). The Q^* scale and the theoretical curve have a 10% uncertainty due to the 10% uncertainty in the determination of the capacitance of the detector.

observed effect be attributed to some conventional charging mechanism—thus invalidating the electrogravitational interpretation of the results? The *a priori* probability that a conventional effect would give the correct magnitude and ω dependence would seem to be rather low; but it is not exactly zero.

Because of the simplicity of this experiment, there seem to be but a few conventional candidate explanations for the observed effect. Several colleagues and I have only been able to identify five plausible candidates:

(1) Quasistatic magnetic fields (e.g., the Earth's field), which might induce currents that could lead to potentials that might give results like those observed. This hypothesis was tested by applying relatively large static magnetic fields (employing permanent magnets) to the apparatus during spin-up/spin-down. The results were unaffected by the application of such magnetic fields.

(2) AC line frequency electromagnetic fields with varying intensity, which might lead to charging of the detection circuit (as mentioned above). Initially, when an external shunt capacitor was placed in the circuit and an AC electric drive motor was utilized, problems of this sort were encountered. Substitution of a pneumatic motor for the electric motor, thus eliminating all varying intensity AC fields, removed this problem. The nonvarying intensity AC fields present, if this mechanism contributes to the observed effect, would manifest themselves as constant zero drift when the sample is not being spun. For the duration of sample spin-up/spin-down (typically $<$ 60 s), such zero drift was entirely negligible ($< 1\%$ of the observed effect).

(3) Frictional deposition of charge on the sample due to its motion through the gas in the case. In this case, once deposited on the sample, such charge would only slowly leak off to ground—it would *not* disappear when the sample is spun down. As can be seen in Fig. 5, where both spin-up and spin down are shown, the induced charge disappears during spin down. Therefore, the frictional charging hypothesis must be rejected.

(4) Piezoelectric effects due to stressing of the dielectric segment of the driveshaft that might lead to a change in the net charge in the detector. Two types of stressing forces may be present: lateral forces due to imbalance in the sample/driveshaft assembly, and torsion due to frictional drag in the guide bearings.[46] Several arguments that eliminate these scenarios can be adduced. The simplest is that prior to dynamic balancing of the sample/driveshaft assembly, these effects should both be more pronounced than after such balancing. Dynamic balancing of the sample/driveshaft assembly was found to *enhance* the observed effect slightly, not reduce it, as would be expected on the piezoelectric hypothesis.

(5) Thermal effects in the detection circuit might produce emf's in the course of the execution of the experiment. Junction (thermocouple) effects can be ignored as the emf's produced are typically on the order of microvolts per degree C. A change in the difference of the temperatures of the detector and inner shield might produce a changing emf (Thomson effect). Direct measurement of the thermal differences present (employing a sensitive thermistor bridge) revealed that thermal effects would produce positive voltages, as opposed to the negative voltages observed, were they sig-

nificant. Moreover, thermal effects, because of their relatively long time constants, would not disappear virtually instantaneously during spin-down. Thus, the ω^2 dependence of Q^* during spindown evident in Fig. 5 would not be present for thermal effects.

Notwithstanding the above considerations, it is possible that the observed effect may be due to conventional causes. Treating the apparatus employed to obtain the results reported here as a prototype, Louis Shen and I are presently designing a second generation device that should be more accurate and precise, as well as ensure that the results of this experiment truly are what they seem to be.[47] If no conventional explanation can be found for the impact and spun cylinder experiments, however, and the electrogravitational interpretation of the results is accepted, at the very least, modification of Maxwell's equations will be required to accommodate Eq. (15) above. The required modification is:

$$\rho_c \rightarrow \rho_c + \beta\rho_m|a| \tag{18}$$

where ρ_c is the normal electric charge density. Thus Maxwell's equations become

$$\begin{aligned}
\nabla \cdot \mathbf{D} &= 4\pi(\rho_c + \beta\rho_m|a|) \\
\nabla \cdot \mathbf{B} &= 0 \\
\nabla \times \mathbf{H} - \frac{1}{c}\frac{\partial \mathbf{D}}{\partial t} &= \frac{4\pi}{c}(\rho_c + \beta\rho_m|a|)\,\mathrm{v} \\
\nabla \times \mathbf{E} + \frac{1}{c}\frac{\partial \mathbf{B}}{\partial t} &= 0
\end{aligned}$$

where the symbols have their usual connotations and Gaussian units are employed. The presence of the electrogravitational terms in Eq. (19) raises several questions of theoretical interest. These issues, however, exceed the scope of this paper.

In conclusion I should note that although putative electrogravitational effects should be discernible in experiments expressly designed to illuminate them, in virtually all other circumstances, if they occur, they are so minute as to be undetectable at present. Straightforward calculations reveal that they will not be found in astrophysical situations (e.g., planetary and stellar magnetism), or in other currently observed phenomena. To my knowledge, there is only one other experiment that approaches the precision required to see electrogravitational induction effects: the muon magnetic anomaly experiment recently performed a CERN.[48] In this experiment, muons and antimuons are given centrepital accelerations of about 1.8×10^{18} cm/sec^2 in a storage ring. The charge shift predicted by Eq. (14) for this situation leads to a shift in the muon magnetic moment, and if the muon charge is presumed invariant, a shift in the muon g factor and magnetic anomaly, $\alpha\,[\equiv \frac{1}{2}(g - 2)]$, results. This shift ($\Delta\alpha$) is easily calculated to be

$$\Delta\alpha = \Delta e/e = \beta m\mu a/e \tag{20}$$

Since the magnetic moments of the muons and antimuons are determined for identical circumstances of acceleration, one might reasonably expect the predicted charge shift to be of the same sign for both. Thus, one expects a splitting of the muon/antimuon magnetic anomaly that is just twice the right-hand side of Eq. (20). Taking $\beta = 3.3 \times 10^{-11}$ and the muon rest mass, this is computed to be 3.4×10^{-8}, and, with the electrogravitational splitting included, the predicted muon magnetic anomaly is

$$\alpha \begin{cases} = (1165938 \pm 8.3) \times 10^{-9} \\ = (1165904 \pm 8.3) \times 10^{-9} \end{cases}$$

The observed values are

$$\alpha_{\mu-} = (1165937 \pm 12) \times 10^{-9}$$
$$\alpha_{\mu+} = (1165911 \pm 11) \times 10^{-9}$$

Although the observed splitting does not quite satisfy conventional criteria for statistical significance, the results are, nonetheless, suggestive.

Acknowledgments

I am indebted to Alan Sweedler for many lengthy discussions and encouragement that catalyzed the spun cylinder experiment. Louis Shen has provided invaluable and extensive collaboration in the execution of this experiment. R. J. Crowley and M. Shapiro have also provided helpful comments on various aspects of this work. The greatest debt I owe, however, is to Wolfgang Yourgrau, mentor and friend, but for whom what little I have done would never have been accomplished. NOTE (added in proof): Since this paper was written results that supersede those reported for the spinning cylinder experiment described in the last section of this paper have been obtained.[49]

References and Notes

1. D. Z. Freedman and P. van Nieuwenhuizen, "Supergravity and the Unification of the Laws of Physics," *Sci. Am.* **238**, 126 (1978).
2. Charles Babbage, *The Ninth Bridgewater Treatise*, 2nd ed. (John Murray, London, 1838), p. 179.
3. Today Mossotti is almost exclusively known for his contribution to the theory of the atomic polarizability of dielectrics embodied in the "Claussius–Mossotti" equation.
4. There are a number of works that deal with the germane issues in physics in the 19th century. Particularly helpful are E. Whittaker, *A History of the Theories of Aether and Electricity*, 2 Vols. (Thomas Nelson and Sons, London, 1951 and 1953); M. B. Hesse, *Forces and Fields* (New York Philosophical Library, 1961); W. H. Brock, editor, *The Atomic Debates* (Leicester University Press, Liecester, 1967); D. M. Knight, *Atoms and Elements*, 2nd ed. (Hutchinson, London, 1970); A. Thackray, *Atoms and Powers* (Harvard University Press, Cambridge, Massachusetts, 1970); and T. H. Levere, *Affinity and Matter* (Clarendon Press, Oxford, 1971). In the extensive journal literature, see especially B. G. Doran, "Origins and Consolidation of Field Theory in Nineteenth-Century Britain: . . .," *Hist. Stud. Phys. Sci.* **6**, 133 (1975); A. J. Rocke, "Atoms and Equivalents: . . .," *Hist. Stud. Phys. Sci.* **9**, 225 (1978); and M. Crosland and C. Smith, "The Transmission of Physics from France to Britain," *Hist. Stud. Phys. Sci.* **9**, 1 (1978).

5. The origins of this view, at least in part, are to be found in the *naturphilosophie* of the German romantics. But they also seem to have had indigenous roots as well. See P. M. Heimann, "Conversion of Forces and the Conservation of Energy," *Centaurus* **18,** 147 (1974).
6. See Y. Elkana, *The Discovery of the Conservation of Energy* (Harvard University Press, Cambridge, Massachusetts, 1974), and Heimann, Ref. 5.
7. O. F. Mossotti, *Sur les forces qui regissent la constitution intérieur des corps* (Turin, 1836). Mossotti's essay was translated into English and published in *Taylor's Scientific Memoirs* **1,** 448 (1839).
8. Reference 7, trans., p. 451.
9. P. S. Laplace, *Mécanique Celeste* (Chelsea Reprint, New York, 1966–69), Vol. 5, pp. 117–121. Laplace's argument deals with pressure/density relations in the caloric, which he supposed to possess a short-range mutual repulsion. It is, therefore, not obvious that Mossotti's claimed analogy is correct.
10. I cannot resist the temptation to remark here that it was almost exactly 100 years later that H. Yukawa [*Phys. Math. Soc. Jpn, Proc.* **17,** 48 (1935)] constructed a nuclear force law that differs only trivially from that which follows from Mossotti's theory. In most discussions of the Yukawa interaction it is implied that his force law requires the transient violation of energy conservation. In light of Mossotti's theory, it is easily seen that this is not true.
11. These views, which can be found throughout Faraday's work of the late 1830s, led to to an illuminating correspondence with R. Hare that is published in L. Pearce Williams, editor, *The Selected Correspondence of Michael Faraday* (Cambridge University Press, Cambridge, 1971), Vol. 1, 350–358, 360–370, and 382–392.
12. Faraday to W. Whewell, 13 December 1836, in *Correspondence,* Vol. 1, 306–307.
13. C. Babbage to Faraday, 20 January 1837, in *Correspondence,* Vol. 1, 308.
14. G.-L. Lesage, "Essai sur l'origine des forces mortes" (ms., Univ. of Geneva Library, 1749). Lesage's theory is predicated on the assumption that space is filled with high-velocity particles that are slightly attenuated by matter. Gravitation is the consequence of the "shadow" in the sea of these particles cast by matter. Shadow theories of gravitation are still independently rediscovered with high frequency and monotonous regularity. (R. H. Dicke, personal communication.)
15. See the reviews of this lecture in *Phil. Mag.* **10,** 317 (1837), and *Lit. Gazette,* 4 Feb. 1837, p. 72.
16. See Ref. 2.
17. Reference 2, p. 183. Note that Babbage has completely missed the main point of Mossotti's force law—namely, the point of stable equilibrium.
18. Reference 2, p. 181.
19. T. Exley, "Remarks on M. Mossotti's theory of Physics, . . .," *Phil. Mag.* **11,** 496 (1837).
20. S. Earnshaw, "On the Nature of the Molecular Forces which Regulate the Constitution of the Luminiferous Ether," *Trans. Camb. Phil. Soc.* **7,** 97 (1839–42).
21. There is also reason to suspect that Earnshaw may have discussed Mossotti's theory with Exley, for Exley remarked that, "A very learned and universally admired professor of Cambridge, who honoured my paper with some judicious remarks, . . ." (Note 19, p. 497).
22. This is wrong. If an ether particle of nonvanishing radius is located at and displaced an infinitesimal distance from, the equilibrium point, the equation that actually applies is Poisson's equation. This, by the way, vitiates the remainder of Earnshaw's argument.
23. In the 1840s, when others had abandoned Mossotti's theory, Faraday favorably referred to to it in his famous papers, "A Speculation Touching Electric Conduction and the Nature of Matter" [*Phil. Mag.* **24,** 136 (1844)] and "Thoughts on Ray-Vibrations" [*Phil. Mag.* **28,** #188 (1846)]. And more than a decade later, shortly before his last major series of experiments, he again brought up Mossotti's theory in his paper, "On the Conservation of Force" [*Phil. Mag.* **13,** 225 (1857)].
24. The results of his experiments, conducted in 1849, were published in the *Phil. Trans.* of 1851. The paper is reprinted in M. Faraday, *Experimental Researches in Electricity* (Taylor and Francis, London, 1855), Vol. 3, p. 161. A more detailed and candid discussion of these experiments can be found in Faraday's *Diary* (G. Bell and Sons, London, 1934), Vol. 5, pp. 150–189 and 199–202. The only discussions of these experiments of any length of which I am aware are L. Pearce Williams, *Michael Faraday* (Chapman and Hall, London, 1965), pp. 465–470, and W. Berkson, *Fields of Force* (John Wiley and Sons, New York, 1974), pp. 112–116. Several articles that shed light on Faraday's thought (although they do not address electrogravitational induction) are T. H. Levere, "Faraday, Matter, and Natural Theology— . . .," *Brit. J. Hist. Sci.* **4,** 95 (1968), P. M. Heimann, "Faraday's Theories of

Matter and Electricity," *Brit. J. Hist. Sci.* **5,** 235 (1971); and J. E. McGuire, "Forces, Powers, Aethers, and Fields," *Boston Stud. Phil. Sci.* **14,** 119 (1974).

25. *Expt, Res. Elec.* **3,** 1–2 (1855).
26. *Diary*, Vol. 5, p. 150.
27. Maxwell was one of the few to realize that Faraday had put his finger on a key problem in the issue of polarity. As he later remarked to Faraday, "You have also seen that the great mystery is not how like bodies repel and unlike attract but how like bodies attract (by gravation [*sic.*])." *Correspondence*, Vol. 2, p. 881.
28. *Diary*, Vol. 5, p. 150.
29. *Diary*, Vol. 5, p. 151.
30. Recalling that inertial motion is the state of gravitational free-fall in general relativity theory, this view is subsumed under the proposition that gravitational radiation (as we would say today) is accompanied by electromagnetic radiation.
31. *Diary*, Vol, 5, p. 152.
32. *Diary*, Vol. 5, p. 152.
33. *Diary*, Vol. 5, p. 156.
34. *Expt. Res. Elec.* **3,** 168.
35. J. C. F. Zöllner, *Erklärung der Universellen Gravitation* (L. Staackmann, Leipzig, 1882).
36. H. A. Lorentz, "Considerations on Gravitation," *Proc. R. Acad. Sci., Amsterdam* **2,** 559 (1900).
37. Faraday, *Diary*, Vol. 7, pp. 334–379.
38. A. Schuster, "Recent Total Solar Eclipses," *Proc. R. Institution* **13,** 273 (1890–92), and "A Critical Examination of the Possible Causes of Terrestrial Magnetism," *Proc. Phys. Soc.* **24,** 121 (1912).
39. Kelvin (Wm. Thomson), in a presidential address to the Royal Society, *Proc. R. Soc.* **52,** 304 (1892).
40. H. A. Wilson, "An Experiment on the Origin of the Earth's Magnetic Field," *Proc. R. Soc. London Ser. A* **104,** 451 (1923).
41. W. F. G. Swann and A. Longacre, "An Attempt to Detect a Magnetic Field as the Result of the Rotation of a Copper Sphere at High Speed," *J. Franklin Institute* **206,** 421 (1928).
42. P. M. S. Blackett, "The Magnetic Field of Massive Rotating Bodies," *Nature* **159,** 658 (1947).
43. P. M. S. Blackett, "A Negative Experiment Relating to Magnetism and the Earth's Rotation," *Phil. Trans. R. Soc. London Ser. A* **245,** 309 (1952); see also: C. W. F. Everitt, "Gravitation, Relativity and Precise Experimentation," in *Proc. First Marcel Grossmann Meeting on General Relativity* (ed. Ruffini, North Holland, 1977), p. 545.
44. J. F. Woodward, "An Experimental Re-examination of Faradayan Electrogravitational Induction," *Gen. Rel. and Grav.*, **12,** 1055 (1980).
45. This is the Faradayan equivalent of the Schuster–Wilson–Blackett hypothesis that rotating matter generates a magnetic field.
46. I might mention here that this is a nonnegligible problem. The guide bearings must be free-running, as any stiffness produces torsion that generates an effect comparable in magnitude to that sought. In the apparatus used, however, the torsionally induced charge was found to have the opposite sign (that is, positive) from that of the observed effect. Thus, it could be separated from the observed effect
47. The chief modifications will be the use of gas bearings and an improved sample geometry.
48. J. Bailey, *et al.*, "Final Report on the CERN Muon Storage Ring . . .," *Nucl. Phys. B* **150,** 1 (1979).
49. J. F. Woodward, "Electrogravitational Induction and Rotation," *Found. of Phys.* **12,** 467 (1982).

Subject Index

Abundance of light elements, 317
Aliya Chadasha, 9
Al Muntada, 23
Analysis, age of, 535
Astrophysics, 488

Baer *-semigroups, 121
Berlin, 5
Biology and physics, 525, 532
Boykott Tag, 7
Bures distance between states, 277

Carcinogenesis, 578
Carcinogens, germinal tissues as targets for, 579
Casuistry, 666
Celestial mechanics, Borelli's, 698
Chance, 789
Chromodynamics, 38, 265-267
Clifford algebras, 69
Colloquia, International, Second and Third, at Denver, ix
Complementarity, 653
Consciousness, 779
Continuum, 405
Cosmic evolutionary effects, 226, 233
Cosmological models, 148, 149
Cosmologically significant facts, 309, 320
Cosmological unity, new, 782
Cosmology, 490
  aporias of, 145, 148
  status and perspectives of, 223
Cosmos, pico- and mega-, connections between, 37
CPT invariance, 87, 91
Crete, 21
Cyprus, 25

Determinacy, 77
Determinism, 679

Dirac
  aether, 325
  algebra, 67
  equation, 73
  fields, nonlinear classical, 363, 364
    confinement of, 373
    and point particles, 374
  -von Neumann axioms, 110
Dissipative dynamics, deterministic, 357
Division ring, 118
Duality, wave-particle, induced superfluorescence and, 433

Eddington's numbers, 40
Einstein, 683
Einstein nonlocality, 517
Electrogravity experiments, Faraday's, 894
Electroweak interaction in scattering, 411
Elementary particles as microuniverses, 377
Emergent properties, 719
Entropy, increase of, in the Carleman model, 495
Ergodic principle, 651
Ethics and science, 659

Falsification, notational, 831
Forum, 23
Foundations of Physics, viii, 28

Galaxies
  clumpy irregular, 133
  Markarian, 134
Gauge theories
  hadronic states in, 271
  nature of, 266
Global frame of reference, 823
God
  as a cosmic guest-worker, 765
  Einstein, Spinoza, and, 683

Grassmann algebras, 59, 63, 69
Groups, supersymmetry, 59, 63

Hagalgal, 23
Haganah, 25
Hidden variables, 114
Hilbert space axiom, 109
Histadrut, 9
History of science and refutation, 835
HOGOA, 9
Hopping, long-range, in disordered systems, 451
Hubble's law, 313

Imperatives, 663
Indistinguishability of microparticles, 607
Induced superfluorescence, 433
Infinity and the natural world, 421
Injection well problem, 566
Interactions among bodies, 679
International Journal of Theoretical Physics, 28
Irreversibility, 679

Jerusalem Radio, 17
Jovilabe, 707
Jupiter, satellites of, 689

Kanazawa Institute of Technology, 27
Klein paradox, 503
Knowledge, canonical, 757
Kreipe, K., General, abduction of, 21

Life, nature and origin of, 805
Light, finite velocity of, 706
Logical positivism, 537, 541-545

Mach–Einstein doctrine, 45
Mach's principle, 238
Macrosystems
  dynamics of, 260
  objective description of, 243
    embedding of, in an extrapolated quantum mechanics, 254
  preparations and measurements of, 248
  registration and preparation procedures for, 249
    time translations of, 251
  trajectories of, 249
Man as the self-sensitizing animal, 777
Mathematics, precision and approximation, 397
Matter, scientific conceptions of, 877
Measurement problem, 91
Metaphysics, 673, 676
Microwave background radiation, 236, 314
Mind
  reality as, 769, 779
  theories of, 774
"Miner's canaries"
  decision paradigms and, 565
  model organisms and, 564
Mossotti, theory of, 886

Negentropy, 101
Numbers, hypercomplex, 60

Objective description of macrosystems, 243, 246
Office of Strategic Services (OSS), 19
Olbers paradox, 310
Orient
  ceasing of, 15
  launching of, 10

Palestine, 25
Phase space
  distributions, 461
  relativistic, 67
Philosophy
  and physics, 671
  of physics, 678
  of science, 870
Physical principles, social exemplifications of, 647
Physical world, nature of, 873
Physics and philosophy, 671
Piron's theorem, 116
Planarians, 549
  acute toxicity studies on, 574
  chemical induction of tumorigenesis and teratogenesis in, 581
Positivism, critique of, 673
Precognition, 103
Predictability, breakdown of, 437
Protomathematics, 402
Psychoanalysis, 537, 540
Psychokinesis, 103
Public Information Office of the British Government of Palestine, 16

Quantum logic, 111
Quantum mechanics (theory)
  extrapolated, of many particles, 243, 253
  fresh look at, 759
  interpretation of, 87, 102, 103
    realistic, 345
  new, 345
  relativistic, 169
    stochastic interpretation of, 325
Quantum systems
  perception of, 457
  states of, 458
Quark
  confinement, 265, 271, 387
  models, 268

Reality as mind, a model of, 769, 779
Red-shift, 229
Reductionism, 525
  epistemological, 530
Refutations, final, 830
Relativity
  conformal, 382
  general, extensions of, 213
  projective, 379
  and our world view, 733
Relativistic quantum equations, stochastic interpretation of, 325, 338
Reversibility, 679

Scattering
  cross section, 177
  polarized
    electron–electron, 411, 417
    electron–positron, 411, 412
  theory, relativistic, 175
School of Higher Studies, 21
Schrödinger equations, new type of, 354
Science
  beyond science, 753
  integrated vs. disintegrated view of, 809
  world views in, 813, 821, 861, 862
Scientific inquiries, queries about, 847
  different kinds of, 866
  model of, 848
  preparations of, 850
  systematics of, 851
Scientific method, 753
Scientific world views, 816
Semantics, philosophical, 625
Seneca's ethics, 659
Sensitivity, 779
Singularities, nature of, 437
Solitary waves, 365
  stability of, 372
Solitons, 365
Sozialistische Arbeiterpartei (SAP), 5
Space–time structure, 517
Spinoza, Benedict de, 683
Spontaneous emission, 53, 55
Star-algebras, 120
Statistical inference
  and entropy, 286
  by minimal Bures distance, 275
Stochastic electrodynamics, 469
Stochastic equations, covariant, 330
Stochastic interpretation of relativistic quantum equations, 325
Strong interactions, theory of, 265
Sturmabteilung (SA), 7
Superfluorescence, induced, and wave-particle duality, 433
Superselection variables formalism, 350

Teleological principles, 655
Terrestrial kilometric radiation, cyclotron–soliton model of, 129
Thermodynamics
  fundamental dilemmas in, 289
  principles of, 648
  second law of, 293
  space–time structure and, 517
  state concept of, 300
  zeroth and first laws of, 304
Time
  physical, 678
  in relativistic physics and theology, 745
  reversal, 100
Topology, algebraic, 209
Total views, belief in the possibility of, 828
Toxicity studies on planarians, 574
Toxicology, predictive, 549
Transphyletic animal similarities, 549, 551

Uncertainty principle, toxicology's, 571
Unified theory, 37
Unitary understanding of nature, early attempts at a, 885

Vacuum and the continuum, 406
Vacuum tunneling, 189, 190
Values, primary, 661
V-E Day, 23
Vienna, 537
Vienna Circle, 537, 541-543
Von Humboldt University, 4

Wave function, collapse of, 94
Wave mechanics, ideas underlying, 83
Wave-particle duality, induced superfluorescence and, 433
Weimar Republic, 4
Weltbühne, 12
Wigner distribution function, 460
Wright, Thomas, the wronging of, 593

Yourgrau, Wolfgang, 1

Zero-point field, 471
Zitterbewegung, 53

NAME INDEX

Aaronson, M., 322
Abelein, M., 587
Abercrombie, M., 590
Abrahamsson, B., 29
Abramowitz, A. A., 588
Abriksov, A. A., 274
Adam, C., 716
Adams, J. C., 590
Aerts, D., 361
Agarwal, G. S., 467
Agassi, J., 841
Aharonov, Y., 107, 273
Aichelburg, P. C., 29
Aiton, E. J., 605
Aldrich, H. S., 23, 30
Alexander, H. G., 768
Alexander, S., 732
Alfvén, H., 167
Allen, C. W., 242
Alpher, R. A., 323
Alumets, J., 589
Alvarez, A., 376
Ambartsumian, V. A., 242
Ambrose, A., 645
Ames, B. N., 591
Amme, R., 29
Amrein, W. O., 188
Anders, T., 411, 420
Anderson, A. D., 591
Anderson, D. L. T., 376
Anderson, P. A. V., 589
Anderson, P. W., 456
Angelasanto, F. A., 590
Anscombe, G. E. E., 645
Anton-Tay, F., 588
Araki, H., 287
Arcidiacono, G., 396
Ariotti, P. E., 716
Aronow, L., 591
Arp, H. C., 241, 323
Aspect, A., 106, 343
Audretsch, J., 167
Austin, J. L., 843
Avery, T. L., 588
Avez, A., 375
Avis, S. J., 210
Axelrod, J., 589
Ayala, F. J., 525, 534
Ayer, A. J., 645

Babbage, C., 905, 906
Badurek, G., 361
Bahcall, J. N., 242
Baierlein, R., 308
Bailey, J., 907
Bailly, J.-S., 717
Bais, F. A., 396
Bakamjian, B., 187
Baldwin, R. L., 591
Ballentine, L. E., 493
Band, W., 34, 308
Bar-Hillel, Y., 449
Barnes, D. W., 449
Barnothy, J. M., 323
Barnothy, M. F., 323
Barut, A. O., 53
Basscha, J., 716
Bastin, T., 492
Bateman, H., 396
Baumrin, B., 841
Bauspiess, W., 361
Beauregard, O. Costa de, 87, 105-107, 493
Becker, G. F., 605
Bekenstein, J. D., 521
Bell, J., 432
Bell, J. S., 106, 361
Bene, G. J., 361
Benoist, R. W., 34, 286, 287
Benvenuti, P., 144
Beran, H., 534
Berenblum, I., 591
Berendsohn, W., 5, 7, 30
Bergman, H. L., 591
Bergmann, P. G., 59, 66, 107
Berkeley, G., 645, 874
Berkson, W., 906
Bernamont, J., 492
Bertalanffy, L. von, 871
Berthold, W., 7, 29
Bessel, F. W., 842
Best, J. B., 549, 586-590
Betchaku, T., 590
Beth, E. W., 449
Bethe, H., 323
Bialynicki-Birula, I., 376
Birau, N., 589
Birkhoff, I. G., 126
Bjorken, J. D., 344
Bjorklund, G. C., 106
Blackett, P. M. S., 907
Blanchard, Ph., 449
Blaquire, A., 375
Blatt, J. M., 308
Bogoliubov, N. N., 449
Bohm, D., 67, 76, 126, 273, 343, 436
Bohr, N., 107, 409, 467, 658, 880
Bolton, J. G., 241
Boltzmann, L., 409, 501, 623
Bolzano, B., 409
Bondi, H., 77, 167, 432, 843
Bone, S., 808
Bonelli, M. L. R., 715
Bonse, V., 361
Borelli, A., 715

Borland, R. E., 456
Born, M., 323, 751
Borzeszkowski, H.-H. von, 34, 38, 51
Bose, S. W., 623
Bourbaki, N., 449
Bourne, G. H., 589
Bower, J., 590
Boyer, T. H., 492
Boyland, E., 591
Bracken, A. J., 58
Bracket, J., 590
Braffort, P., 492
Brandt, W., 5
Breck, A. D., viii, ix, 29, 32-34, 221, 449, 586, 752
Breit, G., 29
Brinkley, S., 589
Brock, W. H., 905
Broglie, L. de, 83, 86, 343, 436
Bronstein, A. A., 589
Bronson, F. H., 588
Brookes, P., 591
Brouwer, J. L., 409
Brown, G., 589
Brown, H., 716
Bruce, H. M., 588
Bryan, W. R., 591
Bub, J., 126
Bubenik, G., 589
Buber, M., 9, 23
Budny, R., 420
Bullock, T. H., 587
Bunge, M., 843, 871
Burbidge, G., 241, 323
Burbidge, M., 323
Bures, D., 287
Butcher, H., 241
Butler, S., 605
Butt, D. K., 106
Butterfield, H., 843

Cajori, F., 751
Caldirola, P., 396
Callan, C. G., 210
Calvin, M., 591
Campbell, D. T., 843
Cantor, G., 409
Canuto, V., 323
Capek, M., 751
Capra, F., 107
Carleman, T., 501
Carnap, R., 542, 843
Carpenter, K., 587
Cartan, E., 343, 396
Casati, G., 396
Casini, C., 144
Cassini, J. D., 717
Cassirer, E., 871
Castell, L., 58
Castlenuovo, G., 396
Castorina, P., 396
Caralieri, E., 591
Cercignani, C., 501
Cesari, L., 501
Cetto, A. M., 343, 492
Chandler, C., 591
Chandrasekhar, S., 493
Charlier, C. V. L., 322
Charon, J. E., 396
Chenkman, J. B. S., 591
Cherry, C., 658
Chesseaux, J. P. L. de, 322
Chissik, S. S., 263
Christopher, J. P., 591
Chu, F. Y. Y., 376
Churchman, C. W., 871
Clark, D. A., 449
Clark, R. W., 751, 752
Clarke, S., 768
Clauser, J. F., 106, 436, 492, 521
Claverie, P., 492
Clifford, A., 126
Cocconi, G., 396
Cochran, A. A., 107
Cohen, E. G. D., 501
Cohen, I. B., 716
Cohen, P. J., 449
Cohen, R. S., 432, 843, 872
Cole, K. D., 132
Collins, R. E., 188
Colodny, R. G., 841, 843
Conant, J. B., 843
Cook, J. L., 188
Cooke, J. H., 188
Cookson, M. J., 591
Coors, J., 29
Coors, W., 29
Coplestone, V. F., 752
Coppola, J. A., 588
Corben, H. C., 396
Corciovei, A., 501
Corey, B. E., 323
Courant, R., 843
Craig, W. L., 432
Crell, B., 501
Crosland, M., 905
Crowley, R. J., 33
Csiszar, I., 501
Cufaro-Petroni, N., 106, 325, 343, 436
Cunningham, E., 396

Dakin, H. D., 808
D'Alembert, J., 605
Danbechis, I., 361
Dashen, R. F., 210
Dauben, J. W., 431
Davies, E. B., 126, 308, 361, 521
Davies, P. C. W., 107
Davis, W. J., 587
Debye, P., 623
Decomps, B., 623
De Finis, F., 396
De Graeve, G. M., 591
Dehnen, H., 145, 167
De Lacaille, N. L., 717
Delambre, J. B., 716
Denbigh, K., 732, 751
Derrick, G., 376
Descartes, R., 107, 716, 874
Deser, S., 210, 211
D'Espagnat, B., 106, 107
Destefano, F., 449
Dethier, V., 588
Dewhurst, K. E., 588
DeWitt, C. M., 75
Dezell, H. E., 588
Dicke, R. H., 323, 906
Dicus, D., 420
Diner, S., 492
Di Paolo, J. A., 591
Dirac, P. A. M., 58, 105, 106, 126, 167, 187, 242, 343, 396, 467, 623
Dobzhansky, T., 534, 587
Dold, A., 210, 211
Dollard, C., 188
Donner, K. J., 242
Doran, B. G., 905
Dosch, H. G., 515
Drake, S., 715
Drell, S. D., 344
Dreyer, J. L. E., 604
Dubois, F., 587, 589, 590
Dudley, H. V., 808
Duff, M. J., 211
Dugas, R., 343
Du Hamel, J.-B., 716
Duhem, P., 107, 843
Dummett, M., 752
Dunford, N., 126
Durand, E., 493
Dyson, F. J., 106, 107

Eakin, R. M., 587
Earnshaw, S., 906
Eberhard, P., 106
Eccles, J., 467
Eckart, C., 343
Eddington, A. S., 242, 307, 396, 409
Edelman, G. M., 534
Eibe, T., 717
Eigen, M., 763
Einasto, J., 242
Einstein, A., 3, 5, 51, 58, 106, 107, 322, 343, 361, 396, 409, 436, 467, 493, 623, 751, 763, 880
Ekers, R. D., 241
Eleftheriou, B. F., 588
El Kabir, D. J., 588
Elkana, Y., 905
Ellis, G., 167
Elsasser, W., 732
Emch, G. G., 126, 263
Emerson, R. W., 658
Engels, F., 242
Epstein, P. S., 658
Evans, G., 752
Everett, H., III, 521
Everitt, C. W. F., 907
Ewer, D. W., 589
Exley, T., 906

Fanchi, J. R., 188
Fantappić, L. 396
Faraday, M., 906, 907
Farquhar, I. E., 308
Farrier, D. S., 590, 591
Farrukh, M. O., 449
Federmann, S., 2
Feigl, H., 27, 107
Feldman, B., 29
Fenech, C., 343
Fermi, E., 492, 493
Fernandez, H. L., 590
Fernbach, S., 274
Ferris, T., 842
Feyerabend, P. K., 432, 841-843, 872
Feynman, R. P., 106, 187, 344, 492, 750, 751, 762
Findlay-Freundlich, E., 323
Finkelstein, R., 375
Fistere, J., 20, 22, 30
Fitchard, E. E., 344, 436
Fitzwater, H., 29
Flamsteed, J., 717
Flandern, T. van, 323
Flato, M., 343, 361
Fleck, L., 871
Fliche, H. A., 322
Flügge, S., 501
Fock, V., 107
Fokker, A. D., 106
Foldy, L. L., 187
Fontana, F., 716
Fontenelle, B. le B. de, 605
Ford, V. K., Jr., 242
Foster, J., 591
Foulis, D., 126, 361
Fournier D'Albe, E. E., 322
Fox, J. P., 590
Fraenkel, A. A., 432, 449
Fraenkel, G. S., 586
Franken, P. A., 493
Fraser, A. C., 645
Freedman, D. Z., 905
Freedman, S. J., 106, 493
Frege, G., 409
French, A. P., 623
Frescura, F. A. M., 75
Freud, S., 537-541
Fritzsch, H., 274
Fronsdal, C., 375
Fry, R. J. M., 591
Fuchs, K., 409
Fürnberg, L., 13
Furry, W., 515
Fuxe, K., 588

Galilei, G., 715, 716
Gal-Or, B., 308
Gamow, G., vii, 323
Ganong, W. F., 588
Garber, J., 24
Garber, T., 24, 26
Garcia, L., 376
Gardner, F. F., 242
Garuccio, A., 343, 436
Gascoyne, P. R. C., 808
Gasperini, M., 396
Gassendi, P., 716, 717
Gauss, J. K. F., 593
Gell-Mann, M., 274, 344
Gentner, W., 762
George, C., 308
Georgi, H., 762
Gilbert, C. W., 590
Ginsberg, B. L., 274
Ginsberg, K., 20
Giovannini, N., 361
Gisin, N., 361
Giuliani, G., 588
Glashow, S., 762
Glauber, R. J., 467
Gleason, A., 126
Goddard, P., 274
Goe, G., 604
Gold, T., 167, 432, 521
Goldberg, S. J., 211
Goldman, A., 29
Goldstein, A., 591
Goldstein, F., 13
Goldstone, J., 274
Goodman, A. V., 587
Gorenstein, M. W., 322
Gorini, V., 308
Gottfried, K., 308
Gould, R. T., 716
Grad, H., 501
Grant, R., 716
Grasso, M., 589
Gray, J., 605
Greechie, R., 126
Green, R. F., 241
Greenberg, O. W., 273
Greenwell, B., 29
Gregory, R., 843
Greim, H., 591
Gross, D. J., 210, 274
Gross, G., 29
Grota, L., 589
Grover, P. L., 591
Grünbaum, A., 843
Gudder, S. P., 109, 126, 286, 287
Guerra, F., 313
Gullvaag, I., 841, 843, 844
Gunn, D. L., 586
Gunn, J. E., 167
Gurnett, D. A., 132
Gürsey, F., 66
Gyftopoulos, E. P., 308

Haag, R., 126
Haavisto, J., 396
Habermas, J., 841, 843
Hagedorn, R., 492
Hakanson, R., 589
Haken, H., 493
Halbwachs, F., 343
Hall, J., 106
Halley, E., 716, 717
Halliwell, J. O., 716
Hand, S., 587
Handle, F., 871
Handler, P., 590
Hanson, N. R., 871
Hardy, G. H., 501
Hardy, H., 241

Hare, R., 906
Harkins, W. D., 242
Harrington, L. W., 590
Harriott, T., 716
Harris, G. W., 588
Hart, B. L., 30
Hatsopoulos, G. N., 308
Hawking, S. W., 167, 521
Hecht, M. K., 587
Hehl, F., 221
Heidelberger, C., 591
Heidmann, J., 133, 144, 322
Heim, K., 751
Heimann, P. M., 905, 906
Heine, H., 604
Heisenberg, W., 273, 409, 492
Heitler, W., 58
Heller, A., 449
Hempel, G., 732, 871
Henderson, T., 842
Henin, F., 308
Hérigone, P., 716
Herman, R. C., 323
Herschel, W., 604, 605
Hertz, H., 47, 409
Hesse, M. B., 871, 905
Hevelius, J., 716
Hewitt, E., 126
Hiatt, H., 591
Hickson, P., 242
Highnam, K. C., 588
Hilbert, D., 409
Hiley, B. J., 67, 75, 76
Hill, L., 588
Hillian, P., 344
Hilton, P. J., 76
Himmelfarb, G., 605
Hintikka, J., 843
Hirsch, R., 30
Hisaw, F. L., 588
Hitler, A., 6
Hjelmslev, J., 409
Hlavatý, V., 51
Hokfeldt, T., 588
Hollaender, A., 591
Holme, C., 16, 29, 30
Holstein, M., 590
Holton, G. J., 687, 843, 871
Hönl, H., 145, 167
Hooke, R., 716
Hooker, C., 881
Horkheimer, M., 843
Hormander, L., 188
Horwitz, L. P., 169, 187, 188
Hoskin, M. A., 604, 605
Houson, C., 871
Howell, W., 587
Hoyle, F., 167, 323, 432, 493
Hsieh, S. H., 323
Hubble, E., 241, 322
Hubner, K., 871
Huchra, J., 322
Hudson, R., 126
Hull, D., 534
Humberd, P., 715
Hume, D., 645
Humphreys, W. C., 871
Huneeus-Cox, F., 590
Huygens, C., 715-717
Hylleraas, E. A., 515
Hyman, L., 586

Iliopoulos, J., 762
Infeld, L., 51
Ingarden, R. S., 308
Ingraham, R. L., 396
Isham, C. J., 189, 210, 211
Ishii, K., 456
Israel, W., 521
Istomin, Ya. N., 132
Ivanenko, D. D., 213, 221, 396
Ivanov, V. P., 589

Jaakkola, T., 33, 223, 241, 242, 322
Jachmann, W., 411, 420
Jackiw, R., 210
Jaki, S. L., 593, 604, 605
James, V. H. T., 588
Jammer, M., 492
Jancel, R., 308
Jánossy, L., 34
Jauch, J. M., 107, 126, 361
Jauncey, D. L., 241
Javoy, M., 322
Jaynes, E. T., 308
Jeans, J., 751
Jensen, S., 871
Jenssen, J. H. D., 515
Johnson, J. H., 716
Jonsson, G., 588
Jordan, P., 26, 167, 396
Jordan, T. F., 467
Josephson, B., 107

Kaasik, A., 242
Kaestner, A. G., 603-605
Kahler, E., 76
Kalman, S. M., 591
Kalter, H., 591
Kano, Y., 467
Kant, I., 431, 645, 875
Kaper, H., 501
Karachentsev, I. O., 241
Karoji, H., 322, 323
Kastler, A., 607, 623
Kastler, D., 126
Kato, T., 361
Katz, A., 308
Kaus, P., 375
Kazemier, B. H., 842
Keenan, L., 589
Keep, C. T. S., 30
Keisler, J., 449
Keleman, P. J., 449
Kellermann, K. I., 241
Kelvin, W. T., 907
Kenk, R., 589
Kepler, J., 715
Kerner, E. H., 187
Khalatnikov, I. M., 274
Khinchin, A. I., 658
Kierkegaard, S., 842
Kim, C., 107
Kimball, F., 588
King, E. C., 587
King, M., 187
Kipper, A., 241
Kishida, Y., 588
Klein, D. C., 589
Klein, F., 409
Klein, O., 396, 515
Kleinholz, L. H., 588
Klotz, I. M., 763
Klug, J., 716
Kneale, W., 842
Knight, D. M., 905
Knock, H., 30
Koestler, A., 534, 842
Komar, A., 66, 188
Koopowitz, H., 586, 589
Kopylov, A. I., 241
Körner, S., 842
Kossakowski, A., 308, 361
Kotin, P., 591
Koyre, A., 716
Kozlovsky, B., 242
Kramers, H. A., 76
Krechet, V., 221
Kreipe, K., 21, 22
Kreutzer, E., 587
Kron, R. G., 241
Kubler, O., 521
Kuck, A., 449
Kuhn, T. S., 106, 763, 842, 871

Kumar, A., 376
Kummer, G., 376
Kung, C., 589
Küster, K., 17, 18, 20
Kyburg, H. E., 843
Kyburg, H. E., Jr., 842, 843

Lakatos, I., 432, 842, 871, 872
Lalande, J., 717
Lamb, W. E., 493
Lambert, J. H., 605
Landau, L. D., 107, 274
Landé, A., 76, 105, 308
Landford, O. E. 1, III, 501
Lange, C. S., 590, 591
Langford, C. H., 645
Laplace, P. S., 906
Larson, R. B., 241
Lasker-Schüler, E., 13, 19
Lavie, Y., 188
Lawrence, J. K., 396
Lawvere, F. W., 449
Lax, M., 58
Lazerowitz, M., 625
Lazlo, E., 871
Leaf, G. K., 501
Lebowitz, Y., 107
Lebsack, M. E., 591
LeDenmat, G., 322
Lee, T. D., 107
Lee, V. J., 343
Lehr, W., 343, 436
Leibniz, G. W. von, 768, 875
Leigh-Fermor, P., 21, 29, 30
Leighton, R. B., 750
Leland, C. G., 604
Lelevier, R., 375
Lender, Th., 589, 590
Lenk, R., 308
Lentz, T. L., 587, 590
Lenz, J. W., 843
Lenzen, V. F., 751
Lesage, G.-L., 906
L'Estrange, R., 669
Lethieullier, S., 605
Leutwyler, H., 274
Levere, T. H., 905, 906
Levy, A., 449
Levy, M., 210
Lewis, C. I., 645
Lewis, J., 126
Liao, P. F., 106
Liebscher, D.-E., 34, 38
Lindahl, R., 126
Lindblad, G., 308
Lindell, E. A., 32
Lindholm, S., 871
Lindsay, R. B., 647, 658
Littlewood, J. E., 501
Livingston, D., 31
Llewellyn-Smith, H., 420
Locke, J., 874
Loeb, J., 586
Loewy, E., 29, 30
Loher, W., 588
Lonegan, B., 842
Longacre, A., 907
Longair, M. S., 241
Lorentz, H. A., 907
Lorenzo, A. Y. de, 589
Lougain, M. S., 521
Lowe, J., 106
Lüders, G., 106
Ludwig, G., 243, 263, 361
Lukash, V. N., 167
Luxemburg, W. A., 449
Lyngholm, C., 32
Lyttleton, R. A., 242

Maccarrone, G. D., 396
MacBride, S. E., 589
MacGregor, J. T., 590
Mach, E., 322
Mack, J. M., 449
Mackey, G. W., 126, 361
MacLane, I., 449
MacMillan, W., 432
MacRae, E. K., 587
Magnes, J. L., 9, 19
Maiani, C., 762
Mandel, L., 106, 436
Mandelbrote, B. M., 588, 658
Mandelstam, S., 31, 265
Mann, K., 30
Mann, T., 20
Mannheim, K., 842
Makhankov, V. G., 376
Marchand, J.-P., 34, 83, 126, 275, 286, 287
Marchildon, L., 66
Margenau, H., viii, 28, 29, 307, 308, 449, 658, 659, 669, 751
Maric, Z., 323
Mark, H., 274
Markov, M. A., 396
Marsden, J. R., 589
Marshall, T. W., 343, 492
Martin, Ph. A., 188
Martini, L., 588
Marx, K., 537
Marzollo, A., 375
Maserick, P., 126
Mates, B., 843
Mattuck, R. D., 107
Maupertuis, P. de, 605
Maxwell, G., 107
Maxwell, J. C., 907
Mayants, L. S., 33
Mayr, S., 716
Mazia, D., 587
McCann, J., 590, 591
McDonald, A., 420
McDowell, J., 752
McElroy, B., 107
McGarvey, M., 588
McGuire, J. E., 906
McKeon, R. M., 715
McLauglin, W., 376
Mehlberg, H., 107
Mehra, J., 308
Mehta, C. L., 467
Merat, P., 33
Mercier, A., 38, 51, 671, 682
Merlivat, L., 322
Meselson, M., 591
Meyer, K., 717
Mezger, P. G., 242
Michel, J., 126
Mielnik, B., 361
Mignani, R., 396
Mikaelian, K. O., 322
Mikhailovskii, A. B., 132
Miley, G. K., 241
Miller, E. C., 591
Miller, J. A., 591
Mills, R. L., 273
Milne, D. K., 242
Misner, C. W., 449, 492
Misra, B., 188
Mitchell, V., 29
Mittelstaedt, P., 361
Moles, M., 33, 241, 322, 323, 343
Molotkov, V. V., 187
Momaev, S. G., 167
Montucla, J. E., 716
Moore, C. E., 876
Moore, J. W., 589
Morita, M., 587-590
Morse, M. P., 589
Mos, L., 841
Moscona, A. A., 590
Mosher, R. E., 211
Moss, W. S., 21, 30
Mossotti, O. F., 905, 906
Mostepanenko, V. M., 167
Motte, A., 751

Mould, J., 322
Moyal, J. E., 467
Müller, F., 515
Muller, R. A., 322
Musgrave, A., 871
Mycielski, J., 376

Nachmansohn, D., 587
Naess, A., 683, 841, 842, 844
Nagel, E., 534, 732
Nagy, B. Sz., 263, 501
Naimark, M., 126
Nambu, Y., 274
Nanopoulos, D. V., 420
Narahashi, T., 589
Narlikar, J. V., 323
Nash, L. K., 843
Ne'eman, Y., 221
Nellhaus, G., 29
Nelson, E., 343, 436, 449
Neuberg, C., 808
Neumann, H., 263, 361
Neumann, J. von, 107, 126, 127, 308
Newton, I., 716, 717, 751
Newton, R. G., 188
Newton, T. D., 188
Nichols, W. W., 590
Nicoll, J. F., 323
Nielsen, H. B., 274
Nienaber, C. J. M., 29
Nietzsche, F., 842
Nieuwenhuizen, P. van, 905
Niiniluoto, I., 872
Niles, L., 589
Nishichenko, V. P., 276
Noel, J., 587
North, J. D., 242, 689
Nottale, L., 34, 322, 323
Notting, B. E., 871

Obukhov, Y., 221
Ochs, W., 287
O'Connel, D. J. K., 242
O'Connell, R. F., 106
O'Connor, D. J., 719
Oemler, A., 241
Oficjalski, J., 376
Olbers, W., 322, 604
Olesen, P., 274
Oliverio, V. T., 591
Olson, R. E., 717
Omnes, R., 167
Oppenheimer, R., 26
Ornstein, R., 467
Ossietzki, C. von, 12
Osterbrock, D. E., 242
Ostriker, J. P., 24 899
Oteri, L., 107
Oudemans, J. A. C., 716

Paassen, P. van, 13
Pagnini, P., 716
Paly, T. de, 501
Pantaleo, M., 221, 396
Papandrea, D. N., 588
Pareto, V., 658
Park, J. L., 34, 289, 307, 308, 343, 436
Parker, R. A., 242
Parkes, A. S., 588
Paul, A. M., 841
Paul, H., 58
Pauli, W., 106, 273, 343
Pauliny-Toth, I. I. K., 241
Pausic, M., 396
Payen, R., 107
Peacocke, A. R., 733, 752
Pearle, P. M., 188
Pearson, K. B., 587
Pecker, J.-C., 33, 34, 241, 309, 322, 323
Pedersen, K. J., 587
Peebles, P. J. E., 242, 323
Peet, J., 17, 29, 30
Peierls, R., 308
Peimbert, M., 242
Peña-Auerbach, L. de la, 343, 492
Penrose, R., 75, 167, 521
Penzias, A. A., 323
Peraino, G., 591
Percival, E. C., 308
Perry, M. J., 396
Pethig, R., 808
Petviashvili, V. I., 132
Pfleegor, R. L., 106, 436
Phillips, J., 591
Phillips, T. E., 590
Piaget, J., 871
Pietschmann, H., 753, 762
Pigon, A., 587, 588
Pilsudski, J., 8
Pine-Coffin, R. S., 752
Pirani, F., 750
Piron, C., 126, 187, 188, 345, 361
Pisano, S., 29
Planck, M., 409, 623, 879
Podolsky, B., 106, 107, 343, 361, 467, 763
Poincaré, H., 409
Pokhotelov, O. A., 132
Polanyi, M., 732, 751, 842
Polder, D., 436
Politzer, H. D., 274
Polya, G., 501
Pomeranchuk, I., 274
Pool, J., 126
Popper, K. R., vii, 27, 432, 467, 534, 536, 544, 546, 548, 732, 762, 842, 844, 872
Poulson, R. E., 590
Pound, J., 717
Powers, R., 126
Prescott, C. Y., 420
Preston, G., 126
Price, W. C., 263
Prickard, A. O., 716
Prigogine, I., 308
Prosperetti, A., 396
Puharitch, A., 107
Pullman, A., 591
Pullman, B., 591
Pullman, M. E., 591
Putnam, H. W., 752

Quine, W. O., 844

Raabe, A., 344
Rabin, Y., 188
Racah, G., 106
Radnitzky, G., 842, 844
Raine, D. J., 396
Ralph, C. L., 589
Rañada, A. F., 363, 375, 376
Rañada, M. F., 376
Randall, C. H., 361
Ranft, J., 492
Rauch, H., 361
Raw, C. J. G., 31, 32, 501
Rayleigh, Lord, 623
Rebbi, C., 210, 274
Recami, E., 377, 396
Redford, J. B., 587
Reed, M., 188
Reichenbach, V. H., 751
Reines, F., 287
Remmer, H., 591
Renninger, W., 107
Renyi, A., 501
Reuse, F., 188
Rheita, A. M. Schyrlaeus, de, 715
Riccioli, G., 716
Richards, P. L., 242, 322
Rideau, G., 343
Riedel, V., 30
Riegel, V., 587
Riesz, F., 263, 501
Rigaud, S. P., 716

Rindler, W., 521
Ringhoffer, K., 420
Riser, N. W., 589
Rizzati, P., 396
Robert, F., 322
Roberts, A. P., 323
Robinson, A., 449
Roche, J., 716
Rocke, A. J., 905
Roemer, O., 717
Rogers, C. R., 763
Rohrlich, F., 107, 187, 188
Roll, P. G., 323
Roller, D. H. D., 843
Roman, P., 396
Rompe, R., 397, 409
Roos, D. K., 396
Rosen, J., 396
Rosen, N., 106, 107, 343, 361, 375, 467, 763
Rosenfeld, L., 106, 308
Rosenvold, R., 587
Rothe, K. W., 436
Royce, J. R., 841
Rubenstein, I., 586
Rubin, V. C., 242
Ruch, E., 501
Rude, S., 589
Ruderman, M., 375
Rudnick, D., 587
Ruelle, D., 308
Ruffini, R., 907
Ruggiero, P., 343
Ruskai, M. B., 308
Russell, B., 27, 432, 536, 537, 543, 645, 750, 876
Russell, K., 591
Russell, R. J., 396
Rutherford, E., 843
Rutherford, W., 682
Rutter, J. W., 211
Rüttiman, G., 127
Rybakov, Yu. P., 376

Saar, E., 242
Sabbata, V. de, 396
Saimi, Y., 589
Sakai, S., 287
Salam, A., 107, 274, 396, 762
Salecker, H., 411, 420
Salmon, W. C., 844
Sandage, A., 241, 242, 322
Sands, M., 750
Sardanashvily, G., 221
Sauter, F., 515
Schäfer, G., 167
Scharrer, B., 589
Schilling, H. K., 751
Schilpp, P. A., 436, 467, 687, 751, 843
Schlegel, R., 421, 432, 751
Schlick, M., 536
Schloat, W., 589
Schmidt, H., 107
Schmidt, M., 241
Schönberg, M., 76
Schopenhauer, A., 875
Schrödinger, E., vii, 58, 167, 188, 732, 763
Schroeder, P., 29
Schuster, A., 907
Schuurmans, M. F. H., 436
Schwartz, J., 126
Schweikert, U., 13, 29
Schwinger, J., 106
Scott, A. C., 376
Scott, W. R., 589
Scully, M. O., 493
Segal, I. E., 127, 241, 323, 467
Selby-Bigge, L. A., 645
Selleri, F., 343, 433, 436
Seneca, L. A., 669
Sexl, R. U., 29, 167, 765
Seydewitz, M., 6, 30
Shah, K. T., 436, 449
Shastri, A. R., 211
Shavachka, A. B., 376
Shieve, W. C., 188
Shimkin, M. B., 591
Shimony, A., 106, 107, 376, 521, 763
Shirley, J. W., 715
Shklovskii, I. S., 242
Shribanowitz, N., 436
Shubik, P., 591
Simmel, G., 842
Simmons, R. F., Jr., 289, 308
Simon, B., 188
Simpson, G. G., 534
Sims, P., 591
Sinha, K. P., 343, 396
Sivaram, C., 343, 396
Skagerstam, B. K., 308
Skinner, Q. D., 590
Skolimowski, H., 769
Skyrme, T. H., 376
Slipher, V. M., 322
Smetanin, E., 221
Smith, B. H., 590
Smith, C., 905
Smith, N. K., 431, 645
Smith, R. B., 587
Smith, R. H., 30
Smith, R. S., 587
Smoot, G. F., 322
Smuts, J., 23
Smythies, J. R., 534
Snow, C. P., ix, 872
Soffer, A., 188
Soler, M., 376
Sommerfeld, A., 623, 687
Souriau, J.-M., 322, 343
Spanier, E. H., 210
Spinner, H. F., 871
Spinrad, H., 241, 242
Spitaleri, C., 396
Spohn, H., 501
Stachow, E. W., 361
Staffeldt, E., 591
Stapp, H. P., 106
Starobinskii, A. A., 167
Stebbins, G. L., 586
Steinberg, M. S., 590
Stene, W. C., 587
Stephen, M. J., 451
Stone, M., 127
Strathdee, J., 396
Steegmüller, W., 872
Stuart, A., 76
Stueckelberg, E. C. G., 187, 188
Sudarshan, E. C. G., 308, 343, 457, 467
Sundler, F., 589
Suppe, F., 872
Suppes, P., 449, 843
Surdin, M., 449, 492, 493
Susskind, L., 274
Svartholm, N., 274
Svengrowski, P., 211
Swann, W. F. G., 907
Sweeney, D. C., 589
Swift, J., 395
Switzer, R. M., 210
Synge, J. L., 789
Szamosi, G., 396
Szczepanski, A., 436
Szekeres, G., 396
Szent-Györgyi, A., 805, 808

Tammann, G. A., 322
Tangora, M. C., 211
Tannery, P., 716
Taroni, A., 492
Tarski, J., 449
Tartar, L., 501
Taton, R., 716
Taylor, J. P., 188
Teerikorpi, P., 241, 242
Tennessen, H., 809, 841-845

ter Haar, D., 129, 322
Tezletski, Ya. P., 343
Thackray, A., 905
Thirring, W., 501
Thomas, L. H., 187
Thompson, R. C., 106, 493
Thonnard, N., 242
Thorell, J., 589
Thorn, C. B., 274
Thorne, K. S., 449, 492
Tierny, M., 449
Tifft, W. G., 241
Tinsley, L. H., 187
Todorov, I. T., 187
Tolman, R. C., 242, 322, 343, 492
Tomonaga, S. I., 106
Törnebohm, H., 847, 872
Torrance, T. F., 751
Toulmin, J., 20
Toulmin, S., 872
Trautman, A., 221
Treder, H.-J., 34, 37, 38, 51, 397, 409
Tremaine, S., 241
Trinkhaus, J. P., 590
Tsui, H., 589
Tuomela, R., 872
Turnbull, H. W., 716
Tyler, A., 587
Tzara, C., 492

Uhlenbeck, G. E., 501
Uhlmann, A., 287, 501
Unruh, W., 521
Urbantke, H., 167
Uson, J., 376

Vallentin, H., 17
van der Merwe, A., viii, x, 29, 32–34, 38, 51, 76, 83, 287, 308, 397, 467, 492, 501, 682,
van Nieuwenhuizen, P., 396
Varadarajan, V., 127
Vaucouleurs, G. de, 242, 322
Vazquez, L., 376
Veneziano, G., 274
Vigier, J.-P., 33, 34, 106, 241, 322, 323, 325, 343, 344, 433, 436
Vigoureux, J. M., 107
Vinnikov, Ya. A., 589
Virgona, J. E., 590
Virieux-Reymond, A., 395
Voros, A., 449
Vowinckel, C., 589
Vrehen, Q. H. F., 436
Vuysje, D., 842

Wadan, K., 411
Wakano, M., 376
Wald, R. M., 521
Walker, E. H., 107
Walsby, H., 842
Walter, H.-A., 30
Walther, H., 436
Warburg, E., 409
Wartofsky, M. W., 432, 872
Watanabe, S., 107
Watson, J., 591
Wedberg, A., 842
Wehl, H., 376, 396, 409
Wehrl, A., 34, 287, 495, 501
Weidler, J. F., 717
Weinberg, S., 274, 396, 432, 493, 521, 762
Weingard, R., 432
Weingarten, P., 842
Weisskopf, V., 106
Weizsäcker, C.-F. von, 58, 107
Wells, M. R., 591
Welsh, J. H., 587
Weltsch, R., 8
Wendell, G., 29
Wenzl, A., 751
Wergeland, H., 503
Wess, J., 274
Weyssenhoff, J. V., 344
Wheeler, J. A., 75, 107, 449, 492, 521
Whewell, W., 906
Whiston, W., 717
White, S., 241
Whitehead, A. N., 752
Whitehead, G. W., 210
Whitney, H., 210
Whitrow, G. J., 432
Whittaker, E., 905
Wiener, P. P., 106
Wigner, E., 106, 107, 188, 467, 732
Wilczek, F., 274, 322
Wilfing, A., 361
Wilkinson, D. T., 323, 396
Williams, J. G., 211
Williams, L. D., 587
Williams, L. P., 906
Wilson, H. A., 907
Wilson, J., 586
Wilson, R. W., 323
Wilson, Th., 242
Winkler, R., 763
Winsten, J. A., 591
Wisdom, J. O., 873
Wison, A. R., 106
Witmer, E. E., 107
Witt, B. de, 167
Wittgenstein, L., 536, 541-545, 645
Wohl, G., 29
Wolf, E., 467
Wolf, F., 9
Wolff, E., 587, 589
Woodward, J. F., 32, 33, 38, 885, 907
Woody, D. P., 242, 322
Works, C., 32
Wraith, G. C., 449
Wright, T., 593-605
Wu, T. T., 274
Wurtman, R. J., 588
Wylie, S., 76
Wyss, W., 286, 287

Yang, C.-N., 107, 273, 274
Yosida, K., 308
Yourgrau, B., 26, 27, 29
Yourgrau, J., 3
Yourgrau, M., 27
Yourgrau, P., 26, 29
Yourgrau, T., 11, 26, 29
Yourgrau, Th., vii, 24, 26, 27, 29
Yourgrau, W., vii, ix, 1-34, 37, 38, 51, 76, 170, 189, 213, 221, 224, 241, 265, 275, 287, 289, 308, 322, 431, 432, 437, 449, 455, 458, 467, 492, 495, 501, 549, 586, 659, 671, 682, 715, 719, 752, 841, 845, 873, 883, 905
Yukawa, H., 906

Zeh, H. D., 517, 521
Zeilinger, A., 361
Zerbe, J., 127
Zermelo, E., 409
Zierler, N., 127
Zipf, G. K., 658
Zöllner, J. C. F., 907
Zuckerman, H., 126
Zwanzig, R., 521
Zwanziger, M., 58
Zweig, A., 5, 10, 19, 30
Zwicky, F., 241, 323